Department for Economic and Social
Information and Policy Analysis
Statistical Division

Département de l'information économique
et sociale et de l'analyse des politiques
Division de statistique

Statistical Yearbook
Fortieth issue

Annuaire statistique
Quarantième édition

1993
Data available as of
31 December 1994

Données disponibles
au 31 décembre 1994

United Nations / Nations Unies   New York, 1995

Note

The designations employed and the presentation of material in this publication do not imply the expression of any opinion whatsoever on the part of the Secretariat of the United Nations concerning the legal status of any country, territory, city or area or of its authorities, or concerning the delimitation of its frontiers or boundaries.

In general, statistics contained in the present publication are those available to the United Nations Secretariat up to end-1994 and refer to mid-1993 or earlier. They therefore reflect country nomenclature in use in mid-1992.

The term "country" as used in this publication also refers, as appropriate, to territories or areas.

The designations "developed" and "developing" are intended for statistical convenience and do not necessarily express a judgment about the stage reached by a particular country or area in the development process.

Symbols of United Nations documents are composed of capital letters combined with figures.

Note

Les appellations employées dans la présente publication et la présentation des données qui y figurent n'impliquent de la part du Secrétariat de l'Organisation des Nations Unies aucune prise de position quant au statut juridique des pays, territoires, villes ou zones ou de leurs autorités, ni quant au tracé de leurs frontières ou limites.

En règle générale, les statistiques contenues dans la présente publication sont celles dont disposait le Secrétariat de l'Organisation des Nations Unies jusqu'à la fin de 1994 et portent sur la période finissant à la moitié de 1993. Elles reflètent donc la nomenclature des pays en vigueur à l'époque.

Le terme "pays", tel qu'il est utilisé ci-après, peut également désigner des territoires ou des zones.

Les appellations "développées" et "en développement" sont employées à des fins exclusivement statistiques et n'expriment pas nécessairement un jugement quant au niveau de développement atteint par tel pays ou telle région.

Les cotes des documents de l'Organisation des Nations Unies se composent de lettres majuscules et de chiffres.

ST/ESA/STAT/SER.Ś/16

UNITED NATIONS PUBLICATION
Sales No. E/F.95.XVII.1

PUBLICATION DES NATIONS UNIES
Numéro de vente : E/F.95.XVII.1

ISBN 92-1-061163-2

Inquiries should be directed to:

SALES SECTION
PUBLISHING DIVISION
UNITED NATIONS
NEW YORK 10017
USA

Adresser toutes demandes de renseignements à la :

SECTION DES VENTES
DIVISION DES PUBLICATIONS
NATIONS UNIES
NEW YORK 10017
USA

Preface

This is the fortieth issue of the United Nations *Statistical Yearbook*, prepared by the Statistical Division, Department for Economic and Social Information and Policy Analysis of the United Nations Secretariat, since 1948. The present issue contains series covering, in general, 1983–1992 or 1984–1993, using statistics available to the Statistical Division up to 31 December 1994.

The *Yearbook* is based on data compiled by the Statistical Division from over 40 different international and national sources. These include the United Nations Statistical Division in the fields of national accounts, industry, energy, transport and international trade; the United Nations Statistical Division and Population Division in the field of demographic statistics; and data provided by over 20 offices of the United Nations system and international organizations in other specialized fields.

United Nations agencies and other international organizations which furnished data are listed under "Statistical sources and references" at the end of the *Yearbook*. Acknowledgement is gratefully made to them for the generous cooperation of national and international statistical services throughout the world in providing data.

The present issue of the *Yearbook* is the fourth to reflect a phased programme of major changes in its organization and presentation, undertaken in 1990, which until then was relatively unchanged since the first issue, in 1948. This programme of changes was adopted in response to the continuing long-term expansion of international data available and demanded by users, in terms of country and subject-matter coverage and level of detail, and the more recent impact of new electronic data processing technologies on data compilation and typesetting. One result of this process has been to reduce the total number of tables from 140 in the 37th issue to 83 in the present issue. An index is continued in the present issue.

Recognizing the tremendous worldwide growth in recent years in the use of microcomputers and the corresponding interest in obtaining statistics in machine-readable form for further study and analysis by users, the thirty-eighth issue of the *Yearbook* was published for the first time in compact disk (CD-ROM) for IBM-compatible microcomputers, in addition to the traditional book form followed by the thirty-ninth issue in 1994.[1] The present issue will also be published in CD-ROM in the fall of 1995. Ad hoc or standing orders for the *Yearbook* in both hard copy and CD-ROM are available from United Nations Publications sales offices in New York and Geneva. A full list of machine-readable products in statistics available from the United Nations Statistical Division may be obtained on request to the Statistical Division at the United Nations Secretariat, New York. The Division has also prepared an inventory of over 100 international statistical databases with some form of public access, *StatBase Locator on Disk—UNSTAT's Guide to International Computerized Statistical Databases*.[2]

Préface

La présente édition est la quarantième de l'*Annuaire statistique des Nations Unies*, établi depuis 1948 par la Division de statistique du Département de l'information économique et sociale et de l'analyse des politiques du Secrétariat de l'Organisation des Nations Unies. Elle contient des séries qui portent d'une manière générale sur la période 1983-1992 ou 1984-1993 et pour lesquelles ont été utilisées les informations dont disposait la Division de statistique au 31 décembre 1994.

L'*Annuaire* est établi à partir des données que la Division de statistique a recueillies auprès de plus de 40 sources différentes, internationales et nationales. Ces sources sont : la Division de statistique du Secrétariat de l'Organisation des Nations Unies pour ce qui concerne les comptabilités nationales, l'industrie, l'énergie, les transports et le commerce international; la Division de statistique et la Division de la population du Secrétariat de l'Organisation des Nations Unies pour les statistiques démographiques; et plus de 20 bureaux du système des Nations Unies et d'organisations internationales pour les autres domaines spécialisés.

Les institutions spécialisées des Nations Unies et autres organisations internationales qui ont fourni des données sont énumérées dans la section "Sources et références statistiques" figurant à la fin de l'ouvrage. Les auteurs de l'*Annuaire statistique* remercient de leur généreuse coopération les offices nationaux et internationaux du monde entier qui ont fourni des données.

La présente édition de l'*Annuaire* est la quatrième à tenir compte des importantes transformations qui, depuis 1990, ont été apportées par étapes successives à son organisation et à sa présentation, lesquelles étaient restées pratiquement inchangées depuis la première édition parue en 1948. Ces modifications ont été adoptées en réponse à l'expansion continue et à long terme des données internationales disponibles et demandées par les utilisateurs, par pays et par sujet ainsi qu'avec une précision toujours plus grande, et compte tenu de l'impact plus récent des nouvelles techniques informatiques sur la compilation des données et la composition automatique des textes. Ce processus a ainsi permis de ramener le nombre de tableau de 140 dans la trente-septième édition à 83 dans l'édition actuelle. L'index (en anglais seulement) fourni pour la première fois dans l'édition précédente figure également dans celle-ci.

En raison de l'expansion extraordinaire que la micro-informatique a connue ces dernières années et de l'intérêt croissant que suscite la présentation de statistiques sur des supports lisibles en machine et exploitables directement par l'utilisateur, la trente-huitième édition de l'*Annuaire* a été publiée pour la première fois sur disque compact (CD/ROM) pour micro-ordinateurs IBM et compatibles, outre l'édition imprimée habituelle suivie par la trente-neuvième édition en 1994.[1] La présente édition sera également publiée sur CD/ROM à l'automne 1995. Les commandes individuelles et les abonnements à l'*Annuaire statistique* (édition imprimée ou sur CD/ROM) peuvent être adressées aux bureaux de vente des publications des Nations Unies à New York et à Genève. La Division de statistique

The organization of the *Yearbook*, described in the Introduction below in more detail, consists of four parts. Part One, World and Region Summary, consists of key world and regional aggregates and totals and is essentially unchanged from previous issues. In the remaining parts, the main subjectmatter is mainly presented according to countries or areas, with in some cases world and regions aggregates also shown. Parts two, three and four cover, respectively, population and social topics, national economic activity and international economic relations. The organization of the population and social topics generally follows the arrangement of subjectmatter in the United Nations framework for integration of demographic and social statistics (FSDS) [52]*; economic activity is taken up according to the classes of the United Nations International Standard Industrial Classification of All Economic Activities (ISIC) [42]; and tables on international economic relations cover merchandise trade, international tourism (a major factor in international trade in services and balance of payments) and financial transactions including development assistance. Each chapter includes brief technical notes on statistical sources and methods for the tables in that chapter. Complete references to sources and related methodological publications are provided at the end of the *Yearbook* in the section "Statistical sources and references".

Annex I provides complete information on country and area nomenclature, and regional and other groupings used in the *Yearbook*, and annex II on conversion coefficients and factors used in various tables. Other symbols and conventions used in the *Yearbook* are shown in the section "Explanatory notes", preceding the Introduction.

Important changes in the last issue of the *Yearbook* included extensive revisions to the list of basic commodity tables in part three, reviewed in consultation with FAO. A substantial number of basic commodities was deleted and replaced by commodities of more contemporary importance in agriculture and manufacturing. New tables on telecommunications and national accounts were added to improve and update coverage in these key areas. The complete list of tables added and omitted from the last issue of the *Yearbook* is given in annex III.

du Secrétariat de l'Organisation des Nations Unies à New York fournit sur demande la liste complète de produits statistiques disponibles sur supports lisibles en machine. La Division publie également *StatBase Locator on Disk — UNSTAT's Guide to International Computerized Databases*,[2] inventaire de plus de 100 bases de données statistiques internationales accessibles au public.

Le plan de l'*Annuaire*, qui est décrit ci-après de manière plus détaillée dans l'introduction, comprend quatre parties. La première partie, "Aperçu mondial et régional", qui se compose des principaux agrégats et totaux aux niveaux mondial et régional, est reprise presque sans changement des éditions précédentes; les trois autres sont consacrées à la population et aux questions sociales (deuxième partie), à l'activité économique nationale (troisième partie) et aux relations économiques internationales (quatrième partie). L'Organisation de la deuxième partie, "Population et questions sociales", suit généralement le plan adopté par l'ONU pour intégrer les statistiques économiques et sociales [52]*; dans la troisième partie, l'activité économique est présentée conformément aux catégories adoptées par l'ONU dans la *Classification internationale type, par industrie, de toutes les branches d'activité économique* [42]; les tableaux de la quatrième partie, consacrée aux relations économiques internationales, portent sur le commerce des marchandises, le tourisme international (élément essentiel du secteur international des services et balance des paiements) et les opérations financières, y compris l'aide au développement. Chaque chapitre comprend une brève note technique sur les sources et méthodes statistiques utilisées pour les tableaux du chapitre. On trouvera à la fin de l'*Annuaire*, dans la section "Sources et références statistiques", des références complètes aux sources et publications méthodologiques connexes.

L'annexe I donne des renseignements complets sur la nomenclature des pays et des zones et sur la façon dont ceux-ci ont été regroupés pour former les régions et autres entités géographiques utilisées dans l'*Annuaire*; l'annexe II fournit des renseignements sur les coefficients et facteurs de conversion employés dans les différents tableaux. Les divers symboles et conventions utilisés dans l'*Annuaire* sont présentés à la section "Notes explicatives" qui précède l'introduction.

Entre autres modifications importantes apportées à l'édition antérieure de l'*Annuaire*, la liste de produits de base figurant dans la troisième partie a été révisée et largement remaniée en consultation avec la FAO. De nombreuses inscriptions étaient remplacées par des produits de base présentant une plus grande importance pour les secteurs agricole et manufacturier contemporains. De nouveaux tableaux sur les télécommunications et les comptabilités nationales étaient ajoutés afin d'améliorer et

---

Another modification introduced in the last issue and continued in this *Yearbook* is to exclude tables for which new data are not available but to retain them in the table of contents for publication in a later issue as new data are compiled and published by the collecting agency. However, all of the tables from the 39th issue will be retained in the CD-ROM of the 40th issue.

The new format adopted for typesetting tables in the *Yearbook* was developed in cooperation with the Graphic Presentation Unit of the United Nations Secretariat, to meet new challenges and design opportunities offered by microcomputer "desktop publishing" technologies. Since the disappearance in the 1970s of mechanical typesetting, photocomposition techniques and equipment and in turn table design for statistics have become much more flexible. As a result, it is no longer possible to expect data suppliers for the *Yearbook* to follow a standard design format, nor economically feasible to reset the tables, which are provided in a wide variety of tabulation and database formats, except using microcomputer techniques. This issue of the *Yearbook* continues to make extensive use of microcomputer database, spreadsheet and typesetting technologies, and an "open" table design, which is easier to photocompose than enclosed designs using horizontal and vertical rules and boxes common to the mechanical era.

As described more fully in the Introduction below, every attempt has been made to ensure that the series contained in the *Yearbook* are sufficiently comparable to provide a reliable general description of economic and social topics throughout the world. Nevertheless, the reader should carefully consult the footnotes and technical notes for any given table for explanations of general limitations of series presented and specific limitations affecting particular data items. Complete information concerning the definitions and concepts used and limitations of the data are provided in the section "Statistical sources and references" at the end of the *Yearbook*. Readers interested in more detailed figures than those shown in the present publication, and in further information on the full range of internationally assembled statistics in specialized fields, should also consult the specialized publications listed in that section.

Of course much remains to be done to bring the *Yearbook* fully up to date in its scope, coverage, timeliness and design, and in its technical notes. The process is inevitably an evolutionary one. Comments on the present *Yearbook* and its future evolution are welcome and should be addressed to the Director, United Nations Statistical Division, New York 10017 USA.

d'actualiser la portée des statistiques dans ces secteurs essentiels. La liste complète des tableaux ajoutés et supprimés depuis la dernière édition de l'*Annuaire* figure à l'annexe III.

Autre modification, les tableaux pour lesquels on ne dispose d'aucune donnée nouvelle ont été omis de l'édition antérieure ainsi que la présente édition, mais ils figurent toujours dans la table des matières et seront repris dans une prochaine édition à mesure que des données nouvelles seront dépouillées et publiées par l'office statistique d'origine. Tous les tableaux qui figuraient dans la trente-neuvième édition seront cependant repris dans la quarantième édition publiée sur CD/ROM.

La nouvelle présentation adoptée pour la composition automatique des tableaux de l'*Annuaire* a été mise au point en coopération avec le Groupe de la présentation graphique du Secrétariat de l'ONU, afin de satisfaire à de nouvelles tâches et de répondre aux possibilités offertes par les techniques de publication assistée par micro-ordinateur. Depuis la disparition de la composition mécanique dans les années 70, les techniques et le matériel de photocomposition — donc la conception des tableaux statistiques — peuvent varier beaucoup plus librement. En conséquence, il ne faut plus s'attendre à ce que les fournisseurs de données suivent un mode de présentation type et il n'est plus rentable de recomposer les tableaux qui utilisent des présentations et des modèles de bases de données très diverses, sinon en recourant à la micro-informatique. La présente édition de l'*Annuaire* continue à utiliser largement une base de données, un tableur et des techniques de composition empruntés à la micro-informatique, et à adopter un mode de tabulation "ouvert" qui convient mieux à la photocomposition que les modes fermés faisant appel à des lignes horizontales et verticales et à des encadrés, procédé courant à l'époque de la composition mécanique.

Comme il est précisé ci-après dans l'introduction, aucun effort n'a été épargné afin que les séries figurant dans l'*Annuaire* soient suffisamment comparables pour fournir une description générale fiable de la situation économique et sociale dans le monde entier. Néanmoins, le lecteur devra consulter avec soins les renvois individuels et les notes techniques de chaque tableau pour y trouver l'explication des limites générales imposées aux séries présentées et des limites particulières propres à certains types de données. On trouvera à la fin de l'ouvrage, dans la section "Sources et références statistiques", des renseignements complets concernant les définitions et concepts utilisés et les limites des données. Les lecteurs qui souhaitent avoir des chiffres plus détaillés que ceux figurant dans le présent volume ou qui désirent se procurer des renseignements sur la gamme complète des statistiques qui ont été compilées à l'échelon international dans tel ou tel domaine particulier devraient consulter les publications énumérées dans la section "Sources et références statistiques".

Il reste sans doute beaucoup à faire pour mettre l'*Annuaire* pleinement à jour en ce qui concerne sa portée, son actualité et sa conception générale, ainsi que ses notes techniques. Il s'agit là inévitablement d'un processus évolutif. Les observations sur la présente édition de l'*Annuaire* et les modifications suggérées pour l'avenir seront reçues avec intérêt et doivent être adressées au Directeur de la Division de statistique de l'ONU, New York, N.Y. 10017 (États-Unis d'Amérique).

---

[1] Statistical Yearbook, thirty-ninth issue, CD-ROM (United Nations publication, Sales No. E.95.XVII.5).

[2] United Nations publication, Sales No. E.94.XVII.8 (issued on one 3 1/2 diskette for IBM-compatible microcomputers).

---

[1] L'Annuaire statistique, la trente-neuvième édition sur CD/ROM (Publication des Nations Unies, numéro de vente E.95.XVII.5).

[2] Publication des Nations Unies, numéro de vente E.94.XVII.8 (sur une disquette de 3,5 inches pour micro-ordinateurs IBM et compatibles).

Contents

Part One
World and Region Summary

Part Two
Population and Social Statistics

Part Three
Economic Activity

Table des matières

Première partie
Aperçu mondial et régional

Deuxième partie
Statistiques démographiques et sociales

Troisième partie
Activité économique

Part Four
International Economic Relations

Quatrième partie
Relations économiques internationales

List of tables

Liste des tableaux

Part Three
Economic Activity

Troisième partie
Activité économique

Part Four
International Economic Relations

Quatrième partie
Relations économiques internationales

---

* This symbol identifies tables presented in the previous (39th or 38th)
   issue of the *Statistical Yearbook* but not contained in the present issue
   because of insufficient new data. These tables will be updated in future
   issues of the *Yearbook* when new data become available.

* Ce symbole indique les tableaux publiés dans l'édition précédente (39ᵉ
   ou 38ᵉ édition) de l'*Annuaire statistique* mais qui n'ont pas été repris
   dans la présente édition fautes de données nouvelles suffisantes. Ces
   tableaux seront actualisés dans les futures livraisons de l'*Annuaire* à
   mesure que des données nouvelles deviendront disponibles.

Explanatory notes

The metric system of weights and measures has been employed throughout the *Statistical Yearbook*. For conversion coefficients and factors, see annex II.

Certain tables contain global aggregates designated variously as "total" or "world". Where a figure represents the summation of the country series shown in the table but is not considered comprehensive for the world, it is labelled "total". Where, however, an aggregate is considered to represent substantially complete world coverage, it is labelled "world". As a rule, allowance has been made in the "world" figures for any gaps that may exist in the country series shown.

In some cases, the comparability of the statistics is affected by geographical changes. As a general rule, the data relate to a given country or area within its present de facto boundaries. Where statistically important, attention is called to changes in territory by means of a footnote. The reader is referred to annex I, concerning country and area nomenclature, where changes in designation are listed.

Numbers in brackets refer to numbered entries in the section "Statistical sources and references" at the end of this book.

In general, statistics presented in the present publication are based on information available to the Statistical Division of the United Nations Secretariat up to 31 December 1994.

Symbols and conventions used in the tables

A point (.) is used to indicate decimals.

A hyphen (-) between years, e.g., 1984-1985, indicates the full period involved, including the beginning and end years; a slash (/) indicates a financial year, school year or crop year, e.g., 1984/85.

"Δ p.a." (change per annum) is used to indicate annual rate of change.

Not applicable or not separately reported	..
Data not available	...
Magnitude zero	-
Magnitude zero, or less than half of unit employed	0 or 0.0
Provisional or estimated figure	*
United Nations estimate	x
Marked break in series	#

Details and percentages in tables do not necessarily add to totals because of rounding.

Notes explicatives

Le système métrique de poids et mesures a été utilisé dans tout l'*Annuaire statistique*. On trouvera à l'annexe II les coefficients et facteurs de conversion.

Certains tableaux contiennent des agrégats globaux désignés par la mention "total" ou "monde", selon les cas. Quand un chiffre représente l'addition des chiffres correspondant à chacun des pays qui figurent dans le tableau, mais ne paraît pas recouvrir le monde entier, il porte la mention "total". Si un agrégat semble correspondre au total mondial, ou très peu s'en faut, il porte la mention "monde". En règle générale, les chiffres portant la mention "monde" s'entendent compte tenu des lacunes qui peuvent exister dans la série de pays indiqués.

Dans certains cas, les changements géographiques intervenus influent sur la comparabilité des statistiques. En règle générale, les données renvoient au pays ou zone en question dans ses frontières actuelles effectives. Une note appelle l'attention sur les changements territoriaux, si cela importe du point de vue statistique. Le lecteur est renvoyé à l'annexe A (nomenclature des pays et zones et groupements régionaux) où il trouvera une liste des changements de désignation.

Les chiffres figurant entre crochets se réfèrent aux entrées numérotées dans la liste des sources et références statistiques à la fin de l'ouvrage.

En général, les statistiques qui figurent dans la présente publication sont fondées sur les informations dont disposait la Division de statistique du Secrétariat de l'ONU au 31 décembre 1994.

Signes et conventions employés dans les tableaux

Les décimales sont précédées d'un point (.).

Un tiret (-) entre des années, par exemple "1984-1985", indique que la période est embrassée dans sa totalité, y compris la première et la dernière année; une barre oblique (/) renvoie à un exercice financier, à une année scolaire ou à une campagne agricole, par exemple "1984/85".

Le symbole "Δ p.a." signifie qu'il s'agit du taux annuel de variation.

Non applicable ou non communiqué séparément	..
Données non disponibles	...
Néant	-
Valeur nulle, ou inférieure à la moitié de la dernière unité retenue	0 ou 0.0
Chiffre provisoire ou estimatif	*
Estimation des Nations Unies	x
Discontinuité notable dans la série	#

Les chiffres étant arrondis, les totaux ne correspondent pas toujours à la somme exacte des éléments ou pourcentages figurant dans les tableaux.

Introduction

This is the fortieth issue of the United Nations *Statistical Yearbook*, prepared by the Statistical Division, Department for Economic and Social Information and Policy Analysis, of the United Nations Secretariat. It contains series covering, in general, 1983-1992 or 1984-1993, based on statistics available to the Statistical Division up to 31 December 1994.

The major purpose of the *Statistical Yearbook* is to provide in a single volume a comprehensive compilation of internationally-available statistics on social and economic conditions and activities in the world, at world, regional and national levels, covering roughly a ten-year period.

Most of the statistics presented in the *Yearbook* are extracted from more detailed, specialized publications prepared by the Statistical Division and by many other international statistical services. Thus, while the specialized publications concentrate on monitoring topics and trends in particular social and economic fields, the *Statistical Yearbook* tables provide data for a more comprehensive, overall description of social and economic structures, conditions, changes and activities. The objective has been to collect, systematize and coordinate the most essential components of comparable statistical information which can give a broad and, to the extent feasible, a consistent picture of social and economic processes at world, regional and national levels.

More specifically, the *Statistical Yearbook* provides systematic information on a wide range of social and economic issues which are of concern in the United Nations system and among the governments and peoples of the world. A particular value of the *Yearbook*, but also its greatest challenge, is that these issues are extensively interrelated. Meaningful social and economic analysis of these issues requires systematization and coordination of the data across many fields. These issues include:

— General economic growth and related economic conditions;

— Economic situation in developing countries and progress towards the objectives adopted for the United Nations development decades;

— Population and urbanization, and their growth and impact;

— Employment, inflation and wages;

— Production of energy and development of new energy sources;

— Expansion of trade;

— Supply of food and alleviation of hunger;

— Financial situation and external payments and receipts;

— Education, training and eradication of illiteracy;

— Improvement in general living conditions;

— Assistance provided to developing countries for social and economic development purposes.

Introduction

La présente édition est la quarantième de l'*Annuaire statistique* des Nations Unies, établi par la Division de statistique du Département de l'information économique et sociale et de l'analyse des politiques du Secrétariat de l'Organisation des Nations Unies. Elle contient des séries de données qui portent d'une manière générale sur les années 1983 à 1992 ou 1984 à 1993, et pour lesquelles ont été utilisées les informations dont disposait la Division de statistique au 31 décembre 1994.

L'*Annuaire statistique* a principalement pour objet de présenter en un seul volume un inventaire complet de statistiques disponibles sur le plan international et concernant la situation et les activités sociales et économiques dans le monde, aux échelons mondial, régional et national, pour une période d'environ 10 ans.

Une bonne partie des données qui figurent dans l'*Annuaire* existent sous une forme plus détaillée dans les publications spécialisées établies par la Division de statistique et par bien d'autres services statistiques internationaux. Alors que les publications spécialisées suivent essentiellement l'évolution dans certains domaines socio-économiques précis, l'*Annuaire statistique* présente les données de manière à fournir une description plus globale et exhaustive des structures, conditions, transformations et activités socio-économiques. On a cherché à recueillir, systématiser et coordonner les principaux éléments de renseignements statistiques comparables, de manière à dresser un tableau général et autant que possible cohérent des processus socio-économiques en cours aux échelons mondial, régional et national.

Plus précisément, l'*Annuaire statistique* a pour objet de présenter des renseignements systématiques sur toutes sortes de questions socio-économiques qui sont liées aux préoccupations actuelles du système des Nations Unies ainsi que des gouvernements et des peuples du monde. Le principal avantage de l'*Annuaire* — et aussi la principale difficulté à surmonter — tient à ce que ces questions sont étroitement interdépendantes. Pour en faire une analyse économique et sociale utile, il est essentiel de systématiser et de coordonner les données se rapportant à de nombreux domaines différents. Ces questions sont notamment les suivantes :

— La croissance économique générale et les aspects connexes de l'économie;

— La situation économique dans les pays en développement et les progrès accomplis vers la réalisation des objectifs des décennies des Nations Unies pour le développement;

— La population et l'urbanisation, leur croissance et leur impact;

— L'emploi, l'inflation et les salaires;

— La production d'énergie et la mise en valeur des énergies nouvelles;

— L'expansion des échanges;

— Les approvisionnements alimentaires et la lutte contre la faim;

Organization of the *Yearbook*

The contents of the *Statistical Yearbook* are planned to serve a general readership. The *Yearbook* endeavours to provide information for various bodies of the United Nations system as well as for other international organizations, for governments and non-governmental organizations, for national statistical, economic and social policy bodies, for scientific and educational institutions, for libraries and for the public. Data published in the *Statistical Yearbook* are also of interest to companies and enterprises and to agencies engaged in marketing research.

The 83 tables of the *Yearbook* are grouped into four broad parts:
— World and Region Summary (chapter I and tables 1-7);
— Population and Social Statistics (chapters II-V and tables 8-20);
— Economic Activity (chapters VI-XIV and tables 21-70);
— International Economic Relations (chapters XV-XIX and tables 71-83).

These four parts present data at two levels of aggregation. The more aggregated information shown in Part One provides an overall picture of development at the world and region levels. More specific and detailed information for analysis concerning individual countries or areas in the three following parts. Each of these is divided into more specific chapters, by topic, and each chapter includes a section, "Technical notes". These notes provide brief descriptions of major statistical concepts, definitions and classifications required for interpretation and analysis of the data. Systematic information on the methodology used for computation of figures can also be found in the publications on methodology of the United Nations and its agencies, listed in the section "Statistical sources and references" at the end of the *Yearbook*. Additional general information on statistical methodology is provided in the section below on "Comparability of statistics" and in the explanatory notes following the Introduction.

More specifically, Part One, World and Region Summary, comprises 7 tables highlighting the principal trends in the world as a whole as well as in regions and in the major economic and social sectors. It contains global totals of important aggregate statistics needed for the analysis of economic growth, the structure of the world economy, major changes in world population, expansion of external merchandise trade, world production and consumption of energy. The global totals are, as a rule, subdivided into major geographical areas.

Part Two, Population and Social Statistics, comprises 13 tables which contain more detailed statistical series on social conditions and levels of living, for example, data on education and cultural activities.

— La situation financière, les paiements extérieurs et les recettes extérieures;
— L'éducation, la formation et l'élimination de l'analphabétisme;
— L'assistance fournie aux pays en développement à des fins socio-économiques.

Présentation de l'*Annuaire*

Le contenu de l'*Annuaire statistique* a été préparé à l'intention de tous les lecteurs intéressés. Les renseignements fournis devraient pouvoir être utilisés par les divers organismes du système des Nations Unies ainsi que par d'autres organisations internationales, par les gouvernements et les organisations non gouvernementales, par les organismes nationaux de statistique et de politique économique et sociale, par les institutions scientifiques et les établissements d'enseignement, les bibliothèques et les particuliers. Les données publiées dans l'*Annuaire statistique* peuvent également intéresser les sociétés et entreprises, et les organismes spécialisés dans les études de marché.

Les 83 tableaux de l'*Annuaire* sont groupés en quatre parties :
— Aperçu mondial et régional (chap. I et tableaux 1 à 7);
— Statistiques démographiques et sociales (chap. II à V et tableaux 8 à 20);
— Activité économique (chap. VI à XIV et tableaux 21 à 70);
— Relations économiques internationales (chap. XV à XIX et tableaux 71 à 83).

Ces quatre parties présentent les données à deux niveaux d'agrégation : les valeurs les plus agrégées qui figurent dans la première partie donnent un tableau global du développement à l'échelon mondial et régional, tandis que les trois autres parties contiennent des renseignements plus précis et détaillés qui se prêtent mieux à une analyse par pays ou par zones. Chacune de ces trois parties est divisée en chapitres portant sur des sujets donnés, et chaque chapitre comprend une section intitulée "Notes techniques" où l'on trouve une brève description des principales notions, définitions et classifications statistiques nécessaires pour interpréter et analyser les données. Les méthodes de calcul utilisées sont également décrites de façon systématique dans les publications se référant à la méthodologie des Nations Unies et de leurs organismes, énumérées à la fin de l'*Annuaire* dans la section "Sources et références statistiques". Le lecteur trouvera un complément d'informations générales ci-après dans la section intitulée "Comparabilité des statistiques", ainsi que dans les notes explicatives qui suivent l'introduction.

Plus spécialement, la première partie, intitulée "Aperçu mondial et régional", comprend sept tableaux présentant les principales tendances dans le monde et dans les régions ainsi que dans les principaux secteurs économiques et sociaux. Elle fournit des chiffres mondiaux pour les principaux agrégats

Part Three, Economic Activity, provides data in 23 tables of statistics on national accounts, index numbers of industrial production, interest rates, labour force, wages and prices, transport, energy, science and technology; and in 27 tables on production in the major branches of the economy (using, in general, the international standard industrial classification, ISIC), namely agriculture, hunting, forestry and fishing; mining and quarrying; manufacturing; and transport and communications. In an innovation in the general approach of the *Yearbook*, consumption data are now being combined with the production data in tables on specific commodities, where feasible.

Part Four, International Economic Relations, comprises 13 tables on international merchandise trade, balance of payments, tourism, finance and development assistance. It focuses on the growth and structure of exports and imports by countries or areas, international tourism, balance of payments and development assistance provided by multilateral and bilateral agencies to developing countries.

An index (in English only) is provided at the end of the *Yearbook*.

Annexes and regional groupings of countries or areas

The annexes to the *Statistical Yearbook* and the section "Explanatory notes", preceding the Introduction, provide additional essential information on the *Yearbook*'s contents and presentation of data.

Annex I provides information on countries or areas covered in the *Yearbook* tables and on their grouping into geographical regions. The geographical groupings shown in the *Yearbook* are generally based on continental regions unless otherwise indicated. However, strict consistency in this regard is impossible. A wide range of classifications is used for different purposes in the various international agencies and other sources of statistics for the *Yearbook*. These classifications vary in response to administrative and analytical requirements.

Neither is there a common agreement in the United Nations system concerning the terms "developed" and "developing", when referring to the stage of development reached by any given country or area and its corresponding classification in one or the other grouping. Thus, the *Yearbook* refers more generally to "developed" or "developing" regions on the basis of conventional practice. Following this practice, "developed regions" comprises northern America, Europe and the former USSR, Australia, Japan and New Zealand, while all of Africa and the remainder of the Americas, Asia and Oceania comprise the "developing regions". These designations are intended for statistical convenience and do not necessarily express a judgement about the stage reached by a particular country or area in the development process.

statistiques nécessaires pour analyser la croissance économique, la structure de l'économie mondiale, les principaux changements dans la population mondiale, l'expansion du commerce extérieur de marchandises, la production et la consommation mondiales d'énergie. En règle générale, les chiffres mondiaux sont ventilés par grandes régions géographiques.

La deuxième partie, intitulée "Statistiques démographiques et sociales", comporte 13 tableaux où figurent des séries plus détaillées concernant les conditions sociales et les niveaux de vie, notamment des données sur l'éducation et les activités culturelles.

La troisième partie, intitulée "Activité économique", présente en 23 tableaux des statistiques concernant les comptes nationaux, les nombres indices relatifs à la production industrielle, les taux d'intérêt, la population active, les prix et les salaires, le transport, l'énergie, les sciences et les techniques; et en 27 tableaux des données sur la production des principales branches d'activité économique (en utilisant en général la *Classification internationale type, par industrie, de toutes les branches d'activité économique*) : agriculture, chasse, sylviculture et pêche; mines et carrières; industries manufacturières; transports et communications. Une innovation a été introduite dans la présentation générale de l'*Annuaire* en ce sens que les tableaux traitant de certains produits de base associent autant que possible les données relatives à la consommation aux valeurs concernant la production.

La quatrième partie, intitulée "Relations économiques internationales", comprend 13 tableaux relatifs au commerce international de marchandises, aux balances des paiements, au tourisme, aux finances et à l'aide au développement. Elle est consacrée essentiellement à la croissance et à la structure des exportations et des importations par pays et par zone, au tourisme international, aux balances des paiements et à l'aide au développement fournie aux pays en développement par les organismes multilatéraux et bilatéraux.

Un index (en anglais seulement) figure à la fin de l'*Annuaire*.

Annexes et groupements régionaux des pays et zones

Les annexes à l'*Annuaire statistique* et la section intitulée "Notes explicatives" offrent d'importantes informations complémentaires quant à la teneur et à la présentation des données figurant dans le présent ouvrage.

L'annexe I donne des renseignements sur les pays ou zones couverts par les tableaux de l'*Annuaire* et sur leur regroupement en régions géographiques. Sauf indication contraire, les groupements géographiques figurant dans l'*Annuaire* sont généralement fondés sur les régions continentales, mais une présentation absolument systématique est impossible à cet égard car les diverses institutions internationales et autres sources de statistiques employées pour la confection de l'*Annuaire* emploient, selon l'objet de

Annex II provides detailed information on conversion coeffients and factors used in various tables, and annexe III provides listings of tables added and omitted in the present edition of the *Yearbook*.

Comparability of statistics

One major aim of the *Statistical Yearbook* is to present series which are as nearly comparable accross countries as the available statistics permit. Considerable efforts are also made among the international suppliers of data and by the staff of the *Yearbook* to ensure the compatibility of various series by coordinating time periods, base years, prices chosen for valuation and so on. This is indispensable in relating various bodies of data to each other and to facilitate analysis across different sectors. Thus, for example, relating data on economic output to those on employment makes it possible to derive some trends in the field of productivity; relating data on exports and imports to those on national product allows an evaluation of the relative importance of external trade in different countries and reveals changes in the role of trade over time.

In general, the data presented reflect the methodological recommendations of the United Nations Statistical Commission, issued in various United Nations publications, and of other international bodies concerned with statistics. Publications containing these recommendations and guidelines are listed in the section "Statistical sources and references" at the end of the *Yearbook*. Use of international recommendations not only promotes international comparability of the data but also ensures a degree of compatibility regarding the underlying concepts, definitions and classifications relating to different series. However, much work remains to be done in this area and, for this reason, some tables can serve only as a first source of data, which require further adjustment before being used for more in-depth analytical studies. Although, on the whole, a significant degree of comparability has been achieved in international statistics, there are many limitations, for a variety of reasons.

One common cause of non-comparability of economic data is different valuations of statistical aggregates such as national income, wages and salaries, output of industries and so forth. Conversion of these and similar series originally expressed in national prices into a common currency, for example into United States dollars, through the use of exchange rates is not always satisfactory owing to frequent wide fluctuations in market rates and differences between official rates and rates which would be indicated by unofficial markets or purchasing power parities. For this reason, data on national income in United States dollars which are published in the *Yearbook* are subject to certain distortions and can be used as only a rough approximation of the relative magnitudes involved.

l'exercice, des classifications fort différentes en réponse à diverses exigences d'ordre administratif ou analytique.

Il n'existe pas non plus dans le système des Nations Unies de définition commune des termes "développé" et "en développement" pour décrire le niveau atteint en la matière par un pays ou une zone donnés ni pour les classifier dans l'un ou l'autre de ces groupes. Ainsi, dans l'*Annuaire*, on s'en remet à l'usage pour qualifier les régions de "développées" ou "en développement". Selon cet usage, les régions développées sont le continent américain au nord du Mexique, l'Europe et l'ancienne URSS, l'Australie, le Japon et la Nouvelle-Zélande, alors que toute l'Afrique et le reste des Amériques, l'Asie et l'Océanie constituent les régions en développement. Ces appellations sont utilisées pour plus de commodité dans la présentation des statistiques et n'impliquent pas nécessairement un jugement quant au stade de développement auquel est parvenu tel pays ou telle zone.

L'annexe II fournit des renseignements sur les coefficients et facteurs de conversion employés dans les différents tableaux, et l'annexe III contient les listes de tableaux qui ont été ajoutés ou omis dans la présente édition de l'*Annuaire*.

Comparabilité des statistiques

L'*Annuaire statistique* a principalement pour objet de présenter des statistiques aussi comparables d'un pays à l'autre que les données le permettent. Les sources internationales de données et les auteurs de l'*Annuaire* ont réalisés des efforts considérables pour faire en sorte que les diverses séries soient compatibles en harmonisant les périodes de référence, les années de base, les prix utilisés pour les évaluations, etc. Cette démarche est indispensable si l'on veut rapprocher divers ensembles de données pour faciliter l'analyse intersectorielle de l'économie. Ainsi, en liant les données concernant la production à celles de l'emploi, on parvient à dégager certaines tendances dans le domaine de la productivité; de même, en associant les données concernant les exportations et importations aux valeurs du produit national, on obtient une évaluation de l'importance relative des échanges extérieurs dans différents pays et de l'évolution du rôle joué par le commerce.

De façon générale, les données sont présentées selon les recommandations méthodologiques formulées par la Commission de statistique de l'ONU et par les autres organisations internationales qui s'intéressent aux statistiques. Les titres des publications contenant ces recommandations et lignes directrices figurent à la fin de l'ouvrage dans la section intitulée "Sources et références statistiques". Le respect des recommandations internationales tend non seulement à promouvoir la comparabilité des données à l'échelon international, mais elle assure également une certaine comparabilité entre les concepts, les définitions et classifications utilisés. Mais comme il reste encore beaucoup à faire dans ce domaine, les données présentées dans certains

The use of different kinds of sources for obtaining data is another cause of incomparability. This is true, for example, in the case of employment and unemployment, where data are collected from such non-comparable sources as sample surveys, social insurance statistics and establishment surveys.

Non-comparability of data may also result from differences in the institutional patterns of countries. Certain variations in social and economic organization and institutions may have an impact on the comparability of the data even if the underlying concepts and definitions are identical.

These and other causes of non-comparability of the data are briefly explained in the technical notes to each chapter.

Statistical sources and reliability and timeliness of data

Statistics and indicators have been compiled mainly from official national and international sources, as these are more authoritative and comprehensive, more generally available as time series and more comparable among countries than other sources. In a few cases, official sources are supplemented by other sources and estimates, where these have been subjected to professional scrutiny and debate and are consistent with other independent sources. The comprehensive international data sources used for most of the tables are presented in the list of "Statistical soruces and references" at the end of the *Yearbook*.

Users of international statistics are often concerned about the apparent lack of timeliness in the available data. Unfortunately, most international data are only available with a delay of at least one to three years after the latest year to which they refer. The reasons for the delay are that the data must first be processed by the national statistical services at the country level, then forwarded to the international statistical services and processed again to ensure as much consistency across countries and over time as possible.

tableaux n'ont qu'une valeur indicative et nécessiteront des ajustements plus poussés avant de pouvoir servir à des analyses approfondies. Bien que l'on soit parvenu, dans l'ensemble, à un degré de comparabilité appréciable en matière de statistiques internationales, diverses raisons expliquent que subsistent encore de nombreuses limitations.

Une cause commune de non-comparabilité des données réside dans la diversité des méthodes d'évaluation employées pour comptabiliser des agrégats tels que le revenu national, les salaires et traitements, la production des différentes branches d'activité industrielle, etc. Il n'est pas toujours satisfaisant de ramener la valeur des séries de ce type — exprimée à l'origine en prix nationaux — à une monnaie commune (par exemple le dollar des États-Unis) car les taux de change du marché connaissent fréquemment de fortes fluctuations tandis que les taux officiels ne coïncident pas avec ceux des marchés officieux ni avec les parités réelles de pouvoir d'achat. C'est pourquoi les données relatives au revenu national, qui sont publiées dans l'*Annuaire* en dollars des États-Unis, souffrent de certaines distorsions et ne peuvent servir qu'à donner une idée approximative des ordres de grandeur relatifs.

Le recours à des sources diverses pour la collecte des données est une autre facteur qui limite la comparabilité, en particulier dans les secteurs de l'emploi et du chômage où les statistiques sont obtenues par des moyens aussi peu comparables que les sondages, le dépouillement des registres d'assurances sociales et les enquêtes auprès des entreprises.

Dans certains cas, les données ne sont pas comparables en raison de différences entre les structures institutionnelles des pays. Certaines variations dans l'organisation et les institutions économiques et sociales peuvent affecter la comparabilité des données même si les concepts et définitions sont fondamentalement identiques.

Ces causes de non-comparabilité des données sont parmi celles qui sont brièvement expliquées dans les notes techniques de chaque chapitre.

Origine, fiabilité et actualité des données

Les statistiques et les indicateurs sont fondés essentiellement sur des données provenant de sources officielles nationales et internationales; c'est en effet la meilleure source si l'on veut des données fiables, complètes et comparables et si l'on a besoin de séries chronologiques. Dans quelques cas, les données officielles sont complétées par des informations et des estimations provenant d'autres sources qui ont été examinées par des spécialistes et confirmées par des sources indépendantes. On trouvera à la fin de l'*Annuaire* la liste des "Sources statistiques et références", qui récapitule les sources des données internationales utilisées pour la plupart des tableaux.

Les utilisateurs des statistiques internationales se plaignent souvent du fait que les données disponibles ne sont pas actualisées. Malheureusement, la plupart des données internationales ne sont disponibles qu'avec un délai de deux ou trois ans après la dernière année à laquelle elles se rapportent. S'il en est ainsi, c'est parce que les données sont d'abord traitées par les services statistiques nationaux avant d'être transmises aux services statistiques internationaux, qui les traitent à nouveau pour assurer la plus grande comparabilité possible entre les pays et entre les périodes.

Part One
World and Region Summary

I
World and region summary (tables 1-7)

This part of the *Statistical Yearbook* presents selected aggregate series on principal economic and social topics for the world as a whole and major regions. The topics include population and surface area, agricultural and industrial production, energy, commodity prices, motor vehicles in use, external trade and government financial reserves. More detailed data on individual countries and areas are provided in the subsequent parts of the present *Yearbook*. These comprise Part Two, Population and Social Statistics; Part Three, Economic Activity; and Part Four, International Economic Relations.

Regional totals may contain incomparabilities between series owing to differences in definitions of regions and lack of data for particular regional components. General information on regional groupings is provided in annex I of the *Yearbook*. Supplementary information on regional groupings used in specific series is provided as necessary in table footnotes and in the technical notes at the end of chapter I.

Première partie
Aperçu mondial et régional

I
Aperçu mondial et régional (tableaux 1 à 7)

Cette partie de l'*Annuaire statistique* présente, pour le monde entier et ses principales subdivisions, un choix d'agrégats ayant trait à des questions économiques et sociales essentielles : population et superficie, production agricole et industrielle, énergie, prix des produits de base, véhicules automobiles en circulation, commerce extérieur et réserves financières publiques. Des statistiques plus détaillées pour divers pays ou zones figurent dans les parties ultérieures de l'*Annuaire*, c'est-à-dire dans les deuxième, troisième et quatrième parties intitulées respectivement : population et statistiques sociales, activités économiques et relations économiques internationales.

Les totaux régionaux peuvent présenter des incomparabilités entre les séries en raison de différences dans la définition des régions et de l'absence de données sur tel ou tel élément régional. A l'annexe I de l'*Annuaire*, on trouvera des renseignements généraux sur les groupements régionaux. Des informations complémentaires sur les groupements régionaux pour certaines séries bien précises sont fournies, lorsqu'il y a lieu, dans les notes figurant au bas des tableaux et dans les notes techniques à la fin du chapitre I.

1
Selected series of world statistics
Séries principales de statistiques mondiales
Population, production, external trade and finance
Population, production, commerce extérieur et finances

Series Séries	Unit or base Unité ou base	1985	1986	1987	1988	1989	1990	1991	1992	1993
World population [1] **Population mondiale** [1]	million	4853	4938	5024	5112	5201	5292	5385	5480	5572

Agriculture, forestry and fishing production • Production agricole, forestière et de la pêche
Index numbers • Indices

All commodities Tous produits	1979–81	114	115	116	119	123	126	127	129	128
Food Produits alimentaires	1979–81	114	116	116	118	123	126	127	129	129
Crops Culture	1979–81	115	115	116	116	121	125	124	128	126
Cereals Céréales	1979–81	117	118	114	112	122	127	122	130	124
Livestock products Produits de l'élevage	1979–81	112	115	118	121	121	126	128	129	129

Quantities • Quantités

Oilcrops Huile cultures	million t.	65	65	68	69	72	76	78	79	79
Meat Viande	million t.	151	157	163	169	172	177	182	185	186
Roundwood Bois rond	million m³	3178	3266	3342	3404	3466	3511	3422	3477	...
Fish catches Prises de poissons	million t.	86.3	92.8	94.4	99.1	100.3	97.6	97.1	98.1	...

Industrial production • Production industrielle
Index numbers [2] • Indices [2]

All commodities Tous produits	1980	108.4	109.4	113.1	119.5	123.8	123.6	122.4	122.4	123.9
Mining Mines	1980	87.1	89.2	90.4	95.8	100.2	99.8	97.3	101.3	106.9
Manufacturing Manufactures	1980	111.9	112.9	117.1	123.9	128.1	127.6	126.2	125.1	125.4

Quantities • Quantités

Coal Houille	million t.	3236	3328	3423	3505	3581	3516	3462	3534	3470
Lignite and brown coal [3] Lignite et charbon brun [3]	million t.	1173	1196	1225	1243	1256	1191	1055	979	988
Crude petroleum [4] Petrole brut [4]	million t.	2641	2770	2772	2891	2935	3299	2976	2991	2877
Natural gas Gaz naturel	pétajoules	61160	62541	67202	70426	73078	75003	76551	77708	77815
Pig–iron and ferro–alloys Fonte et ferro–alliages	million t.	500	497	505	542	557	556	492	533	...
Fabrics • Tissus Cellulosic and non cellulosic fibres Cellulosiques et non cellulosiques	million m²	20693	20900	20355	21361	22320	22225	21084	21625	...
Cotton and wool Coton et laines	million m²	76945	93083	84776	94895	103476	91175	95622	99478	...
Leather footwear Chaussures de cuir	million pairs	4361	4480	4227	4579	4528	4540	4111	4399	...
Sulphuric acid Acide sulfurique	million t.	133	131	135	143	141	136	125	127	...
Soap Savons	million t.	20	21	21	22	23	23	21	20	...
Refrigerators Réfrigérateurs	million	43	45	50	56	55	53	48	50	...
Washing machines Machines et appareils à laver	million	40	42	45	48	46	45	38	37	...

1
Selected series of world statistics
Population, production, external trade and finance [*cont.*]
Séries principales de statistiques mondiales
Population, production, commerce extérieur et finaces [*suite*]

Series Séries	Unit or base Unité ou base	1985	1986	1987	1988	1989	1990	1991	1992	1993
Machine tools · Machines outils										
Drilling and boring machines										
Perceuses	1000	163	158	152	160	155	145	118	101	...
Lathes										
Tours	1000	150	142	136	144	151	149	127	106	...
Lorries · Camions										
Assembled										
Assemblés	1000	456	408	416	518	581	642	623	648	...
Produced										
Fabriqués	1000	11954	11840	12433	13503	12973	11480	11217	13890	...
Aluminium										
Aluminium	1000 t.	19666	19710	21301	22505	23172	21846	20484	21295	...
Cement										
Ciment	million t.	950	996	1044	1110	1146	1154	1110	1189	...
Electricity [5]	billion									
Electricité [5]	milliard kWh	9747	10055	10587	11117	11483	11788	12013	12130	12261
Fertilizers [6]										
Engrais [6]	million t.	136.2	143.5	152.4	152.2	158.3	152.9	147.6	144.2	138.3
Sugar, raw										
Sucre, brut	million t.	98.4	100.2	103.9	104.7	107.1	110.8	111.9	116.7	110.3
Woodpulp · Pâte de bois										
Sawnwood										
Sciages	million m³	468	484	505	508	507	505	370	377	...
Motor vehicles · Véhicules automobiles										
Passenger										
Tourisme	million	31.99	32.43	32.82	33.85	35.12	35.32	33.76	34.25	...
Commercial										
Utilitaires	million	12.52	12.44	12.90	13.86	13.53	12.44	12.01	12.61	...

Transport · Transports

Motor vehicles in use · Véhicules automobiles en service

		1985	1986	1987	1988	1989	1990	1991	1992	1993
Passenger cars										
Voitures de tourisme	1000	373667	393352	395129	407959	422240	468525	450596	451928	...
Commercial vehicles										
Véhicules utilitaires	1000	115165	120860	124037	129593	133832	139857	140969	141930	...

External trade · Commerce extérieur

Value, billion US$ · Valeur, milliard $E=U

		1985	1986	1987	1988	1989	1990	1991	1992	1993
Imports, c.i.f.										
Importations c.i.f.		2028.2	2199.9	2559.6	2915.0	3161.4	3566.6	3546.1	3804.4	3708.3
Exports, f.o.b.										
Exportations f.o.b.		1926.7	2129.4	2490.8	2827.0	3043.4	3436.9	3421.4	3656.5	3634.6

Quantum: index of exports · Quantum : indice des exportations

		1985	1986	1987	1988	1989	1990	1991	1992	1993
All commodities										
Tous produits	1990	74	78	82	88	95	100	104	112	116
Manufactures										
Produits manufacturés	1990	71	74	80	93	92	100	100	103	99

Unit value: index of exports [7] · Valeur unitaire : indice des exportations [7]

		1985	1986	1987	1988	1989	1990	1991	1992	1993
All commodities										
Tous produits	1990	73	77	85	91	91	100	98	99	93
Manufactures										
Produits manufacturés	1990	66	77	86	93	92	100	100	103	98

Primary commodities: price indexes [78] · Produits de base : indices des prix [78]

		1985	1986	1987	1988	1989	1990	1991	1992	1993
All commodities										
Tous produits	1980	85	63	66	70	72	79	69	68	62
Food										
Produits alimentaires	1980	66	74	75	87	87	88	85	87	82
Non−food: of agricultural origin										
Non alimentaires: d'origine agricole	1980	74	75	86	100	98	100	93	91	84
Minerals										
Minéraux	1980	93	57	59	59	62	71	59	57	51

1

Selected series of world statistics
Population, production, external trade and finance [*cont.*]
Séries principales de statistiques mondiales
Population, production, commerce extérieur et finaces [*suite*]

Series Séries	Unit or base Unité ou base	1985	1986	1987	1988	1989	1990	1991	1992	1993

Finance • Finances

International reserves minus gold, billion US $ [9] • Réserves internationales moins l'or, milliard $E−U [9]

All countries										
Tous les pays		405.2	419.0	507.8	542.7	590.9	637.6	671.7	693.1	757.2
Position in IMF										
Disponibilité au FMI		38.7	35.3	31.5	28.3	25.5	23.8	25.9	33.9	32.8
Foreign exchange										
Devise étrangères		348.3	364.1	456.1	494.2	545.0	593.5	625.3	646.3	710.1
SDR (special drawing rights)										
DTS (droits de tirage spéciaux)		18.2	19.5	20.2	20.2	20.5	20.4	20.6	12.9	14.6

Source:
Statistical Division of the United Nations Secretariat (New York),
Food and Agriculture Oganization of the United Nations (Rome),
Motor Vehicle Manufacturers' Association (Detroit) and
International Monetary Fund (Washington, DC).

1 Annual data: mid−year estimates.
2 Excluding Albania, China, Democratic People's Republic of
   Korea, Viet Nam, former Czechoslovakia, former USSR and
   Yugoslavia.
3 Excluding peat for fuel.
4 Excluding natural gas liquids.
5 Electricity (hydro, geothermal, thermal, nuclear) generated by
   establishments for public or private use.
6 1 July − 30 June.
7 Indexes computed in US dollars.
8 Export price indexes.
9 End of period.

Source:
Division de statistique du Sécretariat des Nations Unies (New York),
Organisation des Nations Unies pour l'alimentation et l'agriculture
(Rome), "Motor Vehicles Manufacturers' Association" (Detroit) et
Fonds monétaire internationale (Washington, DC).

1 Données annuelles : estimations au milieu de l'année.
2 Non compris l'Albanie, la Chine, la République populaire
   démocratique de Corée, le Viet Nam, l'ancienne Tchèchoslovaquie,
   l'ancienne URSS et la Yougoslavie.
3 Non compris la tourbe combustible.
4 Non compris le gaz naturel liquéfié.
5 L'électricité (hydraulique, géothermique, thermique, nucléaire)
   produite par des entreprises d'utilisation publique ou privée.
6 1 juillet − 30 juin.
7 Indice calculé en dollars des Etats−Unis.
8 Indice des prix à l'exportation.
9 Fin de la période.

2
Population, rate of increase, birth and death rates, surface area and density
Population, taux d'accroissement, taux de natalité et taux de mortalité, superficie et densité

Macro regions and regions / Grandes régions et régions	Mid–year estimates (millions) / Estimations au milieu de l'année (millions)								Annual rate of increase Taux d'accroissement annuel % 1990–95	Birth rate Taux de natalité (0/000) 1990–95	Death rate Taux de mortalité (0/000) 1990–95	Surface area (km²) Superficie (km²) (000's) 1993	Density[1] Densité[1] 1993
	1950	1960	1970	1975	1980	1985	1990	1993					
World Monde	2520	3021	3697	4077	4444	4846	5285	5544	1.6	25	9	135641	41
Africa Afrique	224	282	364	414	476	549	633	689	2.8	42	14	30306	23
Eastern Africa Afrique orientale	66	83	110	125	145	168	196	214	3.0	46	16	6356	34
Middle Africa Afrique centrale	26	32	40	45	52	61	70	77	3.1	46	15	6613	12
Northern Africa Afrique septentrionale	53	67	85	96	110	126	143	154	2.3	31	9	8525	18
Southern Africa Afrique méridionale	16	20	25	29	33	38	41	45	2.3	32	9	2675	17
Western Africa Afrique occidentale	63	80	104	118	135	156	181	198	3.0	46	16	6138	32
Northern America[2] Amérique septentrionale[2]	166	199	226	239	252	265	278	287	1.0	16	9	21517	13
Latin America Amérique latine	166	217	283	320	358	398	440	465	1.8	26	7	20533	23
Caribbean Caraïbes	17	20	25	27	29	31	34	35	1.3	24	8	235	149
Central America Amérique centrale	37	49	67	78	89	110	113	121	2.2	30	6	2480	49
South America Amérique du Sud	112	147	191	214	240	267	293	300	1.7	25	7	17819	17
Asia[3] Asie[3]	1403	1703	2147	2406	2642	2904	3186	3350	1.6	25	8	31764	105
Eastern Asia Asie orientale	671	792	987	1097	1179	1259	1352	1397	1.0	18	7	11762	119
South Central Asia[4] Asie central méridionale[4]	499	621	788	886	990	1113	1243	1325	2.1	31	10	10776	123
South Eastern Asia Asie mériodionale orientale	182	225	287	324	360	401	442	467	1.8	27	8	4495	104
Western Asia[3] Asie occidentale[3]	50	66	86	99	113	131	149	160	2.4	32	7	4731	34
Europe[3] Europe[3]	549	605	656	676	693	706	722	726	0.2	12	11	22986	32
Eastern Europe Europe orientale	221	254	276	286	295	303	310	310	-0.1	12	12	18813	16
Northern Europe Europe septentrionale	78	82	87	89	90	91	92	93	0.3	14	11	1749	53
Southern Europe Europe mériodionale	109	118	128	132	138	141	143	144	0.1	11	10	1316	109
Western Europe Europe occidentale	141	152	165	169	170	172	176	179	0.6	12	11	1107	162
Oceania[2] Océanie[2]	12.6	15.7	19.3	21.4	22.7	24.5	26.4	27.7	1.5	19	8	8537	3
Australia and New Zealand Australie et Nouvelle Zélande	10.1	12.6	15.4	17.0	17.7	18.9	20.2	21.1	1.4	15	8	7984	3
Melanesia Mélanésie	2.1	2.6	3.3	3.7	4.2	4.7	5.2	5.6	2.2	32	9	541	10
Micronesia Micronésie	0.2	0.2	0.2	0.3	0.3	0.4	0.4	0.5	2.3	33	6	3	167
Polynesia Polynésie	0.2	0.3	0.4	0.4	0.5	0.5	0.5	0.6	1.5	31	6	9	67

2

Population, rate of increase, birth and death rates, surface area and density [*cont.*]

Population, taux d'accroissement, taux de natalité et taux de mortalité, superficie et densité [*suite*]

Source:
Demographic statistics database of the Statistical Division of the United Nations Secretariat.

1   Population per square kilometre of surface area. Figures are merely the quotients of population divided by surface area and are not to be considered as either reflecting density in the urban sense or as indicating the supporting power of a territory's land and resources.
2   Hawaii, a state of the United States of America, is included in Northern America rather than Oceania.
3   The European portion of Turkey is included in Western Asia rather than Europe.
4   Central Asia and Southern Asia.

Source:
Base de données pour les statistiques demographique du Bureau de statistique du Secrétariat de l'ONU.

1   Habitants per kilomètre carré. Il s'agit simplement du quotient calculé en divisasnt la population par la superficie et n'est pas considéré comme indiquant la densité au sens urbain du mot ni l'effectif de population que les terres et les ressources du territoire sont capables de nourrir.
2   Hawaii, un Etat des Etats—Unis d'Amérique, est compris en Amérique. septentrionale plutôt qu'en Océanie.
3   La partie européenne de la Turquie est comprise en Asie Occidentale plutôt qu'en Europe.
4   Asie centrale et Asie méridionale.

3
Index numbers of total agricultural and food production
Indices de la production agricole totale et de la production alimentaire

1979-1981 = 100

Region Région	1984	1985	1986	1987	1988	1989	1990	1991	1992	1993
A. Total agricultural production · Production agricole totale										
World *Monde*	**111**	**114**	**115**	**116**	**118**	**123**	**126**	**127**	**130**	**128**
Africa Afrique	102	111	116	115	122	126	128	134	131	135
North America Amérique du Nord	103	108	104	102	96	104	109	109	115	105
South America Amérique du sud	108	115	111	118	123	125	126	129	133	132
Asia Asie	120	123	126	128	135	139	146	150	155	158
Europe Europe	110	108	109	109	109	110	109	110	106	104
Oceania Océanie	106	109	110	109	113	110	112	112	120	118
former USSR† ancienne URSS†	110	109	116	116	117	121	120	106	..	..
B. Food production · Production alimentaire										
World *Monde*	**111**	**114**	**116**	**116**	**118**	**123**	**126**	**126**	**131**	**129**
Africa Afrique	102	111	116	116	123	127	129	135	132	137
North America Amérique du Nord	103	109	105	103	96	106	109	109	116	106
South America Amérique du sud	108	115	113	118	125	128	129	132	137	136
Asia Asie	119	122	126	128	134	139	146	149	155	158
Europe Europe	110	107	109	109	109	110	109	110	106	105
Oceania Océanie	106	107	108	106	110	109	111	110	120	120
former USSR† ancienne URSS†	111	110	117	118	118	123	122	108	..	..

Source:
Food and Agriculture Organization of the United Nations (Rome).

Source:
Organisation des Nations Unies pour l'alimentation et l'agriculture (Rome).

† For detailed descriptions of data pertaining to former Czechoslovakia, Germany, SFR Yugoslavia and former USSR, see Annex I - Country or area nomenclature, regional and other groupings.

† Pour les descriptions en détails des données relatives à l'ancienne Tchécoslovaquie, l'Allemagne, la Rfs Yougoslavie et l'ancienne URSS, voir l'Annexe I - Nomenclature des pays ou zones, groupements régionaux et autres groupements.

4

Index numbers of per capita total agricultural and food production
Indices de la production agricole totale et de la production alimentaire par habitant

1979-1981 = 100

Region Région	1984	1985	1986	1987	1988	1989	1990	1991	1992	1993
A. Per capita total agricultural production • Production agricole totale par habitant										
World *Monde*	**104**	**105**	**104**	**103**	**103**	**105**	**106**	**105**	**106**	**103**
Africa Afrique	91	96	97	94	97	97	95	97	92	92
North America Amérique du Nord	97	101	96	93	87	92	95	94	98	88
South America Amérique du sud	99	103	98	102	104	104	103	104	105	102
Asia Asie	112	112	112	112	116	117	121	122	124	124
Europe Europe	109	106	107	107	106	107	106	106	104	102
Oceania Océanie	100	101	100	98	100	95	96	95	99	96
former USSR† ancienne URSS†	106	105	110	109	108	111	110	97	..	..
B. Per capita food production • Production alimentaire par habitant										
World *Monde*	**104**	**104**	**104**	**103**	**103**	**105**	**106**	**104**	**106**	**103**
Africa Afrique	91	96	98	95	98	98	96	98	93	93
North America Amérique du Nord	98	102	97	94	86	94	95	94	99	89
South America Amérique du sud	100	103	99	102	106	106	105	106	108	106
Asia Asie	111	111	113	112	115	117	121	121	124	124
Europe Europe	109	106	107	107	106	107	106	106	104	102
Oceania Océanie	100	100	99	95	98	95	95	93	99	98
former USSR† ancienne URSS†	107	105	111	111	110	114	112	98	..	..

Source:
Food and Agriculture Organization of the United Nations (Rome).

† For detailed descriptions of data pertaining to former Czechoslovakia, Germany, SFR Yugoslavia and former USSR, see Annex I - Country or area nomenclature, regional and other groupings.

Source:
Organisation des Nations Unies pour l'alimentation et l'agriculture (Rome).

† Pour les descriptions en détails des données relatives à l'ancienne Tchécoslovaquie, l'Allemagne, la Rfs Yougoslavie et l'ancienne URSS, voir l'Annexe I - Nomenclature des pays ou zones, groupements régionaux et autres groupements.

5
# Index numbers of industrial production: world and regions
# Indices de la production industrielle : monde et régions
1980=100

Regions and industry [ISIC] Régions et industrie [CITI]	Weight(%) Pond.(%)	1985	1986	1987	1988	1989	1990	1991	1992	1993
World · Monde										
Total industry [2–4] **Total, industrie [2–4]**	**100.0**	**107**	**109**	**113**	**120**	**124**	**124**	**122**	**122**	**124**
Total mining [2] **Total, industries extractives [2]**	**16.9**	**87**	**89**	**90**	**96**	**100**	**100**	**97**	**101**	**107**
Coal Houille	1.7	106	107	106	107	107	98	97	93	90
Petroleum, gas Pétrole, gaz	12.6	80	82	83	88	93	93	90	96	104
Metal Minerais métalliques	1.4	110	113	116	128	134	137	137	138	135
Total manufacturing [3] **Total, industries manufacturières [3]**	**76.3**	**110**	**113**	**117**	**124**	**128**	**128**	**126**	**125**	**125**
Light industry Industrie légère	26.6	107	110	114	118	120	120	119	119	118
Heavy industry Industrie lourde	49.7	112	114	119	127	132	132	130	128	129
Selected manufacturing **Industries manufacturières déterminées**										
Food, beverages, tobacco Industries alimentaires, boissons, tabac	9.6	101	113	115	119	121	123	125	126	126
Textiles Textiles	3.9	101	104	107	107	108	103	101	100	98
Apparel, leather, footwear Articles d'habillement, cuir et chaussures	3.0	99	101	102	101	102	96	92	90	88
Wood products, furniture Bois, meubles	3.3	103	107	113	118	119	118	115	116	118
Paper, printing, publishing Papier, imprimerie, édition	5.5	113	119	125	132	135	138	138	138	137
Chemicals and related products Produits chimiques et alliés	11.2	114	117	123	130	134	133	133	136	138
Non–metallic mineral products Produits minéraux non métalliques	3.4	99	102	105	110	114	112	110	110	110
Basic metals Métallurgie de base	6.0	98	96	100	108	110	107	104	101	101
Metal products Ouvrages en métaux	29.2	116	118	123	132	139	139	137	133	134
Electricity, gas, water [4] **Electricité, gaz et eau [4]**	**6.5**	**118**	**120**	**125**	**130**	**135**	**138**	**143**	**145**	**150**

A. Developed regions · Régions developpées

	Weight(%) Pond.(%)	1985	1986	1987	1988	1989	1990	1991	1992	1993
Total industry [2–4] **Total, industrie [2–4]**	**100.0**	**109**	**110**	**114**	**120**	**124**	**123**	**121**	**119**	**119**
Total mining [2] **Total, industries extractives [2]**	**8.8**	**106**	**103**	**105**	**107**	**107**	**106**	**106**	**106**	**107**
Coal Houille	2.0	104	104	104	104	103	94	91	85	81
Petroleum, gas Pétrole, gaz	4.6	107	101	103	101	98	99	102	106	111
Metal Minerais métalliques	1.2	111	114	120	135	144	148	148	148	147
Total manufacturing [3] **Total, industries manufacturières [3]**	**83.7**	**109**	**111**	**114**	**121**	**125**	**124**	**121**	**119**	**118**
Light industry Industrie légère	27.4	105	107	111	114	117	115	113	113	111
Heavy industry Industrie lourde	56.3	111	112	116	124	130	128	125	122	122
Selected manufacturing **Industries manufacturières déterminées**										
Food, beverages, tobacco Industries alimentaires, boissons, tabac	9.5	106	108	111	114	116	116	117	117	116

5

Index numbers of industrial production: world and regions [*cont.*]

Indices de la production industrielle : monde et régions [*suite*]

1980=100

Region̄s and industry [ISIC] Région̄s et industrie [CITI]	Weight(%) Pond.(%)	1985	1986	1987	1988	1989	1990	1991	1992	1993
Textiles Textiles	3.6	98	99	101	101	103	95	91	89	86
Apparel, leather, footwear Articles d'habillement, cuir et chaussures	3.1	96	97	96	95	95	88	83	80	77
Wood products, furniture Bois, meubles	3.7	102	106	112	117	118	116	112	113	114
Paper, printing, publishing Papier, imprimerie, édition	6.3	111	117	123	129	133	135	135	134	132
Chemicals and related products Produits chimiques et alliés	11.5	111	113	118	125	129	126	125	127	127
Non−metallic mineral products Produits minéraux non métalliques	3.6	96	98	100	106	109	106	101	99	99
Basic metals Métallurgie de base	6.8	96	93	96	104	106	103	98	95	94
Metal products Ouvrages en métaux	34.4	116	117	121	131	137	136	133	129	128
Electricity, gas, water [4] **Electricité, gaz et eau [4]**	**7.5**	**115**	**116**	**120**	**124**	**128**	**130**	**134**	**135**	**139**

North America · Amérique du Nord

	Weight(%)	1985	1986	1987	1988	1989	1990	1991	1992	1993
Total industry [2−4] **Total, industrie [2−4]**	**100.0**	**110**	**111**	**116**	**123**	**125**	**126**	**123**	**125**	**130**
Total mining [2] **Total, industries extractives [2]**	**12.4**	**97**	**90**	**92**	**96**	**95**	**96**	**96**	**94**	**93**
Coal Houille	1.6	110	109	114	120	121	129	125	119	118
Petroleum, gas Pétrole, gaz	8.7	92	82	82	83	80	81	81	80	81
Metal Minerais métalliques	1.2	107	107	119	133	139	140	142	144	139
Total manufacturing [3] **Total, industries manufacturières [3]**	**79.4**	**113**	**115**	**120**	**128**	**131**	**132**	**128**	**130**	**136**
Light industry Industrie légère	24.1	114	118	124	128	131	132	129	130	128
Heavy industry Industrie lourde	55.3	112	114	119	127	131	132	127	130	139
Selected manufacturing **Industries manufacturières déterminées**										
Food, beverages, tobacco Industries alimentaires, boissons, tabac	7.5	110	112	115	118	120	122	123	123	121
Textiles Textiles	2.4	99	106	112	112	114	112	112	117	119
Apparel, leather, footwear Articles d'habillement, cuir et chaussures	2.6	91	94	97	98	100	95	90	90	87
Wood products, furniture Bois, meubles	3.9	128	136	145	150	150	147	136	141	148
Paper, printing, publishing Papier, imprimerie, édition	8.0	118	126	134	138	142	145	144	142	136
Chemicals and related products Produits chimiques et alliés	10.6	114	119	126	133	137	139	138	144	145
Non−metallic mineral products Produits minéraux non métalliques	2.6	108	111	113	116	116	113	101	103	109
Basic metals Métallurgie de base	6.1	91	85	92	102	100	99	92	94	99
Metal products Ouvrages en métaux	34.5	117	118	122	133	138	139	134	136	149
Electricity, gas, water [4] **Electricité, gaz et eau [4]**	**8.2**	**109**	**107**	**109**	**115**	**117**	**117**	**117**	**118**	**125**

Europe · Europe

	Weight(%)	1985	1986	1987	1988	1989	1990	1991	1992	1993
Total industry [2−4] **Total, industrie [2−4]**	**100.0**	**106**	**108**	**110**	**115**	**118**	**114**	**111**	**109**	**107**

5

Index numbers of industrial production: world and regions [cont.]

Indices de la production industrielle : monde et régions [suite]

1980=100

Region s and industry [ISIC] Région s et industrie [CITI]	Weight(%) Pond.(%)	1985	1986	1987	1988	1989	1990	1991	1992	1993
Total mining [2] **Total, industries extractives [2]**	**7.0**	**115**	**113**	**115**	**111**	**110**	**104**	**105**	**108**	**110**
Coal Houille	2.6	100	101	97	94	93	76	73	68	60
Petroleum, gas Pétrole, gaz	3.3	132	129	135	129	126	127	135	145	159
Metal Minerais métalliques	0.3	91	84	78	78	80	79	71	68	59
Total manufacturing [3] **Total, industries manufacturières [3]**	**86.2**	**104**	**106**	**109**	**114**	**118**	**113**	**110**	**107**	**104**
Light industry Industrie légère	30.2	101	103	106	109	112	108	106	105	103
Heavy industry Industrie lourde	55.9	106	108	110	116	121	116	112	108	104
Selected manufacturing **Industries manufacturières déterminées**										
Food, beverages, tobacco Industries alimentaires, boissons, tabac	10.6	106	108	110	113	115	114	115	115	115
Textiles Textiles	4.4	97	98	99	100	102	90	85	81	77
Apparel, leather, footwear Articles d'habillement, cuir et chaussures	3.8	98	98	95	93	92	84	78	74	71
Wood products, furniture Bois, meubles	3.7	90	92	97	102	106	102	103	103	101
Paper, printing, publishing Papier, imprimerie, édition	5.2	106	110	116	122	127	128	127	127	127
Chemicals and related products Produits chimiques et alliés	12.7	109	111	115	122	126	118	116	117	116
Non−metallic mineral products Produits minéraux non métalliques	4.1	91	94	96	102	106	102	98	96	93
Basic metals Métallurgie de base	5.9	98	96	96	103	105	97	92	88	83
Metal products Ouvrages en métaux	34.1	108	110	113	119	125	120	115	110	104
Electricity, gas, water [4] **Electricité, gaz et eau [4]**	**6.8**	**116**	**119**	**124**	**127**	**130**	**131**	**137**	**138**	**139**

European Union [+] • Union européenne [+]

	Weight(%) Pond.(%)	1985	1986	1987	1988	1989	1990	1991	1992	1993
Total industry [2−4] **Total, industrie [2−4]**	**100.0**	**104**	**106**	**108**	**112**	**116**	**113**	**112**	**110**	**106**
Total mining [2] **Total, industries extractives [2]**	**6.9**	**113**	**111**	**110**	**104**	**96**	**90**	**90**	**91**	**92**
Coal Houille	2.7	99	100	95	92	90	76	72	67	59
Petroleum, gas Pétrole, gaz	3.2	132	126	130	118	102	101	105	111	123
Metal Minerais métalliques	0.2	87	75	64	64	71	72	69	66	52
Total manufacturing [3] **Total, industries manufacturières [3]**	**86.0**	**103**	**104**	**107**	**112**	**117**	**113**	**112**	**110**	**105**
Light industry Industrie légère	29.5	101	103	105	109	112	110	109	108	106
Heavy industry Industrie lourde	56.5	104	105	107	114	120	115	113	111	105
Selected manufacturing **Industries manufacturières déterminées**										
Food, beverages, tobacco Industries alimentaires, boissons, tabac	10.6	107	108	111	114	117	119	122	122	121
Textiles Textiles	4.2	98	98	99	99	101	93	90	87	81
Apparel, leather, footwear Articles d'habillement, cuir et chaussures	3.9	96	95	92	89	88	82	77	73	71

5
Index numbers of industrial production: world and regions [*cont.*]
Indices de la production industrielle : monde et régions [*suite*]
1980=100

Region s and industry [ISIC] Région s et industrie [CITI]	Weight(%) Pond.(%)	1985	1986	1987	1988	1989	1990	1991	1992	1993
Wood products, furniture										
Bois, meubles	3.4	87	89	94	100	103	100	101	100	98
Paper, printing, publishing										
Papier, imprimerie, édition	5.0	102	106	112	119	124	126	126	126	125
Chemicals and related products										
Produits chimiques et alliés	13.0	108	110	113	120	124	118	117	118	116
Non−metallic mineral products										
Produits minéraux non métalliques	4.2	89	91	93	99	104	102	99	98	95
Basic metals										
Métallurgie de base	5.9	95	92	93	100	103	97	94	90	85
Metal products										
Ouvrages en métaux	34.6	106	108	110	116	124	119	117	113	104
Electricity, gas, water [4]										
Electricité, gaz et eau [4]	**7.1**	**114**	**118**	**122**	**124**	**128**	**130**	**136**	**136**	**138**

European Free Trade Association [+]
Association européenne de libre échange [+]

Total industry [2−4]										
Total, industrie [2−4]	**100.0**	**112**	**114**	**118**	**123**	**130**	**134**	**134**	**136**	**139**
Total mining [2]										
Total, industries extractives [2]	**7.6**	**133**	**141**	**155**	**169**	**211**	**222**	**244**	**270**	**286**
Coal										
Houille	0.1	200	190	171	122	165	155	165	173	127
Petroleum, gas										
Pétrole, gaz	6.4	137	146	162	179	229	241	268	298	318
Metal										
Minerais métalliques	0.5	106	104	106	104	101	99	94	93	95
Total manufacturing [3]										
Total, industries manufacturières [3]	**84.0**	**109**	**110**	**113**	**118**	**122**	**125**	**123**	**122**	**123**
Light industry										
Industrie légère	30.9	103	105	106	108	110	112	110	108	108
Heavy industry										
Industrie lourde	53.1	112	114	117	123	128	133	130	130	132
Selected manufacturing										
Industries manufacturières déterminées										
Food, beverages, tobacco										
Industries alimentaires, boissons, tabac	10.4	105	106	107	109	112	115	118	117	119
Textiles										
Textiles	2.7	93	93	91	91	90	91	87	84	78
Apparel, leather, footwear										
Articles d'habillement, cuir et chaussures	2.9	92	92	86	79	74	72	69	62	57
Wood products, furniture										
Bois, meubles	6.9	94	97	100	103	109	112	107	104	104
Paper, printing, publishing										
Papier, imprimerie, édition	11.4	117	120	127	132	134	135	133	134	137
Chemicals and related products										
Produits chimiques et alliés	9.8	120	119	126	136	144	147	145	148	154
Non−metallic mineral products										
Produits minéraux non métalliques	3.5	104	104	109	109	112	118	112	109	104
Basic metals										
Métallurgie de base	5.0	111	108	111	120	122	120	115	116	117
Metal products										
Ouvrages en métaux	30.4	111	114	114	120	125	132	129	127	128
Electricity, gas, water [4]										
Electricité, gaz et eau [4]	**8.4**	**125**	**125**	**136**	**139**	**142**	**147**	**151**	**156**	**163**

B. Developing regions · Régions en voie de développement

Total industry [2−4]										
Total, industrie [2−4]	**100.0**	**98**	**106**	**110**	**117**	**123**	**127**	**128**	**134**	**142**
Total mining [2]										
Total, industries extractives [2]	**48.4**	**74**	**80**	**80**	**88**	**96**	**96**	**91**	**98**	**107**
Coal										
Houille	0.4	140	149	157	163	176	182	203	226	247

5

Index numbers of industrial production: world and regions [*cont.*]
Indices de la production industrielle : monde et régions [*suite*]
1980=100

Regions and industry [ISIC] Régions et industrie [CITI]	Weight(%) Pond.(%)	1985	1986	1987	1988	1989	1990	1991	1992	1993
Petroleum, gas Pétrole, gaz	43.7	69	75	76	83	91	91	86	92	102
Metal Minerais métalliques	2.6	110	111	109	116	116	116	118	120	115
Total manufacturing [3] **Total, industries manufacturières [3]**	**47.8**	**120**	**128**	**136**	**142**	**146**	**153**	**158**	**164**	**171**
Light industry Industrie légère	23.1	118	124	128	132	136	140	145	149	153
Heavy industry Industrie lourde	24.7	121	132	144	151	156	165	171	178	188
Selected manufacturing **Industries manufacturières déterminées**										
Food, beverages, tobacco Industries alimentaires, boissons, tabac	10.3	126	128	131	136	141	147	152	158	162
Textiles Textiles	4.7	113	120	125	123	124	127	131	133	136
Apparel, leather, footwear Articles d'habillement, cuir et chaussures	2.3	113	121	127	131	137	136	137	138	139
Wood products, furniture Bois, meubles	1.8	108	113	117	122	126	132	139	142	146
Paper, printing, publishing Papier, imprimerie, édition	2.3	133	142	152	158	163	172	180	186	195
Chemicals and related products Produits chimiques et alliés	9.7	127	136	147	155	158	165	167	176	185
Non−metallic mineral products Produits minéraux non métalliques	2.8	113	118	127	130	138	142	152	159	167
Basic metals Métallurgie de base	3.0	115	125	136	139	143	149	151	154	164
Metal products Ouvrages en métaux	9.7	119	134	148	157	162	175	186	192	204
Electricity, gas, water [4] **Electricité, gaz et eau [4]**	**3.8**	**142**	**150**	**163**	**176**	**187**	**199**	**211**	**221**	**234**

Latin America and Caribbean • Amérique latine et Caraïbes

	Weight(%) Pond.(%)	1985	1986	1987	1988	1989	1990	1991	1992	1993
Total industry [2−4] **Total, industrie [2−4]**	**100.0**	**104**	**109**	**112**	**113**	**114**	**115**	**118**	**118**	**121**
Total mining [2] **Total, industries extractives [2]**	**16.8**	**115**	**114**	**113**	**119**	**121**	**125**	**124**	**125**	**126**
Coal Houille	0.2	146	142	141	171	164	156	154	132	125
Petroleum, gas Pétrole, gaz	12.0	116	114	112	118	119	123	122	123	124
Metal Minerais métalliques	3.4	120	120	126	132	136	142	148	153	162
Total manufacturing [3] **Total, industries manufacturières [3]**	**77.9**	**101**	**106**	**110**	**109**	**110**	**109**	**112**	**113**	**116**
Light industry Industrie légère	36.8	105	110	111	111	112	114	117	119	119
Heavy industry Industrie lourde	41.1	96	103	108	107	108	105	108	108	112
Selected manufacturing **Industries manufacturières déterminées**										
Food, beverages, tobacco Industries alimentaires, boissons, tabac	18.2	110	113	114	114	116	120	124	126	125
Textiles Textiles	5.2	94	102	102	100	99	96	98	96	95
Apparel, leather, footwear Articles d'habillement, cuir et chaussures	3.7	96	98	94	93	95	91	90	90	92
Wood products, furniture Bois, meubles	3.0	98	102	104	104	104	107	117	123	126
Paper, printing, publishing Papier, imprimerie, édition	4.4	126	133	139	142	141	145	152	152	160

5

Index numbers of industrial production: world and regions [cont.]

Indices de la production industrielle : monde et régions [suite]

1980=100

Regions and industry [ISIC] Régions et industrie [CITI]	Weight(%) Pond.(%)	1985	1986	1987	1988	1989	1990	1991	1992	1993
Chemicals and related products										
Produits chimiques et alliés	14.0	108	114	120	118	118	119	121	122	125
Non−metallic mineral products										
Produits minéraux non métalliques	4.8	88	94	101	97	98	94	99	101	104
Basic metals										
Métallurgie de base	5.6	94	101	107	107	110	106	104	105	109
Metal products										
Ouvrages en métaux	17.2	88	97	100	99	100	95	98	96	102
Electricity, gas, water [4]										
Electricité, gaz et eau [4]	**5.3**	**128**	**137**	**148**	**157**	**163**	**168**	**176**	**177**	**183**

Asia · Asie

	Weight(%) Pond.(%)	1985	1986	1987	1988	1989	1990	1991	1992	1993
Total industry [2−4]										
Total, industrie [2−4]	**100.0**	**107**	**112**	**118**	**130**	**138**	**144**	**146**	**147**	**152**
Total mining [2]										
Total, industries extractives [2]	**32.5**	**64**	**73**	**75**	**85**	**94**	**92**	**84**	**93**	**105**
Coal										
Houille	0.5	124	131	132	130	140	142	160	181	198
Petroleum, gas										
Pétrole, gaz	30.0	61	70	71	80	90	87	77	86	99
Metal										
Minerais métalliques	0.7	82	83	85	91	97	91	89	87	78
Total manufacturing [3]										
Total, industries manufacturières [3]	**61.7**	**127**	**131**	**138**	**151**	**160**	**170**	**175**	**172**	**173**
Light industry										
Industrie légère	20.4	114	118	122	129	133	139	142	145	147
Heavy industry										
Industrie lourde	41.3	134	137	146	162	172	185	192	186	185
Selected manufacturing										
Industries manufacturières déterminées										
Food, beverages, tobacco										
Industries alimentaires, boissons, tabac	6.6	125	126	131	140	145	151	157	162	166
Textiles										
Textiles	4.2	111	114	119	118	120	123	126	128	128
Apparel, leather, footwear										
Articles d'habillement, cuir et chaussures	1.7	119	129	141	144	149	149	152	152	151
Wood products, furniture										
Bois, meubles	2.1	92	94	99	103	105	109	109	108	107
Paper, printing, publishing [1]										
Papier, imprimerie, édition [1]	4.0	120	126	136	145	156	165	170	170	170
Chemicals and related products										
Produits chimiques et alliés	9.5	126	132	142	155	161	171	174	181	187
Non−metallic mineral products										
Produits minéraux non métalliques	2.9	114	114	120	130	138	147	155	156	159
Basic metals										
Métallurgie de base	6.8	108	107	112	121	124	131	134	127	129
Metal products										
Ouvrages en métaux	22.7	146	151	161	182	195	210	220	208	204
Electricity, gas, water [4]										
Electricité, gaz et eau [4]	**5.8**	**130**	**133**	**141**	**151**	**160**	**173**	**183**	**190**	**198**

Oceania [2] · Océanie [2]

	Weight(%) Pond.(%)	1985	1986	1987	1988	1989	1990	1991	1992	1993
Total industry [2−4]										
Total, industrie [2−4]	**100.0**	**113**	**118**	**126**	**137**	**142**	**143**	**144**	**147**	**156**
Total mining [2]										
Total, industries extractives [2]	**15.9**	**156**	**179**	**200**	**237**	**269**	**287**	**299**	**303**	**309**
Coal										
Houille	4.7	112	114	136	142	154	153	158	146	160
Petroleum, gas										
Pétrole, gaz	2.5	139	138	137	124	140	142	140	151	157
Metal										
Minerais métalliques	7.3	202	251	282	363	418	458	481	494	496

5

Index numbers of industrial production: world and regions [*cont.*]

Indices de la production industrielle : monde et régions [*suite*]

1980=100

Region s and industry [ISIC] Région s et industrie [CITI]	Weight(%) Pond.(%)	1985	1986	1987	1988	1989	1990	1991	1992	1993
Total manufacturing [3]										
Total, industries manufacturières [3]	**73.2**	**102**	**103**	**108**	**115**	**113**	**110**	**108**	**112**	**121**
Light industry										
Industrie légère	31.9	103	104	109	114	110	108	106	109	117
Heavy industry										
Industrie lourde	41.3	101	102	108	115	115	113	110	114	125
Selected manufacturing										
Industries manufacturières déterminées										
Food, beverages, tobacco										
Industries alimentaires, boissons, tabac	13.7	101	104	109	111	112	114	117	126	136
Textiles										
Textiles	2.8	107	112	111	113	108	109	113	101	85
Apparel, leather, footwear										
Articles d'habillement, cuir et chaussures	3.1	116	107	108	111	99	92	86	88	92
Wood products, furniture										
Bois, meubles	4.2	111	101	113	118	106	96	93	88	92
Paper, printing, publishing										
Papier, imprimerie, édition	6.8	112	116	124	133	125	121	116	111	127
Chemicals and related products										
Produits chimiques et alliés	8.8	92	99	103	112	110	112	108	112	124
Non-metallic mineral products										
Produits minéraux non métalliques	3.5	111	104	108	117	114	102	96	105	116
Basic metals										
Métallurgie de base	8.3	105	106	110	116	128	132	133	133	141
Metal products										
Ouvrages en métaux	21.3	95	96	103	111	108	103	98	105	117
Electricity, gas, water [4]										
Electricité, gaz et eau [4]	**10.9**	**127**	**131**	**136**	**143**	**149**	**152**	**155**	**158**	**160**

Source:
Industrial statistics database of the Statistical Division
of the United Nations Secretariat.

Source :
Base de données de statistiques industrielles de la Division de statistique
du Secrétariat de l'ONU.

+   For Member States of this grouping, see Annex I —
    Other groupings.

+   Les Etats membres de ce groupement, voir annexe I —
    Autres groupements.

§   All series exclude Albania, China, Democratic
    People's Republic of Korea, Viet Nam, former Czechoslovakia,
    former USSR and Yugoslavia. Series for "Developed regions"
    include North America (Canada and the United States), Europe
    (excluding former Czechoslovakia and the European countries
    of the former USSR), Australia, Israel, Japan, New Zealand
    and South Africa. Series for "Developing regions" exclude
    Australia, Israel, Japan, New Zealand, South Africa and the Asian
    countries of the former USSR.

§   Aucune série ne comprend l'Albanie, la Chine,
    la République populaire démocratique de Corée, Viet Nam,
    l'ancienne Tchécoslovaquie, l'ancienne URSS et la Yougoslavie.
    Les "régions developpées" comprennent l'Amérique du nord
    (le Canada et les Etats-Unis), l'Europe (non compris l'ancienne
    Tchécoslovaquie et les pays européens de l'ancienne URSS),
    l'Australie, Israël, le Japon, la Nouvelle-Zélande et l'Afrique du sud.
    Les séries pour les "Régions en voie de développement" excluent
    l'Australie, Israël, le Japon, la Nouvelle-Zélande, l'Afrique du sud
    et les pays asiatiques de l'ancienne URSS.

1   Excluding printing and publishing.
2   Including Australia and New Zealand.

1   Non compris l'imprimerie et l'édition.
2   Y compris l'Australie et la Nouvelle-Zélande.

Table 6 follows overleaf
Le tableau 6 est présenté au verso

6
Production, trade and consumption of commercial energy
Production, commerce et consommation d'énergie commerciale
Thousand metric tons of coal equivalent and kilograms per capita
Milliers de tonnes métriques d'équivalent houille et kilogrammes par habitant

Regions	Year	Primary energy production – Production d'energie primaire					Changes in stocks Variations des stocks	Imports Importations	Exports Exportations
		Total Totale	Solids Solides	Liquids Liquides	Gas Gaz	Electricity Electricité			
World	1980	9 412 983	2 660 769	4 419 285	1 844 089	488 840	77 232	3 236 884	3 235 353
	1990	11 491 187	3 261 311	4 590 734	2 586 795	1 052 347	150 894	3 557 565	3 529 054
	1991	11 441 822	3 170 596	4 566 206	2 615 320	1 089 700	42 950	3 571 926	3 516 491
	1992	11 484 729	3 210 064	4 577 334	2 587 179	1 110 152	51 830	4 146 267	4 060 238
Africa	1980	577 211	98 232	436 325	35 227	7 427	6 277	69 920	425 314
	1990	703 444	139 493	461 565	92 179	10 206	3 120	72 374	460 825
	1991	736 187	142 719	485 575	97 527	10 365	2 114	72 551	488 969
	1992	722 012	139 822	474 804	97 136	10 250	−3 120	76 110	486 577
America, North	1980	2 608 282	659 448	972 674	792 627	183 532	23 245	750 517	347 489
	1990	2 979 788	830 208	963 881	840 951	344 747	58 513	778 512	435 174
	1991	2 998 634	807 006	982 558	844 630	364 440	−2 523	762 465	473 213
	1992	3 001 698	800 716	973 143	863 795	364 045	−13 855	800 890	482 250
America, South	1980	338 030	9 210	260 467	43 759	24 593	−1 650	92 837	161 474
	1990	466 532	27 000	315 789	76 792	46 950	4 450	82 496	204 672
	1991	494 317	30 690	333 255	81 040	49 331	2 370	84 482	226 071
	1992	499 078	28 024	340 230	80 692	50 131	4 254	86 817	224 281
Asia	1980	2 461 006	619 155	1 653 384	115 251	73 217	19 986	761 730	1 447 021
	1990	3 250 565	1 083 702	1 681 705	312 250	172 908	66 013	1 041 979	1 357 143
	1991	3 274 194	1 117 926	1 662 512	313 487	180 269	35 522	1 090 254	1 343 189
	1992	3 741 999	1 262 193	1 828 612	461 716	189 478	32 438	1 279 504	1 574 765
Europe	1980	1 311 800	641 619	200 004	325 052	145 126	31 176	1 518 579	459 435
	1990	1 511 367	549 493	306 734	291 935	363 205	6 700	1 527 573	579 108
	1991	1 485 721	487 561	319 480	308 734	369 947	5 248	1 529 046	583 768
	1992	3 277 814	819 751	922 991	1 046 242	488 830	32 733	1 871 449	1 166 524
Oceania	1980	120 948	70 817	31 283	13 587	5 260	−2 920	31 150	40 408
	1990	227 174	144 702	42 215	33 338	6 918	4 653	29 299	102 770
	1991	235 102	150 721	42 118	34 578	7 685	219	28 543	115 783
	1992	242 128	159 558	37 554	37 597	7 418	−620	31 396	125 841
former USSR †	1980	1 995 707	562 288	865 148	518 586	49 685	1 120	12 064	354 213
	1990	2 352 318	486 714	818 844	939 347	107 413	7 445	25 234	389 362
	1991	2 217 666	433 974	740 707	935 323	107 661	...	4 484	285 498
	1992	..	..	..	..	..		..	..

Source:
Energy statistics database of the Statistical Division of the United Nations Secretariat.

Source:
Base de données pour les statistiques énergétiques de la Division de statistique du Secrétariat de l'ONU.

† For detailed descriptions of data pertaining to former Czechoslovakia, Germany, SFR Yugoslavia and former USSR, see annex I – Country or area nomenclature, regional and other groupings.

† Pour les descriptions en détails des données relatives à l'ancienne Tchécoslovaquie, l'Allemagne, la Rfs Yougoslavie et l'ancienne URSS, voir l'Annexe I – Nomenclature des pays ou zones, groupements régionaux et autres groupements.

| Bunkers – Soutes | | | Consumption – Consommation | | | | | | | |
Air Avion	Sea Maritime	Unallocated Nondistribué	Per capita Par habitant	Total Totale	Solids Solides	Liquids Liquides	Gas Gaz	Electricity Electricité	Année	Régions
50 077	159 062	354 272	1 982	8 773 870	2 650 706	3 796 114	1 838 349	488 701	1980	*Monde*
53 238	137 516	313 435	2 045	10 864 614	3 238 538	4 010 774	2 563 176	1 052 126	1990	
51 447	136 869	314 640	2 032	10 951 350	3 187 080	4 058 886	2 615 118	1 090 267	1991	
54 566	140 681	362 394	2 003	10 961 286	3 226 270	4 028 142	2 595 977	1 110 896	1992	
3 577	9 424	12 530	435	190 008	74 220	84 366	24 019	7 403	1980	Afrique
3 268	5 693	21 018	448	281 894	101 148	120 543	50 152	10 051	1990	
3 162	5 654	19 256	447	289 583	103 928	124 188	51 185	10 282	1991	
3 233	4 975	18 817	432	287 640	101 226	127 544	48 751	10 119	1992	
11 288	55 145	96 541	7 616	2 825 091	571 841	1 273 489	796 230	183 530	1980	Amérique du Nord
2 113	30 995	35 478	7 529	3 196 027	708 008	1 320 649	822 600	344 771	1990	
1 944	32 887	32 795	7 486	3 222 783	703 336	1 304 463	850 240	364 744	1991	
2 015	32 788	32 052	7 483	3 267 338	707 863	1 320 551	874 582	364 342	1992	
1 524	3 159	18 489	1 033	247 870	15 503	163 966	43 826	24 575	1980	Amérique du Sud
1 338	3 599	21 927	1 068	313 041	24 799	164 593	76 873	46 776	1990	
1 383	3 961	19 604	1 090	325 410	26 792	168 594	80 794	49 230	1991	
1 341	3 762	23 490	1 082	328 765	23 919	174 242	80 644	49 961	1992	
10 649	44 305	118 588	607	1 582 189	699 951	695 707	113 133	73 397	1980	Asie
12 015	43 488	154 750	844	2 659 136	1 214 668	950 481	321 167	172 820	1990	
17 080	41 536	163 388	866	2 763 734	1 269 895	996 552	317 040	180 246	1991	
18 791	44 582	200 416	950	3 150 512	1 384 928	1 133 729	440 832	191 024	1992	
20 473	44 331	44 786	4 604	2 230 179	716 175	971 678	395 237	147 088	1980	Europe
31 057	51 878	23 903	4 704	2 346 293	671 736	853 084	453 753	367 720	1990	
24 473	51 101	17 691	4 644	2 332 487	613 376	872 148	474 168	372 794	1991	
25 592	52 812	98 786	5 205	3 772 816	948 076	1 215 105	1 121 603	488 032	1992	
2 565	2 611	2 865	4 697	106 569	40 062	47 727	13 520	5 260	1980	Océanie
3 447	1 771	−5 558	5 653	149 390	56 118	56 745	29 609	6 918	1990	
3 406	1 637	−7 446	5 588	150 048	57 790	56 307	28 266	7 685	1991	
3 594	1 668	−11 167	5 655	154 208	60 260	56 964	29 565	7 418	1992	
..	..	60 473	5 995	1 591 965	532 954	559 180	452 383	47 448	1980	ancienne URSS †
..	..	61 919	6 632	1 918 827	462 061	544 673	809 021	103 072	1990	
..	..	69 353	6 415	1 867 299	411 963	536 627	813 424	105 286	1991	
..	..	..	..	..	..	..	..	..	1992	

7

Total exports and imports: index numbers
Exportations et importations totales: indices

Quantum, unit value and terms of trade (1980 = 100)

Quantum, valeur unitaire et termes de l'échange (1980 =100)

Regions[1] Régions[1]	1975	1985	1986	1987	1988	1989	1990	1991	1992	1993
A. Exports • Exportations										
Quantum indices[2] • Indices du quantum[2]										
Total	77	113	120	127	135	147	154	160	171	183
Developed economies **Econ. développées**	73	120	122	127	136	147	154	158	165	166
North America Amérique du Nord	73	98	100	109	126	135	143	151	160	168
Europe **Europe**	74	124	126	131	138	151	158	160	165	167
EU+ UE+	75	124	126	131	137	151	157	160	166	166
EFTA AELE	74	126	129	134	141	151	158	159	163	171
Africa[3] Afrique[3]	61	110	115	106	108	118	122	127	164	184
Asia Asie	65	141	141	142	147	154	161	165	167	165
Oceania Océanie	77	132	131	141	143	150	159	183	193	199
Developing economies **Econ. en dévelop.**	89	98	115	125	135	146	153	165	185	218
America Amérique	83	127	135	139	130	147	151	165	181	192
Europe[4] Europe[4]	66	112	111	117	112	116	111	...	...	...
Africa Afrique	83	83	97	93	98	101	110	109	121	131
Asia **Asie**	87	93	114	129	146	157	165	181	205	224
Middle East Moyen-Orient	109	52	71	71	84	86	78	75	82	92
Other Asia Autres pays d'Asie	58	146	170	203	232	255	277	317	359	390
Unit value indices in US dollars[5] • Indices de la valeur unitaire en dollars E-U[5]										
Total	56	85	88	98	104	105	115	112	113	108
Developed economies **Econ. développées**	63	85	97	109	116	116	127	125	128	122
North America Amérique du Nord	68	107	106	108	117	121	122	121	120	119
Europe **Europe**	61	77	93	107	112	111	128	124	127	116
EU+ UE+	60	77	93	107	112	111	128	124	128	118
EFTA AELE	62	78	92	107	113	111	128	122	122	108
Africa[3] Afrique[3]	60	67	77	94	97	101	108	107	109	105
Asia Asie	66	95	113	124	138	137	136	146	156	168

7

Total exports and imports: index numbers
Quantum, unit value and terms of trade (1980 = 100) [*cont.*]
Exportations et importations totales: indices
Quantum, valeur unitaire et termes de l'échange (1980 =100) [*suite*]

Regions[1] Régions[1]	1975	1985	1986	1987	1988	1989	1990	1991	1992	1993
Oceania Océanie	65	79	78	87	106	111	112	102	99	97
Developing economies Econ. en dévelop.	**43**	**84**	**67**	**75**	**78**	**81**	**88**	**86**	**85**	**85**
America Amérique	50	78	61	67	81	78	83	75	70	70
Europe[4] Europe[4]	57	106	103	109	126	129	143	...	...	...
Africa Afrique	40	77	52	59	55	60	69	65	64	58
Asia **Asie**	**42**	**88**	**71**	**79**	**80**	**84**	**92**	**91**	**92**	**92**
Middle East Moyen-Orient	36	90	53	59	50	56	71	65	67	65
Other Asia Autres pays d'Asie	57	87	81	88	95	96	99	99	99	100

B. Imports · Importations
Quantum indices[2] · Indices du quantum[2]

	1975	1985	1986	1987	1988	1989	1990	1991	1992	1993
Total	**74**	**114**	**123**	**131**	**143**	**154**	**160**	**167**	**179**	**181**
Developed economies Econ. développées	**74**	**117**	**126**	**135**	**144**	**154**	**160**	**163**	**170**	**170**
North America Amérique du Nord	75	138	150	157	165	163	163	162	176	194
Europe Europe	**72**	**111**	**119**	**129**	**137**	**151**	**159**	**163**	**168**	**163**
EU+ UE+	71	110	119	129	137	151	160	166	172	166
EFTA AELE	78	115	124	131	138	147	152	147	143	142
Africa[3] Afrique[3]	77	69	72	72	86	86	79	79	83	93
Asia Asie	81	110	121	132	152	163	173	179	179	188
Oceania Océanie	82	132	127	128	140	168	162	160	182	189
Developing economies Econ. en dévelop.	**74**	**105**	**114**	**121**	**140**	**152**	**159**	**179**	**209**	**212**
Europe[4] Europe[4]	74	69	74	75	69	76	88	...	...	...

Unit value indices in US dollars[5] · Indices de la valeur unitaire en dollars E-U[5]

	1975	1985	1986	1987	1988	1989	1990	1991	1992	1993
Total	**58**	**86**	**87**	**95**	**100**	**102**	**111**	**109**	**109**	**104**
Developed economies Econ. développées	**58**	**85**	**88**	**97**	**102**	**105**	**115**	**113**	**113**	**107**
North America Amérique du Nord	53	100	97	102	108	116	120	119	118	116
Europe Europe	**59**	**78**	**86**	**97**	**102**	**102**	**116**	**113**	**115**	**105**
EU+ UE+	59	78	86	97	102	102	116	113	115	105

7

Total exports and imports: index numbers
Quantum, unit value and terms of trade (1980 = 100) [*cont.*]
Exportations et importations totales: indices
Quantum, valeur unitaire et termes de l'échange (1980 =100) [*suite*]

Regions[1] Régions[1]	1975	1985	1986	1987	1988	1989	1990	1991	1992	1993
EFTA AELE	58	74	87	100	106	104	118	115	117	105
Africa[3] Afrique[3]	58	81	87	105	109	106	116	120	121	116
Asia Asie	52	84	75	82	88	92	96	95	94	93
Oceania Océanie	62	88	92	101	113	110	114	115	113	113
Developing economies[6] **Econ. en dévelop.**[6]	**58**	**89**	**84**	**90**	**93**	**94**	**101**	**99**	**99**	**99**
Europe[4] Europe[4]	57	117	106	111	126	129	142	...	...	...

C. Terms of trade [7] • Termes de l'échange[7]

	1975	1985	1986	1987	1988	1989	1990	1991	1992	1993
Developed economies **Econ. développées**	...	**101**	**110**	**112**	**113**	**111**	**111**	**111**	**113**	**114**
North America Amérique du Nord	...	107	109	106	109	104	102	102	102	103
Europe **Europe**	...	**99**	**108**	**110**	**110**	**109**	**110**	**109**	**111**	**111**
EU+ UE+	...	98	108	110	110	109	110	110	112	112
EFTA AELE	...	105	106	107	107	107	108	107	104	102
Africa[3] Afrique[3]	...	83	88	89	89	95	94	90	90	90
Asia Asie	...	113	150	152	156	150	141	153	165	180
Oceania Océanie	...	90	85	85	94	101	98	89	87	86
Developing economies **Econ. en dévelop.**	...	**94**	**80**	**83**	**84**	**86**	**87**	**87**	**86**	**86**
Europe[4] Europe[4]	...	91	97	98	99	100	100	...	...	...

Source:
Trade statistics database of the Statistical Division of the
United Nations Secretariat.
+   For Member States of this grouping, see
    Annex I - Other groupings.

1   For the composition of the regions see table 71.

2   Quantum indices are derived from value data and unit value
    indices. They are base period weighted.

3   South African Customs Union.
4   Socialist Federal Republic of Yugoslavia only.

5   Regional aggregates are current period weighted.

6   Indices, except those for Europe, are based on estimates
    prepared by the International Monetary Fund.

7   Unit value index of exports divided by unit value index of
    imports.

Source:
Base de données pour les statistiques du commerce extérieur
de la Division de statistique du Secrétariat de l'ONU.
+   Les Etats membres de ce groupement, voir
    annexe I - Autres groupements.

1   Pour la composition des régions, voir le tableau 71 du
    présent numéro.
2   Les indices du quantum sont calculés à partir des chiffres
    de la valeur et des indices de la valeur unitaire. Ils sont
    à coéfficients de pondération correspondant à la période en
    base.
3   L'Union Douanière d'Afrique Méridionale.
4   La République fédérative socialiste de Yougoslavie
    seulement.
5   Les totaux régionaux sont à coéfficients de pondération
    correspondant à la période en cours.
6   Le calcul des indices, sauf ceux pour l'Europe, sont basés
    sur les estimations preparées par le Fonds monétaire
    international.
7   Indices de la valeur unitaire des exportations divisé par
    l'indice de la valeur unitaire des importations.

Technical notes, tables 1-7

Table 1: These series of world aggregates on population and production have been compiled from statistical publications of the United Nations and the specialized agencies.[4, 5, 6, 11, 14, 19, 20, 21, 22] Reference should be made to these sources for details of compilation and coverage.

Table 2 presents for the world and regions estimates of population size, rates of population increase, crude birth and death rates, surface area and population density. Unless otherwise specified all figures are estimates of the order of magnitude and are subject to a substantial margin of error.

All population estimates and rates presented in this table were prepared by the Population Division of the United Nations Secretariat and published in *World Population Prospects 1994*.[26]

The average annual percentage rates of population growth were calculated by the Population Division of the United Nations Secretariat, using an exponential rate of increase formula.

Crude birth and crude death rates are expressed in terms of the average annual number of births and deaths respectively, per 1,000 mid-year population. These rates are estimated.

Surface area totals were obtained by summing the figures for individual countries or areas.

Density is the number of persons in the 1992 total population per square kilometre of total surface area.

The scheme of regionalization used for the purpose of making these estimates is presented in annex I. Although some continental totals are given, and all can be derived, the basic scheme presents eight macro regions that are so drawn as to obtain greater homogeneity in sizes of population, types of demographic circumstances and accuracy of demographic statistics.

Tables 3-4: The index numbers in table 3 refer to agricultural production, which is defined to include both crop and livestock products. Seeds and feed are excluded. The index numbers of food refer to commodities which are considered edible and contain nutrients. Coffee, tea and other inedible commodities are excluded.

The index numbers of agricultural output and food production in table 3 are calculated by the Laspeyres formula with the base year period 1979-1981. The latter is provided in order to diminish the impact of annual fluctuations in agricultural output during base years on the indices for the period. Production quantities of each commodity are weighted by 1979-1981 average national producer prices and summed for each year. The index numbers are based on production data for a calendar year. As in the past, the series include a large number of estimates made by FAO in cases where figures are not available from official country sources.

Notes techniques, tableaux 1-7

Tableau 1 : Ces séries d'agrégats mondiaux sur la population et la production ont été établies à partir de publications statistiques des Nations Unies et d'institutions spécialisées [4, 5, 6, 11, 14, 19, 20, 21, 22]. On doit se référer à ces sources pour tous renseignements détaillés sur les méthodes de calcul et la portée des statistiques.

Le *Tableau 2* présente les estimations mondiales et régionales de la population, des taux d'accroissement de la population, des taux bruts de natalité et de mortalité, de la superficie et de la densité de population. Sauf indication contraire, tous les chiffres sont des estimations de l'ordre de grandeur et comportent une assez grande marge d'erreur.

Toutes les estimations de la population et tous les taux présentés dans ce tableau ont été établis par la Division de la population du Secrétariat des Nations Unies et publiés dans "World Population Prospects 1994" [26].

Les pourcentages annuels moyens de l'accroissement de la population ont été calculés par la Division de la population du Secrétariat des Nations Unies, sur la base d'une formule de taux d'accroissement exponentiel.

Les taux bruts de natalité et de mortalité sont exprimés, respectivement, sur la base du nombre annuel moyen de naissances et de décès par tranche de 1.000 habitants en milieu d'année. Ces taux sont estimatifs.

On a déterminé les superficies totales en additionnant les chiffres correspondant aux différents pays ou régions.

La densité est le nombre de personnes de la population totale de 1992 par kilomètre carré de la superficie totale.

Le schéma de régionalisation utilisé aux fins de l'établissement de ces estimations est présenté à l'Annexe I. Bien que les totaux de certains continents soient donnés et que tous puissent être déterminés, le schéma de base présente huit grandes régions qui sont établies de manière à obtenir une plus grande homogénéité en ce qui concerne l'ampleur des populations, les types de conditions démographiques et la précisions des statistiques démographiques.

Tableaux 3-4 : Les indices du Tableau 3 se rapportent à la production agricole, qui est définie comme comprenant à la fois les produits de l'agriculture et de l'élevage. Les semences et les aliments pour les animaux sont exclus de cette définition. Les indices de la production alimentaire se rapportent aux produits considérés comme comestibles et contenant des éléments nutritifs. Le café, le thé et les produits non comestibles sont exclus.

Les indices de la production agricole et de la production alimentaire présentés au Tableau 3 sont calculés selon la formule de Laspeyres avec les années 1979-1981 comme période de référence, cela afin de limiter l'incidence, sur les indices correspondant à la période considérée, des fluctuations annuelles de la production agricole enregistrée pendant les années de référence. Les chiffres de production de chaque produit sont pondérés par les prix nationaux moyens à la production pour la période 1979-81 et additionnés pour chaque année. Les indices sont fondés sur les données de production

Index numbers for the world and regions are computed in a similar way to the country index numbers except that, instead of using different commodity prices for each country group, "international commodity prices" derived from the Gheary-Khamis formula are used for all country groupings. This method assigns a single "price" to each commodity.

The indexes in table 4 are calculated as a ratio between the index numbers of total agricultural and food production in table 3 described above and the corresponding index numbers of population.

For further information on the series presented in these tables, see the production yearbook published by FAO.[6]

Starting with this issue, the classification of countries by economic classes and regions has been discontinued; only world and continental totals are shown. For consistency of the time series, the totals for Asia and Europe do not include the estimates for the independent republics of the USSR, which previously were considered collectively as a separate region.

Table 5: The indices of industrial production are classified according to divisions, major groups or combinations of major groups of the International Standard Industrial Classification of All Economic Activities (ISIC) for mining, manufacturing and electricity, gas and water.[42] The indices indicate trends in value added in constant US dollars. The measure of value added used is the national accounts concept, which is defined as the gross value of output less the cost of materials, supplies, fuels and electricity consumed and services received.

Each series is compiled by use of the Laspeyres formula, that is, the indices are base-weighted arithmetic means. The weight base year is 1980 and value added, generally at factor values, is used in weighting. For countries using the System of National Accounts, the estimates of value added used for weighting purposes are derived, in most cases, from the results of industrial censuses or other inquiries around 1980. These census results are adjusted to ISIC where necessary and the estimates of value added in units of national currency are then converted to United States dollars.

Value added estimates for eastern Europe and former USSR are made using official data supplied by the statistical authorities or from data extracted from national publications (number of persons engaged in industrial activity, average wages and salaries, cost structure, industrial origin of net material product etc.). The value added estimates are computed separately for their major components, "compensation of employees, operating surplus and consumption of fixed capital", and adjusted to ISIC major groups. Efforts are made to ensure consistency with the international standards. Estimates in national currencies have been converted to United States dollars by using official exchange rates.

de l'année. Comme dans le passé, les séries comprennent un grand nombre d'estimations établies par la FAO lorsqu'elle n'avait pu obtenir de chiffres de sources officielles dans les pays eux-mêmes.

Les indices pour le monde et les régions sont calculés de la même façon que les indices par pays, mais au lieu d'appliquer des prix différents aux produits de base pour chaque groupe de pays, on a utilisé des "prix internationaux" établis d'après la formule de Gheary-Khamis pour tous les groupes de pays. Cette méthode attribue un seul "prix" à chaque produit de base.

Les indices du Tableau 4 sont calculés comme ratio entre les indices de la production alimentaire et de la production agricole totale du Tableau 3 décrits ci-dessus et les indices de population correspondants.

Pour tout renseignement complémentaire sur les séries présentées dans ces tableaux, voir l'Annuaire publié par la FAO [6].

A partir de cette édition de l'Annuaire, les pays ne sont plus classés par catégories économiques et régions, et seuls figurent désormais dans les tableaux les totaux mondiaux et par continent. Pour assurer la cohérence des séries chronologiques, les données des républiques indépendantes de l'URSS, qui constituait une région en soi dans l'ancienne classification, ne figurent pas dans les totaux continentaux de l'Asie et de l'Europe.

Tableau 5 : Les indices de la production industrielle sont classés par catégorie, classe ou groupement de classes de la Classification internationale type, par industrie, de toutes les branches d'activité économique (CITI) et portent sur les industries extractives et manufacturières et les industries de l'électricité, du gaz et de l'eau [42]. Ces indices mesurent les variations de la valeur ajoutée en dollars constants des Etats-Unis. La mesure de la valeur ajoutée utilisée est celle de la comptabilité nationale, qui se définit comme la valeur brute de la production moins le coût des matières premières et des fournitures, des combustibles et de l'électricité consommés ainsi que des services reçus.

Chaque série d'indices est calculée à l'aide de la formule de Laspeyres, c'est-à-dire sous forme de moyennes arithmé-tiques pondérées. L'année de base pour la pondération est 1980 et la valeur ajoutée utilisée dans la pondération est généralement calculée au coût des facteurs. Pour les pays utilisant le Système de comptabilité nationale, les estimations de la valeur ajoutée utilisées pour les pondérations sont tirées, le plus souvent, des résultats de recensements industriels ou d'autres enquêtes effectuées autour de 1980. Ces données sont alignées sur la CITI en cas de besoin et les estimations de la valeur ajoutée, exprimées en monnaie nationale, sont ensuite converties en dollars des Etats-Unis.

Les estimations de la valeur ajoutée pour l'Europe de l'Est et l'ancienne URSS sont faites sur la base des données officielles fournies par les services de statistiques ou de données extraites de publications nationales (nombre de personnes travaillant dans une branche d'activité, salaires et traitements moyens, structure des coûts, produit matériel net par branche d'activité, etc.). Les estimations de la valeur ajoutée sont calculées séparément par grande composante,

The elementary series used in compiling indices for ISIC major groups are, in general, indices for the corresponding category, or its sub-divisions, compiled by national statistical authorities. Adjustments are made, when necessary, to align the national industrial classification with ISIC. In the case of SNA economies, most of the national indices used as elementary series indicate trends in value added at constant prices, in most instances, at factor values. In the case of eastern Europe and former USSR, the indices that serve as elementary series measure the movement in gross industrial output valued at constant enterprise prices, that is, excluding turnover taxes.

The indices for divisions, major groups or combinations of major groups of the ISIC are calculated in three main stages. First, the indices for both the developed and developing SNA economies are calculated. Secondly, these two sets of indices are combined to yield the indices for all market economies. Finally, these indices, in turn are combined with similarly derived indices for Eastern Europe and the former USSR to yield indices for the world as a whole.

Table 6: For description of the series in table 6, see technical notes to table 66 of chapter XIII.

Table 7: For description of the series in table 7, see technical notes to chapter XV. The composition of the regions is presented in table 71.

"rémunération des employés, excédent d'exploitation et consommation de capital fixe", et ajustées aux grandes catégories de la CITI. On s'efforce d'assurer la conformité aux normes internationales. Les estimations en monnaie nationale ont été converties en dollars des Etats-Unis aux taux de change officiels.

Les séries élémentaires utilisées pour établir les indices pour les principales catégories de la CITI sont généralement les indices de la catégorie correspondante, ou de ses subdivisions, calculés par les services nationaux de statistiques. Le cas échéant, des ajustements sont effectués pour aligner la classification nationale par branche d'activité sur la CITI. Dans le cas des économies SCN (Système de comptabilité nationale), la plupart des indices nationaux utilisés comme séries élémentaires indiquent les tendances de la valeur ajoutée aux prix constants, dans la plupart des cas, aux coûts des facteurs. Dans le cas de l'Europe de l'Est et de l'ancienne URSS, les indices qui servent de séries élémentaires mesurent le mouvement de la production industrielle brute évaluée à prix constants par l'entreprise, c'est-à-dire abstraction faite de l'impôt sur le chiffre d'affaires.

Les indices correspondant aux catégories et classes ou groupements de classes de la CITI sont calculés en trois étapes. Premièrement, on calcule les indices pour les économies SCN développées et en développement. Deuxièmement, on combine ces deux séries d'indices pour déterminer les indices correspondant à l'ensemble des économies de marché. Enfin, on combine ces indices à des indices équivalents pour l'Europe de l'Est et l'ancienne URSS pour déterminer les indices pour l'ensemble du monde.

Tableau 6 : On trouvera une description de la série de statistiques du Tableau 6 dans les Notes techniques du Tableau 66 du Chapitre XIII.

Tableau 7 : On trouvera une description de la série de statistiques du Tableau 7 dans les notes techniques du Chapitre XV. La composition des régions est présentée au Tableau 71.

Part Two
Population and Social Statistics

II
**Population and human settlements
(tables 8 and 9)**
III
Education and literacy (tables 10-12)
IV
**Health and child-bearing; nutrition
(table 13)**
V
**Culture and communications
(tables 14-20)**

Part Two of the *Yearbook* presents statistical series on a wide range of population and social topics for all countries or areas of the world for which data are available. These include population and population growth, surface area and density; education at first, second and third levels; AIDS cases; book production, newspapers, television and radio, telefax stations and telephones.

Deuxième partie
Population et statistiques sociales

II
**Population et établissements humains
(tableaux 8 et 9)**
III
**Education et alphabétisation
(tableaux 10 à 12)**
IV
**Santé et maternité; nutrition
(tableau 13)**
V
**Culture et communications
(tableaux 14 à 20)**

La deuxième partie de l'*Annuaire* présente, pour tous les pays ou zones du monde pour lesquels des données sont disponibles, des séries statistiques intéressant une large gamme de questions démographiques et sociales : population et croissance démographique, superficie et densité; enseignement des premier, second et troisième degrés; cas de SIDA; production de livres, journaux, télévision et radio, postes de télécopie et téléphones.

8
Population by sex, rate of population increase, surface area and density
Population selon le sexe, taux d'accroissement de la population, superficie et densité

Country or area Pays ou zone	Date	Latest Census Dernier recensement Both sexes Les deux sexes	Male Masculin	Female Féminin	Midyear estimates (thousands) Estimations au milieu de l'année (milliers) 1990	1993	Type[1] 1993	Annual rate of increase Taux d'accrois- sement annuel % 1990-93	Surface area (km²) Superficie (km²) 1993	Density Densité 1993[2]
Africa · Afrique										
Algeria[3] Algérie[3]	20−III−87	23 033 942	...	...	25 012	x26 722	A6 c1	2.2	2 381 741	11
Angola[4] Angola[4]	15−XII−70	5 646 166	2 943 974	2 702 192	10 020	x10 276	A23c1	0.8	1 246 700	8
Benin Bénin	15−II−92	*4 855 349	*2 365 574	*2 489 775	4 739	*5 215	A14c3	3.2	112 622	46
Botswana Botswana	14−VIII−91	1 326 796	634 400	692 396	1 300	*1 443	A2 c1	3.5	581 730	2
British Indian Territory[5] Territoire britannique de l'océan indien[5]	(⁶)	(⁶)	(⁶)	(⁶)	x2	x2	D28d	0.0	78	26
Burkina Faso Burkina Faso	10−XII−85	7 964 705	3 833 237	4 131 468	9 001	*9 682	A8 c3	2.4	274 000	35
Burundi Burundi	16−VIII−90	5 139 073	2 473 599	2 665 474	5 458	*5 958	A14c3	2.9	27 834	214
Cameroon Cameroun	IV−87	*10 493 655	...	...	x11 526	x12 522	A17c3	2.8	475 442	26
Cape Verde Cap−Vert	23−VI−90	341 491	161 494	179 997	341	x370	A13c1	2.7	4 033	92
Central African Republic République centrafricaine	8−XII−88	2 463 616	1 210 734	1 252 882	x2 927	x3 156	A5c3	2.5	622 984	5
Chad Tchad	8−IV−93	*6 158 992	*2 950 415	*3 208 577	5 687	*6 098	B30c3	2.3	1 284 000	5
Comoros Comores	15−IX−91	*446 817[7]	*221 152[7]	*225 665[7]	x543	x607	A13c3	3.7	2 235	272
Congo Congo	22−XII−84	1 843 421	...	...	x2 232	x2 443	A9 c3	3.0	342 000	7
Côte d'Ivoire Côte d'Ivoire	1−III−88	10 815 694	5 527 343	5 288 351	x11 974	x13 316	A5 c3	3.5	322 463	41
Djibouti Djibouti	1960−61	81 200	...	...	x517	x557	A33d	2.5	23 200	24
Egypt Egypte	17−XI−86	48 254 238	24 709 274	23 544 964	52 691	*56 489	A7c1	2.3	1 001 449	56
Equatorial Guinea[8] Guinée équatoriale[8]	4−VII−83	300 000	144 760	155 240	348	x379	A10c3	2.8	28 051	14
Eritrea Erythrée	9−V−84	2 748 304	1 374 452	1 373 852	x3 082	x3 345	A9 c3	2.7	...	...
Ethiopia Ethiopie	9−V−84	39 868 572	20 062 490	19 806 082	x47 423	x51 859	A9 c3	3.0	...	...
Gabon Gabon	31−VII−93	*1 011 710	*498 710	*513 000	x1 146	x1 248	A33c3	2.8	267 668	5
Gambia Gambie	13−IV−93	*1 025 867	*514 530	*511 337	x923	x1 042	A10c1	4.0	11 295	92
Ghana Ghana	11−III−84	12 296 081	6 063 848	6 232 233	x15 020	x16 446	A9 c1	3.0	238 533	69
Guinea[9] Guinée[9]	4−II−83	4 533 240	...	...	x5 755	x6 306	A10c3	3.0	245 857	26
Guinea−Bissau Guinée−Bissau	16−IV−79	753 313	362 589	390 724	x964	x1 028	A14c1	2.1	36 125	28
Kenya Kenya	24−VIII−89	21 443 636	10 628 368	10 815 268	24032[10]	*28 113	A4 c2	(¹¹)	580 367	48
Lesotho Lesotho	12−IV−86	*1 447 000	...	...	x1 792	x1 943	A7 c3	2.7	30 355	64
Liberia Libéria	1−II−84	*2 101 628	...	...	2 407	*2 640	A9 c3	3.1	111 369	24
Libyan Arab Jamahiriya Jamahiriya arabe libyenne	31−VII−84	*3 637 488	*1 950 152	*1 687 336	4 151[10]	*4 700	A9 c3	(¹¹)	1 759 540	3

8
Population by sex, rate of population increase, surface area and density [*cont.*]
Population selon le sexe, taux d'accroissement de la population,
superficie et densité [*suite*]

Country or area Pays ou zone	Date	Latest Census Dernier recensement Both sexes Les deux sexes	Male Masculin	Female Féminin	Midyear estimates (thousands) Estimations au milieu de l'année (milliers) 1990	1993	Type[1] 1993	Annual rate of increase Taux d'accrois- sement annuel % 1990–93	Surface area (km[2]) Superficie (km[2]) 1993	Density Densité 1993[2]
Madagascar Madagascar	1–VIII–93	*12 092 157	*5 991 171	*6 100 986	11 197	x13 854[10]	A19c3	([11])	587 041	24
Malawi Malawi	1–IX–87	7 988 507	3 867 136	4 121 371	8 289	*9 135	A6 c3	3.2	118 484	77
Mali Mali	1–IV–87	7 696 348[3]	3 760 711[3]	3 935 637[3]	8 156[10]	x10 135	A6 c3	([11])	1 240 192	8
Mauritania Mauritanie	5–IV–88	1 864 236[12]	923 175[12]	941 061[12]	x2 003	*2 148	A5 c3	2.3	1 025 520	2
Mauritius Maurice	1–VII–90	1 056 660	527 760	528 900	1 071	x1 091	A3 b1	0.6	2 040	535
Island of Mauritius Ile Maurice	1–VII–90	1 022 456	510 676	511 780	1 037	*1 098	A3 b1	1.9	1 865	589
Rodrigues Rodrigues	1–VII–90	34 204	17 084	17 120	34	...	..	...	104	...
Others[13] Autres[13]	30–VI–72	366	272	94	...	...	..	...	71	...
Morocco Maroc	3–IX–82	20 419 555	...	...	24 487	*26 069	A11c2	2.1	446 550	58
Mozambique[9] Mozambique[9]	1–VIII–80	11 673 725	5 670 484	6 003 241	14 151	*15 583	A13c1	3.2	801 590	19
Namibia Namibie	21–X–91	1 409 920	686 327	723 593	x1 349	x1 461	A2 c3	2.7	824 292	2
Niger Niger	20–V–88	7 248 100	3 590 070	3 658 030	x7 731	*8 361	A5 c3	2.6	1 267 000	7
Nigeria Nigéria	26–XI–91	*88 514 501	*44 544 531	*43 969 970	x96 154	x105 264	A30c2	3.0	923 768	114
Réunion[3] Réunion[3]	15–III–90	597 828	294 256	303 572	601	*632	A3 b3	1.7	2 510	252
Rwanda Rwanda	15–VIII–91	*7 142 755	...	...	7 181	x7 554	A2 c3	1.7	26 338	287
St. Helena ex. dep. Sainte–Hélène sans dép.	22–II–87	5 644	2 769	2 875	6	6	A6 b3	0.6	122	53
Ascension Ascension	31–XII–78	849	608	241	...	...	..	...	88	...
Tristan da Cunha Tristan da Cunha	22–II–87	296	139	157	...	...	..	0.6	...	...
Sao Tome and Principe Sao Tomé–et–Principe	4–VIII–91	116 998	57 837	59 161	115	*122	A12c1	2.0	964	127
Senegal Sénégal	27–V–88	6 896 808	3 353 599	3 543 209	x7 327	x7 902	A5 c3	2.5	196 722	40
Seychelles Seychelles	17–VIII–87	68 598	34 125	34 473	70	*72	A6 b2	1.3	455	159
Sierra Leone[9] Sierra Leone[9]	15–XII–85	3 515 812	1 746 055	1 769 757	x3 999	x4 297	A8 c1	2.4	71 740	60
Somalia Somalie	1986–1987	7 114 431	3 741 664	3 372 767	x8 677	x8 954	A18c3	1.0	637 657	14
South Africa[9] Afrique du Sud[9]	7–III–91	30 986 920[14]	15 479 528[14]	15 507 392[14]	x37 066	x39 659	A2 c1	2.3	1 221 037	32
Sudan Soudan	15–IV–93	*24 940 683	*12 518 638	*12 422 045	25 752	*28 129	A10c3	2.9	2 505 813	11
Swaziland Swaziland	25–VIII–86	681 059	321 579	359 480	768	x809	A7 c1	1.8	17 364	47
Togo Togo	22–XI–81	2 703 250	...	...	x3 531	x3 885	A12c1	3.2	56 785	68
Tunisia Tunisie	30–III–84	6 966 173	3 547 315	3 418 858	8 074	x8 570	A9 c1	2.0	163 610	52
Uganda Ouganda	12–I–91	16 671 705	8 185 747	8 485 958	x17 949	x19 940	A13c1	3.5	241 038	83
United Rep. of Tanzania Rép.–Unie de Tanzanie	28–VIII–88	*23 174 336	*11 327 511	*11 846 825	25 635	x28 019	A15c3	3.0	883 749	32

8
Population by sex, rate of population increase, surface area and density [cont.]
Population selon le sexe, taux d'accroissement de la population,
superficie et densité [suite]

Country or area Pays ou zone	Date	Latest Census Dernier recensement Both sexes Les deux sexes	Male Masculin	Female Féminin	Midyear estimates (thousands) Estimations au milieu de l'année (milliers) 1990	1993	Type[1] 1993	Annual rate of increase Taux d'accrois- sement annuel % 1990-93	Surface area (km²) Superficie (km²) 1993	Density Densité 1993[2]
Tanganyika Tanganyika	28–VIII–88	*22 533 758	*11 012 647	*11 521 111	24 972	...	..	...	881 289	...
Zanzibar Zanzibar	28–VIII–88	*640 578	*314 864	*325 714	663	...	..	...	2 460	...
Western Sahara[15] Sahara occidental[15]	31–XII–70	76 425	43 981	32 444	x230	x261	A23c1	4.2	266 000	1
Zaïre Zaïre	1–VII–84	29 916 800	14 543 800	15 373 000	35 562	x41 231	A9 c3	4.9	2 344 858	18
Zambia Zambie	20–VIII–90	*7 818 447	*3 843 364	*3 975 083	8 073	x8 936	A13c1	3.4	752 618	12
Zimbabwe Zimbabwe	18–VIII–92	*10 401 767	*5 075 549	*5 326 218	9 369	x10 739	A11c1	4.5	390 757	27
America, North · Amérique du Nord										
Anguilla Anguilla	10–IV–84	6 987	...	...	x7	9	A9 b1	9.1	96	96
Antigua and Barbuda Antigua–et–Barbuda	28–V–91	62 922	...	...	x64	x65	A23b1	0.5	442	147
Aruba[3] Aruba[3]	6–X–91	66 687	32 821	33 866	x67	x69	A12b1	1.0	193	358
Bahamas Bahamas	1–V–90	255 095	124 992	130 103	255	*269	A3 b1	1.7	13 878	19
Barbados Barbade	2–V–90	*257 082	...	...	257	*264	A3 b1	0.8	430	613
Belize Belize	12–V–91	189 774	96 289	93 485	189	*205	A13c1	2.7	22 696	9
Bermuda Bermudes	20–V–91	74 837	...	...	61[16]	63[16]	A13b1	1.3	53	1 189
British Virgin Islands Iles Vierges britanniques	12–V–80	11 697	...	...	x16	x18	A13b1	3.9	153	118
Canada[3] Canada[3]	4–VI–91	27 296 859	...	...	26 584	*28 755	A2 b1	2.6	9 970 610	3
Cayman Islands[3] Iles Caïmanes[3]	15–X–89	25 355	12 372	12 983	26	x29	A4 c1	3.6	264	110
Costa Rica[3] Costa Rica[3]	10–VI–84	2 416 809	1 208 216	1 208 593	2 994	*3 199	A9 b2	2.2	51 100	63
Cuba Cuba	11–IX–81	9 723 605	4 914 873	4 808 732	10 625	*10 905	A12b1	0.9	110 861	98
Dominica Dominique	12–V–91	71 183	35 471	35 712	x71	x71	A2 b1	0.0	751	95
Dominican Republic Rép. dominicaine	12–XII–81	5 545 741	2 793 884	2 751 857	7 170	*7 608	A12c1	2.0	48 734	156
El Salvador El Salvador	6–X–92	*5 047 925	*2 423 004	*2 624 921	x5 172	x5 517	A22b1	2.2	21 041	262
Greenland[3] Groenland[3]	26–X–76	49 630	26 856	22 774	56	*55	A17a1	–0.2	2 175 600	–
Grenada[17] Grenade[17]	30–IV–81	89 088	42 943	46 145	x91	x92	A12b1	0.4	344	267
Guadeloupe[3 18] Guadeloupe[3 18]	9–III–82	327 002	160 112	166 890	385	x413	A3 b1	2.3	1 705	242
Guatemala[9] Guatemala[9]	26–III–81	6 054 227	3 015 826	3 038 401	9 198	*10 030	A12b2	2.9	108 889	92
Haiti[3] Haïti[3]	30–VIII–82	5 053 792	2 448 370	2 605 422	6 486	*6 903	A11c3	2.1	27 750	249
Honduras Honduras	V–88	4 248 561	2 110 106	2 138 455	5 105	*5 595	A19c1	3.1	112 088	50
Jamaica Jamaïque	7–IV–91	*2 366 067	...	...	2 415	x2 411	A2 b1	–0.1	10 990	219
Martinique[3] Martinique[3]	15–III–90	*359 579	173 878	185 701	362	x371	A3 b1	0.8	1 102	337

8

Population by sex, rate of population increase, surface area and density [cont.]
Population selon le sexe, taux d'accroissement de la population,
superficie et densité [suite]

Country or area Pays ou zone	Date	Latest Census Dernier recensement Both sexes Les deux sexes	Male Masculin	Female Féminin	Midyear estimates (thousands) Estimations au milieu de l'année (milliers) 1990	1993	Type[1] 1993	Annual rate of increase Taux d'accrois- sement annuel % 1990-93	Surface area (km²) Superficie (km²) 1993	Density Densité 1993[2]
Mexico[3] Mexique[3]	12−III−90	*81 140 922	*39 878 536	*41 262 386	86 154	*91 261	A13c1	1.9	1 958 201	47
Montserrat Montserrat	12−V−80	11 932	...	...	x11	x11	A13b1	0.0	102	108
Netherlands Antilles[3 9 19] Antilles néerlandaises[3 9 19]	27−I−92	*189 474	*90 707	*98 767	190	x195	A12c1	0.8	800	244
Nicaragua[3] Nicaragua[3]	20−IV−71	1 877 952	921 543	956 409	3 871	*4 265	A22b3	3.2	130 000	33
Panama Panama	13−V−90	2 329 329	1 178 790	1 150 539	2 418	*2 563	A3 c1	1.9	75 517	34
Puerto Rico[3 20] Porto Rico[3 20]	1−IV−90	*3 522 039	...	...	3 528	*3 620	A3 b1	0.9	8 897	407
Saint Kitts and Nevis Saint−Kitts−et−Nevis	12−V−80	44 224	...	...	x42	x42	A13b1	0.0	261	161
Saint Lucia Sainte−Lucie	12−V−91	133 308	...	...	x133	x139	A13b1	1.5	622	223
St. Pierre and Miquelon Saint−Pierre et Miquelon	9−III−82	6 037	2 981	3 056	x6	x6	A11d	0.0	242	25
St. Vincent and Grenadines[21] St.−Vincent−et−Grenadines[21]	12−V−91	106 499	53 165	53 334	x107	x110	A13b1	0.9	388	284
Trinidad and Tobago Trinité−et−Tobago	2−V−90	1 234 388	618 050	616 338	1 227	*1 260	A3 b1	0.9	5 130	246
Turks and Caicos Islands Iles Turques et Caïques	31−V−90	12 350	6 289	6 061	x12	x13	A13d	2.7	430	30
United States[22] Etats−Unis[22]	1−IV−90	248 709 873	121 239 418	127 470 455	249 924	*258 233	A3 b1	1.1	9 363 520	28
United States Virgin[3 20] Iles Vierges américaines[3 20]	1−IV−90	101 809	49 210	52 599	102	x104	A3 c1	0.7	347	300
America, South · Amérique du Sud										
Argentina Argentine	15−V−91	32 615 528	...	...	32 547	*33 778	A2 c1	1.2	2 780 400	12
Bolivia Bolivie	3−VI−92	6 420 792	3 171 265	3 249 527	6 573	*7 065	A1 c3	2.4	1 098 581	6
Brazil[23] Brésil[23]	1−IX−91	146 917 459[3]	72 536 142[3]	74 381 317[3]	144 541	*151 534	A2 c1	1.6	8 511 965	18
Chile Chili	22−IV−92	13 348 401	6 553 254	6 795 147	13 173	*13 813	A1 b1	1.6	756 626	18
Colombia[24] Colombie[24]	15−X−85	27 837 932	13 777 700	14 060 232	32 300	*33 951	A8 b3	1.7	1 138 914	30
Ecuador[25] Equateur[25]	25−XI−90	9 648 189	4 796 412	4 851 777	10 264	*10 981	A3 b3	2.3	283 561	39
Falkland Is. (Malvinas)[26 27] Iles Falkland (Malvinas)[26 27]	5−III−91	2 050	1 095	955	x2	x2	A2 d	0.0	12 173	−
French Guyana[3] Guyane française[3]	15−III−90	114 808	59 798	55 010	x117	x135	A3 c1	4.8	90 000	2
Guyana Guyana	12−V−80	758 619	375 841	382 778	x796	x816	A13b1	0.8	214 969	4
Paraguay Paraguay	26−VIII−92	*4 123 550	*2 069 673	*2 053 877	4 277	*4 643	A11c2	2.7	406 752	11
Peru[9 23] Pérou[9 23]	12−VII−81	17 005 210	8 489 867	8 515 343	21 550	*22 454	A12c2	1.4	1 285 216	17
Suriname Suriname	1−VII−80	352 041	173 083	178 958	404	x414	A13c2	0.9	163 265	3
Uruguay[9] Uruguay[9]	23−X−85	2 955 241	1 439 021	1 516 220	3 094	*3 149	A8 b3	0.6	177 414	18
Venezuela[23] Venezuela[23]	20−X−90	18 105 265	9 019 757	9 085 508	19 325	*20 712	A3 c1	2.3	912 050	23
Asia · Asie										
Afghanistan Afghanistan	23−VI−79	13 051 358[28]	6 712 377[28]	6 338 981[28]	28 16121	x17 691	A14c3	([11])	652 090	27

8
Population by sex, rate of population increase, surface area and density [cont.]
Population selon le sexe, taux d'accroissement de la population,
superficie et densité [suite]

Country or area Pays ou zone	Date	Latest Census Dernier recensement			Midyear estimates (thousands) Estimations au milieu de l'année (milliers)			Annual rate of increase Taux d'accrois- sement	Surface area (km²) Superficie	Density
		Both sexes Les deux sexes	Male Masculin	Female Féminin	1990	1993	Type[1] 1993	annuel % 1990-93	(km²) 1993	Densité 1993[2]
Armenia Arménie	12–I–89	3 304 776[3]	1 619 308[3]	1 685 468[3]	3 545	*3 732	A4 b1	1.7	29 800	125
Azerbaijan Azerbaïdjan	12–I–89	7 021 178[3]	3 423 793[3]	3 597 385[3]	7 153	*7 392	A4 b1	1.1	86 600	85
Bahrain Bahreïn	16–XI–91	508 037	294 346	213 691	486	*539	A2 c1	3.5	694	777
Bangladesh Bangladesh	12–III–91	*104 766 143	*53 918 319	*50 847 824	x108 118	x115 203	A2 c1	2.1	143 998	800
Bhutan Bhoutan	XI–69	1 034 774	...	...	x1 544	x1 596	A24c3	1.1	47 000	34
Brunei Darussalam[9 29] Brunéi Darussalam[9 29]	7–VIII–91	260 482	137 616	122 866	253	x274	A2 c2	2.6	5 765	48
Cambodia[30] Cambodge[30]	17–IV–62	5 728 771	2 862 939	2 865 832	8 568	*9 308	A31c3	2.8	181 035	51
China[31] Chine[31]	1–VII–90	1 160 044 618	...	...	x1155305	x1 196 360	A3 c3	1.2	9 596 961	125
Cyprus[3] Chypre[3]	30–IX–76	612 851	306 144	306 707	702	x726	A17b2	1.1	9 251	78
East Timor Timor oriental	31–X–90	747 750	386 939	360 811	x740	x785	A3 c1	2.0	14 874	53
Georgia Géorgie	12–I–89	5 400 841[3]	2 562 040[3]	2 838 801[3]	5 464	x5 446	A4 b1	-0.1	69 700	78
Hong Kong[32] Hong–kong[32]	15–III–91	5 522 281	2 811 991	2 710 290	5 705	*5 919	A7 b2	1.2	1 075[33]	5 506
India[34] Inde[34]	1–III–91	846 302 688	439 230 458	407 072 230	827 050	x901 459	A2 c1	2.9	3 287 590	274
Indonesia[35] Indonésie[35]	31–X–90	179 378 946	89 463 545	89 915 401	179 830	*189 136	A3 c1	1.7	1 904 569	99
Iran, Islamic Republic of Iran, Rép. islamique d'	1–X–91	55 837 163	28 768 450	27 068 713	54 496	64 169[10]	A7 c1	([11])	1 633 188	39
Iraq Iraq	17–X–87	16 335 199	8 395 889	7 939 310	17 373	19 454	A6 c1	3.8	438 317	44
Israel[3 36] Israël[3 36]	4–VI–83	4 037 620	2 011 590	2 026 030	4 660	*5 256	A10b1	4.0	21 056	250
Japan[37] Japon[37]	1–X–90	123 611 167	60 696 724	62 914 443	123 537	*123 653	A3 b1	0.0	377 801	327
Jordan[38] Jordanie[38]	10–XI–79	2 100 019[39]	1 086 591[39]	1 013 428[39]	x4 259	x4 936	A14b3	4.9	97 740	51
Kazakhstan Kazakhstan	12–I–89	16 536 511	8 012 985	8 523 526	16 670	*16 956	A4 b1	0.6	2 717 300	6
Korea, Dem. People's Rep. Corée, Rép. pop. dém. de	1–V–44	...	...	...	x21 774	x23 048	D30c3	1.9	120 538	191
Korea, Republic of[9 40] Corée, Rép. de[9 40]	1–XI–90	43 410 899	21 782 154	21 628 745	42 869	*44 056	A3 c1	0.9	99 263	444
Kuwait Koweït	21–IV–85	1 697 301	965 297	732 004	2 125	*1 433[10]	A8 c1	([11])	17 818	80
Kyrgyzstan Kirghizistan	12–I–89	4 257 755[3]	2 077 623[3]	2 180 132[3]	4 395	*4 528	A4 b1	1.0	198 500	23
Lao People's Dem. Rep. Rép. dém. populaire Lao	1–III–85	3 584 803	1 757 115	1 827 688	x4 202	x4 605	A8 c3	3 1	236 800	19
Lebanon[41] Liban[41]	15–XI–70	2 126 325[42]	1 080 015[42]	1 046 310[42]	x2 555	x2 806	B23c3	3.1	10 400	270
Macau[43] Macao[43]	30–VIII–91	*385 089	...	...	335	*388	A12c1	4.9	18	21 560
Malaysia Malaisie	14–VIII–91	*17 566 982	*8 861 124	*8 705 858	17 764	*19 239	A13c2	2.7	329 758	58
Maldives Maldives	8–III–90	213 215	109 336	103 879	x216	*238	A8 c1	3.3	298	800
Mongolia Mongolie	5–I–89	2 043 400	...	...	x2 177	x2 318	A4c1	2.1	1 566 500	1

8

Population by sex, rate of population increase, surface area and density [cont.]
Population selon le sexe, taux d'accroissement de la population,
superficie et densité [suite]

Country or area Pays ou zone	Date	Latest Census Dernier recensement Both sexes Les deux sexes	Male Masculin	Female Féminin	Midyear estimates (thousands) Estimations au milieu de l'année (milliers) 1990	1993	Type[1] 1993	Annual rate of increase Taux d'accroissement annuel % 1990–93	Surface area (km²) Superficie (km²) 1993	Density Densité 1993[2]
Myanmar Myanmar	31–III–83	35 307 913[3]	17 518 255[3]	17 789 658[3]	x41 813	x44 596	A10c2	2.1	676 578	66
Nepal[3] Népal[3]	22–VI–91	18 462 081	9 220 914	9 241 167	18 916	x20 812	A2 c1	3.2	140 797	148
Oman Oman	1–XII–93	*2 017 591	...	...	2 000	x1 992[10]	..	([11])	212 457	9
Pakistan[44] Pakistan[44]	1–III–81	84 253 644	44 232 677	40 020 967	112 049	*122 802	A12c1	3.1	796 095	154
Palestine[45] Palestine[45]	18–XI–31	1 035 821	524 268[46]	509 028[46]	...	...	..	...	...	...
Gaza Strip[47] Zone de Gaza[47]	14–IX–67	356 261	172 511	183 750	...	...	..	...	378	...
Philippines[3] Philippines[3]	1–V–90	60 559 116	30 443 187	30 115 929	61 480	*65 649	A13c2	2.2	300 000	219
Qatar Qatar	16–III–86	369 079	247 852	121 227	486	*559	A7 c3	4.7	11 000	51
Saudi Arabia Arabie saoudite	27–IX–92	*16 929 294	*9 466 541	*7 462 753	14 870	x17 119	A1 c3	4.7	2 149 690	8
Singapore[48] Singapour[48]	30–VI–90	2 705 115	1 370 059	1 335 056	2 705	*2 874	A3 b2	2.0	618	4 650
Sri Lanka Sri Lanka	17–III–81	14 846 750	7 568 253	7 278 497	16 993	*17 619	A12c1	1.2	65 610	269
Syrian Arab Republic[49] Rép. arabe syrienne[49]	7–IX–81	9 046 144	4 621 852	4 424 292	12 116	*13 393	A12c1	3.3	185 180	72
Tajikistan Tadjikistan	12–I–89	5 092 603[3]	2 530 245[3]	2 562 358[3]	5 303	x5 767	A4 b1	2.8	143 100	40
Thailand[3] Thaïlande[3]	1–IV–90	*54 532 300	*27 031 200	*27 501 100	56 082	*58 584	A13c1	1.5	513 115	114
Turkey Turquie	21–X–90	56 473 035	28 607 047	27 865 988	56 098	60 227	A3 c1	2.4	774 815	78
Turkmenistan Turkménistan	12–I–89	3 522 717[3]	1 735 179[3]	1 787 538[3]	3 670	x3 921	A4 b1	2.2	488 100	8
United Arab Emirates[50] Emirats arabes unis[50]	15–XII–80	1 043 225	720 360	322 865	x1 671	*1 206	A13c3	([11])	83 600	14
Uzbekistan Ouzbékistan	12–I–89	19 810 077[3]	9 784 156[3]	10 025 921[3]	20 531	x21 860	A4 b1	2.1	447 400	49 *
Viet Nam Viet Nam	1–IV–89	*64 411 713	*31 336 568	*33 075 145	66 233	x71 324	A4 c3	2.5	331 689	215
Yemen Yémen	...	...	...	...	11 279	*12 302	..	2.9	...	...
former Dem. Yemen ancienne Yémen dém.	29–III–88	2 345 266	1 184 359	1 160 907	2 460	*2 929	A5 c3	5.8	332 968	9
former Yemen Arab Rep.[3] anc. Yémen rép. arabe[3]	1–II–86	9 274 173	4 647 310	4 626 863	x9 196	x10 283	A7 c3	3.7	195 000	53
Europe · Europe										
Albania Albanie	12–IV–89	*3 182 400	*1 638 900	*1 543 500	3 256	*3 500	A4 b1	2.4	28 748	122
Andorra Andorre	XI–54	5 664	...	...	53	x61	A39c3	4.9	453	135
Austria[3] Autriche[3]	12–V–81	7 555 338	3 572 426	3 982 912	7 718	*7 988	A12b1	1.1	83 853	95
Belarus Bélarus	12–I–89	10 151 806[3]	4 749 324[3]	5 402 482[3]	10 260	x10 188	A4 b1	−0.2	207 600	49
Belgium[3] Belgique[3]	1–III–81	9 848 647	4 810 349	5 038 298	9 967	*10 010	A2 b1	0.1	30 519	328
Bosnia–Herzegovina[3] Bosnie–Herzégovine[3]	31–III–91	4 365 639	...	...	x4 308	x3 707	A2 b1	−5.0	51 129	73
Bulgaria Bulgarie	4–XII–85	8 948 388	4 430 061	4 518 327	8 991	*8 472	A8 b1	−2.0	110 912[51]	76

8
Population by sex, rate of population increase, surface area and density [*cont.*]
Population selon le sexe, taux d'accroissement de la population,
superficie et densité [*suite*]

Country or area Pays ou zone	Date	Latest Census Dernier recensement Both sexes Les deux sexes	Male Masculin	Female Féminin	Midyear estimates (thousands) Estimations au milieu de l'année (milliers) 1990	1993	Type[1] 1993	Annual rate of increase Taux d'accrois- sement annuel % 1990–93	Surface area (km²) Superficie (km²) 1993	Density Densité 1993[2]
Channel Islands Iles Anglo–Normandes	23–III–86	135 694	65 610	70 084	x142	x146	A7 b1	0.9	195	749
Guernsey[52] Guernesey[52]	21–IV–91	58 867	28 297	30 570	60	58		–0.9	78	744
Jersey Jersey	10–III–91	84 082	40 862	43 220	...	...		...	116	...
Croatia Croatie	31–III–91	4 784 265	...	...	4 778	x4 511	A2 b1	...	56 538	80
Czech Republic Rép. tchéque	3–III–91	10 302 215	4 999 935	5 302 280	10 363	*10 328	A2 b1	–0.1	78 864	131
Denmark[3 53] Danemark[3 53]	1–I–81	5 123 989	2 528 225	2 595 764	5 140	*5 189	A2 a1	0.3	43 077	120
Estonia Estonie	12–I–89	1 565 662[3]	731 392[3]	834 270[3]	1 571	*1 517	A4 b1	–1.2	45 100	34
Faeroe Islands[3] Iles Féroe[3]	22–IX–77	41 969	21 997	19 972	x47	x47	A16b1	–0.3	1 399	34
Finland[3] Finlande[3]	31–XII–90	4 998 478	2 426 204	2 572 274	4 986	*5 067	A3 b1	0.5	338 145	15
France[54 55] France[54 55]	5–III–90	56 634 299[56]	27 553 788[56]	29 080 511[56]	56 735	*57 379	A3 b1	0.4	551 500	104
Germany † Allemagne †	...	...	...	...	79 365	*81 187	..	0.8	356 733	228
Federal Rep. of Germany[3] Rép. féd. d'Allemagne[3]	25–V–87	61 077 042	29 322 923	31 754 119	63 253	*61 241	A6 b1	–1.1	248 647	246
former German Dem Rep.[3] ancienne R. d. allemande[3]	31–XII–81	16 705 635	7 849 112	8 856 523	16 247	x16 204	A12b1	–0.1	108 333	150
Gibraltar[57] Gibraltar[57]	9–XI–81	29 616	14 992	14 624	31	x28	A2 b1	–3.2	6	4 667
Greece Grèce	5–IV–81	*10 269 074[58]	...	...	10089[59]	*10 305[59]	A12b2	0.7	131 990	78
Holy See Saint–Siège	30–IV–48	890	548	342	x1	x1	D4 d	0.0	0[60]	...
Hungary Hongrie	1–I–90	10 374 823	4 984 904	5 389 919	10 365	*10 294	A3 b1	–0.2	93 032	111
Iceland[3] Islande[3]	1–XII–70	204 930	103 621	101 309	255	x263	A23a1	1.1	103 000	3
Ireland Irlande	21–IV–91	3 525 719	1 753 418	1 772 301	3 503	*3 563	A7 b2	0.6	70 284	51
Isle of Man Ile de Man	14–IV–91	69 788	33 693	36 095	69	*71	A2 b1	0.8	572	123
Italy Italie	20–X–91	59 103 833	...	...	57 661[3]	* 57057[3]	A12b1	–0.4	301 268	189
Latvia Lettonie	12–I–89	2 666 567[3]	1 238 806[3]	1 427 761[3]	2 671	*2 586	A4 b1	–1.1	64 600	40
Liechtenstein Liechtenstein	2–XII–80	25 215	...	...	x29	x30	A13b1	1.1	160	188
Lithuania Lithuanie	12–I–89	3 674 802[3]	1 738 953[3]	1 935 849[3]	3 722	*3 730	A4 b1	0.1	65 200	57
Luxembourg[3] Luxembourg[3]	31–III–81	364 602	177 869	186 733	382	x380	A12b2	–0.2	2 586	147
Malta[61] Malte[61]	16–XI–85	345 418	169 832	175 586	354	x361	A8 b2	0.6	316	1 142
Monaco[3] Monaco[3]	4–III–82	27 063	12 598	14 465	x30	x31	A11c1	1.1	1[62]	31000
Netherlands[3] Pays–Bas[3]	28–II–71	13 060 115	...	...	14 952	*15 298	A22a1	0.8	40 844	375
Norway[3] Norvège[3]	3–XI–90	4 247 546	2 099 881	2 147 665	4 241	*4 312	A3 a1	0.6	323 877	13
Poland[63] Pologne[63]	6–XII–88	37 878 641	18 464 373	19 414 268	38 119	*38 505	A5 b1	0.3	323 250	119

8

Population by sex, rate of population increase, surface area and density [cont.]
Population selon le sexe, taux d'accroissement de la population,
superficie et densité [suite]

Country or area Pays ou zone		Latest Census Dernier recensement			Midyear estimates (thousands) Estimations au milieu de l'année (milliers)			Annual rate of increase Taux d'accrois- sement	Surface area (km²)	Density
	Date	Both sexes Les deux sexes	Male Masculin	Female Féminin	1990	1993	Type[1] 1993	annuel % 1990–93	Superficie (km²) 1993	Densité 1993[2]
Portugal[64]										
Portugal[64]	15–IV–91	*9 853 896	...	...	9 868[10]	*9 864	A2 b1	([11])	92 389	107
Republic of Moldova										
Moldova, Rép. de	12–I–89	4 337 592	2 058 160	2 279 432	4 364	*4 356	A4 b1	−0.1	33 700	129
Romania										
Roumanie	7–I–92	22 810 035	11 213 763	11 596 272	23 207	*22 755	A1 b2	−0.7	238 391	95
Russian Federation										
Fédération de Russie	12–I–89	147 021 869[3]	68 713 869[3]	78 308 000[3]	147 913	x147 760	A4 b1	−0.0	17075400	9
San Marino										
Saint–Marin	30–XI–76	19 149	9 654	9 495	23	x24	A17a2	1.2	61	393
Slovakia										
Slovaquie	3–III–91	5 274 335	2 574 061	2 700 274	5 298	*5 318	A2 b1	0.1	49 012	108
Slovenia[3]										
Slovénie[3]	31–III–91	1 965 986	952 611	1 013 375	1 998	*1 991	A2 b1	−0.1	20 256	98
Spain[65]										
Espagne[65]	1–III–91	39 433 942	19 338 083	20 095 859	38 959	*39 141	A2 c1	0.2	505 992	77
Svalbard and Jan Mayen Isl.[66]										
Svalbard et Ile Jan–Mayen[66]	1–XI–60	3 431	2 545	886	...	...		...	62 422	...
Sweden[3]										
Suède[3]	1–IX–90	8 587 353	4 242 351	4 345 002	8 559	*8 712	A3 a1	0.6	449 964	19
Switzerland[3]										
Suisse[3]	4–XII–90	6 873 687	3 390 446	3 483 241	6 712	*6 938	A3 b1	1.1	41 293	168
TFYR of Macedonia[3]										
L'ex–R.y. Macédoine[3]	31–III–91	2 033 964	1 027 352	1 006 612	2 028	x2 119	A2 b1	1.5	25 713	82
Ukraine										
Ukraine	12–I–89	51 452 034[3]	23 745 108[3]	27 706 926[3]	51 839	*52 179	A4 b1	0.2	603 700	86
United Kingdom[67]										
Royaume–Uni[67]	21–IV–91	*56 352 200	...	...	57 561	*58 191	A12b1	0.4	244 100	238
Yugoslavia[3]										
Yougoslavie[3]	31–III–91	*10 337 504	...	...	10 529	*10 485	A2 b1	−0.1	102 173	103
Oceania · Océanie										
American Samoa[3 20]										
Samoa américaines[3 20]	1–IV–90	46 773	...	...	39[10]	x51	A3 b1	([11])	199	256
Australia										
Australie	30–VI–91	16 850 540	8 362 815	8 487 725	17 065	*17 661	A7 b1	1.1	7 713 364	2
Christmas Island										
Iles Christmas	30–VI–81	2 871	1 918	953	...	...	..	...	135	...
Cocos (Keeling) Islands										
Iles des Cocos (Keeling)	30–VI–81	555	298	257	...	...	...	...	14	...
Cook Islands[68]										
Iles Cook[68]	1–XII–86	17 614	9 188	8 426	18	x19	A9 b1	1.4	236	81
Fiji										
Fidji	31–VIII–86	715 375	362 568	352 807	731	x758	A7 b1	1.2	18 274	41
French Polynesia[69]										
Polynésie française[69]	6–IX–88	188 814	98 345	90 469	197	x211	A5 c1	2.3	4 000	53
Guam[3 20]										
Guam[3 20]	1–IV–90	133 152	70 945	62 207	x134	x144	A3 b1	2.4	549	262
Kiribati[70]										
Kiribati[70]	9–V–85	63 883	...	...	x72	x76	A8 c1	1.8	726	105
Marshall Islands										
Iles Marshall	13–XI–88	43 380	...	...	46	*52	A5 c1	3.9	181	287
Micronesia, Federated States of										
Micronésie, Etats fédérés de	1985–89	100 749	...	...	x108	x118	A8 c1	3.0	702	168
Nauru										
Nauru	22–I–77	7 254	...	...	x10	x10	A16c1	0.0	21	476
New Caledonia[71]										
Nouvelle–Calédonie[71]	4–IV–89	164 173	83 862	80 311	170	*179	A4 c1	1.8	18 575	10
New Zealand[72]										
Nouvelle–Zélande[72]	5–III–91	*3 434 952	...	...	3 363	*3 451	A2 b1	0.9	270 534	13

8

Population by sex, rate of population increase, surface area and density [*cont.*]
Population selon le sexe, taux d'accroissement de la population,
superficie et densité [*suite*]

Country or area Pays ou zone	Date	Latest Census Dernier recensement Both sexes Les deux sexes	Male Masculin	Female Féminin	Midyear estimates (thousands) Estimations au milieu de l'année (milliers) 1990	1993	Type[1] 1993	Annual rate of increase Taux d'accrois- sement annuel % 1990-93	Surface area (km²) Superficie (km²) 1993	Density Densité 1993[2]
Niue Nioué	29–X–86	2 531	...	...	x2	x2	A2 d	0.0	260	8
Norfolk Island Ile Norfolk	30–VI–86	2 367	1 170	1 197	...	...	..	...	36	...
Northern Mariana Islands Iles Mariannes du Nord	1990	43 345	...	...	26[10]	x47	A3 c1	(¹¹)	464	101
Palau Palaos	1990	15 122	...	...	x15	x16	A3 c1	2.2	459	35
Papua New Guinea[73] Papouasie–Nouv.–Guinée[73]	22–IX–80	3 010 727	1 575 672	1 435 055	3 699	*3 922	A13c3	2.0	462 840	8
Pitcairn Pitcairn	31–XII–91	66	...	...	...	...	A2 b1	...	5	...
Samoa Samoa	3–XI–81	156 349	81 027	75 322	164	x167	A12c1	0.6	2 831	59
Solomon Islands[74] Iles Salomon[74]	23–XI–86	285 176	147 972	137 204	x320	x354	A7 c1	3.4	28 896	12
Tokelau Tokélau	1–X–82	1 552	751	801	x2	x2	A11b1	0.0	12	167
Tonga Tonga	28–XI–86	94 649	47 611	47 038	97	x98	A7 c1	0.5	747	131
Tuvalu Tuvalu	27–V–79	7 300	...	...	x9	x9	A14c1	0.0	26	346
Vanuatu Vanuatu	16–V–89	142 419	73 384	69 035	147	x161	A4 c1	3.1	12 189	13
Wallis and Futuna Islands Iles Wallis et Futuna	15–II–83	12 408	6 266	6 142	x14	x14	A3 c1	0.0	200	70

Source:
Demographic statistics database of the Statistical Division of
the United Nations Secretariat.

Source:
Base de données pour les statistiques démographique de la
Division de statistique du Secrétariat de l'ONU.

† For detailed descriptions of data pertaining to former
  Czechoslovakia, Germany, SFR Yugoslavia and former USSR,
  see Annex I – Country or area nomenclature, regional and
  other groupings.

* Provisional.
x Estimate prepared by the Population Division of the United
  Nations.
1 For explanation of code, see technical notes to this chapter.
2 Population per squre kilometre of surface area in 1993.
  Figures are merely the quotients of population divided by
  surface area and are not to be considered either as
  reflecting density in the urban sense or as indicating the
  supporting power of a territory's land and resources.
3 De jure population.
4 Including the enclave of Cabinda.
5 Comprising Chagos Archipelago (formerly dependency of
  Mauritius).
6 Census of Chagos Archipelago taken 30 June 1962 gave
  total population of 747 persons.
7 Excluding Mayotte.
8 Comprising Bioko (which includes Pagalu)
  and Rio Muni (which includes Corisco and Elobeys).
9 Mid–year estimates have been adjusted for
  under–enumeration, estimated as follows:

† Pour les descriptions en détails des données relatives à l'ancienne
  Tchécoslovaquie, l'Allemagne, la Rfs Yougoslavie et l'ancienne
  URSS, voir l'Annexe I – Nomenclautre des pays ou zones,
  groupements régionaux et autres groupements.

* Données provisoires.
x Estimation établie par la Division de la population de
  l'Organisation des Nations Unies.
1 Pour l'explication du code, voir la remarque générale concerment
  ce chapître.
2 Nombre d'habîtants au kilomètre carré en 1993. Il s'agit
  simplement du quotient du chiffre de la population divisé
  par celui de la superficie: il ne faut pas y voir d'indication de la densité au
  sens urbain du terme ni de population que les terres et les ressources
  du territoire sont capables de nourrir.
3 Population de droit.
4 Y compris l'enclave de Cabinda.
5 Comprend l'archipel de Chagos (ancienne dépendance de
  Maurice).
6 Le recensement de la population de l'archipel de Chagos au 30 juin 1962
  a donnée comme population total 747 personnes.
7 Non compris Mayotte.
8 Comprend Bioko (qui comprend Pagalu) et Rio Muni (qui
  comprend Corisco et Elobeys).
9 Les estimations au milieu de l'année tiennent compte d'un ajustement
  destiné à compenser les lacunes du dénombrement. Les données de
  recensement ne tiennent pas compte de cet ajustement. En voici le détail:

8

Population by sex, rate of population increase, surface area and density [*cont.*]
Population selon le sexe, taux d'accroissement de la population,
superficie et densité [*suite*]

	Percentage adjustment	Adjusted census total		Adjustement (en pourcentage)	Chiffre de recensement ajusté
Brunei Darussalam	1.06	...	Brunéi Darussalam	1,06	...
Guatemala	13.7	...	Guatemala	13,7	...
Guinea	...	...	Guinée	...	...
Korea, Republic of	1.9	...	Corée, Rép. de	1,9	...
Mozambique	3.8	...	Mozambique	3,8	...
Netherlands Antilles	2.0	...	Antilles néerlandaises	2,0	...
Peru	...	...	Pérou	...	...
Sierra Leone	10.0	*3 002 426	Sierra Leone	10,0	*3 002 426
South Africa	...	...	Afrique du Sud	...	...
Uruguay	2.6	...	Uruguay	2,6	...

10 Estimate not in accord with the latest census and/or the latest estimate.

11 Rate not computed because of apparent lack of comparability between estimates shown for 1990 and 1993.

12 Including an estimate of 224,095 for nomad population.

13 Comprising the islands of Agalega and St. Brandon.

14 Excluding Bophuthatswana, Ciskei, Transkei and Venda.

15 Comprising the Northern Region (former Saguia el Hamra) and Southern Region (former Rio de Oro).

16 De jure population, but excluding persons residing in institutions.

17 Including Carriacou and other dependencies in the Grenadines.

18 Including dependencies: Marie−Galante, la Désirade, les Saintes, Petite−Terre, St. Barthélemy and French part of St. Martin.

19 Comprising Bonaire, Curaçao, Saba, St. Eustatius and Dutch part of St. Martin.

20 Including armed forces in the area.

21 Including Bequia and other islands in the Grenadines.

22 De jure population, but excluding civilian citizens absent from country for extended period of time. Census figures also exclude armed forces overseas.

23 Excluding Indian jungle population.

24 Mid−year estimates for 24 October.

25 Excluding nomadic Indian tribes.

26 Excluding dependencies, of which South Georgia (area 3,755 km^2) had an estimated population of 499 in 1964 (494 males, 5 females). The other dependencies namely, the South Sandwich group (surface area 337 km^2) and a number of smaller islands, are presumed to be uninhabited.

27 A dispute exists between the governments of Argentina and the United Kingdom of Great Britain and Northern Ireland concerning sovereignty over the Falkland Islands (Malvinas).

28 Excluding nomad population.

29 Excluding transients afloat.

30 Excluding foreign diplomatic personnel and their dependants.

31 This total population of China, as given in the communiqué of the State Statistical Bureau releasing the major figures of the census, includes a population of 6,130,000 for Hong Kong and Macau.

32 Comprising Hong Kong island, Kowloon and the New (leased) Territories.

33 Land area only. Total including ocean area within administrative boundaries is 2,916 km^2.

34 Including data for the Indian−held part of Jammu and Kashmir, the final status of which has not yet been determined.

35 Figures provided by Indonesia including East Timor,

10 L'estimation ne s'accorde avec le dernier recensement, et/ou avec la dernière estimation.

11 On n'a pas calculé le taux parce que les estimations pour 1990 et 1993 ne paraissent pas comparables.

12 Y compris une estimation de 224 095 personnes pour la population nomade

13 Y compris les îles Agalega et Saint−Brandon.

14 Non compris Bophuthatswana, Ciskei, Transkei et Venda.

15 Comprend la région septentrionale (ancien Saguia−el−Hamura) et la région méridionale (ancien Rio de Oro).

16 Population de droit, mais non compris les personnes dans les institutions.

17 Y compris Carriacou et les autres dépendances du groupe des îles Grenadines.

18 Y compris les dépendances: Marie−Galante, la Désirade, les Désirade, les Saintes, Petite−Terre, Saint−Barthélemy et la partie française de Saint−Martin.

19 Comprend Bonaire, Curaçao, Saba, Saint−Eustache et la partie néederlandaise de Saint−Martin.

20 Y compris les militaires en garnison sur le territoire.

21 Y compris Bequia et des autres îles dans les Grenadines.

22 Population de droit, mais non compris les civils hors du pays pendant une période prolongée. Les chiffres de recensement ne comprennent pas également les militaires à l'étranger.

23 Non compris les Indiens de la jungle.

24 Estimations au milieu de l'années pour le 24 Octobre.

25 Non compris les tribus d'Indiens nomades.

26 Non compris les dépendances, parmi lesquelles figure la Georgie du Sud (3 755 km^2) avec une population estimée à 499 personnes en 1964 (494 du sexe masculin et 5 du sexe féminin). Les autres dépendances, c'est−à−dire le groupe des Sandwich de Sud (superficie: 337 km^2) et certaines petite−îles, sont présumées inhabitées.

27 La souveraineté sur les îles Falkland (Malvinas) fait l'objet d'un différend entre le Gouvernement argentin et le Gouvernement du Royaume−Uni de Grande−Bretagne et d'Irlande du Nord.

28 Non compris la population nomade.

29 Non compris les personnes de passage à bord des navires.

30 Non compris le personnel diplomatique étranger et les membres de leur famille les accompagnant.

31 Le chiffre indiqué pour la population totale de la Chine, qui figure dans le communiqué du Bureau du statistique de l'Etat publiant les principaux chiffres du recensement, comprennent la population de Hong−kong et Macao qui s'élève à 6 130 000 personnes.

32 Comprend les îles de Hong−kong, Kowloon et les Nouveaux Territoires (à bail).

33 Superficie terrestre seulement. La superficie totale, qui comprend la zone maritime se trouvant à l'intérieur des limites administratives, est de

8
Population by sex, rate of population increase, surface area and density [*cont.*]
Population selon le sexe, taux d'accroissement de la population,
superficie et densité [*suite*]

shown separately.
36 Including data for East Jerusalem and Israeli
résidents in certain other territories under
occupation by Israeli military forces since June 1967.
37 Comprising Hokkaido, Honshu, Shikoku, Kyushu. Excluding
diplomatic personnel outside the country and foreign
military and civilian personnel and their dependants
stationed in the area.
38 Including military and diplomatic personnel and their families
abroad, numbering 933 at 1961 census, but excluding foreign
military and diplomatic personnel and their families in the
country, numbering 389 at 1961 census. Also including
registered Palestinian refugees number 654,092 and 722,687 at
30 June 1963 and 31 May 1967, respectively.
39 Excluding data for Jordanian territory under
occupation since June 1967 by Israeli military forces.
40 Excluding alien armed forces, civilian aliens employed
by armed forces, foreign diplomatic personnel and their
dependants and Korean diplomatic personnel and their
dependants outside the country.
41 Excluding Palestinian refugees in camps.
42 Based on results of sample survey.
43 Comprising Macau City and islands of Taipa and Coloane.
44 Excluding data for Jammu and Kashmir, the final
status of which has not yet been determined,
Junagardh, Manavadar, Gilgit and Baltistan.
45 Former mandated territory administered by the United
Kingdom until 1948.
46 Excluding United Kingdom armed forces, numbering 2,507.
47 Comprising that part of Palestine under Egyptian
administration following the Armistice of 1949 until
June 1967, when it was occupied by Israeli military forces.
48 Excluding transients afloat and non—locally domiciled military
and civilian services personnel and their dependants and visitors,
numbering 5,553, 5,187 and 8,895 respectively at 1980 census.
49 Including Palestinian refugees numbering 193,000 on 1 July 1977.
50 Comprising 7 sheikdoms of Abu Dhabi, Dubai, Sharjah,
Ajaman, Umm al Qaiwain, Ras al Khaimah and Fujairah,
and the area lying within the modified Riyadh line as
announced in October 1955.
51 Excluding surface area of frontier rivers.
52 Including dependencies: Alderey, Brechou, Herm,
Jethou, Lithou and Sark Island.
53 Excluding Faeroe Islands and Greenland, shown separately.
54 Excluding Overseas Departments, namely French Guiana,
Guadeloupe, Martinique and Réunion, shown separately.
55 De jure population, but excluding diplomatic personnel
outside the country and including foreign diplomatic
personnel not living in embassies or consulates.
56 Excluding military personnel stationed outside the
country who do not have a personal residence in France.
57 Excluding armed forces.
58 Including armed forces stationed outside the country,
but including alien armed forces stationed in the area.
59 Including armed forces stationed outside the country,
but excluding alien armed forces stationed in the area.
60 Surface area is 0.44 km².
61 Including Gozo and Comino Islands and civilian
nationals temporarily outside the country.
62 Surface area is 1.49 km².
63 Excluding civilian aliens within the country, but
including civilian nationals temporarily outside the country.
64 Including the Azores and Madeira Islands.
65 Including the Balearic and Canary Islands, and Alhucemas,
Ceuta, Chafarinas, Melilla and Penon de Vélez de la Gomera.
66 Inhabited only during the winter season. Census data are for
total population while estimates refer to Norwegian population

2 916 km².
34 Y compris les données pour la partie du Jammu et du Cacehmire occupée
par l'Inde dont le statut définitif n'a pas encore été déterminé.
35 Les chiffres fournis par l'Indonesie comprennent le Timor
oriental, qui fait l'objet d'une rubrique distincte.
36 Y compris les données pour Jérusalem—Est et les résidents israéliens dans
certains autres territoires occupés depuis juin 1967 par les forces
armées israéliennes.
37 Comprend Hokkaido, Honshu, Shikoku, Kyushu. Non compris le
personnel diplomatique hors du pays, les militaires et agents civils étrangers
en poste sur le territoire et les membres de leur famille les accompagnant.
38 Y compris les militaires et le personnel diplomatique à l'étranger et
les membres de leur famille les accompagnant, au nombre de 933
personnes au recensement de 1961, mais non compris les militaires et
le personnel diplomatique étrangers sur le territoire et les membres
de leur famille les accompagnant, au nombre de 389 personnes au
recensement de 1961. Y compris également les réfugiés de Palestine
immatriculés: 654 092 au 30 juin 1963 et 722 687 au 31 mai 1967.
39 Non compris les données pour le territoire jordanien
occupé depuis juin 1967 par les forces armées israéliennes.
40 Non compris les militaires étrangers, les civils étrangers employés par les
armées, le personnel diplomatique étranger et les membres de leur famille
les accompagnant et le personnel diplomatique coréen hors du pays et les
membres de leur familles les accompagnant.
41 Non compris les réfugiés de Palestine dans les camps.
42 D'après les résultats d'une enquête par sondage.
43 Comprend la ville de Macao et les îles de Taipa et de Colowane.
44 Non compris les données pour le Jammu et le Cachemire, dont le status
définitif n'a pas encore été déterminé, le Junagardh, le Manavadar,
le Gilgit et le Baltistan.
45 Ancien territoire sous mandat administré par le
Royaume—Uni jusqu'à 1948.
46 Non compris les forces armées du Royaume—Uni au nombre de 2 507
personnes.
47 Comprend la partie de la Palestine administrée par l'Egypt depuis
l'armistice de 1949 jusqu'en juin 1967, date laquelle elle a été occupée par
les forces armées israéliennes.
48 Non compris les personnes de passage à bord de navires,
les militaires et agents civils non résidents et les membres de leur
famille les accompagnant, et les visiteurs, soit: 5 553, 5 187 et 8 895
personnes respectivement au recensement de 1980.
49 Y compris les réfugiés de Palestine au nombre de
193 000 au 1er juillet 1977.
50 Comprend les sept cheikhats de Abou Dhabi, Dabai, Ghârdja, Adjmân,
Oumm—al—Quiwaïn, Ras al Khaïma et Foudjaïra, ainsi que la zone
délimitée par la ligne de Riad modifiée comme il a été announcé
en octobre 1955.
51 Non compris la surface des cours d'eau frontières.
52 Y compris les dépendances: Aurigny, Brecqhou, Herm,
Jethou, Lihou et l'île de Sercq.
53 Non compris les îles Féroé et le Groenland, qui
font l'objet de rubriques distinctes.
54 Non compris les départements d'outre—mer, c'est—à—dire la Guyane
française, la Guadeloupe, la Martinique et la Réunion, qui font l'objet de
rubriques distinctes.
55 Population de droit, non compris le personnel diplomatique hors du
pays et y compris le personnel diplomatique étranger qui ne vit
pas dans les ambassades ou les consulats.
56 Non compris les militaires en garnison hors du pays et
sans résidence personnelle en France.
57 Non compris les militaires.
58 Y compris les militaires en garnison hors du pays, mais non compris les
militaires étrangers en garnison sur le territoire.
territoire.
59 Y compris les militaires en garnison hors du pays, mais y compris les
militaires étrangers en garnison sur le territoire.
territoire.

8

Population by sex, rate of population increase, surface area and density [*cont.*]
Population selon le sexe, taux d'accroissement de la population,
superficie et densité [*suite*]

only. Included also in the de jure population of Norway.

67 Excluding Channel Islands and Isle of Man, shown separately.

68 Excluding Niue, shown separately, which is part of Cook Islands, but because of remoteness is administered separately.

69 Comprising Austral, Gambier, Marquesas, Rapa, Society and Tuamotu Islands.

70 Including Christmas, Fanning, Ocean and Washington Islands.

71 Including the islands of Huon, Chesterfield, Loyalty, Walpole and Belep Archipelago.

72 Including Campbell and Kermadec Islands (population 20 in 1961, surface area 148 km^2) as well as Antipodes, Auckland, Bounty, Snares, Solander and Three Kings island, all of which are uninhabited. Excluding diplomatic personnel and armed forces outside the country, the latter numbering 1,936 at 1966 census; also excluding alien armed forces within the country.

73 Comprising eastern part of New Guinea, the Bismarck Archipelago, Bougainville and Buka of Solomon Islands group and about 600 smaller islands.

74 Comprising the Solomon islands group (except Bougainville and Buka which are included with Papua New Guinea shown separately), Ontong, Java, Rennel and Santa Cruz Islands.

60 Superficie: 0,44 km^2.

61 Y compris les îles de Gozo et de Comino et les civils nationaux temporairement hors du pays.

62 Superficie: 1,49 km^2.

63 Non compris les civils étrangers dans le pays, mais y compris les civils nationaux temporairement hors du pays.

64 Y compris les Açores et Madère.

65 Y compris les Baléares et les Canaries, Al Hoceima, Ceuta, les îles Zaffarines, Melilla et Penon de Vélez de la Gomera.

66 N'est habitée pendant la saison d'hiver. Les données de recensement se rapportent à la populatio totale, mais les estimations ne concernent que la population norvégienne, comprise également dans la population de droit de la Norvège.

67 Non compris les îles Anglo–Normandes et l'île de Man, qui font l'objet de rubriques distinctes.

68 Non compris Nioué, qui fait l'objet d'une rubrique distincte et qui fait partie des îles Cook, mais qui, en raison de son éloignement, est administrée séparément.

69 Comprend les îles Australes, Gambier, Marquises, Rapa, de la Societé et Tuamotou.

70 Y compris les îles Christmas, Fanning, Océan et Washington.

71 Y compris les îles Huon, Chesterfield, Loyauté et Walpole, et l'archipel Belep.

72 Y compris les îles Campbell et Kermadec (20 habitants en 1961, superficie: 148 km^2) ainsi que les îles Antipodes, Auckland, Bounty, Snares, Solander et Three Kings, qui sont toutes inhabitées. Non compris le personnel diplomatique et les militaires hors du pays, ces derniers au nombre de 1936 au recensement de 1966; non compris également les militaires étrangers dans le pays.

73 Comprend l'est de la Nouvelle–Guinée, l'archipel Bismarck, Bougainville et Buka (ces deux dernières du groupe des Salomon) et environ 600 îlots.

74 Comprend les îles Salomon (à l'exception de Bougainville et de Buka dont la population est comprise dans celle de Papouasie–Nouvelle Guinée qui font l'objet d'une rubrique distincte), ainsi que les îles Ontong, Java, Rennel et Santa Cruz.

9

Population in urban and rural areas, rates of growth and largest city population
Population urbaine, population rurale, taux d'accroissement et population de la ville la plus peuplée

Country or area Pays ou zone	Year Année	Population estimates Estimations de la population		Growth rate p.a. (%) [1] Taux d'accroissement p.a. (%) [1]		Largest city population Ville la plus peuplée		
		R%	U%	Pop.r.	Pop.u.	Number (000s) Nombre (000s)	%u	%t
Africa · Afrique								
Algeria	1990	48.3	51.7	0.9	4.3	3033	23.5	12.2
Algérie	1995	44.2	55.8	0.5	3.8	3702	23.8	13.3
Angola	1990	71.7	28.3	1.8	5.7	1642	63.1	17.9
Angola	1995	67.8	32.2	2.6	6.3	2207	61.9	19.9
Benin	1990	71.0	29.0	2.4	4.5	169[2]	12.6	3.6
Bénin	1995	68.7	31.3	2.5	4.6	...	...	...
Botswana	1990	76.9	23.1	2.3	7.5	109	36.9	8.5
Botswana	1995	71.9	28.1	1.7	7.0	...	...	...
Burkina Faso	1990	82.2	17.9	1.1	11.6	681	42.4	7.6
Burkina Faso	1995	72.8	27.2	0.3	11.2	...	...	...
Burundi	1990	93.7	6.3	2.7	6.6	234	67.8	4.3
Burundi	1995	92.5	7.5	2.7	6.6	...	...	...
Cameroon	1990	59.7	40.3	1.4	5.3	1001	21.6	8.7
Cameroun	1995	55.1	44.9	1.2	4.9	1322	22.3	10.0
Cape Verde	1990	55.9	44.2	−1.7	7.7	62	41.1	18.2
Cap−Vert	1995	45.7	54.3	−1.3	6.9	...	...	...
Central African Rep.	1990	62.5	37.5	2.0	3.1	474	43.2	16.2
Rép. centrafricaine	1995	60.8	39.3	1.9	3.4	...	...	...
Chad	1990	79.5	20.5	1.8	2.9	613	53.9	11.0
Tchad	1995	78.6	21.4	2.5	3.6	...	...	...
Comoros	1990	72.2	27.8	2.9	5.4	24	15.9	4.4
Comores	1995	69.3	30.7	2.9	5.7	...	...	...
Congo	1990	46.5	53.5	0.5	5.4	793	66.4	35.6
Congo	1995	41.2	58.8	0.6	4.9	1009	66.3	39.0
Côte d'Ivoire	1990	59.6	40.4	2.8	5.2	2168	44.8	18.1
Côte d'Ivoire	1995	56.4	43.6	2.4	5.0	2797	45.0	19.6
Djibouti	1990	19.3	80.7	2.7	6.3	417	100.0	80.7
Djibouti	1995	17.2	82.8	−0.1	2.7	...	...	...
Egypt	1990	56.1	43.9	2.5	2.5	8633	34.9	15.3
Egypte	1995	55.2	44.8	1.9	2.6	9656	34.3	15.3
Equatorial Guinea	1990	64.3	35.8	0.7	6.1	30	23.8	8.5
Guinée équatoriale	1995	57.8	42.2	0.4	5.9	...	...	...
Eritrea	1990	84.2	15.8	2.4	3.9	359	73.7	11.6
Erythrée	1995	82.8	17.2	2.4	4.4	...	...	...
Ethiopia	1990	87.7	12.3	2.7	4.2	1808	31.1	3.8
Ethiopie	1995	86.6	13.4	2.7	4.7	2209	30.0	4.0
Gabon	1990	54.3	45.7	1.3	5.3	286	54.7	25.0
Gabon	1995	50.0	50.0	1.2	4.7	...	...	...
Gambia	1990	77.4	22.7	3.7	6.6	209	100.0	22.6
Gambie	1995	74.5	25.5	3.1	6.2	...	...	...
Ghana	1990	66.0	34.0	2.6	4.2	1405	27.5	9.4
Ghana	1995	63.7	36.3	2.3	4.3	1687	26.6	9.7
Guinea	1990	74.2	25.8	1.9	5.8	1127	76.0	19.6
Guinée	1995	70.4	29.6	2.0	5.8	1508	76.1	22.5
Guinea−Bissau	1990	80.1	19.9	1.6	3.9	71	37.2	7.4
Guinée−Bissau	1995	77.8	22.2	1.6	4.4	...	...	...
Kenya	1990	76.4	23.6	2.4	7.1	1519	27.3	6.4
Kenya	1995	72.3	27.7	2.5	6.8	2079	26.6	7.4
Lesotho	1990	80.6	19.4	1.9	6.5	170	48.9	9.5
Lesotho	1995	76.9	23.1	1.8	6.2	...	...	...
Liberia	1990	57.9	42.1	2.2	4.5	670	61.8	26.0
Libéria	1995	55.1	45.0	2.3	4.6	...	...	...
Libyan Arab Jamah.	1990	17.6	82.4	−1.6	5.0	2595	69.3	57.1
Jamah. arabe libyenne	1995	14.0	86.0	−1.1	4.3	3272	70.4	60.5
Madagascar	1990	76.2	23.8	2.6	6.0	690	23.1	5.5
Madagascar	1995	72.9	27.1	2.3	5.8	...	...	...
Malawi	1990	88.2	11.8	4.8	7.7	310[2]	28.0	3.3
Malawi	1995	86.5	13.5	3.1	6.2	...	...	...

9
Population in urban and rural areas, rates of growth and largest city population [*cont.*]
Population urbaine, population rurale, taux d'accroissement et population de la ville la plus peuplée [*suite*]

Country or area Pays ou zone	Year Année	Population estimates Estimations de la population		Growth rate p.a. (%) [1] Taux d'accroissement p.a. (%) [1]		Largest city population Ville la plus peuplée Number (000s) Nombre (000s)		
		R%	U%	Pop.r.	Pop.u.		%u	%t
Mali	1990	76.2	23.8	2.3	5.5	738	33.7	8.0
Mali	1995	73.1	27.0	2.3	5.7	...	...	...
Mauritania	1990	53.2	46.8	−0.5	6.6	707	75.5	35.3
Mauritanie	1995	46.2	53.8	−0.3	5.4			
Mauritius	1990	59.5	40.5	1.1	0.4	158	36.9	14.9
Maurice	1995	59.4	40.6	1.1	1.2	...	...	...
Morocco	1990	53.9	46.1	1.4	3.2	2815	25.1	11.6
Maroc	1995	51.6	48.4	1.2	3.1	3289	25.2	12.2
Mozambique	1990	73.3	26.8	−1.0	7.3	1561	41.1	11.0
Mozambique	1995	65.8	34.3	0.3	7.4	2227	40.6	13.9
Namibia	1990	68.1	31.9	1.3	6.2	149	34.7	11.0
Namibie	1995	62.6	37.4	1.0	5.9	...	...	...
Niger	1990	84.8	15.2	2.8	5.1	447	38.0	5.8
Niger	1995	83.0	17.0	2.9	5.6	...	...	...
Nigeria	1990	64.8	35.2	1.7	5.4	7742	22.9	8.1
Nigéria	1995	60.7	39.3	1.7	5.2	10287	23.4	9.2
Réunion	1990	36.2	63.9	−0.4	3.3	123	31.9	20.4
Réunion	1995	32.2	67.8	−0.7	2.7			
Rwanda	1990	94.4	5.6	2.8	4.5	219	56.0	3.1
Rwanda	1995	93.9	6.1	2.5	4.2	...	...	...
Sao Tome and Principe	1990	57.7	42.3	0.8	4.7	50	100.0	42.0
Sao Tomé−et−Principe	1995	53.3	46.7	0.6	4.2	...	...	...
Senegal	1990	60.2	39.8	2.1	3.8	1613	55.3	22.0
Sénégal	1995	57.8	42.3	1.7	3.7	1986	56.6	23.9
Seychelles	1990	50.2	49.8	−0.6	3.4	35	100.0	50.0
Seychelles	1995	45.5	54.5	−0.9	2.9	...	...	...
Sierra Leone	1990	67.8	32.2	1.1	4.8	649	50.4	16.2
Sierra Leone	1995	63.8	36.2	1.2	4.8	...	...	...
Somalia	1990	75.8	24.2	1.7	2.8	779	37.1	9.0
Somalie	1995	74.3	25.8	0.9	2.5	982	41.2	10.6
South Africa	1990	50.8	49.2	2.0	2.7	2294	12.6	6.2
Afrique du Sud	1995	49.2	50.8	1.6	2.9	2671	12.7	6.4
Saint Helena	1990	76.0	24.1	0.3	2.6	1	100.0	24.1
Sainte−Hélène	1995	73.1	26.9	0.1	3.1	...	...	...
Sudan	1990	77.5	22.6	2.3	4.1	1944	35.1	7.9
Soudan	1995	75.4	24.6	2.1	4.4	2429	35.1	8.7
Swaziland	1990	73.6	26.4	1.5	6.5	47	24.0	6.3
Swaziland	1995	68.8	31.2	1.4	6.2			
Togo	1990	71.6	28.5	2.5	4.5	513	51.0	14.5
Togo	1995	69.2	30.8	2.5	4.8	...	...	...
Tunisia	1990	45.1	54.9	1.3	2.9	1741	39.2	21.5
Tunisie	1995	42.8	57.3	0.9	2.8	2037	40.0	22.9
Uganda	1990	88.8	11.2	3.2	5.8	754	37.7	4.2
Ouganda	1995	87.5	12.5	3.1	5.8	954	35.7	4.5
United Rep.Tanzania	1990	79.2	20.8	2.4	6.6	1436	27.0	5.6
Rép. Unie de Tanzanie	1995	75.7	24.4	2.0	6.1	1734	24.0	5.8
Western Sahara	1990	43.5	56.5	2.8	5.7	98	75.4	42.6
Sahara occidental	1995	40.1	59.9	2.5	5.3	...	...	...
Zaire	1990	71.9	28.1	3.3	3.4	3455	32.9	9.2
Zaïre	1995	70.9	29.1	2.9	3.9	4214	33.0	9.6
Zambia	1990	58.0	42.0	3.1	4.0	979	28.6	12.0
Zambie	1995	57.0	43.1	2.6	3.5	1327	32.6	14.0
Zimbabwe	1990	71.5	28.5	2.4	5.8	854	30.2	8.6
Zimbabwe	1995	67.9	32.1	1.5	5.0	1044	28.9	9.3
America, North · Amerique du Nord								
Antigua and Barbuda	1990	64.6	35.4	0.4	0.8	23	100.0	35.9
Antigua−et−Barbuda	1995	64.2	35.8	0.5	0.8	...	...	...
Bahamas	1990	16.4	83.6	−2.5	2.7	173	80.8	67.6
Bahamas	1995	13.5	86.5	−2.4	2.2	...	...	...
Barbados	1990	55.2	44.8	−0.5	1.4	115	100.0	44.7
Barbade	1995	52.6	47.4	−0.6	1.5	...	...	...
Belize	1990	52.4	47.7	2.9	2.2	5	5.6	2.6
Belize	1995	53.2	46.8	3.0	2.3	...	...	...

9
Population in urban and rural areas, rates of growth and largest city population [*cont.*]
Population urbaine, population rurale, taux d'accroissement et population de la ville la plus peuplée [*suite*]

Country or area Pays ou zone	Year Année	Population estimates Estimations de la population		Growth rate p.a. (%) [1] Taux d'accroissement p.a. (%) [1]		Largest city population Ville la plus peuplée Number (000s)		
		R%	U%	Pop.r.	Pop.u.	Nombre (000s)	%u	%t
Bermuda	1990	0.0	100.0	0.0	1.7	1	1.6	1.6
Bermudes	1995	0.0	100.0	0.0	0.7	...	...	...
Canada	1990	23.4	76.6	1.2	1.4	3770	17.7	13.6
Canada	1995	23.3	76.7	1.1	1.2	4483	19.8	15.2
Cayman Islands	1990	0.0	100.0	0.0	3.9	13	50.0	50.0
Iles Caïmanes	1995	0.0	100.0	0.0	3.5	...	...	...
Costa Rica	1990	52.9	47.1	2.0	3.7	760	53.2	25.0
Costa Rica	1995	50.3	49.7	1.4	3.5	879	51.7	25.7
Cuba	1990	26.4	73.6	−0.9	1.7	2124	27.2	20.0
Cuba	1995	24.0	76.0	−1.1	1.5	2241	26.7	20.3
Dominican Republic	1990	39.6	60.4	−0.1	3.8	2199	51.2	30.9
Rép. dominicaine	1995	35.4	64.6	−0.3	3.3	2580	51.1	33.0
El Salvador	1990	56.1	43.9	1.3	2.3	597	26.3	11.5
El Salvador	1995	55.0	45.1	1.8	2.7	...	...	...
Greenland	1990	21.7	78.3	−0.2	1.4	23	52.3	41.1
Groënland	1995	20.2	79.8	−0.7	1.1	...	...	...
Guadeloupe	1990	1.5	98.5	−21.4	2.6	26	6.8	6.6
Guadeloupe	1995	0.6	99.4	−17.5	2.0	...	...	...
Guatemala	1990	60.6	39.5	2.4	3.6	842	23.2	9.2
Guatemala	1995	58.5	41.5	2.2	3.9	946	21.5	8.9
Haiti	1990	71.4	28.6	1.3	3.9	1041	56.1	16.0
Haïti	1995	68.4	31.6	1.2	4.0	1266	55.9	17.6
Honduras	1990	59.3	40.7	2.1	4.6	677	34.1	13.9
Honduras	1995	56.1	43.9	1.8	4.5	...	...	...
Jamaica	1990	48.6	51.5	−0.5	1.4	582	47.8	24.6
Jamaïque	1995	46.3	53.7	−0.3	1.5	...	...	...
Martinique	1990	9.5	90.5	−6.8	2.2	102	31.3	28.3
Martinique	1995	6.7	93.3	−5.9	1.6	...	...	...
Mexico	1990	27.4	72.6	0.2	3.1	15085	24.6	17.9
Mexique	1995	24.7	75.3	−0.0	2.8	15643	22.2	16.7
Montserrat	1990	87.6	12.4	−0.8	0.9	2	...	18.2
Montserrat	1995	86.2	13.8	−0.7	1.8	...	...	...
Netherlands Antilles	1990	31.6	68.4	0.5	1.0	115	88.5	60.5
Antilles néerlandaises	1995	30.5	69.5	0.2	1.2	...	...	...
Nicaragua	1990	40.2	59.8	1.1	3.7	964	43.9	26.2
Nicaragua	1995	37.1	62.9	2.1	4.8	1195	42.9	27.0
Panama	1990	48.3	51.7	1.6	2.4	826	66.6	34.4
Panama	1995	46.7	53.3	1.2	2.4	948	67.7	36.0
Puerto Rico	1990	28.7	71.3	−0.5	1.6	1092	43.4	30.9
Porto Rico	1995	26.6	73.4	−0.7	1.4	1101	40.8	30.0
Saint Lucia	1990	54.0	46.0	0.7	2.3	61	100.0	45.9
Sainte−Lucie	1995	51.9	48.1	0.6	2.2	...	...	...
Saint Kitts and Nevis	1990	60.4	39.6	−1.2	0.6	17	100.0	39.6
Saint−Kitts−et−Nevis	1995	57.6	42.4	−1.2	1.1	...	...	...
Saint Pierre and Miquelon	1990	9.2	90.9	−0.6	1.0	6	100.0	90.9
Saint−Pierre−et−Miquelon	1995	8.5	91.5	−0.8	0.9	...	...	...
St. Vincent−Grenadines	1990	59.1	41.0	−1.3	4.5	26	59.1	24.3
St. Vincent−Grenadines	1995	53.0	47.0	−1.3	3.6	...	...	...
Trinidad and Tobago	1990	30.9	69.1	−0.6	2.1	51	6.0	4.1
Trinité−et−Tobago	1995	28.2	71.8	−0.8	1.9	...	...	...
Turks and Caicos Islands	1990	57.4	42.6	4.0	4.8	5	100.0	42.6
Iles Turques et Caiques	1995	55.5	44.5	3.2	4.7	...	...	...
United States	1990	24.8	75.2	0.4	1.1	16056	8.5	6.4
Etats−Unis	1995	23.8	76.2	0.2	1.3	16329	8.1	6.2
United States Virgin Is.	1990	55.5	44.5	0.5	0.5	12	26.7	11.8
Iles Vierges américaines	1995	54.7	45.3	0.3	0.9	...	...	...
America, South · Amerique du Sud								
Argentina	1990	13.5	86.5	−1.0	1.8	10623	37.7	32.6
Argentine	1995	11.9	88.1	−1.3	1.6	10990	36.1	31.8
Bolivia	1990	44.2	55.8	−0.0	4.1	1041	28.4	15.8
Bolivie	1995	39.2	60.8	0.0	4.1	1246	27.7	16.8
Brazil	1990	25.4	74.6	−1.0	3.0	14847	13.4	10.0
Brésil	1995	21.8	78.3	−1.4	2.7	16417	13.0	10.2

9

Population in urban and rural areas, rates of growth and largest city population [*cont.*]
Population urbaine, population rurale, taux d'accroissement et population de la ville la plus peuplée [*suite*]

Country or area Pays ou zone	Year Année	Population estimates Estimations de la population		Growth rate p.a. (%) [1] Taux d'accroissement p.a. (%) [1]		Largest city population Ville la plus peuplée		
		R%	U%	Pop.r.	Pop.u.	Number (000s) Nombre (000s)	%u	%t
Chile	1990	16.7	83.3	1.0	1.9	4588	41.9	34.9
Chili	1995	16.1	83.9	0.9	1.8	5065	42.3	35.5
Colombia	1990	30.0	70.0	−0.1	2.7	4851	21.5	15.0
Colombie	1995	27.3	72.7	−0.3	2.4	5614	22.0	16.0
Ecuador	1990	45.2	54.8	0.8	3.8	1492	26.5	14.5
Equateur	1995	41.6	58.4	0.5	3.5	1717	25.6	15.0
Falkland Is. (Malvinas)	1990	25.5	74.5	−6.3	3.6	2	100.0	74.5
Iles Falkland (Malvinas)	1995	15.9	84.1	−8.9	3.0	...	...	...
French Guiana	1990	25.4	74.6	3.6	5.7	41	47.1	35.0
Guyane française	1995	23.5	76.5	3.0	5.0	...	...	...
Guyana	1990	66.4	33.6	−0.4	1.3	236	88.1	29.6
Guyana	1995	63.8	36.2	0.1	2.4	...	...	...
Paraguay	1990	51.2	48.9	1.7	4.7	630	29.9	14.6
Paraguay	1995	47.3	52.7	1.2	4.3	...	...	...
Peru	1990	30.2	69.8	0.4	2.8	6475	43.0	30.0
Pérou	1995	27.8	72.2	0.3	2.6	7452	43.4	31.3
Suriname	1990	52.5	47.5	0.5	2.0	100	52.6	25.0
Suriname	1995	49.6	50.4	−0.0	2.3	...	...	...
Uruguay	1990	11.1	88.9	−2.3	1.0	1287	46.8	41.6
Uruguay	1995	9.7	90.3	−2.1	0.9	1326	46.1	41.6
Venezuela	1990	9.6	90.4	−3.1	3.3	2773	15.7	14.2
Venezuela	1995	7.2	92.8	−3.5	2.8	2959	14.6	13.6
Asia · Asie								
Afghanistan	1990	81.8	18.3	0.4	2.2	1565	57.0	10.4
Afghanistan	1995	80.0	20.0	5.4	7.7	2034	50.5	10.1
Armenia	1990	32.5	67.5	0.2	1.0	1210	53.5	36.1
Arménie	1995	31.3	68.7	0.6	1.8	1305	52.8	36.3
Azerbaijan	1990	45.6	54.4	1.0	1.6	1751	45.2	24.6
Azerbaïdjan	1995	44.2	55.8	0.6	1.7	1853	43.9	24.5
Bahrain	1990	12.5	87.5	−1.4	4.2	133	31.0	27.1
Bahreïn	1995	9.7	90.3	−2.2	3.4	...	...	...
Bangladesh	1990	84.3	15.7	1.3	5.0	5877	34.7	5.4
Bangladesh	1995	81.7	18.3	1.5	5.3	7832	35.5	6.5
Bhutan	1990	94.7	5.3	2.1	5.6	17	20.5	1.1
Bhoutan	1995	93.6	6.4	1.0	4.8	...	...	...
Brunei Darussalam	1990	42.3	57.8	2.6	2.6	64	43.0	24.9
Brunéi Darussalam	1995	42.2	57.8	2.0	2.1	...	...	...
Cambodia	1990	82.4	17.6	2.5	6.6	369	23.7	4.2
Cambodge	1995	79.3	20.7	2.2	6.2	...	...	...
China	1990	73.8	26.2	0.6	4.5	13452	4.4	1.2
Chine	1995	69.8	30.3	−0.0	4.0	15082	4.1	1.2
Cyprus	1990	48.6	51.4	0.0	2.2	162	44.9	23.1
Chypre	1995	45.9	54.1	−0.0	2.1	...	...	...
East Timor	1990	92.2	7.8	2.4	1.5	58	100.0	7.8
Timor oriental	1995	92.5	7.5	2.0	1.0	...	...	...
Gaza Strip (Palestine)	1990	6.4	93.6	−0.1	4.1	...	...	...
Zone de Gaza (Palestine)	1995	5.5	94.5	2.1	5.0	...	...	...
Georgia	1990	44.0	56.0	−0.4	1.4	1277	42.1	23.6
Géorgie	1995	41.5	58.5	−1.0	1.0	1353	42.4	24.8
Hong Kong	1990	5.9	94.1	−2.9	1.2	5369	100.0	94.1
Hong−kong	1995	5.0	95.0	−2.8	0.8	5574	100.0	95.0
India	1990	74.5	25.5	1.7	3.0	12223	5.6	1.4
Inde	1995	73.2	26.8	1.6	2.9	15093	6.0	1.6
Indonesia	1990	69.4	30.6	0.5	4.9	9250	16.5	5.1
Indonésie	1995	64.6	35.4	0.1	4.5	11500	16.4	5.8
Iran, Islamic Rep. of	1990	43.7	56.3	2.4	4.8	6351	19.2	10.8
Iran, Rép. islamique d'	1995	41.0	59.0	1.3	3.6	6830	17.2	10.2
Iraq	1990	28.2	71.8	1.2	4.2	4044	31.1	22.4
Iraq	1995	25.4	74.6	0.4	3.2	4478	29.4	21.9
Israel	1990	9.7	90.3	1.0	2.0	1790	42.6	38.4
Israël	1995	9.4	90.6	3.1	3.9	1921	37.7	34.1
Japan	1990	22.8	77.2	0.0	0.6	25013	26.2	20.3
Japon	1995	22.4	77.6	−0.2	0.4	26836	27.6	21.5

9

Population in urban and rural areas, rates of growth and largest city population [*cont.*]
Population urbaine, population rurale, taux d'accroissement et population de la ville la plus peuplée [*suite*]

Country or area Pays ou zone	Year Année	Population estimates Estimations de la population		Growth rate p.a. (%) [1] Taux d'accroissement p.a. (%) [1]		Largest city population Ville la plus peuplée Number (000s) Nombre (000s)		
		R%	U%	Pop.r.	Pop.u.		%u	%t
Jordan	1990	32.0	68.0	−0.2	3.3	955	33.0	22.4
Jordanie	1995	28.5	71.5	2.6	5.9	1187	30.5	21.8
Kazakhstan	1990	42.4	57.6	0.3	1.7	1158	12.1	7.0
Kazakhstan	1995	40.3	59.7	−0.5	1.2	1262	12.4	7.4
Korea, Dem. P. R.	1990	40.2	59.8	1.3	2.2	2230	17.1	10.2
Corée, R. p. dém. de	1995	38.8	61.3	1.2	2.4	2470	16.9	10.3
Korea, Republic of	1990	26.2	73.9	−4.9	3.6	10558	33.4	24.6
Corée, République de	1995	18.7	81.3	−5.7	2.9	11641	31.8	25.9
Kuwait	1990	4.2	95.9	−3.6	4.8	1090	53.1	50.9
Koweït	1995	3.0	97.0	−13.0	−6.3	1090	72.7	70.5
Kyrgyzstan	1990	61.8	38.2	1.8	1.8	629	37.8	14.4
Kirghizistan	1995	61.1	38.9	1.4	2.1	...	...	...
Lao People's Dem. Rep.	1990	81.4	18.6	2.5	6.3	415	53.1	9.9
Rép. dém. pop. lao	1995	78.3	21.7	2.2	6.1	...	...	...
Lebanon	1990	16.2	83.8	−6.0	0.3	1563	73.0	61.2
Liban	1995	12.8	87.2	−1.4	4.1	...	...	...
Macau	1990	1.3	98.7	0.7	3.8	338	100.0	98.8
Macao	1995	1.2	98.8	1.3	3.7	...	...	...
Malaysia	1990	50.2	49.8	1.2	4.3	1122	12.6	6.3
Malaisie	1995	46.3	53.7	0.8	3.9	1238	11.5	6.2
Maldives	1990	74.1	25.9	3.2	3.4	56	100.0	25.9
Maldives	1995	73.2	26.8	3.1	4.0	...	...	...
Mongolia	1990	42.0	58.0	1.3	3.7	573	45.4	26.3
Mongolie	1995	39.1	60.9	0.6	3.0	...	...	...
Myanmar	1990	75.3	24.8	2.0	2.8	3302	31.9	7.9
Myanmar	1995	73.8	26.2	1.8	3.3	3851	31.6	8.3
Nepal	1990	89.1	10.9	2.0	7.4	363	17.3	1.9
Népal	1995	86.3	13.7	2.0	7.1	...	...	...
Oman	1990	89.0	11.0	4.1	8.2	53	27.5	3.0
Oman	1995	86.8	13.2	3.7	7.8	...	...	...
Pakistan	1990	68.0	32.0	2.8	4.9	7965	20.4	6.5
Pakistan	1995	65.3	34.7	2.0	4.4	9863	20.2	7.0
Philippines	1990	51.2	48.8	−0.0	4.6	7968	26.9	13.1
Philippines	1995	45.8	54.2	−0.1	4.2	9280	25.3	13.7
Qatar	1990	10.1	89.9	2.5	6.6	293	67.2	60.4
Qatar	1995	8.6	91.4	−0.7	2.9	...	...	...
Saudi Arabia	1990	22.7	77.3	1.3	5.9	1975	15.9	12.3
Arabie saoudite	1995	19.8	80.2	−0.6	2.9	2576	18.0	14.4
Singapore	1990	0.0	100.0	0.0	1.1	2705	100.0	100.0
Singapour	1995	0.0	100.0	0.0	1.0	2848	100.0	100.0
Sri Lanka	1990	78.6	21.4	1.3	1.6	616	16.7	3.6
Sri Lanka	1995	77.6	22.4	1.0	2.2	...	...	...
Syrian Arab Republic	1990	49.8	50.2	2.8	4.3	1790	28.9	14.5
Rép. arabe syrienne	1995	47.6	52.4	2.6	4.3	2052	26.7	14.0
Tajikistan	1990	67.8	32.2	3.3	2.3	607	35.7	11.5
Tadjikistan	1995	67.8	32.2	2.9	2.9	...	...	...
Thailand	1990	81.3	18.7	1.5	2.6	5894	56.6	10.6
Thaïlande	1995	80.0	20.1	0.8	2.5	6566	55.7	11.2
Turkey	1990	39.1	60.9	−1.8	5.2	6507	19.0	11.6
Turquie	1995	31.2	68.8	−2.5	4.4	7817	18.4	12.6
Turkmenistan	1990	55.1	44.9	2.9	2.0	412	25.1	11.3
Turkménistan	1995	55.1	44.9	2.3	2.3	...	...	...
Uzbekistan	1990	59.4	40.6	2.4	2.4	2109	25.5	10.3
Ouzbékistan	1995	58.7	41.3	2.0	2.6	2288	24.3	10.0
United Arab Emirates	1990	19.0	81.0	−0.1	4.9	605[2]	44.7	36.2
Emirats arabes unis	1995	16.0	84.0	−0.9	3.4	...	...	...
Viet Nam	1990	80.1	19.9	2.1	2.5	3237	24.4	4.9
Viet Nam	1995	79.2	20.8	2.0	3.1	3555	23.0	4.8
Yemen	1990	71.1	28.9	2.1	6.7	500[2]	15.3	4.4
Yémen	1995	66.4	33.6	3.6	8.0	...	...	...
Europe · Europe								
Albania	1990	64.3	35.8	1.8	2.7	245	20.8	7.4
Albanie	1995	62.7	37.3	0.4	1.8	...	...	...

9

Population in urban and rural areas, rates of growth and largest city population [cont.]
Population urbaine, population rurale, taux d'accroissement et population de la ville la plus peuplée [suite]

Country or area Pays ou zone	Year Année	Population estimates Estimations de la population		Growth rate p.a. (%) [1] Taux d'accroissement p.a. (%) [1]		Largest city population Ville la plus peuplée		
		R%	U%	Pop.r.	Pop.u.	Number (000s) Nombre (000s)	%u	%t
Andorra	1990	37.5	62.5	4.6	2.5	19	59.4	36.5
Andorre	1995	37.5	62.5	5.5	5.5	...	...	...
Austria	1990	44.6	55.4	0.3	0.4	2055	48.2	26.7
Autriche	1995	44.5	55.5	0.6	0.7	2060	46.6	25.9
Belarus	1990	33.2	66.9	−2.3	2.1	1648	24.1	16.1
Bélarus	1995	28.9	71.2	−2.9	1.1	1766	24.5	17.4
Belgium	1990	3.5	96.5	−2.6	0.3	1148	12.0	11.5
Belgique	1995	3.0	97.0	−2.5	0.4	1122	11.4	11.1
Bosnia & Herzegovina	1990	55.4	44.6	−0.7	3.0	410	21.4	9.5
Bosnie−Herzégovine	1995	51.0	49.0	−6.1	−2.5	...	...	...
Bulgaria	1990	32.3	67.7	−1.8	1.0	1313	21.6	14.6
Bulgarie	1995	29.3	70.7	−2.4	0.4	1384	22.3	15.8
Channel Islands	1990	70.6	29.5	1.5	0.2	28	66.7	19.7
Iles Anglo−Normandes	1995	70.6	29.5	0.8	0.8	...	...	...
Croatia	1990	40.2	59.8	−2.1	1.9	700	25.9	15.5
Croatie	1995	35.6	64.4	−2.6	1.4	...	...	...
Czech Republic	1990	35.1	64.9	−0.3	0.2	1210	18.1	11.7
République tchèque	1995	34.6	65.4	−0.3	0.1	1225	18.2	11.9
Denmark	1990	15.2	84.8	−0.5	0.2	1345	30.9	26.2
Danemark	1995	14.8	85.2	−0.4	0.3	1326	30.0	25.6
Estonia	1990	28.2	71.8	−0.2	0.8	498	44.0	31.6
Estonie	1995	26.9	73.1	−1.5	−0.2	...	...	...
Faeroe Islands	1990	69.5	30.5	0.1	1.6	14	100.0	30.5
Iles Féroé	1995	67.1	32.9	−0.4	1.7	...	...	...
Finland	1990	38.6	61.4	−0.5	0.9	872	28.5	17.5
Finlande	1995	36.8	63.2	−0.4	1.0	1059	32.8	20.7
France	1990	27.3	72.7	0.9	0.4	9334	22.7	16.5
France	1995	27.2	72.8	0.4	0.5	9469	22.4	16.3
Germany †	1990	14.7	85.3	−1.3	0.7	6353	9.4	8.0
Allemagne †	1995	13.4	86.6	−1.2	0.8	6481	9.2	7.9
Gibraltar	1990	0.0	100.0	0.0	−0.2	28	100.0	100.0
Gibraltar	1995	0.0	100.0	0.0	0.0	...	...	...
Greece	1990	37.4	62.6	−0.7	1.4	3492	54.5	34.1
Grèce	1995	34.8	65.2	−1.0	1.2	3693	54.2	35.3
Holy See	1990	0.0	100.0	0.0	0.0	1	100.0	100.0
Saint−Siège	1995	0.0	100.0	0.0	0.0	...	...	...
Hungary	1990	38.0	62.1	−1.7	0.4	2017	31.4	19.5
Hongrie	1995	35.3	64.7	−1.9	0.3	2017	30.8	19.9
Iceland	1990	9.4	90.6	−1.2	1.3	145	62.8	56.9
Islande	1995	8.4	91.6	−1.1	1.3	...	...	...
Ireland	1990	43.1	56.9	−0.6	−0.1	916	46.0	26.2
Irlande	1995	42.5	57.5	0.0	0.5	911	44.6	25.6
Isle of Man	1990	26.1	73.9	0.4	1.7	21	41.2	30.4
Ile de Man	1995	24.7	75.3	0.3	1.8	...	...	...
Italy	1990	33.3	66.7	0.2	0.1	4603	12.1	8.1
Italie	1995	33.4	66.6	0.1	0.0	4251	11.2	7.4
Latvia	1990	28.8	71.2	−0.6	0.9	921	48.4	34.5
Lettonie	1995	27.2	72.9	−2.0	−0.4	924	49.6	36.1
Liechtenstein	1990	79.8	20.2	1.2	1.9	6	100.0	20.7
Liechtenstein	1995	78.6	21.4	1.1	2.5	...	...	...
Lithuania	1990	31.2	68.8	−1.6	1.8	598	23.4	16.1
Lituanie	1995	27.9	72.1	−2.3	0.9	...	...	...
Luxembourg	1990	13.7	86.3	−3.6	1.6	76	23.1	19.9
Luxembourg	1995	10.9	89.1	−3.3	1.9	...	...	...
Malta	1990	12.4	87.6	−2.6	1.1	98	31.6	27.7
Malte	1995	10.7	89.3	−2.3	1.1	...	...	...
Monaco	1990	0.0	100.0	0.0	1.2	30	100.0	100.0
Monaco	1995	0.0	100.0	0.0	1.2	...	...	...
Netherlands	1990	11.3	88.7	0.3	0.7	1053	7.9	7.1
Pays−Bas	1995	11.0	89.0	0.2	0.8	1109	8.0	7.2
Norway	1990	27.7	72.3	−0.2	0.7	681	22.2	16.1
Norvège	1995	26.8	73.2	−0.2	0.7	...	...	...
Poland	1990	37.5	62.5	−0.6	1.2	3449	14.5	9.1
Pologne	1995	35.3	64.7	−1.1	0.9	3552	14.3	9.3

9

Population in urban and rural areas, rates of growth and largest city population [cont.]

Population urbaine, population rurale, taux d'accroissement et population de la ville la plus peuplée [suite]

Country or area Pays ou zone	Year Année	Population estimates Estimations de la population		Growth rate p.a. (%) [1] Taux d'accroissement p.a. (%) [1]		Largest city population Ville la plus peuplée		
		R%	U%	Pop.r.	Pop.u.	Number (000s) Nombre (000s)	%u	%t
Portugal	1990	66.5	33.5	−0.7	1.2	1658	50.2	16.8
Portugal	1995	64.4	35.6	−0.7	1.1	1863	53.3	19.0
Republic of Moldova	1990	52.2	47.8	−0.8	2.4	685	32.9	15.7
République de Moldova	1995	48.3	51.7	−1.3	1.9	...	...	...
Romania	1990	46.7	53.3	−0.5	1.2	2047	16.6	8.8
Roumanie	1995	44.6	55.4	−1.3	0.5	2090	16.5	9.2
Russian Federation	1990	26.0	74.0	−0.9	1.2	9048	8.3	6.1
Fédération de Russie	1995	24.0	76.0	−1.7	0.4	9233	8.3	6.3
San Marino	1990	8.4	91.6	−8.0	1.8	4	19.0	17.4
Saint−Marin	1995	5.8	94.2	−6.1	2.1	...	...	...
Slovakia	1990	43.5	56.5	−0.6	1.3	438	14.8	8.3
Slovaquie	1995	41.2	58.8	−0.7	1.2	...	...	...
Slovenia	1990	41.0	59.0	−2.0	2.2	260	23.0	13.6
Slovénie	1995	36.5	63.6	−2.1	1.8	...	...	...
Spain	1990	24.7	75.4	−0.5	0.7	4172	14.1	10.6
Espagne	1995	23.6	76.5	−0.7	0.5	4072	13.4	10.3
Sweden	1990	16.9	83.1	0.5	0.5	1490	21.0	17.4
Suède	1995	16.9	83.1	0.5	0.5	1545	21.2	17.6
Switzerland	1990	40.5	59.6	0.3	1.3	826	20.3	12.1
Suisse	1995	39.2	60.8	0.4	1.5	897	20.5	12.5
TFYR Macedonia	1990	42.2	57.8	0.3	2.0	445	37.6	21.7
L'ex−R.y. Macédoine	1995	40.1	59.9	0.1	1.8	...	...	...
Ukraine	1990	32.5	67.5	−1.4	1.2	2638	7.6	5.1
Ukraine	1995	29.7	70.3	−1.9	0.7	2809	7.8	5.5
United Kingdom	1990	10.9	89.1	−0.1	0.3	7335	14.3	12.8
Royaume−Uni	1995	10.5	89.5	−0.4	0.4	7335	14.1	12.6
Yugoslavia, SFR †	1990	46.9	53.1	−0.8	2.0	1278	23.7	12.6
Yougoslavie, Rfs †	1995	43.5	56.5	−0.2	2.6	1405	22.9	13.0
Oceania · Océanie								
American Samoa	1990	52.2	47.8	2.8	4.4	...	...	...
Samoa américaines	1995	50.2	49.8	2.3	3.8	...	...	...
Australia	1990	14.9	85.1	2.0	1.5	3524	24.5	20.9
Australie	1995	15.3	84.7	1.9	1.3	3590	23.4	19.9
Cook Islands	1990	42.3	57.7	0.8	2.6	2	18.2	11.1
Iles Cook	1995	39.6	60.4	−0.4	1.8	...	...	...
Fiji	1990	60.7	39.3	0.5	1.2	151	53.0	20.8
Fidji	1995	59.3	40.7	1.1	2.2	...	...	...
French Polynesia	1990	43.6	56.4	2.9	2.2	108	97.3	54.8
Polynésie française	1995	43.6	56.4	2.2	2.2	...	...	...
Guam	1990	62.0	38.0	2.5	1.9	2	3.9	1.5
Guam	1995	61.9	38.2	2.2	2.4	...	...	...
Kiribati	1990	65.4	34.6	1.3	2.3	25	100.0	34.7
Kiribati	1995	64.3	35.7	1.4	2.4	...	...	...
Marshall Islands	1990	34.3	65.7	0.7	4.0	22	73.3	47.8
Iles Marshall	1995	30.9	69.1	0.9	4.0	...	...	...
Federated States of Micronesia	1990	73.7	26.4	2.4	3.4	...	...	...
Etats fédérés de Micron	1995	72.0	28.0	2.3	4.0	...	...	...
Nauru	1990	0.0	100.0	0.0	2.4	10	100.0	100.0
Nauru	1995	0.0	100.0	0.0	2.6	...	...	...
New Caledonia	1990	40.1	59.9	0.6	2.3	100	100.0	59.5
Nouvelle−Calédonie	1995	37.9	62.1	0.4	2.3	...	...	...
New Zealand	1990	15.2	84.8	−0.8	1.0	877	30.8	26.1
Nouvelle−Zélande	1995	13.9	86.1	−0.4	1.5	945	30.7	26.4
Niue	1990	69.2	30.9	−1.8	−1.9	1	100.0	30.9
Nioué	1995	70.6	29.4	−1.7	−3.1	...	...	...
Northern Mariana Islands	1990	47.0	53.0	16.3	16.3	12	52.2	27.9
Iles Mariannes du Nord	1995	46.3	53.7	1.5	2.0	...	...	...
Palau	1990	30.9	69.1	1.0	2.6	11	100.0	69.1
Palaos	1995	29.4	70.6	1.2	2.7	...	...	...
Papua New Guinea	1990	85.0	15.0	2.0	3.6	195	33.9	5.1
Papouasie−Nvl−Guinée	1995	84.0	16.1	2.0	3.6	...	...	...
Pitcairn	1990	100.0	0.0	0.0	0.0	...	...	...
Pitcairn	1995	100.0	0.0	0.0	0.0	...	...	...

9
Population in urban and rural areas, rates of growth and largest city population [*cont.*]
Population urbaine, population rurale, taux d'accroissement et population de la ville la plus peuplée [*suite*]

Country or area Pays ou zone	Year Année	Population estimates Estimations de la population		Growth rate p.a. (%) [1] Taux d'accroissement p.a. (%) [1]		Largest city population Ville la plus peuplée		
		R%	U%	Pop.r.	Pop.u.	Number (000s) Nombre (000s)	%u	%t
Samoa	1990	79.0	21.0	0.3	0.2	33	97.1	20.4
Samoa	1995	79.0	21.0	1.1	1.1	...	...	...
Solomon Islands	1990	85.4	14.6	2.9	6.6	39	83.0	12.2
Iles Salomon	1995	82.9	17.1	2.7	6.5	...	...	...
Tokelau	1990	100.0	0.0	−0.7	0.0	...	...	...
Tokélaou	1995	100.0	0.0	0.0	0.0	...	...	...
Tonga	1990	64.9	35.1	−0.7	4.9	34	100.0	35.4
Tonga	1995	58.9	41.1	−1.6	3.5	...	...	...
Tuvalu	1990	59.6	40.4	0.6	5.3	4	100.0	40.4
Tuvalu	1995	53.8	46.2	−0.6	4.1	...	...	...
Vanuatu	1990	81.5	18.5	2.4	2.8	20	71.4	13.4
Vanuatu	1995	80.7	19.3	2.3	3.4	...	...	...
Wallis and Futuna Islands	1990	100.0	0.0	2.0	0.0	1	0.0	7.1
Iles Wallis et Futuna	1995	100.0	0.0	1.1	0.0	...	...	...

Source:
World Urbanization Prospects, 1994 (United Nations forthcoming publication).

†   For detailed descriptions of data pertaining to former Czechoslovakia, Germany, SFR Yugoslavia and former USSR, see Annex I – Country or area nomenclature, regional and other groupings.

1   Annual rates of growth calculated for periods 1985−1990 and 1990−1995.
2   Data refer to capital city which is not the largest city.

Source:
"World Urbanization Prospects 1994," (publication des Nations Unies).

†   Pour les descriptions en détails des données relatives à l'ancienne Tchécoslovaquie, l'Allemagne, la Rfs Yougoslavie et l'ancienne URSS, voir l'Annexe I – Nomenclature des pays ou zones, groupements régionaux et autres groupements.

1   Ces taux d'accroissement annuel ont été calculés pour les périods 1985−1990 et 1990−1995.
2   Les données se rapportent à la capitale qui n'est pas la ville la plus peuplée.

Technical notes, tables 8 and 9

Table 8 is based on detailed data on population and its growth and distribution published in the United Nations *Demographic Yearbook* [19], which also provides a comprehensive description of methods of evaluation and limitations of data. A brief explanation of the quality code used for total population estimates in table 8 is given below.

For "Type of estimate", the code indicates the method by which official estimates of population for the year 1993 were prepared, so far as could be ascertained. The letters A–D indicate the nature of the base figure; numerals to capital letters indicate the lapse of time since the establishment of the base figure; letters a–d indicate the method of time adjustment by which the base figure is brought up to date; numerals to small letters indicate the quality of time adjustment.

The details of the code classification for this column are given below:

Nature of base data (capital letter)
A.   Complete census of individuals.
B.   Sample survey.
C.   Partial census or partial registration of individuals.
D.   Conjecture.
...   Nature of base data not determined.

Recency of base data (subscript numeral following capital letter)
Numeral indicates time elapsed (in years) since establishment of base figure.

Method of time adjustment (lower-case letter)
a.   Adjustment by continuous population register.
b.   Adjustment based on calculated balance of births, deaths and migration.
c.   Adjustment of assumed rate of population increase.
d.   No adjustment: base figure held constant at least two consecutive years.
...   Method of time adjustment not determined.

Quality of adjustment for types a and b (numeral following letter a or b)
1.   Population balance adequately accounted for.
2.   Adequacy of accounting for population balance not determined but assumed to be adequate.
3.   Population balance not adequately accounted for.

Quality of adjustment for type c (numeral following letter c)
1.   Two or more censuses taken at decennial intervals or less.
2.   Two or more censuses taken, but latest interval exceeds a decennium.
3.   One or no census taken.

Unless otherwise indicated, figures refer to de facto (present-in-area) population for present territory; surface area estimates include inland waters.

Notes techniques, tableaux 8 et 9

Le *Tableau 8* : est fondé sur des données détaillées sur la population, sa croissance et sa distribution, publiées dans l'*Annuaire démographique des Nations Unies* [19], qui offre également une description complète des méthodes d'évaluation et une indication des limites des données. On trouvera ci-après une brève explication du code de qualité utilisé pour les estimations de la population totale présentées au Tableau 8.

Pour le "Type d'estimation", le code indique, dans la mesure où elle a pu être déterminée, la méthode selon laquelle les estimations officielles de la population ont été établies pour l'année 1993. Les lettres A-D indiquent la nature du chiffre de base; les nombres placés à la droite de ces lettres indiquent le temps écoulé depuis l'établissement du chiffre de base; les lettres a-d indiquent la méthode d'actualisation du chiffre de base; les numéros qui suivent ces lettres en petits caractères indiquent la qualité de l'actualisation.

Le détail de la classification codée est donné ci-dessous :

Nature des données de base (majuscules)
A.   Recensement complet de la population.
B.   Enquête par échantillonnage.
C.   Recensement partiel ou enregistrement partiel de la population.
D.   Conjecture.
...   La nature des données de base n'est pas déterminée.

Actualité relative des données (nombre en indice suivant la lettre majuscule)

Le nombre indique le temps écoulé (en années) depuis l'établissement du chiffre de base.

Méthode d'actualisation (lettre minuscule)
a.   Actualisation par enregistrement continu de la population.
b.   Actualisation basée sur le calcul de la balance des naissances, des décès et des migrations.
c.   Actualisatin du taux présumé d'accroissement de la population.
d.   Absence d'actualisation : le chiffre de base est maintenu constant pendant au moins deux années consécutives.
...   Méthode d'actualisation non déterminée.

Qualité de l'actualisation pour les types a et b (chiffre suivant la lettre a ou b)
1.   Balance démographique convenablement établie.
2.   La qualité de l'établissement de la balance démographique n'est pas déterminée, mais on suppose qu'elle est convenable.
3.   Balance démographique non convenablement établie.

In *table 9*, statistics of urban and rural population and population in the largest urban agglomeration by each country or area are from estimates by the Population Division of the United Nations Secretariat [27, 28]. As noted above, because of national differences in the specific characteristics that distinguish urban from rural areas, there are no internationally agreed definitions of urban and rural. In most countries, the distinction is mainly based on size of locality. For the latest available census definition of urban areas in such country or area, reference should be made to the Demographic Yearbook 1992 [19].

For this table, data for largest city refer to the urban agglomeration. An urban agglomeration comprises the city or town proper and also the suburban fringe or thickly settled territory lying outside, but adjacent to, its boundaries.

Annual rates of change in urban and rural population are computed as average annual percentage changes using mid-year population estimates.

Qualité de l'actualisation pour le type c (définie par le chiffre suivant la lettre c)
1. Deux recensements ou plus ont été effectués en dix ans ou moins.
2. Deux recensements ou plus ont été effectués, mais à plus de dix ans d'intervalle.
3. Un recensement effectué ou aucun.

Sauf indication contraire, les chiffres se rapportent à la population effectivement présente sur le territoire tel qu'il est actuellement défini; les estimations de superficie comprennent les étendues d'eau intérieures.

Au *Tableau 9*, les statistiques de la population urbaine, de la population rurale et de la population de la ville la plus peuplée de chaque pays ou zone sont tirées d'estimations de la Division de la population du Secrétariat des Nations Unies [27, 28]. Comme il a été indiqué précédemment, il n'existe pas de définition reconnue à l'échelle internationale des zones urbaines et rurales parce que les caractéristiques retenues pour distinguer ces deux types de zone diffèrent d'un pays à un autre. Dans la plupart des pays, cette distinction est essentiellement fonction de la taille des agglomérations. Pour la définition la plus récente des zones urbaines utilisée dans une région ou un pays donné, se reporter à *l'Annuaire démographique 1992* [19].

Sur ce tableau, les données concernant la ville la plus peuplée se rapportent à l'agglomération urbaine. L'agglomération urbaine comprend la ville proprement dite et ses faubourgs ou banlieues, et tout territoire à forte densité de population situé à sa périphérie.

Les taux annuels de variation des populations urbaines et rurales se calculent sur la base de la variation annuelle moyenne en pourcentage déterminée à partir des estimations de la population en milieu d'année.

10
Education at the first, second and third levels
Enseignement des premier, second et troisième degrés
Number of students and percentage female
Nombre d'étudiants et étudiantes feminines en pourcentage

Country or area Pays ou zone	Years Années	First level Premier degré Total	%F	Years Années	Second level Second degré Total	%F	Years Années	Third level Troisième degré Total	%F
Africa · Afrique									
Algeria	1980	3 118 827	42	1980	1 028 294	39	1980[1]	79 351	26
Algérie	1985	3 481 288	44	1985	1 823 392	42	1985[1]	132 057	31
	1990	4 189 152	45	1990[2]	2 175 580	43	1990	285 930	...
	1992	4 436 363	45	1992	2 305 198	45	1991	298 117	...
Angola	1980	1 300 673	...	1980	190 702	...	1980	2 333	...
Angola	1985	974 498	45	1985[2]	178 910	...	1985	5 034	...
	1990	990 155	*48	1990	186 499	...	1990	6 534	...
Benin	1980	379 926	32	1975	43 123	...	1975	2 118	15
Bénin	1985	444 163	34	1980	89 969	...	1980	4 822	...
	1990[3]	457 140	...	1985	107 172	29	1985	9 063	16
	1991[3]	505 970	...	...	...	...	1990	10 873	*13
Botswana	1980	171 914	55	1980	20 969	55	1980	1 078	35
Botswana	1985	223 608	52	1985	36 144	53	1985	1 938	...
	1990	283 516	52	1990[2]	61 767	52	...	...	...
	1992	301 482	51	1992	81 316	53	1992[4]	6 409	45
Burkina Faso	1980	201 595	37	1980	27 539	33	1975	1 067	20
Burkina Faso	1985	351 807	37	1985	53 565	34	1980	1 644	22
	1990	504 414	38	1990	98 929	34	1985	4 085	23
	1991	530 013	39	1992	115 753	35	1990	5 425	23
Burundi	1981	206 408	38	1980	19 013	*32	1980	1 879	25
Burundi	1985	385 936	42	1985	25 939	34	1985	2 783	24
	1990	633 203	46	1990	44 207	37	1990	3 592	27
	1992	651 086	45	1992	55 713	39	1992	4 256	26
Cameroon	1975	1 122 900	45	1975	143 812	33	...	...	...
Cameroun	1980	1 379 205	45	1980	234 090	35	1980[5]	11 686	...
	1985	1 705 319	46	1985	343 720	38	...	...	...
	1990	1 964 146	46	1990	500 272	41	1990	33 177	...
Cape Verde	1975	64 794	*48	...	...	...	...	...	...
Cap–Vert	1980	57 587	49	1980	3 341	...	...	...	...
	1985	57 909	49	...	...	...	...	...	...
	1990	69 832	...	1989	7 866	50	...	...	...
Central African Rep.	1975	221 432	36	1975	23 011	18	1975	669	...
Rép. centrafricaine	1980	246 174	37	1980	45 211	26	1980	1 719	8
	1985	309 656	39	1985	59 273	27	1985	2 651	11
	1989	323 661	39	1989	49 147	29	...	...	...
Chad	1975	*212 983	*26	1975	16 391	...	1975	547	5
Tchad	1985	337 616	28	1986	44 379	16	1984	1 643	9
	1991	591 417	32	1989	58 570	18	1988[6]	2 983	...
Comoros	1980	59 709	41	1980	13 798	34	...	...	...
Comores	1985	66 084	43	1985	21 056	39	...	...	...
	1990	72 824	42	...	...	...	1989	248	15
	1991	75 577	45	1991	15 878	39	1991	223	28
Congo	1975	319 101	47	1975	102 110	36	1980	7 255	15
Congo	1980	390 676	48	1980	187 585	41	1985	10 684	16
	1985	475 805	49	1985	222 633	44	1990	10 671	18
	1990	502 918	46	1990	183 023	42	1991	12 045	19
Côte d'Ivoire	1980	1 024 585	40	1975[7]	119 482	...	1975	7 174	17
Côte d'Ivoire	1985	1 214 511	41	1980[7]	221 940	30	1980	19 633	...
	1990	1 414 865	41	...	...	...	1986	23 642	18
	1991	1 447 785	42	...	...	...	...	...	...
Djibouti	1980	16 841	...	1980	5 133	...	...	...	...
Djibouti	1985	25 212	41	1985	7 041	39	...	...	...
	1990	31 706	41	1990	9 513	...	...	...	...
	1993	33 005	43	1992	9 740	43	...	...	...
Egypt	1980	4 662 816	40	1980	2 929 168	37	1975	480 016	30
Egypte	1985	6 214 250	43	1985	3 826 601	40	1980	715 701	32
	1990[2]	6 964 306	44	1990[9]	5 507 257	43	1985	854 584	30
	1991[8]	6 541 725	45	1991[9]	5 284 174	45	1990[10]	708 417	35
Equatorial Guinea	1975	*39 000	...	1975	4 523	17	...	...	...
Guinée équatoriale	1980	44 499	...	1979	2 729	...	...	...	...
	1983	61 532	...	1982	4 368	...	1990	578	13

10
Education at the first, second and third levels
Number of students and percentage female [*cont.*]
Enseignement des premier, second et troisième degrés
Nombre d'étudiants et étudiantes feminines en pourcentage [*suite*]

Country or area Pays ou zone	Years Années	First level Premier degré Total	%F	Years Années	Second level Second degré Total	%F	Years Années	Third level Troisième degré Total	%F
Ethiopia	1980	2 130 716	35	...	...	...	1980	14 368	...
Ethiopie	1985	2 448 778	39	1985	666 169	...	1985	27 338	18
	1990	2 466 375	40	1990	865 886	43	1990	34 076	18
	1992	1 855 894	41	1992	720 779	46	1991[11]	26 218	19
Gabon	1975	128 552	49	1975	22 542	35	1975	1 014	20
Gabon	1980	155 081	49	1980	29 406	40	...	...	...
	1985	183 607	49	1985	44 124	42	1988	4 007	31
	1991	210 000	50	1991	51 348	54	...	...	...
Gambia	1980[12]	43 432	35	1980	9 657	30	...	...	...
Gambie	1985[12]	69 017	39	1984	15 913	30	...	...	...
	1990	86 307	*41	1990	20 400	33	...	...	...
	1992	97 262	41	1992	25 929	35	...	...	...
Ghana	1980[3]	1 377 734	44	1980	693 159	38	1975	9 079	16
Ghana	1985[3]	1 505 819	...	1985[13]	749 980	...	...	...	...
	1990	1 945 422	45	1989[13]	829 518	39	...	...	...
	1991[3]	1 796 490	45	...	...	...	1990	...	...
Guinea	1980	257 547	33	1980	98 305	28	1975	12 411	18
Guinée	1985	276 438	32	1985[2]	92 754	26	...	...	...
	1990	346 807	32	1990	85 942	24	1985	8 801	14
	1992	421 869	32	1992	106 811	25	...	...	...
Guinea−Bissau	1976	84 793	33	1975	2 153	31	...	...	...
Guinée−Bissau	1980	74 539	32	1980	4 757	20	...	...	...
	1986	77 004	35	1986	6 450	25	...	...	...
	1988	79 035	36	1988	6 330	32	1988	404	6
Kenya	1975	2 881 155	46	1975	240 969	35	...	...	...
Kenya	1980	3 926 629	47	1980	428 023	41	1980	12 986	...
	1985[2]	4 702 414	48	1985[2]	457 767	38	1985	21 756	26
	1990	5 392 319	49	1988	563 440	41	...	...	...
Lesotho	1980	244 838	59	1975	16 476	56	1980	1 889	...
Lesotho	1985	314 003	56	1980	25 292	60	1984	2 043	61
	1990	351 632	55	1985	37 343	60	1991	4 164	53
	1992	362 657	54	1992	53 485	59	1992	5 359	58
Liberia	1975	104 056	34	1975	34 151	25	1975	2 404	22
Libérie	1980	147 216	35	1980	54 623	28	1979	3 789	28
	1984	132 889	*35	...	...	...	...	...	...
	1986[3]	80 048	...	...	...	...	1987	5 095	23
Libyan Arab Jamahiriya	1975	556 169	46	1975	166 122	34	1975	13 427	18
Jamah. arabe libyenne	1980	662 843	47	1980	296 197	40	1980	20 166	25
	1985[2]	1 011 952	47	1985[2]	143 113	47	1985	30 000	...
	1991	1 238 986	48	1991	215 508	56	1991	72 899	46
Madagascar	1980	1 723 779	49	1979[14]	233 578	...	1980	22 632	...
Madagascar	1984	1 625 216	48	1988	345 302	48	1985	38 310	38
	1990	1 570 721	49	1990	340 191	49	1990	35 824	45
	1993	1 490 317	49	1993	312 939	49	1992	42 681	...
Malawi	1975	641 709	40	1975	15 018	27	1975	2 198	14
Malawi	1980	809 862	41	1980	18 653	28	1980	3 476	31
	1985	942 539	43	1985	25 737	32	1985	3 928	29
	1990	1 400 682	45	1990	32 275	34	1989	5 594	28
Mali	1980	291 159	36	1975	55 444	26	1975	2 936	10
Mali	1985	292 395	37	1979	78 707	...	1980	1 631	11
	1990	340 573	37	1986	63 768	30	1985	6 768	13
	1991	375 131	37	1990	78 523	32	1990	6 703	14
Mauritania	1980	90 530	35	1980	22 102	20	...	...	...
Mauritanie	1985	140 871	40	1985	35 955	...	1985	4 526	...
	1990[3]	166 036	42	1991	39 821	*34	1988	5 808	13
	1992	219 258	44	1992	43 034	33	1991	5 850	15
Mauritius	1980	128 758	49	1975	65 113	44	1975	1 096	14
Maurice	1985	140 714	49	1980	81 926	48	1980	1 038	31
	1990	137 491	49	1985	72 551	47	1985	1 161	36
	1991	135 233	49	1990	79 229	50	...	...	...
Morocco	1980	2 172 289	37	1980	797 110	38	1975	45 322	19
Maroc	1985	2 279 887	38	1985	1 201 858	39	1980	112 405	...
	1990[2]	2 483 691	40	1990[2]	1 123 531	41	1985	181 087	32
	1992	2 727 833	41	1992	1 207 734	41	...	...	...

10

Education at the first, second and third levels
Number of students and percentage female [*cont.*]
Enseignement des premier, second et troisième degrés
Nombre d'étudiants et étudiantes feminines en pourcentage [*suite*]

Country or area Pays ou zone	Years Années	First level Premier degré Total	%F	Years Années	Second level Second degré Total	%F	Years Années	Third level Troisième degré Total	%F
Mozambique	1980[15]	1 387 192	43	1981	107 849	28	1976	906	34
Mozambique	1985	1 248 074	44	1985[2]	151 888	31	1980	1 000	...
	1990[2]	*1 206 278	*41	1990	160 177	36	1985	1 442	23
	1992	1 199 476	43	1992	159 202	38	1992	3 482	26
Namibia	1986	294 985	...	1986	49 571	...		...	...
Namibie	1990	313 970	52	1990	62 354	56	1991	4 157	64
	1992	349 167	50	1992	84 581	55	...	...	...
Niger	1975	142 182	35	1975	14 462	28	1975	541	10
Niger	1980	228 855	35	1980	38 861	29	1980	1 435	20
	1985	275 902	36	...	...	...	1986	3 317	18
	1990	368 732	36	1990	76 758	29	1989	4 506	15
Nigeria	1980	13 760 030	43	1980	2 345 604	35	1975	44 964	...
Nigéria	1985	12 914 870	44	1985	3 088 711	43	1980	150 072	...
	1990	13 776 854	44	1990[2]	3 123 277	42	1985	266 679	27
	1991	14 805 937	44	1991	3 600 620	55	1989	335 824	24
Réunion	1975	95 810	...	1975	50 467	55	...	...	...
Réunion	1980	79 143	...	1980	62 613	...	...	...	...
	1985	73 985	48	1985	69 863	54	...	...	...
	1988	73 747	...	1986	69 585	53	...	...	...
Rwanda	1980	704 924	48	1980	20 672	46	1975	1 108	...
Rwanda	1985	836 877	49	1985[2]	46 998	42	1980	1 243	10
	1990	1 100 437	50	1990	70 400	43	1985	1 987	14
	1991[2]	1 104 902	50	1991	94 586	44	1989	3 389	19
St. Helena	1975	755	48	1975	524	54	...	...	...
Sainte−Hélène	1980	717	51	1980	638	48	1980	36	25
	1985	582	55	1985	513	49	...	...	...
Sao Tome and Principe	1975	14 290	...	1975	4 010	...	...	...	...
Sao Tomé−et−Principe	1980	16 376	...	1980	3 815	...	...	...	...
	1986	17 010	48	1986	5 255	...	...	...	...
	1989	19 822	47	...	...	...	...	...	...
Senegal	1980	419 748	40	1980	95 604	33	1976	8 921	20
Sénégal	1985	583 890	40	1985	130 338	33	1980	13 626	18
	1990	708 448	...	1990	181 170	...	1985	13 354	...
	1991	725 496	42	1991	191 431	35	1991	21 562	24
Seychelles	1980	14 468	51	1980	924	*47	1976	142	93
Seychelles	1985	14 368	49	1985[2]	3 975	50	1980	144	89
	1990	14 362	...	1990	4 396	...	...	...	...
	1993[2]	9 873	49	1993[2]	9 111	50	...	...	...
Sierra Leone	1975	205 910	40	1975	50 478	32	1975	1 642	16
Sierra Leone	1980	315 145	42	1980	68 199	...	1980	2 166	...
	1985	421 689	...	1985	94 717	...	1985	5 690	...
	1990	367 426	41	1990	102 474	37	1990	4 742	...
Somalia	1975	197 706	35	1975	31 857	24	1975	2 040	11
Somalie	1980	271 704	36	1980	43 841	27	...	...	...
	1985	196 496	34	1985	45 686	35	1986	15 672	20
South Africa	1986[16]	4 737 367	49	1989	2 572 226	54	1989[16]	421 152	46
Afrique du Sud	1990	6 951 777	50	1990	2 743 184	54	1990[16]	439 007	44
	1992[16]	5 643 707	49	...	...	...	1992[16]	490 112	46
Sudan	1980	1 464 227	40	1980	384 194	37	1975	21 342	16
Soudan	1985	1 738 341	40	1985	556 587	42	1980	28 788	27
	1990	2 042 743	43	1990	731 624	43	1985	37 367	37
	1991	2 168 180	43	1991	718 298	44	1989	60 134	40
Swaziland	1980	112 019	50	1975	16 876	...	1980	1 875	40
Swaziland	1985	139 345	50	1980	23 665	...	1985	2 732	...
	1990	166 454	50	1985	31 109	...	1990	3 198	43
	1992	180 285	49	...	...	...	1992	3 023	42
Togo	1975	362 895	35	1975	64 404	24	1975	2 353	14
Togo	1980	506 356	38	1981	130 366	25	1980	4 750	15
	1985	462 858	38	1985	97 120	24	1986	6 223	...
	1990	651 962	39	1990	125 545	25	1989	7 826	13
Tunisia	1980	1 054 027	42	1980	293 351	37	1980	31 827	30
Tunisie	1985	1 291 490	45	1985	457 630	40	1985	41 594	36
	1990	1 405 665	46	1990	564 540	43	1990	68 535	39
	1993	1 476 329	47	1992	639 403	45	1992	87 780	41

10
Education at the first, second and third levels
Number of students and percentage female [cont.]
Enseignement des premier, second et troisième degrés
Nombre d'étudiants et étudiantes feminines en pourcentage [suite]

Country or area Pays ou zone	Years Années	First level Premier degré Total	%F	Years Années	Second level Second degré Total	%F	Years Années	Third level Troisième degré Total	%F
Uganda	1975[17]	973 604	40	1975[18]	55 263	26	1980	5 856	23
Ouganda	1980[17]	1 292 377	*43	1980[18]	86 560	...	1985	10 103	23
	1985[17]	2 117 000	...	1985[18]	179 185	...	1990	17 578	28
	1988[17]	2 632 764	...	1988[18]	260 069	...	1992	21 489	29
United Republic of Tanzania	1981[19]	3 538 183	48	1980[19]	78 715	...	1975	3 064	14
Rép.–Unie de Tanzanie	1985[19]	3 169 759	50	1985[19]	92 945	...	...	...	...
	1990[19]	3 379 000	50	1990[19]	167 150	42	1985[4]	4 863	15
	1992[19]	3 603 488	49	1992[19]	189 827	44	1989	5 254	...
Zaire	1980	4 195 699	42	1975	511 481	26	1975	24 853	...
Zaïre	1985	4 650 756	39	1980	861 774	27	1980	28 493	...
	1990	4 562 430	43	1985	959 934	30	1985	40 878	...
	1992	4 870 933	43	1987	1 066 351	32	1988	61 422	...
Zambia	1975	872 392	45	1975	77 672	...	1975	8 403	*14
Zambie	1980	1 041 938	47	1980	102 019	35	...	...	...
	1985	1 348 318	47	1985[2]	140 743	...	1986	14 492	...
	1990	1 461 206	...	...	...	...	1990	15 343	...
Zimbabwe	1980	1 235 036	...	1975	70 005	42	1980	8 339	...
Zimbabwe	1985	2 214 963	48	1980	74 746	...	1985	30 843	...
	1990	2 116 414	50	1985	482 000	...	1990	49 361	...
	1993	2 376 048	48	1991	710 619	44	1992	61 553	27
America, North · Amerique du Nord									
Antigua and Barbuda Antigua–et–Barbuda	1991	9 298	49	1991	5 845	50	...	...	...
Bahamas	1975	31 707	50	1980	28 136	...	1976	5 660	...
Bahamas	1980	32 854	...	1985	27 604	52	1980	4 093	...
	1985	32 848	49	1991	29 559	50	1985	4 531	...
	1992[20]	31 601	50	1992	29 863	50	1987	5 305	68
Barbados	1975	*32 884	*50	1975	29 025	52	...	...	...
Barbade	1980	31 147	50	1980	28 818	50	1980	4 033	54
	1984	30 161	48	1984	28 695	50	1988	4 244	59
	1991	26 662	49	1989	24 004	47	1991	6 888	55
Belize	1980[17]	34 615	...	1982	6 308	*52	...	...	...
Belize	1985[17]	39 212	48	1985	7 048	54	...	...	...
	1990[17]	46 023	*48	1990	7 904	53	...	...	...
	1991[17]	46 874	48	1991	8 901	55	...	...	...
Bermuda	1975	6 808	48	...	...	...	1980	608	51
Bermudes	1980	5 934	49	...	...	...	1982[21]	2 664	...
	1984	5 398	50	...	...	...	...	...	...
British Virgin Islands	1975	2 096	50	1975	821	56	...	...	...
Iles Vierges brit.	1980	1 974	50	1980	791	57	...	...	...
	1984	2 069	48	1983	1 323	58	...	...	...
Canada	1980	2 184 919	49	1980	2 323 228	49	1980[22]	1 172 750	50
Canada	1985	2 254 887	48	1985	2 250 941	49	1985[22]	1 639 410	45
	1990	2 371 558	48	1990	2 292 735	49	1990[22]	1 916 801	54
	1992	2 438 436	48	1992	2 392 064	49	1991[22]	1 942 814	54
Costa Rica	1980	348 674	49	1980	135 830	53	1980[23]	55 593	...
Costa Rica	1985	362 877	48	1985	112 531	52	1985[23]	63 771	...
	1990	435 025	49	1990	130 553	50	1990[23]	74 681	...
	1993	484 958	49	1993	160 343	51	1991[23]	80 442	...
Cuba	1980	1 468 538	48	1980	1 146 414	50	1980	151 733	48
Cuba	1985	1 077 213	47	1985	1 156 555	51	1985	235 224	54
	1990	887 737	48	1990	1 002 338	52	1990	242 434	57
	1992	942 431	49	1992	819 712	52	1992	198 474	58
Dominica	1980	14 815	50	1975	6 487	59	1980	63	22
Dominique	1985	12 340	48	1983	7 562	...	1984	60	67
	1990	12 836	49	1985	7 370	54	1990	430	41
	1992	12 795	49	...	...	...	1991	658	55
Dominican Republic	1975	911 142	...	1975	260 133	...	...	...	...
République dominicaine	1980	1 105 730	...	1980	356 091	...	1985	123 748	...
	1985	1 219 681	50	1985	463 511	...	...	...	...
	1989[3]	1 032 055	49	...	...	...	...	...	...
El Salvador	1975	759 460	48	1980	73 030	48	1975	28 281	33
El Salvador	1980	834 101	49	1984	85 081	53	1980	16 838	31
	1984	883 329	50	1991	94 268	55	1985	70 499	44
	1992	1 028 877	50	1992	105 093	53	1990	78 211	33

10
Education at the first, second and third levels
Number of students and percentage female [cont.]
Enseignement des premier, second et troisième degrés
Nombre d'étudiants et étudiantes feminines en pourcentage [suite]

Country or area Pays ou zone	Years Années	First level Premier degré Total	%F	Years Années	Second level Second degré Total	%F	Years Années	Third level Troisième degré Total	%F
Grenada	1975	21 195	48	1975	10 197	...	...	...	...
Grenade	1980	18 076	48	1980	8 626	59	...	...	...
	1985	20 808	...	1985	6 341	...	...	...	...
	1989	21 616	...	1987	6 497	...	...	...	...
Guadeloupe	1980	53 581	...	1980	49 398	...	...	...	...
Guadeloupe	1985	42 734	...	1985	51 634	53	...	...	...
	1990	38 531	49	1990	49 846	53	...	...	...
	1991	38 255	49	1991	50 556	52	...	...	...
Guatemala	1975	627 126	45	1975	99 233	46	1975[24]	22 881	23
Guatemala	1980	803 404	45	1980	171 903	45.	1980	50 890	...
	1985	1 016 474	45	1985	204 049	...	1985[24]	48 283	...
	1991	1 249 413	46	1991	294 907	...	1986[24]	51 860	...
Haiti	1975	487 135	...	1980	99 894	...	1975	2 881	24
Haïti	1980[25]	642 391	46	1985	143 758	...	1980	4 671	30
	1985[25]	872 500	47	...	...	...	1985	6 288	26
	1990	555 433	48	...	...	...	...	...	...
Honduras	1975	460 744	49	1975	56 705	...	1975	11 907	34
Honduras	1980	601 337	50	1980	127 293	50	1980	25 825	38
	1985	765 809	50	1985	184 112	...	1985	36 620	...
	1991	908 446	50	1991	194 083	55	...	...	...
Jamaica	1976	367 525	50	1975	216 248	54	1980	13 999	...
Jamaïque	1980	359 488	50	1980	248 001	53	1985	10 969	...
	1985	340 059	...	1985	237 713	52	1990	16 018	...
	1990[3]	323 378	50	1990	225 240	52	1991	23 220	...
Martinique	1980	47 382	...	1980	47 745	...	...	...	...
Martinique	1986	33 492	48	1985	47 500	...	...	...	...
	1990	32 744	...	1991	46 373	52	...	...	...
	1992	32 585	49	1992	43 928	54	...	...	...
Mexico	1980	14 666 257	49	1980	4 741 850	47	1975	562 056	...
Mexique	1985	15 124 160	49	1985	6 549 105	48	1980	929 865	33
	1990	14 401 588	49	1990	6 704 297	50	1985	1 207 779	...
	1992	14 425 669	48	1992	6 782 886	49	1990	1 310 835	...
Montserrat	1975	2 635	48	1975	535	...	...	...	...
Monteserrat	1980	1 846	48	1980	887	...	...	...	...
	1981	1 725	50	...	...	...	...	...	...
Netherlands Antilles	1980	32 856	49	...	...	...	...	...	...
Antilles néerlandaises	1982	32 380	49	...	...	...	...	...	...
Nicaragua	1980	472 167	51	1980	139 743	53	1980	35 268	...
Nicaragua	1985	561 551	52	1985	128 499	67	1985	29 001	56
	1990	632 882	51	1990	168 888	58	1990	30 733	52
	1992	703 854	50	...	...	...	1991	31 499	49
Panama	1980	337 522	48	1975	133 181	52	1980	40 369	55
Panama	1985	340 135	48	1980	171 273	52	1985	55 303	58
	1990	351 021	48	1985	184 536	52	1990	53 235	...
	1991	349 858	...	1990	195 903	51	1991	58 625	...
Puerto Rico	...	...	...	...	...	...	1975	97 517	53
Porto Rico	...	...	...	...	...	...	1980	131 184	59
St. Kitts and Nevis	1975	8 804	48	...	...	...	1979	40	70
St. Kitts–Nevis	1980	7 149	49	1985	4 197	*49	1985	212	42
	1985	7 810	42	1991	4 396	51	1991	325	37
	1991	7 236	48	...	...	...	1992	394	55
Saint Lucia	1980	29 605	51	1980	4 485	55	1976	298	53
Sainte–Lucie	1985	32 817	49	1985	6 833	61	1980	301	52
	1990	33 006	49	1990	8 230	*59	1986	367	55
	1992[3]	31 868	...	1991	9 419	61	1987	389	54
St. Pierre and Miquelon	1980	747	49	1980	748	52	...	...	...
St. Pierre et Miquelon	1985	558	47	1985	821	53	...	...	...
	1989	556	...	1986	800	53	...	...	...
St. Vincent and the Grenadines	1975[26]	21 854	49	1975	5 084	58	1985	736	69
Saint–Vincent–et–Grenadines	1980[26]	24 158	...	1981	8 058	59	1986	795	64
	1985[26]	24 561	49	1985	6 782	59	1989	677	68
	1990	22 030	49	1990	10 719	55	...	...	...

10
Education at the first, second and third levels
Number of students and percentage female [cont.]
Enseignement des premier, second et troisième degrés
Nombre d'étudiants et étudiantes feminines en pourcentage [suite]

Country or area Pays ou zone	Years Années	First level Premier degré Total	%F	Years Années	Second level Second degré Total	%F	Years Années	Third level Troisième degré Total	%F
Trinidad and Tobago	1980[17]	167 039	50	1975	66 583	49	1975	4 940	38
Trinité-et-Tobago	1985[17]	168 308	50	1980	89 272	...	1980	5 649	43[27]
	1990[17]	193 992	49	1985	95 302	...	1985	6 582	38
	1991[17]	196 333	49	1988	99 741	50	1988	7161	38
Turks and Caicos Is.	1975[3]	1 764	...	1975	671	...	...	...	...
Iles Turques et Caïques	1980[3]	1 483	48	1980	691	...	...	...	...
	1984[3]	1 429	49	1984	707	...	...	...	...
United States	1980[2]	27 449 000	49	1980	14 556 000	50	1980	12 096 895	51
Etats-Unis	1985	26 870 000	49	1985	13 977 000	49	1985	12 247 055	52
	1990	29 262 830	48	1990	12 436 123	49	1990	13 710 150	54
	1991	29 600 000	...	1991	13 200 000	48	1992	14 422 975	...
US Virgin Islands	...	...	...	1975	*10 590	...	1975	2 069	63
Iles Vierges américaines	1980	21 738	...	1980	6 737	...	1980	2 148	71
	1985	20 548	...	1985	7 948	...	1985	2 602	72
	1990[3]	15 256	51	1990[3]	5 848	59	1990	2 466	75
America, South · Amerique du Sud									
Argentina	1975	3 571 180	49	1975	1 243 058	52	1975	596 736	48
Argentine	1980	3 917 449	49	1981	1 366 444	53	1980	491 473	50
	1985	4 589 291	49	1985	1 800 049	52	1985	846 145	53
	1991	4 874 306	...	1991	2 160 410	...	1991	1 077 212	...
Bolivia	1975	859 413	45	1975	130 029	...	...	...	...
Bolivie	1980	978 250	47	1980	170 710	43	...	...	...
	1986	1 204 534	47	1986	209 293	46	...	...	...
	1990	1 278 775	47	1990	219 232	46	...	...	...
Brazil	1980	22 598 254	49	1980	2 819 182	54	1980	1 409 243	48[28]
Brésil	1985	24 769 736	...	1985	3 016 175	...	1986	1 451 191	...
	1990	28 943 619	...	1990	3 498 777	...	1990[28]	1 540 080	52
	1991	28 742 471	...	1991	3 558 946	...	1991[28]	1 565 056	53
Chile	1980	2 185 459	49	1980	538 309	53	1980	145 497	43
Chili	1985	2 062 344	49	1985	667 797	52	1985	197 437	43
	1990	1 991 178	49	1990	719 819	51	1990	255 358	...
	1992	2 034 839	49	1992	675 073	51	1991	286 962	...
Colombia	1980	4 168 200	50	1980	1 733 192	50	1975	176 098	36
Colombie	1985	4 039 533	50	1985	1 934 032	50	1980	271 630	45
	1990	4 246 658	...	1991	2 377 947	54	1985	391 490	49
	1992	4 525 959	49	1992	2 686 515	54	1991	510 649	51
Ecuador	1975	1 216 233	49	1975	383 624	48	1975	170 173	...
Equateur	1980	1 534 258	49	1980	591 969	...	1980	269 775	...
	1985	1 738 549	49	1985	730 226	50	1984	280 594	...
	1988	1 827 920	...	1987	771 928	50	1990	206 541	...
Falkland Islands (Malvinas)	1975	206	48	1975	126	44	...	...	...
Iles Falkland (Malvinas)	1980	223	58	1980	90	56	...	...	...
French Guiana	1975	7 594	...	1975	5 534	52	...	...	...
Guyane française	1980	9 276	...	1980	7 421	...	...	...	...
	1983	9 780	...	1990	10 722	...	...	...	...
	1990	14 256	...	...	...	...	...	...	...
Guyana	1975	130 240	49	1975	71 327	51	1975	2 852	35
Guyana	1980	130 832	49	1985	76 546	51	1980	2 465	44
	1985	113 857	49	1986	76 012	51	1985	2 328	48
	1988	118 015	*49	...	...	...	1989	4 665	43
Paraguay	1980	518 968	48	1980	118 828	...	1975	18 302	...
Paraguay	1985	570 775	48	1985	150 736	...	1980	26 915	...
	1990	687 331	48	1990	163 734	50	1985	32 090	...
	1992	749 336	48	1992	192 775	50	1990	32 884	...
Peru	1980	3 161 375	48	1980	1 203 116	45	1980	306 353	35
Pérou	1985	3 711 592	48	1985	1 427 261	47	1985	452 462	...
	1990	3 855 282	...	1990	1 697 943	...	1990	681 801	...
	1992	3 853 098	...	1992	1 703 997	...	1992	777 918	...
Suriname	1975	80 171	48	1975	30 603	55	...	...	...
Suriname	1980	74 538	...	1980	24 027	...	1980	2 378	...
	1986	62 633	48	1984	34 608	...	1985	2 751	54
	1988	65 798	49	1988	34 248	54	1990	4 319	53

10
Education at the first, second and third levels
Number of students and percentage female [*cont.*]
Enseignement des premier, second et troisième degrés
Nombre d'étudiants et étudiantes feminines en pourcentage [*suite*]

Country or area Pays ou zone	Years Années	First level Premier degré Total	%F	Years Années	Second level Second degré Total	%F	Years Années	Third level Troisième degré Total	%F
Uruguay	1980	331 247	49	1980	148 294	53	...	...	...
Uruguay	1985	356 002	49	1985	213 774	...	1989	69 428	...
	1990	346 416	49	1990	265 947	...	1990	71 612	...
	1992	338 020	49	1992	272 622	...	1992	68 227	...
Venezuela	1980[2]	3 158 466	...	1980	222 267	58	1975	213 542	...
Venezuela	1985	3 539 890	50	1985	268 580	56	1980	307 133	...
	1990	4 052 947	50	1990	281 419	57	1985	443 064	41
	1991	4 190 047	50	1991	289 430	57	1990	550 030	...
Asia · Asie									
Afghanistan	1975	784 568	15	1975	93 497	11	1975	*12 256	*14
Afghanistan	1980	1 115 993	18	1980	136 898	...	1982	19 652	...
	1985	580 499	31	...	...	...	1986	22 306	14
	1989	726 287	33	...	...	...	1990	24 333	31
Armenia	...	...	...	...	...	...	1980	58 100	...
Arménie	...	...	...	...	...	...	1985	54 800	...
	...	...	...	...	...	...	1990	68 400	...
Azerbaijan	...	...	...	...	...	...	1980	107 000	...
Azerbaïdjan	...	...	...	...	...	...	1985	105 900	...
	...	...	...	...	...	...	1990	105 100	...
	1992	267 946	...	1992	891 839	...	1992	100 985	38
Bahrain	1980	48 451	46	1980	26 528	46	1980	1 908	41
Bahreïn	1985	57 330	49	1985	38 577	48	1985	4 180	60
	1990	66 597	49	1990	47 005	50	1990	6 868	56
	1992	68 898	49	1992	51 413	50	1992	7 763	57
Bangladesh	1975	8 349 834	34	1976	2 183 413	...	1976	158 604	13
Bangladesh	1980	8 240 169	37	1980	2 659 208	24	1980	240 181	14
	1985	8 920 293	40	1985	3 125 219	28	1985	457 862	19
	1990	11 939 949	45	1990	3 592 995	*33	1990	434 309	16
Bhutan	1976	16 671	27	...	...	...	...	...	...
Bhoutan	1980	29 899	...	...	...	...	1980	322	22
	1985	45 395	34	...	...	...	...	...	...
	1993[2]	56 773	43	...	...	...	...	...	...
Brunei Darussalam	1975	30 109	48	1975	14 614	48	1980	143	50
Brunéi Darussalam	1980	30 513	48	1980	17 441	50	1986	601	50
	1985	34 815	...	1985	20 642	...	1987	945	51
	1992	39 782	47	1992	26 836	51	...	...	...
China	1980	146 270 000	45	1980	56 778 008	39	1980[4]	1 161 440	23
Chine	1985	133 701 800	45	1985	50 926 400	40	1985[4]	1 778 608	30
	1990	122 413 800	46	1990	51 054 100	42	1990[4]	2 146 853	33
	1992	122 012 800	47	1992	53 544 000	43	1992[4]	2 270 772	33
Cyprus	1980[29]	48 701	49	1980[29]	47 599	49	1980	1 940	42
Chypre	1985[29]	50 990	48	1985[29]	46 159	49	1985	3 134	48
	1990[29]	62 962	48	1990[29]	44 614	49	1990	6 554	52
	1992[29]	64 313	48	1992[29]	51 641	49	1992	6 263	49
Georgia	...	...	...	...	...	...	1980	85 800	...
Géorgie	...	...	...	...	...	...	1985	88 500	...
	...	...	...	...	...	...	1990	103 900	...
Hong Kong	1980	540 260	48	1975	368 655	47	1975	44 482	25
Hong–kong	1985	534 903	48	1980	468 975	49	1980	38 153	26
	1990	526 700	...	1985	450 367	50	1984	76 844	35
	1991	517 100	...	1987	458 444	49	1992	88 950	42
India	1980	73 873 184	39	1975	23 638 666	30	1975[30]	4 615 992	23
Inde	1985	87 440 514	40	1980	32 748 397	32	1980[30]	3 545 318	26
	1990	99 118 320	41	1985	44 484 544	33	1985[30]	4 470 844	30
	1992	105 370 216	43	1992	63 204 943	36	1988[30]	4 528 956	32
Indonesia	1975	17 776 617	45	1975	3 570 080	38	1975	278 200	...
Indonésie	1980	25 537 053	46	1980	5 721 815	36	1981	565 501	31
	1985	29 897 115	48	1985	*9 479 086	...	1984	980 162	32
	1992	29 598 790	48	1992	10 863 435	45	1992	1 973 094	14
Iran, Islamic Rep. of	1980	4 799 000	...	1980	2 718 461	37	1975[3]	151 905	28
Iran, Rép. islamique d'	1985	6 788 323	44	...	...	...	1985[3]	184 442	29
	1990	9 369 646	46	1990	5 084 832	41	...	...	...
	1992	9 937 369	47	1992	6 322 988	43	...	...	...

10

Education at the first, second and third levels
Number of students and percentage female [*cont.*]
Enseignement des premier, second et troisième degrés
Nombre d'étudiants et étudiantes feminines en pourcentage [*suite*]

Country or area Pays ou zone	Years Années	First level Premier degré Total	%F	Years Années	Second level Second degré Total	%F	Years Années	Third level Troisième degré Total	%F
Iraq Iraq	1980	2 615 910	46	1975	525 255	29	1975	86 111	33
	1985	2 816 326	45	1980	1 033 418	32	1980	106 709	32
	1990	3 328 212	44	1985	1 190 833	35	1985	169 665	36
	1992	2 857 467	45	1992	1 144 938	38	1988	209 818	38
Israel Israël	1980	621 912	...	1980	199 859	...	1976	85 081	46
	1985	699 476	49	1985	251 466	51	1980	97 097	51
	1990	724 502	49	1990	309 098	51	1985	116 062	47
	1992	763 511	49	1992	334 290	51	...	...	...
Japan Japon	1980	11 826 573	49	1980	9 557 563	49	1975	2 248 903	32
	1985	11 095 372	49	1985	11 058 133	49	1980	2 412 117	33
	1990	9 373 295	49	1990	11 025 720	49	1985	2 347 463	35
	1992	8 947 226	49	1991	10 676 866	49	1991	2 899 143	40
Jordan Jordanie	1980	454 391	48	1980	266 368	45	1980	36 549	46
	1985	530 906	48	1985	335 835	48	1985	53 753	45
	1990[2]	926 445	48	1990[2]	100 953	47	1990	80 442	48
	1992	1 014 295	49	1992	113 910	50	1992	88 506	49
Kazakhstan Kazakhstan	...	...	...	...	...	...	1980	260 000	...
	...	...	...	...	...	...	1985	273 400	...
	1993	1 227 130	49		...	...	1990	287 400	...
Korea, Dem. People's Rep. Corée, Rép. pop. dém. de	1976	2 561 674	49	...	...	...	...	...	...
	1987	1 543 000	49	...	...	...	1987	390 000	34
Korea, Republic of Corée, République de	1980	5 658 002	49	1980	4 285 889	45	1980[31]	647 505	...
	1985	4 856 752	49	1985	4 934 975	47	1985	1 455 759	30
	1990	4 868 520	49	1990	4 559 557	48	1990	1 691 429	32
	1993	4 336 252	48	1993	4 479 463	48	1992	1 858 568	34
Kuwait Koweït	1980	148 983	48	1980	181 882	46	1975	8 104	57
	1985	172 975	49	1985	239 581	47	1980	13 630	57
	1990	124 996	48	1991	167 331	49	1985	23 678	54
	1992	122 930	49	1992	177 675	49	1991	28 399	61
Kyrgyzstan Kirghizistan	...	...	...	...	...	...	1980	55 400	...
	...	...	...	...	...	...	1985	58 200	...
	...	...	...	...	...	...	1990	58 800	...
Lao People's Dem. Rep. République dém. pop. lao	1975	317 126	...	1976	48 669	33	...	...	...
	1980	479 291	45	1980	90 435	39	1980	1 408	31
	1985	523 347	45	1985	113 630	41	1985	5 382	36
	1991	580 792	44	1991	125 702	39	1989	4 730	32
Lebanon Liban	1980	405 402	...	1980	287 310	...	1980	79 073	36
	1986	399 029	...	...	...	...	1985	79 500	...
	1991	345 662	48	...	...	...	1991	85 495	48
Macau Macao	1988	31 917	...	...	...	...	1987	7 662	...
	1990	34 972	48	1989	16 687	52	1989	8 824	36
	1991	37 872	48	1991	18 978	52	...	...	...
Malaysia Malaisie	1980	2 008 973	49	1980	1 083 818	48	1980	57 650	39
	1985	2 199 096	49	1985	1 294 990	49	1985	93 249	44
	1990	2 455 522	49	1990	1 420 173	50	1990	121 412	45[32]
	1992	2 652 397	49	1992	1 566 790	51	...	...	...
Maldives Maldives	1980	30 621	...	1980	998	...	...	...	...
	1986	39 775	...	1983	2 756	...	...	...	...
	1992	45 333	49	1992	16 087	49	...	...	...
Mongolia Mongolie	1980	145 200	49	1980	245 600	...	1981	38 200	63
	1985	153 100	...	...	...	...	1985	40 099	...
	1990	166 200	50	...	...	...	1990	31 006	...
	1991	154 600	...	...	...	...	1991	28 209	...
Myanmar Myanmar	1975	3 475 749	48	1975	933 486	...	1975	56 083	...
	1980	4 148 342	48	1980	1 066 300	...	1981	165 000	...
	1985	4 710 616	48	1985	1 283 586	...	1987	202 381	...
	1990	5 384 539	49	1987	1 358 788	...	...	...	...
Nepal Népal	1980	1 067 912	28	1980	512 434	20	1980[3]	34 094	22
	1985[2]	1 812 098	30	1985[2]	496 921	23	1985[3]	54 452	...
	1990	2 788 644	36	1990	708 663	29	1990	93 753	23
	1992	3 034 710	38	1992	855 137	32	1991	110 239	24
Oman Oman	1980	91 895	34	1980	16 776	24	1980	18	–
	1985	177 541	44	1985	48 096	32	1985	990	38
	1990	262 989	47	1990	102 021	44	1990	5 962	44
	1992	289 911	47	1992	140 761	46	1991	7 322	49

10

Education at the first, second and third levels
Number of students and percentage female [*cont.*]
Enseignement des premier, second et troisième degrés
Nombre d'étudiants et étudiantes feminines en pourcentage [*suite*]

Country or area Pays ou zone	Years Années	First level Premier degré Total	%F	Years Années	Second level Second degré Total	%F	Years Années	Third level Troisième degré Total	%F
Pakistan	1975[15]	5 236 203	30	1975	1 935 849	23	1975	127 932	24
Pakistan	1980[15]	5 473 578	33	1980	2 165 832	26	1979	156 558	27
	1985[15]	7 094 059	33	1985	2 933 422	27	1985	267 742	26
	1990[15]	8 855 997	34	1990	3 983 462	29	1989	304 922	28
Palestine · Palestine									
Gaza Strip	1986	109 521	...	1986	59 241	...	1986	5 313	...
Zone de Gaza	1991	127 257	...	1991	73 940	...	1991	4 711	44
Philippines	1981	8 033 642	49	1980	2 928 525	53	1980	1 276 016	53
Philippines	1985	8 925 959	49	1985	3 214 159	50	1985	1 402 000	...
	1990	10 427 077	...	1990	4 033 597	...	1990	1 709 486	...
	1992	10 679 748	...	1992	4 421 649	...	1991	1 656 815	59
Qatar	1980	30 078	48	1980	15 901	48	1980	2 269	62
Qatar	1985	40 636	48	1985	22 574	50	1985	5 344	62
	1990	48 650	48	1990	30 031	50	1990	6 485	69
	1992	49 059	48	1992	35 013	50	1992	7 283	71
Saudi Arabia	1980	926 531	39	1980	348 996	38	1980	62 074	28
Arabie saoudite	1985	1 344 076	43	1985	603 127	38	1985	113 529	39
	1990	1 876 916	46	1990	892 585	44	1990	153 967	43
	1992	2 025 948	47	1992	1 073 361	44	1991	163 688	40
Singapore	1980	291 649	48	1975	183 364	49	1975	22 607	40
Singapour	1985	278 060	47	1980	180 817	50	1980	23 256	39
	1990	257 932	...	...	...	...	1983	35 192	42
	1991	260 286	...	...	...	...			
Sri Lanka	1980	2 081 391	48	...	...	...	1975	15 426	36
Sri Lanka	1985	2 242 645	48	1985	1 462 794	52	1980	42 694	43
	1990[2]	2 112 023	48	1990[2]	2 081 842	51	1985	59 377	40
	1992	2 059 203	48	1992	2 185 277	51		...	...
Syrian Arab Republic	1980	1 555 921	43	1980	604 327	37	1980	140 180	29
Rép. arabe syrienne	1985	2 029 752	46	1985	870 383	40	1985	179 473	35
	1990	2 452 086	46	1990	914 250	41	1990	221 628	39
	1992	2 573 181	47	1992	916 950	44	1992	194 371	38
Tajikistan	...	...	...	...	...	...	1980	54 800	...
Tadjikistan	1992	519 100	48	1992	736 700	48	1985	55 100	...
	1993	570 300	49	1993	652 700	47	1990	68 800	...
Thailand	1980	7 392 563	48	1975	1 193 741	44	1975	130 965	40
Thaïlande	1985	7 150 489	...	1980	1 919 967	...		...	...
	1990	6 464 853	49	...	...	...	1985	1 026 952	...
	1992	6 813 151	...	1990	2 397 262	48	1992	1 156 174	53
Turkey	1980	5 656 494	45	1980	2 217 909	...	1980	246 183	...
Turquie	1985	6 635 858	47	1985	2 927 692	35	1985	469 992	32
	1990	6 861 711	47	1990	3 808 142	37	1990	749 921	34
	1991	6 878 923	47	1991	3 987 423	38	1992	915 765	35
Turkmenistan	...	...	...	...	...	...	1980	35 800	...
Turkménistan	...	...	...	...	...	...	1985	38 800	...
	...	...	...	...	...	...	1990	41 800	...
United Arab Emirates	1980	88 617	48	1980	32 362	45	1980	2 861	49
Emirats arabes unis	1985	152 125	48	1985	62 082	48	1985	7 772	58
	1990	228 980	48	1990	107 881	50	1990	10 196	70
	1992	238 469	48	1992	129 683	51	1991	10 405	75
Uzbekistan	...	...	...	...	...	...	1980	278 100	...
Ouzbékistan	...	...	...	...	...	...	1985	285 500	...
	...	...	...	...	...	...	1990	340 900	...
Viet Nam	1980	7 887 439	47	1976	3 200 912	49	1975	80 323	39
Viet Nam	1985	8 125 836	48	...	...	...	1980[33]	114 701	24
	1990	8 862 292	...	1989	3 651 719	...	...	...	...
	1992	9 476 441	...	1990	...	...	...	...	...
Yemen [34]	1975	254 651	11	1975	24 606	13	1975	2 408	10
Yémen [34]	1980	435 913	13	1980	41 155	13	1980	4 519	11
	1985	981 127	20	1985	146 133	11	1985	12 589	...
	1990	1 291 372	24	1990	420 697	15	1988	23 457	...
Europe · Europe									
Albania	1975	579 303	47	1975	110 519	46	1980	14 568	50
Albanie	1980	552 651	47	1980	163 866	45	1985	21 995	45
	1985	543 775	48	1985	177 679	45	1990	22 059	52
	1990	551 294	48	1990	205 774	45	1992	22 835	52

10
Education at the first, second and third levels
Number of students and percentage female [cont.]
Enseignement des premier, second et troisième degrés
Nombre d'étudiants et étudiantes feminines en pourcentage [suite]

Country or area Pays ou zone	Years Années	First level Premier degré Total	%F	Years Années	Second level Second degré Total	%F	Years Années	Third level Troisième degré Total	%F
Austria	1980	400 397	49	1980	937 484	46	1980	136 774	42
Autriche	1985	343 823	48	1985	847 188	47	1985	173 215	45
	1990	370 210	49	1990	746 272	47	1990	205 767	46
	1992	382 663	49	1992	768 176	47	1992	221 389	47
Belarus	1980	750 300	...	1980	759 700	...	1980	177 000	...
Bélarus	1985	796 600	...	1985	716 700	...	1985	181 900	...
	1990[2]	614 800	...	1990[2]	968 200	...	1990	188 600	...
	1992	635 100	...	1992	970 300	...	1992	187 700	51
Belgium	1980	842 117	49	1980	835 524	50	1975	159 660	41
Belgique	1985	730 288	49	1985	824 997	49	1980	196 153	44
	1990	719 372	49	1990	769 438	49	1985	247 499	46
	1991	711 521	49	1991	765 672	49	1990	276 248	48
Bulgaria	1980	994 018	49	1980	314 753	48	1980	101 359	56
Bulgarie	1985	1 080 979	48	1985	374 565	49	1985	113 795	55
	1990	960 681	48	1990	391 550	50	1990	188 479	51
	1992	877 189	48	1992	374 514	50	1992	195 447	57
Croatia	...	...	...	...	...	...	1980[28]	64 966	...
Croatie	...	...	...	...	...	...	1985[28]	55 886	...
	...	...	...	...	...	...	1990[28]	72 342	...
	1992[35]	436 755	49	1992[35]	190 926	51	1992[28]	77 689	48
former Czechoslovakia †	1980	1 904 476	49	1980	780 811	48	1980	197 041	42
ancienne Tchécoslovaquie †	1985[2]	2 074 403	49	1985	744 059	51	1985	169 344	43
	1990	1 924 001	49	1990	864 215	51	1990	190 409	44
	1991	1 898 470	49	1991	848 721	50	1991	177 110	46
Czech Republic République tchèque	1992	1 160 510	50	1992	549 266	50	1992	116 560	44
Denmark	1980	434 635	49	1980	498 944	49	1980	106 241	49
Danemark	1985	402 707	49	1985	487 526	49	1985	116 319	49
	1990	340 267	49	1990	464 555	49	1990	142 968	52
	1991	327 024	49	1991	455 639	49	1991	150 159	53
Estonia	...	...	...	...	...	...	1980	25 500	...
Estonie	...	...	...	...	...	...	1985	23 500	...
	...	...	...	...	...	...	1990	25 900	...
	1992	119 409	49	1992	121 798	51	1992	24 768	51
Finland	1980	373 347	49	1980	449 322	52	1980	123 165	48
Finlande	1985	379 339	49	1985	424 076	53	1985	127 976	49
	1990	389 067	49	1990	426 864	53	1990	165 714	52
	1992	392 754	49	1992	463 121	54	1992	188 162	53
France	1980	4 610 361	48	1980	5 013 666	53	1980	1 076 717	...
France	1985	4 115 846	48	1985	5 371 593	51	1985	1 278 581	50
	1990	4 149 143	48	1990	5 521 862	50	1990	1 698 938	53
	1992	4 060 408	48	1992	5 573 582	50	1992	1 951 994	54
Germany †	1990	3 431 385	...	1990	7 398 100	...	...	...	...
Germany †	1992	3 470 000	49	1991	7 500 078	48	...	...	...
Federal Republic of Germany	1980	2 783 867	49	1980[36]	6 561 297	50	1980	1 223 221	41
Rép. féd. d'Allemagne	1985	2 271 546	49	1985	7 101 250	48	1985	1 550 211	42
	1990	2 561 267	49	1990	5 972 607	48	1990	1 799 394	40
	1991	2 590 082	49	1991	5 992 998	48	1991	1 867 491	41
former German Dem. Rep.	1980	852 109	48	1980	1 895 579	48	1975	386 000	...
ancienne Rép. dém. allemande	1985	859 830	48	1985	1 519 152	48	1980	400 799	58
	1990	870 118	...	1990	1 425 404	...	1985	432 672	54
	1991	847 970	49	1991	1 507 080	48	1988	438 930	52
Gibraltar	1975	2 808	50	1975	1 629	51	...	...	...
Gibraltar	1980	2 750	50	1980	1 811	50	...	...	...
	1984	2 830	48	1984	1 806	49	...	...	...
Greece	1975	935 730	48	1975	661 796	43	1975	117 246	37
Grèce	1980	900 641	48	1980	740 058	46	1980	121 116	41
	1985	887 735	48	1985	813 534	48	1985	181 901	49
	1989	834 688	48	1989	843 732	48	1990	195 213	50
Holy See [37]	...	...	...	...	...	...	1980	9 104	39
Saint–Siège [37]	...	...	...	...	...	...	1985	9 775	33
	...	...	...	...	...	...	1990	10 938	32
	...	...	...	...	...	...	1992	12 253	32

10
Education at the first, second and third levels
Number of students and percentage female [cont.]
Enseignement des premier, second et troisième degrés
Nombre d'étudiants et étudiantes feminines en pourcentage [suite]

Country or area Pays ou zone	Years Années	First level Premier degré Total	%F	Years Années	Second level Second degré Total	%F	Years Années	Third level Troisième degré Total	%F
Hungary	1980	1 162 203	49	1980	357 334	46	1980	101 166	50
Hongrie	1985	1 297 818	49	1985	422 323	49	1985	99 344	54
	1990	1 130 656	49	1990	514 076	49	1990	102 387	50
	1991	1 081 213	49	1991	531 051	49	1992	117 460	51
Iceland	1975	26 418	49	1975	*25 853	*45	1980	3 633	60
Islande	1980	24 736	...	1980	26 643	47	1985	4 724	55
	1985	24 603	49	1985	27 559	47	1990	5 225	...
	1989	25 525	...	1989	29 059	...	1991	6 161	59
Ireland	1980	419 998	49	1980	300 601	52	1980	54 746	41
Irlande	1985	420 236	49	1985	338 256	51	1985	70 301	43
	1990	416 747	49	1990	345 941	51	1990	90 296	46
	1991	408 567	49	1991	352 408	51	1991	101 108	47
Italy	1980	4 422 888	49	1980	5 307 989	48	1980	1 117 742	43
Italie	1985	3 703 108	49	1985	5 361 579	49	1985	1 185 304	46
	1990	3 055 883	49	1990	5 117 897	49	1990	1 452 286	48
	1992	2 959 564	49	1992	4 892 194	49	1992	1 615 150	51
Latvia	...	...	...	...	...	...	1980	47 200	...
Lettonie	...	...	...	...	...	...	1985	43 900	...
	...	...	...	...	...	...	1990	46 000	...
	1992	133 846	49	1992	242 644	51	1992	41 138	53
Lithuania	...	...	...	...	...	...	1980	71 000	...
Lituanie	...	...	...	...	...	...	1985	65 300	...
	1992	207 522	48	1992	337 890	50	1990	65 600	...
Luxembourg	1975	29 430	49	1975	22 652	49	1975[38]	483	42
Luxembourg	1980	24 628	49	1980	27 487	...	1980[38]	748	35
	1985	22 003	49	1985	25 656	48	1985[38]	759	34
	1990	23 465	51	1987	22 496	49	...	...	...
Malta	1975	29 834	48	1975	32 409	46	1975	1 425	28
Malte	1980	33 063	48	1980	25 501	45	1980	947	24
	1985	36 240	47	1985	27 779	48	1985	1 474	33
	1990	36 899	48	1990	32 544	47	1990	3 123	44
Monaco	1976	1 145	...	1980[3]	2 065	...	...	...	...
Monaco	1980[3]	1 017	46	1982	3 132	...	...	...	...
	1990	1 773	51	1990	2 785	49	...	...	...
	1991	1 761	49	1991	2 858	49	...	...	...
Netherlands	1980	1 333 342	49	1980[39]	1 391 485	48	1980	360 033	40
Pays−Bas	1985	1 109 590	49	1985	1 620 011	48	1985	404 866	41
	1990	1 082 022	50	1990	1 401 739	47	1990	478 869	44
	1992	1 046 192	50	1992	1 369 507	48	1991	493 563	45
Norway	1980	390 186	49	1980	360 776	50	1980	79 117	48
Norvège	1985	335 373	49	1985	387 990	50	1985	94 658	52
	1990	309 432	49	1990	370 779	50	1990	142 521	53
	1992	307 461	49	1992	380 916	50	1992	166 499	54
Poland	1980	4 167 313	49	1980	1 673 869	50	1980	589 134	56
Pologne	1985	4 801 307	48	1985	1 567 641	51	1985	454 190	56
	1990	5 189 118	49	1990	1 887 667	50	1990	544 893	56
	1992	5 231 769	49	1992	2 030 842	50	1992	584 177	56
Portugal	1980	1 240 307	48	1980	398 320	...	1980	92 152	48
Portugal	1985	1 235 312	48	1985	580 248	...	1985[40]	103 585	54
	1990	1 019 794	48	1990	670 035	53	1990	185 762	56
	1991	1 004 848	48	1991	778 432	...	1991	190 856	60
Republic of Moldova	...	...	...	...	...	...	1980	51 300	...
République de Moldova	...	...	...	...	...	...	1985	53 200	...
	1991	306 933	49	1991	448 404	...	1990	54 700	...
Romania	1980	3 236 808	49	1980	871 257	47	1975	164 567	45
Roumanie	1985	3 030 666	49	1985	1 537 548	46	1980	192 769	43
	1990[2]	1 253 480	49	1990[2]	2 837 948	49	1985	159 798	45
	1992	1 201 229	49	1992	2 451 624	49	1992	235 669	47
Russian Federation	...	...	...	...	...	...	1980	3 045 800	...
Fédération de Russie	...	...	...	...	...	...	1985	2 966 100	...
	...	...	...	...	...	...	1990	2 861 000	...
	1992	11 872 357	49	1992	9 443 327	51	1992	2 638 000	50

10
Education at the first, second and third levels
Number of students and percentage female [cont.]
Enseignement des premier, second et troisième degrés
Nombre d'étudiants et étudiantes feminines en pourcentage [suite]

Country or area Pays ou zone	Years Années	First level Premier degré Total	%F	Years Années	Second level Second degré Total	%F	Years Années	Third level Troisième degré Total	%F
San Marino	1980	1 509	49	1980	1 219	49	...	...	...
Saint—Marin	1985	1 411	49	1985	1 248	48	...	...	...
	1990	1 212	47	1990	1 182	52	...	...	...
	1992	1 190	48	1992	1 159	48	...	...	...
Slovakia Slovakia	1992	350 604	49	1992	657 010	50	1992	66 002	48
Slovenia Slevenia	1992	104 441	49	1992	211 426	49	1992	39 264	54
Spain	1980	3 609 623	49	1980	3 976 747	50	1980	697 789	44
Espagne	1985	3 483 948	48	1985	4 555 541	51	1985	935 126	49
	1990	2 820 497	48	1990	4 755 322	50	1990	1 222 089	51
	1991	2 662 490	48	1991	4 773 349	51	1991	1 301 748	51
Sweden	1980	666 679	49	1980	606 833	51	1980	171 356	...
Suède	1985	612 704	...	1985	624 835	...	1985	183 697	...
	1990	578 359	49	1990	588 474	50	1990	192 596	54
	1992	594 891	49	1992	602 703	49	1991	207 265	54
Switzerland	1980	450 942	49	1980	661 315	46	1980	85 127	30
Suisse	1985	376 512	49	1985	634 750	46	1985	110 111	32
	1990	404 154	49	1990	567 396	47	1990	137 486	35
	1992	420 089	49	1992	561 470	47	1992	146 266	36
Ukraine	1980	3 595 200	...	1980	3 406 400	...	1980	880 400	...
Ukraine	1985	3 739 900	...	1985	3 401 100	...	1985	853 100	...
	1990[2]	3 990 500	...	1990	3 407 500	...	1990	894 700	50[28]
	1992	4 102 100	49	1992	3 281 500	...	1991	890 192	50[28]
United Kingdom	1975	5 725 167	49	1975	5 154 371	49	1980	827 146	37
Royaume—Uni	1980	4 910 724	49	1980	5 341 849	50	1985	1 032 491	46
	1985	4 296 000	49	1985	4 877 000	50	1990	1 258 188	48
	1990	4 532 500	49	1990	4 335 600	50	1991	1 385 072	49
Yugoslavia	1990	466 692	49	1990	788 170	49	...	...	...
Yougoslavie	1992	473 902	49	1992	831 506	49	1992	143 268	53
Yugoslavia, SFR †	1975	1 494 825	49	1975	2 264 101	47	1975	394 992	40
Yougoslavie, Rfs †	1980	1 431 582	48	1980	2 426 164	47	1980	411 995	45
	1985	1 448 562	48	1985	2 352 985	47	1985	350 334	46
	1990	1 392 789	48	1990	2 344 331	48	1990	327 092	51
Oceania · Océanie									
American Samoa	1975	6 052	...	1976	*2 602	...	1975	689	49
Samoa américaine	1981	6 744	48	1980	3 000	...	1980	976	56
	1985	7 704	47	1985	3 342	47	1985	758	52
	1991	7 884	48	1991	3 643	46	1988	909	54
Australia	1980	1 718 352	49	1980	1 100 468	50	1980	323 716	45
Australie	1985	1 542 101	49	1985	1 278 272	49	1985	370 048	48
	1990	1 583 024	49	1990	1 278 163	50	1990	485 075	53
	1992	1 623 012	49	1992	1 294 596	49	1992	559 365	53
Cook Islands	1975	5 339		...	...	...	...	...	...
Iles Cook	1983	2 909	...	...	...	...	1980	360	45
	1985	2 713	...	...	...	...			
	1988	2 376		...	...	...	...		
Fiji	1980	116 139	49	1980	49 963	51	1975	1 653	29
Fidji	1985	127 286	49	1985	45 093	50	1980	1 666	...
	1990	143 552	...	1991	61 614	48	1985	2 313	38
	1992	145 630	49	1992	66 890	49	1991	7 908	...
French Polynesia	1981	29 012	48	1981	13 306	55	1980	27	26
Polynésie française	1984	27 401	48	1986	17 878	55	...	...	...
	1990	28 270	48	1990	20 311	53	...	...	...
	1992	29 132	48	1992	22 366	55	1991	301	50
Guam	1975	*20 215	...	1975	*13 242	...	1975	3 800	43
Guam	1980	18 093	...	1980	14 935	...	...	...	...
	1885	16 783	...	1985	14 557	...	1986	7 052	53
	1988	15 516	...	1988	16 017	...	...	...	...
Kiribati	1980	13 235	49	1980	2 440	46	...	...	...
Kiribati	1985	13 440	49	1985	2 196	50	...	...	...
	1990	14 709	50	1990	3 003	49	...	...	...
	1992	16 020	49	1992	3 357	52	...	...	...
Nauru	1974	1 392	47	...	...	...	...	...	...
Nauru	1985	1 451	47	1985	482	50	...	...	...

10
Education at the first, second and third levels
Number of students and percentage female [cont.]
Enseignement des premier, second et troisième degrés
Nombre d'étudiants et étudiantes feminines en pourcentage [suite]

Country or area Pays ou zone	Years Années	First level Premier degré Total	%F	Years Années	Second level Second degré Total	%F	Years Années	Third level Troisième degré Total	%F
New Caledonia	1981	26 779	48	1980	11 945	54	1975	178	35
Nouvelle−Calédonie	1985	22 517	48	1985	18 351	52	1980	438	39
	1990	22 958	48	1990	20 673	52	1985	761	44
	1991	22 325	48	1991	21 908	52	...	...	...
New Zealand	1980	381 262	49	1980	352 427	49	1980	76 643	41
Nouvelle−zélande	1985	329 337	49	1985	354 080	50	1985	95 793	46
	1990	318 568	48	1990	340 915	49	1990	111 504	52
	1992	317 286	48	1992	350 112	49	1992	146 215	54
Niue	1975	1 122	48	1975	271	49	...	...	...
Nioué	1980	666	...	1980	397	...	...	...	...
	1985	503	...	1985	321	...	...	...	...
	1991[2]	371	...	1991[2]	302	53	...	...	...
Palau [41]	1975	30 285	48	1975	7 951	44	...	...	...
Palaos [41]	1980	30 159	...	1980	6 885	...	1980	2 129	24
	...	...	...	1981	6 872	...	...	...	...
Papua New Guinea	1980	299 823	*41	1981	49 334	...	...	...	...
Papouasie−Nouv.−Guinée	1987	384 367	44	1987	63 391	35	1980	*5 040	*22
	1990	415 195	44	1990	65 643	38	1985	5 068	23
	1992	443 552	45	1992	69 596	39	1986	6 397	24
Samoa	1975	32 642	51	1975	15 943	*51	1975	249	*5
Samoa	1980	33 012	48	1980	19 785	49	1981	694	7
	1986	31 412	48	1986	20 604	...	1983[42]	562	47
	1989	37 833	48	...	...	...	...	...	...
Solomon Islands	1980	28 870	41	1980	4 030	...	...	...	...
Iles Salomon	1985	38 716	...	...	...	...	...	...	...
	1990	47 598	44	1990	5 636	37	...	...	...
	1992	53 320	44	1992	6 666	37	...	...	...
Tokelau	1976	386	52	1979	277	47	...	...	...
Tokélaou	1981	434	49	...	...	...	...	...	...
	1983	411	48	1983	488	50	...	...	...
	1991	361	50	...	...	...	...	...	...
Tonga	1980	19 012	...	1981	16 566	47	...	...	...
Tonga	1985	17 019	48	1985	15 232	51	1985	705	56
	1990	16 522	48	1990	14 749	48	...	...	...
	1992	16 658	48	1992	13 902	48	...	...	...
Tuvalu	1976	1 540	...	1976	248	...	...	...	...
Tuvalu	1980	1 327	...	1980	248	...	...	...	...
	1986	1 280	49	...	...	...	...	...	...
	1990[3]	1 485	48	1990	345	52	...	...	...
Vanuatu	1980	23 264	46	1975	1 505	42	...	...	...
Vanuatu	1985	22 897	...	1980	2 426	42	...	...	...
	1990[3]	24 471	47	...	...	...	...	...	...
	1992[3]	26 267	47	1991	4 184	43	...	...	...
former USSR · ancienne URSS									
former USSR †	1975	21 366 200	...	1975	23 171 700	...	1975	4 853 958	50
ancienne URSS †	1980	21 713 900	...	1980	20 274 500	...	1980	5 235 200	...
	1985	23 585 000	...	1985	20 513 200	...	1985	5 147 200	...
	1990	25 633 000	49	1990	21 090 000	...	1990	5 253 088	49

Source:
United Nations Educational, Scientific and Cultural Organization (Paris).

† For detailed descriptions of data pertaining to former Czechoslovakia, Germany, SFR Yugoslavia and former USSR, see Annex I − Country or area nomenclature, regional and other groupings.

1 Data refer to institutions under the authority of the Ministry of Education.
2 Due to a change in classification, data are not comparable with those of previous years.
3 Data refer to public education only.
4 Data refer to full−time only.
5 Data exclude 'Ecole Nationale d'Administration et de

Source:
Organisation des Nations Unies pour l'éducation, la science et la culture (Paris).

† Pour les descriptions en détails des données relatives à l'ancienne Tchécoslovaquie, l'Allemagne, la Rfs Yougoslavie et l'ancienne URSS, voir l'Annexe I − Nomenclature des pays ou zones, groupements régionaux et autres groupements.

1 Les données se rapportent aux institutions relevant du Ministère de l'Education.
2 Suite à un changement de classification, les données ne sont pas comparables à celles des années antérieures.
3 Les données se réfèrent à l'enseignement public seulement.
4 Les données se réfèrent à l'enseignement à plein temps seulement.
5 Les données excluent l'Ecole Nationale d'Administration et de

10
Education at the first, second and third levels
Number of students and percentage female [*cont.*]
Enseignement des premier, second et troisième degrés
Nombre d'étudiants et étudiantes feminines en pourcentage [*suite*]

Magistrature'.

6   Data do not include the University of Law.
7   Data for vocational education refer to schools attached to the Ministry of Education only.
8   Data do not include Al Azhar.
9   Data for general education do not include Al Azhar for 1991 and data for teacher training do not include Al Azhar from 1988.
10   Data exclude private institutions.
11   Data do not include Asmara University and Kotebe college.
12   Data do not include Action aid schools.
13   For general and vocational education, data refer to public education only.
14   For general education, data refer to public education only.
15   Including education preceding the first level.
16   Excluding Transkei, Bophuthatswana, Venda and Ciskei.
17   Data refer to government–maintained and aided schools only.
18   For general education, data refer to government–maintained and aided schools only.
19   Data refer to Tanzania mainland only.
20   Data include a part of education at the second level.
21   Data include adult education.
22   Data include trade and vocational programmes.
23   Data refer only to institutions recognized by the National Council for Higher Education.
24   Data refer to the University of San Carlos only.
25   Data include infant classes.
26   Data include secondary classes attached to primary schools.
27   Data refer to nationals only.
28   Data exclude post–graduate students.
29   Not including Turkish schools.
30   Revised data.
31   Do not include Air and Correspondence Courses.
32   Data do not include polytechnic institutes.
33   Data include correspondence courses.
34   Excluding former Democratic Yemen.
35   Data do not include schools in the war areas.
36   Data do not include technical education consisting of both on the job training and school education.
37   Data refer to students enrolled in higher institutions under the authority of the Holy See.
38   Data refer to students enrolled in institutions located in Luxembourg. At university level, the majority of students pursue their studies in the following countries: Austria, Belgium, France, Germany and Switzerland.
39   Data do not include apprenticeships and health care training.
40   Excluding the University of Porto.
41   Including data for Federated States of Micronesia, Marshall Is. and Northern Marianna Is.
42   Data exclude school of agriculture.

Magistrature.

6   Les données ne comprennent pas l'Université des Sciences juridiques.
7   Les données pour l'enseignement technique se réfèrent aux écoles rattachées au Ministère de l'Education seulement.
8   Les données ne comprennent pas Al Azhar.
9   Les données pour l'enseignement général ne comprennent pas Al Azhar pour 1991 et les données pour l'enseignement normal ne comprennent pas Al Azhar à partir de 1988.
10   Les données excluent les institutions privées.
11   Les données ne comprennent pas l'Université Asmara et le collège Kotebe.
12   Les données n'incluent pas les écoles de Action aid.
13   Pour l'enseignement général et technique, les données se réfèrent à l'enseignement public seulement.
14   Pour l'enseignement général, les données se réfèrent à l'enseignement public seulement.
15   Y compris l'enseignement précédant le premier degré.
16   Non compris Transkei, Bophuthatswana, Venda and Ciskei.
17   Les données se réfèrent aux écoles publiques et subventionnées seulement.
18   Pour l'enseignement général, les données se réfèrent aux écoles publiques et subventionnées seulement.
19   Les données se réfèrent à la Tanzanie continentale seulement.
20   Les données incluent une partie de l'enseignement du second degré.
21   Les données incluent l'éducation des adultes.
22   Les données incluent les programmes d'enseignement techniques et commerciaux.
23   Les données se réfèrent seulement aux institutions reconnues par le Conseil National pour l'Education supérieure.
24   Les données se réfèrent à l'Université de San Carlos seulement.
25   Les données incluent les classes enfantines.
26   Les données incluent les classes secondaires rattachées aux écoles primaires.
27   Les données se réfèrent aux étudiants nationales seulement.
28   Les données excluent le niveau universitaire supérieur.
29   Non compris les écoles turques.
30   Données révisées.
31   Les données n'incluent pas 'Air and Correspondence Courses'.
32   Les données n'incluent pas les instituts polytechniques.
33   Les données incluent les cours par correspondance.
34   Non compris l'ancien Yémen démocratique.
35   Les données n'incluent pas les écoles dans les zones de guerre.
36   Les données n'incluent pas l'enseignement technique dispensé à la fois dans les institutions scolaires et auprés des entreprises.
37   Les données se réfèrent aux étudiants dans les institutions du troisième degré sous l'autorité du Saint–Siège.
38   Les données se réfèrent seulement aux étudiants inscrits dans les institutions du Luxembourg. La plus grande partie des étudiants luxembourgeois poursuivent leurs études universitaires dans les pays suivants : Allemagne, Autriche, Belgique, France, et Suisse.
39   Les données ne comprennent pas l'apprentissage et les programmes relatifs à la santé.
40   Non compris l'Université de Porto.
41   Y compris les données pour les Etats fédérés de Micronésie, les îles Marshall et les îles Mariannes du Nord.
42   Les données excluent l'école d'agriculture.

11
Public expenditure on education at current market prices
Dépenses publiques afférentes à l'enseignement aux prix courants du marché

Country or area Pays ou zone (Currency unit · Unité monétaire)	Years Années	Total educational expenditure Dépenses totales d'éducation			Current educational expenditure Dépenses ordinaires d'éducation		
		Amount Montant (000 000)	% of GNP % du PNB	% total gov't exp. % dép. totales gouv.	Amount Montant (000 000)	% of GNP % du PNB	% current gov't exp. % dép. ordinaires gouv.
Africa · Afrique							
Algeria	1980	12 355	7.8	24.3	8 259	5.2	29.8
Algérie	1985	24 248	8.4	20.7	16 814	5.9	26.2
(dinar)	1990[1]	29 504	5.7	21.1	24 953	4.8	29.7
	1992[1]	53 516	5.7	16.3	...	...	...
Angola	1985	9 643	...	10.8	9 419	...	14.0
Angola	1987	12 854	...	13.8	11 567	...	15.5
(kwansa)	1990[2]	12 076	...	10.7	10 856	...	...
Benin Bénin							
(CFA franc · franc CFA)	1980	...	...	...	12 426	4.2	36.8
Botswana	1979	37	7.0	16.1	27	5.3	21.4
Botswana	1985	111	6.9	15.4	88	5.5	18.6
(pula)	1990[2]	437	7.6	17.0	311	5.4	21.1
	1992[2]	574	7.5	16.0	462	6.1	21.3
Burkina Faso	1980	7 994	2.6	19.8	7 436	2.4	21.1
Burkina Faso	1985	12 901	2.3	21.0	12 292	2.2	22.4
(CFA franc · franc CFA)	1989	18 780	2.7	17.5	18 727	2.7	21.9
	1992[3]	11 934	1.6	...	11 002	1.4	...
Burundi	1979	2 066	3.0	17.5	1 796	2.6	...
Burundi	1985[2]	3 467	2.5	15.5	3 212	2.4	17.1
(franc)	1990[2]	6 570	3.4	16.7	6 370	3.3	19.4
	1992[2]	8 361	3.7	11.9	8 023	3.6	24.2
Cameroon	1980	45 099	3.2	20.3	36 653	2.6	...
Cameroun	1985[2]	109 344	3.0	14.8	90 045	2.5	20.9
(CFA franc · franc CFA)	1990[2]	107 968	3.4	19.6	97 948	3.1	26.9
	1991[2]	91 889	3.1	16.9	81 719	2.8	21.6
Cape Verde	1985	341	3.6	...	325	3.5	15.2
Cap–Vert	1987	493	3.8	14.8	472	3.6	15.3
(escudo)	1991	903	4.1	19.9	890	4.1	20.0
Central African Rep.	1979	5 670	3.8	20.9	5 510	3.6	...
Rép. centrafricaine	1986	9 553	2.8	...	9 313	2.8	25.6
(CFA franc · franc CFA)	1990	9 862	2.8	...	9 622	2.8	...
Chad Tchad							
(CFA franc · franc CFA)	1991	8 284	2.3	...	8 212	2.3	...
Comoros [2]	1985	...	...	...	2 105	4.1	23.1
Comores [2]	1990	...	...	...	2 666	4.0	24.3
(CFA franc · franc CFA)	1992	...	...	...	2 829	4.1	22.0
Congo	1980	22 942	7.0	23.6	21 517	6.6	24.1
Congo	1984	44 442	5.1	9.8	41 033	4.7	14.7
(CFA franc · franc CFA)	1990	37 899	5.6	14.4	36 906	5.5	18.0
	1991	57 092	...	...	56 537	...	19.3
Côte d'Ivoire	1980	147 478	7.2	22.6	123 196	6.0	36.4
Côte d'Ivoire	1985	...	...	...	179 447	6.3	...
(CFA franc · franc CFA)	1992	...	...	...	153 004	6.7	...
Djibouti	1979	1 368	...	11.5	1 046	...	9.6
Djibouti	1985[2]	1 690	...	7.5	1 690	...	...
(franc)	1990	2 614	...	10.5	2 614	...	10.5
	1991	2 872	...	11.1	2 872	...	11.1
Egypt [4]	1981	919	5.7	9.4	722	4.5	10.1
Egypte [4]	1985	1 878	6.3	...	1 775	5.9	10.8
(pound · livre)	1990[2]	2 664	3.8	...	2 406	3.4	...
	1991	4 557	5.0	9.7	3 941	4.3	11.0
Equatorial Guinea Guinée équatoriale							
(CFA francs · francs CFA)	1988	620	1.7	3.9	527	1.5	3.5
Ethiopia	1980	279	3.3	10.4	222	2.6	12.7
Ethiopie	1985	420	4.3	9.5	354	3.6	14.3
(birr)	1990	600	4.9	9.4	494	4.0	11.1

11
Public expenditure on education at current market prices [cont.]
Dépenses publiques afférentes à l'enseignement aux prix courants du marché [suite]

Country or area Pays ou zone (Currency unit · Unité monétaire)	Years Années	Total educational expenditure Dépenses totales d'éducation			Current educational expenditure Dépenses ordinaires d'éducation		
		Amount Montant (000 000)	% of GNP % du PNB	% total gov't exp. % dép. totales gouv.	Amount Montant (000 000)	% of GNP % du PNB	% current gov't exp. % dép. ordinaires gouv.
Gabon Gabon (CFA franc · franc CFA)	1980	22 204	2.7	...	16 055	2.0	...
	1985	69 500	4.5	9.4	47 500	3.1	21.7
	1987	53 372	5.7	...	47 774	5.1	...
	1992[3]	41 529	2.9	...	34 407	2.4	...
Gambia Gambie (dalasi)	1980	13	3.3	...	11	2.9	...
	1985	31	3.2	...	25	2.6	16.4
	1990	95	3.8	11.0	73	2.9	11.6
	1991	78	2.7	12.9	75	2.6	13.2
Ghana Ghana (cedi)	1980	1 319	3.1	17.1	...	...	...
	1985[2]	8 675	2.6	19.0	...	...	...
	1990	61 900	3.1	24.3	53 664	2.7	27.1
Guinea Guinée (franc)	1984	1 491	...	15.3	1 486	...	17.2
	1990	23 483	1.3	...	...	...	...
	1992	61 123	2.2	...	...	...	...
Guinea—Bissau Guinée—Bissau (peso)	1980	...	...	...	208	4.0	...
	1984	...	...	...	539	3.2	11.2
	1987	2 533	2.8	...	2 473	2.8	...
Kenya Kenya (shilling)	1980	3 526	6.8	18.1	3 247	6.2	23.6
	1985	6 171	6.4	...	5 789	6.0	...
	1990	12 955	7.0	16.7	11 000	5.9	19.0
Lesotho Lesotho (maloti)	1980	25	5.1	14.8	20	4.1	19.0
	1984	35	3.7	...	29	3.1	15.1
	1990	99	3.8	12.2	81	3.1	17.4
	1992	204	6.4	...	160	5.0	...
Liberia Libéria (dollar)	1980	62	5.7	24.3	53	4.9	27.0
Libyan Arab Jamahiriya Jamah. arabe libyenne (dinar)	1980	356	3.4	...	224	2.1	...
	1985	575	7.1	19.8	457	5.7	38.1
	1986	636	9.6	20.8	506	7.7	37.1
Madagascar Madagascar (franc)	1980	36 896	4.4	...	31 548	3.7	...
	1985	52 182	2.9	...	49 806	2.8	...
	1990[3]	67 038	1.5	...	64 964	1.5	...
Malawi Malawi (kwacha)	1980	31	3.4	8.4	24	2.6	...
	1985	64	3.5	9.6	46	2.5	...
	1990	165	3.4	10.3	117	2.4	9.8
Mali Mali (CFA franc · franc CFA)	1980	12 903	3.8	30.8	12 752	3.7	32.1
	1985	17 184	3.1	...	17 048	3.1	...
	1987	18 693	3.2	17.3	18 285	3.1	18.5
Mauritania Mauritanie (ouguiya)	1980	...	...	...	1 546	5.0	...
	1985	...	...	...	3 973	8.2	33.2
	1988	...	...	...	3 188	4.8	22.0
Mauritius Maurice (rupee · roupie)	1980	454	5.3	11.6	408	4.7	15.5
	1985	598	3.8	9.8	555	3.5	12.4
	1990	1 384	3.7	11.8	1 287	3.4	14.0
Morocco[2] Maroc[2] (dirham)	1980	4 367	6.1	18.5	3 529	4.9	23.3
	1985	7 697	6.3	22.9	6 079	5.0	28.6
	1990	11 220	5.5	26.1	10 187	4.9	33.6
	1992	13 564	5.8	26.7	11 944	5.1	32.0
Mozambique[5] Mozambique[5] (metical)	1980	2 900	3.8	12.1	2 500	3.2	17.7
	1985	4 400	3.1	10.6	4 100	2.9	12.3
	1990	72 264	6.2	12.0	46 064	4.0	17.5
Namibia Namibie (rand)	1981	25	1.6	...	...	...	...
Niger Niger (CFA franc · franc CFA)	1980	16 533	3.1	22.9	7 763	1.5	16.8
	1989[12]	19 873	2.9	9.0	15 545	2.3	13.6
	1991[12]	20 143	3.1	10.8	19 493	3.0	18.4
Nigeria Nigéria (naira)	1981	3 153	6.4	24.7	2 523	5.1	...
	1985[6]	814	1.2	8.7	699	1.0	19.7
	1986[6]	1 170	1.7	12.0	728	1.1	19.0

11
Public expenditure on education at current market prices [*cont.*]
Dépenses publiques afférentes à l'enseignement aux prix courants du marché [*suite*]

Country or area Pays ou zone (Currency unit • Unité monétaire)	Years Années	Total educational expenditure Dépenses totales d'éducation			Current educational expenditure Dépenses ordinaires d'éducation		
		Amount Montant (000 000)	% of GNP % du PNB	% total gov't exp. % dép. totales gouv.	Amount Montant (000 000)	% of GNP % du PNB	% current gov't exp. % dép. ordinaires gouv.
Réunion	1980	1 314	15.6	...	1 264	15.0	...
Réunion	1985	2 416	...	...	2 292	...	...
(franc)	1990	3 375	...	...	3 343	...	...
	1992[1]	4 854	...	...	4 814	...	...
Rwanda	1980	2 880	2.7	21.6	2 439	2.3	21.5
Rwanda	1984	4 997	3.2	25.1	4 887	3.1	28.1
(franc)	1989	7 222	3.8	25.4	6 793	3.6	29.1
Sao Tome and Principe							
Sao Tomé et Principe	1981	91	8.0	...	...	...	...
(dobra)	1986	100	4.3	18.8	...	...	...
Senegal	1980	...	...	...	27 485	4.5	23.5
Sénégal	1985	...	...	...	46 118	4.2	24.4
(CFA franc • franc CFA)	1990	...	...	...	60 467	3.9	26.8
	1991	...	...	...	61 690	3.9	27.4
Seychelles	1980	52	5.8	14.4	50	5.5	14.0
Seychelles	1985	125	10.7	21.3	120	10.3	21.9
(rupee • roupie)	1990	153	8.1	14.8	153	8.1	18.5
	1992	140	7.2	11.3	...	...	...
Sierra Leone	1980	43	3.8	11.8	41	3.6	14.5
Sierra Leone	1985	112	2.4	12.4	106	2.3	15.5
(leone)	1989	604	1.4	...	577	1.3	...
Somalia [1]	1980	169	1.0	8.7	154	0.9	...
Somalie [1]	1985	371	0.5	4.1	274	0.3	...
(shilling)	1986	434	0.4	2.8	290	0.3	...
South Africa							
Afrique du Sud	1986	8 108	6.0	...	6 844	5.1	...
(rand)	1990[7]	17 153	6.8	...	15 265	6.0	...
Sudan							
Soudan	1980	187	4.8	9.1	172	4.4	12.6
(pound • livre)	1985[1]	...	...	...	580	4.0	15.0
Swaziland	1980	26	6.1	...	20	4.7	23.1
Swaziland	1985	52	5.9	20.3	44	5.0	25.9
(lilangeni)	1989	101	6.1	22.5	88	5.3	25.1
Togo	1980	13 049	5.6	19.4	12 575	5.4	21.0
Togo	1985	15 880	5.0	19.4	15 028	4.7	19.2
(CFA franc • franc CFA)	1990	24 420	5.6	...	22 720	5.2	...
Tunisia	1980	185	5.4	16.4	162	4.7	23.5
Tunisie	1985	389	5.9	14.1	351	5.3	20.8
(dinar)	1990	648	6.1	14.3	569	5.4	17.7
	1992[2]	789	5.9	...	701	5.2	...
Uganda [2]	1980	15	1.2	11.3	14	1.1	12.8
Ouganda [2]	1984	288	2.8	...	205	2.0	...
(shilling)	1990	20 188	1.3	11.5	18 527	1.2	15.1
	1991	35 026	1.7	15.0	33 012	1.6	16.5
United Rep. of Tanzania	1981	2 203	4.5	12.1	1 848	3.8	15.9
Rep. Unie de Tanzania	1985	4 234	3.5	14.0	3 643	3.0	15.6
(shilling)	1990	23 426	5.0	11.4	20 599	4.4	...
Zaire	1980	1 015	2.6	24.2	998	2.6	25.3
Zaïre	1985	3 291	1.0	7.3	3 239	1.0	7.3
(zaire • zaïre)	1988	15 006	1.0	6.4	14 357	0.9	6.4
Zambia	1980	127	4.5	7.6	120	4.2	11.1
Zambie	1985	293	4.6	13.4	272	4.3	14.3
(kwacha)	1990	2 737	2.3	8.7	2 382	2.0	8.7
Zimbabwe	1980	224	6.6	13.7	218	6.4	14.1
Zimbabwe	1986	726	9.1	15.0	721	9.0	...
(dollar)	1990	1 661	10.4	...	...	...	...
	1992[1]	2 040	7.4	...	...	...	...
America, North • Amérique du Nord							
Antigua and Barbuda	1980	9	3.0	...	9	3.0	...
Antigua−et−Barbuda	1984	12	2.7	...	11	2.6	...
(dollar)	1988	...	...	...	31	3.7	...

11
Public expenditure on education at current market prices [cont.]
Dépenses publiques afférentes à l'enseignement aux prix courants du marché [suite]

Country or area Pays ou zone (Currency unit · Unité monétaire)	Years Années	Total educational expenditure Dépenses totales d'éducation			Current educational expenditure Dépenses ordinaires d'éducation		
		Amount Montant (000 000)	% of GNP % du PNB	% total gov't exp. % dép. totales gouv.	Amount Montant (000 000)	% of GNP % du PNB	% current gov't exp. % dép. ordinaires gouv.
Bahamas	1980	...	...	...	53	4.4	22.1
Bahamas	1984	...	...	...	77	4.1	21.1
(dollar)	1990	...	...	...	107	3.6	18.9
	1991	...	...	...	104	3.3	17.4
Barbados	1980	109	6.5	20.5	90	5.4	...
Barbade	1984	139	6.1	...	123	5.4	...
(dollar)	1990	269	7.9	...	218	6.4	...
Belize							
Belize	1986	...	...	...	18	4.1	...
(dollar)	1991	48	5.8	15.5	38	4.7	21.8
Bermuda	1980	26	4.1	...	26	4.0	...
Bermudes	1984	33	3.2	18.4	31	3.0	19.9
(dollar)	1990	53	...	14.5	49	...	15.7
	1991	60	...	...	54	...	...
British Virgin Islands							
Iles Vierges brit.	1980	...	...	...	2	...	13.3
(US dollars · dollars E–U)	1984	4	...	15.3	4	...	16.9
Canada	1980	20 833	7.0	16.3	19 295	6.4	...
Canada	1985	30 287	6.6	11.9	28 202	6.1	...
(dollar)	1990	43 487	6.8	14.2	40 288	6.3	...
	1992	49 955	7.6	14.3	46 514	7.1	...
Cayman Islands Iles Caïmanes							
(dollar)	1980	5	...	...	4	...	...
Costa Rica	1980	3 069	7.8	22.2	2 802	7.1	26.7
Costa Rica	1985	8 181	4.5	22.7	7 787	4.2	26.2
(colon)	1990	22 907	4.6	20.8	22 188	4.4	26.3
	1992	37 625	4.5	20.2	...	...	...
Cuba [8]	1980	1 267	7.2	...	1 135	6.4	...
Cuba [8]	1985	1 690	6.3	...	1 588	5.9	...
(peso)	1990	1 748	...	12.3	1 627	...	14.4
Dominica							
Dominique	1986	17	5.9	16.7	16	5.7	18.5
(dollar)	1989	22	5.8	10.6	20	5.2	19.9
Dominican Republic	1980	139	2.2	16.0	104[2]	1.6	...
Rép. dominicaine	1985	234	1.8	14.0	204[2]	1.6	...
(peso)	1992	1 502	1.6	8.9	970[2]	1.0	...
El Salvador	1980	340	3.9	17.1	320	3.7	22.9
El Salvador	1984	336	3.0	12.5	293	2.6	16.3
(colon)	1990	722	1.8	...	715	1.8	...
	1992	884	1.6	...	883	1.6	...
Grenada							
Grenade	1985	...	...	...	14	4.6	...
(dollar)	1986	...	...	...	16	4.7	...
Guadeloupe	1980	862	14.3	...	818	13.6	...
Guadeloupe	1987	1 419	...	...	1 324	...	...
(franc)	1990	1 973	...	...	1 952	...	...
	1992[1]	2 663	...	...	2 627	...	...
Guatemala	1979	131	1.9	16.6	117	1.7	23.4
Guatemala	1984	163	1.8	12.4	160	1.7	23.1
(quetzal)	1990[2]	468	1.4	11.8	...	...	...
	1991[2]	562	1.2	...	...	...	...
Haiti	1980	107	1.5	14.9	86	1.2	17.2
Haïti	1985	118	1.2	16.5	117	1.2	16.7
(gourde)	1990	216	1.8	20.0	216	1.8	20.1
Honduras	1980	155	3.2	14.2	141	2.9	19.5
Honduras	1985	290	4.2	13.8	286	4.1	...
(lempira)	1989	416	4.2	15.9	404	4.1	...
	1991	621	4.0	...	606	3.9	...
Jamaica	1980[2]	304	7.0	13.1	303	7.0	19.4
Jamaïque	1985	550	5.8	12.1	515	5.4	15.8
(dollar)	1990[2]	1 472	5.6	12.8	1 276	4.9	17.3
	1992	2 633	4.1	10.2	2 360	3.7	16.2

11
Public expenditure on education at current market prices [cont.]
Dépenses publiques afférentes à l'enseignement aux prix courants du marché [suite]

Country or area Pays ou zone (Currency unit · Unité monétaire)	Years Années	Total educational expenditure Dépenses totales d'éducation			Current educational expenditure Dépenses ordinaires d'éducation		
		Amount Montant (000 000)	% of GNP % du PNB	% total gov't exp. % dép. totales gouv.	Amount Montant (000 000)	% of GNP % du PNB	% current gov't exp. % dép. ordinaires gouv.
Martinique Martinique (franc)	1980 1990 1992[1]	931 2 013 2 741	15.3	... ...	887 1 994 2 716	14.5	... ...
Mexico Mexique (peso)	1980 1985 1990 1992	204 326 1 767 324 26 640 958 51 549 597	4.7 3.9 4.0 5.2	... ...	129 797[2] 1 229 314[2] 16 617 522[2] 32 780 581[2]	3.0 2.7 2.5 3.3	... ...
Montserrat Montserrat (dollar)	1988	6	...	9.3	6	...	18.9
Netherlands Antilles Antilles néerlandaises (guilder · florin)	1980 1985 1989 1992	43 57 69 171	3.5 2.8	13.6 14.4	42 57 54 ...	3.4 2.8	20.2 16.3 18.8 ...
Nicaragua Nicaragua (córdoba)	1980 1985[2] 1989[2] 1992[3]	0.662 6.409 399 895 275	3.4 6.8 3.3 4.1	10.4 10.2 ... 10.6	0.580 6.196 383 109 ...	3.0 6.6 3.1 ...	... 12.2 ... ...
Panama Panama (balboa)	1980 1985 1990 1992	166 237 248 329	4.8 4.8 5.2 5.5	19.0 18.7 20.9 18.9	156 231 241 307	4.5 4.7 5.1 5.2	19.8 19.9 22.2 21.9
St. Kitts and Nevis Saint−Kitts−et−Nevis (dollar)	1980 1985 1989[1] 1991[1]	7 12 12 12	5.2 5.8 3.2 2.8	9.4 18.5 12.0 11.6	7 12 11 12	5.2 5.7 3.1 2.7	13.6 19.1 12.9 11.7
Saint Lucia Sainte lucie (dollar)	1980 1986 1990 1991	... 38	... 5.5	... ...	19 36 54 55	6.9 5.2 5.1 4.9	... ...
St. Pierre−Miquelon St. Pierre−Miquelon (franc)	1980 1985 1990[2] 1992[2]	18 41 54 57	... ...	... ...	17 37 41 54	... ...	... ...
St. Vincent and the Grenadines Saint−Vincent−et−Grenadines (dollar)	1986 1990[1]	19 34	5.8 6.7	11.6 13.8	18 26	5.4 5.0	16.8 17.2
Trinidad and Tobago Trinité−et−Tobago (dollar)	1980 1985 1990	564 1 042 788	4.0 6.1 4.0	11.5 ... 11.6	431 912 727	3.0 5.3 3.7	17.6 ... 13.5
Turks and Caicos Is. Iles Turques et Caïques (US dollars · dollars E−U)	1980	1.018	...	...	0.832	...	...
United States Etats−Unis (dollar)	1980[9] 1985 1990	182 849 199 372 292 944	6.7 4.9 5.3	... 15.5 12.3	... 182 875 265 074	... 4.5 4.8	... 16.3 12.3
US Virgin Islands Iles Vierges américaine (US dollars · dollars E−U)	1980 1984	58 74	7.9 7.5	... ...	57 ...	7.8 ...	... ...
America, South · Amérique du Sud							
Argentina Argentine (austral)	1980 1985[2] 1990[2] 1992[10]	1.018 740 7 356 595 6 962	2.7 1.5 1.1 3.1	15.1 8.6 10.9 15.7	0.860 667 7 062 974 ...	2.3 1.3 1.1 ...	18.8 8.9 12.4 ...
Bolivia Bolivie (boliviano)	1980 1985 1990[2 11] 1991[2 11]	0.0051 43 363 467	4.4 2.1 2.7 2.7	25.3	0.0049	4.2	27.0 ... ...
Brazil Brésil (cruzado)	1980 1985 1989	0.431 50 56 101	3.6 3.8 4.6	... ...	... ...	... ...	... ...

11
Public expenditure on education at current market prices [cont.]
Dépenses publiques afférentes à l'enseignement aux prix courants du marché [suite]

Country or area Pays ou zone (Currency unit · Unité monétaire)	Years Années	Total educational expenditure Dépenses totales d'éducation			Current educational expenditure Dépenses ordinaires d'éducation		
		Amount Montant (000 000)	% of GNP % du PNB	% total gov't exp. % dép. totales gouv.	Amount Montant (000 000)	% of GNP % du PNB	% current gov't exp. % dép. ordinaires gouv.
Chile	1980	47 961	4.6	11.9	45 504	4.4	13.4
Chili	1985	101 493	4.4	15.3	...	...	...
(peso)	1990[12]	232 516	2.7	10.4	225 620	2.6	...
	1992	407 645	2.9	12.9	392 411	2.8	15.0
Colombia [2]	1980	29 240	1.9	14.3	27 286	1.7	19.9
Colombie [2]	1985	136 570	2.9	...	127 908	2.7	...
(peso)	1990	526 686	2.8	12.4	...	...	...
	1992	996 567	3.1	...	...	...	...
Ecuador	1980	15 580	5.6	33.3	14 649	5.3	36.0
Equateur	1985	38 009	3.7	20.6	35 611	3.5	25.8
(sucre)	1991	298 126	2.6	17.5	260 332	2.3	23.9
	1992	492 252	2.7	19.2	416 411	2.3	24.0
French Guiana	1980	159	17.6	...	152	16.9	...
Guyane français	1990	501	...	...	490	...	...
(franc)	1992[1]	774	...	...	755	...	...
Guyana	1979	121	9.7	14.0	89	7.2	15.2
Guyana	1985	162	9.8	10.4	135	8.2	13.0
(dollar)	1990	542	4.7	...	435	3.8	5.2
Paraguay	1980	8 793	1.5	16.4	6 267	1.1	...
Paraguay	1985	20 662	1.5	16.7	16 822	1.2	18.8
(guarani)	1990[2]	74 387	1.2	9.1	71 134	1.1	...
	1991[2]	154 328	1.9	10.3	151 577	1.8	...
Peru	1980	176	3.1	15.2	166	2.9	18.5
Pérou	1985	5 042	2.9	15.7	4 855	2.8	17.9
(inti)	1990[13]	89	1.5	14.7	...	...	...
Suriname	1980	105	6.7	22.5	105	6.7	22.8
Suriname	1985	158	9.4	...	...	...	...
(guilder · florin)	1990	250	8.3	...	249	8.2	...
Uruguay	1980	2 035	2.3	10.0	1 927	2.2	...
Uruguay	1985	12 565	2.8	9.3	12 068	2.7	9.3
(peso)	1990	289 354	3.1	15.9	265 660	2.8	16.7
	1992	959 087	2.8	15.4	873 317	2.6	15.5
Venezuela	1980	13 162	4.4	14.7	12 524	4.2	24.3
Venezuela	1985	23 068	5.1	20.3	...	...	...
(bolivar)	1991[2]	136 571	4.6	17.0	...	...	...
	1992	211 659	5.2	23.5	...	...	...
Asia · Asie							
Afghanistan	1980	3 205	2.0	12.7	2 886	1.8	14.4
Afghanistan (afghani)	1990	5 667	...	...	5 282	...	...
Armenia Arménie (rouble)	1990	707	...	20.5	...	...	...
Azerbaijan	1987	939	...	29.3	...	...	...
Azerbaïdjan	1990	1 132	...	24.2	...	...	...
(rouble)	1991	2 066	...	24.7	...	...	...
Bahrain	1980	33	2.9	10.3	28	2.5	14.7
Bahreïn	1985	52	4.0	10.4	49	3.8	14.6
(dinar)	1990[1]	65	4.8	...	61	4.5	...
	1992[1]	72	...	12.3	68	...	14.4
Bangladesh [2]	1980	3 009	1.5	7.8	2 010	1.0	13.6
Bangladesh [2]	1985	7 782	1.9	9.7	6 005	1.5	15.3
(taka)	1990	14 942	2.0	10.3	11 820	1.6	14.4
	1992	20 996	2.3	8.7	16 744	1.9	10.4
Bhutan Bhoutan (ngultrum)	1988	125	3.4	...	...	...	...
Brunei Darussalam	1980	129	1.2	11.8	115	1.1	12.5
Brunéi Darussalam	1986	243	4.6	...	217	4.2	...
(dollar)	1990	253	...	...	229	...	...
	1992[2]	258	...	...	239	...	...

11
Public expenditure on education at current market prices [*cont.*]
Dépenses publiques afférentes à l'enseignement aux prix courants du marché [*suite*]

Country or area Pays ou zone (Currency unit · Unité monétaire)	Years Années	Total educational expenditure Dépenses totales d'éducation			Current educational expenditure Dépenses ordinaires d'éducation		
		Amount Montant (000 000)	% of GNP % du PNB	% total gov't exp. % dép. totales gouv.	Amount Montant (000 000)	% of GNP % du PNB	% current gov't exp. % dép. ordinaires gouv.
China	1980	11 319	1.9	9.3	10 263	1.7	...
Chine	1985	22 489	2.0	12.2	19 770	1.7	...
(yuan)	1990	43 386	1.8	12.8	40 423	1.7	...
	1992	53 874	1.7	12.2	48 976	1.5	13.5
Cyprus [14]	1980	27	3.5	12.9	25	3.3	16.2
Chypre [14]	1985	55	3.7	12.2	53	3.6	13.4
(pound · livre)	1990	89	3.5	11.3	84	3.3	12.3
	1992	120	4.0	12.5	112	3.7	13.1
Hong Kong	1980	3 446	...	14.6	3 036	...	25.5
Hong−kong	1984	6 991	...	18.7	6 190	...	23.0
(dollar)	1990	16 566	...	17.4	...	...	...
	1991	19 552	...	18.1	16 915	...	...
India	1980	37 924	2.8	10.0	37 462	2.7	...
Inde	1985	87 257	3.4	9.4	85 198	3.3	13.0
(rupee · roupie)	1990	207 897	4.0	10.9	205 330	3.9	13.7
	1991	232 425	3.9	11.9	228 878	3.8	13.4
Indonesia	1980	808 087	1.7	8.9	...	...	...
Indonésie	1988[2]	1 238 315	0.9	4.3	1 095 493	0.8	5.5
(rupiah)	1992	5 479 346	2.2	9.4	3 585 711	1.4	10.5
Iran, Islamic Rep. of	1980	498 268	7.5	15.7	440 298	6.6	20.1
Iran, Rép. islamique d'	1985	575 519	3.6	17.2	510 081	3.2	20.6
(rial)	1990	1 493 896	4.1	22.4	1 232 097	3.4	28.8
	1992	3 142 954	4.6	28.2	2 582 561	3.8	31.4
Iraq	1980	418	3.0	...	...	...	...
Iraq	1985	551	4.0	...	...	...	...
(dinar)	1988	690	5.1	...	625	4.6	...
	1991	804	...	...	711	...	...
Israel	1980	9	7.9	7.3	8	7.3	8.9
Israël	1985	1 876	6.4	8.6	1 720	5.9	8.3
(shekel)	1990	6 239	5.8	10.5	5 713	5.3	10.4
Japan	1980	13 908 111	5.8	19.6	9 416 591[15]	3.9	...
Japon	1985	16 142 654	5.0	17.9	15 280 808[9]	4.8	...
(yen)	1989	18 911 000	4.7	16.5	...	...	...
Jordan	1980	64	...	11.3	51	...	15.0
Jordanie	1985	105	5.5	13.0	92	4.8	18.9
(dinar)	1990[1]	101	4.3	8.5	92	3.9	...
	1991[1]	101	4.0	9.2	96	3.8	10.7
Kazakstan	1985	2 174	...	18.9	...	...	...
Kazakstan	1990	3 001	...	17.6	...	...	...
(tenge)	1991	6 252	...	19.1	...	...	...
Korea, Republic of	1980	1 374 736	3.7	...	1 158 967	3.2	...
Corée, République de	1985	3 530 101	4.5	...	2 811 861	3.6	...
(won)	1990	6 159 073	3.6	...	5 495 204	3.2	...
	1992	10 019 080	4.4	14.8	7 996 446	3.5	15.3
Kuwait	1980	219	2.4	8.1	204	2.3	11.2
Koweït	1986	360	4.9	12.6	344	4.7	...
(dinar)	1990[2]	216	2.9	...	210	2.8	...
	1991[2]	279	6.0	...	258	5.5	...
Kyrgyzstan	1987	600	...	25.9	...	...	...
Kirghizistan	1990	720	...	22.4	...	...	...
(rouble)	1991	1 190	...	24.1	...	...	...
Lao People's Dem. Rep.	1980	24	...	1.3	...	...	...
Rep. dém. pop. lao	1985	463	...	4.5	...	...	...
(kip)	1988	2 782	1.2	...	2 097	0.9	...
Lebanon [2]	1980	511	...	13.2	...	...	...
Liban [2]	1985	1 639	...	16.8	...	...	...
(pound · livre)	1992	206 603	2.0	12.5	204 123	2.0	...
Malaysia	1980	3 104	6.0	14.7	2 575	5.0	18.4
Malaysie	1985	4 754	6.6	16.3	4 062	5.6	...
(ringgit)	1990[2]	6 033	5.5	18.3	4 664	4.2	19.3
	1992[2]	7 702	5.5	16.9	6 656	4.8	19.6

11
Public expenditure on education at current market prices [*cont.*]
Dépenses publiques afférentes à l'enseignement aux prix courants du marché [*suite*]

Country or area Pays ou zone (Currency unit · Unité monétaire)	Years Années	Total educational expenditure Dépenses totales d'éducation			Current educational expenditure Dépenses ordinaires d'éducation		
		Amount Montant (000 000)	% of GNP % du PNB	% total gov't exp. % dép. totales gouv.	Amount Montant (000 000)	% of GNP % du PNB	% current gov't exp. % dép. ordinaires gouv.
Maldives Maldives (rufiyaa)	1986	24	...	*7.2	20	...	...
	1990	79	...	*10.0	...	...	...
	1992	220	...	*16.0	...	...	...
Mongolia Mongolie (tugrik)	1985	716	7.8	...	...	...	...
	1990	882	8.6	...	...	...	...
	1991	1 598	8.5	...	...	...	...
Myanmar [2] Myanmar [2] (kyat)	1980	660	1.7	...	584	1.5	...
	1985	1 084	2.0	...	841	1.5	...
	1989	2 948	2.4	...	2 699	2.2	...
Nepal [16] Népal [16] (rupee · roupie)	1980	430	1.8	10.5	...	...	...
	1985	1 241	2.9	10.8	...	...	...
	1990	2 105	2.4	8.6	...	...	...
	1991[2]	2 079	2.0	7.8	...	...	...
Oman Oman (rial)	1980	38	2.1	...	31	1.7	...
	1986	107	4.3	10.8	90	3.7	...
	1990	128	3.5	11.1	117	3.2	18.8
	1992	151	3.8	16.2	135	3.4	19.3
Pakistan Pakistan (rupee · roupie)	1980	4 619	2.0	5.0	3 379	1.5	5.2
	1985	12 645	2.5	...	9 390	1.9	...
	1990[17]	23 570	2.7	...	19 500	2.2	...
	1991[17]	27 790	2.7	...	24 090	2.3	...
Philippines Philippines (peso)	1980	4 191	1.7	9.1	4 023	1.7	13.0
	1985	7 524	1.4	7.4	7 026	1.3	10.0
	1990	31 067	2.9	10.1	28 713	2.7	11.1
	1991	37 033	2.9	10.5	32 872	2.6	11.1
Qatar Qatar (riyal)	1980	792	2.6	7.2	598	2.0	7.8
	1985	1 012	4.1	...	767	3.1	...
	1990	929	3.5	...	904	3.4	...
	1992	957	3.0	...	896	2.9	...
Saudi Arabia Arabie saoudite (riyal)	1980	21 294	5.5	8.7	13 526	3.5	...
	1985	23 540	6.7	12.0	19 283	5.5	...
	1990	25 460	6.2	17.8	24 033	5.9	...
	1992	30 800	6.8	17.0	29 428	6.5	...
Singapore Singapour (dollar)	1980	686	2.8	7.3	587	2.4	10.3
	1985	1 776	4.4	...	1 388	3.4	...
	1988	1 718	3.4	...	1 523	3.0	...
Sri Lanka Sri Lanka (rupee · roupie)	1980[2]	1 799	2.7	7.7	1 535	2.3	13.5
	1985[2]	4 183	2.6	6.9	3 530	2.2	11.4
	1990	8 621	2.7	8.1	7 024	2.2	10.7
	1992	13 883	3.3	8.8	10 898	2.5	10.6
Syrian Arab Republic Rép. arabe syrienne (pound · livre)	1980	2 347	4.6	8.1	1 272[1]	2.5	...
	1985	5 060	6.1	11.8	2 799[1]	3.4	...
	1990	10 720	4.2	17.3	...	...	...
	1991	12 025	4.2	14.2	...	...	...
Thailand Thaïlande (baht)	1980	22 489	3.4	20.6	15 867	2.4	19.1
	1985	39 367	3.9	18.5	33 830	3.4	19.2
	1990	77 420	3.6	20.0	64 702	3.0	21.0
	1991	87 976	3.6	19.1	72 182	2.9	19.9
Turkey Turquie (lira · livre)	1980	117 744	2.8	10.5	98 593	2.3	...
	1985	627 104	2.3	...	523 102	1.9	...
	1990[1]	8 506 541	3.1	...	7 580 817	2.7	...
	1992[1]	30 357 203	4.0	...	27 895 167	3.6	...
Turkmenistan Turkménistan (manat)	1985	431	...	28.0	...	...	...
	1990	655	...	21.0	...	...	...
	1991	1 159	...	19.7	...	...	...
United Arab Emirates Emirats arabes unis (dirham)	1980	1 460	1.3	...	1 153	1.0	...
	1985	1 738	1.7	10.4	1 637	1.6	10.6
	1990	2 280	1.7	14.6	2 174	1.6	14.3
	1992	2 637	1.9	15.2	2 460	1.8	15.1
Yemen Yémen (rial)	1981	1 108	...	15.8	...	...	...
	1985	2 039	4.1	21.8	1 789	3.6	27.4
	1986	2 505	4.6	23.5	2 228	4.1	30.2

11
Public expenditure on education at current market prices [*cont.*]
Dépenses publiques afférentes à l'enseignement aux prix courants du marché [*suite*]

Country or area Pays ou zone (Currency unit · Unité monétaire)	Years Années	Total educational expenditure Dépenses totales d'éducation			Current educational expenditure Dépenses ordinaires d'éducation		
		Amount Montant (000 000)	% of GNP % du PNB	% total gov't exp. % dép. totales gouv.	Amount Montant (000 000)	% of GNP % du PNB	% current gov't exp. % dép. ordinaires gouv.
Europe · Europe							
Albania	1980	767	...	10.3	...	...	...
Albanie	1987	946	...	11.2	...	...	...
(lek)	1990	984	...	...	...	...	...
Andorra							
Andorre							
(peseta)	1986	951	...	15.8	800	...	24.4
Austria	1980	55 016	5.6	8.0	46 955	4.8	8.4
Autriche	1985	78 639	5.9	7.9	70 847	5.3	8.6
(schilling)	1990	97 301	5.4	7.6	89 858	5.0	8.8
	1992	117 519	5.8	7.7	103 693	5.1	8.6
Belarus	1980	1 253	...	...	1 051	...	...
Bélarus	1985	1 499	...	...	1 264	...	...
(rouble)	1990	2 093	...	...	1 758	...	...
	1992	60 520	...	19.3	49 528	...	20.0
Belgium [2]	1980	208 469	6.1	16.3	206 227	6.0	18.6
Belgique [2]	1985	287 388	6.2	15.2	272 843	5.8	16.0
(franc)	1990	325 282	5.1	...	321 427	5.1	...
	1992	361 584	5.2		358 648	5.1	
Bulgaria	1980	1 145	4.5	...	1 098	4.3	...
Bulgarie	1985	1 784	5.5	...	1 598	4.9	...
(lev)	1990	2 357	5.6	...	2 183	5.2	...
	1992	11 729	6.4	...	11 211	6.2	...
formar Czechoslovakia †	1980	23 181	4.0		21 802	3.8	
anc. Tchécoslovaquie †	1985	28 201	4.2	7.9	26 959	4.0	8.6
(koruna · couronne)	1990	37 323	4.6	8.2	35 482	4.4	8.7
	1991[2]	33 546	3.5	6.6	30 816	3.2	6.8
Czech Republic							
Rép. tchèque							
(koruna · couronne)	1992	36 915	5.0	...	33 631	4.5	...
Denmark	1980	25 020	6.9	9.5	22 188	6.1	9.0
Danemark	1985	42 672	7.2	...	...	...	...
(krone · couronne)	1989	55 448	7.5	13.0	52 270	7.1	12.9
	1991	58 960	7.4	11.8	54 896	6.9	11.5
Estonia							
Estonie							
(kroon · couronne)	1992	795	...	31.4	729	...	33.9
Finland	1980	10 036	5.3	...	9 181	4.9	12.0
Finlande	1985	17 682	5.4	11.8	16 356	5.0	11.9
(markka)	1990	28 770	5.8	11.9	26 757	5.4	12.1
	1992	33 086	7.4	11.6	37 191	7.1	11.9
France [18]	1980	142 099	5.0	...	131 441	4.7	...
France [18]	1985	269 191	5.8	...	254 433	5.4	...
(franc)	1990	351 867	5.4	...	327 427	5.1	...
	1992	393 004	5.7	...	362 695	5.2	...
Germany † · Allemagne †							
Federal Republic of Germany	1980	70 099	4.7	9.5	60 558	4.1	...
Rép. féd. d'Allemagne	1985	83 691	4.6	9.2	75 566	4.1	9.4
(deutsche mark)	1990	98 412	4.0	8.6	88 499	3.6	8.7
former German Dem. Republic	1980	...	...	...	9 836	...	...
ancienne Rép. dém. allemande	1985	...	...	...	12 404	...	...
(DDR mark)	1988	15 462	...	...	14 097	...	...
Gibraltar							
Gibraltar	1980	5	9.0	...	3	5.5	...
(pound · livre)	1984	5	6.0	...	5	6.0	...
Greece	1979	31 754	2.2	8.4	29 951	2.0	9.6
Grèce	1985	133 091	2.9	7.5	126 749	2.8	8.5
(drachma · drachme)	1989	265 351	3.1	...	248 040	2.9	...
Hungary	1980	33 099	4.7	5.2	27 516	3.9	6.4
Hongrie	1985	54 061	5.5	6.4	48 125	4.9	7.4
(forint)	1990	122 120	6.1	7.8	110 382	5.5	8.6
	1992	193 772	7.2	7.7	178 954	6.7	8.6

11
Public expenditure on education at current market prices [*cont.*]
Dépenses publiques afférentes à l'enseignement aux prix courants du marché [*suite*]

Country or area Pays ou zone (Currency unit · Unité monétaire)	Years Années	Total educational expenditure Dépenses totales d'éducation			Current educational expenditure Dépenses ordinaires d'éducation		
		Amount Montant (000 000)	% of GNP % du PNB	% total gov't exp. % dép. totales gouv.	Amount Montant (000 000)	% of GNP % du PNB	% current gov't exp. % dép. ordinaires gouv.
Iceland	1980	699	4.6	14.0	...		...
Islande	1985	5 684	5.0	13.8	...	...	...
(krona · couronne)	1990	19 747	5.8	...	14 584	4.3	...
Ireland	1980	595	6.6	...	515	5.7	...
Irlande	1985	1 058	6.7	8.9	963	6.1	10.5
(pound · livre)	1990	1 372	6.0	10.2	1 303	5.7	12.1
	1991	1 479	6.1	9.7	1 416	5.8	12.2
Italy	1979	13 633 022	4.4	11.1	11 791 069	3.8	10.7
Italie	1985	40 533 126	5.0	8.3	36 859 217	4.6	10.0
(lira · lire)	1990[1][12]	40 846 282	3.2	...	40 400 136	3.1	...
	1992	80 268 000	5.4	...	77 347 968[19]	5.2	...
Latvia Lettonie (lat)	1993	87	...	16.1	...	...	...
Lithuania Lituanie (lita)	1992	179	...	22.1	171	...	23.5
Luxembourg	1980	9 792	6.1	14.9	9 305	5.8	19.8
Luxembourg	1986	12 269	4.0	15.7	10 712	3.5	15.4
(franc)	1989	16 363	4.3	...	13 440	3.5	...
Malta	1980	13	3.0	7.8	12	2.9	9.7
Malte	1985	17	3.4	7.7	17	3.3	9.1
(lira · lire)	1990	32	4.0	8.3	30	3.8	10.9
Monaco	1980	...	...	...	43	...	...
Monaco	1989	124	...	5.3	113	...	7.8
(franc)	1992	153	...	5.6	140	...	7.7
Netherlands	1980	26 606	7.8	...	23 079	6.7	...
Pays−Bas	1985	28 298	6.7	...	24 796	5.8	...
(guilder · florin)	1990	32 232	6.3	...	29 005	5.6	...
	1991	33 277	6.2	...	30 231	5.6	...
Norway	1980	19 731	7.2	13.7	16 448	6.0	14.4
Norvège	1985	31 680	6.5	14.6	27 979	5.7	15.3
(krone · couronne)	1990	51 119	7.9	14.6	44 109	6.9	14.5
	1992	59 201	8.7	14.1	51 132	7.5	15.8
Poland	1980	...	...	...	79 984	3.3	7.0
Pologne	1985	497 497	4.9	12.2	405 597	4.0	11.6
(zloty)	1990	28 249 871	5.0	14.6	...	...	...
	1992	62 442 000	5.6	14.0	58 467 000	5.2	14.2
Portugal	1980	53 234	4.4	...	45 443	3.8	...
Portugal	1985	152 886	4.6	...	135 612	4.1	...
(escudo)	1990	412 481	4.8	...	378 252	4.4	...
Republic of Modova République de Moldova	1990	794	...	19.3	...	...	...
(rouble)	1992	14 073	...	26.4	12 923	...	28.6
Romania	1980	19 930	3.3	6.7	17 691	2.9	...
Roumanie	1985	17 941	2.2	...	17 345	2.1	...
(leu)	1990	24 270	2.9	7.3	23 881	2.8	9.0
	1992	216 465	3.6	14.2	208 881	3.5	15.5
San Marino	1980	7 252	...	7.5	6 263	...	9.5
Saint−Marin	1984	11 470	...	10.7	10 474	...	10.3
(lira · lire)	1990	22 048	...	...	21 723	...	...
	1992	30 915	...	...	29 221	...	...
Slovakia Slovaquie (koruna · couronne)	1992	19 829	7.0	...	17 575	6.2	...
Slovenia	1991	16 603	4.8	16.1	15 266	4.4	...
Slovénie (tolar)	1992	55 828	5.7	23.2	52 595	5.3	24.6
Spain	1979	342 376	2.6	...	295 443	2.3	...
Espagne	1985	917 076	3.3	...	821 118	2.9	...
(peseta)	1990	2 181 935	4.4	9.4	1 936 203	3.9	...
	1992	2 680 764	4.6	9.3	2 423 169	4.2	...

11
Public expenditure on education at current market prices [*cont.*]
Dépenses publiques afférentes à l'enseignement aux prix courants du marché [*suite*]

Country or area Pays ou zone (Currency unit · Unité monétaire)	Years Années	Total educational expenditure Dépenses totales d'éducation			Current educational expenditure Dépenses ordinaires d'éducation		
		Amount Montant (000 000)	% of GNP % du PNB	% total gov't exp. % dép. totales gouv.	Amount Montant (000 000)	% of GNP % du PNB	% current gov't exp. % dép. ordinaires gouv.
Sweden	1980	47 322	9.0	14.1	40 886	7.7	...
Suède	1985	65 001	7.7	12.6	57 703	6.8	...
(krona · couronne)	1990	101 363	7.7	13.8	93 083	7.0	...
	1992	116 298	8.8	12.7	112 573	8.7	...
Switzerland	1980	8 873	5.0	18.8	7 937	4.5	20.0
Suisse	1985	11 696	4.8	18.6	10 638	4.4	19.9
(franc)	1990	16 215	5.0	...	14 395	4.4	...
	1991	18 106	5.2	18.8	16 061	4.7	19.5
Ukraine	1980	5 927	...	24.5	5 114	...	...
Ukraine	1985	6 721	...	21.1	5 708	...	...
(rouble)	1990	8 606	...	27.4	6 903	...	25.4
	1992	327 968	...	...	280 690	...	...
United Kingdom	1980	12 856	5.6	13.9	12 094	5.2	...
Royaume–Uni	1985	17 501	4.9	...	16 764	4.7	...
(pound · livre)	1990	26 677	4.9	...	25 318	4.7	...
	1991	29 534	5.2	...	28 045	5.0	...
Yugoslavia, SFR †	1980	8	4.7	32.5	7	4.0	...
Yugoslavie, Rfs †	1985	42	3.4	...	39	3.1	...
(dinar)	1990	60 318	6.1	...	55 911	5.7	...
Yugoslavia Yugoslavie (dinar)	1992	0.292	...	...	0.274	...	
Oceania · Océanie							
American Samoa	1981	11	8.3	16.0	11	8.1	17.9
Samoa américaines	1986	23	...	...	23	...	...
(US dollars · dollars E–U)	1988	26	...	23.7	...	...	...
Australia	1980	7 592	5.5	14.8	6 899	5.0	16.8
Australie	1985	12 925	5.6	12.8	11 848	5.1	14.2
(dollar)	1990	19 364	5.3	14.8	17 889	4.9	14.8
	1991	20 417	5.5	14.1	18 983	5.1	14.6
Cook Islands	1981	3	...	13.1	3	...	...
Iles Cook	1986	5	...	9.5	4	...	9.5
(NZ dollar)	1991	8	...	12.4	8	...	12.7
Fiji	1981	61	5.9	11.3	58	5.5	22.4
Fidji	1986	85	6.0	...	83	5.9	...
(dollar)	1990[2]	94	4.7	...	93	4.7	...
	1992[2]	128	5.6	18.6	124	5.4	...
French Polynesia	1984	1 030	9.8	...	935	8.9	...
Polynésia française	1990[2]	1 205	...	...	1 148	...	...
(CFP franc · franc CFP)	1992[12]	1 785	...	...	1 708	...	...
Guam	1981	49	8.0	...	48	7.9	...
Guam (US dollars · dollars E–U)	1985[1]	60	8.5	...	59	8.3	...
Kiribati	1980	3	...	...	3	...	17.4
Kiribati	1985	3	6.7	18.5	3	6.7	...
(dollar)	1990	4	6.0	18.3	4	6.0	...
	1991	4	6.5	14.8	4	6.5	...
New Caledonia	1981	628	11.8	...	596	11.3	...
Nouvelle–Calédonie	1985	1 007	13.5	...	919	12.3	...
(CFP franc · franc CFP)	1990[2]	1 137	...	...	1 040	...	...
	1992[12]	1 501	...	...	1 458	...	...
New Zealand	1980	1 302	5.8	23.1	1 171	5.2	27.9
Nouvelle–Zélande	1985	2 028	4.7	18.4	1 849	4.3	25.5
(dollar)	1990	4 451	6.4	...	4 252	6.1	...
	1992	5 308	7.2	...	5 107	6.9	...
Niue	1980	0.777	...	13.2	0.755	...	15.9
Nioué	1986	1.394	...	...	1.283	...	...
(NZ dollar)	1988	1.366	...	10.8	1.166	...	10.1
	1991	1.454	...	10.2	1.164	...	10.8
Norfolk Island Ile Norfolk (dollar)	1980	0.425	...	15.1	0.425	...	17.3

11
Public expenditure on education at current market prices [*cont.*]
Dépenses publiques afférentes à l'enseignement aux prix courants du marché [*suite*]

Country or area Pays ou zone (Currency unit · Unité monétaire)	Years Années	Total educational expenditure Dépenses totales d'éducation			Current educational expenditure Dépenses ordinaires d'éducation		
		Amount Montant (000 000)	% of GNP % du PNB	% total gov't exp. % dép. totales gouv.	Amount Montant (000 000)	% of GNP % du PNB	% current gov't exp. % dép. ordinaires gouv.
Palau							
Palaos							
(US dollars · dollars E–U)	1981	26	...	...	26	...	...
Samoa							
Samoa							
(tala)	1990	15	4.2	10.7	14	3.9	15.8
Solomon Islands	1980	5	5.6	11.2	4	4.3	14.1
Iles Salomon	1984	10	4.7	12.4	4	1.9	8.0
(dollar)	1991	24	4.2	7.9	24	4.2	13.5
Tokelau							
Tokélaou							
(NZ dollar)	1981	0.669	...	...	0.630	...	...
Tonga	1980	1.558	...	11.6	1.339	...	13.3
Tonga	1985	3.659	4.4	16.1	3.659	4.4	...
(pa'anga)	1986	4.330	4.2	...	4.330	4.2	...
Tuvalu							
Tuvalu	1989	0.741	...	18.5	0.741	...	...
(australian dollar)	1990	0.854	...	16.2	...	...	...
Vanuatu	1987	924	7.4	24.6	924	7.4	25.4
Vanuatu	1990	831	4.3	...	831	4.3	19.2
(vatu)	1991	929	4.5	...	929	4.5	18.8
former USSR · ancienne URSS							
former USSR † [20]	1980	33 026	7.3	11.2	28 099	6.2	...
ancienne URSS † [20]	1985	39 866	7.0	...	33 319	5.9	...
(rouble)	1990	57 608	8.2	...	46 030	6.5	...

Source:
United Nations Educational, Scientific and Cultural Organization
(Paris).

† For detailed descriptions of data pertaining to former
Czechoslovakia, Germany, SFR Yugoslavia and former
USSR, see Annex I – Country or area nomenclature,
regional and other groupings.

1 Expenditure on third level education is not included.
2 Data refer to expenditure of the Ministry of Education only.
3 Data refer to expenditure of the Ministry of Primary and
Secondary Education only (Nicaragua: expressed in gold
cordobas).
4 Expenditure relating to Al–Azhar is not included.
5 Data include foreign aid received for education.
6 Data refer to expenditure of the Federal government only.
7 Data do not include expenditure for Transkei, Bophuthatswana,
Venda and Ciskei.
8 Expenditure on education is calculated as percentage of
global social product.
9 Data refer to public and private expenditure on education.
10 Figures are in pesos (1 austral = 1000 pesos).
11 Expenditure on universities is not included.
12 Data refer to expenditure of education of the central
government only.
13 Data are expressed in new Soles.
14 Expenditure of the Office of Greek Education only.
15 Data do not include public subsidies to private education.
16 Data refer to regular and development expenditure.
17 Data do not include expenditure on education by other ministries
which are not directly related to education.
18 Metropolitan France.
19 Data include capital expenditure on universities.
20 Expenditure on education is calculated as percentage of net
material product.

Source:
Organisation des Nations Unies pour l'éducation, la science et
la culture (Paris).

† Pour les descriptions en détails des données relatives à l'ancienne
Tchécoslovaquie, l'Allemagne, la Rfs Yougoslavie et l'ancienne
URSS, voir l'Annexe I – Nomenclature des pays ou zones,
groupements régionaux et autres groupements.

1 Les dépenses relatives à l'enseignement du troisième degré ne sont pas
incluses.
2 Les données se réfèrent aux dépenses du Ministère de l'Education seulement.
3 Les données de réfèrent aux dépenses du Ministère des enseignements
primaire et secondaire seulement (Nicaragua: exprimées en cordobas or).
4 Les dépenses relatives à Al–Azhar ne sont pas incluses.
5 Les données comprennent l'aide étrangère reçue pour la éducation.
6 Les dépenses se réfèrent aux dépenses du gouvernement fédéral seulement.
7 Les données ne comprennent pas Transkei, Bophuthatswana, Venda et Ciskei.
8 Les dépenses de l'enseignement sont calculés en pourcentage du produit
social global.
9 Les données se réfèrent aux dépenses publiques et privées afférentes à
l'enseignement.
10 Les chiffres sont exprimés en pesos (1 austral = 1 000 pesos).
11 Les dépenses des universités ne sont pas incluses.
12 Les données se réfèrent aux dépenses de l'éducation du gouvernement central
seulement.
13 Les données sont exprimées en nouveaux Soles.
14 Dépenses du bureau grec de l'éducation seulement.
15 Les données ne comprennent pas les subventions publiques à l'enseignement
privé.
16 Les données se réfèrent aux dépenses ordinaires et de developpement.
17 Les données ne comprennent pas les dépenses d'éducation effectuées
par d'autres ministères qui ne sont pas en rapport direct avec l'enseignement.
18 France métropolitaine.
19 Les données comprennent les dépenses en capital des universités.
20 Les dépenses de l'enseignement sont calculés en pourcentage du produit
matériel net.

12
Illiterate population by sex
Population analphabète selon le sexe

Country or area Pays ou zone	Year Année	Age group Groupe d'âge	Illiterate population Population analphabète			Percentage of illiterates Percentage d'analphabètes		
			Total	M	F	Total	M	F
Africa · Afrique								
Algeria Algérie	1987	15+	6373688	2320756	4052932	50.4	36.6	64.2
Angola [1] Angola [1]	1985	15+	...	...	...	59.0	51.0	...
Benin Bénin	1979	15+	1418051	563351	854700	83.5	74.8	90.5
Botswana [1] Botswana [1]	1990	15+	174500	49600	124900	26.4	16.3	34.9
Burkina Faso Burkina Faso	1975	15+	2803440	1272593	1530847	91.2	85.3	96.7
Burundi [1] Burundi [1]	1982	10+	...	...	...	66.2	57.2	74.3
Cameroon Cameroun	1976	15+	2360088	863884	1496204	58.8	45.4	70.9
Cape Verde Cap–Vert	1990	15+	69930	21363	48567	37.1	25.2	46.7
Central African Rep. République centrafricaine	1975	15+	841995	336544	505451	81.8	70.4	91.6
Chad [1] Tchad [1]	1990	15+	2280300	916800	1364400	70.2	57.8	82.1
Comoros Comores	1980	15+	88780	36429	52351	52.1	44.0	60.0
Congo [1] Congo [1]	1990	15+	484700	163500	321100	43.4	30.0	56.1
Côte d'Ivoire [1] Côte d'Ivoire [1]	1990	15+	2941000	1080600	1860400	46.2	33.1	59.8
Egypt [2] Egypte [2]	1986	15+	14644904	5706276	8938628	54.2	41.6	67.2
Equatorial Guinea Guinée équatoriale	1983	15+	58847	16288	42559	38.0	22.6	51.5
Ethiopia Ethiopie	1984	15+	13533624	5840560	7693064	75.7	67.3	83.6
Gabon [1] Gabon [1]	1990	15+	311400	103000	208400	39.3	26.5	51.5
Gambia [1] Gambie [1]	1990	15+	349900	143600	206300	72.8	61.0	84.0
Ghana [1] Ghana [1]	1990	15+	3258200	1215100	2043100	39.7	30.0	49.0
Guinea [1] Guinée [1]	1990	15+	2947000	1237000	1710000	76.0	65.1	86.6
Guinea–Bissau Guinée–Bissau	1979	15+	342393	130922	211471	80.0	66.7	91.4
Kenya [1] Kenya [1]	1990	15+	3728300	1207400	2520900	31.0	20.2	41.5
Liberia [1] Libéria [1]	1990	15+	839000	352800	486200	60.5	50.2	71.2
Libyan Arab Jamahiriya [1] Jamahiriya arabe libyenne [1]	1990	15+	890300	324500	565700	36.2	24.6	49.6
Madagascar [1] Madagascar [1]	1990	15+	1304500	395300	909200	19.8	12.3	27.1
Malawi Malawi	1987	15+	2214440	706326	1508114	51.5	34.7	66.5
Mali Mali	1988	6+	...	...	...	81.2	73.6	88.6
Mauritania Mauritanie	1988	15+	667342	268955	398387	64.9	53.9	75.4
Mauritius Maurice	1990	15+	149383	54748	94635	20.1	14.8	25.3
Morocco Maroc	1982	15+	8119233	3187079	4932154	69.7	56.3	82.5
Mozambique Mozambique	1980	15+	4557751	1650952	2906799	72.8	56.0	87.8

12
Illiterate population by sex [cont.]
Population analphabète selon le sexe [suite]

Country or area Pays ou zone	Year Année	Age group Groupe d'âge	Illiterate population Population analphabète			Percentage of illiterates Percentage d'analphabètes		
			Total	M	F	Total	M	F
Niger [1] Niger [1]	1990	15+	2683000	1099400	1583600	71.6	59.6	83.2
Nigeria [1] Nigéria [1]	1990	15+	28722600	10758200	17964400	49.3	37.7	60.5
Réunion Réunion	1982	15+	73220	38861	34359	21.4	23.5	19.5
Rwanda Rwanda	1978	15+	1619117	620852	998265	61.8	49.2	73.4
Sao Tome and Principe Sao Tomé−et−Principe	1981	15+	22080	6755	15325	42.6	26.8	57.6
Senegal Sénégal	1988	15+	2652915	1090771	1562144	73.1	63.1	82.1
Seychelles Seychelles	1971	15+	12494	6465	6029	42.3	44.4	40.2
Sierra Leone [1] Sierra Leone [1]	1990	15+	1829500	776800	1052800	79.3	69.3	88.7
Somalia [1] Somalie [1]	1990	15+	3002500	1153700	1848800	75.9	63.9	86.0
South Africa [3] Afrique du Sud [3]	1980	15+	3711776	1796523	1915253	23.8	22.5	25.2
St. Helena Sainte−Hélène	1987	20+	104	65	39	2.7	3.3	2.1
Sudan [1] Soudan [1]	1990	15+	10061100	3947700	6113500	72.9	57.3	88.3
Swaziland Swaziland	1986	15+	116464	48722	67742	32.7	30.1	34.8
Togo Togo	1981	15+	927712	328497	599215	68.6	53.3	81.5
Tunisia Tunisie	1989	15+	2095943	762085	133858	42.7	30.8	54.8
Uganda Ouganda	1991	15+	3855388	1348282	2507106	43.9	31.8	55.2
United Rep. Tanzania [1] République−Unie de Tanzanie [1]	1986	15+	1366721	489830	878891	9.6	...	...
Zaire [1] Zaïre [1]	1990	15+	5465900	1563300	3902700	28.2	16.4	39.3
Zambia [1] Zambie [1]	1990	15+	1170100	398000	772100	27.2	19.2	34.7
Zimbabwe [4] Zimbabwe [4]	1982	15+	852120	292790	559330	22.2	15.8	28.1
America, North · Amérique du Nord								
Barbados [5] Barbade [5]	1970	15+	1093	493	600	0.7	0.7	0.7
Belize [5] Belize [5]	1970	15+	5353	2656	2699	8.8	8.8	8.8
Bermuda [5] Bermudes [5]	1970	15+	586	391	195	1.6	2.1	1.1
British Virgin Islds [5] Iles Vierges britanniques [5]	1970	15+	100	61	39	1.7	1.9	1.5
Canada Canada	1986	15+	659745	...	...	3.4	...	...
Cayman Islands [5] Iles Caïmanes [5]	1970	15+	152	70	82	2.5	2.5	2.4
Costa Rica Costa Rica	1984	15+	112946	55431	57515	7.4	7.3	7.4
Cuba [6] Cuba [6]	1981	10+	...	...	...	3.8	3.8	3.8
Dominica [5] Dominica [5]	1970	15+	2083	944	1139	5.9	6.0	5.8
Dominican Republic République dominicaine	1981	15+	1031629	518236	513809	26.0	26.0	26.0
El Salvador El Salvador	1980	15+	818100	...	...	32.7	...	...
Grenada [5] Grenade [5]	1970	15+	1070	424	646	2.2	2.0	2.4

12
Illiterate population by sex [*cont.*]
Population analphabète selon le sexe [*suite*]

Country or area Pays ou zone	Year Année	Age group Groupe d'âge	Illiterate population Population analphabète			Percentage of illiterates Percentage d'analphabètes		
			Total	M	F	Total	M	F
Guadeloupe Guadeloupe	1982	15+	22359	11231	11128	10.0	10.4	9.6
Guatemala [1] Guatemala [1]	1990	15+	2253200	931900	1321300	44.9	36.9	52.9
Haiti Haïti	1982	15+	2004791	926751	1078040	65.2	62.7	67.5
Honduras [1] Honduras [1]	1990	15+	766000	348800	417200	26.9	24.5	29.4
Jamaica Jamaïques	1987	15+	278578	173683	104895	18.2	23.1	13.5
Martinique Martinique	1982	15+	16814	8824	7990	7.2	8.0	6.6
Mexico Méxique	1990	15+	6161662	2305113	3856549	12.4	9.6	15.0
Montserrat [5] Montserrat [5]	1970	15+	231	100	131	3.4	3.2	3.4
Netherlands Antilles Antilles néerlandaises	1981	15+	10236	4497	5739	6.2	5.8	6.6
Nicaragua [7] Nicaragua [7]	1971	15+	410755	193475	217277	42.5	42.0	42.9
Panama Panama	1990	15+	168644	80700	87944	11.2	10.6	11.7
Puerto Rico Porto Rico	1980	15+	239095	107372	131723	10.9	10.3	11.5
St. Kitts−and−Nevis [5] Saint−Kitts−et−Nevis [5]	1980	15+	674	337	337	2.7	2.9	2.5
Saint Lucia [5] Sainte−Lucie [5]	1970	15+	9195	4251	4944	18.3	19.2	17.6
St. Vincent−and−Grenadines [5] Saint−Vincent−et−Grenadines [5]	1970	15+	1839	779	1060	4.4	4.2	4.5
Trinidad and Tobago Trinité−et−Tobago	1990	15+	25910	9159	16751	3.1	2.0	4.4
Turks and Caicos Is. [5] Iles Turques et Caïques [5]	1970	15+	56	18	38	1.9	1.4	2.3
United States Etats−Unis	1979	14+	...	...	...	0.5	...	...
America, South · Amérique du Sud								
Argentina Argentine	1980	15+	1184964	543174	641790	6.1	5.7	6.4
Bolivia Bolivie	1992	15+	744846	213713	531133	19.9	11.8	27.5
Brazil Brésil	1991	15+	19294646	9300503	9994143	20.1	19.9	20.3
Chile Chili	1992	15+	537744	247531	290213	5.7	5.4	6.0
Colombia Colombie	1985	15+	2271338	1076907	1194431	11.9	11.6	12.2
Ecuador [8] Equateur [8]	1990	15+	690802	273501	417301	12.7	9.6	16.1
French Guiana Guyane française	1982	15+	8372	4321	4051	17.0	16.4	17.7
Guyana [1][5] Guyane [1][5]	1990	15+	24500	8800	15700	3.6	2.5	4.6
Paraguay Paraguay	1982	15+	219120	84340	134780	12.5	9.7	15.2
Peru [9] Pérou [9]	1981	15+	1799458	485486	1313972	18.1	9.9	26.1
Suriname [1] Suriname [1]	1990	15+	13400	6300	7100	5.1	4.9	5.3
Uruguay Uruguay	1985	15+	108400	57300	51100	5.0	5.6	4.5
Venezuela Venezuela	1990	15+	1130567	509864	620703	10.0	9.1	10.8

12
Illiterate population by sex [cont.]
Population analphabète selon le sexe [suite]

Country or area Pays ou zone	Year Année	Age group Groupe d'âge	Illiterate population Population analphabète			Percentage of illiterates Percentage d'analphabètes		
			Total	M	F	Total	M	F
Asia · Asie								
Afghanistan [10] Afghanistan [10]	1979	15+	5832988	2583581	3249407	81.8	69.7	95.0
Armenia Arménie	1989	15+	...	...	...	1.2	0.6	1.9
Azerbaijan Azerbaïdjan	1989	15+	...	...	...	2.7	1.1	4.1
Bahrain Bahreïn	1991	15+	55300	24196	31104	15.9	11.4	23.0
Bangladesh Bangladesh	1981	15+	32923083	14501583	18421500	70.8	60.3	82.0
Brunei Darussalam Brunéi Darussalam	1991	15+	20809	6887	13922	12.2	7.5	17.5
China Chine	1990	15+	181609097	54359731	127249366	22.2	13.0	31.9
Cyprus [1] Chypre [1]	1987	15+	...	...	...	6.0	2.0	9.0
Georgia Géorgie	1989	15+	...	...	...	1.0	0.5	1.5
Hong Kong [5] Hong−kong [5]	1971	15+	571840	126152	445688	22.7	9.9	35.9
India Inde	1981	15+	238097747	93899834	144197913	59.2	45.2	74.3
Indonesia Indonésie	1990	15+	20899440	6553716	14345714	18.4	11.7	24.7
Iran, Islamic Rep. of Iran, Rép. islamique d'	1991	15+	10652344	4113811	6538533	34.3	25.6	43.6
Iraq [1] Iraq [1]	1985	15−45	...	...	...	10.7	9.8	12.5
Israel Israël	1992	15+	183200	50500	132700	5.1	2.9	7.3
Jordan Jordanie	1991	15+	373610	105950	267660	16.8	9.2	24.9
Kazakhstan Kazakhstan	1989	15+	276835	49301	227534	2.5	0.9	3.9
Korea, Republic of [1][11] Corée, République de [1][11]	1990	15+	1185300	134300	1050900	3.7	0.9	6.5
Kuwait Koweït	1985	15+	273513	141082	132431	25.5	21.8	31.2
Kyrgyzstan Kirghizistan	1989	15+	...	...	...	3.0	1.4	4.5
Lao People's Dem. Rep. [1] Rép. dém. pop. lao [1]	1985	15−45	...	...	...	16.1	8.0	24.2
Lebanon [1] Liban [1]	1990	15+	382300	110900	271300	19.9	12.2	26.9
Macau Macao	1970	15+	31917	11894	20023	20.6	15.2	26.1
Malaysia Malaisie	1980	15+	2399790	791000	1608790	30.4	20.4	40.3
Maldives Maldives	1985	15+	8568	4565	4003	8.7	8.8	8.5
Myanmar Myanmar	1983	15+	4492769	1460457	3032312	21.4	14.2	28.3
Nepal Népal	1981	15+	6998148	3053083	3945065	79.4	68.3	90.8
Pakistan Pakistan	1981	15+	34713824	16051771	18662053	74.3	64.6	85.2
Philippines Philippines	1990	15+	2349731	1095697	1254034	6.4	6.0	6.8
Qatar Qatar	1986	15+	64891	45253	19638	24.3	23.2	27.5
Saudi Arabia Arabie saoudite	1982	15+	...	...	...	48.9	28.9	69.2
Singapore Singapour	1990	15+	226677	51307	175370	10.9	4.9	17.0

12
Illiterate population by sex [*cont.*]
Population analphabète selon le sexe [*suite*]

Country or area Pays ou zone	Year Année	Age group Groupe d'âge	Illiterate population Population analphabète			Percentage of illiterates Percentage d'analphabètes		
			Total	M	F	Total	M	F
Sri Lanka Sri Lanka	1981	15+	1271984	424424	847560	13.2	8.7	18.0
Syrian Arab Republic [12] République arabe syrienne [12]	1981	15+	1982265	601390	1380875	44.4	26.4	63.0
Tajikistan Tadjikistan	1989	15+	66973	17189	49784	2.3	1.2	3.4
Thailand Thaïlande	1990	15+	2572127	833682	1738445	6.7	4.4	8.8
Turkey Turquie	1990	15+	7615973	1870245	5745728	20.8	10.1	31.5
Turkmenistan Turkménistan	1989	15+	...	...	...	2.3	1.2	3.4
United Arab Emirates Emirats arabes unis	1985	15+	269983	185397	84586	28.8	27.7	31.3
Uzbekistan Ouzbékistan	1989	15+	...	...	...	2.8	1.5	4.0
Viet Nam Viet Nam	1979	15+	4846849	1340445	3506404	16.0	9.5	21.7
Yemen [1][13] Yémen [1][13]	1990	15+	2558600	876000	1682600	61.5	46.7	73.7
Europe · Europe								
Belarus Bélarus	1989	15+	165406	21917	143489	2.1	0.6	3.4
Bulgaria Bulgarie	1992	15+	147389	44123	103266	2.1	1.3	2.9
Croatia Croatie	1991	15+	124882	22507	102375	3.3	1.2	5.2
Estonia Estonie	1989	15+	3329	687	2642	0.3	0.1	0.4
Greece Grèce	1991	15+	389067	90049	299018	4.8	2.3	7.0
Hungary Hongrie	1980	15+	95542	27756	67786	1.1	0.7	1.5
Italy Italie	1981	15+	1572556	539781	1032775	3.5	2.5	4.5
Latvia Lettonie	1989	15+	11476	2327	9149	0.5	0.2	0.8
Liechtenstein Liechtenstein	1981	10+	68	33	35	0.3	0.3	0.3
Lithuania Lituanie	1989	15+	44308	10436	33872	1.6	0.8	2.2
Malta Malte	1985	20+	33740	16802	16938	14.3	14.8	13.9
Poland Pologne	1978	15+	334586	92609	241977	1.2	0.7	1.7
Portugal Portugal	1981	15+	1506206	524461	981745	20.6	15.2	25.4
Republic of Moldova République de Moldova	1989	15+	113193	20078	93115	3.6	1.4	5.6
Romania Roumanie	1992	15+	577376	125372	452004	3.3	1.5	5.0
Russian Federation Fédération de Russie	1989	15+	2274572	279490	1995082	2.0	0.5	3.2
San Marino San Marino	1976	10+	640	260	380	3.9	3.2	4.7
Spain Espagne	1986	15+	1260789	360483	900306	4.2	2.5	5.8
Ukraine Ukraine	1989	15+	...	...	...	1.6	0.5	2.6
Yugoslavia Yougoslavie	1991	15+	463291	79258	384033	6.7	2.4	10.8
Yugoslavia, SFR † Yougoslavie, Rfs †	1981	15+	1764042	370558	1393484	10.4	4.5	16.1
Oceania · Océanie								
American Samoa Samoa américaines	1980	15+	507	240	267	2.7	2.5	2.8

12
Illiterate population by sex [*cont.*]
Population analphabète selon le sexe [*suite*]

Country or area Pays ou zone	Year Année	Age group Groupe d'âge	Illiterate population Population analphabète			Percentage of illiterates Percentage d'analphabètes		
			Total	M	F	Total	M	F
Fiji Fidji	1986	15+	56203	21633	34570	12.8	9.8	15.8
Guam Guam	1990	15+	1004	511	493	1.0	1.0	1.0
New Caledonia Nouvelle−Calédonie	1976	15+	7133	3370	3763	8.7	7.8	9.7
Palau [14] Palaos [14]	1980	15+	5798	2454	3344	8.1	6.7	9.5
Papua New Guinea [1] Papouasie−Nouvelle−Guinée [1]	1990	15+	1119000	426500	692500	48.0	35.1	62.2
Samoa Samoa	1971	15+	1581	819	762	2.2	2.2	2.1
Tonga Tonga	1976	15+	193	81	112	0.4	0.3	0.5
Vanuatu Vanuatu	1979	15+	28647	13823	14824	47.1	42.7	52.2
former USSR · ancienne URSS								
former USSR ancienne URSS	1989	15+	4282023	644964	3637059	2.0	0.7	3.2

Source:
United Nations Educational, Scientific and Cultural Organization (Paris).

† For detailed description of data pertaining to former Czechoslovakia, Germany, SFR Yugoslavia and former USSR, see Annex I − Country or area nomenclature, regional and other groupings.

1 Estimates.
2 Egyptian population only.
3 Not including Bothuthatswana, Transkei and Veda.
4 Based on a 10% sample of census returns.
5 Persons with no schooling are defined as illiterates.
6 Excluding functionally and physically handicapped.
7 In 1980, after the National Literacy Campaign, the Ministry of Education estimated that the illiteracy rate of the population aged 10 years and over had been reduced to 12.96%.
8 Excluding nomadic Indian tribes.
9 Excluding Indian jungle population.
10 Excluding nomadic population.
11 Based on a sample survey.
12 National population only.
13 Excluding former Democratic Yemen.
14 Including data for Federated States of Micronesia, Marshall Is. and Northern Mariana Is.

Source:
Organisation des Nations Unies pour l'éducation, la science et la culture (Paris).

† Pour les descriptions en détails des données relatives à l'ancienne Tchécoslovaquie, l'Allemagne, la Rfs Yougoslavie et l'ancienne USSR, voir l'Annexe I − Nomenclature des pays ou zones, groupements régionaux et autres groupements.

1 Estimation.
2 Population egyptienne seulement.
3 Non compris le Bothuthatswana, le Transkei and Veda.
4 D'après un échantillon portant sur 10% des bulletins de recensement.
5 Les personnes sans scolarité ont été considérées comme étant analphabètes.
6 Non compris les handicapés physiques et fonctionnels.
7 En 1980, à la fin de la compagne nationale d'alphabétisation, le Ministère de l'Education a estimé que le taux d'analphabétisation de la population âgée de 10 ans et plus a été réduit à 12.96%.
8 Non compris les tribus indiennes nomades.
9 Non compris les indiens de la jungle.
10 Non compris les populations nomades.
11 Basées sur une enquête par sondage.
12 Population nationale seulement.
13 Non compris l'ancien Yémen démocratique.
14 Y compris les données pour les Etats fédérés de Micronésie, les îles Marshall et les îles Mariannes du Nord.

Technical notes, tables 10-12

Detailed data on education accompanied by explanatory notes can be found in the UNESCO *Statistical Yearbook*. [29] Brief notes pertaining to major items of statistical information on education shown in the present edition of the United Nations *Statistical Yearbook* are given below.

The tables included in this chapter cover basic data on education at the first, second and third levels. Definitions for the different levels and types of education given below are based on the "Revised recommendations concerning the international standardization of educational statistics" adopted by the General Conference of UNESCO at its twentieth session (1978). [29, 46]

Table 10: Data on education at the first level refer to education whose main function is to provide basic instruction in the tools of learning (for example at elementary and primary schools). Its length may vary from 4 to 9 years, depending on the organization of the school system in each country. Unless otherwise stated, data cover both public and private schools. Figures on teachers refer to both full-time and part-time but exclude classes organized for adults or for handicapped children.

Data on education at the second level refer to education which is based upon at least four years of previous instruction at first level and provides general or specialized instruction, or both (for example at middle and secondary schools, high schools, teachers' training schools at this level, schools of a vocational or technical nature). Unless otherwise stated data cover both public and private schools. In most cases data include part-time teachers.

Data on education at the third level refer to education which requires as a minimum condition of admission the successful completion of education at the second level or proof of equivalent knowledge or experience. It can be given in different types of institutions such as universities, teacher-training institutes and technical institutes. The figures include, as a rule, both full-time and part-time teachers and students. Correspondence courses are not normally included. Figures on teachers include auxiliary teachers (assistants, demonstrators and the like) but exclude staff with no teaching duties (such as administrators and laboratory technicians).

Table 11: Data on total expenditure refer to public expenditure on public education plus subsidies for private education. Total expenditures cover both current and capital expenditure.

Current expenditures include expenditures on administration, emoluments of teachers and supporting teaching staff, school books and other teaching materials, scholarships, welfare services and maintenance of school buildings.

Capital expenditures include outlays on purchases of land, building construction expenditures and so forth. This item also includes loan transactions.

Notes techniques, tableaux 10-12

On trouvera des données détaillées sur l'éducation accompagnées de notes explicatives dans l'*Annuaire statistique* de l'UNESCO [29]. Des notes sommaires concernant les principaux éléments d'information statistique sur l'éducation figurant dans la présente édition de l'*Annuaire statistique* des Nations Unies sont présentées ci-dessous.

Les tableaux de ce chapitre contiennent des données de base sur l'enseignement aux premier, second et troisième degrés. Les définitions des différents niveaux et types d'enseignement données ci-dessous sont fondées sur les "recommandations révisées concernant la normalisation internationale des statistiques de l'éducation" adoptées par la Conférence générale de l'UNESCO à sa vingtième session (1978) [29, 46].

Tableau 10 : Les données relatives à l'enseignement du premier degré portent sur l'enseignement dont la fonction principale est d'offrir les premiers éléments de l'instruction (par exemple, dans les écoles élémentaires ou primaires). Sa durée peut varier de quatre à neuf ans, selon l'organisation du système scolaire de chaque pays. Sauf indication contraire, les données portent à la fois sur les établissements d'enseignement publics et privés. Les chiffres relatifs aux enseignants désignent à la fois ceux qui enseignent à plein temps et à temps partiel mais ne comprennent pas les classes organisées à l'intention des adultes ou des enfants handicapés.

Les données relatives à l'enseignement secondaire se rapportent à un enseignement fondé sur au moins quatre années d'enseignement préalable au niveau primaire et donnant une formation générale ou spécialisée, ou les deux (par exemple, dans les écoles moyennes et secondaires, les lycées, les écoles normales de ce niveau, les écoles professionnelles ou techniques).

Les données relatives à l'enseignement du troisième degré se rapportent à un enseignement qui exige comme condition minimum d'admission l'achèvement avec succès d'un enseignement secondaire complet ou la preuve de connaissances ou d'une expérience équivalentes. Il peut être dispensé dans différents types d'établissement tels que les universités, les instituts de formation pédagogique et les instituts techniques. En règle générale, les chiffres portent à la fois sur les enseignants et les étudiants à plein temps et à temps partiel. Normalement, les cours par correspondance ne sont pas compris. Les données relatives aux enseignants englobent les professeurs auxiliaires (assistants, maîtres de travaux pratiques, etc.), mais pas le personnel n'exerçant pas de fonctions d'enseignement (tel que les administrateurs et les techniciens de laboratoire).

Tableau 11 : Les données relatives aux dépenses totales se rapportent aux dépenses publiques consacrées à l'enseignement public et aux subventions à l'enseignement privé. Les totaux englobent à la fois les dépenses ordinaires et les dépenses d'équipement.

Data include, unless otherwise indicated, educational expenditure at every level of administration. In general, the data do not include development assistance expenditures on education. Data on gross national product (GNP) used to derive the ratio of total public expenditure on education to GNP are World Bank estimates. For certain countries with centrally planned economies, the concept of material product rather than GNP was employed.

Table 12: Except where otherwise stated, data refer to population 15 years of age and over.

Ability to both read and write a simple sentence on everyday life is used as the criterion of literacy; hence semi-literates (persons who can read but not write) are included with illiterates. Persons for whom literacy is not known are excluded from calculations; consequently the percentage of illiteracy for a given country is based on the number of reported illiterates, divided by the total number of reported literates and illiterates. The data are based on population censuses or surveys.

Les dépenses ordinaires comprennent les dépenses d'administration, les émoluments du personnel enseignant et auxiliaire, les manuels scolaires et autres matériels didactiques, les bourses d'études, les services sociaux et l'entretien des bâtiments scolaires.

Les dépenses d'équipement comprennent les dépenses consacrées à l'achat de terrains, à la construction de bâtiments, etc. Les transactions de prêt sont également incluses dans cette rubrique.

Sauf indication contraire, les données comprennent les dépenses effectuées à tous les niveaux administratifs. En règle générale, elles ne comprennent pas les dépenses d'enseignement financées au titre de l'aide au développement. Les données relatives au produit national brut (PNB), utilisées pour déterminer le ratio du volume total de dépenses publiques consacrées à l'éducation au PNB, sont des estimations de la Banque mondiale. Pour certains pays à économie planifiée, on a utilisé la notion de produit matériel à la place du PNB.

Tableau 12 : Sauf indication contraire, les données se réfèrent à la population âgée de 15 ans et plus.

On utilise l'aptitude à lire et à écrire une phrase simple sur la vie quotidienne comme critère d'alphabétisme; par conséquent, les semi-alphabètes (c'est-à-dire les personnes qui savent lire, mais non écrire) sont assimiliés aux analphabètes. Les personnes dont on ne sait pas si elles savent lire ou écrire sont exclues de ces calculs; par conséquent, le pourcentage d'analphabétisme d'un pays donné est fondé sur le nombre d'analphabètes connus divisé par le total des alphabètes et analphabètes connus. Ces données sont fondées sur les recensements ou enquêtes de population.

13
Estimates of total AIDS cases and cumulative HIV infections and number of reported AIDS cases
Estimations du nombre total de cas de SIDA et du nombre cumulé de personnes infectées par le VIH et nombre de cas de SIDA rapportés

A. Estimated cumulative HIV infections to mid 1994
Estimations du nombre cumulé de personnes infectées par le VIH jusqu'à mi-1994

Regions Régions	Total	Ratio F/M
World **Monde**	more than 16 million plus de 16 millions	about 1 to 1.5 à peu près 1:1,5
Sub-Saharan Africa **Afrique subsaharienne**	more than 10 million plus de 10 millions	approx. 1 to 0.9 à peu près 1:0,9
Latin America and Caribbean **Amérique latine et Caraïbes**	about 2 million à peu près 2 millions	about 1 to 4 à peu près 1:4
South-eastern and southern Asia **Asie du Sud-Est et Asie du Sud**	more than 2.5 million plus de 2,5 millions	approx. 1 to 2.5 à peu près 1:2,5
Northern America; western Europe; Australasia **Amérique du Nord, Europe occidentale, Australasie**	more than 1.5 million plus de 1,5 million	approx. 1 to 5.5 à peu près 1:5,5
Eastern Europe and former USSR; eastern and western Asia and northern Africa; Oceania **Europe de l'Est et l'ancienne URSS, Asie de l'Est, Asie de l'Ouest et Afrique du Nord, Océanie**	more than 200,000 plus de 200 000	about 1 to 5 à peu près 1:5

B. Reported and estimated AIDS cases to end 1994
Cas de SIDA : nombre de cas rapportés jusqu'à la fin 1994, et estimations

Regions Régions	Cases reported to World Health Organization (cumulative) Cas rapportés à l'Organisation mondiale de la santé (cumulatif)	Estimated cases Estimations
World **Monde**	1 025 073	4 500 000+
Africa Afrique	347 713	3 150 000+
Americas Amériques	526 682	810 000+
Asia Asie	17 057	270 000
Europe Europe	127 886	180 000
Oceania Océanie	5 735	45 000

13 C. Reported AIDS cases, 1993 • Cas de SIDA rapportés, 1993

Country or area Pays ou zone	Total reported to Dec. 1991 Total des cas rapportés jusqu'en déc. 1991	New cases reported for 1992 Nouveaux cas rapportés en 1992	New cases reported for 1993 Nouveaux cas rapportés en 1993	Cumulative total to Dec. 1993 Total cumulatif en déc. 1993
Africa • Afrique				
Algeria Algérie	98	0	40	138
Angola Angola	421	147	138	706
Benin Bénin	247	218	277	742
Botswana Botswana	250	189	976	1415
Burkina Faso Burkina Faso	1307	2886	0	4193
Burundi Burundi	5180	1583	799	7562
Cameroon Cameroun	863	1347	862	3072
Cape Verde Cap-Vert	52	13	78	143
Central African Rep. Rép. centrafricaine	3314	416	0	3730
Chad Tchad	224	363	1010	1597
Comoros Comores	3	0	0	3
Congo Congo	3482	1785	1126	6393
Côte d'Ivoire Côte d'Ivoire	10792	3863	4015	18670
Djibouti Djibouti	165	144	144	453
Egypt Egypte	39	23	29	91
Equatorial Guinea Guinée équatoriale	7	12	24	43
Eritrea Erythrée	150	219	300	669
Ethiopia Ethiopie	1627	3256	5124	10007
Gabon Gabon	215	177	80	472
Gambia Gambie	180	86	11	277
Ghana Ghana	7679	2606	2018	12303
Guinea Guinée	440	208	328	976
Guinea-Bissau Guinée-Bissau	172	116	92	380
Kenya Kenya	24423	2762	2941	30126
Lesotho Lesotho	51	131	297	479
Liberia Libéria	24	4	163	191
Libyan Arab Jamahiriya Jamah. arabe libyenne	7	3	2	12
Madagascar Madagascar	3	0	6	9
Malawi Malawi	22300	4655	4916	31871
Mali Mali	853	460	672	1985
Mauritania Mauritanie	26	14	10	50
Mauritius Maurice	10	5	3	18

13 C. Reported AIDS cases, 1993 [*cont.*] • Cas de SIDA rapportés, 1993 [*suite*]

Country or area Pays ou zone	Total reported to Dec. 1991 Total des cas rapportés jusqu'en déc. 1991	New cases reported for 1992 Nouveaux cas rapportés en 1992	New cases reported for 1993 Nouveaux cas rapportés en 1993	Cumulative total to Dec. 1993 Total cumulatif en déc. 1993
Morocco Maroc	98	30	44	172
Mozambique Mozambique	340	322	164	826
Namibia Namibie	1572	2050	1479	5101
Niger Niger	505	304	112	921
Nigeria Nigéria	325	247	409	981
Réunion Réunion	49	16	0	65
Rwanda Rwanda	6578	2908	1220	10706
Sao Tome and Principe Sao Tomé–et–Principe	9	2	13	24
Senegal Sénégal	525	323	349	1197
Seychelles Seychelles	0	1	1	2
Sierra Leone Sierra Leone	51	32	12	95
Somalia Somalie	13	0	0	13
South Africa Afrique du Sud	1123	768	1567	3458
Sudan Soudan	508	184	191	883
Swaziland Swaziland	92	156	165	413
Togo Togo	1278	864	1330	3472
Tunisia Tunisie	105	25	36	166
Uganda Ouganda	30190	4421	9264	43875
United Rep. Tanzania Rép. Unie de Tanzanie	34140	4579	0	38719
Zaire Zaïre	20388	1771	588	22747
Zambia Zambie	5848	1276	22610	29734
Zimbabwe Zimbabwe	10551	8180	9174	27905
America, North • Amérique du Nord				
Anguilla Anguilla	5	0	0	5
Antigua and Barbuda Antigua–et–Barbuda	14	13	7	34
Bahamas Bahamas	838	254	297	1389
Barbados Barbade	252	78	88	418
Belize Belize	46	13	24	83
Bermuda Bermudes	191	17	15	223
Canada Canada	7164	1460	1290	9914
Cayman Islands Iles Caïmanes	11	4	0	15
Costa Rica Costa Rica	332	125	113	570
Cuba Cuba	110	68	80	258

13 C. Reported AIDS cases, 1993 [*cont.*] · Cas de SIDA rapportés, 1993 [*suite*]

Country or area Pays ou zone	Total reported to Dec. 1991 Total des cas rapportés jusqu' en déc. 1991	New cases reported for 1992 Nouveaux cas rapportés en 1992	New cases reported for 1993 Nouveaux cas rapportés en 1993	Cumulative total to Dec. 1993 Total cumulatif en déc. 1993
Dominica Dominique	12	0	14	26
Dominican Republic Rép. dominicaine	1723	326	308	2357
El Salvador El Salvador	315	114	177	606
Grenada Grenade	31	4	21	56
Guadeloupe Guadeloupe	305	48	17	370
Guatemala Guatemala	268	94	118	480
Haiti Haïti	4161	806	0	4967
Honduras Honduras	1658	744	965	3367
Jamaica Jamaïque	334	99	237	670
Martinique Martinique	198	42	26	266
Mexico Mexique	9072	3220	5085	17377
Montserrat Montserrat	8	0	0	8
Netherlands Antilles Antilles néerlandaises	107	10	49	166
Nicaragua Nicaragua	24	6	17	47
Panama Panama	358	121	179	658
Saint Kitts and Nevis Saint-Kitts-et-Nevis	33	4	3	40
Saint Lucia Sainte-Lucie	33	8	12	53
St. Vincent and the Grenadines St. Vincent-et-Grenadines	41	5	6	52
Trinidad and Tobago Trinité-et-Tobago	1030	260	243	1533
Turks and Caicos Islands Iles Turques et Caiques	21	4	14	39
United States Etats-Unis	231811	72061	84562	388434

America, South · Amérique du Sud

Country or area Pays ou zone	Total reported to Dec. 1991	New cases reported for 1992	New cases reported for 1993	Cumulative total to Dec. 1993
Argentina Argentine	1818	1081	1336	4235
Bolivia Bolivie	45	18	20	83
Brazil Brésil	30316	11357	11544	53217
Chile Chili	552	190	178	920
Colombia Colombie	2787	921	519	4227
Ecuador Equateur	194	66	85	345
French Guiana Guyane française	275	67	17	359
Guyana Guyana	230	160	107	497
Paraguay Paraguay	23	17	30	70
Peru Pérou	562	250	235	1047
Suriname Suriname	106	28	35	169

13 C. Reported AIDS cases, 1993 [*cont.*] · Cas de SIDA rapportés, 1993 [*suite*]

Country or area Pays ou zone	Total reported to Dec. 1991 Total des cas rapportés jusqu' en déc. 1991	New cases reported for 1992 Nouveaux cas rapportés en 1992	New cases reported for 1993 Nouveaux cas rapportés en 1993	Cumulative total to Dec. 1993 Total cumulatif en déc. 1993
Uruguay Uruguay	245	90	103	438
Venezuela Venezuela	2308	760	690	3758
Asia · Asie				
Afghanistan Afghanistan	0	0	0	0
Armenia Arménie	0	2	0	2
Azerbaijan Azerbaïdjan	0	0	0	0
Bahrain Bahreïn	7	4	4	15
Bangladesh Bangladesh	1	0	0	1
Bhutan Bhoutan	0	0	0	0
Brunei Darussalam Brunéi Darussalam	3	0	1	4
Cambodia Cambodge	0	0	0	0
China Chine	8	5	23	36
Cyprus Chypre	23	1	6	30
Georgia Géorgie	0	2	0	2
Hong Kong Hong-kong	59	14	19	92
India Inde	102	140	252	494
Indonesia Indonésie	24	10	17	51
Iran, Islamic Rep. of Iran, Rép. islamique d'	44	16	32	92
Iraq Iraq	7	6	11	24
Israel Israël	169	53	50	272
Japan Japon	448	90	147	685
Jordan Jordanie	17	7	7	31
Kazakhstan Kazakhstan	0	0	0	0
Korea, Dem.People's Rep. Corée, Rép. pop. dém. de	0	0	0	0
Korea, Republic of Corée, République de	6	2	8	16
Kuwait Koweït	6	2	2	10
Kyrgyzstan Kirghizistan	0	0	0	0
Lao People's Dem. Rep. Rép. dém. populaire lao	1	0	4	5
Lebanon Liban	31	6	32	69
Macau Macao	2	2	2	6
Malaysia Malaisie	29	40	27	96
Maldives Maldives	0	0	0	0
Mongolia Mongolie	0	0	0	0

13 C. Reported AIDS cases, 1993 [cont.] • Cas de SIDA rapportés, 1993 [suite]

Country or area Pays ou zone	Total reported to Dec. 1991 Total des cas rapportés jusqu' en déc. 1991	New cases reported for 1992 Nouveaux cas rapportés en 1992	New cases reported for 1993 Nouveaux cas rapportés en 1993	Cumulative total to Dec. 1993 Total cumulatif en déc. 1993
Myanmar Myanmar	6	41	142	189
Nepal Népal	9	3	12	24
Oman Oman	24	5	2	31
Pakistan Pakistan	18	8	12	38
Philippines Philippines	78	17	32	127
Qatar Qatar	33	1	7	41
Saudi Arabia Arabie saoudite	44	6	12	62
Singapore Singapour	35	18	22	75
Sri Lanka Sri Lanka	11	9	13	33
Syrian Arab Republic Rép. arabe syrienne	20	3	3	26
Tajikistan Tadjikistan	0	0	0	0
Thailand Thaïlande	606	1489	6098	8193
Turkey Turquie	63	29	30	122
Turkmenistan Turkménistan	0	1	0	1
United Arab Emirates Emirats arabes unis	8	0	0	8
Uzbekistan Ouzbékistan	0	1	1	2
Viet Nam Viet Nam	0	0	89	89
Yemen Yémen	1	3	4	8
Europe • Europe				
Albania Albanie	0	0	0	0
Austria Autriche	691	198	216	1105
Belarus Bélarus	2	6	2	10
Belgium Belgique	1047	251	223	1521
Bulgaria Bulgarie	12	6	6	24
Croatia Croatie	39	8	13	60
Czech Republic République tchèque	23	9	15	47
Denmark Danemark	925	194	236	1355
Estonia Estonie	0	1	1	2
Finland Finlande	89	17	36	142
France France	17720	5146	5805	28671
Germany † Allemagne †	7079	1679	1887	10645
Greece Grèce	559	162	170	891
Hungary Hongrie	79	33	36	148

13 C. Reported AIDS cases, 1993 [*cont.*] • Cas de SIDA rapportés, 1993 [*suite*]

Country or area Pays ou zone	Total reported to Dec. 1991 Total des cas rapportés jusqu' en déc. 1991	New cases reported for 1992 Nouveaux cas rapportés en 1992	New cases reported for 1993 Nouveaux cas rapportés en 1993	Cumulative total to Dec. 1993 Total cumulatif en déc. 1993
Iceland Islande	22	3	6	31
Ireland Irlande	239	69	74	382
Italy Italie	11607	4150	4573	20330
Latvia Lettonie	3	1	3	7
Lithuania Lituanie	2	1	2	5
Luxembourg Luxembourg	45	12	20	77
Malta Malte	22	4	3	29
Monaco Monaco	7	10	7	24
Netherlands Pays—Bas	1943	533	435	2911
Norway Norvège	252	54	62	368
Poland Pologne	85	40	42	167
Portugal Portugal	831	382	462	1675
Republic of Moldova République de Moldova	2	0	2	4
Romania Roumanie	1705	530	402	2637
Russian Federation Fédération de Russie	57	54	21	132
San Marino Saint—Marin	1	0	0	1
Slovakia Slovaquie	3	1	3	7
Slovenia Slovénie	20	4	7	31
Spain Espagne	11342	5361	5576	22279
Sweden Suède	645	127	177	949
Switzerland Suisse	2223	646	679	3548
Ukraine Ukraine	8	4	10	22
United Kingdom Royaume—Uni	5425	1482	1608	8515
Yugoslavia, SFR † Yougoslavie, Rfs †	180	88	54	322
Oceania • Océanie				
Australia Australie	3348	730	707	4785
Cook Islands Iles Cook	0	0	0	0
Fiji Fidji	4	1	1	6
French Polynesia Polynésie française	30	3	9	42
Guam Guam	12	2	4	18
Kiribati Kiribati	0	1	0	1
Marshall Islands Iles Marshall	2	0	0	2
Micronesia, Federated States of Micronésie, Etats fédérés de	2	0	0	2

13 C. Reported AIDS cases, 1993 [cont.] · Cas de SIDA rapportés, 1993 [suite]

Country or area Pays ou zone	Total reported to Dec. 1991 Total des cas rapportés jusqu' en déc. 1991	New cases reported for 1992 Nouveaux cas rapportés en 1992	New cases reported for 1993 Nouveaux cas rapportés en 1993	Cumulative total to Dec. 1993 Total cumulatif en déc. 1993
Nauru Nauru	0	0	0	0
New Caledonia Nouvelle-Calédonie	16	9	6	31
New Zealand Nouvelle-Zélande	303	56	70	429
Niue Nioué	0	0	0	0
Northern Mariana Islands Iles Mariannes du Nord	4	0	0	4
Palau Palaos	0	0	1	1
Papua New Guinea Papouasie-Nvl-Guinée	59	8	8	75
Samoa Samoa	1	0	0	1
Solomon Islands Iles Salomon	0	0	0	0
Tonga Tonga	3	1	1	5
Tokelau Tokélaou	0	0	0	0
Tuvalu Tuvalu	0	0	0	0
Vanuatu Vanuatu	0	0	0	0
Wallis and Futana Islands Iles Wallis et Futana	0	0	0	0

Source:
World Health Organization (Geneva), Global Programme on AIDS, Surveillance, Forecasting and Impact Assessement Unit.

Source:
Organisation mondiale de la santé (Gèneve), Programme mondial de lutte contre le SIDA, equipe de surveillance, prévision et étude d'impact.

†   For detailed descriptions of data pertaining to former Czechoslovakia, Germany, SFR Yugoslavia and former USSR, see Annex I — Country or area nomenclature, regional and other groupings.

†   Pour les descriptions en détails des données relatives à l'ancienne Tchécoslovaquie, l'Allemagne, la Rfs Yougoslavie et l'ancienne URSS, voir l'Annexe I — Nomenclature des pays ou zones, groupements régionaux et autres groupements.

Technical notes, table 13

Table 13: Data on acquired immunodeficiency syndrome (AIDS) have been compiled and estimated by WHO from official national reports and special studies (unpublished).

Notes techniques, tableau 13

Tableau 13 : Les données sur le syndrome d'immuno-déficience acquise (SIDA) ont été compilées et estimées par l'OMS sur la base des rapports officiels soumis par les pays et d'études spéciales (non publiées).

14

Book production: number of titles by UDC classes
Production de livres : nombre de titres classés d'après la CDU

Country or area Pays ou zone	Year Année	Total	Gener– alities Géné– ralités	Philo– sophy Philo– sophie	Reli– gion	Social sciences Sciences sociales	Philo– logy Philo– logie	Pure sciences Sciences pures	Applied sciences Sciences appli– quées	Arts Beaux arts	Litera– ture Litté– rature	Geogr/ History Géogr., histoire
Africa · Afrique												
Algeria Algérie	1992	506	18	10	42	153	28	71	77	19	54	34
Angola Angola	1986	14	–	–	–	–	–	–	–	–	14	–
Benin [1] [2] Bénin [1] [2]	1992	647	10	4	–	534	7	12	77	–	–	3
Botswana [1] Botswana [1]	1991	158	8	–	–	125	1	10	11	–	–	3
Burundi Burundi	1986	54	–	–	10	28	–	–	16	–	–	–
Cape Verde Cap–Vert	1989	10	–	–	–	2	–	–	1	–	6	1
Egypt Egypte	1988	1451	19	47	191	305	65	73	159	65	403	124
Equatorial Guinea [1] Guinée équatoriale [1]	1988	17	–	–	–	8	3	1	3	–	–	2
Ethiopia Ethiopie	1991	240	36	–	23	66	11	8	44	3	31	18
Gambia [1] Gambie [1]	1991	21	–	–	–	15	–	–	5	–	–	1
Kenya [2] Kenya [2]	1991	239	7	–	84	48	26	13	34	1	19	7
Libyan Arab Jamahiriya [1] Jamahiriya arabe libyenne [1]	1988	121	6	7	4	41	6	20	6	–	27	4
Madagascar [1] Madagascar [1]	1992	85	–	2	33	25	–	1	13	1	7	3
Malawi Malawi	1989	141	11	–	31	53	2	7	22	–	11	4
Mauritius Maurice	1992	80	1	1	3	21	13	5	5	1	20	10
Nambia [1] Namibie [1]	1991	193	12	–	4	120	–	–	20	2	24	11
Niger [1] Niger [1]	1991	5	–	–	–	–	–	–	–	–	5	–
Nigeria Nigéria	1992	1562	55	14	142	733	104	71	196	30	148	69
Réunion Réunion	1992	69	1	–	–	20	–	4	5	12	14	13
Rwanda Rwanda	1987	207	18	5	42	72	9	20	31	3	4	3
South Africa Afrique du Sud	1992	4738	176	41	661	1203	225	312	808	210	870	232
Tunisia [2] Tunisie [2]	1992	1165	130	100	35	194	–	68	108	20	310	200
United Rep. Tanzania [1] Rép.–Unie de Tanzanie [1]	1990	172	–	1	18	45	2	7	40	3	47	9
Zimbabwe Zimbabwe	1992	232	6	–	15	107	15	3	48	7	24	7
America, North · Amérique du Nord												
Bahamas [2] Bahamas [2]	1991	15 [3]	2	–	–	4	–	–	–	1	2	3
Barbados [1] Barbade [1]	1990	77	3	1	2	39	–	–	7	14	8	3
Belize Belize	1991	134 [3]	2	–	–	36	1	–	10	1	–	2
Costa Rica Costa Rica	1990	244	5	1	16	111	5	13	34	8	35	16
Cuba [4] Cuba [4]	1991	1017	97	35	5	212	28	48	270	79	211	32
El Salvador [2] El Salvador [2]	1988	15	–	–	–	8	–	1	–	–	4	2

14
Book production: number of titles by UDC classes [*cont.*]
Production de livres : nombre de titres classés d'après la CDU [*suite*]

Country or area Pays ou zone	Year Année	Total	Gener-alities Géné-ralités	Philo-sophy Philo-sophie	Reli-gion	Social sciences Sciences sociales	Philo-logy Philo-logie	Pure sciences Sciences pures	Applied sciences Sciences appli-quées	Arts Beaux arts	Litera-ture Litté-rature	Geogr/ History Géogr., histoire
Haïti Haïti	1989	271	9	4	6	104	5	8	21	18	81	15
Mexico Mexique	1990	2608	281	120	145	943	65	192	179	126	315	242
Nicaragua [5] Nicaragua [5]	1987	41	–	–	–	15	–	–	3	2	21	–
Saint Kitts and Nevis [6] Saint Kitts–et–Nevis [6]	1988	3	–	–	–	3	–	–	–	–	–	–
Saint Pierre and Miquelon [1] Saint–Pierre–et–Miquelon [1]	1991	3	–	–	–	2	–	–	1	–	–	–
United States [7] Etats–Unis [7]	1992	49276 [3]	2153	1806	2540	10147	617	2729	8144	2851	8816	4329
America, South · Amérique du Sud												
Argentina Argentine	1992	5628	144	338	407	1571	53	108	572	358	1712	365
Bolivia Bolivie	1988	447	25	7	26	171	10	24	56	25	56	47
Chile Chili	1992	1820	13	59	127	587	34	58	183	43	548	168
Colombia [2] Colombie [2]	1991	1481	141	28	88	570	43	40	243	52	216	60
Ecuador Equateur	1991	717	30	7	8	197	5	38	33	30	288	81
Guyana Guyana	1989	46	1	–	–	29	–	–	4	3	4	5
Peru Pérou	1991	1063	36	23	34	389	37	63	194	76	108	103
Uruguay Uruguay	1991	1143	6	68	35	340	17	81	205	33	276	82
Venezuela [1] Venezuela [1]	1992	3879	86	130	121	935	79	275	843	386	680	344
Asia · Asie												
Afghanistan Afghanistan	1990	2795	165	25	170	1045	30	125	680	95	200	260
Azerbaijan Azerbaïdjan	1992	599	9	15	9	131	25	36	63	13	241	57
Bangladesh [1 2] Bangladesh [1 2]	1988	1209	63	21	60	457	27	66	107	14	300	94
Brunei Darussalam [8] Brunéi Darussalam [8]	1990	25	2	–	–	13	2	–	–	–	8	–
China [2 9] Chine [2 9]	1990	73923	2588	1206	...	36231	2403	3087	12196	5727	7756	2729
Cyprus Chypre	1992	900	37	11	69	210	41	19	197	79	191	46
India Inde	1991	14438	214	449	807	3204	197	397	3864	331	3746	1229
Indonesia [1] Indonésie [1]	1992	6303 [3]	191	148	892	1192	294	186	788	64	231	246
Iran, Islamic Rep. of [1] Iran, Rép. islamique d' [1]	1991	5018	191	225	947	499	482	229	657	316	1120	352
Japan [10] Japon [10]	1987	36346	1190	1156	630	8015	1523	1924	6875	5046	8170	1817
Jordan [1 2] Jordanie [1 2]	1992	790 [3]	–	3	64	299	1	109	5	29	251	17
Kazakhstan Kazakhstan	1992	1226	–	21	16	293	85	117	313	23	304	54
Korea, Republic of Corée, République de	1992	27889	3044	659	2063	2635	1741	1491	2880	5954	6270	1152
Kuwait [11] Koweït [11]	1992	196	17	5	17	18	–	102	6	15	13	3
Lao People's Dem. Rep. [1] Rép. dém. populaire Lao [1]	1990	109	–	–	3	9	25	1	12	–	58	1
Malaysia Malaisie	1991	3748	85	22	291	906	635	360	349	135	757	208

14
Book production: number of titles by UDC classes [*cont.*]
Production de livres : nombre de titres classés d'après la CDU [*suite*]

Country or area Pays ou zone	Year Année	Total	Gener– alities Géné– ralités	Philo– sophy Philo– sophie	Reli– gion	Social sciences Sciences sociales	Philo– logy Philo– logie	Pure sciences Sciences pures	Applied sciences Sciences appli– quées	Arts Beaux arts	Litera– ture Litté– rature	Geogr/ History Géogr., histoire
Mongolia Mongolie	1990	717	49	–	–	300	2	45	103	–	218	–
Nepal [9] Népal [9]	1989	122	–	4	7	19	25	18	27	3	13	6
Oman [2] Oman [2]	1992	24	1	–	3	–	9	–	1	–	–	10
Pakistan [1] Pakistan [1]	1992	70 [3]	10	–	–	14	–	14	4	–	6	4
Philippines Philippines	1991	825	84	15	40	294	38	27	100	18	183	26
Qatar [2] Qatar [2]	1992	372	23	14	31	186	19	36	24	11	17	11
Sri Lanka Sri Lanka	1992	4225	263	46	312	2209	179	71	400	75	531	139
Syrian Arab Republic [2] République arabe syrienne [2]	1992	598	144	24	62	79	13	1	89	25	112	49
Thailand Thaïlande	1992	7626	413	189	302	2341	234	591	2235	443	457	421
Turkey Turquie	1992	6549 [3]	159	204	526	1591	176	140	811	236	1371	463
Turkmenistan Turkménistan	1992	565	47	5	6	147	21	72	74	18	136	39
United Arab Emirates [12] Emirats arabes unis [12]	1992	302	10	3	46	20	85	99	9	3	–	27
Europe · Europe												
Albania Albanie	1991	381	12	–	4	28	35	74	127	20	63	18
Andorra [2] Andorre [2]	1992	56	3	–	1	15	1	3	4	7	10	12
Austria Autriche	1991	3786 [3]	81	96	132	993	126	156	509	527	437	358
Belarus Bélarus	1992	2364	112	41	27	665	123	140	674	80	418	84
Belgium Belgique	1991	13913	300	311	686	2847	478	817	2228	1325	3696	1225
Bulgaria Bulgarie	1992	4773	127	151	146	814	1	264	729	145	2191	205
Croatia Croatie	1990	2239	...	...	...	...	...	...	...	...	...	...
Czech Republic République Tchèque	1992	6743	330	178	244	1045	353	684	1107	423	1981	398
former Czechoslovakia † ancienne Tchécoslovaquie †	1991	9362	1383	126	226	1378	540	738	2383	589	1492	507
Denmark Danemark	1992	11761	271	510	316	2225	361	932	2723	773	2505	1145
Estonia Estonie	1992	1557	63	42	73	320	–	161	314	87	437	60
Finland Finlande	1992	11033	291	181	456	2603	403	1080	2774	761	1816	668
France France	1992	45379 [3]	755	1642	1402	13062	1040	2038	5094	3525	11659	5162
Germany † [13] Allemagne † [13]	1992	67277	5914	3132	3595	15104	...	2141	9441	5734	13426	8790
Federal Republic of Germany Rép. féd. d'Allemagne	1989	65980	5511	2890	3709	14568	2894	1958	9048	5296	11806	8300
former German Dem. Rep. [14] anc. Rép. dém. allemande [14]	1989	6018 [3]	87	103	281	702	287	458	910	448	1399	287
Greece Grèce	1991	4066	143	158	289	638	159	157	260	246	1633	383
Holy See Saint–Siège	1992	205	5	37	117	31	7	–	–	–	–	8
Hungary Hongrie	1992	8536	249	237	312	1432	439	557	1711	551	2452	596
Iceland Islande	1991	1576 [3]	21	38	41	293	109	70	141	106	533	173

14
Book production: number of titles by UDC classes [cont.]
Production de livres : nombre de titres classés d'après la CDU [suite]

Country or area Pays ou zone	Year Année	Total	Generalities Généralités	Philosophy Philosophie	Religion	Social sciences Sciences sociales	Philology Philologie	Pure sciences Sciences pures	Applied sciences Sciences appliquées	Arts Beaux arts	Literature Littérature	Geogr/ History Géogr., histoire
Italy Italie	1992	29351	880	1678	1752	6223	612	1042	3613	3620	7143	2788
Latvia Lettonie	1992	1509	67	50	74	169	72	64	283	66	361	77
Lithuania Lituanie	1992	2361	179	71	66	416	130	189	559	127	500	124
Luxembourg [13] Luxembourg [13]	1992	417	4	2	9	136	...	11	33	99	77	46
Malta Malte	1992	395	9	7	76	155	17	6	17	22	34	52
Monaco Monaco	1990	41	3	–	1	1	–	1	8	22	2	3
Netherlands [2] Pays–Bas [2]	1992	11844	71	628	833	1701	233	358	2502	874	3251	1393
Norway [15] Norvège [15]	1992	4881	196	139	193	1096	107	270	658	378	1432	412
Poland Pologne	1992	10727	211	333	514	1801	423	1000	2147	631	2691	976
Portugal Portugal	1991	6430 [3]	97	114	85	1031	31	115	348	220	1341	355
Republic of Moldova République de Moldova	1992	802	24	13	12	174	27	57	204	20	225	46
Romania Roumanie	1992	3662	71	83	133	254	201	382	758	118	1442	220
Russian Federation Fédération de Russie	1992	28716	3354	480	585	4550	722	2727	7681	690	6975	952
Slovakia Slovaquie	1992	3308	75	94	197	577	147	325	850	168	734	141
Slovenia Slovénie	1992	2136	94	60	83	517	86	234	327	235	354	146
Spain Espagne	1992	41816	1342	1447	1760	7638	1631	2512	5873	3116	12098	4399
Sweden Suède	1992	12813	305	346	404	2451	405	912	2696	842	3222	1230
Switzerland Suisse	1992	14663 [3]	282	508	846	3188	230	1344	3285	1406	1839	876
TFYR Macedonia L'ex–R.y. Macédoine	1990	559	6	6	3	162	–	13	29	41	279	20
Ukraine Ukraine	1992	4410	7	70	94	837	144	513	1439	69	1050	187
United Kingdom Royaume–Uni	1992	86573	2385	2888	3252	17286	2434	9490	14695	6294	17601	10248
Yugoslavia Yougoslavie	1992	2618	43	42	71	775	3	110	400	308	784	82
Yugoslavia, SFR † Yougoslavie, Rfs †	1990	9797	357	197	297	2651	12	472	1488	925	2950	448
Oceania · Océanie												
Australia Australie	1987	10963	390	76	391	4000	624	596	1766	994	1074	1052
New Caledonia Nouvelle–Calédonie	1987	14	–	–	–	4	2	–	1	3	–	4
Papua New Guinea Papouassie–Nouvelle–Guinée	1991	122	8	–	19	64	1	6	16	4	–	4
former USSR · ancienne URSS												
former USSR † ancienne URSS †	1989	76711 [3]	2244	1596	299	19676	1933	6425	27458	2655	11692	2219

Source:
United Nations Educational, Scientific and Cultural Organization (Paris).

Source:
Organisation des Nations Unies pour l'éducation, la science et la culture (Paris).

† For detailed description of data pertaining to former Czechoslovakia, Germany, SFR Yugoslavia and former USSR, see Annex–Country or area nomenclature, regional and other groupings.

† Pour les descriptions en détails des données relatives à l'ancienne Tchécoslovaquie, l'Allemagne, la Rfs Yougoslavie et l'ancienne URSS, voir l'Annexe I–Nomenclature des pays ou zones, groupements régionaux et autres groupments.

14
Book production: number of titles by UDC classes [*cont.*]
Production de livres : nombre de titres classés d'après la CDU [*suite*]

1	All first editions.
2	Data do not include pamphlets.
3	Including titles for which a class breakdown is not available.
4	Data refer only to the titles published by the Ministry of Culture.
5	Data on pamphlets refer to the first editions of school text books and children's books.
6	Data refer only to first editions of school textbooks.
7	Data do not include pamphlets, school textbooks, government publications and university theses.
8	Data do not include pamphlets and refer only to school textbooks(6), and to first editions of children's books (9) and of government publications(10).
9	Works of UDC class 3 (religion) are distributed among other classes without specification.
10	Data do not include pamphlets and government publications and refer only to first editions.
11	Data refer to govervment publications only.
12	Data refer to school textbooks only.
13	Data on philology are included with those on literature.
14	Data include only books and pamphlets shown in series A of the German National Bibliography (publications of the book market); those of the series B (publications outside the book market) and C (books published by universities) are excluded.
15	Data do not include schools textbooks and government publications.

1　Tous les ouvrages recensés sont des premières éditions.
2　Les données ne comprennent pas les brochures.
3　Y compris les titres pour lesquels aucune répartition par catégorie n'est disponible.
4　Les données se réfèrent seulement aux titres publiés par le Ministère de la Culture.
5　Les données dans les brochures se réfèrent aux premières éditions des manuels scolaires et les livres pour enfants.
6　Les données se réfèrent seulement aux premières éditions des manuels scolaires.
7　Les données ne comprennent pas les brochures, les manuels scolaires, les publications officielles et les thèses universitaires.
8　Les données ne comprennent pas les brochures et se réfèrent seulement aux manuels scolaires (6) et aux premières éditions des livres pour enfants (9) et des publications officielles (10).
9　Les ouvrages de la catégorie 3 de la CDU (religion) sont distribués sans spécification, entre les autres catégories.
10　Les données ne comprennent ni les brochures ni publications officielles, et ne se réfèrent qu'aux premières éditions.
11　Les données ne se réfèrent qu'aux publications officielles.
12　Les données se réfèrent aux manuels scolaires seulement.
13　Les données relatives à la philologie sont comprises avec celles de la littérature.
14　Les données se réfèrent seulement aux livres et brochures de la série A de la Bibliographie nationale allemande (publications en vente dans le commerce). Les données relatives aux séries B (publications non vendues dans le commerce) et C (livres publiés par les universités) ne sont pas prises en compte.
15　Les données n'incluent pas les manuels scolaires et les publications officielles.

15
Book production: number of titles by language of publication
Production de livres : nombre de titres par langue de publication

Country or area / Pays ou zone	Year / Année	Total	National language / Langue nationale	Foreign languages / Langues étrangères					Others / Autres	Two or more languages / Deux langues ou plus
				English / Anglais	French / Français	German / Allemand	Spanish / Espagnol	Russian / Russe		
Africa · Afrique										
Algeria / Algérie	1992	506	258	3	245	–	–	–	–	–
Botswana [1] / Botswana [1]	1991	158	158	–	–	–	–	–	–	–
Burundi / Burundi	1986	54	54	–	...[2]	–	–	–	–	–
Cape Verde / Cap–Vert	1989	10	10	–	–	–	–	–	–	–
Egypt / Egypte	1988	1451	1344	83	23	–	–	–	1	–
Equatorial Guinea [1] / Guinée équatoriale [1]	1988	17	17	–	–	–	–	–	–	–
Ethiopia / Ethiopie	1991	240	206	–	–	–	–	–	34	–
Gambia / Gambie	1991	21	19	...[2]	–	–	–	–	–	–
Libyan Arab Jamahiriya [1] / Jamah. arabe libyenne [1]	1988	121	121	–	–	–	–	–	–	–
Madagascar [1] / Madagascar [1]	1992	85	82	–	...[2]	–	–	–	–	3
Malawi / Malawi	1989	141	87	...[2]	–	–	–	–	13	–
Mauritius / Maurice	1992	80	80	...[2]	...[2]	–	–	–	–	–
Nambia / Namibie	1991	193	193	...[2]	–	–	–	–	–	–
Nigeria / Nigéria	1992	1562	1554	...[2]	6	–	–	–	2	–
Réunion / Réunion	1992	69	68	1	...[2]	–	–	–	–	–
Rwanda / Rwanda	1987	207	191	–	...[2]	–	–	–	11	5
South Africa / Afrique du Sud	1992	4738	4363	...	1	12	2	–	66	294
Zimbabwe / Zimbabwe	1989	337	333	...[2]	–	–	–	–	2	2
America, North · Amérique du Nord										
Barbados [1] / Barbade [1]	1990	77	77	...[2]	–	–	–	–	–	–
Belize / Belize	1991	134	132	...[2]	–	–	–	–	2	–
Costa Rica / Costa Rica	1990	244	243	1	–	–	–	...	–	–
Cuba / Cuba	1989	2199	2040	104	3	2	...[2]	17	–	33
El Salvador [3] / El Salvador [3]	1988	15	15	–	–	–	...[2]	–	–	–
Haiti / Haïti	1989	271	258	–	...[2]	–	–	–	23	–
St. Kitts and Nevis [4] / Saint Kitts–et–Nevis [4]	1989	3	3	...[2]	–	–	–	–	–	–
America, South · Amérique du Sud										
Argentina / Argentine	1992	5628	5611	...	...	...	...	...	17	–
Colombia [3] / Colombie [3]	1991	1481	1481	–	–	–	...[2]	–	–	–
Peru / Pérou	1988	481	472	6	1	–	...[2]	–	2	–
Uruguay / Uruguay	1989	805	801	4	–	–	...[2]	–	–	–

15
Book production: number of titles by language of publication [*cont.*]
Production de livres : nombre de titres par langue de publication [*suite*]

Country or area Pays ou zone	Year Année	Total	National language Langue nationale	Foreign languages Langues étrangères						Two or more languages Deux langues ou plus
				English Anglais	French Français	German Allemand	Spanish Espagnol	Russian Russe	Others Autres	
Asia · Asie										
Afghanistan Afghanistan	1990	2795	2500	100	–	–	–	100	95	–
Azerbaijan Azerbaïdjan	1992	599	576	...	...	...	...	...	12	11
Bangladesh[5] Bangladesh[5]	1988	1209	1209	...[2]	–	–	–	–	–	–
Brunei Darussalam[5] Brunéi Darussalam[5]	1990	25	19	–	–	–	–	–	–	–
Cyprus[1] Chypre[1]	1992	900	563	95	16	19	–	–	53	–
Indonesia[3] Indonésie[3]	1988	1687	1627	20	–	–	–	–	–	40
Jordan Jordanie	1992	790	780	10	–	–	–	–	–	–
Korea, Republic of Corée, République de	1992	27889	27750	138	–	–	–	–	1	–
Kuwait Koweït	1988	793	793	–	–	–	–	–	–	–
Lao People's Dem. Rep. Rép. dém. populaire Lao	1990	109	109	–	–	–	–	–	–	–
Malaysia Malaysie	1991	3748	2249	1015	–	–	–	–	329	155
Nepal[6] Népal[6]	1989	122	86	8	–	–	–	–	28	–
Philippines Philippines	1991	825	605	...	–	–	–	–	140	18
Sri Lanka Sri Lanka	1991	2535	1599	632	–	–	1	–	5	298
Thailand Thaïlande	1992	7626	7120	506	–	–	–	–	–	–
Turkey Turquie	1992	6549	6266	98	39	23	–	–	56	67
United Arab Emirates[6] Emirats arabes unis[6]	1992	302	260	42	–	–	–	–	–	–
Europe · Europe										
Albania Albanie	1991	381	216	9	1	2	–	3	2	1
Andorra Andorre	1992	56	51	–	2	–	3	–	–	–
Austria Autriche	1989	9462	9108	274	31	...[2]	12	4	32	1
Belarus Bélarus	1992	2364	2258	...	...	...	...	...	106	–
Belgium[3] Belgique[3]	1991	13913	12564	684	...[2]	...[2]	60	9	128	468
Bulgaria Bulgarie	1992	4773	4492	77	23	24	17	25	28	87
Czech Republic République Tchèque	1992	6743	6038	173	25	127	10	3	61	306
Denmark Danemark	1992	11761	9820	1252	32	137	18	1	125	376
Estonia Estonie	1992	1557	1078	101	1	28	–	191	16	142
Finland Finlande	1992	11033	9298	1536	12	81	6	10	51	39
France France	1992	45379	35483	459	...[2]	113	59	16	239	923
Holy See Saint−Siège	1992	205	89	70	9	1	11	–	4	21
Hungary Hongrie	1992	8536	8034	239	17	105	6	5	57	73
Italy Italie	1992	29351	27345	620	177	161	35	–	162	851

15
Book production: number of titles by language of publication [*cont.*]
Production de livres : nombre de titres par langue de publication [*suite*]

Country or area Pays ou zone	Year Année	Total	National language Langue nationale	Foreign languages Langues étrangères English Anglais	French Français	German Allemand	Spanish Espagnol	Russian Russe	Others Autres	Two or more languages Deux langues ou plus
Latvia Lettonie	1992	1509	1025	54	–	22	–	358	50	–
Lithuania Lituanie	1992	2361	2028	70	3	20	–	162	63	15
Luxembourg Luxembourg	1989	520	383	–	...[2]	...[2]	–	–	39	98
Malta Malte	1992	395	372	...[2]	4	6	–	–	11	2
Netherlands[3] Pays–Bas[3]	1992	11844	9649	1531	46	195	–	–	79	–
Norway[7] Norvège[7]	1992	4881	4310	435	9	10	–	–	117	–
Poland Pologne	1992	10727	10148	400	33	47	6	51	10	32
Portugal Portugal	1991	6430	3906	1157	742	72	168	15	220	150
Republic of Moldova République de Moldova	1992	802	754	4	8	1	3	...	2	30
Romania Roumanie	1990	2178	1867	49	23	62	–	–	177	–
Russian Federation Fédération de Russie	1992	28716	26853	595	45	53	12	...	977	181
Slovakia Slovaquie	1992	3308	2592	120	10	45	2	16	523	–
Slovenia Slovénie	1992	2136	1851	53	4	31	1	–	37	159
Spain Espagne	1992	41816	40001	622	230	108	...	–	330	525
Sweden Suède	1992	12813	10507	1676	27	49	10	12	146	386
Switzerland Suisse	1992	14663	12323	1907	...[2]	...[2]	–	–	433	...
Ukraine Ukraine	1992	4410	4245	42	9	12	–	...	–	102
Yugoslavia Yugoslavie	1992	2618	2094	136	39	38	–	20	43	248
Yugoslavia, SFR † Yugoslavie, Rfs †	1990	9797	8246	160	20	49	–	15	499	808
Oceania · Océanie										
Australia Australie	1987	10963	10723	...[2]	3	4	10	3	86	134
former USSR · ancienne URSS										
former USSR † ancienne URSS †	1989	76711	72981	1021	349	371	380	...[2]	1356	253

<table>
<tr><td>

Source:
United Nations Educational, Scientific and Cultural Organization (Paris).

† For detailed description of data pertaining to former Czechoslovakia, Germany, SFR Yugoslavia and former USSR, see Annex I– Country or area nomenclature, regional and other groupings.

1 All first editions.
2 Data are included in column "National language".
3 Data do not include pamphlets.
4 Data refer only to the first editions of school textbooks.
5 Language breakdown refers only to ten titles of first editions of school textbooks, children's books and university theses.
6 Data refer only to school textbooks.
7 Data do not include schools textbooks.

</td><td>

Source:
Organisation des Nations Unies pour l'éducation, la science et la culture (Paris).

† Pour les descriptions en détails des données relatives à l'ancienne Tchécoslovaquie, l'Allemagne, la Rfs Yugoslavie et l'ancienne URSS, voir l'Annexe I–Nomenclature des pays ou zones, groupements régionaux et autres groupements.

1 Tous les ouvrages recensés sont des premières éditions.
2 Les chiffres sont inclus dans la colonne "langue nationale".
3 Les données ne tiennent pas compte des brochures.
4 Les données se réfèrent seulement aux premières éditions des manuels scolaires.
5 La répartition par langue ne se réfère qu'aux 10 titres des premières éditions des manuels scolaires, des livres pour enfants et thèses universitaires.
6 Les données se réfèrent seulement aux manuels scolaires.
7 Les manuels scolaires ne sont pas inclus dans les données.

</td></tr>
</table>

16
Daily newspapers
Journaux quotidiens

Country or area	Number Nombre				Circulation Diffusion (estimation) Total (000)				Per 1000 inhabitants Pour 1000 habitants			
Pays ou zone	1975	1980	1985	1992	1975	1980	1985	1992	1975	1980	1985	1992
Africa · Afrique												
Algeria Algérie	4	4	5	5	285	448	570	1000	18	24	26	38
Angola Angola	4	4	4	4	*85	143	103	*116	*14	20	13	*12
Benin Bénin	1	1	1	1	1	1	1	12	0.3	0.3	0.3	2
Botswana Botswana	1	1	1	1	14	19	18	40	19	21	17	30
Burkina Faso Burkina Faso	1	1	2	1	2	2	4	*3	0.2	0.2	0.4	*0.3
Burundi Burundi	1	1	1	1	2	1	2	20	0.5	0.2	0.4	3
Cameroon Cameroun	2	2	1	1	25	65	35	*50	3	8	4	*4
Central African Rep. Rép. centrafricaine	–	–	–	1	–	–	–	*2	–	–	–	*1
Chad Tchad	4	1	1	1	...	1	1	2	...	0.2	0.2	0.4
Congo Congo	3	1	1	6	1	3	8	*19	0.7	2	4	*8
Côte d'Ivoire Côte d'Ivoire	3	2	1	1	*55	81	90	90	*8	10	9	7
Egypt Egypte	12	12	12	16	1095	1701	2383	*2426	30	42	51	*44
Equatorial Guinea Guinée équatoriale	2	2	2	1	*1	*2	*2	*1	*5	*7	*5	*3
Ethiopia Ethiopie	3	3	3	4	44	40	41	*70	1	1	1	*1
Gabon Gabon	1	1	1	1	3	15	20	*20	5	19	20	*16
Gambia Gambie	–	–	6	2	–	–	4	*2	–	–	6	*2
Ghana Ghana	4	5	5	4	500	*500	*510	280	51	*47	*40	18
Guinea–Bissau Guinée–Bissau	1	1	1	1	6	6	6	6	10	8	7	6
Kenya Kenya	3	3	4	5	134	216	283	*354	10	13	14	*14
Lesotho Lesotho	1	3	4	2	1	44	47	14	1	33	30	8
Liberia Libéria	3	3	5	8	13	11	*28	*35	8	6	*13	*13
Libyan Arab Jamahiriya Jamah. arabe libyenne	2	3	3	4	41	55	* 65	* 71	17	* 18	* 17	* 15
Madagascar Madagascar	*12	6	7	7	*55	55	67	48	*7	6	7	4
Malawi Malawi	2	*2	1	1	18	*20	15	25	3	*3	2	2
Mali Mali	1	2	2	2	3	*4	*10	*41	0.5	*1	*1	*4
Mauritania Mauritanie	–	–	–	1	–	–	–	*1	–	–	–	*0.5
Mauritius Maurice	12	10	7	6	82	80	*70	80	91	83	*69	73
Morocco Maroc	7	11	14	14	*250	*270	*320	*335	*14	*14	*15	*13
Mozambique Mozambique	2	2	2	2	42	54	81	81	4	4	6	5
Namibia Namibie	3	4	3	4	*16	27	21	209	*17	25	17	136

16
Daily newspapers [cont.]
Journaux quotidiens [suite]

Country or area	Number Nombre				Circulation Diffusion (estimation) Total (000)				Per 1000 inhabitants Pour 1000 habitants			
Pays ou zone	1975	1980	1985	1992	1975	1980	1985	1992	1975	1980	1985	1992
Niger Niger	2	1	1	1	3	3	4	5	1	1	1	1
Nigeria Nigéria	12	16	19	26	*650	*1100	*1400	*1800	*10	*14	*15	*16
Réunion Réunion	2	3	2	3	27	56	49	55	56	111	89	88
Rwanda Rwanda	1	1	1	1	0.2	0.3	0.3	*0.5	0.0	0.1	0.1	*0.1
Senegal Sénégal	1	1	3	1	25	35	53	50	5	6	8	6
Seychelles Seychelles	2	1	1	1	4	3	3	3	61	48	48	44
Sierra Leone Sierra Leone	2	1	1	1	30	10	10	10	10	3	3	2
Somalia Somalie	1	2	2	1	*3	*5	*7	*9	*1	*1	*1	*1
South Africa Afrique du Sud	*20	*24	24	20	*1000	*1400	1440	1248	*39	*47	43	31
Sudan Soudan	4	6	5	5	*30	105	*250	*620	*2	6	*11	*23
Swaziland Swaziland	1	1	2	3	5	9	10	*12	9	15	14	*15
Togo Togo	1	3	2	2	7	*16	*11	*12	3	*6	*4	*3
Tunisia Tunisie	4	5	6	9	190	272	*280	*410	34	43	*39	*49
Uganda Ouganda	*3	1	1	6	*46	25	25	*80	*4	2	2	*4
United Rep. of Tanzania Rép.−Unie de Tanzanie	3	3	2	3	70	208	101	*220	4	11	5	*8
Zaire Zaïre	6	5	4	9	*50	*60	*50	*112	*2	*2	*2	*3
Zambia Zambie	2	2	2	2	106	110	95	70	22	19	14	8
Zimbabwe Zimbabwe	3	2	3	2	116	133	203	195	19	19	24	18

America, North · Amérique du Nord

Country or area Pays ou zone	1975	1980	1985	1992	1975	1980	1985	1992	1975	1980	1985	1992
Antigua and Barbuda Antigua−et−Barbuda	1	1	1	−	4	6	6	−	67	98	95	−
Bahamas Bahamas	2	3	3	3	31	33	39	35	166	157	167	133
Barbados Barbade	1	2	2	2	24	39	40	41	98	156	158	160
Belize Belize	1	1	1	−	4	3	3	−	31	21	18	−
Bermuda Bermudes	1	1	1	1	11	14	18	16	215	257	321	258
Canada Canada	121	123	117	106[1]	*4900	5425	5566	5815[1]	*216	225	221	215[1]
Cayman Islands Iles Caïmanes	−	−	−	1	−	−	−	8	−	−	−	259
Costa Rica Costa Rica	6	4	6	4	174	251	*280	322	88	110	*106	101
Cuba Cuba	15	17	17	17	*600	1050	1207	1315	*64	108	120	122
Dominican Republic République dominicaine	10	7	7	11	*200	220	216	*265	*40	39	34	*35
El Salvador El Salvador	10	7	4	8	211	291	243	*485	52	64	51	*90
Grenada Grenade	1	1	−	−	*3	*4	−	−	*36	*45	−	−
Guadeloupe Guadeloupe	2	1	2	1	24	32	*34	35	74	99	*96	88
Guatemala Guatemala	10	9	9	5	249	*200	*250	180	41	*29	*31	18

16
Daily newspapers [cont.]
Journaux quotidiens [suite]

Country or area	Number Nombre				Circulation Diffusion (estimation) Total (000)				Per 1000 inhabitants Pour 1000 habitants			
Pays ou zone	1975	1980	1985	1992	1975	1980	1985	1992	1975	1980	1985	1992
Haiti Haïti	7	4	5	4	93	*36	*50	45	19	*7	*9	7
Honduras Honduras	8	6	7	4	*120	212	293	*159	*39	58	67	*29
Jamaica Jamaïque	3	3	4	3	131	109	*138	*160	65	51	*60	*65
Martinique Martinique	2	1	1	1	27	28	32	32	82	86	94	86
Mexico Mexique	216	317	332	292	5499	8322	9964	10231	93	124	132	116
Netherlands Antilles Antilles néerlandaises	5	8	6	6	55	*52	*54	53	328	*304	*309	301
Nicaragua Nicaragua	7	3	3	3	*100	136	*160	90	*41	49	*50	23
Panama Panama	6	5	7	8	131	*110	245	223	75	*56	112	89
Puerto Rico Porto Rico	4	4	5	3	430	512	599	507	144	160	177	141
Trinidad and Tobago Trinité−et−Tobago	3	4	4	4	135	*155	173	175	134	*143	149	138
United States Etats−Unis	1775	1745	1676	1586[1]	60655	62200	62800	60700[1]	281	273	263	240[1]
US Virgin Islands Iles Vierges américaines	3	3	3	2	17	17	21	22	179	168	198	206
America, South · Amérique du Sud												
Argentina Argentine	164	220	218	190	*3000	*4000	*3940	4780	*115	*142	*130	*144
Bolivia Bolivie	14	14	14	16	199	226	*290	*390	41	40	*46	*52
Brazil Brésil	289	343	322	373	4653	5482	6534	*8500	43	45	48	*55
Chile Chili	47	34	*38	45	...	...	...	...	...	...	...	...
Colombia Colombie	40	36	*46	46	*1450	*1400	*1800	*2100	*61	*53	*61	*63
Ecuador Equateur	29	18	26	36	332	558	*800	688	47	69	*86	62
French Guiana Guyane française	1	1	1	1	2	1	1	2	26	14	12	15
Guyana Guyana	2	1	2	2	50	58	78	80	68	76	99	99
Paraguay Paraguay	8	5	6	5	*140	*160	*170	168	*52	*51	*46	37
Peru Pérou	49	66	70	59	1377	*1400	*1600	*1590	91	*81	*82	*71
Suriname Suriname	7	4	5	3	*35	*45	*55	*43	*96	*128	*144	*98
Uruguay Uruguay	30	24	25	32	*800	*700	*680	*750	*283	*240	*226	*240
Venezuela Venezuela	49	66	55	82	*1300	2937	*2700	*4200	*103	196	*157	*208
Asia · Asie												
Afghanistan Afghanistan	15	13	13	16	*80	*90	*110	*206	*5	*6	*8	*11
Armenia Arménie	..	..	..	7	..	..	..	*84	..	..	..	...
Azerbaijan Azerbadjan	..	..	..	6	..	..	..	427	..	..	..	...
Bahrain Bahreïn	−	3	2	3	−	*14	19	*43	−	*40	44	*81
Bangladesh Bangladesh	30	44	60	51	*250	274	591	*710	*3	3	6	*6
Brunei Darussalam Brunéi Darussalam	−	−	−	1	−	−	−	20	−	−	−	74

16
Daily newspapers [cont.]
Journaux quotidiens [suite]

Country or area Pays ou zone	Number Nombre				Circulation Diffusion (estimation) Total (000)				Per 1000 inhabitants Pour 1000 habitants			
	1975	1980	1985	1992	1975	1980	1985	1992	1975	1980	1985	1992
China Chine	...	50	70	74	...	34375	*39000	...	...	35	*36	...
Cyprus Chypre	12	12	10	9	*80	*80	83	77	*131	*127	125	108
Hong Kong Hong–kong	82	...	46	49	...	...	*4100	*4500	...	...	*751	*819
India Inde	835	1173	1802	*2300	9383	14531	19804	*27500	15	21	26	*31
Indonesia Indonésie	*60	84	97	68	*2000	2281	3010	4591	*15	15	18	24
Iran, Islamic Rep. of Iran, Rép. islamique d'	19	*45	15	13	*700	*970	*1250	*1250	*21	*25	*26	*20
Iraq Iraq	7	5	6	6	*230	*340	*600	*660	*21	*26	*39	*34
Israel Israël	23	36	21	31	*850	*1000	*1100	*1240	*246	*258	*260	*242
Japan Japon	176	151	124	121	60782	66258	68296	71690	545	567	565	576
Jordan Jordanie	4	4	4	4	58	66	155	250	22	23	45	58
Korea, Dem.People's Rep. Corée, Rép. pop. dém. de	11	11	11	11	*3500	*4000	*4500	*5000	*211	*219	*226	*221
Korea, Republic of Corée, République de	36	30	35	63	6010	8000	*10000	*18000	170	210	*245	*407
Kuwait Koweït	8	8	8	9	180	305	380	*410	179	222	221	*244
Lao People's Dem. Rep. République dém. pop. lao	8	3	3	3	*15	*14	*13	*14	*5	*4	*4	*3
Lebanon Liban	33	14	13	16	*300	*290	*300	*500	*108	*109	*112	*176
Macau Macao	6	6	9	9	*45	*70	*250	*250	*173	*217	*638	*510
Malaysia Malaisie	31	40	32	39	1038	*810	*1500	*2200	85	*59	*96	*117
Maldives Maldives	1	2	2	2	*1	1	2	3	*7	6	8	13
Mongolia Mongolie	1	2	2	3	112	177	177	*208	77	106	93	*90
Myanmar Myanmar	7	7	7	2	319	*350	511	324	10	*10	14	7
Nepal Népal	29	28	28	25	*110	*120	*130	*140	*8	*8	*8	*7
Oman Oman	–	–	3	4	–	–	51	79	–	–	40	48
Pakistan Pakistan	102	106	118	274	*993	*1032	1149	809	*13	*12	11	6
Philippines Philippines	15	22	15	43	850	2000	2170	*3200	20	41	39	*49
Qatar Qatar	1	3	4	4	*20	*30	60	*70	*117	*131	168	*155
Saudi Arabia Arabie saoudite	12	11	13	13	*215	*350	*450	729	*30	*37	*36	46
Singapore Singapour	10	12	10	10	449	690	924	*930	198	286	361	*336
Sri Lanka Sri Lanka	15	21	17	10	480	450	390	* 480	35	30	24	* 27
Syrian Arab Republic Rép. arabe syrienne	6	7	7	11	*77	*114	*163	*290	*10	*13	*16	*22
Tajikistan Tadjikistan	..	..	..	9	..	..	..	36137	..	..	..	...
Thailand Thaïlande	22	27	32	35	2540	2680	4350	*4150	61	57	85	*74
Turkey Turquie	*500	*400	366	399[2]	*2000	*2500	3020	4000[2]	*50	*56	60	71[2]
United Arab Emirates Emirats arabes unis	2	9	13	11	10	152	290	*335	20	149	215	*201

16
Daily newspapers [*cont.*]
Journaux quotidiens [*suite*]

Country or area	Number Nombre				Circulation Diffusion (estimation) Total (000)				Per 1000 inhabitants Pour 1000 habitants			
Pays ou zone	1975	1980	1985	1992	1975	1980	1985	1992	1975	1980	1985	1992
Uzbekistan Ouzbékistan	..	..	..	12	..	..	..	452	..	..	...	
Viet Nam Viet Nam	2	4	4	4	*230	*520	*560	*570	*5	*10	*9	*8
Yemen Yémen	..	..	..	4	..	..	..	236	..	..	..	19
former Democratic Yemen ancienne Yémen dém.	4	3	3	..	* 14	* 14	* 15	..	* 8	* 8	* 7	..
former Yemen Arab Rep. ancienne Yémen rép. arabe	*3	*3	1	..	*70	*84	110	..	*13	*13	*14	..
Europe · Europe												
Albania Albanie	2	2	2	4	115	145	135	*165	47	54	46	*50
Andorra Andorre	–	–	–	3	–	–	–	*4	–	–	–	*83
Austria Autriche	30	30	33	27	2405	2651	2729	3108	317	351	361	400
Belarus Bélarus	25	27	28	10	2214	2343	2446	1899	236	244	245	...
Belgium Belgique	30	26	24	33	2340	2289	2171	*3100	239	232	220	*310
Bosnia & Herzegovina [3] Bosnie−Herzégovine [3]	..	..	..	2	..	..	..	518	..	..	..	...
Bulgaria Bulgarie	14	14	17	46	2109	2244	2626	1464	242	253	293	164
Croatia Croatie	..	..	8	9[2]	..	..	...	...	..	..	...	...
former Czechoslovakia † ancienne Tchécoslovaquie †	29	30	30	..	4436	4798	5124	..	300	313	331	..
Czech Republic République tchéque	..	..	..	55	..	..	..	*6000	..	..	..	...
Denmark Danemark	49	48	47	42	1723	1874	1855	1710	341	366	362	332
Estonia Estonie	..	..	..	15	..	..	..	..	..	..	..	...
Finland Finlande	60	58	65	58[4]	2100	2414	2661	2578[4]	446	505	543	515[4]
France France	95	90	92	77[1]	*11000	10332	10670	11695[1]	*209	192	193	205[1]
Germany † Allemagne †	..	..	..	355[1]	..	..	..	26425[1]	..	..	..	331[1]
Federal Republic of Germany Rép. féd. d'Allemagne	350	329	319	..	20200	20611	21108	..	327	335	346	..
former German Dem. Rep. ancienne Rép. dém. allemande	40	39	39	..	7946	8777	9320	..	472	524	560	..
Gibraltar Gibraltar	1	1	1	2	3	2	3	*4	100	80	103	*135
Greece Grèce	108	128	140	145	921	*1160	*1350	*1400	102	*120	*136	*137
Holy See Saint−Siège	1	1	1	1	70	70	70	70	..	..	..	..
Hungary Hongrie	27	27	28	28	2455	2648	2717	2896	233	247	255	275
Iceland Islande	5	6	6	5	93	125	113	*135	427	548	467	*519
Ireland Irlande	7	7	7	8	693	779	685	652	218	229	193	187
Italy Italie	78	82	72	78	6497	4775	5511	6068	117	85	96	105
Latvia Lettonie	..	..	..	17	..	..	..	517	..	..	..	193
Liechtenstein Liechtenstein	2	2	2	2	12	14	14	20	483	519	504	700
Lithuania Lituanie	..	..	..	18	..	..	..	836	..	..	..	223

16
Daily newspapers [*cont.*]
Journaux quotidiens [*suite*]

Country or area	Number Nombre				Circulation Diffusion (estimation) Total (000)				Per 1000 inhabitants Pour 1000 habitants			
Pays ou zone	1975	1980	1985	1992	1975	1980	1985	1992	1975	1980	1985	1992
Luxembourg Luxembourg	7	5	4	5	130	135	140	145	359	371	381	384
Malta Malte	6	5	4	3	*55	*60	*56	54	*181	*185	*163	150
Monaco Monaco	2	2	2	1	11	10	*10	8	420	396	*370	296
Netherlands Pays−Bas	84	84	88	...	4194	4612	4496	*4600	307	326	310	*303
Norway Norvège	80	85	82	82	1657	1892	2120	2600	414	463	510	606
Poland Pologne	44	43	45	72	8429	8407	7714	6085	248	236	207	158
Portugal Portugal	30	28	25	25	*612	*480	413	465	*67	*49	42	47
Republic of Moldova République de Moldova	..	..	..	5	..	..	..	*205	..	..	..	*45
Romania Roumanie	20	35	36	76	*3015	*4024	*3601	*7500	*142	*181	*158	*322
Russian Federation Fédération Russe	..	..	..	339	..	..	..	57367	..	..	..	...
San Marino Saint−Marin	4	3	..	..	1	1	..	..	53	48	..	..
Slovakia Slovaquie	..	..	..	21	..	..	..	1680	..	..	..	...
Slovenia Slovénie	3	3	3	6	196	198	216	308	...	...	...	...
Spain Espagne	115	111	102	148	3491	3487	3078	*4100	98	93	80	*105
Sweden Suède	112	114	115	104	4413	4386	4389	4419	539	528	526	511
Switzerland Suisse	95	89	97	83[5]	2574	2483	3213	2635[5]	406	393	497	387[5]
TFYR Macedonia L'ex−R.y.Macédoine	..	..	..	2[2]	..	..	..	*55[2]	..	..	..	...
Ukraine Ukraine	..	..	..	90	..	..	..	6083	..	..	..	...
United Kingdom Royaume−Uni	109	113	104	101	24805	23472	22495	*22100	441	417	397	*383
Yugoslavia, SFR † Yougoslavie, Rfs †	26	27	27	..	1896	2649	2451	..	89	119	106	..
Yugoslavia Yougoslavie	11	12	12	10	350	537	425	544	...	...	...	...
Oceania · Océanie												
American Samoa Samoa américaines	1	2	3	−	4	10	*9	−	120	319	*226	−
Australia Australie	70	62	62	69	*5336	*4700	*4300	*4600	*392	*320	*273	*261
Cook Islands Iles Cook	1	1	1	1	1	2	2	2	42	111	111	118
Fiji Fidji	1	3	3	1	20	64	68	27	35	102	97	37
French Polynesia Polynésie française	4	2	3	4	11	*13	23	*24	85	*86	132	*116
Guam Guam	1	1	1	1	18	18	18	25	193	169	151	179
New Caledonia Nouvelle−Calédonie	2	1	1	3	18	15	19	*23	135	107	123	*133
New Zealand Nouvelle−Zélande	40	32	33	31	*900	1059	1075	*1050	*292	340	331	*304
Papua New Guinea Papouasie−Nouv.−Guinée	1	1	2	2	*20	27	45	64	*7	9	13	16
Tonga Tonga	−	−	−	1	−	−	−	7	−	−	−	72

16
Daily newspapers [*cont.*]
Journaux quotidiens [*suite*]

Country or area	Number Nombre				Circulation Diffusion (estimation)							
					Total (000)				Per 1000 inhabitants Pour 1000 habitants			
Pays ou zone	1975	1980	1985	1992	1975	1980	1985	1992	1975	1980	1985	1992
former USSR · ancienne URSS												
former USSR † ancienne URSS †	691	713	727	..	100928	*109089	120027	..	408	* 423	445	..

Source:
United Nations Educational, Scientific and Cultural Organization (Paris).

† For detaied descriptions of data pertaining to former Czechoslovakia, Germany, SFR Yugoslavia and former USSR, see Annex I — Country or area nomenclature, regional and other groupings.

1 Data refer to 1991.
2 Data refer to 1990.
3 Data shown for 1992 refer only to the territory that is under the control of the Government of the Republic of Bosnia and Herzegovina.
4 Data shown for 1992 do not include newspapers that are distributed free of charge.
5 Data shown for 1992 refer only to newspapers purchased and do not include satellites publications.

Source:
Organisation des Nations Unies pour l'éducation, la science et la culture (Paris).

† Pour les descriptions en détails des données relatives à l'ancienne Tchécoslovaquie, l'Allemagne, la Rfs Yougoslavie et l'ancienne URSS, voir l'Annexe I — Nomenclature des pays ou zones, groupements régionaux et autres groupements.

1 Les données se réfèrent à 1991.
2 Les données se réfèrent à 1990.
3 Les données présentées pour 1992 se réfèrent seulement au territoire qui est sous le contrôle du governement de la République de Bosnie—Herzégovine.
4 Les données pour 1992 n'incluent pas les journaux qui sont distribués gratuitement.
5 Les données préséntées pour 1992 se réfèrent seulement aux journaux payants et n'incluent pas les éditions satellites.

17
Non-daily newspapers and periodicals
Journaux non quotidiens et périodiques

Country or area Pays ou zone	Year Année	Non-daily newspapers Journaux non quotidiens			Periodicals Périodiques		
		Number Nombre	Circulation Diffusion Total (000)	Per 1000 inhabitants Pour 1000 habitants	Number Nombre	Circulation Diffusion Total (000)	Per 1000 inhabitants Pour 1000 habitants
Africa · Afrique							
Algeria Algérie	1990	37	1409	56	48	803	32
Angola Angola	1988	*2	*7	*1	...	...	...
Benin Bénin	1988	*3	*20	*5	...	...	...
Botswana Botswana	1992	4	61	46	14	177	135
Burkina Faso Burkina Faso	1990	10	14	2	37	24	3
Burundi Burundi	1988	*3	*22	*4	...	...	...
Cameroon Cameroun	1988	25	315	29	54	...	...
Cape Verde Cap–Vert	1988	3	7	20	...	...	...
Central African Republic République centrafricaine	1988	*2	*3	*1	...	...	...
Chad Tchad	1988	1	1	0.2	10	...	...
Congo Congo	1990	3	139	62	3	34	15
Côte d'Ivoire Côte d'Ivoire	1988	*5	*172	*15	...	...	...
Djibouti Djibouti	1988	2	*7	*17	7	6	15
Egypt Egypte	1991	35	1502	28	266	1815	34
Ethiopia Ethiopie	1988	4	40	1	3	14	0.3
Gabon Gabon	1988	1	20	18	...	...	...
Gambia Gambie	1990	6	*7	*8	10	885	1028
Ghana Ghana	1990	87	1111	74	121	774	52
Guinea Guinée	1990	*1	*1	*2	*3	*5	*1
Guinea–Bissau Guinée–Bissau	1988	*1	*2	*2	...	...	...
Kenya Kenya	1988	*8	*300	*14	...	...	...
Lesotho Lesotho	1988	*3	*45	*27	...	...	...
Liberia Libéria	1988	*8	*25	*10	...	...	...
Libyan Arab Jamahiriya Jamahiriya arabe libyenne	1988	*1	*15	*4	...	...	...
Madagascar Madagascar	1992	37	*168	*13	63	*191	*15
Malawi Malawi	1992	4	133	13	...	...	...
Mauritius Maurice	1992	25	...	...	62	...	...
Morocco Maroc	1988	*5	*35	*1	...	...	...
Mozambique Mozambique	1988	2	*85	*6	3	1828	132
Namibia Namibie	1990	18	71	49	...	...	...

17
Non–daily newspapers and periodicals [*cont.*]
Journaux non quotidiens et périodiques [*suite*]

Country or area Pays ou zone	Year Année	Non–daily newspapers Journaux non quotidiens			Periodicals Périodiques		
		Number Nombre	Circulation Diffusion		Number Nombre	Circulation Diffusion	
			Total (000)	Per 1000 inhabitants Pour 1000 habitants		Total (000)	Per 1000 inhabitants Pour 1000 habitants
Niger Niger	1988	1	5	1	...	...	...
Nigeria Nigéria	1988	11	45	0.4	92	495	5
Reunion Réunion	1988	*4	*20	*34	...	...	...
Rwanda Rwanda	1990	15	*155	*22	15	101	14
St. Helena Saint–Hélène	1988	1	2	250	...	...	...
Sao Tome and Principe Sao Tomé–et–Principe	1988	*2	*2	*18	...	...	...
Senegal Sénégal	1988	10	63	9	...	...	...
Seychelles Seychelles	1988	*4	*9	*136	...	...	...
Sierra Leone Sierra Leone	1988	*6	*65	*16	...	...	...
Somalia Somalie	1988	*4	*13	*2	...	...	...
South Africa Afrique du Sud	1991	10	1527	39	11	2149	55
Sudan Soudan	1988	*10	*135	*6	10	136	6
Swaziland Swaziland	1988	*1	*7	*10	...	...	...
Togo Togo	1988	*1	*5	*2	...	...	...
Tunisia Tunisie	1988	*9	*244	*32	...	...	...
Uganda [1] Uganda [1]	1990	*5	*70	*4	26	158	10
United Republic of Tanzania Rép. Unie de Tanzanie	1988	*9	*450	*19	...	...	...
Zaire Zaïre	1990	77	...	...	...	...	...
Zambia Zambie	1988	*1	*72	*9	...	...	...
Zimbabwe [1] Zimbabwe [1]	1990	16	428	43	28	680	68
America, North · Amérique du Nord							
Antigua and Barbuda Antigua–et–Barbuda	1988	*5	*10	*154	...	...	...
Bahamas Bahamas	1988	*2	*13	*53	...	...	...
Barbados [2] Barbade [2]	1990	4	*95	*370	52	...	...
Belize Belize	1990	7	37	196	...	...	...
Bermuda Bermudes	1990	3	35	566	...	...	...
British Virgin Islands Iles Vierges brit.	1990	2	4	250	...	...	...
Canada [3] Canada [3]	1992	1627	22043	805	1400	39510	1356
Cayman Islands Iles Caïmanes	1988	*1	*4	*160	...	...	...
Costa Rica Costa Rica	1991	12	106	34	...	...	...
Cuba [3] Cuba [3]	1990	4	36	3	160	2797	264

17
Non–daily newspapers and periodicals [cont.]
Journaux non quotidiens et périodiques [suite]

Country or area Pays ou zone	Year Année	Non–daily newspapers Journaux non quotidiens			Periodicals Périodiques		
			Circulation Diffusion			Circulation Diffusion	
		Number Nombre	Total (000)	Per 1000 inhabitants Pour 1000 habitants	Number Nombre	Total (000)	Per 1000 inhabitants Pour 1000 habitants
Dominica Dominique	1990	2	*7	*97	...	...	...
El Salvador El Salvador	1988	*3	*12	*2	...	...	...
Greenland Groenland	1988	*3	*15	*269	...	...	...
Grenada Grenade	1990	3	...	...	...	...	...
Guadeloupe Guadeloupe	1988	*9	*28	*74	...	...	...
Guatemala Guatemala	1988	*1	*7	*1	–	–	–
Haiti Haïti	1988	*4	*16	*3	...	...	...
Honduras Honduras	1988	*1	*5	*1	...	...	...
Martinique Martinique	1988	*7	*28	*80	...	...	...
Mexico Mexique	1992	56	1258	14	182	28016	318
Montserrat Montserrat	1990	2	2	200	...	...	...
Netherlands Antilles Antilles néerlandaises	1988	*1	*3	*17	...	...	...
Nicaragua Nicaragua	1988	*8	*140	*40	...	...	...
Panama Panama	1988	3	*50	*22	8	...	...
Puerto Rico Porto Rico	1988	*4	*106	*31	...	...	...
Saint Kitts and Nevis Saint Kitts–et–Nevis	1992	2	6	143	9	43	1023
Saint Lucia Sainte Lucie	1990	3	18	135	...	...	...
Saint Pierre and Miquelon Saint–Pierre et Miquelon	1990	1	2	317	...	...	...
Saint Vincent and the Grenadines Saint Vincent–et–Grenadines	1988	2	11	105	...	...	...
Trinidad and Tobago Trinité–et–Tobago	1990	5	125	101	...	...	...
Turks and Caicos Islands Iles Turques et Caïques	1990	1	10	833	...	...	...
United States Etats–Unis	1992	8855	...	...	...	...	...
United States Virgin Islands Iles Vierges américaines	1988	*2	*4	*33	...	...	...
America, South · Amérique du Sud							
Argentina Argentine	1988	*7	*320	*10	...	...	...
Bolivia Bolivie	1988	*8	*16	*2	...	...	...
Brazil Brésil	1988	*1450	*5000	*35	...	...	...
Chile Chili	1992	48	102	7	417	*3450	*254
Colombia Colombie	1988	*6	*260	*8	...	...	...
Falkland Islands (Malvinas) Iles Falkland (Malvinas)	1988	*2	*1	*500	...	...	...
French Guiana Guyane française	1988	*2	*7	*74	...	...	...
Guyana Guyana	1988	*6	*84	*106	...	...	...

17
Non–daily newspapers and periodicals [*cont.*]
Journaux non quotidiens et périodiques [*suite*]

Country or area Pays ou zone	Year Année	Non–daily newspapers Journaux non quotidiens			Periodicals Périodiques		
		Number Nombre	Circulation Diffusion		Number Nombre	Circulation Diffusion	
			Total (000)	Per 1000 inhabitants Pour 1000 habitants		Total (000)	Per 1000 inhabitants Pour 1000 habitants
Paraguay Paraguay	1988	*2	*16	*4	...	...	...
Peru Pérou	1988	*12	*374	*18	45	90	4
Suriname Suriname	1988	*2	*10	*25	...	...	...
Uruguay Uruguay	1988	*95	...	...	...	...	...
Venezuela Venezuela	1988	*45	*450	*24	...	...	...
Asia · Asie							
Afghanistan Afghanistan	1988	37	223	15	105	...	...
Armenia Arménie	1992	57	*200	...	40	5064	...
Azerbaijan Azerbaïdjan	1992	273	3476	...	49	801	...
Bahrain Bahreïn	1988	*6	*65	*137	...	...	...
Bangladesh Bangladesh	1988	490	916	8	...	...	...
Bhutan Bhoutan	1988	*3	*24	*16	...	...	...
Brunei Darussalam Brunéi Darussalam	1992	2	57	211	15	132	487
China [4] Chine [4]	1992	875	134409	113	6486	205060	173
Cyprus Chypre	1992	30	133	186	...	...	...
Hong Kong [5] Hong–kong [5]	1992	17	...	...	598	...	...
India Inde	1988	*8000	*13000	*16	...	...	...
Indonesia Indonésie	1992	92	3501	18	117	3985	21
Iran, Islamic Rep. of Iran, Rép. islamique d'	1990	50	...	...	318	6166	106
Iraq Iraq	1988	*12	*465	*27	...	...	...
Israel Israël	1988	*80	...	...	...	...	...
Japan [1] Japon [1]	1992	*14	*7963	*64	2926	...	...
Jordan Jordanie	1990	6	122	30	31	43	11
Korea, Dem. People's Rep. of Corée, Rép. pop. dém. de	1988	*2	...	...	...	...	...
Kuwait Koweït	1988	59	420	204	...	...	...
Lao People's Dem. Rep. Rép. dém. populaire Lao	1988	*4	*20	*5	...	...	...
Lebanon Liban	1988	*15	*240	*89	...	...	...
Macau [1] Macao [1]	1992	*3	...	...	16	...	...
Malaysia Malaisie	1988	*19	*4500	*265	...	...	...
Maldives Maldives	1988	*4	*5	*25	...	...	...
Mongolia Mongolie	1990	55	1133	517	45	6361	2904

17
Non-daily newspapers and periodicals [cont.]
Journaux non quotidiens et périodiques [suite]

Country or area Pays ou zone	Year Année	Non-daily newspapers Journaux non quotidiens			Periodicals Périodiques		
			Circulation Diffusion			Circulation Diffusion	
		Number Nombre	Total (000)	Per 1000 inhabitants Pour 1000 habitants	Number Nombre	Total (000)	Per 1000 inhabitants Pour 1000 habitants
Myanmar Myanmar	1988	*4	*246	*6	...	...	...
Nepal Népal	1988	*9	*43	*2	...	...	...
Oman Oman	1992	5	...	...	15	...	...
Pakistan [6] Pakistan [6]	1991	719	1957	16	...	...	...
Philippines Philippines	1990	306	*610	*10	1570	*9468	*152
Qatar [1] Qatar [1]	1990	1	5	12	190	120	281
Saudi Arabia Arabie saoudite	1990	6	...	...	...	...	...
Singapore Singapour	1988	*7	*470	*177	...	...	...
Sri Lanka Sri Lanka	1992	80		...	...	...	...
Tajikistan Tadjikistan	1992	110	56853	...	26	18841	...
Thailand Thaïlande	1990	302	...	...	1293	...	...
Turkey Turquie	1990	872	.1500	27	1325	1325	24
United Arab Emirates [1] Emirats arabes unis [1]	1990	1	30	20	80	922	614
Uzbekistan Ouzbékistan	1992	43	1279	...	61	1598	...

Yemen
Yémen
  Former Democratic Yemen
  Ancienne Yémem dém.
	1988	*5	*43	...	...	...	...

Former Yemen Arab Republic Ancienne Yémem rép. arabe	1988	*7	*40	...	...	...	...
Europe · Europe							
Albania Albanie	1989	42	65	20	143	3477	1086
Andorra Andorre	1988	*4	*7	*152	...	...	...
Austria Autriche	1992	117	...	...	2524	...	...
Belarus Bélarus	1992	338	4749	...	155	3765	...
Belgium Belgique	1992	3	...	...	13706	...	...
Bulgaria [7] Bulgarie [7]	1992	871	8992	1004	745	3097	346
Croatia Croatie	1990	563	109699	...	352	6357	...
former Czechoslovakia † ancienne Tchécoslovaquie †	1990	64	849	54	2513	60061	3835
Denmark Danemark	1992	11	1490	289	205	7838	1520
Estonia Estonie	1992	169	...	...	250	2044	1292
Faeroe Islands Iles Féroé	1992	7	6	128	...	...	...
Finland Finlande	1990	382	...	...	5711	...	...
France France	1991	227	3068	54	2672	120018	2107

17
Non–daily newspapers and periodicals [*cont.*]
Journaux non quotidiens et périodiques [*suite*]

Country or area Pays ou zone	Year Année	Non–daily newspapers Journaux non quotidiens			Periodicals Périodiques		
		Number Nombre	Circulation Diffusion Total (000)	Per 1000 inhabitants Pour 1000 habitants	Number Nombre	Circulation Diffusion Total (000)	Per 1000 inhabitants Pour 1000 habitants
Germany † Allemagne †	1991	34	4871	61	8740	386907	4844
Federal Republic of Germany [8] Rép. féd. d'Allemagne [8]	1990	34	4562	...	7831	309041	5049
former German Dem. Rep. l'ancienne Rép. dém. allemande	1988	30	9431	...	1209	23872	1459
Gibraltar Gibraltar	1988	*5	*6	*200	...	...	...
Greece Grèce	1988	1051	...	...	309	...	...
Holy See Saint–Siège	1992	...	...	...	48	102	...
Hungary Hongrie	1991	279	4603	437	1203	14927	1417
Iceland Islande	1992	72	...	...	598	...	...
Ireland Irlande	1988	54	1935	548	...	...	...
Italy Italie	1992	230	1277	22	10064	85071	1472
Latvia Lettonie	1992	186	72454	27045	170	1912	714
Liechtenstein Liechtenstein	1991	3	3	100	...	...	...
Lithuania Lituanie	1992	395	2851	759	237	2602	693
Luxembourg [1] Luxembourg [1]	1990	1	3	8	508	...	...
Malta Malte	1992	8	...	...	359	...	...
Monaco Monaco	1992	...	...	...	3	38	1407
Netherlands Pays–Bas	1990	...	...	...	367	19283	1290
Norway Norvège	1992	66	358	83	7010	...	...
Poland Pologne	1992	57	1856	48	2950	54703	1424
Portugal Portugal	1990	328	...	...	*937	*6359	*644
Republic of Moldova République de Moldova	1992	212	*1269	*281	68	351	78
Romania [3] Roumanie [3]	1990	24	* 812	*35	1379	...	...
Russian Federation Fédération de Russie	1992	4498	86677	585	2592	918218	6200
San Marino Saint–Marin	1992	6	13	565	18	10	435
Slovakia Slovaquie	1992	262	2091	397	424	8725	1656
Slovenia Slovénie	1992	153	...	...	482	...	...
Spain Espagne	1988	85	*3100	*80	...	...	...
Sweden [9] Suède [9]	1992	70	414	48	46	4947	572
Switzerland [10] Suisse [10]	1991	636	...	...	*3079	...	...
TFYR Macedonia L'ex–R.y. Macédoine	1991	*112	*27562	...	*74	*347	...
Ukraine Ukraine	1992	1605	18194	*346	321	3491	*66

17
Non–daily newspapers and periodicals [*cont.*]
Journaux non quotidiens et périodiques [*suite*]

Country or area Pays ou zone	Year Année	Non–daily newspapers Journaux non quotidiens			Periodicals Périodiques		
		Number Nombre	Circulation Diffusion Total (000)	Per 1000 inhabitants Pour 1000 habitants	Number Nombre	Circulation Diffusion Total (000)	Per 1000 inhabitants Pour 1000 habitants
United Kingdom Royaume–Uni	1988	818	29047	509	...	...	...
Yugoslavia Yugoslavie	1992	599	5088	...	397	747	
Yugoslavia, SFR † Yugoslavie, Rfs †	1990	2229	21756	914	1361	4197	176
Oceania · Océanie							
American Samoa Samoa américaines	1990	2	5	102	...	...	...
Australia Australie	1988	*460	*17204	*1040	...	...	
Cook Islands Iles Cook	1988	*1	*2	*88			
Fiji Fidji	1988	*7	*99	*137	...	...	...
French Polynesia Polynésie française	1988	*1	*4	*19			
Guam Guam	1988	*4	*26	*202	...	...	
Kiribati Kiribati	1988	*2	*4	*58	...	...	...
New Caledonia Nouvelle–Calédonie	1988	*1	*5	*31	...	...	...
New Zealand Nouvelle–Zélande	1988	139	*1100	*330			
Niue Nioué	1990	1	2	900	9	10	4900
Norfolk Island Ile Norfolk	1990	1	1	...	...	...	...
Papua New Guinea Papouasie–Nouv.–Guinée	1988	*4	*78	*21	...	...	
Samoa Samoa	1988	*5	*23	*146	...	...	
Solomon Islands Iles Salomon	1988	4	12	40	...	...	
Tokelau Tokélaou	1988	*1	*2	...			
Tonga Tonga	1988	*2	*9	*96			
Tuvalu Tuvalu	1990	1	3	25	...	...	
Vanuatu Vanuatu	1990	1	2	12	...	...	...
former USSR · ancienne URSS							
former USSR † ancienne URSS †	1989	8092	92710	332	5228	5085133	18203

Source:
United Nations Educational, Scientific and Cultural Organization
(Paris).

† For detailed descriptions of data pertaining to former
Czechoslovakia, Germany, SFR Yugoslavia and former USSR,
see Annex I – Country or area nomenclature, regional and
other groupings.

1 Data on non–dailies refer to 1988.
2 Data on periodicals refer to 1988.
3 Data on non–dailies refer to 1989.
4 Data on non–dailies include daily newspapers.

Source:
Organisation des Nations Unies pour l'éducation, la science et
la culture (Paris).

† Pour les descriptions en détails des données relatives à
l'ancienne Tchécoslovaquie, l'Allemagne, la Rfs Yugoslavie et
l'ancienne URSS, voir l'Annexe I – Nomenclature des pays ou
zones, groupements régionaux et autres groupements.

1 Les données pour les non–quotidiens se réfèrent à 1988.
2 Les données relatives aux périodiques se réfèrent à 1988.
3 Les données pour les non–quotidiens se réfèrent à 1989.
4 Les données relatives aux non–quotidiens comprennent les
quotidiens.

17
Non–daily newspapers and periodicals [*cont.*]
Journaux non quotidiens et périodiques [*suite*]

5   Data on periodicals refer only to periodicals for the general public.
6   Data on non–dailies include periodicals.
7   Data on non–daily newspapers include regional editions.
8   Data on periodicals refer to 1989.
9   Data on periodicals refer to 1988 and to periodicals for the general public only.
10 Data on non–dailies include the district newspapers and the home bulletins. Figures on periodicals refer to 1988.

5   Les données relatives aux périodiques se réfèrent seulement aux périodiques destinés au grand public.
6   Les données relatives aux non–quotidiens comprennent les periodiques.
7   Les données relatives aux journaux non–quotidiens comprennent les éditions régionales.
8   Les données relatives aux périodiques se réfèrent à 1989.
9   Les données relatives aux périodiques se réfèrent à 1988 et aux périodiques destinés au grand public seulement.
10 Les données relatives aux non–quotidiens comprennent les journaux de quartier et les bulletins locaux. Les chiffres pour les périodiques se réfèrent à 1988.

18
Television and radio receivers
Postes récepteurs de télévision et de radio

Country or area Pays ou zone	Code §	Number (000) Nombre (000)				Per 1000 inhabitants Pour 1000 habitants			
		1975	1980	1985	1992	1975	1980	1985	1992
Africa · Afrique									
Algeria	T	500	975	1500	2000	31	52	69	76
Algérie	R	3000	3700	4800	6160	187	197	220	234
Angola	T	–	30	37	62	–	4	5	6
Angola	R	118	145	217	282	19	21	27	29
Benin	T	–	5	15	25	–	1	4	5
Bénin	R	150	230	300	442	49	66	75	90
Botswana	T	–	–	–	22	–	–	–	17
Botswana	R	57	75	115	160	75	83	108	122
Burkina Faso	T	9	20	37	50	2	3	5	5
Burkina Faso	R	100	125	150	255	16	18	19	27
Burundi	T	–	–	0	5	–	–	0	1
Burundi	R	100	160	250	360	27	39	53	62
Cameroon	T	–	–	–	288	–	–	–	24
Cameroun	R	232	760	1200	1775	31	88	120	146
Cape Verde	T	–	–	–	1	–	–	–	3
Cap–Vert	R	31	41	50	63	112	142	154	164
Central African Rep.	T	–	1	5	14	–	0	2	5
Rép. centrafricaine	R	70	120	150	220	34	52	57	69
Chad	T	–	–	–	8	–	–	–	1
Tchad	R	500	750	1150	1425	124	168	229	244
Comoros	T	–	–	–	0	–	–	–	0
Comores	R	36	44	56	75	114	120	123	128
Congo	T	3	4	5	14	2	2	3	6
Congo	R	81	100	138	270	56	60	72	114
Côte d'Ivoire	T	110	310	500	765	16	38	50	59
Côte d'Ivoire	R	300	1000	1300	1835	44	122	131	142
Djibouti	T	3	5	12	25	12	16	32	54
Djibouti	R	12	21	30	41	49	69	79	88
Egypt	T	620	1400	3860	6500	17	34	83	119
Egypte	R	4900	6000	12000	18000	135	147	258	328
Equatorial Guinea	T	0	1	2	4	2	5	7	10
Guinée équatoriale	R	78	87	128	155	347	401	410	420
Ethiopia	T	20	30	70	145	1	1	2	3
Ethiopie	R	1000	3000	8000	9900	29	77	186	187
Gabon	T	7	9	22	46	11	12	22	37
Gabon	R	80	105	132	177	126	130	134	143
Gambia									
Gambie	R	59	73	105	155	108	114	141	171
Ghana	T	33	57	150	250	3	5	12	16
Ghana	R	1060	1700	2500	4285	108	158	195	269
Guinea	T	–	6	8	43	–	1	2	7
Guinée	R	110	135	180	257	27	30	36	42
Guinea–Bissau									
Guinée–Bissau	R	10	25	30	40	16	31	34	40
Kenya	T	38	62	100	245	3	4	5	10
Kenya	R	400	650	1600	2200	29	39	81	87
Lesotho	T	–	–	1	11	–	–	0	6
Lesotho	R	22	33	44	60	19	25	29	33
Liberia	T	9	21	35	51	6	11	16	18
Libéria	R	264	335	475	622	164	179	216	226
Libyan Arab Jamahiriya	T	85	168	235	485	35	61	62	100
Jamah. arabe libyenne	R	110	200	800	1100	45	66	211	226
Madagascar	T	8	45	100	260	1	5	10	20
Madagascar	R	720	1600	1950	2565	95	182	190	200
Malawi									
Malawi	R	170	260	1500	2285	32	42	204	221
Mali	T	–	–	1	11	–	–	0	1
Mali	R	81	105	250	430	13	15	32	44
Mauritania	T	–	–	1	50	–	–	0	23
Mauritanie	R	82	150	250	309	60	97	142	144
Mauritius	T	50	92	140	239	56	95	137	218
Maurice	R	162	260	335	395	182	269	328	360

18
Television and radio receivers [cont.]
Postes récepteurs de télévision et de radio [suite]

Country or area Pays ou zone	Code §	Number (000) Nombre (000)				Per 1000 inhabitants Pour 1000 habitants			
		1975	1980	1985	1992	1975	1980	1985	1992
Morocco	T	535	890	1370	1950	31	46	62	74
Maroc	R	1400	3000	3850	5527	81	155	175	210
Mozambique	T	1	2	7	44	0	0	1	3
Mozambique	R	200	254	450	700	19	21	33	47
Namibia	T	–	5	16	32	–	5	13	21
Namibie	R	...	...	150	195	...	...	121	127
Niger	T	0	5	12	38	0	1	2	5
Niger	R	120	250	300	500	25	45	45	61
Nigeria	T	100	550	1000	3800	2	7	11	33
Nigéria	R	4500	7000	14500	20000	68	89	158	173
Réunion	T	41	81	89	102	85	160	161	164
Réunion	R	91	100	122	151	188	198	222	242
Rwanda									
Rwanda	R	65	175	335	485	15	34	56	64
Saint Helena									
Sainte–Hélène	R	1	1	2	2	180	280	315	330
Sao Tome and Principe									
Sao Tomé–et–Principe	R	17	23	27	33	210	245	255	269
Senegal	T	2	8	200	282	0	1	31	37
Sénégal	R	290	360	700	890	60	65	110	115
Seychelles	T	–	–	2	6	–	–	30	85
Seychelles	R	16	21	25	34	271	339	373	472
Sierra Leone	T	8	20	30	45	3	6	8	10
Sierra Leone	R	280	450	775	980	96	138	211	224
Somalia	T	–	–	1	113	–	–	0	12
Somalie	R	68	112	200	350	12	17	25	38
South Africa	T	100	2010	3000	3900	4	68	89	98
Afrique du Sud	R	2400	8000	10000	12100	93	271	298	304
Sudan	T	100	800	1100	2060	6	43	50	77
Soudan	R	2000	3500	5375	6670	125	187	246	250
Swaziland	T	–	1	8	16	–	2	11	20
Swaziland	R	55	81	101	129	114	143	153	163
Togo	T	1	10	15	24	0	4	5	6
Togo	R	350	530	630	795	153	203	208	211
Tunisia	T	191	300	400	670	34	47	55	80
Tunisie	R	775	1000	1185	1680	138	157	163	200
Uganda	T	59	72	90	193	5	6	6	10
Ouganda	R	250	400	1250	2040	22	30	83	109
United Rep. of Tanzania	T	4	7	8	45	0	0	0	2
Rép.–Unie de Tanzanie	R	232	290	365	640	15	16	17	25
Western Sahara	T	2	2	4	5	15	18	21	21
Sahara occidental	R	15	22	33	46	128	169	179	184
Zaire	T	7	10	13	55	0	0	0	1
Zaïre	R	1000	1500	2800	3870	43	56	88	97
Zambia	T	23	60	90	225	5	10	13	26
Zambie	R	100	135	500	705	21	24	73	82
Zimbabwe	T	60	73	178	280	10	10	21	27
Zimbabwe	R	180	240	500	890	29	34	60	84
America, North · Amérique du Nord									
Anguilla									
Anguilla	R	...	...	...	2	...	...	...	310
Antigua and Barbuda	T	15	16	19	24	238	262	302	356
Antigua–et–Barbuda	R	15	17	21	28	238	279	333	417
Bahamas	T	–	31	51	60	–	148	219	225
Bahamas	R	90	102	120	143	484	486	515	542
Barbados	T	46	52	60	72	187	209	237	280
Barbade	R	94	135	200	227	382	542	791	876
Belize	T	–	–	–	33	–	–	–	166
Belize	R	59	71	88	115	457	486	530	584
Bermuda	T	20	30	45	57	377	556	804	923
Bermudes	R	50	60	68	78	943	1111	1214	1260
British Virgin Islands	T	1	2	3	4	64	167	214	221
Iles Vierges brit.	R	5	6	7	9	455	458	486	502

18
Television and radio receivers [cont.]
Postes récepteurs de télévision et de radio [suite]

Country or area Pays ou zone	Code §	Number (000) Nombre (000)				Per 1000 inhabitants Pour 1000 habitants			
		1975	1980	1985	1992	1975	1980	1985	1992
Canada	T	9200	10617	14028	17515	405	441	557	640
Canada	R	16250	17734	23237	28200	715	737	923	1030
Cayman Islands	T	–	2	4	6	–	106	190	193
Iles Caïmanes	R	4	12	19	27	286	706	910	934
Costa Rica	T	128	155	200	450	65	68	76	141
Costa Rica	R	151	190	650	823	77	83	246	258
Cuba	T	595	1273	1940	1750	64	132	192	162
Cuba	R	1805	2914	3282	3732	194	301	326	345
Dominica	T	–	–	–	5	–	–	–	72
Dominica	R	6	31	38	42	81	413	521	589
Dominican Republic	T	180	400	500	650	36	70	78	87
République dominicaine	R	770	900	1020	1280	153	158	159	171
El Salvador	T	135	300	350	501	33	66	74	93
El Salvador	R	1100	1550	1900	2230	269	343	401	413
Greenland	T	3	4	7	11	60	70	132	200
Groënland	R	13	15	19	23	260	300	358	413
Grenada	T	–	–	–	30	–	–	–	331
Grenade	R	22	35	45	54	239	393	500	598
Guadeloupe	T	18	37	77	106	55	113	217	265
Guadeloupe	R	35	50	80	92	106	153	225	230
Guatemala	T	110	175	207	510	18	25	26	52
Guatemala	R	300	350	450	645	50	51	57	66
Haiti	T	13	16	21	32	3	3	4	5
Haïti	R	93	105	140	320	19	20	24	47
Honduras	T	34	65	280	400	11	18	64	73
Honduras	R	142	500	1600	2115	46	137	365	387
Jamaica	T	110	170	215	330	55	80	93	134
Jamaïque	R	550	800	920	1040	273	375	398	421
Martinique	T	22	38	44	50	67	117	129	136
Martinique	R	57	62	67	74	173	190	196	201
Mexico	T	2700	3820	8500	13100	46	57	112	149
Mexique	R	6900	9000	15000	22500	117	134	198	255
Montserrat	T	–	–	–	2	–	–	–	148
Montserrat	R	5	6	6	6	413	458	545	578
Netherlands Antilles	T	35	43	58	64	211	251	331	363
Antilles néerlandaises	R	132	175	190	205	795	1023	1086	1165
Nicaragua	T	83	160	190	260	34	57	59	66
Nicaragua	R	400	670	802	1037	165	239	248	262
Panama	T	185	225	350	420	106	115	161	167
Panama	R	260	300	400	564	149	153	183	224
Puerto Rico	T	630	725	850	952	210	226	252	265
Porto Rico	R	1760	2000	2300	2565	588	624	681	714
Saint Kitts and Nevis	T	4	4	5	9	89	91	116	206
Saint–Kitts–et–Nevis	R	...	...	21	27	...	...	488	648
Saint Lucia	T	2	9	17	26	16	80	135	190
Sainte–Lucie	R	70	81	92	104	648	704	738	759
Saint Pierre and Miquelon	T	2	3	4	4	283	533	600	653
Saint–Pierre–et–Miquelon	R	2	4	4	4	350	583	633	699
Saint Vincent and Grenadines	T	4	5	6	16	46	53	59	144
Saint–Vincent–et–Grenadines	R	32	42	55	76	344	429	539	698
Trinidad and Tobago	T	105	210	320	400	104	194	276	316
Trinité–et–Tobago	R	225	300	500	625	222	277	431	494
Turks and Caicos Is.									
Iles Turques et Caïques	R	3	4	5	7	500	506	510	513
United States	T	121000	155800	190000	208000	560	684	797	815
Etats–Unis	R	401000	454500	500000	540500	1857	1996	2097	2118
US Virgin Islands	T	30	50	59	65	316	510	551	609
Iles Vierges américaines	R	75	82	93	105	789	837	869	981
America, South · Amérique du Sud									
Argentina	T	4000	5140	6500	7300	154	182	214	221
Argentine	R	9890	12000	18000	22600	380	425	593	683
Bolivia	T	45	300	420	775	9	54	66	103
Bolivie	R	1150	2800	3675	4610	235	502	579	613

18
Television and radio receivers [*cont.*]
Postes récepteurs de télévision et de radio [*suite*]

Country or area Pays ou zone	Code §	Number (000) Nombre (000)				Per 1000 inhabitants Pour 1000 habitants			
		1975	1980	1985	1992	1975	1980	1985	1992
Brazil	T	8400	15000	25000	32000	78	124	184	208
Brésil	R	16980	38000	49000	59500	157	313	361	386
Chile	T	700	1225	1750	2850	68	110	144	210
Chili	R	1700	3250	4000	4680	164	292	330	344
Colombia	T	1600	2250	2750	3900	67	85	93	117
Colombie	R	2808	3300	4000	5900	118	124	136	177
Ecuador	T	252	500	600	940	36	62	64	85
Equateur	R	2000	2425	2850	3510	284	299	306	318
Falkland Islands (Malvinas) Iles Falkland (Malvinas)	R	1	1	1	1	300	400	500	503
French Guiana	T	6	13	17	22	107	188	201	212
Guyane française	R	14	30	59	79	241	435	705	733
Guyana	T	–	–	–	32	–	–	–	40
Guyana	R	266	310	355	398	362	408	449	493
Paraguay	T	54	68	85	370	20	22	23	82
Paraguay	R	185	350	600	775	69	111	162	171
Peru	T	610	895	1500	2200	40	52	77	98
Pérou	R	2050	2750	4000	5700	135	159	206	254
Suriname	T	34	40	45	58	93	114	117	132
Suriname	R	110	189	230	280	302	537	601	639
Uruguay	T	351	368	500	725	124	126	166	232
Uruguay	R	1500	1630	1760	1890	530	559	585	604
Venezuela	T	1284	1710	2250	3300	101	114	131	163
Venezuela	R	4775	5900	7000	9040	377	393	408	448
Asia · Asie									
Afghanistan	T	–	45	100	160	–	3	7	8
Afghanistan	R	750	1200	1450	2045	49	75	100	107
Bahrain	T	30	90	170	222	110	259	396	416
Bahreïn	R	85	125	210	287	313	360	490	538
Bangladesh	T	25	80	261	570	0	1	3	5
Bangladesh	R	700	1500	4000	5189	9	17	40	44
Bhutan Bhoutan	R	4	7	18	26	4	6	13	16
Brunei Darussalam	T	14	26	45	64	87	135	199	237
Brunéi Darussalam	R	24	41	55	73	149	212	243	270
Cambodia	T	30	35	52	73	4	5	7	8
Cambodge	R	110	600	800	985	15	92	109	112
China	T	1185	4000	10000	37000	1	4	9	31
Chine	R	15000	55000	120000	216500	16	55	112	182
Cyprus	T	59	85	92	107	97	135	138	149
Chypre	R	115	162	190	210	189	258	286	293
East Timor Timor oriental	R	3	4	5	6	5	7	7	7
Hong Kong	T	837	1114	1275	1630	190	221	234	281
Hong−kong	R	2200	2550	3250	3875	500	506	596	668
India	T	515	3000	10000	33000	1	4	13	37
Inde	R	17228	26000	50000	70000	28	38	65	80
Indonesia	T	300	3000	6438	11500	2	20	38	60
Indonésie	R	5010	15000	21500	28100	37	99	128	147
Iran, Islamic Rep. of	T	1500	2000	2600	3900	45	51	53	63
Iran, Rép. islamique d'	R	2200	6400	10000	14300	66	163	204	232
Iraq	T	415	650	900	1400	38	50	59	73
Iraq	R	1252	2100	3000	4165	114	161	196	216
Israel	T	680	900	1100	1390	197	232	260	271
Israël	R	595	950	1800	2415	172	245	425	471
Japan	T	40000	62976	70000	76500	359	539	579	614
Japon	R	58026	79200	95000	113000	520	678	786	908
Jordan	T	120	171	240	350	46	59	70	82
Jordanie	R	450	550	791	1100	173	188	232	256
Korea, Dem.People's Rep.	T	100	130	200	400	6	7	10	18
Corée, Rép. pop. dém. de	R	750	1500	2000	2750	45	82	101	122
Korea, Republic of	T	2500	6300	7721	9300	71	165	189	211
Corée, République de	R	13509	20000	38605	44250	383	525	946	1002

18
Television and radio receivers [*cont.*]
Postes récepteurs de télévision et de radio [*suite*]

Country or area Pays ou zone	Code §	Number (000) Nombre (000)				Per 1000 inhabitants Pour 1000 habitants			
		1975	1980	1985	1992	1975	1980	1985	1992
Kuwait	T	150	353	450	610	149	257	262	310
Koweït	R	203	390	535	720	202	284	311	365
Lao People's Dem. Rep.	T	–	–	–	28	–	–	–	6
République dém. pop. lao	R	150	350	430	560	50	109	120	125
Lebanon	T	410	750	800	920	148	201	300	324
Liban	R	1321	2000	2050	2370	477	749	768	835
Macau	T	–	–	–	34	–	–	–	69
Macao	R	61	80	99	135	235	248	253	276
Malaysia	T	452	1200	1800	2820	37	87	115	150
Malaisie	R	1420	5650	6600	8080	116	411	421	430
Maldives	T	–	1	3	6	–	7	17	25
Maldives	R	3	7	19	27	20	44	104	119
Mongolia	T	4	6	45	93	2	3	24	40
Mongolie	R	105	160	200	306	73	96	105	132
Myanmar	T	–	1	20	88	–	0	1	2
Myanmar	R	662	774	2500	3580	22	23	67	82
Nepal	T	–	–	20	45	–	–	1	2
Népal	R	113	300	450	690	9	20	26	34
Oman	T	3	35	900	1195	3	35	713	730
Oman	R	15	300	800	1043	20	304	633	637
Pakistan	T	380	938	1304	2300	5	11	13	18
Pakistan	R	4000	5500	8500	11300	54	64	84	91
Philippines	T	756	1050	1500	2900	18	22	27	45
Philippines	R	1800	2100	5000	9030	42	43	90	139
Qatar	T	20	76	120	205	117	331	335	452
Qatar	R	50	90	150	201	292	393	419	444
Saudi Arabia	T	800	2100	3050	4260	110	224	246	268
Arabie saoudite	R	950	2500	3550	4840	131	267	287	304
Singapore	T	421	750	850	1050	186	311	332	379
Singapour	R	725	900	1550	1790	320	373	606	646
Sri Lanka	T	–	35	450	865	–	2	28	49
Sri Lanka	R	700	1454	2551	3525	51	98	158	200
Syrian Arab Republic	T	224	385	600	810	30	44	58	61
Rép. arabe syrienne	R	1400	1700	2200	3392	188	195	213	255
Thailand	T	500	1000	4122	6400	12	21	81	114
Thaïlande	R	4000	6550	8000	10750	97	140	156	192
Turkey	T	1050	3500	8000	10250	26	79	159	176
Turquie	R	4200	5000	7000	9425	105	113	139	161
United Arab Emirates	T	25	89	130	185	50	88	96	111
Emirats arabes unis	R	52	240	380	545	103	236	282	326
Viet Nam	T	...	...	2000	2900	...	...	33	42
Viet Nam	R	3500	5000	6000	7215	73	93	100	104
Yemen	T	..	..	..	350	..	..	..	28
Yémen	R	..	..	..	350	..	..	..	28
former Dem. Yemen	T	31	35	41	..	19	19	19	..
ancienne Yémen dém.	R	90	118	150	..	54	63	70	..
former Yemen Arab Rep.	T	–	5	28	..	–	1	4	..
ancienne Yémen rép. arabe	R	87	110	150	..	16	17	20	..
Europe · Europe									
Albania	T	5	96	232	290	2	36	78	88
Albanie	R	175	400	493	583	72	150	166	176
Andorra	T	2	4	6	20	74	114	133	428
Andorre	R	5	7	9	10	185	189	196	223
Austria	T	2550	2950	3260	3735	336	391	431	480
Autriche	R	3600	3830	4180	4800	475	507	553	617
Belarus	T	1600	2100	2500	2770	170	219	250	...
Bélarus	R	1450	2145	2900	3180	154	223	290	...
Belgium	T	3315	3815	3950	4530	338	387	401	453
Belgique	R	6200	7200	7450	7690	633	731	756	769
Bulgaria	T	1960	2150	2220	2300	225	243	248	257
Bulgarie	R	3150	3500	3750	3980	361	395	419	445
Croatia	T	..	940	976	1520	..	...	...	...
Croatie	R	..	1100	1125	1350	..	...	...	...

18
Television and radio receivers [*cont.*]
Postes récepteurs de télévision et de radio [*suite*]

Country or area Pays ou zone	Code §	Number (000) Nombre (000)				Per 1000 inhabitants Pour 1000 habitants			
		1975	1980	1985	1992	1975	1980	1985	1992
former Czechoslovakia †	T	4987	5900	6080	..	337	385	392	..
ancienne Tchècoslovaquie †	R	7425	7800	8050	..	502	509	519	..
Czech Republic	T	..	..	..	4900	..	..	..	...
Réplublic Tchèque	R	..	..	..	6200	..	..	..	...
Denmark	T	2110	2550	2675	2770	417	498	522	537
Danemark	R	4625	4750	4875	5330	914	927	952	1033
Estonia	T	..	..	..	555	..	..	..	351
Estonie	R	..	..	..	710	..	..	..	449
Faeroe Islands	T	—	4	9	12	—	93	196	247
Iles Féroé	R	15	18	22	24	366	407	478	514
Finland	T	1658	1980	2300	2530	352	414	469	505
Finlande	R	2300	4000	4830	4995	488	837	985	997
France	T	15000	19000	21500	23300	285	353	390	408
France	R	30000	39900	48000	50850	569	741	870	889
Germany †	T	..	..	..	44800	..	..	..	558
Allemagne †	R	..	..	..	71000	..	..	..	885
Federal Republic of Germany	T	25000	27000	29500	..	404	439	483	..
Rép. féd. d'Allemagne	R	52500	55000	57500	..	849	893	942	..
former German Dem. Rep.	T	7835	8600	12300	..	465	514	739	..
ancienne Rép. dém. allemande	R	8650	8970	10250	..	513	536	616	..
Gibraltar	T	6	7	8	10	200	232	272	316
Gibraltar	R	30	33	34	36	1000	1087	1158	1173
Greece	T	1155	1650	1896	2050	128	171	191	201
Grèce	R	2500	3310	4000	4290	276	343	403	421
Hungary	T	2850	3320	4250	4350	271	310	399	414
Hongrie	R	4100	5340	6144	6295	389	499	577	599
Iceland	T	56	65	75	83	257	285	311	319
Islande	R	140	162	185	205	642	711	768	788
Ireland	T	600	785	910	1060	189	231	256	304
Irlande	R	907	1275	2050	2220	285	375	577	637
Italy	T	15000	22000	23600	24350	271	390	413	421
Italie	R	32500	34000	37000	45734	586	602	648	791
Latvia	T	..	..	..	1200	..	..	..	448
Lettonie	R	..	..	..	1600	..	..	..	597
Liechtenstein	T	5	7	9	10	188	269	322	345
Liechtenstein	R	7	13	18	19	308	500	656	686
Lithuania	T	..	..	..	1410	..	..	..	375
Lituanie	R	..	..	..	1427	..	..	..	380
Luxembourg	T	87	90	92	101	240	247	251	267
Luxembourg	R	179	200	228	240	494	549	621	635
Malta	T	176	202	235	267	579	624	683	744
Malte	R	141	165	178	189	464	509	517	526
Monaco	T	16	17	19	22	640	654	704	820
Monaco	R	20	26	29	30	800	1004	1056	1126
Netherlands	T	4230	5650	6700	7400	310	399	463	488
Pays—Bas	R	7710	9200	12000	13755	565	650	829	907
Norway	T	1252	1430	1640	1820	312	350	395	424
Norvège	R	2575	2700	3230	3410	643	661	778	795
Poland	T	7120	8750	10400	11350	209	246	280	295
Pologne	R	9700	10615	12250	16700	285	298	329	435
Portugal	T	860	1540	1765	1860	95	158	178	188
Portugal	R	1530	1660	2000	2260	168	170	202	229
Romania	T	2950	4085	4375	4580	139	184	193	196
Roumanie	R	3725	3930	4100	4640	175	177	180	199
Russian Federation	T	..	..	..	54850	..	..	..	370
Fédération de Russie	R	..	..	..	48500	..	..	..	327
San Marino	T	4	6	7	8	211	300	314	352
Saint—Marin	R	6	10	12	14	316	476	527	590
Slovakia	T	..	..	..	2500	..	..	..	...
Slovaquie	R	..	..	..	3000	..	..	..	569
Slovenia	T	360	460	500	570	...	...	...	...
Slovénie	R	450	500	650	720	...	...	...	* 367
Spain	T	6640	9505	10400	15700	187	253	270	402
Espagne	R	9050	9700	11300	12200	254	258	294	312

18
Television and radio receivers [cont.]
Postes récepteurs de télévision et de radio [suite]

Country or area / Pays ou zone	Code §	Number (000) / Nombre (000)				Per 1000 inhabitants / Pour 1000 habitants			
		1975	1980	1985	1992	1975	1980	1985	1992
Sweden	T	3750	3830	3875	4060	458	461	464	469
Suède	R	4200	7000	7250	7590	513	842	868	877
Switzerland	T	2050	2300	2550	2770	323	364	394	407
Suisse	R	4500	5140	5420	5740	710	813	838	843
The Former Yugoslav Rep. of Macedonia	T	220	262	310	338	...	...	...	...
L'ex–Rép. Yougoslave de Macédoine	R	..	..	..	369	..	..	..	...
Ukraine	T	11035	12722	15190	17300	...	...	...	...
Ukraine	R	23539	28941	36434	41500	...	...	...	...
United Kingdom	T	20200	22600	24500	25100	359	401	433	435
Royaume–Uni	R	39000	53500	57000	66100	694	950	1007	1146
Yugoslavia	T	..	..	..	1800	..	..	..	...
Yougoslavie	R	..	..	..	2000	..	..	..	...
Yugoslavia, SFR †	T	3385	4245	4500	..	159	190	195	..
Yougoslavie, Rfs †	R	4415	5000	5580	..	207	224	241	..
Oceania · Océanie									
American Samoa	T	5	6	7	11	167	173	179	216
Samoa américaines	R	26	32	38	50	867	991	974	1007
Australia	T	4549	5600	7000	8480	334	381	444	482
Australie	R	13900	16000	19500	22400	1020	1089	1237	1273
Cook Islands	T	–	–	–	3	–	–	–	188
Iles Cook	R	7	8	10	13	368	417	578	740
Fiji	T	–	–	–	12	–	–	–	16
Fidji	R	200	300	370	455	347	473	529	616
French Polynesia	T	14	18	27	35	108	121	155	170
Polynésie française	R	65	78	93	116	500	517	532	558
Guam	T	50	63	76	92	526	589	633	658
Guam	R	84	100	150	195	883	935	1250	1403
Kiribati									
Kiribati	R	8	12	13	15	148	203	204	207
Nauru									
Nauru	R	3	4	5	6	400	488	588	576
New Caledonia	T	15	25	40	47	113	179	260	269
Nouvelle–Calédonie	R	55	69	80	97	414	489	519	562
New Zealand	T	958	1035	1397	1530	311	332	430	443
Nouvelle–zélande	R	2670	2755	2950	3215	866	885	909	931
Niue									
Nioué	R	1	1	1	1	163	317	358	563
Norfolk Island	T	–	–	–	1	–	–	–	* 1
Ile Norfolk	R	1	2	2	2	635	1000	1000	* 1000
Palau [1]	T	3	6	8	9	25	44	51	* 52
Palaos [1]	R	72	85	98	135	595	625	636	...
Papua New Guinea	T	–	–	–	10	–	–	–	3
Papouasie–Nouv.–Guinée	R	110	180	230	298	40	58	66	73
Samoa	T	–	3	5	6	–	16	32	40
Samoa	R	23	32	68	76	152	206	433	481
Solomon Islands	T	–	–	–	2	–	–	–	6
Iles Salomon	R	9	20	27	41	47	88	100	120
Tokelau									
Tokélaou	R	...	...	1	1	...	...	500	...
Tonga	T	–	–	–	1	–	–	–	10
Tonga	R	11	20	30	54	125	217	319	555
Tuvalu									
Tuvalu	R	1	2	2	3	175	206	211	230
Vanuatu	T	–	–	–	2	–	–	–	10
Vanuatu	R	13	23	36	45	130	198	267	287
former USSR · ancienne URSS									
former USSR †	T	55200	76500	82400	..	223	296	305	..
ancienne URSS †	R	122477	130000	182800	..	495	504	678	..

18
Television and radio receivers [*cont.*]
Postes récepteurs de télévision et de radio [*suite*]

Source:
United Nations Educational, Scientific and Cultural Organization
(Paris).

† For detailed descriptions of data pertaining to former
Czechoslavakia, Germany, SFR Yugoslavia and former USSR,
see Annex I − − Country or nomenclature, regional and other
groupings.

§ T: Estimated number of television receivers in use.
R: Estimated number of radio receivers in use.

1 Including data for Federated States of Micronesia,
Marshall Islands and Northern Mariana Islands.

Source:
Organisation des Nations Unies pour l'éducation, la science et la culture
(Paris).

† Pour les descriptions en détails des données relatives à l'ancienne
Tchécoslovaquie, l'Allemagne, la Rfs Yougoslavie et l'ancienne USSR,
voir l'Annexe I − − Nomenclature des pays ou zones, groupements
régionaux et autres groupements.

§ T: Estimation du nombre de récepteurs de télévision en service.
R: Estimation du nombre de récepteurs de radio en service.

1 Y compris les données pour les Etats fédérés de Micronésie,
les îles Marshall et les îles Mariannes du Nord.

19
Telefax stations and mobile cellular telephone subscribers
Postes de télécopie et abonnés au téléphone mobile cellulaire

Country or area	Telefax stations Postes de télécopie					Mobile cellular telephone subscribers Abonnés au téléphone mobile cellulaire				
Pays ou zone	1989	1990	1991	1992	1993	1989	1990	1991	1992	1993
Africa · Afrique										
Algeria Algérie	...	...	...	7000	...	0	470	4781	4781	4781
Angola Angola	...	...	...	...	...	0	0	0	0	1100
Benin Bénin	74	144	184	300	...	0	0	0	0	0
Botswana [1] Botswana [1]	...	820	...	235	...	0	0	0	0	0
Burkina Faso Burkina Faso	...	...	...	...	...	0	0	0	0	0
Burundi Burundi	...	600	...	...	...	0	0	0	0	320
Cameroon Cameroun	...	...	...	...	...	0	0	0	0	...
Cape Verde Cap−Vert	...	...	300	...	...	0	0	0	0	0
Central African Rep. Rép. centrafricaine	...	...	...	...	...	0	0	0	0	0
Chad Tchad	...	72	93	128	...	0	0	0	0	0
Comoros Comores	...	27	57	76	...	0	0	0	0	0
Congo Congo	110	...	...	110	...	0	0	0	0	0
Cote d'Ivoire Côte d' Ivoire	...	...	...	...	...	0	0	0	0	0
Djibouti Djibouti	...	101	128	124	...	0	0	0	0	0
Egypt [2] Egypte [2]	...	7620	11695	15629	...	3619	4000	5240	6944	7550
Equatorial Guinea Guinée équatoriale	...	...	...	...	...	0	0	0	0	...
Ethiopia [2] Ethiopie [2]	111	233	328	521	...	0	0	0	0	0
Gabon Gabon	...	191	...	...	...	0	0	0	280	1200
Gambia [1] Gambie [1]	115	190	361	365	...	0	0	0	204	400
Ghana Ghana	230	1921	2677	3817	...	0	0	0	400	1742
Guinea Guinée	...	...	...	...	...	0	0	0	0	42
Guinea−Bissau Guinée−Bissau	...	...	300	...	...	...	0	0	0	...
Kenya [2] Kenya [2]	...	203	...	203	...	0	0	0	1100	1162
Lesotho Lesotho	...	283	338	447	...	0	0	0	0	...
Liberia Libéria	...	...	...	...	...	0	0	0	...	...
Libyan Arab Jamahiriya Jamah. arabe libyenne	...	...	...	...	...	0	0	0	0	...
Madagascar Madagascar	...	...	...	...	...	0	0	0	0	0
Malawi Malawi	298	342	411	680	...	0	0	0	0	0
Mali Mali	80	...	...	...	...	0	0	0	0	0
Mauritania Mauritanie	...	300	...	...	...	0	0	0	0	0
Mauritius Maurice	497	...	1400	1800	...	...	2200	2500	2912	4037
Mayotte Mayotte	...	100	...	...	...	...	0	0	0	...

19
Telefax stations and mobile cellular telephone subscribers [*cont.*]
Postes de télécopie et abonnés au téléphone mobile cellulaire [*suite*]

Country or area Pays ou zone	Telefax stations Postes de télécopie					Mobile cellular telephone subscribers Abonnés au téléphone mobile cellulaire				
	1989	1990	1991	1992	1993	1989	1990	1991	1992	1993
Morocco Maroc	...	...	...	5000	...	700	904	1500	3228	6725
Mozambique Mozambique	782	...	...	1237		0	0	0	0	...
Namibia [1] Namibie [1]	...	...	...	...		0	0	0	0	0
Niger Niger	...	150	207	268	...	0	0	0	0	0
Nigeria Nigéria	...	...	...	...		...	0	0	0	9049
Reunion Réunion	672	1476	1906	...		...	...	2735		
Rwanda Rwanda	...	300	321	481		0	0	0	0	...
Sao Tome and Principe Sao Tomé−et−Principe	...	...	31	91		0	0	0	0	
Senegal Sénégal	...	...	...	...		0	28	0	0	0
Seychelles [1] Seychelles [1]	...	275	347	439	502	0	0	0	0	0
Sierra Leone Sierra Leone	...	23	48	152		0	0	0	0	0
Somalia Somalie	...	...	...	...		...	0	0	0	...
South Africa [1] Afrique du Sud [1]	...	...	...	50000	...	3980	5680	7100	12510	40000
Sudan Soudan	...	...	258	...		0	0	0	0	...
Swaziland [1] Swaziland [1]	...	341	688	793		0	0	0	0	0
United Rep.Tanzania Rép. Unie de Tanzanie	500	600	900	2000	...	0	0	0	0	0
Togo Togo	82	335	430	590		0	0	0	0	0
Tunisia Tunisie	...	25000	5000	10000	...	...	953	1239	1889	2269
Uganda [2] Ouganda [2]	142	698	...	1600	...	0	0	0	0	0
Zaire Zaïre	...	...	...	...		0	0	0	0	...
Zambia [1] Zambie [1]	324	460	460	460	460	0	0	0	0	0
Zimbabwe [2] Zimbabwe [2]	...	1487	2000	2000		0	0	0	0	0
America, North · Amérique du Nord										
Antigua and Barbuda [1] Antigua−et−Barbuda [1]	...	350	...	...	...	...	...	...	...	...
Bahamas Bahamas	490	522	569	524		...	1922	2020	2600	...
Barbados [1] Barbade [1]	600	1300	1500	1503		...	0	486	796	...
Belize [1] Belize [1]	...	...	253	381	...	0	0	0	0	200
Bermuda [1] Bermudes [1]	...	...	...	...	...	...	1112	1440	1936	3400
Canada Canada	200000	300000	450000	500000	...	370000	583000	786000	1022754	1326390
Costa Rica Costa Rica	...	1707	2138	2177	...	0	0	0	3008	4533
Cuba Cuba	...	207	300	392		0	0	...	234	...
Dominica Dominique	...	220	238	290		...	0	0	0	...
Dominican Republic Rép. dominicaine	...	...	2187	2061		...	3166	5605	7190	11020

19
Telefax stations and mobile cellular telephone subscribers [*cont.*]
Postes de télecopie et abonnés au téléphone mobile cellulaire [*suite*]

Country or area	Telefax stations Postes de télécopie					Mobile cellular telephone subscribers Abonnés au téléphone mobile cellulaire				
Pays ou zone	1989	1990	1991	1992	1993	1989	1990	1991	1992	1993
El Salvador										
El Salvador	...	3500	...	...	...	0	0	0	0	1632
Grenada										
Grenade	...	150	217	272	...	0	150	147	181	282
Guadeloupe										
Guadeloupe	...	291	...	...	...	...	814	...	...	...
Guatemala										
Guatemala	...	1000	3000	4000	...	0	293	1221	2141	2990
Haiti										
Haïti	...	...	...	...	...	0	0	0	...	
Honduras										
Honduras	...	...	...	...	...	0	0	0	0	0
Jamaica [1]										
Jamaïque [1]	...	...	...	1567	...	0	0	2512	7910	13986
Martinique										
Martinique	...	690	...	...	...	...	802	...	...	...
Mexico										
Mexique	...	...	...	150000	...	8500	34944	170080	311510	385341
Netherlands Antilles										
Antilles néerlandaises	...	...	...	...	...	0	...	...	...	...
Nicaragua										
Nicaragua	...	...	...	...	...	0	0	0	0	...
Panama										
Panama	...	...	...	...	...	0	0	0	0	0
Puerto Rico										
Porto Rico	...	...	...	...	...	...	20369	33410	51115	68410
Saint Kitts and Nevis [1]										
Saint−Kitts−et−Nevis [1]	...	...	...	...	...	0	...	...	...	...
Saint Lucia [1]										
Sainte−Lucie [1]	...	...	...	...	...	0	...	...	...	...
St. Vincent and Grenadines [1]										
St. Vincent−et−Grenadines [1]	...	...	244	336	...	0	0	...	70	83
Trinidad and Tobago										
Trinité−et−Tobago	...	1806	...	1480	...	0	0	426	1277	1679
United States										
Etats−Unis	1500000	2500000	4500000	6000000	...	3508944[3]	5283055[3]	7557148[3]	11032753[3]	16009461[3]
U.S. Virgin Islands										
Iles Vierges américaines	...	...	...	...	...	...	2000	...	2000	...
America, South · Amérique du Sud										
Argentina [4]										
Argentine [4]	...	...	...	25000	...	2300	12000	25000	46590	125780
Bolivia										
Bolivie	...	...	...	...	...	0	0	500	1556	2651
Brazil										
Brésil	55000	90000	125000	160000	...	0	667	6700	30729	180186
Chile										
Chili	...	...	...	12457	...	4886	13921	18679	64438	85186
Colombia										
Colombie	26000	35000	44000	53000	...	0	0	0	0	0
Ecuador										
Equateur	...	...	...	25000	...	0	0	0	0	0
French Guiana										
Guyane française	109	185	...	...	...	0	0	0	...	
Guyana										
Guyana	...	195	...	...	...	0	0	0	841	1029
Paraguay										
Paraguay	478	1189	1493	1691	...	0	0	0	1500	1500
Peru										
Pérou	1680	1814	1948	2082	...	0	3200	4788	27072	36900
Suriname										
Suriname	137	200	244	250	...	...	0	0	0	1400
Uruguay										
Uruguay	...	...	...	7000	...	0	0	0	1712	4969
Venezuela										
Venezuela	...	...	...	15000	...	3685	7422	14190	78560	182600

19
Telefax stations and mobile cellular telephone subscribers [*cont.*]
Postes de télécopie et abonnés au téléphone mobile cellulaire [*suite*]

Country or area / Pays ou zone	Telefax stations / Postes de télécopie					Mobile cellular telephone subscribers / Abonnés au téléphone mobile cellulaire				
	1989	1990	1991	1992	1993	1989	1990	1991	1992	1993
Asia · Asie										
Afghanistan / Afghanistan	...	...	...	...	...	0	0	0	0	...
Armenia / Arménie	...	...	...	...	...	0	0	0		...
Azerbaijan / Azerbaïdjan	...	...	...	...	...	0	0	0	0	100
Bahrain / Bahreïn	1838	2800	3500	3798	...	4251	6900	7354	9683	11360
Bangladesh [2] / Bangladesh [2]	...	...	...	1440	...	0	0	0	250	500
Bhutan / Bhoutan	...	...	...	130		...	0	0	0	0
Brunei Darussalam / Brunéi Darussalam	1051	1145	1221	1233		...	...	3025	4103	8304
Cambodia / Cambodge	...	...	...	...		0	0	0		
China / Chine	35000	39268	57899	89271	...	9805	18319	47544	176943	638000
Cyprus / Chypre	1854	3000	...	6000		1348	3156	5131	9739	15288
Korea, Dem. People's Rep. / Corée, Rép. pop. dém. de	...	...	...	...	...	0	0	0	0	
Georgia / Géorgie	...	...	...	...	...	...	0	0	0	0
Hong Kong [1] / Hong–kong [1]	88550	111239	138616	196872	233546	98530	141400	202000	233324	285600
India / Inde	1500	5000	10000	35000	...	0	0	0	0	0
Indonesia / Indonésie	7141	15000	20000	35000	...	12928	18096	24528	35546	53438
Iran, Islamic Rep. of [1] / Iran, Rép. islamique d' [1]	400	13000	...	22015		0	0	0	0	
Iraq / Iraq	...	...	...	...		0	0	0	0	...
Israel / Israël	26750	35000	55000	70000	...	...	15240	23000	36104	64484
Japan [1] / Japon [1]	3500000	4000000	5000000	5500000	...	489558	868078	1378108	1712545	2008000
Jordan / Jordanie	...	...	...	6000		...	1439	1462	1462	...
Kazakhstan / Kazakhstan	...	...	...	...		0	0	0	0	
Korea, Republic of / Corée, République de	...	...	...	300000		...	129402	166108	271868	384700
Kuwait / Koweït	...	...	527			...	20735	...	51000	60100
Kyrgyzstan / Kirghizistan	...	...	...	...		0	0	0	0	
Lao People's Dem. Rep. / Rép. dém. populaire lao			150	200		0	0	0	290	340
Lebanon / Liban	...	...	2000	3000		0	0	0	3800	...
Macau / Macao	2643	3548	4336	5608	...	...	3197	6596	12414	18048
Malaysia / Malaisie	24864	40000	45000	46000	...	39419	86620	139920	213120	309030
Maldives / Maldives	...	235	...	...		0	0	0	0	0
Mongolia / Mongolie	...	...	...	...		0	0	0	0	0
Myanmar / Myanmar	...	...	...	...		0	0	0		...
Nepal / Népal	140	...	...	...		0	0	0	0	0
Oman / Oman	...	...	...	1560	...	2098	2730	3672	4721	5616

19
Telefax stations and mobile cellular telephone subscribers [*cont.*]
Postes de télécopie et abonnés au téléphone mobile cellulaire [*suite*]

Country or area Pays ou zone	Telefax stations Postes de télécopie					Mobile cellular telephone subscribers Abonnés au téléphone mobile cellulaire				
	1989	1990	1991	1992	1993	1989	1990	1991	1992	1993
Pakistan Pakistan	...	2300	4100	5300	...	...	2000	8500	13500	16000
Philippines Philippines	...	...	...	20000	...	...	9708	11083	52000	76880
Qatar Qatar	700	950	2800	4000		...	3811	4057	4233	4289
Saudi Arabia Arabie saoudite	...	...	...	50000		...	14851	15331	15828	19000
Singapore [1] Singapour [1]	27299	33269	39354	48343	...	26295	51000	81900	120000	179000
Sri Lanka Sri Lanka	...	...	5890	8100			1010	1973	4000	13000
Syrian Arab Republic Rép. arabe syrienne	...	...	...	...		0	0	0	0	0
Tajikistan Tadjikistan	...	...	...	...			0	0	0	0
Thailand [4] Thaïlande [4]	5487	6324	7452	50000		...	66278	210000	248716	436000
Turkey Turquie	17913	32882	48311	62876		15606	31809	47828	61395	84187
Turkmenistan Turkménistan	...	...	...	...			0	0	0	500
United Arab Emirates Emirats arabes unis	11269	15562	19494	23866		24901	33528	42953	48861	70516
Uzbekistan Ouzbékistan	...	...	...	...			0	0	0	500
Viet Nam Viet Nam	...	...	...	2100	4894	0	0	0	800	4060
Yemen Yémen	...	778	805	782	...	0	0	0	1550	...
Europe · Europe										
Albania Albanie	...	300	500	600	...	0	0	0	0	0
Andorra Andorre	...	...	...	1280		...	...	...	770	780
Aruba Aruba	...	...	...	500		0	0	0	20	...
Austria Autriche	70000	85000	100000	130000	...	50721	73698	115402	172453	220976
Belarus Bélarus	...	...	...	1514		0	0	0	0	324
Belgium Belgique	50000	67000	93320	150000	...	30791	42880	51420	61460	66929
Bulgaria Bulgarie	...	...	...	7500	...	0	0	0	0	1000
Croatia Croatie	1307	3320	7067	9079	14322	...	240	2019	6320	11239
Czech Republic République tchèque	...	4552	16059	29418	45520	...	0	1242	4651	11151
Denmark Danemark	60000	100000	150000	170000	...	123792	148220	175943	210195	359323
Estonia Estonie	...	...	...	1459		0	0	570	4300	8500
Faeroe Islands Iles Féroé	...	...	...	1300		...	...	1429	1718	1605
Finland Finlande	26091	75000	92000	105000	...	157969	226000	287000	353161	466131
France France	350000	580000	600000	630000	...	188500	291000	375300	436000	584000
Germany † Allemagne †	411095	696229	946216	1172700	1296000	163619	272609	532251	969000	1769962
former German D. R. anc. R. d. allemande	256	...	...	...	...	0	...	...	...	...
Gibraltar Gibraltar	...	122	135	160		...	0	0	0	0
Greece Grèce	5562	9200	12082	13273	...	0	0	0	0	17000

19
Telefax stations and mobile cellular telephone subscribers [*cont.*]
Postes de télécopie et abonnés au téléphone mobile cellulaire [*suite*]

Country or area Pays ou zone	Telefax stations Postes de télécopie					Mobile cellular telephone subscribers Abonnés au téléphone mobile cellulaire				
	1989	1990	1991	1992	1993	1989	1990	1991	1992	1993
Greenland Groënland	...	...	...	1153	...	...	...	...	171	438
Hungary Hongrie	4661	9693	14580	24721	...	0	2645	8477	23292	45712
Iceland Islande [1]	1343	1343	...	4000	...	7893	10010	12889	15251	17409
Ireland [1] Irlande [1]	35000	53000	60000	75000	...	13579	25000	32000	44000	61100
Italy Italie	135169	170450	190987	230000	...	66070	266000	568000	783000	1207000
Latvia Lettonie	...	...	...	596	...	...	...	...	1027	4507
Liechtenstein Liechtenstein	...	...	...	...	960	...	...	...	...	...
Lithuania Lituanie	...	...	444	3000	...	0	0	0	267	1239
Luxembourg Luxembourg	2000	3500	4000	5000	...	427	824	1130	1139	5082
Malta Malte	857	1456	1900	2224	2794	0	...	2280	3500	5300
Netherlands Pays-Bas	110000	180000	325000	372750	...	56000	79000	115000	166000	216000
Norway Norvège	65000	85000	107000	120000	...	167651	196828	227733	280000	375000
Poland Pologne	1500	5824	13156	25068	...	0	0	0	2195	10700
Portugal Portugal	4878	10500	17700	26761	...	2782	6500	12600	37262	101231
Republic of Moldova République de Moldova	...	35	122	245	...	...	100	100	160	170
Romania Roumanie	...	4000	5000	10231	...	0	0	0	0	800
Russian Federation Fédération de Russie	...	...	...	100000	...	...	0	300	6000	10000
San Marino Saint-Marin	...	...	...	1800	2500	...	...	...	900	1300
Slovakia Slovaquie	...	2514	7330	16202	...	0	0	119	1537	3125
Slovenia Slovénie	...	73	135	7578	10480	...	...	523	3500	6500
Spain Espagne	99242	144310	176061	195029	...	29783	54700	108451	180296	257281
Sweden Suède	115000	170000	230000	300000	...	349000	461200	568200	656000	774500
Switzerland Suisse	55000	83000	118000	135000	169000	72735	125047	174557	215061	257703
TFYR Macedonia L'ex-R.y. Macédonie	...	...	...	1408	...	...	0	0	0	0
Ukraine Ukraine	...	682	944	1465	...	...	0	0	0	65
United Kingdom [1] Royaume-Uni [1]	556000	750000	900000	1005000	1200000	975000	1114000	1230000	1496000	2215000
Yugoslavia, SFR † Yougoslavie, Rfs †	...	...	...	7871	...	...	...	0	0	0
Oceania · Océanie										
Australia [2] Australie [2]	190000	280000	350000	400000	...	94529	184943	291459	497000	760000
Fiji Fidji	...	1513	2079	2197	...	0	0	0	0	0
French Polynesia Polynésie française	...	800	405	710	...	...	0	0	0	0
Guam Guam	...	...	...	...	...	0	0	867	1301	1907
Kiribati Kiribati	...	38	76	96	...	0	0	0	0	0
Marshall Islands Iles Marshall	...	50	80	120	...	...	0	0	0	0

19
Telefax stations and mobile cellular telephone subscribers [*cont.*]
Postes de télecopie et abonnés au téléphone mobile cellulaire [*suite*]

Country or area	Telefax stations Postes de télécopie					Mobile cellular telephone subscribers Abonnés au téléphone mobile cellulaire				
Pays ou zone	1989	1990	1991	1992	1993	1989	1990	1991	1992	1993
Micronesia, Fed. States of Micronésie, Etats féd. de	...	251	299	...	...	0	0	0	0	0
New Caledonia Nouvelle−Calédonie	450	1000	1200	1350	...	...	0	0	0	0
New Zealand [1] Nouvelle−Zélande [1]	18000	28000	35000	40000	...	28900	54100	72300	100200	143800
Papua New Guinea Papouasie−Nvl−Guinée	600	629	1018	1085	...	0	0	0	0	0
Samoa Samoa	...	...	...	100	...	...	0	0	0	...
Solomon Islands [1] Iles Salomon [1]	...	126	410	440	...	0	0	0	0	...
Tonga Tonga	...	83	127	134	...	...	0	0	0	...
Vanuatu Vanuatu	...	...	...	...	...	0	0	0	0	...

Source:
International Telecommunications Union (Geneva).

† For detailed descriptions of data pertaining to former Czechoslovakia, Germany, SFR Yugoslavia and former USSR, see Annex I − Country or area nomenclature, regional and other groupings.

1 Year beginning 1 April.
2 Year ending 30 June.
3 Source: Cellular Telecommunications Industry Association.
4 Year ending 30 September.

Source:
Union internationale des télecommunications (Genève).

† Pour les descriptions en détails des données relatives à l'ancienne Tchécoslovaquie, l'Allemagne, la Rfs Yougoslavie et l'ancienne URSS, voir l'Annexe I − Nomenclature des pays ou zones, groupements régionaux et autres groupements.

1 L'année commençant le 1er avril.
2 L'année finissant le 30 juin.
3 Source: "Cellular Telecommunications Industry Association".
4 L'année finissant le 30 septembre.

20
Telephones
Téléphones
Number in use and per 100 inhabitants
Nombre en service et par 100 habitants

Country or area Pays ou zone	Number (000) Nombre (000)					Per 100 inhabitants Par 100 habitants				
	1989	1990	1991	1992	1993	1989	1990	1991	1992	1993
Africa · Afrique										
Algeria Algérie	750	794	883	962	1 068	3.1	3.2	3.4	3.7	4.0
Angola Angola	67	70	72	49	53	0.8	0.8	0.8	0.5	0.5
Benin Bénin	14	15	15	16	20	0.3	0.3	0.3	0.3	0.4
Botswana[1] Botswana[1]	22	26	33	36	43	1.8	2.1	2.5	2.7	3.1
Burkina Faso Burkina Faso	14	16	18	20	22	0.2	0.2	0.2	0.2	0.2
Burundi Burundi	8	8	10	13	16	0.1	0.1	0.2	0.2	0.3
Cameroon Cameroun	38	40	42	55	57	0.3	0.3	0.4	0.4	0.5
Cape Verde Cap-Vert	7	8	9	11	15	1.9	2.2	2.4	2.9	3.8
Central African Rep. Rép. centrafricaine	5	5	5	6	7	0.2	0.2	0.2	0.2	0.2
Chad Tchad	4	4	4	4	5	0.1	0.1	0.1	0.1	0.1
Comoros Comores	3	3	4	4	4	0.6	0.7	0.7	0.8	0.8
Congo Congo	16	16	17	18	19	0.7	0.7	0.7	0.7	0.8
Côte d'Ivoire Côte d'Ivoire	68	73	79	86	94	0.6	0.6	0.6	0.7	0.7
Djibouti Djibouti	5	6	6	7	7	1.0	1.1	1.2	1.2	1.3
Egypt[2] Egypte[2]	1 554	1 718	1 927	2 145	2 375	3.0	3.3	3.6	3.9	4.3
Equatorial Guinea Guinée équatoriale	1	1	1	1	...	0.3	0.3	0.3	0.3	...
Eritrea Erythrée	..	..	..	..	20	..	..	..	..	0.6
Ethiopia[2] Ethiopie[2]	116	125	133	127	132	0.2	0.2	0.3	0.2	0.2
Gabon Gabon	19	21	26	28	30	1.7	1.8	2.3	2.3	2.4
Gambia[13] Gambie[13]	8	6	10	14	16	1.0	0.7	1.0	1.4	1.6
Ghana Ghana	43	44	47	48	49	0.3	0.3	0.3	0.3	0.3
Guinea Guinée	11	11	12	11	12	0.2	0.2	0.2	0.2	0.2
Guinea-Bissau Guinée-Bissau	6	6	6	8	9	0.6	0.6	0.6	0.8	0.8
Kenya[2] Kenya[2]	169	183	200	207	215	0.7	0.8	0.8	0.8	0.8

20
Telephones
Number in use and per 100 inhabitants [cont.]
Téléphones
Nombre en service et par 100 habitants [suite]

Country or area Pays ou zone	Number (000) Nombre (000)					Per 100 inhabitants Par 100 habitants				
	1989	1990	1991	1992	1993	1989	1990	1991	1992	1993
Lesotho Lesotho	12	10	10	10	...	0.7	0.6	0.6	0.6	...
Liberia Libéria	9	...	3	5	5	0.4	...	0.1	0.2	0.2
Libyan Arab Jamah. Jamah. arabe libyenne	216	220	233	236	240	4.9	4.8	5.0	4.8	4.8
Madagascar Madagascar	28	32	36	37	35	0.2	0.3	0.3	0.3	0.3
Malawi Malawi	26	27	29	31	33	0.3	0.3	0.3	0.3	0.4
Mali Mali	10	11	12	13	14	0.1	0.1	0.1	0.1	0.1
Mauritania Mauritanie	5	6	6	7	8	0.3	0.3	0.3	0.3	0.4
Mauritius Maurice	53	56	64	79	107	4.9	5.2	5.9	7.2	9.6
Morocco Maroc	335	402	497	654	821	1.4	1.6	1.9	2.5	3.1
Mozambique Mozambique	41	47	53	56	62	0.3	0.3	0.3	0.3	0.4
Namibia[1] Namibie[1]	51	53	57	61	70	3.6	3.7	3.8	4.0	4.5
Niger Niger	9	9	10	10	11	0.1	0.1	0.1	0.1	0.1
Nigeria Nigéria	250	289	294	321	342	0.3	0.3	0.3	0.3	0.3
Réunion Réunion	151	162	175	188	199	25.8	27.4	29.1	30.7	32.0
Rwanda Rwanda	9	10	11	12	...	0.1	0.1	0.2	0.2	...
Saint Helena Sainte-Hélène	1	1	1	1	2	...	...	15.3	18.7	21.9
Sao Tome and Principe Sao Tomé-et-Principe	2	2	2	2	2	2.0	1.9	1.9	1.9	1.9
Senegal Sénégal	36	44	48	58	64	0.5	0.6	0.6	0.7	0.8
Seychelles[1] Seychelles[1]	8	9	9	10	11	11.3	12.8	13.6	14.6	16.2
Sierra Leone Sierra Leone	13	13	14	14	14	0.3	0.3	0.3	0.3	0.3
Somalia Somalie	15	15	15	15	15	0.2	0.2	0.2	0.2	0.2
South Africa[1] Afrique du Sud[1]	3 080	3 315	3 419	3 524	3 660	8.3	8.7	8.8	8.9	9.0
Sudan Soudan	60	62	64	64	64	0.2	0.2	0.2	0.2	0.2
Swaziland[1] Swaziland[1]	13	14	15	15	16	1.6	1.7	1.8	1.8	1.8
Togo Togo	10	11	11	15	17	0.3	0.3	0.3	0.4	0.4

20

Telephones
Number in use and per 100 inhabitants [*cont.*]
Téléphones
Nombre en service et par 100 habitants [*suite*]

Country or area Pays ou zone	Number (000) Nombre (000)					Per 100 inhabitants Par 100 habitants				
	1989	1990	1991	1992	1993	1989	1990	1991	1992	1993
Tunisia Tunisie	277	303	337	375	421	3.5	3.8	4.1	4.5	4.9
Uganda[2] Ouganda[2]	27	28	28	30	21	0.2	0.2	0.2	0.2	0.1
United Rep.Tanzania Rép. Unie de Tanzanie	71	73	76	81	85	0.3	0.3	0.3	0.3	0.3
Western Sahara Sahara occidental	4	4	4	7	...	2.5	2.6	2.6	4.0	...
Zaire Zaïre	32	34	35	36	36	0.1	0.1	0.1	0.1	0.1
Zambia[1] Zambie[1]	66	65	69	76	78	0.9	0.8	0.9	0.9	0.9
Zimbabwe[2] Zimbabwe[2]	121	124	126	127	128	1.3	1.3	1.3	1.2	1.2
America, North · Amérique du Nord										
Anguilla Anguilla	...	...	3	3	3	...	...	35.6	38.7	38.0
Antigua and Barbuda[1] Antigua-et-Barbuda[1]	13	16	14	17	19	20.1	24.6	20.9	25.5	28.9
Aruba Aruba	17	19	20	20	...	26.9	28.4	30.3	30.4	...
Bahamas Bahamas	62	70	62	80	...	24.7	27.4	23.9	30.3	...
Barbados[1] Barbade[1]	78	83	78	80	83	30.4	32.4	30.1	30.9	31.8
Belize[1] Belize[1]	15	17	21	25	29	8.3	9.2	11.0	12.5	14.0
Bermuda[1] Bermudes[1]	36	37	39	40	42	60.9	61.7	63.9	64.7	67.9
British Virgin Islands Iles Vierges britanniques	...	...	...	4	...	...	...	...	23.5	...
Canada[4] Canada[4]	14 648	15 296	15 815	16 247	16 471	55.8	57.5	58.6	59.2	59.2
Cayman Islands[1] Iles Caïmanes[1]	11	12	12	13	14	...	...	43.4	44.0	44.9
Costa Rica Costa Rica	272	281	305	327	364	9.2	9.3	9.8	10.2	11.1
Cuba Cuba	304	337	339	344	...	2.9	3.2	3.2	3.2	...
Dominica Dominique	10	12	12	14	...	13.8	16.2	17.2	19.1	...
Dominican Republic Rép. dominicaine	284	341	411	479	552	4.1	4.8	5.6	6.5	7.4
El Salvador El Salvador	120	125	130	165	174	2.4	2.4	2.5	3.1	3.2
Greenland Groënland	17	17	17	17	18	29.8	29.9	30.3	30.6	31.7
Grenada Grenade	12	15	16	19	20	13.2	16.7	17.8	20.3	22.1

20
Telephones
Number in use and per 100 inhabitants [cont.]
Téléphones
Nombre en service et par 100 habitants [suite]

Country or area Pays ou zone	Number (000) Nombre (000)					Per 100 inhabitants Par 100 habitants				
	1989	1990	1991	1992	1993	1989	1990	1991	1992	1993
Guadeloupe Guadeloupe	107	118	127	139	149	28.0	30.3	32.1	34.6	36.7
Guatemala Guatemala	159	190	202	214	231	1.8	2.1	2.1	2.2	2.3
Haiti Haïti	42	45	45	45	45	0.7	0.7	0.7	0.7	0.7
Honduras Honduras	79	88	94	105	117	1.6	1.7	1.8	1.9	2.1
Jamaica[1] Jamaïque[1]	89	105	132	167	255	3.8	4.5	5.5	7.0	10.6
Martinique Martinique	115	122	132	141	150	32.2	33.9	36.5	38.5	40.5
Mexico Mexique	4 702	5 355	6 025	6 754	7 621	5.9	6.6	7.2	7.9	8.8
Montserrat Montserrat	3	4	4	4	...	...	...	36.4	39.4	...
Netherlands Antilles Antilles néerlandaises	47	47	49	49	50	25.2	24.8	25.3	25.3	25.5
Nicaragua Nicaragua	46	46	50	54	67	1.3	1.3	1.3	1.4	1.7
Panama Panama	201	216	229	243	261	8.5	8.9	9.3	9.6	10.2
Puerto Rico[5] Porto Rico[5]	816	1 002	1 060	1 134	1 207	23.3	28.4	29.9	31.7	33.5
Saint Kitts and Nevis[1] Saint-Kitts-et-Nevis[1]	8	10	10	12	12	19.3	23.1	24.3	27.7	29.6
Saint Lucia[1] Sainte-Lucie [1]	13	17	17	20	24	8.5	11.3	11.2	13.1	15.4
Saint Pierre and Miquelon Saint-Pierre-et-Miquelon	3	3	3	3	3	...	...	54.5	57.0	...
St. Vincent-Grenadines[1] St. Vincent-Grenadines[1]	9	13	15	16	16	8.5	12.3	13.5	14.3	14.8
Trinidad and Tobago Trinité-et-Tobago	158	165	174	180	192	13.0	13.3	13.9	14.2	15.0
Turks and Caicos Islands Iles Turques et Caiques	...	...	...	3	3	...	...	...	24.6	24.9
United States[4] Etats-Unis[4]	132 683	136 337	139 658	144 057	148 084	53.3	54.6	55.3	56.5	57.4
United States Virgin Is. Iles Vierges américaines	46	47	51	54	59	45.0	46.3	51.6	54.8	59.3
America, South · Amérique du Sud										
Argentina[6] Argentine[6]	3 322	3 087	3 199	3 682	4 115	10.4	9.6	9.8	11.1	12.3
Bolivia Bolivie	170	184	185	193	234	2.4	2.6	2.5	2.6	3.0
Brazil Brésil	8 853	9 409	10 076	10 841	11 744	6.0	6.3	6.7	6.9	7.5
Chile Chili	646	860	1 056	1 283	1 520	5.0	6.5	7.9	9.4	11.0

20
Telephones
Number in use and per 100 inhabitants [*cont.*]
Téléphones
Nombre en service et par 100 habitants [*suite*]

Country or area Pays ou zone	Number (000) Nombre (000)					Per 100 inhabitants Par 100 habitants				
	1989	1990	1991	1992	1993	1989	1990	1991	1992	1993
Colombia Colombie	2 177	2 415	2 633	2 822	3 828	6.9	7.5	8.0	8.4	11.3
Ecuador Equateur	454	491	491	531	598	4.4	4.7	4.6	4.8	5.3
French Guiana Guyane française	28	30	34	36	38	25.2	30.4	33.6	30.4	29.8
Guyana Guyana	16	13	16	28	41	2.0	1.6	2.0	3.5	5.1
Paraguay Paraguay	103	112	120	128	142	2.5	2.6	2.7	2.8	3.1
Peru Pérou	531	565	577	614	670	2.5	2.6	2.6	2.7	2.9
Suriname Suriname	34	37	41	44	47	8.3	9.1	10.0	10.8	11.6
Uruguay Uruguay	376	415	451	492	530	12.2	13.4	14.5	15.7	16.8
Venezuela Venezuela	1 465	1 488	1 599	1 832	2 083	7.8	7.7	8.1	9.0	9.9
Asia · Asie										
Afghanistan Afghanistan	35	36	37	29	29	0.2	0.2	0.2	0.1	0.1
Armenia Arménie	530	560	572	578	583	15.2	15.8	15.8	15.7	15.6
Azerbaijan Azerbaïdjan	590	620	627	657	647	8.3	8.7	8.7	8.9	8.7
Bahrain Bahreïn	88	94	101	113	124	17.9	18.7	19.5	21.2	22.9
Bangladesh[2] Bangladesh [2]	192	242	250	256	268	0.2	0.2	0.2	0.2	0.2
Bhutan Bhoutan	2	2	3	3	4	0.1	0.1	0.2	0.2	0.2
Brunei Darussalam Brunéi Darussalam	30	35	39	48	55	12.1	13.6	14.8	17.6	19.7
Cambodia Cambodge	5	5	5	5	6	0.1	0.1	0.1	0.1	0.1
China Chine	5 680	6 850	8 451	11 469	17 332	0.5	0.6	0.7	1.0	1.5
Cyprus Chypre	223	246	269	291	311	32.1	35.0	44.6	47.0	49.4
Georgia Géorgie	510	540	557	573	571	9.4	9.9	10.2	10.5	10.5
Hong Kong[1] Hong-kong[1]	2 345	2 475	2 642	2 820	2 992	41.2	43.4	45.9	48.5	51.0
India Inde	4 589	5 075	5 810	6 797	8 037	0.6	0.6	0.7	0.8	0.9
Indonesia Indonésie	864	1 066	1 246	1 503	1 713	0.5	0.6	0.7	0.8	0.9
Iran, Islamic Rep. of[1] Iran, Rép. islamique d'[1]	2 029	2 199	2 456	2 998	3 598	3.8	3.9	4.3	5.0	5.9

20
Telephones
Number in use and per 100 inhabitants [cont.]
Téléphones
Nombre en service et par 100 habitants [suite]

Country or area Pays ou zone	Number (000) Nombre (000)					Per 100 inhabitants Par 100 habitants				
	1989	1990	1991	1992	1993	1989	1990	1991	1992	1993
Iraq Iraq	675	675	675	675	675	3.9	3.7	3.6	3.5	3.4
Israel Israël	1 534	1 626	1 703	1 804	1 958	33.9	34.9	34.3	35.2	37.1
Japan[17] Japon[17]	52 454	54 528	56 260	57 652	58 459	42.6	44.1	45.4	46.4	46.8
Jordan Jordanie	238	246	265	278	288	7.8	7.5	7.2	7.0	7.0
Kazakhstan Kazakhstan	1 226	1 333	1 425	1 490	1 559	7.4	8.0	8.4	8.7	9.1
Korea, Dem. P. R. Corée, R. p. dém. de	750	780	800	1 089	...	3.5	3.6	3.6	4.8	...
Korea, Republic of Corée, République de	11 792	13 276	14 573	15 593	16 633	27.8	31.0	33.7	35.7	37.8
Kuwait Koweït	311	331	322	346	358	15.2	15.5	22.1	24.5	24.5
Kyrgyzstan Kirghizistan	288	314	332	339	367	6.7	7.1	7.5	7.5	8.1
Lao People's Dem. Rep. Rép. dém. pop. lao	6	7	7	9	9	0.2	0.2	0.2	0.2	0.2
Lebanon Liban	300	300	310	350	...	8.4	8.3	8.4	9.3	...
Macau Macao	79	93	107	121	134	23.7	27.1	30.1	32.3	34.4
Malaysia Malaisie	1 388	1 586	1 817	2 092	2 411	8.0	8.9	10.0	11.2	12.6
Maldives Maldives	5	6	8	9	10	2.5	2.9	3.4	3.7	4.2
Mongolia Mongolie	57	66	68	69	66	2.7	3.0	3.0	3.0	2.8
Nepal Népal	45	57	65	69	72	0.2	0.3	0.3	0.3	0.4
Oman Oman	91	105	117	130	148	6.2	6.9	7.4	7.9	8.6
Pakistan Pakistan	792	843	1 116	1 244	1 605	0.7	0.8	1.0	1.0	1.3
Philippines Philippines	591	610	648	661	860	1.0	1.0	1.0	1.0	1.3
Qatar Qatar	89	92	96	105	111	21.1	18.9	19.4	20.6	21.4
Saudi Arabia Arabie saoudite	1 125	1 234	1 466	1 568	1 575	7.4	7.8	9.0	9.3	9.1
Singapore[1] Singapour[1]	998	1 054	1 101	1 169	1 246	37.7	39.0	39.9	41.5	43.4
Sri Lanka Sri Lanka	66	71	76	80	158	0.2	0.2	0.2	0.2	0.0
Syrian Arab Republic Rép. arabe syrienne	489	496	506	513	550	4.2	4.1	4.0	4.0	4.1
Tajikistan Tadjikistan	230	240	258	268	260	4.4	4.5	4.7	4.8	4.6

20
Telephones
Number in use and per 100 inhabitants [*cont.*]
Téléphones
Nombre en service et par 100 habitants [*suite*]

Country or area Pays ou zone	Number (000) Nombre (000)					Per 100 inhabitants Par 100 habitants				
	1989	1990	1991	1992	1993	1989	1990	1991	1992	1993
Thailand[6 8] Thaïlande[6 8]	1 158	1 325	1 553	1 790	2 185	2.1	2.4	2.7	3.1	3.7
Turkey Turquie	5 862	6 861	8 152	9 410	10 936	10.6	12.1	14.1	16.0	18.4
Turkmenistan Turkménistan	210	220	237	249	265	5.9	6.0	6.3	6.5	6.7
United Arab Emirates Emirats arabes unis	378	430	438	492	553	24.6	27.1	26.7	29.2	32.1
Uzbekistan Ouzbékistan	1 350	1 403	1 458	1 440	1 452	6.7	6.8	7.0	6.7	6.6
Viet Nam Viet Nam	80	99	137	153	260	0.1	0.1	0.2	0.2	0.4
Yemen Yémen	93	125	131	143	162	0.9	1.1	1.0	1.1	1.2
Europe · Europe										
Albania Albanie	39	40	42	45	49	1.2	1.2	1.3	1.3	1.4
Andorra Andorre	19	22	23	25	27	38.6	41.8	40.4	41.8	42.2
Austria Autriche	3 103	3 223	3 344	3 466	3 579	40.7	41.8	42.7	44.0	45.1
Belarus Bélarus	1 464	1 574	1 673	1 744	1 814	14.3	15.3	16.3	16.9	17.6
Belgium Belgique	3 712	3 913	4 096	4 264	4 396	37.3	39.3	40.9	42.5	43.7
Bosnia & Herzegovina Bosnie-Herzégovine	..	..	..	600	...	..	..	..	13.7	...
Bulgaria Bulgarie	1 994	2 175	2 205	2 340	2 300	22.2	25.2	25.7	27.5	26.3
Croatia Croatie	754	823	891	955	1 027	15.9	17.3	18.6	19.9	21.5
Czech Republic République tchèque	1 557	1 624	1 707	1 819	1 961	15.0	15.8	16.6	17.6	19.0
Denmark[9] Danemark[9]	2 848	2 911	2 951	3 005	3 060	55.5	56.6	57.3	58.1	58.9
Estonia Estonie	310	320	332	335	358	19.7	20.4	21.2	21.5	23.2
Faeroe Islands Iles Féroé	22	23	24	24	23	45.8	48.1	50.9	50.7	49.5
Finland Finlande	2 582	2 670	2 718	2 742	2 761	52.0	53.5	54.0	54.4	54.4
France[10] France[10]	26 942	28 085	29 100	30 100	30 900	47.8	49.5	51.0	52.5	53.6
Germany † Allemagne†	28 848	31 887	33 560	35 421	36 900	46.5	40.1	42.0	44.0	45.7
Gibraltar Gibraltar	10	11	12	13	14	32.4	35.8	41.1	45.3	48.7
Greece Grèce	3 788	3 949	4 190	4 497	4 744	37.7	38.9	40.8	43.7	45.7

20
Telephones
Number in use and per 100 inhabitants [*cont.*]
Téléphones
Nombre en service et par 100 habitants [*suite*]

Country or area Pays ou zone	Number (000) Nombre (000)					Per 100 inhabitants Par 100 habitants				
	1989	1990	1991	1992	1993	1989	1990	1991	1992	1993
Hungary Hongrie	916	996	1 128	1 291	1 498	8.7	9.4	10.9	12.5	14.6
Iceland[11] Islande[11]	126	131	136	140	144	49.9	51.2	52.5	53.7	54.4
Ireland[1] Irlande[1]	916	983	1 048	1 113	1 170	26.1	28.1	29.7	31.4	32.8
Italy Italie	21 266	22 350	23 071	23 709	24 176	37.0	38.8	39.9	40.9	41.8
Latvia Lettonie	600	620	643	652	694	22.4	23.2	24.2	24.7	26.8
Liechtenstein Liechtenstein	16	17	17	18	19	51.7	55.1	62.5	61.5	62.4
Lithuania[3] Lituanie[3]	734	781	814	832	858	19.9	20.9	21.7	22.2	22.9
Luxembourg Luxembourg	176	184	192	206	215	46.8	48.3	49.7	52.9	54.1
Malta Malte	122	128	139	150	158	34.6	36.0	38.6	41.4	43.0
Monaco Monaco	23	24	26	29	30	76.7	81.3	92.9	104.0	107.6
Netherlands Pays-Bas	6 690	6 940	7 175	7 395	7 630	45.1	46.4	47.6	48.7	49.9
Norway Norvège	2 070	2 132	2 198	2 268	2 335	49.0	50.3	51.6	52.9	54.2
Poland Pologne	3 124	3 293	3 565	3 938	4 419	8.2	8.6	9.3	10.3	11.5
Portugal Portugal	2 077	2 379	2 694	3 014	3 260	21.2	24.3	27.4	30.6	31.1
Republic of Moldova République de Moldova	427	462	496	511	524	9.8	10.6	11.4	11.7	12.0
Romania Roumanie	2 338	2 366	2 443	2 574	2 624	10.1	10.2	10.6	11.3	11.5
Russian Federation Fédération de Russie	19 300	20 700	22 296	22 849	23 397	13.1	14.0	15.0	15.4	15.8
San Marino Saint-Marin	10	10	11	14	14	48.3	51.2	54.0	67.5	61.3
Slovakia Slovaquie	669	711	759	821	893	12.7	13.4	14.4	15.5	16.7
Slovenia Slovénie	390	422	459	494	516	19.8	21.1	22.9	24.7	25.9
Spain Espagne	11 797	12 603	13 264	13 792	14 253	30.3	32.3	34.0	35.3	36.4
Sweden Suède	5 716	5 849	5 911	5 919	5 903	67.4	68.3	68.6	68.2	67.8
Switzerland Suisse	3 785	3 943	4 081	4 185	4 266	56.9	58.7	60.1	60.6	61.1
TFYR Macedonia L'ex-R.y. Macédoine	270	286	290	312	324	12.8	13.4	13.5	14.4	14.8
Ukraine Ukraine	6 684	7 028	7 344	7 578	7 820	12.9	13.6	14.1	14.5	15.0

20
Telephones
Number in use and per 100 inhabitants [cont.]
Téléphones
Nombre en service et par 100 habitants [suite]

Country or area Pays ou zone	Number (000) Nombre (000)					Per 100 inhabitants Par 100 habitants				
	1989	1990	1991	1992	1993	1989	1990	1991	1992	1993
United Kingdom[1] Royaume-Uni[1]	25 017	25 817	26 357	27 370	28 681	43.7	45.0	45.7	47.3	49.4
Yugoslavia Yougoslavie	1 585	1 682	1 782	1 873	1 923	15.3	16.1	16.9	17.7	18.0
Oceania · Océanie										
American Samoa Samoa américaines	5	6	6	6	...	...	14.1	14.1	15.4	...
Australia[2][12] Australie[2][12]	7 420	7 787	8 046	8 257	8 540	44.2	45.7	46.5	47.1	48.2
Cook Islands Iles Cook	3	3	3	4	4	14.0	15.0	16.9	21.5	25.1
Micronesia,Federated States of Micron, Etats fédérés de	2	2	3	3	6	2.1	2.4	2.6	2.8	5.6
Fiji Fidji	39	42	46	50	54	5.4	5.8	6.2	6.6	7.1
French Polynesia Polynésie française	34	38	41	44	45	17.7	19.3	20.4	21.0	21.1
Guam Guam	36	39	43	46	66	27.1	29.4	31.8	33.3	46.4
Kiribati Kiribati	1	1	1	1	2	1.5	1.7	1.7	1.8	2.3
Marshall Islands Iles Marshall	...	1	1	1	2	...	1.1	1.5	1.7	4.4
Nauru Nauru	1	1	1	1	...	...	...	13.3	12.0	...
New Caledonia Nouvelle-Calédonie	27	28	31	35	39	16.2	16.9	18.4	20.1	21.9
New Zealand[1] Nouvelle-Zélande[1]	1 444	1 469	1 493	1 534	1 593	43.4	43.7	43.8	44.9	46.0
Papua New Guinea Papouasie-Nvl-Guinée	31	30	34	36	40	0.8	0.8	0.9	0.9	1.0
Solomon Islands[1][13] Iles Salomon[1][13]	4	4	5	5	5	1.2	1.4	1.4	1.5	1.5
Tonga Tonga	4	4	5	5	6	4.4	4.8	5.6	6.0	6.4
Tuvalu Tuvalu	0	0	0	...	...	1.3	...	1.3	...	...
Vanuatu Vanuatu	2	3	3	4	4	1.7	1.8	2.0	2.3	2.5
Wallis and Futuna Islands Iles Wallis et Futuna	0	0	...	1	1	1.9	1.4	...	3.5	6.3

Source:
International Telecommunications Union (Geneva).

† For detailed descriptions of data pertaining to
former Czechoslovakia, Germany, SFR Yugoslavia and former
USSR, see Annex I - Country or area nomenclature, regional
and other groupings.

Source:
Union internationale des télécommunications (Genève).

† Pour les descriptions en détails des données
relatives à l'ancienne Tchécoslovaquie, l'Allemagne, la Rfs
Yougoslavie et l'ancienne URSS, voir l'Annexe I -
Nomenclature des pays ou zones, groupements régionaux et
autres groupements.

20

Telephones
Number in use and per 100 inhabitants [cont.]

Téléphones
Nombre en service et par 100 habitants [suite]

1 Year beginning 1 April.	1 L'année commençant le 1er avril.
2 Year ending 30 June.	2 L'année finissant le 30 juin.
3 Excluding public call offices.	3 Cabines publiques exclues.
4 Access lines.	4 Lignes d'accès.
5 Switched access lines.	5 Lignes d'accès par communication.
6 Year ending 30 September.	6 L'année finissant le 30 septembre.
7 Telephone subscriptions.	7 Abonnements téléphoniques.
8 Main telephone stations.	8 Postes téléphoniques principales.
9 Number of subscribers; including ISDN from 1992.	9 Nombre d'abonnés; RNIS inclu à partir de 1992.
10 Including ISDN from 1993.	10 RNIS inclu à partir de 1993.
11 Including PABX.	11 PABX inclu.
12 Telephone services in operation.	12 Les services téléphoniques en exploitation.
13 Billable lines.	13 Lignes payables.

Technical notes, tables 14-20

Tables 14-18 are compiled from the UNESCO *Statistical Yearbook.* [29] The technical notes in the UNESCO *Yearbook* concerning these tables are summarized below.

Table 14: Data on books by subject groups cover printed books and pamphlets and, unless otherwise stated, refer to first editions and re-editions of originals and translations. The grouping by subject follows the Universal Decimal Classification (UDC).

Table 15: Unless otherwise stated, the data refer to the total number of titles (first editions and re-editions of original works and translations) by language of publication.

Table 16: For the purposes of this table, a daily general interest newspaper is defined as a publication devoted primarily to recording general news. It is considered to be "daily" if it appears at least four times a week.

It is known or believed that no daily general-interest newspapers are published in the following 35 countries and territories: *Africa*: Cape Verde, Comoros, Djibouti, Guinea, St. Helena, Sao Tome and Principe, Western Sahara; *America, North*: Anguilla, Antigua and Barbuda, Belize, British Virgin Islands, Dominica, Greenland, Grenada, Montserrat, Panama-former Canal Zone, St. Kitts and Nevis, St. Lucia, St. Pierre and Miquelon, St. Vincent and the Grenadines, Turks and Caicos Islands. *America, South*: Falkland Islands (Malvinas). *Asia*: Bhutan, Cambodia, East Timor; *Europe*: Faeroe Islands; *Oceania*: American Samoa, Kiribati, Nauru, Niue, Norfolk Island, Palau, Samoa, Solomon Islands, Tokelau, Tuvalu, Vanuatu.

Table 17: For the purposes of this table a non-daily general interest newspaper is defined as a publication which is devoted primarily to recording general news and which is published three times a week or less. Under the category of periodicals are included publications of periodical issue, other than newspapers, containing information of a general or of a specialized nature.

Table 18: Data show the estimated number of television receivers in use (indicated by T) and the estimated number of radio receivers (indicated by R) as well as receivers per 1,000 inhabitants. The figures refer to 31 December of the year stated. In these tables the term "receivers" relates to all types of receivers for broadcasts to the general public, including receivers connected to a redistribution system (wired receivers).

Table 19: The number of telefax stations refers to all types of private equipment (e.g. Group 2, Group 3) connected to the PSTN. Some operators report only the equipment sold, leased or registered by them and therefore the actual number may be higher.

Notes techniques, tableaux 14-20

Tableaux 14-18 ont été établis à partir de l'*Annuaire statistique* de l'UNESCO [29]. Les notes techniques de l'*Annuaire* de l'UNESCO concernant ces tableaux sont résumées ci-dessous.

Tableau 14 : Les données concernant la production de livres par groupes de sujets se rapportent aux livres et brochures imprimés, sauf indication contraire, aux premières éditions et aux rééditions d'originaux et aux traductions. Les sujets sont groupés selon la Classification décimale universelle (CDU).

Tableau 15 : Sauf indication contraire, les données se rapportent au nombre total de titres (premières éditions et rééditions d'ouvrages originaux et de traductions) par langue de publication.

Tableau 16 : Dans ce tableau, par "journal quotidien d'information générale", on entend une publication qui a essentiellement pour objet de rendre compte des événements courants. Il est considéré comme "quotidien" s'il paraît au moins quatre fois par semaine.

On sait ou l'on croit savoir qu'il ne paraît aucun journal quotidien d'information générale dans les 35 pays ou territoires suivants : *Afrique* : Cap-Vert, Comores, Djibouti, Guinée, Sahara occidental, Sainte-Hélène, Sao Tomé-et-Principe; *Amérique du Nord* : Anguilla, Antigua-et-Barbuda, Bélize, Dominique, Grenade, Groenland, Iles turques et caïques, Iles vierges britanniques, Montserrat, Panama-ancienne zone du Canal, Saint Kitts-et-Nevis, Sainte-Lucie, Saint-Pierre-et-Miquelon, Saint-Vincent-et-Grenadines; *Amérique du Sud* : Iles Falkland (Malvinas), *Asie* : Bhoutan, Cambodge, Timor oriental; *Europe* : Iles Faeroe; *Océanie* : Ile Norfolk, Palaos, Iles Salomon, Kirabiti, Nauru, Niue, Samoa, Samoa américaines, Tokelau, Tuvalu, Vanuatu.

Tableau 17 : Aux fins de ce tableau, par "journal non quotidien d'information générale", on entend une publication qui a essentiellement pour objet de rendre compte des événements courants et qui est publié trois fois par semaine ou moins. La catégorie périodique comprend les publications périodiques autres que les journaux, contenant des informations de caractère général ou spécialisé.

Tableau 18 : Les données de ces tableaux indiquent le nombre estimatif de récepteurs de télévision en usage (indiqués par un T), et le nombre estimatif de récepteurs de radio (indiqués par R) ainsi que les récepteurs pour 1000 habitants. Les chiffres se rapportent au 31 décembre de l'année indiquée. Dans ces tableaux, le terme "récepteurs" désigne tous les types de récepteur permettant de capter les émissions destinées au grand public, y compris les récepteurs reliés (par fil) à un réseau de redistribution.

The number of mobile cellular subscribers refers to users of portable telephones subscribing to an automatic public mobile telephone service using cellular technology which provides access to the PSTN.

Table 20: This table shows the number of main lines and the main lines per 100 inhabitants for the years indicated. Main telephone lines refer to the telephone lines connecting a customer's equipment to the PSTN and which have a dedicated port on a telephone exchange. Main telephone lines per 100 inhabitants is calculated by dividing the number of main lines by the population and multipying by 100.

Tableau 19 : Le nombre de postes de télécopie désigne tous les types d'équipements privés (par exemple, Groupes 2 et 3) reliés au RTPC. Certains exploitants signalent uniquement les équipements qu'ils ont vendus, loués ou enregistrés; par conséquent, leur nombre réel peut être plus élevé.

Les abonnés mobiles désignent les utilisateurs de téléphones portatifs abonnés à un service automatique public de téléphones mobiles cellulaires ayant accès au RTPC.

Tableau 20 : Ce tableau indique le nombre de lignes principales et les lignes principales par 100 habitants pour les années indiquées. Les lignes principales sont des lignes téléphoniques qui relient l'equipement terminal de l'abonné au RTPC et qui possèdent un accès individualisé aux équipements d'un central téléphonique.

Les lignes principales par 100 habitants se calculent en divisant le nombre de lignes principales par la population et en multipliant par 100.

Part Three
Economic Activity

VI
**National accounts and industrial production
(tables 21-26)**
VII
Financial statistics (tables 27 and 28)
VIII
Labour force (tables 29 and 30)
IX
Wages and prices (tables 31-33)
X
**Agriculture, hunting, forestry and fishing
(tables 34-40)**
XI
Manufacturing (tables 41-60)
XII
Transport (tables 61-65)
XIII
Energy (tables 66 and 67)
XIV
**Science and technology; intellectual property
(tables 68-70)**

Part Three of the *Yearbook* presents statistical series on economic production and consumption for a wide range of economic activities, and other basic series on major economic topics, for all countries or areas of the world for which data are available. Included are basic tables on national accounts, finance, labour force, wages and prices, a wide range of agricultural, mined and manufactured commodities, transport, energy, science and technology, and intellectual property. In most cases, tables present statistics on production; in a few cases, data are presented on stocks and consumption.

International economic topics such as external trade are covered in Part Four.

Troisième partie
Activité économique

VI
**Comptabilités nationales et production
industrielle (tableaux 21 à 26)**
VII
Statistiques financières (tableaux 27 et 28)
VIII
Main-d'oeuvre (tableaux 29 et 30)
IX
Salaires et prix (tableaux 31 à 33)
X
**Agriculture, chasse, forêts et pêche
(tableaux 34 à 40)**
XI
Industries manufacturières (tableaux 41 à 60)
XII
Transports (tableaux 61 à 65)
XIII
Energie (tableaux 66 et 67)
XIV
**Science et technologie; Propriété intellectuelle
(tableaux 68 à 70)**

La troisième partie de l'*Annuaire* présente, pour une large gamme d'activités économiques, des séries statistiques sur la production économique et la consommation, et, pour tous les pays ou zones du monde pour lesquels des données sont disponibles, d'autres séries fondamentales ayant trait à des questions économiques importantes. Y figurent des tableaux de base consacrés à la comptabilité nationale, aux finances, à la main-d'oeuvre, aux salaires et aux prix, à un large éventail de produits agricoles, miniers et manufacturés, aux transports, à l'énergie, science et technologie, et à la propriété intellectuelle. On y trouve, dans la plupart des cas, des statistiques sur la production et parfois des données relatives aux stocks et à la consommation.

Les questions économiques internationales comme le commerce extérieur sont traitées dans la quatrième partie.

21
Gross domestic product: total and per capita
Produit intérieur brut : total et par habitant

In US dollars (millions) [1] at current and constant 1990 prices; per capita US$;
real rates of growth
En monnaie dollars de E−u (millions) [1] aux prix courants et constants de 1990; par habitant en dollars de E−u ;
taux de l'accroissement réels

Country or area Pays ou zone	1984	1985	1986	1987	1988	1989	1990	1991	1992
Afghanistan Afghanistan									
At current prices	6 616	7 192	7 272	7 710	8 479	13 799	18 969	29 390	41 845
Per capita	450	495	501	524	562	879	1 146	1 662	2 195
At constant prices	24 785	24 851	25 595	22 968	21 067	19 580	18 969	18 756	17 953
Growth rates	1.8	0.3	3.0	−10.3	−8.3	−7.1	−3.1	−1.1	−4.3
Albania Albanie									
At current prices	2 358	2 408	2 483	2 464	2 549	3 046	2 170	1 014	653
Per capita	813	813	821	798	810	951	668	308	197
At constant prices	2 088	2 125	2 244	2 226	2 195	2 411	2 170	1 569	1 417
Growth rates	−1.3	1.8	5.6	−0.8	−1.4	9.8	−10.0	−27.7	−9.7
Algeria Algérie									
At current prices	52 951	57 995	63 069	64 475	56 569	53 024	53 549	38 402	45 925
Per capita	2 501	2 662	2 815	2 800	2 391	2 182	2 145	1 498	1 743
At constant prices	50 872	52 153	52 929	52 340	50 623	51 889	53 549	54 754	56 150
Growth rates	6.8	2.5	1.5	−1.1	−3.3	2.5	3.2	2.3	2.6
Andorra Andorre									
At current prices	202	222	317	408	498	548	712	766	836
Per capita	4 695	4 943	6 895	8 870	10 595	11 657	15 145	16 289	17 781
At constant prices	487	519	557	596	656	685	712	732	738
Growth rates	10.0	6.5	7.3	7.1	10.0	4.5	3.9	2.9	0.8
Angola Angola									
At current prices	5 908	6 865	6 443	7 424	8 015	9 327	10 400	5 954	183
Per capita	761	861	787	883	928	1 048	1 131	625	18
At constant prices	8 627	8 929	9 212	9 857	10 407	10 447	10 400	10 867	11 182
Growth rates	4.5	3.5	3.2	7.0	5.6	0.4	−0.4	4.5	2.9
Anguilla Anguilla									
At current prices	16	20	25	32	41	51	58	59	62
Per capita	2 225	2 791	3 608	4 501	5 793	7 235	8 272	8 391	7 734
At constant prices	30	33	38	42	48	52	58	55	55
Growth rates	4.4	11.2	14.2	11.6	13.6	9.7	10.9	−5.9	1.0
Antigua and Barbuda Antigua−et−Barbuda									
At current prices	173	200	238	277	341	370	412	423	439
Per capita	2 754	3 179	3 715	4 326	5 328	5 685	6 333	6 404	6 646
At constant prices	274	295	320	348	374	398	412	429	436
Growth rates	7.5	7.7	8.4	8.7	7.6	6.3	3.5	4.3	1.7
Argentina Argentine									
At current prices	116 914	88 123	105 876	108 826	126 885	76 691	141 350	189 700	228 788
Per capita	3 907	2 905	3 445	3 495	4 024	2 402	4 373	5 799	6 912
At constant prices	149 324	139 446	149 654	153 521	150 628	141 260	141 350	153 941	167 267
Growth rates	1.8	−6.6	7.3	2.6	−1.9	−6.2	0.1	8.9	8.7
Armenia Arménie									
At current prices	9 408	9 531	11 564	13 017	13 269	15 136	14 554	9 086	306
Per capita	2 922	2 854	3 414	3 786	3 843	4 347	4 365	2 661	88
At constant prices	12 448	13 261	13 118	13 148	12 936	15 243	14 554	12 220	6 391
Growth rates	13.0	6.5	−1.1	0.2	−1.6	17.8	−4.5	−16.0	−47.7
Australia Australie									
At current prices	190 220	168 349	177 445	209 055	266 300	293 143	296 317	300 953	294 123
Per capita	12 251	10 683	11 085	12 848	16 095	17 429	17 343	17 351	16 715
At constant prices	247 054	256 680	262 860	276 027	289 319	297 852	296 317	297 364	304 803
Growth rates	5.1	3.9	2.4	5.0	4.8	2.9	−0.5	0.4	2.5
Austria Autriche									
At current prices	63 810	65 173	93 175	117 171	126 869	126 334	158 168	163 990	184 489
Per capita	8 457	8 623	12 294	15 401	16 602	16 456	20 509	21 176	23 725
At constant prices	133 243	136 523	138 138	140 428	146 124	151 722	158 168	162 441	165 121
Growth rates	1.4	2.5	1.2	1.7	4.1	3.8	4.2	2.7	1.6
Azerbaijan Azerbaïdjan									
At current prices	16 960	16 905	19 485	23 016	23 718	25 017	22 072	15 257	1 298
Per capita	2 585	2 538	2 883	3 352	3 398	3 531	3 094	2 120	178
At constant prices	21 774	22 777	21 443	22 711	22 502	24 468	22 072	19 023	12 422
Growth rates	8.6	4.6	−5.9	5.9	−0.9	8.7	−9.8	−13.8	−34.7

21

Gross domestic product: total and per capita

In US dollars (millions) [1] at current and constant 1990 prices; per capita US$;

real rates of growth [cont.]

Produit intérieur brut : total et par habitant

En monnaie dollars de E−u (millions) [1] aux prix courants et constants de 1990; par habitant;

taux de l'accroissement réels [suite]

Country or area Pays ou zone	1984	1985	1986	1987	1988	1989	1990	1991	1992
Bahamas Bahamas									
At current prices	1 824	2 110	2 370	2 625	2 913	3 006	3 134	3 090	3 059
Per capita	7 967	9 057	9 956	10 848	11 841	11 976	12 290	11 885	11 587
At constant prices	2 329	2 644	2 740	2 866	2 932	2 990	3 134	3 034	3 064
Growth rates	2.9	13.5	3.6	4.6	2.3	2.0	4.8	−3.2	1.0
Bahrain Bahrein									
At current prices	3 906	3 705	3 187	3 170	3 359	3 584	3 903	4 250	4 364
Per capita	9 457	8 635	7 161	6 891	7 086	7 344	7 759	8 204	8 188
At constant prices	3 564	3 492	3 548	3 504	3 762	3 856	3 903	4 075	4 156
Growth rates	4.9	−2.0	1.6	−1.2	7.3	2.5	1.2	4.4	2.0
Bangladesh Bangladesh									
At current prices	16 050	16 654	17 733	19 294	20 786	22 856	24 137	24 683	24 784
Per capita	163	165	172	182	192	206	212	212	208
At constant prices	19 093	19 921	20 754	21 354	21 892	23 343	24 137	25 157	26 280
Growth rates	3.0	4.3	4.2	2.9	2.5	6.6	3.4	4.2	4.5
Barbados Barbados									
At current prices	1 145	1 198	1 316	1 449	1 541	1 698	1 711	1 687	1 574
Per capita	4 544	4 737	5 180	5 704	6 044	6 632	6 656	6 539	6 078
At constant prices	1 411	1 420	1 566	1 672	1 731	1 790	1 711	1 662	1 596
Growth rates	2.4	0.7	10.3	6.8	3.5	3.4	−4.4	−2.8	−4.0
Belarus Bélarus									
At current prices	37 733	34 200	44 289	50 829	53 781	59 369	63 213	48 114	4 731
Per capita	3 797	3 429	4 404	5 027	5 290	5 804	6 162	4 681	460
At constant prices	52 353	54 431	57 755	58 170	59 417	64 111	63 213	61 317	55 431
Growth rates	5.0	4.0	6.1	0.7	2.1	7.9	−1.4	−3.0	−9.6
Belgium Belgique									
At current prices	76 709	79 841	111 721	139 636	151 518	153 608	192 303	196 874	219 307
Per capita	7 788	8 099	11 315	14 112	15 272	15 444	19 294	19 717	21 935
At constant prices	163 492	164 826	167 216	170 644	179 153	185 943	192 303	195 933	198 708
Growth rates	2.2	0.8	1.5	2.0	5.0	3.8	3.4	1.9	1.4
Belize Belize									
At current prices	211	209	228	277	315	363	396	430	468
Per capita	1 302	1 260	1 333	1 580	1 749	1 962	2 098	2 219	2 364
At constant prices	252	254	266	297	323	363	396	413	443
Growth rates	2.0	1.0	4.6	11.6	9.0	12.2	9.3	4.2	7.2
Benin Bénin									
At current prices	1 051	1 046	1 336	1 562	1 620	1 528	1 877	1 931	2 137
Per capita	272	263	326	370	372	341	406	405	435
At constant prices	1 646	1 770	1 808	1 781	1 835	1 822	1 877	1 915	1 962
Growth rates	7.9	7.5	2.1	−1.5	3.0	−0.7	3.0	2.0	2.5
Bermuda Bermudes									
At current prices	1 040	1 174	1 297	1 415	1 502	1 592	1 635	1 680	1 698
Per capita	18 900	20 955	22 746	24 398	25 449	26 540	26 802	27 095	27 381
At constant prices	1 465	1 563	1 621	1 687	1 694	1 695	1 635	1 632	1 596
Growth rates	−2.9	6.7	3.7	4.1	0.4	0.0	−3.5	−0.2	−2.2
Bhutan Bhoutan									
At current prices	185	193	222	278	283	270	283	246	245
Per capita	137	140	158	193	192	179	184	156	152
At constant prices	189	196	216	254	257	269	283	299	312
Growth rates	7.0	3.7	10.2	17.8	1.0	4.7	5.3	5.4	4.5
Bolivia Bolivie									
At current prices	7 406	6 487	4 648	4 953	4 834	5 479	5 529	6 058	6 311
Per capita	1 198	1 023	715	743	708	783	771	825	839
At constant prices	5 136	5 087	4 960	5 089	5 240	5 388	5 529	5 758	5 919
Growth rates	−0.6	−1.0	−2.5	2.6	3.0	2.8	2.6	4.1	2.8
Bosnia & Herzegovina Bosnie−Herzégovine									
At current prices	5 526	5 672	7 930	9 080	7 983	10 514	13 032	17 631	...
Per capita	1 293	1 314	1 819	2 064	1 797	2 346	2 886	3 639	...
At constant prices	16 485	16 733	17 370	17 085	16 707	16 969	13 032	9 131	6 392
Growth rates	2.2	1.5	3.8	−1.6	−2.2	1.6	−23.2	−29.9	−30.0
Botswana Botswana									
At current prices	1 083	968	1 296	1 675	2 090	2 719	3 296	3 467	3 943
Per capita	1 047	907	1 176	1 475	1 788	2 260	2 662	2 720	3 003

21

Gross domestic product: total and per capita

In US dollars (millions) [1] at current and constant 1990 prices; per capita US$;

real rates of growth [*cont.*]

Produit intérieur brut : total et par habitant

En monnaie dollars de E−u (millions) [1] aux prix courants et constants de 1990; par habitant;

taux de l'accroissement réels [*suite*]

Country or area Pays ou zone	1984	1985	1986	1987	1988	1989	1990	1991	1992
At constant prices	1 977	2 119	2 278	2 481	2 859	3 136	3 296	3 569	3 737
Growth rates	11.5	7.2	7.5	8.9	15.3	9.7	5.1	8.3	4.7
Brazil Brésil									
At current prices	208 874	223 065	268 082	294 311	328 495	447 473	473 697	401 091	389 572
Per capita	1 574	1 645	1 938	2 086	2 285	3 056	3 178	2 646	2 528
At constant prices	395 429	428 382	461 334	494 287	494 287	494 287	473 697	478 149	473 846
Growth rates	0.0	8.3	7.7	7.1	0.0	0.0	−4.2	0.9	−0.9
British Virgin Islands Iles Vierges britanniques									
At current prices	86	90	98	117	131	156	163	166	175
Per capita	6 638	6 420	6 987	7 777	8 756	9 763	10 164	9 778	10 316
At constant prices	113	114	119	136	151	164	163	156	160
Growth rates	5.6	1.4	3.8	14.7	11.3	8.6	−1.0	−4.0	2.3
Brunei Darussalam Brunéi Darussalam									
At current prices	3 769	3 482	2 314	2 753	2 689	2 996	3 590	3 816	3 919
Per capita	17 132	15 407	9 972	11 518	10 977	11 936	13 969	14 456	14 516
At constant prices	3 574	3 521	3 425	3 494	3 532	3 494	3 590	3 721	3 683
Growth rates	−0.1	−1.5	−2.7	2.0	1.1	−1.1	2.7	3.6	−1.0
Bulgaria Bulgarie									
At current prices	31 991	31 646	36 621	41 990	46 199	47 118	20 726	7 367	9 583
Per capita	3 576	3 532	4 081	4 672	5 136	5 237	2 305	821	1 070
At constant prices	21 837	22 226	23 412	24 597	25 190	25 116	20 726	19 294	17 712
Growth rates	4.6	1.8	5.3	5.1	2.4	−0.3	−17.5	−6.9	−8.2
Burkina Faso Burkina Faso									
At current prices	894	1 045	1 454	1 788	1 900	1 854	2 245	2 221	2 455
Per capita	116	133	180	215	223	212	250	240	258
At constant prices	1 844	2 034	2 115	2 146	2 190	2 201	2 245	2 289	2 343
Growth rates	−2.1	10.3	4.0	1.5	2.0	0.5	2.0	2.0	2.4
Burundi Burundi									
At current prices	1 006	1 171	1 234	1 162	1 089	1 132	1 141	1 170	1 056
Per capita	218	247	252	231	210	212	208	207	181
At constant prices	868	969	1 006	1 048	1 086	1 102	1 141	1 198	1 228
Growth rates	−0.1	11.7	3.8	4.1	3.7	1.4	3.5	5.0	2.5
Cambodia Cambodge									
At current prices	685	650	620	625	765	792	801	862	923
Per capita	96	89	82	81	97	97	96	101	105
At constant prices	763	732	693	684	765	792	801	862	922
Growth rates	−2.5	−4.1	−5.3	−1.3	11.7	3.5	1.2	7.6	7.0
Cameroon Cameroun									
At current prices	7 312	8 545	11 941	13 326	12 657	11 160	13 363	12 788	14 052
Per capita	755	857	1 164	1 261	1 164	997	1 160	1 079	1 152
At constant prices	13 520	14 728	15 793	15 007	13 923	13 088	13 363	13 111	13 373
Growth rates	7.8	8.9	7.2	−5.0	−7.2	−6.0	2.1	−1.9	2.0
Canada Canada									
At current prices	340 754	347 374	360 868	412 351	488 210	544 888	569 432	583 668	563 769
Per capita	13 660	13 795	14 188	16 041	18 779	20 712	21 376	21 623	20 600
At constant prices	473 613	496 033	512 227	533 435	559 915	572 534	569 432	559 343	563 548
Growth rates	6.4	4.7	3.3	4.1	5.0	2.3	−0.5	−1.8	0.8
Cape Verde Cap−Vert									
At current prices	136	143	194	248	286	282	328	341	385
Per capita	431	441	586	734	828	798	905	913	1 002
At constant prices	234	254	261	280	302	317	328	346	363
Growth rates	3.7	8.5	2.7	7.6	7.6	4.9	3.7	5.3	5.1
Cayman Islands Iles Caïmanes									
At current prices	203	220	242	283	345	395	492	513	540
Per capita	10 167	10 476	10 985	12 283	13 800	15 192	18 210	18 333	18 632
At constant prices	287	297	320	349	399	434	492	498	510
Growth rates	7.4	3.5	7.8	9.3	14.2	8.8	13.2	1.3	2.3
Central African Rep. Rép. centrafricaine									
At current prices	787	865	1 122	1 201	1 265	1 237	1 495	1 471	1 631
Per capita	307	329	415	433	444	422	497	476	514
At constant prices	1 325	1 370	1 427	1 384	1 422	1 478	1 495	1 492	1 507
Growth rates	7.2	3.3	4.2	−3.0	2.7	4.0	1.1	−0.2	1.0
Chad Tchad									
At current prices	626	780	817	819	1 044	1 019	1 219	1 290	1 418

21

Gross domestic product: total and per capita
In US dollars (millions) [1] at current and constant 1990 prices; per capita US$;
real rates of growth [cont.]
Produit intérieur brut : total et par habitant
En monnaie dollars de E−u (millions) [1] aux prix courants et constants de 1990; par habitant;
taux de l'accroissement réels [suite]

Country or area Pays ou zone	1984	1985	1986	1987	1988	1989	1990	1991	1992
Per capita	127	155	160	157	196	188	220	227	243
At constant prices	912	1 112	1 066	1 040	1 200	1 252	1 219	1 200	1 212
Growth rates	2.1	21.9	−4.1	−2.4	15.3	4.4	−2.6	−1.6	1.0
Chile Chili									
At current prices	18 656	16 463	17 714	20 682	24 150	28 082	30 167	33 977	41 203
Per capita	1 566	1 358	1 437	1 650	1 894	2 167	2 290	2 538	3 030
At constant prices	21 595	22 020	23 252	24 785	26 597	29 298	30 167	31 996	35 285
Growth rates	5.9	2.0	5.6	6.6	7.3	10.2	3.0	6.1	10.3
China Chine									
At current prices	298 621	290 360	280 584	303 780	378 120	424 876	369 648	379 231	441 791
Per capita	288	276	263	280	344	381	326	330	378
At constant prices	225 414	252 736	276 908	301 261	329 673	346 449	369 648	393 893	444 311
Growth rates	13.8	12.1	9.6	8.8	9.4	5.1	6.7	6.6	12.8
Colombia Colombie									
At current prices	38 256	34 896	34 943	36 371	39 212	39 540	40 274	41 452	43 438
Per capita	1 324	1 184	1 163	1 188	1 257	1 246	1 247	1 261	1 300
At constant prices	31 214	32 184	34 058	35 887	37 345	38 620	40 274	41 119	42 600
Growth rates	3.4	3.1	5.8	5.4	4.1	3.4	4.3	2.1	3.6
Comoros Comores									
At current prices	107	114	162	196	207	199	244	245	273
Per capita	245	252	345	403	411	379	449	435	466
At constant prices	225	231	235	239	246	242	244	248	252
Growth rates	4.2	2.7	2.1	1.6	2.7	−1.6	0.8	1.8	1.4
Congo Congo									
At current prices	2 194	2 161	1 849	2 298	2 212	2 425	2 840	2 909	3 382
Per capita	1 175	1 124	934	1 127	1 054	1 121	1 274	1 266	1 428
At constant prices	2 959	2 925	2 723	2 729	2 777	2 826	2 840	2 772	2 802
Growth rates	7.2	−1.2	−6.9	0.2	1.8	1.8	0.5	−2.4	1.1
Cook Islands Iles Cook									
At current prices	25	26	33	46	52	49	52	55	54
Per capita	1 391	1 438	1 950	2 697	3 054	2 904	3 051	3 255	3 161
At constant prices	35	38	41	42	50	54	52	54	54
Growth rates	15.3	8.8	8.2	1.8	19.0	7.1	−3.3	3.9	1.0
Costa Rica Costa Rica									
At current prices	3 660	3 923	4 404	4 531	4 572	5 175	5 686	5 523	6 311
Per capita	1 426	1 485	1 620	1 620	1 590	1 751	1 874	1 774	1 977
At constant prices	4 509	4 541	4 792	5 021	5 193	5 486	5 686	5 744	6 157
Growth rates	8.0	0.7	5.5	4.8	3.4	5.6	3.7	1.0	7.2
Côte d'Ivoire Côte d'Ivoire									
At current prices	6 841	6 978	9 369	10 375	10 321	9 370	11 959	12 360	13 807
Per capita	715	702	908	969	928	812	998	994	1 069
At constant prices	12 006	12 595	13 022	12 815	12 583	12 458	11 959	12 163	12 418
Growth rates	−2.0	4.9	3.4	−1.6	−1.8	−1.0	−4.0	1.7	2.1
Croatia Croatie									
At current prices	11 025	11 162	15 451	17 969	16 006	20 436	25 323	34 259	25 934
Per capita	2 373	2 397	3 311	3 845	3 419	4 362	5 358	7 232	5 319
At constant prices	32 881	32 924	33 835	33 802	33 487	32 974	25 323	17 744	13 432
Growth rates	2.1	0.1	2.8	−0.1	−0.9	−1.5	−23.2	−29.9	−24.3
Cuba Cuba									
At current prices	18 898	17 968	19 090	19 262	20 752	20 218	20 874	18 738	16 585
Per capita	1 892	1 783	1 876	1 873	1 997	1 925	1 968	1 749	1 534
At constant prices	21 533	21 876	22 248	21 180	21 644	21 542	20 874	15 656	13 464
Growth rates	7.9	1.6	1.7	−4.8	2.2	−0.5	−3.1	−25.0	−14.0
Cyprus Chypre									
At current prices	2 279	2 430	3 094	3 701	4 276	4 568	5 511	5 743	6 639
Per capita	3 463	3 653	4 598	5 442	6 215	6 573	7 850	8 101	9 273
At constant prices	3 787	3 965	4 116	4 404	4 781	5 176	5 511	5 567	6 042
Growth rates	8.8	4.7	3.8	7.0	8.5	8.3	6.5	1.0	8.5
former Czechoslovakia † anc. Tchécoslovaquie †									
At current prices	39 693	39 501	46 343	52 163	51 530	50 464	45 625	34 342	..
Per capita	2 565	2 548	2 984	3 352	3 304	3 229	2 913	2 188	..
At constant prices	41 587	42 756	43 933	45 067	45 976	46 559	45 625	36 866	..
Growth rates	3.1	2.8	2.8	2.6	2.0	1.3	−2.0	−19.2	..

21
Gross domestic product: total and per capita
In US dollars (millions) [1] at current and constant 1990 prices; per capita US$;
real rates of growth [*cont.*]
  Produit intérieur brut : total et par habitant
  En monnaie dollars de E−u (millions) [1] aux prix courants et constants de 1990; par habitant;
  taux de l'accroissement réels [*suite*]

Country or area Pays ou zone	1984	1985	1986	1987	1988	1989	1990	1991	1992
Czech Republic République tchèque									
At current prices	27 863	27 637	32 221	36 163	35 755	34 855	31 606	24 309	27 293
Per capita	2 697	2 674	3 116	3 494	3 453	3 364	3 050	2 359	2 623
At constant prices	29 039	29 215	29 824	29 993	30 608	31 995	31 606	27 113	25 193
Growth rates	..	0.6	2.1	0.6	2.1	4.5	−1.2	−14.2	−7.1
Denmark Danemark									
At current prices	54 580	58 048	82 375	102 326	108 743	104 959	129 745	130 287	142 496
Per capita	10 654	11 333	16 079	19 966	21 206	20 448	25 242	25 308	27 626
At constant prices	115 285	120 230	124 610	124 976	126 432	127 144	129 745	131 338	132 974
Growth rates	4.4	4.3	3.6	0.3	1.2	0.6	2.0	1.2	1.2
Djibouti Djibouti									
At current prices	401	417	431	460	497	509	552	554	578
Per capita	1 096	1 096	1 097	1 136	1 194	1 192	1 255	1 222	1 238
At constant prices	450	457	463	493	523	531	552	546	563
Growth rates	−0.6	1.6	1.2	6.5	6.1	1.4	4.0	−1.1	3.0
Dominica Dominica									
At current prices	90	99	112	126	146	157	167	177	187
Per capita	1 214	1 350	1 535	1 721	1 995	2 174	2 323	2 463	2 594
At constant prices	126	127	136	147	160	159	167	170	175
Growth rates	5.5	1.3	7.1	7.6	8.7	−0.4	5.3	1.8	2.7
Dominican Republic Rép. dominicaine									
At current prices	10 355	4 488	5 434	5 081	4 638	6 687	7 103	7 202	8 155
Per capita	1 652	700	828	757	675	953	991	984	1 092
At constant prices	6 611	6 440	6 644	7 167	7 216	7 509	7 103	7 061	7 617
Growth rates	0.3	−2.6	3.2	7.9	0.7	4.1	−5.4	−0.6	7.9
Ecuador Equateur									
At current prices	12 995	15 958	11 266	10 527	10 012	9 823	10 629	11 612	12 626
Per capita	1 433	1 714	1 179	1 074	997	954	1 008	1 075	1 142
At constant prices	9 206	9 605	9 903	9 310	10 290	10 316	10 629	11 147	11 535
Growth rates	4.2	4.3	3.1	−6.0	10.5	0.3	3.0	4.9	3.5
Egypt Egypte									
At current prices	44 638	52 311	58 314	62 932	87 299	65 260	43 871	33 166	40 898
Per capita	984	1 125	1 223	1 288	1 745	1 274	837	618	746
At constant prices	36 288	38 027	39 017	39 994	41 553	42 799	43 871	44 878	45 012
Growth rates	6.0	4.8	2.6	2.5	3.9	3.0	2.5	2.3	0.3
El Salvador ElSalvador									
At current prices	4 663	5 732	4 075	4 628	5 473	6 446	5 113	5 915	5 982
Per capita	995	1 210	848	948	1 101	1 272	989	1 120	1 109
At constant prices	4 570	4 660	4 690	4 816	4 893	4 945	5 113	5 294	5 566
Growth rates	2.3	2.0	0.6	2.7	1.6	1.0	3.4	3.5	5.1
Equatorial Guinea Guinée équatoriale									
At current prices	68	85	107	131	144	132	163	165	185
Per capita	230	272	332	393	423	384	463	457	502
At constant prices	137	147	142	152	160	156	163	168	173
Growth rates	2.3	7.3	−3.8	7.5	5.3	−2.8	4.4	3.4	2.9
Eritrea Erythrée									
At current prices	300	297	325	339	356	378	376	413	173
Per capita	115	111	118	119	121	125	120	128	53
At constant prices	329	304	326	360	374	382	376	394	376
Growth rates	−0.1	−8.7	0.7	5.0	5.0	8.1	−6.5	−13.6	−32.0
Estonia Estonie									
At current prices	6 412	5 825	7 276	8 537	9 488	10 242	11 977	10 180	1 152
Per capita	4 207	3 792	4 701	5 476	6 043	6 491	7 566	6 427	728
At constant prices	11 686	10 670	10 747	11 280	11 848	12 806	11 977	10 346	7 036
Growth rates	−3.6	−7.0	6.7	9.0	3.6	2.9	−0.4	1.5	−5.2
Ethiopia Ethiopie									
At current prices	4 831	4 778	5 233	5 409	5 685	6 034	6 005	6 592	2 740
Per capita	115	111	118	119	121	125	121	128	52
At constant prices	5 229	4 865	5 190	5 655	5 859	6 031	6 005	6 096	5 779
Growth rates	−2.3	−7.6	7.2	9.4	3.8	2.3	−1.5	4.8	−2.6
Ethiopia excluding Eritrea Ethiopie non compris Erythrée									
At current prices	4 531	4 481	4 908	5 069	5 328	5 656	5 628	6 179	2 639
Per capita	115	111	118	119	121	125	121	128	53

21
Gross domestic product: total and per capita
In US dollars (millions) [1] at current and constant 1990 prices; per capita US$;
real rates of growth [cont.]

Produit intérieur brut : total et par habitant
En monnaie dollars de E−u (millions) [1] aux prix courants et constants de 1990; par habitant;
taux de l'accroissement réels [suite]

Country or area Pays ou zone	1984	1985	1986	1987	1988	1989	1990	1991	1992
At constant prices	4 970	4 590	4 918	5 379	5 584	5 714	5 628	5 897	5 744
Growth rates	−4.5	−7.6	7.1	10.4	3.8	2.3	−1.5	4.8	−4.6
Fiji Fidji									
At current prices	1 178	1 141	1 290	1 178	1 110	1 255	1 228	1 311	1 407
Per capita	1 709	1 632	1 825	1 652	1 548	1 740	1 691	1 791	1 904
At constant prices	1 056	1 014	1 083	1 013	1 031	1 168	1 228	1 235	1 280
Growth rates	8.3	−3.9	6.8	−6.5	1.8	13.2	5.1	0.6	3.6
Finland Finlande									
At current prices	51 307	54 048	70 533	89 088	105 561	115 803	137 544	124 541	108 954
Per capita	10 516	11 026	14 330	18 038	21 308	23 310	27 608	24 933	21 756
At constant prices	112 625	116 417	119 178	124 065	130 149	137 527	137 544	127 826	123 005
Growth rates	3.0	3.4	2.4	4.1	4.9	5.7	0.0	−7.1	−3.8
France France									
At current prices	499 126	523 098	731 912	887 859	962 765	965 358	1 192 217	1 199 291	1 323 721
Per capita	9 094	9 482	13 193	15 912	17 154	17 105	21 020	21 053	23 149
At constant prices	999 309	1 018 107	1 043 744	1 067 234	1 115 243	1 162 665	1 192 217	1 200 806	1 215 251
Growth rates	1.3	1.9	2.5	2.3	4.5	4.3	2.5	0.7	1.2
French Polynesia Polynesie française									
At current prices	1 270	1 388	2 120	2 346	2 475	2 428	3 007	3 036	3 324
Per capita	7 515	7 978	11 841	12 821	13 164	12 581	15 189	14 955	16 060
At constant prices	2 241	2 361	2 560	2 743	2 808	2 893	3 007	3 125	3 157
Growth rates	5.5	5.4	8.4	7.1	2.4	3.0	4.0	3.9	1.0
Gabon Gabon									
At current prices	3 561	3 663	3 468	3 396	3 403	3 662	4 431	4 438	4 864
Per capita	3 748	3 719	3 404	3 225	3 131	3 263	3 823	3 708	3 932
At constant prices	6 018	6 078	5 166	4 346	4 200	4 305	4 431	4 515	4 607
Growth rates	2.0	1.0	−15.0	−15.9	−3.4	2.5	2.9	1.9	2.0
Gambia Gambie									
At current prices	172	201	156	210	244	256	298	306	339
Per capita	238	270	204	266	300	306	346	346	374
At constant prices	245	249	259	267	271	283	298	304	318
Growth rates	−8.2	1.6	4.1	2.8	1.7	4.3	5.2	2.3	4.5
Georgia Géorgie									
At current prices	16 045	17 122	19 583	21 774	23 401	23 168	22 523	10 914	789
Per capita	3 095	3 282	3 730	4 123	4 358	4 252	4 125	1 996	144
At constant prices	20 400	22 983	21 751	21 513	22 142	22 599	22 523	17 953	13 465
Growth rates	12.2	12.7	−5.4	−1.1	2.9	2.1	−0.3	−20.3	− 25.0
Germany † Allemagne †									
At current prices	..	..	..	..	..	..	..	1 696 836	1 938 657
Per capita	..	..	..	..	..	..	..	21 243	24 157
At constant prices	..	..	..	..	..	..	..	1 696 836	1 732 088
Growth rates	..	..	..	..	..	..	..	..	2.1
F. R. Germany R.F. Allemagne									
At current prices	615 211	619 287	886 618	1 107 422	1 193 475	1 183 213	1 501 021	..	..
Per capita	10 074	10 148	14 526	18 125	19 504	19 310	24 477	..	..
At constant prices	1 246 781	1 272 092	1 301 929	1 321 166	1 370 358	1 420 027	1 501 021	..	..
Growth rates	5.7	5.1	4.2	3.5	2.8	2.2	−25.1	..	..
former German D. R. anc. R.d. allemande									
At current prices	90 804	93 722	127 063	158 373	167 621	154 516	141 308	..	..
Per capita	5 441	5 631	7 665	9 604	10 223	9 473	8 696	..	..
At constant prices	158 514	166 540	173 562	179 582	184 598	188 611	141 308	..	..
Growth rates	2.8	2.0	2.3	1.5	3.7	3.6	5.7	..	..
Ghana Ghana									
At current prices	7 730	6 346	5 733	4 853	5 195	5 249	6 226	6 413	6 357
Per capita	624	494	432	354	367	360	415	414	398
At constant prices	4 685	4 923	5 180	5 429	5 732	6 025	6 226	6 463	6 792
Growth rates	8.6	5.1	5.2	4.8	5.6	5.1	3.3	3.8	5.1
Greece Grèce									
At current prices	33 762	33 433	39 397	46 212	53 056	54 042	66 679	70 574	78 260
Per capita	3 413	3 366	3 949	4 613	5 277	5 356	6 587	6 951	7 686
At constant prices	59 516	61 374	62 369	61 949	64 472	66 713	66 679	67 885	68 522
Growth rates	2.8	3.1	1.6	−0.7	4.1	3.5	−0.1	1.8	0.9
Grenada Grenade									
At current prices	102	115	130	150	166	182	200	210	217

21

Gross domestic product: total and per capita
In US dollars (millions) [1] at current and constant 1990 prices; per capita US$;
real rates of growth [cont.]

Produit intérieur brut : total et par habitant
En monnaie dollars de E−u (millions) [1] aux prix courants et constants de 1990; par habitant;
taux de l'accroissement réels [suite]

Country or area Pays ou zone	1984	1985	1986	1987	1988	1989	1990	1991	1992
Per capita	1 131	1 281	1 442	1 671	1 847	2 000	2 202	2 309	2 380
At constant prices	140	152	158	166	177	188	200	206	208
Growth rates	4.7	8.2	3.9	5.1	6.8	6.2	6.7	3.0	0.6
Guadeloupe Guadeloupe									
At current prices	1 035	1 091	1 582	2 045	2 187	2 204	2 763	2 636	2 962
Per capita	2 975	3 073	4 370	5 543	5 815	5 755	7 084	6 674	7 405
At constant prices	2 159	2 142	2 219	2 474	2 563	2 679	2 763	2 653	2 714
Growth rates	0.6	−0.8	3.6	11.5	3.6	4.5	3.1	−4.0	2.3
Guatemala Guatemala									
At current prices	9 470	11 180	8 470	7 084	7 843	8 410	7 650	9 406	10 434
Per capita	1 224	1 404	1 034	840	903	941	832	994	1 071
At constant prices	6 666	6 627	6 636	6 871	7 138	7 420	7 650	7 930	8 309
Growth rates	0.5	−0.6	0.1	3.5	3.9	3.9	3.1	3.7	4.8
Guinea Guinée									
At current prices	17 347	25 085	2 427	2 080	2 384	2 432	2 764	3 016	3 000
Per capita	3 568	5 030	474	395	440	435	480	508	490
At constant prices	2 065	2 145	2 250	2 388	2 519	2 640	2 764	2 893	3 041
Growth rates	−2.5	3.9	4.9	6.1	5.5	4.8	4.7	4.7	5.1
Guinea−Bissau Guinée−Bissau									
At current prices	264	248	230	165	155	198	235	251	134
Per capita	307	284	259	182	167	210	243	255	133
At constant prices	191	184	193	203	217	227	235	242	249
Growth rates	7.0	−3.5	4.6	5.6	6.9	4.5	3.3	3.0	2.9
Guyana Guyana									
At current prices	444	462	520	349	414	256	256	219	239
Per capita	564	585	656	441	522	323	322	274	296
At constant prices	300	303	305	306	296	285	256	272	293
Growth rates	2.2	1.0	0.4	0.3	−3.3	−3.5	−10.1	6.0	7.7
Haiti Haïti									
At current prices	1 816	2 009	2 238	1 968	1 967	1 862	2 487	1 888	1 584
Per capita	316	343	374	323	316	293	384	285	235
At constant prices	2 461	2 476	2 463	2 445	2 464	2 491	2 487	2 504	2 233
Growth rates	0.3	0.6	−0.5	−0.7	0.8	1.1	−0.1	0.7	−10.8
Honduras Honduras									
At current prices	3 231	3 504	3 798	4 064	4 469	4 891	5 902	2 797	2 928
Per capita	763	799	838	868	925	982	1 149	528	536
At constant prices	4 893	5 066	5 208	5 477	5 760	5 895	5 902	6 076	6 422
Growth rates	2.8	3.5	2.8	5.2	5.2	2.3	0.1	2.9	5.7
Hong Kong Hong−kong									
At current prices	31 705	33 513	38 255	47 139	55 554	63 995	71 839	82 837	96 088
Per capita	5 882	6 143	6 935	8 465	9 889	11 299	12 583	14 396	16 567
At constant prices	49 017	49 111	54 558	62 484	67 662	69 581	71 839	74 777	78 775
Growth rates	9.8	0.2	11.1	14.5	8.3	2.8	3.2	4.1	5.3
Hungary Hongrie									
At current prices	20 812	21 078	24 276	26 680	28 571	29 168	33 056	31 391	35 512
Per capita	1 951	1 979	2 284	2 515	2 697	2 759	3 132	2 981	3 378
At constant prices	32 288	32 206	32 700	34 025	34 003	34 253	33 056	29 084	27 775
Growth rates	2.7	−0.3	1.5	4.1	−0.1	0.7	−3.5	−12.0	−4.5
Iceland Islande									
At current prices	2 762	2 871	3 843	5 336	5 891	5 355	6 024	6 490	6 613
Per capita	11 558	11 914	15 750	21 604	23 657	21 251	23 622	25 252	25 436
At constant prices	5 000	5 182	5 519	6 003	5 982	5 993	6 024	6 085	5 854
Growth rates	4.1	3.6	6.5	8.8	−0.3	0.2	0.5	1.0	−3.8
India Inde									
At current prices	203 593	212 016	232 296	257 060	284 970	279 789	303 282	268 006	269 508
Per capita	271	277	297	322	350	337	358	311	306
At constant prices	213 301	224 965	235 920	247 172	272 314	287 074	303 282	307 252	319 542
Growth rates	3.7	5.5	4.9	4.8	10.2	5.4	5.6	1.3	4.0
Indonesia Indonésie									
At current prices	87 612	87 339	80 061	75 932	84 300	94 449	106 859	116 476	128 281
Per capita	534	522	469	436	475	522	580	620	671
At constant prices	77 085	78 983	83 624	87 743	92 814	99 736	106 859	113 909	121 085
Growth rates	7.0	2.5	5.9	4.9	5.8	7.5	7.1	6.6	6.3

21

Gross domestic product: total and per capita
In US dollars (millions) [1] at current and constant 1990 prices; per capita US$;
real rates of growth [cont.]

Produit intérieur brut : total et par habitant
En monnaie dollars de E−u (millions) [1]   aux prix courants et constants de 1990; par habitant;
taux de l'accroissement réels [suite]

Country or area Pays ou zone	1984	1985	1986	1987	1988	1989	1990	1991	1992
Iran, Islamic Rep. of Iran, Rép. islamique d'									
At current prices	164 434	173 253	206 031	279 163	324 752	385 877	538 106	606 426	834 907
Per capita	3 506	3 542	4 049	5 286	5 938	6 829	9 235	10 116	13 561
At constant prices	592 950	594 329	504 579	510 437	466 129	481 587	538 106	562 859	600 007
Growth rates	0.9	0.2	−15.1	1.2	−8.7	3.3	11.7	4.6	6.6
Iraq Iraq									
At current prices	47 982	49 819	48 434	57 558	64 413	67 607	74 909	64 115	68 603
Per capita	3 237	3 252	3 059	3 516	3 806	3 864	4 143	3 433	3 556
At constant prices	61 790	61 780	66 807	79 753	82 187	74 897	74 909	25 379	26 141
Growth rates	0.2	0.0	8.1	19.4	3.1	−8.9	0.0	−66.1	3.0
Ireland Irlande									
At current prices	18 616	19 549	26 625	30 107	32 358	35 953	44 752	45 152	50 492
Per capita	5 262	5 504	7 490	8 488	9 159	10 223	12 775	12 926	14 484
At constant prices	34 586	35 653	35 501	37 094	38 619	41 049	44 752	45 907	48 161
Growth rates	4.4	3.1	−0.4	4.5	4.1	6.3	9.0	2.6	4.9
Israel Israël									
At current prices	27 480	25 867	32 245	38 680	46 810	45 794	54 683	62 687	69 380
Per capita	6 593	6 111	7 516	8 908	10 619	10 156	11 734	12 869	13 522
At constant prices	43 042	44 585	46 562	49 879	51 112	51 345	54 683	58 877	63 154
Growth rates	1.2	3.6	4.4	7.1	2.5	0.5	6.5	7.7	7.3
Italy Italie									
At current prices	413 077	424 512	603 634	759 049	838 842	869 807	1 094 765	1 150 526	1 223 625
Per capita	7 245	7 429	10 542	13 229	14 592	15 106	18 986	19 930	21 177
At constant prices	917 893	941 762	969 276	999 670	1 040 305	1 070 881	1 094 765	1 110 503	1 120 826
Growth rates	2.7	2.6	2.9	3.1	4.1	2.9	2.2	1.4	0.9
Jamaica Jamaïque									
At current prices	2 373	2 015	2 503	2 982	3 487	3 968	4 151	3 497	3 052
Per capita	1 041	872	1 071	1 264	1 465	1 654	1 715	1 431	1 236
At constant prices	3 471	3 310	3 365	3 615	3 720	3 961	4 151	4 161	4 223
Growth rates	−0.9	−4.6	1.7	7.4	2.9	6.5	4.8	0.2	1.5
Japan Japon									
At current prices	1 265 333	1 343 251	1 985 574	2 408 912	2 898 385	2 871 825	2 932 088	3 346 411	3 658 382
Per capita	10 532	11 116	16 347	19 740	23 650	23 338	23 734	26 983	29 387
At constant prices	2 242 453	2 354 213	2 416 074	2 515 269	2 671 417	2 797 374	2 932 088	3 050 777	3 092 219
Growth rates	4.3	5.0	2.6	4.1	6.2	4.7	4.8	4.0	1.4
Jordan Jordanie									
At current prices	4 864	4 818	5 830	6 166	5 925	4 454	3 869	4 041	4 744
Per capita	1 474	1 414	1 657	1 697	1 579	1 149	965	974	1 106
At constant prices	3 772	3 902	4 274	4 419	4 398	3 803	3 869	3 937	4 383
Growth rates	8.6	3.5	9.5	3.4	−0.5	−13.5	1.7	1.8	11.3
Kazakhstan Kazakhstan									
At current prices	36 424	37 103	46 403	51 527	58 194	61 894	69 670	46 286	10 355
Per capita	2 313	2 328	2 876	3 150	3 526	3 726	4 162	2 740	607
At constant prices	47 184	50 089	50 868	50 153	54 780	60 301	69 670	63 689	62 797
Growth rates	8.3	6.2	1.6	−1.4	9.2	10.1	15.5	−8.6	−1.4
Kenya Kenya									
At current prices	6 197	6 138	7 240	7 971	8 510	8 259	8 675	8 261	8 366
Per capita	323	309	352	374	386	362	368	339	332
At constant prices	6 325	6 597	7 068	7 488	7 952	8 326	8 675	8 799	8 837
Growth rates	1.8	4.3	7.1	5.9	6.2	4.7	4.2	1.4	0.4
Kiribati Kiribati									
At current prices	25	20	20	22	30	33	37	39	39
Per capita	399	306	315	328	440	478	514	544	528
At constant prices	33	30	29	29	35	37	37	37	38
Growth rates	2.2	−9.3	−0.6	−1.2	21.0	4.1	−0.3	1.9	2.5
Korea, Dem. P. R. Corée, R.p. dém.									
At constant prices	15 161	16 615	17 237	18 380	19 418	20 561	21 712	20 583	19 554
Growth rates	9.8	9.6	3.7	6.6	5.6	5.9	5.6	−5.2	−5.0
Korea, Republic of Corée, République de									
At current prices	90 131	92 925	105 991	131 816	174 940	212 970	244 043	283 904	296 839
Per capita	2 238	2 277	2 564	3 147	4 124	4 962	5 626	6 483	6 721
At constant prices	140 310	150 036	168 633	188 928	210 618	223 581	244 043	264 698	277 373
Growth rates	9.4	6.9	12.4	12.0	11.5	6.2	9.2	8.5	4.8

21

Gross domestic product: total and per capita
In US dollars (millions) [1] at current and constant 1990 prices; per capita US$;
real rates of growth [cont.]
Produit intérieur brut : total et par habitant
En monnaie dollars de E−u (millions) [1] aux prix courants et constants de 1990; par habitant;
taux de l'accroissement réels [suite]

Country or area Pays ou zone	1984	1985	1986	1987	1988	1989	1990	1991	1992
Kuwait Koweït									
At current prices	21 706	21 429	17 728	20 513	18 620	24 018	18 423	11 177	21 703
Per capita	13 276	12 458	9 708	10 541	9 048	11 276	8 597	5 358	11 017
At constant prices	25 905	24 796	26 732	25 824	23 229	26 318	18 423	10 685	20 300
Growth rates	5.2	−4.3	7.8	−3.4	−10.1	13.3	−30.0	−42.0	90.0
Kyrgyzstan Kirghizistan									
At current prices	8 076	7 301	8 651	10 177	11 294	12 779	12 613	9 886	799
Per capita	2 057	1 821	2 112	2 429	2 649	2 953	2 870	2 218	177
At constant prices	10 431	9 877	9 578	10 043	10 705	12 463	12 613	11 300	9 865
Growth rates	10.2	−5.3	−3.0	4.9	6.6	16.4	1.2	−10.4	−12.7
Lao People's Dem. Rep. Rép. dé. pop. lao									
At current prices	1 851	2 496	1 867	1 138	630	773	865	1 042	1 201
Per capita	529	695	505	298	160	190	206	240	269
At constant prices	668	702	736	728	715	811	865	900	962
Growth rates	11.4	5.1	4.8	−1.1	−1.8	13.4	6.7	4.0	7.0
Latvia Lettonie									
At current prices	10 717	9 934	12 306	13 898	15 354	16 415	18 751	16 380	1 231
Per capita	4 128	3 802	4 677	5 245	5 757	6 125	6 978	6 098	460
At constant prices	15 747	15 713	16 439	16 685	17 719	19 035	18 751	17 200	11 392
Growth rates	4.7	−0.2	4.6	1.5	6.2	7.4	−1.5	−8.3	−33.8
Lebanon Liban									
At current prices	2 811	2 420	2 987	3 680	2 840	3 033	3 325	4 043	4 246
Per capita	1 056	907	1 117	1 374	1 056	1 120	1 214	1 452	1 496
At constant prices	2 845	2 446	3 016	3 717	2 872	3 064	3 325	4 088	4 287
Growth rates	−12.7	−14.0	23.3	23.2	−22.7	6.7	8.5	22.9	4.9
Lesotho Lesotho									
At current prices	316	252	278	370	454	497	584	631	711
Per capita	211	163	176	228	273	291	334	352	387
At constant prices	397	411	419	440	498	556	584	594	598
Growth rates	8.5	3.5	2.0	5.1	13.0	11.8	4.9	1.7	0.8
Liberia Libéria									
At current prices	1 070	1 055	1 037	1 090	1 158	1 194	1 111	1 037	973
Per capita	502	480	457	465	480	479	431	390	354
At constant prices	1 185	1 161	1 148	1 168	1 201	1 235	1 111	1 000	854
Growth rates	−0.6	−2.0	−1.2	1.8	2.8	2.8	−10.0	−10.0	−14.6
Libyan Arab Jamah. Jamah. arabe libyenne									
At current prices	27 072	27 963	21 544	25 768	25 273	26 632	30 848	32 426	29 838
Per capita	7 450	7 386	5 473	6 308	5 968	6 071	6 787	6 887	6 121
At constant prices	29 118	31 542	28 798	27 875	28 073	28 857	30 848	32 391	32 876
Growth rates	−5.0	8.3	−8.7	−3.2	0.7	2.8	6.9	5.0	1.5
Liechtenstein Liechtenstein									
At current prices	509	529	807	1 049	1 162	1 124	1 432	1 462	1 529
Per capita	18 862	19 596	29 900	38 853	41 491	40 149	51 126	52 224	54 607
At constant prices	1 041	1 123	1 203	1 262	1 347	1 399	1 432	1 431	1 430
Growth rates	3.8	7.9	7.2	4.9	6.8	3.9	2.3	0.0	−0.1
Lithuania Lituanie									
At current prices	12 950	13 569	17 180	20 011	22 235	23 967	23 029	21 727	2 575
Per capita	3 644	3 783	4 747	5 479	6 037	6 460	6 172	5 800	686
At constant prices	17 615	19 311	19 644	20 195	22 157	24 693	23 029	19 912	12 943
Growth rates	8.0	9.6	1.7	2.8	9.7	11.4	−6.7	−13.5	−35.0
Luxembourg Luxembourg									
At current prices	3 352	3 457	4 999	6 095	6 805	7 178	8 989	9 336	11 848
Per capita	9 157	9 419	13 584	16 517	18 392	19 347	24 100	24 896	31 343
At constant prices	6 961	7 164	7 506	7 727	8 168	8 712	8 989	9 265	9 438
Growth rates	6.2	2.9	4.8	2.9	5.7	6.7	3.2	3.1	1.9
Madagascar Madagascar									
At current prices	2 374	2 345	3 258	2 566	2 442	2 498	3 080	2 673	3 120
Per capita	239	229	308	235	217	215	256	215	243
At constant prices	2 648	2 708	2 729	2 776	2 871	2 988	3 080	2 889	2 924
Growth rates	2.2	2.3	0.8	1.7	3.4	4.1	3.1	−6.2	1.2
Malawi Malawi									
At current prices	1 194	1 122	1 176	1 228	1 369	1 644	2 145	2 192	1 907
Per capita	170	153	152	150	158	180	224	220	184

21

Gross domestic product: total and per capita
In US dollars (millions) [1] at current and constant 1990 prices; per capita US$;
real rates of growth [*cont.*]

Produit intérieur brut : total et par habitant
En monnaie dollars de E−u (millions) [1] aux prix courants et constants de 1990; par habitant;
taux de l'accroissement réels [*suite*]

Country or area Pays ou zone	1984	1985	1986	1987	1988	1989	1990	1991	1992
At constant prices	1 718	1 853	1 860	1 905	1 954	2 052	2 145	2 299	2 099
Growth rates	3.8	7.8	0.4	2.4	2.6	5.0	4.6	7.2	−8.7
Malaysia Malaisie									
At current prices	33 938	31 200	27 735	31 602	34 696	37 852	42 822	47 111	58 014
Per capita	2 223	1 990	1 722	1 910	2 042	2 170	2 393	2 568	3 087
At constant prices	31 119	30 784	31 124	32 802	35 733	39 022	42 822	46 528	50 160
Growth rates	7.8	−1.1	1.1	5.4	8.9	9.2	9.7	8.7	7.8
Maldives Maldives									
At current prices	77	86	98	94	116	120	143	156	178
Per capita	433	472	516	482	578	580	673	708	782
At constant prices	77	88	95	104	113	123	143	154	164
Growth rates	13.0	13.8	8.6	8.9	8.7	9.3	16.2	7.6	6.3
Mali Mali									
At current prices	1 061	1 158	1 690	1 999	2 067	2 073	2 510	2 451	2 786
Per capita	138	146	207	238	239	232	272	258	284
At constant prices	1 849	2 006	2 329	2 274	2 280	2 451	2 510	2 505	2 747
Growth rates	0.7	8.5	16.1	−2.4	0.2	7.5	2.4	−0.2	9.7
Malta Malte									
At current prices	1 001	1 018	1 304	1 592	1 835	1 924	2 316	2 467	2 705
Per capita	2 936	2 959	3 759	4 560	5 242	5 466	6 544	6 929	7 536
At constant prices	1 672	1 715	1 782	1 855	2 011	2 176	2 316	2 446	2 553
Growth rates	0.9	2.6	3.9	4.1	8.4	8.2	6.5	5.6	4.4
Marshall Islands Iles Marshall									
At current prices	44	43	55	62	69	71	77	72	79
Per capita	1 224	1 128	1 402	1 501	1 643	1 615	1 664	1 528	1 618
At constant prices	48	47	57	65	71	72	77	69	69
Growth rates	2.7	−2.5	22.4	13.2	9.3	0.8	7.0	−9.8	0.1
Martinique Martinique									
At current prices	1 250	1 389	1 996	2 622	2 836	2 835	3 553	3 399	3 842
Per capita	3 708	4 074	5 802	7 534	8 058	7 963	9 869	9 337	10 440
At constant prices	2 621	2 739	2 862	3 072	3 259	3 452	3 553	3 412	3 490
Growth rates	4.5	4.5	4.5	7.3	6.1	5.9	2.9	−4.0	2.3
Mauritania Mauritanie									
At current prices	723	683	803	910	965	1 005	1 152	1 204	1 224
Per capita	420	387	443	488	504	510	569	578	571
At constant prices	927	958	1 013	1 037	1 058	1 113	1 152	1 190	1 235
Growth rates	−2.1	3.4	5.8	2.3	2.1	5.1	3.5	3.3	3.8
Mauritius Maurice									
At current prices	1 041	1 076	1 463	1 831	2 069	2 116	2 556	2 731	3 036
Per capita	1 030	1 055	1 419	1 757	1 965	1 988	2 378	2 515	2 765
At constant prices	1 654	1 767	1 940	2 137	2 281	2 385	2 556	2 670	2 825
Growth rates	4.8	6.9	9.7	10.2	6.8	4.6	7.2	4.5	5.8
Mexico Mexique									
At current prices	175 604	184 498	129 446	140 264	171 770	206 223	244 047	286 631	329 302
Per capita	2 378	2 441	1 674	1 773	2 124	2 494	2 889	3 321	3 736
At constant prices	222 015	227 752	219 235	223 308	226 086	233 631	244 047	252 888	259 971
Growth rates	3.6	2.6	−3.7	1.9	1.2	3.3	4.5	3.6	2.8
Micronesia, Fed. States of Micron, Etats fédérés de									
At current prices	119	109	119	144	184	215	242	269	273
Per capita	1 414	1 249	1 327	1 545	1 920	2 169	2 351	2 516	2 484
At constant prices	144	156	167	182	199	223	242	251	254
Growth rates	2.1	8.3	6.7	9.0	9.1	12.1	8.8	3.7	1.0
Monaco Monaco									
At current prices	246	256	356	430	463	462	569	588	646
Per capita	9 094	9 482	13 193	15 912	17 154	17 107	21 064	20 992	23 082
At constant prices	493	499	509	517	538	557	569	591	596
Growth rates	0.8	1.4	2.0	1.7	3.9	3.7	2.0	4.0	0.8
Mongolia Mongolie									
At current prices	2 585	2 493	2 891	3 337	3 552	3 577	2 441	662	1 107
Per capita	1 392	1 306	1 473	1 653	1 713	1 679	1 114	294	479
At constant prices	1 987	2 110	2 324	2 439	2 584	2 537	2 441	2 094	1 935
Growth rates	6.4	6.2	10.1	4.9	6.0	−1.8	−3.8	−14.2	−7.6
Montserrat Montserrat									
At current prices	35	37	42	56	63	70	67	65	70

21

Gross domestic product: total and per capita
In US dollars (millions) [1] at current and constant 1990 prices; per capita US$;
real rates of growth [*cont.*]

Produit intérieur brut : total et par habitant
En monnaie dollars de E−u (millions) [1] aux prix courants et constants de 1990; par habitant;
taux de l'accroissement réels [*suite*]

Country or area Pays ou zone	1984	1985	1986	1987	1988	1989	1990	1991	1992
Per capita	3 152	3 370	3 842	5 051	5 741	6 364	6 061	5 941	6 376
At constant prices	41	43	45	48	52	58	67	51	53
Growth rates	1.7	4.7	5.2	6.6	9.3	11.5	14.8	−23.6	4.1
Morocco Maroc									
At current prices	12 750	12 871	16 947	18 703	22 079	22 848	25 940	27 754	28 401
Per capita	594	584	750	806	927	935	1 035	1 080	1 079
At constant prices	19 744	20 993	22 694	22 130	24 463	25 070	25 940	27 545	26 425
Growth rates	4.3	6.3	8.1	−2.5	10.5	2.5	3.5	6.2	−4.1
Mozambique Mozambique									
At current prices	2 568	3 639	3 645	1 317	1 108	1 144	1 318	1 061	962
Per capita	193	269	266	95	80	82	93	73	65
At constant prices	1 208	1 102	1 119	1 173	1 226	1 293	1 318	1 359	1 340
Growth rates	3.0	−8.8	1.6	4.8	4.5	5.5	1.9	3.1	−1.4
Myanmar Myanmar									
At current prices	6 392	6 608	8 052	10 325	11 923	18 593	23 971	28 415	37 830
Per capita	174	176	210	263	298	454	573	665	866
At constant prices	25 963	26 703	26 421	25 362	22 483	23 314	23 971	23 725	26 313
Growth rates	4.9	2.9	−1.1	−4.0	−11.4	3.7	2.8	−1.0	10.9
Namibia Namibie									
At current prices	1 502	1 280	1 429	1 724	1 935	1 905	2 183	2 278	2 456
Per capita	1 254	1 037	1 123	1 314	1 431	1 366	1 517	1 533	1 601
At constant prices	1 879	1 889	1 959	2 003	2 128	2 119	2 183	2 294	2 374
Growth rates	0.6	0.6	3.7	2.3	6.2	−0.4	3.0	5.1	3.5
Nepal Népal									
At current prices	2 393	2 434	2 375	2 715	2 962	2 859	3 099	2 790	2 954
Per capita	144	142	135	150	159	150	158	139	144
At constant prices	2 223	2 359	2 461	2 558	2 747	2 871	3 099	3 240	3 307
Growth rates	9.7	6.1	4.3	3.9	7.4	4.5	7.9	4.6	2.1
Netherlands Pays−Bas									
At current prices	126 866	128 079	178 633	217 495	231 413	228 542	283 525	289 822	320 284
Per capita	8 803	8 843	12 265	14 846	15 695	15 398	18 974	19 259	21 130
At constant prices	237 600	243 863	250 568	253 528	260 157	272 332	283 525	289 483	293 542
Growth rates	3.2	2.6	2.7	1.2	2.6	4.7	4.1	2.1	1.4
Netherlands Antilles Antilles néerlandaises									
At current prices	1 072	1 092	1 076	1 190	1 324	1 397	1 559	1 522	1 578
Per capita	6 123	6 240	6 146	6 803	7 566	7 981	8 907	8 699	9 019
At constant prices	1 308	1 281	1 246	1 329	1 427	1 449	1 559	1 497	1 531
Growth rates	−3.0	−2.1	−2.8	6.7	7.4	1.5	7.6	−4.0	2.3
New Caledonia Nouvelle−Calédonie									
At current prices	796	855	1 201	1 491	2 073	2 185	2 512	2 610	2 903
Per capita	5 272	5 551	7 651	9 316	12 793	13 243	14 954	15 354	16 783
At constant prices	1 607	1 679	1 663	1 757	2 365	2 624	2 512	2 611	2 637
Growth rates	1.8	4.5	−1.0	5.7	34.6	11.0	−4.3	3.9	1.0
New Zealand Nouvelle−Zélande									
At current prices	22 862	22 650	28 861	36 638	43 560	42 754	43 940	42 501	41 470
Per capita	7 109	6 976	8 810	11 086	13 069	12 717	12 954	12 416	12 003
At constant prices	42 254	42 646	43 743	44 058	43 548	44 096	43 940	43 234	43 928
Growth rates	5.0	0.9	2.6	0.7	−1.2	1.3	−0.4	−1.6	1.6
Nicaragua Nicaragua									
At current prices	3 063	4 274	6 504	3 414	1 022	861	1 465	1 378	1 466
Per capita	974	1 324	1 965	1 008	295	242	399	362	371
At constant prices	1 833	1 758	1 741	1 728	1 493	1 468	1 465	1 459	1 471
Growth rates	−1.6	−4.1	−1.0	−0.7	−13.6	−1.7	−0.2	−0.4	0.8
Niger Niger									
At current prices	1 461	1 440	1 905	2 233	2 277	2 171	2 506	2 464	2 712
Per capita	228	218	279	317	314	290	324	309	329
At constant prices	2 152	2 274	2 561	2 333	2 376	2 437	2 506	2 556	2 605
Growth rates	−18.2	5.7	12.6	−8.9	1.8	2.6	2.8	2.0	1.9
Nigeria Nigéria									
At current prices	82 931	80 934	41 631	27 113	32 013	30 522	32 426	32 778	29 640
Per capita	931	880	438	276	315	291	299	292	256
At constant prices	22 779	24 989	25 618	25 438	27 956	29 969	32 426	33 983	35 852
Growth rates	−4.8	9.7	2.5	−0.7	9.9	7.2	8.2	4.8	5.5

21

Gross domestic product: total and per capita
In US dollars (millions) [1] at current and constant 1990 prices; per capita US$;
real rates of growth [cont.]

Produit intérieur brut : total et par habitant
En monnaie dollars de E−u (millions) [1] aux prix courants et constants de 1990; par habitant;
taux de l'accroissement réels [suite]

Country or area Pays ou zone	1984	1985	1986	1987	1988	1989	1990	1991	1992
Norway Norvège									
At current prices	55 445	58 182	69 471	83 337	89 501	89 997	105 524	105 923	112 906
Per capita	13 399	14 010	16 664	19 904	21 279	21 296	24 853	24 824	26 331
At constant prices	92 714	97 604	101 684	103 710	103 185	103 797	105 524	107 192	110 726
Growth rates	5.7	5.3	4.2	2.0	−0.5	0.6	1.7	1.6	3.3
Oman Oman									
At current prices	8 705	8 983	7 283	7 809	7 610	8 402	10 535	10 236	11 543
Per capita	7 206	7 112	5 534	5 713	5 363	5 716	6 913	6 479	7 051
At constant prices	7 922	9 012	9 312	8 970	9 484	9 798	10 535	11 504	12 284
Growth rates	16.7	13.8	3.3	−3.7	5.7	3.3	7.5	9.2	6.8
Pakistan Pakistan									
At current prices	33 615	32 304	34 387	38 818	42 756	41 704	47 095	50 517	56 379
Per capita	345	321	330	361	385	364	399	416	452
At constant prices	33 671	35 524	37 816	40 699	42 718	44 652	47 095	50 764	54 673
Growth rates	7.6	5.5	6.5	7.6	5.0	4.5	5.5	7.8	7.7
Panama Panama									
At current prices	4 610	4 948	5 191	5 363	4 605	4 639	5 009	5 496	6 001
Per capita	2 160	2 270	2 331	2 359	1 983	1 958	2 072	2 229	2 386
At constant prices	5 140	5 383	5 562	5 697	4 808	4 788	5 009	5 488	5 958
Growth rates	−0.4	4.7	3.3	2.4	−15.6	−0.4	4.6	9.6	8.6
Papua New Guinea Papouasie−Nvl−Guinée									
At current prices	2 553	2 403	2 649	3 144	3 657	3 559	3 220	3 789	4 291
Per capita	755	694	749	868	988	940	831	956	1 058
At constant prices	2 910	3 014	3 184	3 272	3 367	3 319	3 220	3 527	3 830
Growth rates	−1.0	3.6	5.6	2.8	2.9	−1.4	−3.0	9.5	8.6
Paraguay Paraguay									
At current prices	5 326	4 545	5 407	4 534	6 035	4 363	5 265	5 694	5 873
Per capita	1 488	1 231	1 420	1 156	1 494	1 050	1 231	1 295	1 300
At constant prices	4 186	4 352	4 352	4 540	4 829	5 109	5 265	5 395	5 486
Growth rates	3.1	4.0	0.0	4.3	6.4	5.8	3.0	2.5	1.7
Peru Pérou									
At current prices	20 867	18 040	26 713	44 014	38 372	43 179	37 472	45 504	44 703
Per capita	1 099	929	1 346	2 172	1 855	2 045	1 739	2 069	1 991
At constant prices	40 235	41 144	44 943	48 754	44 679	39 475	37 472	38 347	37 312
Growth rates	4.8	2.3	9.2	8.5	−8.4	−11.6	−5.1	2.3	−2.7
Philippines Philippines									
At current prices	31 408	30 735	29 868	33 307	38 043	42 562	44 050	45 272	52 623
Per capita	582	555	526	572	638	697	706	709	807
At constant prices	37 891	35 122	36 322	38 059	40 439	42 891	44 050	43 763	43 748
Growth rates	−7.3	−7.3	3.4	4.8	6.3	6.1	2.7	−0.7	0.0
Poland Pologne									
At current prices	74 416	69 226	72 038	62 299	67 099	80 151	62 265	77 943	90 497
Per capita	2 016	1 861	1 923	1 654	1 772	2 107	1 631	2 035	2 356
At constant prices	61 282	63 512	66 195	67 491	70 281	70 393	62 265	57 537	58 408
Growth rates	5.6	3.6	4.2	2.0	4.1	0.2	−11.5	−7.6	1.5
Portugal Portugal									
At current prices	19 234	20 682	29 550	36 731	41 700	45 283	59 680	68 614	84 195
Per capita	1 942	2 088	2 984	3 713	4 219	4 586	6 048	6 955	8 534
At constant prices	46 434	47 737	49 714	52 325	54 377	57 180	59 680	60 961	61 628
Growth rates	−1.9	2.8	4.1	5.3	3.9	5.2	4.4	2.1	1.1
Puerto Rico Porto Rico									
At current prices	20 289	21 969	23 878	26 178	28 267	30 604	32 591	33 969	35 797
Per capita	6 065	6 505	7 004	7 612	8 148	8 744	9 233	9 537	9 960
At constant prices	24 200	26 179	27 464	29 255	30 699	31 859	32 591	33 645	34 733
Growth rates	2.1	8.2	4.9	6.5	4.9	3.8	2.3	3.2	3.2
Qatar Qatar									
At current prices	6 870	6 153	5 053	5 446	6 038	6 488	7 360	6 884	7 473
Per capita	20 570	17 188	13 403	13 894	14 946	15 634	17 238	15 644	16 497
At constant prices	6 363	6 221	6 448	6 507	6 811	7 169	7 360	7 302	7 594
Growth rates	2.8	−2.2	3.7	0.9	4.7	5.3	2.7	−0.8	4.0
Republic of Moldova République de Moldova									
At current prices	11 789	10 071	12 896	14 808	16 194	17 892	19 041	13 943	1 121
Per capita	2 824	2 437	3 031	3 452	3 748	4 114	4 363	3 194	257

21

Gross domestic product: total and per capita
In US dollars (millions) [1] at current and constant 1990 prices; per capita US$;
real rates of growth [cont.]

Produit intérieur brut : total et par habitant
En monnaie dollars de E−u (millions) [1] aux prix courants et constants de 1990; par habitant;
taux de l'accroissement réels [suite]

Country or area Pays ou zone	1984	1985	1986	1987	1988	1989	1990	1991	1992
At constant prices	14 988	13 526	14 230	14 845	15 547	17 602	19 041	16 643	11 856
Growth rates	5.0	−9.8	5.2	4.3	4.7	13.2	8.2	−12.6	−28.8
Réunion Réunion									
At current prices	1 776	1 898	2 723	3 643	3 962	4 015	5 162	5 226	5 758
Per capita	3 289	3 451	4 862	6 381	6 808	6 770	8 547	8 511	9 228
At constant prices	4 072	4 214	4 416	4 620	4 805	4 964	5 162	5 344	5 504
Growth rates	2.0	3.5	4.8	4.6	4.0	3.3	4.0	3.5	3.0
Romania Roumanie									
At current prices	38 447	47 806	52 046	58 200	60 177	53 612	38 244	28 785	19 426
Per capita	1 699	2 104	2 280	2 538	2 613	2 318	1 648	1 237	833
At constant prices	42 455	42 307	43 188	43 439	43 115	40 519	38 244	33 305	28 785
Growth rates	5.7	−0.3	2.1	0.6	−0.7	−6.0	−5.6	−12.9	−13.6
Russian Federation Fédération de Russie									
At current prices	554 668	559 412	690 409	795 290	888 303	913 876	966 967	742 914	123 295
Per capita	3 890	3 896	4 771	5 452	6 046	6 187	6 521	4 994	827
At constant prices	971 571	1 015 480	993 824	998 102	1 042 637	1 046 961	966 967	958 043	890 980
Growth rates	2.3	4.5	−2.1	0.4	4.5	0.4	−7.6	−0.9	−7.0
Rwanda Rwanda									
At current prices	1 588	1 715	1 928	2 152	2 327	2 378	2 335	1 701	1 628
Per capita	275	288	314	339	354	350	332	234	216
At constant prices	2 112	2 205	2 304	2 307	2 308	2 383	2 335	2 360	2 394
Growth rates	−4.2	4.4	4.5	0.1	0.0	3.2	−2.0	1.1	1.4
Saint Kitts and Nevis Saint−Kitts−et−Nevis									
At current prices	62	67	83	93	108	117	126	121	131
Per capita	1 440	1 552	1 938	2 167	2 578	2 797	2 999	2 880	3 114
At constant prices	90	95	101	109	116	122	126	131	135
Growth rates	9.0	5.6	6.2	7.4	7.0	5.1	3.0	3.7	3.6
Saint Lucia Sainte−Lucie									
At current prices	151	167	183	191	217	241	257	275	302
Per capita	1 239	1 343	1 450	1 491	1 682	1 842	1 932	2 037	2 206
At constant prices	184	195	206	211	236	247	257	261	278
Growth rates	5.0	6.0	5.8	2.1	12.1	4.6	3.9	1.6	6.6
St. Vincent−Grenadines St. Vincent−Grenadines									
At current prices	103	113	127	142	161	174	194	178	193
Per capita	1 017	1 106	1 236	1 366	1 530	1 640	1 817	1 648	1 771
At constant prices	131	137	147	156	170	182	194	203	213
Growth rates	5.7	4.4	7.3	6.3	8.7	7.0	7.0	4.6	4.7
Samoa Samoa									
At current prices	99	85	90	100	119	109	113	120	123
Per capita	628	543	573	634	755	693	713	759	777
At constant prices	108	115	115	117	116	118	113	112	106
Growth rates	1.3	6.0	0.6	1.0	−0.2	1.3	−4.5	−0.5	−5.0
San Marino San Marino									
At current prices	159	163	242	304	336	347	435	457	485
Per capita	7 245	7 429	10 542	13 229	14 592	15 096	18 916	19 857	21 099
At constant prices	337	346	377	391	413	425	435	442	476
Growth rates	4.3	2.9	8.9	3.7	5.7	2.9	2.3	1.6	7.6
Sao Tome and Principe Sao Tomé−et−Principe									
At current prices	36	37	64	55	49	46	54	42	27
Per capita	354	349	595	499	433	397	454	349	218
At constant prices	51	51	51	50	51	52	54	56	57
Growth rates	−4.1	−1.6	1.0	−1.5	2.0	1.5	3.8	3.5	2.0
Saudi Arabia Arabie saoudite									
At current prices	106 096	97 017	84 780	72 387	73 552	76 142	82 996	89 030	94 366
Per capita	8 995	7 837	6 563	5 394	5 293	5 297	5 581	5 786	5 927
At constant prices	70 783	65 099	64 580	62 071	68 479	70 968	82 996	87 973	91 499
Growth rates	−3.9	−8.0	−0.8	−3.9	10.3	3.6	16.9	6.0	4.0
Senegal Sénégal									
At current prices	2 337	2 578	3 762	4 599	4 980	4 625	5 528	5 608	6 285
Per capita	377	404	574	682	718	649	754	745	812
At constant prices	4 688	4 865	5 087	5 290	5 558	5 473	5 528	5 672	5 820
Growth rates	−4.0	3.8	4.6	4.0	5.1	−1.5	1.0	2.6	2.6
Seychelles Seychelles									
At current prices	151	169	209	249	284	308	373	373	409

21

Gross domestic product: total and per capita
In US dollars (millions) [1] at current and constant 1990 prices; per capita US$;
real rates of growth [*cont.*]

Produit intérieur brut : total et par habitant
En monnaie dollars de E−u (millions) [1] aux prix courants et constants de 1990; par habitant;
taux de l'accroissement réels [*suite*]

Country or area Pays ou zone	1984	1985	1986	1987	1988	1989	1990	1991	1992
Per capita	2 293	2 521	3 071	3 613	4 113	4 405	5 254	5 260	5 684
At constant prices	256	283	286	299	315	347	373	381	395
Growth rates	8.0	10.3	1.2	4.4	5.3	10.3	7.5	2.2	3.5
Sierra Leone Sierra Leone									
At current prices	1 088	939	465	579	894	735	547	508	741
Per capita	304	256	124	150	227	182	132	119	169
At constant prices	449	487	482	508	522	534	547	533	501
Growth rates	−4.5	8.5	−1.0	5.5	2.7	2.3	2.5	−2.6	−6.0
Singapore Singapour									
At current prices	18 775	17 693	17 757	20 245	24 845	29 146	35 001	40 201	46 025
Per capita	7 424	6 917	6 861	7 733	9 379	10 880	12 915	14 677	16 621
At constant prices	24 278	23 880	24 321	26 609	29 574	32 306	35 001	37 353	39 520
Growth rates	8.3	−1.6	1.8	9.4	11.1	9.2	8.3	6.7	5.8
Slovakia Slovaquie									
At current prices	11 626	11 859	14 123	15 847	15 669	15 560	13 568	9 878	11 104
Per capita	2 267	2 297	2 720	3 034	2 984	2 949	2 561	1 872	2 085
At constant prices	12 165	12 666	13 182	13 518	13 771	13 918	13 568	11 595	11 315
Growth rates	0.0	4.1	4.1	2.5	1.9	1.1	−2.5	−14.5	−2.4
Slovenia Slovénie									
At current prices	7 508	7 678	10 662	12 280	10 775	12 102	17 304	17 856	16 695
Per capita	3 914	3 989	5 521	6 339	5 545	6 212	8 860	9 096	8 298
At constant prices	18 216	18 424	18 994	18 792	18 474	18 150	17 304	15 692	14 672
Growth rates	2.1	1.1	3.1	−1.1	−1.7	−1.8	−4.7	−9.3	−6.5
Solomon Islands Iles Salomon									
At current prices	174	160	145	146	176	179	177	195	207
Per capita	667	593	520	506	589	576	554	589	606
At constant prices	141	145	144	148	155	166	177	184	200
Growth rates	7.2	2.8	−0.7	2.3	5.4	6.9	6.7	4.0	8.2
Somalia Somalie									
At current prices	3 110	2 211	1 650	1 612	1 636	1 127	645	373	335
Per capita	405	281	205	197	197	133	74	42	36
At constant prices	558	612	598	659	659	659	645	621	546
Growth rates	3.3	9.5	−2.1	10.1	0.1	0.0	−2.2	−3.7	−12.0
South Africa Afrique du Sud									
At current prices	74 563	56 196	62 656	80 849	87 600	89 214	102 155	108 076	114 775
Per capita	2 276	1 673	1 819	2 290	2 422	2 408	2 691	2 780	2 882
At constant prices	95 437	94 281	94 298	96 279	100 322	102 627	102 155	101 757	99 620
Growth rates	5.1	−1.2	0.0	2.1	4.2	2.3	−0.5	−0.4	−2.1
Spain Espagne									
At current prices	158 743	165 849	230 803	292 664	344 739	380 182	491 761	527 294	574 530
Per capita	4 140	4 311	5 980	7 562	8 886	9 778	12 623	13 511	14 697
At constant prices	385 151	395 213	407 860	430 868	453 107	474 575	491 761	502 692	506 702
Growth rates	1.5	2.6	3.2	5.6	5.2	4.7	3.6	2.2	0.8
Sri Lanka Sri Lanka									
At current prices	5 792	5 808	6 154	6 413	6 878	6 886	7 935	8 936	9 583
Per capita	365	361	377	387	410	405	461	512	542
At constant prices	6 404	6 725	7 013	7 126	7 324	7 469	7 935	8 316	8 674
Growth rates	5.1	5.0	4.3	1.6	2.8	2.0	6.2	4.8	4.3
Sudan Soudan									
At current prices	8 715	6 081	8 804	10 386	10 638	16 592	26 617	39 594	4 953
Per capita	412	279	392	449	447	677	1 056	1 528	186
At constant prices	25 876	24 504	25 462	26 022	25 164	27 026	26 617	26 804	29 594
Growth rates	−4.3	−5.3	3.9	2.2	−3.3	7.4	−1.5	0.7	10.4
Suriname Suriname									
At current prices	968	978	998	1 098	1 301	1 520	1 728	2 077	2 807
Per capita	2 582	2 555	2 560	2 758	3 204	3 680	4 096	4 830	6 408
At constant prices	1 595	1 627	1 639	1 537	1 658	1 728	1 728	1 782	1 863
Growth rates	−1.9	2.0	0.8	−6.2	7.8	4.2	0.0	3.1	4.5
Swaziland Swaziland									
At current prices	507	367	463	559	576	672	751	821	955
Per capita	793	558	684	804	807	918	1 000	1 065	1 205
At constant prices	558	580	632	616	656	699	751	770	792
Growth rates	8.0	3.9	9.0	−2.5	6.6	6.5	7.5	2.5	2.8

21

Gross domestic product: total and per capita
In US dollars (millions) [1] at current and constant 1990 prices; per capita US$;
real rates of growth [*cont.*]

Produit intérieur brut : total et par habitant
En monnaie dollars de E−u (millions) [1] aux prix courants et constants de 1990; par habitant;
taux de l'accroissement réels [*suite*]

Country or area Pays ou zone	1984	1985	1986	1987	1988	1989	1990	1991	1992
Sweden Suède									
At current prices	96 022	100 626	132 732	160 802	181 236	190 217	227 900	236 946	244 774
Per capita	11 525	12 051	15 837	19 093	21 400	22 329	26 605	27 520	28 291
At constant prices	201 079	205 545	210 105	216 090	220 987	226 222	227 900	223 734	219 975
Growth rates	4.0	2.2	2.2	2.8	2.3	2.4	0.7	−1.8	−1.7
Switzerland Suisse									
At current prices	90 736	92 772	135 277	170 792	183 428	177 493	226 022	231 998	242 583
Per capita	14 114	14 339	20 767	26 027	27 746	26 647	33 674	34 304	35 606
At constant prices	189 932	196 975	202 621	206 733	212 729	220 944	226 022	225 957	225 822
Growth rates	1.8	3.7	2.9	2.0	2.9	3.9	2.3	0.0	−0.1
Syrian Arab Republic Rép. arabe syrienne									
At current prices	19 195	21 204	25 461	32 538	16 574	18 610	23 905	28 170	33 050
Per capita	1 921	2 049	2 375	2 930	1 441	1 561	1 935	2 200	2 489
At constant prices	20 951	22 232	21 133	21 536	24 393	22 208	23 905	26 681	29 254
Growth rates	−4.1	6.1	−4.9	1.9	13.3	−9.0	7.6	11.6	9.6
Tajikistan Tadjikistan									
At current prices	7 153	7 301	8 367	9 551	10 972	10 703	10 961	7 683	577
Per capita	1 615	1 597	1 979	1 954	2 177	2 067	2 067	1 412	103
At constant prices	9 338	10 014	9 401	9 555	10 522	10 498	10 961	10 580	7 935
Growth rates	10.3	7.2	−6.1	1.6	10.1	−0.2	4.4	−3.5	−25.0
Thailand Thaïlande									
At current prices	41 798	38 900	43 097	50 535	61 667	72 231	85 288	98 343	110 421
Per capita	830	760	830	960	1 157	1 338	1 560	1 775	1 967
At constant prices	49 681	51 990	54 867	60 090	68 074	76 441	85 288	92 010	98 819
Growth rates	5.8	4.6	5.5	9.5	13.3	12.3	11.6	7.9	7.4
TFYR Macedonia L'ex−R.y. Macédoine									
At current prices	2 468	2 477	3 573	4 105	3 572	4 720	5 851	7 917	6 753
Per capita	1 262	1 258	1 803	2 058	1 780	2 339	2 885	3 883	3 285
At constant prices	6 248	6 203	6 640	6 555	6 342	6 464	5 851	5 269	4 495
Growth rates	3.1	−0.7	7.0	−1.3	−3.2	1.9	−9.5	−10.0	−14.7
Togo Togo									
At current prices	698	740	1 050	1 232	1 383	1 348	1 672	1 741	2 103
Per capita	237	244	336	383	417	394	473	478	559
At constant prices	1 346	1 373	1 428	1 518	1 612	1 671	1 672	1 741	1 792
Growth rates	7.0	2.0	4.0	6.3	6.2	3.6	0.0	4.2	2.9
Tonga Tonga									
At current prices	65	56	67	74	88	97	101	122	124
Per capita	696	594	711	783	924	1 013	1 053	1 262	1 280
At constant prices	92	97	95	99	102	103	101	106	108
Growth rates	2.5	5.4	−2.3	4.3	2.5	1.6	−2.0	5.3	1.9
Trinidad and Tobago Trinité−et−Tobago									
At current prices	7 756	7 266	4 794	4 798	4 497	4 323	5 116	5 275	5 442
Per capita	6 780	6 264	4 077	4 028	3 726	3 538	4 139	4 217	4 302
At constant prices	5 976	5 730	5 542	5 289	5 082	5 040	5 116	5 241	5 158
Growth rates	−5.8	−4.1	−3.3	−4.6	−3.9	−0.8	1.5	2.5	−1.6
Tunisia Tunisie									
At current prices	8 031	8 275	8 821	9 604	10 031	10 004	12 299	12 940	15 552
Per capita	1 133	1 140	1 188	1 267	1 296	1 267	1 526	1 573	1 851
At constant prices	9 937	10 500	10 332	10 929	11 089	11 430	12 299	12 776	13 829
Growth rates	5.7	5.7	−1.6	5.8	1.5	3.1	7.6	3.9	8.2
Turkey Turquie									
At current prices	60 530	67 819	76 651	88 353	92 041	108 415	149 972	148 197	154 479
Per capita	1 232	1 347	1 488	1 679	1 713	1 977	2 679	2 592	2 647
At constant prices	108 618	113 461	122 290	133 420	136 189	137 438	149 972	151 491	159 815
Growth rates	7.4	4.5	7.8	9.1	2.1	0.9	9.1	1.0	5.5
Turkmenistan Turkménistan									
At current prices	6 857	6 648	8 291	9 916	10 847	11 051	11 411	8 400	1 471
Per capita	2 175	2 059	2 501	2 910	3 105	3 089	3 111	2 231	381
At constant prices	9 022	9 143	9 295	9 919	10 422	10 828	11 411	11 284	10 268
Growth rates	8.0	1.3	1.7	6.7	5.1	3.9	5.4	−1.1	−9.0
Tuvalu Tuvalu									
At current prices	5	5	5	6	8	8	8	9	9
Per capita	550	559	543	647	702	690	673	751	713

21

Gross domestic product: total and per capita
In US dollars (millions) [1] at current and constant 1990 prices; per capita US$;
real rates of growth [cont.]

Produit intérieur brut : total et par habitant
En monnaie dollars de E−u (millions) [1] aux prix courants et constants de 1990; par habitant;
taux de l'accroissement réels [suite]

Country or area Pays ou zone	1984	1985	1986	1987	1988	1989	1990	1991	1992
At constant prices	7	7	7	7	8	8	8	8	9
Growth rates	7.3	−2.0	−0.6	10.0	10.4	−4.3	2.5	4.8	1.0
Uganda Ouganda									
At current prices	2 306	3 200	3 617	4 072	4 630	4 463	3 161	2 596	2 851
Per capita	158	213	234	255	281	262	180	143	153
At constant prices	2 426	2 427	2 449	2 617	2 828	3 037	3 161	3 279	3 376
Growth rates	−4.3	0.0	0.9	6.9	8.0	7.4	4.1	3.7	2.9
Ukraine Ukraine									
At current prices	159 337	150 941	184 203	213 972	234 267	245 774	247 447	168 800	18 664
Per capita	3 139	2 964	3 605	4 172	4 544	4 749	4 769	3 243	358
At constant prices	220 132	222 317	226 681	231 287	230 075	244 055	247 447	233 020	219 738
Growth rates	3.1	1.0	2.0	2.0	−0.5	6.1	1.4	−5.8	−5.7
former USSR † ancienne URSS †									
At current prices	930 626	914 118	1 126 234	1 295 133	1 442 175	1 503 987	1 535 135	..	..
Per capita	3 383	3 294	4 022	4 583	5 060	5 235	5 306	..	..
At constant prices	1 412 122	1 436 831	1 473 496	1 498 205	1 559 578	1 596 509	1 535 135	..	..
Growth rates	2.7	1.7	2.6	1.7	4.1	2.4	−3.8	..	..
United Arab Emirates Emirats arabes unis									
At current prices	27 743	27 081	21 674	23 799	23 728	27 506	33 780	34 323	34 666
Per capita	21 439	20 075	15 448	16 379	15 808	17 792	21 259	21 057	20 758
At constant prices	30 803	30 067	24 365	25 890	25 269	28 688	33 780	33 014	33 344
Growth rates	4.3	−2.4	−19.0	6.3	−2.4	13.5	17.8	−2.3	1.0
United Kingdom Royaume−Uni									
At current prices	434 281	461 961	562 510	691 432	715 956	843 198	980 490	977 133	1 049 030
Per capita	7 684	8 159	9 913	12 151	12 544	14 729	17 078	16 976	18 182
At constant prices	804 907	833 080	869 083	910 472	955 818	976 644	980 490	958 188	953 742
Growth rates	2.5	3.5	4.3	4.8	5.0	2.2	0.4	−2.3	−0.5
United Rep.Tanzania Rép. Unie de Tanzanie									
At current prices	5 813	6 904	4 882	3 532	3 336	2 835	2 538	3 150	2 720
Per capita	275	315	216	151	137	113	98	117	98
At constant prices	2 044	2 098	2 166	2 276	2 373	2 452	2 538	2 634	2 745
Growth rates	3.4	2.6	3.3	5.1	4.2	3.3	3.5	3.8	4.2
United States Etats−Unis									
At current prices	3 763 467	4 016 649	4 230 784	4 496 574	4 853 962	5 204 509	5 489 600	5 654 400	5 953 300
Per capita	15 925	16 844	17 580	18 513	19 799	21 026	21 961	22 392	23 332
At constant prices	4 654 942	4 801 471	4 934 970	5 087 200	5 287 618	5 431 743	5 489 600	5 445 729	5 575 429
Growth rates	6.7	3.1	2.8	3.1	3.9	2.7	1.1	−0.8	2.4
Uruguay Uruguay									
At current prices	4 829	4 719	5 859	7 329	7 583	7 992	8 282	9 479	11 026
Per capita	1 615	1 569	1 936	2 408	2 478	2 597	2 677	3 046	3 523
At constant prices	6 797	6 897	7 508	8 103	8 103	8 207	8 282	8 436	9 060
Growth rates	−1.1	1.5	8.9	7.9	0.0	1.3	0.9	1.9	7.4
Vanuatu Vanuatu									
At current prices	124	118	115	122	144	141	154	179	180
Per capita	964	889	844	878	1 005	966	1 024	1 171	1 149
At constant prices	140	141	138	139	140	146	154	160	160
Growth rates	6.9	1.1	−2.0	0.4	0.6	4.5	5.2	4.1	0.0
Venezuela Venezuela									
At current prices	59 856	61 965	60 519	48 029	60 226	43 551	48 598	53 462	60 435
Per capita	3 575	3 610	3 440	2 665	3 263	2 306	2 515	2 707	2 994
At constant prices	42 680	42 763	45 547	47 178	49 924	45 646	48 598	53 327	56 961
Growth rates	−1.4	0.2	6.5	3.6	5.8	−8.6	6.5	9.7	6.8
Viet Nam Viet Nam									
At current prices	...	...	...	...	...	6 154	6 491	7 383	9 074
Per capita	...	...	...	...	...	94	97	108	131
At constant prices	4 786	5 082	5 231	5 438	5 718	6 179	6 491	6 880	7 451
Growth rates	8.4	6.2	2.9	4.0	5.2	8.0	5.1	6.0	8.3
Yemen Yémen									
At current prices	..	..	..	..	..	6 266	7 873	7 777	8 960
Per capita	..	..	..	..	..	556	674	643	715
At constant prices	..	..	..	..	..	8 176	7 873	7 566	8 449
Growth rates	..	..	..	..	..	..	−3.7	−3.9	11.7
former Dem. Yemen ancienne Yémen dém.									
At current prices	1 124	1 107	994	1 304	1 369	..	..	..	..

21
Gross domestic product: total and per capita
In US dollars (millions) [1] at current and constant 1990 prices; per capita US$;
real rates of growth [*cont.*]
Produit intérieur brut : total et par habitant
En monnaie dollars de E−u (millions) [1]   aux prix courants et constants de 1990; par habitant;
taux de l'accroissement réels [*suite*]

Country or area Pays ou zone	1984	1985	1986	1987	1988	1989	1990	1991	1992
Per capita	541	518	452	575	585	..	..	..	..
At constant prices	1 493	1 449	1 269	1 311	1 369	..	..	..	..
Growth rates	5.9	−3.0	−12.4	3.3	4.5	..	..	..	..
former Yemen Arab Rep. anc. Yémen rép. arabe									
At current prices	3 353	4 181	3 903	4 229	5 491	..	..	..	..
Per capita	457	549	493	515	644	..	..	..	..
At constant prices	3 387	4 156	4 398	4 821	5 491	..	..	..	..
Growth rates	2.4	22.7	5.8	9.6	13.9	..	..	..	..
Yugoslavia Yougoslavie									
At current prices	17 208	17 448	24 369	27 902	24 760	32 715	40 524	54 826	40 463
Per capita	1 698	1 709	2 371	2 698	2 378	3 124	3 849	5 182	3 840
At constant prices	42 864	42 984	44 569	43 836	42 858	43 331	40 524	37 511	27 684
Growth rates	1.6	0.3	3.7	−1.6	−2.2	1.1	−6.5	−7.4	−26.2
Yugoslavia, SFR † Yougoslavie, Rfs †									
At current prices	43 538	44 238	61 705	71 017	62 764	81 848	101 413	..	..
Per capita	1 896	1 913	2 651	3 031	2 662	3 453	4 259	..	..
At constant prices	93 426	94 365	98 277	100 103	98 329	99 790	101 413	..	..
Growth rates	1.5	1.0	4.1	1.9	−1.8	1.5	1.6	..	..
Zaire Zaïre									
At current prices	2 760	2 953	3 390	3 238	3 732	3 804	3 677	3 594	3 714
Per capita	90	93	104	96	107	105	98	93	93
At constant prices	3 181	3 263	3 361	3 428	3 497	3 568	3 677	3 574	3 617
Growth rates	3.0	2.6	3.0	2.0	2.0	2.1	3.0	−2.8	1.2
Zambia Zambie									
At current prices	3 468	3 463	2 413	2 423	3 759	4 551	4 260	3 799	2 225
Per capita	524	505	339	329	493	577	524	453	258
At constant prices	3 896	3 895	3 999	4 109	4 189	4 229	4 260	4 177	3 989
Growth rates	−0.6	0.0	2.7	2.8	1.9	1.0	0.7	−2.0	−4.5
Zimbabwe Zimbabwe									
At current prices	5 148	4 525	4 978	5 258	6 356	6 428	6 726	6 194	5 002
Per capita	636	541	574	586	684	668	676	603	473
At constant prices	5 223	5 586	5 733	5 662	6 192	6 596	6 726	6 856	6 019
Growth rates	−1.9	6.9	2.6	−1.2	9.4	6.5	2.0	1.9	−12.2

Source:
National accounts database of the Statistical Division of
the United Nations Secretariat.

† For detailed descriptions of data pertaining to
former Czechoslovakia, Germany, SFR Yugoslavia and former
USSR, see Annex I − Country or area nomenclature, regional
and other groupings.

1   The exchange rates used for the conversion of national currency data
into US dollars are the average market rates as published by the
International Monetary Fund in the International Financial Statistics.
Official exchange rates were used only when a free market rate was
not available. For non−members of the fund, the conversion rates
used are the average of United Nations operational rates of exchange.
It should be noted that the use of IMF rates may distort the US dollar
income level figures in a number of countries. Therefore,
comparability of data both across countries and over time is distorted.

Source:
Base de données sur les comptes nationaux de la Division de
statistique du Secrétariat de l'ONU.

† Pour les descriptions en détails des données relatives
à l'ancienne Tchécoslovaquie, l'Allemagne, la Rfs
Yougoslavie et l'ancienne URSS, voir l'Annexe I −
Nomenclature des pays ou zones, groupements régionaux
et autres groupements.

1 Les taux de change utilisés pour la conversion en dollars des États−
Unis des données libellées en monnaie nationale sont les taux
moyens du marché publiés par le Fonds monétaire international dans
les "Statistiques financières internationales". Le taux de
change officiel n'a été utilisé qu'en l'absence d'un taux du
marché libre. Pour les pays qui ne sont pas membres du Fonds, les
taux de change utilisés sont la moyenne des taux appliqués pour les
opérations des Nations Unies. Il est à noter que l'utilisation des taux
du FMI risque de fausser, dans plusieurs pays, les chiffres du revenu
exprimés en dollars des États−Unis. La comparabilité des données,
entre pays, et à des dates différentes, s'en trouve donc faussée.

22
Expenditure on gross domestic product at current prices
Dépenses imputées au produit intérieur brut aux prix courants

Percentage distribution
Répartition en pourcentage

Country or area Pays ou zone	Year Année	GDP at current prices (Million nat. cur.) PIB aux prix courants (Mil. monnaie nat.)	% of GDP — en % du PIB					
			Govt. final cons. exp. Consom. finale des admin. publiques	Private final cons. exp. Consom. finale privée	Increase in stocks Variation des stocks	Gross fixed capital form. Formation brute de capital fixe	Exports of goods/ services Exportations de biens et services	Imports of goods/ services Importations de biens et services
Albania	1985	16856	9.3	58.6	0.5	33.2	-1.6[1]	...
Albanie	1989	18674	8.8	61.0	0.4	31.3	-1.5[1]	...
	1990	16812	10.2	72.7	-10.1	34.6	-7.4[1]	...
Algeria	1985	291597	15.8	47.9	1.4	31.8	23.5	20.4
Algérie	1988	334607	19.5	52.4	1.8	26.5	14.9	15.1
	1989	403460	17.5	52.1	3.1	26.9	19.3	18.9
Angola	1985	205400	31.0	47.0	0.7	17.1	32.9	28.7
Angola	1989	278866	28.9	48.2	0.9	11.2	33.8	23.1
	1990	308062	28.5	44.7	0.6	11.1	38.9	23.8
Anguilla	1985	53	14.9	40.3	0.0	43.2	68.0	66.4
Anguilla	1990	156	12.8	29.0	0.0	63.7	70.6	76.1
	1991	159	14.8	...	...	...	...	...
Antigua and Barbuda	1985	541	18.3	71.5	0.0	28.0	75.7	93.5
Antigua−et−Barbuda	1986	642	18.9	69.7	0.0	36.1	75.2	99.9
	1992	1010	...	...	...	...	...	...
Argentina	1985	5	76.9[2]	...	...	17.6[3]	11.7	6.3
Argentine	1991	180898	83.8[2]	...	...	14.6[3]	7.7	6.1
	1992	226638	84.8[2]	...	...	16.7[3]	6.6	8.1
Australia [4]	1985	239970[5]	18.7	59.9	0.6	24.7[6]	16.1	19.2
Australie [4]	1991	389247[5]	18.3	62.4	-0.5	19.6[6]	17.7	17.4
	1992	405860[5]	18.3	62.6	-0.1	19.5[6]	18.4	19.0
Austria	1985	1348425	18.9	57.5	0.8[5]	22.6	40.7	40.6
Autriche	1991	1922548	18.2	55.3	0.4[5]	25.1	40.1	39.2
	1992	2035606	18.4	55.2	0.1[5]	24.9	39.6	38.3
Bahamas	1990	3134	13.7	74.1	0.8	21.4	48.4	58.7
Bahamas	1991	3090	14.3	77.1	-0.4	20.8	45.8	57.0
	1992	3059	14.6	74.2	0.2	20.8	45.9	55.2
Bahrain	1985	1393	22.5	31.5	-0.7	35.0	100.3	88.6
Bahrëin	1989	1348	26.5	37.5	2.2	28.6	99.6	94.3
	1990	1468	25.8	35.9	4.0	27.3	118.7	102.3
Bangladesh [4]	1985	466227	12.5	84.3	0.0	12.5	7.3	16.6
Bangladesh [4]	1991	906502	13.8	80.4	0.0	12.1	10.0	16.3
	1992	968804	13.8	78.9	0.0	12.7	11.2	16.7
Barbados	1985	2410	18.9	58.0	0.3	15.1	67.7	60.1
Barbade	1991	3393	18.9	66.7	0.9	16.2	47.4	50.1
	1992	3171	20.2	62.7	-1.5	11.0	50.0	42.4
Belarus	1990	42051	23.7	46.7	4.7	22.7	47.3	45.0
Bélarus	1991	84182	21.1	45.3	7.1	22.9	38.1	34.6
	1992	913968	15.1	51.1	6.6	25.7	60.0	58.5
Belgium	1985	4740756	17.1	65.5	-0.7	15.6	76.9	74.4
Belgique	1991	6705447	14.8	63.1	-0.1	19.4	72.3	69.5
	1992	7032336	14.8	63.2	-0.1	19.2	69.3	66.4
Belize	1985	418	22.8	51.4	4.2	17.4	93.7	89.5
Belize	1991	861	19.4	66.3	1.7	28.5	70.7	86.6
	1992	936	18.7	64.4	1.2	29.7	68.7	82.8
Benin	1985	469778	12.5	79.6	2.4	13.4	33.3	41.3
Bénin	1990	502300	13.2	80.4	0.8	13.4	20.4	28.2
	1991	535500	12.0	82.6	0.9	13.6	22.0	31.1
Bermuda [6]	1985	1174	11.7	66.6	0.0	17.7	59.8	55.8
Bermudes [6]	1991	1680	12.9	69.4	0.0	13.4	56.9	52.6
	1992	1698	12.8	70.5	0.0	13.9	58.9	56.1
Bhutan	1985	2392	23.5	63.0	3.4	41.9	15.4	47.2
Bhoutan	1990	4961	19.6	59.8	-0.7	33.7	29.9	42.3
	1991	5569	18.7	59.9	-1.3	35.0	31.3	43.7

22
Expenditure on gross domestic product at current prices
Percentage distribution [cont.]
Dépenses imputées au produit intérieur brut aux prix courants
Répartition en pourcentage [suite]

Country or area Pays ou zone	Year Année	GDP at current prices (Million nat. cur.) PIB aux prix courants (Mil. monnaie nat.)	Govt. final cons. exp. Consom. finale des admin. publiques	Private final cons. exp. Consom. finale privée	Increase in stocks Variations des stocks	Gross fixed capital form. Formation de brute capital fixe	Exports of goods/ services Exportations de biens et services	Imports of goods/ services Importations de biens et services
Bolivia	1985	2867	8.3	62.5	5.2	11.7	29.0	16.7
Bolivie	1990	17542	11.5	70.5	−1.8	11.0	29.2	20.3
	1991	21690	11.5	71.5	1.3	12.0	27.1	23.2
Botswana [4]	1985	2421	22.0	37.3	−2.7	18.9	71.8	47.3
Botswana [4]	1987	3796	27.7	29.4	−21.2	28.5	82.2	46.6
	1988	5472	22.1	22.6	1.0	40.8	67.7	54.2
Brazil	1985	1	9.9	65.7	2.2	16.9	12.2	7.1
Brésil	1991	164486	14.5	64.6	...	19.0	8.5	6.6
	1992	1846813	15.2	62.4	...	19.1	9.7	6.3
British Virgin Islands	1985	90	18.8	74.5	2.5	37.5	97.1	130.5
Iles Vierges brit.	1988	131	20.1	68.9	2.7	34.3	107.6	133.5
	1989	156	20.7	64.9	2.6	31.6	104.8	124.6
Brunei Darussalam	1982	9126	10.0	...	−0.0	12.4	...	...
Brunéi Darussalam	1983	8163	11.3	...	−0.0	9.9	...	...
	1984	8069	31.1	...	0.0	6.5	...	...
Bulgaria	1985	32595	8.5	61.5	5.8	26.4	−2.1 [1]	...
Bulgarie	1991	131058	5.4	70.4	9.0	18.9	−3.7 [1]	...
	1992	195000	9.7	78.3	...	20.5 [3]	−8.6 [1]	...
Burkina Faso	1984	390565	19.7	78.9	0.8	23.3	22.6	45.3
Burkina Faso	1985	469313	15.5	87.8	3.3	24.2	16.9	47.6
	1986	503500	...	...	...	...	...	...
Burundi	1985	141347	16.1	77.5	0.1	14.2	9.9	17.7
Burundi	1991	211898	17.0	83.9	−0.5	18.1	10.0	28.5
	1992	226384	15.6	82.9	0.4	21.1	9.0	29.0
Cameroon [4]	1985	4106200	11.3	62.6	0.7	24.8	21.5	20.9
Cameroun [4]	1989	3420900	10.8	69.2	0.0	17.0	21.4	18.4
	1990	3423600	10.2	68.8	0.0	16.4	21.1	16.5
Canada	1985	474339 [5]	20.1	57.2	0.5	19.9	28.4	26.0
Canada	1991	668708 [5]	21.3	60.3	−0.1	20.0	24.7	25.7
	1992	681427 [5]	21.7	60.6	−0.4	19.1	26.7	27.3
Cape Verde	1985	13081	21.0	87.7	−1.9	45.5	22.1	74.4
Cap−Vert	1987	17984	20.4	84.2	1.8	39.2	16.6	62.2
	1988	20640	19.2	86.5	−2.1	37.4	15.5	56.5
Cayman Islands	1985	264 [5]	16.3	62.9	...	20.1	69.7	70.1
Iles Caïmanes	1990	590 [5]	14.2	62.5	...	21.4	64.1	58.5
	1991	616 [5]	15.1	62.5	...	21.8	58.9	52.8
Central African Rep.	1985	388545	15.1	79.1	3.0	12.4	22.0	31.6
Rép. centrafricaine	1987	360942	17.5	80.6	−0.2	12.9	17.8	28.6
	1988	376748	16.1	80.7	0.7	9.8	17.7	25.1
Chile	1985	2651937	13.4	67.0	0.3	16.8	28.1	25.7
Chili	1991	11870595	9.7	63.5	0.5	21.7	33.3	28.7
	1992	14939958	9.7	64.6	0.0	23.7	30.9	28.9
Colombia	1985	4966000	10.7	69.4	1.5	17.5	14.4	13.6
Colombie	1991	26107000	10.3	66.4	1.4	14.6	22.6	15.3
	1992	33143000	11.0	69.9	1.5	15.7	18.9	17.1
Congo	1985	970850	16.4	41.6	1.8	28.5	56.8	45.1
Congo	1988	658964	21.1	60.1	−1.0	19.6	40.6	40.4
	1989	773524	18.7	52.8	−0.5	16.4	47.6	35.0
Costa Rica	1985	197920	15.8	60.1	6.6	19.3	30.7	32.5
Costa Rica	1991	689848	16.2	59.5	5.1	19.7	38.5	39.1
	1992	878198	15.8	61.0	7.2	20.9	38.7	43.6
Côte d'Ivoire	1985	3134800	14.1	58.6	1.2	11.8	46.8	32.4
Côte d'Ivoire	1990	2939000	17.0	71.8	−1.8	8.5	31.7	27.1
	1991	2960000	16.5	73.0	−1.1	8.6	30.0	27.0
Cyprus	1985	1482 [5]	14.1	63.8	3.2	27.2	48.7	58.8
Chypre	1991	2666 [5]	18.5	65.8	1.5	24.4	47.2	57.3
	1992	3081 [5]	19.2	63.6	2.8	25.7	49.9	61.0
Czech Republic	1985	473700	...	...	2.0	25.6	...	...
République tchèque	1991	716593	18.8	...	6.7	23.1	57.7	50.7
	1992	771300	20.7	...	−1.6	25.1	60.2	58.7

22

Expenditure on gross domestic product at current prices
Percentage distribution *[cont.]*
Dépenses imputées au produit intérieur brut aux prix courants
Répartition en pourcentage *[suite]*

Country or area Pays ou zone	Year Année	GDP at current prices (Million nat. cur.) PIB aux prix courants (Mil. monnaie nat.)	% of GDP − en % du PIB					
			Govt. final cons. exp. Consom. finale des admin. publiques	Private final cons. exp. Consom. finale privée	Increase in stocks Variations des stocks	Gross fixed capital form. Formation de brute capital fixe	Exports of goods/ services Exportations de biens et services	Imports of goods/ services Importations de biens et services
Denmark	1985	615072	25.3	54.8	0.8	18.7	36.7	36.3
Danemark	1991	827379	25.4	52.0	−0.3	16.4	37.4	30.9
	1992	853981	25.4	51.8	−0.0	15.2	37.1	29.4
Djibouti	1980	60313	33.6	64.2	1.5	12.9	39.0	51.2
Djibouti	1981	67193	30.8	66.6	−0.3	13.0	44.7	54.8
Dominica	1985	266	22.5	72.4	0.0	28.5	36.5	60.0
Dominique	1990	452	20.3	64.1	1.1	39.7	50.1	75.3
	1991	479	20.0	71.4	1.1	40.2	46.4	79.1
Dominican Republic	1985	13972	8.0	76.7	0.4	20.4	29.3	34.7
Rép. dominicaine	1991	100070	4.1	82.5	0.1	16.9	24.6	28.1
	1992	112368	4.8	80.9	0.1	21.0	24.1	31.0
Ecuador	1985	1109940	11.5	64.5	2.1	16.1	26.8	20.9
Equateur	1991	12201436	7.7	69.2	2.7	19.2	31.4	30.2
	1992	19452299	7.2	68.1	2.4	19.1	31.0	27.9
Egypt [4]	1986	51946	12.8	67.0	2.9	27.4	12.5	22.6
Egypte [4]	1990	110143	10.0	76.3	−0.6	22.4	28.1	36.2
	1991	136190	8.9	80.8	0.1	17.9	29.5	37.2
El Salvador	1985	14331	15.5	81.2	−1.2	12.0	22.3	29.9
El Salvador	1991	47792	11.0	87.5	0.4	13.5	14.8	27.1
	1992	54853	10.6	89.0	0.5	15.7	13.6	29.3
Equatorial Guinea	1985	38067	15.5	80.3	−4.5	12.0	27.3	30.5
Guinée équatoriale	1990	44349	15.3	53.2	−3.1	34.6	59.7	59.7
	1991	46429	14.4	75.9	−2.3	18.4	28.4	34.7
Estonia	1985	495	16.2	64.1	2.6	30.4	55.0	67.9
Estonie	# 1991	1832	13.2	57.6	4.9	19.5	31.9	27.0
	1992	14255	12.5	59.5	6.9	19.3	55.4	53.5
Ethiopia [7]	1985	9924	19.7	75.9	...	15.5	11.5	22.6
Ethiopie [7]	1991	13332	28.2	69.9	...	10.7	10.9	19.6
	1992	13508	16.0	86.7	...	9.0	7.7	19.4
Fiji	1985 [5]	1316	19.1	63.7	0.9	18.2	44.4	44.8
Fidji	1989 [5]	1861	15.9	65.0	1.3	12.5	61.5	57.1
	1990	2090	15.4	64.8	1.2	18.1	64.0	63.6
Finland	1985	331628 [5]	20.2	54.5	−0.1	23.9	29.6	28.5
Finlande	1991	490868 [5]	24.2	56.0	−1.9	22.4	22.3	22.9
	1992	475674 [5]	24.9	57.1	−1.0	18.4	27.0	25.6
France	1985	4700143	19.4	61.1	−0.4	19.3	23.9	23.2
France	1991	6746923	18.3	60.2	0.3	20.9	22.7	22.4
	1992	6987221	18.6	60.5	−0.3	20.0	23.1	21.8
French Guyana	1985	3173	41.7	68.5	0.1	37.9	53.6	101.8
Guyane française	1988	4929	36.1	67.5	1.1	35.0	75.5	115.2
	1989	5684	35.7	67.5	1.6	44.5	71.0	120.2
French Polynesia	1985	226772	41.9	52.8	0.4	35.3	8.8	39.2
Polynésie française	1989	281667	40.9	59.4	0.2	23.2	8.8	32.5
	1990	297754	40.4	60.0	0.2	21.2	9.2	30.9
Gabon	1985	1645800	18.6	30.5	...	37.3 [3]	56.9	43.2
Gabon	1988	1013600	21.8	48.1	...	36.2 [3]	37.3	43.4
	1989	1168066	18.4	48.4	...	23.3 [3]	50.3	40.3
Gambia [4]	1985	1085	14.9	...	...	7.1	43.4	46.8
Gambie [4]	1990	2630	14.5	...	...	5.0	...	...
	1991	2948	12.5	...	...	5.3	...	...
Germany † · Allemagne †								
F. R. Germany	1985	1823180	20.1	56.9	0.1	19.5	32.5	29.0
R.f. Allemagne	1991	2635000	17.7	54.2	0.3	21.4	33.9	27.5
	1992	2794200	17.9	54.0	−0.2	21.2	33.5	26.5
Ghana	1985	343048	9.4	83.0	0.0	9.5	9.7	11.6
Ghana	1991	2574800	11.4	83.9	0.1	12.7	15.7	23.7
	1992	3008800	13.3	84.6	0.1	12.8	16.0	26.8
Greece	1985	4617816 [5]	20.4	65.5	2.2	19.1	21.2	32.8
Grèce	1991	12802352 [5]	19.9	71.4	1.8	18.6	22.7	33.2
	1992	14846938 [5]	19.7	71.7	1.8	18.0	23.2	33.0

22
Expenditure on gross domestic product at current prices
Percentage distribution *[cont.]*
Dépenses imputées au produit intérieur brut aux prix courants
Répartition en pourcentage *[suite]*

Country or area Pays ou zone	Year Année	GDP at current prices (Million nat. cur.) PIB aux prix courants (Mil. monnaie nat.)	% of GDP – en % du PIB					
			Govt. final cons. exp. Consom. finale des admin. publiques	Private final cons. exp. Consom. finale privée	Increase in stocks Variations des stocks	Gross fixed capital form. Formation de brute capital fixe	Exports of goods/ services Exportations de biens et services	Imports of goods/ services Importations de biens et services
Grenada	1985	311	21.9	77.2	−1.7	31.2	47.9	76.4
Grenade	1991	567	18.8	69.2	3.7	40.0	45.4	77.1
	1992	578	19.9	66.5	2.1	32.4	38.6	59.4
Guadeloupe	1985	9802	34.6	95.2	0.3	21.8	7.4	59.3
Guadeloupe	1988	13025	31.3	90.7	−0.8	27.5	8.0	56.7
	1989	14062	31.6	90.2	0.8	29.2	5.6	57.4
Guatemala	1985	11180	7.0	83.1	0.5	11.0	18.5	20.1
Guatemala	1991	47302	5.7	83.9	2.1	12.2	17.7	21.6
	1992	53949	6.5	85.1	2.8	15.5	17.6	27.4
Guinea−Bissau	1990	510094	11.4	100.9	0.9	13.8	12.0	39.0
Guinée−Bissau	1991	854985	12.6	100.6	0.9	10.4	13.4	38.0
	1992	1530010	10.7	111.1	...	...	8.2	56.5
Guyana	1985	1964	35.6	53.7	0.0	20.9	53.1	63.3
Guyana	\1991	38966	11.8	5.5	0.0	35.3	...	...
	1992	46734	13.7	5.0	0.0	53.7	...	...
Haiti	1985	10047	94.3[2]	...	...	16.7	27.0	38.0
Haïti	1991	14755	95.0[2]	...	...	13.0	13.6	21.5
	1992	14999	96.9[2]	...	...	8.1	7.3	12.3
Honduras	1985	6977	15.0	72.1	0.2	17.9	25.1	30.4
Honduras	1991	16314	10.8	67.5	5.7	19.0	34.5	37.6
	1992	18772	11.6	66.5	3.6	22.4	32.2	36.3
Hong Kong	1985	261070	7.6	64.2	0.6	21.9	107.6	101.9
Hong−kong	1991	642930	8.0	60.8	0.6	27.6	138.0	135.1
	1992	745407	8.6	60.3	1.7	27.6	143.3	141.5
Hungary	1985	1033658	19.5	53.4	2.5	22.5	42.2	40.1
Hongrie	# 1991	2346035	80.8[2]	...	0.4	21.7	35.6	38.4
	1992	2805012	81.7[2]	...	−0.9	20.4	32.4	33.6
Iceland	1985	119996	17.5	62.2	−0.8	20.4	41.3	40.6
Islande	1991	384053	20.0	61.5	0.3	18.9	33.1	33.9
	1992	381750	20.7	61.9	0.0	17.0	32.4	31.9
India [6]	1985	2622430[5]	11.1	67.4	3.2	20.7	5.7	8.3
Inde [6]	1991	6156600[5]	11.3	62.3	2.0	22.0	9.2	9.2
	1992	7055700	11.1	60.6	3.7	21.3	...	...
Indonesia	1985	96997000	11.2	59.0	5.0	23.1	22.2	20.4
Indonésie	1991	227502000	9.1	55.1	6.6	28.9	27.4	27.8
	1992	260786000	9.5	52.7	7.2	28.4	29.1	26.9
Iran, Islamic Republic of [8]	1985	15775000[5]	15.5	61.0	3.6	17.5	7.9	8.0
Iran, Rép. islamique d' [8]	1991	50107000[5]	10.7	63.2	11.6	21.6	14.8	19.5
	1992	67811000[5]	11.8	60.9	9.5	23.1	14.2	16.9
Iraq	1985	15494	28.6	52.3	−4.1	27.8	24.4	28.9
Iraq	1990	23297	26.4	50.5	−4.2	26.7	18.5	17.8
	1991	19940	35.3	48.2	2.6	16.5	2.7	5.3
Ireland	1985	17790	18.6	59.6	0.9	19.0	60.4	58.4
Irlande	# 1991	27949	15.9	58.6	2.1	16.8	60.4	53.9
	1992	29609	16.1	58.3	−0.5	16.1	63.1	53.1
Israel	1985	30455	34.3	54.3	0.3	17.5	39.6	46.2
Israël	1991	143656	28.0	57.2	0.9	22.7	27.7	36.4
	1992	171552	26.3	57.4	1.3	21.9	28.7	35.7
Italy	1985	810580000	16.4	61.4	1.8	20.7	22.8	23.2
Italie	1991	1429453000	17.5	61.9	0.7	19.7	19.6	19.4
	1992	1504323000	17.6	62.9	0.3	19.1	20.0	19.9
Jamaica	1985	11203	15.5	69.4	2.3	23.0	58.2	68.5
Jamaïque	1991	44158	12.6	60.9	0.5	26.8	56.9	57.6
	1992	72540	9.5	59.7	0.3	28.3	70.4	68.2
Japan	1985	320419000	9.6	58.9	0.7	27.5	14.5	11.1
Japon	1991	451297000	9.1	56.5	0.7	31.8	10.4	8.5
	1992	463850000	9.3	57.1	0.3	30.8	10.2	7.8
Jordan	1985	1970	26.8	90.6	2.1	19.6	37.2	76.3
Jordanie	1991	2778	26.5	91.4	2.1	21.9	43.1	85.0
	1992	3257	23.7	92.8	1.8	30.1	43.0	91.3

22

Expenditure on gross domestic product at current prices
Percentage distribution *[cont.]*
Dépenses imputées au produit intérieur brut aux prix courants
Répartition en pourcentage *[suite]*

Country or area Pays ou zone	Year Année	GDP at current prices (Million nat. cur.) PIB aux prix courants (Mil. monnaie nat.)	Govt. final cons. exp. Consom. finale des admin. publiques	Private final cons. exp. Consom. finale privée	Increase in stocks Variations des stocks	Gross fixed capital form. Formation de brute capital fixe	Exports of goods/ services Exportations de biens et services	Imports of goods/ services Importations de biens et services
Kenya	1985	5042	17.5	58.0	8.1	17.5	25.3	26.3
Kenya	1991	11044	17.1	62.9	2.0	19.3	27.4	28.7
	1992	12904	16.3	68.5	0.3	16.4	26.8	28.4
Korea, Republic of	1985	80847000[5]	10.1	59.2	1.0	28.2	34.6	33.3
Corée, Rép. de	1991	208201000[5]	10.7	52.7	1.2	37.8	29.2	31.7
	1992	231727000[5]	11.3	53.4	0.2	35.6	29.9	31.0
Kuwait	1985	6450	22.4	47.8	−0.9	19.8	53.7	42.7
Koweït	1991	3184	163.6	67.0	4.9	50.8	15.4	201.6
	1992	6367	37.8	39.5	2.0	33.6	35.8	48.8
Kyrgyzstan	1985	31	21.6	58.0	6.2	30.5	−16.4	...
Kirghizistan	1991	87[5]	16.9	55.0	14.7	17.0	37.8	39.2
	1992	772[5]	10.2	51.1	41.6	12.3	34.2	45.7
Latvia	1990	12488	8.6	52.7	17.1	23.0	47.7	49.0
Lettonie	1991	28665	10.3	46.2	27.6	6.2	35.2	25.5
	1992	200911	12.5	39.4	30.1	11.2	79.9	73.1
Lesotho	1985	551	24.7	152.0	−0.2	49.6	12.7	138.8
Lesotho	1991	1739	18.1	132.5	−0.4	71.8	14.4	136.4
	1992	2026	18.8	131.3	−0.4	70.3	18.7	138.7
Liberia	1985	1055[5]	13.0	67.0	0.7	12.0	44.6	30.6
Libéria	1988	1158[5]	11.8	63.3	0.3	10.0	39.0	27.8
	1989	1194[5]	11.9	55.0	0.3	8.1	43.7	23.1
Libyan Arab Jamah. Jamah. arabe libyenne	1985	8277	31.7	37.6	0.4	19.7	37.4	26.7
Lithuania	1985	9190	22.4	56.9	2.0	32.3	61.6	75.3
Lithuanie	1991	38187	12.1	50.8	3.4	24.4	32.2	22.9
	1992	332830	13.0	67.8	−5.1	15.2	32.4	23.2
Luxembourg	1985	205255	15.7	58.7	2.6	17.7	108.6	103.3
Luxembourg	1991	318804	17.1	57.3	2.4	29.0	94.3	100.0
	1992	339450	17.1	56.3	2.5	27.7	89.1	92.7
Madagascar	1985	1553400	13.5	77.8	0.0	14.0	14.5	19.8
Madagascar	1991	4906400	8.6	92.2	0.0	8.2	17.3	26.2
	1992	5584500	8.2	90.0	0.0	11.6	15.6	25.3
Malawi	1985	1929	17.9	74.1[9]	...	13.5	24.4	29.9
Malawi	1991	6145	16.1	83.0[9]	...	10.2	15.6	24.9
	1992	6873	18.1	84.1[9]	...	15.7	22.0	39.9
Malaysia	1985	77470	15.3	52.0	−2.3	29.8	54.9	49.8
Malaisie	1991	129559	14.2	54.7	1.3	35.6	81.4	87.3
	1992	147784	13.1	51.5	−0.6	34.3	78.0	76.2
Maldives Maldives	1985	596	...	444.5	...	...	27.5	57.0
Mali	1985	520200	20.2	83.9	−2.4	19.2	18.5	39.4
Mali	1991	691400	15.3	85.1	−2.4	20.0	17.5	35.5
	1992	737400	14.2	82.2	2.7	17.6	17.8	34.6
Malta	1985	476	17.7	70.0	1.7[5]	26.4	72.5	88.3
Malte	1991	807	18.2	61.3	1.9[5]	29.6	87.0	98.1
	1992	872	18.8	60.8	−0.2[5]	27.6	92.2	99.3
Martinique	1985	12484	33.7	87.4	−0.5	16.7	11.7	48.9
Martinique	1988	16896	29.8	84.6	1.3	23.8	7.4	46.9
	1989	18086	29.9	83.7	1.6	23.6	7.9	46.8
Mauritania	1985	52665	22.3	67.3	3.9	28.5	58.8	80.8
Mauritanie	1987	67216	13.6	82.6	1.7	20.8	48.3	67.0
	1988	72635	14.2	79.6	1.4	17.0	49.1	61.3
Mauritius	1985	16618	11.5	66.9	4.8	18.7	53.5	55.4
Maurice	1991	42635	11.7	64.5	−1.4	29.0	65.3	69.3
	1992	47647	11.7	63.3	−0.1	28.6	62.5	65.9
Mexico	1985	47392000	9.2	64.5	2.1	19.1	15.4	10.3
Méxique	1991	865166000	9.0	71.8	2.9	19.5	13.8	17.0
	1992	1019156000	10.1	72.2	2.5	20.8	12.6	18.1
Montserrat	1985	100	20.3	96.3	1.5	24.7	11.7	54.4
Montserrat	1986	114	18.7	89.5	2.8	33.0	10.1	53.9

22

Expenditure on gross domestic product at current prices
Percentage distribution *[cont.]*
Dépenses imputées au produit intérieur brut aux prix courants
Répartition en pourcentage *[suite]*

Country or area Pays ou zone	Year Année	GDP at current prices (Million nat. cur.) PIB aux prix courants (Mil. monnaie nat.)	Govt. final cons. exp. Consom. finale des admin. publiques	Private final cons. exp. Consom. finale privée	Increase in stocks Variations des stocks	Gross fixed capital form. Formation de brute capital fixe	Exports of goods/ services Exportations de biens et services	Imports of goods/ services Importations de biens et services
Morocco	1985	129510	15.8	65.2	4.0	23.1	24.9	33.1
Maroc	1991	241650	15.2	66.9	0.5	22.3	22.2	27.1
	1992	242490	16.4	66.7	−0.1	23.0	22.5	28.6
Mozambique	1985	147000	15.6	87.1	...	6.8[3]	−10.2[1]	...
Mozambique	1991	1967000	19.2	90.4	...	46.8[3]	22.6	73.0
	1992	2764000	21.6	93.5	...	43.5[3]	26.7	85.3
Myanmar[6]	1985	55989	88.5[2]	...	0.1	15.4	4.6	8.6
Myanmar[6]	1991	178553	85.6[2]	...	0.6	15.1	1.6	3.0
	1992	230935	86.9[2]	...	0.7	13.4	1.6	2.7
Namibia	1985	2854	27.9	46.5[10]	0.2	13.3	65.9	53.8
Namibie	1991	6203	29.1	68.9[10]	−0.3	11.2	61.2	70.0
	1992	7027	31.6	66.8[10]	−0.9	12.4	57.4	67.3
Nepal[14]	1985	44417	9.8	76.1	1.8	21.1	12.1	21.0
Népal[14]	1991	103948	11.6	81.0	2.2	18.9	13.9	27.7
	1992	126186	10.0	80.6	3.2	18.7	19.5	32.0
Netherlands	1985	425350	15.7	59.5	0.3	19.7	60.8	56.0
Pays−Bas	1991	541880	14.4	59.6	0.6	20.5	54.3	49.4
	1992	563220	14.5	60.3	0.4	20.4	52.3	47.8
Netherlands Antilles Antilles néerlandaises	1985	1966	31.9	70.0	0.8	15.0	66.6	84.2
New Caledonia	1985	139650[5]	37.8	54.8	−1.4	15.4	33.4	39.7
Nouvelle−Calédonie	1989	253475[5]	27.6	51.8	1.8	21.3	31.7	34.5
	1990	250427[5]	32.6	57.3	−1.1	24.4	22.0	35.4
New Zealand[6]	1985	45435[5]	16.2	60.9	0.6	25.8	30.7	33.7
Nouvelle−Zélande[6]	1991	73379[5]	16.9	62.5	0.7	16.7	29.3	26.4
	1992	77067[5]	16.1	61.4	2.1	17.3	30.8	28.6
Nicaragua	1985	115404	35.7	48.2	2.4	20.7	14.8	21.8
Nicaragua	1986	435742	35.4	55.8	3.1	13.8	12.8	20.8
	1987	2389500	24.7	58.1	3.0	7.9	22.1	15.7
Niger	1985	647100	15.0	78.8	1.0	14.3	21.0	30.1
Niger	1989	692600	18.0	73.0	−0.1	12.3	18.6	21.8
	1990	682300	17.2	74.1	1.1	11.7	16.8	20.9
Nigeria	1985	72355	10.1	74.7	−0.6	7.7	16.7	8.7
Nigéria	1991	324794	3.9	68.4	0.1	11.2	39.9	23.5
	1992	553154	3.7	73.7[9]	...	10.7	35.6	23.6
Norway	1985	500199	18.5	49.1	2.2	22.0	47.1	38.9
Norvège	1991	686687	21.5	50.9	0.3	18.5	44.8	36.0
	1992	701652	22.4	52.0	−0.9	19.1	43.2	35.9
Oman	1985	3454	27.2	32.6[9]	...	27.6	49.9	37.2
Oman	1991	3917	35.6	37.9[9]	...	16.9	47.8	38.2
	1992	4417	39.3	33.4[9]	...	17.0	48.4	38.1
Pakistan[4]	1985	514532	12.8	76.3	1.7	17.0	12.3	20.1
Pakistan[4]	1991	1211245	12.8	70.2	1.5	18.5	17.3	20.4
	1992	1359338	12.7	69.7	1.6	18.8	17.0	19.7
Panama	1985	4948	21.1	61.0	−0.4	15.6	36.8	34.2
Panama	1991	5496	17.4	61.4	1.2	16.8	38.8	35.8
	1992	6001	19.2	59.4	1.7	21.1	37.1	38.5
Papua New Guinea	1985	2403	23.8	66.6	1.4	18.6	42.5	52.9
Papouasie−Nouv.−Guinée	1991	3606	22.4	60.1	−0.6	28.0	42.3	52.2
	1992	4140	22.5	57.9	0.0	23.8	45.2	49.3
Paraguay	1985	1393890	6.5	76.7	1.3	20.7	21.4	26.6
Paraguay	1991	8280772	6.6	78.3	1.1	23.7	24.7	34.4
	1992	9670838	6.5	78.7	1.0	21.9	21.6	29.8
Peru	1985	197[11]	11.3	62.4	0.2	21.4	20.0	15.3
Pérou	1991	33107	3.4	81.6	1.2	16.3	9.5	12.0
	1992	52452	4.0	80.9	1.0	17.6	10.3	13.8
Philippines	1985	571883[5]	7.6	73.6	−2.1	17.5	24.0	21.9
Philippines	1991	1244037[5]	10.2	73.7	0.2	19.7	30.0	31.5
	1992	1342515[5]	9.7	75.8	0.2	20.9	29.4	33.4

22

Expenditure on gross domestic product at current prices
Percentage distribution *[cont.]*

Dépenses imputées au produit intérieur brut aux prix courants
Répartition en pourcentage *[suite]*

Country or area Pays ou zone	Year Année	GDP at current prices (Million nat. cur.) PIB aux prix courants (Mil. monnaie nat.)	% of GDP − en % du PIB					
			Govt. final cons. exp. Consom. finale des admin. publiques	Private final cons. exp. Consom. finale privée	Increase in stocks Variations des stocks	Gross fixed capital form. Formation de brute capital fixe	Exports of goods/ services Exportations de biens et services	Imports of goods/ services Importations de biens et services
Poland	1985	10445000	18.1	52.7	6.5	21.2	18.2	16.9
Pologne	# 1991	824330000	20.4	58.3	2.7	18.8	23.1	22.8
	1992	1142429000	19.3	64.1	0.9	16.7	21.2	22.3
Portugal	1985	3523945	15.5	67.9	−1.2	21.8	37.3	41.4
Portugal	1991	9913400	17.8	63.3	2.3	26.0	31.9	41.3
	1992	11366000	18.3	62.9	2.3	26.2	29.1	38.8
Puerto Rico [4]	1985	21969	14.7	71.7	−0.3	10.6	66.6	63.3
Porto Rico [4]	1991	34004	14.1	62.9	0.9	14.8	74.7	67.5
	1992	35834	14.2	62.9	1.3	15.7	68.7	62.9
Qatar	1985	22398	35.2	25.1	0.1	17.7	51.4	29.5
Qatar	1991	25056	35.7	31.4	1.7	17.4	47.4	33.5
	1992	27202	33.0	30.8	1.0	18.0	51.2	34.0
Republic of Moldova	1985	9	...	...	...	...	−3.2 [1]	...
Rép. de Moldova	1990	13	...	...	...	...	−2.9 [1]	...
	1991	23	...	...	...	...	−6.8 [1]	...
Réunion	1985	17054	32.8	86.1	−2.5	21.9	6.4	44.7
Réunion	1988	23604	27.6	79.9	1.7	29.9	4.2	43.4
	1989	25613	27.9	82.7	1.8	27.8	4.1	44.3
Romania	1985	817400 [5]	10.2	50.0	2.9	30.1	4.1	...
Roumanie	1991	2198900 [5]	15.1	60.0	14.8	14.3	−4.2	...
	1992	5982300 [5]	15.0	61.7	16.2	14.9	−7.7	...
Rwanda	1985	173700	11.3	80.5	1.7	15.6	10.8	19.9
Rwanda	1991	212900	21.6	82.2	−1.5	11.9	9.8	23.9
	1992	217300	25.1	76.8	−0.0	14.9	7.4	24.2
Saint Kitts−Nevis	1982	158	22.5	74.1	0.0	34.9	51.0	82.5
Saint−Kitts−et−Nevis	1983	154	23.1	91.0	0.0	32.4	51.2	97.6
	1984	167	22.4	77.5	0.0	31.9	55.5	87.3
Saint Vincent−Grenadines	1985	305	19.8	57.7	3.3	25.0	73.0	78.8
St.−Vincent−et−Grenad.	1989	469	19.6	74.4	2.7	26.7	63.8	87.2
	1990	525	17.8	67.2	1.4	30.3	70.6	87.3
Sao Tome and Principe	1986	2478	30.3	76.1	0.9	13.6	...	50.8
Sao Tomé−et−Principe	1987	3003	24.8	63.1	1.1	15.4	...	43.1
	1988	4221	21.2	71.8	0.0	15.7	...	66.8
Saudi Arabia [4]	1985	313941	36.4	50.5	−3.4 [5]	24.3	36.0	43.9
Arabie saoudite [4]	1991	431920	38.2	39.1	1.2 [5]	18.2	45.7	42.3
	1992	455130	32.7	40.4	2.0 [5]	21.7	43.1	39.9
Senegal	1985	1158100	16.8	84.7	−1.7	11.5	29.7	40.9
Sénégal	1990	1583000	14.1	77.1	0.0	13.2	25.2	29.6
	1991	1623300	12.9	78.6	0.0	13.6	23.6	28.8
Seychelles	1985	1205	34.6	58.2	0.0	22.7	48.2	63.7
Seychelles	1991	1980	28.2	50.7	1.0	21.3	48.1	49.3
	1992	2221	30.4	50.6	0.3	20.9	45.8	48.0
Sierra Leone [4]	1985	7481	6.7	83.4	1.5	9.8	11.4	12.7
Sierra Leone [4]	1989	82837	6.6	84.7	0.5	13.5	19.7	25.1
	1990	150175	10.4	77.9	1.8	10.1	25.4	25.7
Singapore	1985	38924 [5]	14.3	45.1	0.3	42.2	−2.4	...
Singapour	1991	69452 [5]	10.4	43.7	−1.1	39.4	8.1	...
	1992	74975 [5]	9.8	43.1	0.4	40.4	6.2	...
Slovakia	1985	203258	19.6	49.9	2.3	29.4	−1.1 [1]	...
Slovaquie	1991	280053 [5]	20.6	51.2	6.9	28.3	46.3	49.3
	1992	301800	24.8	54.0	−3.8	28.5	93.1	87.3
Solomon Islands	1985	237	28.1	64.2	5.1	21.1	51.1	69.6
Iles Salomon	1987	293	36.3	63.1	2.7	20.4	55.9	78.4
	1988	367	31.4	68.6	2.7	30.0	52.4	85.0
Somalia	1985	87290	10.6	90.5 [10]	2.9	8.9	4.2	17.0
Somalie	1986	118781	9.7	89.1 [10]	1.0	16.8	5.8	22.4
	1987	169608	11.1	88.8 [10]	4.8	16.8	5.9	27.2
South Africa	1985	123126 [5]	17.3	53.7	−3.0	23.3	32.2	23.2
Afrique du Sud	1991	297895 [5]	20.8	60.2	−1.9	18.0	24.9	19.7
	1992	327068 [5]	21.3	62.2	−0.9	15.9	23.9	20.0

22
Expenditure on gross domestic product at current prices
Percentage distribution [cont.]
Dépenses imputées au produit intérieur brut aux prix courants
Répartition en pourcentage [suite]

Country or area Pays ou zone	Year Année	GDP at current prices (Million nat. cur.) PIB aux prix courants (Mil. monnaie nat.)	% of GDP − en % du PIB					
			Govt. final cons. exp. Consom. finale des admin. publiques	Private final cons. exp. Consom. finale privée	Increase in stocks Variations des stocks	Gross fixed capital form. Formation de brute capital fixe	Exports of goods/ services Exportations de biens et services	Imports of goods/ services Importations de biens et services
Spain	1985	28200900	14.7	64.1	0.0	19.2	22.7	20.8
Espagne	1991	54820600	16.1	62.4	0.7	23.9	17.2	20.3
	1992	58852000	16.8	63.2	0.9	22.0	17.6	20.4
Sri Lanka	1985	157763[5]	12.2	74.9	0.1	23.9	26.9	39.6
Sri Lanka	1991	369720[5]	13.7	73.3	0.3	23.4	28.9	39.1
	1992	421755[5]	12.8	73.1	0.2	22.9	32.0	41.4
Suriname	1985	1747	33.7	55.0	−3.4	17.9	36.7	39.9
Suriname	1991	3708	29.3	52.1	4.3	18.0	17.7	21.4
	1992	5010	24.7	52.4	4.6	18.6	13.0	13.3
Swaziland [12]	1985	803	22.8	72.5	1.6	24.6	56.4	78.0
Swaziland [12]	1990	2340	17.6	51.9	0.7	20.0	75.5	65.8
	1991	2413	19.7	60.0	0.9	18.3	79.4	78.3
Sweden	1985	866601	27.9	51.2	−0.1	19.3	35.3	33.6
Suède	1991	1447327	27.2	53.3	−1.5	19.4	27.9	26.4
	1992	1439835	27.8	53.9	−0.4	17.0	27.9	26.2
Switzerland	1985	227950	13.3	61.9	0.6	23.8	39.1	38.6
Suisse	1991	331075	13.9	57.8	1.4	25.6	35.3	33.9
	1992	339470	14.4	58.5	−0.2	23.7	36.0	32.5
Syrian Arab Rep.	1985	83225	23.8	65.7	...	24.1	12.0	25.4
Rép. arabe syrienne	1991	316204	15.0	74.8	...	17.7	24.0	31.6
	1992	370990	16.2	71.6	...	23.6	26.3	37.7
Thailand	1985	1056496[5]	13.5	62.2	1.1	27.2	23.2	25.9
Thaïlande	1990	2182100[5]	9.5	57.3	0.9	40.2	34.2	41.7
	1991	2509427[5]	9.4	56.6	0.8	41.8	35.3	42.4
Togo	1985	332500	14.2	66.0	5.2	22.9	48.3	56.7
Togo	1986	363600	14.4	69.0	5.3	23.8	35.6	48.2
Trinidad and Tobago	1985	18071	22.7	54.4	−1.5	20.2	32.6	28.4
Trinité−et−Tobago	1991	22380	17.1	61.0	−0.1	13.6	41.4	33.0
	1992	22222	18.1	59.0	0.4	11.7	41.2	30.4
Tunisia	1985	6910	16.5	63.0	−0.2	26.8	32.6	38.7
Tunisie	1988	8605	16.1	64.6	−0.8	19.5	42.5	42.0
	1989	9497	16.9	64.9	1.6	21.1	42.8	47.4
Turkey	1985	35400000	8.9	71.3	0.4	21.6	16.3	18.4
Turquie	1991	618249000[5]	12.7	65.9	−1.0	22.9	15.0	17.0
	1992	1061642000[5]	13.3	63.9	−0.3	22.7	15.6	17.9
Ukraine	1985	128300	...	...	...	...	...	...
Ukraine	1990	164800	...	...	...	...	...	...
	1991	230800	22.6	70.1	−23.3	26.0	33.9	31.1
United Arab Emirates	1985	99416	19.7	28.5	0.5	24.6	58.0	31.2
Emirats arabes unis	1991	124500	16.9	41.4	1.1	20.7	67.6	47.7
	1992	128400	17.8	45.5	1.2	23.2	69.1	56.8
United Kingdom	1985	356172	21.1	60.6	0.2	17.0	28.8	27.8
Royaume−Uni	1991	571782	21.7	63.5	−0.9	16.9	23.6	24.8
	1992	594183	22.3	64.1	−0.3	15.6	23.7	25.2
United Rep. of Tanzania	1985	120621	15.4	77.2	1.7	14.0	6.2	14.5
Rép.−Unie de Tanzanie	1990	494999	14.3	74.0	3.6	43.9	20.8	56.7
	1991	690421	13.5	76.3	2.9	38.0	17.7	48.3
United States	1985	4016649	18.1	64.7	0.7	19.5	7.4	10.4
Etats−Unis	1991	5654300	18.2	66.9	−0.1	15.6	10.5	11.0
	1992	5953300	17.7	67.1	0.1	15.7	10.6	11.3
Uruguay	1985	479	14.4	68.5	1.7	9.6	26.8	21.1
Uruguay	1991	20271	13.5	69.9	1.6	11.9	23.1	19.9
	1992	35346	12.8	72.6	0.7	13.1	22.4	21.5
former USSR †	1985	777000[13]	20.5	47.5	...	...	...	...
ancienne URSS †	1987	825000[13]	20.8	47.1	...	...	...	...
	1988	875400[13]	21.5	47.0	...	...	...	...
Vanuatu	1985	12534[5]	35.9	56.6	5.5	22.7	51.0	72.3
Vanuatu	1989	16367[5]	29.8	64.4	3.7	33.4	36.7	64.7
	1990	17899[5]	28.2	62.9	2.7	40.8	46.4	76.6

22
Expenditure on gross domestic product at current prices
Percentage distribution *[cont.]*
Dépenses imputées au produit intérieur brut aux prix courants
Répartition en pourcentage *[suite]*

Country or area Pays ou zone	Year Année	GDP at current prices (Million nat. cur.) PIB aux prix courants (Mil. monnaie nat.)	% of GDP − en % du PIB					
			Govt. final cons. exp. Consom. finale des admin. publiques	Private final cons. exp. Consom. finale privée	Increase in stocks Variations des stocks	Gross fixed capital form. Formation de brute capital fixe	Exports of goods/ services Exportations de biens et services	Imports of goods/ services Importations de biens et services
Venezuela	1985	464741	10.4	61.8	1.2	17.3	25.0	15.8
Venezuela	1991	3037492	9.7	66.5	0.5	18.2	31.4	26.2
	1992	4132307	8.9	69.9	2.6	20.6	26.2	28.1
Viet Nam	1989	24308000	17.3	82.9	1.7	9.9	27.6	39.4
Viet Nam	# 1990	38166000⁵	17.0	80.9	...	...	−7.7¹	
	1991	69959000⁵	14.9	80.4	...	...	−5.6¹	...
Yemen	1989	61406	26.8	78.7	0.2	18.5	16.6	40.8
Yémen	1990	77159	27.4	75.2	0.9	15.5	13.6	32.6
former Yemen Arab Rep.	1985	30939	17.9	95.0	−0.2	14.7	3.8	31.2
anc. Yémen rép. arabe	1986	37472	18.4	90.7	0.1	13.2	3.1	25.5
former Dem. Yemen	1985	369	53.0	94.5	...	43.1³	10.9	101.4
ancienne Yémen dém	1986	311	58.0	105.3	...	34.9³	7.3	105.4
Yugoslavia, SFR †	1985	1195⁵	13.9	49.8	17.5	21.8	22.1	23.2
Yougoslavie, SFR †	1989	235395⁵	14.4	47.4	28.0	14.5	25.3	29.2
	1990	1147787	17.6	66.1	7.3	14.7	23.7	29.4
Zambia	1985	9351	66.0	18.0	...	15.5	29.3	28.9
Zambie	1990	123487	54.7	14.1	...	30.7	34.3	33.8
	1991	234504	60.4	14.3	...	24.6	26.4	25.7
Zimbabwe	1985	7295	21.5	57.5	4.3	15.5	28.8	27.6
Zimbabwe	1988	11441	23.7	49.3	4.0	17.8	30.0	24.8
	1989	13794	23.6	54.5	1.6	17.4	30.0	27.0

Source:
National accounts database of the Statistical Division of the
United Nations Secretariat.

†   For detailed descriptions of data pertaining to
     former Czechoslovakia, Germany, SFR Yugoslavia and former
     USSR, see annex I − Country or area nomenclature, regional
     and other groupings.

1   Net exports.
2   Including private final consumption expenditure.
3   Gross capital formation.
4   Fiscal year beginning 1 July.
5   Including a statistical discrepancy.
6   Fiscal year beginning 1 April.
7   Fiscal year beginning 7 July.
8   Fiscal year beginning 21 March.
9   Including increase in stocks.
10   Obtained as a residual.
11   Thousand.
12   Fiscal year ending 30 June.
13   Gross national product.
14   Fiscal year ending 15 July.

Source:
Base de données sur les comptes nationaux de la division
de statistique du Secrétariat de l'ONU.

†   Pour les descriptions en détails des données relatives
     à l'ancienne Tchécoslovaquie, l'Allemagne, la Rfs
     Yougoslavie et l'ancienne URSS, voir l'Annexe I −
     Nomenclature des pays ou zones, groupements
     régionaux et autres groupements.

1   Exportations nettes.
2   Y compris dépenses consommation final privée.
3   Formation brute de capital.
4   L'année fiscale commençant le 1er juillet.
5   Y compris   divergence statistique.
6   L'année fiscale commençant le 1er avril.
7   L'année fiscale finissant le 7 juillet.
8   L'année fiscale commençant le 21 mars.
9   Y compris les variations des stocks.
10   Données résiduelles.
11   Milliers.
12   L'année fiscale finissant le 30 juin.
13   Produit nationale brut.
14   L'année fiscale commençant 15 juillet.

23
Gross domestic product by kind of economic activity at current prices
Produit intérieur brut par genre d'activité economique aux prix courants

Percentage distribution
Répartition en pourcentage

Country or area Pays ou zone	Year Année	GDP at current prices (Mil. nat.cur.) PIB aux prix courants (Mil. mon.nat.)	% of GDP – en % du PIB Agriculture, hunting, forestry & fishing Agriculture, chasse, sylviculture et pêche	Mining & quarrying Industries extractives	Manufac- turing Manufac- turières	Electricity, gas and water Electricité, gaz et eau	Construc- tion Construc- tion	Wholesale/ retail trade, restaurants and hotels Commerce, restaurants, hôtels	Transport, storage & commu- nication Transports, entrepôts, communi- cations	Other activities Autres activités
Angola	1985	205400	13.4	28.3	9.6	0.2	4.8	12.9[1]	4.8	25.9[1]
Angola	1989	278866	19.1	29.4	6.1	0.2	3.2	11.2[1]	3.0	27.8[1]
	1990	308062	17.9	32.7	5.0	0.1	2.9	10.7[1]	3.2	27.6[1]
Anguilla	1985	53	5.3	1.0	0.9	1.3	13.4	32.1	12.9	33.0
Anguilla	1990	156	3.9	0.4	0.6	1.0	19.8	33.6	10.5	30.3
	1991	159	2.5	0.7	0.7	1.6	17.1	34.2	11.4	31.8
Antigua and Barbuda	1985	541	4.3	0.9	3.8	3.2	6.6	22.5	15.0	43.8
Antigua−et−Barbuda	1986	642	3.8	1.5	3.3	3.1	7.9	20.8	13.8	45.7
Argentina	1985	5	7.6	2.0	29.6	2.0	5.7	16.5	0.5	31.7
Argentine	1991	180898	6.9	2.1	24.4	1.6	4.7	15.9	5.2	39.5
	1992	226638	6.0	1.8	21.9	1.7	5.3	15.4	5.2	42.7
Australia [2]	1985	239970	4.0	6.5	17.2	3.5	7.6	17.0[1]	7.0	37.2[1]
Australie [2]	1991	389247	3.1	4.2	14.4	3.5	7.1	17.5[1]	7.5	42.7[1]
	1992	405860	3.2	4.3	14.7	3.5	6.9	17.4[1]	7.5	42.5[1]
Austria	1985	1348425	3.3	0.4	26.9	3.0	6.6	15.9	5.8	38.0
Autriche	1991	1922548	2.7	0.2	25.6	2.8	7.2	16.5	6.3	38.6
	1992	2035606	2.4	0.2	25.1	2.7	7.4	16.6	6.5	39.0
Bahamas	1990	3134	2.6	2.7	...	2.1	3.5	29.6	7.3	52.1
Bahamas	1991	3090	3.4	3.3	...	2.5	3.4	25.9	6.6	54.9
	1992	3059	2.9	3.4	...	2.9	3.0	23.0	7.4	57.3
Bahrain	1985	1393	1.2	28.6	10.0	1.8	9.6	8.6	11.9	28.3
Bahrëin	1989	1348	1.2	17.5	17.8	2.1	6.5	10.2	11.3	33.4
	1990	1468	1.0	22.1	17.2	2.0	6.4	10.1	9.6	31.7
Bangladesh [2]	1985	466227	40.4	0.0	9.3	0.6	5.6	9.3	11.7	23.1
Bangladesh [2]	1991	906502	34.5	0.0	9.1	1.5	5.9	8.6	12.0	28.3
	1992	968804	33.1	0.0	9.4	1.7	5.9	8.1	11.7	30.0
Barbados	1985	2410	5.7	1.2	9.6	3.0	4.9	28.1	7.7	39.9
Barbade	1991	3393	4.8	0.5	6.8	2.9	4.8	26.3	7.5	46.4
	1992	3171	4.6	0.5	6.4	3.3	3.6	25.7	7.8	48.0
Belarus	1990	42051	23.5	0.2	37.5	...	8.0	3.5	7.0	20.3
Bélarus	1991	84182	20.7	0.1	40.0	...	7.7	4.8	5.9	20.8
	1992	913968	21.6	0.1	35.8	...	7.0	3.8	8.9	22.9
Belgium	1985	4740756[3]	2.2	0.2	23.5	2.5	5.2	16.1	7.9	42.1
Belgique	1991	6705447[3]	1.9	...	...	...	5.5	18.1	8.2	42.5
	1992	7032336[3]	1.6	...	...	...	5.5	18.4	8.0	43.0
Belize	1985	418	17.8	0.3	14.6	2.2	4.7	13.5	8.4	38.5
Belize	1991	861	16.9	0.7	13.0	2.2	6.9	15.3	9.8	35.2
	1992	936	17.1	0.6	11.7	2.5	6.8	15.0	10.1	36.2
Benin	1985	469778	32.0	4.5	7.5	0.7	3.2	15.9[1]	8.0	28.2[1]
Bénin	1988	482434	34.8	0.9	8.3	0.9	3.1	17.5[1]	7.6	27.0[1]
	1989	487525	36.3	0.9	8.8	0.8	3.1	16.8[1]	7.5	25.7[1]
Bhutan	1985	2392	51.7	0.8	5.4	0.3	12.1	8.5	4.4	16.8
Bhoutan	1991	5569	41.9	1.7	9.0	7.5	6.6	6.9	6.9	19.5
	1992	6337	41.4	3.1	9.1	7.3	6.8	7.0	7.1	18.2
Bolivia	1985	2867	29.0	11.8	12.0	0.9	4.2	10.8	10.6	20.6
Bolivie	1986	8924	27.5	10.0	13.3	1.1	2.9	10.9	11.8	22.7
	1988	11361	18.4	7.5	15.6	1.2	3.2	13.9	10.8	29.4
Botswana [2]	1985	2421	5.5	46.8	5.1	2.4	4.0	6.8	2.7	26.7
Botswana [2]	1990	6995	5.2	42.3	4.3	2.3	5.6	13.9	2.5	23.9
	1991	7810	5.1	39.3	4.4	2.3	5.9	15.2	2.8	25.1
Brazil	1985	1	10.5	3.0	30.7	2.1	5.5	8.2	4.8	35.1
Brésil	1988	86	9.2	1.7	28.1	2.5	7.3	7.3	4.9	39.0
	1989	1266	7.8	1.4	27.0	2.2	8.4	7.1	5.1	40.9
British Virgin Islands	1985	90	4.1	0.2	2.6	3.2	6.6	27.4	9.7	46.4
Iles Vierges brit.	1988	131	3.4	0.2	2.8	3.6	5.6	26.1	11.0	47.3
	1989	156	3.0	0.2	2.8	3.3	6.0	25.3	13.6	45.7

23
Gross domestic product by kind of economic activity at current prices
Percentage distribution [cont.]
Produit intérieur brut par genre d'activité économique aux prix courants
Répartition en pourcentage [suite]

% of GDP − en % du PIB

Country or area Pays ou zone	Year Année	GDP at current prices (Mil. nat.cur.) PIB aux prix courants (Mil. mon.nat.)	Agriculture, hunting, forestry & fishing Agriculture, chasse, sylviculture et pêche	Mining & quarrying Industries extractives	Manufac− turing Manufac− turières	Electricity, gas and water Electricité, gaz et eau	Construc− tion Construc− tion	Wholesale/ retail trade, restaurants and hotels Commerce, restaurants, hôtels	Transport, storage & commu− nication Transports, entrepôts, communi− cations	Other activities Autres activités
Brunei Darussalam Brunéi Darussalam	1985 1991 1992	7752 6604 6372	1.2 2.6 3.0	59.2 38.6 41.7	10.1 8.3 ...	0.4 0.9 1.0	2.1 4.4 4.9	10.6 12.8 12.3	1.8 4.5 4.9	14.6 27.9 32.3
Bulgaria Bulgarie	1985 1991 1992	32595 131058 195000	11.9 17.0 11.5	54.4[4] 43.7[4] 39.5[4]	... ...	... ...	8.1 4.4 5.2	5.7 7.5 8.1	6.1 7.3 5.9	13.8 21.6 29.9
Burkina Faso Burkina Faso	1985	455882	46.9	0.1	11.2	0.7	1.2	10.0	6.8	23.2
Burundi Burundi	1985 1989 1990	141347 179549 196656	54.8 46.0 51.2	0.5[5] 1.1[5] 0.8[5]	12.1 18.1 16.4	... ...	4.0 3.2 3.4	11.4 10.6 4.8	2.2 3.2 3.0	15.0 17.8 20.4
Cameroon [2] Cameroun [2]	1985 1989 1990	4106200 3420900 3423600	21.6 24.2 25.0	12.6 9.4 9.1	12.0 14.2 13.6	1.1 1.7 1.7	6.8 4.2 3.8	16.0 15.2 15.4	6.1 7.0 7.3	23.8 24.2 24.1
Canada Canada	1985 1989 1990	474340 645146 664414	2.8 2.4 2.4	5.7 3.2 3.2	17.2 16.9 16.0	3.1 2.8 2.7	5.5 6.3 6.2	12.2 12.7 12.6	6.3 5.6 5.6	47.2 50.1 51.2
Cape Verde Cap−Vert	1985 1987 1988	13081 17984 20640	15.6 20.1 20.2	0.9 0.6 0.6	5.5 5.7 5.4	0.1 1.0 1.0	10.7 10.5 10.8	27.6 24.3 24.8	14.0 12.8 12.6	25.5 25.0 24.5
Cayman Islands Iles Caïmanes	1985 1990 1991	264 590 616	0.4 0.3 0.3	0.4 0.3 0.3	2.3 1.5 1.5	2.3 3.1 3.1	9.5 9.5 8.9	20.5 24.1 22.4	12.5 10.7 10.6	52.3[6] 50.7[6] 52.9[6]
Central African Rep. Rép. centrafricaine	1983 1984 1985	251055 275730 311430	39.6 39.6 42.0	2.5 2.8 2.5	7.6 7.9 7.4	0.5 0.9 0.8	2.0 2.6 2.6	20.5 21.1 21.8	4.1 4.2 4.2	23.2 22.0 21.0
Chile Chili	1985 1987 1988	2651937 4540556 5917879	7.4 8.8 8.5	13.2 11.4 15.5	15.7 17.5 18.1	2.6 2.5 2.5	4.9 4.8 5.1	13.4 15.4 13.4	6.0 6.1 6.0	36.8 33.5 30.9
Colombia Colombie	1985 1991 1992	4966000 26107000 33143000	17.0 17.0 15.7	4.2 8.2 7.1	21.4 20.4 19.4	2.2 2.6 2.7	6.9 5.0 5.7	14.1 14.4 15.5	8.2 9.6 10.4	26.2 22.6 23.5
Congo Congo	1985 1988 1989	970850 658964 773524	7.5 13.9 13.0	41.0 16.8 27.9	5.6 8.6 7.0	1.2 1.9 1.8	6.1 2.6 1.8	11.4 16.3 14.4	7.3 11.0 9.1	20.0 28.9 24.9
Cook Islands Iles Cook	1985 1986	52 63	14.3 12.7	0.1 0.1	4.6 5.1	0.1 1.1	2.6 3.7	26.1 23.0	10.9 12.2	41.1 42.1
Costa Rica Costa Rica	1985 1991 1992	197920 689848 878198	18.9 17.3 16.1	22.1 19.9 20.3	... ...	3.2 3.5 4.0	3.6 2.8 2.6	20.5 20.2 20.9	4.9 5.2 5.3	26.9 31.0 30.8
Côte d'Ivoire Côte d'Ivoire	1981 1982 1984	2291401 2486544 2855779	28.7 26.2 27.7	1.2 2.1 2.9	10.0 12.1 12.0	1.8 2.1 1.3	5.6 5.1 2.2	16.8 18.4 ...	7.9 7.1 7.5	28.0 26.9 ...
Cyprus Chypre	1985 1991 1992	1481 2666 3081	7.5 6.2 5.8	0.5 0.3 0.2	15.7 14.3 13.4	2.2 2.1 2.0	10.0 10.2 10.0	18.7 20.0 21.2	9.3 8.7 8.6	36.0 38.1 39.0
Czech Republic République tchèque	1989 1990 1991	524565 567322 716593	8.9 7.3 5.6	41.8[4] 39.3[4] 43.9[4]	... ...	... ...	7.8 8.1 6.3	11.4 13.2 10.5	4.8 4.3 4.1	25.2 27.8 29.7
Denmark Danemark	1985 1991 1992	615072 827379 853982	4.9 3.5 3.1	1.0 0.8 0.9	17.1 16.2 16.6	1.1 1.6 1.4	5.0 4.6 4.6	13.2 12.2 12.4	7.0 8.0 7.9	50.7 53.2 53.1
Djibouti Djibouti	1981 1982 1983	56818 59383 59997	3.7 4.3 4.3	... ...	8.3 8.1 8.2	2.0 2.6 3.2	7.0 7.5 7.6	18.6 16.3 15.7	9.9 9.7 9.8	50.4 51.6 51.2
Dominica Dominique	1985 1990 1991	266 452 479	23.4 21.5 21.2	0.5 0.7 0.8	5.4 5.9 5.7	2.3 2.5 2.7	5.6 6.2 6.1	8.9 10.8 11.0	11.0 13.3 13.5	42.7 39.3 39.0

23
Gross domestic product by kind of economic activity at current prices
Percentage distribution *[cont.]*
Produit intérieur brut par genre d'activité économique aux prix courants
Répartition en pourcentage *[suite]*

			% of GDP — en % du PIB							
Country or area Pays ou zone	Year Année	GDP at current prices (Mil. nat.cur.) PIB aux prix courants (Mil. mon.nat.)	Agriculture, hunting, forestry & fishing Agriculture, chasse, sylviculture et pêche	Mining & quarrying Industries extractives	Manufac- turing Manufac- turières	Electricity, gas and water Electricité, gaz et eau	Construc- tion Construc- tion	Wholesale/ retail trade, restaurants and hotels Commerce, restaurants, hôtels	Transport, storage & commu- nication Transports, entrepôts, communi- cations	Other activities Autres activités
Dominican Republic	1982	7965	17.7	2.6	18.3[7]	1.0	6.7	17.0	5.3[1]	31.4[1]
Rép. dominicaine	1983	8623	17.2	2.7	17.7[7]	0.9	7.8	16.8	5.4[1]	31.6[1]
	1984	10355	18.5	2.4	16.5[7]	0.9	8.5	17.2	5.0[1]	31.0[1]
Ecuador	1985	1109940	13.3	17.1	18.9	0.3	4.4	15.6	8.6	21.8
Equateur	1991	12201436	14.5	11.3	20.8	−0.1	4.0	22.2	8.6	18.7
	1992	19452299	13.2	12.3	22.4	−0.1	4.1	21.5	7.6	19.1
Egypt [2]	1989	87741	18.0	4.5	16.7	1.2[8]	5.1	20.3	8.5	25.6
Egypte [2]	1990	110143	16.2	10.0	15.8	1.3[8]	4.7	19.7	9.8	22.4
	1991	136190	15.2	9.8	15.7	1.5[8]	4.5	19.6	10.4	23.4
El Salvador	1985	14331	18.2	0.1	16.4	2.3	3.0	27.2	4.3[1]	28.4[1]
El Salvador	1991	47792	10.2	0.2	18.7	2.3	2.7	35.0	4.8[1]	26.1[1]
	1992	54853	9.4	0.2	18.9	2.3	2.8	35.8	4.8[1]	25.7[1]
Equatorial Guinea	1985	38067	57.0	...	1.8	2.6	3.2	8.4	2.9	24.1
Guinée équatoriale	1990	44350	51.7	...	1.3	3.3	3.7	7.4	2.1	30.6
	1991	46429	50.2	...	1.3	2.9	2.8	7.1	1.8	33.8
Estonia	1985	495	20.7		39.0	...	8.0	7.5	5.6	19.3
Estonie	# 1991	1832	16.9	1.6	32.7	1.7	6.1	8.0	6.2	26.8
	1992	14255	12.5	2.4	28.6	4.3	4.6	14.5	11.9	21.1
Ethiopia [9]	1985	9924	38.9	0.2	10.1	0.7	4.0	9.7	6.2[1]	30.1[1]
Ethiopie [9]	1991	13332	37.8	0.3	9.5	1.4	3.0	8.6	6.6[1]	32.9[1]
	1992	13508	46.7	0.3	8.4	1.2	2.6	9.5	5.0[1]	26.3[1]
Fiji	1985	1316	16.4	1.1	8.4	3.0	4.9	16.0	9.3	41.0
Fidji	1988	1588	17.6	3.9	8.6	3.3	3.8	17.8	10.3	34.8
	1989	1861	17.5	3.0	9.4	3.0	3.5	20.4	9.1	34.1
Finland	1985	331628	7.3	0.4	22.8	2.5	7.0	11.8	7.1	41.0
Finlande	1991	490868	4.9	0.4	18.3	2.3	7.5	10.6	7.4	48.7
	1992	475674	4.4	0.4	19.5	2.3	5.5	10.7	7.6	49.7
France	1985	4700143	3.9	0.9	22.0	2.5	5.2	14.6	6.1	44.9
France	1991	6746923	3.0	0.5	20.8	2.3	5.2	15.2	5.9	47.1
	1992	6987221	2.8	0.5	20.5	2.4	5.3	15.1	6.0	47.5
French Guiana	1985	3173	6.4	4.9	...	−0.5	10.6	12.8	12.2	53.7
Guyane française	1988	4929	8.9	5.2	...	0.7	9.7	11.7	14.3	49.4
	1989	5684	10.6	7.0	...	1.3	13.4	11.8	6.8	49.0
French Polynesia	1985	226772	3.8	...	8.5	0.8	12.0	...	5.8	27.7
Polynésie française	1989	281667	4.4	...	7.2	1.5	6.6	22.8	...[10]	57.5[10]
	1990	297754	4.7	...	7.3	1.7	6.1	22.7	...[10]	57.6[10]
Gabon	1985	1645800	6.2	46.2	5.3[11]	1.8	6.3	6.6	4.8	22.8
Gabon	1988	1013600	10.7	21.5	6.9[11]	2.8	5.0	13.8	8.6	30.8
	1989	1168066	10.0	31.2	5.5[11]	2.4	5.3	12.0	7.9	25.7
Gambia [2]	1985	1085	25.5	0.0	6.4	0.6	3.7	34.1	9.4	20.2
Gambie [2]	1990	2630	23.1	0.0	6.1	0.8	4.5	39.0	10.0	16.5
	1991	2948	22.9	0.0	5.9	0.9	5.0	38.3	10.6	16.4
Germany † · Allemagne †										
F.R. Germany	1985	1823180	1.8	0.9	31.7[12]	2.8	5.2	9.8	5.8	42.1
R.f. Allemagne	1991	2635000	1.3	0.5	30.0[12]	2.3	5.4	10.1	5.5	45.0
	1992	2794200	1.2	...	28.7[12]	...	5.8	...	5.4	28.9
Ghana	1985	343048	44.9	1.1	11.5	1.2	2.9	24.8	5.3	8.3
Ghana	1986	550515	44.4	1.6	11.2	1.6	2.4	22.8	4.6	11.4
Greece	1985	4617816	15.5	1.9	16.3	2.3	5.7	11.7	6.7[1]	39.9[1]
Grèce	1991	12802353	14.4	1.2	13.6	2.3	6.1	10.8	6.0[1]	45.6[1]
	1992	14846939	12.7	1.1	13.1	2.3	5.7	11.1	5.8[1]	48.3[1]
Grenada	1985	311	14.2	0.3	4.0	2.1	6.2	15.7	10.1	47.4
Grenade	1990	541	13.2	0.3	4.2	2.4	8.2	15.2	11.1	45.4
	1991	567	12.1	0.3	4.3	2.5	8.5	15.9	11.7	44.6
Guadeloupe	1985	9802	9.8	4.7[13]	...	−0.5	4.7	16.7	5.6	58.9
Guadeloupe	1988	13025	10.1	5.3[13]	...	0.7	5.0	16.1	5.9	56.9
	1989	14062	8.4	5.4[13]	...	0.3	6.8	17.8	5.5	55.9

23

Gross domestic product by kind of economic activity at current prices
Percentage distribution [cont.]
Produit intérieur brut par genre d'activité économique aux prix courants
Répartition en pourcentage [suite]

% of GDP — en % du PIB

Country or area Pays ou zone	Year Année	GDP at current prices (Mil. nat.cur.) PIB aux prix courants (Mil. mon.nat.)	Agriculture, hunting, forestry & fishing Agriculture, chasse, sylviculture et pêche	Mining & quarrying Industries extractives	Manufac- turing Manufac- turières	Electricity, gas and water Electricité, gaz et eau	Construc- tion Construc- tion	Wholesale/ retail trade, restaurants and hotels Commerce, restaurants, hôtels	Transport, storage & commu- nication Transports, entrepôts, communi- cations	Other activities Autres activités
Guinea–Bissau	1989	358875	44.6	7.9 [4]	...	...	9.7	25.7	3.6	8.5
Guinée–Bissau	1990	510094	44.6	8.2 [4]	...	...	10.0	25.7	3.7	7.8
	1991	854985	44.7	8.5 [4]	...	...	8.4	25.8	3.9	8.7
Guyana	1985	1964	25.8	2.5	8.1 [5]	...	6.1	6.8	6.5	44.1
Guyana	1991	38966	40.5	13.5	3.7 [5]	...	3.0	4.5	5.4	29.4
	1992	46734	43.8	9.7	3.8 [5]	...	3.0	4.3	4.9	30.4
Honduras	1985	6977	19.0	2.0	13.3	1.6	5.2	12.1	5.8 [1]	41.0 [1]
Honduras	1991	16314	19.5	1.3	14.5	3.0	4.6	9.6	5.6 [1]	42.0 [1]
	1992	18772	17.4	1.6	15.3	2.8	5.7	9.4	5.6 [1]	42.2 [1]
Hong Kong	1985	261070	0.5	0.1	20.3	2.6	4.6	20.2	7.5	44.1
Hong–kong	1989	499157	0.3	0.0	18.0	2.2	5.0	22.8	8.5	43.2
	1990	558859	0.3	0.0	16.5	2.3	5.3	23.4	9.0	43.2
Hungary	1985	1033658	16.1	5.7	26.4	3.3	7.2	9.4	7.5	24.4
Hongrie	# 1991	2346035	8.6	3.0	21.0	4.0	4.8	16.1	7.8	34.7
	1992	2805083	7.2	2.4	20.9	4.3	5.2	14.9	7.0	38.0
Iceland	1985	119996	9.5	...	14.4	4.7	6.1	8.5	6.1	50.8
Islande	1990	354375	10.0	...	13.5	3.3	6.4	10.3	6.0	50.4
	1991	384054	9.9	...	13.5	3.1	6.2	10.6	5.9	50.8
India [14]	1985	2622430	29.4	2.4	15.9	1.9	4.9	11.8	5.4	28.2
Inde [14]	1990	5308650	28.3	2.0	16.5	1.9	5.1	11.1	6.4	28.6
	1991	6095000	28.6	1.9	15.8	2.0	5.1	10.9	6.8	29.0
Indonesia	1985	96997000	23.2	14.0	16.0	0.4	5.5	15.9	6.3 [1]	18.8 [1]
Indonésie	1991	227503000	19.6	13.8	20.9	0.8	5.9	16.2	6.1 [1]	16.8 [1]
	1992	260787000	19.2	11.9	21.7	0.8	6.2	16.4	6.5 [1]	17.4 [1]
Iran, Islamic Rep. of [15]	1985	15775000	19.7	10.1	8.2	0.9	7.4	14.5	7.5	31.7
Iran, Rép. islamique d' [15]	1989	27787000	24.0	6.6	10.5	1.1	4.6	19.9	6.4	26.8
	1990	36645000	23.0	10.8	12.0	1.1	3.9	17.9	7.2	24.1
Iraq	1985	15494	13.9	22.5	9.6	1.3	9.1	12.5	5.0	26.2
Iraq	1990	23297	19.8	14.3	8.8	1.1	7.3	14.8	9.0	24.9
	1991	19940	30.3	0.7	6.4	0.8	4.1	18.1	13.3	26.3
Ireland	1985	17790	8.8	30.2 [4]	...	...	5.4	10.8	4.8	40.0
Irlande	# 1990	26983	7.4	30.2 [4]	...	...	4.6	12.0	5.0	40.9
	1991	27949	7.0	30.7 [4]	...	...	4.8	11.2	4.9	41.4
Israel	1985	21493	4.9	22.6 [13]	...	2.0	4.3	13.8	7.6	44.9 [16]
Israël	1991	100167	2.4	21.5 [13]	...	2.3	7.8	9.6	7.5	49.0 [16]
	1992	117988	2.6	21.7 [13]	...	2.4	8.0	10.4	8.1	46.8 [16]
Italy	1985	810580000 [3]	4.5	24.2 [13]	...	4.6	6.3	19.1 [11]	5.4	35.9
Italie	1991	1429453000 [3]	3.3	21.1 [13]	...	5.4	5.9	18.5 [11]	5.8	40.0
	1992	1504323000 [3]	3.1	20.5 [13]	...	5.7	5.8	18.4 [11]	6.0	40.3
Jamaica	1985	11203	6.0	5.1	20.0	3.2	9.1	23.9	8.3	24.4
Jamaïque	1991	44158	7.0	10.9	19.1	1.8	12.6	21.0	8.0	19.5
	1992	72540	8.0	9.4	19.6	2.5	12.9	23.6	7.7	16.2
Japan	1985	320419000	3.2	0.3	29.5	3.2	7.9	13.4	6.6 [1]	35.9 [1]
Japon	1991	451297000	2.3	0.3	29.1	2.7	10.0	12.8	6.3 [1]	36.4 [1]
	1992	463850000	2.2	0.3	27.9	2.8	10.1	12.8	6.3 [1]	37.6 [1]
Jordan	1985	1970	4.3	3.3	10.4	2.0	7.9	14.7	13.5	43.9
Jordanie	1991	2778	6.3	4.5	12.4	2.2	4.5	9.2	13.2	47.8
	1992	3257	6.3	3.6	13.1	2.2	4.7	8.3	13.1	48.8
Kenya	1985	5043	28.6	0.2	10.3	1.5	4.7	10.3	5.9	38.4
Kenya	1991	11044	23.6	0.3	10.6	1.3	5.2	10.3	6.2	42.6
	1992	12904	23.0	0.2	10.2	1.2	5.0	11.0	7.0	42.5
Korea, Republic of	1985	80847000	12.8	1.0	30.3	2.8	7.7	12.2	7.6	25.6
Corée, Rép. de	1991	208201000	8.0	0.4	28.2	2.0	15.5	10.2	7.0	28.7
	1992	231727000	7.6	0.4	27.3	2.1	15.2	10.1	7.2	30.1
Kuwait	1985	6450	0.6	49.4	5.9	−2.4	4.0	8.8	4.2	29.4
Kowëit	1991	3184	0.3	14.1	7.1	−3.1	6.0	16.6	2.7	56.3
	1992	6367	0.3	42.7	14.6	−1.5	2.0	7.7	2.0	32.1

23

Gross domestic product by kind of economic activity at current prices
Percentage distribution *[cont.]*
Produit intérieur brut par genre d'activité économique aux prix courants
Répartition en pourcentage *[suite]*

% of GDP − en % du PIB

Country or area Pays ou zone	Year Année	GDP at current prices (Mil. nat.cur.) PIB aux prix courants (Mil. mon.nat.)	Agriculture, hunting, forestry & fishing Agriculture, chasse, sylviculture et pêche	Mining & quarrying Industries extractives	Manufac-turing Manufac-turières	Electricity, gas and water Electricité, gaz et eau	Construc-tion Construc-tion	Wholesale/retail trade, restaurants and hotels Commerce, restaurants, hôtels	Transport, storage & communication Transports, entrepôts, communi-cations	Other activities Autres activités
Kyrgyzstan	1990	42	32.5	26.7[4]	...	...	7.4	5.3	5.7	22.4
Kirghizistan	1991	87	39.2	21.0[4]	...	...	6.6	5.5	4.9	22.8
	1992	772	37.1	34.3[4]	...	...	3.2	4.8	3.4	17.1
Latvia	1990	12488	21.1	0.2	33.2	1.7	9.4	6.6	10.5	17.4
Lettonie	1991	28665	22.5	0.2	34.7	2.2	5.6	11.4	7.2	16.3
	1992	200911	16.5	0.1	26.4	1.4	4.7	13.6	16.6	20.8
Lesotho	1985	551	17.6	0.3	9.1[7]	0.7	14.2	9.4	2.6	46.1
Lesotho	1991	1739	10.5	0.2	12.0[7]	1.0	22.1	8.5	2.7	43.0
	1992	2026	8.7	0.2	12.5[7]	1.6	22.6	8.8	2.1	43.4
Liberia	1985	1055	33.3	12.2	6.1	1.6	3.1	5.7	6.8	31.3
Libéria	1988	1158	35.6	9.9	6.9	1.6	2.5	5.5	6.8	31.1
	1989	1194	34.4	10.2	6.8	1.6	2.2	5.3	6.6	32.8
Libyan Arab Jamah. Jaman. arabe libyenne	1985	8277	3.4	40.4[17]	4.4	1.2	11.1	6.8	4.8	27.8
Lithuania	1985	9190	27.8	36.9[4]	...	...	11.5	4.5	8.1	11.3
Lithuanie	1991	38187	...	...	...	...	...	...	...	8.2
	1992	332830	15.9	42.3[4]	...	...	5.5	8.7	6.1	21.4
Luxembourg	1985	205255[3]	2.6	0.1	30.0	2.4	5.4	16.6	5.4	37.6
Luxembourg	1990	300409[3]	1.9	0.3	25.8	1.7	7.2	16.5	6.8	39.8
	1991	318804[3]	1.4	0.3	24.2	1.7	7.5	16.4	6.9	41.5
Madagascar Madagascar	1985	1553400	42.0	16.4[4,19]	...	...	...	...	...	...
Malawi	1985	1655	32.3[18]	...	16.3	1.4	2.0	11.9	4.4	31.5
Malawi	1986	1793	32.6[18]	...	19.1	1.1	1.9	10.9	3.3	31.1
Malaysia Malaisie	1983	69941	18.7	13.9	19.1	1.5	5.5	11.6	5.2	24.4
Maldives	1985	613	28.8	1.9	5.5[5]	...	8.0	15.9	22.1	17.9
Maldives	1986	697	30.5	1.7	5.5[5]	...	7.6	14.8	22.8	17.0
Mali	1985	520200	38.2	2.7	7.0[7]	4.2[19]	...	17.5	5.2	25.2
Mali	1991	691400	44.1	1.6	6.6[7]	4.1[19]	...	19.3	4.7	19.4
	1992	737400	45.2	1.5	6.7[7]	4.2[19]	...	18.5	4.8	19.1
Malta	1985	476	4.1	4.4[19]	26.7	4.7	...	14.0	5.0[1]	41.2[1]
Malte	1990	735	3.1	3.1[19]	23.9	7.3	...	12.8	5.1[1]	44.8[1]
	1991	807	2.9	3.3[19]	23.3	7.2	...	12.6	5.2[1]	45.5[1]
Martinique	1985	12484	7.8	6.0[13]	...	2.0	3.7	16.8	4.8	58.8
Martinique	1988	16896	6.2	7.5[13]	...	2.6	4.4	18.9	5.6	54.8
	1989	18086	5.9	7.4[13]	...	2.6	4.5	19.1	5.6	54.9
Mauritania	1985	52665	20.0	10.8	11.5[5]	...	7.0	13.1	6.1	31.6
Mauritanie	1988	72635	29.0	6.9	11.7[5]	...	5.6	11.8	4.6	30.4
	1989	83520	30.9	9.4	9.3[5]	...	5.8	...	4.5	22.7
Mauritius	1985	16618	12.8	0.1	17.2	2.4	4.7	11.0	9.1	42.7
Maurice	1990	37990	10.3	0.1	19.6	1.3	5.8	14.4	9.2	39.3
	1991	42750	9.5	0.1	19.5	1.8	6.1	14.3	9.8	39.0
Mexico	1985	47392000	9.1	4.6	23.4[20]	0.9	4.4	28.1	6.7	22.8
Mexique	1991	865166000	7.7	2.1	22.3[20]	1.5	4.2	24.8	8.8	28.6
	1992	1019156000	7.0	2.1	21.2[20]	1.5	4.8	24.0	9.3	30.2
Montserrat	1982	81	4.1	1.4	4.9	2.6	10.2	20.0	8.0	48.8
Montserrat	1983	87	3.8	0.8	6.1	3.5	6.6	18.6	9.8	50.8
	1984	94	4.2	1.2	6.0	3.5	6.6	17.1	10.4	51.1
Morocco	1985	129510	16.6	4.3	18.6	4.8[21]	5.7	14.1	6.4	29.5
Maroc	1991	241650	19.9	2.3	17.7	6.3[21]	5.0	12.7	5.8	30.3
	1992	242490	14.9	2.2	18.6	7.0[21]	5.0	13.5	6.3	32.7
Myanmar[14]	1985	55989	48.2	1.0	9.9	0.5[8]	1.7	23.9	4.0[1]	10.9[1]
Myanmar[14]	1991	178553	57.8	0.6	7.3	0.2[8]	2.2	22.6	2.7[1]	6.7[1]
	1992	230935	59.8	0.6	7.4	0.2[8]	1.8	22.5	2.1[1]	5.6[1]
Namibia	1985	2854	9.2	32.6	4.3	1.7	2.5	9.9	4.9	35.0
Namibie	1991	6203	11.0	17.7	4.9	1.8	2.0	10.9	5.9	45.9
	1992	7027	9.8	17.2	5.2	1.7	1.9	10.4	6.0	47.7

23
Gross domestic product by kind of economic activity at current prices
Percentage distribution *[cont.]*
Produit intérieur brut par genre d'activité économique aux prix courants
Répartition en pourcentage *[suite]*

% of GDP — en % du PIB

Country or area Pays ou zone	Year Année	GDP at current prices (Mil. nat.cur.) PIB aux prix courants (Mil. mon.nat.)	Agriculture, hunting, forestry & fishing Agriculture, chasse, sylviculture et pêche	Mining & quarrying Industries extractives	Manufacturing Manufacturières	Electricity, gas and water Electricité, gaz et eau	Construction Construction	Wholesale/retail trade, restaurants and hotels Commerce, restaurants, hôtels	Transport, storage & communication Transports, entrepôts, communications	Other activities Autres activités
Nepal [22] Népal [22]	1985	44417	53.9	0.3	4.5	0.4	8.1	4.1	6.2	22.5
	1991	103948	50.1	0.1	6.1	0.6	7.8	5.7	5.7	23.9
	1992	126186	48.7	0.1	7.4	0.8	8.1	6.0	6.1	22.8
Netherlands Pays-Bas	1985	425350	3.9	8.3	17.7	1.8	4.6	13.3 [23]	6.0	44.4
	1991	541880	3.9	3.4	18.4	1.7	5.1	14.8 [23]	6.4	46.3
	1992	563220	3.6	2.9	18.3	1.6	5.3	14.8 [23]	6.4	47.1
Netherlands Antilles Antilles néerlandaises	1985	1966	0.9	...	7.1	3.4	8.0	23.1	10.8	46.8
New Caledonia Nouvelle-Calédonie	1985	139650	1.8	15.6	4.7	2.3	3.4	23.8	3.4 [1]	45.0 [1]
	1989	253475	1.8	23.6	4.4	2.3	4.6	21.9	4.5 [1]	36.9 [1]
	1990	250427	2.0	10.4	6.4	2.5	5.7	23.0	5.6 [1]	44.5 [1]
New Zealand [14] Nouvelle-Zélande [14]	1985	45435	8.8	1.3	21.1	3.2	5.9	19.0	7.8	32.9
	1989	71435	8.4	1.2	18.0	2.9	4.6	15.5	8.0	41.3
	1990	73601	7.2	1.5	17.4	2.8	4.2	16.8	8.0	42.0
Niger Niger	1985	647141	36.7	8.1	7.1	2.2	3.5	15.8	4.2	22.4
	1986	643362	36.0	7.4	7.7	2.5	4.5	14.1	4.1	23.7
	1987	649846	33.7	7.6	8.7	2.8	5.1	13.7	4.1	24.3
Nigeria Nigéria	1985	72355	36.8	17.3	8.6	0.7	2.1	13.4	5.4	15.7
	1990	260637	32.4	33.3	5.5	0.5	1.7	14.0	2.2	10.6
	1991	324794	30.4	37.2	5.7	0.4	1.5	13.1	2.0	9.8
Norway Norvège	1985	500201	3.0	18.2	14.0	3.7	5.4	10.8	7.6	37.3
	1990	660552	3.1	13.4	13.7	4.1	4.4	11.1	9.8	40.4
	1991	686686	2.9	13.3	13.5	4.0	3.9	11.0	10.4	41.0
Oman Oman	1985	3454	2.7	48.8	2.4	1.1	7.0	12.4	2.9	22.8
	1991	3917	3.7	42.6	4.3	1.6	3.9	13.8	3.7	26.3
	1992	4417	3.3	42.7	4.3	1.5	4.0	13.9	3.6	26.6
Pakistan [2] Pakistan [2]	1985	514532	25.0	0.6	14.7	2.1	3.7	14.1	8.0 [1]	31.7 [1]
	1991	1211245	23.4	0.6	15.5	3.1	3.6	14.6	8.3 [1]	31.0 [1]
	1992	1359338	22.4	0.6	15.5	3.3	3.8	14.4	9.2 [1]	30.8 [1]
Panama Panama	1985	4948	10.1	0.1	8.5	4.3	4.6	14.0	12.9	45.4
	1991	5496	11.1	0.1	8.2	4.0	3.7	12.8	13.4	46.6
	1992	6001	10.8	0.2	8.4	3.6	5.0	12.9	13.8	45.4
Papua New Guinea Papua New Guinea	1985	2403	33.3	10.0	10.9	1.9	3.9	9.9	5.0 [1]	25.2 [1]
	1990	3076	29.0	14.7	9.0	1.7	5.0	9.6	6.2 [1]	24.8 [1]
	1991	3606	26.0	17.0	9.6	1.6	6.2	9.9	6.7 [1]	22.9 [1]
Paraguay Paraguay	1985	1393890	28.9	0.4	16.2	2.2	5.9	25.9	4.2 [1]	16.3 [1]
	1988	3319124	29.6	0.5	16.8	2.5	4.9	27.5	4.0 [1]	14.2 [1]
	1989	4608400	29.6	0.5	17.0	2.4	5.6	27.4	3.8 [1]	13.7 [1]
Peru Pérou	1985	197 [24]	9.4	9.9	24.3	1.2	7.1	18.0	6.2	24.0
	1991	33107	6.7	2.2	28.1	0.6	8.5	19.1	5.6	29.1
	1992	52452	6.5	2.2	26.2	0.8	9.3	19.0	5.7	30.3
Philippines Philippines	1985	571883	24.6	2.1	25.2	2.8	5.1	15.7	5.5	19.1
	1991	1244037	21.2	1.4	25.4	2.5	5.0	15.2	5.9	23.4
	1992	1342515	21.6	1.2	24.6	2.5	5.0	15.3	5.8	24.0
Poland Pologne	1990	591518000	8.4	44.9 [4]	...	...	9.2	12.7	4.9	19.8
	1991	824330000	6.9	40.2 [4]	...	...	10.2	13.1	5.6	24.0
Portugal Portugal	1985	3523945	8.0	30.4 [13]	...	3.5	5.7	22.4	7.7	22.4
	1989	7130260	6.2	28.8 [13]	...	2.9	6.3	19.6	5.8	30.3
	1990	8507434	5.8	27.9 [13]	...	3.1	6.9	19.8	5.4	31.2
Puerto Rico [2] Porto Rico [2]	1985	21969	1.7	0.1	38.9	3.1	1.4 [25]	16.2	5.6	33.1
	1991	34004	1.2	0.1	39.4	3.0	2.2 [25]	15.9	5.3	32.9
	1992	35834	1.1	0.1	39.4	3.0	2.2 [25]	16.0	5.2	32.9
Qatar Qatar	1985	22398	1.0	42.8	7.9	0.9	5.9	5.3	2.0	34.3
	1991	25056	0.9	32.5	13.3	1.2	4.1	6.7	3.1	38.1
	1992	27202	0.9	35.8	12.7	1.1	4.1	6.7	2.9	35.7
Réunion Réunion	1985	17054	4.1	7.6 [13]	...	4.1	5.5	16.0	4.7	58.0
	1988	23604	4.9	7.8 [13]	...	4.1	6.5	18.0	5.2	53.5
	1989	25613	3.6	8.2 [13]	...	4.5	5.8	19.2	4.8	53.9

23
Gross domestic product by kind of economic activity at current prices
Percentage distribution [cont.]
Produit intérieur brut par genre d'activité économique aux prix courants
Répartition en pourcentage [suite]

% of GDP − en % du PIB

Country or area Pays ou zone	Year Année	GDP at current prices (Mil. nat.cur.) PIB aux prix courants (Mil. mon.nat.)	Agriculture, hunting, forestry & fishing Agriculture, chasse, sylviculture et pêche	Mining & quarrying Industries extractives	Manufac-turing Manufac-turières	Electricity, gas and water Electricité, gaz et eau	Construc-tion Construc-tion	Wholesale/retail trade, restaurants and hotels Commerce, restaurants, hôtels	Transport, storage & commu-nication Transports, entrepôts, communi-cations	Other activities Autres activités
Romania	1985	817400	14.0	54.5[4]	...	...	7.1	4.9	6.8	12.8
Roumanie	# 1991	2198900	18.7	43.3[4]	...	...	4.7	14.1	6.3	12.8
	1992	5982300	18.9	44.7[4]	...	...	4.4	13.2	6.4	12.4
Rwanda	1985	173700	41.8	0.3	13.7	0.5	8.7	12.2	4.9	17.8
Rwanda	1988	177940	37.9	0.2	14.0	0.7	6.9	12.7	7.1	20.5
	1989	190220	39.8	0.4	13.1	0.5	6.8	12.8	6.8	19.8
Saint Kitts−Nevis	1985	172[26]	9.6	0.3	12.8	1.0	8.8	20.8	12.3	34.5
Saint−Kitts−et−Nevis	1988	279[26]	10.0	0.3	15.9	1.0	10.3	20.6	14.5	27.4
	1989	302[26]	9.2	0.4	15.7	0.9	11.8	20.6	14.4	27.0
Saint Lucia	1985	389[26]	15.0	0.6	8.5	3.9	6.9	22.7	10.4	32.0
Sainte−Lucie	1986	427[26]	16.6	0.6	8.0	3.9	7.5	22.1	9.9	31.6
	1987	446[26]	14.4	0.6	7.9	4.2	8.1	23.4	9.6	31.9
Saint Vincent−Grenadines	1985	305	16.3	0.2	9.6	3.3	6.4	11.1	15.3	38.0
St.−Vincent−et−Gren.	1989	469	14.5	0.2	9.2	4.0	7.6	11.1	16.9	36.5
	1990	525	16.0	0.2	7.3	4.1	8.1	11.8	17.8	34.7
Sao Tome and Principe	1986	2478	26.6	...	2.1	0.3	3.1	17.7	4.8	45.4
Sao Tomé−et−Principe	1987	3003	29.4	...	1.2	1.3	3.5	15.9	5.2	43.4
	1988	4221	28.9	...	1.6	0.9	3.8	16.9	3.7	44.2
Saudi Arabia [2]	1985	313941	4.4	28.7	7.8	0.1	12.3	9.6	7.6	29.5
Arabie saoudite [2]	1987	275453	6.6	23.6	8.7	0.2	12.1	10.1	8.0	30.6
	1988	281971	7.3	22.5	8.8	0.3	11.2	9.5	8.1	32.4
Senegal	1985	1158100	18.7	0.3	12.8	1.8	2.8	...	...	...
Sénégal	1990	1583000	20.1	0.7	13.3	1.9	2.8	...	...	...
	1991	1623300	19.4	0.6	13.4	1.9	2.9	...	...	...
Seychelles	1985	1205	5.8	0.0	9.7	2.6	6.1	24.7	14.1	37.1
Seychelles	1990	1967	4.8	0.0	10.1	1.4	4.8	27.0	12.0	39.9
	1991	1980	4.8	0.0	10.8	2.1	5.1	25.0	14.9	37.2
Sierra Leone [2]	1985	7481	39.1	17.5	4.3	0.2	2.3	14.0	8.1	14.5
Sierra Leone [2]	1989	82837	36.9	7.0	7.0	0.2	1.9	24.7	10.7	11.6
	1990	150175	35.0	9.4	8.7	0.1	1.2	20.1	8.9	16.7
Singapore	1985	38924	0.7	0.3	23.6	2.0	10.6	17.0	13.4	32.3
Singapour	1991	69452	0.2	0.1	29.1	1.9	6.7	16.7	13.4	31.8
	1992	74975	0.2	0.1	27.9	1.8	7.8	15.9	13.7	32.7
Slovenia	1985	178	5.1	1.1	36.5	2.8	5.6	9.6	7.3	32.0
Slovénie	1991	350649	5.5	1.1	32.8	4.6	3.7	10.4	6.1	35.9
	1992	1003783	5.0	1.7	29.7	2.3	4.1	12.9	6.5	37.8
Solomon Islands	1985	236	45.6	−0.6	3.4	0.9	3.8	9.4	4.6	32.9
Iles Salomon	1986	250	43.3	−1.0	4.0	1.0	4.6	7.5	5.2	35.4
Somalia	1985	87290	63.6	0.3	4.7	0.1	2.2	9.7	6.5	12.9
Somalie	1986	118781	59.3	0.3	5.3	0.2	2.5	9.8	6.9	15.7
	1987	169608	62.5	0.3	4.9	−0.5	2.8	10.3	6.5	13.2
South Africa	1985	123126	5.3	13.5	21.1	3.9	3.4	10.8	8.1	34.0
Afrique du Sud	1991	297895	4.4	9.1	22.3	3.9	2.8	12.1	7.6	37.8
	1992	327068	3.5	8.7	22.5	3.9	2.7	12.4	7.5	38.8
Spain	1985	28200900	5.9	1.0	26.7	2.9	6.7	20.2	5.6	31.0
Espagne	1991	54820600	4.1	...	18.8	...	9.3	...	...	...
	1992	58852000	3.5	...	17.5	...	8.6	...	...	...
Sri Lanka	1985	157763	24.4[27]	0.8	16.6[27]	1.9	7.6	18.4	11.0	19.3
Sri Lanka	1991	369720	22.2[27]	1.1	17.0[27]	1.8	6.7	19.8	10.1	21.4
	1992	421755	21.1[27]	1.0	17.1[27]	1.8	6.5	21.0	10.2	21.3
Suriname	1985	1747	8.2	5.4	11.8	4.2	5.6	15.3	6.5	43.0
Suriname	1991	3708	10.9	2.4	10.7	3.9	6.6	19.2	5.3	41.1
	1992	5010	13.7	2.1	10.2	3.0	7.0	21.7	5.2	37.2
Swaziland [28]	1985	803	16.9	2.2	13.3	2.7	3.1	10.4	5.4	45.9
Swaziland [28]	1987	1118	13.7	1.7	21.1	2.7	2.8	10.1	6.6	41.3
	1988	1325	13.8	1.2	20.0	2.6	2.8	10.7	6.2	42.7

23

Gross domestic product by kind of economic activity at current prices
Percentage distribution *[cont.]*
Produit intérieur brut par genre d'activité économique aux prix courants
Répartition en pourcentage *[suite]*

% of GDP − en % du PIB

Country or area Pays ou zone	Year Année	GDP at current prices (Mil. nat.cur.) PIB aux prix courants (Mil. mon.nat.)	Agriculture, hunting, forestry & fishing Agriculture, chasse, sylviculture et pêche	Mining & quarrying Industries extractives	Manufac- turing Manufac- turières	Electricity, gas and water Electricité, gaz et eau	Construc- tion Construc- tion	Wholesale/ retail trade, restaurants and hotels Commerce, restaurants, hôtels	Transport, storage & commu- nication Transports, entrepôts, communi- cations	Other activities Autres activités
Sweden	1985	866601	3.3	0.4	21.5	2.6	5.9	10.7	5.4	50.1
Suède	1991	1447327	2.2	0.2	17.9	2.8	6.8	9.8	6.1	54.3
	1992	1439835	2.1	0.2	17.6	3.0	6.6	9.7	6.2	54.6
Switzerland	1985	227950	3.6	0.0	25.7	2.2	7.6	19.3	6.5	35.1
Suisse	1990	313990	3.1	...	24.4	1.9	8.4	18.0	5.9	37.5
Syrian Arab Rep.	1985	83225	21.0	7.1	7.7	0.2	6.8	22.2	9.8	25.0
Rép. arabe syrienne	1991	316204	29.8	12.6	5.8	1.6	3.8	21.0	9.4	16.0
	1992	370990	29.8	11.1	4.2	1.4	3.8	23.7	9.1	17.0
Thailand	1985	1056496	15.8	2.5	21.9	2.4	5.1	23.9	7.4	21.0
Thaïlande	1990	2182100	12.8	1.6	27.3	2.2	6.1	23.6	7.2	19.2
	1991	2509427	12.8	1.6	28.2	2.1	6.8	22.9	7.0	18.7
Trinidad and Tobago	1985	18071	3.0	22.1	8.7	1.0	12.1	14.2	8.8	30.1
Trinité−et−Tobago	1991	22380	2.4	16.0	14.2	1.2	9.6	18.1	8.8	29.8
	1992	22222	2.5	13.5	13.6	1.6	9.6	18.9	9.0	31.3
Tunisia	1985	6910	15.2	10.1	11.8	1.7	6.1	18.5	5.2	31.3
Tunisie	1988	8605	11.8	7.9	14.1	1.7	4.4	20.5	6.6	33.0
	1989	9497	12.1	8.0	14.5	1.6	4.7	20.4	6.6	32.1
Turkey	1985	35401000	19.0	1.6	18.1	1.8	5.8	19.1	14.1	20.5
Turquie	1991	618249000	15.7	1.9	22.1	2.1	6.7	18.9	12.3	20.3
	1992	1061642000	15.4	1.7	21.7	2.5	6.5	19.1	12.6	20.6
Uganda	1985	24746[26]	61.6	0.1	3.3	0.4	2.5	12.8	2.6	16.7
Ouganda	1991	2004054[26]	54.5	0.3	4.2	0.7	6.0	13.0	4.4	16.9
	1992	3399090[26]	57.1	0.3	3.9	0.8	5.9	12.6	3.9	15.5
United Arab Emirates	1985	99416	1.4	45.3	9.3	2.2	8.9	8.8	4.2	19.9
Emirats arabes unis	1989	100976	1.9	38.7	8.6	2.2	9.5	10.6	5.6	23.0
	1990	124008	1.6	46.7	7.5	1.9	8.1	9.1	4.8	20.4
United Kingdom	1985	356172	1.7	6.2	21.8	2.4	5.2	11.7	6.3	44.7
Royaume−Uni	1991	571782	1.6	1.8	19.7	2.4	5.7	12.6	7.1	49.3
	1992	594183	1.6	1.7	19.3	2.3	5.4	12.2	7.0	50.6
United Rep. of Tanzania	1985	120621	50.8	0.2	5.5[7]	0.9	1.7	11.8	5.8	23.3
Rép.−Unie de Tanzanie	1990	494999	47.2	1.0	3.7[7]	1.5	2.6	11.4	7.3	25.3
	1991	690421	52.0	1.0	3.0[7]	1.2	2.1	12.1	6.8	21.9
United States	1985	4016600	2.1	3.3	20.0	3.2	4.5	16.5	6.2	44.1
Etats Unis	# 1990	5489600	2.1	1.9	18.8	2.9	4.4	15.8	5.9	48.2
	1991	5654400	2.0	1.6	18.3	3.0	4.0	15.9	6.0	49.2
Uruguay	1985	479	13.6	0.2	29.4	3.4	3.0	12.5	6.0	31.9
Uruguay	1990	9698	11.5	0.2	26.2	2.8	3.1	12.2	6.5	37.5
	1991	19135	9.8	0.2	25.1	2.8	3.8	12.5	6.4	39.4
Vanuatu	1985	12534	29.5	...	3.8	1.6	2.7	32.3	7.0	23.2
Vanuatu	1988	15006	19.5	...	4.7	1.4	5.7	33.1	8.1	27.5
	1989	16367	19.2	...	5.4	1.6	5.8	32.1	8.4	27.6
Venezuela	1985	464741	5.8	13.5[29]	21.9[30]	1.5[21]	6.1	15.9	5.3	30.0
Venezuela	1991	3037492	5.5	18.2[29]	19.6[30]	2.1[21]	5.4	18.6	5.7	24.9
	1992	4132307	5.3	14.8[29]	18.3[30]	2.0[21]	6.5	20.1	7.0	26.0
Viet Nam	1989	24308000	40.5	19.7[4]	...	...	3.6	11.3	2.5	22.5
Viet Nam	# 1990	38166000	38.6	19.9[4]	...	...	3.8	12.1	3.2	22.4
	1991	69959000	40.8	20.3[4]	...	...	3.3	11.7	3.8	20.1
Yemen	1989	61406	23.9	6.4	9.5	1.9	4.6	12.7	8.5	32.5
Yémen	1990	77159	20.9	9.1	8.5	1.8	...	12.4	7.8	11.5
Yemen, former Arab Rep.	1985	30939	26.0	0.8	10.9	0.8	5.0	13.5	11.3	31.8
anc. Yémen rép. arabe	1986	37472	28.5	1.4	12.3	0.9	3.4	13.1	11.0	29.5
former Dem. Yemen	1985	382	10.3	7.6[4]	...	...	10.6	12.0	8.3	51.2
ancienne Yémen dém.	1987	391	11.9	7.3[4]	...	...	8.2	12.9	8.2	51.4
	1988	411	12.8	7.3[4]	...	...	9.5	11.4	8.5	50.5
Yugoslavia, SFR †	1985	1195	11.5	2.8	34.5	2.2	6.9	10.9	7.4	23.7
Yougoslavie, SFR †	1989	235395	10.8	2.3	39.5	1.6	6.1	6.3	10.0	23.3
	1990	1147787	10.8	2.1	26.2	1.5	6.6	7.2	10.3	35.2

23

Gross domestic product by kind of economic activity at current prices
Percentage distribution *[cont.]*
Produit intérieur brut par genre d'activité économique aux prix courants
Répartition en pourcentage *[suite]*

Country or area Pays ou zone	Year Année	GDP at current prices (Mil. nat.cur.) PIB aux prix courants (Mil. mon.nat.)	% of GDP − en % du PIB							
			Agriculture, hunting, forestry & fishing Agriculture, chasse, sylviculture et pêche	Mining & quarrying Industries extractives	Manufacturing Manufacturières	Electricity, gas and water Electricité, gaz et eau	Construction Construction	Wholesale/ retail trade, restaurants and hotels Commerce, restaurants, hôtels	Transport, storage & communication Transports, entrepôts, communications	Other activities Autres activités
Zambia Zambie	1985	9351	8.7	24.9	18.9	1.2	4.0	9.1	5.8	27.5
	1990	123487	11.5	20.5	20.7	0.5	5.1	10.6	5.6	25.6
	1991	234504	12.0	14.4	26.3	0.8	4.7	10.2	6.7	24.8
Zimbabwe Zimbabwe	1985	7295	18.0	4.6	20.4	2.0 [21]	2.1	10.7	5.9	36.4
	1990	16674	14.3	4.1	24.8	2.4 [21]	2.2	9.4	6.4	36.5
	1991	21818	17.0	4.3	25.6	2.4 [21]	2.3	8.7	5.7	34.0

Source:
National accounts database of the Statistical Division of the United Nations Secretariat.

† For detailed descriptions of data pertaining to former Czechoslovakia, Germany, SFR Yugoslavia and former USSR, see annex I − Country or area nomenclature, regional and other groupings.

1 Restaurants and hotels are included in "other activities".
2 Fiscal year beginning 1 July.
3 The breakdown by kind of economic activity used in this table is according to the classification NACE/CLIO.
4 Including manufacturing and electricity, gas and water.
5 Including electricity, gas and water.
6 Excluding banks and insurance companies registered in Cayman Islands but with no physical presence in the Islands.
7 Including handicrafts.
8 Electricity only. Gas and water are included in "other activities".
9 Fiscal year ending 7 July.
10 Trasnport, storage and communication are included in "other activities".
11 Including repair services.
12 Including quarrying.
13 Including manufacturing.
14 Fiscal year beginning 1 April.
15 Fiscal year beginning 21 March.
16 Including a statistical discrepancy.
17 Including gas and oil production.
18 Including non−monetary output.
19 Including construction.
20 Including basic petroleum manufacturing.
21 Excluding gas.
22 Fiscal year ending 15 July.
23 Including real estate brokers.
24 Thousand.
25 Contract construction only.
26 Gross domestic product in factor values.
27 Processing of tea and rubber is included in the agricultural sector.
28 Fiscal year ending 30 June.
29 Including crude petroleum and natural gas production.
30 Including petroleum refining.

Source:
Base de données sur les comptes nationaux de la division de statistique du Secrétariat de l'ONU.

† Pour les descriptions en détails des données relatives a l'ancienne Tchécoslovaquie, l'Allemagne, la Rfs Yougoslavie et l'ancienne URSS, voir l'Annexe I − Nomenclature des pays ou zones, groupements regionaux et autres groupements.

1 Restaurants et hôtels sont incluses dans "autres activités".
2 L'année fiscale commençant le 1er juillet.
3 La ventilation par branche utilisée est conforme à la nomenclature NACE/CLIO.
4 Y compris les industries manufacturières, l'électricité, le gaz et l'eau.
5 Y compris l'électricité, le gaz et l'eau.
6 Non compris les banques et les compagnies d'assurances enregistrées aux îles Caïmanes, mais sans présence matérielle dans ce territoire.
7 Y compris l'artisanat.
8 Seulement électricité. Le gaz et l'eau sont incluses dans autres activités".
9 L'année fiscale finissant le 7 juillet.
10 Transports, entrepôts et communications sont incluses dans "autres activités".
11 Y compris les services de répartion.
12 Y compris les carrières.
13 Y compris les industries manufacturières.
14 L'année fiscale commençant le 1er avril.
15 L'année fiscale commençant le 21 mars.
16 Y compris erreurs et omissions.
17 Y compris la production de gaz et de pétrole.
18 Y compris la production non commercialisée.
19 Y compris construction.
20 Y compris la production de pétrole.
21 Non compris le gaz.
22 L'année fiscale finissant le 15 juillet.
23 Y compris les agences immobilières.
24 Milliers.
25 Construction sous contrat seulement.
26 Produit intérieur brut au coût de facteurs.
27 La transformation du thé et du caoutchouc est incluse dans le secteur agricole.
28 L'année fiscale finissant le 30 juin.
29 Y compris la production de pétrole brut et de gas naturel.
30 Y compris le raffinage du pétrole.

24
Government final consumption expenditure by function at current prices
Consommation finale des administrations publiques par fonction aux prix courants

Percentage distribution
Répartition en pourcentage

Country or area	Year	Govt. final consumption expenditures Consommation finale des administrations (M. nat'l. curr.)	General public services Adminis— tration publique générale	Defence Defence	Public order & safety Sûreté publique	Education Enseigne— ment	Health Santé	Social services Services sociaux	Economic services Services écono— miques	Other functions Autres fonctions
Anguilla	1985	8	22.8	...	12.4	23.9	14.8	1.8	8.9	15.5
Anguilla	1990	20	19.3	...	12.3	23.6	15.7	2.4	13.9	12.8
	1991	23	25.9	...	11.6	22.2	15.3	2.5	10.7	11.9
Antigua and Barbuda [1]	1985	81	25.1	2.2	11.4	14.0	11.5	6.6	20.9	8.2
Antigua—et—Barbuda [1]	1986	108	26.1	2.3	11.6	14.7	10.0	7.2	21.2	6.9
Australia [2]	1985	44755	13.0	13.9	6.6	23.2	17.3	3.5	16.6	5.9
Australie [2]	1991	71324	14.5	12.1	7.2	21.3	17.5	5.5	16.2	5.7
	1992	74344	14.7	12.2	7.1	21.3	16.9	5.7	15.9	6.1
Austria	1985	254999	16.5	6.6	4.8	22.0	23.5	18.4	6.2	2.1
Autriche	1991	349632	17.0	5.2	4.5	22.2	25.8	18.4	4.8	2.1
	1992	374952	...	4.9	...	...	...	...	...	...
Bahamas	1985	295	12.2	3.4	12.5	27.5	20.0	2.4	20.7	1.0
Bahamas	# 1991	443	16.5	4.1	13.5	22.3	18.7	4.7	19.0	1.4
	1992	448	17.9	4.0	13.4	23.4	17.9	5.1	17.0	1.6
Belarus	1990	9947	...	56.9	2.3	13.1	7.3[3]	...	4.4	16.0
Bélarus	1991	17803	...	46.8	9.5	15.6	13.6[3]	...	3.6	10.9
	1992	138083	...	16.1	18.9	26.3	20.9[3]	...	4.5	13.3
Belgium	1985	808931	15.0	15.3	8.6	37.8	2.9	...	7.8	6.7
Belgique	1986	839302	15.0	15.4	9.4	36.9	2.9	5.9	8.0	6.6
Belize [4]	1989	229	12.7	4.3	5.3	16.5	7.9	0.7	40.3	12.4
Belize [4]	1990	279	12.7	3.4	7.9	15.3	6.8	3.3	37.5	13.2
	1991	321	16.6	3.4	7.0	16.8	6.6	4.0	32.6	13.1
Bermuda [4]	1985	138	36.9	2.0	...	21.3	3.7	3.6	27.1	5.3
Bermudes [4]	1991	217	34.6	1.8	...	22.7	4.1	3.8	27.2	5.8
	1992	218	34.9	1.5	...	22.6	4.2	3.8	27.4	5.5
British Virgin Islands [1]	1985	17	23.3	...	10.8	23.6	14.9	1.8	20.7	4.9
Iles Vierges brit. [1]	1986	19	21.8	...	11.4	22.7	14.5	2.4	22.1	5.1
	1987	21	23.3	...	11.0	21.1	15.8	2.6	20.2	6.0
Brunei Darussalam	1982	914	22.9	41.4	5.5	14.2	5.0	0.2	5.0	5.8
Brunéi Darussalam	1983	922	24.6	35.3	6.0	15.2	5.7	0.2	5.5	7.5
	1984	2512	69.3	12.8	2.7	6.5	2.6	0.1	2.5	3.4
Cameroon	1985	465500	29.9	11.6[5]	...	17.6	5.6	0.6	11.3	23.3
Cameroun	1987	391000	28.3	14.7[5]	...	21.6	6.1	0.8	7.3	21.0
	1988	378400	35.5	12.4[5]	...	21.5	6.0	0.9	6.2	17.6
Cayman Islands	1985	45	31.1	...	15.6	15.6	15.6	2.2	17.8	2.2
Iles Caïmanes	1990	94	26.6	...	14.9	14.9	16.0	4.3	19.1	4.3
	1991	103	27.2	...	14.6	14.6	14.6	5.8	20.4	2.9
Colombia	1985	531264	31.3	11.6	...	28.6	9.8	6.3	10.9	1.7
Colombie	1990	2076459	30.8	12.1	...	26.3	9.8	7.2	12.1	1.6
	1991	2684541	29.7	11.5	...	26.9	9.4	7.4	12.8	2.2
Cyprus	1985	209	14.0	6.0	14.3	23.3	11.8	14.1	11.6	4.9
Chypre	1991	494	11.4	26.5[6]	9.9	17.7	9.2	12.8	8.4	4.2
	1992	591	10.4	32.3[6]	8.8	16.3	9.5	11.6	7.0	4.2
Denmark	1985	155481	11.7	8.4[5]	...	23.6	19.7	22.4	8.8	5.3
Danemark	1989	196546	11.8	7.9[5]	...	23.8	19.8	24.1	7.8	4.8
	1990	202504	11.3	8.0[5]	...	23.7	19.8	24.3	8.0	4.9
Ecuador	1985	137000	10.8	18.3	5.5	31.4	6.3	5.1	11.7	10.9
Equateur	1990	777131	13.0	14.5	7.0	27.5	4.8	6.1	13.9	13.1
	1991	1009334	12.9	14.9	7.2	27.8	4.6	6.2	14.9	11.5
Estonia										
Estonie	1992	1775	12.2	2.5	9.9	40.1	2.4	3.6	22.8	6.5
Fiji	1985	252	28.6	6.0	...	27.8	12.3	0.4	22.6	2.4
Fidji	1988	264	33.0	10.6	...	23.9	10.6	0.4	18.9	2.7
	1989	304	28.9	12.2	...	25.7	10.9	0.3	18.8	3.3
Finland	1985	66967	9.5	7.7	5.8	25.2	22.8	14.6	7.0	7.4
Finlande	1991	118719	9.4	6.9	5.5	25.0	22.3	16.4	7.7	6.7
	1992	118480	9.2	8.0	5.5	25.0	22.1	15.9	7.8	6.6

24
Government final consumption expenditure by function at current prices
Percentage distribution [cont.]
Consommation finale des administrations publiques par fonction aux prix courants
Répartition en pourcentage [suite]

Country or area	Year	Govt. final consumption expenditures Consommation finale des administrations (M. nat'l. curr.)	General public services Administration publique générale	Defence Défense	Public order & safety Sûreté publique	Education Enseignement	Health Santé	Social services Services sociaux	Economic services Services économiques	Other functions Autres fonctions
France	1985	910315	13.0	16.5	4.5	25.7	16.5	8.1	6.6	9.1
France	1989	1106075	13.0	16.3	4.6	25.3	17.1	7.6	6.2	9.9
	1990	1165996	11.9	16.7	4.6	25.8	17.0	7.8	6.3	9.9
Gambia [1] [2]	1985	265	22.6	...	...	10.3[7]	5.6	0.1	45.8	15.7
Gambie [1] [2]	1990	819	22.0	...	...	12.9[7]	6.4	0.1	18.6	40.1
	1991	804	22.2	...	...	12.6[7]	5.7	0.1	24.1	35.3
Germany † · Allemagne †										
F.R. Germany	1985	365720	10.2	13.6	7.7	19.8	30.1	10.2	4.5	3.9
R.f. Allemagne	1989	418820	10.4	12.8	7.9	19.1	30.1	11.3	4.3	4.0
	1990	444070	10.4	11.9	7.9	18.9	30.7	12.0	4.2	3.9
Greece	1985	942	39.2	30.8	...	14.6	9.9	1.6	3.9[8]	...
Grèce	1990	2221	41.4	25.3	...	15.7	11.9	1.5	4.3[8]	...
	1991	2548	42.6	24.7	...	15.6	11.7	1.3	4.1[8]	...
Guinea−Bissau	1986	6423	44.0	...	...	18.4	11.8	0.9	22.7	2.2
Guinée−Bissau	1987	10776	44.0	...	...	18.4	11.8	1.2	22.7	1.9
Honduras	1985	1046	37.4	18.0	...	25.7	9.1	...	...	13.6
Honduras	1988	1468	29.7	18.0	...	27.5	11.9	...	...	12.9
	1989	1607	25.6	17.2	...	27.9	11.9	...	...	17.4
Iceland	1985	20953	6.8	0.0	7.1	20.0	34.5	5.6	11.6	14.4
Islande	1991	76918	8.3	0.0	6.5	19.9	34.5	7.4	8.8	14.5
	1992	78930	8.1	0.0	6.4	20.1	33.6	7.8	8.7	15.3
India [1] [4]	1985	241870	22.2[5]	35.8	...	14.7	7.2	2.7	13.6	3.8
Inde [1] [4]	1989	450920	22.7[5]	36.5	...	14.5	6.4	3.1	13.6	3.2
	1990	512210	23.0[5]	34.3	...	15.1	6.9	3.4	13.4	3.9
Iran, Islamic Rep. of [9]	1985	2443000	2.3	38.0	6.8	20.1	6.6	7.5	8.4	10.2
Iran, Rép. islamique d' [9]	1989	3294000	3.9	33.9	4.7	24.5	8.4	8.3	6.1	10.1
	1990	4054000	3.3	27.7	5.6	25.9	7.4	9.6	7.2	13.3
Israel [4]	1985	11662	5.9	54.4	3.5	15.9	8.0	1.4	2.1	8.8
Israël [4]	1986	15220	6.6	50.6	4.0	17.1	8.6	1.5	2.3	9.4
	1987	20599	6.2	53.3	3.9	16.1	7.9	1.5	2.1	9.0
Italy	1985	133265000	15.2	12.6	9.4	28.1	18.7	4.3	6.7	4.9
Italie	1991	249773000	16.4	10.8	9.8	27.1	21.2	4.2	6.1	4.5
	1992	263137000	16.1	10.6	9.9	27.6	21.0	4.2	6.1	·4.5
Japan	1985	31038000	26.2[5]	9.5	...	36.3	3.9	5.6	10.0	8.5
Japon	1991	41671000	27.0[5]	9.9	...	34.1	4.5	6.0	9.0	.9.6
	1992	43672000	27.0[5]	9.7	...	33.7	4.6	6.1	9.1	9.8
Jordan	1985	411	63.6[5]	...	...	15.9	5.2	0.6	4.6	5.4
Jordanie	1986	461	65.5[5]	...	...	15.7	5.4	0.5	4.7	3.8
	1987	460	63.5[5]	...	...	16.9	5.7	0.5	5.1	4.2
Kenya	1985	868	17.6	14.5	...	37.8	10.9	19.1[8]	...	...
Kenya	1991	1890	24.0	12.3	...	39.4	9.6	14.7[8]	...	...
	1992	2102	24.1	9.8	...	40.4	10.5	15.2[8]	...	...
Korea, Republic of	1985	8136000	14.8	43.8	9.9	22.0	1.1	2.2	3.3	2.8
Corée, Rép. de	1991	22212000	15.9	34.7	11.3	24.2	1.3	4.8	4.5	3.3
	1992	26299000	16.1	32.5	11.4	25.4	1.3	5.0	4.8	3.5
Kuwait	1985	1445	51.3[5]	...	...	23.3	12.6	2.1	3.7	7.1
Koweït	1986	1404	48.0[5]	...	...	24.9	13.3	2.3	4.1	7.4
	1987	1371	45.5[5]	...	...	25.4	13.8	2.5	4.1	8.8
Lesotho	1985	136	22.7	...	31.6	4.1	11.4[3]	...	23.3	6.8
Lesotho	1991	314	25.1	...	29.8	4.2	10.7[3]	...	22.4	7.8
	1992	381	25.0	...	29.4	4.2	10.6[3]	...	22.9	7.9
Malaysia	1985	11844	16.3	20.3	9.4	27.6	9.3	4.5	12.5[8]	...
Malaisie	1991	18391	12.8	23.6	9.3	28.8	9.1	4.6	11.7[8]	...
	1992	19304	12.7	22.4	9.6	29.5	9.4	4.6	11.8[8]	...
Maldives	1985	121	29.9	15.0[5]	...	14.5	7.9	5.6	11.8	15.3
Maldives	1986	139	32.3	16.3[5]	...	16.2	8.3	5.3	5.6	15.9
Malta	1985	84	12.8	17.4	...	23.1	25.0	2.5	3.3	15.8
Malte	1990	129	12.6	13.8	...	25.2	25.4	6.7	3.1	13.1
	1991	147	14.6	12.6	...	29.2	23.9	5.5	3.1	11.2

24
Government final consumption expenditure by function at current prices
Percentage distribution *[cont.]*
Consommation finale des administrations publiques par fonction aux prix courants
Répartition en pourcentage *[suite]*

Country or area	Year	Govt. final consumption expenditures Consommation finale des administrations (M. nat'l. curr.)	General public services Administration publique générale	Defence Défence	Public order & safety Sûreté publique	Education Enseignement	Health Santé	Social services Services sociaux	Economic services Services économiques	Other functions Autres fonctions
Mauritius	1985	1915	15.6	2.0	12.9	23.0	17.5	2.9	20.5	5.6
Maurice	1990	4455	16.3	3.2	14.0	21.0	17.1	2.5	18.8	7.0
	1991	4930	16.0	3.5	15.0	20.8	16.1	2.6	17.4	8.6
Montserrat										
Montserrat	1985	20	18.7	0.3	11.3	22.4	15.3	4.8	24.5	2.7
Nepal [10]	1985	4445	13.0	12.4	10.3	28.1	5.3	0.0	24.1	3.3
Népal [10]	1987	6571	14.9	11.5	7.2	24.3	7.7	0.4	30.5	3.5
	1988	7303	14.4	7.0	10.4	22.7	3.6	1.6	34.9	5.2
Netherlands	1985	67670	...	17.6	...	31.8	...	4.6	...	45.9 [11]
Pays-Bas	1988	71160	...	18.2	...	30.7	...	4.9	...	46.2 [11]
	1989	72490	...	18.0	...	30.3	...	4.9	...	46.8 [11]
Norway	1985	92653	7.5	15.6	4.0	26.0	23.2	8.8	11.5	3.3
Norvège	1990	139115	7.9	15.9	4.2	25.7	22.4	10.0	10.5	3.4
	1991	147478	8.1	15.1	4.3	25.7	22.6	10.5	10.2	3.5
Oman	1985	938	...	82.5 [5]	...	9.1	4.9	...	3.6	...
Oman	1991	1395	...	79.8 [5]	...	12.5	5.2	...	2.4	...
	1992	1735	...	82.5 [5]	...	10.8	4.7	...	2.1	...
Pakistan [2]	1985	65662	59.1	...	7.5	9.8	4.5	3.0	12.9	3.1
Pakistan [2]	1990	145575	57.2	...	7.4	10.4	4.5	3.3	14.3	2.8
	1991	155569	56.9	...	7.6	10.5	4.5	3.3	14.4	2.8
Panama	1985	1044	31.0 [12]	...	...	24.8	7.6	20.8	10.9	5.0
Panama	1991	958	30.2 [12]	...	...	27.9	11.8	18.2	6.5	5.3
	1992	1151	37.1 [12]	...	...	25.3	10.7	17.1	5.0	5.3
Peru	1985	22 [13]	57.4 [12]	...	...	24.7	8.3	1.9	6.0	1.6
Pérou	1987	94 [13]	55.5 [12]	...	...	27.3	7.1	2.9	5.3	2.0
	1988	475 [13]	54.1 [12]	...	...	26.1	7.6	2.7	6.7	2.7
Portugal	1985	546899	12.4	16.2	9.7	24.7	22.5	4.8	6.5	3.2
Portugal	1986	678841	12.2	17.1	9.6	25.3	21.7	4.7	6.4	3.0
Saint Vincent-Grenadines	1985	63	9.3	0.0	13.0	26.6	18.7	6.9	24.3	1.3
St.-Vincent-et-Grenad.	1988	91	14.9	0.0	11.0	25.7	19.1	5.4	22.8	1.1
	1989	94	10.7	0.0	11.3	26.1	22.4	6.5	21.5	1.5
Seychelles	1989	475	9.4	11.5	4.9	29.7	11.8	2.6	17.1	13.0
Seychelles	1990	544	9.5	10.2	5.1	29.2	12.8	3.5	18.3	11.5
	1991	558	10.3	11.0	5.7	26.4	13.7	5.2	17.0	10.7
Sierra Leone [1] [2]	1985	900	15.1	3.9	...	10.1	4.3	0.3	38.7	27.4
Sierra Leone [1] [2]	1989	7620	14.4	7.3	...	10.2	5.0	1.4	45.9	15.9
	1990	32337	11.9	5.6	...	6.2	2.1	0.6	27.9	45.6
Slovenia	1990	34226	39.6	0.7	...	21.3	27.6	7.6	...	3.2
Slovénie	1991	65845	35.5	7.6	...	21.0	25.6	7.3	...	3.0
	1992	213669	34.1	7.6	...	19.8	28.6	7.0	...	2.9
Spain	1985	4151700	3.9	13.3	7.5	17.9	23.6	10.9	5.9	17.0
Espagne	1988	5924400	5.5	12.3	7.6	18.8	23.2	11.0	6.0	15.5
	1989	6831300	5.8	11.5	7.4	18.6	24.5	10.3	6.6	15.4
Sri Lanka	1985	19170	32.0	24.3	...	16.5	8.7	11.6	6.2	0.9
Sri Lanka	1991	50766	19.8	23.0	...	12.5	8.0	23.7	12.3	0.6
	1992	53965	20.2	25.9	...	13.7	8.4	19.5	11.6	0.6
Sweden	1985	241754	9.0	9.6	4.6	20.5	24.0	17.0	8.0	7.3
Suède	1991	394394	9.2	10.1	5.3	18.9	23.1	19.2	7.5	6.8
	1992	400284	0.0	0.0	...	...	...	...	...	...
Thailand	1985	142923	19.8	42.3 [5]	...	27.1	7.1	0.5	2.1	1.1
Thaïlande	1990	206824	19.9	36.5 [5]	...	30.5	8.2	0.6	2.3	2.0
	1991	237078	20.5	36.0 [5]	...	30.1	8.7	0.6	2.1	2.0
Tonga [14]	1985	21	20.7	4.2	7.0	13.6	12.2	1.9	31.9	8.5
Tonga [14]	1986	27	22.7	3.7	7.1	13.8	11.9	1.9	29.0	9.7
	1987	32	26.9	3.4	6.9	13.1	10.3	1.9	26.3	11.3
Trinidad and Tobago	1985	4109	14.7	...	13.8	19.5	17.0	0.5	28.3	6.3
Trinité-et-Tobago	1990	3487	22.6	...	13.5	18.9	15.4	0.5	22.8	6.4
	1991	3838	19.6	...	13.9	18.3	15.2	0.4	24.7	7.9

24
Government final consumption expenditure by function at current prices
Percentage distribution [cont.]
Consommation finale des administrations publiques par fonction aux prix courants
Répartition en pourcentage [suite]

Country or area	Year	Govt. final consumption expenditures Consommation finale des administrations (M. nat'l. curr.)	Per cent — Pourcentage							
			General public services Administration publique générale	Defence Défence	Public order & safety Sûreté publique	Education Enseignement	Health Santé	Social services Services sociaux	Economic services Services économiques	Other functions Autres fonctions
Ukraine	1989	26600	0.0	52.3	...	20.3	15.0[3]	...	4.1	8.3
Ukraine	1991	52100	0.0	41.3	...	25.9	20.3[3]	...	3.6	8.8
United Kingdom	1985	75296	5.0	23.7	7.7	19.9	22.3	6.8	6.3	8.4
Royaume—Uni	1991	124245	5.6	19.8	9.2	20.1	23.6	7.6	5.6	8.4
	1992	132418	5.8	18.0	9.6	20.3	25.1	7.6	5.5	8.1
United Rep. of Tanzania [1][2]	1985	27440	21.4	13.3	7.8	6.5	4.8	0.5	23.6	22.1
Rép.—Unie de Tanzanie [1][2]	1990	134691	23.4	8.6	7.8	6.6	4.9	0.4	15.9	32.3
	1991	207000	17.2	6.2	6.2	6.9	4.9	0.4	22.1	35.9
Vanuatu	1985	4501	15.3	5.5[5]	...	17.4	9.2[3]	...	14.8	37.8
Vanuatu	1988	4969	17.1	6.8[5]	...	21.1	8.4[3]	...	18.3	28.4
	1989	4881	15.9	7.7[5]	...	17.5	8.1[3]	...	24.6	26.3
Venezuela	1985	48547	...	...	...	29.1	11.2	...	...	...
Venezuela	1991	293214	...	...	...	21.5	10.7	...	...	...
	1992	366047	...	...	...	20.4	11.0	...	...	...
Zimbabwe	1985	1559	47.8[5]	...	...	28.7	10.5	...	9.5	3.5
Zimbabwe	1990	7425	28.1	13.1	4.9	19.9	6.5	3.1	20.8	3.5
	1991	7788	16.7	14.3	6.1	26.7	7.4	4.1	23.0	4.5

Source:
National accounts database of the Statistical Division of the
United Nations Secretariat.

† For detailed descriptions of data pertaining to
former Czechoslovakia, Germany, SFR Yugoslavia and former
USSR, see annex I — Country or area nomenclature, regional
and other groupings.

1 Central government estimates only (India: incl. state government).
2 Fiscal year beginning 1 July.
3 Including social services.
4 Fiscal year beginning 1 April.
5 Including public order and safety (Jordan, Kuwait, Zimbabwe:
defence).
6 Including military expenditure of government.
7 Including recreational, cultural and religious affairs.
8 Including "other functions" (Kenya: including economic services).
9 Fiscal year beginning 21 March.
10 Fiscal year ending 15 July.
11 Including general public services, health services and
economic services.
12 Including defence.
13 Thousand.
14 Fiscal year ending 30 June.

Source:
Base de données sur les comptes nationaux de la division
de statistique du Secrétariat de l'ONU.

† Pour les descriptions en détails des données relatives
à l'ancienne Tchécoslovaquie, l'Allemagne, la Rfs
Yougoslavie et l'ancienne URSS, voir l'Annexe I —
Nomenclature des pays ou zones, groupements régionaux
et autres groupements.

1 Administration centrale seulement (Inde : y compris
administration des États).
2 L'année fiscale commençant le 1er juillet.
3 Services sociaux compris.
4 L'année fiscale commençant le 1er avril.
5 Y compris l'ordre public et la sécurité (Jordanie, Koweït,
Zimbabwe : défense).
6 Y compris les dépenses militaires de l'État.
7 Loisirs, affaires culturelles et religieuses.
8 Y compris les "autres fonctions" (Kenya : services
économique compris).
9 L'année fiscale commençant le 21 mars.
10 L'année fiscale finissant le 15 juillet.
11 Y compris l'administration publique générale, les soins de
santé et les services économiques.
12 Défense comprise.
13 Milliers.
14 L'année fiscale finissant le 30 juin.

25
Private final consumption expenditure by type and purpose at current prices
Consommation finale privée par catégorie de dépenses et par fonction aux prix courants

Percentage distribution
Répartition en pourcentage

Per cent — Pourcentage

Country or area Pays ou zone	Year Année	Private final consump. expend. Consommation finale privée (M. nat.curr.)	Food, beverages, & tobacco Alimentation boissons et tabac	Clothing/ footwear Articles d'habille— ment et chaussures	Gross rent fuel and power Loyers bruts, chauffuge et éclairage	Furniture household equip. and operation Muebles articles de ménager et dépenses d'entretien courant de la maison	Medical and health expenses Soins med. et dépenses de santé	Transport and communi— cation Transports et communi— cations	Recreation entertain— ment and education services Loisirs, spectacles et ensei— gnement	Other functions Autres fonctions
Australia [1] Australie [1]	1985	143738	22.4	6.7[2]	19.5	7.5	6.4	14.8	9.0	13.8
	1991	242750	20.6	5.7[2]	20.6	6.6	7.2	15.0	9.7	14.7
	1992	253952	20.7	5.5[2]	20.4	6.6	7.3	14.9	9.8	14.7
Austria Autriche	1985	775529	23.5	11.0	19.6	7.1	5.0	17.2	6.1	10.6
	1991	1064082	21.0	9.8	19.0	8.1	5.7	17.9	8.0	10.5
	1992	1124248	20.7	9.4	19.0	8.2	6.0	18.0	8.0	10.7
Belgium Belgique	1985	3106060	21.5	7.5	18.8	9.8	10.4	12.5	6.0	13.4
	1991	4227944	18.6	7.9	16.7	10.8	11.4	13.1	6.5	15.0
	1992	4443873	17.9	7.7	16.8	10.6	11.7	13.2	6.3	15.9
Bolivia Bolivie	1985	1793	52.0	3.6	7.8	5.2	2.1	14.8	3.4	11.0
	1986	6214	47.7	3.3	8.0	6.7	2.5	16.5	4.2	11.1
	1988	8591	40.1	5.1	12.5	9.7	2.1	17.7	3.0	9.9
Canada Canada	1985	271099	17.4	6.1	22.2	9.1	4.2	15.9	10.5	14.5
	1991	403506	16.0	5.3	23.7	8.7	4.4	14.4	10.8	16.6
	1992	412766	15.6	5.2	24.1	8.7	4.6	14.2	11.0	16.7
Cape Verde Cap—Vert	1985	11470	60.7	3.2	13.8	7.7	0.8	8.3	5.9	−0.4
	1987	15134	60.4	2.9	13.6	7.2	0.6	10.1	5.8	−0.5
	1988	17848	62.6	2.5	13.5	6.9	0.5	8.8	5.6	−0.5
Colombia Colombie	1985	3446000	37.1	6.4	12.3	5.5	5.9	14.0	5.5	13.2
	1991	17348000	34.6	4.5	10.5	5.8	6.2	18.0	5.7	14.7
	1992	23183000	34.2	4.5	10.0	5.7	6.4	18.5	5.4	15.3
Cyprus Chypre	1985	946	36.6	13.2	11.3	14.4	3.6	19.8	9.7	−7.1
	1991	1754	33.4	13.6	9.5	13.2	4.0	21.5	10.1	−4.3
	1992	1961	34.0	13.6	9.3	14.3	4.2	21.3	10.6	−6.5
Denmark Danemark	1985	337215	23.2	5.9	25.1	6.8	1.8	17.5	9.6	9.9
	1991	429961	20.8	5.5	27.8	6.5	2.2	15.2	10.4	11.6
	1992	442198	21.0	5.3	27.9	6.2	2.3	15.2	10.1	12.1
Ecuador Equateur	1985	715659	38.4	10.9	7.2	6.0	3.8	12.5[3]	...	21.2
	1991	8440031	39.0	10.0	5.2	7.4	4.2	12.7[3]	...	21.4
	1992	13255745	38.6	9.6	5.0	7.1	4.6	12.7[3]	...	22.5
Fiji Fidji	1985	838	32.6	4.8	13.6	8.4	1.8	13.1	4.3	21.5
	1990	1292	31.2	8.1	12.9	7.7	2.0	12.0	3.9	22.2
	1991	1421	30.9	7.8	13.1	7.8	2.0	11.6	3.6	23.2
Finland Finlande	1985	180887	24.9	5.3	18.0	6.7	3.7	16.2	8.7	16.4
	1991	274709	22.3	5.1	19.5	6.0	4.6	14.3	9.1	19.0
	1992	271505	22.6	4.7	21.4	5.8	4.8	13.5	9.0	18.2
France France	1985	2871097	20.6	7.0[4]	19.0	8.3	8.6	16.8	7.0	12.6
	1991	4061968	19.2	6.3[4]	19.7	7.8	9.7	16.3	7.6	13.4
	1992	4226992	18.8	6.2[4]	20.2	7.7	9.9	16.2	7.6	13.3
Germany † · Allemagne † F.R. Germany R.f. Allemagne	1985	1036530	22.5[5]	8.2	21.8	8.4	3.2	14.9[3]	9.6	11.3
	1991	1428310	20.8[5]	7.9	20.2	9.1	3.3	17.6[3]	10.2	10.8
	1992	1510030	19.9[5]	7.6	20.5	9.0	3.4	17.3[3]	10.2	12.1
Greece Grèce	1985	3025492	41.1	9.0	11.4	8.8	3.8	14.6	6.1	5.0
	1991	9136782	38.5	8.8	12.7	8.1	3.7	15.2	5.9	7.2
	1992	10648168	38.1	8.3	13.1	7.9	4.0	15.5	5.8	7.3

25
Private final consumption expenditure by type and purpose at current prices
Percentage distribution *[cont.]*
Consommation finale privée par catégorie de dépenses et par fonction aux prix courants
Répartition en pourcentage *[suite]*

Per cent — Pourcentage

Country or area Pays ou zone	Year Année	Private final consump. exp. Consomma- tion finale privée (M. nat.curr.)	Food, beverages, & tobacco Alimentation boissons et tabac	Clothing/ footwear Articles d'habille- ment et chaussures	Gross rent fuel and power Loyers bruts, chauffuge et éclairage	Furniture household equip. and operation Muebles articles de ménager et dépenses d'entretien courant de la maison	Medical and health expenses Soins med. et dépenses de santé	Transport and communi- cation Transports et communi- cation	Recreation entertain- ment and education services Loisirs, spectacles et ensei- gnement	Other functions Autres fonctions
Honduras	1985	5033	45.2	9.1	22.5	8.3	7.0	3.0	2.4	2.5
Honduras	1986	5421	45.1	9.1	22.5	8.3	7.0	3.0	2.4	2.6
Hong Kong	1985	167698	21.5	17.2[4]	15.2	10.8	6.0	7.5	8.9	12.8
Hong kong	1991	390913	16.1	20.0[4]	14.5	12.2	5.6	8.7	8.4	14.4
	1992	449342	15.0	21.3[4]	14.8	12.8	5.6	9.9	7.9	12.7
Hungary	1985	552294[5]	47.5[5]	9.9	9.5	9.1	0.7	10.0	6.7	6.6
Hungary	# 1990	1046338[5]	44.0[5]	7.9	10.6	8.6	0.9	12.9	6.4	8.7
	1991	1303792[5]	40.9[5]	7.4	11.8	8.7	1.5	15.2	6.6	7.8
Iceland	1985	74689	25.7	9.5	16.0	10.0	1.5	15.5	8.6	13.2
Islande	1991	236276	24.6	7.9	14.9	8.4	1.8	15.4	11.3	15.6
	1992	236214	25.0	7.7	15.5	7.9	2.2	14.6	11.3	15.8
India [6]	1985	1768520	55.3	11.6	12.1	4.4	2.9	7.5	3.0	3.2
Inde [6]	1990	3340070	53.1	10.7	10.8	4.6	2.5	10.7	3.8	3.7
	1991	3875880	54.3	10.0	9.9	4.3	2.3	11.5	3.8	3.8
Iran, Islamic Rep. of	1985	9627000	43.2[5]	9.3	24.4	6.1	4.3	6.3	1.6	5.0
Iran, Rép. islamique d'	1989	18448000	46.5[5]	12.0	23.4	5.5	3.4	4.6	1.4	3.1
	1990	24071000	42.4[5]	11.7	24.8	6.4	3.8	5.0	1.7	4.2
Ireland	1985	10598	40.5	7.3	11.8	6.9	3.6	13.1	9.6	7.1
Irlande	# 1991	16371	35.6	7.1	13.0	7.2	4.0	13.4	12.0	7.7
	1992	17268	35.6	7.0	12.6	7.2	4.1	13.0	12.3	8.2
Israel	1985	16552	28.6	6.1	22.8	9.5	5.4	11.6	7.0	8.9
Israël	1991	82177	25.0	4.6	23.9	10.2	5.9	11.0	6.8	12.5
	1992	98513	24.7	4.8	23.5	10.1	6.0	12.3	7.2	11.4
Italy	1985	498048000	25.3	10.3	15.0	9.1	5.6	12.6	8.4	13.7
Italie	1991	884753000	20.3	9.9	15.7	9.5	6.8	12.1	8.9	16.8
	1992	946937000	20.0	9.8	15.9	9.4	6.8	12.2	8.9	17.0
Jamaica	1985	7772	50.5	4.5	15.3	7.0	2.9	17.3	3.1	-0.5
Jamaïque	1987	9849	52.1	6.0	14.5	6.9	3.4	15.8	2.9	-1.4
	1988	11388	49.6	5.8	13.1	6.8	3.5	15.3	2.8	3.2
Japan	1985	188760000	22.0	6.6	18.6	6.1	10.4	9.6	9.7	17.1
Japon	1991	255084000	20.1	6.3	19.2	6.1	10.5	9.8	9.9	18.1
	1992	264778000	19.6	6.0	19.7	5.9	10.8	9.5	10.1	18.4
Jordan	1985	1415	40.4	5.7	6.5	4.8	4.1	5.9	6.4	26.3
Jordanie	1986	1238	40.7	5.6	6.5	4.9	4.1	5.9	6.4	25.9
Korea, Republic of	1985	47875000	39.4	5.1	11.1	5.1	6.2	10.1	10.8	12.3
Corée, République de	1991	109655000	34.7	4.4	11.1	6.0	7.3	10.9	11.2	14.4
	1992	123746000	33.6	4.2	11.3	5.8	7.3	11.3	11.3	15.2
Luxembourg	1985	120523	24.3	6.9	21.9	9.6	7.0	17.7	3.6	9.1
Luxembourg	1990	166543	19.9	6.3	20.3	11.1	7.7	18.0	4.5	12.3
	1991	182597	19.1	6.0	20.4	11.1	7.6	19.7	4.3	11.8
Malta	1985	333	42.6	9.9	6.9	10.1	3.8	17.4	6.4	3.1
Malte	1990	461	39.7	9.3	6.9	11.4	4.5	20.6	9.2	-1.5
	1991	495	39.6	9.6	7.0	12.5	4.4	20.8	9.5	-3.4
Mexico	1985	30575000	38.5	9.7	8.4	13.3	4.0	9.2	5.5	11.3
Mexique	1991	621208000	34.8	7.3	12.4	11.0	4.1	11.5	5.0	14.0
	1992	736137000	33.7	7.1	12.8	10.4	4.2	12.2	5.2	14.4
Netherlands	1985	252910	16.6	7.2	18.7	6.4	12.5	12.4	9.5	16.8
Pays—Bas	1991	323000	14.9	6.9	18.2	7.1	12.7	12.8	10.4	17.0
	1992	339520	14.7	6.6	18.3	6.9	12.9	13.3	10.0	17.3
New Zealand [6]	1985	27712	17.4	6.7	18.6	8.2	4.9	18.8	8.1	17.4
Nouvelle—Zélande [6]	1989	43952	17.6	5.5	21.4	7.1	5.8	16.5	7.9	18.2
	1990	46076	18.4	5.1	22.0	6.8	6.3	15.5	7.7	18.2
Norway	1985	245439	25.1	7.8	16.6	7.8	3.8	16.8	8.2	13.9
Norvège	1990	336065	25.3	6.8	19.1	6.9	4.7	13.1	8.9	15.2
	1991	349705	25.4	6.8	19.4	6.7	5.2	12.4	9.1	14.9

25

Private final consumption expenditure by type and purpose at current prices
Percentage distribution *[cont.]*

Consommation finale privée par catégorie de dépenses et par fonction aux prix courants
Répartition en pourcentage *[suite]*

Per cent — Pourcentage

Country or area Pays ou zone	Year Année	Private final consump. exp. Consommation finale privée (M. nat.curr.)	Food, beverages, & tobacco Alimentation boissons et tabac	Clothing/ footwear Articles d'habillement et chaussures	Gross rent fuel and power Loyers bruts, chauffuge et éclairage	Furniture household equip. and operation Muebles articles de ménager et dépenses d'entretien courant de la maison	Medical and health expenses Soins med. et dépenses de santé	Transport and communication Transports et communication	Recreation entertainment and education services Loisirs, spectacles et enseignement	Other functions Autres fonctions
Peru	1985	123[7]	38.2	8.4	2.2	11.8	4.0	9.3	9.2	16.8
Pérou	1990	4768	31.0	9.0	1.1	10.6	7.4	7.9	13.9	...
	1991	27008	...	7.5	2.4	7.5	4.0	7.4	11.3	...
Philippines	1985	420832	59.2	3.8	3.7	13.3	...	4.9	...	15.1
Philippines	1991	916367	58.1	3.7	4.2	13.2	...	5.3	...	15.5
	1992	1017001	57.7	3.7	4.1	13.5	...	5.0	...	16.0
Portugal	1985	2393244	39.5	9.8	5.2	9.1	4.8	15.5	6.1	9.9
Portugal	1986	2876199	38.6	10.7	5.2	9.0	4.7	16.0	6.0	9.9
Puerto Rico [1]	1985	15746	26.4	8.6	15.4	7.3	6.4	15.7	6.6	13.6
Porto Rico [1]	1991	21398	23.0	8.2	15.6	7.1	9.0	13.7	7.6	15.8
	1992	22539	22.1	7.9	15.5	7.6	9.6	14.3	7.5	15.5
Sierra Leone [1]	1985	6238	64.5	3.0	14.9	2.7	0.9	6.9	1.8	5.2
Sierra Leone [1]	1986	16267	66.7	2.7	13.9	2.2	1.1	8.1	1.4	4.0
Singapore	1985	17553	29.2	8.9	12.4	11.5	4.0	15.3	14.0	4.7
Singapour	1991	30361	22.2	8.5	11.5	10.5	5.5	16.4	17.6	7.9
	1992	32289	21.9	8.3	12.0	10.4	5.4	17.0	17.0	8.0
South Africa	1985	66167	34.0	7.0	12.1	10.4	4.2	16.7[8]	6.0	9.5
Afrique du Sud	1991	179283	37.6	7.5	9.4	10.1	4.9	14.9[8]	6.3	9.3
	1992	203407	38.3	7.2	9.5	9.9	4.8	14.5[8]	6.2	9.4
Spain	1985	18080000	26.6	9.2	15.5	7.2	3.8·	14.4	7.0	16.4
Espagne	1991	34213000	21.9	9.2	13.2	6.9	4.3	15.8	6.9	21.8
	1992	37175700	21.3	9.0	13.1	6.8	4.5	16.3	6.9	22.1
Sri Lanka	1985	118101	54.4	6.8	5.7	4.5	1.5	15.8	4.7	6.6
Sri Lanka	1991	270927	56.7	5.9	4.4	4.8	1.5	14.1	3.8	8.8
	1992	308350	55.6	7.9	4.3	3.4	1.5	15.2	3.0	9.1
Sweden	1985	443671	23.4	7.2	25.6	6.0	2.5	15.3	9.5	10.6
Suède	1991	771310	19.6	6.6	27.8	6.0	2.6	16.0	9.4	11.9
	1992	776025	18.9	6.3	30.0	5.5	3.0	15.4	9.3	11.7
Switzerland	1985	141015	27.4	4.5	19.7	5.1	8.6	10.9	9.4	14.4
Suisse	1991	191205	25.8	4.1	19.0	4.6	9.9	11.1	9.8	15.5
	1992	198680	25.2	4.0	19.5	4.4	10.6	11.1	9.6	15.7
Thailand	1985	657365	38.7	11.3	11.5	8.0	6.5	11.2	4.6	8.2
Thaïlande	1990	1249651	32.9	12.0	8.8	10.5	8.6	14.1	4.9	8.2
	1991	1420297	32.1	12.0	8.3	11.3	8.3	13.3	4.9	9.7
United Kingdom	1985	215972	24.7	6.9	20.0	6.5	1.2	16.9	9.3	14.5
Royaume−Umi	1991	363137	21.3	5.8	18.3	6.4	1.5	16.5	9.8	20.5
	1992	380695	21.0	5.6	18.9	6.4	1.6	16.4	9.9	20.3
United States	1985	2598435	13.5	6.2	19.3	6.1	14.0	16.3	9.1	15.4
Etats−Unis	1991	3779900	12.4	6.0	18.6	5.9	17.2	13.6	10.2	16.0
	1992	3997100	12.0	6.1	18.4	5.9	17.6	13.7	10.3	16.0
Vanuatu	1985	7091	45.6	7.0	8.3	2.9	...	18.3	...	11.2
Vanuatu	1988	9562	46.6	5.5	7.4	2.7	...	22.5	. ...	9.1
	1989	10545	46.0	4.9	7.9	2.8	...	22.9	...	8.9
Venezuela	1985	287321	32.1	10.3	14.1	3.5	2.4	7.2	3.1	27.4
Venezuela	1991	2021222	39.7	8.0	9.0	3.9	2.1	6.1	2.4	28.9
	1992	2887169	38.8	7.7	8.2	4.4	2.2	7.1	2.3	29.2
Zimbabwe	1985	3894	34.6	11.4	16.9	7.7	4.2	1.9[9]	6.5	17.0
Zimbabwe	1986	4521	31.8	10.6	15.5	11.6	5.2	1.3[9]	5.8	18.2
	1987	4387	29.7	10.2	15.1	12.7	7.0	1.0[9]	6.5	17.8

25

Private final consumption expenditure by type and purpose at current prices
Percentage distribution *[cont.]*
Consommation finale privée par catégorie de dépenses et par fonction aux prix courants
Répartition en pourcentage *[suite]*

Source:
National accounts database of the Statistical Division of the
United Nations Secretariat.

† For detailed descriptions of data pertaining to
former Czechoslovakia, Germany, SFR Yugoslavia and former
USSR, see annex I – Country or area nomenclature, regional
and other groupings.

1 Fiscal year beginning 1 July.
2 Including drapery.
3 Including fuel.
4 Including personal effects.
5 Including expenditures in restaurants, cafes and hotels
(Germany: excluding hotels; Hungary: including hospitals,
medical institutions and schools).
6 Fiscal year beginning 1 April.
7 Thousand.
8 Incluidng packaged tours.
9 Including personal transport equipment only.
Communication is included in "Other functions".

Source:
Base de données sur les comptes nationaux de la Division de
statistique du Secrétariat de l'ONU.

† Pour les descriptions en détails des données relatives
à l'ancienne Tchécoslovaquie, l'Allemagne, la Rfs
Yougoslavie et l'ancienne URSS, voir l'Annexe I –
Nomenclature des pays ou zones, groupements régionaux et
autres groupements.

1 L'année fiscale commençant le 1er juillet.
2 Y compris les tissus d'ameublement.
3 Y compris le carburant.
4 Y compris les effets personnels.
5 Y compris les dépenses faites dans les restaurants, les cafés
et hôtels (Allemagne : hôtels exclus; Hondrie : hôpitaux,
établissements médicaux et écoles compris).
6 L'année fiscale commençant le 1er avril.
7 Milliers.
8 Voyages organisés compris.
9 Y compris seulement le matériel de transport individuel.
Les communications sont incluses dans "les autres fonctions".

26
# Index numbers of industrial production
# Indices de la production industrielle
1980=100

Country or area and industry [SITC] Pays ou zone et industrie [CITI]	1984	1985	1986	1987	1988	1989	1990	1991	1992	1993
Africa · Afrique										
Algeria Algérie										
Total industry [2−4]										
Total, industrie [2−4]	152	157	164	166	167	165	169	167	163	164
Total mining [2]										
Total, industries extractives [2]	117	124	132	128	122	124	127	116	122	109
Total manufacturing [3]										
Total, industries manufacturières [3]	174	179	189	186	188	182	183	176	164	163
Food, beverages, tobacco										
Aliments, boissons, tabac	141	145	160	165	168	171	173	171	162	170
Textiles										
Textiles	165	166	162	154	138	140	147	143	149	140
Petroleum products										
Produits pétroliers	192	186	206	200	207	206	213	213	216	210
Non−metallic mineral products										
Produits minéraux non métalliques	136	141	147	155	158	153	151	150	157	148
Electricity [4]										
Electricité [4]	155	174	186	199	216	226	239	260	274	294
Côte d'Ivoire Côte d'Ivoire										
Total industry [2−4]										
Total, industrie [2−4]	117	112	120	120	116	114	108	105	104	...
Total mining [2]										
Total, industries extractives [2]	115	111	103	88	56	14	12	10	8	...
Total manufacturing [3]										
Total, industries manufacturières [3]	118	113	123	121	118	115	124	121	121	...
Food, beverages, tobacco										
Aliments, boissons, tabac	88	114	129	133	84	130	121	128	135	...
Textiles and clothing										
Textiles, habillement	96	122	120	126	140	145	118	112	109	...
Chemicals and petroleum										
Produits chimiques et pétrole	99	109	112	117	129	117	119	116	114	...
Metal products										
Produits métalliques	70	92	124	94	87	89	71	61	65	...
Electricity, gas and water										
Electricité, gaz et eau	103	110	120	124	136	133	122	114	118	...
Egypt Egypte										
Total industry [2−4]										
Total, industrie [2−4]	139	152	139	144	146	142	136	144	137	141
Total mining [2]										
Total, industries extractives [2]	130	146	142	142	145	141	143	146	150	154
Total manufacturing [3]										
Total, industries manufacturières [3]	142	152	136	142	143	139	130	138	126	129
Food, beverages, tobacco										
Aliments, boissons, tabac	174	210	237	203	197	233	185	187	190	227
Textiles										
Textiles	138	145	149	130	108	112	114	122	112	105
Chemicals, coal, petroleum products										
Produits chimiques, houillers, pétroliers	154	168	143	184	185	169	171	180	165	165
Basic metals										
Métaux de base	119	134	148	169	169	182	179	201	162	160
Metal products										
Produits métalliques	172	206	179	185	190	182	162	158	142	142
Electricity [4]										
Electricité [4]	153	185	167	178	198	189	190	222	227	234
Ethiopia[17] Ethiopie[17]										
Total industry [2−4]										
Total, industrie [2−4]	136	143	152	157	151	142	103	...	...	...
Food, beverages, tobacco										
Aliments, boissons, tabac	153	165	167	170	166	151	122	...	...	...

26
Index numbers of industrial production [*cont.*]
Indices de la production industrielle [*suite*]
1980=100

Country or area and industry [SITC] Pays ou zone et industrie [CITI]	1984	1985	1986	1987	1988	1989	1990	1991	1992	1993
Textiles										
Textiles	108	116	127	135	115	124	78	...	...	...
Chemicals										
Produits chimiques	184	156	238	188	152	122	62	...	...	...
Metal products										
Produits métalliques	137	130	139	144	108	92	57	...	...	...
Ghana Ghana										
Total industry [2−4] [1]										
Total, industrie [2−4] [1]	**60**	**74**	**83**	**87**	**94**	**103**	**111**	...	...	...
Total mining [2]										
Total, industries extractives [2]	**78**	**88**	**82**	**88**	**94**	**114**	**134**	...	...	...
Total manufacturing [3]										
Total, industries manufacturières [3]	**57**	**71**	**79**	**82**	**90**	**101**	**106**	...	...	...
Food, beverages, tobacco										
Aliments, boissons, tabac	65	74	82	94	101	94	101	...	...	...
Chemical and rubber products										
Produits chimiques ou en caoutchouc	116	92	110	150	194	177	166	...	...	...
Non−metallic mineral products										
Produits minéraux non métalliques	81	122	91	95	141	192	225	...	...	...
Electricity [4]										
Electricité [4]	**85**	**103**	**153**	**160**	**161**	**99**	**109**	...	...	...
Kenya Kenya										
Total mining [2]										
Total, industries extractives [2]	**159**	**187**	**169**	**210**	**234**	**297**	**283**	**271**	**248**	**293**
Total manufacturing [3] [2]										
Total, industries manufacturières [3] [2]	**115**	**120**	**128**	**135**	**143**	**141**	**149**	**154**	**156**	**159**
Food, beverages, tobacco										
Aliments, boissons, tabac	113	119	130	143	151	154	154	153	155	155
Textiles										
Textiles	103	108	116	119	122	122	138	132	132	152
Chemicals, coal, petroleum products										
Produits chimiques, houillers, pétroliers	117	121	130	140	156	156	178	198	189	186
Metal products										
Produits métalliques	93	92	92	94	106	106	110	116	108	99
Malawi Malawi										
Total industry [2−4]										
Total, industrie [2−4]	**97**	**100**	**102**	**97**	**103**	**112**	**127**	**134**	**132**	**125**
Total manufacturing [3]										
Total, industries manufacturières [3]	**98**	**101**	**102**	**95**	**102**	**110**	**126**	**132**	**129**	**120**
Food, beverages, tobacco										
Aliments, boissons, tabac	138	139	149	151	148	176	189	174	185	188
Textiles [3]										
Textiles [3]	117	122	115	104	108	112	129	198	168	134
Electricity and water [4]										
Electricité et eau [4]	**120**	**123**	**132**	**144**	**146**	**157**	**175**	**185**	**197**	**200**
Morocco Maroc										
Total industry [2−4]										
Total, industrie [2−4]	**113**	**117**	**121**	**124**	**136**	**131**	**144**	**143**	**147**	**145**
Total mining [2] [4]										
Total, industries extractives [2] [4]	**121**	**120**	**118**	**117**	**134**	**101**	**118**	**103**	**108**	**107**
Total manufacturing [3] [5]										
Total, industries manufacturières [3] [5]	**110**	**116**	**121**	**124**	**135**	**137**	**149**	**152**	**155**	**153**
Food, beverages, tobacco										
Aliments, boissons, tabac	107	105	110	95	101	100	102	107	107	113
Textiles										
Textiles	102	117	128	143	152	151	157	162	167	161
Chemicals and petroleum products										
Produits chimiques et pétroliers	118	122	122	109	119	119	125	121	132	130
Basic metals										
Métaux de base	95	81	67	124	146	169	167	178	172	165

26
Index numbers of industrial production [*cont.*]
Indices de la production industrielle [*suite*]
1980=100

Country or area and industry [SITC] Pays ou zone et industrie [CITI]	1984	1985	1986	1987	1988	1989	1990	1991	1992	1993
Metal products Produits métalliques	77	80	77	97	101	110	117	122	121	114
Electricity [4] [6] Electricité [4] [6]	119	127	136	142	148	160	168	170	177	181
Nigeria Nigéria										
Total industry [2−4] Total, industrie [2−4]	106	121	119	141	157	151	158	168	173	160
Total mining [2] Total, industries extractives [2]	116	125	122	110	121	136	144	150	156	146
Total manufacturing [3] Total, industries manufacturières [3]	116	139	134	179	202	214	226	248	254	226
Electricity [4] Electricité [4]	118	135	163	160	170	223	168	169	180	191
Senegal Sénégal										
Total industry [2−4] Total, industrie [2−4]	115	118	106	118	124	112	118	102	110	103
Total mining [2] Total, industries extractives [2]	120	125	130	115	149	147	130	100	133	103
Total manufacturing [3] [17] Total, manufactures [3] [17]	113	117	101	117	119	105	115	101	104	101
Food, beverages, tobacco Aliments, boissons, tabac	104	88	73	114	145	130	135	82	112	110
Textiles Textiles	125	150	91	124	140	73	89	80	81	72
Chemicals, coal, petroleum products [8] Produits chimiques, houillers, pétroliers [8]	79	85	83	98	95	80	94	106	98	99
Electricity and water [4] Electricité et eau [4]	118	120	128	148	144	140	139	140	157	151
South Africa Afrique du Sud										
Total mining [2] Total, industries extractives [2]	104	105	102	97	100	98	96	95	96	98
Total manufacturing [3] Total, industries manufacturières [3]	108	105	104	108	113	114	112	107	104	104
Food, beverages, tobacco Aliments, boissons, tabac	115	117	118	121	122	127	131	128	130	123
Textiles Textiles	97	91	94	92	91	88	78	74	71	72
Chemicals Produits chimiques	106	104	106	104	112	115	118	116	111	111
Basic metals Métaux de base	89	92	89	88	100	106	99	89	82	80
Metal products Produits métalliques	104	96	91	95	103	104	100	96	92	90
Tunisia Tunisie										
Total industry [2−4] Total, industrie [2−4]	112	112	114	114	118	120	120	124	125	122
Total mining [2] Total, industries extractives [2]	93	80	98	106	107	117	108	108	103	88
Total manufacturing [3] Total, industries manufacturières [3]	118	121	123	125	134	136	142	139	142	144
Food, beverages, tobacco Aliments, boissons, tabac	118	125	132	133	143	144	147	150	157	157
Textiles Textiles	112	118	112	127	134	134	128	112	115	121
Chemicals and petroleum products Produits chimiques et pétroliers	114	114	127	134	150	157	163	166	165	168
Basic metals Métaux de base	100	96	112	107	107	105	108	111	107	105
Metal products Produits métalliques	133	135	103	94	101	109	125	125	120	118
Electricity and water [4] Electricité et eau [4]	124	124	122	118	117	118	112	124	126	117

26
Index numbers of industrial production [*cont.*]
Indices de la production industrielle [*suite*]
1980=100

Country or area and industry [SITC] Pays ou zone et industrie [CITI]	1984	1985	1986	1987	1988	1989	1990	1991	1992	1993
Zambia Zambie										
Total industry [2−4]										
Total, industrie [2−4]	**98**	**99**	**97**	**96**	**97**	**96**	**96**	**93**	**98**	...
Total mining [2] [9]										
Total, industries extractives [2] [9]	**90**	**87**	**86**	**84**	**81**	**82**	**79**	**72**	**80**	...
Total manufacturing [3] [10]										
Total, manufactures [3] [10]	**109**	**115**	**113**	**117**	**123**	**122**	**125**	**125**	**127**	...
Food, beverages, tobacco										
Aliments, boissons, tabac	102	102	100	104	113	108	125	128	159	...
Textiles and clothing										
Textiles, habillement	120	152	133	114	147	156	161	145	134	...
Basic metals										
Métaux de base	82	96	90	92	92	67	50	50	57	...
Electricity and water [4]										
Electricité et eau [4]	106	109	106	91	91	73	84	94	82	...
Zimbabwe Zimbabwe										
Total industry [2−4]										
Total, industrie [2−4]	**100**	**108**	**113**	**118**	**122**	**129**	**134**	**137**	**127**	**118**
Total mining [2] [9]										
Total, industries extractives [2] [9]	**97**	**97**	**99**	**103**	**102**	**107**	**108**	**109**	**107**	**104**
Total manufacturing [3] [10]										
Total, manufactures [3] [10]	**101**	**112**	**115**	**118**	**124**	**131**	**139**	**143**	**130**	**119**
Food, beverages, tobacco										
Aliments, boissons, tabac	106	106	112	121	124	124	138	142	144	124
Textiles										
Textiles	124	175	190	196	203	208	217	226	176	192
Chemicals and petroleum products										
Produits chimiques et pétroliers	113	122	122	119	131	146	159	159	138	129
Basic metals and metal products										
Métaux de base et produits métalliques	93	100	98	93	101	113	117	118	107	82
Electricity and water [4]										
Electricité et eau [4]	**97**	**111**	**132**	**171**	**177**	**208**	**202**	**186**	**180**	**159**
America, North · Amérique du Nord										
Barbados Barbade										
Total industry [2−4]										
Total, industrie [2−4]	**100**	**96**	**102**	**97**	**102**	**106**	**110**	**108**	**101**	**97**
Total mining [2]										
Total, industries extractives [2]	**154**	**166**	**156**	**144**	**135**	**127**	**134**	**128**	**128**	**129**
Total manufacturing [3]										
Total, industries manufacturières [3]	**92**	**87**	**93**	**87**	**93**	**97**	**101**	**98**	**90**	**86**
Food, beverages, tobacco										
Aliments, boissons, tabac	84	85	89	95	93	94	104	110	105	110
Wearing apparel										
Habillement	114	96	77	90	93	72	60	48	31	26
Chemicals and petroleum products										
Produits chimiques et pétroliers	89	88	99	93	105	116	120	114	95	99
Electricity and gas										
Electricité et gaz	**124**	**132**	**139**	**145**	**157**	**162**	**164**	**168**	**170**	**170**
Canada Canada										
Total industry [2−4]										
Total, industrie [2−4]	**109**	**115**	**115**	**120**	**125**	**128**	**128**	**126**	**127**	**130**
Total mining [2]										
Total, industries extractives [2]	**112**	**117**	**113**	**123**	**134**	**131**	**130**	**132**	**135**	**140**
Total manufacturing [3]										
Total, industries manufacturières [3]	**106**	**112**	**113**	**118**	**124**	**126**	**121**	**113**	**113**	**119**
Food, beverages, tobacco										
Aliments, boissons, tabac	101	106	104	105	106	103	103	101	103	104
Textiles										
Textiles	81	104	115	116	117	116	109	104	105	111
Paper and paper products										
Papier, produits en papier	98	97	102	108	109	104	101	98	97	101
Chemicals, coal, petroleum products										
Produits chimiques, houillers, pétroliers	114	114	116	125	133	139	141	132	134	138

26
Index numbers of industrial production [*cont.*]
Indices de la production industrielle [*suite*]
1980=100

Country or area and industry [SITC] Pays ou zone et industrie [CITI]	1984	1985	1986	1987	1988	1989	1990	1991	1992	1993
Basic metals										
Métaux de base	104	111	107	119	125	121	114	115	117	129
Metal products										
Produits métalliques	110	119	120	125	138	144	138	126	127	137
Electricity, gas and water [4]										
Electricité, gaz, eau [4]	112	120	123	127	129	128	122	127	129	131
Costa Rica [11] Costa Rica [11]										
Total industry [2−4]										
Total, industrie [2−4]	103	104	111	118	120	125	150	151	165	177
Total mining [2]										
Total, industries extractives [2]	87	90	88	84	87	98	97	106	129	...
Total manufacturing [3]										
Total, industries manufacturières [3]	99	101	108	114	117	121	124	127	140	149
Electricity, gas and water [4]										
Electricité, gaz et eau [4]	139	129	137	147	151	159	169	177	187	198
El Salvador [13] El Salvador [13]										
Total industry [2−4]										
Total, industrie [2−4]	87	91	93	96	98	101	104	109	116	...
Total mining [2]										
Total, industries extractives [2]	103	103	103	103	128	128	128	128	154	...
Total manufacturing [3]										
Total, industries manufacturières [3]	85	88	90	93	96	98	101	106	112	...
Food, beverages, tobacco										
Aliments, boissons, tabac	95	102	103	105	109	111	114	120	127	...
Textiles										
Textiles	55	47	56	59	53	57	59	62	67	...
Chemical products and petroleum										
Produits chimiques et pétroliers	100	91	94	96	96	104	106	112	117	...
Basic metals										
Métaux de base	91	103	120	122	124	122	134	134	145	...
Metal products										
Produits métalliques	63	67	68	71	72	70	74	77	83	...
Electricity, gas and water [4]										
Electricité, gaz et eau [4]	102	107	109	111	113	114	121	129	138	...
Honduras [14] Honduras [14]										
Total industry [2−4]										
Total, industrie [2−4]	113	116	119	123	132	138	140	142	151	160
Total mining [2]										
Total, industries extractives [2]	132	135	126	77	105	118	108	114	126	130
Total manufacturing [3]										
Total, industries manufacturières [3]	109	110	115	122	128	133	134	136	145	154
Food, beverages, tobacco										
Aliments, boissons, tabac	124	129	130	135	141	157	198	249	285	...
Textiles										
Textiles	141	165	158	141	149	170	232	324	372	...
Chemicals										
Produits chimiques	119	120	120	119	129	151	211	310	365	...
Electricity [4]										
Electricité [4]	129	148	158	185	208	217	246	248	250	267
Mexico Méxique										
Total industry [2−4] [15]										
Total, industrie [2−4] [15]	105	112	108	113	115	122	128	132	135	135
Total mining [2]										
Total, industries extractives [2]	128	129	126	131	131	131	136	137	139	139
Total manufacturing [3] [16]										
Total, manufactures [3] [16]	101	108	104	108	111	118	124	129	132	131
Food, beverages, tobacco										
Aliments, boissons, tabac	109	115	117	118	117	124	128	133	135	135
Textiles [3]										
Textiles [3]	93	98	92	92	96	97	95	91	91	88
Chemicals, coal, petroleum products [8]										
Produits chimiques, houillers, pétroliers [8]	113	119	116	121	123	132	140	148	153	148
Basic metals										

26
Index numbers of industrial production [*cont.*]
Indices de la production industrielle [*suite*]
1980=100

Country or area and industry [SITC] Pays ou zone et industrie [CITI]	1984	1985	1986	1987	1988	1989	1990	1991	1992	1993
Métaux de base	100	97	89	102	108	113	120	117	116	124
Metal products										
Produits métalliques	88	101	94	103	113	125	132	146	146	145
Electricity [4]										
Electricité [4]	126	135	140	148	156	169	178	185	195	202
Panama Panama										
Total mining [2]										
Total, industries extractives [2]	83	79	81	87	51	48	62	104	144	177
Total manufacturing [3]										
Total, industries manufacturières [3]	105	110	113	121	92	97	109	117	128	137
Food, beverages, tobacco										
Aliments, boissons, tabac	105	105	105	112	97	104	111	116	122	128
Textiles										
Textiles	74	93	97	96	68	95	110	112	99	100
Non−metallic mineral products										
Produits minéraux non−métalliques	88	89	108	116	46	45	62	96	120	152
Basic metals										
Métaux de base	58	62	70	107	36	34	58	85	98	122
Metal products										
Produits métalliques	104	120	128	145	76	75	104	99	105	120
Trinidad and Tobago Trinité−et−Tobago										
Total industry [2−4] [12] [68]										
Total, industrie [2−4] [12] [68]	103	99	121	125	120	121	124	138	151	141
Total manufacturing [3] [12]										
Total, industries manufacturières [3][12]	98	94	116	119	114	115	118	132	144	135
Food, beverages, tobacco										
Aliments, boissons, tabac	98	85	88	89	87	89	90	91	91	86
Textiles										
Textiles	55	43	53	38	29	33	42	38	39	24
Chemicals and petroleum products										
Produits chimiques et pétroliers	56	64	67	70	72	70	71	69	72	70
Metal products										
Produits métalliques	75	56	64	54	50	46	45	61	66	56
Electricity [4]										
Electricité [4]	154	154	171	181	183	181	188	198	210	208
United States Etats−Unis										
Total industry [2−4]										
Total, industrie [2−4]	110	112	113	119	124	126	126	124	128	133
Total mining [2]										
Total, industries extractives [2]	102	99	92	91	92	91	93	92	90	88
Total manufacturing [3]										
Total, industries manufacturières [3]	113	116	120	127	133	135	135	132	137	143
Food, beverages, tobacco										
Aliments, boissons, tabac	108	111	113	116	120	122	125	126	126	124
Textiles										
Textiles	100	99	105	112	112	114	112	112	118	119
Paper and paper products										
Papier, produits en papier	115	116	124	131	135	135	138	138	140	147
Chemicals, coal, petroleum products										
Produits chimiques, houillers, pétroliers	106	109	115	121	127	130	133	133	138	139
Basic metals										
Métaux de base	91	89	83	90	99	98	97	89	92	96
Metal products										
Produits métalliques	114	117	118	122	132	138	139	135	137	149
Electricity [4]										
Electricité [4]	101	104	100	104	109	112	113	114	114	121
America, South · Amérique du Sud										
Argentina Argentine										
Total industry [2−4] [1]										
Total, industrie [2−4] [1]	95	86	96	98	93	87	89	99	...	...
Total mining [2]										
Total, industries extractives [2]	97	94	87	93	100	103	109	103	...	...

26
Index numbers of industrial production [*cont.*]
Indices de la production industrielle [*suite*]
1980=100

Country or area and industry [SITC] Pays ou zone et industrie [CITI]	1984	1985	1986	1987	1988	1989	1990	1991	1992	1993
Total manufacturing [3]										
Total, industries manufacturières [3]	**94**	**85**	**95**	**96**	**92**	**85**	**87**	**97**	...	...
Food, beverages, tobacco										
Aliments, boissons, tabac	107	105	113	114	112	110	113	125	...	...
Textiles										
Textiles	107	91	110	104	106	100	103	120	...	...
Chemicals, coal, petroleum products										
Produits chimiques, houillers, pétroliers	99	89	101	104	103	94	102	109	...	...
Basic metals										
Métaux de base	100	77	99	110	114	103	109	112	...	...
Metal products										
Produits métalliques	82	74	77	82	75	63	59	70	...	...
Electricity and gas [4]										
Electricité et gaz [4]	**122**	**125**	**129**	**136**	**126**	**120**	**131**	**134**	...	...
Bolivia Bolivie										
Total industry [2−4]										
Total, industrie [2−4]	**76**	**67**	**65**	**66**	**73**	**79**	**87**	**92**	**95**	**101**
Total mining [2]										
Total, industries extractives [2]	**72**	**59**	**42**	**43**	**62**	**73**	**85**	**84**	**85**	**95**
Total manufacturing [3]										
Total, industries manufacturières [3]	**71**	**64**	**67**	**69**	**72**	**76**	**82**	**88**	**91**	**95**
Electricity, gas and water [4]										
Electricité, gaz et eau [4]	**141**	**133**	**134**	**126**	**132**	**135**	**144**	**156**	**168**	**185**
Brazil Brésil										
Total industry [2−4]										
Total, industrie [2−4]	**91**	**99**	**110**	**111**	**107**	**110**	**100**	**98**	**94**	**101**
Total mining [2]										
Total, industries extractives [2]	**158**	**176**	**182**	**181**	**182**	**189**	**194**	**196**	**197**	**198**
Total manufacturing [3] [18]										
Total, manufactures [3] [18]	**89**	**97**	**108**	**109**	**105**	**108**	**98**	**96**	**92**	**99**
Food, beverages, tobacco										
Aliments, boissons, tabac	104	106	108	114	112	115	117	124	122	124
Textiles										
Textiles	78	89	101	100	94	94	85	87	83	83
Chemicals and petroleum products										
Produits chimiques et pétroliers	114	121	123	130	126	125	115	106	106	110
Basic metals [19]										
Métaux de base [19]	89	95	106	107	104	109	95	90	89	96
Chile Chili										
Total mining [2] [20] [21]										
Total, industries extractives [2] [20] [21]	**125**	**129**	**130**	**131**	**137**	**148**	**150**	**167**	**177**	**184**
Total manufacturing [3] [22]										
Total, manufactures [3] [22]	**98**	**98**	**106**	**110**	**119**	**129**	**128**	**136**	**153**	**157**
Food, beverages, tobacco										
Aliments, boissons, tabac	119	116	125	121	128	140	134	136	157	163
Textiles										
Textiles	98	105	124	130	124	126	118	130	125	120
Chemicals and petroleum products										
Produits chimiques et pétroliers	95	93	103	108	120	134	133	140	154	162
Basic metals										
Métaux de base	110	107	111	114	122	129	131	128	139	136
Metal products										
Produits métalliques	64	68	75	86	92	104	105	106	134	149
Colombia Colombie										
Total industry [2−4]										
Total, industrie [2−4]	**110**	**110**	**127**	**139**	**141**	**145**	**151**	**150**	**158**	**162**
Total mining [2]										
Total, industries extractives [2]	**161**	**153**	**246**	**287**	**283**	**299**	**291**	**280**	**285**	**284**
Total manufacturing [3]										
Total, industries manufacturières [3]	**102**	**104**	**112**	**120**	**122**	**125**	**133**	**133**	**142**	**146**
Food, beverages, tobacco										
Aliments, boissons, tabac	106	114	116	119	118	118	130	123	130	136
Textiles										

26
Index numbers of industrial production [*cont.*]
Indices de la production industrielle [*suite*]
1980=100

Country or area and industry [SITC] Pays ou zone et industrie [CITI]	1984	1985	1986	1987	1988	1989	1990	1991	1992	1993
Textiles	100	104	115	125	118	112	115	115	123	121
Chemicals, coal, petroleum products Produits chimiques, houillers, pétroliers	125	134	152	164	163	165	168	174	157	163
Basic metals [10] Métaux de base [10]	106	106	116	134	143	141	145	140	156	163
Metal products Produits métalliques	95	87	96	104	117	109	115	105	119	140
Electricity, gas and water [4] **Electricité, gaz et eau [4]**	**119**	**117**	**118**	**126**	**137**	**139**	**142**	**147**	**151**	**155**
Ecuador Equateur **Total manufacturing [3]** **Total, industries manufacturières [3]**	**107**	**113**	**117**	**120**	**126**	**131**	**140**	**158**	**166**	**169**
Food, beverages, tobacco Aliments, boissons, tabac	92	96	99	102	102	105	110	121	121	119
Textiles Textiles	94	98	95	91	94	96	100	100	89	77
Chemicals, coal, petroleum products Produits chimiques, houillers, pétroliers	114	123	128	128	132	128	145	185	189	206
Basic metals Métaux de base	187	150	142	129	146	129	132	167	167	184
Metal products Produits métalliques	106	122	136	136	149	163	174	205	235	251
Paraguay Paraguay **Total manufacturing [3]** **Total, industries manufacturières [3]**	**98**	**102**	**101**	**106**	**114**	**148**	**102**	**78**	**81**	**75**
Food, beverages, tobacco Aliments, boissons, tabac	109	111	116	123	111	148	152	112	124	110
Textiles Textiles	114	169	115	95	196	215	224	106	77	104
Chemicals, coal, petroleum products Produits chimiques, houillers, pétroliers	66	74	67	76	87	78	81	65	62	84
Basic metals Métaux de base	251	271	254	560	622	615	509	220	217	244
Metal products Produits métalliques	184	212	192	190	181	168	145	144	140	113
Peru Pérou **Total manufacturing [3]** **Total, industries manufacturières [3]**	**84**	**89**	**105**	**120**	**104**	**84**	**81**	**86**	**81**	**...**
Food, beverages, tobacco Aliments, boissons, tabac	92	94	114	132	118	94	94	99	93	99
Textiles Textiles	98	113	121	134	126	116	100	97	84	90
Chemicals, coal, petroleum products Produits chimiques, houillers, pétroliers	79	83	103	124	111	78	78	84	79	89
Basic metals Métaux de base	94	99	96	100	78	86	75	89	89	94
Metal products Produits métalliques	45	54	81	101	70	42	45	50	43	40
Uruguay Uruguay **Total manufacturing [3] [23]** **Total, manufactures [3] [23]**	**76**	**75**	**84**	**93**	**91**	**89**	**98**	**88**	**78**	**73**
Food, beverages, tobacco Aliments, boissons, tabac	88	95	96	97	101	103	103	104	110	105
Textiles [3] Textiles [3]	86	85	99	106	101	96	100	108	109	98
Chemicals, coal, petroleum products Produits chimiques, houillers, pétroliers	77	75	83	97	100	99	102	100	96	76
Metal products Produits métalliques	46	48	62	85	75	69	69	68	64	60
Venezuela Venezuela **Total industry [2−4]** **Total, industrie [2−4]**	**156**	**181**	**217**	**310**	**392**	**...**	**...**	**...**	**...**	**...**

26
Index numbers of industrial production [*cont.*]
Indices de la production industrielle [*suite*]
1980=100

Country or area and industry [SITC] Pays ou zone et industrie [CITI]	1984	1985	1986	1987	1988	1989	1990	1991	1992	1993
Total mining [2]										
Total, industries extractives [2]	**91**	**118**	**177**	**319**	**454**	...	...	...	...	...
Total manufacturing [3] [24]										
Total, manufactures [3] [24]	**152**	**176**	**218**	**311**	**393**	...	...	...	...	...
Food, beverages, tobacco										
Aliments, boissons, tabac	161	192	219	282	350	624		...	...	...
Textiles										
Textiles	166	185	216	331	451	613		...	...	...
Chemicals, coal, petroleum products										
Produits chimiques, houillers, pétroliers	117	150	203	299	406	493		...	...	...
Basic metals [10]										
Métaux de base [10]	124	119	173	272	386	608		...	...	...
Metal products										
Produits métalliques	121	135	170	255	311	332		...		
Asia · Asie										
Bangladesh [25] **Bangladesh** [25]										
Total industry [2−4]										
Total, industrie [2−4]	**124**	**127**	**146**	**148**	**153**	**172**	**176**	**194**	**219**	**240**
Total mining [2]										
Total, industries extractives [2]	**178**	**199**	**236**	**277**	**295**	**315**	**325**	**359**	**394**	**386**
Total manufacturing [3]										
Total, industries manufacturières [3]	**122**	**126**	**144**	**145**	**148**	**167**	**171**	**189**	**214**	**236**
Food, beverages, tobacco										
Aliments, boissons, tabac	104	108	124	127	129	156	158	151	145	163
Textiles										
Textiles	88	77	89	88	87	91	80	80	82	78
Basic metals [10]										
Métaux de base [10]	70	55	50	45	51	57	44	34	36	64
Electricity [4]										
Electricité [4]	**172**	**180**	**210**	**246**	**285**	**304**	**326**	**350**	**363**	**379**
Cyprus Chypre										
Total industry [2−4]										
Total, industrie [2−4]	**118**	**116**	**119**	**131**	**142**	**147**	**154**	**156**	**162**	**151**
Total mining [2]										
Total, industries extractives [2]	**65**	**71**	**64**	**68**	**66**	**55**	**55**	**54**	**58**	**67**
Total manufacturing [3] [26]										
Total, manufactures [3] [26]	**118**	**115**	**118**	**130**	**138**	**144**	**151**	**152**	**156**	**140**
Food, beverages, tobacco										
Aliments, boissons, tabac	126	126	125	137	146	157	168	168	178	164
Textiles										
Textiles	101	95	94	102	124	113	125	118	136	124
Chemicals										
Produits chimiques	125	140	156	166	221	234	212	201	223	217
Metal products										
Produits métalliques	128	134	123	141	157	143	141	147	157	164
Electricity [4]										
Total, industries manufacturières [3]	**118**	**123**	**134**	**150**	**167**	**181**	**198**	**196**	**228**	**248**
India [25] **Inde** [25]										
Total industry [2−4]										
Total, industrie [2−4]	**128**	**140**	**149**	**165**	**178**	**188**	**209**	**213**	**221**	**223**
Total mining [2]										
Total, industries extractives [2]	**157**	**164**	**176**	**184**	**194**	**210**	**216**	**221**	**223**	**227**
Total manufacturing [3]										
Total, industries manufacturières [3]	**122**	**135**	**143**	**160**	**173**	**181**	**205**	**206**	**214**	**214**
Food, beverages, tobacco										
Aliments, boissons, tabac	120	125	129	129	136	137	153	160	162	156
Textiles										
Textiles	99	109	107	114	107	108	122	132	138	149
Chemicals and petroleum products										
Produits chimiques et du pétroliers	138	149	172	197	227	238	254	254	268	283
Basic metals										
Métaux de base	122	133	140	156	169	167	180	195	194	203

26
Index numbers of industrial production [*cont.*]
Indices de la production industrielle [*suite*]
1980=100

Country or area and industry [SITC] Pays ou zone et industrie [CITI]	1984	1985	1986	1987	1988	1989	1990	1991	1992	1993
Metal products										
Produits métalliques	128	146	157	194	209	227	279	264	275	255
Electricity [4]										
Électricité [4]	138	149	164	178	192	216	233	253	266	284
Indonesia Indonésie										
Total mining [2]										
Total, industries extractives [2]	87	82	86	77	82	88	91	100	97	97
Total manufacturing [3]										
Total, industries manufacturières [3]	121	130	145	162	185	208	236	262	291	324
Food, beverages, tobacco										
Aliments, boissons, tabac	114	116	132	147	168	177	204	182	186	221
Textiles										
Textiles	100	97	127	134	163	184	221	233	253	252
Chemicals										
Produits chimiques	145	169	175	163	144	174	187	184	222	282
Iran, Islamic Rep. of [27] Iran, Rép. islamique d' [27]										
Total manufacturing [3] [28] [29]										
Total, manufactures [3] [28] [29]	151	152	121	113	104	109	130	160	170	...
Food, beverages, tobacco										
Aliments, boissons, tabac	120	125	110	104	98	97	111	121	136	...
Textiles										
Textiles	146	148	135	118	105	96	103	117	123	...
Chemicals										
Produits chimiques	140	141	140	131	123	137	186	189	195	...
Non−metallic mineral products										
Produits minéraux non−métalliques	141	145	138	144	136	142	162	180	192	...
Israel Israël										
Total industry [2−4]										
Total, industrie [2−4]	116	120	124	130	126	124	132	141	153	...
Total mining [2] [20]										
Total, industries extractives [2] [20]	103	108	118	126	115	118	125	136	142	...
Total manufacturing [3]										
Total, industries manufacturières [3]	116	120	124	130	126	124	132	141	153	...
Food and beverages										
Aliments et boissons	128	125	144	161	161	157	159	161	...	...
Textiles										
Textiles	103	99	98	99	92	92	97	104	...	...
Chemicals and petroleum products										
Produits chimiques et du pétroliers	128	128	131	146	148	156	163	170	...	...
Basic metals										
Métaux de base	99	85	83	86	90	85	106	118	...	...
Metal products										
Produits métalliques	118	123	111	122	118	114	120	128	...	...
Japan Japon										
Total industry [2−4]										
Total, industrie [2−4]	114	118	118	122	134	142	148	150	141	134
Total mining [2]										
Total, industries extractives [2]	94	94	94	86	82	79	75	77	76	75
Total manufacturing [3] [31]										
Total, manufactures [3] [31]	114	119	118	122	134	142	148	150	141	135
Food and beverages										
Aliments et boissons	101	101	102	103	106	108	108	109	109	108
Textiles										
Textiles	99	98	94	92	93	92	90	88	85	76
Chemicals and petroleum products										
Produits chimiques et du pétroliers	107	107	107	111	120	127	134	136	136	134
Basic metals										
Métaux de base	99	101	96	100	108	111	115	116	107	104
Metal products										
Produits métalliques	134	142	143	149	169	182	193	200	181	172
Electricity and gas [4]										
Électricité et gaz [4]	116	120	120	126	132	139	149	155	158	159

26
Index numbers of industrial production [cont.]
Indices de la production industrielle [suite]
1980=100

Country or area and industry [SITC] Pays ou zone et industrie [CITI]	1984	1985	1986	1987	1988	1989	1990	1991	1992	1993
Jordan Jordanie										
Total industry [2−4]										
Total, industrie [2−4]	**152**	**155**	**157**	**172**	**158**	**166**	**172**	**167**	**178**	**180**
Total mining [2] [67]										
Total, industries extractives [2] [67]	**159**	**155**	**160**	**175**	**144**	**170**	**147**	**114**	**110**	**108**
Total manufacturing [3]										
Total, industries manufacturières [3]	**145**	**145**	**150**	**168**	**152**	**160**	**167**	**161**	**175**	**178**
Food and beverages										
Aliments et boissons	112	108	80	84	95	93	98	94	133	137
Textiles										
Textiles	112	149	132	149	156	133	142	123	106	117
Chemicals and petroleum products										
Produits chimiques et du pétroliers	140	135	139	145	135	154	165	149	162	156
Basic metals										
Métaux de base	130	155	162	172	152	138	137	153	183	144
Electricity and gas [4]										
Electricité et gaz [4]	197	216	266	313	290	261	330	341	408	445
Korea, Rep. of Corée, Rép. de										
Total industry [2−4]										
Total, industrie [2−4]	**158**	**165**	**199**	**238**	**269**	**278**	**302**	**329**	**346**	**366**
Total mining [2] [32]										
Total, industries extractives [2] [32]	**106**	**116**	**124**	**126**	**126**	**112**	**102**	**102**	**88**	**82**
Total manufacturing [3]										
Total, industries manufacturières [3]	**161**	**168**	**205**	**246**	**279**	**288**	**314**	**342**	**360**	**380**
Food, beverages, tobacco										
Aliments, boissons, tabac	140	150	164	183	205	219	232	251	257	261
Textiles										
Textiles	124	124	149	166	171	166	165	162	156	142
Chemicals and petroleum products										
Produits chimiques et du pétroliers	139	146	164	192	237	259	295	346	396	394
Basic metals										
Métaux de base	181	189	210	239	259	282	316	350	366	407
Metal products										
Produits métalliques	261	268	342	429	504	497	571	628	665	685
Electricity [4]										
Electricité [4]	**144**	**156**	**174**	**198**	**230**	**254**	**289**	**322**	**360**	**401**
Malaysia Malaisie										
Total industry [2−4]										
Total, industrie [2−4]	**142**	**138**	**152**	**164**	**188**	**210**	**236**	**262**	**285**	**312**
Total mining [2]										
Total, industries extractives [2]	**163**	**163**	**189**	**194**	**218**	**235**	**247**	**259**	**266**	**268**
Total manufacturing [3] [33]										
Total, manufactures [3] [33]	**130**	**122**	**130**	**147**	**169**	**193**	**223**	**254**	**281**	**317**
Food, beverages, tobacco										
Aliments, boissons, tabac	115	123	129	136	145	164	175	170	178	190
Textiles										
Textiles	132	128	144	170	189	218	228	228	239	241
Chemicals										
Produits chimiques	105	111	126	143	148	154	157	202	187	197
Basic metals										
Métaux de base	165	131	102	125	166	198	226	256	305	344
Metal products										
Produits métalliques	181	149	183	218	269	317	418	528	592	704
Electricity [4]										
Electricité [4]	133	143	157	169	188	210	239	270	308	347
Mongolia Mongolie										
Total industry [2−4]										
Total, industrie [2−4]	**145**	**156**	**168**	**176**	**182**	**187**	**183**	**156**	...	...
Food, beverages, tobacco										
Aliments, boissons, tabac	140	137	142	144	148	152	139	152	...	...
Textiles										
Textiles	201	227	240	246	244	276	248	197	...	...

26
Index numbers of industrial production [*cont.*]
Indices de la production industrielle [*suite*]
1980=100

Country or area and industry [SITC] Pays ou zone et industrie [CITI]	1984	1985	1986	1987	1988	1989	1990	1991	1992	1993
Chemicals Produits chimiques	126	130	150	162	192	198	187	162	...	...
Metal products Produits métalliques	140	159	165	184	186	186	158	146	...	...
Electricity [4] **Electricité [4]**	**144**	**181**	**202**	**214**	**217**	**224**	**236**	**211**	...	...
Pakistan [25] Pakistan [25]										
Total industry [2−4] **Total, industrie [2−4]**	**144**	**160**	**170**	**186**	**191**	**204**	**218**	**232**	**238**	...
Total mining [2] **Total, industries extractives [2]**	**147**	**183**	**187**	**212**	**218**	**250**	**275**	**278**	**278**	**274**
Total manufacturing [3] **Total, industries manufacturières [3]**	**143**	**154**	**165**	**179**	**183**	**192**	**206**	**218**	**228**	...
Philippines Philippines										
Total industry [2−4] **Total, industrie [2−4]**	**193**	**231**	**258**	**296**	**358**	**408**	**451**	**515**	**540**	**571**
Total mining [2] **Total, industries extractives [2]**	**75**	**85**	**71**	**87**	**92**	**125**	**157**	**195**	**262**	**197**
Total manufacturing [3] [2] **Total, industries manufacturières [3] [2]**	**207**	**241**	**291**	**333**	**406**	**459**	**506**	**577**	**597**	**643**
Food, beverages, tobacco Aliments, boissons, tabac	233	272	264	288	343	379	411	468	504	521
Textiles Textiles	178	166	221	256	274	288	289	322	317	338
Non−metallic mineral products Produits minéraux non−métalliques	172	173	182	213	256	310	355	468	474	544
Metal products Produits métalliques	231	357	598	669	826	953	1072	1198	1326	1461
Electricity [4] **Electricité [4]**	**126**	**126**	**130**	**141**	**148**	**152**	**154**	**158**	**159**	**165**
Singapore Singapour										
Total manufacturing [3] **Total, industries manufacturières [3]**	**116**	**107**	**116**	**136**	**162**	**178**	**196**	**206**	**211**	**232**
Food, beverages, tobacco Aliments, boissons, tabac	87	90	93	96	112	132	132	140	150	154
Textiles Textiles	42	30	30	35	37	39	36	39	34	30
Chemicals, coal, petroleum products Produits chimiques, houillers, pétroliers	137	136	145	143	149	169	198	216	211	232
Basic metals Métaux de base	125	123	124	129	136	139	151	144	150	163
Metal products Produits métalliques	119	103	108	132	158	182	181	196	196	198
Sri Lanka Sri Lanka										
Total manufacturing [3] [34] **Total, manufactures [3] [34]**	**100**	**144**	**155**	**141**	**138**	**147**	**175**	**224**	**223**	...
Food, beverages, tobacco Aliments, boissons, tabac	59	96	115	102	92	109	114	171	147	...
Textiles and clothing Textiles, habillement	168	177	188	218	282	364	487	661	918	...
Chemicals and rubber products Produits chimiques et en caoutchouc	100	131	116	104	108	84	115	119	122	...
Basic metals Métaux de base	37	61	84	73	102	94	142	159	122	...
Metal products Produits métalliques	76	60	55	80	78	102	132	115	120	...
Syrian Arab Rep. Rép. arabe syrienne										
Total industry [2−4] **Total, industrie [2−4]**	**143**	**136**	**141**	**151**	**151**	**170**	**186**	**199**	**212**	**216**
Total mining [2] **Total, industries extractives [2]**	**98**	**101**	**125**	**143**	**181**	**229**	**270**	**310**	**338**	**343**
Total manufacturing [3] **Total, industries manufacturières [3]**	**178**	**163**	**161**	**165**	**147**	**153**	**163**	**165**	**174**	**179**

26
Index numbers of industrial production [*cont.*]
Indices de la production industrielle [*suite*]
1980=100

Country or area and industry [SITC] Pays ou zone et industrie [CITI]	1984	1985	1986	1987	1988	1989	1990	1991	1992	1993
Food, beverages, tobacco										
Aliments, boissons, tabac	175	147	150	131	121	123	136	148	150	142
Textiles										
Textiles	141	118	133	118	92	102	100	93	107	108
Non−metallic mineral products										
Produits minéraux non−métalliques	191	191	191	174	145	171	134	141	155	171
Electricity and water [4]										
Electricité et eau [4]	**170**	**183**	**172**	**191**	**224**	**237**	**258**	**276**	**255**	**289**
Turkey Turquie										
Total industry [2−4]										
Total, industrie [2−4]	**141**	**149**	**166**	**184**	**187**	**194**	**212**	**217**	**228**	**247**
Total mining [2]										
Total, industries extractives [2]	**124**	**138**	**153**	**161**	**152**	**171**	**182**	**200**	**194**	**177**
Total manufacturing [3]										
Total, industries manufacturières [3]	**145**	**152**	**168**	**186**	**188**	**191**	**210**	**214**	**223**	**243**
Food, beverages, tobacco										
Aliments, boissons, tabac	158	154	151	165	170	179	188	203	200	217
Textiles										
Textiles	162	173	190	209	216	226	231	210	221	232
Chemicals, coal, petroleum products										
Produits chimiques, houillers, pétroliers	170	176	211	249	244	251	261	256	261	267
Basic metals										
Métaux de base	168	188	220	244	241	247	288	264	279	308
Metal products										
Produits métalliques	195	206	225	245	233	228	298	330	366	441
Electricity, gas and water [4]										
Electricité, gaz et eau [4]	**147**	**162**	**170**	**190**	**206**	**222**	**246**	**256**	**287**	**316**
Europe · Europe										
Austria Autriche										
Total industry [2−4]										
Total, industrie [2−4]	**104**	**109**	**110**	**112**	**116**	**123**	**132**	**134**	**133**	**130**
Total mining [2] [35]										
Total, industries extractives [2] [35]	**113**	**110**	**108**	**112**	**101**	**100**	**104**	**94**	**84**	**77**
Total manufacturing [3]										
Total, industries manufacturières [3]	**104**	**109**	**110**	**110**	**116**	**124**	**133**	**136**	**135**	**131**
Food, beverages, tobacco										
Aliments, boissons, tabac	109	112	115	116	118	124	135	143	145	146
Textiles										
Textiles	89	91	91	86	90	94	98	99	96	85
Chemicals and petroleum products										
Produits chimiques et pétroliers	121	116	113	118	131	136	138	139	140	136
Basic metals										
Métaux de base	106	107	102	102	112	118	115	108	105	101
Metal products										
Produits métalliques	106	118	123	119	127	140	157	162	160	156
Electricity [4]										
Electricité [4]	**106**	**110**	**110**	**123**	**119**	**123**	**125**	**128**	**126**	**131**
Belgium Belgique										
Total industry [2−4]										
Total, industrie [2−4]	**102**	**104**	**105**	**107**	**114**	**118**	**122**	**119**	**119**	**113**
Total mining [2] [37]										
Total, industries extractives [2] [37]	**90**	**84**	**75**	**65**	**56**	**45**	**32**	**23**	**15**	**7**
Total manufacturing [3] [31]										
Total, manufactures [3] [31]	**102**	**104**	**105**	**108**	**115**	**120**	**125**	**123**	**124**	**118**
Food, beverages, tobacco										
Aliments, boissons, tabac	115	118	120	124	127	118	139	143	144	145
Textiles										
Textiles	104	104	105	105	106	107	105	95	100	90
Chemicals and petroleum products										
Produits chimiques et pétroliers	109	112	117	119	126	118	132	132	144	139
Basic metals										
Métaux de base	93	92	86	88	99	105	101	99	84	81

26
Index numbers of industrial production [*cont.*]
Indices de la production industrielle [*suite*]
1980=100

Country or area and industry [SITC] Pays ou zone et industrie [CITI]	1984	1985	1986	1987	1988	1989	1990	1991	1992	1993
Metal products										
Produits métalliques	101	105	106	106	113	114	122	121	115	110
Electricity [4]										
Electricité [4]	**102**	**107**	**110**	**118**	**122**	**126**	**133**	**135**	**135**	**132**
Bulgaria Bulgarie										
Total industry [2−4]										
Total, industrie [2−4]	**120**	**124**	**129**	**134**	**141**	**139**	**116**	**90**	**76**	**74**
Total mining [2] [9] [38]										
Total, industries extractives [2] [9] [38]	**106**	**105**	**109**	**109**	**108**	**108**	**96**	**74**	**74**	**77**
Total manufacturing [3] [10] [39]										
Total, manufactures [3] [10] [39]	**120**	**124**	**130**	**135**	**142**	**140**	**116**	**89**	**74**	**64**
Food, beverages, tobacco										
Aliments, boissons, tabac	104	107	107	107	110	111	100	80	...	...
Textiles										
Textiles	115	112	114	121	125	132	130	89	...	...
Chemicals										
Produits chimiques	149	153	164	162	167	168	142	147	...	...
Basic metals										
Métaux de base	118	117	120	120	123	119	70	38	...	...
Metal products										
Produits métalliques	137	153	167	183	200	196	149	106	...	...
Electricity and steam [4]										
Electricité et vapeur [4]	**132**	**124**	**127**	**133**	**140**	**137**	**124**	**116**	**102**	**103**
former Czechoslovakia † ancienne Tchécoslovaquie †										
Total industry [2−4]										
Total, industrie [2−4]	**111**	**115**	**119**	**121**	**124**	**125**	**120**	**90**	**79**	...
Total mining [2]										
Total, industries extractives [2]	**102**	**101**	**101**	**102**	**103**	**100**	**91**	**77**	**69**	...
Total manufacturing [3] [39]										
Total, manufactures [3] [39]	**111**	**116**	**119**	**122**	**125**	**126**	**121**	**89**	**74**	...
Food, beverages, tobacco										
Aliments, boissons, tabac	107	107	109	110	110	113	111	96	87	...
Textiles										
Textiles	108	111	114	115	118	120	120	78	67	...
Chemicals, coal, petroleum products										
Produits chimiques, houillers, pétroliers	107	112	117	120	123	123	128	94	88	...
Basic metals										
Métaux de base	102	104	105	107	108	108	107	81	69	...
Metal products										
Produits métalliques	131	139	146	152	156	156	150	102	75	...
Electricity, gas and water [4]										
Electricité, gaz et eau [4]	**108**	**112**	**118**	**120**	**122**	**124**	**123**	**118**	**119**	...
Denmark Danemark										
Total industry [2−4]										
Total, industrie [2−4]	**116**	**121**	**129**	**125**	**127**	**130**	**131**	**133**	**136**	**132**
Total mining [2]										
Total, industries extractives [2]	**100**	**110**	**138**	**122**	**124**	**132**	**116**	**113**	**130**	**116**
Total manufacturing [3] [40]										
Total, manufactures [3] [40]	**117**	**122**	**131**	**127**	**129**	**133**	**133**	**135**	**138**	**134**
Food, beverages, tobacco										
Aliments, boissons, tabac	113	117	122	120	123	122	126	135	134	134
Textiles										
Textiles	117	116	120	116	111	113	106	106	105	91
Chemicals, coal, petroleum products										
Produits chimiques, houillers, pétroliers	121	127	134	134	137	137	138	139	148	148
Basic metals										
Métaux de base	101	97	93	87	91	103	96	93	92	79
Metal products										
Produits métalliques	116	125	136	129	131	138	139	141	143	137
Electricity [4]										
Electricité [4]	**83**	**107**	**113**	**108**	**103**	**84**	**103**	**143**	**120**	**125**

26
Index numbers of industrial production [*cont.*]
Indices de la production industrielle [*suite*]
1980=100

Country or area and industry [SITC] Pays ou zone et industrie [CITI]	1984	1985	1986	1987	1988	1989	1990	1991	1992	1993
Finland Finlande										
Total industry [2−4]										
Total, industrie [2−4]	**112**	**116**	**118**	**124**	**129**	**133**	**132**	**120**	**122**	**128**
Total mining [2]										
Total, industries extractives [2]	**119**	**123**	**127**	**124**	**134**	**146**	**143**	**132**	**127**	**123**
Total manufacturing [3]										
Total, industries manufacturières [3]	**112**	**116**	**118**	**124**	**128**	**133**	**132**	**118**	**121**	**127**
Food, beverages, tobacco										
Aliments, boissons, tabac	109	110	114	116	120	122	122	121	121	125
Textiles										
Textiles	84	79	75	78	73	70	64	52	53	55
Paper and paper products										
Papier, produits en papier	114	113	116	121	129	130	132	128	133	144
Chemicals, coal, petroleum products										
Produits chimiques, houillers, pétroliers	107	110	109	118	125	129	132	128	130	129
Basic metals										
Métaux de base	116	120	121	125	132	136	139	138	154	168
Metal products										
Produits métalliques	121	132	135	146	152	163	161	134	142	155
Electricity, gas and water [4]										
Electricité, gaz et eau [4]	**111**	**121**	**121**	**130**	**133**	**134**	**136**	**141**	**141**	**147**
France France										
Total industry [2−4]										
Total, industrie [2−4]	**100**	**101**	**101**	**103**	**108**	**112**	**114**	**114**	**114**	**111**
Total mining [2]										
Total, industries extractives [2]	**90**	**86**	**81**	**81**	**77**	**75**	**72**	**72**	**67**	**64**
Total manufacturing [3] [43]										
Total, manufactures [3] [43]	**97**	**97**	**98**	**99**	**104**	**108**	**110**	**108**	**108**	**104**
Food, beverages, tobacco										
Aliments, boissons, tabac	106	106	105	108	111	113	118	120	122	122
Textiles										
Textiles	89	86	84	81	80	80	77	72	71	66
Chemicals, coal, petroleum products										
Produits chimiques, houillers, pétroliers	108	109	109	112	119	125	126	128	134	135
Basic metals										
Métaux de base	84	82	80	80	86	88	88	85	84	76
Metal products										
Produits métalliques	93	92	92	94	101	107	110	107	103	95
Electricity and gas [4]										
Electricité et gaz [4]	**122**	**130**	**136**	**142**	**145**	**150**	**154**	**168**	**171**	**175**
Germany † Allemagne †										
Germany, Fed. Rep. of Allemagne, Rép. féd. d'										
Total industry [2−4]										
Total, industrie [2−4] [47]	**99**	**104**	**107**	**107**	**111**	**117**	**123**	**126**	**124**	**115**
Total mining [2] [47]										
Total, industries extractives [2] [47]	**92**	**93**	**91**	**87**	**85**	**84**	**84**	**83**	**81**	**72**
Total manufacturing [3] [47]										
Total, manufactures [3] [47]	**99**	**104**	**107**	**107**	**112**	**118**	**124**	**128**	**125**	**116**
Food, beverages, tobacco										
Aliments, boissons, tabac	102	105	108	109	113	116	129	138	137	136
Textiles										
Textiles	91	94	95	94	92	93	95	94	87	77
Chemicals, coal, petroleum products										
Produits chimiques, houillers, pétroliers	103	104	103	104	111	112	115	117	120	118
Basic metals										
Métaux de base	93	97	92	90	99	102	98	97	91	83
Metal products										
Produits métalliques	101	111	116	117	120	130	137	141	136	121
Electricity and gas [4]										
Electricité et gaz [4]	**107**	**112**	**111**	**115**	**118**	**121**	**125**	**128**	**129**	**127**
former German Dem. Rep. ancienne Rép. dém. allemande										
Total industry [2−4]										
Total, industrie [2−4]	**114**	**119**	**123**	**126**	**130**	**133**	**94**	**...**	**...**	**...**

26
Index numbers of industrial production [*cont.*]
Indices de la production industrielle [*suite*]
1980=100

Country or area and industry [SITC] Pays ou zone et industrie [CITI]	1984	1985	1986	1987	1988	1989	1990	1991	1992	1993
Total mining [2] [44] **Total, industries extractives** [2] [44]	**115**	**118**	**120**	**119**	**120**	**116**	**82**	...	...	...
Total manufacturing [3] [44] [45] **Total, manufactures** [3] [44] [45]	**114**	**119**	**123**	**127**	**131**	**135**	**95**	...	...	...
Food, beverages, tobacco Aliments, boissons, tabac	107	110	113	113	114	115	66	...	...	...
Textiles Textiles	107	109	111	114	117	119	74	...	...	...
Chemicals, coal, petroleum products Produits chimiques, houillers, pétroliers	108	112	113	114	119	122	65	...	...	...
Basic metals Métaux de base	108	112	113	117	117	116	69	...	...	...
Metal products Produits métalliques	125	134	141	149	158	165	75	...	...	...
Electricity and gas [4] **Electricité et gaz** [4]	114	118	121	122	123	123	93	...	...	...
Greece Grèce										
Total industry [2−4] **Total, industrie** [2−4]	**104**	**107**	**107**	**106**	**111**	**113**	**110**	**109**	**108**	**105**
Total mining [2] [48] **Total, industries extractives** [2] [48]	**178**	**183**	**185**	**182**	**189**	**180**	**174**	**172**	**161**	**150**
Total manufacturing [3] [49] **Total, manufactures** [3] [49]	**98**	**101**	**100**	**98**	**103**	**106**	**103**	**102**	**100**	**97**
Food, beverages, tobacco Aliments, boissons, tabac	110	116	111	102	112	120	114	121	128	128
Textiles Textiles	92	95	102	104	101	99	95	86	79	74
Chemicals, coal, petroleum products Produits chimiques, houillers, pétroliers	110	115	115	117	125	132	133	124	126	126
Basic metals Métaux de base	90	88	86	85	95	96	93	96	92	86
Metal products Produits métalliques	88	84	87	79	83	82	83	85	88	80
Electricity and gas [4] **Electricité et gaz** [4]	**109**	**123**	**126**	**136**	**144**	**151**	**155**	**152**	**163**	**167**
Hungary Hongrie										
Total industry [2−4] **Total, industrie** [2−4]	**109**	**110**	**112**	**116**	**116**	**111**	**102**	**82**	**74**	**76**
Total mining [2] [44] **Total, industries extractives** [2] [44]	**97**	**99**	**99**	**99**	**95**	**88**	**78**	**69**	**56**	**55**
Total manufacturing [3] [44] [50] **Total, manufactures** [3] [44] [50]	**110**	**110**	**112**	**117**	**117**	**111**	**101**	**76**	**63**	**65**
Food, beverages, tobacco Aliments, boissons, tabac	111	109	111	114	111	112	109	106	92	89
Textiles Textiles	100	101	102	104	107	103	85	58	43	38
Chemicals, coal, petroleum products Produits chimiques, houillers, pétroliers	119	125	128	134	142	135	120	96	81	81
Basic metals Métaux de base	106	107	110	111	116	124	107	71	53	49
Metal products Produits métalliques	115	119	123	129	130	132	110	89	64	74
Electricity, gas and water [4] **Electricité, gaz et eau** [4]	**117**	**120**	**123**	**130**	**129**	**130**	**132**	**125**	**104**	**101**
Ireland Irlande										
Total industry [2−4] [51] **Total, industrie** [2−4] [51]	**124**	**128**	**131**	**143**	**158**	**176**	**184**	**190**	**208**	**220**
Total mining [2] **Total, industries extractives** [2]	**108**	**72**	**79**	**85**	**69**	**89**	**84**	**77**	**71**	**83**
Total manufacturing [3] [28] **Total, manufactures** [3] [28]	**125**	**131**	**135**	**149**	**167**	**187**	**196**	**202**	**222**	**234**
Food, beverages, tobacco Aliments, boissons, tabac	110	117	121	132	139	145	149	156	167	173

26
Index numbers of industrial production [*cont.*]
Indices de la production industrielle [*suite*]
1980=100

Country or area and industry [SITC] Pays ou zone et industrie [CITI]	1984	1985	1986	1987	1988	1989	1990	1991	1992	1993
Textiles										
Textiles	96	93	92	95	99	103	111	111	117	119
Chemicals										
Produits chimiques	145	166	165	172	200	240	247	301	353	388
Basic metals										
Métaux de base	121	130	120	120	136	150	158	141	128	134
Metal products										
Produits métalliques	157	158	166	201	244	282	300	291	326	345
Electricity, gas and water [4]										
Electricité, gaz et eau [4]	125	133	127	124	128	135	144	155	160	169
Italy Italie										
Total industry [2−4]										
Total, industrie [2−4]	**95**	**97**	**100**	**104**	**110**	**114**	**114**	**113**	**112**	**110**
Total mining [2]										
Total, industries extractives [2]	**99**	**100**	**105**	**118**	**130**	**128**	**130**	**128**	**124**	**129**
Total manufacturing [3] [52]										
Total, manufactures [3] [52]	**95**	**96**	**100**	**103**	**109**	**113**	**112**	**111**	**110**	**108**
Food, beverages, tobacco										
Aliments, boissons, tabac	101	105	109	115	119	120	121	122	126	126
Textiles										
Textiles	96	97	101	105	106	112	110	111	110	105
Chemicals, coal, petroleum products										
Produits chimiques, houillers, pétroliers	90	90	93	96	98	98	99	94	96	93
Basic metals										
Métaux de base	92	93	93	96	104	108	104	104	104	100
Metal products										
Produits métalliques	97	101	110	113	122	128	127	119	113	106
Electricity and gas [4]										
Electricité et gaz [4]	100	103	107	113	116	121	125	128	130	129
Luxembourg Luxembourg										
Total industry [2−4]										
Total, industrie [2−4]	**113**	**121**	**124**	**122**	**133**	**144**	**143**	**143**	**142**	**139**
Total mining [2]										
Total, industries extractives [2]	**32**	**29**	**29**	**33**	**36**	**40**	**44**	**49**	**48**	**47**
Total manufacturing [3] [53]										
Total, manufactures [3] [53]	**114**	**122**	**124**	**123**	**134**	**144**	**143**	**143**	**142**	**139**
Food, beverages, tobacco										
Aliments, boissons, tabac	123	129	134	130	128	133	138	144	143	135
Chemicals										
Produits chimiques	118	143	204	269	268	308	280	300	434	498
Basic metals [10]										
Métaux de base [10]	94	102	97	90	102	104	100	96	89	89
Metal products										
Produits métalliques	139	151	155	147	161	176	183	190	179	171
Electricity and gas [4]										
Electricité et gaz [4]	121	126	132	138	148	154	158	164	160	163
Malta Malte										
Total industry [2−4] [1]										
Total, industrie [2−4] [1]	**118**	**120**	**127**	**131**	**142**	**167**	**189**	**216**	...	...
Total mining [2]										
Total, industries extractives [2]	**86**	**64**	**84**	**95**	**91**	**170**	**195**	**294**	...	...
Total manufacturing [3]										
Total, industries manufacturières [3]	**113**	**118**	**128**	**132**	**144**	**169**	**191**	**217**	...	...
Food, beverages, tobacco										
Aliments, boissons, tabac	143	146	152	167	178	186	182	226	...	...
Metal products [54]										
Produits métalliques [54]	125	135	153	165	214	277	352	391	...	...
Electricity, gas and water [4]										
Electricité, gaz et eau [4]	125	132	143	160	176	188	202	226	...	...
Netherlands Pays−Bas										
Total industry [2−4]										
Total, industrie [2−4]	**102**	**106**	**106**	**107**	**110**	**114**	**117**	**120**	**120**	**119**

26
Index numbers of industrial production [*cont.*]
Indices de la production industrielle [*suite*]
1980=100

Country or area and industry [SITC] Pays ou zone et industrie [CITI]	1984	1985	1986	1987	1988	1989	1990	1991	1992	1993
Total mining [2]										
Total, industries extractives [2]	**86**	**93**	**87**	**90**	**79**	**82**	**83**	**92**	**92**	**93**
Total manufacturing [3]										
Total, industries manufacturières [3]	**106**	**109**	**112**	**112**	**117**	**122**	**125**	**126**	**128**	**125**
Food, beverages, tobacco										
Aliments, boissons, tabac	108	105	112	112	113	119	125	130	129	126
Textiles										
Textiles	98	100	95	96	98	100	104	99	93	92
Chemicals, coal, petroleum products [55]										
Produits chimiques, houillers, pétroliers [55]	122	130	134	135	147	150	155	151	149	150
Basic metals										
Métaux de base	108	110	108	99	108	114	107	109	110	109
Metal products										
Produits métalliques	105	110	110	112	112	119	126	126	125	120
Electricity, gas and water [4]										
Electricité, gaz et eau [4]	**105**	**108**	**112**	**108**	**111**	**111**	**112**	**118**	**119**	**118**
Norway Norvège										
Total industry [2−4]										
Total, industrie [2−4]	**121**	**125**	**128**	**137**	**141**	**155**	**157**	**161**	**171**	**177**
Total mining [2]										
Total, industries extractives [2]	**132**	**139**	**148**	**165**	**181**	**230**	**242**	**267**	**297**	**317**
Total manufacturing [3]										
Total, industries manufacturières [3]	**108**	**110**	**111**	**112**	**110**	**111**	**111**	**109**	**111**	**113**
Food, beverages, tobacco										
Aliments, boissons, tabac	94	92	92	93	92	93	91	94	94	95
Textiles										
Textiles	72	76	77	72	63	58	59	59	57	55
Paper and paper products										
Papier, produits en papier	133	137	136	135	136	145	143	142	139	148
Chemicals, coal, petroleum products										
Produits chimiques, houillers, pétroliers	184	182	178	189	192	203	230	219	224	234
Basic metals										
Métaux de base	123	123	121	127	136	138	138	136	137	138
Metal products										
Produits métalliques	98	102	104	105	101	102	102	100	106	109
Electricity and gas [4]										
Electricité et gaz [4]	**121**	**118**	**111**	**119**	**125**	**136**	**139**	**127**	**134**	**137**
Poland Pologne										
Total industry [2−4]										
Total, industrie [2−4]	**95**	**99**	**103**	**106**	**111**	**110**	**81**	**70**	**72**	**76**
Total mining [2]										
Total, industries extractives [2]	**102**	**104**	**104**	**106**	**105**	**106**	**78**	**72**	**72**	**69**
Total manufacturing [3] [56]										
Total, manufactures [3] [56]	**94**	**97**	**102**	**105**	**111**	**108**	**80**	**70**	**71**	**78**
Food, beverages, tobacco [57]										
Aliments, boissons, tabac [57]	90	94	98	100	101	94	70	69	75	85
Textiles										
Textiles	83	88	89	90	98	102	61	49	48	53
Chemicals, coal, petroleum products										
Produits chimiques, houillers, pétroliers	95	98	102	106	112	107	79	73	73	79
Basic metals										
Métaux de base	89	89	91	90	93	89	69	54	51	51
Metal products										
Produits métalliques	100	106	114	122	132	131	100	75	75	85
Electricity [4]										
Electricité [4]	**121**	**128**	**136**	**135**	**135**	**134**	**122**	**115**	**108**	**94**
Portugal [58] **Portugal** [58]										
Total industry [2−4]										
Total, industrie [2−4]	**117**	**119**	**128**	**133**	**138**	**144**	**161**	**161**	**158**	**153**
Total mining [2]										
Total, industries extractives [2]	**100**	**109**	**98**	**86**	**95**	**237**	**422**	**475**	**516**	**398**
Total manufacturing [3] [31]										
Total, manufactures [3] [31]	**117**	**118**	**118**	**135**	**138**	**142**	**150**	**147**	**142**	**138**

26
Index numbers of industrial production [*cont.*]
Indices de la production industrielle [*suite*]
1980=100

Country or area and industry [SITC] Pays ou zone et industrie [CITI]	1984	1985	1986	1987	1988	1989	1990	1991	1992	1993
Food, beverages, tobacco										
Aliments, boissons, tabac	98	98	101	107	116	122	131	132	123	126
Chemicals, coal, petroleum products										
Produits chimiques, houillers, pétroliers	110	119	130	136	145	151	162	136	127	117
Basic metals										
Métaux de base	107	104	105	113	121	119	120	107	116	108
Metal products										
Produits métalliques	87	85	85	87	92	92	100	99	96	87
Electricity [4]										
Electricité [4]	127	127	130	132	145	168	184	194	198	199
Romania Roumanie										
Total industry [2–4]										
Total, industrie [2–4]	**116**	**120**	**129**	**133**	**130**	**107**	**83**	**65**	**62**	**64**
Total mining [2] [59]										
Total, industries extractives [2] [59]	**112**	**114**	**118**	**128**	**133**	**134**	**97**	**79**	**78**	**79**
Total manufacturing [3] [60]										
Total, manufactures [3] [60]	**116**	**121**	**130**	**129**	**133**	**130**	**106**	**81**	**61**	**57**
Food, beverages, tobacco [61]										
Aliments, boissons, tabac [61]	107	108	112	121	121	121	106	88	73	63
Textiles										
Textiles	110	114	121	123	125	124	106	92	65	63
Chemicals, coal, petroleum products										
Produits chimiques, houillers, pétroliers	118	119	129	124	130	125	99	64	56	58
Basic metals [10]										
Métaux de base [10]	119	117	127	121	124	121	95	70	49	52
Metal products										
Produits métalliques	119	129	137	141	145	134	127	95	68	69
Electricity and steam [4]										
Electricité et vapeur [4]	**112**	**114**	**124**	**128**	**137**	**135**	**141**	**128**	**118**	**108**
Spain Espagne										
Total industry [2–4]										
Total, industrie [2–4]	**101**	**103**	**107**	**112**	**115**	**120**	**120**	**119**	**116**	**110**
Total mining [2]										
Total, industries extractives [2]	**131**	**130**	**124**	**107**	**102**	**110**	**104**	**99**	**96**	**90**
Total manufacturing [3]										
Total, industries manufacturières [3]	**99**	**101**	**105**	**111**	**114**	**119**	**119**	**117**	**113**	**107**
Food, beverages, tobacco										
Aliments, boissons, tabac	112	117	116	125	129	127	133	136	131	133
Textiles										
Textiles	92	96	104	108	100	105	102	97	91	83
Chemicals, coal, petroleum products										
Produits chimiques, houillers, pétroliers	102	104	107	108	109	115	116	112	112	110
Basic metals										
Métaux de base	105	107	99	98	101	108	104	105	100	98
Metal products										
Produits métalliques	89	91	99	110	120	130	130	127	122	109
Electricity and gas [4]										
Electricité et gaz [4]	**109**	**115**	**117**	**122**	**127**	**133**	**136**	**139**	**139**	**135**
Sweden Suède										
Total industry [2–4] [51]										
Total, industrie [2–4] [51]	**107**	**110**	**110**	**114**	**117**	**120**	**121**	**115**	**113**	**117**
Total mining [2] [62]										
Total, industries extractives [2] [62]	**98**	**103**	**102**	**103**	**100**	**94**	**95**	**99**	**94**	**98**
Total manufacturing [3]										
Total, industries manufacturières [3]	**107**	**110**	**111**	**114**	**117**	**121**	**121**	**114**	**113**	**116**
Food, beverages, tobacco										
Aliments, boissons, tabac	103	104	104	104	106	108	108	105	102	106
Textiles										
Textiles	94	96	94	94	92	88	96	90	77	69
Paper and paper products										
Papier, produits en papier	108	108	111	118	120	120	118	117	118	122
Chemicals, coal, petroleum products										
Produits chimiques, houillers, pétroliers	113	113	111	123	130	130	128	130	134	144

26
Index numbers of industrial production [*cont.*]
Indices de la production industrielle [*suite*]
1980=100

Country or area and industry [SITC] Pays ou zone et industrie [CITI]	1984	1985	1986	1987	1988	1989	1990	1991	1992	1993
Basic metals Métaux de base	108	109	103	105	113	114	103	99	101	106
Metal products Produits métalliques	111	119	118	120	124	131	135	125	121	126
Switzerland Suisse										
Total industry [2−4] **Total, industrie [2−4]**	**97**	**103**	**107**	**108**	**117**	**119**	**122**	**123**	**122**	**122**
Total manufacturing [3] **Total, industries manufacturières [3]**	**97**	**103**	**106**	**107**	**117**	**120**	**123**	**124**	**123**	**122**
Food, beverages, tobacco Aliments, boissons, tabac	101	101	102	104	106	109	111	113	113	114
Textiles Textiles	102	105	107	105	106	104	99	96	94	90
Chemicals, coal, petroleum products Produits chimiques, houillers, pétroliers	115	123	125	129	146	169	171	173	179	193
Basic metals Métaux de base	98	101	103	105	115	116	117	110	110	104
Metal products Produits métalliques	88	94	99	99	106	106	114	121	116	109
Electricity and gas [4] **Electricité et gaz [4]**	102	114	116	121	123	110	112	116	119	123
United Kingdom Royaume Uni										
Total industry [2−4] **Total, industrie [2−4]**	**102**	**108**	**110**	**115**	**120**	**123**	**123**	**118**	**118**	**120**
Total mining [2] [63] **Total, industries extractives [2]** [63]	**106**	**117**	**121**	**121**	**111**	**94**	**90**	**92**	**96**	**104**
Total manufacturing [3] [63] **Total, manufactures [3]** [63]	**99**	**102**	**103**	**108**	**116**	**121**	**121**	**114**	**114**	**115**
Food, beverages, tobacco Aliments, boissons, tabac	102	101	102	105	107	108	109	**108**	109	110
Textiles Textiles	94	97	80	84	83	80	78	70	70	70
Chemicals, coal, petroleum products Produits chimiques, houillers, pétroliers	108	115	118	123	129	135	134	138	140	144
Basic metals Métaux de base	100	100	99	105	117	119	112	100	96	96
Metal products Produits métalliques	98	102	88	89	96	104	104	97	95	95
Electricity, gas and water [4] **Electricité, gaz et eau [4]**	**89**	**108**	**119**	**122**	**122**	**122**	**125**	**132**	**132**	**140**
Yugoslavia, SFR † Yougoslavie, Rfs †										
Total industry [2−4] **Total, industrie [2−4]**	**111**	**114**	**119**	**120**	**119**	**120**	**107**	**98**	...	...
Total mining [2] **Total, industries extractives [2]**	**109**	**115**	**116**	**117**	**116**	**113**	**108**	**102**	...	...
Total manufacturing [3] **Total, industries manufacturières [3]**	**112**	**114**	**117**	**119**	**117**	**120**	**106**	**94**	...	...
Food, beverages, tobacco Aliments, boissons, tabac	110	108	110	113	109	83	108	90	...	...
Textiles Textiles	107	109	116	117	115	112	92	65	...	...
Chemicals, coal, petroleum products Produits chimiques, houillers, pétroliers	117	118	126	128	134	131	118	84	...	...
Basic metals Métaux de base	122	127	129	124	128	131	113	78	...	...
Metal products Produits métalliques	117	122	131	131	129	132	110	65	...	...
Electricity and gas [4] **Electricité et gaz [4]**	**116**	**119**	**124**	**128**	**132**	**131**	**131**	**131**	...	...

26
Index numbers of industrial production [*cont.*]
Indices de la production industrielle [*suite*]
1980=100

Country or area and industry [SITC] Pays ou zone et industrie [CITI]	1984	1985	1986	1987	1988	1989	1990	1991	1992	1993
Oceania · Océanie										
Australia [25] Australie [25]										
Total industry [2−4]										
Total, industrie [2−4]	**108**	**110**	**111**	**119**	**126**	**127**	**127**	**126**	**130**	**137**
Total mining [2]										
Total, industries extractives [2]	**125**	**139**	**132**	**149**	**154**	**168**	**176**	**180**	**182**	**184**
Total manufacturing [3]										
Total, industries manufacturières [3]	**100**	**100**	**102**	**109**	**116**	**113**	**111**	**108**	**113**	**122**
Food, beverages, tobacco										
Aliments, boissons, tabac	101	101	104	109	111	112	114	117	126	136
Textiles										
Textiles	108	107	112	111	113	108	109	113	101	85
Chemicals, coal, petroleum products										
Produits chimiques, houillers, pétroliers	107	104	109	118	120	119	124	124	125	136
Basic metals										
Métaux de base	105	105	106	110	116	128	132	133	133	141
Metal products										
Produits métalliques	93	95	96	103	111	108	103	98	105	117
Electricity and gas [4]										
Electricité et gaz [4]	**120**	**124**	**127**	**133**	**140**	**147**	**150**	**152**	**155**	**158**
Fiji Fidji										
Total industry [2−4]										
Total, industrie [2−4]	**116**	**103**	**123**	**110**	**117**	**128**	**138**	**141**	**144**	**152**
Total mining [2]										
Total, industries extractives [2]	**195**	**241**	**369**	**370**	**552**	**545**	**532**	**355**	**478**	**489**
Total manufacturing [3]										
Total, industries manufacturières [3]	**113**	**98**	**118**	**102**	**103**	**114**	**126**	**133**	**129**	**139**
Food, beverages, tobacco										
Aliments, boissons, tabac	120	95	126	117	112	128	124	128	132	132
Electricity [4]										
Electricité [4]	**124**	**129**	**138**	**136**	**147**	**157**	**166**	**172**	**186**	**192**
former USSR · ancienne URSS										
former USSR † ancienne URSS †										
Total industry [2−4] [64]										
Total, industrie [2−4] [64]	**115**	**119**	**125**	**129**	**134**	**136**	**135**	**124**	...	...
Total mining [2] [65]										
Total, industries extractives [2] [65]	**105**	**107**	**111**	**113**	**116**	**115**	**111**	**99**	...	...
Total manufacturing [3] [45]										
Total, manufactures [3] [45]	**116**	**120**	**126**	**131**	**136**	**139**	**138**	**126**	...	...
Food, beverages, tobacco [66]										
Aliments, boissons, tabac [66]	115	112	111	115	119	123	124	127	...	...
Textiles										
Textiles	103	106	108	110	114	117	115	110	...	...
Chemicals, coal, petroleum products										
Produits chimiques, houillers, pétroliers	112	115	120	124	129	131	130	127	...	...
Basic metals										
Métaux de base	110	113	118	121	125	125	122	113	...	...
Metal products										
Produits métalliques	126	135	144	152	160	164	165	170	...	...
Electricity and steam [4]										
Electricité et vapeur [4]	**116**	**120**	**123**	**129**	**132**	**133**	**135**	**134**	...	...

Source:
Industrial statistics database of the Statistical Division of the
United Nations Secretariat.

† For detailed descriptions of data pertaining to former
Czechoslovakia, Germany, SFR Yugoslavia and former USSR,
see Annex I − Country or area nomenclature, regional and
other groupings.

1 Calculated by the Statistical Division of the United Nations.
2 Excluding basic metals.

Source:
Base de données de statistiques industrielles de la Division de statistique
du Secrétariat de l'ONU.

† Pour les descriptions en détails des données relatives
à l'ancienne Tchécoslovaquie, l'Allemagne, la Rfs Yougoslavie et
et l'ancienne URSS, voir l'Annexe I − Nomenclature des pays
ou zones, groupements régionaux et autres groupements.

1 Calculé par la Division de Statistique de l'Organisation des
Nations Unies.

26
Index numbers of industrial production [*cont.*]
Indices de la production industrielle [*suite*]
1980=100

3   Including clothing and footwear.
4   Excluding coal, stone quarrying, clay and sand pits.
5   Excluding wearing apparel except footwear, furniture, printing and publishing, petroleum refineries and some miscellaneous manufacturing products.
6   Including coal mining and petroleum refineries.
7   Including food, beverages, tobacco, textiles, shoes, chemicals, non–metallic mineral products, beds, mattresses and some metal products only.
8   Including rubber and plastic products.
9   Including non–ferrous metal basic industries.
10   Excluding non–ferrous metal basic industries.
11   Index numbers of gross domestic product (at constant prices of 1966).
12   Including oil and sugar.
13   Index numbers of gross domestic product (at constant prices of 1962).
14   Index numbers of gross domestic product (at constant prices of 1978).
15   Including construction.
16   Excluding furniture.
17   Index numbers of gross domestic product (at constant prices of 1978/79).
18   Excluding leather and leather products, wood products, furniture, printing and publishing and miscellaneous manufacturing industries.
19   Manufacture of fabricated metal products, except machinery and equipment, is included in basic metals.
20   Excluding the extraction of natural gas.
21   Including the smelting and refining of copper.
22   Excluding slaughtering, sawmills and the smelting and refining of copper.
23   Including mining.
24   Excluding footwear: including cold storage and production of motion pictures.
25   Figures relate to 12 months beginning 1 July of year stated.
26   Excluding coal, petroleum products and basic metals.
27   Figures relate to 12 months beginning 21 March of year stated.
28   Excluding petroleum refineries.
29   Excluding tobacco industry.
30   Extraction of non–metallic minerals only.
31   Excluding printing and publishing.
32   Excluding saltern.
33   Excluding made–up textiles and other wearing apparel, leather, fur and leather products, footwear, furniture and fixtures, printing and publishing, non–electrical machinery and miscellaneous industries.
34   Excluding machinery and transport equipment.
35   Inlcuding magnesite products.
36   Including stone quarrying, clay and sand pits, excluding sawmills, printing and publishing, and coal products.
37   Excluding metal mining.
38   Including logging and fishing.
39   Excluding publishing.
40   Excluding shipbuilding and repairing.
41   Excluding iron foundries.
42   Including iron foundries.
43   Excluding clothing.
44   Stone quarrying, clay and sand pits are included in manufacturing. Coal briquetting is included in mining.
45   Including commercial fishing. Excluding publishing.
46   Including fishing.
47   Stone quarrying, clay and sand pits are included in manufacturing.
48   Including magnesite roasting.
49   Including car repairs.
50   Including waterworks and gasworks, excluding publishing.
51   Excluding electricity and gas.

2   Non compris les métaux de base.
3   Y compris l'industrie d'habillement et des chaussures.
4   Non compris l'extraction du charbon, l'extraction de la pierre à bâtir, de l'argile et du sable.
5   Non compris l'industrie d'articles d'habillement, à l'exclusion des chaussures, l'industrie de meuble, l'imprimerie et l'édition, raffineries de pétrole et quelques produits de l'industrie manufacturière diverse.
6   Y compris l'extraction du charbon et les raffineries de pétrole.
7   Y compris aliments, boissons, tabac, textiles, chaussures, produits chimiques, produits minéraux non–métalliques, lits, sommiers et quelques produits métalliques seulement.
8   Y compris l'industrie du caoutchouc et articles en matière plastique.
9   Y compris la métallurgie de base de métaux non–ferreux.
10   Non compris la métallurgie de base de métaux non–ferreux.
11   Indices du produit intérieur brut (aux prix constants de 1966).
12   Y compris le pétrole et le sucre.
13   Indices du produit intérieur brut (aux prix constants de 1962).
14   Indices du produit intérieur brut (aux prix constants de 1978).
15   Y compris la construction.
16   Non compris l'industrie du meuble.
17   Indices du produit intérieur brut (aux prix constants de 1978/79).
18   Non compris le cuir et les articles en cuir, les produits de l'industrie du bois, les meubles, l'imprimerie et l'édition et industries manufacturières diverses.
19   Fabrication d'ouvrages en métaux, à l'exclusion des machines et du matériel est comprise dans les industries des métaux de base.
20   Non compris l'extraction du gaz naturel.
21   Y compris la fonderie et l'affinage du cuivre.
22   Non compris l'abattage du bétail, les scieries, la fonderie et l'affinage du cuivre.
23   Y compris les industries extractives.
24   Non compris les chaussures, y compris les entrepôts frigorifiques et la production cinématographique.
25   Les chiffres se rapportent à 12 mois commençant le 1er juillet de l'année indiquée.
26   Non compris les produits dérivés du charbon, des produits pétroliers et des métaux de base.
27   Les chiffres se rapportent à 12 mois commençant le 21 mars de l'année indiquée.
28   Non compris les raffineries de pétrole.
29   Non compris l'industrie du tabac.
30   Extraction des minéraux non–métalliques seulement.
31   Non compris l'imprimerie et l'édition.
32   Non compris les salines.
33   Non compris la confection d'ouvrages en tissue et autres industries d'habillement, l'industrie du cuir, de la fourrure et des articles en cuir, la fabrication des chaussures, l'industrie des meubles, l'imprimerie, l'édition, la construction de machines non électriques et les industries diverses.
34   Non compris la construction de machines et du matériel de transport.
35   Y compris les articles de magnésite.
36   Y compris l'extraction de la pierre à bâtir, de l'argile et du sable, non compris les scieries, l'imprimerie, l'édition et les produits du charbon.
37   Non compris l'extraction des minerais métalliques.
38   Y compris l'exploitation forestière et la pêche.
39   Non compris l'édition.
40   Non compris la construction navale et la réparation des navires.
41   Non compris les fonderies de fer.
42   Y compris les fonderies de fer.
43   Non compris les articles d'habillement.
44   L'extraction de la pierre à bâtir, de l'argile et du sable est comprise dans les industries manufacturières. La fabrication des briquettes de charbon est comprise dans les industries extractives.
45   Y compris la prise de poissons. Non compris l'édition.
46   Y compris la pêche.

26
Index numbers of industrial production [*cont.*]
Indices de la production industrielle [*suite*]
1980=100

52 Excluding some groups of miscellaneous manufacturing.
53 Excluding paper and paper products and miscellaneous manufacturing industries.
54 Including basic metals.
55 Including plastic products.
56 Including also fishing, film laboratories, repair services and gas production.
57 Including fishing, extraction of salt, Excluding animal fodder.
58 Annual figures are monthly averages.
59 Excluding stone quarrying, clay and sand pits, extraction of salt and chemical minerals.
60 Including fishing, stone quarrying, clay and sand pits, gas works, the extraction of salt and chemical minerals. Excluding publishing.
61 Including fishing and the extraction of salt.
62 Excluding clay and sand pits.
63 These figures do not include an adjustment to correct mixed sales and production data to true production index numbers; consequently, they are not strictly comparable with the United Kingdom's output of production industries' series.
64 Including logging, motion picture production, cleaning and dyeing.
65 Excluding oil and natural gas drilling, prospecting and preparing sites for the extraction of minerals.
66 Including commercial fishing and the processing and cold storage of fish and fish products by factory—type vessels.
67 Phosphate only.
68 Excluding mining.

47 L'extraction de la pierre à bâtir, de l'argile et du sable est comprise dans les industries manufacturières.
48 Y compris le rôtissage du magnésite.
49 Y compris la réparation des véhicules automobiles.
50 Y compris les usines des eaux et les usines à gaz, non compris l'édition.
51 Non compirs l'électricité et le gaz.
52 Non compris quelques groupes d'industries manufacturières diverses.
53 Non compris le papier et les ouvrages en papier et les industries manufacturières diverses.
54 Y compris les métaux de base.
55 Y compris articles en matière plastique.
56 Y compris également la pêche, les laboratoires de films, les services de réparation et les usines à gaz.
57 Y compris la pêche, l'extraction du sel. Non compris la préparation des fourrages.
58 Les chiffres annuels sont des moyennes mensuelles.
59 Non compris l'extraction de la pierre à bâtir, de l'argile et du sable, l'extraction du sel et l'extraction des minéraux destinés à l'industrie chimique.
60 Y compris la pêche, l'extraction de la pierre à bâtir, de l'argile et du sable, les usines à gaz, l'extraction du sel et des minéraux chimiques. Non compris l'édition.
61 Y compris la pêche et l'extraction du sel.
62 Non compris l'extraction de l'argile et du sable.
63 Ces chiffres ne comprennent pas un ajustement corrigeant les ventes mixtes et les données de production aux indices réels de production. En conséquence, ils ne sont pas strictement comparables avec ceux du Royaume Uni relatifs au rendement de la production des industries.
64 Y compris l'exploitation forestière, la production cinématographique, le nettoyage et la teinture.
65 L'exploitation des puits de pétrole et des puits de gaz naturel, la prospection et la préparation du terrain avant l'extraction des minéraux ne sont pas comprises.
66 Y compris la pêche commerciale, le traitement des poissons et des produits poissonniers et les entrepôts frigorifiques dans les usines flottantes.
67 Le phosphate seulement.
68 Non compris les mines.

Technical Notes, tables 21-26

Detailed internationally comparable data on national accounts are compiled and published annually by the Statistical Division, Department for Economic and Social Information and Policy Analysis, of the United Nations Secretariat. Data of the national accounts aggregates for countries or areas are based on the concepts and definitions contained in *A System of National Accounts* [49], Studies in Methods, Series F, No. 2, Rev. 3. A summary of the conceptual framework, classifications and definitions of trasactions is found in the annual United Nations publication, *National Accounts Statistics: Main Aggregates and Detailed Tables* [24].

The national accounts data shown in this publication offer in the form of analytical tables a summary of some selected principal national accounts aggregates based on official detailed national accounts data of 184 countries and areas. Every effort has been made to present the estimates of the various countries or areas in a form designed to facilitate international comparability. Differences in concept, scope, coverage and classification are footnoted. Detailed footnotes identifying these differences are also available in the national accounts yearbook mentioned above. Such differences should be taken into account if misleading comparisons among countries or areas are to be avoided.

Table 21 shows total and per capita gross domestic product (GDP) expressed in United States dollars at current prices and at constant 1990 prices and its corresponding rates of growth. The table is designed to facilitate international comparisons of levels of income generated in production. In order to have a comparable coverage for the years 1984 - 1992 for as many countries as possible, the official GDP national currency data are supplemented by estimates prepared by the Statistical Division, based on a variety of data derived from national and international sources. National currency data are converted to United States dollars using the average market rates as published by the International Monetary Fund in the *International Financial Statistics* [11]. Official exchange rates are used only when market rate is not available. For non-members of the Fund, the conversion rates are the average of United Nations operational rates of exchange. It should be noted that the conversion introduces deficiencies of comparability over time and between countries which should be taken into account when using the data. The comparability over time is distorted when there are large and discrepant fluctuations in the exchange rates vis-a-vis domestic inflation. Per capita GDP data are likewise affected by the same distortions resulting from exchange rate conversion. Per capita takes the growth of the population into account as well.

Notes techniques, tableaux 21 à 26

La Division de statistique du Département de l'information économique et sociale et de l'analyse des politiques du Secrétariat de l'Organisation des Nations Unies établit et publie chaque année des données détaillées sur les comptes nationaux se prêtant à des comparaisons internationales. Les données relatives aux agrégats des différents pays et territoires sont établies à l'aide des concepts et des définitions figurant dans le *Système de comptabilité nationale* [49], études méthodologiques, série F, No 2, Rev. 3. On trouvera un résumé de l'appareil conceptuel des classifications et des définitions des transactions dans *National Accounts Statistics: Main Aggregates and Detailed Tables* [24], publication annuelle des Nations Unies.

Les comptes nationaux figurant dans cette publication présentent, sous forme de tableaux analytiques, un résumé de certains agrégats importants calculés à partir des comptes nationaux détaillés de 184 pays et territoires. Tout a été fait pour présenter des estimations relatives aux divers pays ou territoires sous une forme facilitant les comparaisons internationales. Les différences de définition, de portée, de couverture et de classification sont indiquées dans les notes en bas de page. Des notes détaillées précisant ces différences figurent également dans l'annuaire des comptes nationaux mentionné plus haut.

En raison de ces différences, l'interprétation des comparaisons entre pays et territoires doit être prudente.

Le *tableau 21* donne le produit intérieur brut (PIB) total et par habitant, exprimé en dollars des États-Unis à prix courants et aux prix constants de 1990, ainsi que les taux de croissance correspondants. Le tableau est conçu pour faciliter les comparaisons internationales du revenu engendré par la production. À l'aide des données les plus diverses provenant de sources nationales et internationales, ia Division de statistique établit des estimations destinées à compléter les chiffres officiels du PIB exprimé dans la monnaie nationale de façon à rendre possible une comparaison d'autant de pays que possible pour les années 1984 à 1992. Les données libellées en monnaie nationale sont converties en dollars des États-Unis à l'aide des taux moyens du marché publiés par le Fonds monétaire international dans les *Statistiques financières internationales* [11]. Le taux de change officiel n'est utilisé que si l'on ne dispose pas d'un taux du marché. Pour les pays qui ne sont pas membres du Fonds, les taux de change utilisés sont la moyenne des taux de change retenus pour les opérations des Nations Unies. Il est à noter que cette opération de conversion fausse la comparaison des données dans le temps et entre pays et toute interprétation des données doit donc en tenir compte. Les comparaisons dans le temps sont faussées quand l'évolution des taux de change s'écarte nettement de celle de l'inflation intérieure. De même, le PIB par habitant

The GDP constant price series based partly on data officially provided by countries and partly based on estimates made by the Statistical Division are transformed into index numbers and rebased to 1990=100. The resulting data are then converted into US dollars at the rate prevailing in the base year 1990. The growth rates are based on the estimates of GDP at constant 1990 prices. The rate of the year in question is obtained by dividing the GDP of that year by the GDP of the preceeding year.

Table 22 features the distribution of GDP by expenditure breakdown at current prices. It shows what portion of income is spent by the government and by the private sector on consumption, what is spent on investment and what revenues are obtained from exports after deducting the expenditure on imports. The percentages are derived from official data of countries as reported to the United Nations and published in the annual national accounts yearbook.

Table 23 shows the distribution of GDP originating from each industry component based on the *International Standard Industrial Classification of All Economic Activities* [42]. This table reflects the economic structure of production in the country. The percentages are based on official GDP estimates broken down by kind of economic activity at current prices: agriculture, hunting, forestry and fishing; mining and quarrying; manufacturing; electricity, gas and water; construction; wholesale, retail trade, restaurants and hotel; transport, storage and communication; and other activities comprised of financial and community services, producers of government services, other producers and import duties and taxes.

Table 24 presents the distribution of total government final consumption expenditure by function at current prices. The breakdown by function includes: general public services; defence; public order and safety; education; health; social serives; economic services; and other function which includes housing, community amenities, recreational, cultural and religious affairs. The government expenditure is equal to the service produced by general government for its own use. These services are not sold, they are valued in the GDP at their cost to the government.

Table 25 shows the distribution of total private final consumption expenditure by type and purpose at current prices. Private consumption expenditure measures the expenditure of all resident non-government units which includes all housholds and private non-profit institutions serving households. The percentage shares include: food, beverages and tobacco; clothing and footwear; gross rent, fuel and power; furniture, furnishings and household equipment; medical care and health expenses; transport and communication; recreational, entertainment, education and

est susceptible d'être affecté par les mêmes distorsions liées au taux de change. Les données exprimées par habitant tiennent compte aussi de l'accroissement de la population.

La série de statistiques du PIB à prix constants est fondée en partie sur des données officiellement communiquées par les pays et en partie sur des estimations effectuées par la Division de statistique; les données permettent de calculer des indices, la base 100 étant retenue pour 1990. Les données ainsi obtenues sont alors converties en dollars des États-Unis au taux de change de l'année de base (1990). Les taux de croissance sont calculés à partir des estimations du PIB aux prix constants de 1990. Le taux de croissance de l'année en question est obtenu en divisant le PIB de l'année par celui de l'année précédente.

Le *tableau 22* donne la répartition du PIB par catégorie de dépenses aux prix courants. Il indique quelle est la fraction du revenu national qui est consacrée à la consommation par les administrations et par le secteur privé, quelle est celle qui est affectée à l'investissement, et quelles recettes proviennent des exportations une fois les importations déduites. Les pourcentages sont calculés à partir des données officiellement communiquées par les pays à l'Organisation des Nations Unies et sont publiés dans l'annuaire des comptes nationaux.

Le *tableau 23* donne la répartition du PIB par secteur industriel, selon le classement proposé par la *Classification internationale type, par industrie, de toutes les branches d'activité économique (CITI) [42]*. Ce tableau donne donc la structure économique de la production dans chaque pays. Les pourcentages sont établis à partir des estimations officielles du PIB, ventilées entre les diverses branches d'activité économique aux prix courants : agriculture, chasse, forêts et pêche; industries minières et extractives; industrie manufacturière; électricité, gaz et eau; construction; commerce de gros et de détail, restaurants et hôtels; transports, entrepôts et communications; autres activités, y compris les services financiers, les services communautaires, les services fournis par les administrations, divers autres services et les droits et taxes d'importation.

Le *tableau 24* donne la répartition des dépenses de consommation finale des administrations, par fonction, aux prix courants. La répartition par fonction est la suivante : administration publique générale; défense; ordre public et sécurité; éducation; santé; services sociaux; services économiques; et autres fonctions incluant le logement, les aménagements collectifs, les équipements de loisir et les activités culturelles et religieuses. Les dépenses des administrations sont considérées comme égales aux services produits par l'administration pour son propre usage. Ces services ne sont pas vendus et ils sont évalués, dans le PIB, à leur coût pour l'administration.

cultural services; and other fuctions which include
miscellaneous goods and services, purchases abroad by
resident households deducting the expenditure of non-
resident in the domestic market and the expenditure of
private non-profit insitutions serving households.

Table 26: Detailed descriptions of national practices in
the compilation of production index numbers are given in
the United Nations *1977 Supplement to the Statistical
Yearbook and Monthly Bulletin of Statistics.* [23] Some
differences in national practice in compilation, as well as
major deviations from ISIC in the scope of the indexes, are
indicated in the footnotes to this table.

Le *tableau 25* donne la répartition des dépenses totales
de consommation finale privée par type et par objet aux prix
courants. Les dépenses privées de consommation mesurent
donc les dépenses de toutes les entités résidentes autres que
les administrations, y compris tous les ménages et les entités
privées à but non lucratif fournissant des services aux
ménages. La répartition en pourcentage distingue les
rubriques suivantes : aliments, boissons et tabac, articles
d'habillement et chaussures, loyer brut, combustible et
électricité, mobilier et équipement des ménages, soins
médicaux et dépenses de santé, services de loisir, éducatifs et
culturels, autres fonctions, y compris les biens et services
divers, achats à l'étranger effectués par les ménages
résidents, moins les dépenses des non-résidents sur le marché
intérieur, et dépenses des institutions privées à but non
lucratif fournissant des services aux ménages.

Tableau 26 : Des descriptions détaillées des pratiques
nationales employées pour la compilation des indices de
production sont données dans le *Supplément 1977 à
l'Annuaire statistique et au Bulletin mensuel de statistique
des Nations Unies* [23]. Certaines différences dans la
méthode nationale de compilation, ainsi que les écarts
importants par rapport à la CITI dans la portée des indices,
sont indiqués dans les notes figurant au bas de ce tableau.

27
Rates of discount of central banks
Taux d'escompte des banques centrales
Per cent per annum, end of period
Pour cent par année, fin de la période

Country or area Pays ou zone	1985	1986	1987	1988	1989	1990	1991	1992	1993	1994
Aruba Aruba	...	9.50	9.50	9.50	9.50	9.50	9.50	9.50	9.50	9.50
Australia Australie	15.98	16.92	14.95	13.20	17.23	15.24	10.99	6.96	5.83	5.75
Austria Autriche	4.00	4.00	3.00	4.00	6.50	6.50	8.00	8.00	5.25	...
Bahamas Bahamas	8.50	7.50	7.50	9.00	9.00	9.00	9.00	7.50	7.00	6.50
Bangladesh Bangladesh	11.25	10.75	10.75	10.75	10.75	9.75	9.25	8.50	6.00	5.50
Barbados Barbade	13.00	8.00	8.00	8.00	13.50	13.50	18.00	12.00	8.00	9.50
Belgium Belgique	9.75	8.00	7.00	7.75	10.25	10.50	8.50	7.75	5.25	4.50
Belize Belize	20.00	12.00	12.00	10.00	12.00	12.00	12.00	12.00	12.00	12.00
Benin Bénin	10.50	8.50	8.50	9.50	11.00	11.00	11.00	12.50	10.50	10.00
Botswana Botswana	9.00	9.00	8.50	6.50	6.50	8.50	12.00	14.25	14.25	...
Brazil Brésil	380.00	89.00	401.00	2 282.00	38 341.00	1 083.00	2 494.00	1 489.00	5 757.00	56.00
Burkina Faso Burkina Faso	10.50	8.50	8.50	9.50	11.00	11.00	11.00	12.50	10.50	10.00
Burundi Burundi	7.00	5.00	7.00	7.00	...	...	...	...	...	...
Cameroon Cameroun	9.00	8.00	8.00	9.50	10.00	11.00	10.75	...	...	...
Canada Canada	9.49	8.49	8.66	11.17	12.47	11.78	7.67	7.36	4.11	7.00
Central African Rep. Rép. centrafricaine	9.00	8.00	8.00	9.50	10.00	11.00	10.75	...	...	...
Chad Tchad	9.00	8.00	8.00	9.50	10.00	11.00	10.75	...	...	...
Colombia Colombie	27.00	33.80	34.80	34.30	36.90	# 46.50	45.00	34.40	33.50	...
Comoros Comores	10.00	10.00	8.50	8.50	...	...	...	...	...	...
Congo Congo	9.00	8.00	8.00	9.50	10.00	11.00	10.75	...	...	...
Costa Rica Costa Rica	28.00	27.50	31.38	31.50	31.61	37.80	42.50	29.00	35.00	37.75
Côte d'Ivoire Côte d'Ivoire	10.50	8.50	8.50	9.50	11.00	11.00	11.00	12.50	10.50	10.00
Cyprus Chypre	6.00	6.00	6.00	6.00	6.50	6.50	6.50	6.50	6.50	...
former Czechoslovakia† anc. Tchécoslovaquie†	...	...	...	...	...	8.50	9.50	9.50	..	..
Czech Republic République tchèque	..	..	..	..	..	..	..	..	8.00	8.50

27
Rates of discount of central banks
Per cent per annum, end of period [*cont.*]
Taux d'escompte des banques centrales
Pour cent par année, fin de la période [*suite*]

Country or area Pays ou zone	1985	1986	1987	1988	1989	1990	1991	1992	1993	1994
Denmark Danemark	7.00	7.00	7.00	7.00	7.00	8.50	9.50	9.50	6.25	5.00
Ecuador Equateur	23.00	23.00	23.00	23.00	32.00	35.00	49.00	49.00	33.57	44.88
Egypt Egypte	13.00	13.00	13.00	13.00	14.00	14.00	20.00	18.40	16.50	14.00
Equatorial Guinea Guinée équatoriale	9.00	8.00	8.00	9.50	10.00	11.00	10.75	...	...	...
Ethiopia Ethiopie	6.00	6.00	3.00	3.00	3.00	3.00	3.00	5.25	12.00	12.00
Fiji Fidji	11.00	8.00	11.00	11.00	8.00	8.00	8.00	6.00	6.00	6.00
Finland Finlande	9.00	7.00	7.00	8.00	8.50	8.50	8.50	9.50	5.50	5.25
France France	9.50	9.50	9.50	9.50	9.50	9.50	9.50	9.50	9.50	9.50
Gabon Gabon	9.00	8.00	8.00	9.50	10.00	11.00	10.75	12.00	...	...
Gambia Gambie	15.00	20.00	21.00	19.00	15.00	16.50	15.50	17.50	13.50	13.50
Germany † Allemagne† F. R. Germany R. f. Allemagne	.. 4.00	.. 3.50	.. 2.50	.. 3.50	.. 6.00	.. 6.00	8.00 ..	8.25 ..	5.75 ..	4.50 ..
Ghana Ghana	18.50	20.50	23.50	26.00	26.00	33.00	20.00	30.00	35.00	33.00
Greece Grèce	20.50	20.50	20.50	19.00	19.00	19.00	19.00	19.00	21.50	20.50
Guatemala Guatemala	9.00	9.00	9.00	9.00	13.00	18.50	16.50	...	...	...
Guinea-Bissau Guinée-Bissau	...	...	...	...	...	42.00	42.00	45.50	41.00	26.00
Guyana Guyana	14.00	14.00	14.00	14.00	35.00	30.00	32.50	24.30	17.00	20.30
Honduras Honduras	24.00	24.00	24.00	24.00	24.00	28.20	30.10	26.10	...	...
Hungary Hongrie	10.50	9.50	10.00	10.50	14.00	20.00	26.00	20.00	22.00	25.00
Iceland Islande	30.00	21.00	49.20	24.10	38.40	21.00	21.00	# 10.50	5.50	4.70
India Inde	10.00	10.00	10.00	10.00	10.00	10.00	12.00	12.00	12.00	12.00
Ireland Irlande	10.25	13.25	9.25	8.00	12.00	11.25	10.75	...	7.00	6.25
Israel Israël	79.60	31.40	26.80	30.90	15.00	13.00	14.20	10.40	9.80	17.00
Italy Italie	15.00	12.00	12.00	12.50	13.50	12.50	12.00	12.00	8.00	7.50
Jamaica Jamaïque	21.00	21.00	21.00	21.00	21.00	21.00	...	...	...	...

27
Rates of discount of central banks
Per cent per annum, end of period [*cont.*]
Taux d'escompte des banques centrales
Pour cent par année, fin de la période [*suite*]

Country or area Pays ou zone	1985	1986	1987	1988	1989	1990	1991	1992	1993	1994
Japan Japon	5.00	3.00	2.50	2.50	4.25	6.00	4.50	3.25	1.75	1.75
Jordan Jordanie	6.25	6.25	6.25	6.25	8.00	8.50	8.50	8.50	8.50	8.50
Kenya Kenya	12.50	12.50	12.50	16.02	16.50	19.43	20.27	20.46	45.50	21.50
Korea, Republic of Corée, République de	5.00	7.00	7.00	8.00	7.00	7.00	7.00	7.00	5.00	5.00
Kuwait Koweït	6.00	6.00	6.00	7.50	7.50	...	7.50	7.50	5.80	7.00
Lebanon Liban	19.70	21.85	21.85	21.84	21.84	21.84	18.04	16.00	20.22	16.49
Lesotho Lesotho	12.00	9.50	9.00	15.50	17.00	15.75	18.00	15.00	13.50	13.50
Libyan Arab Jamah. Jamah. arabe libyenne	5.00	5.00	5.00	5.00	...	...	...	...	...	...
Madagascar Madagascar	11.50	11.50	11.50	11.50	...	...	...	...	...	...
Malawi Malawi	11.00	11.00	14.00	11.00	11.00	14.00	13.00	20.00	25.00	...
Malaysia Malaisie	4.13	3.89	3.20	3.33	4.44	6.79	7.38	...	...	...
Mali Mali	10.50	8.50	8.50	9.50	11.00	11.00	11.00	12.50	10.50	10.00
Malta Malte	6.00	6.00	5.50	5.50	5.50	5.50	5.50	5.50	5.50	5.50
Mauritania Mauritanie	6.50	6.50	6.50	6.50	7.00	7.00	7.00	7.00	...	...
Mauritius Maurice	11.00	11.00	10.00	10.00	12.00	12.00	11.30	8.30	8.30	13.80
Mongolia Mongolie	...	...	...	...	...	...	...	...	628.80	435.00
Morocco Maroc	8.50	8.50	8.50	8.50	...	...	...	...	8.00	...
Namibia Namibie	...	...	...	...	...	...	20.50	16.50	14.50	15.50
Nepal Népal	15.00	11.00	11.00	11.00	11.00	11.00	13.00	13.00	...	...
Netherlands Pays-Bas	5.00	4.50	3.75	4.50	7.00	7.25	8.50	7.75	5.00	...
Netherlands Antilles Antilles néerlandaises	8.00	8.00	6.00	6.00	6.00	6.00	6.00	6.00	6.00	5.00
New Zealand Nouvelle-Zélande	19.80	24.60	18.55	15.10	15.00	13.25	8.30	9.15	5.70	9.75
Nicaragua Nicaragua	...	...	...	12 874.60	311.00	10.00	15.00	15.00	11.80	10.50
Niger Niger	10.50	8.50	8.50	8.50	11.00	11.00	11.00	12.50	10.50	10.00
Nigeria Nigéria	10.00	10.00	12.75	12.75	18.50	18.50	15.50	17.50	26.00	...

27
Rates of discount of central banks
Per cent per annum, end of period [cont.]
Taux d'escompte des banques centrales
Pour cent par année, fin de la période [suite]

Country or area Pays ou zone	1985	1986	1987	1988	1989	1990	1991	1992	1993	1994
Norway Norvège	10.70	14.80	13.80	12.00	11.00	10.50	10.00	11.00	7.00	6.75
Pakistan Pakistan	10.00	10.00	10.00	10.00	10.00	10.00	10.00	10.00	...	...
Papua New Guinea Papouasie-Nvl-Guinée	9.75	11.40	8.80	10.80	9.55	9.30	9.30	7.12	# 6.39	...
Paraguay Paraguay	...	...	...	10.00	21.00	30.00	18.00	18.00	18.00	18.00
Peru Pérou	42.60	36.10	29.80	748.00	865.60	289.60	67.70	48.50	28.60	16.10
Philippines Philippines	12.75	10.00	10.00	10.00	12.00	14.00	14.00	14.30	9.40	8.30
Poland Pologne	4.00	4.00	4.00	6.00	140.00	55.00	40.00	38.00	35.00	33.00
Portugal Portugal	23.50	17.00	14.96	13.71	14.33	14.50	14.50	14.50	13.71	11.63
Rwanda Rwanda	9.00	9.00	9.00	9.00	9.00	14.00	14.00	11.00	11.00	...
Senegal Sénégal	10.50	8.50	8.50	9.50	11.00	11.00	11.00	12.50	10.50	10.00
Somalia Somalie	12.00	12.00	12.00	45.00	45.00	...	...	...	...	...
South Africa Afrique du Sud	13.00	9.50	9.50	14.50	18.00	18.00	17.00	14.00	12.00	13.00
Spain Espagne	10.50	11.84	13.50	12.40	14.52	14.71	12.50	13.25	9.00	7.38
Sri Lanka Sri Lanka	11.00	11.00	10.00	10.00	14.00	15.00	17.00	17.00	17.00	17.00
Swaziland Swaziland	12.50	9.50	9.00	11.00	12.00	12.00	13.00	12.00	11.00	12.00
Sweden Suède	10.50	7.50	7.50	8.50	10.50	11.50	8.00	# 10.00	5.00	7.00
Switzerland Suisse	4.00	4.00	2.50	3.50	6.00	6.00	7.00	6.00	4.00	3.50
Syrian Arab Republic Rép. arabe syrienne	5.00	5.00	5.00	5.00	...	...	...	...	...	...
Thailand Thaïlande	11.00	8.00	8.00	8.00	8.00	12.00	11.00	11.00	9.00	9.50
Togo Togo	10.50	8.50	8.50	9.50	11.00	11.00	11.00	12.50	10.50	10.00
Trinidad and Tobago Trinité-et-Tobago	7.50	5.97	7.50	9.50	9.50	9.50	11.50	13.00	13.00	13.00
Tunisia Tunisie	9.25	9.25	9.25	9.25	...	...	...	...	...	...
Turkey Turquie	52.00	48.00	45.00	54.00	54.00	45.00	45.00	...	...	...
Uganda Ouganda	24.00	36.00	31.00	45.00	55.00	50.00	46.00	41.00	24.00	15.00
United Rep.Tanzania Rép. Unie de Tanzanie	4.25	6.50	11.31	12.67	15.17	...	...	...	14.50	...

27
Rates of discount of central banks
Per cent per annum, end of period [*cont.*]
Taux d'escompte des banques centrales
Pour cent par année, fin de la période [*suite*]

Country or area Pays ou zone	1985	1986	1987	1988	1989	1990	1991	1992	1993	1994
United States Etats-Unis	7.50	5.50	6.00	6.50	7.00	6.50	3.50	3.00	3.00	4.75
Uruguay Uruguay	145.10	138.40	143.40	154.50	219.60	251.60	219.00	162.40	164.30	182.30
Venezuela Venezuela	8.00	8.00	8.00	8.00	45.00	43.00	43.00	52.20	71.25	48.00
Yugoslavia, SFR† Yougoslavie, Rfs†	61.00	56.00	131.00	372.00	8 187.00	30.00	40.00	..	..	..
Zaire Zaïre	26.00	26.00	29.00	37.00	50.00	45.00	55.00	55.00	95.00	145.00
Zambia Zambie	25.00	30.00	15.00	15.00	...	...	...	47.00	72.50	...
Zimbabwe Zimbabwe	9.00	9.00	9.00	9.00	9.00	10.25	20.00	...	28.50	29.50

Source:
International Monetary Fund (Washington, DC).

† For detailed descriptions of data pertaining to
former Czechoslovakia, Germany, SFR Yugoslavia and former
USSR, see Annex I - Country or area nomenclature, regional
and other groupings.

Source:
Fonds monétaire international (Washington, DC).

† Pour les descriptions en détails des données
relatives à l'ancienne Tchécoslovaquie, l'Allemagne, la Rfs
Yougoslavie et l'ancienne URSS, voir l'Annexe I -
Nomenclature des pays ou zones, groupements régionaux et
autres groupements.

28
Money market rates
Taux de l'argent hors banque
Treasury bill and call money rates: per cent per annum

Taux d'intérêt des bons du Trésor et de l'argent au jour le jour : pour cent par année

Country or area Pays ou zone	1985	1986	1987	1988	1989	1990	1991	1992	1993	1994
Antigua and Barbuda Antigua-et-Barbuda										
Treasury bill										
Bons du Trésor	7.00	7.00	7.00	7.00	7.00	7.00	7.00	7.00	7.00	7.00
Australia Australie										
Treasury bill										
Bons du Trésor	15.34	15.39	12.80	12.14	16.80	14.15	9.96	6.26	5.00	5.69
Call money										
Argent au jour le jour	14.70	15.75	13.06	11.90	16.75	14.81	10.47	6.44	5.11	5.18
Austria Autriche										
Call money										
Argent au jour le jour	6.11	5.19	4.35	4.59	7.46	8.53	9.10	9.35	7.22	...
Bahamas Bahamas										
Treasury bill										
Bons du Trésor	5.90	3.47	2.40	4.46	5.21	5.85	6.49	5.32	3.96	1.88
Bahrain Bahreïn										
Treasury bill										
Bons du Trésor	...	...	6.40	7.40	9.10	...	5.90	3.80	3.30	4.80
Call money										
Argent au jour le jour	...	7.20	7.10	7.90	9.20	8.50	6.30	4.00	3.50	5.20
Barbados Barbade										
Treasury bill										
Bons du Trésor	0.55	4.42	4.84	4.75	4.90	7.07	9.34	10.88	5.44	7.26
Belgium Belgique										
Call money										
Argent au jour le jour	8.27	6.64	5.66	5.04	7.00	8.29	# 9.38	9.38	8.21	5.72
Belize Belize										
Treasury bill										
Bons du Trésor	12.76	10.80	8.80	8.32	7.36	7.37	6.71	5.37	4.59	4.27
Benin Bénin										
Call money										
Argent au jour le jour	10.66	8.58	8.37	8.72	10.07	10.98	10.94	11.44	...	...
Brazil Brésil										
Treasury bill										
Bons du Trésor	215.00	151.00	195.00	483.00	382.00	...	...	...	...	...
Burkina Faso Burkina Faso										
Call money										
Argent au jour le jour	10.67	8.58	8.37	8.72	10.07	10.98	10.94	11.44	...	...
Canada Canada										
Treasury bill										
Bons du Trésor	9.43	8.97	8.15	9.48	12.05	12.81	8.73	6.59	4.84	5.54
Call money										
Argent au jour le jour	9.84	8.16	8.50	10.35	12.06	11.62	7.40	6.79	3.79	5.54
Côte d'Ivoire Côte d'Ivoire										
Call money										
Argent au jour le jour	10.66	8.58	8.37	8.72	10.07	10.98	10.94	11.44	...	...
Denmark Danemark										
Call money										
Argent au jour le jour	10.33	9.22	10.20	8.52	9.66	10.97	9.78	11.35	10.73	6.21
Dominica Dominique										
Treasury bill										
Bons du Trésor	6.50	6.50	6.50	6.50	6.50	6.50	6.50	6.50	6.40	6.40
Ethiopia Ethiopie										
Treasury bill										
Bons du Trésor	3.00	3.00	3.00	3.00	3.00	3.00	3.00	5.25	12.00	12.00

28
Money market rates
Treasury bill and call money rates: per cent per annum [cont.]
Taux de l'argent hors banque
Taux d'intérêt des bons du Trésor et de l'argent au jour le jour : pour cent par année [suite]

Country or area Pays ou zone	1985	1986	1987	1988	1989	1990	1991	1992	1993	1994
Fiji Fidji										
Treasury bill										
Bons du Trésor	7.03	6.36	9.76	1.78	2.74	4.40	5.61	3.65	2.91	2.69
Call money										
Argent au jour le jour	6.61	6.55	9.02	1.49	2.34	2.92	4.28	3.06	2.91	4.10
Finland Finlande										
Call money										
Argent au jour le jour	13.46	11.90	10.03	9.97	12.56	14.00	13.08	13.25	7.77	5.35
France France										
Call money										
Argent au jour le jour	9.93	7.74	7.98	7.52	9.07	9.85	9.49	10.35	8.75	5.69
Germany † Allemagne†										
Treasury bill										
Bons du Trésor	..	..	..	..	..	..	8.27	8.32	6.22	5.05
Call money										
Argent au jour le jour	..	..	..	..	..	..	8.84	9.42	7.49	5.35
F. R. Germany R. f. Allemagne										
Treasury bill										
Bons du Trésor	5.04	3.86	3.28	3.62	6.28	8.13	..	..	..	..
Call money										
Argent au jour le jour	5.19	4.57	3.72	4.01	6.59	7.92	..	..	..	..
Ghana Ghana										
Treasury bill										
Bons du Trésor	17.13	18.47	21.71	19.76	19.84	21.78	29.23	19.38	30.95	27.72
Greece Grèce										
Treasury bill										
Bons du Trésor	...	17.00	17.30	16.30	16.50	18.50	18.80	17.70	18.20	...
Grenada Grenade										
Treasury bill										
Bons du Trésor	6.50	6.50	6.50	6.50	6.50	6.50	6.50	6.50	6.50	6.50
Guyana Guyana										
Treasury bill										
Bons du Trésor	12.80	12.80	11.30	11.00	15.20	30.00	30.90	25.70	16.80	17.70
Hungary Hongrie										
Treasury bill										
Bons du Trésor	...	...	...	...	...	...	...	21.60	16.70	25.30
Iceland Islande										
Treasury bill										
Bons du Trésor	...	...	...	26.39	23.00	12.92	14.25	9.96	8.88	4.88
Call money										
Argent au jour le jour	0.00	0.00	31.52	34.49	21.57	12.73	14.85	12.37	8.61	4.96
India Inde										
Call money										
Argent au jour le jour	10.00	9.97	9.83	...	...	15.57	19.35	15.23	8.64	...
Indonesia Indonésie										
Call money										
Argent au jour le jour	10.33	13.00	14.51	15.00	12.57	14.37	15.12	12.14	...	...
Ireland Irlande										
Treasury bill										
Bons du Trésor	11.78	11.85	10.70	7.81	9.70	10.90	10.12	...	...	...
Call money										
Argent au jour le jour	11.87	12.28	10.84	7.84	9.55	11.10	10.45	15.11	10.49	5.66
Israel Israël										
Treasury bill										
Bons du Trésor	210.10	19.90	20.00	16.00	12.90	15.10	14.50	11.80	10.50	...

28
Money market rates
Treasury bill and call money rates: per cent per annum [*cont.*]
Taux de l'argent hors banque
Taux d'intérêt des bons du Trésor et de l'argent au jour le jour : pour cent par année [*suite*]

Country or area Pays ou zone	1985	1986	1987	1988	1989	1990	1991	1992	1993	1994
Italy Italie										
Treasury bill										
Bons du Trésor	13.71	11.40	10.73	11.13	12.58	12.38	12.54	14.32	10.58	9.17
Call money										
Argent au jour le jour	15.25	13.41	11.51	11.29	12.69	12.38	12.18	13.97	10.20	...
Jamaica Jamaïque										
Treasury bill										
Bons du Trésor	19.03	20.88	18.16	18.50	19.10	26.21	25.56	34.36	28.85	42.98
Japan Japon										
Call money										
Argent au jour le jour	6.46	4.79	3.51	3.62	4.87	7.24	7.46	4.58	# 3.06	2.20
Kenya Kenya										
Treasury bill										
Bons du Trésor	13.90	13.22	12.86	13.48	13.86	14.78	16.59	16.53	49.80	23.32
Korea, Republic of Corée, République de										
Call money										
Argent au jour le jour	9.30	9.70	8.90	9.60	13.30	14.00	17.00	14.30	12.10	12.50
Kuwait Koweït										
Treasury bill										
Bons du Trésor	5.68	5.69	5.48	6.01	8.28	...	...	...	...	...
Call money										
Argent au jour le jour	...	7.53	6.08	6.12	8.70	...	...	...	7.43	6.27
Lebanon Liban										
Treasury bill										
Bons du Trésor	14.96	18.67	26.91	25.17	18.84	18.84	17.47	22.40	18.27	15.09
Lesotho Lesotho										
Treasury bill										
Bons du Trésor	17.60	11.21	10.75	11.42	15.75	16.33	15.75	14.20	...	# 9.44
Libyan Arab Jamah. Jamah. arabe libyenne										
Call money										
Argent au jour le jour	4.00	4.00	4.00	4.00	...	...	...	...		
Malawi Malawi										
Treasury bill										
Bons du Trésor	12.31	12.75	14.25	15.75	15.75	12.92	11.50	15.62	23.54	...
Malaysia Malaisie										
Call money										
Argent au jour le jour	7.57	8.02	2.85	3.22	4.72	6.81	7.83	...	3.25	...
Maldives Maldives										
Call money										
Argent au jour le jour	9.00	9.00	8.67	8.50	7.33	7.00	...	...	...	...
Mali Mali										
Call money										
Argent au jour le jour	10.66	8.58	8.37	8.72	10.07	10.98	10.94	11.43	...	...
Mauritius Maurice										
Call money										
Argent au jour le jour	11.20	11.10	10.30	...	...	13.30	12.20	9.00	7.70	10.70
Mexico Mexique										
Treasury bill										
Bons du Trésor	63.20	...	103.07	# 69.15	44.99	34.76	19.28	15.62	15.03	14.10
Call money										
Argent au jour le jour	62.43	88.01	95.59	69.01	# 47.43	37.36	23.58	18.87	17.38	16.47
Morocco Maroc										
Treasury bill										
Bons du Trésor	10.00	10.50	10.50	10.50	10.50	9.50	9.50	...	...	...

28
Money market rates
Treasury bill and call money rates: per cent per annum [*cont.*]
Taux de l'argent hors banque
Taux d'intérêt des bons du Trésor et de l'argent au jour le jour : pour cent par année [*suite*]

Country or area Pays ou zone	1985	1986	1987	1988	1989	1990	1991	1992	1993	1994
Call money Argent au jour le jour	9.41	9.44	...	...	...	...	...	...	...	...
Namibia Namibie Treasury bill Bons du Trésor	...	...	...	...	...	...	...	13.88	12.16	11.35
Nepal Népal Treasury bill Bons du Trésor	5.00	5.00	5.00	5.00	5.62	7.93	8.80	9.00	...	...
Netherlands Pays-Bas Call money Argent au jour le jour	6.30	5.83	5.16	4.48	6.99	8.29	9.01	9.27	7.10	5.14
Netherlands Antilles Antilles néerlandaises Treasury bill Bons du Trésor	7.21	7.34	6.36	5.79	5.96	6.10	...	...	4.83	4.48
New Zealand Nouvelle-Zélande Treasury bill Bons du Trésor	...	19.97	20.49	...	13.51	13.78	9.74	6.72	6.21	6.69
Niger Niger Call money Argent au jour le jour	10.66	8.58	8.37	8.72	10.07	10.98	10.94	11.44	...	...
Pakistan Pakistan Call money Argent au jour le jour	8.13	6.59	6.25	6.32	6.30	7.29	7.64	7.51	11.00	8.36
Papua New Guinea Papouasie-Nvl-Guinée Treasury bill Bons du Trésor	10.40	12.32	10.44	10.12	10.50	11.40	10.33	8.88	6.25	...
Philippines Philippines Treasury bill Bons du Trésor	26.73	16.08	11.51	14.67	18.65	23.67	21.48	16.02	12.45	12.71
Poland Pologne Treasury bill Bons du Trésor	...	...	...	...	...	...	...	44.00	33.20	28.80
Call money Argent au jour le jour	...	...	...	...	...	...	49.90	# 29.50	24.50	23.30
Portugal Portugal Treasury bill Bons du Trésor	20.90	15.56	13.89	12.96	...	13.52	14.19	12.88	...	...
Call money Argent au jour le jour	20.17	14.52	13.69	12.34	12.84	13.73	15.81	# 17.48	13.25	10.62
Saint Kitts and Nevis Saint-Kitts-et-Nevis Treasury bill Bons du Trésor	6.50	6.50	6.50	6.50	6.50	6.50	6.50	6.50	6.50	6.50
Saint Lucia Sainte-Lucie Treasury bill Bons du Trésor	7.00	7.00	7.00	7.00	7.00	7.00	7.00	7.00	7.00	7.00
St. Vincent-Grenadines St. Vincent-Grenadines Treasury bill Bons du Trésor	6.50	6.50	6.50	6.50	6.50	6.50	6.50	6.50	6.50	6.50
Senegal Sénégal Call money Argent au jour le jour	10.66	8.58	8.37	8.72	10.07	10.98	10.94	11.43	...	...
Seychelles Seychelles Treasury bill Bons du Trésor	12.44	12.91	15.15	13.90	13.41	13.17	13.40	13.40	13.25	11.78

28
Money market rates
Treasury bill and call money rates: per cent per annum [*cont.*]
Taux de l'argent hors banque
Taux d'intérêt des bons du Trésor et de l'argent au jour le jour : pour cent par année [*suite*]

Country or area Pays ou zone	1985	1986	1987	1988	1989	1990	1991	1992	1993	1994
Sierra Leone Sierra Leone										
Treasury bill Bons du Trésor	12.00	14.50	16.50	18.00	22.00	47.50	50.67	78.63	28.64	12.19
Singapore Singapour										
Call money Argent au jour le jour	5.38	4.27	3.89	4.30	5.34	6.61	4.76	2.74	2.50	3.59
Solomon Islands Iles Salomon										
Treasury bill Bons du Trésor	9.58	12.00	11.33	11.00	11.00	11.00	13.71	13.50	12.15	...
South Africa Afrique du Sud										
Treasury bill Bons du Trésor	17.56	10.43	8.71	12.03	16.84	17.80	16.68	13.77	11.31	...
Call money Argent au jour le jour	18.21	10.92	9.50	13.90	18.77	19.46	17.02	14.11	...	...
Spain Espagne										
Treasury bill Bons du Trésor	10.90	8.63	8.03	# 10.79	13.56	14.17	12.45	12.44	10.53	8.11
Call money Argent au jour le jour	11.60	11.50	16.07	11.30	14.39	14.76	13.20	13.01	...	...
Sri Lanka Sri Lanka										
Treasury bill Bons du Trésor	13.39	10.47	7.30	13.59	14.81	14.08	13.75	16.19	16.52	12.68
Call money Argent au jour le jour	14.56	12.95	13.14	18.65	22.19	21.56	25.42	21.63	25.65	18.54
Swaziland Swaziland										
Treasury bill Bons du Trésor	16.47	9.76	5.96	7.28	10.16	11.14	12.67	12.34	8.25	8.35
Call money Argent au jour le jour	...	...	...	...	8.39	10.50	10.61	10.25	9.73	7.01
Sweden Suède										
Treasury bill Bons du Trésor	14.17	9.83	9.39	10.08	11.50	13.66	11.59	12.85	8.35	7.40
Call money Argent au jour le jour	13.85	10.15	9.16	10.08	11.52	13.45	11.81	18.42	9.08	7.36
Switzerland Suisse										
Treasury bill Bons du Trésor	4.15	3.54	3.18	3.01	6.60	8.32	7.74	7.76	4.75	3.97
Call money Argent au jour le jour	3.75	3.17	2.51	2.22	6.50	8.33	7.73	7.47	4.94	3.85
Thailand Thailande										
Treasury bill Bons du Trésor	11.02	6.76	3.63	5.08	...	...	...	...	...	...
Call money Argent au jour le jour	13.48	8.07	5.91	8.66	9.82	12.73	10.58	7.07	...	...
Togo Togo										
Call money Argent au jour le jour	10.66	8.58	8.37	8.72	10.07	10.98	10.94	11.44	...	...
Trinidad and Tobago Trinité-et-Tobago										
Treasury bill Bons du Trésor	3.47	3.99	4.63	...	7.13	7.50	7.67	9.26	9.45	...
Tunisia Tunisie										
Call money Argent au jour le jour	10.28	9.95	10.00	9.15	9.40	11.53	11.79	11.73	10.48	8.81
Turkey Turquie										
Treasury bill Bons du Trésor	...	...	41.92	54.56	48.01	43.46	67.01	72.17	...	...

28
Money market rates
Treasury bill and call money rates: per cent per annum [cont.]
Taux de l'argent hors banque
Taux d'intérêt des bons du Trésor et de l'argent au jour le jour : pour cent par année [suite]

Country or area Pays ou zone	1985	1986	1987	1988	1989	1990	1991	1992	1993	1994
Call money Argent au jour le jour	...	...	39.82	60.62	40.66	51.90	72.74	65.35	62.83	136.47
Uganda Ouganda										
Treasury bill Bons du Trésor	22.00	30.67	30.50	33.00	42.17	41.00	34.17	...	# 21.30	12.52
United Kingdom Royaume-Uni										
Treasury bill Bons du Trésor	11.55	10.36	9.25	9.78	13.05	14.08	10.96	8.93	5.18	5.21
Call money Argent au jour le jour	10.78	10.68	9.66	10.31	13.88	14.68	11.75	9.55	5.46	4.76
United States Etats-Unis										
Treasury bill Bons du Trésor	7.49	5.97	5.83	6.67	8.11	7.51	5.41	3.46	3.02	4.27
Call money Argent au jour le jour	8.10	6.80	6.66	7.61	9.22	8.10	5.70	3.52	3.02	4.20
Vanuatu Vanuatu										
Call money Argent au jour le jour	7.00	6.96	6.50	7.50	7.08	7.00	7.00	5.92	6.00	6.00
Yugoslavia, SFR† Yougoslavie, Rfs†										
Call money Argent au jour le jour	...	64.00	93.30	423.30	4 150.80	...	...	...	...	...
Zambia Zambie										
Treasury bill Bons du Trésor	13.21	24.25	16.50	15.17	18.50	25.92	...	...	124.02	...
Zimbabwe Zimbabwe										
Treasury bill Bons du Trésor	8.48	8.71	8.73	8.38	8.17	8.39	13.34	...	33.04	29.22
Call money Argent au jour le jour	8.80	9.10	9.30	9.08	8.73	8.68	17.36	...	34.18	...

Source:
International Monetary Fund (Washington, DC).

† For detailed descriptions of data pertaining to
former Czechoslovakia, Germany, SFR Yugoslavia and former
USSR, see Annex I - Country or area nomenclature, regional
and other groupings.

Source:
Fonds monétaire international (Washington, DC).

† Pour les descriptions en détails des données
relatives à l'ancienne Tchécoslovaquie, l'Allemagne, la Rfs
Yougoslavie et l'ancienne URSS, voir l'Annexe I -
Nomenclature des pays ou zones, groupements régionaux et
autres groupements.

Technical notes, tables 27 and 28

Detailed information and current figures relating to tables 27 and 28 are contained in *International Financial Statistics*, published monthly by the International Monetary Fund [11] and in the United Nations *Monthly Bulletin of Statistics*. [23]

Table 27: Rates shown represent those rates at which the central bank either discounts or makes advances against eligible commercial paper and/or government securities for commercial banks or brokers. For countries with more than one rate applicable to such discounts or advances, the rate shown is the one at which the largest proportion of central bank credit operations is understood to be transacted.

Table 28: Treasury bill rates represent the average tender rates per annum on new issues of bills (ordinarily 3-month issues) offered by the treasury during the period. Call money rates are the average rates per annum on loans which were available on demand in the open market during the period.

Notes techniques, tableaux 27 et 28

Les informations détaillées et les chiffres courants concernant les tableaux 27 et 28 figurent dans les *Statistiques financières internationales* publiées chaque mois par le Fonds monétaire international [11] et dans le *Bulletin mensuel de statistique* des Nations Unies [23].

Tableau 27 : Les taux indiqués représentent les taux pratiqués par la banque centrale à l'escompte, ou pour avance de fonds, dans toute transaction portant sur des effets de commerce ou des obligations de l'Etat détenus par les banques commerciales ou des courtiers. Pour les pays où il existe plus d'un taux applicable à de telles transactions, le tableau indique le taux que la Banque centrale semble pratiquer pour la plupart de ses opérations de crédit.

Tableau 28 : Les taux d'intérêt des bons du Trésor représentent les taux moyens annuels d'adjudication des nouvelles émissions de bons (généralement à trois mois) offertes par le Trésor au cours de la période considérée. Les taux d'intérêt de l'argent au jour le jour représentent les taux moyens annuels des prêts pouvant être obtenus sans préavis sur l'"open market" pendant la période considérée.

Table 29 follows overleaf
Le tableau 29 est présenté au verso

29
Employment by industry
Emploi par industrie

		Total employment (000s) Emploi total (000s)		Persons employed, by branch of economic activity (000s) Personnes employées, par branches d'activité économique (000s)							
				Agriculture, hunting, forestry and fishing Agriculture, chasse sylviculture, pêche		Mining and quarrying Industries extractives		Manufacturing Industries manufacturières		Electricity, gas, water Electricité, gaz, eau	
Country or area Pays ou zone	Year Année	M	F	M	F	M	F	M	F	M	F
Albania[1 2] Albanie[1 2]	1980	397.2	262.4	75.0	62.6	...	...	126.7[3]	102.0[3]	...	...
	1991	494.1	356.4	107.2	88.0	...	...	151.2[3]	141.1[3]	...	...
Angola[4] Angola[4]	1983	355.5	...	84.1	...	...	...	57.8	...	...	...
	1986	367.6	...	75.5	...	...	...	66.1	...	...	...
Argentina Argentine	1982	3740.6	...	4.6	...	2.5	...	846.3	...	16.9	...
	1984	3897.0	...	10.2	...	2.2	...	962.2	...	21.2	...
Australia[5 6] Australie[5 6]	1980	3997.8	2286.5	313.0	94.3	76.8	6.2	937.3	309.3	119.5	10.1
	1993	4420.5	3259.1	284.7	121.6	80.7	8.2	791.8	301.3	82.5	13.0
Austria[16] Autriche[16]	1980	1703.9	1136.8	28.5	12.8	24.1	2.6	607.5	313.2	27.3	4.9
	1993[9]	2090.6	1517.5	126.4	117.6	7.7	0.9	660.6	246.5	31.3	4.0
Bahamas Bahamas	1986	53.2	44.2	4.1	0.8	0.1	...	2.5	2.5	1.4	0.4
	1989	59.8	52.7	3.8	1.1	0.3	0.1	2.0	2.2	1.3	0.2
Bahrain[7 40] Bahreïn[7 40]	1987	80.3	5.7	2.2	0.1	0.1	0.0	16.5	0.5	2.6	0.0
	1993	99.5	10.5	0.9	0.3	0.7	0.4	21.5	3.2	3.2	0.5
Bangladesh Bangladesh	1983	25547.0	2429.0	16231.0	216.0	46.0	...	1785.0	698.0	68.0	...
	1989	29386.0	20761.0	17735.0	14836.0	82.0	6.0	2491.0	4484.0	14.0	3.0
Barbados[5 6] Barbade[5 6]	1981	57.4	42.8	5.6	3.8	...	...	6.4	7.7	1.1	0.1
	1993	51.2	44.2	3.2	2.0	...	...	4.9	5.1	1.1	0.2
Belarus[4] Bélarus[4]	1984	5091.7	...	1305.0	...	1534.4	...	1416.8	...	24.5	...
	1993	4792.5	...	1048.8	...	1429.6	...	1311.8	...	38.1	...
Belgium[1 6 9] Belgique[1 6 9]	1980	2409.7	1290.5	90.2	25.4	27.8	0.5	702.2	219.1	30.3	2.4
	1992	2236.0	1517.1	68.5	26.4	6.3	0.3	576.1	175.6	25.7	3.4
Benin[4] Bénin[4]	1980	66.2	...	5.0	...	0.1	...	6.5	...	0.7	...
	1985	80.8	...	6.1	...	0.4	...	8.3	...	1.1	...
Bermuda[9 10] Bermudes[9 10]	1980	16.9	12.8	0.2	0.0	0.1	0.0	0.7	0.4	0.3	0.1
	1993	16.9	16.5	0.4	0.0	0.1	0.0	0.7	0.4	0.4	0.1
Bolivia[5 6] Bolivie[5 6]	1980	1316.7	402.9	686.1	113.5	70.7	5.0	107.7	69.4	6.3	0.6
	1990	1375.2	468.2	743.8	129.6	43.1	4.4	79.5	50.8	8.2	0.8
Botswana[5 9] Botswana[5 9]	1980	64.0	19.4	3.7	0.5	6.9	0.3	4.6	0.9	1.4	0.0
	1992	145.6	81.9	4.2	1.8	7.3	0.4	16.1	9.5	2.3	0.3
Brazil[5 6 9] Brésil[5 6 9]	1981	31266.0	14199.0	10495.0	2805.0	674.0[16 20]	78.0	5117.0	1693.0	...	...
	1990	40018.0	22083.0	11235.0	2945.0	764.0[16 20]	96.0	6774.0	2637.0	...	...
British Virgin Islands[4 10] Iles Vierges brit.[4 10]	1983	4.6	...	0.1	...	0.0	...	0.2	...	0.1	...
	1987	6.6	...	0.1	...	0.0	...	0.4	...	0.2	...
Brunei Darussalam[4 9 12] Brunéi Darussalam[4 9 12]	1980	27.1	...	0.5	...	4.8	...	2.2	...	...	...
	1986	30.0	...	0.7	...	4.8	...	2.7	...	...	...
Bulgaria[4 13] Bulgarie[4 13]	1980	4363.9	...	1062.7	...	108.6	...	1401.0	...	25.4	...
	1992[38]	3112.9	...	558.9	...	...	...	1026.8[3]	...	...	...
Burundi[4 9 14] Burundi[4 9 14]	1980	38.2	...	...	...	1.5	...	3.9	...	...	...
	1991	44.7	...	6.6	...	0.3	...	6.2	...	0.9	...
Canada[5 6] Canada[5 6]	1980	6459.0	4249.0	450.0	134.0	172.0	23.0	1543.0[15]	568.0[15]	105.0[16]	19.0[16]
	1993	6753.0	5630.0	395.0	155.0	129.0	20.0	1297.0[15]	503.0[15]	108.0[16]	34.0[16]
Central African Rep.[4 40] Rép. centrafricaine[4 40]	1980	14.5	...	4.7	...	...	...	2.1	...	0.6	...
	1990	13.0	...	2.3	...	...	...	4.0	...	0.8	...
Chad[10] Tchad[10]	1986	8.4	0.5	1.6	0.1	...	...	0.5	0.0	0.5	0.0
	1991	11.8	0.7	1.3	0.1	0.2	0.0	5.2	0.1	0.0	0.0
Chile[5 6 17] Chili[5 6 17]	1980	2297.9	959.3	503.0	26.9	69.8	2.0	374.8	149.1	22.9	1.5
	1993	3375.2	1610.5	727.4	97.9	88.3	3.9	594.5	240.7	23.4	3.5
China[9 18] Chine[9 18]	1985	60795.0	29100.0	5055.0	2776.0	...	...	25221.0[3]	12924.0[3]	...	...
	1989	66867.0	34220.0	5076.0	2867.0	6218.0	1589.0	20292.0	13144.0	1089.0	395.0
Colombia[6 9] Colombie[6 9]	1980	1977.6	1224.6	37.1	6.2	11.9	1.0	520.1	300.9	20.3	3.7
	1992	2819.5	2021.7	53.0	15.4	15.0	4.3	660.8	477.9	27.1	5.4
Costa Rica[5 6 9] Costa Rica[5 6 9]	1980	548.3	176.4	188.5	10.4	...	...	83.0[8]	34.9[8]	...	...
	1992	735.7	307.3	234.6	16.6	1.5	...	123.7	73.5	11.5	1.5
Côte d'Ivoire[4 41] Côte d'Ivoire[4 41]	1980	470.2	...	71.0	...	...	...	73.5[3]	...	...	...
	1990	385.0	...	53.3	...	...	...	60.0[3]	...	...	...
Croatia Croatie	1981	857.49	560.1	52.1	14.5	11.1	1.5	295.7	196.8	22.6	4.5
	1992	639.64	519.4	46.1	15.5	6.8	1.2	226.6	162.9	22.5	5.2
Cuba[2] Cuba[2]	1980	1784.6	822.0	437.3[21]	74.1[21]	...	...	420.4[3]	145.3[3]	11.1	2.6
	1988[38]	2135.6	1309.8	519.4	142.7	29.9	5.3	452.2	230.9	29.2	9.7

Construction Construction		Trade, restaurants and hotels Commerce, restaurants, hôtels		Transport, storage, communications Transports, entrepôts, communications		Finance, insurance, real est.,bus. services Services financières, immob.,et apparentées		Community, social and personal services Services à collectivité services soc. et pers.	
M	F	M	F	M	F	M	F	M	F
65.7	6.5	23.5	24.5	23.4	4.4	10.6	4.8	72.3	57.6
69.7	6.4	27.7	29.6	32.1	6.4	12.8	13.9	93.4	71.0
19.8	...	30.0	...	24.1	...	1.1	...	132.7	...
21.6	...	31.9	...	27.4	...	1.6	...	135.2	...
230.9	...	659.6	...	190.5	...	238.7	...	1137.2	...
219.3	...	635.0	...	218.5	...	252.5	...	1101.1	...
439.1	47.8	813.6	680.7	378.2	82.1	286.5	231.2	628.3	822.3
470.1	75.6	1038.1	917.3	373.3	115.3	442.0	414.1	833.3	1281.9
240.1	27.6	212.6	287.3	171.6	35.3	81.6	81.3	310.6	371.8
291.6	25.8	282.1	423.3	186.3	43.6	130.7	131.1	367.2	515.5
8.0	0.2	15.3	16.2	5.2	2.7	3.5	4.5	13.0	0.8
9.6	0.3	16.7	19.6	5.8	3.1	3.6	4.9	15.8	1.2
30.0	0.3	12.9	0.8	4.1	0.4	4.4	1.5	7.5	2.0
32.3	0.3	17.7	1.3	8.1	2.0	4.8	1.6	9.3	1.9
477.0	9.0	3115.0	140.0	1077.0	11.0	133.0	3.0	2080.0	214.0
610.0	51.0	3909.0	220.0	1268.0	9.0	229.0	8.0	1606.0	188.0
6.1[8]	0.2[8]	12.6	10.8	3.6	1.2	1.7	2.2	20.3	16.8
6.6[8]	0.3[8]	7.0	7.6	3.4	1.2	2.3	3.2	18.4	19.7
350.6	...	317.7	...	429.3	...	27.6	...	1022.3	...
369.1	...	405.3	...	328.9	...	33.6	...	1072.7	...
274.9	12.3	300.6	282.2	238.9	36.1	141.7	84.3	603.1	628.2
229.6	15.7	318.0	316.2	211.3	45.8	196.8	144.8	603.7	788.8
7.4	...	8.0	...	8.8	...	2.2	...	27.6	...
5.5	...	9.1	...	9.0	...	2.5	...	39.0	...
1.7	0.1	6.1	5.3	1.6	0.6	1.4	2.0	4.6	4.3
1.6	0.1	5.3	5.4	1.6	0.7	2.2	3.2	4.7	6.4
93.8	0.8	58.5	68.7	89.9	3.0	11.2	2.6	192.5	139.4
49.3	0.5	68.5	81.7	130.4	5.1	12.5	3.2	239.9	192.0
13.1	0.3	6.2	4.3	3.2	0.3	3.3	1.1	21.6	11.6
28.9	4.8	18.4	22.5	7.5	2.7	11.4	6.2	49.4	33.8
3592.0	71.0	3301.0[43]	1387.0[43]	1619.0	149.0	825.0	398.0	5643.0[11]	7617.0[11]
3726.0	97.0	5060.0[43]	2916.0[43]	2246.0	194.0	1151.0	565.0	9062.0[11]	12633.0[11]
0.5	...	0.4	...	0.4	...	0.2	...	2.6	...
0.5	...	0.6	...	0.4	...	0.3	...	4.1	...
9.6	...	4.5	...	1.7	...	1.3	...	2.7	...
9.2	...	6.4	...	1.4	...	2.0	...	2.8	...
328.7	...	370.1	...	295.3	...	54.4	...	717.7	...
193.3	...	328.4	...	241.6[48]	...	52.4	...	681.5	...
5.3	...	...	...	1.2	...	...	...	...	...
2.4	...	3.8	...	2.9	...	1.9	...	17.7	...
570.0	55.0	1274.0	1137.0	613.0	170.0	492.0	526.0	1241.0	1618.0
585.0	75.0	1514.0	1398.0	559.0	208.0	699.0	767.0	1468.0	2469.0
0.8	...	4.8	...	1.2	...	0.0	...	0.2	...
1.3	...	4.3	...	1.5	...	0.3	...	...	...
1.1	0.0	1.2	0.0	1.2	0.1	0.1	0.1	1.9	0.1
1.0	0.0	0.1	0.1	0.9	0.1	0.3	0.0	2.1	0.3
148.4	3.4	365.3	224.6	194.3	16.6	74.1	26.8	539.8	507.0
391.0	12.0	517.7	407.8	315.6	39.2	184.1	104.8	532.8	700.6
4602.0	1192.0	5028.0	3407.0	4623.0	1225.0	599.0	331.0	14837.0[19]	7021.0[19]
4550.0	1233.0	5639.0	4167.0	4947.0	1453.0	864.0	492.0	17396.0[19]	8637.0[19]
218.2	11.8	475.0	285.9	176.2	16.2	136.0	81.5	380.9	516.9
288.3	16.3	680.0	552.7	266.2	32.1	216.0	130.0	604.4	781.3
55.7	0.5	90.8[19]	40.5[19]	44.3[20]	3.2[20]	...	...	84.5	86.5
60.3	1.0	105.0	67.8	44.1	4.7	29.8	8.0	119.1	131.8
50.0	...	38.0	...	65.8	...	...	...	171.9[19]	...
17.4	...	25.9	...	53.5	...	...	...	174.9[19]	...
135.2	14.0	101.3	121.8	107.9	24.0	20.4	24.1	111.2	159.0
69.4	10.2	77.1	105.7	80.7	23.9	16.7	25.4	93.8	169.6
243.4	24.2	190.9	135.5	163.4	32.9	19.1	15.0	299.0	392.4
280.1	42.5	223.3	213.2	187.4	54.8	33.5[44]	29.6[44]	380.6	581.1

29
Employment by industry [*cont.*]
Emploi par industrie [*suite*]

Persons employed, by branch of economic activity (000s)
Personnes employées, par branches d'activité économique (000s)

Country or area Pays ou zone	Year Année	Total employment (000s) Emploi total (000s) M	F	Agriculture, hunting, forestry and fishing Agriculture, chasse sylviculture, pêche M	F	Mining and quarrying Industries extractives M	F	Manufacturing Industries manufacturières M	F	Electricity, gas, water Electricité, gaz, eau M	F
Cyprus[5]	1980	122.8	69.7	18.3	18.6	1.6	0.1	21.5	18.4	1.4	0.1
Chypre[5]	1993	163.8*	106.6*	19.3	15.7	0.7		25.0	20.5	1.4	0.1
former Czechoslovakia †	1980[1]	4389.0	3772.0	592.0	443.0	171.0	35.0	1459.0	1274.0	76.0	31.0
anc. Tchécoslovaquie †	1990[13]	4317.0	3932.0	576.0	367.0	219.0	41.0	1349.0	1247.0	89.0	35.0
Denmark[6][22]	1981	1307.8	1060.9	137.6	36.7	2.1	0.2	357.5	147.6	14.1	2.0
Danemark[6][22]	1991	1429.6	1217.6	112.3	37.0	2.1	0.4	357.6	176.3	17.2	2.6
Ecuador[6][9][39]	1987[4]	1147.5	...	28.2[8]	...	...	...	243.4	...	...	...
Ecuador[6][9][39]	1993	1636.4	1014.4	169.4	24.6	16.6	1.9	302.7	159.8	16.0	3.0
Egypt[6][9]	1980	9106.3	692.8	4086.2	65.7	17.8	2.1	1364.2	74.8	78.5	4.7
Egypte[6][9]	1991	10971.7	2855.1	3220.5	1111.7	37.5	0.7	1772.7	354.3	119.5	14.8
El Salvador[6][39]	1980	926.8	460.3	472.8	46.9	2.9	0.5	127.0	96.3	7.6	0.4
El Salvador[6][39]	1992[38]	1134.4	647.2	542.2	96.1	2.3	0.3	167.4	141.3	9.7	1.9
Estonia[4][13]	1980[2]	788.0	...	109.7		...		267.5[3]		...	
Estonie[4][13]	1993	516.5		58.5		13.4		127.3[3]		11.7	
Fiji[49]	1980	80.5	...	2.6		1.1		15.4		2.3	
Fidji[49]	1993	94.3	...	2.0		1.9		23.5		2.7	
Finland[16]	1980	1271.0	1088.0	186.0	127.0	...	...	395.0[3]	233.0[3]		
Finlande[16]	1993[38]	1071.0	993.0	117.0	56.0	4.0	...	265.0	131.0	18.0	5.0
France[1]	1980	13261.5	8485.8	1197.8	623.5	...		5833.4[3]	1879.4[3]	...	...
France[1]	1992	12641.1	9643.5	777.7	372.1	...		4784.6[3]	1555.2[3]	...	...
French Polynesia[4]	1982	34.0	...	0.6	...	0.2		1.6		0.3	...
Polynésie française[4]	1989	41.7	...	0.9		0.1		2.2		0.4	...
Gambia[7][9]	1983	25.0	4.3	2.2	0.3	...		3.1	0.9	1.4	0.0
Gambie[7][9]	1987	21.1	5.0	1.4	0.6			1.9	0.5	0.8	0.1
Germany † · Allemagne †											
F. R. Germany[6][9]	1980	16782.0	10092.0	732.0	706.0	335.0	26.0	6386.0	2747.0	218.0	36.0
R. f. Allemagne[6][9]	1992	17845.0	12249.0	601.0	440.0	213.0	17.0	6561.0	2694.0	229.0	50.0
former German D. R.[6][9]	1980	3847.7	3909.9	201.8	135.6	...		1988.5[3]	1507.7[3]	...	
anc. R. d. allemande[6][9]	1992	3778.0	3068.0	213.0	125.0	91.0	31.0	1048.0	514.0	87.0	34.0
Ghana[49]	1980	337.2	...	54.9	...	23.8	...	35.1	...	6.6	...
Ghana[49]	1989	214.9	...	18.2	...	14.5	...	27.9	...	6.1	...
Gibraltar[9][23]	1980	8.9	2.9	...	...	...	...	2.6	0.5	0.2	0.0
Gibraltar[9][23]	1992	9.9	4.8	...	...	...		0.4	0.1	0.3	0.0
Greece[16][17]	1981	2423.5	1107.5	622.4	461.1	18.8	0.4	486.0[15]	194.8[15]	26.1	4.2
Grèce[16][17]	1992	2403.2	1281.3	468.7	338.0	17.6	0.7	487.3[15]	211.4[15]	30.9	5.8
Guam[9][13]	1982	18.2	12.5	0.1	...	...	...	0.9	0.3	...	...
Guam[9][13]	1987	25.8	18.3	0.2	...	...		1.0	0.7	...	...
Guatemala[4][13]	1980	755.5	...	373.5	...	3.7	...	83.1	...	13.5[16]	...
Guatemala[4][13]	1993	823.2		214.6	...	2.4	...	136.7	...	11.1[16]	...
Haiti[5][9]	1980	999.2	954.4	812.4	507.2	1.0	0.2	57.6	74.5	1.5	0.1
Haïti[5][9]	1988	1123.3	665.1	854.9	329.9	9.0	8.6	66.3	49.2	2.2	0.7
Honduras[6]	1985	162.1	118.6	2.5	0.3	0.6	0.5	32.2	22.4	2.6	0.4
Honduras[6]	1992	1152.0	522.7[39]	607.6	32.3[39]	5.4	358.0[39]	125.4	125.0[39]	6.8	1.1[39]
Hong Kong[5][6]	1980	1458.0	779.8	22.5	9.5	0.6	...	516.0	427.1	10.7	0.9
Hong-kong[5][6]	1993	1777.1	1039.0	13.7	3.8	0.3	...	376.4	224.0	17.4	1.9
Hungary	1980	2782.4	2261.7	679.1	433.4	...	...	927.3[3]	748.3[3]	...	...
Hongrie	1992	2185.2	2131.3	364.8	204.7	...	...	718.2[3]	599.6[3]	...	...
Iceland[4][6]	1980	105.9	...	14.0	...	...	...	25.9[8]	...	0.9	...
Islande[4][6]	1990	124.6	...	13.1	...	...	...	23.3[8]	...	1.1	...
India[9][27][28]	1980	19603.0	2702.0	829.0	461.0	836.0	86.0	5309.0	563.0	679.0	16.0
Inde[9][27][28]	1989	22417.0	3545.0	975.0	467.0	975.0	78.0	5670.0	575.0	880.0	28.0
Indonesia[16]	1980	34618.5	16934.6	19828.4	9179.5	329.0	60.4	2537.5	2112.5	60.3	6.1
Indonésie[16]	1992	47644.6	30459.5	25629.5	17224.0	489.9	104.9	4187.1	3660.5	157.5	15.8
Ireland[6][9]	1983	777.6	346.4	167.6	21.8	9.1	0.5	160.4	57.6	13.8	1.3
Irlande[6][9]	1991	748.1	377.0	141.9	12.3	6.1	0.5	157.8	63.4	12.5	1.5
Israel[5][6][29]	1980	801.9	452.2	60.7	18.6	...	...	229.9[8]	64.3[8]	11.2	1.7
Israël[5][6][29]	1993	1033.1	718.1	48.5	13.5	...		271.1[8]	100.2[8]	14.9	2.6
Italy[16]	1980	14135.0	6540.0	1851.0	1048.0	201.0[20]	19.0[20]	3715.0	1724.0	...	...
Italie[16]	1992	14021.0	7588.0	1105.0	644.0	212.0[20]	26.0[20]	3179.0	1500.0	...	...
Jamaica[5][6]	1980	424.7	273.8	198.3	62.3	7.0	1.2	53.6	19.3	...	...
Jamaïque[5][6]	1990	514.2	379.3	174.4	58.4	6.4	0.8	96.3	39.8	...	...

Construction Construction		Trade, restaurants and hotels Commerce, restaurants, hôtels		Transport, storage, communications Transports, entrepôts, communications		Finance, insurance, real est.,bus. services Services financières, immob.,et apparentées		Community, social and personal services Services à collectivité services soc. et pers.	
M	F	M	F	M	F	M	F	M	F
20.3	1.7	20.7	13.0	7.9	1.6	5.2	3.0	22.2	12.4
21.9	1.4	37.3	30.5	12.1	4.1	10.3	9.1	33.1	24.5
583.0	96.0	250.0	647.0	344.0	170.0	128.0	126.0	777.0	938.0
548.0	99.0	228.0	700.0	359.0	192.0	140.0	147.0	795.0	1091.0
155.9	16.0	164.7	171.9	125.2	38.0	83.7	73.7	256.7	567.4
147.9	19.0	196.7	184.4	132.1	49.7	132.0	106.4	322.4	631.0
91.8	...	282.6	...	67.5[20]	...	71.5	...	361.8	...
156.6	3.8	395.7	390.6	140.5	15.6	81.7	37.7	356.6	377.4
419.5	6.1	836.4	47.9	475.7	27.6	106.5	20.3	1600.3	381.5
1024.1	110.6	1247.6	230.1	984.9	99.8	350.1	72.0	2213.5	860.1
56.1	0.4	76.1	183.6	54.5	2.7	10.3	5.7	119.4	123.7
79.2	1.8	123.6	184.0	56.5	4.5	12.2	7.8	139.3	209.4
71.5	...	72.1	...	75.2	...	4.0	...	176.8	...
33.2	...	66.6	...	47.4	...	29.1	...	129.5	...
9.0	...	13.4	...	8.1	...	4.4	...	24.1	...
6.8	...	13.7	...	10.2	...	5.9	...	27.6	...
160.0	16.0	137.0	191.0	135.0	48.0	44.0	83.0	207.0	386.0
113.0	12.0	132.0	169.0	116.0	42.0	84.0	95.0	219.0	480.0
...	...	6230.3[19]	5982.8[19]	...	...	...	...	...	...
...	...	7078.8[19]	7716.2[19]	...	...	...	...	...	...
4.7	...	7.4	...	2.3	...	1.5	...	15.4	...
5.3	...	9.5	...	9.7	...	2.8	...	15.3	...
1.9	0.0	2.5	0.4	5.5	0.3	0.4	0.1	8.0	2.2
2.9	0.0	3.7	0.8	2.7	0.4	0.8	0.3	7.0	2.3
1972.0	188.0	1651.0	2127.0	1201.0	326.0	814.0	735.0	3474.0	3202.0
1815.0	247.0	1899.0	2578.0	1261.0	451.0	1342.0	1218.0	3923.0	4555.0
484.3	92.7	219.5[19]	632.5[19]	407.9	230.8	...	...	545.8[11 24]	1310.6[11 24]
688.0	81.0	340.0	538.0	359.0	186.0	171.0	208.0	781.0[11]	1351.0[11]
22.4	...	22.1	...	17.0	...	11.0	...	144.3	...
10.2	...	11.1	...	6.4	...	11.5	...	109.0	...
2.3	0.1	1.4	0.8	0.6	0.1	0.2	0.2	1.3	1.0
3.0	0.2	1.8	1.4	0.5	0.1	0.6	0.9	2.5	1.7
290.5	2.1	359.7	168.9	247.4	26.4	78.1	38.6	292.9	210.9
242.5	3.7	423.9	263.4	220.7	29.5	120.1	80.5	391.5	348.1
1.6	0.1	2.9[43]	2.6[43]	1.3[20]	0.3[20]	0.5	0.8	2.3[11]	1.8[11]
4.2	0.2	4.0[43]	4.0[43]	1.5[20]	0.7[20]	0.8	1.2	4.0[11]	3.7[11]
27.7	...	58.9[19]	...	21.6	...	...	...	173.5	...
26.4	...	102.6[19]	...	25.2	...	...	...	304.2	...
21.9	0.3	28.0	293.2	14.0	1.4	3.2	0.5	59.6	77.0
19.1	3.3	65.2	195.6	14.6	2.0	2.2	1.3	62.9	52.5
13.9	0.5	38.1	30.6	11.3	1.8	7.8	4.6	52.8	57.9
70.0	2.6[39]	130.8	151.1[39]	45.7	6.3[39]	19.6	10.4[39]	140.1	193.5[39]
157.0	10.1	319.1	129.4	147.5	15.7	64.9	38.0	219.3	149.0
214.1	9.6	462.0	338.4	266.2	51.4	158.7	110.6	268.3	299.2
327.0	71.2	173.8	314.2	303.1	100.5	...	...	372.1[19 26]	594.1[19 26]
211.2	60.1	205.4	412.3	250.4	123.6	...	...	435.2[19 26]	791.0[19 26]
10.7	...	14.2	...	7.7	...	5.7	...	26.8	...
12.4	...	18.1	...	8.4	...	10.0	...	38.2	...
1077.0	64.0	364.0	20.0	2646.0	76.0	825.0	72.0	7048.0	1344.0
1157.0	58.0	404.0	31.0	2895.0	130.0	1204.0	152.0	8280.0	2025.0
1625.1	42.1	3495.5	3226.7	1453.6	22.5	252.6	51.4	4958.2	2229.3
2259.4	104.0	5604.6	5495.7	2439.7	72.5	376.9	184.9	6447.5	3525.5
84.2	3.3	110.3	75.8	56.5	13.2	44.5	33.3	128.9	137.6
76.3	3.5	113.9	85.7	52.3	12.9	51.0	43.5	133.7	151.3
74.1	5.1	94.9	50.8	70.5	15.5	52.5	50.4	202.2	242.5
111.7	6.6	147.4	103.3	81.2	24.9	96.6	87.3	255.1	376.9
1972.0	69.0	2533.0	1265.0	1005.0	129.0	369.0	153.0	2489.0	2133.0
1824.0	110.0	2893.0	1723.0	971.0	180.0	638.0	441.0	3200.0	2962.0
24.3[15]	0.7[15]	...	...	24.7[16 20]	7.5[16 20]	29.3[11]	58.3[11]	84.7	123.0
57.2[15]	1.8[15]	...	...	32.3[16 20]	9.3[16 20]	49.2[11]	95.4[11]	95.4	171.6

29
Employment by industry [cont.]
Emploi par industrie [suite]

Country or area Pays ou zone	Year Année	Total employment (000s) Emploi total (000s)		Agriculture, hunting, forestry and fishing Agriculture, chasse sylviculture, pêche		Mining and quarrying Industries extractives		Manufacturing Industries manufacturières		Electricity, gas, water Electricité, gaz, eau	
		M	F	M	F	M	F	M	F	M	F
Japan[16] Japon[16]	1980	33940.0	21420.0	2940.0	2830.0	100.0	10.0	8400.0	5270.0	260.0	40.0
	1993	38400.0	26100.0	2070.0	1760.0	50.0	10.0	9450.0	5850.0	300.0	50.0
Jordan[7 9 23] Jordanie[7 9 23]	1980	96.2	20.4	...	...	5.9	0.0	11.9	1.3	1.9	0.0
	1991	181.3	55.8	...	...	7.6	0.2	33.9	4.3	4.6	0.2
Kazakhstan[4 25] Kazakhstan[4 25]	1984	7064.0	...	1612.0	...	221.0	...	1224.0		63.0	...
	1992	7356.0	...	1762.0	...	257.0		1160.0	...	83.0	...
Kenya[9 30] Kenya[9 30]	1980	829.0	176.8	186.4	44.9	2.2	0.1	128.2	13.1	9.6	0.5
	1991	1117.1	324.6	207.6	64.4	3.5	0.9	167.0	21.9	19.2	3.2
Korea, Republic of[5 6] Corée, Rép. de[5 6]	1980	8462.0	5222.0	2620.0	2034.0	113.0	11.0	1800.0	1155.0	39.0	5.0
	1993[38]	11493.0	7710.0	1534.0	1311.0	50.0	3.0	2811.0	1773.0	56.0	9.0
Latvia Lettonie	1990[4]	1409.0	...	245.0		4.0		373.0		14.0	
	1992	703.0	642.0	177.0	92.0	2.0	1.0	169.0	152.0	11.0	5.0
Lithuania Lituanie	1984	875.1	967.3	236.3	140.0	...	...	274.8[3]	286.1[3]	147.0	28.6
	1992	873.8	981.4	203.8	149.3	3.9	2.1	233.5	265.4	21.3	9.2
Luxembourg[1] Luxembourg[1]	1983	106.0	51.9	4.9	2.6	0.2	...	34.1	5.2	1.3	0.1
	1990	125.0	64.6	4.3	1.9	0.2	...	33.1	4.1	1.2	0.2
Macao[5 6] Macao[5 6]	1989[9]	95.4	66.6	0.9	0.1			23.9	35.5	1.5	0.4
	1992	98.8	70.4	0.2	0.5	0.0		18.6	29.6	0.8	0.1
Madagascar[4 40] Madagascar[4 40]	1986	259.0	...	68.0	...	4.5	...	45.0[8]	...	9.0	...
	1991	285.9	...	76.3	...	4.5	...	93.1[8]	...	11.5	...
Malawi Malawi	1980	328.2	42.2	157.2	25.2	0.6	...	38.0	2.0	3.8	0.3
	1991	487.9	106.5	243.4	77.6	0.8	0.0	70.3	6.5	6.3	0.6
Malaysia[5 6] Malaisie[5 6]	1980	3185.8	1601.6	1079.5	701.1	40.1	6.3	457.6	294.1	62.1	3.6
	1990	4310.7	2374.3	1137.8	599.8	32.7	4.2	697.3	635.5	42.2	4.4
Malta[9] Malte[9]	1980[1]	87.2	31.1	6.4	0.8	1.2	...	23.7[15]	14.5[15]	1.2[16]	
	1991	96.0	34.3	2.9	0.4	0.6	...	25.0[15]	10.9[15]	1.8[16]	0.1[16]
Mauritius[7 9] Maurice[7 9]	1980	145.5	52.1	42.7[32]	15.4[32]	0.1	0.1	16.0	20.4	4.3	0.1
	1992	180.7	101.7	32.7[32]	11.7[32]	0.1	0.1	44.2	65.1	3.3	0.1
Mexico Mexique	1988[4]	28128.0	...	6616.0	...	260.0	...	5548.0	...	130.0	...
	1992	22748.0	10085.0	7721.0	1122.0	165.0	6.0	3372.0	1706.0	87.0	12.0
Montserrat[4 5] Montserrat[4 5]	1980	4.6	...	0.5	...	0.0	...	0.5	...	0.1	...
	1987	5.2	...	0.5	...	0.0	...	0.6		0.1	
Morocco[6 39] Maroc[6 39]	1991	2590.9	809.1	109.2	23.2	39.1	1.9	634.8	378.0	28.6	2.7
	1992	2733.5	760.8	104.0	21.1	43.9	1.5	548.4	340.9	30.5	2.5
Mozambique[4 25] Mozambique[4 25]	1987	192.7		15.1	...	5.0	...	111.6		3.8	...
	1988	201.6		16.9	...	4.9	...	117.0		2.9	...
Myanmar[4 5] Myanmar[4 5]	1980	13208.0	...	8864.0	...	68.0	...	1009.0	...	16.0	...
	1992	15737.0	...	10867.0	...	79.0	...	1132.0	...	17.0	...
Netherlands[16] Pays−Bas[16]	1981	3518.0	1590.0	229.0	42.0	8.0	1.0	865.0	177.0	43.0	4.0
	1993[49]	3771.0	2154.0	180.0	53.0	8.0	...	862.0	202.0	36.0	6.0
Netherlands Antilles[4 5] Antilles néerlandaises[4 5]	1980	76.0	...	0.3	...	0.4	...	10.1	...	1.2	...
	1989[33 38]	43.8	...	...	...	0.4[34]	...	4.2	...	0.9	...
New Caledonia[4] Nouvelle−Calédonie[4]	1985	33.4	...	1.6	...	1.0	...	4.6[20]	...	...	
	1989	44.4	...	2.1	...	1.1	...	4.6		0.6	
New Zealand[5 6] Nouvelle−Zélande[5 6]	1986	904.0	641.0	116.0	49.0	6.0	0.0	217.0	101.0	14.0	2.0
	1993	838.0	657.0	108.0	49.0	4.0	...	175.0	80.0	9.0	2.0
Nicaragua[4] Nicaragua[4]	1980	146.4	...	4.5	...	3.4	...	30.1	...	3.4	...
	1992	214.7	...	28.0	...	1.3	...	33.4	...	6.1	...
Niger Niger	1980	25.1	0.9	1.6	...	5.8	0.0	2.6	0.0	1.2	0.0
	1991	22.1	2.0	1.8	0.1	3.0	0.1	1.7	0.1	4.2	0.3
Norway[16] Norvège[16]	1980	1133.0	775.0	112.0	47.0	10.0	1.0	291.0	92.0	17.0	2.0
	1993[38]	1086.0	918.0	82.0	29.0	20.0	5.0	216.0	76.0	18.0	4.0
Pakistan[5 6 47] Pakistan[5 6 47]	1985	24360.0	2601.0	11677.0	1954.0	46.0		3390.0	296.0	183.0	3.0
	1992	26878.0	3875.0	12034.0	2558.0	46.0	...	3244.0	517.0	252.0	3.0
Panama[5 6 9] Panama[5 6 9]	1982	399.1	162.0	148.6	8.9	0.3	...	38.1	16.2	6.6	1.0
	1992	560.5	234.6	202.0	7.2	1.7	0.4	53.7	21.8	7.8	2.0
Paraguay[5 6] Paraguay[5 6]	1982	156.6	104.5	2.3	0.4	0.3	0.1	25.4	13.1	3.8	0.4
	1993	326.5	243.2	8.0	1.2	0.1	...	68.2	32.5	3.5	0.6
Peru[16 45] Pérou[16 45]	1987	1233.3	827.8	10.5	5.9	13.0	0.6	314.9	138.5	5.5	0.4
	1992	1500.1	910.4	17.1	2.4	5.7	2.3	302.7	110.4	9.4	2.4
Philippines[1 5 6 9] Philippines[1 5 6 9]	1980	11083.0	6070.0	6629.0	2265.0	117.0	13.0	979.0	871.0	47.0	8.0
	1993[38]	15468.0	8975.0	8263.0	2931.0	119.0	11.0	1321.0	1134.0	90.0	16.0

Construction Construction		Trade, restaurants and hotels Commerce, restaurants, hôtels		Transport, storage, communications Transports, entrepôts, communications		Finance, insurance, real est.,bus. services Services financières, immob.,et apparentées		Community, social and personal services Services à collectivité services soc. et pers.	
M	F	M	F	M	F	M	F	M	F
4720.0	770.0	6720.0	5760.0	3090.0	410.0	1830.0	1350.0	5810.0[11]	4940.0[11]
5370.0	1030.0	7300.0	7180.0	3300.0	640.0	3020.0	2450.0	7360.0[11]	7020.0[11]
3.4	0.1	7.1[19 43]	0.3[19 43]	5.9	0.8	4.4	0.8	55.8	17.1
3.1	0.1	12.1[19 43]	1.0[19 43]	19.5	1.9	8.6	3.1	91.8	45.0
619.0	...	445.0	...	821.0	...	42.0[46]	...	1604.0	...
681.0	...	437.0	...	665.0	...	46.0[46]	...	1906.0	...
60.7	2.4	63.2	7.3	48.9	6.3	33.4	6.3	296.3	95.8
68.3	4.2	97.1	19.6	65.6	10.6	51.4	14.9	437.4	185.0
770.0	72.0	1350.0	1275.0	564.0	56.0	229.0	103.0	978.0	511.0
1518.0	162.0	2380.0	2450.0	904.0	103.0	789.0	566.0	1451.0	1333.0
136.0	...	170.0	...	106.0	...	88.0	...	83.0	...
78.0	11.0	62.0	125.0	70.0	35.0	34.0	41.0	41.0	31.0
33.6	117.5	102.1	45.7	1.2	8.1	77.9	336.0	2.2	5.3
144.0	25.4	63.9	163.5	80.7	43.3	4.5	28.7	116.1	292.2
14.0	0.8	17.1	16.8	8.9	1.6	8.7	6.6	16.8	18.2
17.7	1.1	20.9	19.3	10.6	2.3	9.1	7.7	27.9	28.0
13.3	0.9	20.0	11.6	6.2	0.9	3.7	2.9	24.8	13.4
14.3	1.4	23.5	17.6	7.8	1.7	4.7	4.3	28.6	15.2
27.8	...	...	...	24.2	...	33.4[42]	...	151.8	...
28.4	...	...	...	26.1	...	37.6[42]	...	169.2	...
32.6	0.2	23.9	3.2	16.6	0.9	11.2	1.0	44.5[31]	9.5[31]
44.1	0.5	26.3	3.0	20.9	1.4	15.2	1.7	60.6[31]	15.1[31]
257.8	15.6	499.5	192.9	193.0	16.6	...	...	578.4	371.2
404.2	19.7	761.1	456.7	267.2	34.6	163.9	94.6	804.1	524.7
5.0	...	10.2[43]	4.4[43]	7.1	0.9	...	...	32.5[11]	10.4[11]
5.8	0.1	9.9[43]	3.7[43]	8.4	1.1	2.9[19]	2.1[19]	38.9[11]	15.9[11]
7.3	0.1	7.2	2.0	7.7	0.4	3.3	1.2	50.5	12.4
10.4	0.2	14.6	4.4	12.0	1.5	6.9	2.9	52.6	15.7
1528.0	...	5430.0	...	1062.0	...	395.0	...	6882.0	...
1816.0	63.0	3646.0	3246.0	1243.0	119.0	657.0	423.0	3837.0	3368.0
0.7	...	0.6	...	0.2	...	0.2	...	1.8	...
0.9	...	0.7	...	0.3	...	0.2	...	2.0	...
261.1	3.0	649.2	63.3	171.8	9.1	52.7	22.9	639.3[15]	304.0
278.1	3.8	698.6	62.8	189.8	10.3	52.8	23.3	779.8[15]	293.0
19.6	...	5.5	...	29.8	...	...	...	2.1	...
21.5	...	6.4	...	29.3	...	...	...	2.7	...
195.0	...	1262.0	...	443.0	...	772.0	...	579.0	...
188.0	...	1396.0	...	388.0	...	1205.0	...	465.0	...
447.0	25.0	564.0	347.0	279.0	43.0	257.0	145.0	791.0	792.0
365.0	24.0	553.0	404.0	302.0	77.0	420.0	249.0	992.0	1105.0
6.1	...	23.8	...	3.9	...	4.1	...	25.7	...
3.9	...	10.7	...	3.8	...	4.3	...	15.5	...
3.1	...	5.6	...	1.6	...	2.8	...	9.2	...
5.5	...	7.4	...	1.9	...	3.3	...	11.6	...
92.0	11.0	140.0	156.0	78.0	31.0	64.0	70.0	172.0	218.0
72.0	9.0	162.0	155.0	66.0	25.0	77.0	72.0	164.0	265.0
6.8	...	16.5	...	8.1	...	8.5	...	61.7	...
6.6	...	21.5	...	8.8	...	11.2	...	96.3	...
7.7	0.2	2.7	0.1	1.6	0.1	1.2	0.1	0.8	0.3
2.3	0.0	2.1	0.3	3.0	0.5	0.6	0.2	3.5	0.5
134.0	8.0	147.0	173.0	136.0	38.0	56.0	45.0	227.0	366.0
107.0	9.0	163.0	186.0	111.0	47.0	88.0	65.0	278.0	497.0
1502.0	8.0	3074.0	37.0	1394.0	8.0	226.0	11.0	2701.0	284.0
1987.0	49.0	3952.0	120.0	1593.0	22.0	268.0	9.0	3484.0	597.0
37.0	1.5	45.5	28.4	28.4	6.1	15.8	8.6	65.8	88.6
42.1	0.9	104.4	52.9	40.8	6.5	21.7	13.9	85.4	128.9
25.6	0.3	32.5	28.7	12.9	2.0	10.9	4.6	42.6	54.7
41.8	...	79.4	73.3	24.0	3.6	29.1	15.3	72.2	116.5
112.8	7.1	309.4	318.7	118.4	14.2	73.4	24.1	275.4	318.3
133.9	2.0	410.6	410.4	153.8	12.8	108.4	36.1	358.5	331.5
590.0	10.0	601.0	1197.0	695.0	31.0	204.0	104.0	1221.0[11]	1570.0[11]
1080.0	22.0	1168.0	2247.0	1299.0	60.0	298.0	198.0	1821.0[11]	2352.0[11]

29
Employment by industry [cont.]
Emploi par industrie [suite]

| | | Persons employed, by branch of economic activity (000s) Personnes employées, par branches d'activité économique (000s) | | | | | | | | |
| | | Total employment (000s) Emploi total (000s) | | Agriculture, hunting, forestry and fishing Agriculture, chasse sylviculture, pêche | | Mining and quarrying Industries extractives | | Manufacturing Industries manufacturières | | Electricity, gas, water Electricité, gaz, eau | |
Country or area Pays ou zone	Year Année	M	F	M	F	M	F	M	F	M	F
Poland[4,5]	1981	18005.7	...	5389.7	...	539.0	...	5089.3	...	154.4	...
Pologne[4,5]	1991	16560.3	...	4415.2	...	480.8	...	4026.9	...	147.2	...
Portugal[1,6,35]	1980[17]	2419.0	1542.0	540.0	542.0	22.0	1.0	651.0	378.0	17.0	2.0
Portugal[1,6,35]	1993	2486.0	1971.0	262.7	252.9	19.0	0.7	598.2	444.4	25.8	6.3
Puerto Rico[5,6]	1980	487.0	273.0	39.0	1.0	...	...	76.0	65.0	12.0	1.0
Porto Rico[5,6]	1993	593.0	411.0	32.0	2.0	1.0	...	94.0	74.0	13.0	3.0
Rep. of Moldova[4,5]	1981	2069.0	...	826.0	...	0.0	...	409.0	...	11.0	...
Rép. de Moldova[4,5]	1992	2050.0	...	820.0	...	0.0	...	394.0	...	14.0	...
Romania[5]	1980[2]	5803.3	4546.8	1298.6	1789.0	226.8	31.9	1969.4	1406.3	36.7	7.6
Roumanie[5]	1992	5570.2	4887.8	1605.0	1843.8	227.6	44.4	1568.3	1296.9	127.4	36.3
Russian Federation[4]	1990	75324.7	...	10499.1	...	1235.9	...	20181.9	...	595.1	...
Fédération de Russie[4]	1992	72071.1	...	11078.9	...	1252.8	...	18652.8	...	669.3	...
San Marino[6,9]	1980	6.2	3.4	0.3	0.2	...	...	2.3	1.5	0.0	0.0
Saint-Marin[6,9]	1993	8.6	5.6	0.2	0.1	...	...	3.2	1.5	...	...
Seychelles[4,10]	1980	17.9	...	1.9	...	...	...	1.1[3]	...	0.7	...
Seychelles[4,10]	1989	22.3	...	2.2	...	...	...	2.5[3]	...	...	...
Sierra Leone[4,7,9,27]	1980	69.9	...	6.4	...	6.9	...	7.4	...	1.8	...
Sierra Leone[4,7,9,27]	1988	70.2	...	7.3	...	5.8	...	8.6	...	2.7	...
Singapore[6,9]	1980	695.0	373.9	10.6	3.5	1.3	0.3	168.8	143.9	8.6	1.1
Singapour[6,9]	1993	952.3	639.7	3.6	0.3	0.3	...	247.8	181.7	5.9	1.6
Slovakia[9]	1984	1309.0	1082.0	222.0	147.0	23.0	5.0	445.0	324.0	25.0	7.0
Slovaquie[9]	1992[37]	1264.0	911.0	169.0	87.0	24.0	4.0	353.0	236.0	32.0	9.0
Slovenia	1983	496.2	412.3	41.3	44.6	11.2	1.1	217.9	164.9	9.1	2.5
Slovénie	1993[1]	402.7	363.4	32.7	18.5	8.7	0.7	173.9	127.2	9.3	2.4
Solomon Islands[4,9,25]	1980	19.8	...	6.2	...	...	...	2.2[8]	...	0.3[36]	...
Iles Salomon[4,9,25]	1992	26.8	...	6.3	...	...	...	2.0[8]	...	0.3[36]	...
South Africa[4,9]	1980	4812.9	...	...	...	709.0	...	1421.4	...	45.3	...
Afrique du Sud[4,9]	1993	4950.5	...	...	...	561.7	...	1400.5	...	42.5	...
Spain[1,6]	1980	8258.3	3298.8	1618.9	607.8	89.6	2.2	2265.8	675.4	77.8	3.7
Espagne[1,6]	1993[38]	7850.3	3987.3	869.7	328.1	55.6	2.6	1859.4	542.3	73.5	6.5
Sri lanka[7]	1980	715.3	363.1	273.1	263.9	4.8	0.4	124.3	55.8	5.0	0.1
Sri lanka[7]	1991	449.0	350.4	165.7	168.5	2.6	0.4	96.6	131.5	11.7	1.0
Sudan	1982	353.4	...	45.3	...	3.3	...	67.4	...	22.0	...
Soudan	1992	244.0	...	11.8	...	0.1	...	10.9	...	34.1	...
Suriname[17]	1983[4]	68.5	...	...	...	5.1	...	5.5	...	1.3	...
Suriname[17]	1993	53.0	34.2	2.5	0.6	2.3	0.2	6.8	1.0	1.4	0.2
Swaziland[9]	1980	56.1	19.0	22.7	7.3	2.6	0.0	6.9	2.4	1.1	0.1
Swaziland[9]	1986	55.0	21.4	18.1	5.0	2.3	0.1	7.5	3.4	1.3	0.1
Sweden[16]	1980[22]	2327.0	1906.0	178.0	59.0	14.0	1.0	750.0	276.0	31.0	6.0
Suède[16]	1993[38]	2026.0	1938.0	100.0	36.0	10.0	1.0	527.0	199.0	28.0	7.0
Switzerland[5]	1980	2021.0	1145.0	160.0	58.0	...	...	943.0[3]	264.0[3]	...	...
Suisse[5]	1993	2095.0	1294.0	135.0	57.0	...	...	607.0[3]	219.0[3]	...	...
Syrian Arab Republic[5,6,9]	1983	1830.8	294.8	497.5	153.8	7.5	0.1	254.2	27.1	21.1	0.9
Rép. arabe syrienne[5,6,9]	1991	2710.3	539.6	625.0	292.0	6.7	...	421.5	34.6	7.9	0.6
TFYR Macedonia[22]	1984[2]	319.0	160.0	31.0	11.0	12.0	1.0	102.0	66.0	8.0	1.0
L'ex-R.y. Macédoine[22]	1993	263.0	158.0	26.0	8.0	9.0	1.0	88.0	60.0	10.0	2.0
Thailand[5,6]	1980[22]	11866.3	10657.5	8048.7	7893.9	27.7	8.8	1036.0[15]	752.7[15]	54.2[16]	5.5[16]
Thaïlande[5,6]	1991[17]	16850.9	14287.5	10089.6	8687.7	41.5	12.8	1717.5[15]	1747.5[15]	92.3[16]	17.9[16]
Togo[4]	1980	46.0	...	1.2	...	3.4	...	6.7	...	1.7	...
Togo[4]	1987	63.9	...	5.7	...	2.6	...	5.1	...	2.4	...
Trinidad and Tobago[6]	1980	270.2	117.7	29.5	10.1	...	...	44.5[8]	17.7[8]	...	...
Trinité-et-Tobago[6]	1993[38]	261.8	142.8	37.7	8.0	13.1	1.9	28.1	12.2	6.2	0.9
Turkey[5,6,9]	1982[39]	4681.0	647.0	173.0	55.0	54.0	2.0	1208.0	192.0	14.0	1.0
Turquie[5,6,9]	1993	13348.0	5990.0	4029.0	4408.0	149.0	2.0	2258.0	645.0	105.0	6.0
former USSR †[4]	1980	125626.0	...	25236.0	...	...	...	36891.0[3]	...	...	...
ancienne URSS †[4]	1990	124971.0	...	22761.0	...	...	...	35400.0[3]	...	...	...
United Kingdom[6,9,10]	1980	15248.0	10078.0	523.0	131.0	408.0	32.0	4961.0	2055.0	282.0	71.0
Royaume-Uni[6,9,10]	1993	13741.0	11576.0	427.0	120.0	...	...	4936.0[3]	1629.0[3]	...	...
United Rep. of Tanzania[9]	1980	398.3	73.9	72.3	5.9	4.3	0.4	70.3	9.6	12.6	1.2
Rép.-Unie de Tanzanie[9]	1984	448.7	86.2	67.6	6.8	3.7	0.5	82.0	9.1	18.6	1.8
United States[5,6]	1980	57186.0	42117.0	2842.0	687.0	845.0	134.0	15036.0	6906.0	1153.0[16]	252.0[16]
Etats-Unis[5,6]	1993	64700.0	54606.0	2582.0	676.0	561.0	108.0	13249.0	6309.0	1252.0[16]	345.0[16]

Construction Construction		Trade, restaurants and hotels Commerce, restaurants, hôtels		Transport, storage, communications Transports, entrepôts, communications		Finance, insurance, real est.,bus. services Services financières, immob.,et apparentées		Community, social and personal services Services à collectivité services soc. et pers.	
M	F	M	F	M	F	M	F	M	F
1414.6	...	1581.5	...	1416.8	...	371.5	...	2434.3	...
1203.8	...	1750.5	...	999.7	...	341.7	...	3099.3	...
363.0	9.0	282.0	184.0	135.0	25.0	59.0	21.0	345.0	379.0
349.5	15.8	498.1	369.1	160.9	47.0	182.1	117.9	389.7	717.6
45.0	...	101.0	39.0	30.0	5.0	13.0	9.0	170.0[11]	151.0[11]
54.0	2.0	128.0	74.0	30.0	7.0	16.0	17.0	225.0[11]	232.0[11]
115.0	...	112.0	...	141.0	...	8.0	...	375.0	...
120.0	...	111.0	...	102.0	...	9.0	...	422.0	...
761.7	95.9	263.6	356.3	607.5	102.2	11.6	18.3	627.4	739.3
502.9	76.3	408.6	520.6	503.6	145.0	242.5	255.7	384.3	668.8
8168.1	...	5085.8	...	5818.2	...	401.6	...	19607.1	...
7246.6	...	4914.2	...	5631.8	...	493.6	...	18641.8	...
0.9	0.0	0.7	0.7	0.1	0.0	0.1	0.0	1.7	1.0
1.1	0.0	1.2	1.3	0.2	0.1	0.2	0.2	2.4	2.4
2.7[8]	...	3.4	...	2.2	...	0.8	...	2.7	...
1.7[8]	...	4.4	...	3.1	...	0.7	...	4.6	...
7.8	...	5.6	...	7.9	...	1.9	...	24.2	...
7.3	...	3.3	...	7.7	...	1.8	...	25.7	...
52.9	5.3	158.6	86.1	101.7	20.8	41.9	37.1	147.1	75.7
91.1	11.0	216.2	147.4	131.0	35.8	88.1	85.3	167.6	176.5
209.0	32.0	73.0	188.0	111.0	44.0	40.0	38.0	162.0	296.0
161.0	37.0	143.0	113.0	109.0	52.0	105.0	65.0	168.0	308.0
61.7	8.2	36.5	66.1	43.0	11.9	12.9	15.0	62.6	98.0
29.8	4.5	29.7	55.2	32.5	10.4	14.3	19.3	71.8	105.2
1.6	...	2.0	...	1.5	...	0.3	...	5.8	...
1.1	...	3.2	...	1.4	...	1.2	...	11.1	...
364.1	...	756.3	...	452.5	...	122.0	...	942.3[31]	...
374.5	...	764.4	...	303.1	...	191.5	...	1312.4[31]	...
1017.7	20.5	1185.5	824.8	607.6	57.5	315.8	81.4	1073.0	1023.6
1046.1	42.3	1596.5	1095.5	606.5	88.4	554.5	372.2	1188.6	1509.3
88.9	9.6	80.3	18.0	88.4	3.2	29.9	7.3	20.9	4.7
15.9	3.5	57.8	18.6	47.9	4.6	35.0	14.9	15.8	7.3
23.7	...	12.9	...	69.3	...	11.4	...	58.1	...
2.7	...	3.5	...	40.8	...	23.3	...	116.9	...
3.0	...	6.7	...	3.0	...	1.5	...	42.3	...
4.3	0.4	7.2	0.6	4.1	0.2	1.2	2.0	20.7	22.6
5.9	0.2	3.5	2.4	2.9	0.4	1.6	0.8	9.0	5.5
5.1	0.2	4.3	3.1	4.9	0.7	2.2	1.2	9.2	7.6
260.0	26.0	281.0	300.0	214.0	81.0	155.0	128.0	444.0	1028.0
216.0	19.0	289.0	277.0	190.0	87.0	207.0	161.0	454.0	1148.0
...	...	919.0[19]	823.0[19]	...	...	...	...	...	...
277.0	22.0	324.0	366.0	161.0	52.0	226.0	145.0	366.0	433.0
306.5	4.1	208.8	5.7	126.1	4.3	15.6	3.9	393.3	94.9
334.3	6.4	369.0	9.3	158.4	8.6	20.2	4.5	767.4	183.7
52.0	3.0	34.0	25.0	21.0	3.0	7.0	5.0	39.0	43.0
34.0	3.0	26.0	23.0	18.0	3.0	6.0	6.0	35.0	50.0
374.2	61.6	881.3[19]	1034.5[19]	425.7	30.2	...	...	1017.6[11]	869.2[11]
998.1	180.1	1648.3[19 43]	1828.8[19 43]	739.0	94.9	...	...	1515.0[11]	1710.6[11]
6.2	...	...	...	2.1	...	7.1[11]	...	17.5	...
5.3	...	8.2[19]	...	4.2	...	...	...	30.5	...
73.6[20]	7.4[20]	39.5	39.7	29.2	3.2	...	...	53.0[19]	39.6[19]
39.7	4.6	35.5	35.6	25.2	4.8	15.4	12.5	60.5	62.2
397.0	3.0	1105.0	48.0	362.0	20.0	169.0	59.0	1185.0	264.0
1119.0	18.0	2290.0	186.0	892.0	63.0	359.0	112.0	2147.0	552.0
1240.0	...	9694.0	...	11958.0	...	649.0	...	28522.0	...
2550.0	...	9800.0	...	10225.0	...	720.0	...	31425.0	...
1500.0	117.0	2282.0	2508.0	1289.0	291.0	1033.0	831.0	2662.0	4027.0
...	...	8125.0[19]	9809.0[19]	...	...	...	...	...	...
23.0	0.9	27.7	4.0	49.4	2.8	10.1	3.0	128.7	46.2
19.7	1.4	35.4	2.6	52.6	2.8	11.3	4.7	157.7	56.5
5717.0	498.0	10838.0	9353.0	3725.0	1395.0	3800.0	4551.0	13228.0[11]	18341.0[11]
6603.0	617.0	13202.0	11566.0	4810.0	2074.0	6129.0	6862.0	16314.0[11]	26046.0[11]

29
Employment by industry [*cont.*]
Emploi par industrie [*suite*]

| | | Total employment (000s) Emploi total (000s) | | Persons employed, by branch of economic activity (000s) Personnes employées, par branches d'activité économique (000s) | | | | | | | |
| | | | | Agriculture, hunting, forestry and fishing Agriculture, chasse sylviculture, pêche | | Mining and quarrying Industries extractives | | Manufacturing Industries manufacturières | | Electricity, gas, water Electricité, gaz, eau | |
Country or area Pays ou zone	Year Année	M	F	M	F	M	F	M	F	M	F
US Virgin Islands[4]	1980	37.5	...	0.2	...	...	...	3.2	...	...	...
Iles Vierges amé.[4]	1993	48.9	...	0.2	...	0.0	...	2.9	...	...	...
Uruguay[1 6 39]	1984[17]	569.7	362.9	...	...	35.6[20 34]	5.5[20 34]	117.8	72.1		
Uruguay[1 6 39]	1992	676.2	470.2	47.2	4.7	1.6	0.1	150.5	91.4	13.2	3.3
Venezuela[5 6 17]	1980	3061.2	1183.9	609.1	28.3	57.0	5.8	483.8	190.9	42.5	7.5
Venezuela[5 6 17]	1993	4815.6	3219.3	719.8	36.3	60.4	8.8	794.9	295.5	47.2	9.7
Yugoslavia,SFR † [22 37]	1980	3663.0	2019.0	223.0	55.0	115.0	11.0	1322.0	746.0	92.0	16.0
Yugoslavie,Rfs † [22 37]	1989	4012.0	2685.0	253.0	85.0	129.0	15.0	1570.0	1025.0	125.0	23.0
Zambia[4 17]	1980	379.3	...	32.6		63.1	...	47.8		8.0	...
Zambie[4 17]	1989	359.6	...	37.2		54.2	...	50.9		8.7	...
Zimbabwe[30]	1980	838.5	171.4	242.0	85.0	65.0	1.2	147.7	11.7	6.5	0.2
Zimbabwe[30]	1993	989.0	237.8	227.2	90.2	46.0	1.7	171.7	13.7	7.5	0.3

Source:
International Labour Office (Geneva).

Source:
Bureau international du travail (Genève).

† For detailed descriptions of data pertaining to former Czechoslovakia, Germany, SFR Yugoslavia and former USSR, see Annex I – Country or area nomenclature, regional and other groupings.

1 Including armed forces (Philippines: living in private households; France, Greece, Italy, Portugal, Spain, Uruguay: excluding compulsory military service).
2 State sector (Estonia, Romania, TFYR Macedonia: including cooperative sector).
3 Including data for mining and quarrying, electricity, gas and water France, Switzerland (1980), U.K.: and construction).
4 Both sexes.
5 Civilian labour force employed.
6 Persons aged 15 years and over (Egypt: 6 years and over; Bolivia, Brazil, El Salvador, Honduras, Indonesia, Pakistan, Paraguay (1993), Syrian A.R.: 10 years and over; Colombia, Costa Rica, Ecuador, Paraguay (1982), Portugal (1980), Turkey: 12 years and over; Denmark, Finland: 15 to 74 years; Greece, Italy, Jamaica, Macau, Peru, Portugal (1993), San Marino, Uruguay: 14 years and over; Iceland, Puerto Rico, Spain, United Kingdom, United States: 16 years and over; Malaysia, Netherlands: 15 to 64 years; Norway: 16 to 74 years; Sweden: 16 to 64 years; Thailand: 11 years and over prior to 1989; 13 years and over beginning 1989).
7 Establishments with 5 or more persons employed (Bahrain, Mauritius: 10 or more persons; Sierra Leone: 6 or more persons).
8 Including mining and quarrying (Barbados, Madagascar: quarrying only).
9 One month of each year.
10 Excluding unpaid family workers and employees in private domestic services (Bermuda, British Virgin Island, Chad: unpaid family workers only).
11 Including data for restaurants and hotels (Guam, Japan, Puerto Rico: hotels only; Malta: and business services; Jamaica, Togo: and trade; Brazil: and storage, excluding sanitary services and international bodies).
12 Excluding government and personal services.
13 Excluding armed forces.
14 Bujumbara.
15 Including repair and installation services (Greece, Morocco: repairs only).
16 Including sanitary services.
17 One quarter of each year.
18 State owned enterprises.

† Pour les descriptions en détails des données relatives à l'ancienne Tchécoslovaquie, l'Allemagne, la Rfs Yougoslavie et l'ancienne URSS, voir l'Annexe I – Nomenclature des pays ou zones, groupements régionaux et autres groupements.

1 Y compris les forces armées (Philippines: vivant en ménages privés; Espagne, France, Grèce, Italie, Portugal: non compris les militaires du contigent).
2 Secteur d'état (Estonie, L'ex–R.y.Macédoine, Roumanie: y compris secteur coopératives).
3 Y compris les données concernant les industries extractives, l'électricité, le gaz at l'eau (France, Royaume–Uni, Suisse (1980): et la construction).
4 Les deux sexes.
5 Main–d'oeuvre civile occupée.
6 Personnes âgée 15 ans et plus (Egypte:6 ans et plus; Bolivie, Brésil, El Salvador, Honduras, Indonésie, Pakistan, Paraguay (1993), Rép. arabe syrienne: 10 ans et plus; Colombie, Costa Rica, Ecuador, Paraguay (1982), Portugal (1980), Turquie: 12 ans et plus; Danemark, Finlande: de 15 à 74 ans; Grèce, Italie, Jamaïque, Macao, Pérou, Saint–Marin, Uruguay: 14 ans et plus; Islande, Porto Rico, Espagne, Royaume–Uni, Etats–Unis: 16 ans et plus; Malaisie, Pays–Bas: de 15 à 64 ans; Norvège: de 16 à 74 ans; Suède: de 16 à 64 ans; Thaïlande: 11 ans et plus avant 1989; 13 ans et plus commençant 1989.
7 Etablissements occupant cinq personnes et plus (Bahreïn, Maurice: 10 et plus personnes; Sierra Leone: 6 et plus personnes).
8 Y compris les industries extractives (Barbade, Madagascar: les carrières seulement).
9 Un mois de chaque année.
10 Non compris les travailleurs familiaux non rémunérés et les personnes occupées à des services domestiques privés (Bermudes, Iles Vierges brit., Tchad: travailleurs familiaux non rémunérés seulement).
11 Y compris les données concernant les restaurants et les hôtels (Guam, Japon, Porto Rico: les hôtels seulement; Malte: et les services entreprises; Jamaïque, Togo: et le commerce; Brésil: et les entrepôts, non compris les services sanitaires et les organismes internationaux).
12 Non compris les services gouvernementaux et personnels.
13 Non compris les forces armées.
14 Bujumbara.
15 Y compris les services de réparation et d'installation (Grèce, Maroc: les services de réparation seulement).
16 Y compris les services sanitaires.
17 Un trimestre de chaque année.
18 Entreprises d'Etat.
19 Y compris les données concernant les banques, les assurances, les

Construction Construction		Trade, restaurants and hotels Commerce, restaurants, hôtels		Transport, storage, communications Transports, entrepôts, communications		Finance, insurance, real est.,bus. services Services financières, immob.,et apparentées		Community, social and personal services Services à collectivité services soc. et pers.	
M	F	M	F	M	F	M	F	M	F
3.5[8]	...	10.5	...	2.1[20]	...	2.2	...	2.5	...
5.4[8]	...	14.9	...	2.7[20]	...	3.3	...	5.7	...
50.0	0.7	109.8	61.0	56.7	8.0	32.8	15.2	167.0	200.3
76.6	1.8	121.5	84.3	56.5	8.6	36.8	27.6	171.8	248.1
373.9	12.8	548.4	254.5	275.9	26.5	115.3	71.6	554.1	585.2
611.5	25.0	989.1	576.6	415.6	46.4	273.5	186.2	900.1	1033.1
580.0	52.0	384.0	400.0	369.0	71.0	80.0	90.0	498.0	579.0
480.0	60.0	408.0	514.0	416.0	105.0	91.0	119.0	540.0	739.0
43.8	...	31.4	...	23.9	...	22.7	...	106.1	...
20.8	...	26.6	...	26.1	...	24.7	...	110.2[31]	...
41.6	0.6	58.2	12.1	42.6	3.0	7.5	5.0	227.4[19]	52.6[19]
86.6	4.1	79.9	14.9	46.2	3.7	56.1	9.8	267.8[19]	99.5[19]

19 Including data for finance, insurance, real estate and business services (Jordan, Malta, Thailand: finance, insurance and real estate only; Guatemala: finance and insurance only; France, Switzerland, United Kingdom: and transport, storage, communication, community, social and personal services; Zimbabwe: business services only; China: real estate and business services only).
20 Including data for electricity, gas and water.
21 Excluding hunting and fishing.
22 Average of less than twelve months.
23 Non−agricultural activities.
24 Excluding repair and installation services and sanitary services.
25 Employees only.
26 Non−material activities.
27 Employees including working proprietors.
28 Public sector and establishments of non−agricultural private sector with ten or more persons employed.
29 Including the residents of East Jerusalem.
30 Excluding small establishments in rural areas.
31 Excluding domestic services.
32 Including data for sugar and tea factories.
33 Curaçao.
34 Including data for agriculture, hunting, forestry and fishing.
35 Including Azores and Madiera.
36 Excluding gas.
37 Socialized sector (Slovakia: prior to 1992).
38 Methodology revised.
39 Urban areas.
40 Private sector.
41 Modern sector.
42 Including trade.
43 Excluding restaurant and hotels (Guam, United States: hotels).
44 Including petroleum and gas extraction.
45 Lima.
46 Excluding real estate and business services.
47 July of preceding year to June of current year.
48 Excluding storage.
49 Persons working or seeking work for less than 12 hours per week are no longer included.

affaires immobilières et les services aux entreprises (Jordanie, Malte, Thaïlande : les banques, les assurances, les affaires immobilières seulement; Guatemala : les banques et les assurances seulement; France, Royaume−Uni, Suisse : et les transports, les entrepôts, les communications, services à collectivité et les services social et personnel; Zimbabwe : les services aux entreprises seulement; Chine : les affaires immobilières et les services aux entreprises seulement).
20 Y compris les données concernant l'électricité, le gaz et l'eau.
21 Non compris la chasse et la pêche.
22 Moyenne de moins de douze mois.
23 Activités non agricoles.
24 Non compris les services de réparation et d'installation et les service sanitaires.
25 Salariés seulement.
26 Activités non matérielles.
27 Salariés y compris les propriétaires exploitants.
28 Secteur public et établissements du secteur privé non agricole occupant dix personnes et plus.
29 Y compris les résidents de Jerusalem Est.
30 Non compris les petites entreprises des zones rurales.
31 Non compris les services domestiques.
32 Y compris les données concernant les fabriques de sucre et de thé.
33 Curaçao.
34 Y compris les données concernant l'agriculture, la chasse, le sylviculture et la pêche.
35 Y compris Açores et Madère.
36 Non compris le gaz.
37 Secteur socialisé (Slovaquie: avant 1992).
38 Méthodologie reviseé.
39 Régions urbaines.
40 Secteur privé.
41 Secteur moderne.
42 Y compris le commerce.
43 Non compris les restaurants et hôtels (Guam, Etats−Unis: hôtels).
44 Y compris l'extraction du pétrole et du gaz.
45 Lima.
46 Non compris les affaires immobilières et les services aux entreprises.
47 Juillet de l'année précédente à juin de l'année en cours.
48 Non compris les entrepôts.
49 Ne sont plus comprises les personnes qui travaillent, ou qui travaillent, ou qui cherchent moins de 12 heures de travail par semaine.

30
Unemployment
Chômage

Number (thousands) and percentage of unemployed
Nombre (milliers) et pourcentage des chômeurs

Country or area Pays ou zone §	1984	1985	1986	1987	1988	1989	1990	1991	1992	1993
Albania Albanie										
MF [IV]	77.5	94.0	92.0	89.4	105.8	113.4	150.7	139.8	...	...
% MF [IV]	5.7	6.7	6.4	6.1	7.0	7.3	9.5	9.1	...	...
Algeria Algérie										
MF [I] [1 2]	...	...	...	...	...	946.0	1156.0	1261.0	1482.0	...
M [I] [1 2]	...	...	...	...	...	876.0	1069.0	1155.0	1348.0	...
F [I] [1 2]	...	...	...	...	...	70.0	87.0	106.0	134.0	...
% MF [I] [1 2]	...	...	.	...	...	17.0	19.7	21.1	23.8	...
American Samoa Samoa américaines										
MF [IV] [2]	1.4	1.5	1.6	...	...	...	...	...	...	...
% MF [IV] [2]	13.1	12.5	13.4	...	...	...	...	...	...	...
Angola Angola										
MF [III] [3 4]	52.0	56.1	69.4	...	...	...	...	...	...	...
M [III] [3 4]	34.8	34.9	43.6	...	...	...	...	...	...	...
F [III] [3 4]	17.3	21.2	25.8	...	...	...	...	...	...	...
% MF [III] [3 4]	14.2	14.9	18.9	...	...	...	...	...	...	...
% M [III] [3 4]	11.8	11.5	14.8	...	...	...	...	...	...	...
% F [III] [3 4]	24.7	28.5	35.1	...	...	...	...	...	...	...
Argentina Argentine										
MF [I] [2 5]	152.1	216.2	177.8[1]	230.5	251.2	322.6	...	...	...	...
M [I] [2 5]	92.8	143.0	107.7[1]	126.8	137.2	195.2	...	...	...	...
F [I] [2 5]	59.3	73.3	70.1[1]	103.7	114.0	127.4	...	...	...	...
% MF [I] [2 5]	3.8	5.3	4.4[1]	5.3	5.9	7.3	...	...	...	...
% M [I] [2 5]	...	...	...	4.5	5.2	7.0	...	...	...	...
% F [I] [2 5]	...	...	...	6.6	7.2	7.7	...	...	...	...
Australia Australie										
MF [I] [2]	# 641.2[6]	602.9[6]	# 613.1	628.9	576.2	# 508.1[7]	584.8[7]	814.5[7]	925.1[7]	939.2[7]
M [I] [2]	# 384.6[6]	355.9[6]	# 351.8	361.0	321.1	# 276.5[7]	332.3[7]	489.5[7]	566.2[7]	574.0[7]
F [I] [2]	# 256.7[6]	246.9[6]	# 261.3	267.9	255.1	#231.6[7]	252.5[7]	325.0[7]	358.9[7]	365.1[7]
% MF [I] [2]	# 9.0[6]	8.3[6]	# 8.1	8.1	7.2	# 6.2[7]	6.9[7]	9.6[7]	10.8[7]	10.9[7]
% M [I] [2]	# 8.7[6]	7.9[6]	# 7.7	7.8	6.8	# 5.7[7]	6.7[7]	9.9[7]	11.4[7]	11.5[7]
% F [I] [2]	# 9.5[6]	8.8[6]	# 8.7	8.6	7.9	# 6.8[7]	7.2[7]	9.2[7]	10.0[7]	10.1[7]
Austria Autriche										
MF [I] [2]	127.7[8]	120.7[8]	105.8[8]	# 130.2	122.0	108.6	114.8	125.4	132.4	...
M [I] [2]	79.8[8]	73.8[8]	64.7[8]	# 74.3	66.5	58.2	63.0	70.9	74.4	...
F [I] [2]	47.9[8]	46.9[8]	41.1[8]	# 55.9	55.5	50.4	51.8	54.5	58.0	...
MF [III] [2]	130.5	139.5	152.0	164.5	158.6	149.2	165.8	185.0	193.1	222.3
M [III] [2]	80.6	84.2	88.9	95.0	89.8	81.0	89.0	99.0	107.2	126.7
F [III] [2]	49.9	55.3	63.1	69.5	68.8	68.2	76.8	86.0	85.9	95.6
% MF [I] [2]	...	...	...	3.8	3.7	3.1	3.2	3.5	3.7	...
% M [I] [2]	...	...	...	3.6	3.3	2.8	3.0	3.3	3.5	...
% F [I] [2]	...	...	...	4.1	4.0	3.6	3.6	3.7	3.8	...
% MF [III] [2]	4.5	4.8	5.2	5.6	5.3	5.0	5.4	5.8	5.9	6.8
% M [III] [2]	4.7	4.9	5.1	5.5	5.1	4.6	4.9	5.3	5.7	6.7
% F [III] [2]	4.3	4.7	5.2	5.7	5.6	5.5	6.0	6.5	6.2	6.9
Bahamas Bahamas										
MF [I] [2]	...	...	13.5	...	13.7	14.9[1]	...	...	20.0	...
M [I] [2]	...	...	5.7	...	5.4	7.4[1]	...	...	9.7	...
F [I] [2]	...	...	7.8	...	8.3	7.5[1]	...	...	10.2	...
% MF [I] [2]	...	...	12.2	...	11.0	11.7[1]	...	...	14.8	...
% M [I] [2]	...	...	9.7	...	8.2	11.0[1]	...	...	13.8	...
% F [I] [2]	...	...	15.0	...	14.2	12.5[1]	...	...	16.0	...
Bahrain Bahreïn										
MF [III] [3 9]	4.1	6.3	6.7	4.0	4.5	3.4	3.0	3.3	3.0	3.6
M [III] [3 9]	...	...	...	3.1	3.6	2.5	2.1	2.4	2.2	2.9
F [III] [3 9]	...	...	...	0.9	0.9	0.9	0.8	0.9	0.9	0.7

30
Unemployment
Number (thousands) and percentage of unemployed [*cont.*]
Chômage
Nombre (milliers) et pourcentage des chômeurs [*suite*]

Country or area Pays ou zone §	1984	1985	1986	1987	1988	1989	1990	1991	1992	1993
Barbados Barbade										
MF [I] [2]	19.2	21.2	20.7	21.4	21.2[10]	17.1[10]	18.6[10]	20.9	28.7	30.9
M [I] [2]	7.9	8.6	8.0	8.4	7.7[10]	5.9[10]	6.6[10]	8.6	13.2	14.0
F [I] [2]	11.3	12.6	12.7	13.0	13.5[10]	11.2[10]	12.0[10]	12.3	15.5	16.9
% MF [I] [2]	17.1	18.7	17.7	17.9	17.4[10]	13.7[10]	15.0[10]	17.1	23.0	24.5
% M [I] [2]	13.0	14.1	13.0	13.3	12.3[10]	9.1[10]	10.3[10]	13.3	20.4	21.5
% F [I] [2]	22.1	24.0	23.0	23.1	22.9[10]	18.7[10]	20.2[10]	21.4	25.7	27.7
Belarus Bélarus										
MF [III] [1]	..	..	..	..	..	..	..	2.3	24.0	66.3
M [III] [1]	..	..	..	..	..	..	..	0.5	4.4	22.3
F [III] [1]	..	..	..	..	..	..	..	1.8	19.6	44.0
% MF [III] [1]	..	..	..	..	..	..	..	...	0.5	1.4
% M [III] [1]	..	..	..	..	..	..	..	...	0.1	0.5
% F [III] [1]	..	..	..	..	..	..	..	...	0.4	0.9
Belgium Belgique										
MF [I] [14]	486.3	455.3	450.2	444.8	396.0	326.1	285.2	282.4	315.9	386.4
M [I] [14]	206.1	185.8	176.3	177.8	167.2	128.0	109.2	110.8	137.1	169.1
F [I] [14]	280.2	269.5	273.9	267.0	228.8	198.1	176.0	171.6	178.8	217.3
MF [III]	595.8	# 558.3	516.8	500.8	459.4	419.3	402.8	429.5	472.9	549.7
M [III]	273.7	# 246.0	217.6	208.9	187.8	167.5	161.3	178.0	199.1	237.5
F [III]	322.1	# 312.3	299.1	292.0	271.6	251.8	241.5	251.5	273.8	312.2
% MF [I] [14]	12.3	11.4	11.3	11.3	10.1	8.3	7.2	7.0	7.7	9.3
% M [I] [14]	8.3	7.5	7.1	7.3	7.0	5.3	4.5	4.5	5.6	6.9
% F [I] [14]	18.8	17.9	17.9	17.6	15.1	13.0	11.4	10.7	10.7	12.7
% MF [III]	14.4	# 13.6	12.6	12.2	11.1	10.1	9.6	10.2	11.2	13.0
% M [III]	10.9	# 9.9	8.9	8.6	7.7	6.9	6.6	7.3	8.1	9.7
% F [III]	19.9	# 19.1	18.1	17.4	16.0	14.7	14.1	14.3	15.5	17.4
Bermuda Bermudes										
MF [III]	0.0	0.0	0.0	0.1	0.1	0.0	0.1	0.2	...	...
Bolivia Boilivie										
MF [I] [11]	85.3	58.2	46.2	78.0	...	96.1	67.3	33.7	42.8	...
M [I] [11]	...	...	...	...	...	51.1	37.2	20.4	27.9	...
F [I] [11]	...	...	...	...	...	45.0	30.0	13.3	14.9	...
MF [IV] [12]	293.0	370.9	415.4	430.7	388.4	443.2	433.4	...	...	...
M [IV] [12]	...	323.2	362.0	365.1	329.3	375.7	366.6	...	...	...
F [IV] [12]	...	47.6	53.4	65.6	59.1	67.5	66.9	...	...	...
% MF [I] [11]	...	...	...	...	...	...	7.3	5.8	5.4	...
% M [I] [11]	...	...	...	...	...	...	6.9	5.6	5.4	...
% F [I] [11]	...	...	...	...	...	...	7.8	6.0	5.5	...
% MF [IV] [12]	15.1	18.0	20.0	20.5	18.0	20.0	19.0	...	...	...
Brazil Brésil										
MF [I] [1 12 13]	2234.3	1875.3	1380.2	2133.0	2319.4	1891.0	2367.5	...	...	...
M [I] [1 12 13]	1437.3	1171.5	854.1	1315.3	1410.3	1244.0	1582.4	...	...	...
F [I] [1 12 13]	797.1	703.8	526.1	817.7	909.1	647.0	785.1	...	...	...
% MF [I] [1 12 13]	4.3	3.4	2.4	3.6	3.8	3.0	3.7	...	...	...
% M [I] [1 12 13]	4.1	3.2	2.3	3.4	3.6	3.1	3.8	...	...	...
% F [I] [1 12 13]	4.6	3.8	2.7	4.0	4.2	2.9	3.4	...	...	...
Bulgaria Bulgarie										
MF [III] [1]	...	...	...	...	...	...	65.1	419.1	576.9	626.1
M [III] [1]	...	...	...	...	...	...	22.7	190.7	274.5	298.4
F [III] [1]	...	...	...	...	...	...	42.4	228.4	302.4	327.7
% MF [III] [1]	...	...	...	...	...	...	1.7	11.1	15.3	16.4
Burkina Faso Burkina Faso										
MF [III] [3 14]	29.0	32.5	32.0[15]	35.3	# 34.6	38.1	42.0	34.8	29.8	27.4
M [III] [3 14]	...	28.8	26.8[15]	30.4	# 29.3	32.4	37.4	30.4	25.9	...
F [III] [3 14]	...	3.8	4.7[15]	4.9	# 5.3	5.7	4.6	4.4	3.9	...
Burundi Burundi										
MF [III] [3 16]	2.6	1.9	6.8	8.2	9.3	11.1	14.5	13.8	7.3	...
M [III] [3 16]	2.2	1.7	6.0	7.4	8.4	9.8	...	9.6	...	...
F [III] [3 16]	0.4	0.3	0.8	0.8	0.9	1.3	...	4.2	...	...
Cameroon Cameroun										
MF [III] [2]	37.8	14.3	25.5	19.2	...	...	...	...	...	...

30
Unemployment
Number (thousands) and percentage of unemployed [*cont.*]
Chômage
Nombre (milliers) et pourcentage des chômeurs [*suite*]

Country or area Pays ou zone §	1984	1985	1986	1987	1988	1989	1990	1991	1992	1993
Canada Canada										
MF [I] [2]	1384.0	1311.0	1215.0	1150.0	1031.0	1018.0	1109.0	1417.0	1556.0	1562.0
M [I] [2]	792.0	739.0	677.0	623.0	546.0	548.0	613.0	818.0	910.0	896.0
F [I] [2]	592.0	572.0	539.0	527.0	485.0	470.0	496.0	599.0	647.0	667.0
% MF [I] [2]	11.2	10.5	9.5	8.8	7.8	7.5	8.1	10.3	11.3	11.2
% M [I] [2]	11.2	10.3	9.3	8.5	7.4	7.3	8.1	10.8	12.0	11.7
% F [I] [2]	11.3	10.7	9.8	9.3	8.3	7.9	8.1	9.7	10.4	10.6
Central African Rep. Rép. centrafricaine										
MF [III] [3,17]	9.1	8.2	˙9.7	8.9[1]	9.1	7.8	7.8	7.7	5.7	5.6
M [III] [3,17]	...	7.5	7.8	8.1[1]	8.3	7.2	7.1	7.2	5.2	5.2
F [III] [3,17]	...	0.7	0.9	0.8[1]	0.8	0.6	0.7	0.5	0.5	0.4
Chile Chili										
MF [I] [2,10]	541.3	# 516.5	374.2	343.5	286.1	249.8	268.9	253.6	217.1	233.6
M [I] [2,10]	351.1	# 346.7	250.0	222.5	177.4	162.4	184.8	168.6	132.1	147.6
F [I] [2,10]	190.1	# 169.8	124.0	120.9	108.6	87.4	84.0	85.0	85.1	86.0
% MF [I] [2,10]	13.9	# 12.1	8.8	7.9	6.3	5.3	5.6	5.3	4.4	4.5
% M [I] [2,10]	...	# 11.7	8.4	7.3	5.6	5.0	5.7	5.1	4.1	4.2
% F [I] [2,10]	...	# 13.4	9.7	9.3	7.8	6.1	5.7	5.8	5.6	5.1
China Chine										
MF [IV] [1,11,18]	2357.0	2385.0	2644.0	2766.0	2960.0	3779.0	3832.0	3522.0	3603.0	...
M [IV] [1,11,18]	752.0	783.0	805.0	953.0	1001.0	1942.0	1313.0	1207.0	1298.0	...
F [IV] [1,11,18]	1207.0	1186.0	1288.0	1398.0	1452.0	1837.0	1814.0	1677.0	1700.0	...
% MF [IV] [1,11,18]	1.9	1.8	2.0	2.0	2.0	2.6	2.5	2.3	2.3	...
% M [IV] [1,11,18]	0.6	0.6	0.6	0.7	0.7	1.3	0.9	0.8	...	...
% F [IV] [1,11,18]	1.0	0.9	1.0	1.0	1.0	1.3	1.2	1.1	...	...
Colombia Colombie										
MF [I] [1,19,20]	463.2	499.9	482.8	429.0	403.0	356.5	491.6	501.6	486.6	...
M [I] [1,19,20]	230.5	228.7	221.1	190.7	175.5	159.5	232.8	216.8	196.6	...
F [I] [1,19,20]	232.7	271.2	261.8	238.3	227.5	197.1	258.8	254.8	290.0	...
Costa Rica Costa Rica										
MF [I] [1,20]	44.4	60.8	56.7	# 54.5	54.9	38.7	49.5	59.1	44.0	...
M [I] [1,20]	32.0	42.5	40.3	# 33.1	31.9	23.4	31.7	35.5	26.4	...
F [I] [1,20]	12.4	18.3	16.5	# 21.4	23.1	15.3	17.8	23.5	17.6	...
% MF [I] [1,20]	5.0	6.8	6.2	# 5.6	5.5	3.8	4.6	5.5	4.1	...
% M [I] [1,20]	5.0	6.5	6.0	# 4.7	4.4	3.2	4.2	4.8	3.5	...
% F [I] [1,20]	5.0	7.9	6.9	# 7.9	8.0	5.3	5.9	7.4	5.4	...
Côte d'Ivoire Côte d'Ivoire										
MF [III] [1,21]	73.5	86.4	92.0	107.8	120.6	128.5	140.3	136.9	114.9	...
M [III] [1,21]	55.8	62.5	...	73.2	81.9	89.4	99.0	...	88.2	...
F [III] [1,21]	17.7	23.9	...	34.6	38.7	39.1	41.3	...	26.7	...
Croatia Croatie										
MF [III]	114.0	120.0	123.0	123.0	135.0	140.0	161.0	254.0	267.0	...
M [III]	44.0	46.0	48.0	48.0	55.0	57.0	70.0	121.0	126.0	...
F [III]	70.0	74.0	75.0	75.0	80.0	83.0	91.0	133.0	141.0	...
% MF [III]	6.0	7.0	7.0	6.0	7.0	7.0	8.0	14.0	15.0	...
% M [III]	4.0	4.0	4.0	4.0	5.0	5.0	6.0	11.0	12.0	...
% F [III]	10.0	10.0	10.0	9.0	10.0	10.0	11.0	16.0	18.0	...
Cyprus Chypre										
MF [III] [4]	8.0	8.3	9.2	8.7	7.4	6.2	5.1	8.3	5.2	7.6
M [III] [4]	4.5	4.6	4.8	4.4	3.7	2.9	2.5	3.8	2.4	3.2
F [III] [4]	3.5	3.7	4.4	4.3	3.7	3.3	2.6	4.5	2.8	4.4
% MF [III] [4]	3.3	3.3	3.7	3.4	2.8	2.3	1.8	3.0	1.8	2.6
% M [III] [4]	2.9	2.9	2.9	2.7	2.2	1.7	1.4	2.2	1.3	2.0
% F [III] [4]	4.1	4.2	5.0	4.7	3.8	3.3	2.5	4.4	2.6	4.1
former Czechoslovakia† anc. Tchécoslovaquie†										
MF [III] [1]	...	...	...	...	...	1.0	77.0	524.0	395.0	...
M [III] [1]	...	...	...	...	...	...	38.0	239.0	187.0	...
F [III] [1]	...	...	...	...	...	...	39.0	285.0	208.0	...
% MF [III] [1]	...	...	...	...	...	...	1.0	6.6	5.1	...
% M [III] [1]	...	...	...	...	...	...	0.9	5.9	4.7	...
% F [III] [1]	...	...	...	...	...	...	1.0	7.3	5.4	...

30
Unemployment
Number (thousands) and percentage of unemployed [*cont.*]
     Chômage
     Nombre (milliers) et pourcentage des chômeurs [*suite*]

Country or area Pays ou zone §	1984	1985	1986	1987	1988	1989	1990	1991	1992	1993
Czech Republic République tchèque										
MF [III] [1]	..	..	..	..	..	..	39.0	222.0	135.0	185.0
M [III] [1]	..	..	..	..	..	..	19.0	95.0	57.0	81.0
F [III] [1]	..	..	..	..	..	..	20.0	127.0	78.0	104.0
% MF [III] [1]	..	..	..	..	..	..	0.7	4.1	2.6	3.5
% M [III] [1]	..	..	..	..	..	..	0.7	3.5	2.2	3.0
% F [III] [1]	..	..	..	..	..	..	0.8	4.8	3.0	4.1
Denmark Danemark										
MF [I] [10 22]	231.4	199.5	153.6	200.0	207.9	253.9	242.4	264.8	...	...
M [I] [10 22]	112.1	88.6	62.0	92.0	99.0	124.3	122.8	129.3	...	...
F [I] [10 22]	119.2	110.9	91.6	108.0	108.9	129.6	119.7	135.5	...	...
MF [III] [23]	276.0	251.8	220.4	221.9	243.9	264.9	271.7	296.1	318.3	348.8
M [III] [23]	129.7	110.9	90.8	96.1	108.8	120.0	124.0	137.2	148.8	168.6
F [III] [23]	146.3	140.9	129.6	25.7	135.1	145.0	147.7	158.9	169.5	180.2
% MF [III] [23]	10.1	9.1	7.9	7.9	8.7	9.5	9.7	10.6	11.4	12.4
% M [III] [23]	8.8	7.5	6.1	6.4	7.3	8.1	8.4	9.2	10.0	11.3
% F [III] [23]	11.7	11.0	10.0	9.6	10.3	11.1	11.3	12.1	12.9	13.7
Ecuador Equateur										
MF [I] [1 11 20]	...	...	...	89.5[24]	155.4	187.0	150.5	237.0	263.2	240.8
M [I] [1 11 20]	...	...	...	39.8[24]	73.0	88.1	67.0	91.9	105.3	108.6
F [I] [1 11 20]	...	...	...	49.7[24]	82.3	98.9	83.2	145.2	157.9	132.3
% MF [I] [1 11 20]	...	...	...	7.2[24]	7.0	7.9	6.1	8.5	8.9	9.4
% M [I] [1 11 20]	...	...	...	5.2[24]	5.1	5.9	4.3	5.4	6.0	6.6
% F [I] [1 11 20]	...	...	...	10.4[24]	10.3	11.1	9.1	13.2	13.2	14.0
Egypt Egypte										
MF [I] [1 25]	756.0	...	...	...	...	1107.9	1346.5	1463.4	...	...
M [I] [1 25]	487.1	...	...	...	...	615.9	602.3	692.2	...	...
F [I] [1 25]	268.9	...	...	...	...	492.0	744.2	771.2	...	...
% MF [I] [1 25]	6.0	...	...	...	...	6.9	8.6	9.6	...	...
% M [I] [1 25]	4.8	...	...	...	...	5.4	5.2	5.9	...	...
% F [I] [1 25]	11.4	...	...	...	...	10.7	17.9	21.3	...	...
El Salvador El Salvador										
MF [I] [12]	...	280.2	# 28.3[26]	...	# 74.1[11]	72.0	97.9	72.5[1]	81.0	...
M [I] [12]	...	128.0	# 14.5[26]	...	# 50.9[11]	46.2	54.8	43.9[1]	47.5	...
F [I] [12]	...	152.3	# 13.8[26]	...	# 23.2[11]	25.8	43.1	28.6[1]	33.5	...
% MF [I] [12]	...	16.9	# 7.9[26]	...	# 9.4[11]	8.4	10.0	7.5[1]	7.9	...
% M [I] [12]	...	12.4	# 7.5[26]	...	# 11.0[11]	10.0	10.1	8.3[1]	8.4	...
% F [I] [12]	...	24.3	# 8.5[26]	...	# 7.1[11]	6.8	9.8	6.6[1]	7.2	...
Estonia Estonie										
MF [III]	..	..	..	..	..	..	..	0.9	14.9	16.3
M [III]	..	..	..	..	..	..	..	0.2	7.5	7.5
F [III]	..	..	..	..	..	..	..	0.6	7.4	8.8
% MF [III]	..	..	..	..	..	..	..	0.1	1.7	1.9
% M [III]	..	..	..	..	..	..	..	0.1	1.6	1.6
% F [III]								0.2	1.8	2.1
Ethiopia Ethiopie										
MF [III] [27 28]	54.7	56.4	52.6	58.2	55.3	51.3	44.2	44.3	70.9	...
M [III] [27 28]	33.2	33.3	30.7	37.9	34.1	28.6	25.8	24.9	52.1	...
F [III] [27 28]	21.5	23.1	21.9	20.3	21.2	22.7	18.4	19.4	18.8	...
Fiji Fidji										
MF [IV] [2]	16.8	18.6	18.2	23.0	23.0	15.0	16.0	15.0	15.8	15.8
% MF [IV] [2]	7.5	8.1	7.5	9.3	9.4	6.1	6.4	5.9	6.0	5.9
Finland Finlande										
MF [I] [29]	133.0	129.0	138.0	130.0	116.0	89.0	88.0	193.0	328.0	444.0
M [I] [29]	72.0	73.0	82.0	78.0	67.0	48.0	54.0	124.0	203.0	259.0
F [I] [29]	61.0	56.0	56.0	53.0	48.0	41.0	34.0	69.0	125.0	184.0
MF [III] [2 30]	124.0	129.0	136.0	130.0	119.0	97.0	94.0	181.0	319.0	436.0
M [III] [2 30]	68.0	71.0	76.0	74.0	65.0	50.0	52.0	108.0	186.0	245.0
F [III] [2 30]	56.0	58.0	60.0	56.0	54.0	47.0	42.0	73.0	133.0	191.0
% MF [I] [29]	5.2	5.0	5.4	5.1	4.5	3.5	3.4	7.6	13.5	17.7
% M [I] [29]	5.4	5.5	6.1	5.8	5.1	3.6	4.0	9.2	15.2	19.5
% F [I] [29]	5.0	4.6	4.6	4.3	4.0	3.3	2.8	5.7	10.5	15.7

30
Unemployment
Number (thousands) and percentage of unemployed [*cont.*]
Chômage
Nombre (milliers) et pourcentage des chômeurs [*suite*]

Country or area Pays ou zone §	1984	1985	1986	1987	1988	1989	1990	1991	1992	1993
France France										
MF [III] [31]	2340.3	2458.4	2516.6	2621.7	2562.9	2531.9	2504.7	2709.1	2911.2	3172.0
M [III] [31]	1226.7	1272.5	1275.1	1297.7	1224.8	1177.4	1148.7	1266.4	1404.6	1603.9
F [III] [31]	1113.6	1185.9	1241.5	1324.0	1338.1	1354.5	1355.9	1442.7	1506.6	1568.1
MF [IV]	2357.0	2473.6	2520.3	2567.2	2456.0	2323.0	2204.9	2348.9	2590.7	2911.4
M [IV]	1103.7	1179.1	1190.5	1169.3	1092.5	998.9	947.8	1031.8	1168.3	1384.0
F [IV]	1253.3	1294.5	1329.9	1398.0	1363.5	1324.1	1257.0	1317.1	1422.5	1527.4
% MF [IV]	9.8	10.2	10.4	10.5	10.0	9.4	8.9	9.4	10.3	11.6
% M [IV]	7.9	8.4	.8.5	8.4	7.8	7.1	6.7	7.3	8.3	9.9
% F [IV]	12.4	12.7	12.9	13.3	12.9	12.4	11.7	12.0	12.9	13.8
French Guiana Guyane française										
MF [III] [3 31 32]	3.7	4.2[1]	3.7	3.4	3.3	3.8	4.4	4.7	6.9	8.1
M [III] [3 31 32]	1.8	2.1[1]	1.8	1.6	1.6	1.9	2.3	2.5	4.0	4.7
F [III] [3 31 32]	1.9	2.1[1]	1.9	1.8	1.7	2.0	2.1	2.2	3.0	3.4
% MF [III] [3 31 32]	12.2	13.5[1]	12.0	11.0	10.6	12.2	13.9	9.7	...	...
% M [III] [3 31 32]	6.0	10.5[1]	9.2	8.4	...	9.4	11.7	8.2	...	...
% F [III] [3 31 32]	16.3	18.5[1]	16.7	15.4	...	16.8	17.6	11.6	...	...
French Polynesia Polynésie française										
MF [III] [4]	0.9	0.9	1.0	0.5	0.7	0.6	0.6	...	...	...
Germany † Allemagne †										
F. R. Germany R. f. Allemagne										
MF [I] [12]	2207.0	2385.0	2290.0	2359.0	2314.0	2147.0	1971.0	1676.0	1788.0	...
M [I] [12]	1110.0	1177.0	1127.0	1236.0	1145.0	1046.0	943.0	852.0	930.0	...
F [I] [12]	1098.0	1209.0	1163.0	1123.0	1169.0	1101.0	1028.0	824.0	858.0	...
MF [III] [29]	2265.6	2304.0	2228.0	2228.8	2241.6	2037.8	1883.1	1689.4	1808.3	2270.3
M [III] [29]	1276.7	1289.1	1200.0	1207.4	1198.8	1069.8	967.7	897.7	982.8	1277.1
F [III] [29]	988.9	1015.0	1028.0	1021.4	1042.8	968.0	915.4	791.7	825.5	993.3
% MF [I] [12]	8.7	9.2	8.8	9.0	8.7	8.0	7.0	6.0	6.0	...
% M [I] [12]	7.1	7.5	7.2	7.8	7.2	6.5	5.7	5.2	5.6	...
% F [I] [12]	11.0	11.8	11.2	10.7	11.0	10.2	8.8	7.1	7.2	...
% MF [III] [29]	9.1	9.3	9.0	8.9	8.7	7.9	7.2	6.3	6.6	8.2
% M [III] [29]	8.5	8.6	8.0	8.0	7.8	6.9	6.3	5.8	6.2	8.0
% F [III] [29]	10.2	10.4	10.5	10.2	10.0	9.4	8.4	7.0	7.2	8.4
former German D. R. anc. R. d. allemande										
MF [III]	...	...	...	...	...	...	...	912.8	1170.3	1148.8
M [III]	...	...	...	...	...	...	...	382.9	429.1	414.5
F [III]	...	...	...	...	...	...	...	530.0	741.1	734.3
% MF [III]	...	...	...	...	...	...	...	10.3	14.8	15.8
% M [III]	...	...	...	...	...	...	...	8.5	10.5	11.0
% F [III]	...	...	...	...	...	...	...	12.3	19.6	21.0
Ghana Ghana										
MF [III] [33]	21.2	24.2	25.8	...	28.7	27.4	30.2	30.7	30.6	...
M [III] [33]	15.4	19.2	22.5	...	26.1	24.7	26.6	27.5	27.5	...
F [III] [33]	5.8	5.1	3.3	...	2.6	2.7	3.2	3.3	3.1	...
% MF [III] [33]	0.4	0.4	0.5	...	...	...	...	...	...	...
% M [III] [33]	0.6	0.7	0.8	...	...	...	...	...	...	...
% F [III] [33]	0.2	0.2	0.1	...	...	...	...	...	...	...
Gibraltar Gibraltar										
MF [III] [34]	0.5	0.5	0.5	0.3	0.5	0.6	...	...	...	...
M [III] [34]	0.3	0.3	0.4	0.2	0.5	0.5	...	...	...	...
F [III] [34]	0.2	0.1	0.1	0.1	0.1	0.1	...	...	...	...
Greece Grèce										
MF [I] [4 10]	315.0	303.9	286.9	286.2	303.4	296.0	280.8	301.1	349.8	...
M [I] [4 10]	152.4	141.9	127.1	127.9	121.5	114.6	107.1	120.8	137.9	...
F [I] [4 10]	162.6	162.0	159.9	158.3	181.9	181.4	173.7	180.3	211.9	175.9
MF [III] [2]	71.2	89.0	110.5	117.9	115.3	133.9	* 140.2	* 173.2	184.7	87.9
M [III] [2]	42.0	51.2	61.0	64.1	58.7	64.6	* 68.1	* 83.9	89.7	87.9
F [III] [2]	29.2	37.8	49.4	53.9	56.6	69.3	* 72.1	* 89.3	95.0	...
% MF [I] [4 10]	8.3	7.8	7.4	7.4	7.7	7.5	7.0	7.7	8.7	...
% M [I] [4 10]	6.1	5.7	5.1	5.1	4.9	4.6	4.3	4.8	5.4	...
% F [I] [4 10]	12.3	11.7	11.6	11.4	12.5	12.4	11.7	12.9	14.2	...
% MF [III] [2]	4.2	5.1	6.1	6.4	6.0	6.5	* 6.4	* 7.3	7.6	7.1

30
Unemployment
Number (thousands) and percentage of unemployed [*cont.*]
Chômage
Nombre (milliers) et pourcentage des chômeurs [*suite*]

Country or area Pays ou zone §	1984	1985	1986	1987	1988	1989	1990	1991	1992	1993
Greenland Groënland										
MF [III]	2.3	2.4	...	...	...	...	...	...	1.9	1.8
M [III]	1.2	1.2	...	...	...	...	...	...	...	...
F [III]	1.1	1.2	...	...	...	...	...	...	...	...
Guadeloupe Guadeloupe										
MF [III] [31]	21.5	24.2	27.2	27.8	29.5	30.8	29.4[1]	34.3[1]	...	...
M [III] [31]	...	11.4	12.2	11.7	11.5	13.5	11.7[1]	13.9[1]	...	...
F [III] [31]	...	12.8	15.0	16.1	18.0	21.3	17.6[1]	20.4[1]	...	...
% MF [III] [31]	20.3	22.0	27.0	23.0	24.0	24.0	17.0	19.9		
% M [III] [31]	...	19.0	22.0	16.0	16.0	16.0	...	...		
% F [III] [31]	...	27.0	33.0	31.0	33.0	34.0	...	...		
Guam Guam										
MF [I] [31]	2.9	2.7	2.2	1.6	1.6	1.1	1.3	1.7	1.8	2.6
M [I] [31]	1.7	1.4	1.2	0.9	...	...	...	...	...	...
F [I] [31]	1.1	1.3	1.0	0.6	...	...	...	...	...	...
% MF [I] [31]	8.4	7.8	6.1	4.4	4.2	2.8	2.8	3.5	3.9	5.5
% M [I] [31]	8.5	5.6	7.1	4.3	...	...	...	...	...	...
% F [I] [31]	8.3	8.8	6.9	4.5	...	...	...	...	...	...
Guatemala Guatemala										
MF [III] [12 35]	4.8	2.7	2.7	2.2	1.9	1.7	1.8	...	...	...
M [III] [12 35]	3.4	1.7	1.8	1.4	1.3	1.2	1.3	...	...	...
F [III] [12 35]	1.4	1.0	0.9	0.8	0.6	0.5	0.5	...	...	...
Guyana Guyana										
MF [III] [36 37]	11.7	13.3	10.6	9.2	...	...	...	...	...	...
M [III] [36 37]	7.2	5.8	5.2	4.8	...	...	...	...	...	...
F [III] [36 37]	4.4	7.5	5.4	4.4	...	...	...	...	...	...
Haiti Haïti										
MF [IV] [12]	...	...	...	...	561.8	...	339.7	...	...	...
M [IV] [12]	...	...	...	...	265.0	...	191.3	...	...	...
F [IV] [12]	...	...	...	...	296.7	...	148.3	...	...	...
Hong Kong Hong–kong										
MF [I] [2]	101.0	# 83.6[38]	76.1	47.5	38.0	30.0	37.0	50.3	54.7	56.9
M [I] [2]	68.8	# 58.6[38]	51.1	29.8	24.1	19.3	23.5	33.7	35.3	36.2
F [I] [2]	32.2	# 25.1[38]	24.9	17.7	13.8	10.7	13.5	16.6	19.4	20.7
% MF [I] [2]	3.9	# 3.2[38]	2.8	1.7	1.4	1.1	1.3	1.8	2.0	2.0
% M [I] [2]	4.2	# 3.5[38]	3.0	1.7	1.4	1.1	1.3	1.9	2.0	2.0
% F [I] [2]	3.4	# 2.6[38]	2.5	1.8	1.4	1.1	1.3	1.6	1.9	2.0
Hungary Hongrie										
MF [III] [1]	...	...	...	...	...	...	79.5	406.1	663.0	632.1
M [III] [1]	...	...	...	...	...	...	49.1	239.0	390.0	376.1
F [III] [1]	...	...	...	...	...	...	30.4	167.1	273.0	256.0
% MF [III] [1]	...	...	...	...	...	...	1.7	8.5	12.3	12.1
% M [III] [1]	...	...	...	...	...	...	1.8	9.2	14.0	14.2
% F [III] [1]	...	...	...	...	...	...	1.4	7.6	10.5	10.1
Iceland Islande										
MF [III] [31]	1.5	1.1	0.8	0.6	0.8	2.1	2.3	1.9	3.9	5.6
M [III] [31]	0.7	0.5	0.4	0.3	0.3	0.9	1.1	1.0	1.9	2.7
F [III] [31]	0.8	0.7	0.4	0.3	0.5	1.2	1.2	0.9	1.9	2.9
% MF [III] [31]	1.2	0.9	0.7	0.4	0.6	1.7	1.8	1.5	3.0	4.3
% M [III] [31]	0.9	0.6	0.5	0.3	0.4	1.2	1.4	1.3	2.6	3.6
% F [III] [31]	1.8	1.3	0.8	0.6	0.9	2.2	2.2	1.7	3.6	5.4
India Inde										
MF [III] [4]	23034.0	24861.0	28261.0	30542.0	# 30050.0[1]	32776.0	34632.0	36300.0	36758.0	...
M [III] [4]	19186.0	20628.0	23476.0	25251.0	# 24590.0[1]	26668.0	27932.0	28992.0	29105.0	...
F [III] [4]	3848.0	4233.0	4785.0	5291.0	# 5461.0[1]	6109.0	6700.0	7308.0	7653.0	...
Indonesia Indonésie										
MF [I] [12]	...	1368.0[39]	1855.0	1843.0	2106.0	2083.0	1952.0	2032.0	2199.0	...
M [I] [12]	...	898.0[39]	1127.0	1148.0	1269.0	1251.0	1155.0	1147.0	1292.0	...
F [I] [12]	...	470.0[39]	728.0	695.0	837.0	832.0	796.0	885.0	907.0	...
MF [III] [3 40]	577.0	785.2	855.0	1017.2	1352.4	1518.5	1238.7	1042.3	...	...
M [III] [3 40]	430.7	552.3	586.3	672.8	900.8	991.7	735.9	...	...	...
F [III] [3 40]	146.3	233.0	268.7	344.4	451.6	526.8	502.9	...	...	...

30
Unemployment
Number (thousands) and percentage of unemployed [cont.]
Chômage
Nombre (milliers) et pourcentage des chômeurs [suite]

Country or area Pays ou zone §	1984	1985	1986	1987	1988	1989	1990	1991	1992	1993
Ireland Irlande										
MF [I] [12]	203.5	226.0	227.4	231.6	218.5	202.2	178.9	208.4	...	...
M [I] [12]	156.4	172.8	173.7	176.3	169.6	157.3	138.1	156.3	...	...
F [I] [12]	47.1	53.2	53.8	55.2	49.0	44.9	40.8	52.1	...	...
MF [III] [31]	214.2	230.6	236.4	247.3	241.4	231.6	224.7	253.9	283.1	294.0
M [III] [31]	159.1	170.2	172.0	176.2	169.7	160.0	152.1	170.5	187.2	194.0
F [III] [31]	55.1	60.4	64.4	71.1	71.7	71.6	72.6	83.5	96.0	101.0
% MF [I] [12]	15.6	17.4	17.4	17.6	16.7	15.6	13.7	15.6	...	...
% M [I] [12]	17.0	18.8	19.0	19.4	18.6	17.6	15.5	17.3	...	...
% F [I] [12]	12.2	13.8	13.7	13.6	12.3	11.3	9.9	12.1	...	...
% MF [III] [31]	16.4	17.7	18.1	18.8	18.4	17.9	17.2	19.0	...	...
% M [III] [31]	17.3	18.5	18.8	19.3	18.4	17.9	17.0	18.8	...	...
% F [III] [31]	14.3	15.7	16.4	17.7	17.9	17.9	17.6	19.5	...	...
Isle of Man Ile de Man										
MF [III]	2.0	2.2	2.1	1.6	0.9	0.6	0.6	1.0	1.4	1.6
M [III]	1.4	1.5	1.4	1.1	0.6	0.4	0.4	0.7	1.0	1.2
F [III]	0.6	0.7	0.7	0.6	0.3	0.2	0.1	0.3	0.4	0.4
% MF [III]	7.4	8.0	7.7	5.9	3.4	2.0	2.1	# 3.0[41]	4.3[41]	5.0
% M [III]	8.3	8.8	8.4	6.3	3.7	2.3	2.6	# 3.9[41]	5.4[41]	6.3
% F [III]	6.0	6.7	6.6	5.3	2.9	1.6	1.3	# 1.9[41]	2.7[41]	3.1
Israel Israël										
MF [I] [42 43]	85.1[4]	# 97.0[2]	104.2[2]	90.1[2]	100.0[2]	142.5[2]	158.0[2]	187.4[2]	207.5[2]	194.9[2]
M [I] [42 43]	46.5[4]	# 56.5[2]	59.1[2]	47.6[2]	53.0[2]	75.7[2]	82.2[2]	89.9[2]	99.7[2]	96.2[2]
F [I] [42 43]	38.6[4]	# 40.1[2]	45.2[2]	42.5[2]	47.0[2]	66.8[2]	75.8[2]	97.5[2]	107.8[2]	98.7[2]
% MF [I] [42 43]	5.9[4]	# 6.7[2]	7.1[2]	6.1[2]	6.4[2]	8.9[2]	9.6[2]	10.6[2]	11.2[2]	10.0[2]
% M [I] [42 43]	5.2[4]	# 6.3[2]	6.5[2]	5.2[2]	5.7[2]	7.9[2]	8.4[2]	8.6[2]	9.2[2]	8.5[2]
% F [I] [42 43]	7.0[4]	# 7.2[2]	7.9[2]	7.3[2]	7.6[2]	10.3[2]	11.3[2]	13.4[2]	13.9[2]	12.1[2]
Italy Italie										
MF [I] [4]	2303.0	2382.0	2611.0	2832.0	2885.0	2865.0	2621.0	2653.0	2799.0	...
M [I] [4]	986.0	1024.0	1115.0	1228.0	1240.0	1220.0	1102.0	1142.0	1226.0	...
F [I] [4]	1317.0	1358.0	1496.0	1604.0	1645.0	1646.0	1519.0	1511.0	1573.0	...
% MF [I] [4]	10.0	10.3	11.1	11.9	12.0	12.0	11.0	10.9	11.5	...
% M [I] [4]	6.6	6.8	7.4	8.1	8.1	8.1	7.3	7.5	8.1	...
% F [I] [4]	16.5	16.7	17.8	18.5	15.4	18.7	17.1	16.8	17.3	...
Jamaica Jamaïque										
MF [I] [4]	266.8	260.8	250.5	224.3	203.3	177.4	166.6	...	...	...
M [I] [4]	87.9	88.4	85.3	77.0	68.0	54.1	52.8	...	...	...
F [I] [4]	179.0	172.4	165.2	147.3	135.3	123.3	113.8	...	...	...
% MF [I] [4]	25.5	25.0	23.6	21.0	18.9	16.8	15.7	...	...	...
% M [I] [4]	15.8	15.7	15.0	13.2	11.9	9.5	9.3	...	...	...
% F [I] [4]	36.6	36.0	33.8	30.4	27.0	25.2	23.1	...	...	...
Japan Japon										
MF [I] [2]	1610.0	1560.0	1670.0	1730.0	1550.0	1420.0	1340.0	1370.0	1420.0	1660.0
M [I] [2]	960.0	930.0	990.0	1040.0	910.0	830.0	770.0	780.0	820.0	950.0
F [I] [2]	650.0	630.0	670.0	690.0	640.0	590.0	570.0	590.0	600.0	710.0
% MF [I] [2]	2.7	2.6	2.8	2.8	2.5	2.3	2.1	2.1	2.2	2.5
% M [I] [2]	2.7	2.6	2.7	2.8	2.5	2.2	2.0	2.0	2.1	2.4
% F [I] [2]	2.8	2.7	2.8	2.8	2.6	2.3	2.2	2.2	2.2	2.6
Kazakhstan Kazakhstan										
MF [III]	..	..	..	..	..	..	..	6.0	70.0	78.1
M [III]	..	..	..	..	..	..	..	2.0	21.0	27.7
F [III]	..	..	..	..	..	..	..	4.0	49.0	50.4
% MF [III]	..	..	..	..	..	..	..	0.1	0.9	1.0
Korea, Republic of Corée, République de										
MF [I] [2]	568.0	622.0	611.0	519.0	435.0	459.0	451.0	436.0	464.0	551.0[44]
M [I] [2]	444.0	480.0	480.0	397.0	315.0	326.0	318.0	287.0	303.0	373.0[44]
F [I] [2]	124.0	141.0	131.0	122.0	120.0	134.0	133.0	150.0	161.0	178.0[44]
% MF [I] [2]	3.8	4.0	3.8	3.1	2.5	2.6	2.4	2.3	2.4	2.8[44]
% M [I] [2]	4.8	5.0	4.9	3.9	3.0	3.0	2.9	2.5	2.6	3.1[44]
% F [I] [2]	2.2	2.4	2.1	1.8	1.7	1.8	1.8	2.0	2.1	2.3[44]

30
Unemployment
Number (thousands) and percentage of unemployed [cont.]
Chômage
Nombre (milliers) et pourcentage des chômeurs [suite]

Country or area Pays ou zone §	1984	1985	1986	1987	1988	1989	1990	1991	1992	1993
Latvia Lettonie										
MF [III][1]	...	...	...	...	...	...	...	...	31.3	76.7
M [III][1]	...	...	...	...	...	...	...	...	12.9	35.9
F [III][1]	...	...	...	...	...	...	...	...	18.4	40.8
% MF [III][1]	...	...	...	...	...	...	...	...	2.3	5.8
% M [III][1]	...	...	...	...	...	...	...	...	1.8	5.2
% F [III][1]	...	...	...	...	...	...	...	...	4.8	6.4
Lithuania Lituanie										
MF [III][1]	...	...	...	...	...	...	...	4.8	66.5	65.5
M [III][1]	...	...	...	...	...	...	...	...	...	33.5
F [III][1]	...	...	...	...	...	...	...	...	...	32.0
Luxembourg Luxembourg										
MF [III][45]	2.7	2.6	2.3	2.7	2.5	2.3	2.1	2.3	2.7	3.5
M [III][45]	1.4	1.3	1.2	1.5	1.5	1.4	1.2	1.4	1.5	2.0
F [III][45]	1.3	1.3	1.1	1.2	1.0	0.9	0.9	0.9	1.2	1.5
% MF [III][45]	1.8	1.7	1.5	1.7	1.6	1.4	1.3	1.4	1.6	...
% M [III][45]	1.4	1.3	1.2	1.5	1.5	1.4	1.1	1.3	1.5	...
% F [III][45]	2.5	2.3	1.8	2.0	1.7	1.5	1.5	1.5	1.9	...
Macau Macao										
MF [I][4]	...	...	...	...	...	6.2[1]	5.3[1]	5.3[1]	3.8	...
M [I][4]	...	...	...	...	...	2.6[1]	2.5[1]	2.6[1]	2.1	...
F [I][4]	...	...	...	...	...	3.6[1]	2.8[1]	2.7[1]	1.7	...
% MF [I][4]	...	...	...	...	...	3.7[1]	3.2[1]	3.0[1]	2.2	...
% M [I][4]	...	...	...	...	...	2.7[1]	2.5[1]	2.5[1]	2.1	...
% F [I][4]	...	...	...	...	...	5.1[1]	4.1[1]	3.7[1]	2.4	...
Madagascar Madagascar										
MF [III][13]	29.0	28.8	24.2	18.4	16.1	15.7	...	...	...	...
Malaysia Malaisie										
MF [III][2]	72.1[1]	80.7[1]	86.9[1]	# 82.5	82.7	75.6	61.2	48.6	45.2	...
M [III][2]	44.6[1]	49.3[1]	...	# ...	48.5	38.9	...	...	...	
F [III][2]	27.5[1]	31.4[1]	...	# ...	27.0	22.3	...	...	...	
Malta Malte										
MF [III][146]	10.4	9.9	8.5	5.6	5.2	4.8	5.1	4.9	5.6	6.1
M [III][146]	8.1	7.8	6.6	4.6	4.2	4.1	4.3	4.0	4.5	5.3
F [III][146]	2.3	2.2	1.9	1.0	1.0	0.8	0.8	1.0	1.1	1.0
% MF [III][146]	...	8.1	6.8	4.4	4.0	3.7	3.8	3.6	4.0	4.5
% M [III][146]	...	8.4	6.9	4.8	4.3	4.1	4.4	4.0	4.4	5.2
% F [III][146]	...	7.2	6.2	3.1	3.1	2.3	2.3	2.6	3.0	2.5
Martinique Martinique										
MF [III][31]	25.6	29.5	33.9	30.7	...	...	...	...	...	...
M [III][131]	13.6	13.6	...	13.1	...	...	...	...	...	...
F [III][131]	15.6	17.2	...	19.4	...	...	...	...	...	...
Mauritius Maurice										
MF [III][247]	70.2	64.8	54.6	46.8	27.7	18.1	12.8	10.6	7.9	...
M [III][247]	52.5	48.8	42.2	36.3	19.9	11.8	7.3	5.2	3.4	...
F [III][247]	17.8	16.0	12.4	10.5	7.8	6.3	5.5	5.4	4.6	...
Mexico Mexique										
% MF [I][2026]	6.0	# 4.4[11]	4.3	3.9	3.6	3.0	2.8	2.6	2.8	3.4
% M [I][2026]	5.3	# 3.5[11]	3.7	3.4	3.0	2.6	2.6	2.5	2.7	3.2
% F [I][2026]	7.6	# 5.7[11]	5.3	4.8	4.7	3.8	3.1	3.0	3.2	3.9
Montserrat Montserrat										
MF [IV]	0.3	0.3	0.2	0.1	...	...	...	...	...	...
% MF [IV]	5.8	5.3	4.2	2.0	...	...	...	...	...	...
Morocco Maroc										
MF [I][211]	...	...	...	...	...	...	601.2	695.5	649.9	...
M [I][211]	...	...	...	...	...	...	401.4	459.3	400.7	...
F [I][211]	...	...	...	...	...	...	199.8	236.2	249.2	...
% MF [I][211]	...	...	...	14.7	13.9	16.3	15.4	17.0	16.0	...
% M [I][211]	...	...	...	13.4	12.8	15.0	13.9	15.1	13.0	...
% F [I][211]	...	...	...	18.5	17.3	19.8	19.6	22.6	25.3	...

30

Unemployment

Number (thousands) and percentage of unemployed [cont.]

Chômage

Nombre (milliers) et pourcentage des chômeurs [suite]

Country or area Pays ou zone §	1984	1985	1986	1987	1988	1989	1990	1991	1992	1993
Myanmar Myanmar										
MF [III] [48]	441.0	338.0[1]	354.4[1]	331.4	312.7	485.8	555.3	559.0	502.6	...
Netherlands Pays—Bas										
MF [I] [29]	...	620.5[2]	...	# 622.0	609.0	558.0	516.0	490.0	...	# 481.0[49]
M [I] [29]	...	362.9[2]	...	# 290.0	291.0	261.0	227.0	226.0	...	# 228.0[49]
F [I] [29]	...	257.6[2]	...	# 332.0	318.0	297.0	288.0	264.0	...	# 253.0[49]
MF [III] [29]	822.4	761.0	710.7	685.5	# 433.0	390.0	346.0	319.0	336.0	415.0
M [III] [29]	555.2	498.0	453.5	428.7	# 278.0	241.0	209.0	187.0	195.0	240.0
F [III] [29]	267.2	263.0	257.2	256.8	# 155.0	149.0	137.0	132.0	141.0	175.0
% MF [I] [29]	...	10.8[2]	...	# 10.0	9.0	8.0	7.5	7.0	...	# 7.5[49]
% M [I] [29]	...	9.6[2]	...	# 7.0	7.0	6.0	5.4	5.3	...	# 5.7[49]
% F [I] [29]	...	13.0[2]	...	# 14.0	13.0	12.0	10.7	9.5	...	# 10.5[49]
% MF [III] [29]	14.1	12.9	12.0	11.5	# 6.5	5.8	5.0	4.5	5.3	6.5
% M [III] [29]	14.6	13.0	11.8	11.1	# 6.8	5.8	5.0	4.4	4.9	6.0
% F [III] [29]	13.3	12.8	12.3	12.1	# 6.1	5.8	5.1	4.7	6.1	7.3
Netherlands Antilles Antilles néerlandaises										
MF [IV] [50]	15.8	17.2[51]	17.9[51]	17.6[51]	13.3	11.7	11.2	9.4	...	...
M [IV] [50]	6.7	7.3[51]	7.8[51]	...	...	...	5.7	4.3		
F [IV] [50]	9.1	9.9[51]	10.1[51]	...	...	...	5.5	5.1		
% MF [IV] [50]	20.8	22.2[51]	...	24.6[51]	24.4	21.0	19.8	16.4		
% M [IV] [50]	15.8	15.9[51]	...	...	...	...	17.2	13.1		
% F [IV] [50]	29.4	31.3[51]	...	...	...	...	23.6	20.9	...	
New Caledonia Nouvelle—Calédonie										
MF [III] [31]	1.1	1.5	1.6	# 4.5	5.0	5.2	5.7	6.2	6.6	6.8
M [III] [31]	0.8	0.9	...	# 2.4	...	...	...	...	...	...
F [III] [31]	0.3	0.4	...	# 2.1	...	...	...	...	...	...
New Zealand Nouvelle—Zélande										
MF [I] [2]	...	...	64.2	66.0	89.0	112.6	124.6	167.4	168.9	157.2
M [I] [2]	...	...	33.0	36.6	51.3	66.0	74.0	99.8	100.7	93.3
F [I] [2]	...	...	31.1	29.4	37.7	46.5	50.5	67.6	68.2	63.9
MF [III] [33 52]	66.5	53.2	67.2	88.1	120.9	153.6	163.8	196.0	216.9	212.7
M [III] [33 52]	41.4	33.5	45.0	60.8	84.6	106.6	112.1	135.9	149.8	143.2
F [III] [33 52]	25.2	19.7	22.2	27.3	36.4	47.0	51.7	60.1	67.1	69.4
% MF [I] [2]	...	...	4.0	4.0	5.6	7.1	7.8	10.3	10.3	9.5
% M [I] [2]	...	...	3.6	3.9	5.6	7.3	8.2	10.9	10.9	10.0
% F [I] [2]	...	...	4.6	4.3	5.6	6.9	7.2	9.6	9.5	8.9
Nicaragua Nicaragua										
MF [IV] [12]	...	34.6	51.8	66.6	71.8	107.2	145.6	194.2	...	...
M [IV] [12]	...	18.6	27.9	35.9	38.9	57.8	78.5	104.8	...	...
F [IV] [12]	...	16.0	23.9	30.7	32.9	49.4	67.1	89.4	...	...
% MF [IV] [12]	...	3.2	4.7	5.8	6.0	8.4	11.1	14.0	...	...
% M [IV] [12]	...	2.6	3.8	4.7	4.9	6.9	6.9	11.3	...	...
% F [IV] [12]	...	4.6	6.5	8.0	8.3	12.0	15.4	19.4	...	...
Niger Niger										
MF [III]	27.4	29.0	27.7	27.2	26.1	24.6	20.9	20.9	...	...
M [III]	26.8	28.2	26.7	25.7	24.8	23.4	19.9	19.9	...	...
F [III]	0.5	0.8	1.0	1.5	1.3	1.3	1.1	1.0	...	...
Nigeria Nigéria										
MF [III] [2]	34.3	28.3	32.5	57.3	60.5	57.6	57.1	60.2	180.4	...
Norway Norvège										
MF [I] [53]	64.0	53.0	42.0	45.0	69.0	106.0	112.0	116.0	126.0	127.0
M [I] [53]	36.0	25.0	18.0	21.0	36.0	61.0	66.0	68.0	76.0	77.0
F [I] [53]	28.0	28.0	24.0	25.0	33.0	45.0	46.0	48.0	50.0	50.0
MF [III] [31]	66.6	51.4	36.2	32.4	49.3	82.9	92.7	100.7	114.4	118.2
M [III] [31]	41.1	29.8	20.2	18.4	30.0	51.6	57.1	62.8	71.0	73.3
F [III] [31]	25.5	21.7	16.0	14.0	19.3	31.3	35.6	37.9	43.3	44.8
% MF [I] [53]	3.2	2.6	2.0	2.1	3.2	4.9	5.2	5.5	5.9	6.0
% M [I] [53]	3.1	2.2	1.5	1.7	3.0	5.1	5.6	5.9	6.5	6.6
% F [I] [53]	3.2	3.1	2.5	2.6	3.4	4.7	4.8	5.0	5.1	5.2
% MF [III] [31]	3.2	2.5	1.8	1.5	2.3	3.8	4.3	4.7	5.4	5.5

30
Unemployment
Number (thousands) and percentage of unemployed [*cont.*]
   Chômage
   Nombre (milliers) et pourcentage des chômeurs [*suite*]

Country or area Pays ou zone §	1984	1985	1986	1987	1988	1989	1990	1991	1992	1993
Pakistan Pakistan										
MF [III] [3 54]	183.6	212.3	224.6	247.6	265.5	251.8	238.8	221.7	204.3	...
MF [IV] [12 27]	1082.0	1042.0	1018.0	903.0	937.0	966.0	996.0	1999.0	2061.0	...
M [IV] [12 27]	1021.0	1005.0	970.0	862.0	907.0	935.0	964.0	1238.0	1276.0	...
F [IV] [12 27]	61.0	37.0	48.0	41.0	30.0	31.0	32.0	761.0	785.0	...
% MF [IV] [12 27]	3.9	3.7	3.6	3.1	3.1	3.1	3.1	6.3	6.3	...
% M [IV] [12 27]	4.2	4.0	3.9	3.3	3.4	3.4	3.4	4.5	4.5	...
% F [IV] [12 27]	1.9	1.5	1.7	1.1	0.9	0.9	0.9	16.8	16.8	...
Panama Panama										
MF [I] [12]	68.8	88.3	75.8	91.1	127.8	133.7	...	138.4	125.4	...
M [I] [12]	38.9	46.3	42.9	48.5	74.9	75.0	...	72.8	62.5	...
F [I] [12]	29.9	42.0	32.9	42.7	52.9	58.7	...	61.6	62.9	...
% MF [I] [12]	10.1	12.3	10.5	11.8	16.3	16.3	...	16.1	13.6	...
% M [I] [12]	8.2	9.5	8.7	9.4	14.0	13.7	...	12.8	10.0	...
% F [I] [12]	14.2	18.5	14.5	16.8	21.4	21.6	...	22.6	21.2	...
Paraguay Paraguay										
MF [I] [20 26]	28.5	22.0	26.7	24.8	22.2	32.0	34.1	26.6	29.1	30.5[12]
M [I] [20 26]	20.8	12.4	15.1	16.1	12.8	19.3	20.3	16.6	20.2	19.1[12]
F [I] [20 26]	7.7	9.6	11.6	8.7	9.4	12.7	13.8	10.0	8.9	11.4[12]
% MF [I] [1 26]	7.3	5.1	6.1	5.5	4.7	6.1	6.6	5.1	5.3	5.1[12]
% M [I] [1 26]	9.1	5.1	6.3	6.5	4.8	6.6	6.6	5.4	6.4	5.5[12]
% F [I] [1 26]	4.8	5.1	5.8	4.3	4.6	5.6	6.5	4.7	3.8	4.5[12]
Peru Pérou										
MF [I] [4 55]	153.1[56]	...	111.7	103.5	...	186.7	...	146.3	251.0	...
M [I] [4 55]	76.2[56]	...	40.9	48.6	...	84.8	...	73.7	120.9	...
F [I] [4 55]	76.9[56]	...	70.8	54.9	...	101.9	...	72.6	130.2	...
% MF [I] [4 55]	...	...	5.3	4.8	...	7.9	...	5.8	9.4	...
% M [I] [4 55]	...	...	3.4	3.8	...	6.0	...	4.8	7.5	...
% F [I] [4 55]	...	...	8.0	6.2	...	10.7	...	7.3	12.5	...
Philippines Philippines										
MF [I] [12]	1465.0[10]	1316.0[10]	1438.0[10]	# 2085.0	1954.0	2009.0	1993.0	2267.0	2263.0	2379.0
M [I] [12]	675.0[10]	644.0[10]	686.0[10]	# 1163.0	1131.0	1101.0	1099.0	1290.0	1303.0	1384.0
F [I] [12]	790.0[10]	672.0[10]	752.0[10]	# 922.0	823.0	908.0	893.0	977.0	959.0	995.0
% MF [I] [12]	7.0[10]	6.1[10]	6.4[10]	# 9.1	8.3	8.4	8.1	9.0	8.6	8.9
% M [I] [12]	5.2[10]	4.8[10]	4.9[10]	# 8.1	7.6	7.3	7.1	8.1	7.9	8.2
% F [I] [12]	10.0[10]	8.2[10]	8.9[10]	# 10.9	9.5	10.3	9.8	10.5	9.9	10.0
Poland Pologne										
MF [III]	...	...	...	...	...	...	591.0	1684.0	2355.0	2737.0
M [III]	...	...	...	...	...	...	290.0	717.0	1103.0	1290.0
F [III]	...	...	...	...	...	...	301.0	967.0	1252.0	1447.0
% MF [III]	...	...	...	...	...	...	3.5	9.7	13.3	15.7
% M [III]	...	...	...	...	...	...	3.2	7.9	11.8	14.3
% F [III]	...	...	...	...	...	...	3.8	11.8	14.9	17.3
Portugal Portugal										
MF [I] [12]	393.9	397.0	393.4	329.0	...	243.4	231.1	207.5	# 194.1[4]	257.5[4]
M [I] [12]	161.2	171.0	176.2	143.4	...	95.1	90.0	77.6	# 90.7[4]	120.0[4]
F [I] [12]	232.7	225.9	217.2	185.6	...	148.3	141.1	129.9	# 103.4[4]	137.5[4]
% MF [I] [12]	8.5	8.5	8.3	7.0	...	5.0	4.7	4.1	# 4.1[4]	5.5[4]
% M [I] [12]	5.9	6.3	6.4	5.2	...	3.4	3.2	2.8	# 3.4[4]	4.6[4]
% F [I] [12]	12.2	11.6	10.9	9.3	...	7.2	6.6	5.8	# 5.0[4]	6.5[4]
Puerto Rico Porto Rico										
MF [I] [30 31]	202.0	216.0	194.0	178.0	165.0	163.0	160.0	186.0	197.0	206.0
M [I] [30 31]	151.0	160.0	145.0	130.0	120.0	118.0	115.0	131.0	139.0	144.0
F [I] [30 31]	51.0	56.0	50.0	48.0	45.0	45.0	45.0	55.0	58.0	62.0
% MF [I] [30 31]	20.7	21.8	18.9	16.8	15.0	14.6	14.1	16.0	16.6	17.0
% M [I] [30 31]	23.7	24.8	22.0	19.4	17.4	16.9	16.2	17.9	19.0	19.5
% F [I] [30 31]	15.1	16.2	13.5	12.4	10.8	10.8	10.7	12.6	12.8	13.2
Rep. of Moldova Rép. de Moldava										
MF [III] [1]	...	...	...	...	...	...	...	0.1	15.0	14.1
M [III] [1]	...	...	...	...	...	...	...	–	5.9	5.5
F [III] [1]	...	...	...	...	...	...	...	0.1	9.1	8.9
% MF [III] [1]	...	...	...	...	...	...	...	–	0.7	0.7

30
Unemployment
Number (thousands) and percentage of unemployed [*cont.*]
Chômage
Nombre (milliers) et pourcentage des chômeurs [*suite*]

Country or area Pays ou zone §	1984	1985	1986	1987	1988	1989	1990	1991	1992	1993
Réunion Réunion										
MF [III][31]	36.4	45.0	51.6	52.8	56.7	59.5	53.7	59.3	80.1	80.2
M [III][31]	24.3	28.2	26.8	25.8	29.2	30.2	28.4	30.4	41.9	43.1
F [III][31]	15.9	19.6	24.5	26.3	29.2	27.3	25.4	28.8	38.1	37.1
% MF [III][31]	...	...	...	...	...	...	23.0	25.4	34.3	34.4
Romania Roumanie										
MF [III][1]	...	...	...	...	...	...	...	337.4	929.0	1164.7
M [III][1]	...	...	...	...	...	...	...	129.0	366.0	479.2
F [III][1]	...	...	...	...	...	...	...	208.4	563.0	685.5
% MF [III][1]	...	...	...	...	...	...	...	3.0	8.4	10.2
% M [III][1]	...	...	...	...	...	...	...	2.2	6.2	8.1
% F [III][1]	...	...	...	...	...	...	...	4.0	10.7	12.6
Russian Federation Fédération de Russie										
MF [III][1]	...	...	...	...	...	...	...	61.9	577.7	...
M [III][1]	...	...	...	...	...	...	...	18.8	160.7	...
F [III][1]	...	...	...	...	...	...	...	43.1	417.0	...
% MF [III][1]	...	...	...	...	...	...	...	0.1	0.1	...
% M [III][1]	...	...	...	...	...	...	...	0.1	0.5	...
% F [III][1]	...	...	...	...	...	...	...	0.1	1.1	...
Saint Helena Sainte–Hélène										
MF [III]	...	...	...	...	...	0.2	0.2	0.2	0.2	0.2
M [III]	...	...	...	...	...	0.0	0.1	0.1	0.1	0.1
F [III]	...	...	...	...	...	0.1	0.1	0.1	0.1	0.1
Saint Pierre and Miquelon Saint–Pierre–et–Miquelon										
MF [III][57]	0.3	0.3	0.3	0.3	0.4	0.3	...	...	...	0.4
M [III][57]	0.1	0.2	0.2	0.2	0.2	0.2	...	...	...	
F [III][57]	0.1	0.1	0.1	0.2	0.2	0.2	...	...	...	
% MF [III][57]	10.3	10.9	11.0	11.7	13.2	11.3	...	...	...	
San Marino Saint–Marin										
MF [IV][14]	0.7	0.7	0.7	0.7	0.7	0.6	0.6	0.5	0.5	0.6
M [IV][14]	0.2	0.2	0.2	0.2	0.2	0.1	0.2	0.1	0.1	0.2
F [IV][14]	0.5	0.5	0.5	0.5	0.5	0.5	0.5	0.3	0.4	0.4
% MF [IV][14]	7.2	6.9	6.9	6.2	6.0	5.3	5.5	4.3	4.2	5.1
% M [IV][14]	3.9	3.0	3.2	2.8	2.7	2.1	2.4	2.3	2.0	2.6
% F [IV][14]	12.0	12.5	12.0	10.9	10.5	9.5	9.7	6.9	7.1	8.1
Senegal Sénégal										
MF [III][3 21 58]	12.8	10.8	10.2	8.1	17.3	8.3	10.4	14.4	12.0	...
M [III][3 21 58]	11.1	9.0	8.4	6.4	15.0	7.1	8.3	13.1	10.0	...
F [III][3 21 58]	1.7	1.8	1.9	1.8	2.4	1.2	2.1	1.3	2.0	...
Seychelles Seychelles										
MF [III][59]	5.6[1]	5.7	...	...	...	...	...	...	...	...
M [III][59]	3.0[1]	2.4	...	...	...	...	...	...	...	...
F [III][59]	2.6[1]	3.3	...	...	...	...	...	...	...	...
% MF [III][59]	20.8[1]	22.5	...	...	...	...	...	...	...	...
% M [III][59]	19.9[1]	16.2	...	...	...	...	...	...	...	...
% F [III][59]	21.0[1]	31.6	...	...	...	...	...	...	...	...
Sierra Leone Sierra Leone										
MF [III][3 20 60]	2.5	2.8	2.7	2.4	...	...	...	...	...	...
Singapore Singapour										
MF [I][12]	35.2	53.2	84.1	62.4	46.2	30.7	25.2	30.0	43.4	43.7
M [I][12]	21.8	34.1	57.5	42.1	32.3	20.2	15.9	18.7	26.4	25.2
F [I][12]	13.3	19.1	26.6	20.3	13.8	10.4	9.3	11.3	17.0	18.5
MF [III][4]	6.2	9.2	9.5	8.1	4.5	2.7	1.7	1.2	1.0	1.0
M [III][4]	3.7	5.6	6.5	5.6	3.2	1.9	1.2	0.8	0.7	0.7
F [III][4]	2.5	3.6	2.9	2.5	1.4	0.8	0.5	0.4	0.3	0.3
% MF [I][12]	2.7	4.1	6.5	4.7	3.3	2.2	1.7	1.9	2.7	2.7
% M [I][12]	2.6	4.2	7.0	5.1	3.8	2.3	1.7	2.0	2.7	2.6
% F [I][12]	2.8	4.1	5.5	4.0	2.6	1.9	1.6	1.8	2.6	2.8

30
Unemployment
Number (thousands) and percentage of unemployed [*cont.*]
Chômage
Nombre (milliers) et pourcentage des chômeurs [*suite*]

Country or area Pays ou zone §	1984	1985	1986	1987	1988	1989	1990	1991	1992	1993
Slovakia Slovaquie										
MF [III]	..	..	..	..	..	..	..	169.0	285.5	323.2
M [III]	..	..	..	..	..	..	..	83.4	141.1	167.2
F [III]	..	..	..	..	..	..	..	85.6	144 4	156.0
% MF [III]	..	..	..	..	..	..	..	6.6	11.4	12.7
% M [III]	..	..	..	..	..	..	..	6.3	11.1	12.5
% F [III]	..	..	..	..	..	..	..	6.9	11.7	12.9
Slovenia Slovénie										
MF [III]	15.3	14.7	14.2	15.2	21.3	28.2	44.6	75.1	102.6	129.1
M [III]	7.0	6.8	6.9	7.8	11.3	14.4	23.2	41.5	57.5	72.5
F [III]	8.3	7.9	7.3	7.4	10.1	13.8	21.4	33.6	45.1	56.6
% MF [III]	...	...	1.4	1.5	2.2	2.9	4.7	8.2	11.5	14.4
% M [III]	...	...	1.3	1.5	2.1	2.8	4.5	8.5	12.1	15.3
% F [III]	...	...	1.7	1.7	2.3	3.1	4.8	7.9	10.8	13.5
South Africa Afrique du Sud										
MF [III][61]	...	...	...	...	122.1	116.5	110.7	247.8	287.8	...
M [III][61]	...	...	...	...	82.7	80.0	77.4	176.6	201.1	...
F [III][61]	...	...	...	...	39.4	36.5	33.3	71.2	86.7	...
Spain Espagne										
MF [I][31]	2728.2	2938.5	2933.0	2937.7	2847.9	2560.8	2441.2	2463.7	2788.5	3481.3
M [I][31]	1792.9	1907.1	1852.1	1641.3	1464.5	1263.0	1166.1	1191.9	1384.5	1836.7
F [I][31]	935.3	1031.5	1080.8	1296.4	1383.4	1297.8	1275.1	1271.8	1404.1	1644.6
MF [III][45]	2475.2	2642.0	2758.7	2924.1	2858.3	2550.3	2350.0	2289.0	2259.9	2537.9
M [III][45]	1529.0	1599.3	1572.2	1525.4	1359.7	1087.5	942.5	910.7	954.2	1193.0
F [III][45]	946.2	1042.7	1186.4	1398.8	1498.6	1462.8	1407.5	1378.3	1305.7	1344.9
% MF [I][31]	20.3	21.6	21.2	20.5	19.5	17.3	16.3	16.4	18.4	22.7
% M [I][31]	19.1	20.2	19.4	17.1	15.2	13.0	12.0	12.3	14.3	19.0
% F [I][31]	23.0	25.0	25.3	27.5	27.7	25.4	24.2	23.8	25.6	29.2
% MF [III][45]	18.4	19.5	20.0	20.4	19.5	17.2	15.7	15.2	14.9	16.6
% M [III][45]	16.3	16.9	16.5	15.9	14.1	11.2	9.7	9.4	9.9	12.3
% F [III][45]	23.5	25.5	28.0	29.7	30.0	28.6	26.7	25.8	23.8	23.9
Sri Lanka Sri Lanka										
MF [I][10 12]	...	840.3	...	...	...	...	1005.1	843.3[62]	817.6[62]	874.1[62]
M [I][10 12]	...	433.2	...	...	...	...	395.8	378.0[62]	408.7[62]	349.3[62]
F [I][10 12]	...	407.0	...	...	...	...	609.2	463.3[62]	409.0[62]	524.8[62]
% MF [I][10 12]	...	...	...	...	...	...	14.4	14.1[62]	14.1[62]	14.7[62]
% M [I][10 12]	...	...	...	...	...	...	9.1	10.0[62]	10.6[62]	9.1[62]
% F [I][10 12]	...	...	...	...	...	...	23.5	21.2[62]	21.0[62]	25.2[62]
Sudan Soudan										
MF [III][3]	63.4	48.8	63.1	...	...	25.5	70.1	# 19.9[63]	5.3[63]	...
M [III][3]	48.9	36.6	48.6	...	...	15.5	44.4	# 10.2[63]	3.7[63]	...
F [III][3]	14.4	12.3	14.5	...	...	9.9	25.7	# 9.7[63]	1.6[63]	...
Suriname Suriname										
MF [III]	12.9	17.0	13.4	# 2.8[64]	3.0	2.4	3.9	3.7	1.4	1.0
M [III]	6.5	8.6	7.3	# 1.3[64]	1.4	1.2	1.2	1.1	0.5	0.4
F [III]	6.5	8.3	6.1	# 1.5[64]	1.6	1.2	2.7	2.6	0.9	0.6
Sweden Suède										
MF [I]	136.0[53]	125.0[53]	# 98.0[45]	92.0[45]	78.0[45]	67.0[45]	75.0[45]	133.0[45]	233.0[45]	# 356.0
M [I]	69.0[53]	65.0[53]	# 51.0[45]	48.0[45]	40.0[45]	34.0[45]	40.0[45]	77.0[45]	145.0[45]	# 219.0
F [I]	67.0[53]	60.0[53]	# 47.0[45]	44.0[45]	38.0[45]	33.0[45]	36.0[45]	55.0[45]	88.0[45]	# 137.0
MF [III][45]	91.9	84.9	84.2	78.0	61.1	56.3	66.4	114.6	214.5	326.1
M [III][45]	48.1	42.3	42.4	38.9	31.2	29.2	35.6	66.4	131.3	193.7
F [III][45]	43.8	42.6	41.9	39.2	29.9	27.1	30.9	48.1	83.2	131.4
% MF [I]	3.1[53]	2.8[53]	# 2.2[45]	2.1[45]	1.8[45]	1.5[45]	1.6[45]	3.0[45]	5.3[45]	# 8.2
% M [I]	3.0[53]	2.8[53]	# 2.2[45]	2.1[45]	1.8[45]	1.5[45]	1.7[45]	3.3[45]	6.3[45]	# 9.7
% F [I]	3.3[53]	2.9[53]	# 2.2[45]	2.1[45]	1.8[45]	1.5[45]	1.6[45]	2.6[45]	4.2[45]	# 6.6
% MF [III][45]	2.8	2.5	2.5	2.3	1.7	1.6	1.9	3.2	5.9	8.7
% M [III][45]	2.8	2.6	2.4	2.2	1.8	1.6	2.0	3.9	...	...
% F [III][45]	2.7	2.6	2.5	2.2	1.7	1.6	1.7	2.7	...	...

30
Unemployment
Number (thousands) and percentage of unemployed [*cont.*]
Chômage
Nombre (milliers) et pourcentage des chômeurs [*suite*]

Country or area Pays ou zone §	1984	1985	1986	1987	1988	1989	1990	1991	1992	1993
Switzerland Suisse										
MF [III] [2]	35.2	30.3	25.7	24.7	22.2	17.5	18.1	39.2	92.3	163.1
M [III] [2]	19.7	16.4	13.4	12.5	11.4	9.1	9.8	22.7	54.7	96.6
F [III] [2]	15.5	13.9	12.3	12.1	10.8	8.4	8.3	16.5	37.6	66.6
% MF [III] [2]	1.1	1.0	0.8	0.8	0.7	0.6	0.6	1.3	2.5	4.5
% M [III] [2]	1.0	0.8	0.7	0.6	0.6	0.5	0.5	1.2	2.5	4.4
% F [III] [2]	1.4	1.2	1.1	1.1	1.0	0.8	0.7	1.5	2.7	4.7
Syrian Arab Republic Rép. arabe syrienne										
MF [I] [1 12]	109.6	...	...	...	...	177.3	...	235.4	...	...
M [I] [1 12]	82.8	...	...	...	...	130.8	...	147.3	...	...
F [I] [1 12]	26.8	...	...	...	...	46.5	...	88.2	...	...
% MF [I] [1 12]	4.7	...	...	...	...	5.8	...	6.8	...	...
% M [I] [1 12]	4.1	...	...	...	...	5.1	...	5.2	...	...
% F [I] [1 12]	8.2	...	...	...	...	9.5	...	14.0	...	...
TFYR Macedonia L'ex–R.y. Macédoine										
MF [III]	128.0	136.0	141.0	141.0	140.0	150.0	156.0	165.0	172.0	175.0
M [III]	57.0	61.0	64.0	66.0	66.0	72.0	76.0	82.0	87.0	89.0
F [III]	71.0	75.0	77.0	75.0	74.0	78.0	80.0	83.0	85.0	86.0
% MF [III]	21.1	21.6	21.7	21.5	21.4	22.6	23.6	26.0	27.8	29.3
% M [III]	15.3	15.8	16.3	16.5	16.9	18.3	19.4	21.7	23.8	25.3
% F [III]	30.6	30.9	30.2	29.1	28.2	28.8	29.7	32.3	33.7	35.2
Thailand Thaïlande										
MF [I] [10 65]	1279.8[66]	994.6[66]	968.7[66]	1721.6[66]	929.2[66]	# 433.1	710.0	869.3	...	...
M [I] [10 65]	541.1[66]	456.2[66]	463.9[66]	672.7[66]	419.3[66]	# 204.5	347.4	350.1	...	...
F [I] [10 65]	738.7[66]	538.4[66]	504.8[66]	1048.9[66]	509.8[66]	# 228.6	362.5	519.1	...	...
% MF [I] [10 65]	2.9[66]	3.7[66]	3.5[66]	5.9[66]	3.1[66]	# 1.4	2.2	2.7	...	...
% M [I] [10 65]	2.5[66]	6.9[66]	3.1[66]	4.3[66]	2.6[66]	# 1.2	2.1	2.0	...	...
% F [I] [10 65]	3.4[66]	4.4[66]	3.9[66]	7.6[66]	3.6[66]	# 1.6	2.4	3.5	...	...
Togo Togo										
MF [III] [21]	4.7	4.1	...	...	...	...	...	...	...	...
M [III] [21]	4.2	3.6	...	...	...	...	...	...	...	...
F [III] [21]	0.5	0.5	...	...	...	...	...	...	...	...
Trinidad and Tobago Trinité–et–Tobago										
MF [I] [2 8]	63.2	72.8	81.2	# 106.6	104.7	103.4	93.6	91.2	99.2	99.9
M [I] [2 8]	37.6	46.4	51.5	# 65.7	66.5	64.8	55.1	49.6	54.3	56.3
F [I] [2 8]	25.6	26.4	29.7	# 40.9	38.2	38.6	38.5	41.5	44.9	43.7
% MF [I] [2 8]	13.5	15.5	17.2	# 22.3	22.0	22.0	20.0	18.5	19.6	19.8
% M [I] [2 8]	12.0	15.0	16.4	# 20.7	21.1	20.8	17.8	15.7	17.0	17.6
% F [I] [2 8]	16.0	17.0	18.9	# 25.3	23.6	24.5	24.2	23.4	23.9	23.4
Tunisia Tunisie										
MF [III] [3 48]	74.3	84.0	80.2	83.7	91.5	105.9	...	...	...	...
M [III] [3 48]	59.7	69.1	65.8	67.3	72.8	80.5	...	...	...	...
F [III] [3 48]	14.6	14.9	14.3	16.4	18.7	25.5	...	...	...	...
Turkey Turquie										
MF [I] [1 20]	855.0[11]	# 2040.0	...	...	1709.0	1802.0	1572.0	1714.0	1656.0	1659.0
M [I] [1 20]	548.0[11]	# 1431.0	...	...	1081.0	1196.0	1085.0	1253.0	1184.0	1191.0
F [I] [1 20]	307.0[11]	# 609.0	...	...	627.0	607.0	487.0	461.0	472.0	467.0
MF [III] [4]	761.0	935.0	1053.0	1124.0	1155.0	1076.0	980.0	859.0	840.0	...
M [III] [4]	639.0	782.0	877.0	929.0	952.0	888.0	809.0	707.0	695.0	...
F [III] [4]	122.0	153.0	177.0	195.0	204.0	189.0	171.0	152.0	145.0	...
% MF [I] [1 20]	11.9[11]	# 11.2	...	...	8.3	8.5	7.4	8.3	7.8	7.9
% M [I] [1 20]	9.0[11]	# 11.3	...	...	7.7	8.4	7.5	8.8	8.1	8.2
% F [I] [1 20]	29.1[11]	# 11.1	...	...	9.6	8.8	7.2	7.1	7.2	7.2
United Kingdom Royaume–Uni										
MF [I] [31]	3206.0	3076.0	3073.0	2979.0	2489.0	2063.0	1949.0	2385.0	2732.0	2891.0
M [I] [31]	1916.0	1862.0	1857.0	1789.0	1465.0	1207.0	1149.0	1494.0	1846.0	1967.0
F [I] [31]	1290.0	1214.0	1217.0	1190.0	1005.0	857.0	800.0	891.0	886.0	924.0
MF [II] [30 67 68 69]	3159.8	3271.2	# 3292.9	2953.4	# 2370.4	# 1798.7	1664.5	2291.9	2778.6	2919.2
M [II] [30 67 68 69]	2197.4	2251.7	# 2254.7	2045.8	# 1650.5	# 1290.8	1232.3	1737.1	2126.0	2236.0
F [II] [30 67 68 69]	962.5	1019.5	# 1036.6	907.6	# 719.9	# 507.9	432.2	554.9	652.6	683.1
% MF [I] [31]	11.8	11.2	11.2	10.7	8.8	7.2	6.8	8.3	9.6	10.2
% M [I] [31]	12.0	11.6	11.6	11.1	9.0	7.4	7.0	9.2	11.5	12.4
% F [I] [31]	11.5	10.7	10.6	10.2	8.4	7.0	6.5	7.2	7.2	7.5

30
Unemployment
Number (thousands) and percentage of unemployed [cont.]
Chômage
Nombre (milliers) et pourcentage des chômeurs [suite]

Country or area / Pays ou zone §	1984	1985	1986	1987	1988	1989	1990	1991	1992	1993
% MF [II] [30 67 68 69]	11.6	11.8	# 11.8	10.6	# 8.4	# 6.3	5.9	8.1	9.9	10.4
% M [II] [30 67 68 69]	13.5	13.7	# 13.8	12.5	# 10.1	# 7.9	7.6	10.7	13.3	...
% F [II] [30 67 68 69]	8.8	9.1	# 9.0	7.8	# 6.1	# 4.2	3.5	4.6	5.4	...
United States Etats–Unis										
MF [I] [31]	8539.0	8312.0	8237.0	7425.0	6701.0	6528.0	6874.0	8426.0	9384.0	8734.0
M [I] [31]	4744.0	4521.0	4530.0	4101.0	3655.0	3525.0	3799.0	4817.0	5380.0	4932.0
F [I] [31]	3794.0	3791.0	3707.0	3324.0	3046.0	3003.0	3075.0	3609.0	4005.0	3801.0
% MF [I] [31]	7.4	7.1	6.9	6.1	5.4	5.2	5.4	6.6	7.3	6.7
% M [I] [31]	7.3	6.9	6.8	6.1	5.3	5.1	5.4	6.9	7.6	7.0
% F [I] [31]	7.6	7.4	7.1	6.2	5.5	5.3	5.4	6.3	6.9	6.5
United States Virgin Is. Iles Vierges américaines										
MF [III] [70]	3.3	2.6	2.1	1.4	1.5	1.7	1.3	1.4	1.1	1.9
% MF [III] [70]	7.6	6.0	4.7	3.0	3.3	3.7	2.8	2.8	3.5	3.5
Uruguay Uruguay										
MF [I] [4 11]	145.3[10]	...	122.0[10]	108.7	104.1	98.4	105.7	111.0	112.8	...
M [I] [4 11]	68.0[10]	...	58.1[10]	48.6	46.3	44.9	50.6	52.6	49.6	...
F [I] [4 11]	77.3[10]	...	63.9[10]	60.1	57.8	53.5	55.1	58.4	63.2	...
% MF [I] [4 11]	...	...	10.7[10]	9.1	8.6	8.0	8.5	9.0	9.0	...
% M [I] [4 11]	...	...	8.5[10]	6.7	6.3	6.1	6.9	7.2	6.9	...
% F [I] [4 11]	...	...	13.9[10]	12.6	11.9	10.8	10.9	11.6	11.9	...
Venezuela Venezuela										
MF [I] [2]	734.9	766.8	667.4	575.3	478.2	621.1	741.7	701.0	579.8	497.8
M [I] [2]	...	577.3	502.2	448.9	367.3	469.5	535.8	478.6	415.1	366.2
F [I] [2]	...	189.6	165.8	124.6	110.6	151.9	205.8	222.5	164.6	131.6
% MF [I] [2]	13.0	13.1	11.0	9.2	7.3	9.2	10.4	9.5	7.5	6.4
% M [I] [2]	...	13.5	11.4	9.9	7.8	9.8	10.9	9.6	8.1	7.1
% F [I] [2]	...	11.6	9.9	7.2	6.1	7.6	9.3	9.4	6.9	5.6
Yugoslavia, SFR† Yougoslavie, Rfs†										
MF [III]	974.8	1039.6	1086.7	1080.6	1131.8	1201.2	1308.5	...	...	...
M [III] [34]	430.0	460.4	482.9	477.4	511.0	550.4	610.2	...	...	...
F [III] [33]	544.7	579.3	603.8	603.1	620.7	650.8	698.3	...	...	...
% MF [III]	13.3	13.8	14.1	13.6	14.1	14.9	16.4	...	...	...
% M [III] [34]	9.8	10.2	10.4	10.2	10.9	11.7	...	...	...	...
% F [III] [33]	18.6	19.1	19.0	18.5	18.7	19.3	...	...	...	...

Source:
International Labour Office (Geneva).

Source:
Bureau international du Travail (Genève).

§   I = Labour force sample surveys and general
household sample surveys.
II = Social insurance statistics.
III = Employment office statistics.
IV = Official estimates.

§   I = Enquêtes par sondage sur la main–d'oeuvre et
enquêtes générales par sondage auprès des ménages
II = Statistiques d'assurances sociales.
III = Statistiques des bureaux de placement.
IV = Evaluations officielles.

†   For detailed descriptions of data pertaining to former
Czechoslovakia, Germany, SFR Yugoslavia and former
USSR, see Annex I – Country or area nomenclature,
regional and other groupings.

†   Pour les descriptions en détails des données relatives
à l'ancienne Tchécoslovaquie, l'Allemagne, la Rfs
Yougoslavie et l'ancienne URSS, voir l'Annexe I –
Nomenclature des pays ou zones, groupements régionaux
et autres groupements.

1   One month of each year (India: beginning 1988).
2   Persons aged 15 years and over.
3   Applications for work (including persons in employment).
4   Persons aged 14 years and over.
5   Gran Buenos Aires.
6   Including unpaid family workers who worked for less
than 15 hours.
7   Beginning 1989, based on the 1991 Census of Population
and Housing.
8   Excluding the unemployed not previously employed (Trinidad
and Tobago: beginning 1987).
9   Private sector.
10 Average of less than twelve months.
11 Urban areas.

1   Un mois de chaque année (Inde: à partir de 1988).
2   Personnes âgées de 15 ans et plus.
3   Demandeurs d'emploi (y compris les pesonnes d'un emploi).
4   Personnes âgées de 14 ans et plus.
5   Gran Buenos Aires.
6   Y compris les travailleurs familiaux non rémunérés qui ont
travaillé moins de 15 heures.
7   A partir de 1989, basées sur le recensement de la population et de
l'habitat de 1991.
8   Non compris les chômeurs n'ayant jamais travaillé
précédemment (Trinité–et–Tobago: à partir de 1987).
9   Secteur privé.
10 Moyenne de moins de douze mois.
11 Les régions urbaines.

30
Unemployment
Number (thousands) and percentage of unemployed [*cont.*]
Chômage
Nombre (milliers) et pourcentage des chômeurs [*suite*]

12 Persons aged 10 years and over.	12 Personnes âgées de 10 ans et plus.
13 Excluding the rural population of the northern region.	13 Non compris la population rurale de la région du Nord.
14 Prior to 1989, three employment offices. Beginning 1989, four employment offices.	14 Avant 1989, trois bureau de placement. A partir de 1989, quatre bureau de placement.
15 Including unemployed not registered by sex.	15 Y compris chômeure non répartis par sexe.
16 Bujumbura.	16 Bujumbura.
17 Bangui.	17 Bangui.
18 Young people aged 16 to 25 years.	18 Jeunes gens de 16 à 25 ans.
19 7 main cities of the country.	19 7 villes principales du pays.
20 Persons aged 12 years and over.	20 Personnes âgées de 12 ans et plus.
21 Persons aged 14 to 55 years.	21 Personnes âgées de 14 à 55 ans.
22 Persons aged 15 to 74 years.	22 Personnes âgées de 15 à 74 ans.
23 Persons aged 16 to 66 years.	23 Personnes âgées de 16 à 66 ans.
24 Quito, Guayaquil and Cuenca.	24 Quito, Guayaquil et Cuenca.
25 Persons aged 6 years and over (Egypt: beginning 1989).	25 Personnes âgées de 6 ans et plus (Egypte: à partir de 1989).
26 Metropolitan area.	26 Région métropolitaine.
27 Year ending in June of the year indicated.	27 Année se terminant en juin de l'année indiquée.
28 Persons aged 18 to 55 years.	28 Personnes âgées de 18 à 55 ans.
29 Persons aged 15 to 64 years.	29 Personnes âgées de 15 à 64 ans.
30 Excluding persons temporarily laid off.	30 Non compris les personnes temporairement mises à pied.
31 Persons aged 16 years and over.	31 Personnes âgées de 16 ans et plus.
32 Cayenne and Kourou.	32 Cayenne et Kourou.
33 Persons aged 15 to 60 years.	33 Personnes âgées de 15 à 60 ans.
34 Persons aged 15 to 65 years.	34 Personnes âgées de 15 à 65 ans.
35 Guatemala city only.	35 Ville de Guatemala seulement.
36 Persons aged 14 to 60 years.	36 Personnes âgées de 14 à 60 ans.
37 Georgetown, New Amsterdam, Anna Regina and Fort Wellington districts.	37 Districts de Georgetown, New Amsterdam, Anna Regina et Fort Wellington.
38 Excluding unpaid family workers who worked for one hour or more (Hong Kong: beginning 1985).	38 Non compris les travailleurs familiaux non rémunérés ayant travaillé une heure ou plus (Hong-kong: à partir de 1985).
39 Intercensal Population Survey results.	39 Résultats de l'enquête intercensitaire de population.
40 Persons aged 10 to 56 years.	40 Personnes âgées de 10 à 56 ans.
41 Beginning 1991, rates calculated on basis of 1991 census.	41 A partir de 1991, taux calculés sur la base du Recensement de 1991.
42 Including the residents of East Jerusalem.	42 Y compris les résidents de Jérusalem—Est.
43 Including persons who did not work in the country during the previous 12 months.	43 Y compris les personnes qui n'ont pas travaillé dans le pays pendant les 12 mois précédents.
44 Data classified according to ISIC, Rev.3.	44 Données classées selon le CITI, Rév.3.
45 Persons aged 16 to 64 years.	45 Personnes âgées de 16 à 64 ans.
46 Persons aged 16 to 61 years.	46 Personnes âgées de 16 à 61 ans.
47 Excluding Rodrigues.	47 Non compris Rodrigues.
48 Persons aged 18 years and over.	48 Personnes âgées de 18 ans et plus.
49 Persons working or seeking work for less than 12 hours per week are no longer included.	49 Ne sont plus comprises les personnes qui travaillent, ou qui cherchent moins de 12 heures de travail par semaine.
50 Curaçao.	50 Curaçao.
51 Excluding Aruba.	51 Non compris Aruba.
52 Including students seeking vacation work.	52 Y compris les étudiants qui cherchent un emploi pendant les vacances.
53 Persons aged 16 to 74 years.	53 Personnes âgées de 16 à 74 ans.
54 Persons aged 18 to 60 years.	54 Personnes âgées de 18 à 60 ans.
55 Lima.	55 Lima.
56 Excluding domestic services.	56 Non compris les services domestiques.
57 Persons aged 16 to 60 years.	57 Personnes âgées de 16 à 60 ans.
58 Dakar.	58 Dakar.
59 Persons aged 15 to 63 years.	59 Personnes âgées de 15 à 63 ans.
60 Excluding persons registered at the Maritime Pool.	60 Non compris les demandeurs d'emploi enregistrés au bureau de placement maritime.
61 Excluding Transkei, Bophuthatswana, Venda, Ciskei; elsewhere persons enumerated at de facto dwelling place.	61 Non compris Transkei, Bophuthatswana, Venda, Ciskei; ailleurs personnes énumérées aux logis de facto.
62 Excluding Northern and Eastern provinces.	62 Non compris les provinces du Nord et de l'Est.
63 Khartoum province.	63 Province de Khartoum.
64 Change in registration system; unemployed must re—register every 3 months.	64 Modification du système d'enregistrement les chômeurs doivent se réinscrire tous les 3 mois.
65 Persons aged 13 years and over.	65 Personnes âgées de 13 ans et plus.
66 Persons aged 11 years and over.	66 Personnes âgées de 11 ans et plus.
67 Claimants at unemployment benefits offices.	67 Demandeurs auprès des bureaux de prestations de chômage.
68 Beginning Sept. 1988, excluding most under 18—year—olds.	68 A partir de sept. 1988, non compris la plupart des moins de 18 ans.
69 Beginning Sept. 1989, excluding some men formerly employed in the coalmining industry.	69 Dès sept. 1989, non compris certains hommes ayant précédemment travaillé dans l'industrie charbonnière.
70 Persons aged 16 to 65 years.	70 Personnes âgées de 16 à 65 ans.

Technical notes, tables 29 and 30

Detailed data on labour force and related topics are published in the ILO *Year Book of Labour Statistics*.[9] The series shown in the *Statistical Yearbook* give an overall picture of the availability and disposition of labour resources and, in conjunction with other macro-economic indicators, can be useful for an overall assessment of economic performance. The ILO *Year Book of Labour Statistics* provides a comprehensive description of the methodology underlying the labour series. Brief definitions of the major categories of labour statistics are given below.

"Employment" is defined to include persons above a specified age who, during a specified period of time, were in one of the following categories:

(a) "Paid employment", comprising persons who perform some work for pay or profit during the reference period or persons with a job but not at work due to temporary absence, such as vacation, strike, education leave;

(b) "Self-employment", comprising employers, own-account workers, members of producers' cooperatives, persons engaged in production of goods and services for own consumption and unpaid family workers;

(c) Members of the armed forces. Students, homemakers and others mainly engaged in non-economic activities during the reference period who, at the same time, were in paid employment or self-employment are considered as employed on the same basis as other categories.

"Unemployment" is defined to include persons above a certain age and who during a specified period of time were:

(a) "Without work", i.e. were not in paid employment or self-employment;

(b) "Currently available for work", i.e. were available for paid employment or self-employment during the reference period; and

(c) "Seeking work", i.e. had taken specific steps in a specified period to find paid employment or self-employment.

The following categories of persons are not considered to be unemployed:

(a) Persons intending to establish their own business or farm, but who had not yet arranged to do so and who were not seeking work for pay or profit;

(b) Former unpaid family workers not at work and not seeking work for pay or profit.

For various reasons, national definitions of employment and unemployment often differ from the recommended international standard definitions and thereby limit international comparability. Intercountry comparisons are also complicated by a variety of types of data collection systems used to obtain information on employed and unemployed persons.

Notes techniques, tableaux 29 et 30

Des données détaillées sur la main-d'oeuvre et des sujets connexes sont publiées dans l'*Annuaire des Statistiques du Travail* du BIT [9]. Les séries indiqués dans l'*Annuaire des Statistiques* donnent un tableau d'ensemble des disponibilités de main-d'oeuvre et de l'emploi de ces ressources et, combinées à d'autres indicateurs économiques, elles peuvent être utiles pour une évaluatation générale de la performance économique. L'*Annuaire des statistiques du Travail* du BIT donne une description complète de la méthodologie employée pour établir les séries sur la main-d'oeuvre. On trouvera ci-dessous quelques brèves définitions des grandes catégories de statistiques du travail.

Le terme "Emploi" désigne les personnes dépassant un âge déterminé qui, au cours d'une période donnée, se trouvaient dans l'une des catégories suivantes :

(a) La catégorie "emploi rémunéré", composée des personnes faisant un certain travail en échange d'une rémunération ou d'un profit pendant la période de référence, ou les personnes ayant un emploi, mais qui ne travaillaient pas en raison d'une absence temporaire (vacances, grève, congé d'études);

(b) La catégorie "emploi indépendant" regroupe les employeurs, les travailleurs indépendants, les membres de coopératives de producteurs et les personnes s'adonnant à la production de biens et de services pour leur propre consommation et la main-d'oeuvre familiale non rémunérée;

(c) Les membres des forces armées, les étudiants, les aides familiales et autres personnes qui s'adonnaient essentiellement à des activités non économiques pendant la période de référence et qui, en même temps, avaient un emploi rémunéré ou indépendant, sont considérés comme employés au même titre que les personnes des autres catégories.

Par "chômeurs", on entend les personnes dépassant un âge déterminé et qui, pendant une période donnée, éaient :

(a) "sans emploi", c'est-à-dire sans emploi rémunéré ou indépendant;

(b) "disponibles", c'est-à-dire qui pouvaient être engagées pour un emploi rémunéré ou pouvaient s'adonner à un emploi indépendant au cours de la période de référence; et

(c) "à la recherche d'un emploi", c'est-à-dire qui avaient pris des mesures précises à un certain moment pour trouver un emploi rémunéré ou un emploi indépendant.

Ne sont pas considérés comme chômeurs :

(a) Les personnes qui, pendant la période de référence, avaient l'intention de créer leur propre entreprise ou exploitation agricole, mais n'avaient pas encore pris les dispositions nécessaires à cet effet et qui n'étaient pas à la recherche d'un emploi en vue d'une rémunération ou d'un profit;

Table 29 presents absolute figures on the distribution of employed persons by major divisions of economic activity. Data are arranged as far as possible according to the major divisions of economic activity of the *International Standard Industrial Classification of All Economic Activities*. [42] The column for total employment includes economic activities not adequately defined. The four main sources of these statistics are: labour force sample surveys and general household sample surveys, social insurance statistics, establishment surveys and official estimates.

Table 30: Figures are presented in absolute numbers and in percentages. Data are normally annual averages of monthly, quarterly or semi-annual data.

The series generally represent the total number of persons wholly unemployed or temporarily laid off. Percentage figures, where given, are calculated by comparing the number of unemployed to the total members of that group of the labour force on which the unemployment data are based.

(b)   Les anciens travailleurs familiaux non rémunérés qui n'avaient pas d'emploi et n'étaient pas à la recherche d'un emploi en vue d'une rémunération ou d'un profit.

Pour diverses raisons, les définitions nationales de l'emploi et du chômage diffèrent souvent des définitions internationales types recommandées, limitant ainsi les possibilités de comparaison entre pays. Ces comparaisons se trouvent en outre compliquées par la diversité des systèmes de collecte de données utilisés pour recueillir des informations sur les personnes employées et les chômeurs.

Le *Tableau 29* présente des chiffres en valeur absolue sur la répartition des personnes employées par branche d'activité économique. Les données sont présentées autant que possible selon les principales divisions de l'activité économique de la *Classification internationale type, par Industrie, de toutes les branches d'activité économique* [42]. La colonne indiquant l'emploi total englobe les activités économiques non convenablement définies. Les quatre sources principales de ces statistiques sont : les enquêtes par sondage sur la main-d'oeuvre et les enquêtes générales par sondage auprès des ménages, les statistiques d'assurances sociales, les enquêtes auprès des établissements et les évaluations officielles.

Tableau 30 : Les chiffres sont présentés en valeur absolue et en pourcentage. Les données sont normalement des moyennes annuelles des données mensuelles, trimestrielles ou semestrielles.

Les séries représentent généralement le nombre total des chômeurs complets ou des personnes temporairement mises à pied. Les données en pourcentage, lorsqu'elles figurent au tableau, sont calculées par comparaison du nombre de chômeurs au nombre total des personnes du groupe de main-d'oeuvre sur lequel sont basées les données relatives au chômage.

31
Earnings in manufacturing
Gains dans les industries manufacturières
By hour, day, week or month
Par heure, jour, semaine ou mois

Country or area and unit Pays ou zone et unité	1984	1985	1986	1987	1988	1989	1990	1991	1992	1993
Albania: lek Albanie : lek										
MF - month mois[1 2 3]	513.0	521.0	528.0	533.0	533.0	542.0	554.0	* 665.0	...	...
Antigua and Barbuda: EC dollar Antigua-et-Barbuda : EC dollar										
MF - week semaine[4]	...	...	...	456.3	502.0	416.0	...	505.0	505.0	
Argentina: peso Argentine : peso										
MF - hour heure[4 5 6]	62.4	# 0.4	0.8	1.7	7.7	186.6	5 383.7	13.0[7]	# 1.5	1.6
Australia: dollar Australie : dollar										
MF - hour heure[8 9]	...	...	...	...	...	...	12.9	13.3	13.7	14.0
M - hour heure[8 9]	...	...	...	...	...	...	13.5	13.8	14.2	14.6
F - hour heure[8 9]	...	...	...	...	...	...	11.1	11.7	12.0	12.4
Austria: schilling Autriche : schilling										
MF - month mois[2 10]	15 453.0	16 395.0	17 116.0	17 646.0	18 318.0	19 130.0	20 496.0	21 547.0	22 784.0	23 758.0
Azerbaijan: rouble Azerbaïdjan : rouble										
MF - month mois[1 2]	178.2	181.9	184.4	188.1	196.8	209.4	218.7	400.2	3 280.2	26 085.0
Bahrain: dinar Bahrein : dinar										
MF - month mois[1 3 12]	...	256.0[11]	...	304.0	248.0	241.0	225.0	220.0	204.0	190.0
Barbados: dollar Barbade : dollar										
MF - week semaine	...	...	...	...	238.5[9]	235.3[9]	251.5[9]	255.7		...
Belarus: rouble Bélarus : rouble										
MF - month mois[1 13]	184.0	191.0	197.0	204.0	226.0	249.0	283.0	# 596.0	5 852.0	68 866.0
Belgium: franc Belgique : franc										
MF - hour heure[9 10]	281.8	293.8	295.3	302.2	309.8	327.1	342.4	363.1	379.6	...
M - hour heure[9 10]	299.7	312.5	314.2	321.2	329.3	348.2	364.0	385.7	403.8	...
F - hour heure[9 10]	223.3	232.3	232.7	239.9	245.3	257.6	271.2	287.7	300.2	...
Bolivia: boliviano Bolivie : boliviano										
MF - month mois[1 4 9]	...	67.0	172.0	279.0	371.0	437.0	493.0	612.0	686.0	* 748.0
Botswana: pula Botswana : pula										
MF - month mois[1 9 14]	211.9	249.2	282.4	285.3	330.9	345.0	383.0	427.0	453.0	...
Brazil: cruzeiro Brésil : cruzeiro										
MF - month mois[1 7 15]	423.8	1 607.2	3 507.0	10 949.0	75 579.0	# 1 110.7	26 076.0	136 699.0	...	...
Bulgaria: lev Bulgarie : lev										
MF - month mois[1 3]	217.1	227.0	239.9	243.8	262.8	281.8	335.8	916.9	2 243.0[2]	
Burundi: franc Burundi : franc										
MF - month mois[1 16 17]	17 229.0	17 421.0	17 532.0	17 600.0	...	...	...	...	...	...
Canada: dollar Canada : dollar										
MF - hour heure[1 18]	11.2	11.6	12.0	12.4	12.9	13.6	14.3	14.9	15.5	15.8
Chile: peso Chili : peso										
MF - month mois[9 10 19]	20 379.0	24 409.0	29 157.0	34 137.0	41 497.0	50 432.0	64 447.0	83 908.0	102 514.0	118 612.0
China: yuan renminbi Chine : yuan renminbi										
MF - month mois[1 2 13]	82.3	# 92.6	106.4	118.1	148.5	166.8	174.2	191.3	231.2	...
Colombia: peso Colombie : peso										
MF - month mois[1 20]	262.5	317.9	393.6	483.1	614.1	788.3	...	...	...	...

31

Earnings in manufacturing
By hour, day, week or month [cont.]
Gains dans les industries manufacturières
Par heure, jour, semaine ou mois [suite]

Country or area and unit Pays ou zone et unité	1984	1985	1986	1987	1988	1989	1990	1991	1992	1993
Costa Rica: colon Costa Rica : colon										
MF - month mois[1,9,21]	7 110.0	8 673.0	9 588.0	# 13 211.0	14 658.0	16 784.0	20 037.0	27 229.0	32 949.0	38 631.0
M - month mois[1,9,21]	7 699.0	9 422.0	10 370.0	# 14 207.0	16 573.0	18 534.0	21 887.0	30 152.0	36 427.0	42 225.0
F - month mois[1,9,21]	5 645.0	6 932.0	7 815.0	# 11 021.0	11 102.0	13 505.0	16 262.0	21 733.0	26 282.0	30 556.0
Croatia: new dinar Croatie : nouveau dinar										
MF - month mois[1]	2.4	4.2	8.7	18.0	48.0	764.0	4 218.0	7 447.0	34 024.0	...
Cuba: peso Cuba : peso										
MF - month mois[1,2,3]	192.0	193.0	# 189.0	183.0	# 183.0	...	...	...	...	...
Cyprus: pound Chypre : livre										
MF - week semaine[9,16,19,22]	48.9	52.4	54.0	58.1	61.9	68.0	74.2	80.3	89.0	101.0
M - week semaine[9,16,19,22]	65.5	70.1	72.5	76.7	80.6	89.4	98.8	105.6	116.4	134.2
F - week semaine[9,16,19,22]	36.4	39.1	40.7	44.3	47.7	52.1	56.8	63.0	69.9	75.8
former Czechoslovakia†: koruna anc. Tchécoslovaquie† : couronne										
MF - month mois[23]	2 795.0	2 854.0	2 903.0	2 958.0	3 018.0	3 085.0	3 184.0	3 716.0	..	..
M - month mois[23]	3 197.0	3 264.0	3 320.0	3 383.0	3 452.0	3 529.0	3 642.0	4 250.0	..	..
F - month mois[23]	2 170.0	2 216.0	2 254.0	2 297.0	2 343.0	2 395.0	2 472.0	2 885.0	..	..
Czech Republic: koruna République tchèque : couronne										
MF - month mois[1]	2 965.0	3 015.0	3 055.0	3 112.0	3 172.0	3 239.0	3 325.0	3 882.0	4 565.0	5 717.0
M - month mois[1]	...	...	...	3 560.0	3 628.0	3 705.0	3 803.0	4 440.0	5 221.0	6 539.0
F - month mois[1]	...	...	...	2 416.0	2 463.0	2 515.0	2 582.0	3 014.0	3 544.0	4 439.0
Denmark: krone Danemark : couronne										
MF - hour heure[11,22,24]	72.3	75.8	78.7	86.8	92.1	96.0	99.9	104.6	108.3	...
M - hour heure[11,22,24]	74.7	78.3	81.5	90.2	95.6	99.6	103.8	108.5	112.2	...
F - hour heure[11,22,24]	64.1	67.1	69.2	75.9	80.7	84.3	87.8	92.1	95.4	...
Ecuador: sucre Equateur : sucre										
MF - hour heure	72.7	90.0	117.3	145.8	223.3	337.0	467.9	669.4	...	...
Egypt: pound Egypte : livre										
MF - week semaine[9,12]	...	29.0	33.0	38.0	* 41.0	* 46.0	* 54.0		...	...
M - week semaine[9,12]	...	30.0	34.0	39.0	* 43.0	* 48.0	* 56.0		...	...
F - week semaine[9,12]	...	22.0	25.0	28.0	* 31.0	* 34.0	* 38.0		...	...
El Salvador: colon El Salvador : colon										
MF - hour heure[25]	3.3	3.3	# 3.2	3.2	3.2	3.3	3.3	...	...	...
M - hour heure[25]	3.3	3.6	# 3.5	3.3	3.3	3.4	3.4	...	...	...
F - hour heure[25]	2.9	2.9	# 3.0	3.0	3.0	3.1	3.2	...	...	...
Estonia: kroons Estonie : couronne										
MF - month mois[1,4,26]	224.6	230.0	237.2	246.2	268.0	292.6	359.8	850.7	# 527.0	* 1 036.0
Fiji: dollar Fidji : dollar										
MF - day jour[4,9]	12.0	12.2	11.8	12.3	12.6	11.4	11.4	...	...	...

31
Earnings in manufacturing
By hour, day, week or month [*cont.*]
Gains dans les industries manufacturières
Par heure, jour, semaine ou mois [*suite*]

Country or area and unit Pays ou zone et unité	1984	1985	1986	1987	1988	1989	1990	1991	1992	1993
Finland: markka Finlande : markka										
MF - hour heure[2]	29.7	32.1	34.0	36.5	39.7	43.5	47.7	50.7	52.3	...
M - hour heure[2]	32.0	34.6	36.6	39.2	42.6	46.7	51.1	53.9	55.4	...
F - hour heure[2]	24.7	26.6	28.3	30.3	32.9	35.8	39.5	42.1	43.4	...
France: franc France : franc										
MF - hour heure[9]	35.7	# 37.8	39.3	41.0	# 41.8	43.4	45.5	47.5	49.4	50.6
M - hour heure[9]	38.1	# 40.2	41.6	43.4	# 44.4	46.2	48.4	50.5	52.4	53.7
F - hour heure[9]	29.9	# 31.8	33.0	34.4	# 35.1	36.3	38.2	39.7	41.2	42.5
French Polynesia: CFP franc Polynésie française : franc CFP										
MF - month mois[4 27]	62 530.0	70 000.0	77 700.0	82 000.0	84 000.0	86 500.0	...	...	...	
Gambia: dalasi Gambie : dalasi										
MF - day jour[1 9 12]	7.0	# 12.3	11.5	11.1	...					
Germany † Allemagne† **F. R. Germany: deutsche mark R. f. Allemagne : deutsche mark**										
MF - hour heure[16]	15.5	16.2	16.8	17.5	18.3	19.1	20.1	21.3	22.5	23.8
M - hour heure[16]	16.5	17.2	17.9	18.6	19.5	20.3	21.3	22.6	23.8	25.0
F - hour heure[16] **former German D. R.: mark anc. R. d. allemande : mark**	12.0	12.5	13.0	13.6	14.2	14.7	15.5	16.5	17.5	18.5
MF - hour heure	...	...	...	...	...	...	...	9.4	11.9	13.9
M - hour heure	...	...	...	...	...	...	...	9.7	12.3	14.4
F - hour heure	...	...	...	...	...	...	...	8.4	10.5	11.9
Ghana: cedi Ghana : cedi										
MF - month mois[19]	3 441.0	5 059.0	8 787.0	15 216.0	21 411.0	* 36 793.0			...	...
Gibraltar: pound Gibraltar : livre										
MF - week semaine[9]	138.8	140.9	150.2	152.7	190.6	190.1	238.5	214.1	239.4	...
M - week semaine[9]	146.8	149.0	157.9	158.7	203.3	194.3	247.7	254.4	254.5	...
F - week semaine[9]	93.3	97.5	108.5	118.6	122.0	137.7	151.7	148.8	161.7	...
Greece: drachma Grèce : drachme										
MF - hour heure[10 12]	262.2	# 314.2	354.1	388.2	459.7	554.0	661.3	772.1	878.2	970.8
M - hour heure[10 12]	292.6	# 347.2	393.2	430.2	509.2	614.5	733.8	854.7	967.5	1 063.6
F - hour heure[10 12]	222.9	# 269.5	302.3	333.7	397.4	481.0	575.3	673.2	765.3	851.4
Guam: dollar Guam : dollar										
MF - hour heure[1 9]	5.5	6.2	6.3	6.0	6.3	7.3	8.1	9.1	9.3	10.1
Hong Kong: dollar Hong-kong : dollar										
MF - day jour[4]	90.1	98.3	106.4	119.4	136.9	157.0	179.5	200.7	218.6	241.7
M - day jour[4]	104.2	115.1	125.9	143.3	166.1	191.7	224.5	249.9	274.8	313.8
F - day jour[4]	84.4	91.2	98.1	108.2	123.7	140.3	155.8	173.6	189.6	206.8
Hungary: forint Hongrie : forint										
MF - month mois[10 12]	4 913.0[28]	5 366.0[28]	5 716.0[28]	6 214.0[28]	# 7 761.0	9 121.0	11 167.0	13 992.0	17 636.0	21 751.0
India: rupee Inde : roupie										
MF - month mois[29]	799.5	740.2	890.7	862.9	913.2	* 886.1	...	...	...	...

31
Earnings in manufacturing
By hour, day, week or month [*cont.*]
Gains dans les industries manufacturières
Par heure, jour, semaine ou mois [*suite*]

Country or area and unit Pays ou zone et unité	1984	1985	1986	1987	1988	1989	1990	1991	1992	1993
Ireland: pound Irlande : livre										
MF - hour heure[9][30]	3.9	# 4.2	4.5	4.7	4.9	5.1	5.4	5.6	5.9	...
M - hour heure[9][22]	4.4	# 4.8	5.1	5.3	5.5	5.7	6.0	6.3	6.6	...
F - hour heure[9][22]	3.0	# 3.2	3.5	3.6	3.8	4.0	4.2	4.4	4.7	...
Israel: new shekel Israël : nouveau shekel										
MF - day jour[1][2][31][32]	8 470.0	# 29.9	47.5	64.9	78.8	95.2	...	...	...	...
Italy: lira Italie : lire										
MF - hour heure[33]	...	...	...	...	...	...	100.0	109.5	115.7	
Jamaica: dollar Jamaïque : dollar										
MF - week semaine[1]	...	...	271.9	309.4	345.9	391.8	450.6[11]	701.0	895.0	
Japan: yen Japon : yen										
MF - month mois[1][16][34]	292 255.0	299 531.0	305 414.0	313 170.0	318 663.0	336 648.0	352 020.0	368 011.0	372 594.0	371 356.0
M - month mois[1][16][34]	356 561.0	367 182.0	373 324.0	381 138.0	393 804.0	414 981.0	436 135.0	450 336.0	454 482.0	...
F - month mois[1][16][34]	152 519.0	154 571.0	158 550.0	163 944.0	164 673.0	173 097.0	180 253.0	193 112.0	198 058.0	...
Jordan: dinar Jordanie : dinar										
MF - day jour[1][9]	3.7	4.0	4.2	4.5	4.2	4.4	4.6	4.6	...	...
Kazakhstan: tenge Kazakhstan : tenge										
MF - month mois[1][35]	195.0	199.0	204.0	209.0	229.0	250.0	277.0	489.0	5 675.0	# 144.0
Kenya: shilling Kenya : shilling										
MF - month mois[1][9][19]	1 836.2	1 928.7	2 078.1	2 293.8	2 469.7	2 797.5	3 064.6	3 324.2	...	...
M - month mois[1][9][19]	1 879.3	2 025.2	2 137.8	2 376.7	2 576.0	2 890.9	3 159.8	3 430.1	...	...
F - month mois[1][9][19]	1 439.7	1 531.0	1 554.3	1 551.1	1 751.7	2 001.7	2 317.7	2 515.8	...	...
Korea, Republic of: won Corée, République de : won										
MF - month mois[1][7][16][19]	245.0	270.0	294.0	329.0	393.0	492.0	591.0	690.0	799.0	# 885.0[36]
M - month mois[1][7][16][19]	317.0	347.0	375.0	413.0	491.0	609.0	724.0	843.0	964.0	# 1 056.0[36]
F - month mois[1][7][16][19]	150.0	163.0	182.0	208.0	250.0	307.0	364.0	428.0	497.0	# 551.0[36]
Latvia: lat Lettonie : lat										
MF - month mois[1][3][35]	...	...	...	...	...	...	1.6	3.2	22.1	# 48.7
M - month mois[1][3][35]	...	...	...	...	...	...	...	...	23.5	...
F - month mois[1][3][35]	...	...	...	...	...	...	...	...	20.2	...
Luxembourg: franc Luxembourg : franc										
MF - hour heure[9][10]	310.7	320.0	332.2	339.2	357.0	374.0	379.0	408.0	...	...
M - hour heure[9][10]	321.9	331.1	344.3	352.4	375.0	393.0	399.0	429.0	...	...
F - hour heure[9][10]	197.7	207.3	209.1	217.2	218.0	234.0	248.0	261.0	...	...
Macau: pataca Macao : pataca										
MF - month mois[1][21]	...	...	...	...	...	1 859.0[9]	2 058.0[9]	2 232.0[9]	2 509.0	...
M - month mois[1][21]	...	...	...	...	...	...	...	...	3 321.0	...
F - month mois[1][21]	...	...	...	...	...	...	...	...	2 222.0	...
Malawi: kwacha Malawi : kwacha										
MF - month mois[1]	72.4	90.2	101.3	125.8	135.7	147.0	176.8	* 137.6	...	...
Mauritius: rupee Maurice : roupie										
MF - day jour[9][10][37][38]	30.8[39]	33.8[39]	34.4[39]	# 37.6	45.5	52.6	60.5	83.3	92.5	95.1

31
Earnings in manufacturing
By hour, day, week or month [cont.]
Gains dans les industries manufacturières
Par heure, jour, semaine ou mois [suite]

Country or area and unit Pays ou zone et unité	1984	1985	1986	1987	1988	1989	1990	1991	1992	1993
Mexico: new peso Mexique : nouveau peso										
MF - hour heure[10 40]	243.0	# 314.0	527.0	1 337.0	2 666.0	3 257.0	4 152.0	5 217.0	7 501.0	# 9.0
Myanmar: kyat Myanmar : kyat										
M - month mois[1 11]	255.7	260.1	282.2	325.7	533.1	729.5	...	631.9	880.5	...
F - month mois[1 11]	246.7	256.8	243.6	261.1	496.9	705.6	...	670.9	866.8	...
Netherlands: guilder Pays-Bas : florin										
MF - hour heure[1 9 30]	18.7	19.0	19.4	19.9	20.3	20.8	21.5	22.3	22.7	...
M - hour heure[19 22]	19.8	20.1	20.5	21.0	21.5	21.9	22.8	23.5	23.9	...
F - hour heure[19 22]	14.6	15.0	15.4	15.6	16.1	16.5	17.1	17.8	18.3	...
New Zealand: dollar Nouvelle-Zélande : dollar										
MF - hour heure[1 41]	7.7	8.4	10.1	11.0	12.0	12.6	13.4	13.9	14.1	14.2
M - hour heure[1 41]	8.4	9.1	10.9	11.9	12.9	13.5	14.4	14.8	15.0	15.1
F - hour heure[1 41]	5.9	6.4	7.8	8.6	9.6	10.2	10.8	11.2	11.6	11.6
Nicaragua: córdoba Nicaragua : córdoba										
MF - hour heure[1 42]	...	...	...	...	...	5.8[7]	590.1[7]	# 4.7	8.0	...
Norway: krone Norvège : couronne										
MF - hour heure[19 22]	57.1	61.5	67.7	78.6	83.0	87.3	92.5	97.3	100.4	103.2
M - hour heure[19 22]	58.6	63.3	69.7	81.0	85.4	89.5	94.6	99.5	102.7	105.4
F - hour heure[19 22]	49.2	52.9	58.4	67.8	72.0	76.5	81.8	86.7	89.2	91.8
Pakistan: rupee Pakistan : roupie										
MF - month mois[1]	713.4	881.6	1 026.4	1 115.4	1 130.9	1 289.7	...	...	...	...
Paraguay: guarani Paraguay : guaraní										
MF - month mois[1]	35 958.0	53 100.0	67 466.0	...	115 897.0	177 464.0	220 548.0	273 537.0	298 682.0	380 096.0
M - month mois[1]	37 359.0	54 216.0	68 493.0	...	118 473.0	170 222.0	234 234.0	292 787.0	342 552.0	407 117.0
F - month mois[1]	29 364.0	47 127.0	62 324.0	...	105 252.0	212 743.0	155 744.0	196 738.0	177 754.0	298 952.0
Peru: new sol Pérou : nouveau sol										
MF - day jour[10 43 44]	12.6[9]	26.7[9]	# 80.6	161.0	758.2	18 636.6	# 0.3	6.0	10.1	
Philippines: peso Philippines : peso										
MF - month mois[1 12 45]	1 642.0	1 951.0	2 183.0	2 537.0	2 995.0	3 441.0	* 4 087.0	...	...	...
Poland: zloty Pologne : zloty										
MF - month mois[1 13 19]	16 717.0	19 901.0	24 076.0	29 382.0	54 708.0	212 170.0	996.0[7]	1 620.0[7]	...	...
Portugal: escudo Portugal : escudo										
MF - hour heure	...	...	...	...	...	281.0	324.0	370.0	419.0	436.0
M - hour heure	...	...	...	...	...	320.0	368.0	419.0	480.0	486.0
F - hour heure	...	...	...	...	...	222.0	254.0	295.0	326.0	390.0
Puerto Rico: dollar Porto Rico : dollar										
MF - hour heure	5.0	5.2	5.3	5.4	5.6	5.8	6.0	6.3	6.7	7.0
Republic of Moldova: rouble République de Moldova : rouble										
MF - month mois[1]	173.2	185.6	182.4	188.2	212.2	244.4	287.6	475.2	3 676.2	...
Romania: leu Roumanie : leu										
MF - month mois[3 28]	2 697.0	2 718.0	2 728.0	2 719.0	2 835.0	2 920.0	3 146.0	6 842.0	# 17 493.0	...
San Marino: lira Saint-Marin : lire										
MF - day jour[1]	60 134.0	65 742.0	72 911.0	76 133.0	...	84 782.0	84 835.0	102 579.0	111 611.0	113 772.0

31
Earnings in manufacturing
By hour, day, week or month [*cont.*]
Gains dans les industries manufacturières
Par heure, jour, semaine ou mois [*suite*]

Country or area and unit Pays ou zone et unité	1984	1985	1986	1987	1988	1989	1990	1991	1992	1993
Seychelles: rupee Seychelles : roupie										
MF - month mois[1 29]	1 856.0	1 956.0	2 003.0	2 075.0	1 863.0[46]	1 975.0[46]	2 187.0[46]	2 259.0[46]	# 2 349.0[46]	...
Sierra Leone: leone Sierra Leone : leone										
MF - week semaine[11 22]	31.9	45.7	79.7	80.5	...	...	...	...	...	...
Singapore: dollar Singapour : dollar										
MF - month mois[1 47]	...	...	975.8	1 008.8	1 115.9	1 242.9	1 395.0	1 551.8	1 686.2	1 817.8
M - month mois[1 47]	...	...	...	...	...	1 623.0	1 797.5	1 970.1	2 127.4	2 266.2
F - month mois[1 47]	...	...	...	...	...	876.2	983.3	1 096.8	1 190.7	1 294.5
Slovakia: koruna Slovaquie : couronne										
MF - month mois[1 3 48]	2 853.0	2 912.0	2 973.0	3 017.0	3 088.0	3 156.0	3 262.0	# 3 776.0	4 370.0	5 234.0
Slovenia: new dinar Slovénie : nouveau dinar										
MF - month mois[1]	4.0	7.0	16.0	35.0	110.0	1 895.0	8 696.0	14 742.0	43 321.0	62 473.0
South Africa: rand Afrique du Sud : rand										
MF - month mois[1 49]	713.0	805.0	911.0	1 045.0	1 215.0	1 438.0	1 660.0	1 890.0	2 195.0	...
Spain: peseta Espagne : peseta										
MF - hour heure[1]	514.0	564.0	620.0	678.0	735.0	780.0	902.0	994.0	1 080.0	1 159.0
Sri Lanka: rupee Sri Lanka : roupie										
MF - hour heure[11]	4.4	5.1	5.5	6.0	6.6	7.5	9.5	11.2	11.8	13.7
M - hour heure[11]	4.9	5.9	5.8	6.4	6.9	8.2	9.8	11.7	12.3	14.1
F - hour heure[11]	3.4	4.3	4.5	4.6	4.8	5.4	8.9	9.4	10.4	12.7
Sudan: pound Soudan : livre										
MF - month mois[10]	...	...	...	...	248.2	...	375.4	...	1 210.3	...
Swaziland: emalangeni Swaziland : emalangeni										
M - month mois[5 9]	806.0	881.0	946.0	1 032.0	941.0	1 040.0	...	...	...	...
F - month mois[5 9]	...	265.0	572.0	577.0	404.0	561.0	...	...	...	...
Sweden: krona Suède : couronne										
MF - hour heure[11 19 22]	54.0	58.6	62.7	67.0	72.2	79.3	87.3	91.7	98.3	98.5
M - hour heure[11 19 22]	55.1	59.8	63.9	68.5	73.8	81.1	89.5	93.8	100.7	100.7
F - hour heure[11 19 22]	49.6	53.7	57.8	61.7	66.4	72.6	79.5	83.7	90.1	90.1
Switzerland: franc Suisse : franc										
M - hour heure[9 16 22]	18.5	19.0	19.9	20.5	21.3	22.1	23.4	25.0	26.2	...
F - hour heure[9 16 22]	12.3	12.8	13.4	13.8	14.3	15.0	15.9	17.0	17.8	...
Thailand: baht Thaïlande : baht										
MF - month mois[1]	2 717.0	2 826.0	2 631.0	...	...	2 995.0	3 366.0	3 734.0	4 016.0	...
M - month mois[1]	...	...	...	...	...	3 726.0	...	...	...	...
F - month mois[1]	...	...	...	...	...	2 453.0	...	...	...	...
Tonga: pa'anga Tonga : pa'anga										
MF - week semaine[1 49]	24.7	26.4	28.6	30.8	33.8	43.0	42.6	51.4	51.3	...
Turkey: lira Turquie : livre										
MF - day jour[19 50]	1 336.3	...	...	...	8 461.0	17 820.5	30 582.2	61 620.4	88 144.3	135 236.0
M - day jour[19 50]	1 372.5	...	...	...	8 642.9	18 679.1	31 231.1	62 140.3	91 204.1	138 746.1
F - day jour[19 50]	1 280.4	...	...	...	6 877.3	13 830.0	25 299.5	56 660.2	84 558.2	122 720.8

31
Earnings in manufacturing
By hour, day, week or month [cont.]
Gains dans les industries manufacturières
Par heure, jour, semaine ou mois [suite]

Country or area and unit Pays ou zone et unité	1984	1985	1986	1987	1988	1989	1990	1991	1992	1993
Ukraine: rouble Ukraine : rouble										
MF - month mois[1 2 13]	196.5	201.5	# 193.5	198.2	215.8	...	...	...	...	...
former USSR†: rouble ancienne URSS† : rouble										
MF - month mois[1 2 13]	198.3	203.3	208.1	213.6	232.2	253.9	...	...	...	...
United Kingdom: pound Royaume-Uni : livre										
MF - hour heure[2 9 51 52]	3.4	3.6	3.9	4.1	4.4	4.8	5.2	5.6	6.0	6.2
M - hour heure[2 9 51 52]	3.6	3.9	4.1	4.4	4.7	5.1	5.5	6.0	6.4	6.6
F - hour heure[2 9 51 52]	2.4	2.6	2.8	3.0	3.2	3.4	3.8	4.1	4.3	4.5
United States: dollar Etats-Unis : dollar										
MF - hour heure[10]	9.2	9.5	9.7	9.9	# 10.2[36]	10.5	10.8	11.2	11.5	11.7
United States Virgin Is.: dollar Iles Vierges américaines : dollar										
MF - hour heure[1]	9.5	9.4	9.6	9.4	9.9	10.9	11.9	12.5	13.7	15.0
Uruguay: new peso Uruguay : nouveau peso										
MF - month mois[1 53]	100.0	169.8	322.2	572.5	949.3	1 751.6	3 453.2	7 371.5	12 564.4	...
Yugoslavia, SFR†: new dinar Yougoslavie, Rfs† : nouveau dinar										
MF - month mois[1 13]	2.2	3.9	8.0	16.6	44.6	773.0	...	...	...	...
Zimbabwe: dollar Zimbabwe : dollar										
MF - month mois[19 54]	382.2	435.4	471.2	527.9	590.3	665.5	796.1	928.1	1 123.3	1 283.1[11]

Source:
International Labour Office (Geneva).

† For detailed descriptions of data pertaining to
former Czechoslovakia, Germany, SFR Yugoslavia and former
USSR, see Annex I - Country or area nomenclature, regional
and other groupings.

1 Employees.
2 Including mining and quarrying (Azerbaizan and Bulgaria
(1992): electricity, gas and water; China: electricity,
water, geological and prospecting activities; prior to 1985,
excluding geological and prospecting activities; Cuba:
prior to 1986, including water; prior to 1988, including
fishing and gas; Finland: electricity; Ukraine: prior to
1986 only; United Kingdom: quarrying only).

3 State sector (Bahrain: private sector only; Bulgaria and
Romania (prior to 1992): including cooperative sector; Cuba:
civilian; Slovakia: prior to 1991 only).
4 Wage rates.
5 Skilled workers only.
6 1984: pesos; one austral is equivalent to 1,000 pesos.
1985-1991: australes; one peso is equivalent to 10,000
australes.
7 Figures in thousands.
8 Full-time adult non-managerial employees.

9 One month of each year.
10 Wage earners.
11 Average of less than twelve months.
12 Establishments with ten or more persons employed only

Source:
Bureau international du Travail (Genève).

† Pour les descriptions en détails des données
relatives à l'ancienne Tchécoslovaquie, l'Allemagne, la Rfs
Yougoslavie et l'ancienne URSS, voir l'Annexe I -
Nomenclature des pays ou zones, groupements régionaux et
autres groupements.

1 Salariés.
2 Y compris les industries extractives (Azerbaïdjan and
Bulgarie (1992): l'électricité, le gaz et l'eau; Chine:
l'électricité, l'eau, les recherches geologiques et de
prospection; avant 1985, non compris les recherches
geologiques et de prospection; Cuba: avant 1986, y compris
l'eau; avant 1988, y compris la pêche, et le gaz; Finlande:
l'électricité; Ukraine: avant 1986 seulement; Royaume-Uni: Y
compris les carrières seulement).
3 Secteur d'etat (Bahreïn: secteur privé; Bulgarie et Roumanie
(avant 1992): y compris secteur de coopératif; Cuba: civils;
Slovaquie: avant 1991 seulement).
4 Taux de salaire.
5 Ouvriers qualifiés seulement.
6 1984: pesos; un austral équivaut à 1000 pesos. 1985-1991:
australes; un peso équivaut à 10000 australes.

7 Données in milliers.
8 Salariés adultes à plein temps, non compris les cadres
dirigeants.
9 Un mois de chaque année.
10 Ouvriers.
11 Moyenne de moins de douze mois.
12 Etablissements occupant dix personnes et plus seulement

31
Earnings in manufacturing
By hour, day, week or month [*cont.*]

Gains dans les industries manufacturières
Par heure, jour, semaine ou mois [*suite*]

(Gambia: 5 or more persons; Hungary: 20 persons or more, prior to 1992 all legal economic units).

13 Socialized sector (Belarus and Poland: prior to 1991).
14 Citizens only.
15 Prior to 1989, former cruzeiros; 1 current cruzeiro = 1 million former cruzeiros.
16 Including family allowances (F. R. Germany: paid directly by the employers; Japan: and mid and end of year bonuses).
17 Bujumbura only.
18 Paid by the hour.
19 Including the value of payments in kind (Sweden: including holidays and sick-leave payments).
20 Index numbers; base: 1980 equals 100.
21 Labour force sample survey.
22 Adults only.
23 State industry.
24 Excluding vacation pay.
25 San Salvador only, prior to 1986.
26 Prior to 1992, Kroon. One kroon is equivalent to 10 roubles.
27 Minimunm rates.
28 Net earnings after deduction of income taxes.
29 Including electricity, gas, water and services (Seychelles (before 1992): electricity and water).
30 Including juveniles.
31 Including payments subject to income tax.
32 Prior to 1985, shekels; one new shekel is equivalent to 1,000 shekels.
33 Index of hourly wage rates (1990=100).
34 Marked break in series: Japan at 1985: sample design revised.
35 Prior to 1993, roubles (Kazakhstan: 1 tenge = 500 roubles; Latvia: 1 lat = 200 roubles).
36 New industrial classification.
37 Excluding sugar and tea factories.
38 Daily rates of pay.
39 Including workers on piece rates of pay.
40 Prior to 1993, pesos; 1 new peso = 1,000 pesos.
41 Establishments with the equivalent of more than 2 full-time paid employees.
42 Prior to 1991, old córdobas; one new córdoba = 1,000 old córdobas.
43 Lima only.
44 Prior to 1990, Intís. One new sol is equivalent to 1 million intís.
45 Computed on the basis of annual wages.
46 Earnings are exempted from income tax.
47 Social security statistics.
48 Excluding enterprises with less than 25 employees. In 1991, excluding enterprises with less than 100 employees.
49 Including employers' non-statutory contributions to certain funds.
50 Insurance statistics.
51 Full-time workers on adult rates of pay.
52 Excluding Northern Ireland.
53 Index of average monthly earnings (Oct.-Dec. 1984=100).
54 All persons engaged.

(Gambie: 5 personnes et plus; Hongrie: 20 personnes et plus, avant 1992 ensemble des unités économiques dotées d'un statut juridique).

13 Secteur socialisé (Bélarus et Pologne: avant 1991).
14 Nationaux seulement.
15 Avant 1989, cruzeiros (anciens); 1 cruzeiro (nouveau) = 1 million de cruzeiros (anciens)
16 Y compris les allocations familiales (Allemagne, Rép. féd. d': payées directement par les employeurs; Japon: et les primes de milieu et de fin d'année).
17 Bujumbura seulement.
18 Rémunérés à l'heure.
19 Y compris la valeur des paiements en nature (Suède: y compris les versements pour les vacances et congés de maladie).
20 Indices; base: 1980 égal à 100.
21 Enquête par sondage sur la main-d'oeuvre.
22 Adultes seulement.
23 Industrie d'Etat.
24 Non compris les versements pour congés payés.
25 San Salvador seulement, avant 1986.
26 Avant 1992, Couronne. Une couronne équivaut à 10 roubles.
27 Taux minima.
28 Gains nets après déduction de l'impôt sur le revenu.
29 Y compris l'électricite, le gaz, l'eau et les services (Seychelles (avant 1992): l'électricite et l'eau).
30 Y compris les jeunes gens.
31 Y compris les versements soumis à l'impôt sur le revenu.
32 Avant 1985, shekels; un nouveaux shekel équivaut à 1000 shekels.
33 Indice des taux de salaire horaire (1990=100).
34 Discontinuité notable dans la série: Japon 1985: plan d'échantillonnage révisé.
35 Avant 1993, roubles (Kazakhstan: 1 tenge = 500 roubles; Lettonie: 1 lat = 200 roubles).
36 Nouvelle classification industrielle.
37 Non compris les fabriques de sucre et de thé.
38 Rémunérés sur la base de taux de salaire journaliers.
39 Y compris les travailleurs aux pièces.
40 Avant 1993, pesos; 1 nouveau peso = 1000 pesos.
41 Etablissements occupant plus de l'équivaut de 2 salariés plein temps.
42 Avant 1991, anciens córdobas; un nouveau córdoba = 1000 anciens córdobas.
43 Lima seulement.
44 Avant de 1990, Intís. Un nouveau sol équivaut à 1 million d'Intís.
45 Calculés sur la base de salaires annuels.
46 Les gains sont exempts de l'impôts sur le revenu.
47 Statistiques de la sécurité sociale.
48 Non compris les entreprises occupant moins de 25 salariés. En 1991, non compris les entreprises occupant moins de 100 salariés.
49 Y compris les cotisations des employeurs certains fonds privés.
50 Statistiques d'assurances.
51 Travailleurs à plein temps rémunérés sur la base de taux de salaire pour adultes.
52 Non compris l'Irlande du Nord.
53 Indices des gains mensuels moyens (oct.-déc. 1984=100).
54 Ensemble de l'effectif occupé.

32
Producers prices and wholesale prices
Prix à la production et des prix de gros
Index numbers: 1980 = 100
Indices : 1980 = 100

Country or area and groups	1987	1988	1989	1990	1991	1992	1993	Pays ou zone et groupes
Argentina								**Argentine**
Domestic supply[1,2,3]	365	1 872	66 123	1 128 670	2 375 572	2 517 955	2 558 363	Offre intérieure[1,2,3]
Domestic production[3]	365	1 859	65 021	1 150 802	2 443 959	2 600 625	2 664 949	Production intérieure[3]
Agricultural products[2,3]	457	2 187	78 419	1 248 401	2 435 898	2 874 159	3 039 688	Produits agricoles[2,3]
Industrial products[2,3]	352	1 815	63 215	1 137 647	2 445 044	2 563 756	2 591 744	Produits industriels[2,3]
Import products[3]	370	1 991	76 632	917 612	1 723 428	1 729 594	1 732 667	Produits importés[3]
Australia								**Australie**
Industrial products[2,3,4]	163	175	186	204	207	208	213	Produits industriels[2,3,4]
Exported goods[5]	...	105	105	100	94	88	89	Produits exportés[5]
Raw materials[3]	...	108	119	124	121	127	126	Matières premières[3]
Austria								**Autriche**
Domestic supply[2,6,7]	111	111	113	116	117	116	116	Offre intérieure[2,6,7]
Agricultural products	109	107	107	115	117	105	102	Produits agricoles
Producers' material[2,6,7]	108	108	112	113	112	111	110	Matériaux de production[2,6,7]
Consumers' goods[2,6]	115	113	115	121	123	124	124	Biens de consommation[2,6]
Capital goods[2,6]	108	107	105	107	108	109	106	Biens d'équipement[2,6]
Bangladesh								**Bangladesh**
Domestic supply[2,6,7]	197	209	226	245	255	264	269	Offre intérieure[2,6,7]
Agricultural products[2,6,8]	213	226	244	265	269	277	281	Produits agricoles[2,6,8]
Industrial products[2,3,6]	168	177	190	206	227	240	245	Produits industriels[2,3,6]
Raw materials[6]	146	150	171	190	196	205	209	Matières premières[6]
Finished goods[6]	155	161	169	188	196	214	220	Produits finis[6]
Belgium								**Belgique**
Domestic supply	126	127	135	136	134	134	133	Offre intérieure
Agricultural products[2]	131	133	140	137	139	143	141	Produits agricoles[2]
Industrial products	125	127	136	136	134	134	132	Produits industriels
Intermediate products	119	120	128	129	124	122	121	Produits intermédiaires
Consumers' goods	135	137	143	143	145	149	147	Biens de consommation
Capital goods	139	141	146	150	155	158	159	Biens d'équipement
Brazil								**Brésil**
Domestic supply[9,10]	748	5 956	83 111	2 378 647	12 004 206	# 100	2 167	Offre intérieure[9,10]
Agricultural products[9,10]	744	5 824	72 408	2 356 680	13 874 940	# 100	2 237	Produits agricoles[9,10]
Industrial products[9,10]	713	5 764	82 271	2 188 515	10 303 545	# 100	2 146	Produits industriels[9,10]
Raw materials[9,10]	608	4 543	55 609	1 067 362	5 689 838	# 100	2 166	Matières premières[9,10]
Producers' material[9,10]	678	5 376	73 031	2 033 744	10 125 359	# 100	2 155	Matériaux de production[9,10]
Consumers' goods[9,10]	797	6 394	93 316	2 689 839	13 854 613	# 100	2 192	Biens de consommation[9,10]
Capital goods[3,10]	773	7 265	99 321	2 883 393	12 071 429	# 100	2 082	Biens d'équipement[3,10]
Canada								**Canada**
Agricultural products[2]	98	104	110	107	100	99	...	Produits agricoles[2]
Industrial products[2,3]	136	142	144	145	143	144	149	Produits industriels[2,3]
Raw materials[11]	123	119	123	128	120	121	128	Matières premières[11]
Intermediate products[12]	110	117	119	118	115	114	118	Produits intermédiaires[12]
Finished goods[12]	114	114	117	120	122	125	129	Produits finis[12]
Chile								**Chili**
Domestic supply	433	459	528	643	782	873	947	Offre intérieure
Domestic production	431	454	528	650	802	908	980	Production intérieure

32
Producers prices and wholesale prices
Index numbers: 1980 = 100 [*cont.*]
Prix à la production et des prix de gros
Indices : 1980 = 100 [*suite*]

Country or area and groups	1987	1988	1989	1990	1991	1992	1993	Pays ou zone et groupes
Agricultural products	424	397	484	600	735	842	898	Produits agricoles
Industrial products[7]	440	497	562	702	860	986	1 077	Produits industriels[7]
Import products	444	483	531	613	689	714	787	Produits importés
Colombia								**Colombie**
Domestic supply[2 6]	314	407	511	664	817	963	1 088	Offre intérieure[2 6]
Domestic production[6]	304	395	495	648	809	967	1 099	Production intérieure[6]
Agricultural products[6]	417	541	696	848	1 101	1 306	1 390	Produits agricoles[6]
Industrial products[6]	356	464	581	760	932	1 100	1 207	Produits industriels[6]
Import products[6]	409	529	673	848	962	1 045	1 149	Produits importés[6]
Exported goods[6]	316	405	389	626	607	640	702	Produits exportés[6]
Raw materials[6]	340	419	519	656	773	944	1 152	Matières premières[6]
Intermediate products[6]	413	547	711	898	1 089	1 252	1 388	Produits intermédiaires[6]
Finished goods[6]	423	541	702	752	965	1 191	1 376	Produits finis[6]
Costa Rica								**Costa Rica**
Domestic supply[7 13]	863	1 053	1 207	1	1 237	...	...	Offre intérieure[7 13]
Croatia								**Croatie**
Industrial products[10]	..	..	..	4	11	100	1 613	Produits industriels[10]
Producers' material[10]	..	..	..	4	10	100	1 650	Matériaux de production[10]
Consumers' goods[10]	..	..	..	5	12	100	1 558	Biens de consommation[10]
Capital goods[10]	..	..	..	4	9	100	1 595	Biens d'équipement[10]
Cyprus								**Chypre**
Industrial products	109	134	142	149	155	160	...	Produits industriels
Denmark								**Danemark**
Domestic supply[2 9]	138	143	151	153	154	153	152	Offre intérieure[2 9]
Domestic production[2 9]	144	149	158	160	162	161	160	Production intérieure[2 9]
Import products[9]	128	133	142	143	143	140	139	Produits importés[9]
Producers' material	133	139	148	150	151	149	149	Matériaux de production
Consumers' goods	146	150	157	158	161	160	156	Biens de consommation
Dominican Republic								**Rép. dominicaine**
Domestic supply[2 3 6]	141	204	329	625	780	836	...	Offre intérieure[2 3 6]
Industrial products[3]	119	221	335	499	500	574	...	Produits industriels[3]
Consumers' goods[3]	144	202	416	335	405	871	...	Biens de consommation[3]
Ecuador								**Equateur**
Agricultural products	473	733	1 264	1 499	3 650	5 663	...	Produits agricoles
Industrial products	395	694	1 221	1 754	2 676	4 096	...	Produits industriels
Egypt								**Egypte**
Domestic supply[6 7]	228	288	366	428	505	570	559	Offre intérieure[6 7]
Raw materials[6]	205	273	369	480	576	705	818	Matières premières[6]
Intermediate products[6]	182	226	294	396	486	566	568	Produits intermédiaires[6]
Finished goods[6]	198	269	327	404	490	559	604	Produits finis[6]
Capital goods	164	211	261	305	418	586	649	Biens d'équipement
El Salvador								**El Salvador**
Domestic supply[6 7 14]	206	216	237	282	...	...	...	Offre intérieure[6 7 14]
Finland								**Finlande**
Domestic supply	138	143	151	156	156	158	163	Offre intérieure
Domestic production	142	150	159	166	166	165	167	Production intérieure

32
Producers prices and wholesale prices
Index numbers: 1980 = 100 [*cont.*]
Prix à la production et des prix de gros
Indices : 1980 = 100 [*suite*]

Country or area and groups	1987	1988	1989	1990	1991	1992	1993	Pays ou zone et groupes
Import products	111	118	123	124	125	134	148	Produits importés
Raw materials	121	130	138	140	137	137	145	Matières premières
Finished goods	136	144	152	155	156	161	168	Produits finis
Consumers' goods	148	152	157	162	166	172	177	Biens de consommation
Capital goods	154	162	173	184	189	190	191	Biens d'équipement
France								**France**
Agricultural products	140	141	151	152	153	141	134	Produits agricoles
Germany †								**Allemagne†**
Domestic supply[15]	..	..	..	..	100	101	102	Offre intérieure[15]
Finished goods[15]	..	..	..	..	100	103	104	Produits finis[15]
Producers' material[15]	..	..	..	..	100	101	101	Matériaux de production[15]
Consumers' goods[15]	..	..	..	..	100	100	104	Biens de consommation[15]
Capital goods[15]	..	..	..	..	100	103	104	Biens d'équipement[15]
F. R. Germany								**R. f. Allemagne**
Domestic production	111	114	119	118	118	119	117	Production intérieure
Agricultural products[16]	95	95	103	99	98	96	88	Produits agricoles[16]
Industrial products	116	117	121	123	126	128	128	Produits industriels
Import products	99	101	105	103	103	100	98	Produits importés
Exported goods	116	119	122	122	122	125	125	Produits exportés
Raw materials	89	92	99	95	91	89	86	Matières premières
Intermediate products	110	113	117	116	116	115	113	Produits intermédiaires
Finished goods[3]	99	119	122	124	128	132	...	Produits finis[3]
Producers' material	117	119	123	125	128	128	...	Matériaux de production
Consumers' goods	112	113	117	119	123	126	...	Biens de consommation
Capital goods	126	129	132	136	140	144	...	Biens d'équipement
Greece								**Grèce**
Domestic supply[17 18 19]	327	360	409	474	553	616	687	Offre intérieure[17 18 19]
Domestic production[18 19]	320	352	398	474	563	631	705	Production intérieure[18 19]
Agricultural products[18 20]	321	353	409	503	619	629	664	Produits agricoles[18 20]
Industrial products[18 19]	320	352	395	468	551	632	715	Produits industriels[18 19]
Import products[17 18 19]	392	428	481	542	621	699	783	Produits importés[17 18 19]
Exported goods[18 19 21]	301	333	387	413	458	487	542	Produits exportés[18 19 21]
Honduras								**Honduras**
Domestic supply	136	138	157	201	...	...	...	Offre intérieure
Domestic production	127	133	151	193	...	...	...	Production intérieure
Agricultural products	113	132	154	199	...	...	...	Produits agricoles
Import products	155	169	215	285	...	...	...	Produits importés
India								**Inde**
Domestic supply	163	173	183	199	222	249	271	Offre intérieure
Agricultural products[22]	183	200	208	220	254	303	323	Produits agricoles[22]
Industrial products[3]	154	163	177	193	213	236	257	Produits industriels[3]
Raw materials	162	176	184	202	239	265	278	Matières premières
Indonesia								**Indonésie**
Domestic supply[6 19]	200	209	228	250	263	277	287	Offre intérieure[6 19]
Domestic production[23]	143	158	169	180	197	211	...	Production intérieure[23]
Agricultural products[6]	212	237	258	278	301	328	366	Produits agricoles[6]

32
Producers prices and wholesale prices
Index numbers: 1980 = 100 [cont.]
Prix à la production et des prix de gros
Indices : 1980 = 100 [suite]

Country or area and groups	1987	1988	1989	1990	1991	1992	1993	Pays ou zone et groupes
Industrial products[3][6]	205	224	238	252	278	312	...	Produits industriels[3][6]
Import products[6][19]	218	226	245	263	278	287	291	Produits importés[6][19]
Exported goods[6]	162	162	180	218	210	218	215	Produits exportés[6]
Iran, Islamic Rep. of								**Iran, Rép. islamique d'**
Domestic production[2][5][8]	...	68	82	100	126	170	192	Production intérieure[2][5][8]
Agricultural products[2][5][8]	...	...	95	100	128	162	171	Produits agricoles[2][5][8]
Industrial products[2][5][8]	...	55	66	100	124	166	188	Produits industriels[2][5][8]
Import products[2][5][8]	...	66	74	100	121	168	193	Produits importés[2][5][8]
Exported goods[2][5][8]	...	85	88	100	153	163	167	Produits exportés[2][5][8]
Raw materials[2][5][8]	...	72	78	100	126	149	155	Matières premières[2][5][8]
Ireland								**Irlande**
Domestic supply[2][17][24]	151	158	166	162	164	165	173	Offre intérieure[2][17][24]
Agricultural products[2][24]	140	155	163	144	140	144	154	Produits agricoles[2][24]
Industrial products[2][3][24]	156	162	170	167	169	171	179	Produits industriels[2][3][24]
Capital goods[6]	158	164	173	178	184	186	208	Biens d'équipement[6]
Israel								**Israël**
Industrial products[3]	85	100	124	134	157	173	187	Produits industriels[3]
Italy								**Italie**
Domestic supply[2][19]	176	184	196	210	221	226	237	Offre intérieure[2][19]
Agricultural products[2][19]	164	169	186	196	214	210	206	Produits agricoles[2][19]
Industrial products[2][19]	177	186	197	212	222	228	242	Produits industriels[2][19]
Producers' material	162	169	180	195	202	209	217	Matériaux de production
Consumers' goods	186	195	208	220	235	245	253	Biens de consommation
Capital goods	206	217	231	244	255	257	273	Biens d'équipement
Japan								**Japon**
Domestic supply[2][6]	88	88	90	92	92	93	93	Offre intérieure[2][6]
Domestic production[6]	93	92	94	95	97	96	94	Production intérieure[6]
Agricultural products[6][8]	99	98	99	100	101	102	101	Produits agricoles[6][8]
Industrial products[6][8]	93	92	94	96	97	97	97	Produits industriels[6][8]
Import products[6][17]	81	86	89	91	90	88	88	Produits importés[6][17]
Exported goods[21]	106	113	114	110	112	111	105	Produits exportés[21]
Raw materials[6]	61	58	63	67	64	69	71	Matières premières[6]
Intermediate products[6]	86	86	88	90	91	90	90	Produits intermédiaires[6]
Finished goods[6]	100	98	99	101	102	103	102	Produits finis[6]
Producers' material[6]	82	82	84	87	87	85	88	Matériaux de production[6]
Consumers' goods[6]	100	98	98	99	101	102	102	Biens de consommation[6]
Capital goods[6]	99	99	101	103	104	104	102	Biens d'équipement[6]
Jordan								**Jordanie**
Domestic supply[2][6]	122	133	178	203	214	224	226	Offre intérieure[2][6]
Korea, Republic of								**Corée, République de**
Domestic supply	127	130	132	138	145	147	150	Offre intérieure
Agricultural products[8][25]	132	155	158	179	198	204	203	Produits agricoles[8][25]
Raw materials	156	164	168	181	198	193	196	Matières premières
Intermediate products	121	121	120	122	126	129	129	Produits intermédiaires
Finished goods	132	139	143	152	162	165	169	Produits finis
Producers' material	124	124	124	126	131	133	134	Matériaux de production

32
Producers prices and wholesale prices
Index numbers: 1980 = 100 [cont.]
Prix à la production et des prix de gros
Indices : 1980 = 100 [suite]

Country or area and groups	1987	1988	1989	1990	1991	1992	1993	Pays ou zone et groupes
Consumers' goods	131	139	142	153	165	168	172	Biens de consommation
Capital goods	125	132	139	142	146	150	153	Biens d'équipement
Kuwait								**Koweït**
Domestic supply[6 9]	111	116	126	...	...	...	...	Offre intérieure [6 9]
Agricultural products[6 9]	107	107	109	...	...	...	...	Produits agricoles [6 9]
Raw materials[6]	88	85	96	...	...	...	...	Matières premières [6]
Intermediate products[6]	102	114	114	...	...	...	...	Produits intermédiaires [6]
Finished goods[6]	116	119	127	...	...	...	...	Produits finis [6]
Producers' material[6]	95	101	110	...	...	...	...	Matériaux de production[6]
Consumers' goods[6]	114	121	130	...	...	...	...	Biens de consommation[6]
Capital goods[6]	124	125	139	...	...	...	...	Biens d'équipement[6]
Luxembourg								**Luxembourg**
Industrial products	137	140	151	148	144	140	138	Produits industriels
Import products	146	150	155	162	161	167	171	Produits importés
Exported goods	135	139	150	146	141	136	133	Produits exportés
Intermediate products	133	136	148	143	136	129	126	Produits intermédiaires
Consumers' goods	151	156	161	161	170	178	182	Biens de consommation
Capital goods	149	151	159	163	170	172	172	Biens d'équipement
Mexico								**Mexique**
Domestic supply[7 26]	4 407	8 784	9 905	12 161	14 478	16 220	17 284	Offre intérieure[7 26]
Agricultural products	3 290	9 120	12 429	17 805	21 865	26 798	28 241	Produits agricoles
Exported goods	3 721	5 938	7 212	9 293	9 210	9 881	9 675	Produits exportés
Raw materials	4 865	10 343	11 878	13 908	16 255	17 215	17 905	Matières premières
Consumers' goods[8 26]	4 738	9 637	10 917	13 528	16 628	18 868	20 241	Biens de consommation[8 26]
Capital goods[8 26 27]	4 085	8 714	9 354	10 914	13 058	14 381	15 610	Biens d'équipement[8 26 27]
Netherlands								**Pays-Bas**
Agricultural products[16 28]	105	106	115	109	110	106	...	Produits agricoles[16 28]
Industrial products	113	115	119	118	120	121	121	Produits industriels
Import products	95	95	103	101	99	95	92	Produits importés
Exported goods[21]	103	103	110	109	108	105	103	Produits exportés [21]
Raw materials	104	106	112	108	107	104	100	Matières premières
Intermediate products	109	112	116	114	115	116	114	Produits intermédiaires
Producers' material	99	99	107	104	102	99	96	Matériaux de production
Consumers' goods	116	117	121	122	125	129	130	Biens de consommation
Capital goods	120	122	126	128	130	131	131	Biens d'équipement
New Zealand								**Nouvelle-Zélande**
Domestic supply[2 29]	213	225	242	251	251	257	264	Offre intérieure[2 29]
Agricultural products[23]	161	164	197	199	176	194	203	Produits agricoles[23]
Industrial products[23]	186	193	206	215	217	222	229	Produits industriels[23]
Norway								**Norvège**
Domestic supply	152	160	169	174	180	180	180	Offre intérieure
Import products	130	133	141	143	140	138	138	Produits importés
Exported goods[30]	...	100	112	116	112	104	103	Produits exportés [30]
Raw materials	138	156	173	172	171	165	159	Matières premières
Finished goods	150	159	166	170	170	167	168	Produits finis
Consumers' goods	166	175	182	190	199	201	...	Biens de consommation

32
Producers prices and wholesale prices
Index numbers: 1980 = 100 [cont.]

Prix à la production et des prix de gros
Indices : 1980 = 100 [suite]

Country or area and groups	1987	1988	1989	1990	1991	1992	1993	Pays ou zone et groupes
Pakistan								**Pakistan**
Domestic supply[2 6 7]	158	166	180	194	219	235	257	Offre intérieure[2 6 7]
Agricultural products[6]	160	169	182	194	215	238	264	Produits agricoles [6]
Industrial products[6]	152	159	176	198	230	241	252	Produits industriels [6]
Raw materials[6]	156	163	177	185	207	224	248	Matières premières[6]
Panama								**Panama**
Domestic supply	126	118	121	132	132	137	132	Offre intérieure
Peru								**Pérou**
Domestic supply[3 5]	243	1 718	# 2	100	363	639	...	Offre intérieure[3 5]
Domestic production[3 5]	255	1 873	# 2	100	407	640	...	Production intérieure[3 5]
Agricultural products[3 5 31]	393	1 688	# 2	100	450	720	1 089	Produits agricoles[3 5 31]
Industrial products[3 8]	216	2 018	# 1	100	395	618	936	Produits industriels [3 8]
Import products[3 5]	194	1 353	# 1	100	393	626	961	Produits importés [3 5]
Capital goods[5]	...	...	11	100	427	1 036	1 036	Biens d'équipement[5]
Philippines [i]								**Philippines**
Domestic supply[2 6 7]	315	354	391	431	...	...	...	Offre intérieure[2 6 7]
Romania								**Roumanie**
Industrial products	.	124	124	153	470	1 316	3 211	Produits industriels
Singapore								**Singapour**
Domestic supply[7]	85	83	86	87	82	79	78	Offre intérieure[7]
Domestic production[2 3 32]	77	76	77	80	76	73	67	Production intérieure[2 3 32]
Agricultural products[2 32]	96	99	99	103	...	...	...	Produits agricoles [2 32]
Import products[17 32]	87	88	88	87	82	81	80	Produits importés [17 32]
Exported goods	80	79	79	79	75	68	78	Produits exportés
Slovakia								**Slovaquie**
Agricultural products[33]	..	77	100	100	105	112	128	Produits agricoles[33]
South Africa								**Afrique du Sud**
Domestic supply[34]	247	279	321	361	402	435	464	Offre intérieure[34]
Domestic production[34]	245	279	320	361	405	442	473	Production intérieure[34]
Agricultural products	232	259	271	282	310	366	386	Produits agricoles
Industrial products[3]	239	273	315	352	392	421	449	Produits industriels [3]
Import products	251	279	324	357	387	403	423	Produits importés
Spain								**Espagne**
Domestic supply[19]	183	190	198	203	205	208	213	Offre intérieure [19]
Producers' material	179	186	193	195	193	193	196	Matériaux de production
Consumers' goods	184	190	198	205	212	219	226	Biens de consommation
Capital goods	197	207	216	225	234	...	...	Biens d'équipement
Sri Lanka								**Sri Lanka**
Domestic supply	181	213	232	284	310	337	...	Offre intérieure
Domestic production	249	217	263	334	373	394	417	Production intérieure
Import products	153	170	193	244	259	267	283	Produits importés
Exported goods	191	245	229	259	279	329	...	Produits exportés
Producers' material	180	199	236	286	313	...	...	Matériaux de production
Consumers' goods	183	220	261	284	310	343	367	Biens de consommation
Capital goods	150	164	190	265	300	...	...	Biens d'équipement
Sweden								**Suède**
Domestic supply[2 19 35]	158	166	179	188	191	188	202	Offre intérieure[2 19 35]

32
Producers prices and wholesale prices
Index numbers: 1980 = 100 [cont.]
Prix à la production et des prix de gros
Indices : 1980 = 100 [suite]

Country or area and groups	1987	1988	1989	1990	1991	1992	1993	Pays ou zone et groupes
Domestic production[2 19 35]	165	175	190	197	201	199	210	Production intérieure[2 19 35]
Import products[19 35]	143	148	157	162	163	160	182	Produits importés[19 35]
Exported goods[19 35]	160	168	180	186	187	183	200	Produits exportés[19 35]
Producers' material	166	175	190	198	201	200	210	Matériaux de production
Switzerland								**Suisse**
Domestic supply[2 7]	108	111	116	117	118	118	118	Offre intérieure[2 7]
Domestic production[2 7]	114	116	120	123	125	125	126	Production intérieure[2 7]
Agricultural products[2]	114	117	116	118	119	115	112	Produits agricoles[2]
Import products[7]	95	98	105	104	102	100	100	Produits importés[7]
Raw materials	106	110	116	116	116	115	114	Matières premières
Consumers' goods	119	120	123	127	129	131	132	Biens de consommation
Syrian Arab Republic								**Rép. arabe syrienne**
Domestic supply	336	491	563	684	787	800	862	Offre intérieure
Raw materials	370	694	801	819	1 147	966	975	Matières premières
Intermediate products	262	321	516	776	861	884	745	Produits intermédiaires
Consumers' goods	298	327	557	702	716	730	793	Biens de consommation
Thailand								**Thaïlande**
Domestic supply[2 9]	115	125	131	136	144	145	144	Offre intérieure[2 9]
Agricultural products	110	122	129	131	147	150	142	Produits agricoles
Industrial products[3]	114	120	124	135	141	140	138	Produits industriels[3]
Exported goods[9]	90	100	101	100	102	104	102	Produits exportés[9]
Raw materials	103	121	127	122	134	134	123	Matières premières
Intermediate products	107	115	119	125	135	135	135	Produits intermédiaires
Finished goods	126	133	139	145	155	157	160	Produits finis
Consumers' goods	124	129	135	141	152	156	158	Biens de consommation
Trinidad and Tobago								**Trinité-et-Tobago**
Domestic supply[23]	122	129	141	143	...	...	...	Offre intérieure[23]
Tunisia								**Tunisie**
Agricultural products[23]	132	136	146	160	170	177	194	Produits agricoles[23]
Industrial products[23]	143	164	178	186	191	192	195	Produits industriels[23]
Turkey								**Turquie**
Domestic supply[2 19 23]	369	621	1 053	1 571	2 441	3 956	6 266	Offre intérieure[2 19 23]
Agricultural products[23]	351	530	961	1 480	2 233	3 634	5 894	Produits agricoles[23]
Industrial products[36]	100	180	1 093	1 586	2 464	3 934	6 159	Produits industriels[36]
United Kingdom								**Royaume-Uni**
Agricultural products[4]	123	125	133	135	139	141	149	Produits agricoles[4]
Industrial products[3 4]	148	155	163	173	181	186	190	Produits industriels[3 4]
Raw materials	131	136	143	143	142	139	146	Matières premières
Finished goods	151	158	165	176	185	191	198	Produits finis
United States								**Etats-Unis**
Domestic supply[2]	114	119	125	130	130	131	132	Offre intérieure[2]
Agricultural products[2]	93	102	108	109	103	101	104	Produits agricoles[2]
Industrial products[2 37]	117	121	127	132	132	133	135	Produits industriels[2 37]
Raw materials	98	101	108	114	106	105	108	Matières premières
Intermediate products	112	119	124	127	127	127	129	Produits intermédiaires
Finished goods	120	123	129	135	138	140	142	Produits finis

32
Producers prices and wholesale prices
Index numbers: 1980 = 100 [cont.]

Prix à la production et des prix de gros
Indices : 1980 = 100 [suite]

Country or area and groups	1987	1988	1989	1990	1991	1992	1993	Pays ou zone et groupes
Consumers' goods	117	120	127	133	136	137	139	Biens de consommation
Capital goods	130	133	138	143	148	150	153	Biens d'équipement
Uruguay								**Uruguay**
Domestic supply[2 27 38]	2 069	3 257	5 641	11 715	...	34 033	44 977	Offre intérieure[2 27 38]
Agricultural products[38]	2 017	3 003	5 227	8 670	13 008	31 487	33 034	Produits agricoles[38]
Industrial products[27 38]	2 081	3 333	5 764	12 102	23 572	36 897	50 351	Produits industriels[27 38]
Venezuela								**Venezuela**
Domestic supply[6 7]	303	362	714	909	1 111	1 373	1 859	Offre intérieure[6 7]
Domestic production[6 27]	291	351	706	920	1 136	1 430	1 942	Production intérieure[6 27]
Agricultural products[6 7]	374	463	654	1 028	1 388	1 730	2 171	Produits agricoles[6 7]
Industrial products[6 7]	292	347	703	882	1 069	1 320	1 799	Produits industriels[6 7]
Import products[6 7]	339	396	750	901	1 074	1 263	1 700	Produits importés[6 7]
Yugoslavia, SFR†								**Yougoslavie, Rfs†**
Domestic supply[10 30]	33	100	1 483	7 800	16 667	# 100	..	Offre intérieure[10 30]
Agricultural products[10 30]	36	100	1 286	7 143	...	# 100	..	Produits agricoles[10 30]
Producers' material[10 30]	32	100	1 318	6 923	15 385	# 100	..	Matériaux de production[10 30]
Consumers' goods[10 30]	33	100	1 500	8 167	16 667	# 100	..	Biens de consommation[10 30]
Capital goods[10 30]	50	100	1 517	8 450	16 667	# 100	..	Biens d'équipement[10 30]
Zambia								**Zambie**
Agricultural products	658	765	1 642	3 170	...	...	...	Produits agricoles
Industrial products	1 194	851	1 567	3 378	6 477	...	...	Produits industriels
Exported goods	631	1 138	1 802	4 046	7 683	...	...	Produits exportés
Producers' material	1 038	1 442	3 043	5 728	...	...	...	Matériaux de production
Consumers' goods	764	713	1 607	3 245	5 904	...	...	Biens de consommation
Capital goods	1 929	1 560	2 616	6 302	1 633	...	...	Biens d'équipement
Zimbabwe								**Zimbabwe**
Domestic production[30]	...	100	...	...	...	289	348	Production intérieure[30]

Source:
Price statistics database of the Statistical Division of the
United Nations Secretariat.

Source:
Base de données pour les statistiques des prix de la
Division de statistique du Secrétariat de l'ONU.

† For detailed descriptions of data pertaining to
former Czechoslovakia, Germany, SFR Yugoslavia and former
USSR, see Annex I - Country or area nomenclature, regional
and other groupings.

† Pour les descriptions en détails des données
relatives à l'ancienne Tchécoslovaquie, l'Allemagne, la Rfs
Yougoslavie et l'ancienne URSS, voir l'Annexe I -
Nomenclature des pays ou zones, groupements régionaux et
autres groupements.

1 Base: 1985 = 100.
2 Including exported products.
3 Manufacturing industry only.
4 Prices relate only to products for sale or transfer to other
   sectors or for use as capital equipment.
5 Base: 1990 = 100 (beginning 1989).
6 Prices are collected from wholesalers.
7 Exclusive of products of mining and quarrying.
8 Including imported products.
9 Agricultural products and products of manufacturing
   industry.
10 Base: 1992 = 100 (Brazil and Yugoslavia, SFR: beginning
   1992).

1 Base : 1985 = 100.
2 Y compris les produits exportés.
3 Industries manufacturières seulement.
4 Uniquement les prix des produits destinés à être vendus ou
   transférés à d'autres secteurs ou à être utilisés comme
   biens d'équipement.
5 Base : 1990 = 100 (à partir de 1989).
6 Prix recueillis auprès des grossistes.
7 Non compris les produits des industries extractives.
8 Y compris les produits importés.
9 Produits agricoles et produits des industries
   manufacturières.
10 Base : 1992 = 100 (Brésil et Yougoslavie, Rfs: à partir de
   1992).

32

Producers prices and wholesale prices
Index numbers: 1980 = 100 [cont.]

Prix à la production et des prix de gros
Indices : 1980 = 100 [suite]

11 Valued at purchasers' values.
12 Base: 1986 = 100.
13 San José.
14 San Salvador.
15 Base: 1991 = 100.
16 Excluding forestry, fishing and hunting.
17 Imports are valued c.i.f.
18 Finished products only.
19 Excluding electricity, gas and water.
20 Including mining and quarrying.
21 Exports are valued f.o.b.
22 Primary articles include food articles, non-food articles and minerals.
23 Base:1983 = 100.
24 Excluding Value Added Tax.
25 Including marine foods.
26 Mexico City.
27 Excluding mining and quarrying.
28 Crop growing production only, excluding live stock production.
29 Output price index.
30 Base: 1988 = 100.
31 Excluding fishing.
32 Not a sub-division of the domestic supply index.
33 Base: 1989 = 100.
34 Excluding gold mining.
35 Excluding agricultural products.
36 Base: 1987 = 100.
37 Excluding foods and feeds production.
38 Montevideo.

11 A la valeur d'acquisition.
12 Base : 1986 = 100.
13 San José.
14 San Salvador.
15 Base : 1991 = 100.
16 Non compris sylviculture, pêche et chasse.
17 Les importations sont évaluées à leur valeur c.a.f.
18 Produits finis uniquement.
19 Non compris l'électricité, le gaz et l'eau.
20 Y compris les industries extractives.
21 Les exportations sont évaluées f.o.b.
22 Les articles premiéres comprennent des articles des produits alimentaires, non alimentaires et des mineraux.
23 Base : 1983 = 100.
24 Non compris taxe sur la valeur ajoutée.
25 Y compris l'alimentation marine.
26 Mexico.
27 Non compris les industries extractives.
28 Cultures uniquement, non compris les produits de l'élevage.

29 Indice des prix de production.
30 Base : 1988 = 100.
31 Non compris la pêche.
32 Pas un élément de l'indice de l'offre intérieure.
33 Base : 1989 = 100.
34 Non compris l'extraction de l'or.
35 Non compris les produits agricoles.
36 Base : 1987 = 100.
37 Non compris les produits alimentaires et d'affouragement.
38 Montevideo.

33
Consumer price index numbers
Indices des prix à la consommation
All items and food; 1980 = 100
Ensemble et aliments; 1980 = 100

Country or area Pays ou zone	1984	1985	1986	1987	1988	1989	1990	1991	1992	1993
Afghanistan[1 2] **Afghanistan**[1 2]	202	219	215	254	324	555	818	1 177	...	...
Food Aliments[2]	214	237	233	289	354	596	1 063	...	...	...
Albania[4] **Albanie**[4]	...	...	...	...	...	98[3]	100[3]	204[3]	442	...
Food Aliments[4]	...	...	...	...	...	97[3]	100[3]	211[3]	467	...
Algeria[5 6] **Algérie**[5 6]	140	155	174	187	198	216	# 118[7]	148	195	236
Food[5 6] Aliments[5 6]	110	123	144	155	161	176	# 123[7]	147	185	230
American Samoa[1 6] **Samoa américaines**[1 6]	103	104	105	110	115	120	130	135	141	142
Food Aliments[6]	102	101	103	105	111	116	124	129	135	133
Angola[4 8] **Angola**[4 8]	...	...	...	...	...	...	...	184	733	10 841
Food Aliments[4 8]	...	...	...	...	...	...	...	194	794	13 307
Anguilla[9] **Anguilla**[9]	...	100	102	105	110	115	121	127	131	...
Food Aliments[9]	...	100	103	104	110	121	128	133	136	...
Antigua and Barbuda[10] **Antigua-et-Barbuda**[10]	171	173	173	180	191	199	213	225	...	...
Food[10] Aliments[10]	170	174	171	182	192	207	220	234	...	...
Argentina[11 12 13] **Argentine**[11 12 13]	1	5	10	23	100[14]	3 295	79 531	216 062	269 861	298 498
Food[11 12 13] Aliments[11 12 13]	1	5	10	23	100[14]	3 187	71 040	185 492	241 460	266 327
Aruba[15] **Aruba**[15]	100	104	105	109	112	117	123	130	135	142
Food[15] Aliments[15]	100	100	103	110	118	128	138	146	150	157
Australia **Australie**	140	149	162	176	189[14]	203	218	225	227[14]	231
Food Aliments	137	145	158	167	180[14]	196	204	211	214[14]	219
Austria **Autriche**	123	127	129	131[14]	133	137	141	146	152	157
Food Aliments	120	123	125	126[14]	127	129	133	138	144	148
Azerbaijan Azerbaïdjan Food Aliments	104	106	110	112	112	113	114	221	2 634	30 546
Bahamas **Bahamas**	128	134	141	149	155	164	171	184	194	199
Food Aliments	126	133	150	157	168	177	190	207	211	212
Bahrain **Bahreïn**	120	# 102[3 16]	100	98	98	100	101	102	102	106[3]
Food Aliments	104	# 100[3 16]	100	96	95	96	96	98	98	100[3]

33
Consumer price index numbers
All items and food; 1980 = 100 [cont.]
Indices des prix à la consommation
Ensemble et aliments; 1980 = 100 [suite]

Country or area Pays ou zone	1984	1985	1986	1987	1988	1989	1990	1991	1992	1993
Bangladesh[17 18 19] **Bangladesh**[17 18 19]	**158**	**175**	**194**	**213**	**233**	**256**	**277**	**296**	**309**	...
Food[17 18 19] Aliments[17 18 19]	155	170	194	214	227	248	262	278	288	...
Barbados **Barbade**	**139**	**145**	**147**	**152**	**159**	**169**	**174**	**185**	**196**	**198**
Food Aliments	132	138	142	148	158	172	179	188	188	188
Belarus **Bélarus**	**105**	**105**	**108**	**109**	**109**	**111**	**116**	**228**	**# 2 437**	**31 444**
Food Aliments	107	109	116	121	121	122	125	236	# 2 726	41 302
Belgium **Belgique**	**134**	**141**	**142**	**145**	**146**	**151**	**156**	**161**	**165**	**169**
Food Aliments	136	141	143	143	143	147	153	156	155	154
Belize **Belize**	**129**	**134**	**135**	**138**	**142**	**145**	**150**	**158**	**162**	**165**
Food Aliments	122	123	123	125	130	135	138	146	150	154
Benin[20 21] **Bénin**[20 21]	...	...	...	...	...	...	...	**100**	**104**	...
Food[6 21] Aliments[6 21]	102	92	86	86	85	82	82	81	# 106[20]	...
Bermuda **Bermudes**	**134**	**139**[14]	**145**	**151**	**159**	**168**	**178**	**185**	**190**	**195**
Food Aliments	128	129[14]	132	140	146	157	166	171	172	175
Bolivia[22 23] **Bolivie**[22 23]	**153**	**18 158**	**68 337**	**78 299**	**90 829**	**104 611**	**122 520**	**148 793**	...	...
Food[22 23] Aliments[22 23]	173	19 688	74 160	81 762	91 117	104 193	123 728	150 457	...	...
Botswana[24] **Botswana**[24]	**133**	**144**	**159**[14]	**174**	**189**	**211**	**234**	**262**	**304**[14]	**348**
Food[24] Aliments[24]	139	152	167[14]	184	200	219	244	273	324[14]	369
Brazil[22 25] **Brésil**[22 25]	**2**	**7**	**17**	**53**	**364**	**4 733**	**142 023**	**# 511**[26]	**5 439**	**109 888**
Food[22 25] Aliments[22 25]	3	8	20	58	429	5 997	163 608	# 495[26]	5 321	106 152
British Virgin Islands **Iles Vierges britanniques**	**127**	**128**	**131**	**132**	**138**	**145**	**152**	**162**	**167**	...
Food Aliments	123	127	129	128	135	141	146	156	159	...
Brunei Darussalam[17] **Brunéi Darussalam**[17]	**121**	**124**	**126**	**128**	**129**	**132**	**134**	**# 102**[26]	**103**	**107**
Food[17] Aliments[17]	128	129	130	131	133	134	134	# 103[26]	103	106
Bulgaria **Bulgarie**	**103**	**105**	**108**	**108**	**110**	**116**	**# 100**[26]	**439**	**787**	**1 450**
Food Aliments	105	108	112	112	114	116	# 100[26]	475	830	1 614
Burkina Faso[27 28] **Burkina Faso**[27 28]	**105**	**112**	**109**	**106**	**110**	**110**	**109**	**112**	**110**	**110**
Food[27 28] Aliments[27 28]	110	113	105	93	103	99	98	107	99	93

33

Consumer price index numbers
All items and food; 1980 = 100 [*cont.*]

Indices des prix à la consommation
Ensemble et aliments; 1980 = 100 [*suite*]

Country or area Pays ou zone	1984	1985	1986	1987	1988	1989	1990	1991	1992	1993
Burundi[29]										
Burundi[29]	**147**	**152**	**155**	**166**	**174**	**194**	**207**	**226**	**236**	**# 102**[30]
Food[29]										
Aliments[29]	145	157	145	144	155	184	198	211	209	# 98[30]
Cameroon[31]										
Cameroun[31]	**162**	**181**	**195**	**220**	**224**	**219**	**223**	**227**	**230**	...
Food[31]										
Aliments[31]	162	160	162	172	173	160[3]	144[3]	143	140	...
Canada										
Canada	**138**	**143**	**149**	**155**	**162**	**170**	**178**	**188**	**191**	**194**
Food										
Aliments	131	135	141	148	151	157	163	171	171	173
Cape Verde[27 32]										
Cap-Vert[27 32]	**112**[1]	**118**[1]	**130**[1]	**136**[1]	**140**[1]	**148**[1]	**164**[1]	**179**[1 33]	**# 125**[7 33]	**138**
Food[27 32]										
Aliments[27 32]	112	111	125	131	135	147	166	185	# 134[7]	146
Cayman Islands										
Iles Caïmanes	**130**	**133**	**137**	**142**	**150**	**159**	**171**	**185**	**191**	...
Food[9]										
Aliments[9]	...	100	103	104	106	111	121	125	124	...
Central African Rep.[1 24 34]										
Rép. centrafricaine[1 24 34]	**133**	**147**	**150**	**140**	**134**	**135**	**135**	**131**	**130**	**126**
Food[24 34]										
Aliments[24 34]	133	147	147	133	127	128	129	124	123	118
Chad[27 35]										
Tchad[27 35]	**125**	**129**	**107**	**101**[3]	**# 100**[13]	**95**	**96**	...	...	...
Food[27 35]										
Aliments[27 35]	135	136	100	86[3]	# 100[13]	92	91	...	...	...
Chile[36]										
Chili[36]	**201**	**262**	**313**	**376**	**431**	**504**	**636**	**774**	**893**	**1 007**
Food[36]										
Aliments[36]	180	231	287	356	403	486	613	771	909	1 008
China										
Chine	**110**	**123**	**131**	**143**	**172**	**200**	**203**	**213**	**232**	...
Food										
Aliments	112	126	136	148	178	209	211	216	...	...
Colombia[37]										
Colombie[37]	**222**	**277**	**328**	**405**	**521**	**656**[14]	**850**	**1 109**	**1 418**	**1 725**
Food[37]										
Aliments[37]	225	296	353	447	593	736[14]	936	1 216	1 577	1 812
Congo[1 38]										
Congo[38]	**162**	**169**	**169**	**171**	**168**	**168**	**173**	**170**	**163**	...
Food[38]										
Aliments[38]	164	173	169	170	160	160	162	156	143	...
Cook Islands[39]										
Iles Cook[39]	**164**	**184**	**202**	**224**	**243**	**# 100**[7]	**105**	**111**	**115**	**124**
Food[39]										
Aliments[39]	162	179	198	219	237	# 100[7]	104	106	110	117
Costa Rica[11 40]										
Costa Rica[11 40]	**387**	**443**	**498**	**582**	**703**	**819**	**975**	**1 255**	**1 528**	**1 677**
Food[11 40]										
Aliments[11 40]	420	472	527	605	727	858	1 014	1 277	1 583	1 758
Côte d'Ivoire[41]										
Côte d'Ivoire[41]	**129**	**131**	**144**[14 37]	**154**	**165**	**167**[3]	**165**[3]	**168**	...	**# 100**[42]
Food[41]										
Aliments[41]	120	122	140[14 37]	156	175	177[3]	153[3]	155	...	# 100[42]

33
Consumer price index numbers
All items and food; 1980 = 100 [cont.]
Indices des prix à la consommation
Ensemble et aliments; 1980 = 100 [suite]

Country or area Pays ou zone	1984	1985	1986	1987	1988	1989	1990	1991	1992	1993
Croatia[22] Croatie[22]	4	7	13	30	88	1 139	7 905	# 224[26]	1 646	26 105
Food[22] Aliments[22]	4	7	14	30	92	1 178	7 602	# 223[26]	1 832	26 709
Cyprus Chypre	131	138	140	144[26]	148	154	161	169	180	189
Food Aliments	136	143	147	151[14]	158	165	173	184	199	203
former Czechoslovakia† anc. Tchécoslovaquie†	108	110	111	111[14]	111	113	124	196	217	..
Food Aliments	111	114	114	114[14]	114	114	127	184	200	..
Czech Republic[30] République tchèque[30]	..	..	..	..	..	..	..	100	111	134
Food[30] Aliments[30]	..	..	..	..	..	..	..	100	109	127
Denmark Danemark	140	146	152	158	165	173	177	182	186	188
Food Aliments	142	148	151	152	158	164	165	166	169	168
Dominica Dominique	126	128	132	139	142	152	157	166	175	178
Food Aliments	119	120	122	133	137	148	145	153	166	169
Dominican Republic[43] Rép. dominicaine[43]	154	212	233	269	389	566	...	...	...	...
Food Aliments	141	202	226	273	418	620	...	...	...	...
Ecuador[24] Equateur[24]	227	290	357	462	731	1 284	1 906	2 835	4 383	6 354
Food[24] Aliments[24]	286	373	460	592	973	1 828	2 697	4 007	6 172	8 754
Egypt Egypte	172	195	240	286	337	409	477	572	649	728
Food Aliments	181	205	255	313	378	473	548	639	693	745
El Salvador[44] El Salvador[44]	162	198	262	327	391	460	571	653	726	860[14]
Food[44] Aliments[44]	169	200	264	331	428	544	684	807	910	1 144[14]
Estonia[26] Estonie[26]	..	..	..	..	..	..	100	302	3 553	6 744
Food[26] Aliments[26]	..	..	..	..	..	..	100	333	3 453	6 004
Ethiopia[45 46] Ethiopie[45 46]	121	144	130	127	136	146	154	209	231	...
Food[46] Aliments[46]	124	155	132	124	133	142	150	211	236	...
Faeroe Islands Iles Féroé	149	156	156	156	161	171	180	187	190	...
Food Aliments	159	162	169	176	181	195	206	217	...	...
Falkland Is. (Malvinas)[1 47] Iles Falkland (Malvinas)[1 47]	138	151	161	165	177	...	# 100[26 33]	105	112	113
Food[47] Aliments[47]	127	135	138	140	158	161	# 100[26]	104	...	...

33
Consumer price index numbers
All items and food; 1980 = 100 [cont.]
Indices des prix à la consommation
Ensemble et aliments; 1980 = 100 [suite]

Country or area Pays ou zone	1984	1985	1986	1987	1988	1989	1990	1991	1992	1993
Fiji										
Fidji	**134**	**140**	**# 102**[9]	**108**	**120**	**128**	**138**	**147**	**154**	**162**
Food										
Aliments	137	148	# 98[9]	104	123	136	147	149	149	159
Finland										
Finlande	**142**	**151**	**156**	**162**	**170**[14]	**181**	**192**	**200**	**205**	**209**[14]
Food										
Aliments	146	157	163	166	170[14]	176	183	187	188	189[14]
France										
France	**149**	**158**	**162**	**167**	**172**	**178**	**184**	**190**	**194**	**199**[14]
Food										
Aliments	151	159	164	167	170	177	184	190	191	191[14]
French Guiana[48]										
Guyane française[48]	**159**	**170**	**174**	**181**	**185**	**193**	**200**	**204**	**210**	...
Food[48]										
Aliments[48]	157	168	170	178	180	188	195	197	200	...
French Polynesia										
Polynésie française[24]	**166**	**184**	**186**	**188**	**192**	**197**[14]	**200**	**202**	**205**	**208**
Food[24]										
Aliments[24]	150	163	158	154	157	164[14]	167	166	167	170
Gabon[49]										
Gabon[49]	**148**	**159**	**169**	**168**	**153**	**164**	**176**[3]	**176**	**159**	**160**
Food[49]										
Aliments[49]	...	157	166	165	143	156	174[3]	172	144	146
Gambia[50]										
Gambie[50]	**159**	**188**	**294**	**364**	**406**	**440**	**493**	**585**	**587**	**625**
Food[50]										
Aliments[50]	158	186	297	367	413	442	504	546	592	644
Germany †[30 51]										
Allemagne†[30 51]	..	..	..	..	..	..	**87**[3]	**100**	**111**	**121**
Food[30 51]										
Aliments[30 51]	..	..	..	..	..	..	93[3]	100	103	104
F. R. Germany										
R. f. Allemagne	**118**	**121**	**121**	**121**	**123**	**126**	**129**	**134**	**139**	**145**
Food										
Aliments	116	117	118	117	117	120	124	127	131	134
former German D. R.										
anc. R. d. allemande	**100**	**100**	**100**	**100**	**100**	**100**	..	..	..	..
Food										
Aliments	100	100	100	100	100	100	..	..	..	..
Ghana										
Ghana	**824**	**909**	**1 132**	**1 583**	**2 080**	**2 605**	**3 575**	**4 219**	**4 644**	**5 827**
Food										
Aliments	779	692	833	1 153	1 547	1 935	2 711	2 955	3 261	4 075
Gibraltar[24]										
Gibraltar[24]	**123**	**130**	**135**	**141**	**146**	**153**	**162**	**174**	**186**	**195**
Food[24]										
Aliments[24]	119	124	128	134	139	146	155	166	174	177
Greece										
Grèce	**215**	**256**	**315**	**367**	**417**	**474**	**570**	**682**	**790**	**903**
Food										
Aliments	220	263	316	356	395	471	570	681	776	875
Greenland[24]										
Groënland[24]	**137**	**147**	**155**	**160**	**172**	**181**	**190**	**198**	**201**	**203**
Food[24]										
Aliments[24]	140	150	150	154	159	169	180	190	192	193

33
Consumer price index numbers
All items and food; 1980 = 100 [cont.]
Indices des prix à la consommation
Ensemble et aliments; 1980 = 100 [suite]

Country or area Pays ou zone	1984	1985	1986	1987	1988	1989	1990	1991	1992	1993
Grenada **Grenade**	**144**	**147**	**148**	**147**	**153**	**161**	**165**	**170**	**176**	**181**
Food Aliments	141	142	145	145	152	165	171	174	175	182
Guadeloupe[44] **Guadeloupe**[44]	**150**	**161**	**165**	**170**	**174**	**178**	**183**	**188**	**194**	**198**[14]
Food[44] Aliments[44]	150	160	165	170	172	174	179	182	185	190[14]
Guam **Guam**	**143**	**148**	**152**	**159**	**167**	**185**	**212**	**233**	**257**	**278**
Food Aliments	157	165	177	196	217	239	283	334	384	453
Guatemala[15 44] **Guatemala**[15 44]	**100**	**119**	**163**	**183**	**202**	**225**	**318**	**424**	**453**	**514**
Food[15 44] Aliments[15 44]	100	121	168	194	221	248	369	483	507	578
Guinea[52 53] **Guinée**[52 53]	...	...	...	**100**	**127**	**163**	**195**	**232**	**272**	**291**
Food[52 53] Aliments[52 53]	...	...	...	100	134	154	199	231	268	293
Guinea-Bissau **Guinée-Bissau** Food[52 54]										
Aliments[52 54]	...	...	52[3]	100	160	290	386	608	...	...
Guyana[44] **Guyana**[44]	**209**[3]	**245**	**264**	**339**	**475**	**795**[3]	...	...	...	...
Food[44] Aliments[44]	259[3]	320	348	452	664	1 132[3]	...	...	...	...
Haiti[11] **Haïti**[11]	**140**	**155**	**160**	**141**	**147**	**158**	**191**	**220**	**251**	...
Food[11] Aliments[11]	135	150	155	126	134	144	177	203	224	...
Honduras **Honduras**	**137**	**139**	**144**	**148**	**156**	**172**	**212**	**283**	**308**	...
Food Aliments	121	123	126	127	138	155	196	282	299	...
Hong Kong **Hong-kong**	**150**	**155**	**159**[14]	**168**	**180**	**199**	**218**	**243**[14]	**266**	**289**
Food Aliments	151	151	153[14]	160	176	198	217	242[14]	263	282
Hungary **Hongrie**	**130**	**139**	**147**	**159**	**184**	**215**	**277**	**374**	**460**	**564**
Food Aliments	128	136	138	151	175	206	279	340	405	524
Iceland[55] **Islande**[55]	**540**	**715**	**867**	**1 030**	**1 291**[14]	**1 564**	**1 796**	**1 919**	**1 990**	**2 069**
Food[55] Aliments[55]	577	794	975	1 138	1 524[14]	1 816	2 047	2 099	2 129	2 178
India[56] **Inde**[56]	**148**	**156**	**170**	**184**	**199**[3]	**216**[14]	**235**	**268**	**300**	**319**
Food[56] Aliments[56]	149	154	169	185	199[3]	215[14]	234	272	308	325
Indonesia **Indonésie**	**152**	**159**	**168**	**184**	**199**	**211**	**# 108**[7]	**118**	**127**	**139**
Food Aliments	146	149	162	180	203	220	# 107[7]	116	125	134

33
Consumer price index numbers
All items and food; 1980 = 100 [*cont.*]
Indices des prix à la consommation
Ensemble et aliments; 1980 = 100 [*suite*]

Country or area Pays ou zone	1984	1985	1986	1987	1988	1989	1990	1991	1992	1993
Iran, Islamic Rep. of[16] **Iran, Rép. islamique d'**[16]	...	**84**	**100**	**129**	**165**	**202**	**218**	**255**	**314**	**376**
Food[16] Aliments[16]	...	79	100	124	141	169	174	213	285	348
Iraq[57] **Iraq**[57]	**191**	**199**	**202**	**230**	**279**	**296**	**# 161**[13]	**462**[3]	...	...
Food[57] Aliments[57]	197	202	203	250	309	331	# 176[13]	640[3]	...	...
Ireland **Irlande**	**169**	**178**	**185**	**191**	**195**	**203**	**210**[14]	**217**	**223**	**227**
Food Aliments	154	160	167	172	177	185	462[14]	468	476	478
Isle of Man **Ile de Man**	**137**	**145**	**149**	**155**	**163**	**174**	**189**	**203**	**212**	**217**
Food Aliments	134	139	145	154	164	177	194	210	219	229
Israel **Israël**	**5 560**	**22 500**	**33 323**[14]	**39 938**	**46 447**[14]	**55 833**	**65 418**	**77 838**	**87 144**	**96 689**
Food Aliments	5 196	21 640	33 499[14]	38 909	46 029[14]	55 484	60 231	68 518	77 389	81 319
Italy **Italie**	**174**	**190**	**202**[14]	**211**	**222**	**236**	**251**	**267**[14]	**280**[58]	**293**
Food Aliments	166	180	190[14]	198	206	219	233	248[14]	260	266
Jamaica **Jamaïque**	**170**	**214**	**246**	**262**	**284**	**# 115**[13]	**139**	**211**	**374**	...
Food Aliments	168	211	249	267	292	# 120[13]	147	227	403	...
Japan **Japon**	**112**	**114**	**115**	**115**	**116**	**119**	**122**	**126**	**129**	**130**
Food Aliments	113	115	115	114	115	117	122	128	128	130
Jordan **Jordanie**	**126**	**130**	**130**[14]	**130**	**138**	**174**	**202**	**219**	**227**	**238**
Food Aliments	118	121	123[14]	121	128	154	186	206	213	213
Kazakhstan[59] **Kazakhstan**[59]	..	..	..	..	..	..	..	3	100	2 265
Food[59] Aliments[59]	..	..	..	..	..	..	..	5	100	2 297
Kenya[37 60] **Kenya**[37 60]	**166**	**187**	**194**	**204**	**221**	**243**	**# 100**[26]	**119**	**155**	**225**
Food[37 60] Aliments[37 60]	164	195	202	209	225	246	# 100[26]	124	167	242
Kiribati[61] **Kiribati**[61]	**125**	**131**	**140**	**149**	**153**	**158**	**166**	**175**	**181**	**192**
Food[61] Aliments[61]	120	125	134	142	148	152	157	164	168	178
Korea, Republic of **Corée, République de**	**138**	**141**	**145**	**149**	**160**	**169**	**184**	**201**	**213**	**223**
Food Aliments	134	139	143	147	162	174	191	215	228	236
Kuwait **Koweït**	**123**	**124**	**126**	**126**	**128**	**133**	**146**[3]	**159**[3]	...	...
Food Aliments	112	112	112	110	111	115	129[3]	142[3]	...	...

33
Consumer price index numbers
All items and food; 1980 = 100 [cont.]
Indices des prix à la consommation
Ensemble et aliments; 1980 = 100 [suite]

Country or area Pays ou zone	1984	1985	1986	1987	1988	1989	1990	1991	1992	1993
Latvia[4] **Lettonie**[4]	..	..	..	..	..	..	..	**200**	**2 099**	**4 390**
Food[4] Aliments[4]	..	..	..	..	..	..	..	184	1 587	3 013
Lesotho **Lesotho**	**165**	**186**	**220**	**246**	**274**	**315**	**351**	**414**	**484**	...
Food Aliments	168	190	223	252	276	307	342	405	500	...
Liberia[62] **Libéria**[62]	**120**	**119**	**123**	**129**	**142**	**150**	**162**[3]	...	...	...
Food[62] Aliments[62]	113	109	108	108	129	141[3]	161[3]	...	...	...
Lithuania **Lituanie**	**105**	**106**	**109**	**111**	**111**	**113**	**# 100**[26]	**316**	**3 546**	**18 093**
Food Aliments	107	110	118	125	124	124	# 100[26]	318	3 295	16 528
Luxembourg[1] **Luxembourg**[1]	**136**	**140**[14]	**140**	**140**[33]	**142**	**147**	**152**	**# 103**[26,58]	**106**	**110**
Food Aliments	138	# 104[15]	107	105	106	110	114	# 103[26]	104	104
Macau[1,15] **Macao**[1,15]	**100**	**102**	**104**	**109**	**117**	**128**	**138**	**151**	**163**	**174**
Food[15] Aliments[15]	100	100	102	106	115	127	138	150	163	174
Madagascar[63] **Madagascar**[63]	**226**	**249**	**286**	**330**	**417**	**454**	**508**	**551**	**631**	**694**
Food[63] Aliments[63]	223	251	297	321	385	420	479	523	610	661
Malawi[1,37,64] **Malawi**[1,37,64]	**167**	**185**	**211**	**264**	**354**	**398**	**444**	**492**	**616**	...
Food[37,64] Aliments[37,64]	153	167	193	245	324	375	423	478	613	...
Malaysia **Malaisie**	**125**	**125**	**126**	**127**	**130**	**134**	**138**[14]	**144**	**151**	**156**
Food Aliments	125	122	122	122	126	131	137[14]	143	153	156
Mali[13] **Mali**[13]	...	...	...	...	**100**	**100**	**101**	**102**	**96**	**95**
Food[13] Aliments[13]	...	...	...	...	100	97	98	102	92	91
Malta[27] **Malte**[27]	**100**	**99**	**101**	**102**	**103**	**104**	**107**	**109**	**# 102**[30]	**106**
Food[27] Aliments[27]	99	98	100	101	101	100	105	106	# 100[30]	...
Martinique **Martinique**	**155**	**167**	**173**	**179**	**183**	**189**	**196**	**202**	**210**	**217**[14]
Food Aliments	153	164	169	177	178	181	188	194	200	205[14]
Mauritius[6] **Maurice**[6]	**114**	**121**	**123**	**124**	**# 109**[52]	**123**	**140**	**149**	**156**[14]	**173**
Food[6] Aliments[6]	117	125	128	131	# 112[52]	127	141	147	# 102[65]	117
Mexico **Mexique**	**679**	**1 072**	**1 995**	**4 626**	**9 907**	**11 889**	**15 058**	**18 470**	**21 335**	**23 415**
Food Aliments	648	1 035	1 921	4 441	9 305	11 192	14 034	16 855	18 748	19 983

33
Consumer price index numbers
All items and food; 1980 = 100 [*cont.*]
Indices des prix à la consommation
Ensemble et aliments; 1980 = 100 [*suite*]

Country or area Pays ou zone	1984	1985	1986	1987	1988	1989	1990	1991	1992	1993
Morocco **Maroc**	**149**	**160**	**174**	**179**	**183**	**189**	**201**	**218**	**229**	**# 129**[7]
Food Aliments	154	166	181	182	183	187	200	217	228	# 133[7]
Myanmar[66] **Myanmar**[66]	**116**	**124**	**135**	**167**	**196**	**249**	**292**	**387**	**472**	**611**
Food[66] Aliments[66]	110	121	128	162	196	252	300	413	505	702
Namibia[26 67] **Namibie**[26 67]	**48**	**55**	**62**	**70**	**78**	**90**	**100**	**112**	**133**	**144**
Food[26 67] Aliments[26 67]	45	47	54	63	72	86	100	105	126	134
Nepal **Népal**	**144**	**155**	**185**	**204**	**223**	**242**[14]	**264**	**306**	**358**	**...**
Food Aliments	143	152	187	210	...	250[14]	268	317	374	...
Netherlands **Pays-Bas**	**120**	**123**	**123**	**122**	**123**	**125**	**128**	**133**	**137**	**140**
Food Aliments	117	118	117	115	115	116	119	123	127	128
Netherlands Antilles[68] **Antilles néerlandaises**[68]	**125**	**126**	**127**	**132**	**135**	**141**	**146**	**152**[14]	**154**	**157**
Food[68] Aliments[68]	125	123	125	132	142	151	163	175[14]	180	185
New Caledonia[69] **Nouvelle-Calédonie**[69]	**160**	**169**	**172**	**173**	**178**	**185**	**189**	**196**	**202**	**207**[14]
Food[69] Aliments[69]	167	176	174	173	178	186	187	193	198	206[14]
New Zealand **Nouvelle-Zélande**	**153**	**176**	**200**	**231**	**246**	**260**	**276**	**283**	**286**	**289**
Food Aliments	145	166	185	209	222	242	259	261	262	265
Nicaragua **Nicaragua**	274	877	6 853	69 256	# 1[13 22]	49[22]	3 694[22]	104 998[22]	139 174[22]	...
Food Aliments	334	1 156	11 413	124 481	# 1[13 22]	41[22]	2 994[22]	85 364[22]	104 875[22]	...
Niger[1 70] **Niger**[1 70]	**145**	**144**	**139**	**130**	**128**	**124**	**124**	**# 96**[7]	**95**	**...**
Food[70] Aliments[70]	142	139	131	116	111	105	105	# 94[7]	92	...
Nigeria[71] **Nigéria**[71]	**224**	**236**	**# 100**[16]	**115**	**162**	**244**	**262**	**297**	**429**	**674**
Food[71] Aliments[71]	242	251	# 100[16]	118	182	277	287	322	471	744
Niue **Nioué**	**163**	**181**	**194**	**208**	**219**	**240**	**255**	**268**	**281**	**285**
Food Aliments	158	176	191	202	210	222	235	244	251	258
Northern Mariana Islands **Iles Marianas du Nord**	**155**	**159**	**167**	**...**	**188**	**181**	**156**	**167**	**182**	**190**
Food Aliments	133	139	143	...	154	146	148	158	166	172
Norway **Norvège**	**146**	**154**	**165**	**180**	**192**	**200**	**209**	**216**	**221**	**226**
Food Aliments	154	164	179	193	205	211	218	221	225	222

33
Consumer price index numbers
All items and food; 1980 = 100 [*cont.*]
Indices des prix à la consommation
Ensemble et aliments; 1980 = 100 [*suite*]

Country or area Pays ou zone	1984	1985	1986	1987	1988	1989	1990	1991	1992	1993
Oman[26 72] **Oman**[26 72]	...	...	...	...	...	...	**100**	**105**	**106**	**107**
Food[26 72] Aliments[26 72]	...	...	...	...	...	...	100	103	102	101
Pakistan[6] **Pakistan**[6]	**113**	**119**	**123**	**129**	**141**	**152**	**165**	**185**	**202**	**221**
Food[6] Aliments[6]	113	118	121	129	144	156	170	188	208	228
Panama[73] **Panama**[73]	**116**	**117**	**117**	**118**	**119**[3]	**119**	**120**	**122**	**124**	**124**
Food[73] Aliments[73]	120	120	121	124	125[3]	125	126	129	134	134
Papua New Guinea **Papouasie-Nvl-Guinée**	**132**	**137**	**145**	**149**	**158**	**165**	**176**	**188**	**196**	**206**
Food Aliments	128	133	137	140	145	151	165	178	184	188
Paraguay[74] **Paraguay**[74]	**166**	**208**	**274**	**334**	**411**	**517**	**715**	**888**	**1 023**	**1 209**
Food[74] Aliments[74]	173	220	316	391	484	583	853	1 026	1 178	1 381
Peru[4 11 22 75] **Pérou**[4 11 22 75]	**13**	**34**	**60**	**112**	**855**	**# 1**[3]	**100**	**510**	**884**	**1 314**
Food[4 11 22 75] Aliments[4 11 22 75]	13	31	58	96	611	# 1[3]	100	448	770	...
Philippines **Philippines**	**198**	**245**	**243**	**251**	**273**	**307**	**350**	**415**	**453**	**487**
Food Aliments	199	245	241	250	276	314	352	406	433	460
Poland **Pologne**	**342**	**393**	**463**	**580**	**929**	**3 261**	**22 362**	**38 082**	**54 458**	**73 681**
Food Aliments	401	448	517	638	986	3 946	24 985	37 727	51 611	68 384
Portugal[1] **Portugal**[1]	**238**	**284**	**318**	**347**	**381**[14]	**429**	**486**	**541**	**# 109**[30]	**116**
Food Aliments	243	286	312	339	371[14]	425	483	530	# 107[30]	110
Puerto Rico[15] **Porto Rico**[15]	**117**	**117**	**117**	**120**	**124**	**128**	**136**	**140**	**143**	**147**
Food[15] Aliments[15]	116	117	118	121	125	131	143	149	157	167
Qatar[24 76] **Qatar**[24 76]	**110**	**112**	**114**	**118**	**124**	...	...	...	...	...
Food[24 76] Aliments[24 76]	114	116	117	118	121	...	...	...	...	...
Republic of Moldova[30] **République de Moldova**[30]	..	..	..	..	..	..	..	**100**	**1 209**	**23 014**
Food[30] Aliments[30]	..	..	..	..	..	..	..	100	1 188	20 382
Réunion[44] **Réunion**[44]	**148**	**158**	**162**	**167**	**169**	**176**	**183**	**190**	**196**	**203**[14]
Food Aliments	143	152	156	158	154	160	163	170	174	180[14]
Romania[26] **Roumanie**[26]	**90**	**90**	**91**	**92**	**94**	**95**	**100**	**275**	**# 310**[30]	**1 105**
Food[26] Aliments[26]	89	90	91	92	94	96	100	299	# 337[30]	1 174

33
Consumer price index numbers
All items and food; 1980 = 100 [cont.]
Indices des prix à la consommation
Ensemble et aliments; 1980 = 100 [suite]

Country or area Pays ou zone	1984	1985	1986	1987	1988	1989	1990	1991	1992	1993
Russian Federation[30] Fédération de Russie[30]	..	..	..	..	..	..	..	100	1 629	15 869
Food[30] Aliments[30]	..	..	..	..	..	..	..	100	1 690	16 750
Rwanda[24 77] Rwanda[24 77]	127	129	127	133	137	138	144	172	188	211
Food[24 77] Aliments[24 77]	160	161	146	158	171	176	185	210	225	...
Saint Helena[24] Sainte-Hélène[24]	136	146	140	145	150	# 105[13]	108	112	118	128
Food[24] Aliments[24]	139	146	144	151	156	# 105[13]	109	111	116	123
Saint Kitts and Nevis[78] Saint-Kitts-et-Nevis[78]	123	126	126	127	128	134	140	146	150	...
Food[78] Aliments[78]	121	121	121	122	122	128	131	138	145	...
Saint Lucia[9] Sainte-Lucie[9]	100[3]	100	102	109	111	115	120	127	134	135
Food Aliments	101[3]	100	103	115	114	118	124	134	141	140
Saint Pierre and Miquelon[24] Saint-Pierre-et-Miquelon[24]	155	168	164	166	172	183	191	...	...	...
Food[24] Aliments[24]	159	172	167	168	175	191	194	...	...	...
St. Vincent-Grenadines[16 79] St. Vincent-Grenadines[16 79]	97	99	100	103	103	106	114	120	125	130
Food[16 79] Aliments[16 79]	99	99	100	104	104	104	112	123	127	130
Samoa[1] Samoa[1]	186	203	213	224	243	259	299	293	318	323
Food Aliments	182	200	214	219	239	247	296	270	300	296
San Marino Saint-Marin	175	190	202[14]	212	222	236	252	270	289	304
Food Aliments	167	184	196[14]	203	211	224	237	250	264	279
Saudi Arabia[24 80] Arabie saoudite[24 80]	100	97	94	92	93	94	96[14]	100	100	101
Food[24 80] Aliments[24 80]	101	99	97	96	96	98	100[14]	107	111	113
Senegal[81] Sénégal[81]	155	175	186	178	175	176	177	173	173	172
Food[81] Aliments[81]	151	165	175	162	161	162	163	158	156	154
Seychelles Seychelles	90	91	91	93	95	96	100	102	105	107
Food Aliments	88	89	89	92	94	95	100	103	104	107
Sierra Leone[82] Sierra Leone[82]	439	776	1 404	3 952	5 189	8 446	17 817	32 578	46 049	...
Food[82] Aliments[82]	409	736	1 307	3 625	4 859	8 069	17 231	31 966	45 405	...
Singapore Singapour	117	117	116	116	118	121	125	129	132	135
Food Aliments	117	116	114	114	115	117	118	120	121	123

33
Consumer price index numbers
All items and food; 1980 = 100 [cont.]
Indices des prix à la consommation
Ensemble et aliments; 1980 = 100 [suite]

Country or area Pays ou zone	1984	1985	1986	1987	1988	1989	1990	1991	1992	1993
Slovakia **Slovaquie**	..	**110**	**111**	**111**	**111**	**113**	**125**	**201**	**221**	**# 272**[30]
Food Aliments	..	114	114	114	114	114	127	187	200	# 130[30]
Slovenia[22] **Slovénie**[22]	**40**	**72**	**141**	**326**	**977**	**13 534**	**88 190**	**189 609**	**582 667**	**774 365**
Food[22] Aliments[22]	45	80	158	343	1 068	14 987	92 454	197 204	602 851	757 784
Solomon Islands[83] **Iles Salomon**[83]	**156**	**171**	**194**	**216**	**252**	**289**	**318**	**362**	**401**	**436**[14]
Food[83] Aliments[83]	165	176	196	212	255	300	326	380	424	461[14]
South Africa **Afrique du Sud**	**166**	**193**	**228**	**265**	**299**	**343**	**392**	**452**	**515**	**565**
Food Aliments	169	188	227	278	322	358	415	496	622	664
Spain **Espagne**	**163**	**178**	**194**	**204**	**214**	**228**	**243**	**258**	**273**	**285**
Food Aliments	163	178	197	207	215	231	247	255	265	267
Sri Lanka[84] **Sri Lanka**[84]	**174**	**176**	**190**	**205**	**234**	**261**	**317**	**356**	**396**	**443**
Food[84] Aliments[84]	176	176	189	205	236	260	321	359	402	447
Sudan[37] **Soudan**[37]	**274**	**398**	**516**	**650**	**965**	**1 604**	**# 385**[85]	**790**	...	...
Food[37] Aliments[37]	287	421	554	679	1 107	1 602	# 392[85]	804	...	...
Suriname[86] **Suriname**[86]	**126**	**140**	**166**	**255**	**274**	**276**	**336**	**423**	**607**	**1 479**
Food[86] Aliments[86]	118	130	163	293	307	311	404	480	743	1 961
Swaziland[26][37] **Swaziland**[26][37]	...	...	...	...	...	**89**	**100**	**102**	**122**	**137**
Food[26][37] Aliments[26][37]	...	...	...	...	...	85	100	114	128	145
Sweden **Suède**	**143**	**154**	**160**	**167**	**177**	**188**	**208**	**227**	**232**	**243**
Food Aliments	161	173	185	191	202	213	229	239	227	228
Switzerland **Suisse**	**119**	**123**	**124**	**126**	**128**	**133**	**140**	**148**	**154**	**159**[14]
Food Aliments	125	129	131	132	135	138	145	152	153	154[14]
Syrian Arab Republic[87] **Rép. arabe syrienne**[87]	**157**	**184**	**251**	**400**	**538**	**599**	**715**	**770**	**844**	**943**
Food[87] Aliments[87]	152	181	256	419	583	633	796	824	852	963
Thailand **Thaïlande**	**124**	**127**	**130**	**133**	**138**	**145**	**154**	**163**	**169**	**175**
Food Aliments	118	115	116	119	125	135	146	156	163	166
TFYR Macedonia[22][24] **L'ex-R.y. Macédoine**[22][24]	**3**	**5**	**9**	**20**[14]	**58**	**794**	**5 532**	**11 662**	**187 903**	**868 110**[14]
Food[22][24] Aliments[22][24]	3	5	10	21[14]	61	824	5 277	10 792	182 704	814 494[14]

33
Consumer price index numbers
All items and food; 1980 = 100 [cont.]
Indices des prix à la consommation
Ensemble et aliments; 1980 = 100 [suite]

Country or area Pays ou zone	1984	1985	1986	1987	1988	1989	1990	1991	1992	1993
Togo[88]										
Togo[88]	**140**	**138**	**144**	**144**	**144**	**142**	**144**	**144**	...	...
Food[88]										
Aliments[88]	143	131	136	135	155	127	130	124	...	...
Tonga[1 15]										
Tonga[1 15]	**100**	**117**	**142**	**149**	**164**	**170**	**189**	**207**	**223**	**225**
Food[15]										
Aliments[15]	100	119	149	154	170	168	178	192	215	207
Trinidad and Tobago										
Trinité-et-Tobago	**169**	**181**	**195**	**216**	**233**	**260**	**289**	**300**	**319**	**353**
Food										
Aliments	180	195	215	257	290	355	416	441	480	570
Tunisia										
Tunisie	**147**	**158**	**167**	**179**	**191**	**205**	**218**	**237**	**249**	**259**
Food										
Aliments	146	160	169	180	194	211	225	245	255	262
Turkey[6]										
Turquie[6]	**195**	**283**	**381**	**100**	**# 174**[52]	**284**	**455**	**754**	**1 283**	**2 131**[52]
Food[6]										
Aliments[6]	198	279	363	100	# 183[52]	310	510	852	1 459	2 386[52]
Tuvalu[89]										
Tuvalu[89]	**137**	**142**	**154**	...	**181**	**186**	**193**	**205**	**193**	**197**
Food[89]										
Aliments[89]	130	137	149	...	177	175	179	190	173	...
Uganda[26 44]										
Ouganda[26 44]	...	...	...	...	...	...	**100**	**128**	**197**	**207**
Food[26 44]										
Aliments[26 44]	...	...	...	...	...	...	100	124	205	197
Ukraine										
Ukraine	**104**	**104**	**105**	**107**	**107**	**109**	...	...	...	...
Food										
Aliments	105	107	112	117	117	118	...	...	...	...
former USSR†[90]										
ancienne URSS†[90]	**104**	**105**	**107**	**108**	**110**	**112**	**118**	..	..	..
Food[90]										
Aliments[90]	107	109	114	119	121	122	125	..	..	..
United Kingdom										
Royaume-Uni	**133**	**142**	**146**	**152**[14]	**160**	**172**	**189**	**200**	**207**	**211**
Food										
Aliments	127	131	136	140[14]	145	153	165	174	178	181
United Rep. Tanzania[91]										
Rép. Unie de Tanzanie[91]	**280**	**374**	**495**	**643**	**843**	**1 061**	**1 271**	**1 554**	**1 897**	**2 342**
Food[91]										
Aliments[91]	288	372	501	654	881	1 095	1 209	1 482	1 798	2 202
United States										
Etats-Unis	**126**	**131**	**133**	**138**	**144**	**151**	**159**	**165**	**170**	**175**
Food										
Aliments	119	122	126	131	136	144	152	158	160	163
Uruguay[92]										
Uruguay[92]	**370**	**637**	**1 123**[14]	**1 836**	**2 978**	**5 374**	**11 422**	**23 069**	**38 862**	**59 885**
Food[92]										
Aliments[92]	365	595	1 141[14]	1 821	2 877	5 194	11 456	21 245	33 996	51 458
Vanuatu[26 44]										
Vanuatu[26 44]	...	**66**	**70**	**81**	**89**	**95**	**100**	**106**	**111**	**115**
Food[26 44]										
Aliments[26 44]	...	71	75	85	95	98	100	103	104	108

33
Consumer price index numbers
All items and food; 1980 = 100 [cont.]
Indices des prix à la consommation
Ensemble et aliments; 1980 = 100 [suite]

Country or area Pays ou zone	1984	1985	1986	1987	1988	1989	1990	1991	1992	1993
Venezuela[11][93] Venezuela[11][93]	152	# 111[15]	124	159	206	380	535	718	943	1 303
Food[11][93] Aliments[11][93]	164	# 122[15]	146	206	289	652	960	1 322	1 725	2 300
former Dem. Yemen[94] anciennce Yémen dém.[94]	128	134	135	139	139	...	...	...	...	...
Food[94] Aliments[94]	120	120	121	126	125	...	...	...	...	...
Yugoslavia[26] Yougoslavie[26]	..	..	..	..	..	..	100	222	20 021	...
Food[26] Aliments[26]	..	..	..	..	..	..	100	200	19 217	...
Yugoslavia, SFR†[22] Yougoslavie, Rfs†[22]	4	7	13	29	85	1 151	7 834	17 078	..	..
Food[22] Aliments[22]	4	7	14	29	87	1 184	7 482	15 271	..	..
Zambia[37] Zambie[37]	184	253	390	568	878	2 009	4 239	8 164	24 280	70 172
Food[37] Aliments[37]	188	256	383	554	877	2 027	4 262	8 139	25 916	75 889
Zimbabwe[37] Zimbabwe[37]	185	201	230	258	278	313	368	457	668	...
Food[37] Aliments[37]	199	212	240	276	300	343	406	509	781	...

Source:
International Labour Office (Geneva).

Source:
Bureau international du Travail (Genève).

† For detailed descriptions of data pertaining to
former Czechoslovakia, Germany, SFR Yugoslavia and former
USSR, see Annex I - Country or area nomenclature, regional
and other groupings.

† Pour les descriptions en détails des données
relatives à l'ancienne Tchécoslovaquie, l'Allemagne, la Rfs
Yougoslavie et l'ancienne URSS, voir l'Annexe I -
Nomenclature des pays ou zones, groupements régionaux et
autres groupements.

1 Excluding "Rent" (Madagascar: Madagascans).
2 Kabul.
3 Average of less than twelve months.
4 Index base: Dec. 1990 = 100 (Peru: beginning 1989).
5 Algiers.
6 Index base: 1982 = 100
7 Index base: 1989 = 100 (Algeria, Indonesia: beginning 1990;
Cook Islands: beginning 1989; Niger: beginning 1991).

8 Luanda.
9 Index base: 1985 = 100 (Fiji: beginning 1986; Saint-Lucia:
beginning 1984).
10 Index base: 1978 = 100.
11 Metropolitan area.
12 Buenos Aires.
13 Index base: 1988 = 100 (Chad, Nicaragua: beginning 1988;
Iraq: beginning 1990; Jamaica, Saint Helena: beginning
1989).
14 Series linked to former series.
15 Index base: 1984 = 100 (Luxembourg: beginning 1987;
Venezuela: beginning 1985).
16 Index base: 1986 = 100 (Bahrain: beginning 1985; Nigeria:
beginning 1986).
17 Government officials.
18 Middle income group.
19 Dhaka.
20 Base: Dec. 1991 = 100.
21 Cotonou.

1 Non compris le groupe "Loyer" (Madagascar: Madagascans).
2 Kaboul.
3 Moyenne de moins de douze mois.
4 Indices base : déc. 1990 = 100 (Pérou : à partir de 1989).
5 Algers.
6 Indices base : 1982 = 100.
7 Indices base : 1989 = 100 (Algérie, Indonésie : à partir de
1990; Iles Cook : à partir de 1989 ; Niger : à partir de
1991).
8 Luanda.
9 Indices base : 1985 = 100 (Fidji : à partir de 1986 ;
Sainte-Lucie : à partir de 1984).
10 Indices base : 1978 = 100.
11 Région métropolitaine.
12 Buenos Aires.
13 Indices base : 1988 = 100(Iraq : à partir de 1990;
Jamaïque,Sainte-Hélène : à partir de 1989; Nicaragua,Tchad :
à partir de 1988).
14 Série enchaînée à la précédente.
15 Indices base : 1984 = 100 (Luxembourg : à partir de 1987 ;
Venezuela : à partir de 1985).
16 Indices base : 1986 = 100 (Bahreïn : à partir de 1985 ;
Nigéria : à partir de 1986).
17 Fonctionnaires.
18 Familles à revenu moyen.
19 Dhaka.
20 Base : déc. 1991 = 100.
21 Cotonou.

33

Consumer price index numbers
All items and food; 1980 = 100 [cont.]

Indices des prix à la consommation
Ensemble et aliments; 1980 = 100 [suite]

22 Due to lack of space, multiply each figure by 100 (Brazil: by 1000; Slovenia: by 10).
23 La Paz.
24 Index base: 1981 = 100.
25 Sao Paulo.
26 Index base: 1990 = 100 (Brazil: beginning 1991; Kenya: beginning 1990; Luxembourg: beginning 1991).
27 Index base: 1983 = 100 (Malta: beginning 1984).
28 Ouagadougou.
29 Bujumbura.
30 Index base: 1991 = 100.
31 Yaounde, Africans.
32 Praia.
33 Including rent (Falkland Is.: beginning 1990; Cape Verde: beginning 1992; Luxembourg: beginning 1987).
34 Bangui.
35 N'Djamena.
36 Santiago.
37 Low income group.
38 Brazzaville, Africans.
39 Rarotonga.
40 San Jose.
41 Abidjan, Africans.
42 Index base: 1993 = 100.
43 Including direct taxes.
44 Urban areas.
45 Excluding Eritrea.
46 Addis Ababa. Excluding "Rent".
47 Stanley.
48 Cayenne.
49 Libreville, Africans.
50 Banjul, Kombo St. Mary.
51 5 new Länder and Berlin (East).
52 Index base: 1987 = 100 (Turkey: beginning 1988).
53 Conakry.
54 Bissau.
55 Reykjavik.
56 Industrial workers.
57 Index base: 1979 = 100.
58 Excluding tobacco (Italy: beginning Feb. 1992; Luxembourg: beginning 1991).
59 Index base: 1992 = 100.
60 Nairobi.
61 Tarawa.
62 Monrovia.
63 Antananarivo, Madagascans.
64 Blantyre.
65 Base July 1991 - June 1992 = 100.
66 Yangon.
67 Windhoek.
68 Curaçao.
69 Nouméa.
70 Niamey, Africans.
71 Rural and urban areas.
72 Muscat.
73 Panama City.
74 Asuncion.
75 Lima.
76 Doha.
77 Kigali.
78 Saint Kitts.
79 St. Vincent.
80 All cities.
81 Dakar.

22 En raison du manque de place, multiplier chaque chiffre par 100 (Brésil : par 1000 ; Slovénie : par 10).
23 La Paz.
24 Indices base: 1981 = 100.
25 Sao Paulo.
26 Indices base : 1990 = 100 (Brésil : partir de 1991 ; Kenya : à partir de 1990 ; Luxembourg : à partir de 1991).
27 Indices base : 1983 = 100 (Matle : à partir de 1984).
28 Ouagadougou.
29 Bujumbura.
30 Indices base : 1991 = 100.
31 Yaoundé, Africains.
32 Praya.
33 Y compris loyer (Iles Falkland : à partir de 1990 ; Cap-Vert : à partir de 1992 ; Luxembourg : à partir de 1987).
34 Bangui.
35 N'Djamena.
36 Santiago.
37 Familles à revenu modique.
38 Brazzaville, Africains.
39 Rarotonga.
40 San José.
41 Abidjan, Africains.
42 Indices base : 1993 = 100.
43 Y compris les impôts directs.
44 Régions urbaines.
45 Non compris Erythrée.
46 Addis Abéba. Non compris le groupe "loyer".
47 Stanley.
48 Cayenne.
49 Libreville, Africains.
50 Banjul, Kombo St. Mary.
51 5 nouveau Länder et Berlin (Est).
52 Indices base : 1987 = 100 (Turquie : à partir de 1988).
53 Conakry.
54 Bissau.
55 Reykjavik.
56 Travailleurs de l'industrie.
57 Indices base : 1979 = 100.
58 Non compris tabac (Italie : à partir de Feb. 1992 ; Luxembourg : à partir de 1991).
59 Indices base : 1992 = 100.
60 Nairobi.
61 Tarawa.
62 Monrovia.
63 Antananarivo, Malgaches.
64 Blantyre.
65 Base juillet 1991 - juin 1992 = 100.
66 Yangon.
67 Windhoek.
68 Curaçao.
69 Nouméa.
70 Niamey, Africains.
71 Régions rurales et urbaines.
72 Muscat.
73 Panamá.
74 Asunción.
75 Lima.
76 Doha.
77 Kigali.
78 Saint Kitts.
79 St. Vincent.
80 Ensemble des villes.
81 Dakar.

33

Consumer price index numbers
All items and food; 1980 = 100 [*cont.*]

Indices des prix à la consommation
Ensemble et aliments; 1980 = 100 [*suite*]

82 Freetown.
83 Honiara.
84 Colombo.
85 Index base: Jan. 1988 = 100 (Sudan: beginning 1990).
86 Paramaribo.
87 Damascus.
88 Lome.
89 Funafuti.
90 Including Belarus and Ukraine shown separately in this
   table.
91 Tanganyika only.
92 Montevideo.
93 Caracas.
94 Aden.

82 Freetown.
83 Honiara.
84 Colombo.
85 Indices base : 1988 = 100 (Soudan : à partir de 1990).
86 Paramaribo.
87 Damas.
88 Lomé.
89 Funafuti.
90 Y compris le Bélarus et l'Ukraine, figurant séparément dans
   ce tableau.
91 Tanganyika seulement.
92 Montevideo.
93 Caracas.
94 Aden.

Technical notes, tables 31-33

In *Table 31*, the series generally relate to the average earnings of wage earners in manufacturing industries. Earnings generally include bonuses, cost of living allowances, taxes, social insurance contributions payable by the employed person and, in some cases, payments in kind, and normally exclude social insurance contributions payable by the employers, family allowances and other social security benefits. The time of year to which the figures refer is not the same for all countries. Unless otherwise stated, the series relate to wage earners of both sexes, irrespective of age.

Some of the series do not conform to the above for one or more of the following reasons: inclusion of salaried employees, inclusion of non-manufacturing industries and use of wage rates instead of earnings.

In the case of countries with widely fluctuating exchange rates or with multiple exchange systems it is advisable to consult table 79 on exchange rates.

For international definitions, further details and current figures, see the International Labour Office *Year Book of Labour Statistics*, *Bulletin of Labour Statistics* and the United Nations *Monthly Bulletin of Statistics*. [9, 23]

In *table 32*, producer prices are prices at which producers sell their output on the domestic market or for export. Wholesale prices, in the strict sense, are prices at which wholesalers sell on the domestic market or for export. In practice, many national wholesale price indexes are a mixture of producer and wholesale prices for domestic goods representing prices for purchases in large quantities from either source. In addition, these indexes may cover the prices of goods imported in quantity for the domestic market either by producers or by retail or wholesale distribution.

Producer or wholesale price indexes normally cover the prices of the characteristic products of agriculture, forestry and fishing, mining and quarrying, manufacturing, and electricity, gas and water supply. Prices are normally measured in terms of transaction prices, including non-deductible indirect taxes less subsidies, in the case of domestically-produced goods and import duties and other non-deductible indirect taxes less subsidies in the case of imported goods.

The Laspeyres index number formula is generally used and, for the purpose of the presentation, the national index numbers have been recalculated, where necessary, on the reference base 1980=100.

The price index numbers for each country are arranged according to the following scheme:

   (a)   Components of supply
       000 Domestic supply
       010 Domestic production for domestic market

Notes techniques, tableaux 31-33

Au *Tableau 31*, les séries se rapportent généralement aux gains moyens des salariés des industries manufacturières. Ces gains, en général, comprennent normalement les primes, les indemnités de vie chère, les impôts, les cotisations des travailleurs à une caisse d'assurance sociale et, dans certains cas, des paiements en nature; ils excluent normalement les contributions de l'employeur à la caisse d'assurance sociale, les allocations familiales et autres prestations de la sécurité sociale. La période de l'année à laquelle se rapportent les chiffres n'est pas la même pour tous les pays. Sauf indication contraire, les séries se rapportent aux salariés des deux sexes et ne tiennent pas compte de l'âge.

Certaines des séries s'écartent des normes indiquées ci-dessus pour une ou plusieurs des raisons suivantes : l'inclusion des employés salariés, l'inclusion des industries non manufacturières et l'utilisation des taux de rémunération au lieu des gains.

Dans le cas des pays à larges fluctuations des cours des changes ou à système de changes multiples, il est recommandé de consulter le tableau 79 sur les cours de changes.

Pour les définitions internationales, plus de détails et pour les chiffres courants, voir l'*Annuaire des Statistiques du Travail*, le *Bulletin des statistiques du travail* du Bureau international du travail et le *Bulletin mensuel de statistique* des Nations Unies [9, 23].

Au *tableau 32*, les prix à la production sont les prix auxquels les producteurs vendent leur production sur le marché intérieur ou à l'exportation. Les prix de gros, au sens strict du terme, sont les prix auxquels les grossistes vendent sur le marché intérieur ou à l'exportation. En pratique, les indices nationaux des prix de gros combinent souvent les prix à la production et les prix de gros de biens nationaux représentant les prix d'achat par grandes quantités au producteur ou au grossiste. En outre, ces indices peuvent s'appliquer aux prix de biens importés en quantités pour être vendus sur le marché intérieur par les producteurs, les détaillants ou les grossistes.

Les indices de prix de gros ou de prix à la production comprennent aussi en général les prix des produits provenant de l'agriculture, de la sylviculture et de la pêche, des industries extractives (mines et carrières), de l'industrie manufacturière ainsi que les prix de l'électricité, de gaz et de l'eau. Les prix sont normalement ceux auxquels s'effectue la transaction, y compris les impôts indirects non déductibles, mais non compris les subventions dans le cas des biens produits dans le pays et y compris les taxes à l'importation et autres impôts indirects non déductibles, mais non compris les subventions dans le cas des biens importés.

On utilise généralement la formule de Laspeyres et, pour la présentation, on a recalculé les indices nationaux, le cas échéant, en prenant comme base de référce 1980=100.

011 Agricultural products
012 Industrial products
020 Import products
110 Exported goods
(b) Stage of processing
210 Raw materials
220 Intermediate products
230 Finished goods
(c) End-use
310 Producers' material
320 Consumers' goods
330 Capital goods

The leading number in each line of the tables identifies the position in the classification scheme of the price index numbers concerned.

Description of the general methods used in compiling the related national indexes is given in the United Nations *1977 Supplement to the Statistical Yearbook and the Monthly Bulletin of Statistics*.[45]

In *table 33*, unless otherwise stated, the index covers all the main classes of expenditure on all items and on food. Monthly data for many of these series and descriptions of them may be found in the United Nations *Monthly Bulletin of Statistics* and the United Nations *1977 Supplement to the Statistical Yearbook and the Monthly Bulletin of Statistics*.[45]

Les indices des prix pour chaque pays sont présentés suivant la classification ci-après :
(a) Eléments de l'offre
000 Offre intérieure
010 Production nationale pour le marché intérieur
011 Produits agricoles
012 Produits industriels
020 Produits importés
110 Produits exportés
(b) Stade de la transformation
210 Matières premières
220 Produits intermédiaires
230 Produits finis
(c) Utilisation finale
310 Biens de production
320 Biens de consommation
330 Biens d'équipement

Le premier chiffre de chacune des entrées du tableau dénote la position de l'indice en question dans la classification.

Les méthodes générales utilisées pour calculer les indices nationaux correspondants sont exposées dans : *1977 Supplément à l'Annuaire statistique et au Bulletin mensuel de statistique* des Nations Unies [45].

Au *tableau 33*, sauf indication contraire, les indices donnés englobent tous les groupes principaux de dépenses pour l'ensemble et les aliments. Les données mensuelles pour plusieurs de ces séries et définitions figurent dans le *Bulletin mensuel de statistique* (ONU) et dans le *1977 Supplément à l'Annuaire statistique et au Bulletin mensuel de statistique* des Nations Unies [45].

34
Agricultural production
Production agricole
Index numbers: 1979-81 = 100
Indices : 1979-81 = 100

Country or area Pays ou zone	All commodities Ensemble des produits					Food Produit alimentaires				
	1989	1990	1991	1992	1993	1989	1990	1991	1992	1993
Africa · Afrique										
Algeria Algérie	150	150	178	182	174	149	150	177	182	173
Angola Angola	103	103	108	112	108	106	107	112	117	112
Benin Bénin	160	166	179	179	185	155	159	170	171	174
Botswana Botswana	111	115	117	112	105	111	115	117	112	105
Burkina Faso Burkina Faso	155	148	180	181	185	153	145	178	179	183
Burundi Burundi	118	127	130	134	127	119	129	132	136	133
Cameroon Cameroun	115	116	115	110	110	115	117	114	111	114
Cape Verde Cap-Vert	188	194	152	114	121	188	194	152	114	121
Central African Rep. Rép. centrafricaine	122	123	126	129	132	122	123	128	132	135
Chad Tchad	130	128	140	136	130	121	118	129	134	131
Comoros Comores	125	126	150	129	139	126	126	150	129	139
Congo Congo	117	118	109	111	115	118	118	109	112	115
Côte d'Ivoire Côte d'Ivoire	136	139	135	134	143	142	143	142	147	156
Egypt Egypte	131	138	144	152	158	142	150	157	165	170
Ethiopia Ethiopie	107	111	110	115	..	108	111	111	116	..
Gabon Gabon	118	121	121	122	122	118	121	121	122	122
Gambia Gambie	132	108	120	104	110	131	108	120	103	109
Ghana Ghana	156	127	158	161	176	158	128	160	162	178
Guinea Guinée	122	128	133	141	150	119	125	131	139	148
Guinea-Bissau Guinée-Bissau	130	136	137	137	142	130	137	137	138	143
Kenya Kenya	149	146	144	139	131	150	147	145	142	129
Lesotho Lesotho	113	123	96	82	99	113	124	94	79	98
Liberia Libéria	122	79	72	76	68	126	93	91	91	87
Libyan Arab Jamah. Jamah. arabe libyenne	111	122	128	120	84	111	122	128	120	84
Madagascar Madagascar	121	122	123	124	128	121	123	125	126	130

34
Agricultural production
Index numbers: 1979-81 = 100 [cont.]
Production agricole
Indices : 1979-81 = 100 [suite]

Country or area Pays ou zone	All commodities Ensemble des produits					Food Produit alimentaires				
	1989	1990	1991	1992	1993	1989	1990	1991	1992	1993
Malawi Malawi	111	110	120	96	126	105	101	111	86	117
Mali Mali	129	126	138	132	138	126	122	134	126	135
Mauritania Mauritanie	119	118	122	113	115	119	118	122	113	115
Mauritius Maurice	109	115	116	122	120	109	116	117	122	121
Morocco Maroc	182	171	198	149	152	181	171	198	149	152
Mozambique Mozambique	101	107	101	84	95	105	111	105	86	98
Namibia Namibie	98	95	100	101	105	100	97	102	104	107
Niger Niger	104	105	125	124	123	104	105	125	124	123
Nigeria Nigéria	149	167	174	189	197	149	167	174	189	197
Réunion Réunion	92	106	117	118	106	92	106	118	119	106
Rwanda Rwanda	113	113	117	118	110	111	110	117	114	109
Sao Tome and Principe Sao Tomé-et-Principe	78	69	77	97	98	79	69	77	98	99
Senegal Sénégal	142	135	144	139	160	142	135	143	139	160
Sierra Leone Sierra Leone	126	126	126	118	115	122	121	121	112	110
Somalia Somalie	126	115	90	76	85	126	115	90	76	85
South Africa Afrique du Sud	110	103	106	86	99	111	104	107	86	102
Sudan Soudan	97	92	118	129	113	95	91	120	132	116
Swaziland Swaziland	129	126	130	114	124	128	126	131	115	121
Togo Togo	135	139	138	140	164	132	136	131	131	156
Tunisia Tunisie	120	144	173	160	169	120	144	174	161	169
Uganda Ouganda	138	143	146	147	155	137	143	146	148	155
United Rep.Tanzania Rép. Unie de Tanzanie	122	125	122	117	116	124	127	123	118	117
Zaire Zaïre	133	138	143	147	151	133	138	143	148	152
Zambia Zambie	145	129	136	118	157	143	128	133	117	153
Zimbabwe Zimbabwe	132	129	130	84	120	125	127	119	66	105

34
Agricultural production
Index numbers: 1979-81 = 100 [cont.]
Production agricole
Indices : 1979-81 = 100 [suite]

Country or area Pays ou zone	All commodities Ensemble des produits					Food Produit alimentaires				
	1989	1990	1991	1992	1993	1989	1990	1991	1992	1993
America, North · Amérique du Nord										
Antigua and Barbuda Antigua-et-Barbuda	104	110	114	117	116	104	109	114	117	117
Bahamas Bahamas	90	98	104	96	96	90	98	104	96	96
Barbados Barbade	75	78	69	66	64	75	78	69	66	64
Belize Belize	110	119	124	125	134	110	119	124	125	134
Canada Canada	113	128	128	124	123	114	129	129	126	124
Costa Rica Costa Rica	131	137	138	145	145	128	138	136	143	149
Cuba Cuba	112	109	112	94	77	111	108	111	94	75
Dominica Dominique	147	151	151	153	153	147	151	151	153	153
Dominican Republic Rép. dominicaine	126	121	125	126	128	129	125	131	136	139
El Salvador El Salvador	79	87	87	94	95	106	107	106	115	116
Grenada Grenade	81	84	82	82	84	81	84	82	82	84
Guadeloupe Guadeloupe	117	98	109	114	120	117	98	109	114	120
Guatemala Guatemala	109	116	117	115	114	133	143	144	144	144
Haiti Haïti	101	96	93	87	85	102	97	94	89	86
Honduras Honduras	123	133	129	135	138	120	126	128	132	138
Jamaica Jamaïque	102	118	119	123	129	102	117	118	123	117
Martinique Martinique	127	137	125	131	126	127	137	125	131	126
Mexico Mexique	105	115	119	117	125	105	115	120	119	128
Netherlands Antilles Antilles néerlandaises	49	39	77	21	54	49	39	77	21	54
Nicaragua Nicaragua	77	72	78	79	76	86	93	86	90	96
Panama Panama	111	114	122	123	120	110	112	120	121	118
Puerto Rico Porto Rico	103	98	102	97	93	102	97	101	96	92
Saint Lucia Sainte-Lucie	166	178	152	179	173	166	178	152	179	173
St. Vincent-Grenadines St. Vincent-Grenadines	158	164	152	168	169	158	164	152	168	169

34
Agricultural production
Index numbers: 1979-81 = 100 [cont.]
Production agricole
Indices : 1979-81 = 100 [suite]

Country or area Pays ou zone	All commodities Ensemble des produits					Food Produit alimentaires				
	1989	1990	1991	1992	1993	1989	1990	1991	1992	1993
Trinidad and Tobago Trinité-et-Tobago	84	101	100	93	104	86	102	102	95	107
United States Etats-Unis	102	106	105	114	104	104	106	104	115	104
America, South · Amérique du Sud										
Argentina Argentine	101	111	115	118	112	101	110	114	119	112
Bolivia Bolivie	122	131	144	139	145	125	134	147	141	147
Brazil Brésil	138	130	133	140	141	142	134	137	146	147
Chile Chili	130	135	138	143	149	131	136	139	144	150
Colombia Colombie	127	136	140	141	142	136	143	143	140	145
Ecuador Equateur	130	143	150	153	159	129	142	149	153	159
Guyana Guyana	94	80	91	97	99	95	81	92	99	100
Paraguay Paraguay	169	177	164	155	170	163	171	155	157	175
Peru Pérou	130	116	119	110	117	132	119	123	114	122
Suriname Suriname	102	92	105	105	107	102	92	105	105	107
Uruguay Uruguay	124	115	116	123	119	125	113	114	124	119
Venezuela Venezuela	131	132	136	141	139	131	132	136	143	141
Asia · Asie										
Afghanistan Afghanistan	76	77	79	75	76	77	78	79	75	76
Bangladesh Bangladesh	124	125	129	130	130	125	127	130	132	131
Bhutan Bhoutan	100	105	107	109	110	99	105	106	108	110
Brunei Darussalam Brunéi Darussalam	170	157	145	140	148	171	157	145	140	149
Cambodia Cambodge	199	194	194	202	209	197	191	189	196	203
China Chine	146	159	167	171	179	145	158	163	170	180
Cyprus Chypre	109	101	77	113	137	109	101	77	113	137
Hong Kong Hong-kong	111	114	113	113	113	111	114	113	113	113
India Inde	147	148	152	159	162	148	149	154	161	163

34
Agricultural production
Index numbers: 1979-81 = 100 [cont.]
Production agricole
Indices : 1979-81 = 100 [suite]

Country or area Pays ou zone	All commodities Ensemble des produits					Food Produit alimentaires				
	1989	1990	1991	1992	1993	1989	1990	1991	1992	1993
Indonesia Indonésie	149	158	161	171	174	152	161	165	176	179
Iran, Islamic Rep. of Iran, Rép. islamique d'	146	169	182	202	207	147	170	184	205	211
Iraq Iraq	132	146	101	119	132	133	147	101	119	133
Israel Israël	111	124	108	113	111	123	137	125	129	128
Japan Japon	99	98	91	96	81	102	101	94	99	84
Jordan Jordanie	134	176	185	219	240	134	178	189	222	244
Korea, Dem. P. R. Corée, R. p. dém. de	130	128	125	116	108	129	127	124	115	106
Korea, Republic of Corée, République de	108	113	105	114	108	110	115	107	116	109
Lao People's Dem. Rep. Rép. dém. pop. lao	146	156	138	154	146	147	156	137	153	145
Lebanon Liban	147	167	184	193	192	150	172	189	199	197
Macau Macao	106	105	87	80	92	106	105	87	80	92
Malaysia Malaisie	177	184	198	204	223	211	224	244	256	285
Maldives Maldives	121	128	122	126	125	121	128	122	126	125
Mongolia Mongolie	116	115	118	102	96	118	117	120	102	97
Myanmar Myanmar	117	118	119	125	137	118	121	122	127	140
Nepal Népal	153	159	159	152	163	155	161	161	155	166
Pakistan Pakistan	150	155	167	166	171	146	150	156	163	170
Philippines Philippines	112	119	116	117	118	111	119	116	117	119
Saudi Arabia Arabie saoudite	547	583	530	584	574	558	595	540	596	585
Singapore Singapour	93	93	73	56	41	94	94	73	56	42
Sri Lanka Sri Lanka	92	101	96	93	98	94	103	97	98	100
Syrian Arab Republic Rép. arabe syrienne	97	125	126	145	135	94	125	123	140	134
Thailand Thaïlande	130	124	132	135	135	127	119	128	126	126
Turkey Turquie	126	133	134	134	134	127	134	136	135	136
Viet Nam Viet Nam	147	152	156	165	175	146	151	155	163	173

34
Agricultural production
Index numbers: 1979-81 = 100 [cont.]
Production agricole
Indices : 1979-81 = 100 [suite]

Country or area	All commodities Ensemble des produits					Food Produit alimentaires				
Pays ou zone	1989	1990	1991	1992	1993	1989	1990	1991	1992	1993
Yemen Yémen	130	117	105	130	137	132	117	105	129	136
Europe · Europe										
Albania Albanie	117	116	89	105	107	119	119	92	109	110
Austria Autriche	111	112	113	110	111	111	112	113	110	111
Belgium-Luxembourg Belgique-Luxembourg	116	110	130	139	144	115	110	130	139	145
Bulgaria Bulgarie	100	92	86	79	63	107	98	91	86	69
former Czechoslovakia† anc. Tchécoslovaquie†	129	126	118	112	..	129	126	119	113	..
Denmark Danemark	126	135	136	120	133	126	135	136	120	133
Finland Finlande	116	122	110	100	107	116	122	110	100	107
France France	104	106	105	112	103	104	106	106	112	103
Germany † Allemagne†	116	117	115	107	109	116	117	115	107	109
Greece Grèce	114	97	115	113	112	110	93	111	106	104
Hungary Hongrie	114	108	115	87	79	114	109	116	87	80
Iceland Islande	85	83	88	90	80	85	84	88	90	81
Ireland Irlande	109	123	127	130	130	109	123	126	130	130
Italy Italie	103	98	107	107	103	102	97	106	107	103
Malta Malte	116	120	128	127	134	116	120	128	127	134
Netherlands Pays-Bas	119	113	118	121	129	119	113	117	121	129
Norway Norvège	107	119	114	106	119	107	119	114	106	119
Poland Pologne	117	119	114	102	108	118	121	116	104	111
Portugal Portugal	127	136	139	119	103	128	136	139	119	103
Romania Roumanie	99	89	89	73	85	100	89	90	73	86
Spain Espagne	120	125	120	122	115	120	125	119	122	115
Sweden Suède	102	108	93	86	97	102	108	93	86	97
Switzerland Suisse	115	111	112	113	109	115	111	112	113	109

34
Agricultural production
Index numbers: 1979-81 = 100 [cont.]
Production agricole
Indices : 1979-81 = 100 [suite]

Country or area	All commodities Ensemble des produits					Food Produit alimentaires				
Pays ou zone	1989	1990	1991	1992	1993	1989	1990	1991	1992	1993
United Kingdom Royaume-Uni	109	111	114	112	103	109	110	114	112	103
Yugoslavia, SFR† Yougoslavie, Rfs†	104	95	99	..	..	104	96	99	..	..
Oceania · Océanie										
Australia Australie	111	116	114	124	130	108	112	107	119	129
Fiji Fidji	115	120	107	112	114	116	120	108	113	114
French Polynesia Polynésie française	94	96	95	92	96	96	98	97	94	97
New Caledonia Nouvelle-Calédonie	97	99	100	104	106	100	101	104	108	111
New Zealand Nouvelle-Zélande	108	102	106	109	106	113	108	114	118	117
Papua New Guinea Papouasie-Nvl-Guinée	129	129	128	132	136	128	130	130	137	139
Samoa Samoa	110	100	89	87	94	110	100	88	87	94
Solomon Islands Iles Salomon	122	129	135	139	134	122	129	135	140	134
Tonga Tonga	89	87	85	86	85	89	87	85	86	85
Vanuatu Vanuatu	95	123	107	102	106	95	124	107	103	107
former USSR† · ancienne URSS†										
former USSR† ancienne URSS†	121	120	106	..	..	123	122	108	..	..

Source:
Food and Agriculture Organization of the United Nations (Rome).

† For detailed descriptions of data pertaining to former Czechoslovakia, Germany, SFR Yugoslavia and former USSR, see Annex I - Country or area nomenclature, regional and other groupings.

Source:
Organisation des Nations Unies pour l'alimentation et l'agriculture (Rome).

† Pour les descriptions en détails des données relatives à l'ancienne Tchécoslovaquie, l'Allemagne, la Rfs Yougoslavie et l'ancienne URSS, voir l'Annexe I - Nomenclature des pays ou zones, groupements régionaux et autres groupements.

35

Cereals
Céréales

Production: thousand metric tons
Production : milliers de tonnes métriques

Country or area Pays ou zone	1984	1985	1986	1987	1988	1989	1990	1991	1992	1993
World **Monde**	**1 788 078**	**1 824 432**	**1 837 138**	**1 769 926**	**1 727 600**	**1 871 076**	**1 947 616**	**1 876 710**	**1 961 449**	**1 894 299**
Africa **Afrique**	**64 478**	**83 465**	**87 918**	**80 281**	**92 888**	**96 361**	**88 167**	**99 379**	**83 952**	**94 702**
Algeria Algérie	1 461	2 919	2 404	2 066	1 038	2 006	1 629	3 812	3 350	2 012
Angola Angola	335 [1]	325 [1]	356	382	352	289	264	386	465	335
Benin Bénin	483	535	497	401	557	565	546	587	609	719
Botswana Botswana	8	19	22	23	106	76	54	45	15	34
Burkina Faso Burkina Faso	1 089	1 583	1 890	1 637	2 101	1 952	1 518	2 455	2 477	2 495
Burundi Burundi	224	256	266	285	288	255	293	299	309	* 300
Cameroon Cameroun	657	815	1 094	715	900	858	816	1 003	905 [1]	970 [1]
Cape Verde Cap-Vert	3	1	12	21	17	10	11	3	5	6 [1]
Central African Rep. Rép. centrafricaine	102	111	178	126	143	124	93	88	79	83 [1]
Chad Tchad	354	704	687	563	782	582	602	931	977	747
Comoros Comores	17	17	17	16	17	19	19	19	19 [1]	19 [1]
Congo Congo	14	16	22	20	21	26	26	26 [1]	27	27 [1]
Côte d'Ivoire Côte d'Ivoire	1 105	1 089	1 049	1 086	1 144	1 193	1 239	1 286	1 340	1 305
Egypt Egypte	8 455	8 561	8 754	9 445	9 764	11 102	13 011	13 839	14 576	14 147
Ethiopia[2] Ethiopie[2]	4 240	4 820	6 504	6 195	6 384	6 355	6 457 [1]	6 305 [1]	7 000 [1]	6 617 [1]
Eritrea Erythrée	..	..	..	..	..	..	..	..	..	* 73
Gabon Gabon	14	14	21	25	25	21	23 [1]	24 [1]	25	25 [1]
Gambia Gambie	87	116	102	92	100	96	90	111	96	102
Ghana Ghana	1 066	921	867	1 057	1 146	1 184	844	1 436	1 254	1 645
Guinea Guinée	700	739	814	801	760	802	859	913	984	956
Guinea-Bissau Guinée-Bissau	159	158 [1]	162	153 [1]	150	148	167	179	169	181
Kenya Kenya	1 728	2 901	3 371	2 866	3 285	3 178	2 729	2 770	2 925	2 112
Lesotho Lesotho	132	167	132	146	234	200	238	133	80	158

35
Cereals
Production: thousand metric tons [*cont.*]
Céréales
Production : milliers de tonnes métriques [*suite*]

Country or area Pays ou zone	1984	1985	1986	1987	1988	1989	1990	1991	1992	1993
Liberia Libéria	298	289	288	298	298	294	100 [1]	109 [1]	102 [1]	71 [1]
Libyan Arab Jamah. Jamah. arabe libyenne	275	235	285	276	283	322	273	298 [1]	313 [1]	303 [1]
Madagascar Madagascar	2 274	2 320	2 385	2 338	2 307	2 542	2 577	2 489	2 617	2 732
Malawi Malawi	1 457	1 423	1 364	1 255	1 489	1 588	1 413	1 680	670	2 137
Mali Mali	1 113	1 719	1 728	1 639	2 197	2 157	1 771	2 415	1 819	2 074
Mauritania Mauritanie	52	108	127	152	174	184	105	105	107	164
Mauritius Maurice	4	5	8	4	4	2	2	2	2	2 [1]
Morocco Maroc	3 759	5 312	7 825	4 337	7 959	7 429	6 276	8 668	2 952	2 930
Mozambique Mozambique	665 [1]	746 [1]	829	569	562	607	734	546	239	766
Namibia Namibie	88 [1]	63	81	69	83	97	99	114	32	81
Niger Niger	1 070	1 849	1 834	1 439	2 389	1 842	1 480	2 384	2 253	1 813
Nigeria Nigéria	10 531	11 911	12 744	12 626	14 560	15 122	13 733	13 018	12 523	14 360
Réunion Réunion	11	12	13	11	13	13	11	13	16	15 [1]
Rwanda Rwanda	296	337	297	292	292	262	308	329	290	199
Sao Tome and Principe [1] Sao Tomé-et-Principe [1]	1	1	1	1	1	1	1	1	1	1
Senegal Sénégal	709	1 249	887	1 054	867	1 067	977	946	856	1 070
Sierra Leone Sierra Leone	561	488	577	517	547	574	562	600	469	* 550
Somalia Somalie	497	514	586	543	601	654	581 [1]	256	202	286 [1]
South Africa Afrique du Sud	7 802	10 870	11 352	11 483	11 347	14 884	11 016	10 810	4 616	12 004
Sudan Soudan	1 440	4 036	3 799	1 699	5 132	1 975	1 716	4 590	5 438	3 151
Swaziland Swaziland	154	178	164	101	118	138	99	142	58	88
Togo Togo	440	371	366	368	504	566	484	465	503	633
Tunisia Tunisie	1 053	2 098	624	1 917	295	641	1 638	2 556	2 199	1 918
Uganda Ouganda	944	1 171	1 058	1 220	1 398	1 636	1 580	1 576	1 792	1 795
United Rep.Tanzania Rép. Unie de Tanzanie	3 142	3 622	3 864	4 029	3 685	4 791	3 842	3 792	3 517	3 908

35

Cereals
Production: thousand metric tons [*cont.*]
Céréales
Production : milliers de tonnes métriques [*suite*]

Country or area Pays ou zone	1984	1985	1986	1987	1988	1989	1990	1991	1992	1993
Zaire Zaïre	1 058	1 141	1 193	1 256	1 332	1 415	1 487	1 538	1 585	1 752
Zambia Zambie	925	1 193	1 319	1 158	2 055	1 967	1 210	1 225	613	1 758
Zimbabwe Zimbabwe	1 423	3 416	3 096	1 505	3 080	2 553	2 560	2 060	481	3 072
America, North **Amérique du Nord**	**386 542**	**428 103**	**401 245**	**361 080**	**268 807**	**359 443**	**400 375**	**362 515**	**435 274**	**343 883**
Bahamas[1] Bahamas[1]	1	1	1	1	1	1	1	1	1	1
Barbados[1] Barbade[1]	2	2	2	2	2	2	2	2	2	2
Belize Belize	22	25	23	28	29	24	22[1]	22[1]	25[1]	26[1]
Canada Canada	42 800	48 239	56 965	51 682	35 788	48 402	56 797	53 850	49 500	52 241
Costa Rica Costa Rica	363	381	342	299	315	246	281	260	272	204
Cuba Cuba	651	620	667	562	585	632	533	462	371	277
Dominican Republic Rép. dominicaine	634	608	560	613	547	554	488	526	579	* 589
El Salvador El Salvador	731	697	631	647	807	802	825	729	919	922
Guatemala Guatemala	1 150	1 283	1 402	1 390	1 389	1 422	1 458	1 297	1 495	1 520
Haiti Haïti	432	445	476	462	469	453	361	400	442	385[1]
Honduras Honduras	608	502	550	534	527	619	685	693	692	765
Jamaica Jamaïque	9	8	7	6	4	3	2	3	4	6
Mexico Mexique	23 707	27 403	23 553	23 636	21 067	21 429	25 570	23 616	26 976	25 825
Nicaragua Nicaragua	451	506	478	455	506	484	488	388	498	566
Panama Panama	263	301	293	304	298	328	345	337	337	327
Puerto Rico Porto Rico	5	5	6	2	2	...	...	...	...	...
St. Vincent-Grenadines St. Vincent-Grenadines	1[1]	1[1]	1[1]	1[1]	1	1	1	1[1]	1[1]	1[1]
Trinidad and Tobago Trinité-et-Tobago	7	7	7	10	10	15	17	18[1]	25	22
United States Etats-Unis	314 705	347 069	315 281	280 447	206 461	284 026	312 499	279 909	353 133	260 205
America, South **Amérique du Sud**	**77 856**	**77 817**	**78 194**	**82 926**	**82 036**	**79 655**	**68 433**	**73 903**	**84 733**	**82 981**
Argentina Argentine	31 796	28 077	26 405	22 873	22 188	18 338	19 798	21 627	25 282	24 756

35
Cereals
Production: thousand metric tons [*cont.*]
Céréales
Production : milliers de tonnes métriques [*suite*]

Country or area Pays ou zone	1984	1985	1986	1987	1988	1989	1990	1991	1992	1993
Bolivia Bolivie	866	973	839	865	834	845	788	1 011	851	1 078
Brazil Brésil	32 711	36 011	37 319	44 112	42 921	43 943	32 490	36 682	44 058	43 044
Chile Chili	2 116	2 360	2 675	2 819	2 800	3 148	2 981	2 864	2 901	2 642
Colombia Colombie	3 258	3 144	,3 067	3 206	3 555	4 007	4 314	3 948	3 674	3 614
Ecuador Equateur	819	832	1 127	1 297	1 489	1 439	1 383	1 444	1 624	1 434
French Guiana Guyane française	5	8	9	13	14	16	22	29	24	* 25
Guyana Guyana	301	281	243	229	229	240	146	253	287	303[1]
Paraguay Paraguay	1 008	1 152	1 143	1 447	1 595	1 550	1 612	698	803	864[1]
Peru Pérou	2 166	1 829	1 891	2 354	2 369	2 443	1 792	1 773	1 507	1 933
Suriname Suriname	302	299	300	272	265	261	196	229	261	260[1]
Uruguay Uruguay	1 082	1 028	925	1 021	1 291	1 488	1 130	1 144	1 538	1 435
Venezuela Venezuela	1 428	1 822	2 250	2 418	2 485	1 937	1 780	2 201	1 923	1 595
Asia Asie	771 152	755 621	775 299	764 619	803 186	832 035	872 904	861 862	888 195	902 883
Afghanistan Afghanistan	* 3 418	* 3 242	3 084	3 394	2 997	2 834	2 705	2 724	2 420[1]	2 540
Armenia Arménie	..	..	..	..	..	..	..	..	342	305
Azerbaijan Azerbaïdjan	..	..	..	..	..	..	..	..	1 256	1 105
Bangladesh Bangladesh	23 256	24 135	24 266	24 304	24 450	27 886	27 747	28 462	28 654	29 250
Bhutan Bhoutan	179	167	164	140	95	95	106	106	106	106[1]
Brunei Darussalam Brunéi Darussalam	3	1	1	1	1	1[1]	1[1]	1[1]	1[1]	1[1]
Cambodia Cambodge	1 308	1 854	2 144	1 853	2 541	2 726	2 588	2 460	2 281	2 560[1]
China Chine	365 851	339 794	352 006	359 159	351 747	367 560	404 332	395 015	400 164	412 262
Cyprus Chypre	93	112	68	126	158	148	109	65	182	194
Gaza Strip (Palestine) Zone de Gaza (Palestine)	2	1	1[1]	1[1]	1[1]	1[1]	1[1]	1[1]	1[1]	1[1]
Georgia Géorgie	..	..	..	..	..	..	..	..	458	397
India Inde	164 478	165 682	164 955	156 114	183 867	199 413	193 919	191 570	201 715	201 479

35

Cereals
Production: thousand metric tons [cont.]
Céréales
Production : milliers de tonnes métriques [suite]

Country or area Pays ou zone	1984	1985	1986	1987	1988	1989	1990	1991	1992	1993
Indonesia Indonésie	43 424	43 362	45 647	45 234	48 328	50 918	51 913	50 944	56 236	* 54 398
Iran, Islamic Rep. of Iran, Rép. islamique d'	10 052	10 728	11 899	12 197	12 103	10 787	13 686	14 447	15 844	17 368
Iraq Iraq	1 100	2 932	2 281	1 728	2 591	1 497	3 455	2 673	3 092	3 283
Israel Israël	171	163	203	326	223	209	303	190	267	229
Japan Japon	16 012	15 856	15 805	14 527	13 867	14 318	14 449	13 065	14 287	10 738
Jordan Jordanie	62	85	43	117	129	82	130	108	148	82
Kazakhstan Kazakhstan	..	..	..	..	..	..	..	..	29 648	21 500
Korea, Dem. P. R. Corée, R. p. dém. de	8 064	8 064[1]	8 509[1]	8 391[1]	8 577[1]	8 494[1]	7 331[1]	7 316[1]	6 733	5 210
Korea, Republic of Corée, République de	9 269	8 808	8 638	8 467	9 164	8 950	8 434	7 866	7 835	7 099
Kuwait Koweït	1	3	3	4[1]	4	4[1]	3[1]	...	2[1]	2[1]
Kyrgyzstan Kirghizistan	..	..	..	..	..	..	..	..	1 516	1 603
Lao People's Dem. Rep. Rép. dém. pop. lao	1 355	1 431	1 491	1 243	1 054	1 448	1 558	1 292	1 561	1 298
Lebanon Liban	26	34	47[1]	70	73	79	82	83	88	81
Malaysia Malaisie	1 594	1 873	1 773	1 730	1 815	1 873	1 995	2 175	2 106	2 137
Mongolia Mongolie	597	884	869	689	814	841	720	597	496	510[1]
Myanmar Myanmar	14 960	15 067	14 861	14 239	13 648	14 278	14 442	13 671	15 341	17 945
Nepal Népal	4 310	4 374	3 999	4 759	5 307	5 672	5 847	5 520	4 919	5 325
Oman Oman	2	1	4	4	4	5	5	5[1]	5[1]	5[1]
Pakistan Pakistan	17 536	17 699	20 866	18 454	19 240	21 018	20 957	21 138	22 113	23 766
Philippines Philippines	11 269	12 728	13 338	12 818	13 399	13 981	14 173	14 329	13 688	14 281
Qatar Qatar	1	2	2	3	3	3	4	4	4	4[1]
Saudi Arabia Arabie saoudite	1 444	2 188	2 461	2 929	3 692	3 932	4 137	4 504	4 667	4 897
Sri Lanka Sri Lanka	2 464	2 702	2 639	2 177	2 524	2 101	2 580	2 430	2 374	2 487[1]
Syrian Arab Republic Rép. arabe syrienne	1 444	2 543	3 167	2 300	5 001	1 404	3 100	3 288	4 361	5 386
Tajikistan Tadjikistan	..	..	..	..	..	..	..	..	278	303

35
Cereals
Production: thousand metric tons [*cont.*]
Céréales
Production : milliers de tonnes métriques [*suite*]

Country or area Pays ou zone	1984	1985	1986	1987	1988	1989	1990	1991	1992	1993
Thailand Thaïlande	24 517	25 615	23 401	21 414	26 167	25 240	21 170	23 872	24 141	22 226
Turkey Turquie	26 314	26 493	29 358	29 282	30 894	23 499	30 201	31 148	29 157	31 749
Turkmenistan Turkménistan	..	..	..	..	..	..	..	..	708	900
United Arab Emirates Emirats arabes unis	6	5	6	6	5	6	8	8	8[1]	8[1]
Uzbekistan Ouzbékistan	..	..	..	..	..	..	..	..	2 252	2 101
Viet Nam Viet Nam	16 042	16 466	16 578	15 669	17 820	19 840	19 901	20 299	22 343	23 105
Yemen Yémen	480	484	700	718	843	864	767	448	811	825
Europe **Europe**	**305 398**	**283 557**	**282 227**	**275 011**	**285 805**	**292 553**	**284 752**	**303 547**	**254 669**	**259 925**
Albania Albanie	1 039	995	1 046	1 026	998	1 033	940	495	418	610
Austria Autriche	5 353	5 551	5 108	4 965	5 359	5 009	5 290	5 045	4 323	* 4 042
Belarus Bélarus	..	..	..	..	..	..	..	..	7 066	7 355[1]
Belgium-Luxembourg Belgique-Luxembourg	2 533	2 225	2 401	2 060	2 352	2 375	2 122	2 217	2 193	2 287
Bosnia & Herzegovina[1] Bosnie-Herzégovine [1]	..	..	..	..	..	..	..	..	1 361	1 256
Bulgaria Bulgarie	9 251	5 384	8 492	7 278	7 820	9 527	8 115	8 974	6 668	5 749
Croatia Croatie	..	..	..	..	..	..	..	..	2 355	2 733
former Czechoslovakia† anc. Tchécoslovaquie†	11 984	11 775	10 805	11 777	11 907	12 047	12 626	11 939	10 196	..
Czech Republic République tchèque	..	..	..	..	..	..	..	..	...	6 475
Denmark Danemark	9 300	7 956	7 969	7 184	8 068	8 795	9 607	9 231	6 954	8 217
Estonia Estonie	..	..	..	..	..	..	..	..	594	675
Finland Finlande	3 647	3 642	3 520	2 183	2 826	3 809	4 296	3 429	2 603	3 341
France France	58 323	55 989	50 387	52 961	56 059	57 615	55 119	60 338	60 574	55 817
Germany † Allemagne†	..	..	..	..	..	..	..	39 268	34 758	36 222
F. R. Germany R. f. Allemagne	26 489	25 915	25 593	23 843	27 112	26 113	25 883	..	..	..
former German D. R. anc. R. d. allemande	11 349	11 640	11 664	11 218	9 820	10 768	1 167	..	..	..
Greece Grèce	5 513	4 492	5 286	5 183	5 624	5 827	4 463	6 224	5 048	4 924

35

Cereals
Production: thousand metric tons [*cont.*]
    Céréales
    Production : milliers de tonnes métriques [*suite*]

Country or area Pays ou zone	1984	1985	1986	1987	1988	1989	1990	1991	1992	1993
Hungary Hongrie	15 731	14 809	14 301	14 168	14 966	15 417	12 561	15 797	9 981	9 040
Ireland Irlande	2 514	2 096	1 955	2 108	2 078	1 919	1 966	1 964	2 016	1 607
Italy Italie	19 928	18 029	18 698	18 400	17 400	17 133	17 411	19 219	19 954	19 549
Latvia Lettonie	..	..	..	..	..	..	..	..	1 127	1 199
Lithuania Lituanie	..	..	..	..	..	..	..	..	2 198	2 240
Malta Malte	9	9	9[1]	9[1]	9	9	9[1]	9[1]	9[1]	9[1]
Netherlands Pays-Bas	1 407	1 132	1 266	1 107	1 222	1 368	1 361	1 254	1 292	1 343
Norway Norvège	1 407	1 302	1 109	1 285	1 066	1 180	1 568	1 482	1 010	1 403
Poland Pologne	24 392	23 741	25 036	26 060	24 504	26 958	28 014	27 812	19 962	23 417
Portugal Portugal	1 469	1 380	1 641	1 727	1 450	1 833	1 388	1 796	1 263	1 307
Republic of Moldova République de Moldova	..	..	..	..	..	..	..	..	1 976	3 255
Romania Roumanie	20 045	19 503	19 725	16 889	19 286	18 379	17 174	19 307	12 288	15 493
Russian Federation Fédération de Russie	...	...	...	...	...	...	...	...	103 795	94 907
Slovakia Slovaquie	..	..	..	..	..	..	..	..	..	3 201
Slovenia Slovénie	..	..	..	..	..	..	..	..	426	404
Spain Espagne	21 032	20 972	16 520	20 697	23 834	19 698	18 762	19 457	14 505	17 319
Sweden Suède	6 897	5 629	5 811	5 170	4 743	5 493	6 380	5 160	3 759	5 394
Switzerland Suisse	1 118	1 054	960	939	1 244	1 411	1 268	1 313	1 213	1 291
TFYR Macedonia L'ex-R.y. Macédoine	..	..	..	..	..	..	..	..	624	497
Ukraine Ukraine	..	..	..	..	..	..	..	..	35 548	42 725
United Kingdom Royaume-Uni	26 614	22 486	24 509	21 698	21 063	22 729	22 569	22 635	22 063	19 317
Yugoslavia Yougoslavie	..	..	..	..	..	..	..	..	6 851	7 663
Yugoslavia, SFR† Yougoslavie, Rfs†	18 051	15 850	18 416	15 077	14 996	16 110	14 166	19 183	..	
Oceania Océanie	**29 755**	**26 169**	**25 253**	**20 987**	**22 678**	**23 027**	**23 876**	**19 544**	**25 864**	**29 354**
Australia Australie	28 600	24 944	24 031	19 959	21 870	22 319	22 967	18 701	25 097	28 549

35
Cereals
Production: thousand metric tons [*cont.*]
Céréales
Production : milliers de tonnes métriques [*suite*]

Country or area Pays ou zone	1984	1985	1986	1987	1988	1989	1990	1991	1992	1993
Fiji Fidji	24	29	27	24	34	33	34	31	24	22
New Caledonia Nouvelle-Calédonie	1	2	2	1	1	1	1	1	1	1[1]
New Zealand Nouvelle-Zélande	1 119	1 184	1 188	1 000	769	671	869	809	738	778
Papua New Guinea Papouasie-Nvl-Guinée	2	2	2	2	3[1]	3[1]	3[1]	3[1]	3[1]	3[1]
Solomon Islands Iles Salomon	7	6	2	...	...	...	...	...	...	...
Vanuatu[1] Vanuatu[1]	1	1	1	1	1	1	1	1	1	1
former USSR† **ancienne URSS†**	**152 897**	**169 700**	**187 001**	**185 021**	**172 201**	**188 002**	**209 109**	**155 960**	..	..

Source:
Food and Agriculture Organization of the United Nations
(Rome).

† For detailed descriptions of data pertaining to
former Czechoslovakia, Germany, SFR Yugoslavia and former
USSR, see Annex I - Country or area nomenclature, regional
and other groupings.

1 FAO estimate.
2 Prior to 1993, including data for Eritrea.

Source:
Organisation des Nations Unies pour l'alimentation et
l'agriculture (Rome).

† Pour les descriptions en détails des données
relatives à l'ancienne Tchécoslovaquie, l'Allemagne, la Rfs
Yougoslavie et l'ancienne URSS, voir l'Annexe I -
Nomenclature des pays ou zones, groupements régionaux et
autres groupements.

1 Estimation de la FAO.
2 Avant de 1993, y compris les données pour Erythrée.

36
Oil crops, in oil equivalent
Cultures d'huile, en équivalent d'huile
Production: thousand metric tons
Production : milliers de tonnes métriques

Country or area Pays ou zone	1984	1985	1986	1987	1988	1989	1990	1991	1992	1993
World *Monde*	59 659	64 778	64 924	67 863	68 823	72 442	75 519	77 489	78 719	79 213
Africa **Afrique**	4 381	4 705	5 178	5 260	5 469	5 489	5 639	5 817	5 632	5 918
Algeria Algérie	43	58	70	67	58	53	51[1]	86[1]	75[1]	63[1]
Angola Angola	60	60	60	59	58	57	58	58	59	59
Benin Bénin	82	89	85	77	87	82	86	92	90	89[1]
Botswana Botswana	1	1	0	0	1	1	1	1[1]	1[1]	1[1]
Burkina Faso Burkina Faso	63	95	94	81	93	88	88	73	77[1]	78[1]
Burundi Burundi	7	7	8	8	8	7	8	8	8	8[1]
Cameroon Cameroun	165	189	192	187	196	232	246	222	214	213
Cape Verde Cap-Vert	1[1]	2	2	2	2	2	2	2	1[1]	1[1]
Central African Rep. Rép. centrafricaine	48	47	65	57	62	64	59	50	47	44
Chad Tchad	44	45	51	54	55	75	63	61	74	72[1]
Comoros Comores	7	7	7	6	6	6	6	7	7[1]	7[1]
Congo Congo	25	25	25	26	26	27	26	26	26	26[1]
Côte d'Ivoire Côte d'Ivoire	275	297	339	351	320	322	348	353	382	398
Egypt Egypte	165	173	172	157	158	141	153	163	178	170
Equatorial Guinea[1] Guinée équatoriale[1]	8	7	7	7	7	7	7	7	7	7
Ethiopia Ethiopie	116	120	116	121	127	133	138	144	145	...
Gabon Gabon	6[1]	6	8	9	10	10	10	11[1]	11[1]	11[1]
Gambia Gambie	37	28	39	42	35	45	28	31	22	25
Ghana Ghana	145	143	170	167	185	199	178	171	193	202
Guinea Guinée	87	85	84	81	87	87	87	87	95	97
Guinea-Bissau Guinée-Bissau	23	19	24	20[1]	18	18	19	18	17	18
Kenya Kenya	17	20	20	22	23	25	25	28	25	25

36
Oil crops, in oil equivalent
Production: thousand metric tons [*cont.*]
Cultures d'huile, en équivalent d'huile
Production : milliers de tonnes métriques [*suite*]

Country or area Pays ou zone	1984	1985	1986	1987	1988	1989	1990	1991	1992	1993
Liberia Libéria	40	40[1]	41[1]	41[1]	41[1]	41[1]	36[1]	31	31	31[1]
Libyan Arab Jamah. Jamah. arabe libyenne	31	33	* 31	17	18	19	20	20[1]	21[1]	21[1]
Madagascar Madagascar	30	31	32	30	31	33	32	31	33	33
Malawi Malawi	39	45	59	56	50	26	17	25	13	26
Mali Mali	66	75	90	89	107	109	114	102	93	100[1]
Mauritania Mauritanie	1	1	1	1	2[1]	2[1]	2[1]	2[1]	2[1]	1[1]
Mauritius Maurice	1	1	1	1	1	1	1	1	1	1[1]
Morocco Maroc	93	103	117	158	163	136	162	134	182	214
Mozambique Mozambique	109	107	110	112	113	114	115	115	105	105
Niger Niger	10	18	20	14	16	9	6	15	19	20
Nigeria Nigéria	1 137	1 212	1 385	1 340	1 448	1 524	1 658	1 700	1 836	1 826
Rwanda Rwanda	6	7	7	7	6	5	6	7[1]	7[1]	7[1]
Sao Tome and Principe Sao Tomé-et-Principe	5	5	5	5	6[1]	5[1]	5[1]	5[1]	5[1]	5[1]
Senegal Sénégal	176	205	281	321	246	283	238	247	200	216
Seychelles [1] Seychelles [1]	2	2	2	2	1	1	1	1	1	1
Sierra Leone Sierra Leone	73	74	76	75	73	71	71	73	85	75
Somalia Somalie	22	30	24	25	25	26	24	19	11[1]	17.
South Africa Afrique du Sud	119	189	173	221	291	270	305	317	131	209
Sudan Soudan	270	230	286	330	370	217	148	158	213	231[1]
Swaziland Swaziland	2	2	3	2[1]	4	5	5[1]	4[1]	3[1]	7[1]
Togo Togo	27	37	37	39	39	36	44	39	38	41
Tunisia Tunisie	107	118	128	112	62	146	184	294	141	179[1]
Uganda Ouganda	64	60	67	66	72	82	97	97	101	108
United Rep.Tanzania Rép. Unie de Tanzanie	115	111	128	129	125	120	123	136	129	128

36
Oil crops, in oil equivalent
Production: thousand metric tons [cont.]
Cultures d'huile, en équivalent d'huile
Production : milliers de tonnes métriques [suite]

Country or area Pays ou zone	1984	1985	1986	1987	1988	1989	1990	1991	1992	1993
Zaire Zaïre	328	341	343	359	382	395	411	419	428	438
Zambia Zambie	29	28	24	27	28	25	24	23	11	33
Zimbabwe Zimbabwe	54	76	67	80	124	106	104	103	39	82
America, North Amérique du Nord	13 796	14 982	13 966	14 214	12 156	13 472	13 972	15 122	15 434	14 196
Canada Canada	1 662	1 774	1 924	1 881	1 885	1 512	1 783	2 045	1 791	2 523
Costa Rica Costa Rica	53	44	53	56	69	81	84	87	86	95
Cuba Cuba	7	7	8	8	8	8	8[1]	8[1]	7[1]	8[1]
Dominica[1] Dominique[1]	2	2	2	2	2	2	1	2	2	2
Dominican Republic Rép. dominicaine	25	20	25	24	29	32	31	31	28	34[1]
El Salvador El Salvador	24	26	20	17	18	18	22	18	17	19
Grenada Grenade	1	1	1[1]	1[1]	1[1]	1[1]	1[1]	1[1]	1[1]	1[1]
Guatemala Guatemala	32	35	35	34	40	34	40	44	53	52
Haiti Haïti	22[1]	22[1]	23[1]	22[1]	21[1]	20[1]	19[1]	17[1]	14	14[1]
Honduras Honduras	63	72	83	84	81	86	90	93	90	93[1]
Jamaica Jamaïque	17	18	24	26	27	12	11	13	16	16[1]
Mexico Mexique	477	505	432	499	421	488	433	410	361	340
Nicaragua Nicaragua	31	24	20	20	17	17	22	23	20	16
Panama Panama	3	3	3	3	3	3	3	3	3	3[1]
Puerto Rico Porto Rico	1	1	1	1	1	1	1	1	1	1
Saint Lucia[1] Sainte-Lucie[1]	4	4	4	4	3	3	4	3	5	5
St. Vincent-Grenadines St. Vincent-Grenadines	3	3	3	3	3	3[1]	3	3[1]	3[1]	3[1]
Trinidad and Tobago[1] Trinité-et-Tobago[1]	7	5	6	6	5	4	5	5	6	6
United States Etats-Unis	11 360	12 413	11 297	11 522	9 521	11 147	11 413	12 315	12 931	10 965
America, South Amérique du Sud	6 668	7 855	7 295	7 078	8 400	8 934	9 370	8 609	9 147	9 424

36
Oil crops, in oil equivalent
Production: thousand metric tons [cont.]
Cultures d'huile, en équivalent d'huile
Production : milliers de tonnes métriques [suite]

Country or area Pays ou zone	1984	1985	1986	1987	1988	1989	1990	1991	1992	1993
Argentina Argentine	2 526	2 878	3 319	2 491	3 341	2 716	3 841	3 940	3 845	3 449
Bolivia Bolivie	19	21	33	28	34	54	55	86	82	114
Brazil Brésil	3 431	4 141	3 074	3 589	3 906	4 914	4 196	3 344	4 091	4 636
Chile Chili	6	26	58	52	64	55	32	36	34	15
Colombia Colombie	185	208	235	235	291	325	369	413	377	408
Ecuador Equateur	96	112	157	174	168	218	207	218	204	205
Guyana Guyana	5	5	6	* 6	6	6[1]	7[1]	7[1]	7[1]	7[1]
Paraguay Paraguay	286	327	244	336	381	428	460	398	344	422
Peru Pérou	37	50	53	49	62	67	58	53	44	44
Suriname Suriname	9	10	8	6	4	5	3	3	4	3[1]
Uruguay Uruguay	16	20	39	33	29	30	27	29	31	28
Venezuela Venezuela	52	56	69	80	114	115	114	82	85	93
Asia Asie	25 436	27 044	27 926	28 642	30 966	31 888	34 466	35 097	36 763	38 389
Afghanistan Afghanistan	34	32	33	29	29	28	26	30	30	30
Azerbaijan Azerbaïdjan	...	...	...	...	...	...	...	...	37	38[1]
Bangladesh Bangladesh	148	161	151	142	145	142	145	146	155	159
Bhutan Bhoutan	1	1[1]	1[1]	1[1]	1	1	1[1]	1[1]	1[1]	1[1]
Cambodia Cambodge	9	13	13	14	13	15	15	19	20	20[1]
China Chine	8 612	9 319	8 966	9 659	8 727	8 300	9 715	10 090	9 810	10 309
Cyprus Chypre	3	3	3	2	4	3	3	2	4	3[1]
Gaza Strip (Palestine) Zone de Gaza (Palestine)	1	1	* 1	* 1	* 1	* 0	* 1	* 0	1[1]	0[1]
Georgia Géorgie	...	...	...	...	...	...	...	...	6	6
India Inde	5 312	4 919	4 769	4 922	6 923	7 195	7 238	7 438	8 486	8 167
Indonesia Indonésie	3 249	3 447	3 682	3 808	4 271	4 413	5 210	5 450	6 001	6 372

36

Oil crops, in oil equivalent
Production: thousand metric tons [cont.]
Cultures d'huile, en équivalent d'huile
Production : milliers de tonnes métriques [suite]

Country or area Pays ou zone	1984	1985	1986	1987	1988	1989	1990	1991	1992	1993
Iran, Islamic Rep. of Iran, Rép. islamique d'	73	74	76	68	81	79	95	85	92	97
Iraq Iraq	9	11	12	12	14	19	36	18	27	32
Israel Israël	39	48	41	32	36	31	35	28	40	30
Japan Japon	69	64	69	76	67	68	59	52	50	32
Jordan Jordanie	11	4	7	5	16	6	14	9	18	9[1]
Kazakhstan Kazakhstan	...	...	...	...	...	...	...	...	75	117
Korea, Dem. P. R. Corée, R. p. dém. de	81	78[1]	80[1]	83[1]	83[1]	78[1]	85[1]	82	76	72
Korea, Republic of Corée, République de	78	75	75	79	88	85	76	64	64	62
Kyrgyzstan Kirghizistan	...	...	...	...	...	...	...	...	* 6	6[1]
Lao People's Dem. Rep. Rép. dém. pop. lao	5	6	3	4	4	5	5	6	7	7
Lebanon Liban	12	8	12	17	11	12	16	13	26	14[1]
Malaysia Malaisie	4 394	4 893	5 356	5 310	5 893	7 069	7 142	7 134	7 421	8 625
Maldives Maldives	1	1	1	1	1	2	* 2	* 2	2	* 2
Myanmar Myanmar	359	439	432	426	378	297	317	329	278	328
Nepal Népal	32	34	34	35	39	41	41	40	39	41
Pakistan Pakistan	475	550	589	618	616	643	700	889	676	604
Philippines Philippines	848	1 182	1 533	1 429	1 368	1 296	1 313	1 242	1 257	1 297
Saudi Arabia Arabie saoudite	2	1	1	2	2	2	2	3	3	3
Singapore Singapour	1	1	1	1	0[1]	0[1]	0[1]	0[1]	0[1]	0[1]
Sri Lanka Sri Lanka	199	279	307	232	197	250	255	220	238	211
Syrian Arab Republic Rép. arabe syrienne	127	107	156	109	180	90	172	131	216	168
Tajikistan Tadjikistan	...	...	...	...	...	...	...	...	* 46	65[1]
Thailand Thaïlande	415	457	485	507	584	658	664	651	679	765
Turkey Turquie	672	659	816	813	953	844	836	677	757	637

36
Oil crops, in oil equivalent
Production: thousand metric tons [*cont.*]
    Cultures d'huile, en équivalent d'huile
    Production : milliers de tonnes métriques [*suite*]

Country or area Pays ou zone	1984	1985	1986	1987	1988	1989	1990	1991	1992	1993
Turkmenistan Turkménistan	...	...	...	...	...	...	...	...	140	125
Uzbekistan Ouzbékistan	...	...	...	...	...	...	...	...	419	433
Viet Nam Viet Nam	143	167	184	200	202	210	211	239	252	264
Yemen Yémen	* 5	* 5	5	4	4	5	5	5	6	7
Europe **Europe**	**5 803**	**6 196**	**6 591**	**8 375**	**7 374**	**7 803**	**7 566**	**8 762**	**7 767**	**7 389**
Albania Albanie	27	26	21	21	15	20	12	10	10	10[1]
Austria Autriche	6	8	10	35	54	60	62	84	93	93[1]
Belarus[1] Bélarus[1]	...	...	...	...	...	...	...	...	44	40
Belgium-Luxembourg Belgique-Luxembourg	7	5	6	8	7	8	11	12	10	* 11
Bosnia & Herzegovina[1] Bosnie-Herzégovine[1]	...	...	...	...	...	...	...	...	2	3
Bulgaria Bulgarie	202	156	210	175	157	195	166	189	250	163
Croatia Croatie	...	...	...	...	...	...	...	...	37	38
former Czechoslovakia† anc. Tchécoslovaquie†	133	127	139	152	169	180	178	230	211	...
Denmark Danemark	167	191	217	195	177	229	278	254	142	151
Estonia[1] Estonie[1]	...	...	...	...	...	...	...	...	1	1
Finland Finlande	30	31	43	31	42	44	41	33	46	45
France France	908	1 147	1 163	2 064	1 837	1 595	1 743	1 909	1 525	1 304
Germany † Allemagne†	343	419	499	585	597	677	764	1 094	984	1 018
Greece Grèce	367	515	430	430	463	474	306	497	470	535
Hungary Hongrie	295	325	411	382	344	353	331	385	338	303
Ireland Irlande	5	5	2	6	4	4	7	8	6	6
Italy Italie	535	934	733	1 233	941	1 147	722	1 285	908	996
Latvia Lettonie	...	...	...	...	...	...	...	...	1	0[1]
Lithuania Lituanie	...	...	...	...	...	...	...	...	4	4[1]

36
Oil crops, in oil equivalent
Production: thousand metric tons [*cont.*]
Cultures d'huile, en équivalent d'huile
Production : milliers de tonnes métriques [*suite*]

Country or area Pays ou zone	1984	1985	1986	1987	1988	1989	1990	1991	1992	1993
Netherlands Pays-Bas	17	13	9	13	11	11	12	11	6	5
Norway Norvège	7	4	4	4	3	3	3	2	3	5
Poland Pologne	332	385	462	421	426	563	434	371	269	212
Portugal Portugal	100	72	98	80	55	94	71	118	59	55
Republic of Moldova République de Moldova	...	...	...	...	...	...	...	...	80	92
Romania Roumanie	451	363	438	388	362	346	264	291	341	308
Russian Federation Fédération de Russie	...	...	...	...	...	...	...	...	1 432	1 323
Slovenia Slovénie	...	...	...	...	...	...	...	...	3	2[1]
Spain Espagne	1 242	835	944	1 289	988	1 051	1 308	1 099	1 266	1 147
Sweden Suède	134	130	131	88	100	148	148	101	99	* 88
Switzerland Suisse	15	14	14	17	18	19	16	18	16	18
TFYR Macedonia L'ex-R.y. Macédoine	...	...	...	...	...	...	...	...	16	8
Ukraine Ukraine	...	...	...	...	...	...	...	...	908	869
United Kingdom Royaume-Uni	324	313	336	478	374	344	465	514	493	479
Yugoslavia Yougoslavie	...	...	...	...	...	...	...	...	163	174
Yugoslavia, SFR† Yougoslavie, Rfs†	158	177	273	280	229	237	227	245	...	...
Oceania **Océanie**	**645**	**759**	**724**	**665**	**662**	**678**	**642**	**731**	**778**	**778**
American Samoa[1] Samoa américaines[1]	1	1	1	1	1	1	1	1	1	1
Australia Australie	168	269	228	188	228	227	176	276	254	248
Cocos (Keeling) Islands[1] Iles des Cocos (Keeling)[1]	1	1	1	1	1	1	1	1	1	1
Cook Islands Iles Cook	1[1]	1[1]	1[1]	1[1]	1[1]	1[1]	1[1]	1[1]	0	1
Fiji Fidji	30	28	34	25	23	25	33	26	32	26[1]
French Polynesia[1] Polynésie française[1]	9	14	14	15	12	12	13	11	11	11
Guam[1] Guam[1]	4	4	4	5	5	5	5	5	5	5

36
Oil crops, in oil equivalent
Production: thousand metric tons [*cont.*]
    Cultures d'huile, en équivalent d'huile
    Production : milliers de tonnes métriques [*suite*]

Country or area Pays ou zone	1984	1985	1986	1987	1988	1989	1990	1991	1992	1993
Kiribati[1] Kiribati[1]	13	9	7	7	14	10	6	9	9	8
New Caledonia[1] Nouvelle-Calédonie[1]	2	2	2	2	1	1	1	1	1	1
New Zealand Nouvelle-Zélande	0[1]	0[1]	3	3	2	1	1	1	1	1[1]
Palau[1] Palaos[1]	21	23	23	23	18	18	18	18	18	18
Papua New Guinea Papouasie-Nvl-Guinée	270	283	282	281	246	270	268	284	332	346
Samoa Samoa	20[1]	23[1]	25	22[1]	21[1]	21[1]	18[1]	12[1]	13[1]	17[1]
Solomon Islands Iles Salomon	48	49	43	40	44	48	49	45	59	55
Tonga[1] Tonga[1]	8	7	10	6	5	4	4	4	4	4
Vanuatu Vanuatu	49	42	45	45	40	32	47	35	34	34[1]
former USSR† ancienne URSS†	2 930	3 236	3 244	3 628	3 796	4 177	3 865	3 351	...	...

Source:
Food and Agriculture Organization of the United Nations
(Rome).

† For detailed descriptions of data pertaining to
former Czechoslovakia, Germany, SFR Yugoslavia and former
USSR, see Annex I - Country or area nomenclature, regional
and other groupings.

1 FAO estimate.

Source:
Organisation des Nations Unies pour l'alimentation et
l'agriculture (Rome).

† Pour les descriptions en détails des données
relatives à l'ancienne Tchécoslovaquie, l'Allemagne, la Rfs
Yougoslavie et l'ancienne URSS, voir l'Annexe I -
Nomenclature des pays ou zones, groupements régionaux et
autres groupements.

1 Estimation de la FAO.

37
Livestock
Cheptel

Thousand head

Milliers de têtes

Country or area	1986	1987	1988	1989	1990	1991	1992	1993	Pays ou zone
World									**Monde**
Cattle	1265 772	1266 265	1259 481	1274 699	1283 140	1283 782	1277 834	1277 794	Bovine
Sheep	1115 158	1137 580	1146 265	1174 494	1182 800	1159 904	1134 196	1110 782	Ovine
Pigs	786 098	799 854	793 831	810 342	817 625	862 900	862 388	870 705	Porcine
Horses	60 163	60 281	60 333	60 723	60 814	61 157	60 608	60 376	Chevaline
Asses	41 361	41 718	42 297	42 676	43 151	43 389	43 889	43 863	Asine
Mules	14 113	14 358	14 500	14 671	14 748	14 864	15 027	15 067	Mulassière
Africa									**Afrique**
Cattle	178 196	176 269	178 965	183 301	188 492	188 923	188 012	188 140	Bovine
Sheep	188 756	193 931	194 237	200 011	202 106	202 978	206 261	206 347	Ovine
Pigs	13 177	13 842	14 354	15 495	16 451	17 588	18 873	20 478	Porcine
Horses	4 211	4 296	4 372	4 419	4 582	4 643	4 678	4 743	Chevaline
Asses	12 116	12 071	12 322	12 253	12 685	13 071	13 330	13 369	Asine
Mules	1 260	1 306	1 300	1 314	1 342	1 374	1 400	1 413	Mulassière
Algeria									**Algérie**
Cattle	1 347	1 416	1 435	1 405	1 393	1 430[1]	1 450[1]	1 460[1]	Bovine
Sheep	15 830	16 148	16 429	17 316	17 698	18 500[1]	18 600[1]	18 800[1]	Ovine
Pigs [1]	5	5	5	5	5	5	6	6	Porcine [1]
Horses	80	84	85	86	81	83[1]	84[1]	84[1]	Chevaline
Asses	340	354	327	322	299	340[1]	340[1]	340[1]	Asine
Mules	131	127	113	101	100	107[1]	107[1]	107[1]	Mulassière
Angola									**Angola**
Cattle	3 400	3 300	3 200	3 100	3 100[1]	3 150	3 200	3 200[1]	Bovine
Sheep	255[1]	250[1]	250[1]	240	240	240	250	250[1]	Ovine
Pigs	750[1]	770[1]	770[1]	800	800	805	810	810[1]	Porcine
Horses [1]	1	1	1	1	1	1	1	1	Chevaline [1]
Asses [1]	5	5	5	5	5	5	5	5	Asine [1]
Benin									**Bénin**
Cattle	895	896	925	943	1 080	1 088	1 095[1]	1 100[1]	Bovine
Sheep	830	831	821	846	869	893	920[1]	940[1]	Ovine
Pigs	392	420	437	459	462	515	530[1]	550[1]	Porcine
Horses [1]	6	6	6	6	6	6	6	6	Chevaline [1]
Asses [1]	1	1	1	1	1	1	1	1	Asine [1]
Botswana									**Botswana**
Cattle	2 332	2 263	2 408	2 543	2 696	2 844	2 800[1]	2 700[1]	Bovine
Sheep	229	240	259	286	317	320[1]	325[1]	325[1]	Ovine
Pigs	11	11	13	15	16	16[1]	16[1]	16[1]	Porcine
Horses	24	24	29	32	34	34[1]	34[1]	34[1]	Chevaline
Asses	143[1]	144[1]	148	151	152[1]	153[1]	153[1]	153[1]	Asine
Mules	3[1]	3[1]	2	2	3[1]	3[1]	3[1]	3[1]	Mulassière
Burkina Faso									**Burkina Faso**
Cattle	3 637	3 711	3 785	3 860	3 937	4 015	4 096	4 178	Bovine
Sheep	4 484	4 619	4 757	4 900	5 050	5 198	5 350	5 500	Ovine
Pigs	441	459	477	496	516	518	530	540	Porcine
Horses	21	21	22	22	22	22	22	23	Chevaline
Asses	380	387	395	403	411	419	427	436[1]	Asine
Burundi									**Burundi**
Cattle	479	422	429	423	432	409	420[1]	420[1]	Bovine
Sheep	329	313	350	327	361	381	390[1]	390[1]	Ovine
Pigs	77	80	115	91	103	92	100[1]	100[1]	Porcine
Cameroon									**Cameroun**
Cattle	4 255	4 362	4 471	4 582	4 697	4 700[1]	4 730[1]	4 867[1]	Bovine
Sheep	2 473	2 597	2 897	3 170	* 3 500	3 550[1]	3 560[1]	3 770[1]	Ovine
Pigs	1 451	1 178	1 237	1 299	1 364	1 370[1]	1 380[1]	1 434[1]	Porcine
Horses	18[1]	15	13	14[1]	14[1]	15[1]	15[1]	15[1]	Chevaline
Asses	34[1]	25	33	35[1]	35[1]	36[1]	36[1]	36[1]	Asine
Cape Verde									**Cap-Vert**
Cattle	12	13	18	19	19	16	19[1]	21[1]	Bovine
Sheep	2	4	5	6	8	4	6[1]	6[1]	Ovine
Pigs	67	80	85	91	99	100[1]	102[1]	105[1]	Porcine
Horses	0	0[1]	1	1	1[1]	0	1[1]	1[1]	Chevaline
Asses	6[1]	6	7	11	11[1]	13	11[1]	11[1]	Asine
Mules	1	1[1]	1	2	2[1]	2	2[1]	2[1]	Mulassière

37
Livestock
Thousand head [*cont.*]
Cheptel
Milliers de têtes [*suite*]

Country or area	1986	1987	1988	1989	1990	1991	1992	1993	Pays ou zone
Central African Rep.									**Rép. centrafricaine**
Cattle	2 216	2 306	2 398	2 495	2 595	2 677	2 700[1]	2 781[1]	Bovine
Sheep	113	117	122	128	134	135[1]	137[1]	142[1]	Ovine
Pigs	360	371	382	393	405	441	460[1]	474[1]	Porcine
Chad									**Tchad**
Cattle	3 886	4 002	4 098	4 197	4 297	4 400	4 507	4 600[1]	Bovine
Sheep	1 800	1 800	1 815	1 870	1 964	1 983	* 2 043	2 100[1]	Ovine
Pigs	11	12	13	13	14	15	14[1]	14[1]	Porcine
Horses	182	186	188	192	195	182	185[1]	184[1]	Chevaline
Asses	227	222	234	259	264	269	270[1]	271[1]	Asine
Comoros									**Comores**
Cattle	49[1]	44	45	47	47[1]	47[1]	47[1]	47[1]	Bovine
Sheep	11[1]	12	12	13	13[1]	14[1]	14[1]	15[1]	Ovine
Asses [1]	4	4	4	4	4	5	5	5	Asine [1]
Congo									**Congo**
Cattle	70	70	69	62	70	64	65[1]	67[1]	Bovine
Sheep	94	97	101	101	104	107	110[1]	111[1]	Ovine
Pigs	44	47	50	44	50[1]	52[1]	55[1]	56[1]	Porcine
Côte d'Ivoire									**Côte d'Ivoire**
Cattle	* 885	917	992	1 049	1 108	1 145	1 180	1 205	Bovine
Sheep	1 050[1]	1 051	1 090	1 115	1 134	1 161	1 190	1 219	Ovine
Pigs	333[1]	335	342	351	360	372	* 382	* 392	Porcine
Djibouti									**Djibouti**
Cattle [1]	120	140	160	180	195	190	190	190	Bovine [1]
Sheep [1]	410	412	414	400	430	470	470	470	Ovine [1]
Asses [1]	8	8	8	8	8	8	8	8	Asine [1]
Egypt									**Egypte**
Cattle	1 855	2 300[1]	2 780	2 721	2 618	2 973	3 286[1]	3 226[1]	Bovine
Sheep	3 683	3 793	3 908	3 481	3 364	3 084	3 385	3 707	Ovine
Pigs	26[1]	22[1]	21[1]	23	24	24	25[1]	26[1]	Porcine
Horses	9	9[1]	10[1]	10[1]	10[1]	10[1]	10[1]	10[1]	Chevaline
Asses	1 879	1 600[1]	1 500[1]	1 158	1 380[1]	1 530	1 550[1]	1 550[1]	Asine
Mules	1	1[1]	1[1]	1[1]	1[1]	1[1]	1[1]	1[1]	Mulassière
Equatorial Guinea									**Guinée équatoriale**
Cattle [1]	4	5	5	5	5	5	5	5	Bovine [1]
Sheep [1]	35	35	35	35	35	36	36	36	Ovine [1]
Pigs [1]	5	5	5	5	5	5	5	5	Porcine [1]
Eritrea									**Erythrée**
Cattle [1]	..	..	..	..	..	..	..	1 550	Bovine [1]
Sheep [1]	..	..	..	..	..	..	..	1 512	Ovine [1]
Ethiopia									**Ethiopie**
Cattle [2]	* 30 000	* 27 000	* 27 000	* 28 900	30 000[1]	30 000[1]	31 000[1]	29 450[1]	Bovine [2]
Sheep	* 23 000	* 24 000	* 24 000	* 24 000	* 22 960	23 000[1]	23 200[1]	21 700[1]	Ovine
Pigs [1]	19	18	18	19	19	20	20	20	Porcine [1]
Horses [1 2]	2 450	2 500	2 550	2 600	2 650	2 700	2 750	2 750	Chevaline [1 2]
Asses [1 2]	4 600	4 700	4 800	4 900	5 000	5 100	5 200	5 200	Asine [1 2]
Mules [1 2]	510	530	550	570	590	610	630	630	Mulassière [1 2]
Gabon									**Gabon**
Cattle	19	21	24	27	27[1]	28[1]	29[1]	30[1]	Bovine
Sheep	146	149	153	157	160[1]	165[1]	170[1]	170[1]	Ovine
Pigs	136	141	147	159	160[1]	162[1]	164[1]	165[1]	Porcine
Gambia									**Gambie**
Cattle	320[1]	350[1]	387	390[1]	400[1]	390[1]	400[1]	400[1]	Bovine
Sheep	154[1]	157[1]	163	140[1]	121	121	121[1]	121[1]	Ovine
Pigs	11[1]	11[1]	11	11[1]	11[1]	11[1]	11[1]	11[1]	Porcine
Horses	16[1]	17[1]	17	18	16	16	16[1]	16[1]	Chevaline
Asses	40[1]	40[1]	40	41	37	31	30[1]	30[1]	Asine
Ghana									**Ghana**
Cattle	1 135	1 170	1 145	1 136	1 145	1 195	1 159	1 200[1]	Bovine
Sheep	1 814	1 989	2 046	2 212	2 224	2 162	2 126	2 200[1]	Ovine
Pigs	469	399	478	559	474	454	413	450[1]	Porcine
Horses	2	2	2	2	1	1	2	2[1]	Chevaline
Asses	13	14	10	11	10	12	13	13[1]	Asine
Guinea									**Guinée**
Cattle	1 500[1]	1 500[1]	1 500[1]	1 436	1 472	1 530[1]	1 612	1 650[1]	Bovine
Sheep	445[1]	440[1]	435[1]	432	420	425[1]	431	435[1]	Ovine

37
Livestock
Thousand head [*cont.*]
Cheptel
Milliers de têtes [*suite*]

Country or area	1986	1987	1988	1989	1990	1991	1992	1993	Pays ou zone
Pigs	30[1]	28[1]	25[1]	24	22	27[1]	32	33[1]	Porcine
Horses	1[1]	2[1]	2[1]	2	2[1]	2[1]	2[1]	2[1]	Chevaline
Asses	2[1]	2[1]	1[1]	1	1[1]	1[1]	1[1]	1[1]	Asine
Guinea—Bissau									**Guinée—Bissau**
Cattle	333	350[1]	370[1]	400	410	425[1]	450	475[1]	Bovine
Sheep	200	210[1]	220[1]	230[1]	242	245[1]	250	255[1]	Ovine
Pigs	286	290[1]	290[1]	290[1]	290[1]	290	300	310[1]	Porcine
Horses	1	1[1]	1[1]	2	2	2	2	2[1]	Chevaline
Asses	4[1]	4[1]	5[1]	5	5	5	5	5[1]	Asine
Kenya									**Kenya**
Cattle	12 600[1]	12 645	13 050	13 457	13 793	13 000[1]	12 000[1]	11 000[1]	Bovine
Sheep	6 500[1]	* 6 040	* 6 167	* 6 325	* 6 516	6 500[1]	6 000[1]	5 500[1]	Ovine
Pigs	92[1]	94	97	100	105	105[1]	105[1]	105[1]	Porcine
Horses [1]	2	2	2	2	2	2	2	2	Chevaline [1]
Lesotho									**Lesotho**
Cattle	525	625	627	583	650	660[1]	600[1]	650[1]	Bovine
Sheep	1 392	1 704	1 650	1 505	1 641	1 676	1 600[1]	1 665[1]	Ovine
Pigs	71	70[1]	72[1]	73[1]	74[1]	75[1]	75[1]	76[1]	Porcine
Horses	127	129	116	104	121[1]	122[1]	122[1]	123[1]	Chevaline
Asses	136	148	131	151	155[1]	158[1]	160[1]	162[1]	Asine
Mules [1]	1	1	1	1	1	1	1	1	Mulassière [1]
Liberia									**Libéria**
Cattle [1]	42	42	42	40	38	37	36	36	Bovine [1]
Sheep [1]	239	240	240	230	220	215	210	210	Ovine [1]
Pigs [1]	130	140	140	130	120	120	120	120	Porcine [1]
Libyan Arab Jamah.									**Jamah. arabe libyenne**
Cattle	140[1]	90	95	102	120	125[1]	130[1]	135[1]	Bovine
Sheep	5 000[1]	4 500	4 500	5 000	5 200	5 500[1]	5 600[1]	5 650[1]	Ovine
Horses	30[1]	20	25	20	20	25[1]	28[1]	29[1]	Chevaline
Asses [1]	61	61	61	62	62	62	63	65	Asine [1]
Madagascar									**Madagascar**
Cattle	10 207	10 220	10 232	10 243	10 254	* 10 265	* 10 276	10 280[1]	Bovine
Sheep	664	670[1]	683	700[1]	* 737	* 721	730[1]	735[1]	Ovine
Pigs	1 412	1 532	1 412	1 400	1 431	* 1 461	* 1 493	1 495[1]	Porcine
Horses	1	0	0	0	0[1]	0[1]	0[1]	0[1]	Chevaline
Malawi									**Malawi**
Cattle	1 011	1 055	860	850[1]	* 836	* 899	* 967	970[1]	Bovine
Sheep	165	170[1]	170[1]	175[1]	* 177	* 186	* 195	195[1]	Ovine
Pigs	282	313	250	240[1]	* 233	* 235	* 238	240[1]	Porcine
Asses	2	1	1	2[1]	2[1]	2[1]	2[1]	2[1]	Asine
Mali									**Mali**
Cattle	4 475	4 589	4 738	4 826	4 996	5 198	5 373	5 554	Bovine
Sheep	* 5 340	* 5 329	5 527	5 771	6 086	6 359	6 658	6 971	Ovine
Pigs	54	56	58	55	56	67	75	84	Porcine
Horses	67	54	56	55	77	83	85	87[1]	Chevaline
Asses	383	348	510	517	575	590	600	610[1]	Asine
Mauritania									**Mauritanie**
Cattle	1 200	1 220	1 260	1 300	1 350	1 400	1 000[1]	1 000[1]	Bovine
Sheep	* 4 100	* 4 200	* 4 500	* 4 800	* 5 100	5 300[1]	5 000[1]	4 800[1]	Ovine
Horses [1]	16	16	17	17	18	18	18	18	Chevaline [1]
Asses [1]	149	149	149	150	151	153	154	155	Asine [1]
Mauritius									**Maurice**
Cattle	32	32	33	34	33	34[1]	33	34[1]	Bovine
Sheep	7	7	7	7	7	6[1]	7	7[1]	Ovine
Pigs	10	13	10	10	13	14	14	14[1]	Porcine
Morocco									**Maroc**
Cattle	2 851	3 178	3 137	3 324	3 346	3 183	3 269	2 924	Bovine
Sheep	14 545	16 136	12 733	13 761	13 514	13 308	17 201	16 302	Ovine
Pigs [1]	8	9	9	9	9	9	9	10	Porcine [1]
Horses	172	198	198	184	194	188	186	180	Chevaline
Asses	785	854	904	918	912	896	940	946	Asine
Mules	494	524	511	515	523	528	536	547	Mulassière
Mozambique									**Mozambique**
Cattle [1]	1 340	1 350	1 360	1 370	1 380	1 370	1 250	1 250	Bovine [1]
Sheep [1]	116	117	119	120	121	118	118	118	Ovine [1]
Pigs [1]	150	155	160	165	170	165	170	170	Porcine [1]

37

Livestock
Thousand head [*cont.*]
Cheptel
Milliers de têtes [*suite*]

Country or area	1986	1987	1988	1989	1990	1991	1992	1993	Pays ou zone
Asses [1]	20	20	20	20	20	20	20	20	Asine [1]
Namibia									**Namibie**
Cattle	1 990	1 835	1 970	2 014	2 087	2 212	2 206	2 300[1]	Bovine
Sheep	2 741	2 811	3 046	3 242	3 328	3 295	2 863	2 900[1]	Ovine
Pigs	19	19[1]	18[1]	20	18	17	15	15[1]	Porcine
Horses	49[1]	49[1]	50[1]	47	52	53[1]	55	55[1]	Chevaline
Asses	68[1]	68[1]	68[1]	68[1]	68[1]	68[1]	* 70	71[1]	Asine
Mules	6[1]	6[1]	6[1]	6[1]	6[1]	6[1]	* 7	7[1]	Mulassière
Niger									**Niger**
Cattle	1 429	1 495	1 563	1 636	1 711	1 790	1 800[1]	1 800[1]	Bovine
Sheep	2 549	2 676	2 900	2 950	3 098	3 253	3 400[1]	3 505[1]	Ovine
Pigs [1]	36	37	37	37	37	38	38	39	Porcine [1]
Horses	74	76	78	80	82[1]	82[1]	82[1]	82[1]	Chevaline
Asses	367	382	397	412	431	449	450[1]	462[1]	Asine
Nigeria									**Nigéria**
Cattle	13 156	13 415	13 759	14 170	14 640	15 140	15 700	16 316	Bovine
Sheep	10 701	11 107	11 575	11 971	12 460	13 000	13 500	14 000[1]	Ovine
Pigs	2 000[1]	2 400[1]	2 600[1]	3 000[1]	3 410	4 263	5 328	6 660	Porcine
Horses	216[1]	214[1]	212[1]	210[1]	208	206[1]	205[1]	204[1]	Chevaline
Asses	800[1]	830[1]	860[1]	900[1]	936	960[1]	1 000[1]	1 000[1]	Asine
Réunion									**Réunion**
Cattle	19	19	19	18	20	22	23	23[1]	Bovine
Sheep	3	3	2	2	2	2	2	2[1]	Ovine
Pigs	70	99	97	84	86	94	92	93[1]	Porcine
Rwanda									**Rwanda**
Cattle	614	583	579	594	582	600[1]	610[1]	610[1]	Bovine
Sheep	349	363	364	393	389	390[1]	395[1]	402[1]	Ovine
Pigs	89	105	126	120	116	120[1]	130[1]	130[1]	Porcine
Saint Helena									**Sainte−Hélène**
Cattle	1[1]	1	1	1	1	1	1	1	Bovine
Sheep	1[1]	1	2	2	2	1	1	1	Ovine
Pigs	1[1]	1	1	1	1	1	1	1	Porcine
Sao Tome and Principe									**Sao Tomé−et−Principe**
Cattle [1]	3	3	3	4	4	4	4	4	Bovine [1]
Sheep [1]	2	2	2	2	2	2	2	2	Ovine [1]
Pigs [1]	3	3	3	3	3	3	3	3	Porcine [1]
Senegal									**Sénégal**
Cattle	2 484	2 543	2 465	2 540	2 622	2 687	2 700[1]	2 750[1]	Bovine
Sheep	3 159	3 326	3 120	3 300[1]	3 400[1]	* 3 800	* 4 200	* 4 400	Ovine
Pigs	310	285	288	291	295	300[1]	310[1]	320[1]	Porcine
Horses	314	338	360	380[1]	440	453	431	498	Chevaline
Asses	255	278	290	300[1]	303	328	362	364	Asine
Seychelles									**Seychelles**
Cattle [1]	2	2	2	2	2	2	2	2	Bovine [1]
Pigs [1]	16	16	17	17	18	18	18	18	Porcine [1]
Sierra Leone									**Sierra Leone**
Cattle	333[1]	333[1]	333	333	333	333	333[1]	333[1]	Bovine
Sheep	280[1]	280[1]	264	267	271	274	275[1]	278[1]	Ovine
Pigs	47	50	50	50	50	50	50	50	Porcine [1]
Somalia									**Somalie**
Cattle	4 571	4 770	4 983	4 990	3 800[1]	2 000[1]	1 000[1]	1 500[1]	Bovine
Sheep	12 274	13 195	14 304	14 350[1]	12 000[1]	9 500[1]	6 000[1]	6 500[1]	Ovine
Pigs [1]	10	10	10	10	10	7	2	6	Porcine [1]
Horses [1]	1	1	1	1	1	1	1	1	Chevaline [1]
Asses [1]	25	25	25	25	25	25	20	23	Asine [1]
Mules [1]	23	23	23	24	24	23	18	20	Mulassière [1]
South Africa									**Afrique du Sud**
Cattle	* 11 750	* 11 799	11 820[1]	11 850[1]	* 13 398	* 13 512	* 13 311	* 13 239	Bovine
Sheep *	29 481	29 753	29 640	30 935	32 665	32 580	32 110	30 000	Ovine *
Pigs	* 1 445	* 1 455	1 460[1]	1 470[1]	1 480[1]	1 490[1]	1 490[1]	1 499[1]	Porcine
Horses [1]	230	230	230	230	230	230	230	230	Chevaline [1]
Asses [1]	210	210	210	210	210	210	210	210	Asine [1]
Mules [1]	14	14	14	14	14	14	14	14	Mulassière [1]
Sudan									**Soudan**
Cattle	19 632	19 738	19 858	20 167	20 583	21 028	* 21 600	21 600[1]	Bovine
Sheep	18 691	18 807	19 207	19 668	20 168	20 700	* 22 600	22 500[1]	Ovine

37
Livestock
Thousand head [*cont.*]
Cheptel
Milliers de têtes [*suite*]

Country or area	1986	1987	1988	1989	1990	1991	1992	1993	Pays ou zone
Horses [1]	20	21	20	21	22	22	23	23	Chevaline [1]
Asses [1]	650	660	650	670	675	680	681	670	Asine [1]
Mules [1]	1	1	1	1	1	1	1	1	Mulassière [1]
Swaziland									**Swaziland**
Cattle	653	641	640	679	716	740	753	753[1]	Bovine
Sheep	28	28	26	25	24	24	23	23[1]	Ovine
Pigs	18	21	18	19	24	28	31	31[1]	Porcine
Horses	2	1	1	1	1	1	1	1[1]	Chevaline
Asses	13	13	12	12	12	12	12	12[1]	Asine
Togo									**Togo**
Cattle	232	235	255	250[1]	243	247	246	246	Bovine
Sheep	1 048	1 094	1 175	1 147	1 144	1 200	1 200	1 200	Ovine
Pigs	236	245	385	433	709	709	800[1]	850[1]	Porcine
Horses	1	2	2[1]	2[1]	2[1]	2[1]	2[1]	2[1]	Chevaline
Asses	3	2	3	4	3	3[1]	3[1]	3[1]	Asine
Tunisia									**Tunisie**
Cattle	624	666	634	626	622	631	* 636	659	Bovine
Sheep	5 409	5 707	5 581	5 548	5 966	6 290	6 400[1]	7 110	Ovine
Pigs	4[1]	4[1]	3[1]	5[1]	7	6[1]	6[1]	6[1]	Porcine
Horses [1]	55	55	55	55	55	56	56	56	Chevaline [1]
Asses [1]	217	218	220	224	226	229	229	230	Asine [1]
Mules [1]	75	75	76	77	78	79	80	81	Mulassière [1]
Uganda									**Ouganda**
Cattle	5 200	3 905	4 260	4 417	4 913	5 000[1]	5 100[1]	5 200[1]	Bovine
Sheep	1 680	1 400[1]	1 300[1]	1 320[1]	1 350[1]	* 1 380	* 1 560	* 1 760	Ovine
Pigs	250	470	452	716	824	850[1]	880[1]	900[1]	Porcine
Asses [1]	17	17	17	17	17	17	17	17	Asine [1]
United Rep.Tanzania									**Rép. Unie de Tanzanie**
Cattle	12 688	12 777	12 866	12 956	13 047	13 138	* 13 217	* 13 296	Bovine
Sheep	3 499	3 512	3 526	3 541	3 557	3 556	* 3 706	* 3 828	Ovine
Pigs	285[1]	290[1]	300[1]	310[1]	320[1]	* 330	* 330	* 335	Porcine
Asses [1]	170	171	172	173	174	175	176	177	Asine [1]
Zaire									**Zaïre**
Cattle	1 340	1 386	1 434	1 484	1 535	1 586	1 600[1]	1 650[1]	Bovine
Sheep	824	849	874	900	927	974	980[1]	985[1]	Ovine
Pigs	802	858	918	982	1 050	1 118	1 120[1]	1 130[1]	Porcine
Zambia									**Zambie**
Cattle	2 520	2 601	2 638	2 672	* 2 878	* 2 984	* 3 095	* 3 204	Bovine
Sheep	* 44	* 47	* 51	55	* 60	* 62	* 63	* 67	Ovine
Pigs	187	196	207	297	295[1]	296[1]	290[1]	293[1]	Porcine
Asses	1[1]	1[1]	1[1]	1	2[1]	2[1]	2[1]	2[1]	Asine
Zimbabwe									**Zimbabwe**
Cattle	5 783	5 918	5 805	5 850	6 218	6 374	4 700	4 000[1]	Bovine
Sheep	512	569	673	569	600	584	550[1]	530[1]	Ovine
Pigs	219	216	238	304	290	305	285[1]	270[1]	Porcine
Horses	* 22	22[1]	23[1]	23[1]	24[1]	24[1]	24[1]	23[1]	Chevaline
Asses [1]	99	100	101	102	103	104	104	103	Asine [1]
Mules [1]	1	1	1	1	1	1	1	1	Mulassière [1]
America, North									**Amérique du Nord**
Cattle	169 672	164 647	162 083	162 630	161 779	161 956	161 467	162 830	**Bovine**
Sheep	17 775	18 401	18 603	18 657	18 840	18 246	18 087	18 089	**Ovine**
Pigs	87 128	86 210	87 629	89 209	85 617	86 717	90 755	93 282	**Porcine**
Horses	14 105	14 112	14 430	14 484	14 297	14 558	14 335	14 340	**Chevaline**
Asses	3 650	3 652	3 658	3 689	3 688	3 687	3 684	3 691	**Asine**
Mules	3 620	3 631	3 645	3 654	3 664	3 675	3 684	3 695	**Mulassière**
Antigua and Barbuda									**Antigua−et−Barbuda**
Cattle [1]	18	18	18	16	16	16	16	16	Bovine [1]
Sheep [1]	13	13	13	13	12	13	13	13	Ovine [1]
Pigs [1]	4	4	4	4	4	4	4	4	Porcine [1]
Horses [1]	1	1	1	1	1	1	1	1	Chevaline [1]
Asses [1]	2	2	2	2	1	2	2	2	Asine [1]
Bahamas									**Bahamas**
Cattle [1]	5	5	5	5	5	6	6	6	Bovine [1]
Sheep [1]	39	40	40	40	40	40	40	40	Ovine [1]
Pigs [1]	15	15	15	15	15	15	15	15	Porcine [1]

37
Livestock
Thousand head [*cont.*]
Cheptel
Milliers de têtes [*suite*]

Country or area	1986	1987	1988	1989	1990	1991	1992	1993	Pays ou zone
Barbados									**Barbade**
Cattle [1]	21	23	24	27	30	30	31	33	Bovine [1]
Sheep [1]	63	65	65	65	65	65	65	66	Ovine [1]
Pigs [1]	44	43	44	44	45	45	45	45	Porcine [1]
Horses [1]	1	1	1	1	1	1	1	1	Chevaline [1]
Asses [1]	2	2	2	2	2	2	2	2	Asine [1]
Mules [1]	2	2	2	2	2	2	2	2	Mulassière [1]
Belize									**Belize**
Cattle	49	* 49	* 50	* 50	* 51	* 51	* 54	* 58	Bovine
Sheep	4	4[1]	4[1]	4[1]	4[1]	4[1]	4[1]	4[1]	Ovine
Pigs	25	25[1]	26[1]	26[1]	26[1]	26[1]	26[1]	26[1]	Porcine
Horses [1]	5	5	5	5	5	5	5	5	Chevaline [1]
Mules [1]	4	4	4	4	4	4	4	4	Mulassière [1]
Bermuda									**Bermudes**
Cattle [1]	1	1	1	1	1	1	1	1	Bovine [1]
Pigs [1]	2	2	2	2	2	2	2	2	Porcine [1]
Horses	0[1]	0	0[1]	1	1[1]	1[1]	1[1]	1[1]	Chevaline
British Virgin Islands									**Iles Vierges britanniques**
Cattle [1]	2	2	2	2	2	2	2	2	Bovine [1]
Sheep	6[1]	6	6[1]	6[1]	6[1]	6[1]	6[1]	6[1]	Ovine [1]
Pigs	2[1]	2	2[1]	2[1]	2[1]	2[1]	2[1]	2[1]	Porcine
Canada									**Canada**
Cattle	10 956	10 667	10 756	10 984	11 220	11 289	11 713	11 786	Bovine
Sheep	490	485	521	560	595	628	654	662	Ovine
Pigs	9 967	9 998	10 801	10 951	10 392	10 172	10 498	10 572	Porcine
Horses [1]	390	400	405	415	415	415	420	420	Chevaline [1]
Mules [1]	4	4	4	4	4	4	4	4	Mulassière [1]
Cayman Islands									**Iles Caïmanes**
Cattle	2	2	2	2	2	1	1	1[1]	Bovine
Costa Rica									**Costa Rica**
Cattle	* 2 306	* 2 294	2 190	* 2 168	* 2 201	* 2 175	* 2 132	* 2 122	Bovine
Sheep	3	3	3	3[1]	3[1]	3[1]	3[1]	3[1]	Ovine
Pigs	* 183	* 160	* 237	237[1]	241[1]	244[1]	245[1]	244[1]	Porcine
Horses [1]	114	114	114	114	114	114	114	114	Chevaline [1]
Asses [1]	7	7	7	7	7	7	7	7	Asine [1]
Mules [1]	5	5	5	5	5	5	5	5	Mulassière [1]
Cuba									**Cuba**
Cattle	5 020	5 007	4 984	4 927	4 920	4 920[1]	4 700[1]	4 500[1]	Bovine
Sheep [1]	380	382	382	385	385	385	350	310	Ovine [1]
Pigs [1]	1 873	1 996	2 125	2 401	2 002	1 903	1 703	1 603	Porcine [1]
Horses	740	718	703	630	629	629[1]	625[1]	580[1]	Chevaline
Asses	4	4	4	5	5	5[1]	5[1]	5[1]	Asine
Mules	30	31	32	31	31	32[1]	32[1]	32[1]	Mulassière
Dominica									**Dominique**
Cattle [1]	9	9	9	9	9	9	9	9	Bovine [1]
Sheep [1]	7	7	8	7	8	8	8	8	Ovine [1]
Pigs [1]	5	5	5	5	5	5	5	5	Porcine [1]
Dominican Republic									**Rép. dominicaine**
Cattle	2 055	2 092	2 129	2 245	* 2 240	* 2 365	* 2 356	* 2 371	Bovine
Sheep	88[1]	* 95	* 100	* 110	115[1]	120[1]	122[1]	128[1]	Ovine
Pigs	368	389	409	429	431[1]	* 769	750[1]	850[1]	Porcine
Horses [1]	300	300	310	310	315	320	320	329	Chevaline [1]
Asses [1]	140	140	142	142	143	143	143	145	Asine [1]
Mules [1]	130	130	132	132	133	133	133	135	Mulassière [1]
El Salvador									**El Salvador**
Cattle	1 050	1 088	1 144	1 176	1 220	1 243	* 1 276	* 1 345	Bovine
Sheep [1]	5	5	5	5	5	5	5	5	Ovine [1]
Pigs	411	418	377	289	317	308	* 315	* 325	Porcine
Horses [1]	92	93	93	94	94	95	95	96	Chevaline [1]
Asses [1]	2	2	2	2	3	3	3	3	Asine [1]
Mules [1]	23	23	23	23	23	23	23	24	Mulassière [1]
Greenland									**Groënland**
Sheep [1]	21	21	21	21	22	22	22	22	Ovine [1]
Grenada									**Grenade**
Cattle [1]	5	4	4	4	4	4	4	4	Bovine [1]
Sheep [1]	13	12	12	11	11	12	12	12	Ovine [1]

37

Livestock
Thousand head [cont.]
Cheptel
Milliers de têtes [suite]

Country or area	1986	1987	1988	1989	1990	1991	1992	1993	Pays ou zone
Pigs [1]	3	3	3	3	3	3	3	3	Porcine [1]
Asses [1]	1	1	1	1	1	1	1	1	Asine [1]
Guadeloupe									**Guadeloupe**
Cattle	82	76	74	74	68	67	64	56	Bovine
Sheep	5[1]	5[1]	6[1]	4	3	4	3	4	Ovine
Pigs	44	41	43	38	32	27[1]	30[1]	30[1]	Porcine
Horses [1]	1	1	1	1	1	1	1	1	Chevaline [1]
Guatemala									**Guatemala**
Cattle	2 022	2 004	2 071	2 047	2 032	2 077	2 097	* 2 055	Bovine
Sheep	530[1]	490[1]	440[1]	425	434	438	430	430[1]	Ovine
Pigs	862	850[1]	820[1]	800	797[1]	769[1]	850[1]	850[1]	Porcine
Horses [1]	100	110	112	112	113	114	114	116	Chevaline [1]
Asses [1]	9	9	9	9	9	9	9	9	Asine [1]
Mules [1]	38	38	38	38	38	38	38	38	Mulassière [1]
Haiti									**Haïti**
Cattle	1 250[1]	1 200[1]	1 150[1]	1 100[1]	1 000[1]	900[1]	800	800[1]	Bovine
Sheep [1]	87	87	87	86	86	86	86	85	Ovine [1]
Pigs [1]	500	450	400	350	300	250	200	200	Porcine [1]
Horses [1]	420	420	430	432	435	430	400	400	Chevaline [1]
Asses [1]	215	216	216	217	215	214	210	210	Asine [1]
Mules [1]	84	84	85	85	83	82	81	80	Mulassière [1]
Honduras									**Honduras**
Cattle	* 2 574	* 2 532	* 2 489	2 424	* 2 424	* 2 388	* 2 351	* 2 315	Bovine
Sheep	* 7	* 7	7[1]	7[1]	8[1]	8[1]	8[1]	8[1]	Ovine
Pigs	* 563	* 567	* 600	* 590	* 590	587	591	596	Porcine
Horses [1]	170	171	171	171	171	172	172	172	Chevaline [1]
Asses [1]	22	22	22	22	22	22	22	22	Asine [1]
Mules [1]	68	69	69	69	69	69	69	69	Mulassière [1]
Jamaica									**Jamaïque**
Cattle [1]	290	290	290	290	310	320	320	330	Bovine [1]
Sheep [1]	3	3	2	2	2	3	2	2	Ovine [1]
Pigs [1]	200	190	195	215	220	150	180	180	Porcine [1]
Horses [1]	4	4	4	4	4	4	4	4	Chevaline [1]
Asses [1]	23	23	23	23	23	23	23	23	Asine [1]
Mules [1]	10	10	10	10	10	10	10	10	Mulassière [1]
Martinique									**Martinique**
Cattle	38	37	35	37	37	37	38[1]	38[1]	Bovine
Sheep [1]	93	95	98	102	110	118	125	135	Ovine [1]
Pigs [1]	48	49	51	51	53	55	60	63	Porcine [1]
Horses [1]	2	2	2	2	2	2	2	2	Chevaline [1]
Mexico									**Mexique**
Cattle	* 32 300	* 31 156	* 31 240	33 068	32 054	31 460	* 30 157	* 30 649	Bovine
Sheep	5 699	5 926	5 761	5 863	5 480[1]	5 000[1]	5 300[1]	5 876	Ovine
Pigs	18 397	18 722	15 884	16 157	15 203	15 786	16 502	16 832	Porcine
Horses [1]	6 140	6 150	6 160	6 170	6 170	6 175	6 180	6 185	Chevaline [1]
Asses [1]	3 183	3 184	3 185	3 186	3 187	3 188	3 189	3 190	Asine [1]
Mules [1]	3 140	3 150	3 160	3 170	3 180	3 190	3 200	3 210	Mulassière [1]
Montserrat									**Montserrat**
Cattle [1]	9	9	9	9	10	10	10	10	Bovine [1]
Sheep [1]	4	4	4	4	5	5	5	5	Ovine [1]
Pigs [1]	1	1	1	1	1	1	1	1	Porcine [1]
Netherlands Antilles									**Antilles néerlandaises**
Cattle [1]	1	1	1	1	1	1	1	1	Bovine [1]
Sheep [1]	4	6	5	6	6	7	7	7	Ovine [1]
Pigs [1]	3	3	3	3	3	2	3	3	Porcine [1]
Asses [1]	3	3	3	3	3	3	3	3	Asine [1]
Nicaragua									**Nicaragua**
Cattle	* 2 110	* 1 885	* 1 700	1 800[1]	1 680[1]	* 1 600	* 1 640	* 1 645	Bovine
Sheep [1]	3	3	4	4	4	4	4	4	Ovine [1]
Pigs	* 750	* 749	* 700	650[1]	600[1]	570[1]	550[1]	530[1]	Porcine
Horses [1]	260	255	250	250	250	250	250	250	Chevaline [1]
Asses [1]	8	8	8	8	8	8	8	8	Asine [1]
Mules [1]	45	45	45	45	45	45	45	45	Mulassière [1]
Panama									**Panama**
Cattle	1 430	1 410	1 423	1 417	1 388	1 399	1 427	1 427[1]	Bovine
Pigs	250	229	211	202	226	256	292	297[1]	Porcine

37

Livestock
Thousand head [*cont.*]
Cheptel
Milliers de têtes [*suite*]

Country or area	1986	1987	1988	1989	1990	1991	1992	1993	Pays ou zone
Horses	138[1]	141[1]	144[1]	147[1]	151[1]	156	155[1]	158[1]	Chevaline
Mules	3[1]	3[1]	3[1]	4[1]	4[1]	4	4[1]	4[1]	Mulassière
Puerto Rico									**Porto Rico**
Cattle	600	580	559	586	601	599	616[1]	552[1]	Bovine
Sheep [1]	7	7	7	7	7	7	8	8	Ovine [1]
Pigs	206	199	195	199	203	209	119[1]	115[1]	Porcine
Horses [1]	21	22	22	22	22	23	23	23	Chevaline [1]
Asses [1]	2	2	2	2	2	2	2	2	Asine [1]
Mules [1]	2	2	2	3	3	3	3	3	Mulassière [1]
Saint Kitts and Nevis									**Saint−Kitts−et−Nevis**
Cattle [1]	5	5	5	5	5	5	5	5	Bovine [1]
Sheep [1]	14	15	15	15	15	15	15	14	Ovine [1]
Pigs [1]	2	2	2	2	2	2	2	2	Porcine [1]
Saint Lucia									**Sainte−Lucie**
Cattle [1]	12	12	12	12	12	12	12	12	Bovine [1]
Sheep [1]	15	15	15	16	16	16	16	16	Ovine [1]
Pigs [1]	12	12	12	12	12	12	12	13	Porcine [1]
Horses [1]	1	1	1	1	1	1	1	1	Chevaline [1]
Asses [1]	1	1	1	1	1	1	1	1	Asine [1]
Mules [1]	1	1	1	1	1	1	1	1	Mulassière [1]
Saint Pierre and Miquelon									**Saint−Pierre−et−Miquelon**
Pigs	1	1	1	0	0	0[1]	0[1]	...	Porcine
Saint Vincent−Grenadines									**Saint Vincent−Grenadines**
Cattle	7	6	7	7	6	6	6[1]	6[1]	Bovine
Sheep	16	15	14	13	13	12	12[1]	12[1]	Ovine
Pigs	9	10[1]	12	11	10	9	9[1]	9[1]	Porcine
Asses [1]	1	1	1	1	1	1	1	1	Asine [1]
Trinidad and Tobago									**Trinité−et−Tobago**
Cattle [1]	57	60	70	65	60	60	55	55	Bovine [1]
Sheep [1]	11	12	13	13	14	14	14	14	Ovine [1]
Pigs [1]	62	72	65	50	54	54	54	48	Porcine [1]
Horses [1]	1	1	1	1	1	1	1	1	Chevaline [1]
Asses [1]	2	2	2	2	2	2	2	2	Asine [1]
Mules [1]	2	2	2	2	2	2	2	2	Mulassière [1]
United States									**Etats−Unis**
Cattle	105 378	102 118	99 622	98 065	98 162	98 896	99 559	100 611	Bovine
Sheep	10 145	10 572	10 945	10 858	11 364	11 200	10 750	10 191	Ovine
Pigs	52 314	51 001	54 384	55 469	53 821	54 477	57 684	59 815	Porcine
Horses [1]	5 203	5 203	5 500	5 600	5 400	5 650	5 450	5 480	Chevaline [1]
Asses [1]	24	24	26	55	53	52	51	56	Asine [1]
Mules [1]	28	28	28	28	28	28	28	28	Mulassière [1]
United States Virgin Is.									**Iles Vierges américaines**
Cattle	11	5	8[1]	8[1]	8[1]	8[1]	8[1]	8[1]	Bovine
Sheep	3	3	3[1]	3[1]	3[1]	3[1]	3[1]	3[1]	Ovine
Pigs	2	3	3[1]	3[1]	3[1]	3[1]	3[1]	3[1]	Porcine
America, South									**Amérique du Sud**
Cattle	253 566	258 219	262 877	266 248	269 900	274 269	277 537	280 221	Bovine
Sheep	102 299	105 000	105 220	105 384	104 456	102 787	101 083	99 860	Ovine
Pigs	51 016	51 045	51 540	51 781	52 329	53 019	51 363	49 699	Porcine
Horses	13 709	13 876	13 988	14 508	14 787	14 932	14 831	14 863	Chevaline
Asses	3 800	3 884	3 919	3 978	4 020	4 050	4 059	4 079	Asine
Mules	3 190	3 237	3 279	3 321	3 356	3 370	3 409	3 434	Mulassière
Argentina									**Argentine**
Cattle	52 537	51 683	* 50 782	* 49 500	* 50 582	* 50 080	* 50 020	50 320[1]	Bovine
Sheep	29 167	28 750	29 167	* 29 345	* 28 571	* 26 500	* 25 706	* 24 500	Ovine
Pigs	* 4 000	3 700[1]	3 400[1]	2 800[1]	2 500[1]	2 600[1]	2 100[1]	2 200[1]	Porcine
Horses *	3 000	3 000	2 900	3 200	3 400	3 400	3 300	3 300	Chevaline *
Asses [1]	90	90	90	90	90	90	90	90	Asine [1]
Mules [1]	165	167	168	170	172	172	173	174	Mulassière [1]
Bolivia									**Bolivie**
Cattle	5 055	5 239	5 402	5 476	5 543	5 607	5 779	5 800	Bovine
Sheep	6 034	7 246	7 505	7 701	7 676	7 342	7 472	7 512	Ovine
Pigs	1 788	1 902	2 019	2 127	2 176	2 177	2 226	* 2 273	Porcine
Horses [1]	310	315	315	320	320	320	322	322	Chevaline [1]
Asses [1]	600	610	620	630	630	630	631	631	Asine [1]
Mules [1]	80	80	80	80	80	80	81	81	Mulassière [1]

37

Livestock
Thousand head [*cont.*]
Cheptel
Milliers de têtes [*suite*]

Country or area	1986	1987	1988	1989	1990	1991	1992	1993	Pays ou zone
Brazil									**Brésil**
Cattle	132 222	135 720	139 599	144 154	147 102	152 136	* 153 000	153 350[1]	Bovine
Sheep	19 660	19 860	20 085	20 041	20 015	20 128	* 19 500	19 701[1]	Ovine
Pigs	32 539	32 480	32 121	33 015	33 623	34 290	* 33 050	* 31 050	Porcine
Horses	5 735	5 855	5 971	6 098	6 122	6 237	6 200[1]	6 310[1]	Chevaline
Asses	1 286	1 295	1 304	1 322	1 343	1 364	1 350[1]	1 364[1]	Asine
Mules	1 921	1 952	1 984	2 009	2 033	2 035	2 060[1]	2 081[1]	Mulassière
Chile									**Chili**
Cattle	3 220	3 371	3 466	3 466	3 336	3 404	3 461	3 557	Bovine
Sheep	5 806	6 470	5 300	4 721	4 887	4 801	4 689	4 629	Ovine
Pigs	1 150	1 300	1 360	1 057	1 125	1 251	1 226	1 288	Porcine
Horses [1]	490	490	490	500	520	530	530	450	Chevaline [1]
Asses [1]	28	28	28	28	28	28	28	28	Asine [1]
Mules [1]	10	10	10	10	10	10	10	10	Mulassière [1]
Colombia									**Colombie**
Cattle	23 593	23 971	24 245	24 415	24 384	24 350	24 772	25 324	Bovine
Sheep	2 538	2 576	2 542	2 545	2 547	2 550	2 553	2 540	Ovine
Pigs	2 440	2 511	2 580	2 600	2 640	2 642	2 644	2 635	Porcine
Horses	1 860	1 898	1 936	1 974	1 975[1]	1 980[1]	2 000[1]	2 000[1]	Chevaline
Asses	672	682	693	703	703[1]	705[1]	705[1]	710[1]	Asine
Mules	600	606	612	618	618[1]	620[1]	620[1]	622[1]	Mulassière
Ecuador									**Equateur**
Cattle	3 765	3 884	3 997	4 177	4 359	4 516	4 682	* 4 819	Bovine
Sheep	1 195	1 293	1 226	1 329	1 420	1 501	1 565	* 1 602	Ovine
Pigs	1 843	1 620	1 922	2 092	2 220	2 327	2 425	* 2 540	Porcine
Horses	372	404	427	460	492	516	512	510[1]	Chevaline
Asses	162	215	221	242	253	259	262	263[1]	Asine
Mules	105	113	116	124	131	141	151	152[1]	Mulassière
Falkland Is. (Malvinas)									**Îles Falkland (Malvinas)**
Cattle	7	6	6	6	5	5	5	5	Bovine
Sheep	699	692	705	745	740	729	713	721	Ovine
Horses	2	2	2	2	2	2	2	1	Chevaline
French Guiana									**Guyane française**
Cattle	17	17	15	16	15[1]	14[1]	13[1]	12[1]	Bovine
Sheep	3[1]	3[1]	4[1]	4	* 4	* 4	* 4	* 4	Ovine
Pigs	10[1]	10	9	8	9	9	9	10	Porcine
Guyana									**Guyana**
Cattle [1]	130	130	140	140	130	140	150	160	Bovine [1]
Sheep [1]	126	126	127	128	129	130	130	130	Ovine [1]
Pigs [1]	60	65	60	50	40	35	30	30	Porcine [1]
Horses [1]	2	2	2	2	2	2	2	2	Chevaline [1]
Asses [1]	1	1	1	1	1	1	1	1	Asine [1]
Paraguay									**Paraguay**
Cattle	7 151	7 374	7 780	8 074	8 254	7 627	7 886	8 074[1]	Bovine
Sheep	388	411	432	451	457	357	365	371[1]	Ovine
Pigs	1 508	1 809	2 108	2 305	2 444	2 580	2 700[1]	2 915[1]	Porcine
Horses	317	323	328	334	335	320	327	330[1]	Chevaline
Asses [1]	30	31	31	31	31	31	31	31	Asine [1]
Mules [1]	13	14	14	14	14	14	14	14	Mulassière [1]
Peru									**Pérou**
Cattle	4 172	4 161	4 174	4 214	4 102	4 042	3 972	3 950	Bovine
Sheep	12 876	13 075	12 922	12 970	12 257	12 226	11 912	11 915	Ovine
Pigs	2 453	2 222	2 376	2 434	2 400	2 417	2 396	2 400	Porcine
Horses [1]	655	655	655	660	660	660	665	665	Chevaline [1]
Asses [1]	490	490	490	490	500	500	520	520	Asine [1]
Mules [1]	220	220	220	220	222	222	224	224	Mulassière [1]
Suriname									**Suriname**
Cattle	67	76	84	89	92	92	96	97[1]	Bovine
Sheep	4	6	8	8	10	9	9	9[1]	Ovine
Pigs	19	19	21	25	32	31	38	36[1]	Porcine
Uruguay									**Uruguay**
Cattle	9 300	9 945	10 331	9 446	8 723	8 889	9 508	10 093	Bovine
Sheep	23 337	24 006	24 689	24 872	25 220	25 986	25 941	* 25 702	Ovine
Pigs	195	220	215	215	215	215[1]	220[1]	223[1]	Porcine
Horses	469	437	466	462	465[1]	470[1]	475[1]	477[1]	Chevaline

37
Livestock
Thousand head [*cont.*]
Cheptel
Milliers de têtes [*suite*]

Country or area	1986	1987	1988	1989	1990	1991	1992	1993	Pays ou zone
Asses [1]	1	1	1	1	1	1	1	1	Asine [1]
Mules [1]	4	4	4	4	4	4	4	4	Mulassière [1]
Venezuela									**Venezuela**
Cattle	12 331	12 641	12 856	13 076	13 272	* 13 368	* 14 192	* 14 660	Bovine
Sheep	467	488	508	523	525[1]	525[1]	525[1]	525[1]	Ovine
Pigs	3 011	3 187	3 349	3 053	2 904	2 445[1]	2 300[1]	2 100[1]	Porcine
Horses	495[1]	495[1]	495[1]	495[1]	495[1]	495[1]	495[1]	495[1]	Chevaline
Asses [1]	440	440	440	440	440	440	440	440	Asine [1]
Mules [1]	72	72	72	72	72	72	72	72	Mulassière [1]
Asia									**Asie**
Cattle	**380 308**	**386 893**	**381 967**	**387 002**	**388 301**	**390 244**	**393 037**	**398 935**	**Bovine**
Sheep	**314 631**	**321 657**	**326 030**	**342 628**	**344 729**	**342 188**	**341 011**	**341 636**	**Ovine**
Pigs	**404 657**	**412 471**	**405 663**	**420 405**	**432 414**	**443 759**	**453 482**	**471 014**	**Porcine**
Horses	**17 184**	**17 109**	**16 767**	**16 617**	**16 447**	**16 306**	**16 207**	**16 075**	**Chevaline**
Asses	**20 395**	**20 732**	**21 077**	**21 460**	**21 487**	**21 322**	**21 583**	**21 516**	**Asine**
Mules	**5 632**	**5 783**	**5 912**	**6 036**	**6 062**	**6 140**	**6 257**	**6 253**	**Mulassière**
Afghanistan									**Afghanistan**
Cattle	1 532	1 500[1]	1 500[1]	1 500[1]	1 500[1]	1 500[1]	1 500[1]	1 500[1]	Bovine
Sheep	10 500	12 000[1]	14 150	14 150	14 170	14 200	14 200[1]	14 200[1]	Ovine
Horses	* 410	400[1]	387[1]	374[1]	362[1]	350	340[1]	320[1]	Chevaline
Asses	* 1 325	1 300	1 280[1]	1 260[1]	1 240[1]	1 220[1]	1 200[1]	1 180[1]	Asine
Mules	* 30	30[1]	29[1]	28[1]	27[1]	26[1]	25[1]	24[1]	Mulassière
Armenia									**Arménie**
Cattle	..	..	..	..	..	..	499	549	Bovine
Sheep	..	..	..	..	..	..	854	850[1]	Ovine
Pigs	..	..	..	..	..	..	84	* 243	Porcine
Horses	..	..	..	..	..	..	9	9[1]	Chevaline
Asses	..	..	..	..	..	..	3	4[1]	Asine
Azerbaijan									**Azerbaïdjan**
Cattle	..	..	..	..	..	..	1 716	* 1 570	Bovine
Sheep	..	..	..	..	..	..	5 088	5 055[1]	Ovine
Pigs *	..	..	..	..	..	..	140	123	Porcine *
Horses [1]	..	..	..	..	..	..	35	33	Chevaline [1]
Asses [1]	..	..	..	..	..	..	5	6	Asine [1]
Bahrain									**Bahreïn**
Cattle [1]	8	9	13	15	15	15	15	16	Bovine [1]
Sheep [1]	7	8	8	8	8	9	9	9	Ovine [1]
Bangladesh									**Bangladesh**
Cattle	22 348	22 567	22 789	23 015	23 244	23 259	23 480	23 923	Bovine
Sheep	739	770	803	837	873	902	940	989	Ovine
Bhutan									**Bhoutan**
Cattle	376	387	393	387	406[1]	413[1]	422[1]	429[1]	Bovine
Sheep	* 16	36	47	48	49[1]	50[1]	52[1]	54[1]	Ovine
Pigs	89	70	66	63	72[1]	73[1]	74[1]	74[1]	Porcine
Horses	24	26	26	26	25[1]	27[1]	28[1]	29[1]	Chevaline
Asses [1]	18	18	18	18	18	18	18	18	Asine [1]
Mules [1]	9	9	9	9	9	9	10	10	Mulassière [1]
Brunei Darussalam									**Brunéi Darussalam**
Cattle [1]	1	1	1	1	1	1	1	1	Bovine [1]
Pigs [1]	18	15	17	15	14	14	14	14	Porcine [1]
Cambodia									**Cambodge**
Cattle	1 705	1 852	1 891	2 095	2 181	2 257	2 468	2 468[1]	Bovine
Pigs	1 161	1 251	1 500	1 737	1 515	1 550	2 043	2 043[1]	Porcine
Horses [1]	13	14	15	16	17	18	19	20	Chevaline [1]
China									**Chine**
Cattle	66 991	70 964	73 963	77 025	76 969	78 635	79 775	82 641	Bovine
Sheep	94 210	99 009	102 656	110 571	113 508	112 816	110 855	109 720	Ovine
Pigs	338 070	344 248	334 862	349 172	360 594	370 975	379 735	393 965	Porcine
Horses	11 081	10 988	10 691	10 541	10 295	10 175	10 095	10 018	Chevaline
Asses	10 415	10 689	10 846	11 052	11 136	11 198	11 158	10 983	Asine
Mules	4 972	5 113	5 248	5 366	5 391	5 494	5 606	5 610	Mulassière
Cyprus									**Chypre**
Cattle	42	44	45	46	49	55	55	56[1]	Bovine
Sheep	325	325	310	300	310	290	295	295[1]	Ovine
Pigs	201	225	266	284	281	278	296	297[1]	Porcine
Horses	1	1	1	1	1[1]	1[1]	1[1]	1[1]	Chevaline

37

Livestock
Thousand head [*cont.*]
Cheptel
Milliers de têtes [*suite*]

Country or area	1986	1987	1988	1989	1990	1991	1992	1993	Pays ou zone
Asses	6	6	6	5	5[1]	5[1]	5[1]	5[1]	Asine
Mules	2	2	2	2	2[1]	2[1]	2[1]	2[1]	Mulassière
Gaza Strip (Palestine)									**Zone de Gaza (Palestine)**
Cattle	3[1]	3	3	3[1]	3[1]	3[1]	3[1]	3[1]	Bovine
Sheep	19[1]	20	24	24[1]	24[1]	24[1]	24[1]	24[1]	Ovine
Georgia									**Géorgie**
Cattle *	..	..	..	..	..	..	1 216	1 130	Bovine *
Sheep	..	..	..	..	..	..	* 1 400	1 350[1]	Ovine
Pigs *	..	..	..	..	..	..	782	688	Porcine *
Horses [1]	..	..	..	..	..	..	20	20	Chevaline [1]
Asses [1]	..	..	..	..	..	..	2	2	Asine [1]
Hong Kong									**Hong−kong**
Cattle	3	2	1	1	2	2	2	2	Bovine
Pigs	372	353	358	350	304	234	175	104	Porcine
Horses	1	1	1	1	1	1	1	1	Chevaline
India									**Inde**
Cattle	* 197 895	199 528	* 190 791	* 190 614	* 191 750	* 193 328	* 192 650	* 192 700	Bovine
Sheep *	46 330	45 750	42 702	43 204	43 706	44 207	44 407	44 608	Ovine *
Pigs [1]	10 200	10 200	10 300	10 300	10 400	10 450	10 500	10 547	Porcine [1]
Horses [1]	920	950	953	955	960	965	970	980	Chevaline [1]
Asses [1]	1 200	1 300	1 328	1 400	1 450	1 500	1 500	1 550	Asine [1]
Mules [1]	132	134	135	138	139	140	141	142	Mulassière [1]
Indonesia									**Indonésie**
Cattle	9 635	9 742	9 776	10 095	10 410	10 665	10 800[1]	11 000[1]	Bovine
Sheep	5 280	5 363	5 825	5 910	6 006	6 108	6 200[1]	6 300[1]	Ovine
Pigs	6 220	6 340	6 484	6 946	7 136	7 612	8 000[1]	8 200[1]	Porcine
Horses	715	658	675	683	683	695	700[1]	705[1]	Chevaline
Iran, Islamic Rep. of									**Iran, Rép. islamique d'**
Cattle	6 691	6 500[1]	6 368	7 918	7 532	6 697	* 6 900	7 000[1]	Bovine
Sheep	42 695	41 000	40 665	45 000	44 581	44 681	* 45 000	45 200[1]	Ovine
Horses	279[1]	276[1]	274[1]	271	255	255	255[1]	255[1]	Chevaline
Asses	2 050[1]	2 000[1]	1 970[1]	2 014	1 860	1 860	1 900[1]	1 900[1]	Asine
Mules	139[1]	138[1]	137[1]	* 136	135[1]	134[1]	134[1]	133[1]	Mulassière
Iraq									**Iraq**
Cattle	1 578	1 750[1]	1 500[1]	1 900[1]	1 535[1]	1 150[1]	1 150[1]	1 200[1]	Bovine
Sheep	8 981	9 000	* 9 000	9 500[1]	8 000[1]	5 800[1]	5 900[1]	6 000[1]	Ovine
Horses [1]	53	55	55	58	60	18	40	45	Chevaline [1]
Asses [1]	400	400	410	415	416	150	355	360	Asine [1]
Mules [1]	25	25	26	26	27	10	22	24	Mulassière [1]
Israel									**Israël**
Cattle	316	325	345	348	342	331	349	357	Bovine
Sheep	262	306	372	394	380	375	360	330	Ovine
Pigs	130	130[1]	130[1]	130[1]	115[1]	100[1]	100[1]	100[1]	Porcine
Horses [1]	4	4	4	4	4	4	4	4	Chevaline [1]
Asses [1]	5	5	5	5	5	5	5	5	Asine [1]
Mules [1]	2	2	2	2	2	2	2	2	Mulassière [1]
Japan									**Japon**
Cattle	4 742	4 694	4 667	4 682	4 760	4 873	4 980	5 024	Bovine
Sheep	26	27	29	30	31	30	29	27	Ovine
Pigs	11 061	11 354	11 725	11 866	11 817	11 335	10 966	10 783	Porcine
Horses	23	22	22	22	23	24	26	27	Chevaline
Jordan									**Jordanie**
Cattle	31	29	30	29	35[1]	40[1]	40[1]	40[1]	Bovine
Sheep	930	1 219	1 279	1 523	1 556	1 900[1]	1 900[1]	1 900[1]	Ovine
Horses	3	3	3	4	4[1]	4[1]	4[1]	4[1]	Chevaline
Asses	19[1]	19	19	19	19[1]	19[1]	19[1]	19[1]	Asine
Mules	3[1]	3	3	3	3[1]	3[1]	3[1]	3[1]	Mulassière
Kazakhstan									**Kazakhstan**
Cattle	..	..	..	..	..	..	9 084	8 313	Bovine
Sheep	..	..	..	..	..	..	33 908	33 000[1]	Ovine
Pigs	..	..	..	..	..	..	2 794	2 459	Porcine
Horses [1]	..	..	..	..	..	..	1 510	1 500	Chevaline [1]
Asses [1]	..	..	..	..	..	..	45	45	Asine [1]
Korea, Dem. P. R.									**Corée, Rép. pop. dém. de**
Cattle [1]	1 150	1 200	1 250	1 280	1 300	1 300	1 300	1 300	Bovine [1]
Sheep [1]	360	368	372	380	385	390	390	390	Ovine [1]

37

Livestock
Thousand head [cont.]
Cheptel
Milliers de têtes [suite]

Country or area	1986	1987	1988	1989	1990	1991	1992	1993	Pays ou zone
Pigs [1]	2 900	3 050	3 100	3 145	3 200	3 300	3 300	3 300	Porcine [1]
Horses [1]	41	42	43	44	44	45	46	46	Chevaline [1]
Asses [1]	3	3	3	3	3	3	3	3	Asine [1]
Mules [1]	2	2	2	2	2	2	2	2	Mulassière [1]
Korea, Republic of									**Corée, République de**
Cattle	2 807	2 386	2 039	2 051	2 126	2 269	2 527	* 2 814	Bovine
Sheep	4	3	3	3	3	3	4	4[1]	Ovine
Pigs	3 347	4 281	4 852	4 801	4 528	5 046	5 463	* 5 928	Porcine
Horses	3	3	4	5	5	5	5	6[1]	Chevaline
Kuwait									**Koweït**
Cattle	23	28	24	20	18[1]	1[1]	5[1]	12[1]	Bovine
Sheep	227	282	280	305	200[1]	50[1]	100[1]	150[1]	Ovine
Horses [1]	3	3	3	4	3	0	0	0	Chevaline [1]
Kyrgyzstan									**Kirghizistan**
Cattle	..	..	..	..	..	..	1 095	* 1 002	Bovine
Sheep	..	..	..	..	..	..	9 200	9 000[1]	Ovine
Pigs	..	..	..	..	..	..	299	* 264	Porcine
Horses [1]	..	..	..	..	..	..	315	310	Chevaline [1]
Asses [1]	..	..	..	..	..	..	10	10	Asine [1]
Lao People's Dem. Rep.									**Rép. dém. pop. lao**
Cattle	646	703	764	817	842	899	993	1 010	Bovine
Pigs	1 280	1 420	1 268	1 350	1 372	1 469	1 561	1 559	Porcine
Horses	42[1]	42[1]	42[1]	43[1]	44[1]	36	29	29[1]	Chevaline
Lebanon									**Liban**
Cattle	55[1]	61	59	59	65[1]	70	73[1]	77[1]	Bovine
Sheep	200[1]	208	204	210	220[1]	238	240[1]	250[1]	Ovine
Pigs	39[1]	49	47	49	45[1]	44	42[1]	40[1]	Porcine
Horses	4[1]	5[1]	5[1]	6[1]	8[1]	10	11[1]	12[1]	Chevaline
Asses	13[1]	14[1]	16[1]	18[1]	19[1]	21	22[1]	23[1]	Asine
Mules	5[1]	6[1]	6[1]	7[1]	7[1]	8	8[1]	8[1]	Mulassière
Macau									**Macao**
Pigs [1]	1	1	1	1	1	1	1	1	Porcine [1]
Malaysia									**Malaisie**
Cattle	607	663	637	661	668	701	718	* 735	Bovine
Sheep	91	130	150	183	205	247	276	308	Ovine
Pigs	2 176	2 261	2 113	2 345	2 678	2 708	2 843	2 983	Porcine
Horses [1]	5	5	5	5	5	5	5	5	Chevaline [1]
Mongolia									**Mongolie**
Cattle	2 408	2 480	2 526	2 541	2 693	2 849	2 822	2 819	Bovine
Sheep	13 249	13 194	13 234	13 451	14 265	15 083	14 721	14 657	Ovine
Pigs	56	80	120	171	192	135	83	49	Porcine
Horses	1 971	2 018	2 047	2 103	2 200	2 262	2 259	2 200	Chevaline
Myanmar									**Myanmar**
Cattle	9 758	9 919	10 091	9 126	9 298	9 382	9 470	9 584	Bovine
Sheep	317	304	313	269	276	279	284	305	Ovine
Pigs	2 986	3 059	3 199	2 449	2 243	2 372	2 514	2 529	Porcine
Horses	133	136	138	119	121	118	116	120	Chevaline
Mules [1]	9	9	9	9	9	9	9	9	Mulassière [1]
Nepal									**Népal**
Cattle	6 372	6 363	6 343	6 285	6 281	6 255	6 246	6 237	Bovine
Sheep	808	837	873	910	892	906	912	911	Ovine
Pigs	456	476	516	548	574	592	599	630	Porcine
Oman									**Oman**
Cattle	130[1]	135[1]	136[1]	136[1]	137	138[1]	140[1]	142[1]	Bovine
Sheep [1]	136	137	138	139	140	143	145	148	Ovine [1]
Asses [1]	24	24	25	25	25	26	26	26	Asine [1]
Pakistan									**Pakistan**
Cattle	16 749	16 951	17 609	17 643	17 677	17 711	17 745	17 779	Bovine
Sheep	23 287	23 868	24 463	25 072	25 698	26 338	26 995	27 668	Ovine
Horses	452	454	378	373	368	363	358	354	Chevaline
Asses	2 857	2 938	3 202	3 309	3 420	3 534	3 653	3 775	Asine
Mules	65	65	71	72	72	73	74	75	Mulassière
Philippines									**Philippines**
Cattle	1 814	1 747	1 700	1 682	1 629	1 680	1 729	1 781	Bovine
Sheep [1]	30	30	30	30	30	30	30	30	Ovine [1]

37
Livestock
Thousand head [*cont.*]
    Cheptel
    Milliers de têtes [*suite*]

Country or area	1986	1987	1988	1989	1990	1991	1992	1993	Pays ou zone
Pigs	7 275	7 038	7 581	7 909	7 990	8 006	8 022	7 954	Porcine
Horses	195	195[1]	200[1]	200[1]	200[1]	200[1]	200[1]	210[1]	Chevaline
Qatar									**Qatar**
Cattle	8	9	10	10	10	9	11	11[1]	Bovine
Sheep	119	121	123	126	130	122	142	145[1]	Ovine
Horses	1	1	1	1	1	1	1	1[1]	Chevaline
Saudi Arabia									**Arabie saoudite**
Cattle	303	195	195	194	193	200	204	210[1]	Bovine
Sheep	7 272	6 812	6 194	6 173	6 383	6 847	7 046	7 100[1]	Ovine
Horses [1]	3	3	3	3	3	3	3	3	Chevaline [1]
Asses [1]	108	107	106	105	104	103	102	100	Asine [1]
Singapore									**Singapour**
Pigs	516	459	321	350[1]	300[1]	250[1]	200[1]	150[1]	Porcine
Sri Lanka									**Sri Lanka**
Cattle	1 783	1 808	1 788	1 820	1 773	1 477	1 568	1 600[1]	Bovine
Sheep	29	28	28	30	26	20	17	19[1]	Ovine
Pigs	86	97	95	94	85	84	91	90[1]	Porcine
Horses [1]	2	2	2	2	2	2	2	2	Chevaline [1]
Syrian Arab Republic									**Rép. arabe syrienne**
Cattle	706	710	763	800	787	771	765	770[1]	Bovine
Sheep	11 669	12 669	13 691	14 011	14 509	15 194	14 665	16 000[1]	Ovine
Pigs [1]	1	1	1	1	1	1	1	1	Porcine [1]
Horses	42	41	40	43	41	39	37	38[1]	Chevaline
Asses	199	184	177	178	168	161	158	160[1]	Asine
Mules	29	28	27	27	26	25	27	27[1]	Mulassière
Tajikistan									**Tadjikistan**
Cattle	..	..	..	..	..	..	1 266	* 1 159	Bovine
Sheep	..	..	..	..	..	..	2 620	2 550[1]	Ovine
Pigs	..	..	..	..	..	..	163	* 143	Porcine
Horses [1]	..	..	..	..	..	..	53	53	Chevaline [1]
Asses [1]	..	..	..	..	..	..	36	36	Asine [1]
Thailand									**Thaïlande**
Cattle	4 879	4 969	5 072	5 285	5 459	5 631	5 815	7 190	Bovine
Sheep	73	95	131	156	162	166	176	136	Ovine
Pigs	4 201	4 209	4 685	4 679	4 728	4 859	4 655	4 800[1]	Porcine
Horses	19	19	19	18	20	20	19	20	Chevaline
Turkey									**Turquie**
Cattle	12 466	12 713	12 713	12 562	12 173	11 377	11 973	11 951	Bovine
Sheep	42 500	43 758	43 796	45 384	43 647	40 553	40 433	39 416	Ovine
Pigs	8	8	7	9	8	12	10	12	Porcine
Horses	604	600	590	557	545	513	496	483	Chevaline
Asses	1 192	1 188	1 153	1 119	1 084	985	944	895	Asine
Mules	206	216	205	208	210	202	192	181	Mulassière
Turkmenistan									**Turkménistan**
Cattle	..	..	..	..	..	..	777	* 711	Bovine
Sheep	..	..	..	..	..	..	5 380	5 405[1]	Ovine
Pigs	..	..	..	..	..	..	237	* 209	Porcine
Horses [1]	..	..	..	..	..	..	20	22	Chevaline [1]
Asses [1]	..	..	..	..	..	..	26	26	Asine [1]
United Arab Emirates									**Emirats arabes unis**
Cattle	41	44	46	46	49	53	55[1]	58[1]	Bovine
Sheep	194	207	222	238	254	272	275[1]	277[1]	Ovine
Uzbekistan									**Ouzbékistan**
Cattle	..	..	..	..	..	..	5 113	5 275	Bovine
Sheep	..	..	..	..	..	..	8 275	8 407	Ovine
Pigs	..	..	..	..	..	..	654	529	Porcine
Horses	..	..	..	..	..	..	113	123	Chevaline
Asses	..	..	..	..	..	..	143	158	Asine
Viet Nam									**Viet Nam**
Cattle	2 598	2 784	2 979	3 127	3 202	3 117	3 136	3 320	Bovine
Pigs	11 808	11 796	12 051	11 643	12 221	12 261	12 194	14 861	Porcine
Horses	133	137	136	133	143	141	134	133	Chevaline
Yemen									**Yémen**
Cattle	1 100	1 120	1 137	1 170	1 175	1 117	1 139	1 163	Bovine
Sheep	* 3 483	3 488	3 602	3 720	3 756	3 568	3 640	3 715	Ovine

37
Livestock
Thousand head [*cont.*]
Cheptel
Milliers de têtes [*suite*]

Country or area	1986	1987	1988	1989	1990	1991	1992	1993	Pays ou zone
Horses [1]	3	3	3	3	3	3	3	3	Chevaline [1]
Asses	547[1]	524[1]	500	500	500	500	500	500	Asine
Europe									**Europe**
Cattle	130 751	127 546	124 698	124 980	124 394	120 277	114 014	109 183	**Bovine**
Sheep	136 596	142 974	144 342	144 935	145 948	142 348	138 764	133 550	**Ovine**
Pigs	147 756	152 098	152 454	150 617	147 151	181 788	173 982	170 200	**Porcine**
Horses	4 638	4 509	4 387	4 303	4 300	4 324	4 345	4 216	**Chevaline**
Asses	1 088	1 069	1 012	987	962	941	898	858	**Asine**
Mules	410	399	363	344	322	304	276	271	**Mulassière**
Albania									**Albanie**
Cattle	618	671	695	699	633	600[1]	500[1]	450[1]	Bovine
Sheep	1 346	1 432	1 525	1 592	1 646	1 600[1]	1 400[1]	1 200[1]	Ovine
Pigs	209	214	196	181	220	170[1]	135[1]	140[1]	Porcine
Horses	* 88	* 94	* 99	* 101	* 105	100[1]	100[1]	105[1]	Chevaline
Asses [1]	52	53	53	53	53	53	53	53	Asine [1]
Mules [1]	22	23	23	23	23	23	23	23	Mulassière [1]
Austria									**Autriche**
Cattle	2 651	2 637	2 590	2 541	2 562	2 534	2 532	2 401	Bovine
Sheep	243	256	261	256	287	309	323	312	Ovine
Pigs	3 926	3 801	3 947	3 874	3 773	3 638	3 629	3 720	Porcine
Horses	45	44	45	44	48	49	57	61	Chevaline
Belarus									**Bélarus**
Cattle	..	..	..	..	..	..	6 577	6 221	Bovine
Sheep *	..	..	..	..	..	..	404	361	Ovine *
Pigs	..	..	..	..	..	..	4 703	4 308	Porcine
Horses	..	..	..	..	..	..	212	215	Chevaline
Asses [1]	..	..	..	..	..	..	8	8	Asine [1]
Belgium−Luxembourg									**Belgique−Luxembourg**
Cattle	3 163	3 190	3 159	3 174	3 257	3 360	3 311	3 303	Bovine
Sheep	166	161	166	162	165	169	163	156[1]	Ovine
Pigs	5 484	5 838	5 958	6 310	6 511	6 496	6 597	6 963	Porcine
Horses	26	24	23	22	21	21	21	21	Chevaline
Bosnia & Herzegovina									**Bosnie−Herzégovine**
Cattle [1]	..	..	..	..	..	..	826	685	Bovine [1]
Sheep [1]	..	..	..	..	..	..	1 287	1 080	Ovine [1]
Pigs [1]	..	..	..	..	..	..	590	550	Porcine [1]
Horses [1]	..	..	..	..	..	..	70	56	Chevaline [1]
Bulgaria									**Bulgarie**
Cattle	1 706	1 678	1 649	1 613	1 575	1 457	1 310	974	Bovine
Sheep	9 724	9 563	8 886	8 609	8 130	7 938	6 703	4 814	Ovine
Pigs	3 912	4 050	4 034	4 119	4 352	4 187	3 141	2 680	Porcine
Horses	120	121	123	122	119	115	114	114	Chevaline
Asses	345	341	333	329	329	329	329	303	Asine
Mules	26	26	25	24	22	19	17	21	Mulassière
Croatia									**Croatie**
Cattle	..	..	..	..	..	..	590	590	Bovine
Sheep	..	..	..	..	..	..	539	524	Ovine
Pigs	..	..	..	..	..	..	1 183	1 262	Porcine
Horses	..	..	..	..	..	..	27	22	Chevaline
Asses	..	..	..	..	..	..	13	12	Asine
Mules [1]	..	..	..	..	..	..	4	3	Mulassière [1]
former Czechoslovakia †									**anc. Tchécoslovaquie †**
Cattle	5 065	5 073	5 044	5 075	5 129	4 923	4 347	..	Bovine
Sheep	1 087	1 104	1 075	1 047	1 051	1 030	886	..	Ovine
Pigs	6 651	6 833	7 235	7 384	7 498	7 090	7 139	..	Porcine
Horses	46	46	45	44	42	39	34	..	Chevaline
Czech Republic									**Tchéque République**
Cattle	..	..	..	..	..	..	..	2 512	Bovine
Sheep	..	..	..	..	..	..	..	254	Ovine
Pigs	..	..	..	..	..	..	..	4 599	Porcine
Horses	..	..	..	..	..	..	..	19	Chevaline
Denmark									**Danemark**
Cattle	2 623	2 490	2 323	2 232	2 241	2 222	2 180	2 115	Bovine
Sheep	52	70	73	86	100	122	102	93	Ovine
Pigs	9 104	9 422	9 048	9 120	9 282	9 767	10 345	10 870	Porcine
Horses	30	29	34	35	38	34	29	19	Chevaline

37
Livestock
Thousand head [*cont.*]
Cheptel
Milliers de têtes [*suite*]

Country or area	1986	1987	1988	1989	1990	1991	1992	1993	Pays ou zone
Estonia									**Estonie**
Cattle	..	..	..	..	..	...	708	* 661	Bovine
Sheep	..	..	..	..	..	...	143	144[1]	Ovine
Pigs	..	..	..	..	..	...	799	* 772	Porcine
Horses	..	..	..	..	..	...	8	8[1]	Chevaline
Faeroe Islands									**Iles Féroé**
Cattle [1]	2	2	2	2	2	2	2	2	Bovine [1]
Sheep	65	66	67[1]	67[1]	67[1]	67[1]	68[1]	68[1]	Ovine
Finland									**Finlande**
Cattle	1 567	1 485	1 434	1 379	1 363	1 315	1 263	1 232	Bovine
Sheep	70	66	63	59	61	57	61	62	Ovine
Pigs	1 211	1 309	1 291	1 327	1 348	1 290	1 357	1 309	Porcine
Horses	37	39	36	40	44	45	49	49	Chevaline
France									**France**
Cattle	23 290	21 967	21 340	21 377	21 394	21 450	20 970	20 328	Bovine
Sheep	12 432	12 044	11 495	11 208	11 209	11 170	10 640	10 380	Ovine
Pigs	11 842	12 419	12 643	12 410	12 276	12 013	12 068	12 564	Porcine
Horses	271	268	266	269	319	322	328	333	Chevaline
Asses [1]	24	24	25	25	25	25	25	25	Asine [1]
Mules	12[1]	12[1]	12[1]	12[1]	13	13	12	12	Mulassière
Germany †									**Allemagne †**
Cattle	..	..	..	..	..	19 488	17 134	* 16 200	Bovine
Sheep	..	..	..	..	..	3 239	2 488	* 2 298	Ovine
Pigs	..	..	..	..	..	30 819	26 063	* 26 466	Porcine
Horses	..	..	..	..	..	491	492[1]	496[1]	Chevaline
F. R. Germany									**R. f. Allemagne**
Cattle	15 627	15 305	14 887	14 659	14 563	..	..	..	Bovine
Sheep	1 296	1 383	1 414	1 464	1 533	..	..	..	Ovine
Pigs	24 282	24 502	23 670	22 589	22 165	..	..	..	Porcine
Horses	370	367	365	375	391[1]	..	..	..	Chevaline
former German D. R.									**anc. R. d. allemande**
Cattle	5 827	5 804	5 721	5 710	5 724	..	..	..	Bovine
Sheep	2 587	2 647	2 656	2 634	2 603	..	..	..	Ovine
Pigs	12 946	12 840	12 503	12 464	12 013	..	..	..	Porcine
Horses	105	105	104	102	93[1]	..	..	..	Chevaline
Greece									**Grèce**
Cattle	722	725	683	677	654	624	616	* 631	Bovine
Sheep	8 342	8 258	8 612	8 670	8 723	8 660	9 694	9 659	Ovine
Pigs	1 009	1 061	998	1 011	1 001	996	1 125	1 040	Porcine
Horses	67	74	57	53	49	45	40[1]	40[1]	Chevaline
Asses	177	188	157	147	137	127	115[1]	110[1]	Asine
Mules	84	89	75	70	65	60	55[1]	50[1]	Mulassière
Hungary									**Hongrie**
Cattle	1 766	1 725	1 664	1 690	1 598	1 571	1 420	1 159	Bovine
Sheep	2 465	2 337	2 336	2 215	2 069	1 865	1 808	1 752	Ovine
Pigs	8 280	8 687	8 216	8 327	7 660	8 000	5 993	5 364	Porcine
Horses	98	95	88	76	75	76	75	75[1]	Chevaline
Asses	4[1]	5	5[1]	4[1]	4[1]	4	4[1]	4[1]	Asine
Iceland									**Islande**
Cattle	71	69	71	73	75	78	75	76[1]	Bovine
Sheep	676	624	587	561	549	511	500	500[1]	Ovine
Pigs [1]	15	18	18	17	19	19	20	21	Porcine [1]
Horses	56	59	64	69	72	74	75	78[1]	Chevaline
Ireland									**Irlande**
Cattle	5 813	5 670	5 633	5 699	5 969	6 101	6 158	6 265	Bovine
Sheep	3 304	3 672	4 301	4 991	5 714	5 864	5 988	6 125	Ovine
Pigs	994	980	960	1 015	1 110	1 249	1 346	1 423	Porcine
Horses	56	55	53	52	54	53[1]	53[1]	53[1]	Chevaline
Asses [1]	17	17	16	16	15	15	14	14	Asine [1]
Mules [1]	1	1	1	1	1	1	1	1	Mulassière [1]
Italy									**Italie**
Cattle	8 908	8 819	8 794	8 737	8 746	8 140	8 004	7 783	Bovine
Sheep	11 293	11 451	11 457	11 569	10 848	10 848	10 435	10 403	Ovine
Pigs	9 169	9 278	9 383	9 359	9 254	8 837	8 549	8 307	Porcine
Horses	248	253	250	256	271	288	316	330[1]	Chevaline

37
Livestock
Thousand head [*cont.*]
Cheptel
Milliers de têtes [*suite*]

Country or area	1986	1987	1988	1989	1990	1991	1992	1993	Pays ou zone
Asses	94	91	86	81	74	51	* 37	30[1]	Asine
Mules	56	52	50	47	43	33	* 23	20[1]	Mulassière
Latvia									**Lettonie**
Cattle	..	..	..	..	..	..	1 383	1 144	Bovine
Sheep	..	..	..	..	..	..	184	165	Ovine
Pigs	..	..	..	..	..	..	1 246	867	Porcine
Horses	..	..	..	..	..	..	30	28	Chevaline
Liechtenstein									**Liechtenstein**
Cattle	6	6	6	6	6	6	6	6[1]	Bovine
Sheep	2	2	2	2	3	3	3	3[1]	Ovine
Pigs	3	3	3	3	3	4	3	3[1]	Porcine
Lithuania									**Lituanie**
Cattle	..	..	..	..	..	..	2 197	1 701	Bovine
Sheep	..	..	..	..	..	..	58	52	Ovine
Pigs	..	..	..	..	..	..	2 180	1 360	Porcine
Horses	..	..	..	..	..	..	80	78[1]	Chevaline
Malta									**Malte**
Cattle	16[1]	18[1]	19[1]	21	21[1]	22[1]	23[1]	24[1]	Bovine
Sheep	5[1]	6[1]	6[1]	6	6[1]	6[1]	6[1]	6[1]	Ovine
Pigs	95[1]	95[1]	98[1]	101	101[1]	102[1]	107[1]	109[1]	Porcine
Horses	1[1]	1[1]	1[1]	1	1[1]	1[1]	1[1]	1[1]	Chevaline
Asses	1[1]	1[1]	1[1]	1	1[1]	1[1]	1[1]	1[1]	Asine
Netherlands									**Pays—Bas**
Cattle	5 123	4 895	4 546	4 772	4 926	5 057	* 4 876	* 4 794	Bovine
Sheep	868	985	* 1 169	* 1 405	* 1 702	* 1 882	* 1 954	* 2 000	Ovine
Pigs	13 481	14 349	* 14 226	* 13 820	* 13 634	* 13 788	* 13 727	* 13 709	Porcine
Horses	63	64	64[1]	64[1]	65[1]	65[1]	65[1]	65[1]	Chevaline
Norway									**Norvège**
Cattle	968	945	932	949	953	974	984	976	Bovine
Sheep	2 340	2 248	2 210	2 183	2 211	2 211	2 363	2 316	Ovine
Pigs	738	779	745	657	710	721	766	745	Porcine
Horses	16	16	17	17	19	21	21	21	Chevaline
Poland									**Pologne**
Cattle	10 919	10 523	10 322	10 733	10 049	8 844	8 221	7 643	Bovine
Sheep	4 991	4 739	4 377	4 409	4 158	3 234	1 870	1 268	Ovine
Pigs	18 949	18 546	19 605	18 835	19 464	21 868	22 086	18 860	Porcine
Horses	1 272	1 141	1 051	973	941	939	900	841	Chevaline
Portugal									**Portugal**
Cattle *	1 310	1 332	1 332	1 356	1 335	1 375	1 416	1 345	Bovine *
Sheep	5 000[1]	* 5 100	* 5 298	* 5 354	* 5 567	* 5 673	* 5 640	* 5 601	Ovine
Pigs	* 3 092	* 2 454	2 455	2 331	2 598	2 664	2 564	2 547	Porcine
Horses [1]	26	26	26	26	26	26	25	25	Chevaline [1]
Asses [1]	175	170	170	170	170	170	170	170	Asine [1]
Mules [1]	85	80	80	80	80	80	80	80	Mulassière [1]
Republic of Moldova									**République de Moldova**
Cattle	..	..	..	..	..	..	1 000	* 909	Bovine
Sheep	..	..	..	..	..	..	1 239	1 070[1]	Ovine
Pigs	..	..	..	..	..	..	1 753	1 600[1]	Porcine
Horses	..	..	..	..	..	..	48	49[1]	Chevaline
Asses [1]	..	..	..	..	..	..	5	5	Asine [1]
Romania									**Roumanie**
Cattle	* 6 692	6 703	6 559	* 6 416	6 291	5 381	4 355	3 683	Bovine
Sheep	17 342	17 219	16 839	16 210	15 435	14 062	13 879	12 079	Ovine
Pigs	13 651	14 095	14 328	14 351	11 671	12 003	10 954	9 852	Porcine
Horses	672	686	693	702	663	670	749	721	Chevaline
Asses [1]	35	36	36	36	35	35	35	34	Asine [1]
Russian Federation									**Fédération de Russie**
Cattle	..	..	..	..	..	..	54 677	52 226	Bovine
Sheep	..	..	..	..	..	..	52 195	48 183	Ovine
Pigs	..	..	..	..	..	..	35 384	31 520	Porcine
Horses	..	..	..	..	..	..	2 590	2 556	Chevaline
Asses	..	..	..	..	..	..	22	22	Asine
Mules [1]	..	..	..	..	..	..	1	1	Mulassière [1]
Slovakia									**Slovaquie**
Cattle	..	..	..	..	..	..	..	1 203	Bovine
Sheep	..	..	..	..	..	..	..	467	Ovine

37
Livestock
Thousand head [*cont.*]
Cheptel
Milliers de têtes [*suite*]

Country or area	1986	1987	1988	1989	1990	1991	1992	1993	Pays ou zone
Pigs	..	..	..	..	..	..	..	2 281	Porcine
Horses [1]	..	..	..	..	..	..	..	13	Chevaline [1]
Slovenia									**Slovénie**
Cattle	..	..	..	..	..	...	484	504	Bovine
Sheep	..	..	..	..	..	...	28	21	Ovine
Pigs	..	..	..	..	..	...	529	602	Porcine
Horses	..	..	..	..	..	...	11	8	Chevaline
Spain									**Espagne**
Cattle	4 953	4 795	5 061	5 185	5 126	5 063	4 976	4 800	Bovine
Sheep	17 894	22 994	23 064	22 739	24 037	24 625	24 615	* 24 800	Ovine
Pigs	15 780	17 303	16 614	16 911	16 002	17 247	18 260	* 18 000	Porcine
Horses	248	248	247[1]	245[1]	260[1]	273[1]	263[1]	263[1]	Chevaline
Asses	140	120[1]	105[1]	100[1]	90[1]	90[1]	90[1]	90[1]	Asine
Mules	117	110[1]	90[1]	80[1]	70[1]	70[1]	60[1]	60[1]	Mulassière
Sweden									**Suède**
Cattle	1 716	1 656	1 662	1 688	1 718	1 707	1 773	1 773[1]	Bovine
Sheep	407	397	395	401	406	419	448	448[1]	Ovine
Pigs	2 439	2 234	2 274	2 264	2 264	2 201	2 279	* 2 390	Porcine
Horses	57[1]	57[1]	58[1]	58[1]	58[1]	54	54[1]	54[1]	Chevaline
Switzerland									**Suisse**
Cattle	1 902	1 858	1 837	1 850	1 855	1 829	1 783	1 745	Bovine
Sheep	365	355	367	371	395	409	415	424	Ovine
Pigs	1 973	1 917	1 941	1 869	1 787	1 723	1 706	1 692	Porcine
Horses	48	48	49	48	45	49	52	54	Chevaline
Asses	2	2	2	2[1]	2[1]	2[1]	2[1]	2[1]	Asine
TFYR Macedonia									**L'ex–R.y. Macédoine**
Cattle	..	..	..	..	..	...	285	285	Bovine
Sheep	..	..	..	..	..	...	2 251	2 351	Ovine
Pigs	..	..	..	..	..	...	173	173	Porcine
Horses	..	..	..	..	..	...	65	39[1]	Chevaline
Ukraine									**Ukraine**
Cattle	..	..	..	..	..	...	23 728	22 457	Bovine
Sheep	..	..	..	..	..	...	7 259	6 597	Ovine
Pigs	..	..	..	..	..	...	17 839	16 175	Porcine
Horses	..	..	..	..	..	...	717	707	Chevaline
Asses [1]	..	..	..	..	..	...	19	19	Asine [1]
United Kingdom									**Royaume–Uni**
Cattle	12 695	12 476	11 855	11 909	11 922	11 641	11 620	11 708	Bovine
Sheep	24 540	25 976	27 820	29 103	29 678	28 944	29 493	29 333	Ovine
Pigs	7 930	7 955	7 915	7 627	7 383	7 520	7 705	7 869	Porcine
Horses [1]	163	165	168	168	169	170	172	173	Chevaline [1]
Asses [1]	10	10	10	10	10	10	10	10	Asine [1]
Yugoslavia									**Yougoslavie**
Cattle *	..	..	..	..	..	...	1 975	1 991	Bovine *
Sheep	..	..	..	..	..	...	2 715	* 2 752	Ovine
Pigs	..	..	..	..	..	...	3 844	* 4 092	Porcine
Horses	..	..	..	..	..	...	89	68[1]	Chevaline
Yugoslavia, SFR †									**Yougoslavie, Rfs †**
Cattle	5 034	5 030	4 881	4 759	4 705	4 514	..	..	Bovine
Sheep	7 693	7 819	7 824	7 564	7 596	7 431	..	..	Ovine
Pigs	7 821	8 459	8 323	7 396	7 231	7 378	..	..	Porcine
Horses	409	384	362	340	314	303	..	..	Chevaline
Asses	* 13	* 13	* 14	* 13	18	29	..	..	Asine
Mules [1]	6	5	5	5	4	4	..	..	Mulassière [1]
Oceania									**Océanie**
Cattle	32 390	30 588	30 591	30 939	31 884	32 470	32 733	33 457	Bovine
Sheep	214 251	213 407	217 049	222 179	228 156	218 407	200 782	189 110	Ovine
Pigs	4 593	4 686	4 786	4 735	4 701	4 601	4 876	4 773	Porcine
Horses	517	478	504	489	480	475	454	429	Chevaline
Asses	12	9	8	9	9	10	10	9	Asine
American Samoa									**Samoa américaines**
Pigs	13[1]	13[1]	13[1]	13	11[1]	11[1]	11[1]	11[1]	Porcine
Australia									**Australie**
Cattle	23 436	21 915	21 851	22 434	23 162	23 662	23 880	24 062	Bovine
Sheep	146 776	149 157	152 443	161 603	170 297	163 238	148 203	138 102	Ovine
Pigs	2 553	2 611	2 706	2 671	2 648	2 531	2 792	2 646	Porcine

37
Livestock
Thousand head [*cont.*]
Cheptel
Milliers de têtes [*suite*]

Country or area	1986	1987	1988	1989	1990	1991	1992	1993	Pays ou zone
Horses	346	313	335	317	310	308	289	272	Chevaline
Asses	5[1]	2	1	2	2	3	3	2	Asine
Cook Islands									**Iles Cook**
Pigs	17	17[1]	16	16	17	17	18	25	Porcine
Fiji									**Fidji**
Cattle	254[1]	261[1]	265[1]	268[1]	273[1]	280	288[1]	295[1]	Bovine
Pigs	78[1]	80[1]	83[1]	86[1]	88[1]	91	93[1]	95[1]	Porcine
Horses [1]	41	41	42	42	43	43	43	43	Chevaline [1]
French Polynesia									**Polynésie française**
Cattle	10	8	8	9	7	6[1]	6[1]	5[1]	Bovine
Pigs	32	32	32	32	32	32[1]	33[1]	34[1]	Porcine
Horses	2	2	2[1]	2[1]	2	2[1]	2[1]	2[1]	Chevaline
Guam									**Guam**
Pigs [1]	14	14	14	5	4	4	4	4	Porcine [1]
Kiribati									**Kiribati**
Pigs	* 10	10[1]	10[1]	9[1]	9[1]	9[1]	9[1]	9[1]	Porcine
Nauru									**Nauru**
Pigs [1]	2	2	2	2	3	3	3	3	Porcine [1]
New Caledonia									**Nouvelle−Calédonie**
Cattle	125[1]	130[1]	136	121	122[1]	125	124[1]	125[1]	Bovine
Sheep	3[1]	2	3[1]	3	4[1]	4	4[1]	4[1]	Ovine
Pigs	35[1]	35	36[1]	36[1]	37[1]	38	39[1]	39[1]	Porcine
Horses	9[1]	10[1]	10[1]	11[1]	11[1]	11	12[1]	12[1]	Chevaline
New Zealand									**Nouvelle−Zélande**
Cattle	8 279	7 999	8 058	7 828	8 034	8 100	8 145	8 675	Bovine
Sheep	67 470	64 244	64 600	60 569	57 852	55 162	52 571	51 000	Ovine
Pigs	435	426	414	411	395	407	411	430	Porcine
Horses	100	94	97	98	94	91	88	80	Chevaline
Niue									**Nioué**
Pigs	1[1]	1[1]	2[1]	2	2	2[1]	2[1]	2[1]	Porcine
Palau [5]									**Palaos** [5]
Cattle [1]	11	12	12	12	13	13	14	14	Bovine [1]
Pigs [1]	28	29	30	30	30	31	31	32	Porcine [1]
Papua New Guinea									**Papouasie−Nvl−Guinée**
Cattle	* 116	105[1]	* 101	101[1]	103[1]	105[1]	105[1]	105[1]	Bovine
Sheep	3	3[1]	3[1]	3[1]	4[1]	4[1]	4[1]	4[1]	Ovine
Pigs [1]	950	960	975	990	1 000	1 000	1 010	1 022	Porcine [1]
Horses	1	1[1]	1[1]	1[1]	1[1]	2[1]	2[1]	2[1]	Chevaline
Samoa									**Samoa**
Cattle	22[1]	24	24	24	24[1]	25[1]	24[1]	25[1]	Bovine
Pigs	175[1]	193	193	193	185[1]	180[1]	175[1]	178[1]	Porcine
Horses	3[1]	3[1]	3[1]	3	3[1]	3[1]	3[1]	3[1]	Chevaline
Asses [1]	7	7	7	7	7	7	7	7	Asine [1]
Solomon Islands									**Iles Salomon**
Cattle	20	14	* 13	13[1]	13[1]	13[1]	13[1]	12[1]	Bovine
Pigs [1]	48	49	50	51	52	53	54	54	Porcine [1]
Tokelau									**Tokélaou**
Pigs [1]	1	1	1	1	1	1	1	1	Porcine [1]
Tonga									**Tonga**
Cattle	9[1]	10[1]	10	12	10[1]	10[1]	10[1]	10[1]	Bovine
Pigs	90[1]	105[1]	120	94	94[1]	94[1]	94[1]	94[1]	Porcine
Horses	11[1]	11[1]	11	11	11[1]	12[1]	12[1]	11[1]	Chevaline
Tuvalu									**Tuvalu**
Pigs	9	8	8	11	11	13	13[1]	13[1]	Porcine
Vanuatu									**Vanuatu**
Cattle	107	110	113	116	123[1]	130	125[1]	128[1]	Bovine
Pigs	76[1]	77	58	58	60	60	59[1]	59[1]	Porcine
Horses	3[1]	3	3[1]	3[1]	3[1]	3[1]	3[1]	3[1]	Chevaline
Wallis and Futuna Islands									**Iles Wallis et Futuna**
Pigs [1]	24	24	24	24	24	25	25	25	Porcine [1]
former USSR †									**ancienne URSS †**
Cattle	120 888	122 103	118 300	119 600	118 388	115 643	..	..	Bovine
Sheep	140 850	142 210	140 783	140 700	138 564	132 949	..	..	Ovine
Pigs	77 772	79 501	77 403	78 100	78 963	75 428	..	..	Porcine

37

Livestock
Thousand head [*cont.*]
Cheptel
Milliers de têtes [*suite*]

Country or area	1986	1987	1988	1989	1990	1991	1992	1993	Pays ou zone
Horses	5 800	5 900	5 885	5 904	5 921	5 919	..	..	Chevaline
Asses	300	300	300	300[1]	300[1]	310[1]	..	..	Asine
Mules	1	1	1[1]	1[1]	1[1]	1[1]	..	..	Mulassière

Source:
Food and Agriculture Organization of the United Nations
(Rome).

† For detailed descriptions of data pertaining to
former Czechoslovakia, Germany, SFR Yugoslavia and former
USSR, see Annex I – Country or area nomenclature, regional
and other groupings.

1 FAO estimate.
2 Prior to 1993, including data for Eritrea.
3 Including data for Federated States of Micronesia, Marshall
Islands, and Northern Mariana Islands.

Source:
Organisation des Nations Unies pour l'alimentation et
l'agriculture (Rome).

† Pour les descriptions en détails des données
relatives à l'ancienne Tchécoslovaquie, l'Allemagne, la Rfs
Yougoslavie et l'ancienne URSS, voir l'Annexe I –
Nomenclature des pays ou zones, groupements régionaux et
autres groupements.

1 Estimation de la FAO.
2 Avant de 1993 y compris données de Erythrée.
3 Y compris les données pour les Etats fédérés de Micronésie,
les îles Marshall et les îles Mariannes du Nord.

38
Roundwood
Bois rond

Production (solid volume of roundwood without bark): million cubic metres
Production (volume solide de bois rond sans écorce) : millions de mètres cubes

Country or area Pays ou zone	1983	1984	1985	1986	1987	1988	1989	1990	1991	1992
World **Monde**	3 040.8	3 149.5	3 178.0	3 265.6	3 342.4	3 404.3	3 464.5	3 510.8	3 422.0	3 476.7
Africa **Afrique**	422.8	433.6	448.2	460.2	473.3	486.4	499.8	514.6	526.4	539.3
Algeria Algérie	1.8[1]	1.8[1]	1.9[1]	1.9[1]	2.0[1]	2.1[1]	2.1[1]	2.2[1]	2.2	2.3
Angola Angola	5.0	5.2	5.3	5.4	5.5	5.7	5.8	6.0	6.2[1]	6.4[1]
Benin Bénin	4.1[1]	4.2	4.3	4.5	4.6	4.7	4.9[1]	5.0	5.2[1]	5.4[1]
Botswana[1] Botswana[1]	1.1	1.1	1.1	1.2	1.2	1.2	1.3	1.3	1.4	1.4
Burkina Faso Burkina Faso	7.3[1]	7.5	7.7	7.9[1]	8.1[1]	8.3[1]	8.5[1]	8.7[1]	9.0[1]	9.3[1]
Burundi Burundi	3.5[1]	3.6[1]	3.7[1]	3.8[1]	3.9[1]	4.0[1]	4.1	4.2	4.4	4.5[1]
Cameroon Cameroun	11.3	11.7	12.2	12.5	12.8	13.0	13.5	14.1	14.3	14.6[1]
Central African Rep. Rép. centrafricaine	3.2	3.3	3.4	3.4	3.5	3.5	3.5	3.5	3.4	3.4[1]
Chad Tchad	3.4[1]	3.5[1]	3.6[1]	3.6[1]	3.7[1]	3.8[1]	3.9[1]	3.9	4.0	4.2[1]
Congo Congo	2.4	2.5	2.6	2.8	3.0	3.4	3.5	3.6	3.6	3.6[1]
Côte d'Ivoire Côte d'Ivoire	12.1	12.3	12.0	12.1	12.0	11.8	12.2	13.3	13.1	13.3
Egypt[1] Egypte[1]	1.9	1.9	2.0	2.0	2.1	2.1	2.2	2.2	2.3	2.4
Equatorial Guinea Guinée équatoriale	0.6	0.5	0.6	0.6[1]	0.6[1]	0.6[1]	0.6[1]	0.6[1]	0.6	0.6
Ethiopia[2] Ethiopie[2]	35.9	36.7[1]	37.5[1]	38.4[1]	39.5[1]	40.6	41.8[1]	43.0	44.3	45.6[1]
Gabon Gabon	3.3	3.6	3.5	3.5	3.5[1]	3.8	3.9	4.2	4.3[1]	4.3[1]
Gambia Gambie	0.8	0.9	0.9	0.9[1]	0.9[1]	0.9[1]	0.9[1]	0.9[1]	0.9[1]	0.9[1]
Ghana Ghana	16.2	16.4	16.5	16.6	16.8	17.0	17.2	17.7	17.1	17.2
Guinea Guinée	3.4[1]	3.4[1]	3.5[1]	3.6	3.7	3.8	3.9	4.0	4.1	4.2
Guinea-Bissau[1] Guinée-Bissau[1]	0.6	0.6	0.6	0.6	0.6	0.6	0.6	0.6	0.6	0.6
Kenya Kenya	27.4	28.5	29.5[1]	30.5[1]	31.6[1]	32.7	33.8	34.9	36.1[1]	37.3[1]
Lesotho[1] Lesotho[1]	0.5	0.5	0.5	0.5	0.6	0.6	0.6	0.6	0.6	0.6
Liberia Libéria	4.1	4.2	4.5	5.3	5.6	5.9	6.0[1]	6.1[1]	5.7	6.1
Libyan Arab Jamah.[1] Jamah. arabe libyenne[1]	0.6	0.6	0.6	0.6	0.6	0.6	0.6	0.6	0.6	0.6

38
Roundwood
Production (solid volume of roundwood without bark): million cubic metres [*cont.*]
Bois rond
Production (volume solide de bois rond sans écorce) : millions de mètres cubes [*suite*]

Country or area Pays ou zone	1983	1984	1985	1986	1987	1988	1989	1990	1991	1992
Madagascar Madagascar	6.7[1]	6.8[1]	7.0[1]	7.2[1]	7.4[1]	7.6[1]	7.9	8.1	8.3[1]	8.6[1]
Malawi Malawi	6.4[1]	6.6	6.9	7.2[1]	7.6[1]	8.1[1]	8.5	9.0	9.4[1]	9.7
Mali Mali	4.5	4.7	4.8	5.0	5.1	5.3	5.4[1]	5.6[1]	5.8[1]	6.0[1]
Morocco Maroc	1.8	2.0[1]	2.2	2.0	2.0[1]	1.9	2.1	2.0	2.5	2.4
Mozambique Mozambique	14.8	15.0	15.2	15.6	15.9	15.9	15.9	15.9	16.0	16.0
Niger[1] Niger[1]	4.0	4.1	4.2	4.4	4.5	4.7	4.8	5.0	5.1	5.3
Nigeria Nigéria	86.7[1]	89.5[1]	92.5	95.4	98.3[1]	101.4[1]	104.5[1]	107.7[1]	111.0[1]	114.3
Rwanda Rwanda	5.3	5.6	5.8	5.8[1]	5.8[1]	5.8[1]	6.6	5.6	5.6	5.7
Senegal Sénégal	3.9	4.0	4.4	4.6	4.6	4.5	4.4	4.9	4.9	4.9
Sierra Leone Sierra Leone	2.6	2.7[1]	2.7	2.8	2.9[1]	2.9[1]	3.0[1]	3.1	3.1	3.2[1]
Somalia[1] Somalie[1]	7.1	7.3	7.5	7.6	7.8	7.9	8.1	8.3	8.5	8.8
South Africa[3] Afrique du Sud[3]	20.6	19.0	19.0[1]	18.6	19.0	19.4	19.4[1]	20.1	19.7	19.7[1]
Sudan Soudan	18.6[1]	19.2[1]	19.7[1]	20.3[1]	20.9[1]	21.5[1]	22.2	22.8	23.4[1]	24.1[1]
Swaziland Swaziland	2.2[1]	2.2[1]	2.2[1]	2.2[1]	2.2[1]	2.2[1]	2.2[1]	2.2[1]	2.3	2.3[1]
Togo Togo	0.7	0.7	0.8	0.8	0.8	0.8	0.9	0.9	1.2	1.3
Tunisia Tunisie	2.7	2.8	2.9	2.9	3.0	3.0	3.1	3.2	3.3	3.3
Uganda Ouganda	11.5[1]	11.8[1]	12.1[1]	12.5[1]	12.9[1]	13.3	13.7	14.2	14.6[1]	15.0[1]
United Rep.Tanzania Rép. Unie de Tanzanie	25.4	26.3	27.2	28.3	29.5	30.5[1]	31.6[1]	32.6[1]	33.8[1]	34.9[1]
Zaire Zaïre	30.2	31.2	35.0	36.1	37.2	38.8	39.5	40.8	42.0	43.2[1]
Zambia Zambie	10.0[1]	10.3[1]	10.8[1]	11.2[1]	11.7[1]	12.2[1]	12.6	13.2	13.7	13.8
Zimbabwe Zimbabwe	6.8	7.1	7.2	7.3	7.6	7.9	7.9[1]	7.9[1]	8.0	8.0[1]
America, North **Amérique du Nord**	**642.3**	**689.7**	**684.6**	**718.8**	**739.7**	**751.0**	**756.7**	**757.6**	**723.4**	**746.6**
Bahamas[1] Bahamas[1]	0.1	0.1	0.1	0.1	0.1	0.1	0.1	0.1	0.1	0.1
Belize Belize	0.1	0.2	0.2	0.2	0.2	0.2[1]	0.2[1]	0.2[1]	0.2[1]	0.2[1]
Canada Canada	157.0	167.5	168.7	177.1	177.1	180.1	177.7	181.2	178.0	186.0[1]

38
Roundwood
Production (solid volume of roundwood without bark): million cubic metres [*cont.*]
Bois rond
Production (volume solide de bois rond sans écorce) : millions de mètres cubes [*suite*]

Country or area Pays ou zone	1983	1984	1985	1986	1987	1988	1989	1990	1991	1992
Costa Rica Costa Rica	3.2	3.4	3.5[1]	3.6[1]	3.9	4.0	4.1	4.1	4.2	4.3[1]
Cuba Cuba	3.3	3.3	3.3	3.4	3.2	3.3	3.1	3.1	3.1[1]	3.1[1]
Dominican Republic Rép. dominicaine	1.0	1.0[1]	1.0[1]	1.0[1]	1.0[1]	1.0[1]	1.0	1.0	1.0[1]	1.0[1]
El Salvador[1] El Salvador[1]	4.0	4.0	4.1	4.1	4.2	4.3	4.4	4.5	4.6	4.7
Guatemala Guatemala	9.5	9.8	10.0	10.2	10.4	10.7	11.0	11.3	11.3[1]	11.3[1]
Haiti[1] Haïti[1]	5.1	5.2	5.3	5.4	5.5	5.6	5.7	5.8	5.9	6.1
Honduras Honduras	4.8	5.1[1]	5.4	5.5	5.8	6.0	6.0	6.1	6.2	6.2
Jamaica Jamaïque	0.1	0.1	0.1[1]	0.2	0.2	0.2	0.2	0.2	0.2	0.2
Mexico Mexique	19.2	20.1	20.7	20.6[1]	21.3[1]	21.4[1]	22.8[1]	22.7	22.6[1]	23.0[1]
Nicaragua Nicaragua	3.4[1]	2.9	3.0	3.0	3.1	3.2	3.2	3.3	3.4	3.6
Panama Panama	1.0[1]	0.9[1]	1.0[1]	0.9[1]	0.9[1]	0.9[1]	1.0[1]	1.0[1]	1.0	1.0
Trinidad and Tobago Trinité-et-Tobago	0.1	0.1	0.1	0.1	0.1[1]	0.1	0.1	0.1	0.1	0.1
United States Etats-Unis	430.4	466.1	458.3	483.5	502.6	509.9	516.0	512.8	481.4	495.8
America, South **Amérique du Sud**	**296.9**	**302.2**	**309.6**	**318.4**	**325.6**	**332.8**	**342.4**	**343.8**	**352.5**	**358.8**
Argentina Argentine	11.5	11.3	11.1	10.8	10.9	11.5	11.3	11.3	11.3[1]	11.9
Bolivia Bolivie	1.3[1]	1.3[1]	1.4[1]	1.4[1]	1.4[1]	1.5	1.5	1.6	1.6[1]	1.6[1]
Brazil Brésil	225.4[1]	230.6	236.0	241.0	245.8	250.8	260.6	259.2	265.7[1]	268.9[1]
Chile Chili	14.0	15.5	16.2	17.2	18.0	19.1	19.3	22.0	24.1	25.8
Colombia Colombie	17.2	17.6	18.1	18.9	19.2	19.5	19.8	20.1	20.3	20.6[1]
Ecuador Equateur	7.9	5.8	6.3	6.6	6.9	7.0	7.2	7.7	7.3	7.5
French Guiana[1] Guyane française[1]	0.3	0.3	0.3	0.3	0.3	0.3	0.3	0.3	0.3	0.3
Guyana Guyana	0.2[1]	0.2	0.2	0.2	0.2	0.2[1]	0.2[1]	0.2	0.2	0.2
Paraguay Paraguay	6.8[1]	7.5	7.7	8.2	8.5	8.4	8.4[1]	8.4[1]	8.5[1]	8.5[1]
Peru Pérou	7.8[1]	7.7	7.7	8.5	8.7	8.8	8.6	7.7	7.8	7.8
Suriname Suriname	0.2	0.2	0.2	0.2	0.2	0.2	0.2	0.1	0.1	0.2

38
Roundwood
Production (solid volume of roundwood without bark): million cubic metres [*cont.*]
Bois rond
Production (volume solide de bois rond sans écorce) : millions de mètres cubes [*suite*]

Country or area Pays ou zone	1983	1984	1985	1986	1987	1988	1989	1990	1991	1992
Uruguay Uruguay	2.9	3.1	3.3	3.7	3.9	3.7	3.6	3.8	3.9	4.1
Venezuela Venezuela	1.4	1.1	1.3	1.4	1.6	1.8	1.6	1.6	1.5	1.5[1]
Asia **Asie**	**950.6**	**975.0**	**984.1**	**1 007.7**	**1 033.2**	**1 054.6**	**1 070.0**	**1 073.8**	**1 090.3**	**1 119.7**
Afghanistan[1] Afghanistan[1]	5.9	5.8	5.8	5.8	5.8	6.0	6.2	6.5	6.8	7.3
Bangladesh Bangladesh	25.8	26.4	27.1	27.7	28.4[1]	29.1[1]	29.7[1]	30.4[1]	31.2[1]	31.9[1]
Bhutan[1] Bhoutan[1]	1.4	1.4	1.4	1.4	1.5	1.5	1.5	1.6	1.6	1.6
Brunei Darussalam Brunéi Darussalam	0.3[1]	0.3[1]	0.3[1]	0.3[1]	0.3[1]	0.3[1]	0.3	0.3	0.3[1]	0.3[1]
Cambodia[1] Cambodge[1]	5.0	5.2	5.3	5.5	5.6	5.7	5.9	6.0	6.4	6.8
China[1] Chine[1]	245.4	258.4	265.3	271.4	276.5	279.6	281.8	280.0	282.3	296.6
Cyprus Chypre	0.1	0.1	0.1	0.1	0.1	0.1	0.1	0.1	0.1	0.1
Hong Kong[1] Hong-kong[1]	0.2	0.2	0.2	0.2	0.2	0.2	0.2	0.2	0.2	0.2
India[1] Inde[1]	237.5	243.1	248.6	253.3	258.0	262.8	267.5	272.4	277.3	282.4
Indonesia Indonésie	151.7	155.6	154.7	161.2	167.7[1]	173.9[1]	178.4[1]	176.4[1]	182.5[1]	185.6[1]
Iran, Islamic Rep. of Iran, Rép. islamique d'	6.7[1]	6.7[1]	6.7[1]	6.7[1]	6.7[1]	6.7	6.7	6.8	6.8	6.8[1]
Iraq[1] Iraq[1]	0.1	0.1	0.1	0.1	0.1	0.1	0.1	0.1	0.1	0.2
Israel Israël	0.1[1]	0.1[1]	0.1[1]	0.1[1]	0.1[1]	0.1[1]	0.1	0.1	0.1[1]	0.1[1]
Japan Japon	31.6	33.1	33.5	32.1	31.3	31.1	30.8	29.6	28.3	28.1[1]
Korea, Dem. P. R.[1] Corée, R. p. dém. de[1]	4.4	4.5	4.5	4.5	4.6	4.6	4.6	4.7	4.7	4.8
Korea, Republic of Corée, République de	8.4	7.7	6.7	6.7[1]	6.9[1]	6.8[1]	6.8	6.5[1]	6.5[1]	6.5[1]
Lao People's Dem. Rep. Rép. dém. pop. lao	3.4	3.5	3.7	3.7	3.8[1]	4.0[1]	4.1[1]	4.3	4.5	4.4
Lebanon Liban	0.5	0.5[1]	0.5[1]	0.5[1]	0.5	0.5	0.5	0.5	0.5	0.5
Malaysia Malaisie	41.2	39.7	37.5	39.0	44.4	48.5	50.8	49.7	50.1	54.0
Mongolia Mongolie	2.4[1]	2.4[1]	2.4[1]	2.4[1]	2.4[1]	2.4[1]	2.4[1]	2.4[1]	2.1	2.0
Myanmar Myanmar	18.9	18.6	19.2	19.5	19.8	20.1	20.5	21.5	22.3	22.7
Nepal Népal	15.5[1]	15.9[1]	16.4[1]	16.8[1]	17.3[1]	17.7[1]	18.2[1]	18.6	19.1	19.6[1]

38
Roundwood
Production (solid volume of roundwood without bark): million cubic metres [cont.]
Bois rond
Production (volume solide de bois rond sans écorce): millions de mètres cubes [suite]

Country or area Pays ou zone	1983	1984	1985	1986	1987	1988	1989	1990	1991	1992
Pakistan Pakistan	19.3[1]	20.1[1]	21.0	21.6[1]	22.2	22.9	24.0	25.7	26.2	26.6
Philippines Philippines	35.5	35.9	35.5	36.1	37.6	38.1	38.3	38.4	38.7	38.7
Sri Lanka Sri Lanka	8.5	8.6	8.5	8.6	8.8[1]	8.9[1]	9.0	9.0	9.1	9.2
Syrian Arab Republic Rép. arabe syrienne	0.1	0.0[1]	0.0[1]	0.0[1]	0.1	0.1[1]	0.1	0.1	0.1	0.1
Thailand Thaïlande	34.9	35.7	36.1	36.7	37.3	37.6	37.0	37.0	37.2	37.6
Turkey Turquie	20.8	19.7	16.3	17.9	16.8	16.8	15.5	15.8	15.3	15.3
Viet Nam Viet Nam	24.7[1]	25.5[1]	26.4[1]	27.2[1]	28.1[1]	28.2[1]	28.6[1]	28.8	29.5	29.6
Yemen[1] Yémen[1]	0.3	0.3	0.3	0.3	0.3	0.3	0.3	0.3	0.3	0.3
Europe **Europe**	**338.0**	**345.7**	**345.4**	**347.3**	**346.6**	**354.3**	**369.8**	**393.4**	**330.2**	**331.4**
Albania Albanie	2.3[1]	2.3[1]	2.3[1]	2.3[1]	2.3[1]	2.3[1]	2.3[1]	2.1	2.6[1]	2.6[1]
Austria Autriche	13.6	14.2	14.2	13.7	13.6	15.0	16.3	16.8	15.6	13.9
Belgium-Luxembourg Belgique-Luxembourg	3.0	3.1	3.3	3.3	3.7	4.0	4.8	5.6	4.8	4.7
Bulgaria Bulgarie	4.8	4.8	4.8	4.5	3.6	3.5	4.2	4.1	3.7	3.7[1]
former Czechoslovakia† anc. Tchécoslovaquie†	18.8	18.9	19.0	18.9	18.7	18.1	18.2	18.2	15.3	15.3
Denmark Danemark	3.0	2.6	2.3	2.3	2.1	2.2	2.1	2.3	2.3	2.3[1]
Finland Finlande	38.4	40.5	41.7	40.7	41.9	44.9	47.1	43.2	34.9	38.7
France France	38.7	38.8	38.9	39.9	41.1	43.0	44.7	45.4	44.2	44.8
Germany † Allemagne†	..	..	..	..	..	..	..	..	41.2	37.3
F. R. Germany R. f. Allemagne	28.2	30.7	30.7[1]	30.3	31.0	32.6	36.9	73.5	..	..
former German D. R. anc. R. d. allemande	10.4	10.6	10.9	10.8	10.6	10.9	11.3	11.3[1]	..	..
Greece Grèce	2.9	2.7	3.0	3.2	2.9	3.1	2.8	2.5	2.5	2.5
Hungary Hongrie	6.4	6.3	6.8	7.0	6.8	6.6	6.6	6.2	5.9	5.2
Ireland Irlande	1.0	1.3	1.3	1.2	1.3	1.4	1.5	1.7	1.7	1.8
Italy Italie	8.4	9.2	9.4	9.6	9.1	9.1	8.8	8.0	8.4	8.4
Netherlands Pays-Bas	0.9	0.9	1.1	1.1	1.2	1.3	1.3	1.4	1.1	1.2

38

Roundwood
Production (solid volume of roundwood without bark): million cubic metres [*cont.*]
Bois rond
Production (volume solide de bois rond sans écorce) : millions de mètres cubes [*suite*]

Country or area Pays ou zone	1983	1984	1985	1986	1987	1988	1989	1990	1991	1992
Norway Norvège	9.6	10.0	9.5	9.9	10.4	11.0	11.5	11.8	11.3	10.9[1]
Poland Pologne	24.7	24.0	23.3	24.3	23.3	22.8	21.4	17.8	17.2	20.5
Portugal Portugal	8.6	8.5	9.4	9.9	9.4	9.4	10.3	11.3	10.9	10.9[1]
Romania Roumanie	22.9	24.1	23.0	20.7	19.9	18.2	16.3	13.3	13.7	13.7[1]
Spain Espagne	15.0	14.6	14.0	15.0	15.5	15.2	18.8	18.2	18.2	17.1
Sweden Suède	53.0	53.0	51.5	52.4	53.1	53.9	55.9	53.5	52.5	53.6
Switzerland Suisse	4.0	5.0	4.2	4.7	4.6	4.6	4.6	6.3	4.6	4.6
United Kingdom Royaume-Uni	4.0	3.9	4.8	5.2	5.5	6.1	6.5	6.4	6.4	6.4
Yugoslavia, SFR† Yougoslavie, Rfs†	15.4	15.9	15.9	16.1	15.2	15.2	15.6	12.8	11.5	11.5[1]
Oceania Océanie	**34.5**	**35.3**	**38.1**	**39.2**	**38.5**	**38.8**	**39.4**	**41.3**	**43.0**	**43.7**
Australia Australie	16.3	18.0	19.7	20.0	19.6[1]	19.8	19.5	20.1	19.5	19.5
Fiji Fidji	0.2	0.2	0.2	0.2	0.3	0.3	0.3	0.3[1]	0.3[1]	0.3[1]
New Zealand Nouvelle-Zélande	9.7	8.8	9.8	10.3	9.7	9.8	10.7	12.0	14.3	15.0
Papua New Guinea Papouasie-Nvl-Guinée	7.6	7.6	7.6	7.9[1]	8.2[1]	8.2[1]	8.2[1]	8.2	8.2[1]	8.2[1]
Samoa[1] Samoa[1]	0.1	0.1	0.1	0.1	0.1	0.1	0.1	0.1	0.1	0.1
Solomon Islands Iles Salomon	0.5	0.5	0.5	0.6	0.4	0.4	0.4	0.5	0.5[1]	0.5[1]
Vanuatu Vanuatu	0.0	0.0[1]	0.0[1]	0.0[1]	0.1[1]	0.1[1]	0.1[1]	0.1[1]	0.1[1]	0.1[1]
former USSR† ancienne URSS†	355.7	367.9	368.0	373.9	385.5	386.5	386.5	386.4	386.4	337.1

Source:
Food and Agriculture Organization of the United Nations (Rome).

† For detailed descriptions of data pertaining to former Czechoslovakia, Germany, SFR Yugoslavia and former USSR, see Annex I - Country or area nomenclature, regional and other groupings.

1 FAO estimate.
2 Including Eritrea.
3 Including data for Namibia.

Source:
Organisation des Nations Unies pour l'alimentation et l'agriculture (Rome).

† Pour les descriptions en détails des données relatives à l'ancienne Tchécoslovaquie, l'Allemagne, la Rfs Yougoslavie et l'ancienne URSS, voir l'Annexe I - Nomenclature des pays ou zones, groupements régionaux et autres groupements.

1 Estimation de la FAO.
2 Y compris Erythrée.
3 Y compris les données pour la Namibie.

39
Fish catches
Quantités pêchées
All fishing areas: thousand metric tons
Toutes les zones de pêche : milliers de tonnes métriques

Country or area Pays ou zone	1983	1984	1985	1986	1987	1988	1989	1990	1991	1992
World *Monde*	77 497.0	83 932.2	86 377.7	92 845.4	94 402.5	99 085.7	100 311.1	97 556.1	97 051.7	98 112.8
Afghanistan Afghanistan	1.5	1.5	1.5	1.5	1.5	1.5	1.5	1.5	1.5	1.5
Albania Albanie	9.0	8.0	11.4	11.9	13.1	14.6	12.0	14.9	13.6[1]	13.2[1]
Algeria Algérie	65.0	65.6	66.1	65.5	94.4	106.7	99.7	91.1	80.1	95.8
American Samoa Samoa américaines	0.1[2]	0.2[2]	0.1[2]	0.1[2]	0.1	0.1	0.1	0.0	0.0	0.0[1]
Angola Angola	110.9	72.7	74.5	57.4	82.9	101.8	111.3	107.0	75.2	74.5
Antigua and Barbuda Antigua-et-Barbuda	1.1	1.5	2.4	2.4[1]	2.4[1]	2.4[1]	2.4[1]	2.2[1]	2.4[1]	2.3[1]
Argentina Argentine	416.4	315.2	406.8	420.7	559.8	493.4	486.6	555.6	640.6	705.3
Armenia Arménie	..	..	..	..	..	..	..	..	4.5	4.5[1]
Aruba Aruba	0.8[1]	0.8[1]	0.8	0.7[1]	0.6[1]	0.6[1]	0.5[1]	0.4[1]	0.4[1]	0.3
Australia[2] Australie[2]	169.2[1]	169.4[1]	161.4	180.3	204.8	213.3	181.8	220.4	227.3	233.9
Austria Autriche	4.7	4.4	4.5	4.6	4.6	5.1	5.0	4.8	4.5	4.1
Azerbaijan Azerbaïdjan	..	..	..	..	..	..	..	..	39.7	39.0[1]
Bahamas Bahamas	5.2	5.3	7.7	5.9	7.1	7.2	8.2	7.5	9.2	9.8
Bahrain Bahreïn	4.8	5.6	7.8	8.1	7.8	6.7	9.2	8.1	7.6	8.0
Bangladesh Bangladesh	726.6	756.0	775.6	796.9	817.0	829.9	843.6	847.8	892.7	966.7
Barbados Barbade	6.5	5.8	3.9	4.2	3.7	9.1	2.5	3.0	2.1	3.3
Belarus Bélarus	..	..	..	..	..	..	..	..	15.5	15.0[1]
Belgium Belgique	48.6	48.2	45.1	39.5	40.4	41.8	39.9	41.6	40.2	37.4
Belize Belize	1.5	1.3	1.4	1.5	1.5	1.5	1.8	1.5	1.6	1.6[1]
Benin Bénin	34.6	35.3[1]	36.4[1]	38.8[1]	41.9	37.3	41.9	41.7[1]	41.0[1]	40.0[1]
Bermuda Bermudes	0.5	0.5	0.7	0.8	0.8	0.8	0.8	0.5	0.4	0.4
Bhutan Bhoutan	1.0	1.0	1.0	1.0	1.0	1.0	1.0	1.0	1.0	1.0
Bolivia Bolivie	4.1[1]	4.1[1]	4.2	3.9	4.3	4.4	6.0	7.4	5.4	5.2
Bosnia & Herzegovina Bosnie-Herzégovine	..	..	..	..	..	..	..	..	...	3.0[1]

39
Fish catches
All fishing areas: thousand metric tons [*cont.*]
Quantités pêchées
Toutes les zones de pêche : milliers de tonnes métriques [*suite*]

Country or area Pays ou zone	1983	1984	1985	1986	1987	1988	1989	1990	1991	1992
Botswana Botswana	1.3	1.5	1.5	1.7	1.9	1.9	1.9	1.9[1]	1.9[1]	1.9[1]
Brazil Brésil	880.9	960.6	967.6	957.6	948.0	830.1	850.0	802.9[1]	800.0[1]	790.0[1]
British Virgin Islands Iles Vierges britanniques	0.9	1.0	1.1	1.2	1.2	1.3	1.4	1.4	1.4	1.4[1]
Brunei Darussalam Brunéi Darussalam	3.1	3.4	4.0	4.1	3.9	2.0	2.3	2.4	1.7	1.7
Bulgaria Bulgarie	121.1	115.5	100.2	109.7	111.7	118.0	103.0	57.1	48.8[1]	35.9[1]
Burkina Faso Burkina Faso	7.1[1]	7.4[1]	7.4[1]	7.6[1]	7.8[1]	7.9[1]	8.0	7.0	7.0	7.5[1]
Burundi Burundi	11.4	11.4	11.4	11.8	12.0	11.7	10.8	17.4	21.0	23.0
Cambodia Cambodge	68.2	62.8	67.6	71.4	79.6	82.2	76.6	105.0	111.1	102.6
Cameroon Cameroun	89.3	87.3	86.0	84.0	82.5	82.5[1]	77.6	77.6[1]	78.0[1]	82.0
Canada Canada	1 349.4	1 284.1	1 453.3	1 512.8	1 565.2	1 610.3	1 572.8	1 624.3	1 534.7	1 251.0
Cape Verde Cap-Vert	11.9	10.7	10.2	6.4	6.9	6.1	8.6	7.7	8.5	8.5[1]
Cayman Islands Iles Caïmanes	0.7	0.4	0.4	0.5	1.1	0.4	0.6[1]	0.8[1]	0.8	0.8
Central African Rep. Rép. centrafricaine	13.0	13.3	13.3	13.2	13.1	13.1	13.1	13.1	13.6	13.3
Chad[1] Tchad[1]	40.0	50.0	55.0	60.0	70.0	60.0	65.0	60.0	60.0	65.0
Channel Islands Iles Anglo-Normandes	3.5	3.7	3.0	2.3	2.4	2.8[1]	2.9	3.0[1]	2.8[1]	2.9
Chile Chili	3 977.0	4 499.3	4 804.4	5 571.6	4 814.6	5 209.9	6 454.2	5 195.4	6 002.8	6 501.8
China Chine	5 213.3	5 926.8	6 778.8	8 000.1	9 346.2	10 358.7	11 220.0	12 095.4	13 135.0	15 007.5
Colombia Colombie	57.5	79.1	71.5	83.4	85.5	89.1	98.3	128.0	109.2	158.9
Comoros Comores	4.8[1]	5.0[1]	5.2[1]	5.3[1]	5.3[1]	5.5[1]	6.8	8.0	6.8[1]	8.0
Congo Congo	35.3	32.8	29.9	31.6	37.9	42.0	45.8	48.2	45.6	40.0
Cook Islands Iles Cook	1.0[1]	1.0[1]	1.0[1]	1.1[1]	1.1[1]	1.1[1]	1.2[1]	1.2[1]	1.1	1.0
Costa Rica Costa Rica	10.6	16.6	20.5	21.0	17.2[1]	17.1[1]	17.4[1]	17.6[1]	17.9	18.1
Côte d'Ivoire Côte d'Ivoire	94.6	88.1	110.7	104.4	101.7	89.5	99.2	104.4	85.1	87.0
Croatia Croatie	..	..	..	..	..	..	..	..	...	31.8
Cuba Cuba	198.4	199.6	219.8	244.4	214.6	231.3	192.1	188.2	165.2	109.4

39
Fish catches
All fishing areas: thousand metric tons [*cont.*]
Quantités pêchées
Toutes les zones de pêche : milliers de tonnes métriques [*suite*]

Country or area Pays ou zone	1983	1984	1985	1986	1987	1988	1989	1990	1991	1992
Cyprus Chypre	2.0	2.3	2.4	2.6	2.6	2.5	2.6	2.7	2.7	2.8
former Czechoslovakia† anc. Tchécoslovaquie†	19.5	19.7	20.0	21.3	20.7	21.2	21.6	22.6	21.9	24.1
Denmark Danemark	1 869.8	1 851.5	1 796.9	1 849.8	1 707.8	1 974.4	1 929.3	1 518.0	1 795.8	1 995.0
Djibouti Djibouti	0.4	0.4	0.4	0.4	0.4	0.5	0.4	0.4	0.4[1]	0.4[1]
Dominica Dominique	0.8[1]	0.7[1]	0.6	0.6	0.7[1]	0.7[1]	0.7[1]	0.6[1]	0.6	0.8
Dominican Republic Rép. dominicaine	15.3	14.7	18.4	17.3	20.4	12.9	21.8	20.1[1]	17.3	13.6
Ecuador Equateur	371.9	882.8	1 087.0	1 003.4	680.1	876.0	739.9	391.2	384.1	347.1
Egypt Egypte	157.3	163.8	215.9	229.1	231.0	284.2	293.6	313.0	299.9	287.1
El Salvador El Salvador	7.6	12.2	16.1	20.5	21.5	11.7	11.6	9.2	11.3	12.6
Equatorial Guinea Guinée équatoriale	2.3[1]	4.0	3.6	4.4	4.0[1]	4.0[1]	4.0[1]	3.7[1]	3.5[1]	3.6[1]
Estonia Estonie	..	..	..	..	..	..	..	..	358.1	132.0
Ethiopia Ethiopie	3.9[1]	4.3[1]	4.0[1]	4.1[1]	4.0[1]	4.1	4.3	5.0	4.3	4.6
Faeroe Islands Iles Féroé	330.0	347.4	374.0	352.2	391.0	359.9[1]	308.6	287.2	246.0	270.8
Falkland Is. (Malvinas) Iles Falkland (Malvinas)	0.0	0.0	0.0	0.0	0.0	2.6	4.6	5.9	1.5	1.8
Fiji Fidji	27.8	27.9	27.6	27.0	35.3	32.4	32.8	35.0	31.1	31.3
Finland Finlande	130.3	134.5	135.2	131.1	106.3	120.9	110.5	95.6	82.5	100.3
France France	804.1	798.1	832.8	874.1	848.9	888.1	908.6[1]	897.4[1]	800.3	800.0[1]
French Guiana Guyane française	2.1	2.2	2.5	3.3	5.3	5.5	6.6	6.7	7.3	7.7
French Polynesia Polynésie française	2.2	2.8	2.3	2.3	2.7	3.1	3.3	3.0	2.6	1.9[1]
Gabon Gabon	19.4[1]	21.0	21.0	20.2	22.2[1]	22.1[1]	20.5[1]	20.0[1]	22.0[1]	21.0[1]
Gambia Gambie	11.7	11.9	10.7	13.3	14.6	13.9	19.8	17.9	23.7	22.7
Georgia Géorgie	..	..	..	..	..	..	..	..	61.0	50.0[1]
Germany † Allemagne†	542.9	553.2	425.8	414.3	398.4	392.4	411.1	387.9	300.3	304.8
Ghana Ghana	250.9	269.1	276.0	319.5	382.0	361.9	361.7	392.8	362.8	426.5
Greece Grèce	99.9	108.2	114.8	124.7	134.6	126.9	139.8	141.7	155.5[1]	171.7

39
Fish catches
All fishing areas: thousand metric tons [*cont.*]
Quantités pêchées
Toutes les zones de pêche : milliers de tonnes métriques [*suite*]

Country or area Pays ou zone	1983	1984	1985	1986	1987	1988	1989	1990	1991	1992
Greenland Groënland	99.3	86.3	94.9	102.9	100.4	120.3	168.8	138.2	113.4	113.3
Grenada Grenade	1.5	1.6	1.7	2.7	2.2	2.0	1.7	1.8	2.0	2.1
Guadeloupe Guadeloupe	8.8	9.0	8.4	8.5	8.6	8.2	8.5	8.6	8.4	8.5[1]
Guam Guam	0.3	0.4	0.5	0.6	0.5	0.6	0.6	0.7	0.8	0.7
Guatemala Guatemala	2.4	3.0	2.8	2.2	2.5	2.9	3.8	7.1	6.8	6.9
Guinea[1] Guinée[1]	26.0	28.0	30.0	32.0	34.0	34.0	34.0	35.0	37.5	37.0
Guinea-Bissau Guinée-Bissau	2.7	2.8	3.7[1]	3.7[1]	4.1[1]	4.7[1]	5.4[1]	5.4[1]	5.0[1]	5.2[1]
Guyana Guyana	35.4	37.2	37.6	37.4	36.8	36.5	35.3	36.9	40.8	41.4
Haiti[1] Haïti[1]	6.5	6.9	6.4	6.0	5.8	5.5	5.5	5.2	5.2	5.0
Honduras Honduras	11.4	8.4	9.7	20.6	23.1	19.9	17.1	15.5	21.0	18.8
Hong Kong Hong-kong	189.3	199.7	198.2	213.6	228.1	238.2	242.5	234.5	230.9	229.5
Hungary Hongrie	43.9	39.0	36.9	36.1	36.8	38.3	35.5	33.9	29.4	29.4
Iceland Islande	839.2	1 535.0	1 680.4	1 658.6	1 632.7	1 757.7	1 502.4	1 508.1	1 050.3	1 577.2
India Inde	2 508.6	2 864.5	2 826.1	2 923.2	2 906.6	3 125.4	3 640.3	3 794.2	4 044.2	4 175.1
Indonesia Indonésie	2 204.9	2 251.9	2 332.7	2 457.0	2 583.9	2 795.2	2 948.4	3 044.2	3 251.8	3 357.7
Iran, Islamic Rep. of Iran, Rép. islamique d'	111.9	115.8	118.5	152.1	211.4	235.4	260.5	271.0	277.4	327.5
Iraq[1] Iraq[1]	22.5	21.0	21.5	20.6	20.5	18.0	18.6	16.0	14.4	23.5
Ireland Irlande	203.4	210.0	230.7	231.6	250.7	257.5	201.7	239.6	261.9	275.4
Isle of Man Ile de Man	8.3	7.7	7.0	5.8	5.6	5.6	5.7	4.1	4.6	4.5
Israel Israël	23.0	24.4	27.3	25.2	28.4	28.2	26.5	23.6	20.7	18.9
Italy Italie	552.7	580.0	591.2	570.4	564.6	579.8	551.3	525.7	561.1	555.9
Jamaica Jamaïque	8.7	9.7	10.5	10.8	10.7	9.7	10.7	10.6[1]	10.6[1]	10.7[1]
Japan Japon	11 254.7	12 021.5	11 409.3	11 976.5	11 857.6	11 966.1	11 173.4	10 354.2	9 301.1	8 460.3
Jordan Jordanie	0.0[1]	0.0[1]	0.0[1]	0.1	0.1	0.1	0.1	0.1	0.0	0.0
Kazakhstan Kazakhstan	..	..	..	..	..	..	..	..	82.7	80.0[1]

39
Fish catches
All fishing areas: thousand metric tons [*cont.*]
Quantités pêchées
Toutes les zones de pêche : milliers de tonnes métriques [*suite*]

Country or area Pays ou zone	1983	1984	1985	1986	1987	1988	1989	1990	1991	1992
Kenya Kenya	98.1	91.0	106.0	119.8	131.2	138.1	146.4	201.8	198.6	198.5
Kiribati Kiribati	20.7	16.4	21.0	27.5	36.1	23.9	34.1	29.4[1]	29.6[1]	30.5[1]
Korea, Dem. P. R.[1] Corée, R. p. dém. de[1]	1 600.0	1 650.0	1 700.0	1 700.2	1 700.3	1 700.0	1 700.1	1 750.1	1 700.1	1 750.1
Korea, Republic of Corée, République de	2 400.3	2 476.8	2 649.9	3 103.4	2 876.6	2 731.5	2 840.6	2 843.1	2 521.2	2 695.6
Kuwait Koweït	8.7	9.6	10.1	7.6	7.7	10.8	7.7	4.5	2.0	7.9
Kyrgyzstan Kirghizistan	..	..	..	..	..	..	..	..	1.2	1.2[1]
Lao People's Dem. Rep. Rép. dém. pop. lao	26.0[1]	26.0[1]	26.0[1]	26.0[1]	28.0[1]	28.0[1]	28.0[1]	28.0[1]	29.0[1]	30.0
Latvia Lettonie	..	..	..	..	..	..	..	..	416.2	158.1
Lebanon Liban	1.4[1]	1.5[1]	1.7[1]	1.9	1.9	1.8	1.8	1.5	1.8	1.8[1]
Liberia Libéria	15.3	14.7	11.5	16.1	18.7	16.1	14.8	6.5	9.6	8.9
Libyan Arab Jamah.[1] Jamah. arabe libyenne[1]	7.7	7.8	9.7	9.7	8.6	9.7	7.8	8.7	8.0	8.4
Lithuania Lituanie	..	..	..	..	..	..	..	..	475.0	192.5
Macau Macao	8.2[1]	11.8[1]	12.4[1]	8.0	3.5	2.5	3.5	2.6	2.3	2.7
Madagascar Madagascar	63.4	67.5	69.2	84.3	92.4	101.1	97.7	104.6	101.2	106.6
Malawi Malawi	67.0	65.1	62.1	72.9	88.6	78.8	70.8	74.1	63.7	64.0[1]
Malaysia Malaisie	740.7	669.6	639.7	621.9	619.3[1]	612.4[1]	609.6[1]	604.0[1]	620.0[1]	640.0[1]
Maldives Maldives	38.6	55.1	61.9	59.3	57.0	71.5	71.2	78.7	80.7	82.2.
Mali Mali	61.3	54.7	54.2	61.0	55.7	55.9	71.8	70.5	68.8	68.5
Malta Malte	1.0	1.2	2.5	1.1	1.0	0.9	0.9	0.8	0.7	0.6
Marshall Islands Iles Marshall	0.2[1]	0.2[1]	0.2[1]	0.2[1]	0.2[1]	0.2[1]	0.2[1]	0.2[1]	0.2	0.2
Martinique Martinique	5.1	5.2	4.6	4.1	3.2	3.1	3.3	3.4	6.4	6.0[1]
Mauritania Mauritanie	81.6	93.8	103.3	98.2	99.4	97.6	92.6[1]	80.5[1]	88.7[1]	85.0[1]
Mauritius Maurice	9.8	10.6	12.4	12.9	18.2	17.2	17.2	14.7	18.9	19.2
Mexico Mexique	1 064.6	1 108.2	1 226.5	1 315.7	1 419.2	1 372.6	1 469.9	1 400.9	1 453.3	1 247.6
Micronesia,Federated States of Micron, Etats fédérés de	1.8	1.8	3.4	3.5	3.6	2.6[1]	2.1[1]	2.1[1]	1.4	1.5[1]

39

Fish catches
All fishing areas: thousand metric tons [*cont.*]
Quantités pêchées
Toutes les zones de pêche : milliers de tonnes métriques [*suite*]

Country or area Pays ou zone	1983	1984	1985	1986	1987	1988	1989	1990	1991	1992
Mongolia Mongolie	0.3	0.4	0.4	0.4	0.4	0.3	0.3	0.1	0.1	0.1 [1]
Montserrat Montserrat	0.1 [1]	0.1 [1]	0.1 [1]	0.1 [1]	0.1	0.1	0.0	0.2	0.1	0.1
Morocco Maroc	453.9	467.5	473.2	595.3	494.1	551.5	520.4	565.6	593.1	548.1
Mozambique Mozambique	42.5	35.8	36.3	34.9	39.1	36.5	37.1 [1]	39.5 [1]	38.9 [1]	37.5 [1]
Myanmar [2] Myanmar [2]	587.6	613.7	648.8	686.5	685.9	704.5	733.8	743.8	769.2	800.0
Namibia Namibie	12.1 [1]	12.6 [1]	13.1 [1]	14.1 [1]	31.6	32.6	20.3	257.1	204.7	293.1
Nauru Nauru	0.1 [1]	0.2 [1]	0.2 [1]	0.2 [1]	0.2 [1]	0.2 [1]	0.2 [1]	0.2 [1]	0.2	0.4
Nepal [2] Népal [2]	4.7	4.9	9.1	9.4	10.7	12.1	12.5	14.5	15.6	16.5
Netherlands Pays-Bas	506.0	432.4	504.2	454.8	446.1	398.9	451.7	459.0	443.1	438.0
Netherlands Antilles Antilles néerlandaises	1.0 [1]	1.0 [1]	1.0	1.1 [1]	1.1 [1]	1.2 [1]	1.2 [1]	1.2 [1]	1.1 [1]	1.2 [1]
New Caledonia Nouvelle-Calédonie	1.3	3.3	2.7	4.0	4.8	3.7	3.3	5.3	4.9	3.5
New Zealand Nouvelle-Zélande	282.3	322.8	305.1	346.0	419.8	552.7	567.6	560.0	609.0	679.3
Nicaragua Nicaragua	4.5	4.3	4.2	2.5	5.0	4.7	4.6	3.1	5.7	6.7
Niger Niger	3.3	3.0	2.0	2.4	2.3	2.5	4.8	3.4	3.2	2.1
Nigeria Nigéria	272.3 [1]	263.2 [1]	244.5	271.5	260.7	279.4 [1]	299.7 [1]	316.3	267.2	318.4
Niue Nioué	0.1	0.1	0.1	0.1	0.1	0.1	0.1	0.1	0.1	0.1
Northern Mariana Islands Iles Marianas du Nord	0.2	0.3	0.2	0.3	0.2	0.2	0.2	0.2	0.1	0.1 [1]
Norway Norvège	2 835.8	2 465.8	2 119.0	1 915.0	1 949.5	1 839.9	1 909.8	1 711.3	2 095.9	2 549.1
Oman Oman	108.8 [2]	105.2 [2]	101.2 [2]	96.4	136.1	165.6	117.7	120.2	117.8	112.3
Pakistan Pakistan	343.4	372.3	408.4	415.7	427.7	445.4	446.2	479.0	515.5	553.1
Palau Palaos	1.3	1.3	1.4 [1 3]	1.4 [1 3]	1.4 [1 3]	1.4 [1 3]	1.6 [1 3]	3.9 [1 3]	4.1	4.1 [1 3]
Panama Panama	168.3	131.1	289.2	131.5	155.5	125.1	190.4	143.7	146.1	143.5
Papua New Guinea [1] Papouasie-Nvl-Guinée [1]	15.5	17.8	25.5	25.3	25.5	26.0	26.0	26.6	25.3	25.7
Paraguay Paraguay	3.5	5.0	7.5	13.0	10.0	10.0	11.0	12.5	13.0	18.0
Peru Pérou	1 569.5	3 319.9	4 138.1	5 616.2	4 587.4	6 641.7	6 853.8	6 875.1	6 949.4	6 842.7

39
Fish catches
All fishing areas: thousand metric tons [*cont.*]
       Quantités pêchées
       Toutes les zones de pêche : milliers de tonnes métriques [*suite*]

Country or area Pays ou zone	1983	1984	1985	1986	1987	1988	1989	1990	1991	1992
Philippines Philippines	1 976.1	1 933.7	1 865.0	1 916.3	1 988.7	2 010.4	2 098.8	2 208.8	2 311.8	2 271.9
Poland Pologne	735.1	719.2	683.5	645.2	670.9	654.9	564.8	473.0	457.4	505.7
Portugal Portugal	249.9	298.0	313.5	408.7	389.6	346.7	332.1	322.0	324.7	295.3
Puerto Rico Porto Rico	2.7	2.4	1.5	1.3	1.3	1.7	2.0	2.1	2.3	1.7
Qatar Qatar	2.1	3.2	2.5	2.0	2.7	3.1	4.4	5.7	8.1	7.8
Republic of Moldova République de Moldova	..	..	..	..	..	..	..	..	5.2	5.0[1]
Réunion Réunion	2.6	2.6	2.2	1.7	1.8	2.1	2.0	2.0	2.3	2.4
Romania Roumanie	242.6	232.2	237.6	271.1	264.4	267.6	224.8	127.7	124.9	95.3
Russian Federation Fédération de Russie	..	..	..	..	..	..	..	..	6 894.2	5 611.2
Rwanda Rwanda	1.2	0.8	0.9	1.5	1.7	1.3	1.5	2.5	3.6	3.7
Saint Helena Sainte-Hélène	0.6	0.7	0.6	0.6	0.7	0.8	1.0	0.8	0.6	0.7
Saint Kitts and Nevis Saint-Kitts-et-Nevis	1.2	1.4[1]	1.6	1.6[1]	1.7[1]	1.7[1]	1.7[1]	1.7[1]	1.8[1]	1.7[1]
Saint Lucia Sainte-Lucie	0.9	0.9	1.1	0.8	0.7	0.8	0.8	0.9	0.9[1]	1.0
Saint Pierre and Miquelon Saint-Pierre-et-Miquelon	10.1	12.3	12.6	23.8	23.7	14.0	18.5	19.1	19.2	13.8
St. Vincent-Grenadines St. Vincent-Grenadines	0.5[1]	0.5[1]	0.5	0.6	0.7	4.6	5.8	8.8	8.0	2.2
Samoa Samoa	3.8	3.7[1]	3.6	3.2	3.1	2.5[1]	1.9	0.6	0.6	1.3
Sao Tome and Principe Sao Tomé-et-Principe	4.0	4.4	4.0	2.8	2.8	2.9	3.1	3.6	3.5[1]	3.6[1]
Saudi Arabia Arabie saoudite	36.0[1]	40.0[1]	43.7	45.5	47.9	47.1	53.4[1]	46.4[1]	43.2	44.0
Senegal Sénégal	263.5	247.3	246.0	255.6	255.0	260.7	287.1	297.9	319.7	326.9
Seychelles Seychelles	3.9	3.8	4.1	4.5	3.9	4.3	4.4	5.4	8.0	6.6[1]
Sierra Leone Sierra Leone	51.1[1]	52.7[1]	53.5[1]	53.3[1]	53.2[1]	53.3[1]	53.9[1]	51.8[1]	51.4	56.2
Singapore Singapour	19.5	26.2	23.9	21.4	16.7	15.2	12.6	13.3	13.1	11.6
Slovenia Slovénie	..	..	..	..	..	..	..	..	...	3.5[1]
Solomon Islands Iles Salomon	47.3	48.9	44.0	55.5	44.6	55.1	57.0	54.8	69.3	60.0[1]
Somalia Somalie	11.5	19.9	16.8	16.9[1]	17.5[1]	18.2[1]	18.2[1]	17.5[1]	16.1[1]	15.3[1]

39

Fish catches
All fishing areas: thousand metric tons [*cont.*]
Quantités pêchées
Toutes les zones de pêche : milliers de tonnes métriques [*suite*]

Country or area Pays ou zone	1983	1984	1985	1986	1987	1988	1989	1990	1991	1992
South Africa Afrique du Sud	935.9	735.2	776.7	820.7	1 426.6	1 302.5	878.5	537.6	501.0	695.3
Spain Espagne	1 412.8	1 440.6	1 482.8	1 489.0	1 525.5	1 593.4	1 560.0[1]	1 400.0[1]	1 320.0[1]	1 330.0[1]
Sri Lanka Sri Lanka	218.8	169.3	179.2	181.5	185.7	197.5	205.3	165.4	198.1	206.2
Sudan Soudan	29.5	29.8	26.3	24.0	27.2	29.2	30.3	31.7	33.3	31.7[1]
Suriname Suriname	3.6	4.1	4.1	3.7	5.2	3.7	6.2	6.5[1]	7.4	10.9
Swaziland Swaziland	0.1[1]	0.1[1]	0.1[1]	0.1[1]	0.1[1]	0.1	0.1[1]	0.1[1]	0.1[1]	0.1[1]
Sweden Suède	269.5	282.2	239.6	215.1	214.5	251.0	257.8	260.1	245.0	314.7
Switzerland Suisse	3.5	4.1	4.5	4.6	4.5	4.2	4.4	4.2	4.8	3.9
Syrian Arab Republic Rép. arabe syrienne	4.4	5.4	5.9	5.3	5.4	5.5	5.1	5.8	5.5[1]	5.4[1]
Tajikistan Tadjikistan	..	..	..	..	..	..	..	..	3.9	3.9[1]
Thailand Thaïlande	2 260.0	2 134.8	2 225.1	2 536.3	2 779.1	2 642.1	2 699.8	2 786.4	2 967.8	2 855.0
TFYR Macedonia L'ex-R.y. Macédoine	..	..	..	..	..	..	..	..	..	2.0[1]
Togo Togo	14.6	14.5	15.5	14.8	15.2	15.5	16.5	15.8	12.5	10.8
Tokelau Tokélaou	0.1[1]	0.2[1]	0.2[1]	0.2[1]	0.2[1]	0.2[1]	0.2[1]	0.2[1]	0.2	0.2
Tonga Tonga	2.4	2.6	2.7	3.0	2.8	2.7	2.7	1.7	2.0	2.3
Trinidad and Tobago Trinité-et-Tobago	4.2	3.6	5.5	4.1	6.6[1]	7.4[1]	8.0[1]	8.4[1]	12.2	15.0
Tunisia Tunisie	68.7	76.6	91.1	92.8	99.3	102.9	95.4	88.8	87.8	88.6
Turkey Turquie	557.3	566.9	578.1	582.9	627.9	676.0	457.1	384.8	363.5	454.3
Turkmenistan Turkménistan	..	..	..	..	..	..	..	..	43.0	40.0[1]
Turks and Caicos Islands Iles Turques et Caiques	1.2	1.2	1.3	1.5	1.3	1.3	1.3	1.0	1.2	1.3
Tuvalu Tuvalu	0.8	0.8	0.3	0.7	0.9	1.4	0.5	0.5	0.5	0.5
Uganda Ouganda	172.0	212.3	160.8	197.6	200.0	214.3	212.2	245.2	254.9	250.0[1]
Ukraine Ukraine	..	..	..	..	..	..	..	..	789.2	453.9
former USSR† ancienne URSS†	9 816.7	10 592.9	10 522.8	11 260.0	11 159.6	11 332.1	11 310.1	10 389.0	..	..
United Arab Emirates Emirats arabes unis	72.7	72.7	72.3	79.3	85.2	89.5	91.2	95.1	92.3	95.0

39
Fish catches
All fishing areas: thousand metric tons [cont.]
Quantités pêchées
Toutes les zones de pêche : milliers de tonnes métriques [suite]

Country or area Pays ou zone	1983	1984	1985	1986	1987	1988	1989	1990	1991	1992
United Kingdom Royaume-Uni	839.9	836.3	891.4	851.2	945.7	937.7	823.9	792.9	820.1	833.9
United Rep.Tanzania Rép. Unie de Tanzanie	239.2	277.3	300.6	309.9	342.3	393.0	377.1	414.0	326.8	331.6
United States Etats-Unis	4 319.2	4 990.9	4 950.9	5 186.4	5 992.0	5 956.2	5 778.1	5 870.4	5 488.7	5 602.9
United States Virgin Is. Iles Vierges américaines	0.6	0.7	0.6	0.9	0.9	0.7	0.8	0.7	0.9	0.9[1]
Uruguay Uruguay	143.4	133.0	138.4	140.7	137.8	107.3	121.7	90.8	143.7	125.7
Uzbekistan Ouzbékistan	..	..	..	..	..	..	..	..	27.4	28.1
Vanuatu Vanuatu	2.5	2.9	3.6	3.2	3.3	3.4[1]	3.3[1]	5.6[1]	3.5[1]	2.7[1]
Venezuela Venezuela	230.4	259.4	263.6	284.2	297.6	285.5	328.9	331.9	351.7	320.6
Viet Nam Viet Nam	757.1	776.3	808.0	824.7	871.4	900.0[1]	930.0[1]	960.0[1]	1 020.0[1]	1 080.3
Wallis and Futuna Islands Iles Wallis et Futuna	0.1[1]	0.1[1]	0.1[1]	0.1[1]	0.1	0.1	0.1	0.1	0.1	0.2
Yemen Yémen	63.7	65.7	71.3	72.7	72.4	73.2[1]	72.9[1]	77.9	83.4	80.7
Yugoslavia, SFR† Yougoslavie, Rfs†	79.7	73.4	75.0	77.5	81.3	71.8	71.7	65.5	35.6[1]	...
Zaire Zaïre	102.0[1]	148.3	148.5	156.5	162.0[1]	162.0[1]	166.0[1]	162.0[1]	160.0[1]	150.0[1]
Zambia Zambie	67.2	64.6	67.7	68.2	63.6	60.6	66.7	64.5	65.9	67.5
Zimbabwe Zimbabwe	13.6	16.4	17.4	18.9	19.2	22.2	24.0	25.8	22.2	22.5[1]

Source:
Food and Agriculture Organization of the United Nations
(Rome).

† For detailed descriptions of data pertaining to
former Czechoslovakia, Germany, SFR Yugoslavia and former
USSR, see Annex I - Country or area nomenclature, regional
and other groupings.

1 FAO estimate.
2 Data refer to a split year period.
3 Including data for Federated States of Micronesia, Marshall
Is. and Northern Mariana Is.

Source:
Organisation des Nations Unies pour l'alimentation et
l'agriculture (Rome).

† Pour les descriptions en détails des données
relatives à l'ancienne Tchécoslovaquie, l'Allemagne, la Rfs
Yougoslavie et l'ancienne URSS, voir l'Annexe I -
Nomenclature des pays ou zones, groupements régionaux et
autres groupements.

1 Estimation de la FAO.
2 Les données se réfèrent à une année fractionnée.
3 Y compris les données pour les Etats fédérés de Micronésie,
les îles Marshall et les îles Mariannes du Nord.

40
Fertilizers
Engrais

Nitrogenous, phosphate and potash: thousand metric tons
Azotés, phosphatés et potassiques : milliers de tonnes métriques

Country or area Pays ou zone	Production Production					Consumption Consommation				
	1988/89	1989/90	1990/91	1991/92	1992/93	1988/89	1989/90	1990/91	1991/92	1992/93
World Monde										
Nitrogenous fertilizers										
Engrais azotés	85 718.2	84 862.1	81 899.9	80 625.4	79 932.7	79 606.1	79 142.2	77 242.3	75 462.5	73 631.0
Phosphate fertilizers										
Engrais phosphatés	41 378.8	39 732.6	38 978.7	38 612.9	34 816.8	37 988.6	37 392.6	36 278.6	35 292.0	31 525.3
Potash fertilizers										
Engrais potassiques	31 157.9	28 327.8	26 710.8	24 980.2	23 510.2	28 042.2	26 885.5	24 526.9	23 572.8	20 775.1
Africa Afrique										
Nitrogenous fertilizers										
Engrais azotés	2 144.2	2 143.0	2 241.3	2 410.3	2 478.3	2 094.2	2 043.9	2 096.5	2 104.5	2 143.7
Phosphate fertilizers										
Engrais phosphatés	2 506.8	2 442.6	2 611.3	2 392.7	2 392.3	1 174.5	1 077.0	1 104.9	1 020.8	1 134.9
Potash fertilizers										
Engrais potassiques	...	...	...	...	...	457.1	488.7	485.4	470.6	501.0
Algeria Algérie										
Nitrogenous fertilizers										
Engrais azotés	105.5	* 88.0	90.4	73.2	88.3	76.0	* 46.3	* 63.2	* 41.0	* 46.6
Phosphate fertilizers										
Engrais phosphatés	70.7	* 45.3	53.7	34.6	* 33.5	60.2	* 47.7	46.3	34.6	* 31.9
Potash fertilizers										
Engrais potassiques	...	...	...	...	...	33.8	* 23.0	* 29.0	* 20.0	* 18.2
Angola Angola										
Nitrogenous fertilizers										
Engrais azotés	...	...	...	...	...	5.2	9.2	* 2.5	* 2.3	* 3.1
Phosphate fertilizers										
Engrais phosphatés	...	...	...	...	...	6.7	* 8.0	* 5.0	* 2.3	* 3.0
Potash fertilizers										
Engrais potassiques	...	...	...	...	...	4.0	6.4	* 2.0	* 2.3	* 3.0
Benin Bénin										
Nitrogenous fertilizers										
Engrais azotés	...	...	...	...	...	2.1	* 1.1	5.1	5.0	6.2
Phosphate fertilizers										
Engrais phosphatés	...	...	...	...	...	2.9	* 1.1	3.4	4.3	5.8
Potash fertilizers										
Engrais potassiques	...	...	...	...	...	1.9	* 1.1	2.5	2.5	3.3
Botswana Botswana										
Nitrogenous fertilizers										
Engrais azotés	...	...	...	...	...	* 0.3	* 0.3	* 0.3	* 0.3	0.4 [1]
Phosphate fertilizers										
Engrais phosphatés	...	...	...	...	...	* 0.4	* 0.4	* 0.4	* 0.4	0.4 [1]
Potash fertilizers										
Engrais potassiques	...	...	...	...	...	* 0.2	* 0.2	* 0.2	* 0.2	0.2 [1]
Burkina Faso Burkina Faso										
Nitrogenous fertilizers										
Engrais azotés	...	...	...	...	...	8.0	8.2	9.2	* 9.0	* 9.3
Phosphate fertilizers										
Engrais phosphatés	...	...	0.1	0.2	0.3	3.9	7.9	7.5	* 6.8	8.0 [1]
Potash fertilizers										
Engrais potassiques	...	...	...	...	...	3.3	4.5	4.4	4.0	4.0
Burundi Burundi										
Nitrogenous fertilizers *										
Engrais azotés *	...	...	...	...	...	1.0	2.0	1.0	0.1	2.1
Phosphate fertilizers *										
Engrais phosphatés *	...	...	...	...	...	0.9	2.4	1.0	0.3	1.8
Potash fertilizers *										
Engrais potassiques *	...	...	...	...	...	0.7	0.3	0.1	0.1	0.7
Cameroon Cameroun										
Nitrogenous fertilizers										
Engrais azotés	...	...	...	...	...	22.6	* 18.2	* 8.1	* 6.0	* 9.4

40

Fertilizers
Nitrogenous, phosphate and potash: thousand metric tons [cont.]
Engrais
Azotés, phosphatés et potassiques : milliers de tonnes métriques [suite]

Country or area	Production Production					Consumption Consommation				
Pays ou zone	1988/89	1989/90	1990/91	1991/92	1992/93	1988/89	1989/90	1990/91	1991/92	1992/93
Phosphate fertilizers										
Engrais phosphatés	...	...	...	...	...	6.2	* 4.4	* 2.7	* 2.8	* 4.0
Potash fertilizers										
Engrais potassiques	...	...	...	...	...	5.6	* 6.0	* 9.8	* 9.3	* 7.5
Central African Rep. Rép. centrafricaine										
Nitrogenous fertilizers										
Engrais azotés	...	...	...	...	...	0.5	0.7	* 0.8	* 0.8	* 0.9
Phosphate fertilizers										
Engrais phosphatés	...	...	...	...	...	0.0	0.0	...	...	* 0.1
Potash fertilizers										
Engrais potassiques	...	...	...	...	...	0.0	0.0	* 0.1	* 0.1	* 0.1
Chad Tchad										
Nitrogenous fertilizers										
Engrais azotés	...	...	...	...	...	2.2	2.0	* 2.0	* 3.4	* 3.1
Phosphate fertilizers										
Engrais phosphatés	...	...	...	...	...	1.0	1.1	* 1.9	* 2.6	* 2.7
Potash fertilizers										
Engrais potassiques	...	...	...	...	...	1.7	1.8	* 1.9	* 2.6	* 2.6
Comoros Comores										
Nitrogenous fertilizers										
Engrais azotés	...	...	...	...	...	...	...	...	* 0.1	0.1 [1]
Congo Congo										
Nitrogenous fertilizers										
Engrais azotés	...	...	...	...	...	* 0.5	0.2	0.6	* 1.3	* 1.1
Phosphate fertilizers										
Engrais phosphatés	...	...	...	...	...	...	0.2	0.4	...	...
Potash fertilizers										
Engrais potassiques	...	...	...	...	...	* 0.2	0.2	0.5	* 0.1	* 0.9
Côte d'Ivoire Côte d'Ivoire										
Nitrogenous fertilizers *										
Engrais azotés *	...	...	...	...	...	8.2	10.0	13.5	21.0	30.0
Phosphate fertilizers										
Engrais phosphatés	* 1.5	* 2.5	* 2.5	* 2.5	* 2.5	* 8.0	* 11.0	* 8.2	* 8.5	9.0 [1]
Potash fertilizers *										
Engrais potassiques *	...	...	...	...	...	25.0	18.0	14.0	14.0	10.0
Egypt Egypte										
Nitrogenous fertilizers										
Engrais azotés	* 676.7	* 678.0	676.1	823.6	824.8	* 799.1	754.1	745.1	* 775.0	* 745.0
Phosphate fertilizers										
Engrais phosphatés	* 202.0	* 217.4	* 194.9	163.1	* 175.0	* 203.7	164.9	184.1	* 150.0	* 115.0
Potash fertilizers										
Engrais potassiques	...	...	...	...	...	* 30.7	* 46.1	35.5	* 38.4	* 22.0
Ethiopia Ethiopie										
Nitrogenous fertilizers * [2]										
Engrais azotés * [2]	...	...	...	...	...	30.2	38.0	42.1	43.0	53.2
Phosphate fertilizers * [2]										
Engrais phosphatés * [2]	...	...	...	...	...	48.3	59.0	69.7	56.0	79.0
Potash fertilizers * [2]										
Engrais potassiques * [2]	...	...	...	...	...	...	...	...	...	0.4
Gabon Gabon										
Nitrogenous fertilizers										
Engrais azotés	...	...	...	...	...	0.3	* 0.4	0.1	* 0.2	* 0.1
Phosphate fertilizers										
Engrais phosphatés	...	...	...	...	...	0.0	* 0.1	* 0.5	* 0.1	* 0.1
Potash fertilizers										
Engrais potassiques	...	...	...	...	...	0.6	* 0.7	* 0.5	* 0.3	* 0.3
Gambia Gambie										
Nitrogenous fertilizers										
Engrais azotés	...	...	...	...	...	0.3	0.7	* 0.2	* 0.3	* 0.2

40

Fertilizers
Nitrogenous, phosphate and potash: thousand metric tons [*cont.*]
Engrais
Azotés, phosphatés et potassiques : milliers de tonnes métriques [*suite*]

Country or area Pays ou zone	Production Production					Consumption Consommation				
	1988/89	1989/90	1990/91	1991/92	1992/93	1988/89	1989/90	1990/91	1991/92	1992/93
Phosphate fertilizers										
Engrais phosphatés	...	...	...	...	...	0.3	0.7	* 0.2	* 0.3	* 0.4
Potash fertilizers										
Engrais potassiques	...	...	...	...	...	0.3	0.7	* 0.2	* 0.3	* 0.2
Ghana Ghana										
Nitrogenous fertilizers										
Engrais azotés	...	...	...	...	...	7.0	5.8	* 8.0	* 7.0	* 7.0
Phosphate fertilizers										
Engrais phosphatés	...	...	...	...	...	2.8	1.2	* 3.0	* 0.2	* 2.4
Potash fertilizers										
Engrais potassiques	...	...	...	...	...	2.8	1.5	* 2.0	* 0.8	* 1.1
Guinea Guinée										
Nitrogenous fertilizers										
Engrais azotés	...	...	...	...	...	0.3	0.4	0.7	1.4	* 2.2
Phosphate fertilizers										
Engrais phosphatés	...	...	...	...	...	0.1	0.1	0.1	0.1	* 0.6
Potash fertilizers										
Engrais potassiques	...	...	...	...	...	0.3	0.3	0.3	0.4	* 0.6
Guinea-Bissau Guinée-Bissau										
Nitrogenous fertilizers										
Engrais azotés	...	...	...	...	* 0.1	0.2	0.4	0.2	* 0.2	0.2[1]
Phosphate fertilizers										
Engrais phosphatés	...	...	...	...	...	0.1	0.3	0.2	0.2	0.1
Potash fertilizers										
Engrais potassiques	...	...	...	...	...	0.0	0.2	0.1	0.1	* 0.1
Kenya Kenya										
Nitrogenous fertilizers										
Engrais azotés	...	...	...	...	...	67.2	* 45.0	* 57.0	* 51.1	* 46.0
Phosphate fertilizers										
Engrais phosphatés	...	...	...	...	...	51.1	* 62.0	* 51.0	* 35.6	* 44.0
Potash fertilizers										
Engrais potassiques	...	...	...	...	...	6.3	* 9.8	* 8.0	* 8.7	* 10.5
Lesotho Lesotho										
Nitrogenous fertilizers *										
Engrais azotés *	...	...	...	...	...	0.4	0.5	0.5	0.8	0.7
Phosphate fertilizers *										
Engrais phosphatés *	...	...	...	...	...	4.0	4.0	4.0	4.0	4.0
Potash fertilizers *										
Engrais potassiques *	...	...	...	...	...	0.1	0.1	0.1	1.1	1.0
Liberia Libéria										
Nitrogenous fertilizers *										
Engrais azotés *	...	...	...	...	...	1.6	0.1	0.1	...	...
Phosphate fertilizers *										
Engrais phosphatés *	...	...	...	...	...	0.1	0.2	0.1	...	...
Potash fertilizers *										
Engrais potassiques *	...	...	...	...	...	1.7	2.4	0.1	...	...
Libyan Arab Jamah. Jamah. arabe libyenne										
Nitrogenous fertilizers										
Engrais azotés	* 144.3	* 124.0	* 89.0	* 172.5	285.7	* 32.5	* 30.0	* 35.0	* 35.0	* 30.0
Phosphate fertilizers										
Engrais phosphatés	...	...	...	...	...	54.2	* 47.0	* 41.0	* 47.5	* 59.0
Potash fertilizers										
Engrais potassiques	...	...	...	...	...	1.2	* 1.8	* 1.6	* 2.3	* 0.4
Madagascar Madagascar										
Nitrogenous fertilizers										
Engrais azotés	...	...	...	...	...	3.8	2.5	4.4	3.7	3.9
Phosphate fertilizers										
Engrais phosphatés	...	...	...	...	...	3.5	1.0	2.5	2.3	1.6

40

Fertilizers
Nitrogenous, phosphate and potash: thousand metric tons [cont.]
Engrais
Azotés, phosphatés et potassiques : milliers de tonnes métriques [suite]

Country or area	Production Production					Consumption Consommation				
Pays ou zone	1988/89	1989/90	1990/91	1991/92	1992/93	1988/89	1989/90	1990/91	1991/92	1992/93
Potash fertilizers										
Engrais potassiques	...	...	...	...	...	3.4	2.7	4.0	3.4	2.1
Malawi Malawi										
Nitrogenous fertilizers										
Engrais azotés	...	...	...	...	...	27.2	* 30.3	* 30.0	* 46.0	* 51.8
Phosphate fertilizers										
Engrais phosphatés	...	...	...	...	...	17.3	* 18.0	* 13.0	* 16.0	* 14.0
Potash fertilizers										
Engrais potassiques	...	...	...	...	...	6.7	* 6.5	* 5.0	* 8.0	* 8.0
Mali Mali										
Nitrogenous fertilizers *										
Engrais azotés *	...	...	...	...	...	9.0	8.7	9.2	9.5	10.7
Phosphate fertilizers *										
Engrais phosphatés *	...	...	...	...	...	3.4	9.1	6.0	5.5	9.0
Potash fertilizers *										
Engrais potassiques *	...	...	...	...	...	1.0	...	...	...	3.0
Mauritania Mauritanie										
Nitrogenous fertilizers *										
Engrais azotés *	...	...	...	...	...	2.0	1.7	1.5	1.5	1.7
Phosphate fertilizers *										
Engrais phosphatés *	...	...	...	...	...	0.6	0.6	0.4	...	...
Potash fertilizers *										
Engrais potassiques *	...	...	...	...	...	0.6	...	...	...	...
Mauritius Maurice										
Nitrogenous fertilizers										
Engrais azotés	11.4	* 9.0	12.0	12.4	14.6	11.0	* 11.5	11.2	11.2	11.3
Phosphate fertilizers										
Engrais phosphatés	...	...	...	...	...	4.1	* 4.7	3.8	3.9	3.6
Potash fertilizers										
Engrais potassiques	...	...	...	...	...	13.1	* 16.0	12.8	12.5	11.8
Morocco Maroc										
Nitrogenous fertilizers										
Engrais azotés	* 232.9	* 273.1	* 344.1	* 346.0	* 296.5	139.6	148.3	144.2	147.5	130.4
Phosphate fertilizers										
Engrais phosphatés	* 969.8	* 935.8	* 1 179.9	* 1 070.0	* 997.2	119.3	116.6	* 125.0	119.2	* 124.0
Potash fertilizers										
Engrais potassiques	...	...	...	...	...	56.4	55.9	56.3	* 45.4	* 66.2
Mozambique Mozambique										
Nitrogenous fertilizers *										
Engrais azotés *	...	...	...	...	...	1.1	1.7	2.2	2.1	3.0
Phosphate fertilizers *										
Engrais phosphatés *	...	...	...	...	...	0.3	0.3	0.2	1.7	1.2
Potash fertilizers *										
Engrais potassiques *	...	...	...	...	...	0.2	0.4	0.2	1.3	0.7
Niger Niger										
Nitrogenous fertilizers										
Engrais azotés	...	...	...	...	...	1.1	1.8	1.4	0.2	* 0.9
Phosphate fertilizers										
Engrais phosphatés	...	...	...	...	...	0.4	0.7	0.7	0.2	* 0.4
Potash fertilizers										
Engrais potassiques	...	...	...	...	...	0.2	0.3	0.2	0.1	* 0.1
Nigeria Nigéria										
Nitrogenous fertilizers										
Engrais azotés	* 243.4	* 272.4	* 284.0	231.0	271.0	* 160.0	197.4	210.0	212.0	270.1
Phosphate fertilizers										
Engrais phosphatés	* 27.4	44.1	50.2	57.9	83.0	* 100.0	93.5	96.1	110.6	166.4
Potash fertilizers										
Engrais potassiques	...	...	...	...	...	* 52.0	87.1	94.2	106.6	129.8

40

Fertilizers
Nitrogenous, phosphate and potash: thousand metric tons [*cont.*]
Engrais
Azotés, phosphatés et potassiques : milliers de tonnes métriques [*suite*]

Country or area / Pays ou zone	Production					Consumption / Consommation				
	1988/89	1989/90	1990/91	1991/92	1992/93	1988/89	1989/90	1990/91	1991/92	1992/93
Réunion Réunion										
Nitrogenous fertilizers Engrais azotés	...	...	...	...	...	5.2	5.7	* 3.6	* 5.0	5.2
Phosphate fertilizers Engrais phosphatés	...	...	...	...	...	3.5	3.7	* 2.7	* 3.0	5.0
Potash fertilizers Engrais potassiques	...	...	...	...	...	4.5	5.3	* 5.8	* 6.0	4.8
Rwanda Rwanda										
Nitrogenous fertilizers Engrais azotés	...	...	...	...	...	0.3	0.3	1.2	0.7	0.4
Phosphate fertilizers Engrais phosphatés	...	...	...	...	...	0.2	0.1	1.0	0.5	0.1
Potash fertilizers Engrais potassiques	...	...	...	...	...	0.1	0.1	0.7	0.4	0.1
Senegal Sénégal										
Nitrogenous fertilizers Engrais azotés	* 12.0	14.4	15.7	26.8	* 25.0	* 8.0	6.2	6.0	8.6	9.0[1]
Phosphate fertilizers Engrais phosphatés	* 35.0	27.5	* 50.0	46.4	* 33.0	* 10.0	3.2	2.7	5.1	* 6.0
Potash fertilizers Engrais potassiques	...	...	...	...	...	* 8.0	3.2	3.4	1.7	* 2.0
Sierra Leone Sierra Leone										
Nitrogenous fertilizers Engrais azotés	...	...	...	...	...	0.2	0.9	* 0.6	* 0.2	* 0.6
Phosphate fertilizers Engrais phosphatés	...	...	...	...	...	0.2	0.4	* 0.4	* 0.2	* 0.4
Potash fertilizers Engrais potassiques	...	...	...	...	...	0.2	0.3	* 0.3	* 0.2	* 0.4
Somalia Somalie										
Nitrogenous fertilizers * Engrais azotés *	...	...	...	...	...	1.3	1.4	1.7	...	...
Phosphate fertilizers * Engrais phosphatés *	...	...	...	...	...	0.5	0.8	0.5	...	...
Potash fertilizers * Engrais potassiques *	...	...	...	...	...	0.4	0.5	0.5	...	...
South Africa Afrique du Sud										
Nitrogenous fertilizers Engrais azotés	* 440.0	* 400.0	* 430.0	* 425.0	* 410.0	380.9	375.2	* 377.6	* 357.8	* 363.0
Phosphate fertilizers Engrais phosphatés	* 404.0	* 375.0	* 380.0	* 339.0	340.0[1]	317.7	273.0	277.0	* 259.0	* 295.0
Potash fertilizers Engrais potassiques	...	...	...	...	...	137.2	127.8	137.0	* 123.0	* 127.0
Sudan Soudan										
Nitrogenous fertilizers Engrais azotés	...	...	...	...	...	45.6	46.9	73.7	76.8	* 59.4
Phosphate fertilizers Engrais phosphatés	...	...	...	...	...	1.8	2.4	15.6	16.4	15.0[1]
Swaziland Swaziland										
Nitrogenous fertilizers Engrais azotés	...	...	...	...	...	7.2	* 7.0	7.1	* 6.0	* 6.0
Phosphate fertilizers Engrais phosphatés	...	...	...	...	...	3.5	* 2.0	2.4	* 2.5	* 2.5
Potash fertilizers Engrais potassiques	...	...	...	...	...	3.1	* 3.0	3.2	* 3.0	* 3.2
Togo Togo										
Nitrogenous fertilizers Engrais azotés	...	...	...	...	...	5.2	* 6.5	* 5.0	5.5	5.9
Phosphate fertilizers Engrais phosphatés	...	...	...	...	...	* 4.0	* 4.5	* 5.0	3.7	3.7

40
Fertilizers
Nitrogenous, phosphate and potash: thousand metric tons [*cont.*]
Engrais
Azotés, phosphatés et potassiques : milliers de tonnes métriques [*suite*]

Country or area Pays ou zone	Production Production					Consumption Consommation				
	1988/89	1989/90	1990/91	1991/92	1992/93	1988/89	1989/90	1990/91	1991/92	1992/93
Potash fertilizers Engrais potassiques	...	...	...	...	...	3.0	* 1.0	* 1.5	2.6	2.6
Tunisia Tunisie										
Nitrogenous fertilizers Engrais azotés	197.5	206.3	202.0	209.4	202.0	* 46.4	* 49.5	* 36.7	* 52.0	54.7
Phosphate fertilizers Engrais phosphatés	746.9	753.5	657.1	631.1	687.1	* 49.9	* 45.9	* 44.2	* 45.0	48.1
Potash fertilizers * Engrais potassiques *	...	...	...	...	...	5.0	8.3	2.0	2.0	5.8
Uganda Ouganda										
Nitrogenous fertilizers Engrais azotés	...	...	...	...	...	0.1	0.3	0.1	* 0.5	* 0.4
Phosphate fertilizers Engrais phosphatés	...	...	...	...	...	0.0	0.1	* 0.1	* 0.3	* 0.1
Potash fertilizers Engrais potassiques	...	...	...	...	...	0.0	0.1	...	* 0.4	* 0.3
United Rep.Tanzania Rép. Unie de Tanzanie										
Nitrogenous fertilizers Engrais azotés	0.7	* 4.1	3.4	2.9	...	27.0	* 28.7	36.7	33.6	33.1
Phosphate fertilizers Engrais phosphatés	0.7	* 3.0	1.5	2.9	...	11.1	* 16.3	11.7	11.3	10.0
Potash fertilizers Engrais potassiques	...	...	...	...	...	3.0	* 4.0	2.9	4.7	4.8
Zaire Zaïre										
Nitrogenous fertilizers Engrais azotés	...	...	...	...	...	1.9	* 4.0	* 4.0	* 4.0	* 1.0
Phosphate fertilizers Engrais phosphatés	...	...	...	...	...	0.5	* 1.4	* 0.5	* 1.8	* 0.2
Potash fertilizers Engrais potassiques	...	...	...	...	...	0.4	* 2.5	* 1.7	* 2.4	* 1.1
Zambia Zambie										
Nitrogenous fertilizers Engrais azotés	* 5.7	* 1.9	* 6.0	4.7	* 3.7	56.2	* 51.8	* 38.0	40.1	* 57.3
Phosphate fertilizers Engrais phosphatés	...	...	...	...	...	20.4	* 19.4	* 15.9	15.4	* 18.2
Potash fertilizers Engrais potassiques	...	...	...	...	...	8.1	* 8.0	* 5.7	7.0	* 9.0
Zimbabwe Zimbabwe										
Nitrogenous fertilizers Engrais azotés	74.1	71.7	88.6	82.7	* 56.6	88.2	82.2	95.0	75.7	* 67.0
Phosphate fertilizers Engrais phosphatés	48.9	38.4	41.4	45.0	* 40.7	47.5	35.7	46.6	40.5	* 39.0
Potash fertilizers Engrais potassiques	...	...	...	...	...	30.1	30.6	35.1	32.3	* 31.0
America, North Amérique du Nord										
Nitrogenous fertilizers **Engrais azotés**	17 509.1	17 185.7	17 745.9	18 343.5	18 388.3	12 715.1	13 252.9	13 369.7	13 376.1	13 370.3
Phosphate fertilizers **Engrais phosphatés**	10 440.5	10 499.5	10 888.5	12 079.6	11 470.4	4 958.7	5 093.5	4 953.7	4 987.6	5 085.4
Potash fertilizers **Engrais potassiques**	9 201.9	7 780.5	8 528.2	7 951.9	8 227.0	5 155.1	5 509.7	5 294.8	5 145.9	5 226.1
Bahamas Bahamas										
Nitrogenous fertilizers * Engrais azotés *	...	...	...	...	...	0.2	0.2	...	0.2	0.2
Phosphate fertilizers * Engrais phosphatés *	...	...	...	...	...	0.2	0.2	...	...	...
Potash fertilizers * Engrais potassiques *	...	...	...	...	...	0.1	0.1	0.1	0.1	0.1

40

Fertilizers
Nitrogenous, phosphate and potash: thousand metric tons [*cont.*]
Engrais
Azotés, phosphatés et potassiques : milliers de tonnes métriques [*suite*]

Country or area	Production Production					Consumption Consommation				
Pays ou zone	1988/89	1989/90	1990/91	1991/92	1992/93	1988/89	1989/90	1990/91	1991/92	1992/93
Barbados Barbade										
Nitrogenous fertilizers										
Engrais azotés	...	...	...	...	...	* 1.5	* 1.5	* 1.5	* 1.5	1.5[1]
Phosphate fertilizers										
Engrais phosphatés	...	...	...	...	...	* 0.3	* 0.1	* 0.2	* 0.2	0.2[1]
Potash fertilizers										
Engrais potassiques	...	...	...	...	...	* 1.3	* 1.4	* 1.0	* 1.0	1.0[1]
Belize Belize										
Nitrogenous fertilizers										
Engrais azotés	...	...	...	...	...	1.8	* 1.9	* 1.5	* 1.5	* 1.7
Phosphate fertilizers										
Engrais phosphatés	...	...	...	...	...	1.0	* 1.2	* 1.7	* 2.0	* 2.4
Potash fertilizers										
Engrais potassiques	...	...	...	...	...	1.6	* 1.8	* 1.9	* 1.7	* 1.9
Bermuda Bermudes										
Nitrogenous fertilizers *										
Engrais azotés *	...	...	...	...	...	0.1	0.1	...	...	0.1
Canada Canada										
Nitrogenous fertilizers										
Engrais azotés	* 3 000.0	2 706.0	2 842.8	2 904.2	2 921.1	1 160.2	1 197.2	1 157.8	1 253.3	1 272.6
Phosphate fertilizers										
Engrais phosphatés	* 463.9	440.0	354.0	437.0	395.5	614.4	609.2	578.2	592.2	598.4
Potash fertilizers										
Engrais potassiques	8 088.9	* 6 773.5	7 520.2	7 013.9	7 289.0	356.1	359.8	337.9	310.3	310.3
Costa Rica Costa Rica										
Nitrogenous fertilizers *										
Engrais azotés *	27.7	24.0	30.2	39.4	41.9	59.0	63.0	55.6	62.4	65.0
Phosphate fertilizers *										
Engrais phosphatés *	...	...	...	...	...	15.9	14.0	15.0	20.0	18.0
Potash fertilizers										
Engrais potassiques	...	...	...	...	...	* 26.0	* 30.0	* 38.0	* 38.0	41.8
Cuba Cuba										
Nitrogenous fertilizers										
Engrais azotés	142.1	145.9	* 140.0	* 120.0	100.0[1]	304.7	* 366.7	* 283.0	* 240.0	* 192.0
Phosphate fertilizers										
Engrais phosphatés	14.6	14.9	* 7.0	* 7.0	* 6.0	80.6	83.0	* 82.0	* 65.0	* 20.7
Potash fertilizers										
Engrais potassiques	...	...	...	...	...	213.4	211.7	* 215.0	* 60.0	* 36.5
Dominica Dominique										
Nitrogenous fertilizers										
Engrais azotés	...	...	...	...	...	* 1.0	* 2.0	* 2.2	* 2.4	2.4[1]
Phosphate fertilizers										
Engrais phosphatés	...	...	...	...	...	* 1.0	* 1.2	* 0.6	* 1.3	1.3[1]
Potash fertilizers										
Engrais potassiques	...	...	...	...	...	* 1.0	* 1.2	* 0.6	* 1.3	1.3[1]
Dominican Republic Rép. dominicaine										
Nitrogenous fertilizers										
Engrais azotés	...	...	...	...	...	25.1	32.8	47.0	* 51.0	* 45.0
Phosphate fertilizers										
Engrais phosphatés	...	...	...	...	...	* 18.1	* 20.0	18.8	* 22.0	* 25.6
Potash fertilizers										
Engrais potassiques	...	...	...	...	...	17.3	* 20.0	* 23.0	* 24.0	* 30.0
El Salvador El Salvador										
Nitrogenous fertilizers										
Engrais azotés	7.7	5.5	11.2	9.2	...	67.8	56.2	53.2	55.4	* 59.4
Phosphate fertilizers										
Engrais phosphatés	0.1	0.1	0.1	0.1	...	24.9	17.9	18.9	19.1	* 15.8
Potash fertilizers										
Engrais potassiques	...	...	...	...	...	4.5	4.0	3.2	3.0	* 3.1

40

Fertilizers
Nitrogenous, phosphate and potash: thousand metric tons [*cont.*]
Engrais
Azotés, phosphatés et potassiques : milliers de tonnes métriques [*suite*]

Country or area Pays ou zone	Production Production					Consumption Consommation				
	1988/89	1989/90	1990/91	1991/92	1992/93	1988/89	1989/90	1990/91	1991/92	1992/93
Guadeloupe Guadeloupe										
Nitrogenous fertilizers										
Engrais azotés	...	...	...	...	...	4.1	* 3.9	* 3.1	* 3.4	* 3.5
Phosphate fertilizers										
Engrais phosphatés	...	...	...	...	...	3.6	* 2.5	* 1.4	* 3.6	* 1.2
Potash fertilizers										
Engrais potassiques	...	...	...	...	...	3.7	* 2.5	* 1.4	* 1.8	* 3.4
Guatemala Guatemala										
Nitrogenous fertilizers *										
Engrais azotés *	7.0	7.0	7.0	8.0	8.5	82.2	90.5	88.0	98.0	100.0
Phosphate fertilizers *										
Engrais phosphatés *	7.0	7.0	7.0	7.0	1.9	30.0	18.0	27.0	27.0	35.0
Potash fertilizers *										
Engrais potassiques *	...	...	...	...	...	15.7	16.0	17.0	18.0	22.0
Haiti Haïti										
Nitrogenous fertilizers *										
Engrais azotés *	...	...	...	...	...	1.4	2.3	0.9	2.7	4.2
Phosphate fertilizers *										
Engrais phosphatés *	...	...	...	...	...	0.3	0.7	...	...	...
Potash fertilizers *										
Engrais potassiques *	...	...	...	...	...	0.5	0.7	0.1	0.1	0.1
Honduras Honduras										
Nitrogenous fertilizers										
Engrais azotés	...	...	...	...	...	21.5	20.3	11.1	* 20.0	* 20.9
Phosphate fertilizers										
Engrais phosphatés	...	...	...	...	...	8.2	6.0	2.1	8.0	10.0[1]
Potash fertilizers										
Engrais potassiques	...	...	...	...	...	10.1	7.2	1.8	7.6	* 8.0
Jamaica Jamaïque										
Nitrogenous fertilizers *										
Engrais azotés *	...	...	...	...	...	16.0	16.0	7.4	9.2	4.2
Phosphate fertilizers										
Engrais phosphatés	...	...	...	...	...	4.9	* 4.1	* 3.6	* 3.8	* 4.1
Potash fertilizers										
Engrais potassiques	...	...	...	...	...	8.4	* 11.0	* 8.1	* 12.7	* 13.0
Martinique Martinique										
Nitrogenous fertilizers										
Engrais azotés	...	...	...	...	...	6.4	* 6.4	* 8.1	* 6.6	7.0
Phosphate fertilizers										
Engrais phosphatés	...	...	...	...	...	4.2	* 4.5	* 5.2	* 2.7	2.2
Potash fertilizers										
Engrais potassiques	...	...	...	...	...	9.8	* 8.0	* 11.8	* 8.0	11.7
Mexico Mexique										
Nitrogenous fertilizers										
Engrais azotés	1 348.6	* 1 497.9	* 1 358.9	* 1 465.0	* 1 346.0	* 1 269.6	* 1 292.8	* 1 346.3	* 1 130.0	* 1 230.0
Phosphate fertilizers										
Engrais phosphatés	375.8	* 447.5	* 383.3	* 488.4	* 383.0	* 394.9	* 354.3	* 373.8	* 380.0	* 298.0
Potash fertilizers										
Engrais potassiques	...	...	...	...	...	78.5	* 92.8	* 78.5	* 73.0	* 88.0
Nicaragua Nicaragua										
Nitrogenous fertilizers										
Engrais azotés	...	...	...	...	...	* 55.0	21.9	31.8	26.6	* 26.0
Phosphate fertilizers										
Engrais phosphatés	...	...	...	...	...	* 12.0	7.6	6.1	6.2	* 0.4
Potash fertilizers										
Engrais potassiques	...	...	...	...	...	* 5.0	5.8	2.1	* 2.0	* 4.9

40

Fertilizers
Nitrogenous, phosphate and potash: thousand metric tons [*cont.*]
Engrais
Azotés, phosphatés et potassiques : milliers de tonnes métriques [*suite*]

Country or area Pays ou zone	Production Production 1988/89	1989/90	1990/91	1991/92	1992/93	Consumption Consommation 1988/89	1989/90	1990/91	1991/92	1992/93
Panama Panama										
Nitrogenous fertilizers										
Engrais azotés	...	...	...	...	...	* 21.1	18.4	21.7	* 16.0	* 19.0
Phosphate fertilizers										
Engrais phosphatés	...	...	...	...	...	* 7.1	4.0	5.1	4.2	3.6
Potash fertilizers										
Engrais potassiques	...	...	...	...	...	* 10.3	10.9	11.9	5.1	* 8.8
Saint Kitts-Nevis Saint-Kitts-et-Nevis										
Nitrogenous fertilizers										
Engrais azotés	...	...	...	...	...	* 0.5	0.4	0.4	* 0.5	* 0.5
Phosphate fertilizers										
Engrais phosphatés	...	...	...	...	...	* 1.0	0.2	0.2	* 0.3	0.3[1]
Potash fertilizers										
Engrais potassiques	...	...	...	...	...	* 1.6	0.3	0.3	* 0.3	0.3[1]
Saint Lucia Sainte-Lucie										
Nitrogenous fertilizers										
Engrais azotés	...	...	...	...	...	* 1.5	2.9	2.4	* 3.0	3.0[1]
Phosphate fertilizers										
Engrais phosphatés	...	...	...	...	...	* 0.1	2.1	1.7	* 1.8	* 1.8
Potash fertilizers										
Engrais potassiques	...	...	...	...	...	* 0.1	2.2	1.7	* 1.7	* 1.7
St. Vincent-Grenadines St. Vincent-Grenadines										
Nitrogenous fertilizers *										
Engrais azotés *	...	...	...	...	...	2.6	1.5	0.5	0.5	0.5
Phosphate fertilizers										
Engrais phosphatés	...	...	...	...	...	* 0.7	* 0.9	* 0.9	* 0.9	1.0[1]
Potash fertilizers *										
Engrais potassiques *	...	...	...	...	...	0.5	0.9	0.9	0.9	0.9
Trinidad and Tobago Trinité-et-Tobago										
Nitrogenous fertilizers										
Engrais azotés	* 248.0	* 223.4	* 231.8	* 240.7	209.8	1.3	* 5.0	* 6.0	* 7.0	* 7.0
Phosphate fertilizers										
Engrais phosphatés	...	...	...	...	...	0.2	* 0.2	* 0.2	0.6	* 0.8
Potash fertilizers										
Engrais potassiques	...	...	...	...	...	0.6	* 1.6	* 1.6	* 1.6	* 2.0
United States Etats-Unis										
Nitrogenous fertilizers										
Engrais azotés	12 728.0	12 576.0	13 124.0	13 557.0	13 761.0	9 609.4	10 047.9	10 239.3	10 383.9	10 303.6
Phosphate fertilizers										
Engrais phosphatés	9 579.0	9 590.0	10 137.0	* 11 140.0	10 684.0	3 734.7	3 941.4	3 810.7	3 826.4	4 044.3
Potash fertilizers										
Engrais potassiques	1 113.0	1 007.0	1 008.0	938.0	938.0	4 389.0	4 719.9	4 537.0	4 573.7	4 635.3
United States Virgin Is. Iles Vierges américaines										
Nitrogenous fertilizers										
Engrais azotés	...	...	...	...	...	* 1.0	* 1.0	* 1.0	1.0[1]	1.0[1]
Phosphate fertilizers										
Engrais phosphatés	...	...	...	...	...	* 0.3	* 0.3	* 0.3	* 0.3	0.3[1]
America, South Amérique du Sud										
Nitrogenous fertilizers										
Engrais azotés	1 415.6	1 430.3	1 372.8	1 351.1	1 218.3	1 860.7	1 791.7	1 734.4	1 669.9	1 785.1
Phosphate fertilizers										
Engrais phosphatés	1 450.0	1 215.0	1 209.6	1 260.9	1 185.2	2 134.2	1 861.3	1 693.4	1 757.1	1 900.1
Potash fertilizers										
Engrais potassiques	55.7	109.4	68.1	138.1	112.6	1 778.0	1 623.8	1 528.2	1 528.3	1 704.6
Argentina Argentine										
Nitrogenous fertilizers										
Engrais azotés	* 51.0	* 50.0	* 40.5	42.1	32.7	* 99.1	* 99.4	* 101.4	* 96.0	* 107.6
Phosphate fertilizers										
Engrais phosphatés	...	...	...	...	...	* 51.2	* 42.7	* 51.5	69.0	92.7

40

Fertilizers
Nitrogenous, phosphate and potash: thousand metric tons [cont.]
Engrais
Azotés, phosphatés et potassiques : milliers de tonnes métriques [suite]

Country or area	Production Production					Consumption Consommation				
Pays ou zone	1988/89	1989/90	1990/91	1991/92	1992/93	1988/89	1989/90	1990/91	1991/92	1992/93
Potash fertilizers										
Engrais potassiques	...	...	...	...	...	* 17.0	* 11.0	* 12.6	20.1	13.1
Bolivia Bolivie										
Nitrogenous fertilizers										
Engrais azotés	...	...	...	...	...	3.1	4.6	2.3	3.6	5.5
Phosphate fertilizers										
Engrais phosphatés	...	...	...	...	...	* 1.3	5.3	2.5	3.5	7.6
Potash fertilizers										
Engrais potassiques	...	...	...	...	...	0.2	0.3	0.3	0.7	0.7
Brazil Brésil										
Nitrogenous fertilizers										
Engrais azotés	705.1	748.5	737.2	704.3	664.6	815.0	823.3	779.3	781.5	865.5
Phosphate fertilizers										
Engrais phosphatés	1 356.9	1 109.4	* 1 090.7	* 1 139.4	1 075.7	1 507.4	1 296.2	* 1 201.6	* 1 247.4	1 346.1
Potash fertilizers										
Engrais potassiques	55.7	109.4	68.1	101.1	77.3	1 406.3	1 263.7	1 183.2	1 206.0	1 372.8
Chile Chili										
Nitrogenous fertilizers										
Engrais azotés	* 130.0	* 130.0	* 126.0	* 119.7	* 126.0	* 160.0	158.3	152.9	* 161.0	* 190.0
Phosphate fertilizers										
Engrais phosphatés	* 4.0	* 4.0	* 5.0	* 4.0	* 4.0	* 139.2	125.9	113.8	* 111.6	* 130.0
Potash fertilizers										
Engrais potassiques	...	...	...	* 37.0	* 35.3	* 25.0	27.7	28.8	* 37.0	40.0[1]
Colombia Colombie										
Nitrogenous fertilizers										
Engrais azotés	105.8	95.7	* 99.7	* 100.0	* 60.0	236.0	* 269.0	* 312.0	* 271.0	280.7
Phosphate fertilizers										
Engrais phosphatés	* 20.7	* 39.0	* 40.0	* 28.0	* 29.0	101.3	* 121.8	* 126.7	* 120.0	123.0
Potash fertilizers										
Engrais potassiques	...	...	...	...	...	129.1	153.5	* 163.8	* 150.0	160.0
Ecuador Equateur										
Nitrogenous fertilizers										
Engrais azotés	...	...	...	...	...	* 40.0	41.1	36.7	* 45.0	* 57.4
Phosphate fertilizers										
Engrais phosphatés	...	...	...	...	...	* 19.0	21.9	* 10.6	* 16.8	* 27.0
Potash fertilizers										
Engrais potassiques	...	...	...	...	...	* 25.8	14.4	19.9	* 24.1	* 30.3
French Guiana Guyane française										
Nitrogenous fertilizers										
Engrais azotés	...	...	...	...	...	0.5	* 0.3	* 0.7	* 0.9	* 0.6
Phosphate fertilizers										
Engrais phosphatés	...	...	...	...	...	0.2	* 0.2	* 0.2	* 0.2	0.3
Potash fertilizers										
Engrais potassiques	...	...	...	...	...	0.2	* 0.2	* 0.2	* 0.2	0.1
Guyana Guyana										
Nitrogenous fertilizers *										
Engrais azotés *	...	...	...	...	...	10.6	12.0	8.0	13.0	10.5
Phosphate fertilizers *										
Engrais phosphatés *	...	...	...	...	...	1.6	1.8	2.2	1.6	0.9
Potash fertilizers *										
Engrais potassiques *	...	...	...	...	...	2.1	2.5	1.8	1.0	0.8
Paraguay Paraguay										
Nitrogenous fertilizers										
Engrais azotés	...	...	...	...	...	2.2	3.5	3.7	4.3	4.8
Phosphate fertilizers										
Engrais phosphatés	...	...	...	...	...	3.8	12.8	11.2	* 13.4	* 14.6
Potash fertilizers										
Engrais potassiques	...	...	...	...	...	2.2	3.4	3.1	2.7	2.5

40

Fertilizers
Nitrogenous, phosphate and potash: thousand metric tons [*cont.*]
Engrais
Azotés, phosphatés et potassiques : milliers de tonnes métriques [*suite*]

Country or area Pays ou zone	Production Production					Consumption Consommation				
	1988/89	1989/90	1990/91	1991/92	1992/93	1988/89	1989/90	1990/91	1991/92	1992/93
Peru Pérou										
Nitrogenous fertilizers Engrais azotés	79.7	36.2	32.4	20.0	10.0	156.1	113.2	* 104.1	58.6	64.9
Phosphate fertilizers Engrais phosphatés	2.9	4.0	4.7	3.0	3.5	37.6	18.8	12.8	11.2	10.8
Potash fertilizers Engrais potassiques	...	...	...	...	...	21.6	19.8	8.3	5.7	4.8
Suriname Suriname										
Nitrogenous fertilizers Engrais azotés	...	...	...	...	...	1.6	1.3	* 0.6	* 0.6	* 2.6
Phosphate fertilizers Engrais phosphatés	...	...	...	...	...	0.2	0.2	* 0.2	* 0.2	* 0.1
Potash fertilizers Engrais potassiques	...	...	...	...	...	0.2	0.2	* 0.2	* 0.2	* 0.2
Uruguay Uruguay										
Nitrogenous fertilizers Engrais azotés	...	...	...	...	...	30.9	24.5	27.7	28.3	30.0[1]
Phosphate fertilizers Engrais phosphatés	15.5	8.6	* 12.0	* 17.0	17.0[1]	35.7	42.3	41.2	46.5	45.0[1]
Potash fertilizers Engrais potassiques	...	...	...	...	...	2.4	3.8	3.0	4.0	* 4.3
Venezuela Venezuela										
Nitrogenous fertilizers Engrais azotés	* 344.0	* 370.0	* 337.0	* 365.0	* 325.0	305.6	* 241.1	* 205.0	* 206.0	165.0[1]
Phosphate fertilizers Engrais phosphatés	* 50.0	* 50.0	* 57.2	* 69.5	* 56.0	235.7	171.3	* 119.0	* 115.6	* 102.0
Potash fertilizers Engrais potassiques	...	...	...	...	...	146.0	123.4	* 103.0	* 76.7	* 75.0
Asia Asie										
Nitrogenous fertilizers Engrais azotés	30 212.5	31 400.8	31 969.8	32 280.1	33 713.2	34 939.7	35 909.7	37 186.8	37 908.3	39 524.7
Phosphate fertilizers Engrais phosphatés	9 682.6	9 041.4	9 656.7	10 462.2	10 233.0	12 254.9	12 763.6	13 832.7	15 161.4	14 466.6
Potash fertilizers Engrais potassiques	2 052.2	2 167.4	2 183.6	2 165.4	2 243.0	4 697.3	4 471.4	5 176.8	5 756.4	5 166.3
Afghanistan Afghanistan										
Nitrogenous fertilizers Engrais azotés	55.2	55.2	* 53.0	* 48.8	* 42.0	49.8	50.0	* 44.1	* 45.0	36.2
Phosphate fertilizers * Engrais phosphatés *	...	...	...	...	...	5.9	5.6	0.4	5.8	6.9
Armenia Arménie										
Nitrogenous fertilizers * Engrais azotés *	..	..	..	...	...	..	..	..	...	15.0
Phosphate fertilizers * Engrais phosphatés *	..	..	..	...	...	..	..	..	...	5.0
Potash fertilizers * Engrais potassiques *	..	..	..	...	...	..	..	..	...	5.0
Azerbaijan Azerbaïdjan										
Nitrogenous fertilizers * Engrais azotés *	..	..	..	...	...	..	..	..	...	50.0
Phosphate fertilizers * Engrais phosphatés *	..	..	..	...	61.0	..	..	..	...	20.0
Potash fertilizers * Engrais potassiques *	..	..	..	...	40.0	..	..	..	...	5.0
Bahrain Bahreïn										
Nitrogenous fertilizers Engrais azotés	...	...	...	...	...	0.2	0.2	0.3	* 0.3	* 0.3
Phosphate fertilizers Engrais phosphatés	...	...	...	...	...	0.1	0.1	0.2	* 0.2	* 0.2

40

Fertilizers
Nitrogenous, phosphate and potash: thousand metric tons [*cont.*]
Engrais
Azotés, phosphatés et potassiques : milliers de tonnes métriques [*suite*]

Country or area Pays ou zone	Production Production					Consumption Consommation				
	1988/89	1989/90	1990/91	1991/92	1992/93	1988/89	1989/90	1990/91	1991/92	1992/93
Potash fertilizers Engrais potassiques	...	...	...	...	...	0.1	0.1	0.2	* 0.2	* 0.2
Bangladesh Bangladesh										
Nitrogenous fertilizers Engrais azotés	667.4	677.8	653.6	756.8	* 924.1	524.0	629.8	608.6	704.6	713.0
Phosphate fertilizers Engrais phosphatés	65.6	67.3	50.9	50.0	* 21.6	191.4	221.1	236.3	216.3	* 211.0
Potash fertilizers Engrais potassiques	...	...	...	...	...	56.5	56.1	88.2	82.2	* 76.0
Bhutan Bhoutan										
Nitrogenous fertilizers Engrais azotés	...	...	...	...	...	* 0.1	* 0.1	* 0.1	* 0.1	0.1[1]
Brunei Darussalam Brunéi Darussalam										
Nitrogenous fertilizers * Engrais azotés *	...	...	...	...	...	0.3	0.1	1.4	1.4	...
Phosphate fertilizers * Engrais phosphatés *	...	...	...	...	...	0.2	0.1	1.4	1.5	1.5
Potash fertilizers * Engrais potassiques *	...	...	...	...	...	0.2	0.2	1.4	1.5	1.5
Cambodia Cambodge										
Nitrogenous fertilizers Engrais azotés	...	...	...	...	...	...	...	* 1.6	4.2	5.0[1]
Phosphate fertilizers Engrais phosphatés	...	* 3.5	* 3.7	...	...	...	* 3.5	* 6.2	* 1.5	1.5[1]
Potash fertilizers * Engrais potassiques *	...	...	...	...	...	0.1	0.3	...	...	0.1
China Chine										
Nitrogenous fertilizers Engrais azotés	13 954.0	14 515.0	14 914.5	15 372.5	15 963.0	18 514.1	* 18 855.3	* 19 449.8	20 020.1	20 403.0
Phosphate fertilizers Engrais phosphatés	3 766.2	3 807.7	4 196.0	4 677.5	4 700.0	5 162.2	* 5 272.1	* 5 878.5	7 326.3	6 758.0
Potash fertilizers Engrais potassiques	53.8	56.0	47.0	97.0	152.0	1 645.7	* 1 300.8	* 1 699.1	2 402.4	1 993.8
Cyprus Chypre										
Nitrogenous fertilizers Engrais azotés	16.5	11.2	...	...	...	12.1	12.5	12.4	12.2	16.0
Phosphate fertilizers Engrais phosphatés	38.2	24.1	...	...	...	8.2	8.3	8.3	8.1	9.8
Potash fertilizers Engrais potassiques	...	...	...	...	...	1.3	1.7	2.0	1.8	2.6
Georgia Géorgie										
Nitrogenous fertilizers * Engrais azotés *	..	..	..	...	76.0	..	..	..	...	60.0
Phosphate fertilizers * Engrais phosphatés *	..	..	..	...	...	..	..	..	...	10.0
India Inde										
Nitrogenous fertilizers Engrais azotés	6 712.4	6 747.4	6 993.1	7 301.5	7 430.6	7 251.0	7 385.9	7 997.2	8 046.3	8 426.1
Phosphate fertilizers Engrais phosphatés	2 289.6	1 834.0	2 088.9	2 596.4	2 355.8	2 757.8	3 052.9	3 258.8	3 355.6	2 907.8
Potash fertilizers Engrais potassiques	...	...	...	...	...	1 068.4	1 168.0	1 328.0	1 360.6	883.9
Indonesia Indonésie										
Nitrogenous fertilizers Engrais azotés	2 032.7	2 369.0	* 2 348.3	* 2 302.4	* 2 279.7	* 1 495.0	* 1 474.0	* 1 496.0	* 1 552.0	* 1 696.3
Phosphate fertilizers Engrais phosphatés	551.5	551.3	588.8	* 500.9	* 437.0	* 610.4	* 598.0	* 585.0	* 584.0	* 593.4

40

Fertilizers

Nitrogenous, phosphate and potash: thousand metric tons [*cont.*]

Engrais

Azotés, phosphatés et potassiques : milliers de tonnes métriques [*suite*]

Country or area Pays ou zone	Production Production					Consumption Consommation				
	1988/89	1989/90	1990/91	1991/92	1992/93	1988/89	1989/90	1990/91	1991/92	1992/93
Potash fertilizers * Engrais potassiques *	...	...	...	...	...	286.8	274.0	306.0	271.7	290.0
Iran, Islamic Rep. of Iran, Rép. islamique d'										
Nitrogenous fertilizers Engrais azotés	115.9	* 308.1	* 376.2	* 451.0	* 571.7	548.4	* 668.3	* 558.1	* 658.6	* 834.0
Phosphate fertilizers Engrais phosphatés	1.8	* 49.7	* 80.5	* 162.0	* 186.8	521.5	* 512.4	* 587.0	* 482.6	* 487.6
Potash fertilizers Engrais potassiques	...	...	...	...	...	5.8	* 0.2	* 15.9	* 15.0	* 49.4
Iraq Iraq										
Nitrogenous fertilizers Engrais azotés	* 333.4	* 450.0	* 409.0	* 95.0	* 130.0	* 146.5	138.4	* 130.0	* 95.0	* 130.0
Phosphate fertilizers Engrais phosphatés	* 396.0	* 415.0	* 207.0	* 30.0	* 80.0	* 62.0	75.1	* 75.0	* 39.7	* 80.0
Potash fertilizers Engrais potassiques	...	...	...	...	...	* 0.7	1.9	* 4.2	...	...
Israel Israël										
Nitrogenous fertilizers Engrais azotés	* 73.3	* 75.0	* 75.0	* 75.0	75.0[1]	51.9	53.8	49.6	41.3	43.4
Phosphate fertilizers Engrais phosphatés	* 145.0	* 180.0	* 200.0	* 185.0	* 210.0	22.3	21.4	20.7	20.0	21.5
Potash fertilizers Engrais potassiques	* 1 218.4	* 1 301.0	* 1 295.0	* 1 250.0	* 1 296.0	33.5	33.9	32.1	32.3	32.9
Japan Japon										
Nitrogenous fertilizers Engrais azotés	977.0	946.0	957.0	931.0	906.0	640.0	641.0	612.0	574.0	572.0
Phosphate fertilizers Engrais phosphatés	490.0	445.0	429.0	425.0	393.0	726.0	728.0	690.0	695.0	699.0
Potash fertilizers Engrais potassiques	...	...	...	...	...	577.0	569.0	537.0	494.0	513.0
Jordan Jordanie										
Nitrogenous fertilizers Engrais azotés	110.7	108.5	107.3	107.8	* 99.7	14.3	* 15.0	* 10.0	14.8	9.7
Phosphate fertilizers Engrais phosphatés	282.9	277.2	274.2	275.5	* 254.8	9.2	* 5.0	4.6	4.6	5.0[1]
Potash fertilizers Engrais potassiques	* 780.0	810.4	841.6	818.4	* 795.0	3.7	* 4.0	* 1.4	1.1	1.4[1]
Kazakhstan Kazakhstan										
Nitrogenous fertilizers * Engrais azotés *	..	..	..	...	253.0	..	..	..	...	150.0
Phosphate fertilizers * Engrais phosphatés *	..	..	..	...	345.0	..	..	..	...	315.0
Potash fertilizers * Engrais potassiques *	..	..	..	...	12.6	..	..	..	...	10.0
Korea, Dem. P. R. Corée, R. p. dém. de										
Nitrogenous fertilizers Engrais azotés	* 660.0	* 660.0	* 660.0	* 660.0	660.0[1]	* 634.0	* 644.0	* 655.4	* 659.0	* 652.5
Phosphate fertilizers * Engrais phosphatés *	137.0	137.0	137.0	137.0	130.0	147.6	163.2	158.9	148.5	130.0
Potash fertilizers * Engrais potassiques *	...	...	...	...	...	30.4	7.5	18.1	3.6	45.0
Korea, Republic of Corée, République de										
Nitrogenous fertilizers Engrais azotés	* 678.8	583.4	554.4	* 570.5	* 623.3	439.0	455.0	465.2	* 473.5	* 486.6
Phosphate fertilizers Engrais phosphatés	488.0	* 393.8	414.4	398.0	* 421.8	194.0	235.1	228.0	242.8	219.1
Potash fertilizers * Engrais potassiques *	...	...	...	...	...	246.0	231.1	222.7	233.7	258.0

40

Fertilizers
Nitrogenous, phosphate and potash: thousand metric tons [*cont.*]
Engrais
Azotés, phosphatés et potassiques : milliers de tonnes métriques [*suite*]

Country or area	Production Production					Consumption Consommation				
Pays ou zone	1988/89	1989/90	1990/91	1991/92	1992/93	1988/89	1989/90	1990/91	1991/92	1992/93
Kuwait Koweït										
Nitrogenous fertilizers										
Engrais azotés	384.5	* 386.4	* 204.0	...	118.4	0.8	* 0.8	* 1.0	...	0.8
Kyrgyzstan Kirghizistan										
Nitrogenous fertilizers *										
Engrais azotés *	..	..	..	...	...	..	..	..	...	25.3
Phosphate fertilizers *										
Engrais phosphatés *	..	..	..	...	...	..	..	..	...	1.7
Potash fertilizers *										
Engrais potassiques *	..	..	..	...	...	..	..	..	...	5.0
Lao People's Dem. Rep. Rép. dém. pop. lao										
Nitrogenous fertilizers										
Engrais azotés	...	...	...	...	...	* 0.1	* 0.1	* 1.1	1.7	3.1
Phosphate fertilizers										
Engrais phosphatés	...	...	...	...	...	* 0.1	* 0.1	* 0.3	0.8	0.2
Potash fertilizers										
Engrais potassiques	...	...	...	...	...	* 0.1	* 0.1	* 0.1	0.0	0.0
Lebanon Liban										
Nitrogenous fertilizers *										
Engrais azotés *	...	...	...	...	...	11.3	14.0	11.0	12.4	14.2
Phosphate fertilizers										
Engrais phosphatés	* 20.0	* 15.0	* 34.0	* 40.0	* 40.0	* 7.0	* 10.0	* 12.0	* 15.0	20.0 [1]
Potash fertilizers *										
Engrais potassiques *	...	...	...	...	...	4.3	3.6	2.4	2.3	3.9
Malaysia Malaisie										
Nitrogenous fertilizers										
Engrais azotés	* 247.0	* 249.0	* 211.0	* 325.0	* 287.4	* 273.0	288.0	* 313.0	* 320.0	* 295.0
Phosphate fertilizers										
Engrais phosphatés	...	...	...	...	...	* 150.0	150.1	* 163.5	* 166.2	* 160.0
Potash fertilizers										
Engrais potassiques	...	...	...	...	...	* 312.0	* 391.0	475.0	* 480.0	* 510.0
Mongolia Mongolie										
Nitrogenous fertilizers										
Engrais azotés	...	...	...	...	...	11.3	10.0	9.1	* 10.0	* 11.0
Phosphate fertilizers										
Engrais phosphatés	...	...	...	...	...	3.2	3.5	3.7	* 4.0	* 3.5
Potash fertilizers										
Engrais potassiques	...	...	...	...	...	5.3	3.6	1.8	* 2.0	* 0.6
Myanmar Myanmar										
Nitrogenous fertilizers										
Engrais azotés	* 111.9	* 88.3	* 60.0	46.7	54.3	86.6	63.7	* 53.0	62.7	55.0
Phosphate fertilizers										
Engrais phosphatés	...	...	...	...	...	18.7	17.3	14.8	14.5	11.4
Potash fertilizers										
Engrais potassiques	...	...	...	...	...	1.4	1.8	2.9	1.3	2.6
Nepal Népal										
Nitrogenous fertilizers										
Engrais azotés	...	...	...	...	...	39.8	49.2	52.0	* 55.0	* 69.4
Phosphate fertilizers										
Engrais phosphatés	...	...	...	...	...	15.3	16.7	19.1	* 22.7	* 21.4
Potash fertilizers										
Engrais potassiques	...	...	...	...	...	1.8	1.3	1.6	1.2	* 1.2
Oman Oman										
Nitrogenous fertilizers										
Engrais azotés	...	...	...	...	...	2.4	2.4	4.8	4.9	5.0 [1]
Phosphate fertilizers										
Engrais phosphatés	...	...	...	...	...	1.4	1.1	1.5	1.5	1.5 [1]
Potash fertilizers										
Engrais potassiques	...	...	...	...	...	1.5	1.8	3.1	1.8	* 1.5

40

Fertilizers
Nitrogenous, phosphate and potash: thousand metric tons [*cont.*]
Engrais
Azotés, phosphatés et potassiques : milliers de tonnes métriques [*suite*]

Country or area Pays ou zone	Production Production					Consumption Consommation				
	1988/89	1989/90	1990/91	1991/92	1992/93	1988/89	1989/90	1990/91	1991/92	1992/93
Pakistan Pakistan										
Nitrogenous fertilizers										
Engrais azotés	1 112.5	1 156.3	1 120.3	1 045.0	1 227.3	1 324.9	1 467.6	1 471.6	1 462.9	1 635.4
Phosphate fertilizers										
Engrais phosphatés	100.6	105.1	104.7	106.0	104.8	390.4	382.5	388.5	393.3	484.2
Potash fertilizers										
Engrais potassiques	...	...	...	...	...	24.5	40.1	32.8	23.3	24.1
Philippines Philippines										
Nitrogenous fertilizers										
Engrais azotés	127.8	127.0	* 120.8	143.2	* 116.0	371.0	376.0	400.6	298.6	* 360.3
Phosphate fertilizers										
Engrais phosphatés	198.5	191.0	199.4	191.9	* 143.9	77.5	84.1	105.0	71.3	* 72.5
Potash fertilizers										
Engrais potassiques	...	...	...	...	...	54.9	77.1	82.5	67.5	* 63.3
Qatar Qatar										
Nitrogenous fertilizers										
Engrais azotés	358.6	359.0	350.1	* 367.5	379.9	* 0.6	1.2	1.4	* 1.5	* 1.5
Saudi Arabia Arabie saoudite										
Nitrogenous fertilizers										
Engrais azotés	417.5	* 428.4	* 584.0	* 658.0	661.7	* 256.2	* 265.0	* 273.0	* 284.0	291.9
Phosphate fertilizers										
Engrais phosphatés	...	...	* 27.0	* 200.0	* 172.2	* 220.4	175.0	* 193.0	* 200.0	218.2
Potash fertilizers										
Engrais potassiques	...	...	...	...	...	* 30.0	* 35.0	* 23.0	* 24.0	28.6
Singapore Singapour										
Nitrogenous fertilizers *										
Engrais azotés *	...	...	...	...	...	2.6	2.6	2.6	2.6	2.6
Phosphate fertilizers *										
Engrais phosphatés *	...	...	...	...	...	0.5	0.5	0.5	0.5	0.5
Potash fertilizers *										
Engrais potassiques *	...	...	...	...	...	2.5	2.5	2.5	2.5	2.5
Sri Lanka Sri Lanka										
Nitrogenous fertilizers										
Engrais azotés	...	...	...	...	...	* 109.0	108.5	92.4	95.8	102.2
Phosphate fertilizers										
Engrais phosphatés	* 6.5	6.8	9.1	5.5	7.2	* 39.1	40.7	29.2	30.9	30.4
Potash fertilizers										
Engrais potassiques	...	...	...	...	...	* 61.0	61.6	49.6	50.5	51.1
Syrian Arab Republic Rép. arabe syrienne										
Nitrogenous fertilizers										
Engrais azotés	81.3	108.1	31.8	* 15.0	62.7	160.6	153.6	184.8	* 180.0	* 196.0
Phosphate fertilizers										
Engrais phosphatés	30.0	14.9	21.3	* 25.0	* 1.7	109.3	91.6	111.9	* 120.0	* 130.0
Potash fertilizers										
Engrais potassiques	...	...	...	...	...	10.5	4.6	6.5	* 8.9	* 10.3
Tajikistan Tadjikistan										
Nitrogenous fertilizers *										
Engrais azotés *	..	..	..	...	83.0	..	..	..	...	71.0
Phosphate fertilizers *										
Engrais phosphatés *	..	..	..	...	...	..	..	..	...	56.0
Potash fertilizers *										
Engrais potassiques *	..	..	..	...	...	..	..	..	...	7.0
Thailand Thaïlande										
Nitrogenous fertilizers										
Engrais azotés	...	...	...	...	...	439.7	494.9	576.5	485.5	* 596.2
Phosphate fertilizers										
Engrais phosphatés	...	...	...	...	...	200.8	188.8	318.3	235.5	* 306.8
Potash fertilizers										
Engrais potassiques	...	...	...	...	...	137.5	117.8	148.9	* 125.0	* 192.0

40

Fertilizers
Nitrogenous, phosphate and potash: thousand metric tons [*cont.*]
Engrais
Azotés, phosphatés et potassiques : milliers de tonnes métriques [*suite*]

Country or area Pays ou zone	Production Production					Consumption Consommation				
	1988/89	1989/90	1990/91	1991/92	1992/93	1988/89	1989/90	1990/91	1991/92	1992/93
Turkey Turquie										
Nitrogenous fertilizers										
Engrais azotés	* 725.2	700.0	947.1	748.9	831.4	1 081.6	1 140.4	1 199.7	1 099.5	1 206.1
Phosphate fertilizers										
Engrais phosphatés	615.4	468.0	524.8	387.9	474.4	490.2	599.7	624.8	619.6	658.1
Potash fertilizers										
Engrais potassiques	...	...	...	...	...	41.8	58.0	63.0	47.5	63.3
Turkmenistan Turkménistan										
Nitrogenous fertilizers *										
Engrais azotés *	..	..	..	...	71.0	..	..	..	...	100.0
Phosphate fertilizers *										
Engrais phosphatés *	..	..	..	...	130.0	..	..	..	...	65.0
Potash fertilizers *										
Engrais potassiques *	..	..	..	...	...	..	..	..	...	10.0
United Arab Emirates Emirats arabes unis										
Nitrogenous fertilizers										
Engrais azotés	* 243.8	266.4	228.3	237.8	232.3	* 2.0	7.8	9.2	14.2	* 14.0
Phosphate fertilizers *										
Engrais phosphatés *	...	...	...	...	...	1.6	1.6	1.2	1.3	1.3
Potash fertilizers *										
Engrais potassiques *	...	...	...	...	...	1.8	2.7	2.1	2.0	2.0
Uzbekistan Ouzbékistan										
Nitrogenous fertilizers *										
Engrais azotés *	..	..	..	...	1 012.0	..	..	..	...	400.0
Phosphate fertilizers *										
Engrais phosphatés *	..	..	..	...	350.0	..	..	..	...	300.0
Potash fertilizers *										
Engrais potassiques *	..	..	..	...	...	..	..	..	...	60.0
Viet Nam Viet Nam										
Nitrogenous fertilizers										
Engrais azotés	15.2	25.3	* 11.0	20.7	* 36.8	* 428.9	424.0	* 419.0	598.6	* 628.8
Phosphate fertilizers										
Engrais phosphatés	* 59.7	55.0	66.0	68.5	* 98.0	* 109.6	* 97.7	103.3	128.8	* 213.2
Potash fertilizers										
Engrais potassiques	...	...	...	...	...	* 50.0	* 20.0	22.2	* 15.9	* 60.0
Yemen Yémen										
Nitrogenous fertilizers										
Engrais azotés	...	...	...	...	...	* 16.6	16.5	19.1	* 16.0	* 12.0
Phosphate fertilizers										
Engrais phosphatés	...	...	...	...	...	* 1.1	0.8	2.8	* 3.0	* 1.1
Potash fertilizers										
Engrais potassiques	...	...	...	...	...	* 0.1	* 0.1	0.5	* 0.6	* 1.5
Europe Europe										
Nitrogenous fertilizers										
Engrais azotés	18 522.3	18 140.6	15 207.7	13 946.2	13 110.9	15 958.4	15 728.4	13 682.3	12 081.8	10 940.9
Phosphate fertilizers										
Engrais phosphatés	7 381.3	6 802.4	5 308.5	4 292.3	3 606.0	7 882.0	7 372.0	6 058.0	4 856.7	4 266.6
Potash fertilizers										
Engrais potassiques	8 547.8	8 037.9	6 895.7	6 193.7	5 837.7	8 646.1	8 148.8	6 642.3	5 335.6	4 658.2
Albania Albanie										
Nitrogenous fertilizers										
Engrais azotés	69.3	79.9	* 73.4	* 22.3	* 5.7	69.3	79.9	* 73.4	* 22.3	* 14.8
Phosphate fertilizers										
Engrais phosphatés	25.0	24.3	* 25.5	* 7.7	* 4.0	25.0	24.3	* 25.5	* 7.7	* 8.7
Potash fertilizers										
Engrais potassiques	...	...	...	...	...	2.2	* 2.3	* 3.1	* 1.3	...

40

Fertilizers
Nitrogenous, phosphate and potash: thousand metric tons [*cont.*]
Engrais
Azotés, phosphatés et potassiques : milliers de tonnes métriques [*suite*]

Country or area	Production					Consumption				
Pays ou zone	1988/89	1989/90	1990/91	1991/92	1992/93	1988/89	1989/90	1990/91	1991/92	1992/93
Austria Autriche										
Nitrogenous fertilizers										
Engrais azotés	* 221.5	* 230.0	* 227.0	* 233.0	* 210.0	140.9	135.6	* 135.0	* 132.0	* 124.0
Phosphate fertilizers										
Engrais phosphatés	* 110.0	* 85.0	* 70.0	* 71.0	* 76.0	78.3	74.4	* 74.0	* 72.0	* 65.0
Potash fertilizers										
Engrais potassiques	...	...	...	...	...	103.0	97.8	* 94.0	* 93.1	* 78.0
Belarus Bélarus										
Nitrogenous fertilizers *										
Engrais azotés *	..	..	..	...	580.0	..	..	..	...	250.0
Phosphate fertilizers *										
Engrais phosphatés *	..	..	..	...	101.0	..	..	..	...	295.0
Potash fertilizers *										
Engrais potassiques *	..	..	..	...	3 311.0	..	..	..	...	850.0
Belgium-Luxembourg Belgique-Luxembourg										
Nitrogenous fertilizers *										
Engrais azotés *	666.0	678.0	725.0	760.0	755.0	196.0	190.5	186.0	182.0	173.0
Phosphate fertilizers *										
Engrais phosphatés *	325.0	332.5	362.0	340.0	340.0	87.0	86.5	78.0	65.0	56.0
Potash fertilizers *										
Engrais potassiques *	...	...	...	...	...	132.0	131.0	120.0	115.0	103.5
Bosnia & Herzegovina Bosnie-Herzégovine										
Nitrogenous fertilizers *										
Engrais azotés *	..	..	..	...	10.0	..	..	..	...	10.0
Bulgaria Bulgarie										
Nitrogenous fertilizers										
Engrais azotés	956.3	926.0	911.1	760.4	678.9	548.0	* 495.0	* 450.0	377.3	225.1
Phosphate fertilizers										
Engrais phosphatés	178.6	168.8	46.6	36.9	37.3	258.2	220.3	132.6	61.1	40.7
Potash fertilizers										
Engrais potassiques	...	...	...	...	...	112.0	91.6	97.0	33.2	21.4
Croatia Croatie										
Nitrogenous fertilizers *										
Engrais azotés *	..	..	..	...	280.0	..	..	..	...	90.0
Phosphate fertilizers *										
Engrais phosphatés *	..	..	..	...	80.0	..	..	..	...	30.0
Potash fertilizers *										
Engrais potassiques *	..	..	..	...	...	..	..	..	...	70.0
former Czechoslovakia† anc. Tchécoslovaquie†										
Nitrogenous fertilizers										
Engrais azotés	585.2	* 675.0	* 604.0	450.6[1]	* 361.0	642.0	704.8	* 588.3	* 289.0	* 226.0
Phosphate fertilizers										
Engrais phosphatés	313.0	* 320.0	311.3	* 47.1	* 40.0	480.4	* 437.0	* 359.0	* 75.0	70.0[1]
Potash fertilizers										
Engrais potassiques	...	...	...	...	...	484.0	460.6	* 355.9	* 65.0	* 55.0
Denmark Danemark										
Nitrogenous fertilizers *										
Engrais azotés *	191.0	200.4	184.9	138.0	120.0	377.0	400.4	394.9	369.5	332.9
Phosphate fertilizers *										
Engrais phosphatés *	130.0	136.2	81.1	75.0	65.0	92.0	94.8	88.6	76.0	67.0
Potash fertilizers *										
Engrais potassiques *	...	...	...	...	...	145.0	155.4	149.7	134.7	132.0
Estonia Estonie										
Nitrogenous fertilizers *										
Engrais azotés *	..	..	..	...	49.0	..	..	..	...	43.0
Phosphate fertilizers *										
Engrais phosphatés *	..	..	..	...	58.0	..	..	..	...	48.0
Potash fertilizers *										
Engrais potassiques *	..	..	..	...	...	..	..	..	...	50.0

40

Fertilizers
Nitrogenous, phosphate and potash: thousand metric tons [*cont.*]
Engrais
Azotés, phosphatés et potassiques : milliers de tonnes métriques [*suite*]

Country or area	Production					Consumption				
Pays ou zone	1988/89	1989/90	1990/91	1991/92	1992/93	1988/89	1989/90	1990/91	1991/92	1992/93
Finland Finlande										
Nitrogenous fertilizers										
Engrais azotés	* 293.4	* 277.6	* 268.0	* 254.0	* 241.0	199.4	231.5	206.8	166.5	170.6
Phosphate fertilizers										
Engrais phosphatés	* 216.3	* 183.3	* 171.9	* 162.3	* 139.0	140.6	143.1	117.2	* 76.0	82.0
Potash fertilizers										
Engrais potassiques	...	...	...	...	...	133.3	140.9	119.1	88.9	90.0
France France										
Nitrogenous fertilizers										
Engrais azotés	1 675.0	1 572.0	1 524.0	1 636.0	1 333.0	2 605.0	2 660.0	2 492.0	2 569.0	2 154.0
Phosphate fertilizers										
Engrais phosphatés	1 116.0	1 025.0	916.0	926.0	693.0	1 460.0	1 494.0	1 349.0	1 253.0	1 029.0
Potash fertilizers										
Engrais potassiques	* 1 411.0	* 1 199.4	* 1 292.4	* 1 129.0	* 1 141.0	1 935.0	1 949.0	1 842.0	1 741.0	1 348.0
Germany † Allemagne†										
Nitrogenous fertilizers										
Engrais azotés	..	..	* 1 165.0	1 095.0	* 1 205.0	..	..	* 1 788.0	* 1 720.0	* 1 680.0
Phosphate fertilizers *										
Engrais phosphatés *	..	..	294.0	250.0	224.0	..	..	609.0	519.0	490.0
Potash fertilizers *										
Engrais potassiques *	..	..	4 462.0	3 902.0	3 525.0	..	..	875.0	729.0	673.0
F. R. Germany R. f. Allemagne										
Nitrogenous fertilizers										
Engrais azotés	918.6	787.8	..	..	..	1 539.9	1 487.2	..	..	..
Phosphate fertilizers										
Engrais phosphatés	339.2	307.2	..	..	..	643.5	594.4	..	..	..
Potash fertilizers										
Engrais potassiques	2 267.9	2 291.4	..	..	..	887.1	791.6	..	..	..
former German D. R. anc. R. d. allemande										
Nitrogenous fertilizers										
Engrais azotés	1 382.2	* 1 133.0	..	..	..	873.2	* 813.0	..	..	..
Phosphate fertilizers										
Engrais phosphatés	299.7	287.5	..	..	..	348.8	356.3	..	..	..
Potash fertilizers										
Engrais potassiques	3 510.4	3 199.8	..	..	..	583.3	595.5	..	..	..
Greece Grèce										
Nitrogenous fertilizers										
Engrais azotés	* 388.0	* 407.0	396.2	379.9	* 297.0	409.2	425.7	426.6	408.3	* 326.0
Phosphate fertilizers										
Engrais phosphatés	* 206.3	* 198.0	* 198.6	* 191.8	* 117.0	176.4	188.8	187.3	* 176.0	* 132.0
Potash fertilizers										
Engrais potassiques	...	...	...	...	...	62.4	73.2	71.3	67.2	53.0
Hungary Hongrie										
Nitrogenous fertilizers										
Engrais azotés	591.4	591.2	421.5	224.5	138.9	646.3	583.0	358.0	159.5	123.2
Phosphate fertilizers										
Engrais phosphatés	* 230.4	220.0	* 109.0	51.0	5.4	322.1	266.0	127.0	24.2	7.3
Potash fertilizers										
Engrais potassiques	...	...	...	...	...	449.1	372.0	186.0	37.4	14.7
Iceland Islande										
Nitrogenous fertilizers										
Engrais azotés	9.6	10.3	10.1	13.0	10.7	11.5	10.4	11.8	12.1	10.8
Phosphate fertilizers										
Engrais phosphatés	...	...	...	...	...	5.9	5.2	5.8	6.1	5.3
Potash fertilizers										
Engrais potassiques	...	...	...	...	...	4.5	4.6	4.8	4.5	3.9
Ireland Irlande										
Nitrogenous fertilizers										
Engrais azotés	* 252.0	* 296.7	* 279.0	* 250.0	* 286.0	* 349.0	* 378.5	* 370.0	358.0	* 336.0

40

Fertilizers
Nitrogenous, phosphate and potash: thousand metric tons [*cont.*]
Engrais
Azotés, phosphatés et potassiques : milliers de tonnes métriques [*suite*]

Country or area Pays ou zone	Production Production					Consumption Consommation				
	1988/89	1989/90	1990/91	1991/92	1992/93	1988/89	1989/90	1990/91	1991/92	1992/93
Phosphate fertilizers										
Engrais phosphatés	...	...	...	...	...	* 148.0	* 145.5	* 138.5	136.2	* 137.0
Potash fertilizers										
Engrais potassiques	...	...	...	...	...	* 193.6	* 182.3	* 183.8	178.1	175.0
Italy Italie										
Nitrogenous fertilizers										
Engrais azotés	1 297.1	1 168.6	862.0	916.9	991.0	924.0	827.3	879.0	906.7	906.0
Phosphate fertilizers										
Engrais phosphatés	412.1	266.0	426.7	389.0	325.0	715.5	607.9	644.6	662.8	596.0
Potash fertilizers										
Engrais potassiques	159.4	130.5	6.0	126.1	13.1	452.5	377.7	424.3	416.4	365.0
Latvia Lettonie										
Nitrogenous fertilizers *										
Engrais azotés *	..	..	..	...	18.3	..	..	..	...	63.0
Phosphate fertilizers *										
Engrais phosphatés *	..	..	..	...	60.0	..	..	..	...	45.0
Potash fertilizers *										
Engrais potassiques *	..	..	..	...	...	..	..	..	...	60.0
Lithuania Lituanie										
Nitrogenous fertilizers *										
Engrais azotés *	..	..	..	...	230.0	..	..	..	...	88.0
Phosphate fertilizers *										
Engrais phosphatés *	..	..	..	...	50.0	..	..	..	...	29.0
Potash fertilizers *										
Engrais potassiques *	..	..	..	...	...	..	..	..	...	50.0
Malta Malte										
Nitrogenous fertilizers										
Engrais azotés	...	...	...	...	...	0.5	0.4	* 0.7	* 0.7	* 1.2
Phosphate fertilizers										
Engrais phosphatés	...	...	...	...	...	0.0	0.0	0.0		...
Potash fertilizers										
Engrais potassiques	...	...	...	...	...	0.1	0.0	0.0	...	...
Netherlands Pays-Bas										
Nitrogenous fertilizers										
Engrais azotés	1 836.9	1 847.9	* 1 875.0	* 1 840.0	* 1 840.0	455.7	412.4	* 390.0	* 370.0	* 380.0
Phosphate fertilizers										
Engrais phosphatés	* 378.7	* 378.4	* 365.0	* 345.0	* 287.0	86.3	76.1	* 74.0	* 75.0	* 73.0
Potash fertilizers										
Engrais potassiques	...	...	...	...	...	104.8	98.3	* 94.6	* 94.0	* 90.0
Norway Norvège										
Nitrogenous fertilizers										
Engrais azotés	* 450.7	* 494.0	517.4	* 512.0	491.0	* 110.1	* 110.4	* 110.0	110.9	109.3
Phosphate fertilizers										
Engrais phosphatés	195.9	229.9	228.7	211.9	211.0	* 39.8	* 36.6	* 34.8	33.9	31.4
Potash fertilizers										
Engrais potassiques	...	...	...	...	...	* 67.9	* 65.5	55.0	62.9	60.2
Poland Pologne										
Nitrogenous fertilizers										
Engrais azotés	1 622.2	1 642.6	1 303.0	1 104.7	1 081.4	1 520.6	* 1 478.6	735.2	619.0	* 683.3
Phosphate fertilizers										
Engrais phosphatés	962.1	946.0	467.0	252.4	329.0	943.7	751.8	410.6	222.5	* 232.3
Potash fertilizers										
Engrais potassiques	...	...	...	...	...	1 160.6	1 003.1	606.7	293.4	276.9
Portugal Portugal										
Nitrogenous fertilizers *										
Engrais azotés *	158.9	133.0	126.0	116.0	102.0	156.5	145.4	150.1	135.0	134.0
Phosphate fertilizers *										
Engrais phosphatés *	60.8	76.7	63.3	39.0	37.0	89.1	81.1	80.3	75.4	77.0
Potash fertilizers *										
Engrais potassiques *	...	...	...	...	...	48.9	47.5	48.0	40.7	47.0

40
Fertilizers
Nitrogenous, phosphate and potash: thousand metric tons [*cont.*]
Engrais
Azotés, phosphatés et potassiques : milliers de tonnes métriques [*suite*]

Country or area	Production Production					Consumption Consommation				
Pays ou zone	1988/89	1989/90	1990/91	1991/92	1992/93	1988/89	1989/90	1990/91	1991/92	1992/93
Republic of Moldova République de Moldova										
Nitrogenous fertilizers *										
Engrais azotés *	..	..	...	...		..	..	..	...	64.7
Phosphate fertilizers *										
Engrais phosphatés *	..	..	..	...	...	..	..	..	...	46.1
Potash fertilizers *										
Engrais potassiques *	..	..	..	...	...	..	..	..	...	24.0
Romania Roumanie										
Nitrogenous fertilizers										
Engrais azotés	2 130.2	2 035.4	* 1 249.1	* 823.0	* 1 085.7	739.0	778.2	* 656.1	* 274.0	* 257.8
Phosphate fertilizers										
Engrais phosphatés	725.3	647.7	* 387.3	* 228.0	* 273.0	395.6	360.0	* 313.1	* 145.0	* 133.0
Potash fertilizers										
Engrais potassiques	...	...	...	...	...	291.8	240.0	* 133.9	* 43.0	* 30.6
Russian Federation Fédération de Russie										
Nitrogenous fertilizers *										
Engrais azotés *	..	..	..	...	5 708.0	..	..	..	...	2 622.6
Phosphate fertilizers *										
Engrais phosphatés *	..	..	..	...	2 840.0	..	..	..	...	1 540.0
Potash fertilizers *										
Engrais potassiques *	..	..	..	...	3 454.0	..	..	..	...	1 348.0
Slovenia Slovénie										
Nitrogenous fertilizers *										
Engrais azotés *	..	..	..	...	...	..	..	..	...	45.7
Phosphate fertilizers										
Engrais phosphatés	..	..	..	...	...	..	..	..	...	12.6
Potash fertilizers *										
Engrais potassiques *	..	..	..	...	...	..	..	..	...	11.8
Spain Espagne										
Nitrogenous fertilizers										
Engrais azotés	935.5	966.7	869.6	862.2	584.0	1 168.4	1 109.4	1 063.1	998.7	765.1
Phosphate fertilizers										
Engrais phosphatés	419.0	368.2	293.0	270.6	181.2	542.2	559.4	534.2	501.7	388.3
Potash fertilizers										
Engrais potassiques	754.2	728.7	642.3	542.6	628.6	383.2	383.3	378.6	381.4	381.1
Sweden Suède										
Nitrogenous fertilizers										
Engrais azotés	127.7	154.4	176.0	147.5	140.0[1]	240.4	221.5	211.7	174.7	* 191.0
Phosphate fertilizers										
Engrais phosphatés	122.1	108.0	108.0	79.6	* 24.0	69.3	69.7	57.8	* 46.0	* 46.0
Potash fertilizers										
Engrais potassiques	...	...	...	...	...	71.9	71.4	58.7	* 61.0	* 61.0
Switzerland Suisse										
Nitrogenous fertilizers										
Engrais azotés	33.2	30.2	26.7	29.0	24.4	71.6	70.3	63.4	63.0	* 62.0
Phosphate fertilizers										
Engrais phosphatés	2.7	2.4	2.5	3.0	2.1	40.0	39.4	38.3	37.0	36.0
Potash fertilizers										
Engrais potassiques	...	...	...	...	...	66.0	67.4	66.2	* 59.0	* 58.0
TFYR Macedonia L'ex-R.y. Macédoine										
Nitrogenous fertilizers *										
Engrais azotés *	..	..	..	...	...	..	..	..	...	3.0
Phosphate fertilizers *										
Engrais phosphatés *	..	..	..	...	...	..	..	..	...	3.0
Potash fertilizers *										
Engrais potassiques *	..	..	..	...	...	..	..	..	...	9.0
Ukraine Ukraine										
Nitrogenous fertilizers *										
Engrais azotés *	..	..	..	...	2 600.0	..	..	..	...	1 247.0

40
Fertilizers
Nitrogenous, phosphate and potash: thousand metric tons [*cont.*]
Engrais
Azotés, phosphatés et potassiques : milliers de tonnes métriques [*suite*]

Country or area Pays ou zone	Production Production 1988/89	1989/90	1990/91	1991/92	1992/93	Consumption Consommation 1988/89	1989/90	1990/91	1991/92	1992/93
Phosphate fertilizers * Engrais phosphatés *	..	..	..	...	1 393.0	..	..	..	...	900.0
Potash fertilizers * Engrais potassiques *	..	..	..		144.3		..	..	...	750.0
United Kingdom Royaume-Uni										
Nitrogenous fertilizers * Engrais azotés *	1 100.0	1 070.0	980.0	930.0	759.0	1 462.0	1 582.0	1 525.0	1 365.0	1 326.0
Phosphate fertilizers * Engrais phosphatés *	248.0	172.0	128.0	90.0	66.0	433.0	428.0	380.0	371.0	368.0
Potash fertilizers * Engrais potassiques *	444.9	488.0	493.0	494.0	530.0	521.0	525.0	465.0	410.0	420.0
Yugoslavia Yougoslavie										
Nitrogenous fertilizers * Engrais azotés *	..	..	..	...	80.0	..	..	..	...	80.0
Phosphate fertilizers * Engrais phosphatés *	..	..	..	...	50.0	..	..	..	...	50.0
Potash fertilizers * Engrais potassiques *	...	..	..	...	...	..	..	..	...	30.0
Yugoslavia, SFR† Yougoslavie, Rfs†										
Nitrogenous fertilizers Engrais azotés	630.4	519.8	* 433.6	* 448.2	..	503.0	443.7	417.3	* 288.3	
Phosphate fertilizers Engrais phosphatés	365.2	319.3	253.0	225.0	..	261.4	231.2	* 199.0	* 139.0	..
Potash fertilizers Engrais potassiques	...	...	...	...	..	250.9	221.8	* 209.6	* 154.3	..
Oceania Océanie										
Nitrogenous fertilizers Engrais azotés	300.2	289.8	278.0	290.0	343.7	451.0	497.6	503.9	543.9	617.2
Phosphate fertilizers Engrais phosphatés	973.3	716.6	437.2	439.7	542.0	1 028.4	1 050.2	806.0	913.9	995.9
Potash fertilizers Engrais potassiques	...	...	...	...	...	264.6	262.1	254.2	245.1	284.9
Australia Australie										
Nitrogenous fertilizers Engrais azotés	* 230.2	* 230.0	215.0	233.0	280.0	* 393.9	* 440.3	* 439.4	* 462.3	* 510.0
Phosphate fertilizers Engrais phosphatés	809.8	* 511.0	* 269.6	229.0	322.0	* 836.0	* 790.5	* 578.9	* 640.0	* 690.0
Potash fertilizers * Engrais potassiques *	...	...	...	...	...	157.3	162.6	145.4	142.1	150.0
Fiji Fidji										
Nitrogenous fertilizers Engrais azotés	...	...	...	...	...	10.1	* 13.2	* 12.0	* 10.0	* 8.5
Phosphate fertilizers Engrais phosphatés	...	...	...	...	...	10.4	2.7	* 3.0	* 2.0	2.5[1]
Potash fertilizers Engrais potassiques	...	...	...	...	...	4.7	7.2	* 8.0	* 2.0	* 2.1
French Polynesia Polynésie française										
Nitrogenous fertilizers * Engrais azotés *	...	...	...	...	...	0.3	0.3	0.3	0.3	0.4
Phosphate fertilizers * Engrais phosphatés *	...	...	...	...	...	0.3	0.3	0.3	0.3	0.4
Potash fertilizers * Engrais potassiques *	...	...	...	...	...	0.3	0.3	0.3	0.2	0.2
New Caledonia Nouvelle-Calédonie										
Nitrogenous fertilizers * Engrais azotés *	...	...	...	...	...	0.5	0.5	1.0	0.4	0.3

40
Fertilizers
Nitrogenous, phosphate and potash: thousand metric tons [cont.]
Engrais
Azotés, phosphatés et potassiques : milliers de tonnes métriques [suite]

Country or area	Production Production					Consumption Consommation				
Pays ou zone	1988/89	1989/90	1990/91	1991/92	1992/93	1988/89	1989/90	1990/91	1991/92	1992/93
Phosphate fertilizers *										
Engrais phosphatés *	...	...	...	...	...	0.5	0.5	0.5	0.5	0.5
Potash fertilizers *										
Engrais potassiques *	...	...	...	...	...	0.2	0.2	0.3	0.3	0.3
New Zealand Nouvelle-Zélande										
Nitrogenous fertilizers										
Engrais azotés	* 70.0	* 59.8	* 63.0	* 57.0	63.7	* 40.0	* 37.0	* 45.8	* 64.4	* 90.0
Phosphate fertilizers										
Engrais phosphatés	* 153.5	* 205.6	* 167.6	* 210.7	* 220.0	* 178.5	253.2	220.3	* 268.9	* 300.0
Potash fertilizers										
Engrais potassiques	...	...	...	...	...	* 97.0	85.7	96.3	* 98.5	* 130.0
Papua New Guinea Papouasie-Nvl-Guinée										
Nitrogenous fertilizers *										
Engrais azotés *	...	...	...	...	...	6.2	6.3	5.4	6.5	8.0
Phosphate fertilizers										
Engrais phosphatés	...	...	...	...	...	* 2.7	* 3.0	* 3.0	* 2.2	2.5 [1]
Potash fertilizers										
Engrais potassiques	...	...	...	...	...	* 5.1	* 6.2	* 3.9	* 2.0	2.3 [1]
former USSR† ancienne URSS†										
Nitrogenous fertilizers										
Engrais azotés	15 614.3	14 272.0	13 084.4	* 12 004.2	..	11 587.0	9 918.0	8 668.7	* 7 778.0	..
Phosphate fertilizers										
Engrais phosphatés	8 944.3	9 015.0	8 866.9	* 7 685.6	..	8 556.0	8 175.0	7 830.0	* 6 594.6	..
Potash fertilizers										
Engrais potassiques	11 300.3	10 232.6	9 035.2	* 8 531.0	..	7 044.0	6 381.0	5 145.2	* 5 032.9	..

Source:
Food and Agriculture Organization of the United Nations
(Rome).

† For detailed descriptions of data pertaining to
former Czechoslovakia, Germany, SFR Yugoslavia and former
USSR, see Annex I - Country or area nomenclature, regional
and other groupings.

1 FAO estimate.
2 Including Eritrea.

Source:
Organisation des Nations Unies pour l'alimentation et
l'agriculture (Rome).

† Pour les descriptions en détails des données
relatives à l'ancienne Tchécoslovaquie, l'Allemagne, la Rfs
Yougoslavie et l'ancienne URSS, voir l'Annexe I -
Nomenclature des pays ou zones, groupements régionaux et
autres groupements.

1 Estimation de la FAO.
2 Y compris Erythrée.

Technical notes, tables 34-40

The series shown on agriculture and fishing have been furnished by the Food and Agriculture Organization of the United Nations (FAO). They refer to the following three topics:

(a)   Long-term trends in growth of agricultural output and food supply;

(b)   Output of principal agricultural commodities and in a few cases consumption;

(c)   Basic means of production.

Agricultural output is defined to include all crop and livestock products except those used for seed and fodder and other intermediate uses in agriculture; for example deductions are made for eggs used for hatching. · Intermediate input of seeds and fodder and similar items refer to both domestically produced and imported commodities.

Detailed data and technical notes are published by FAO in its yearbooks.[4, 6]

The index-numbers of agricultural output and food production are calculated by the Laspeyres formula with the base year period 1979-1981. The latter is provided in order to diminish the impact of annual fluctuations in agricultural output during base years on the indices for the period. Production quantities of each commodity are weighted by 1979-1981 average national producer prices and summed for each year. The index numbers are based on production data for a calendar year. These may differ in some instances from those actually produced and published by the individual countries themselves due to variations in concepts, coverage, weights and methods of calculation. Efforts have been made to estimate these methodological differences to achieve a better international comparability of data. The series include a large amount of estimates made by FAO in cases where no official or semi-official figures are available from the countries.

In *table 34*, The "All commodities index" relates to the production of all crops and livestock products. The "Food Index" includes those commodities which are considered edible and contain nutrients.

In *table 35*, Cereals, production data relate to crops harvested for grain only. Cereals harvested for hay, green feed or used for grazing are excluded.

Table 36: Oil crops, or oil-bearing crops, are those crops yielding seeds, nuts or fruits which are used mainly for the extraction of culinary or industrial oils, excluding essential oils. In this table, data for oil crops represent the total production of oil seeds, oil nuts and oil fruits harvested in the year indicated and expressed in terms of oil equivalent and cake/meal equivalent. That is to say, these figures do not relate to the actual production of vegetable oils and cake/meal, but to the potential

Notes techniques, tableaux 34-40

Les séries présentées sur l'agriculture et la pêche ont été fournies par l'Organisation des Nations Unies pour l'alimentation et l'agriculture (FAO). Elles portent sur les trois aspects suivants :

(a)   Les tendances à long terme de la croissance de la production agricole et des approvisionnements alimentaires;

(b)   La production des principales denrées agricoles et, dans certains cas, la consommation;

(c)   Les moyens essentiels de production.

La production agricole se définit comme comprenant l'ensemble des produits agricoles et des produits de l'élevage à l'exception de ceux utilisés comme semences et comme aliments pour les animaux, et pour les autres utilisations intermédiaires en agriculture; par exemple, on déduit les oeufs utilisés pour la reproduction. L'apport intermédiaire de semences et d'aliments pour les animaux et d'autres éléments similaires se rapportent à la fois à des produits locaux et importés.

Des données détaillées et des notes techniques sont publiées par la FAO dans ses annuaires [4, 6].

Les indices de la production agricole et de la production alimentaire sont calculés selon la formule de Laspeyres avec les années 1979-1981 pour période de base. Le choix d'une période de plusieurs années permet de diminuer l'incidence des fluctuations annuelles de la production agricole pendant les années de base sur les indices pour cette période. Les quantités produites de chaque denrée sont pondérées par les prix nationaux moyens à la production de 1979-1981, et additionnées pour chaque année. Les indices sont fondés sur les données de production d'une année civile. Ils peuvent différer dans certains cas des indices effectivement établis et publiés par les pays eux-mêmes par suite de différences dans les concepts, la couverture, les pondérations et les méthodes de calcul. On s'est efforcé d'estimer ces différences méthodologiques afin de rendre les données plus facilement comparables à l'échelle internationale. Les séries comprennent une grande quantité d'estimations faites par la FAO dans les cas où les pays n'avaient pas fourni de chiffres officiels ou semi-officiels.

Au *Tableau 34*, l'"Indice relatif à l'ensemble des produits" se rapporte à la production de tous les produits de l'agriculture et de l'élevage. L'"Indice des produits alimentaires" comprend les produits considérés comme comestibles et qui contiennent des éléments nutritifs.

Au *Tableau 35*, Céréales : les données sur la production se rapportent uniquement aux céréales récoltées pour le grain; celles cultivées pour le foin, le fourrage vert ou le pâturage en sont exclues.

Tableau 36 : On désigne sous le nom de cultures oléagineuses l'ensemble des cultures produisant des graines, des noix ou des fruits, essentiellement destinées à l'extraction

production if the total amounts produced from all oil crops were processed into oil and cake/meal in producing countries in the same year in which they were harvested. Naturally, the total production of oil crops is never processed into oil in its entirety, since depending on the crop, important quantities are also used for seed, feed and food. However, although oil and cake/meal extraction rates vary from country to country, in this table the same extraction rate for each crop has been applied for all countries. Moreover. it should be borne in mind that the crops harvested during the latter months of the year are generally processed into oil during the following year. In spite of these deficiencies in coverage, extraction rates and time reference the data reported here are useful as they provide a valid indication of year-to-year changes in the size of total oil-crop production. The actual production of vegetable oils in the world is about 80 percent of the production reported here. In addition, about two million tonnes of vegetable oils are produced every year from crops which are not included among those defined above. The most important of these oils are maize-germ oil and rice-bran oil. The actual world production of cake/meal derived from oil crops is also about 80 percent of the production reported in the table.

In *table 37*, Livestock, data refer to livestock numbers grouped into twelve-month periods ending 30 September of the year stated and cover all animals irrespective of their age and place or purpose of their breeding.

In *table 39*, the data cover as far as possible both sea and inland fisheries and are expressed in terms of live weight. They generally include crustaceans and molluscs but exclude seaweed and aquatic mammals such as whales and dolphins. Data include landings by domestic craft in foreign ports and exclude the landings by foreign craft in domestic ports. The flag of the vessel is considered as the paramount indication of the nationality of the catch.

In *table 40*, data generally refer to the fertilizer year 1 July-30 June.

Nitrogenous fertilizers: data refer to the nitrogen content of commercial inorganic fertilizers.

Phosphate fertilizers: data refer to commercial phosphoric acid (P_2O_5) and cover the P_2O_5 of superphosphates, ammonium phosphate and basic slag.

Potash fertilizers: data refer to K_2O content of commercial potash, muriate, nitrate and sulphate of potash, manure salts, kainit and nitrate of soda potash.

d'huiles alimentaires ou industrielles, à l'exclusion des huiles essentielles. Dans ce tableau, les chiffres se rapportent à la production totale de graines, noix et fruits oléagineux récoltés au cours de l'année de référence et sont exprimés en équivalent d'huile et en équivalent de tourteau/farine. En d'autres termes, ces chiffres ne se rapportent pas à la production effective mais à la production potentielle d'huiles végétales et de tourteau/farine dans l'hypothèse où les volumes totaux de produits provenant de toutes les cultures d'oléagineux seraient transformés en huile et en tourteau/farine dans les pays producteurs l'année même où ils ont été récoltés. Bien entendu, la production totale d'oléagineux n'est jamais transformée intégralement en huile, car des quantités importantes qui varient suivant les cultures sont également utilisées pour les semailles, l'alimentation animale et l'alimentation humaine. Toutefois, bien que les taux d'extraction d'huile et de tourteau/farine varient selon les pays, on a appliqué dans ce tableau le même taux à tous les pays pour chaque oléagineux. En outre, il ne faut pas oublier que les produits récoltés au cours des derniers mois de l'année sont généralement transformés en huile dans le courant de l'année suivante.

En dépit de ces imperfections qui concernent le champ d'application, les taux d'extraction et les périodes de référence, les chiffres présentés ici sont utiles, car ils donnent une indication valable des variations de volume que la production totale d'oléagineux enregistre d'une année à l'autre. La production mondiale effective d'huiles végétales atteint 80 pour cent environ de la production indiquée ici. En outre, environ 2 millions de tonnes d'huiles végétales sont produites chaque année à partir de cultures non comprises dans les catégories définies ci-dessus. Les principales sont l'huile de germs de maïs et l'huile de son de riz. La production mondiale effective tourteau/farine d'oléagineux représente environ 80 pour cent de production indiquée dans le tableau.

Au *Tableau 37*, Elevage : les statistiques sur les effectifs du cheptel sont groupées en périodes de 12 mois se terminant le 30 septembre de l'année indiquée et s'entendent de tous les animaux, quels que soient leur âge et l'emplacement ou le but de leur élevage.

Au *Tableau 39*, les données englobent autant que possible la pêche maritime et intérieure, et sont exprimées en poids vif. Elles comprennent, en général, crustacés et mollusques, mais excluent les plantes marines et les mammifères aquatiques (baleines, dauphins, etc.). Les données comprennent les quantités débarquées par des bateaux nationaux dans des ports étrangers et excluent les quantités débarquées par des bateaux étrangers dans des ports nationaux. Le pavillon du navire est considéré comme la principale indication de la nationalité de la prise.

Au *Tableau 40*, Engrais : les données se rapportent en général à une période d'un an comptée du 1er juillet au 30 juin.

Engrais azotés : les données se rapportent à la teneur en azote des engrais commerciaux inorganiques.

Engrais phosphatés : les données se rapportent à l'acide phosphorique (P_2O_5) et englobent la teneur en (P_2O_5) des superphosphates, du phosphate d'ammonium et des scories de déphosphoration.

Engrais potassiques: les données se rapportent à la teneur en K_2O des produits potassiques commerciaux, muriate, nitrate et sulfate de potasse, sels d'engrais, kainite et nitrate de soude potassique.

41
Sugar
Sucre

Production and consumption: thousand metric tons; consumption per capita: kilograms
Production et consommationa : milliers de tonnes métriques ; consommation par habitant : kilogrammes

Country or area Pays ou zone	1984	1985	1986	1987	1988	1989	1990	1991	1992	1993
World Monde										
Production	99 219	98 365	100 193	103 411	104 659	107 124	110 818	112 049	117 125	112 074
Consumption	96 681	97 859	101 316	106 171	106 291	107 234	108 418	108 847	112 391	111 545
Consumption per cap.(kg)	20	20	21	21	21	21	21	20	21	20
Africa · Afrique										
Algeria Algérie										
Production *	7	0	0	0	0	0	0	0	0	0
Consumption *	650	600	585	630	675	710	810	850	880	850
Consumption per cap.(kg)	31	28	26	27	29	29	32	33	33	34
Angola Angola										
Production *	50	50	50	30	30	25	25	30	25	20
Consumption *	105	100	100	77	75	90	95	100	120	120
Consumption per cap.(kg)	12	11	11	8	8	9	10	11	12	12
Benin Bénin										
Production *	3	5	5	5	5	7	5	5	4	5
Consumption *	30	30	30	20	20	15	15	20	25	30
Consumption per cap.(kg)	8	8	7	5	5	3	3	4	5	6
Botswana Botswana										
Consumption	36	37	37	40	45	45	50	55	50	45
Consumption per cap.(kg)	34	34	33	34	37	36	38	41	37	31
Burkina Faso Burkina Faso										
Production *	27	10	10	25	25	20	30	26	27	25
Consumption *	31	35	35	30	30	40	35	30	30	30
Consumption per cap.(kg)	5	5	5	4	4	5	4	3	3	3
Burundi Burundi										
Production	0	0	0	0	3	9	8	14	10	10
Consumption	7	7	12	15	14	15	17	11	12	12
Consumption per cap.(kg)	2	2	2	3	3	3	3	2	2	2
Cameroon Cameroun										
Production	* 59	* 50	* 40	* 28	67	* 35	* 75	* 70	* 70	* 65
Consumption	70	* 60	* 45	* 40	63	* 40	* 75	* 80	* 80	* 80
Consumption per cap.(kg)	7	6	4	4	6	4	7	7	7	6
Cape Verde Cap−Vert										
Consumption *	9	9	10	15	13	12	11	13	14	16
Consumption per cap.(kg)	27	27	30	43	36	32	29	34	37	42
Central African Rep. Rép. centrafricaine										
Consumption *	2	7	10	5	4	3	3	3	3	4
Consumption per cap.(kg)	1	3	4	2	1	1	1	1	1	1
Chad Tchad										
Production	* 15	* 8	* 10	* 20	* 20	* 25	* 25	* 30	* 30	32
Consumption *	30	30	18	20	20	30	50	40	40	40
Consumption per cap.(kg)	6	6	4	4	4	5	9	7	7	6
Comoros Comores										
Consumption	3	3	3	3	3	3	3	4	4	4
Consumption per cap.(kg)	7	7	6	6	6	6	6	7	7	7
Congo Congo										
Production	31	* 25	* 32	* 35	* 40	* 35	* 35	21	30	26
Consumption	* 18	* 20	* 20	* 20	* 20	* 25	* 20	* 17	18	16
Consumption per cap.(kg)	11	12	10	10	10	12	9	7	8	5
Côte d'Ivoire Côte d'Ivoire										
Production	121	* 125	* 120	* 165	* 165	* 160	* 160	* 155	* 160	* 155
Consumption	119	* 125	* 125	* 130	* 155	* 155	* 160	* 160	* 165	* 165
Consumption per cap.(kg)	12	13	12	12	14	13	13	13	13	12
Djibouti Djibouti										
Consumption	8	8	9	9	10	10	10	9	12	10
Consumption per cap.(kg)	24	19	20	23	24	23	23	20	26	20
Egypt Egypte										
Production *	780	900	950	1 000	1 035	947	955	1 060	1 060	1 090
Consumption *	1 600	1 600	1 650	1 650	1 775	1 650	1 725	1 745	1 750	1 675
Consumption per cap.(kg)	34	33	35	34	35	32	33	32	32	30
Ethiopia Ethiopie										
Production	200	191	193	* 195	169	183	* 184	* 161	165	* 185
Consumption	177	144	159	* 160	139	162	* 160	* 160	* 160	* 160
Consumption per cap.(kg)	4	3	4	4	3	3	3	3	3	3

41

Sugar

Production and consumption: thousand metric tons; consumption per capita: kilograms [*cont.*]

Sucre

Production et consommation : milliers de tonnes métriques ; consommation par habitant : kilogrammes [*suite*]

Country or area Pays ou zone	1984	1985	1986	1987	1988	1989	1990	1991	1992	1993
Gabon Gabon										
Production	* 11	* 12	18	* 19	* 20	* 15	* 20	* 22	* 25	* 20
Consumption *	15	18	16	16	17	17	18	18	18	19
Consumption per cap.(kg)	13	16	15	15	15	15	15	14	15	15
Gambia Gambie										
Consumption *	27	30	30	30	35	35	35	35	35	35
Consumption per cap.(kg)	42	47	46	38	43	42	41	40	40	40
Ghana Ghana										
Consumption *	30	30	55	65	80	80	90	95	120	125
Consumption per cap.(kg)	2	2	4	5	6	6	6	6	8	8
Guinea Guinée										
Production *	10	5	5	10	10	20	20	18	18	10
Consumption *	30	35	50	55	40	50	50	55	60	65
Consumption per cap.(kg)	6	6	10	11	8	9	9	9	10	10
Guinea−Bissau Guinée−Bissau										
Consumption	4	3	3	3	4	4	3	3	3	2
Consumption per cap.(kg)	4	3	3	3	4	4	3	3	3	2
Kenya Kenya										
Production	371	* 260	* 200	* 365	* 430	* 440	* 440	* 485	* 380	* 415
Consumption	348	* 420	* 420	* 450	* 460	* 475	* 500	* 515	* 535	* 550
Consumption per cap.(kg)	18	21	20	20	19	19	19	20	20	20
Liberia Libéria										
Production	3	3	3	3	3	3	1	1	1	1
Consumption	10	10	15	15	17	15	11	6	6	6
Consumption per cap.(kg)	5	5	7	7	7	6	5	2	2	2
Libyan Arab Jamah. Jamah. arabe libyenne										
Consumption *	140	150	190	175	160	150	170	180	180	180
Consumption per cap.(kg)	39	42	48	43	38	34	37	38	37	36
Madagascar Madagascar										
Production	79	99	98	107	122	120	118	96	97	104
Consumption	72	84	81	82	79	76	87	87	88	85
Consumption per cap.(kg)	7	8	8	8	7	7	8	8	7	7
Malawi Malawi										
Production	160	154	168	181	187	173	204	210	209	137
Consumption	53	62	73	89	108	107	124	130	152	162
Consumption per cap.(kg)	8	8	10	12	13	13	15	14	17	18
Mali Mali										
Production	10	21	21	17	21	22	* 25	* 30	* 25	* 25
Consumption	40	* 44	47	* 55	* 50	* 70	* 75	* 80	* 80	* 80
Consumption per cap.(kg)	5	5	6	7	6	8	9	8	8	8
Mauritania Mauritanie										
Consumption *	25	25	40	70	65	70	55	60	60	55
Consumption per cap.(kg)	14	13	22	38	34	36	27	29	28	25
Mauritius Maurice										
Production	610	684	748	733	672	602	661	648	681	604
Consumption	40	39	40	40	41	40	41	42	41	39
Consumption per cap.(kg)	39	38	41	38	37	37	38	40	41	36
Morocco Maroc										
Production	441	433	352	* 450	590	469	* 520	518	509	520
Consumption	681	707	725	* 700	756	740	* 775	778	810	830
Consumption per cap.(kg)	30	32	33	30	32	31	32	31	32	33
Mozambique Mozambique										
Production	39	* 60	* 40	* 25	* 40	* 25	* 32	* 25	* 30	* 20
Consumption *	90	90	90	60	45	50	45	45	50	70
Consumption per cap.(kg)	7	6	6	4	3	4	3	3	3	5
Niger Niger										
Consumption *	10	30	30	30	25	20	25	25	23	28
Consumption per cap.(kg)	2	5	4	4	3	3	3	3	3	3
Nigeria Nigéria										
Production *	60	50	45	40	31	55	55	60	50	50
Consumption *	550	600	625	625	425	370	415	480	600	650
Consumption per cap.(kg)	6	6	6	6	4	4	4	4	5	5
Réunion ¹ Réunion ¹										
Production	246	228	244	226	252	171	192	215	227	183
Rwanda Rwanda										
Production	2	2	2	4	5	3	3	4	4	4

41

Sugar
Production and consumption: thousand metric tons; consumption per capita: kilograms [*cont.*]
Sucre
Production et consommation : milliers de tonnes métriques ; consommation par habitant : kilogrammes [*suite*]

Country or area Pays ou zone	1984	1985	1986	1987	1988	1989	1990	1991	1992	1993
Consumption	12	17	19	12	12	10	10	11	12	12
Consumption per cap.(kg)	2	3	3	2	2	2	1	2	2	2
Senegal Sénégal										
Production	47	* 65	73	* 71	* 72	* 70	* 80	* 90	* 90	* 85
Consumption	71	* 75	71	* 80	* 80	* 90	* 120	* 120	* 120	* 120
Consumption per cap.(kg)	12	12	11	12	11	13	16	16	16	15
Sierra Leone Sierra Leone										
Production	6	* 5	* 5	* 6	* 5	* 5	* 5	* 5	* 3	* 4
Consumption *	15	18	16	17	20	16	18	18	19	22
Consumption per cap.(kg)	4	5	4	5	5	4	4	4	4	5
Somalia Somalie										
Production *	45	54	30	35	40	47	35	30	20	20
Consumption *	100	90	90	80	60	50	40	37	43	47
Consumption per cap.(kg)	19	19	11	10	7	6	5	4	5	5
South Africa Afrique du Sud										
Production	2 276[2]	2 540[2]	2 248	2 235	2 470	2 293	2 226	2 462	1 715	1 282
Consumption	1 334	1 368	1 381	1 433	1 417	1 390	1 433	1 382	1 327	1 303
Consumption per cap.(kg)	40	43	40	41	39	38	38	38	34	33
Sudan Soudan										
Production *	360	450	550	525	500	385	425	485	490	520
Consumption *	450	470	550	600	550	440	450	490	480	500
Consumption per cap.(kg)	22	22	25	26	23	18	18	19	18	18
Swaziland Swaziland										
Production	429	396	537	461	464	504	527	517	495	458
Consumption	22	22	* 24	* 35	* 41	* 49	* 47	* 52	* 73	* 103
Consumption per cap.(kg)	38	35	36	50	58	67	61	68	69	77
Togo Togo										
Production	—	—	0	* 4	* 5	* 5	* 5	* 5	* 5	* 5
Consumption *	50	50	50	40	40	35	35	30	30	19
Consumption per cap.(kg)	18	17	16	12	12	10	10	8	8	5
Tunisia Tunisie										
Production	16	17	21	27	26	22	25	22	28	25
Consumption	180	212	189	212	202	192	212	215	246	250
Consumption per cap.(kg)	24	30	25	28	26	24	26	26	28	28
Uganda Ouganda										
Production *	20	20	10	20	40	40	25	45	40	50
Consumption *	21	22	30	70	70	70	35	45	46	55
Consumption per cap.(kg)	1	1	2	4	4	4	2	3	3	3
United Rep.Tanzania Rép. Unie de Tanzanie										
Production *	129	105	100	95	80	100	115	115	105	120
Consumption *	122	126	115	100	85	100	100	115	125	150
Consumption per cap.(kg)	6	6	5	4	4	4	4	4	5	5
Zaire Zaïre										
Production	59	* 65	* 55	* 75	* 75	* 90	* 85	* 88	* 90	* 85
Consumption *	85	90	95	120	120	110	95	95	100	110
Consumption per cap.(kg)	3	3	3	4	4	3	3	3	3	3
Zambia Zambie										
Production	141	143	127	139	146	143	135	134	155	147
Consumption	118	122	105	114	123	114	116	105	111	86
Consumption per cap.(kg)	18	17	16	16	16	15	14	13	13	10
Zimbabwe Zimbabwe										
Production	463	456	507	459	453	502	493	346[3]	9	51
Consumption	223	225	238	253	270	283	297	294	234	229
Consumption per cap.(kg)	28	27	28	29	30	31	32	29	29	22
America, North · Amerique du Nord										
Bahamas Bahamas										
Consumption	9	10	10	12	9	10	11	10	10	10
Consumption per cap.(kg)	39	41	40	48	38	40	42	39	39	37
Barbados Barbade										
Production	98	101	113	84	81	67	70	67	* 55	* 48
Consumption	* 14	14	14	13	14	13	12	11	* 12	* 11
Consumption per cap.(kg)	56	54	54	53	57	52	49	42	46	42
Belize Belize										
Production	109	110	105	88	89	94	108	103	108	108
Consumption	7	6	6	7	7	7	9	9	9	11
Consumption per cap.(kg)	48	39	39	39	40	42	46	47	47	54

41

Sugar
Production and consumption: thousand metric tons; consumption per capita: kilograms [*cont.*]

Sucre
Production et consommation : milliers de tonnes métriques ; consommation par habitant : kilogrammes [*suite*]

Country or area Pays ou zone	1984	1985	1986	1987	1988	1989	1990	1991	1992	1993
Bermuda Bermudes										
Consumption	2	2	2	2	4	2	2	2	2	2
Consumption per cap.(kg)	40	37	37	33	67	30	33	33	27	33
Canada Canada										
Production	110	* 60	* 106	* 147	* 110	* 117	* 140	* 150	* 125	* 124
Consumption	1 072[4]	* 1 050	* 1 100	1 120	* 1 100	* 1 050	* 1 050	* 1 100	* 1 120	* 1 150
Consumption per cap.(kg)	43	41	43	44	42	40	41	41	41	40
Costa Rica Costa Rica										
Production	* 245	* 230	220	* 230	237	* 220	246	279	301	* 305
Consumption	* 150	* 150	164	* 166	* 170	* 168	178	183	186	* 190
Consumption per cap.(kg)	62	60	61	60	59	59	60	60	60	61
Cuba Cuba										
Production	7 783	7 889	7 467	7 232	8 119	7 579	8 445	7 233	7 219	4 246
Consumption [5]	728	887[6]	762	772	746	882	937	956	942	796
Consumption per cap.(kg)	70	65	75	75	72	84	88	89	87	73
Dominican Republic Rép. dominicaine										
Production	1 133	921	895	816	777	693	590	628	593	621
Consumption	258	304	294	351	223	244	201	251	277	290
Consumption per cap.(kg)	42	49	46	52	32	35	29	35	28	38
El Salvador El Salvador										
Production	242	279	292	262	178	196	* 220	186	233	* 250
Consumption	159	159	176	161	173	163	* 165	* 150	116	* 165
Consumption per cap.(kg)	31	33	34	32	34	31	32	28	21	30
Guadeloupe Guadeloupe										
Production	41[1]	53[1]	65[1]	63[1]	76[10]	78[10]	26[10]	53[10]	38[10]	63[10]
Guatemala Guatemala										
Production	555	583	651	639	720	735	939	1 038	1 165	1 226
Consumption	265	279	300	320	332	353	360	348	368	391
Consumption per cap.(kg)	34	35	37	38	38	39	39	37	38	39
Haiti Haïti										
Production	* 41	* 50	* 40	35	* 30	* 30	* 35	* 25	* 20	* 20
Consumption *	53	62	60	55	55	70	85	90	95	85
Consumption per cap.(kg)	10	12	11	10	10	13	15	14	14	12
Honduras Honduras										
Production	207	235	227	* 190	* 180	* 180	* 205	175	* 185	* 185
Consumption	114	120	114	* 120	* 150	* 160	* 175	161	* 170	* 175
Consumption per cap.(kg)	30	27	29	26	31	32	34	31	31	31
Jamaica Jamaïque										
Production	188	209	199	189	222	205	209	234	228	219
Consumption	95	97	102	108	113	125	114	116	117	123
Consumption per cap.(kg)	45	46	44	47	48	53	48	48	48	50
Martinique Martinique										
Production	5[1]	9[1]	8[10]	8[10]	8[10]	7[10]	6[10]	7[10]	6[10]	7[10]
Mexico Mexique										
Production	3 308	3 492	4 068	4 061	3 909	3 570	3 384	3 744	3 745	* 4 360
Consumption	3 343	3 548	3 451	3 657	4 070	4 023	4 424	4 200	4 520	* 4 500
Consumption per cap.(kg)	45	45	43	45	49	48	55	52	54	49
Netherlands Antilles Antilles néerlandaises										
Consumption *	9	8	9	8	8	9	10	11	10	8
Consumption per cap.(kg)	35	30	32	32	32	36	40	48	44	35
Nicaragua Nicaragua										
Production	267	240	256	199	209	160	212	* 225	* 190	* 185
Consumption	154	146	157	151	136	* 150	* 150	* 140	* 135	* 150
Consumption per cap.(kg)	49	45	47	43	38	40	39	35	33	35
Panama Panama										
Production	176	160	139	* 100	* 90	* 110	* 90	* 130	* 130	* 120
Consumption	76	79	80	* 70	* 80	* 85	* 90	* 90	* 95	* 95
Consumption per cap.(kg)	36	36	36	31	35	36	37	36	38	37
Saint Kitts and Nevis Saint—Kitts—et—Nevis										
Production	31	27	28	25	26	25	* 25	20	* 20	* 25
Consumption	2	2	2	2	2	2	* 2	2	* 2	* 2
Consumption per cap.(kg)	42	43	44	51	50	41	40	40	41	41
Trinidad and Tobago Trinité—et—Tobago										
Production	67	80	95	88	94	100	122	104	114	108
Consumption	66	65	63	71	67	60	63	61	55	59
Consumption per cap.(kg)	56	55	52	58	53	48	49	49	48	47

41

Sugar
Production and consumption: thousand metric tons; consumption per capita: kilograms [*cont.*]

Sucre
Production et consommation : milliers de tonnes métriques ; consommation par habitant : kilogrammes [*suite*]

Country or area Pays ou zone	1984	1985	1986	1987	1988	1989	1990	1991	1992	1993
United States Etats—Unis										
Production	5 342	5 415	5 685	6 631	6 429	6 206	5 740	6 477	6 805	7 045
Consumption	7 738	7 290	7 036	7 385	7 420	7 561	7 848	7 887	8 098	8 192
Consumption per cap.(kg)	34	32	29	30	30	30	31	31	31	32
America, South · Amerique du Sud										
Argentina Argentine										
Production	1 545	1 188	1 120	1 063	1 283	1 017	1 351	* 1 560	1 379	1 093
Consumption	1 003	974 [7]	1 093	1 104	895	914	1 070	* 1 140	1 174	* 1 200
Consumption per cap.(kg)	33	31	35	34	27	28	33	35	36	36
Bolivia Bolivie										
Production	198	* 175	* 180	161	162	* 170	* 225	* 230	* 220	* 220
Consumption	195	* 189	* 170	134	* 175	* 170	* 185	* 185	* 190	* 200
Consumption per cap.(kg)	31	29	26	20	25	24	25	24	24	25
Brazil Brésil										
Production	9 259	8 455	7 999	9 266	7 874	* 7 326	* 8 007	9 453	9 925	10 097
Consumption	6 201	6 080	6 589	6 573	6 241	* 7 401	* 6 615	7 276	* 7 349	* 7 575
Consumption per cap.(kg)	47	45	48	47	43	50	44	48	47	48
Chile Chili										
Production	360	351	481	437	443	445	371	360	525	490
Consumption	402	402	440 [8]	467	466	498	508	518	588	598
Consumption per cap.(kg)	36	34	35	37	36	38	39	39	44	44
Colombia Colombie										
Production	1 177	1 367	1 272	1 293	1 364	1 523	1 593	1 633	1 813	1 833
Consumption [9]	983	1 044	1 101	1 208	1 143	1 163	1 195	1 318	1 262	1 158
Consumption per cap.(kg)	33	35	39	42	35	36	36	39	37	33
Ecuador Equateur										
Production	329	* 300	286	341	292	300	334	* 335	387	366
Consumption	319	* 324	353	383	313	327	360	* 375	439	* 361
Consumption per cap.(kg)	38	35	38	40	32	33	35	36	41	33
Guyana Guyana										
Production	256	258	261	234	178	170	134	168	255	255
Consumption	37	31	35	45	38	34	29	26	24	22
Consumption per cap.(kg)	52	40	43	56	47	41	35	32	29	31
Paraguay Paraguay										
Production	* 92	* 80	* 80	* 112	* 112	* 118	89	* 105	* 95	* 110
Consumption	* 78	* 85	* 90	* 100	* 100	* 110	79	* 90	* 91	* 100
Consumption per cap.(kg)	24	22	24	26	25	26	19	21	20	22
Peru Pérou										
Production	605	* 710	585	* 560	571	* 625	* 590	* 580	* 480	* 410
Consumption	620	* 650	733	* 850	* 750	* 725	* 650	* 675	* 700	* 675
Consumption per cap.(kg)	32	33	36	41	35	33	29	31	31	30
Suriname Suriname										
Production *	10	7	10	10	10	10	10	5	5	5
Consumption *	15	16	16	14	15	15	18	20	20	15
Consumption per cap.(kg)	42	43	42	35	37	38	45	49	46	31
Uruguay Uruguay										
Production	* 100	* 90	98	103	63	88	* 84	79	73	* 35
Consumption	* 95	* 95	82	83	78	68	* 75	88	92	* 95
Consumption per cap.(kg)	32	32	27	30	26	22	24	28	30	30
Venezuela Venezuela										
Production	402	495	588	634	521	569	542	567	570	* 565
Consumption	711	705	758	777	853	706	732	731	642	* 720
Consumption per cap.(kg)	42	41	43	43	46	39	37	37	34	35
Asia · Asie										
Afghanistan Afghanistan										
Production	* 3	* 3	0	0	0	0	0	0	0	0
Consumption	100	80	80	80	80	55	39	40	50	50
Consumption per cap.(kg)	6	4	4	4	5	4	2	2	3	2
Armenia Arménie										
Consumption	..	..	..	..	..	..	..	..	* 80	*75
Consumption per cap.(kg)	..	..	..	..	..	..	..	..	2	2
Azerbaijan Azerbaïdjan										
Consumption	..	..	..	..	..	..	..	..	* 225	* 215
Consumption per cap.(kg)	..	..	..	..	..	..	..	..	3	3
Bangladesh Bangladesh										
Production	157	94	* 180	* 200	* 200	* 130	* 190	* 250	* 220	* 220

41

Sugar

Production and consumption: thousand metric tons; consumption per capita: kilograms [cont.]

Sucre

Production et consommation : milliers de tonnes métriques ; consommation par habitant : kilogrammes [suite]

Country or area Pays ou zone	1984	1985	1986	1987	1988	1989	1990	1991	1992	1993
Consumption *	260	330	340	340	300	275	250	275	285	300
Consumption per cap.(kg)	3	3	3	3	3	3	2	2	2	3
Brunei Darussalam Brunéi Darussalam										
Consumption	6	6	6	7	8	8	10	9	8	8
Consumption per cap.(kg)	27	27	30	30	33	32	39	35	28	27
Cambodia Cambodge										
Consumption *	5	5	5	5	5	5	5	5	10	15
Consumption per cap.(kg)	1	1	1	1	1	1	1	1	1	2
China Chine										
Production	* 4 300	* 4 800	* 5 700	* 5 450	* 4 875	* 5 350	* 6 250	6 944	8 864	8 093
Consumption *	5 700	6 350	6 700	7 450	7 600	7 150	7 125	7 515	7 615	7 720
Consumption per cap.(kg)	5	6	6	7	7	6	6	6	6	6
Cyprus Chypre										
Consumption	21	* 20	21	21	17	* 20	* 25	* 27	* 28	* 28
Consumption per cap.(kg)	32	30	30	30	25	29	36	38	39	46
Georgia Géorgie										
Production	..	..	..	..	..	..	..	..		
Consumption	..	..	..	..	..	..	..	..	*15	* 10
Consumption per cap.(kg)	..	..	..	..	..	..	..	..	* 153	* 145
									3	3
Hong Kong Hong-kong										
Consumption *	125	130	135	140	145	150	150	153	155	158
Consumption per cap.(kg)	23	24	24	25	26	26	26	27	27	26
India Inde										
Production	6 635	7 016	7 594	9 215	10 207	* 9 912	12 068	13 113	13 873	* 11 750
Consumption	8 237	8 974	8 694	9 732	10 175	* 10 575	11 075	11 721	12 387	* 12 989
Consumption per cap.(kg)	11	12	12	13	13	13	13	14	14	15
Indonesia Indonésie										
Production	1 759	1 705	2 150	* 2 200	* 2 205	2 171	2 346	2 438	* 2 350	* 2 490
Consumption	1 726	1 794	2 123	* 2 350	* 2 545	* 2 600	* 2 650	2 629	* 2 750	* 2 850
Consumption per cap.(kg)	11	11	13	14	15	15	15	14	14	14
Iran, Islamic Rep. of Iran, Rép. islamique d'										
Production *	* 600	700	600	600	725	600	620	710	875	935
Consumption *	1 250	1 300	1 300	1 300	1 150	1 000	1 200	1 400	1 500	1 500
Consumption per cap.(kg)	29	29	27	26	22	19	22	25	26	26
Iraq Iraq										
Consumption *	600	600	600	600	575	600	450	300	400	385
Consumption per cap.(kg)	40	38	37	37	33	34	25	16	21	19
Israel Israël										
Consumption *	300	250	250	250	250	250	260	270	290	300
Consumption per cap.(kg)	72	59	58	57	56	55	56	55	56	57
Japan Japon										
Production	876	928	953	960	944	998	982	1 005	1 023	861
Consumption	2 747	2 891	2 738	2 690	2 905	2 801	2 833	2 846	2 773	2 678
Consumption per cap.(kg)	23	24	23	22	24	23	23	23	22	22
Jordan Jordanie										
Consumption	* 135	* 135	* 140	* 150	* 160	* 160	* 170	* 175	* 185	171
Consumption per cap.(kg)	40	39	39	40	41	39	40	40	41	36
Kazakhstan Kazakhstan										
Production	..	..	..	..	..	..	..	..		
Consumption	..	..	..	..	..	..	..	..	*105	*100
Consumption per cap.(kg)	..	..	..	..	..	..	..	..	* 525	* 500
									3	3
Korea, Dem. P. R. Corée, R.p. dém. de										
Consumption	120	120	120	120	120	120	120	120	125	125
Consumption per cap.(kg)	6	6	6	6	6	6	6	5	6	5
Korea, Republic of Corée, République de										
Consumption	523	548	643	668	765	839	817	857	860	852
Consumption per cap.(kg)	13	13	15	16	18	20	19	20	18	19
Kuwait Koweït										
Consumption *	65	65	68	65	60	60	60	40	50	45
Consumption per cap.(kg)	36	38	38	35	31	29	28	19	25	23
Kyrgyzstan Kirghizistan										
Production	..	..	..	..	..	..	..	..	* 15	* 10
Consumption	..	..	..	..	..	..	..	..	* 135	* 145
Consumption per cap.(kg)	..	..	..	..	..	..	..	..	3	3
Lao People's Dem. Rep. Rép. dém. pop. lao										
Consumption *	6	6	6	6	6	6	7	8	8	10

41
Sugar
Production and consumption: thousand metric tons; consumption per capita: kilograms [cont.]
Sucre
Production et consommation : milliers de tonnes métriques ; consommation par habitant : kilogrammes [suite]

Country or area Pays ou zone	1984	1985	1986	1987	1988	1989	1990	1991	1992	1993
Consumption per cap.(kg)	1	2	2	2	2	2	2	2	2	2
Lebanon Liban										
Consumption *	70	70	70	95	110	110	110	115	115	110
Consumption per cap.(kg)	27	26	26	35	41	41	40	41	41	38
Macau Macao										
Consumption	3	3	3	3	3	3	3	3	3	3
Consumption per cap.(kg)	9	8	7	10	9	9	9	9	8	8
Malaysia Malaisie										
Production *	75	70	70	90	90	100	110	105	105	105
Consumption	583	* 600	* 615	* 575	* 625	* 650	* 675	* 700	* 720	* 740
Consumption per cap.(kg)	38	38	38	35	37	38	38	39	39	39
Maldives Maldives										
Consumption	5	6	9	9	9	7	9	8	8	8
Consumption per cap.(kg)	29	31	47	45	45	33	41	34	34	33
Mongolia Mongolie										
Consumption	41	42	45	45	50	50	50	55	50	45
Consumption per cap.(kg)	22	22	23	23	25	24	23	24	22	19
Myanmar Myanmar										
Production	* 65	* 59	54	50	27	31	38	35	50	47
Consumption *	65	55	55	42	30	27	34	36	49	50
Consumption per cap.(kg)	2	3	1	1	1	1	1	1	1	1
Nepal Népal										
Production	18	25	15	20	20	15	15	25	20	25
Consumption	35	35	40	35	40	35	40	40	40	41
Consumption per cap.(kg)	2	2	2	2	2	2	2	2	2	2
Pakistan Pakistan										
Production	* 1 355	* 1 450	1 151	* 1 425	1 943	2 052	1 989	2 198	* 2 630	* 2 770
Consumption	* 1 300	* 1 400	* 1 750	* 2 005	1 978	2 089	2 290	2 662	* 2 720	* 2 775
Consumption per cap.(kg)	14	* 15	* 18	19	19	19	20	23	23	23
Philippines Philippines										
Production	2 578	1 665	1 514	1 304	1 495	1 878	1 686	1 847	1 919	2 091
Consumption	1 281	1 340	1 180	1 438	1 225	1 471	1 582	1 512	1 645	1 739
Consumption per cap.(kg)	24	25	21	25	21	25	26	25	26	26
Saudi Arabia Arabie saoudite										
Consumption *	450	400	350	380	400	450	475	475	500	505
Consumption per cap.(kg)	41	41	29	28	29	31	32	31	31	31
Singapore Singapour										
Consumption *	125	130	145	175	200	190	190	195	210	210
Consumption per cap.(kg)	50	51	56	67	76	71	70	71	75	73
Sri Lanka Sri Lanka										
Production	20	* 17	35	* 30	24	* 29	* 30	67	60	69
Consumption *	300	350	315	320	330	345	350	360	360	405
Consumption per cap.(kg)	19	22	20	20	21	21	21	21	23	23
Syrian Arab Republic Rép. arabe syrienne										
Production *	110	50	50	40	40	30	50	75	100	100
Consumption *	330	385	385	385	375	370	395	430	450	460
Consumption per cap.(kg)	33	32	36	35	33	32	33	34	35	34
Tajikistan Tadjikistan										
Consumption	..	..	..	..	..	..	..	..	* 110	* 115
Consumption per cap.(kg)	..	..	..	..	..	..	..	..	2	2
Thailand Thaïlande										
Production	2 550	2 393	2 718	2 532	2 638	4 338	3 542	4 248	5 078	3 825
Consumption	701	721	744	883	886	981	1 105	1 189	1 264	1 368
Consumption per cap.(kg)	14	14	14	16	16	18	20	21	22	23
Turkey Turquie										
Production	1 654	1 398	1 414	1 784	1 414	1 565	* 1 565	2 052	1 790	1 799
Consumption	1 429	1 348	1 483	1 658	1 534	1 641	* 1 775	1 631	1 268	1 389
Consumption per cap.(kg)	30	27	29	32	29	30	32	28	22	23
Turkmenistan Turkménistan										
Consumption	..	..	..	..	..	..	..	..	* 100	* 105
Consumption per cap.(kg)	..	..	..	..	..	..	..	..	3	3
Uzbekistan Ouzbékistan										
Consumption	..	..	..	..	..	..	..	..	* 475	* 490
Consumption per cap.(kg)	..	..	..	..	..	..	..	..	3	2
Viet Nam Viet Nam										
Production *	410	470	440	440	460	465	465	475	495	530

41

Sugar
Production and consumption: thousand metric tons; consumption per capita: kilograms [cont.]
Sucre
Production et consommation : milliers de tonnes métriques ; consommation par habitant : kilogrammes [suite]

Country or area Pays ou zone	1984	1985	1986	1987	1988	1989	1990	1991	1992	1993
Consumption *	450	450	475	500	500	500	520	510	520	570
Consumption per cap.(kg)	4	4	8	8	8	8	8	8	8	8
Yemen Yémen										
Consumption *	300	280	275	285	305	240	250	250	265	275
Consumption per cap.(kg)	70	64	60	57	58	44	47	22	22	22
Europe · Europe										
Albania Albanie										
Production	40	33	*35	*40	*45	25	17	*15	*17	*10
Consumption	50	55	*58	*60	*65	68	70	*68	*65	*85
Consumption per cap.(kg)	17	19	19	20	21	21	22	21	19	25
Austria Autriche										
Production	464	468	307	390	357	390	432	507	436	497
Consumption [18]	*360	348	357	370	375	373	396	425	426	412
Consumption per cap.(kg)	48	46	47	49	49	49	51	54	54	52
Belarus Bélarus										
Production	..	..	..	..	..	..	..	..	117	129
Consumption	..	..	..	..	..	..	..	..	347	425
Consumption per cap.(kg)	..	..	..	..	..	..	..	..	4	4
Belgium [1] Belgique [1]										
Production	910	961	915	915	910	1 312	1 023	1 121	959	...
Bulgaria Bulgarie										
Production *	110	58	40	35	40	76	35	65	35	15
Consumption	455	471	432	482	420	401	289	160	*160	*160
Consumption per cap.(kg)	51	53	48	47	45	32	18	18	18	19
former Czechoslovakia † anc. Tchécoslovaquie †										
Production	*833	*840	*850	*775	*660	*756	717	*790	743	..
Consumption	*800	*800	*800	*750	*750	*800	741	*800	*795	..
Consumption per cap.(kg)	52	52	52	48	48	51	48	51	51	..
Czech Republic République tchèque										
Production	..	..	..	..	..	..	..	..	..	624
Consumption	..	..	..	..	..	..	..	..	..	*485
Consumption per cap.(kg)	..	..	..	..	..	..	..	..	..	5
Denmark [10] Danemark [10]										
Production	595	574	547	422	549	530	572	527	446	372
Estonia Estonie										
Production	..	..	..	..	..	..	..	...	0	*5
Consumption	..	..	..	..	..	..	..	...	*72	*70
Consumption per cap.(kg)	..	..	..	..	..	..	..	...	*5	*5
Finland Finlande										
Production	129	103	133	70	147	173	169	163	153	154[12]
Consumption	203[11]	202[11]	209[11]	207[11]	223[11]	217	189	213	227	242
Consumption per cap.(kg)	39	41	43	42	45	44	38	42	45	48
France [10][13] France [10][13]										
Production	3 956	3 953	3 734	3 973	4 372	4 198	4 595	4 422	4 738	*4 783
Germany † Allemagne †										
Production	..	..	..	..	..	..	..	4 224	4 373	*4 861
F. R. Germany [10] R.f. Allemagne [10]										
Production	3 151	3 454	3 479	2 963	3 004	3 337	3 396	..	..	..
former German D. R. anc. R.d. allemande										
Production	750	798	805	750	658	610	..	..	..	..
Consumption	733	757	746	740	787	782	..	..	..	..
Consumption per cap.(kg)	44	46	45	45	47	47	..	..	..	..
Gibraltar Gibraltar										
Consumption	1	1	1	1	1	2	3	3	3	3
Consumption per cap.(kg)	25	33	33	33	33	67	75	85	78	78
Greece [10] Grèce [10]										
Production	237	345	312	197	235	421	312	*297	*386	*335
Hungary Hongrie										
Production	493	579	510	538	513	630	580	*700	391	248[14]
Consumption	485	518	513	555	488	572	657	*575	608	382
Consumption per cap.(kg)	45	50	49	52	46	54	61	56	59	37
Iceland Islande										
Consumption	11	11	12	*13	*14	*15	*14	*14	*15	*15
Consumption per cap.(kg)	46	48	52	52	56	60	54	54	58	58

41

Sugar
Production and consumption: thousand metric tons; consumption per capita: kilograms [*cont.*]
Sucre
Production et consommation : milliers de tonnes métriques ; consommation par habitant : kilogrammes [*suite*]

Country or area Pays ou zone	1984	1985	1986	1987	1988	1989	1990	1991	1992	1993
Ireland [10] Irlande [10]										
Production	239	189	202	242	212	233	245	*232	*253	*185
Italy [10] Italie [10]										
Production	1 385	1 352	1 868	1 867	1 607	1 879	1 586	1 641	2 032	*1 537
Latvia Lettonie										
Production	..	..	..	..	..	..	..	..	*30	*35
Consumption	..	..	..	..	..	..	..	..	*151	*145
Consumption per cap.(kg)	..	..	..	..	..	..	..	..	*6	*6
Lithuania Lituanie										
Production	..	..	..	..	..	..	..	..	85	*70
Consumption	..	..	..	..	..	..	..	..	116	*117
Consumption per cap.(kg)	..	..	..	..	..	..	..	..	3	*3
Malta Malte										
Consumption	15[15]	17	15	15	15	17	18	13	17	17
Consumption per cap.(kg)	41	52	44	44	43	50	51	37	46	47
Netherlands [10] Pays—Bas [10]										
Production	1 000	1 000	1 325	1 064	1 075	1 240	1 304	*1 137	*1 230	*1 140
Norway Norvège										
Consumption	167	175	170	173	158	168	171	177	178	177
Consumption per cap.(kg)	40	42	41	41	38	40	40	42	41	41
Poland Pologne										
Production	1 933	1 841	*1 881	*1 820	1 823	1 864	2 212	1 618	1 566	2 201
Consumption	2 012	1 690	*1 646	*1 800	1 903	1 634	1 624	1 660	1 618	1 691
Consumption per cap.(kg)	54	45	44	48	50	43	43	43	42	44
Portugal Portugal										
Production [10]	8	9	4	2	1	1	1	*2	*2	4[20]
Consumption *	290	330	...	...	...	...	...	...	...	...
Consumption per cap.(kg)	29	32	...	...	...	...	...	...	...	...
Republic of Moldova République de Moldova										
Production	..	..	..	..	..	..	..	..	*180	*220
Consumption	..	..	..	..	..	..	..	..	*203	*205
Consumption per cap.(kg)	..	..	..	..	..	..	..	..	*5	*5
Russian Federation Fédération de Russie										
Production	..	..	..	..	..	..	..	..	2 437	*2 717
Consumption	..	..	..	..	..	..	..	..	*6 145	*5 850
Consumption per cap.(kg)	..	..	..	..	..	..	..	..	*4	*4
Romania Roumanie										
Production	*600	*540	451	438	363	473	408	344	280	141
Consumption *	700	650	525	495	475	540	555	565	615	475
Consumption per cap.(kg)	31	29	23	22	21	23	24	24	24	21
Slovakia Slovaquie										
Production	..	..	..	..	..	..	..	..	..	*150
Consumption	..	..	..	..	..	..	..	..	..	*190
Consumption per cap.(kg)	..	..	..	..	..	..	..	..	..	4
Spain Espagne										
Production	1 221	1 090	# 1 111[10]	1 093[10]	1 306[10]	953[10]	994[10]	949[10]	1 037[10]	*1 181[10]
Consumption	1 146	*960	...	...	...	...	...	...	...	...
Consumption per cap.(kg)	30	25	...	...	...	...	...	...	...	...
Sweden Suède										
Production [16]	399	350	391	274	385	424	445	266	333	413
Consumption	382	390	387	372	383	377	379	380	382	388
Consumption per cap.(kg)	46	46	47	44	45	44	44	44	44	44
Switzerland Suisse										
Production	131	139	129	123	150	152	160	136	137	*150
Consumption	287[17]	287[17]	289[17]	287[17]	282[17]	291	305	307	311	*315
Consumption per cap.(kg)	43	43	46	43	42	43	46	46	46	45
Ukraine Ukraine										
Production	..	..	..	..	..	..	..	...	3 824	4 160
Consumption	..	..	..	..	..	..	..	...	*2 881	*2 575
Consumption per cap.(kg)	..	..	..	..	..	..	..	...	*6	*5
United Kingdom [10] Royaume—Uni [10]										
Production [19]	1 428	1 315	1 433	1 333	1 417	1 377	1 349	1 326	1 604	1 558

41
Sugar
Production and consumption: thousand metric tons; consumption per capita: kilograms [*cont.*]
Sucre
Production et consommation : milliers de tonnes métriques ; consommation par habitant : kilogrammes [*suite*]

Country or area Pays ou zone	1984	1985	1986	1987	1988	1989	1990	1991	1992	1993
Yugoslavia, SFR † Yougoslavie, Rfs †										
Production	* 930	933	801	* 920	691	930	* 945	* 900	* 500	* 475
Consumption *	900	950	950	950	900	907	950	810	725	585
Consumption per cap.(kg)	39	41	41	41	38	38	40	34	30	29
Oceania · Océanie										
Australia Australie										
Production	3 627	3 439	3 439	3 511	3 759	3 887	3 612	3 195	4 363	4 488
Consumption	750	764	818	817	844	882	864	835	829	909
Consumption per cap.(kg)	48	49	51	50	51	53	51	48	47	51
Fiji Fidji										
Production	484	367	508	426	377	466	378	456	451	451
Consumption	36	36	35	36	38	37	40	44	45	46
Consumption per cap.(kg)	52	51	48	49	53	51	54	59	60	60
New Zealand Nouvelle−Zélande										
Consumption	165	* 170	* 165	* 170	* 170	* 165	* 170	* 165	* 175	* 185
Consumption per cap.(kg)	51	52	51	52	52	50	51	49	51	54
Papua New Guinea Papouasie−Nvl−Guinée										
Production	34	30	10	24	51	30	28	* 35	* 30	* 30
Consumption	28	27	30	30	29	29	27	* 27	* 27	* 30
Consumption per cap.(kg)	9	8	9	7	8	8	7	7	7	8
Samoa Samoa										
Production	1	3	2	2	2	2	2	2	2	2
Consumption	3	3	3	3	3	3	3	3	3	3
Consumption per cap.(kg)	19	20	19	19	19	19	19	19	19	19
former USSR · ancienne URSS										
former USSR † ancienne URSS †										
Production	8 587	8 261	8 696	9 565	8 913	9 533	9 159	6 898	..	..
Consumption *	13 200	12 610	14 050	14 950	14 350	13 150	13 400	11 908	..	..
Consumption per cap.(kg)	48	45	50	53	50	46	46	41	..	..

Source:
International Sugar Organization (London).

Source:
L'organisation internationale du sucre (Londres).

† For detailed descriptions of data pertaining to former Czechoslovakia, Germany, SFR Yugoslavia and former USSR, see Annex I − Country or area nomenclature, regional and other groupings.

† Pour les descriptions en détails des données relatives à l'ancienne Tchécoslovaquie, l'Allemagne, la Rfs Yougoslavie et l'ancienne URSS, voir l'Annexe I − Nomenclature des pays ou zones, groupements régionaux et autres groupments.

1 Official figures communicated directly to the Statistical Division of the United Nations.
2 Excluding high−test molasses to a reported equivalent of raw sugar: 1984 − 6,278 t; 1985 − 74,219 t.
3 3,268 t lost in fire.
4 Including sugar subsequently exported as blends and mixtures.
5 Including non−human consumption: 1984 − 31,119 t; 1985 − 77,946 t; 1986 − 8,831 t; 1987 − 64,646 t; 1988 − 46,834 t; 1989 − 167,173 t; 1990 − 129,608 t; 1991 − 92,446 t.
6 Including 149,293 t destroyed in hurricane kate.
7 Excluding 100,348 t adjustment for loss.
8 Excluding consumption in Free Zone.
9 Including non−human consumption: 1986 − 98,608 t; 1987 − 147,262 t; 1988 − 122,058 t; 1989 − 52,230 t; 1991 − 13,541 t; 1992 − 8,938 t; 1993 − 8,344 t.
10 Source: Food and Agriculture Organization of the United Nations.
11 Including sugar for non−human consumption: 1984 − 16,775 t; 185 − 12,812 t; 1986 − 26,552 t; 1987 − 31,312 t; 1988 − 20,912 t.
12 Of which 1,041 t of sugar produced from imported Estonian beet.
13 Crop year ending 30 September of year stated.
14 In addition, 140,412 t of sugar manufactured from Thick (beet) juice were imported from Poland.

1 Données officielles fournies directement de la Division de statistique des Nations Unies.
2 Á l'exclusion des mélasses interverties, en équivalent de sucre brut : 1984 : 6 278 t; 1985 : 74 219 t.
3 Dont 3 268 tonnes détruites par le feu.
4 Y compris le sucre réexporté sous forme de mélanges.
5 Dont consommation non humaine : 1984 − 31 119 t; 1985 − 77 946 t; 1986 − 8 831 t; 1987 − 64 646 t; 1988 − 46 834 t; 1989 − 167 173 t; 1990 − 129 608 t; 1991 − 92 446 t.
6 Dont 149 293 tonnes détruites par l'ouragan Kate.
7 Non compris un ajustement de 100 348 t pour pertes.
8 Non compris la consommation dans la Zone libre.
9 Dont consommation non humaine : 1986 − 98 608 t; 1987 − 147 262 t; 1988 − 122 058 t; 1989 − 52 230 t; 1991 − 13 541 t; 1992 − 8 938 t; 1993 − 8 344 t.
10 Source: Organisation des Nations Unies pour l'alimentation et l'agriculture.
11 Y compris le sucre pour la consommation non humaine : 1984 − 16 775 t; 1985 − 12 812 t; 1986 − 26 552 t; 1987 − 31 312 t; 1988 − 20 912 t.
12 Dont 1 041 tonnes de sucre à l'aide de betteraves importées d'Estonie.

41
Sugar
Production and consumption: thousand metric tons; consumption per capita: kilograms [*cont.*]
Sucre
Production et consommation : milliers de tonnes métriques ; consommation par habitant : kilogrammes [*suite*]

15 Calculated.
16 Including sales from government stocks: 1984 − 3,602 t; 1985 − 3,181 t; 1986 − 3,775 t; 1988 − 4,324 t; 1989 − 2,771 t.
17 Calculated by Switzerland.
18 Including non−human consumption: 1984 − 9,002t; 1986 − 7,862t; 1987 − 9,989t; 1988 − 12,384t; 1989 − 8,288t; 1990 − 5,387t; 1991 − 11,799t.
19 Sugar produced from home−grown beet.
20 FAO estimate.

13 Production par campagne se terminant au 30 septembre de l'année indiquée.
14 En outre, 140 412 t de sucre fabriqué à partir du jus de betteraves épais importé de Pologne.
15 Données obtenues par le calcul.
16 Dont ventes par prélèvement dans les réserves publiques : 1984 − 3 602 t; 1985 − 3 181 t; 1986 − 3 775 t; 1988 − 4 324 t; 1989 − 2 771 t.
17 Données calculées par la Suisse.
18 Y compris la consommation non humaine : 1984 − 9 002 t; 1986 − 7 862 t; 1987 − 9 989 t; 1988 − 12 384 t; 1989 − 8 288 t; 1990 − 5 387 t; 1991 − 11 799 t.
19 Sucre provenant de betteraves récoltées localement.
20 Estimation de la FAO.

42

Meat

Viande

Production: thousand metric tons
Production : milliers de tonnes métriques

Country or area Pays ou zone	1984	1985	1986	1987	1988	1989	1990	1991	1992	1993
World *Monde*	114 653	118 097	121 031	123 887	128 265	129 848	133 752	135 631	136 428	136 501
Beaf and veal Boeuf et veau	48 904	49 848	51 093	51 759	52 077	52 412	53 948	54 521	53 576	52 739
Pork Porc	57 676	59 996	61 587	63 502	67 193	68 084	70 064	71 155	72 924	73 891
Mutton and lamb Mouton et agneau	8 073	8 253	8 352	8 625	8 996	9 352	9 740	9 955	9 927	9 872
Africa *Afrique*	5 017	5 112	5 056	5 162	5 246	5 438	5 599	5 776	5 901	5 885
Beaf and veal Boeuf et veau	3 342	3 395	3 313	3 285	3 319	3 374	3 482	3 614	3 678	3 662
Pork Porc	406	428	464	502	518	557	598	629	683	689
Mutton and lamb Mouton et agneau	1 269	1 289	1 279	1 375	1 410	1 507	1 519	1 533	1 540	1 534
Algeria *Algérie*	136	153	155	189	200	222	232	247	249	246
Beaf and veal Boeuf et veau	48 [1]	67	66	74	* 81	* 85	* 89	* 94	* 95	97 [1]
Mutton and lamb [1] Mouton et agneau [1]	89	86	88	115	119	* 137	* 142	* 152	154	149
Angola *Angola*	77	78	82	82	83	81	81	81	81	81
Beaf and veal [1] Boeuf et veau [1]	53	54	57	57	57	54	54	54	54	54
Pork [1] Porc [1]	19	20	21	21	21	22	22	22	22	22
Mutton and lamb [1] Mouton et agneau [1]	4	4	4	4	5	5	5	5	5	5
Benin *Bénin*	24	24	23	23	24	24	26	27	28	28
Beaf and veal [1] Boeuf et veau [1]	13	13	13	13	13	14	15	16	16	16
Pork [1] Porc [1]	4	4	4	5	5	5	5	6	6	6
Mutton and lamb [1] Mouton et agneau [1]	6	7	5	5	5	6	6	6	6	6
Botswana *Botswana*	45	50	49	41	40	43	49	51	56	46
Beaf and veal [1] Boeuf et veau [1]	41	46	44	36	34	37	42	44	49	39
Pork [1] Porc [1]	0	0	0	0	0	1	1	1	1	1
Mutton and lamb [1] Mouton et agneau [1]	4	4	4	5	5	6	7	7	7	7
Burkina Faso *Burkina Faso*	51	58	61	64	66	69	73	75	77	79
Beaf and veal [1] Boeuf et veau [1]	27	33	34	36	37	37	37	38	40	41
Pork [1] Porc [1]	5	5	5	5	6	6	6	6	6	6
Mutton and lamb [1] Mouton et agneau [1]	19	21	22	23	24	26	29	30	31	32
Burundi *Burundi*	18	16	19	19	22	20	22	21	22	22
Beaf and veal [1] Boeuf et veau [1]	10	9	11	11	11	11	11	11	11	11

42

Meat
Production: thousand metric tons [cont.]
Viande
Production : milliers de tonnes métriques [suite]

Country or area Pays ou zone	1984	1985	1986	1987	1988	1989	1990	1991	1992	1993
Pork[1]										
Porc[1]	4	3	4	5	7	5	6	5	6	6
Mutton and lamb[1]										
Mouton et agneau[1]	4	4	4	4	4	4	4	4	4	4
Cameroon **Cameroun**	**82**	**84**	**90**	**96**	**106**	**111**	**116**	**117**	**118**	**122**
Beaf and veal										
Boeuf et veau	60	56	53	62	68	71[1]	72[1]	73[1]	73[1]	74[1]
Pork[1]										
Porc[1]	10	14	17	14	15	16	16	17	17	17
Mutton and lamb[1]										
Mouton et agneau[1]	13	15	19	20	22	25	27	28	28	30
Cape Verde **Cap-Vert**	**2**	**2**	**3**	**3**	**3**	**4**	**4**	**4**	**4**	**4**
Pork[1]										
Porc[1]	2	2	2	2	3	3	3	3	3	3
Mutton and lamb[1]										
Mouton et agneau[1]	0	0	0	0	0	0	0	0	0	1
Central African Rep. **Rép. centrafricaine**	**39**	**40**	**41**	**46**	**49**	**51**	**53**	**55**	**56**	**59**
Beaf and veal										
Boeuf et veau	29[1]	30[1]	31[1]	* 35	* 37	* 39	* 40	42[1]	43[1]	45[1]
Pork										
Porc	6[1]	6[1]	6[1]	* 7	* 7	* 7	* 8	* 8	8[1]	8[1]
Mutton and lamb[1]										
Mouton et agneau[1]	4	4	4	* 5	* 5	* 5	* 5	5	5	6
Chad **Tchad**	**41**	**44**	**70**	**69**	**81**	**75**	**80**	**84**	**86**	**88**
Beaf and veal[1]										
Boeuf et veau[1]	26	29	52	51	62	57	63	65	67	68
Mutton and lamb[1]										
Mouton et agneau[1]	14	16	18	18	18	18	17	19	19	20
Comoros **Comores**	**1**	**1**	**1**	**1**	**1**	**1**	**1**	**2**	**2**	**2**
Beaf and veal[1]										
Boeuf et veau[1]	1	1	1	1	1	1	1	1	1	1
Congo **Congo**	**5**	**4**	**5**	**5**	**5**	**4**	**5**	**5**	**5**	**5**
Beaf and veal[1]										
Boeuf et veau[1]	2	2	2	2	2	1	2	2	2	2
Pork[1]										
Porc[1]	1	2	2	2	2	2	2	2	2	3
Mutton and lamb[1]										
Mouton et agneau[1]	1	1	1	1	1	1	1	1	1	1
Côte d'Ivoire **Côte d'Ivoire**	**65**	**66**	**63**	**59**	**59**	**57**	**54**	**56**	**58**	**59**
Beaf and veal[1]										
Boeuf et veau[1]	43	44	41	37	37	34	31	33	34	34
Pork[1]										
Porc[1]	13	13	13	13	14	14	14	15	15	15
Mutton and lamb[1]										
Mouton et agneau[1]	9	9	9	8	8	9	9	8	9	9
Djibouti **Djibouti**	**5**	**5**	**6**	**6**	**6**	**6**	**6**	**7**	**7**	**7**
Beaf and veal[1]										
Boeuf et veau[1]	1	1	2	2	2	2	2	3	3	3
Mutton and lamb[1]										
Mouton et agneau[1]	4	4	4	4	4	4	4	4	4	5

42

Meat
Production: thousand metric tons [*cont.*]
Viande
Production : milliers de tonnes métriques [*suite*]

Country or area Pays ou zone	1984	1985	1986	1987	1988	1989	1990	1991	1992	1993
Egypt										
Egypte	**312**	**317**	**314**	**339**	**368**	**376**	**375**	**404**	**419**	**428**
Beaf and veal[1]										
Boeuf et veau[1]	235	237	230	252	277	287	284	313	321	328
Pork										
Porc	3[1]	3	3	2[1]	2[1]	2[1]	2[1]	3[1]	3[1]	3[1]
Mutton and lamb										
Mouton et agneau	75[1]	77	81	85	89	87[1]	89[1]	88[1]	95[1]	97[1]
Eritrea										
Erythrée	..	..	..	..	..	..	..	..	..	**25**
Beaf and veal[1]										
Boeuf et veau[1]	..	..	..	..	..	..	..	..	..	15
Mutton and lamb[1]										
Mouton et agneau[1]	..	..	..	..	..	..	..	..	..	10
Ethiopia[2]										
Ethiopie[2]	**349**	**366**	**458**	**353**	**353**	**387**	**394**	**395**	**394**	**370**
Beaf and veal[2]										
Boeuf et veau[2]	198[1]	216[1]	310[1]	* 206	* 206	* 237	* 245	* 245	244[1]	230[1]
Pork[1 2]										
Porc[1 2]	1	1	1	1	1	1	1	1	1	1
Mutton and lamb										
Mouton et agneau	150[1]	149[1]	147[1]	* 146[1]	* 146[1]	* 149[1]	* 148[1]	149[1]	149[1]	139
Gabon										
Gabon	**3**	**3**	**3**	**3**	**4**	**4**	**4**	**4**	**4**	**4**
Beaf and veal[1]										
Boeuf et veau[1]	0	0	1	1	1	1	1	1	1	1
Pork[1]										
Porc[1]	2	2	2	2	2	2	2	2	2	2
Mutton and lamb[1]										
Mouton et agneau[1]	1	1	1	1	1	1	1	1	1	1
Gambia										
Gambie	**5**	**6**	**6**	**7**	**7**	**7**	**7**	**7**	**7**	**7**
Beaf and veal[1]										
Boeuf et veau[1]	4	4	5	5	6	6	6	6	6	6
Mutton and lamb[1]										
Mouton et agneau[1]	1	1	1	1	1	1	1	1	1	1
Ghana										
Ghana	**37**	**38**	**39**	**39**	**41**	**44**	**41**	**42**	**40**	**42**
Beaf and veal[1]										
Boeuf et veau	19	20	20	20	20	20	20	21	20	21
Pork[1]										
Porc[1]	9	9	11	9	11	12	11	10	9	10
Mutton and lamb[1]										
Mouton et agneau[1]	9	9	9	10	10	12	11	11	11	11
Guinea										
Guinée	**16**	**16**	**16**	**16**	**16**	**15**	**16**	**16**	**17**	**18**
Beaf and veal[1]										
Boeuf et veau[1]	13	13	13	13	13	13	13	13	14	15
Pork[1]										
Porc[1]	1	1	1	1	1	1	0	1	1	1
Mutton and lamb[1]										
Mouton et agneau[1]	2	2	2	2	2	2	2	3	3	3
Guinea-Bissau										
Guinée-Bissau	**12**	**12**	**12**	**13**	**13**	**13**	**13**	**14**	**14**	**15**
Beaf and veal[1]										
Boeuf et veau[1]	3	3	3	3	3	3	3	3	4	4
Pork[1]										
Porc[1]	8	9	9	9	9	9	9	9	9	10

42

Meat
Production: thousand metric tons [*cont.*]
Viande
Production : milliers de tonnes métriques [*suite*]

Country or area Pays ou zone	1984	1985	1986	1987	1988	1989	1990	1991	1992	1993
Mutton and lamb[1] Mouton et agneau[1]	1	1	1	1	1	1	1	1	1	1
Kenya **Kenya**	**245**	**246**	**250**	**270**	**294**	**286**	**312**	**307**	**289**	**266**
Beaf and veal Boeuf et veau	198[1]	192[1]	199	219	238	228	250[1]	245[1]	230[1]	210[1]
Pork[1] Porc[1]	4	5	5	5	5	5	5	5	5	5
Mutton and lamb[1] Mouton et agneau[1]	42	49	46	46	50	54	56	56	53	51
Lesotho **Lesotho**	**21**	**21**	**22**	**22**	**22**	**23**	**23**	**24**	**23**	**24**
Beaf and veal[1] Boeuf et veau[1]	12	12	12	12	13	13	13	14	13	13
Pork[1] Porc[1]	2	3	3	3	3	3	3	3	3	3
Mutton and lamb[1] Mouton et agneau[1]	6	6	7	7	7	7	7	8	7	7
Liberia **Libéria**	**7**	**7**	**7**	**7**	**7**	**7**	**6**	**6**	**6**	**6**
Beaf and veal[1] Boeuf et veau[1]	2	2	1	1	1	1	1	1	1	1
Pork[1] Porc[1]	4	4	4	4	4	4	4	4	4	4
Mutton and lamb[1] Mouton et agneau[1]	1	1	1	1	1	1	1	1	1	1
Libyan Arab Jamah. **Jamah. arabe libyenne**	**96**	**78**	**74**	**66**	**73**	**58**	**48**	**56**	**52**	**55**
Beaf and veal[1] Boeuf et veau[1]	34	31	32	25	28	22	19	25	20	22
Mutton and lamb[1] Mouton et agneau[1]	62	47	42	42	45	36	29	31	32	33
Madagascar **Madagascar**	**182**	**179**	**181**	**188**	**192**	**194**	**196**	**199**	**200**	**201**
Beaf and veal[1] Boeuf et veau[1]	136	134	134	138	141	142	143	143	143	143
Pork[1] Porc[1]	37	36	38	41	42	44	45	47	48	48
Mutton and lamb[1] Mouton et agneau[1]	9	9	9	9	9	9	9	9	9	9
Malawi **Malawi**	**26**	**21**	**31**	**32**	**30**	**29**	**30**	**30**	**31**	**31**
Beaf and veal[1] Boeuf et veau[1]	15	10	16	16	16	16	17	17	17	18
Pork[1] Porc[1]	7	7	11	13	10	10	9	9	10	10
Mutton and lamb[1] Mouton et agneau[1]	3	4	3	3	3	3	4	4	4	4
Mali **Mali**	**92**	**87**	**96**	**96**	**104**	**112**	**117**	**122**	**128**	**134**
Beaf and veal[1] Boeuf et veau[1]	57	54	62	61	65	69	72	74	78	81
Pork[1] Porc[1]	2	2	2	2	2	2	2	2	2	3
Mutton and lamb[1] Mouton et agneau[1]	34	31	33	33	37	41	44	46	48	50
Mauritania **Mauritanie**	**33**	**34**	**33**	**33**	**34**	**36**	**38**	**41**	**40**	**39**

42

Meat
Production: thousand metric tons [*cont.*]
Viande
Production : milliers de tonnes métriques [*suite*]

Country or area Pays ou zone	1984	1985	1986	1987	1988	1989	1990	1991	1992	1993
Beaf and veal Boeuf et veau	* 16	* 17	* 15	* 15	* 16	* 16	* 17	* 20	20[1]	20[1]
Mutton and lamb[1] Mouton et agneau[1]	17	17	18	18	18	20	21	21	20	20
Mauritius Maurice	**2**	**2**	**2**	**2**	**3**	**3**	**3**	**3**	**4**	**4**
Beaf and veal Boeuf et veau	1	1	1	1	2	2	2	2	2	3[1]
Pork Porc	1	1	1	1	1	1	1[1]	1	1	1[1]
Morocco Maroc	**177**	**152**	**161**	**225**	**198**	**266**	**258**	**270**	**267**	**241**
Beaf and veal Boeuf et veau	* 115	85[1]	94[1]	140[1]	122[1]	150	145	165	160	154
Pork Porc	1	0	0	1	1	1	1	1	1[1]	1[1]
Mutton and lamb Mouton et agneau	61	67	67[1]	84	76[1]	115[1]	112[1]	104[1]	107[1]	86
Mozambique Mozambique	**49**	**50**	**51**	**53**	**54**	**55**	**56**	**62**	**52**	**52**
Beaf and veal[1] Boeuf et veau[1]	36	37	38	38	39	40	41	47	36	36
Pork[1] Porc[1]	10	10	11	11	12	12	12	12	12	12
Mutton and lamb[1] Mouton et agneau[1]	3	3	3	3	3	3	3	3	3	3
Namibia Namibie	**48**	**49**	**51**	**53**	**55**	**60**	**57**	**65**	**67**	**69**
Beaf and veal[1] Boeuf et veau[1]	34	35	35	36	38	41	39	45	46	48
Pork[1] Porc[1]	2	2	2	3	3	3	3	2	2	2
Mutton and lamb[1] Mouton et agneau[1]	12	12	14	14	15	16	16	18	19	19
Niger Niger	**52**	**37**	**53**	**60**	**64**	**67**	**65**	**67**	**68**	**71**
Beaf and veal Boeuf et veau	24[1]	17[1]	24[1]	30[1]	29[1]	30	32[1]	34[1]	35[1]	36[1]
Pork[1] Porc[1]	1	1	1	1	1	1	1	1	1	1
Mutton and lamb[1] Mouton et agneau[1]	27	19	28	29	34	36	31	32	32	33
Nigeria Nigéria	**592**	**621**	**495**	**500**	**483**	**492**	**497**	**533**	**583**	**599**
Beaf and veal Boeuf et veau	426	438	283	267	236	218	204	205	210	219[1]
Pork[1] Porc[1]	48	59	75	90	97	112	128	159	199	202
Mutton and lamb[1] Mouton et agneau[1]	117	124	136	143	150	162	165	168	174	178
Réunion Réunion	**7**	**7**	**7**	**8**	**8**	**9**	**9**	**9**	**9**	**9**
Beaf and veal Boeuf et veau	1	1	1	1	1	1	1	1[1]	1[1]	1[1]
Pork Porc	6	5	6	6	7	7	8	8[1]	8[1]	8[1]
Rwanda Rwanda	**19**	**21**	**20**	**19**	**21**	**21**	**21**	**22**	**22**	**22**

42

Meat
Production: thousand metric tons [*cont.*]
Viande
Production : milliers de tonnes métriques [*suite*]

Country or area Pays ou zone	1984	1985	1986	1987	1988	1989	1990	1991	1992	1993
Beaf and veal[1] Boeuf et veau[1]	14	15	14	12	14	14	14	14	14	14
Pork[1] Porc[1]	2	2	2	2	3	3	3	3	3	3
Mutton and lamb[1] Mouton et agneau[1]	4	4	4	4	4	5	5	5	5	5
Senegal **Sénégal**	**53**	**58**	**66**	**66**	**65**	**68**	**72**	**76**	**78**	**80**
Beaf and veal[1] Boeuf et veau[1]	38	38	40	39	40	41	43	44	44	45
Pork[1] Porc[1]	4	5	7	6	6	7	7	7	7	7
Mutton and lamb[1] Mouton et agneau[1]	11	15	19	20	19	20	23	25	27	28
Seychelles **Seychelles**	**1**	**1**	**1**	**1**	**1**	**1**	**1**	**1**	**1**	**1**
Pork[1] Porc[1]	1	1	1	1	1	1	1	1	1	1
Sierra Leone **Sierra Leone**	**8**	**9**	**8**	**8**	**8**	**9**	**9**	**9**	**9**	**9**
Beaf and veal[1] Boeuf et veau[1]	5	5	5	5	5	5	5	5	5	5
Pork[1] Porc[1]	2	2	2	2	2	2	2	2	2	2
Mutton and lamb[1] Mouton et agneau[1]	1	1	1	1	1	1	1	1	1	1
Somalia **Somalie**	**113**	**117**	**105**	**130**	**137**	**143**	**130**	**95**	**70**	**78**
Beaf and veal[1] Boeuf et veau[1]	40	42	35	44	50	53	44	24	13	20
Mutton and lamb[1] Mouton et agneau[1]	73	75	69	86	87	90	86	71	57	58
South Africa **Afrique du Sud**	**902**	**924**	**878**	**910**	**901**	**890**	**959**	**991**	**1 039**	**1 026**
Beaf and veal Boeuf et veau	620	* 638	* 616	* 628	615[1]	* 603	* 661	* 700	* 745	* 729
Pork Porc	* 118	120[1]	120[1]	122[1]	123[1]	* 121	* 130	* 123	* 129	* 130
Mutton and lamb[1] Mouton et agneau[1]	164	166	142	160	163	167	167	168	165	167
Sudan **Soudan**	**362**	**371**	**337**	**301**	**304**	**312**	**322**	**338**	**359**	**369**
Beaf and veal Boeuf et veau	261	267	248[1]	210[1]	205	211	218	231	250[1]	258[1]
Mutton and lamb Mouton et agneau	101	104	89	92	99	101	104	107	109[1]	111[1]
Swaziland **Swaziland**	**16**	**18**	**19**	**19**	**15**	**16**	**16**	**18**	**17**	**15**
Beaf and veal[1] Boeuf et veau[1]	13	14	15	15	12	12	12	15	13	12
Pork[1] Porc[1]	1	1	1	1	1	1	1	1	1	1
Mutton and lamb[1] Mouton et agneau[1]	3	3	3	3	3	3	3	3	3	3
Togo **Togo**	**11**	**13**	**13**	**14**	**16**	**17**	**22**	**22**	**24**	**24**
Beaf and veal[1] Boeuf et veau[1]	5	5	5	5	5	5	5	5	5	5

42

Meat
Production: thousand metric tons [*cont.*]
Viande
Production : milliers de tonnes métriques [*suite*]

Country or area Pays ou zone	1984	1985	1986	1987	1988	1989	1990	1991	1992	1993
Pork[1] Porc[1]	3	4	3	3	5	6	10	10	11	12
Mutton and lamb[1] Mouton et agneau[1]	3	4	5	5	6	6	7	7	7	7
Tunisia Tunisie	**72**	**80**	**81**	**83**	**73**	**73**	**74**	**77**	**78**	**80**
Beaf and veal Boeuf et veau	34	36	39	38	34	34	34	36	38	38
Mutton and lamb Mouton et agneau	39	44	42	45	39	39	40	41	41	41
Uganda Ouganda	**119**	**123**	**95**	**100**	**101**	**130**	**146**	**149**	**153**	**157**
Beaf and veal Boeuf et veau	88	90	60	59	64[1]	73[1]	81[1]	83[1]	84[1]	86[1]
Pork Porc	* 10	* 11	13[1]	25	23[1]	39[1]	45[1]	46[1]	48[1]	49[1]
Mutton and lamb[1] Mouton et agneau[1]	* 21	* 22	22	16	14	18	20	21	22	23
United Rep.Tanzania Rép. Unie de Tanzanie	**189**	**200**	**208**	**209**	**224**	**233**	**235**	**238**	**241**	**244**
Beaf and veal[1] Boeuf et veau[1]	154	163	170	171	186	194	195	197	199	200
Pork[1] Porc[1]	8	8	8	8	9	9	9	9	9	9
Mutton and lamb[1] Mouton et agneau[1]	27	29	29	30	30	31	31	32	33	34
Zaire Zaïre	**60**	**62**	**65**	**68**	**70**	**74**	**77**	**81**	**81**	**82**
Beaf and veal Boeuf et veau	22	23	24	25	26	26	27[1]	28[1]	29[1]	30[1]
Pork[1] Porc[1]	28	29	31	33	34	36	38	40	40	40
Mutton and lamb[1] Mouton et agneau[1]	10	10	10	10	11	11	12	12	12	12
Zambia Zambie	**35**	**39**	**40**	**42**	**43**	**46**	**48**	**49**	**52**	**53**
Beaf and veal[1] Boeuf et veau[1]	28	32	32	33	34	34	36	37	41	41
Pork[1] Porc[1]	5	6	6	6	7	9	9	10	9	9
Mutton and lamb[1] Mouton et agneau[1]	1	2	2	2	2	2	2	2	2	2
Zimbabwe Zimbabwe	**103**	**99**	**90**	**103**	**99**	**93**	**100**	**101**	**111**	**91**
Beaf and veal[1] Boeuf et veau[1]	88	84	73	84	79	72	78	80	90	74
Pork[1] Porc[1]	9	8	8	10	11	11	11	11	11	8
Mutton and lamb[1] Mouton et agneau[1]	6	6	8	9	9	9	10	10	10	9
America, North Amérique du Nord	**23 091**	**23 112**	**23 047**	**22 715**	**23 311**	**22 929**	**22 459**	**22 905**	**23 645**	**23 648**
Beaf and veal Boeuf et veau	13 415	13 544	14 136	13 695	13 683	13 367	13 119	13 205	13 315	13 380
Pork Porc	9 425	9 321	8 671	8 795	9 395	9 321	9 092	9 451	10 086	10 024

42

Meat
Production: thousand metric tons [*cont.*]
Viande
Production : milliers de tonnes métriques [*suite*]

Country or area Pays ou zone	1984	1985	1986	1987	1988	1989	1990	1991	1992	1993
Mutton and lamb **Mouton et agneau**	**250**	**247**	**241**	**226**	**234**	**241**	**248**	**249**	**244**	**244**
Antigua and Barbuda **Antigua-et-Barbuda**	**1**	**1**	**1**	**1**	**1**	**1**	**1**	**1**	**1**	**1**
Beaf and veal[1] Boeuf et veau[1]	0	1	1	1	1	0	1	1	1	1
Barbados **Barbade**	**6**	**5**	**6**	**5**	**5**	**6**	**6**	**6**	**6**	**6**
Beaf and veal Boeuf et veau	0	0	0	0	0	1	1	1	1	1[1]
Pork[1] Porc[1]	5	5	5	5	5	5	5	5	5	5
Belize **Belize**	**2**	**2**	**2**	**3**	**2**	**2**	**2**	**2**	**2**	**2**
Beaf and veal Boeuf et veau	1	1	1	1	1	1	1[1]	1[1]	1[1]	1[1]
Pork[1] Porc[1]	1	1	1	1	1	1	1	1	1	1
Canada **Canada**	**2 043**	**2 125**	**2 130**	**2 083**	**2 137**	**2 138**	**2 033**	**1 995**	**2 129**	*** 2 141**
Beaf and veal Boeuf et veau	991	1 029	1 028	953	947	952	900	867	910	930
Pork Porc	1 044	1 088	1 094	1 122	1 182	1 177	1 124	1 119	1 209	1 200
Mutton and lamb Mouton et agneau	9	8	8	8	8	9	9	10	10	* 11
Costa Rica **Costa Rica**	**88**	**106**	**102**	**107**	**102**	**100**	**102**	**111**	**100**	**102**
Beaf and veal Boeuf et veau	77	94	* 92	97	* 86	86	87	94	81	82
Pork Porc	11	12	10	10	16	14	14	17	19	20
Cuba **Cuba**	**212**	**220**	**230**	**227**	**227**	**229**	**232**	**232**	**186**	**166**
Beaf and veal Boeuf et veau	142	142	148	140	141	138	141[1]	140[1]	112[1]	100[1]
Pork Porc	68	76	80	85	84	89[1]	89[1]	90[1]	72[1]	65[1]
Mutton and lamb[1] Mouton et agneau[1]	2	2	2	2	2	2	2	2	2	1
Dominican Republic **Rép. dominicaine**	**62**	**78**	**81**	**85**	**99**	**104**	**106**	**125**	**129**	**142**
Beaf and veal Boeuf et veau	54	63	64	67	79	81	82	84	83	86[1]
Pork Porc	7	13[1]	14[1]	17[1]	18[1]	20[1]	21[1]	38[1]	43[1]	53[1]
Mutton and lamb[1] Mouton et agneau[1]	2	2	2	2	2	2	2	2	3	3
El Salvador **El Salvador**	**35**	**35**	**36**	**33**	**36**	**39**	**37**	**33**	**33**	**36**
Beaf and veal Boeuf et veau	22	21	22	19	22	28	27	24	* 23	* 26
Pork[1] Porc[1]	13	14	14	14	13	10	10	9	10	10
Guadeloupe **Guadeloupe**	**8**	**7**	**7**	**7**	**6**	**6**	**6**	**5**	**5**	**5**
Beaf and veal Boeuf et veau	4	4	4	3	3	3	3	3	3	3

42

Meat
Production: thousand metric tons [*cont.*]
Viande
Production : milliers de tonnes métriques [*suite*]

Country or area Pays ou zone	1984	1985	1986	1987	1988	1989	1990	1991	1992	1993
Pork Porc	3	3	3	3	3	* 2	2	* 2	2[1]	2[1]
Guatemala Guatemala	**70**	**69**	**53**	**64**	**66**	**80**	**79**	**70**	**74**	**71**
Beaf and veal Boeuf et veau	50	49	34	46	48	61	61	52	56	53
Pork Porc	16	16	16	14	14	15	14	15	* 14	* 14
Mutton and lamb[1] Mouton et agneau[1]	3	4	4	4	4	4	4	4	4	4
Haiti Haïti	**52**	**52**	**52**	**49**	**47**	**45**	**41**	**36**	**32**	**32**
Beaf and veal[1] Boeuf et veau[1]	32	32	30	29	28	27	24	22	20	20
Pork[1] Porc[1]	16	16	18	16	15	14	13	10	9	9
Mutton and lamb[1] Mouton et agneau[1]	4	4	4	4	4	4	4	4	4	4
Honduras Honduras	**46**	**46**	**38**	**56**	**58**	**59**	**59**	**58**	**58**	**58**
Beaf and veal Boeuf et veau	34	35	27	* 45	* 46	* 46	* 46	45[1]	44[1]	45[1]
Pork Porc	11	10	11	* 11	* 12	* 13	13	13	13	13
Jamaica Jamaïque	**23**	**22**	**22**	**22**	**23**	**23**	**24**	**23**	**23**	**23**
Beaf and veal Boeuf et veau	15	14	15	14	14	13	15	16	16[1]	16[1]
Pork Porc	7	7	6	6	7	8	7	5	6[1]	6[1]
Mutton and lamb[1] Mouton et agneau[1]	2	2	2	2	2	2	2	2	2	2
Martinique Martinique	**6**	**5**	**5**	**5**	**5**	**5**	**5**	**5**	**5**	**5**
Beaf and veal Boeuf et veau	4	3	3	3	3	2	3	3	3[1]	3[1]
Pork[1] Porc[1]	2	2	2	2	2	2	2	2	2	2
Mutton and lamb[1] Mouton et agneau[1]	1	1	1	0	0	0	0	1	1	0
Mexico Mexique	**2 433**	**2 279**	**2 269**	**2 245**	**2 188**	**1 946**	**1 928**	**2 058**	**2 126**	**2 194**
Beaf and veal Boeuf et veau	925	927	1 248	1 273	1 271	1 162	1 114	1 189	1 247	* 1 300
Pork Porc	1 455	1 293	959	915	861	727	757	812	820	830
Mutton and lamb Mouton et agneau	52	59	62	58	* 56	* 57	* 56	* 57	* 59	* 64
Montserrat Montserrat	**1**	**1**	**1**	**1**	**1**	**1**	**1**	**1**	**1**	**1**
Beaf and veal[1] Boeuf et veau[1]	1	1	1	1	1	1	1	1	1	1
Nicaragua Nicaragua	**66**	**58**	**52**	**44**	**44**	*** 63**	**68**	**55**	**57**	**58**
Beaf and veal Boeuf et veau	51	45	37	30	33	50	57	45	48	48
Pork Porc	14[1]	13[1]	14[1]	14	11	* 13	10	9[1]	9[1]	10[1]

42

Meat
Production: thousand metric tons [*cont.*]
Viande
Production : milliers de tonnes métriques [*suite*]

Country or area Pays ou zone	1984	1985	1986	1987	1988	1989	1990	1991	1992	1993
Panama **Panama**	**67**	**71**	**73**	**68**	**64**	**67**	**76**	**72**	**68**	**68**
Beaf and veal Boeuf et veau	57	59	59	58	54	57	64	61	56	57[1]
Pork Porc	10	12	13	10	10	10	12	11	13	11[1]
Puerto Rico **Porto Rico**	**40**	**43**	**51**	**52**	**45**	**45**	**50**	**50**	**39**	**36**
Beaf and veal Boeuf et veau	24	24	25	28	22	20	21	19	21	19
Pork Porc	15	18	25	24	22	25	28	30	18	17
Saint Lucia **Sainte-Lucie**	**1**	**1**	**1**	**1**	**1**	**1**	**1**	**1**	**1**	**1**
Beaf and veal[1] Boeuf et veau[1]	1	1	1	1	1	1	1	1	1	1
Pork[1] Porc[1]	1	1	1	1	1	1	1	1	1	1
St. Vincent-Grenadines **St. Vincent-Grenadines**	**1**	**1**	**1**	**1**	**1**	**1**	**1**	**1**	**1**	**1**
Pork[1] Porc[1]	0	1	1	1	1	1	1	1	1	1
Trinidad and Tobago **Trinité-et-Tobago**	**5**	**5**	**5**	**5**	**5**	**4**	**4**	**4**	**4**	**3**
Beaf and veal Boeuf et veau	1	1	1	1	2	1	1	1	1	1
Pork Porc	4	3	3	3	3	2	2	2	2	2
Mutton and lamb[1] Mouton et agneau[1]	1	1	1	0	0	0	0	0	0	0
United States **Etats-Unis**	**17 818**	**17 873**	**17 824**	**17 547**	**18 145**	**17 963**	**17 595**	**17 956**	**18 561**	**18 489**
Beaf and veal Boeuf et veau	10 927	10 996	11 292	10 884	10 879	10 633	10 465	10 534	10 586	10 586
Pork Porc	6 719	6 715	6 379	6 520	7 114	7 173	6 965	7 258	7 817	7 751
Mutton and lamb Mouton et agneau	172	162	153	143	152	157	165	164	158	152
United States Virgin Is. **Iles Vierges américaines**	**1**	**1**	**1**	**0**	**1**	**1**	**1**	**1**	**1**	**1**
Beaf and veal[1] Boeuf et veau[1]	1	1	1	0	0	0	0	1	1	1
America, South **Amérique du Sud**	**8 345**	**8 823**	**8 629**	**8 941**	**9 560**	**9 946**	**10 207**	**10 301**	**10 466**	**10 501**
Beaf and veal Boeuf et veau	6 466	6 914	6 641	6 738	7 149	7 645	7 804	7 836	7 859	7 932
Pork Porc	1 567	1 593	1 656	1 867	2 066	1 949	2 038	2 106	2 245	2 199
Mutton and lamb Mouton et agneau	312	316	333	336	346	352	365	360	362	370
Argentina **Argentine**	*** 2 889**	**3 145**	**3 087**	**2 996**	**2 917**	**2 898**	**2 898**	**2 971**	**2 876**	**2 956**
Beaf and veal Boeuf et veau	2 553	2 848	2 813	2 700	2 590	2 626	2 635	2 700	* 2 647	* 2 716
Pork Porc	220[1]	193	170	200	237	181	171	180	* 140	* 155
Mutton and lamb[1] Mouton et agneau[1]	* 115	104	104	96	90	91	92	91	* 89	* 85

42
Meat
Production: thousand metric tons [*cont.*]
Viande
Production : milliers de tonnes métriques [*suite*]

Country or area Pays ou zone	1984	1985	1986	1987	1988	1989	1990	1991	1992	1993
Bolivia **Bolivie**	**183**	**195**	**185**	**196**	**211**	**207**	**214**	**214**	**211**	**217**
Beaf and veal Boeuf et veau	116	125	115	121	131	135	130	132	127	132
Pork[1] Porc[1]	53	55	54	57	61	54	65	65	67	68
Mutton and lamb[1] Mouton et agneau[1]	14	15	16	18	19	19	18	16	17	17
Brazil **Brésil**	**2 921**	**3 081**	**2 872**	**3 342**	**3 785**	**3 906**	**4 098**	**4 159**	**4 468**	**4 413**
Beaf and veal Boeuf et veau	2 096	2 223	1 958	2 262	2 581	2 748	2 836	2 885	3 062	* 3 080
Pork Porc	757[1]	780[1]	825[1]	986[1]	* 1 100	* 1 050	* 1 150	* 1 160	* 1 291	* 1 215
Mutton and lamb[1] Mouton et agneau[1]	68	79	89	94	104	108	112	* 114	* 115	* 118
Chile **Chili**	**272**	**259**	**270**	**282**	**315**	**352**	**385**	**376**	**355**	**390**
Beaf and veal Boeuf et veau	197	175	177	175	197	221	242	230	200	* 225
Pork Porc	59	66	75	88	100	113	123	129	138	* 146
Mutton and lamb[1] Mouton et agneau[1]	16	18	17	19	18	17	19	18	17	19
Colombia **Colombie**	**725**	**740**	**788**	**755**	**814**	**943**	**941**	**913**	**818**	**798**
Beaf and veal Boeuf et veau	599	609	652	621	669	805	* 795	* 768	670	651
Pork Porc	114	118	123	* 122	* 131	126	133	133	135	134
Mutton and lamb Mouton et agneau	12	13	* 12	* 13	* 13	12[1]	13[1]	12[1]	13[1]	13[1]
Ecuador **Equateur**	**150**	**159**	**162**	**164**	**165**	**167**	**215**	**218**	**227**	**236**
Beaf and veal Boeuf et veau	83	89	90	92	95	98	140	137	140	143
Pork[1] Porc[1]	62	65	68	68	65	64	71	76	82	87
Mutton and lamb[1] Mouton et agneau[1]	5	6	5	4	4	5	5	5	5	5
Falkland Is. (Malvinas) **Iles Falkland (Malvinas)**	**1**	**1**	**1**	**1**	**1**	**1**	**1**	**1**	**1**	**1**
Mutton and lamb[1] Mouton et agneau[1]	1	1	1	1	1	1	1	1	1	1
French Guiana **Guyane française**	**2**	**2**	**2**	**1**	**1**	**1**	**1**	**1**	**1**	**1**
Beaf and veal[1] Boeuf et veau[1]	0	0	1	1	1	1	1	1	1	0
Pork Porc	1	1	1	1	1	1	1	1	1	1[1]
Guyana **Guyana**	**3**	**3**	**3**	**4**	**4**	**4**	**3**	**4**	**4**	**4**
Beaf and veal Boeuf et veau	2	2	2	2	2	2	2	3	3[1]	3[1]
Pork Porc	1	1	1	1	1	1	1	0	1[1]	1[1]
Mutton and lamb[1] Mouton et agneau[1]	1	1	1	1	1	1	1	1	1	1

42

Meat
Production: thousand metric tons [cont.]
Viande
Production : milliers de tonnes métriques [suite]

Country or area Pays ou zone	1984	1985	1986	1987	1988	1989	1990	1991	1992	1993
Paraguay **Paraguay**	**219**	**225**	**228**	**213**	**244**	*** 290**	*** 311**	*** 320**	*** 336**	**351**
Beaf and veal Boeuf et veau	112	115	118	101	131	* 172	* 189	* 167	* 171	180[1]
Pork[1] Porc[1]	104	107	107	108	109	114	118	150	162	167
Mutton and lamb[1] Mouton et agneau[1]	3	3	3	3	3	3	3	3	3	3
Peru **Pérou**	**202**	**197**	**192**	**217**	**235**	**231**	**232**	**221**	**227**	**223**
Beaf and veal Boeuf et veau	103	101	90	107	117	112	117	109	111	105
Pork * Porc *	72	70	77	83	91	91	82	84	88	91
Mutton and lamb Mouton et agneau	27	26	25	27	28	29	33	28	28	27
Suriname **Suriname**	**3**	**3**	**3**	**3**	**3**	**4**	**4**	**5**	**4**	**4**
Beaf and veal Boeuf et veau	1	1	1	1	1	2	2	3	3	3[1]
Pork Porc	1	2	1	1	1	2	2	2	1	2[1]
Uruguay **Uruguay**	**363**	**395**	**392**	**350**	**402**	**464**	**416**	**434**	**447**	**411**
Beaf and veal Boeuf et veau	301	332	324	279	326	386	333	350	360	* 317
Pork Porc	19	18	17	19	20	20	22	22	22	* 23
Mutton and lamb Mouton et agneau	43	45	52	53	56	58	61	* 63	* 64	* 72
Venezuela **Venezuela**	**413**	**418**	**444**	**417**	**464**	**478**	**489**	**463**	**492**	**496**
Beaf and veal Boeuf et veau	301	294	300	276	307	338	* 382	* 351	* 365	* 377
Pork Porc	105	117	136	133	148	132	* 99	* 103	* 118	* 110
Mutton and lamb Mouton et agneau	7	7	7	8	8	8	8[1]	9[1]	9[1]	9[1]
Asia **Asie**	**28 244**	**30 768**	**32 715**	**33 944**	**36 206**	**38 059**	**40 586**	**42 977**	**45 534**	**47 394**
Beaf and veal **Boeuf et veau**	**5 501**	**5 658**	**6 039**	**6 386**	**6 505**	**6 819**	**7 227**	**7 671**	**8 060**	**8 515**
Pork **Porc**	**19 847**	**22 171**	**23 624**	**24 353**	**26 304**	**27 605**	**29 490**	**31 218**	**33 264**	**34 508**
Mutton and lamb **Mouton et agneau**	**2 896**	**2 938**	**3 052**	**3 206**	**3 397**	**3 635**	**3 869**	**4 088**	**4 210**	**4 372**
Afghanistan **Afghanistan**	**207**	**195**	**183**	**185**	**191**	**197**	**203**	**203**	**203**	**203**
Beaf and veal Boeuf et veau	68[1]	68[1]	68	65[1]	65[1]	65[1]	65[1]	65[1]	65[1]	65[1]
Mutton and lamb Mouton et agneau	139[1]	127[1]	* 115	120[1]	126[1]	132[1]	138[1]	138[1]	138[1]	138[1]
Armenia **Arménie**	..	..	..	..	..	..	..	..	**58**	**44**
Beaf and veal Boeuf et veau	..	..	..	..	..	..	..	..	33	* 27
Pork Porc	..	..	..	..	..	..	..	..	17	* 12

42

Meat
Production: thousand metric tons [*cont.*]
Viande
Production : milliers de tonnes métriques [*suite*]

Country or area Pays ou zone	1984	1985	1986	1987	1988	1989	1990	1991	1992	1993
Mutton and lamb Mouton et agneau	..	..	..	..	..	..	..	..	8	5[1]
Azerbaijan **Azerbaidjan**	..	..	..	..	..	..	..	..	**88**	**80**
Beaf and veal * Boeuf et veau *	..	..	..	..	..	..	..	..	48	42
Pork Porc	..	..		..	..	..	..	..	8[1]	* 8
Mutton and lamb[1] Mouton et agneau[1]	..	..	..	..	..	..	..	..	32	30
Bahrain **Bahreïn**	**6**	**6**	**6**	**7**	**7**	**8**	**8**	**8**	**8**	**8**
Beaf and veal[1] Boeuf et veau[1]	1	1	1	1	1	1	1	1	1	1
Mutton and lamb[1] Mouton et agneau[1]	5	5	5	6	6	7	7	7	7	8
Bangladesh **Bangladesh**	**165**	**188**	**194**	**199**	**205**	**211**	**218**	**223**	**230**	**240**
Beaf and veal Boeuf et veau	125	135	137	138	140	* 141	143	143	145	147
Mutton and lamb Mouton et agneau	40	53	57	61	65	* 70	75	80	86	93
Bhutan **Bhoutan**	**6**	**6**	**6**	**7**	**7**	**7**	**7**	**7**	**7**	**7**
Beaf and veal[1] Boeuf et veau[1]	5	5	5	5	5	5	5	6	6	6
Pork[1] Porc[1]	1	1	1	1	1	1	1	1	1	1
Brunei Darussalam **Brunéi Darussalam**	**1**	**2**	**2**	**2**	**2**	**2**	**2**	**2**	**2**	**2**
Beaf and veal[1] Boeuf et veau[1]	0	2	1	1	1	1	1	1	1	2
Pork[1] Porc[1]	1	1	1	0	1	0	0	0	0	0
Cambodia **Cambodge**	**45**	**52**	**52**	**57**	**60**	**62**	**64**	**66**	**70**	**73**
Beaf and veal[1] Boeuf et veau[1]	22	25	25	26	26	28	29	30	31	33
Pork[1] Porc[1]	23	27	27	32	34	34	35	36	38	40
China **Chine**	**16 180**	**18 442**	**20 043**	**20 802**	**22 852**	**24 185**	**26 149**	**28 374**	**30 526**	**32 060**
Beaf and veal * Boeuf et veau *	415	471	593	796	963	1 078	1 261	1 545	1 808	2 105
Pork Porc	15 179	17 378	18 828	19 287	21 087	22 145	23 820	25 649	27 468	28 640[1]
Mutton and lamb Mouton et agneau	* 586	* 593	* 622	* 719	* 802	* 962	* 1 068	* 1 180	* 1 251	1 316[1]
Cyprus **Chypre**	**32**	**34**	**35**	**38**	**40**	**42**	**44**	**43**	**45**	**45**
Beaf and veal Boeuf et veau	2	3	4	4	4	4	4	5	5	5[1]
Pork Porc	23	24	23	25	27	29	31	32	34	34[1]
Mutton and lamb Mouton et agneau	6	7	8	9	9	8	8	7	7	7[1]
Gaza Strip (Palestine) **Zone de Gaza (Palestine)**	**3**	**2**	**2**	**2**	**2**	**2**	**3**	**2**	**3**	**3**

42

Meat
Production: thousand metric tons [*cont.*]
Viande
Production : milliers de tonnes métriques [*suite*]

Country or area Pays ou zone	1984	1985	1986	1987	1988	1989	1990	1991	1992	1993
Beaf and veal Boeuf et veau	* 1	* 1	* 1	* 1	* 1	* 1	* 1	* 1	1[1]	1[1]
Mutton and lamb Mouton et agneau	* 1	* 1	* 1	* 1	* 1	* 1	* 2	* 1	1[1]	1[1]
Georgia **Géorgie**	..	..	..	..	..	..	..	..	**91**	**86**
Beaf and veal * Boeuf et veau *	..	..	..	..	..	..	..	..	38	36
Pork * Porc *	..	..	..	..	..	..	..	..	47	44
Mutton and lamb Mouton et agneau	..	..	..	..	..	..	..	..	* 6	6[1]
Hong Kong **Hong-kong**	**217**	**221**	**237**	**238**	**240**	**231**	**227**	**206**	**195**	**181**
Beaf and veal Boeuf et veau	37	38	40	40	40	37	39	38	35	32
Pork Porc	180	182	197	197	200[1]	194[1]	188	168	160	149
India **Inde**	**2 791**	**2 844**	**2 880**	**3 124**	**2 980**	**3 073**	**3 262**	**3 348**	**3 386**	**3 460**
Beaf and veal[1] Boeuf et veau[1]	1 912	1 954	1 981	2 206	2 086	2 168	2 319	2 361	2 398	2 458
Pork[1] Porc[1]	354	354	354	355	357	359	361	364	366	367
Mutton and lamb[1] Mouton et agneau[1]	525	536	546	563	538	547	583	623	623	635
Indonesia **Indonésie**	**638**	**672**	**729**	**713**	**731**	**804**	**850**	**891**	**918**	**943**
Beaf and veal Boeuf et veau	216	227	239	* 216[1]	* 192	* 227	* 223	235[1]	243[1]	246[1]
Pork[1] Porc[1]	347	369	413	418	462	495	545	572	589	594
Mutton and lamb[1] Mouton et agneau[1]	75	76	78	* 79	* 77	* 82	* 83	84	86	103
Iran, Islamic Rep. of **Iran, Rép. islamique d'**	**444**	**461**	**483**	**483**	**493**	**526**	**550**	**585**	**630**	**639**
Beaf and veal[1] Boeuf et veau[1]	174	177	180	183	190	* 211	* 220	* 245	* 280	285
Mutton and lamb Mouton et agneau	270[1]	284[1]	303[1]	300[1]	304[1]	* 315	* 330	* 340	* 350	354[1]
Iraq **Iraq**	**87**	**81**	**77**	**85**	**76**	**94**	**72**	**54**	**65**	**67**
Beaf and veal[1] Boeuf et veau[1]	52	49	46	54	44	59	45	33	40	40
Mutton and lamb[1] Mouton et agneau[1]	35	32	31	32	32	36	27	21	25	27
Israel **Israël**	**46**	**45**	**45**	**47**	**51**	**50**	**51**	**53**	**52**	**52**
Beaf and veal Boeuf et veau	34	32	32	33	35	36	36	* 38	* 37	37[1]
Pork Porc	8	9	8	9	9	9	9	9[1]	9[1]	9[1]
Mutton and lamb Mouton et agneau	* 4	* 4	* 5	* 5	* 6	* 6	* 6	* 6	* 6	6[1]
Japan **Japon**	**1 960**	**2 087**	**2 111**	**2 147**	**2 149**	**2 142**	**2 105**	**2 058**	**2 026**	**2 027**
Beaf and veal Boeuf et veau	536	555	559	565	570	548	549	575	592	593

42
Meat
Production: thousand metric tons [*cont.*]
Viande
Production : milliers de tonnes métriques [*suite*]

Country or area Pays ou zone	1984	1985	1986	1987	1988	1989	1990	1991	1992	1993
Pork Porc	1 424	1 532	1 552	1 582	1 579	1 594	1 555	1 483	1 434	1 433
Jordan **Jordanie**	**10**	**10**	**6**	**8**	**8**	**9**	**10**	**16**	**16**	**16**
Beaf and veal Boeuf et veau	2	1	1	1	1	1	* 1	* 1	* 1	2[1]
Mutton and lamb Mouton et agneau	8	9	5	7	7	8	* 9	* 15	* 15	15[1]
Kazakhstan **Kazakhstan**	..	..	..	..	..	..	..	..	**1 086**	**1 023**
Beaf and veal * Boeuf et veau *	..	..	..	..	..	..	..	..	580	551
Pork * Porc *	..	..	..	..	..	..	..	..	260	235
Mutton and lamb[1] Mouton et agneau[1]	..	..	..	..	..	..	..	..	246	237
Korea, Dem. P. R. **Corée, R. p. dém. de**	**180**	**189**	**195**	**206**	**209**	**211**	**213**	**210**	**206**	**183**
Beaf and veal[1] Boeuf et veau[1]	36	38	39	41	42	44	45	45	45	30
Pork[1] Porc[1]	141	148	153	161	164	164	164	161	158	150
Mutton and lamb[1] Mouton et agneau[1]	3	3	3	3	3	4	4	4	4	3
Korea, Republic of **Corée, République de**	**554**	**601**	**531**	**586**	**603**	**606**	**679**	**664**	**891**	**973**
Beaf and veal Boeuf et veau	123	166	208	208	177	125	128	132	* 137	* 175
Pork Porc	429	434	322	377	425	480	550	530	* 752	* 796
Mutton and lamb[1] Mouton et agneau[1]	2	1	1	1	1	1	1	2	2	2
Kuwait **Koweït**	**45**	**35**	**36**	**35**	**35**	**35**	**18**	**9**	**18**	**26**
Beaf and veal Boeuf et veau	5[1]	2	2[1]	2	2	1	1[1]	0[1]	0[1]	1[1]
Mutton and lamb[1] Mouton et agneau[1]	40	* 33	* 33	* 33	* 33	34	17	9	17	25
Kyrgyzstan * **Kirghizistan ***	..	..	..	..	..	..	..	..	**179**	**168**
Beaf and veal * Boeuf et veau *	..	..	..	..	..	..	..	..	80	76
Pork * Porc *	..	..	..	..	..	..	..	..	34	33
Mutton and lamb[1] Mouton et agneau[1]	..	..	..	..	..	..	..	..	65	59
Lao People's Dem. Rep. **Rép. dém. pop. lao**	**82**	**88**	**88**	**90**	**83**	**83**	**85**	**88**	**94**	**94**
Beaf and veal[1] Boeuf et veau[1]	33	36	37	36	33	32	31	33	36	35
Pork[1] Porc[1]	49	52	50	54	50	51	53	54	58	59
Mutton and lamb[1] Mouton et agneau[1]	0	0	0	0	0	0	0	1	1	1
Lebanon **Liban**	**27**	**26**	**23**	**21**	**20**	**15**	**13**	**25**	**25**	**25**
Beaf and veal[1] Boeuf et veau[1]	16	18	15	13	8	7	7	15	15	16

42

Meat
Production: thousand metric tons [*cont.*]
Viande
Production : milliers de tonnes métriques [*suite*]

Country or area Pays ou zone	1984	1985	1986	1987	1988	1989	1990	1991	1992	1993
Pork[1] Porc [1]	0	0	0	1	1	1	1	1	1	1
Mutton and lamb[1] Mouton et agneau[1]	11	8	8	8	11	7	5	9	9	9
Macau **Macao**	**9**	**8**	**8**	**8**	**9**	**9**	**9**	**8**	**8**	**8**
Beaf and veal Boeuf et veau	1	1[1]	1[1]	1[1]	2[1]	2[1]	2[1]	1[1]	1[1]	1[1]
Pork Porc	8	7	7	6	7	7	8[1]	8[1]	8[1]	8[1]
Malaysia **Malaisie**	**160**	**170**	**175**	**191**	**180**	**202**	**235**	**243**	**240**	**238**
Beaf and veal[1] Boeuf et veau[1]	17	16	15	15	17	15	15	18	18	18
Pork[1] Porc[1]	143	153	159	175	163	186	219	225	222	219
Mutton and lamb[1] Mouton et agneau [1]	1	1	1	1	1	1	1	1	1	1
Mongolia **Mongolie**	**196**	**187**	**201**	**196**	**190**	**200**	**206**	**224**	**194**	**163**
Beaf and veal Boeuf et veau	68	68	75	73	73	73	66	84	76	48[1]
Pork Porc	2	2	2	3	4	6	8	4	2	5[1]
Mutton and lamb Mouton et agneau	126	116	123	121	112	122	132	136	116	110[1]
Myanmar **Myanmar**	**188**	**190**	**193**	**196**	**199**	**189**	**193**	**198**	**205**	**209**
Beaf and veal[1] Boeuf et veau[1]	98	99	101	103	105	105	106	109	110	111
Pork[1] Porc[1]	83	83	85	86	87	77	80	83	88	91
Mutton and lamb Mouton et agneau	* 7[1]	8[1]	* 7	* 7	* 7	7	7	7	7	7
Nepal **Népal**	**121**	**125**	**128**	**130**	**135**	**139**	**140**	**142**	**143**	**144**
Beaf and veal[1] Boeuf et veau[1]	86	90	92	93	96	98	98	99	100	100
Pork Porc	7	7	7	8	9	9	10	10	10	10
Mutton and lamb Mouton et agneau	28[1]	28	29	29	31	31	32	32	33	33
Oman **Oman**	**12**	**13**	**13**	**13**	**15**	**18**	**18**	**19**	**18**	**19**
Beaf and veal[1] Boeuf et veau[1]	3	3	3	3	3	3	3	3	3	3
Mutton and lamb[1] Mouton et agneau[1]	8	10	10	10	12	15	15	16	16	16
Pakistan **Pakistan**	**784**	**831**	**881**	**934**	**1 140**	**1 220**	**1 350**	**1 430**	**1 516**	**1 607**
Beaf and veal Boeuf et veau	443	466	490	516	626	667	729	765	803	844
Mutton and lamb Mouton et agneau	341	365	391	418	514	553	621	665	713	763
Philippines **Philippines**	**671**	**503**	**576**	**628**	**705**	**798**	**860**	**849**	**868**	**867**
Beaf and veal Boeuf et veau	* 85	87	96	108	113	* 132	* 125	* 130	* 129	* 126

42

Meat
Production: thousand metric tons [*cont.*]
Viande
Production : milliers de tonnes métriques [*suite*]

Country or area Pays ou zone	1984	1985	1986	1987	1988	1989	1990	1991	1992	1993
Pork Porc	569	397	460	500	566	639	709	691	* 710	* 712
Mutton and lamb[1] Mouton et agneau[1]	17	19	19	20	25	27	27	28	29	29
Qatar **Qatar**	**10**	**7**	**18**	**11**	**13**	**7**	**11**	**14**	**11**	**13**
Mutton and lamb[1] Mouton et agneau[1]	10	7	18	10	13	7	11	14	10	13
Saudi Arabia **Arabie saoudite**	**87**	**94**	**102**	**97**	**107**	**110**	**111**	**106**	**114**	**114**
Beaf and veal Boeuf et veau	26	25	25	* 18	* 24	* 25	* 28	* 27	* 28	* 28
Mutton and lamb Mouton et agneau	61	69	77	* 79	* 83	* 85	* 83	* 79	86[1]	86[1]
Singapore **Singapour**	**70**	**74**	**74**	**79**	**75**	**76**	**77**	**82**	**85**	**86**
Pork Porc	68	72	72	77	74	75	76	81	84	85
Mutton and lamb Mouton et agneau	2	1	1	1	1	1	1	1	1	1
Sri Lanka **Sri Lanka**	**41**	**42**	**36**	**45**	**36**	**36**	**50**	**28**	**41**	**41**
Beaf and veal[1] Boeuf et veau[1]	* 36	* 38	32	41	32	32	46	24	37	38
Pork Porc[1]	* 1	1	1	1	1	1	2	2	2[1]	2[1]
Mutton and lamb[1] Mouton et agneau[1]	* 3	* 3	3	2	2	2	2	2	2	2
Syrian Arab Republic **Rép. arabe syrienne**	**128**	**118**	**126**	**132**	**144**	**143**	**152**	**162**	**146**	**150**
Beef and veal Boeuf et veau	30[1]	30[1]	30[1]	30[1]	31[1]	32[1]	32[1]	33	29	30[1]
Mutton and lamb Mouton et agneau	98[1]	88[1]	95	101[1]	113[1]	112[1]	120	129	118	120[1]
Tajikistan **Tadjikistan**	..	..	..	..	..	..	..	..	63	61
Beaf and veal * Boeuf et veau *	..	..	..	..	..	..	..	..	39	37
Pork Porc	..	..	..	..	..	..	..	..	5[1]	* 6
Mutton and lamb[1] Mouton et agneau[1]	..	..	..	..	..	..	..	..	19	18
Thailand **Thaïlande**	**500**	**602**	**502**	**554**	**562**	**563**	**576**	**612**	**596**	**650**
Beef and veal[1] Boeuf et veau[1]	223	224	226	227	228	227	237	271	252	298
Pork[1] Porc[1]	277	378	276	326	334	335	338	340	343	351
Mutton and lamb[1] Mouton et agneau[1]	1	1	1	1	1	1	1	1	1	1
Turkey * **Turquie** *	**777**	**713**	**851**	**723**	**710**	**756**	**742**	**715**	**704**	**701**
Beaf and veal Boeuf et veau	401	333	466	341	329	381	372	348	* 339	338[1]
Pork[1] Porc[1]	1	1	0	0	0	0	0	0	0	0
Mutton and lamb * Mouton et agneau *	375	380	385	382	380	375	370	367	365	363[1]

42

Meat
Production: thousand metric tons [cont.]
Viande
Production : milliers de tonnes métriques [suite]

Country or area Pays ou zone	1984	1985	1986	1987	1988	1989	1990	1991	1992	1993
Turkmenistan * **Turkménistan ***	..	..	..	..	..	..	..	..	**77**	**77**
Beaf and veal * Boeuf et veau *	..	..	..	..	..	..	..	..	37	35
Pork Porc	..	..	..	..	..	..	..	..	6	8
Mutton and lamb[1] Mouton et agneau[1]	..	..	..	..	..	..	..	..	34	34
United Arab Emirates **Emirats arabes unis**	**27**	**26**	**26**	**31**	**32**	**28**	**31**	**34**	**35**	**35**
Beaf and veal[1] Boeuf et veau[1]	6	5	6	6	5	5	5	6	6	6
Mutton and lamb[1] Mouton et agneau[1]	21	21	21	25	27	23	26	28	29	29
Uzbekistan **Ouzbékistan**	..	..	..	..	..	..	..	..	**427**	**366**
Beaf and veal Boeuf et veau	..	..	..	..	..	..	..	..	323	* 265
Pork Porc	..	..	..	..	..	..	..	..	36	* 46
Mutton and lamb Mouton et agneau	..	..	..	..	..	..	..	..	68	55[1]
Viet Nam **Viet Nam**	**659**	**698**	**762**	**816**	**827**	**885**	**904**	**892**	**908**	**932**
Beaf and veal[1] Boeuf et veau[1]	127	135	134	142	162	168	172	173	174	179
Pork Porc	531	561	625	671	662	714	729	716	730[1]	750[1]
Mutton and lamb[1] Mouton et agneau[1]	2	2	3	3	3	3	3	3	3	3
Yemen **Yémen**	**64**	**65**	**65**	**65**	**67**	**68**	**70**	**67**	**68**	**70**
Beaf and veal Boeuf et veau	29[1]	* 29	30[1]	30[1]	* 30	* 31	* 32	* 31	* 31	* 32
Mutton and lamb Mouton et agneau	35[1]	36	35[1]	35[1]	37	37	38[1]	36[1]	37[1]	38[1]
Europe **Europe**	**32 645**	**32 810**	**33 369**	**33 973**	**33 951**	**33 351**	**34 673**	**34 529**	**33 663**	**33 089**
Beaf and veal **Boeuf et veau**	**11 139**	**11 151**	**11 251**	**11 280**	**10 626**	**10 338**	**11 326**	**11 612**	**11 041**	**10 238**
Pork **Porc**	**20 167**	**20 280**	**20 765**	**21 317**	**21 930**	**21 554**	**21 786**	**21 352**	**21 135**	**21 424**
Mutton and lamb **Mouton et agneau**	**1 340**	**1 379**	**1 352**	**1 376**	**1 395**	**1 459**	**1 561**	**1 565**	**1 487**	**1 427**
Albania **Albanie**	**34**	**37**	**36**	**37**	**42**	**45**	**49**	**53**	**49**	**44**
Beaf and veal Boeuf et veau	15[1]	17	15	17	19	20	22	25	25	20[1]
Pork Porc	10[1]	10	11	11	12	12	14	15	12	13[1]
Mutton and lamb[1] Mouton et agneau[1]	10	* 10	* 9	* 9	* 11	* 13	* 12	* 12	* 12	11
Austria **Autriche**	**695**	**701**	**698**	**705**	**745**	**735**	**747**	**758**	**772**	**764**
Beaf and veal Boeuf et veau	224	221	231	230	219	212	224	236	239	236[1]
Pork Porc	* 467	* 476	* 462	* 470	* 522	* 518	* 517	* 517	* 527	523[1]

42

Meat
Production: thousand metric tons [*cont.*]
Viande
Production : milliers de tonnes métriques [*suite*]

Country or area Pays ou zone	1984	1985	1986	1987	1988	1989	1990	1991	1992	1993
Mutton and lamb Mouton et agneau	* 4[1]	* 3[1]	* 4[1]	* 5[1]	5	5	6	6[1]	6[1]	5[1]
Belarus **Bélarus** Beaf and veal	..	..	..	..	..	..	..	..	824	793
Boeuf et veau Pork	..	..	..	..	..	..	..	..	495	* 459
Porc Mutton and lamb	..	..	..	..	..	..	..	..	323	* 330
Mouton et agneau	..	..	..	..	..	..	..	..	5	4[1]
Belgium-Luxembourg **Belgique-Luxembourg**	**1 062**	**1 059**	**1 081**	**1 121**	**1 137**	**1 143**	**1 114**	**1 305**	**1 316**	**1 314**
Beaf and veal Boeuf et veau	319	326	326	326	317	305	323	381	359	362
Pork Porc	734	725	747	788	813	830	784	915	951	* 945
Mutton and lamb Mouton et agneau	8	8	8	7	7	7	7	8	6	* 7
Bosnia & Herzegovina **Bosnie-Herzégovine** Beaf and veal[1]	..	..	..	..	..	..	..	..	104	91
Boeuf et veau[1] Pork[1]	..	..	..	..	..	..	..	..	43	37
Porc[1] Mutton and lamb[1]	..	..	..	..	..	..	..	..	50	44
Mouton et agneau[1]	..	..	..	..	..	..	..	..	10	10
Bulgaria **Bulgarie** Beaf and veal	**549**	**563**	**591**	**581**	**586**	**608**	**590**	**543**	**506**	**411**
Boeuf et veau Pork	130	134	130	129	122	125	121	108	135	* 98[1]
Porc Mutton and lamb	335	334	372	372	394	412	406	362	311	* 266
Mouton et agneau	84	94	* 89	80	71	72	63	73	60	47[1]
Croatia **Croatie** Beaf and veal	..	..	..	..	..	..	..	..	177	175
Boeuf et veau Pork	..	..	..	..	..	..	..	..	40	38[1]
Porc Mutton and lamb	..	..	..	..	..	..	..	..	* 135	135[1]
Mouton et agneau	..	..	..	..	..	..	..	..	2	2[1]
former Czechoslovakia† **anc. Tchécoslovaquie†** Beaf and veal	**1 231**	**1 247**	**1 254**	**1 281**	**1 331**	**1 352**	**1 326**	**1 190**	**1 162**	...
Boeuf et veau Pork	406	413	412	412	405	407	403	354	319	...
Porc Mutton and lamb	814	823	832	858	915	934	913	827	834	...
Mouton et agneau	11	11	10	11	11	11	10	9	8	...
Czech Republic **République tchèque** Beaf and veal	..	..	..	..	..	..	..	..	..	795
Boeuf et veau Pork	..	..	..	..	..	..	..	..	..	216
Porc Mutton and lamb	..	..	..	..	..	..	..	..	..	576
Mouton et agneau	..	..	..	..	..	..	..	..	..	3

42

Meat
Production: thousand metric tons [*cont.*]
Viande
Production : milliers de tonnes métriques [*suite*]

Country or area Pays ou zone	1984	1985	1986	1987	1988	1989	1990	1991	1992	1993
Denmark **Danemark**	**1 282**	**1 320**	**1 388**	**1 384**	**1 385**	**1 369**	**1 412**	**1 486**	**1 589**	**1 709**
Beaf and veal Boeuf et veau	247	236	243	235	217	205	202	213	217	203
Pork Porc	1 035	1 083	1 144	1 149	1 167	1 163	1 208	1 272	1 370	1 504
Mutton and lamb Mouton et agneau	1	1	1	1	1	1	1	2	2	2
Estonia * **Estonie ***	..	..	..	..	..	..	..	..	155	149
Beaf and veal * Boeuf et veau *	..	..	..	..	..	..	..	..	65	62
Pork * Porc *	..	..	..	..	..	..	..	..	88	84
Mutton and lamb[1] Mouton et agneau[1]	..	..	..	..	..	..	..	..	2	3
Faeroe Islands **Iles Féroé**	**1**	**1**	**1**	**1**	**1**	**1**	**1**	**1**	**1**	**1**
Mutton and lamb[1] Mouton et agneau[1]	0	0	0	1	1	1	1	1	1	1
Finland **Finlande**	**296**	**300**	**300**	**310**	**290**	**291**	**306**	**300**	**295**	**287**
Beaf and veal Boeuf et veau	124	126	125	127	115	110	118	122	117	116
Pork Porc	170	172	174	181	174	180	187	177	176	169
Mutton and lamb Mouton et agneau	1	2	1	1	1	1	1	1	1	1
France **France**	*** 3 860**	**3 743**	**3 770**	**3 874**	**3 857**	**3 693**	**3 977**	**4 129**	**4 245**	**4 216**
Beaf and veal Boeuf et veau	1 991	1 893	1 911	1 963	1 828	1 673	1 912	2 026	2 079	1 902
Pork Porc	1 684	1 662	1 677	1 729	1 852	1 844	1 871	1 918	1 994	2 151
Mutton and lamb Mouton et agneau	* 185	188	182	182	177	176	194	185	172	163
Germany † **Allemagne†**	*** 6 522**	*** 6 656**	*** 6 855**	*** 6 881**	**6 692**	**6 474**	**6 619**	**6 150**	**5 556**	*** 5 314**
Beaf and veal Boeuf et veau	* 1 982	* 1 993	* 2 110	* 2 099	1 988	1 953	2 112	2 182	1 829	* 1 670
Pork Porc	* 4 495	* 4 620	* 4 703	* 4 734	4 663	4 478	4 457	3 918	3 684	3 600[1]
Mutton and lamb Mouton et agneau	* 44[1]	* 43	* 42	* 48	41	43	50	50	44[1]	* 44[1]
Greece **Grèce**	**366**	**349**	**352**	**358**	**358**	**363**	**364**	**362**	**357**	**359**
Beaf and veal Boeuf et veau	91	84	85[1]	85[1]	84[1]	84[1]	83[1]	81	80	80
Pork Porc	148	138	142	141	140	139	140	153	153	154
Mutton and lamb Mouton et agneau	127	126	125	132	134	139	141	* 128	* 124	* 125
Hungary **Hongrie**	**1 263**	**1 165**	**1 118**	**1 171**	**1 137**	**1 135**	**1 137**	**1 059**	**890**	**810**
Beaf and veal Boeuf et veau	132	146	124	129	110	114	114	123	123	* 97
Pork Porc	1 124	1 011	986	1 037	1 022	1 014	1 018	931	764	710[1]

42

Meat
Production: thousand metric tons [*cont.*]
Viande
Production : milliers de tonnes métriques [*suite*]

Country or area Pays ou zone	1984	1985	1986	1987	1988	1989	1990	1991	1992	1993
Mutton and lamb Mouton et agneau	8	9	8	5	5	7	5	6	* 3	* 3
Iceland **Islande**	**17**	**18**	**19**	**18**	**16**	**16**	**15**	**15**	**15**	**15**
Beaf and veal Boeuf et veau	3	3	3	3	3	3	3	3	3	3[1]
Pork Porc	1	2	2	2	2	3	3	3	2	2[1]
Mutton and lamb Mouton et agneau	13	13	13	13	11	10	10	9	9	9[1]
Ireland **Irlande**	**587**	**632**	**693**	**672**	**656**	**639**	**758**	**825**	**862**	**850**
Beaf and veal Boeuf et veau	402	448	509	484	459	432	515	554	565	* 530
Pork Porc	144	135	137	141	148	144	157	179	203	* 220
Mutton and lamb Mouton et agneau	41	48	46	48	50	63	86	92	94	* 100
Italy **Italie**	**2 471**	**2 462**	**2 420**	**2 476**	**2 507**	**2 520**	**2 583**	**2 599**	**2 645**	**2 588**
Beaf and veal Boeuf et veau	1 182	1 205	1 180	1 175	1 165	1 146	1 165	* 1 182	* 1 218	* 1 187
Pork Porc	1 218	1 187	1 172	1 231	1 269	1 295	1 333	1 333	1 342	* 1 320
Mutton and lamb Mouton et agneau	71	70	67	70	73	79	85	85	* 86[1]	* 81[1]
Latvia **Lettonie**	..	..	..	..	..	..	..	..	224	190
Beaf and veal Boeuf et veau	..	..	..	..	..	..	..	..	120	* 98
Pork Porc	..	..	..	..	..	..	..	..	101	88
Mutton and lamb Mouton et agneau	..	..	..	..	..	..	..	..	4	4[1]
Lithuania **Lituanie**	..	..	..	..	..	..	..	..	340	363
Beaf and veal Boeuf et veau	..	..	..	..	..	..	..	..	176	* 183
Pork Porc	..	..	..	..	..	..	..	..	163	* 179
Mutton and lamb Mouton et agneau	..	..	..	..	..	..	..	..	1[1]	1
Malta **Malte**	**7**	**8**	**9**	**9**	**10**	**10**	**10**	**10**	**11**	**11**
Beaf and veal Boeuf et veau	2	1	1[1]	2[1]	2	2	2	2[1]	2[1]	2[1]
Pork Porc	5	7	7[1]	8[1]	8	8	8	8[1]	9[1]	9[1]
Netherlands **Pays-Bas**	**1 831**	**1 933**	**1 994**	**2 062**	**2 151**	**2 106**	**2 197**	**2 232**	**2 237**	**2 372**
Beaf and veal Boeuf et veau	515	511	539	540	507	486	521	* 623	* 635	* 604
Pork Porc	1 306	1 412	1 443	1 511	1 631	1 606	1 661	* 1 591	* 1 584	* 1 749
Mutton and lamb Mouton et agneau	10	11	11	* 11[1]	13[1]	* 14[1]	* 15[1]	* 18[1]	* 18[1]	* 19[1]
Norway **Norvège**	**180**	**183**	**186**	**196**	**191**	**183**	**189**	**189**	**200**	**202**

42

Meat
Production: thousand metric tons [*cont.*]
Viande
Production : milliers de tonnes métriques [*suite*]

Country or area Pays ou zone	1984	1985	1986	1987	1988	1989	1990	1991	1992	1993
Beaf and veal Boeuf et veau	71	74	75	78	77	75	81	80	85	85
Pork Porc	84	84	85	92	90	84	83	85	91	91
Mutton and lamb Mouton et agneau	25	25	26	26	25	24	24	24	25	25
Poland Pologne	**2 011**	**2 236**	**2 511**	**2 495**	**2 544**	**2 513**	**2 609**	**2 643**	**2 602**	*** 2 492**
Beaf and veal Boeuf et veau	687	724	752	735	689	637	725	663	544	* 484
Pork Porc	1 304	1 486	1 728	1 729	1 828	1 854	1 855	1 947	2 036	* 1 995
Mutton and lamb Mouton et agneau	21	26	31	30	26	22	29	33	23	* 13
Portugal Portugal	**308**	**299**	**302**	**330**	**326**	**369**	**424**	**421**	**416**	**428**
Beaf and veal Boeuf et veau	99	100	105	105	115	131	116	128	124	115
Pork Porc	183	174	171	199	183	211	279	263	265	285
Mutton and lamb Mouton et agneau	25	24	26	26	28	27	28	30	27	28[1]
Republic of Moldova République de Moldova	..	..	..	..	..	..	..	..	193	165
Beaf and veal Boeuf et veau	..	..	..	..	..	..	..	..	75	* 67
Pork Porc	..	..	..	..	..	..	..	..	114	95[1]
Mutton and lamb Mouton et agneau	..	..	..	..	..	..	..	..	4	3[1]
Romania Roumanie	**1 396**	**1 244**	**1 299**	**1 296**	**1 209**	**1 024**	**1 284**	**1 316**	**1 195**	**1 154**
Beaf and veal Boeuf et veau	* 297	* 255	* 235	* 218	* 208	* 176	385[1]	385[1]	305[1]	255[1]
Pork Porc	* 1 010	* 909	* 985	* 1 002	* 927	* 773	788	834	789	804[1]
Mutton and lamb Mouton et agneau	* 89	* 80	* 79	* 76	* 74	* 75	112[1]	97[1]	102[1]	95[1]
Russian Federation Fédération de Russie	..	..	..	..	..	..	..	..	6 482	6 222
Beaf and veal Boeuf et veau	..	..	..	..	..	..	..	..	3 500	* 3 400
Pork Porc	..	..	..	..	..	..	..	..	2 700	* 2 550
Mutton and lamb[1] Mouton et agneau[1]	..	..	..	..	..	..	..	..	282	272
Slovakia Slovaquie	..	..	..	..	..	..	..	..	..	**290**
Beaf and veal Boeuf et veau	..	..	..	..	..	..	..	..	..	67[1]
Pork Porc	..	..	..	..	..	..	..	..	..	220[1]
Mutton and lamb Mouton et agneau	..	..	..	..	..	..	..	..	..	3[1]
Slovenia Slovénie	..	..	..	..	..	..	..	..	79	90
Beaf and veal Boeuf et veau	..	..	..	..	..	..	..	..	38	42[1]

42

Meat
Production: thousand metric tons [cont.]
Viande
Production : milliers de tonnes métriques [suite]

Country or area Pays ou zone	1984	1985	1986	1987	1988	1989	1990	1991	1992	1993
Pork Porc	..	..	..	..	..	..	..	* 41	48[1]
Spain **Espagne**	**2 034**	**1 998**	**2 050**	**2 164**	**2 404**	**2 384**	**2 537**	**2 630**	**2 706**	**2 730**
Beaf and veal Boeuf et veau	398	401	440	450	450	459	514	509	539	* 530
Pork Porc	1 429	1 388	1 399	1 489	1 722	1 703	1 789	1 877	1 918	* 1 950
Mutton and lamb Mouton et agneau	207	210	211	225	231	222	234	244	249	* 250[1]
Sweden **Suède**	**484**	**495**	**462**	**427**	**432**	**451**	**441**	**409**	**412**	**435**
Beaf and veal Boeuf et veau	155	158	147	135	127	139	145	137	130	* 135
Pork Porc	324	332	309	288	299	307	291	268	278	* 295
Mutton and lamb Mouton et agneau	5	5	5	5	5	5	5	4	* 5	* 5
Switzerland **Suisse**	**446**	**459**	**461**	**454**	**438**	**441**	**439**	**444**	**435**	**422**
Beaf and veal Boeuf et veau	165	170	171	172	154	157	164	173	165	155
Pork Porc	276	285	286	278	279	280	270	265	264	260
Mutton and lamb Mouton et agneau	4	5	5	5	5	5	5	6	6	6
TFYR Macedonia **L'ex-R.y. Macédoine**	..	..	..	..	..	..	..	..	30	29
Beaf and veal Boeuf et veau	..	..	..	..	..	..	..	..	8	7[1]
Pork[1] Porc[1]	..	..	..	..	..	..	..	..	10	12
Mutton and lamb Mouton et agneau	..	..	..	..	..	..	..	..	12	11[1]
Ukraine **Ukraine**	..	..	..	..	..	..	..	..	2 870	* 2 213
Beaf and veal Boeuf et veau	..	..	..	..	..	..	..	..	1 656	* 1 275
Pork Porc	..	..	..	..	..	..	..	..	1 180	* 908
Mutton and lamb Mouton et agneau	..	..	..	..	..	..	..	..	35	30[1]
United Kingdom **Royaume-Uni**	**2 387**	**2 454**	**2 335**	**2 421**	**2 285**	**2 283**	**2 318**	**2 384**	**2 284**	**2 200**
Beaf and veal Boeuf et veau	1 152	1 179	1 062	1 118	946	978	1 002	1 020	959	856
Pork Porc	947	971	983	1 007	1 017	939	946	979	970	1 011
Mutton and lamb Mouton et agneau	288	304	290	296	322	366	370	385	355	333
Yugoslavia * **Yougoslavie ***	..	..	..	..	..	..	..	..	516	493
Beaf and veal Boeuf et veau	..	..	..	..	..	..	..	..	* 117	105[1]
Pork[1] Porc[1]	..	..	..	..	..	..	..	..	373	365
Mutton and lamb Mouton et agneau	..	..	..	..	..	..	..	..	* 26	* 23

42

Meat
Production: thousand metric tons [*cont.*]
Viande
Production : milliers de tonnes métriques [*suite*]

Country or area Pays ou zone	1984	1985	1986	1987	1988	1989	1990	1991	1992	1993
Yugoslavia, SFR† **Yougoslavie, Rfs†**	**1 327**	**1 248**	**1 187**	**1 252**	**1 224**	**1 202**	**1 230**	**1 075**	..	..
Beaf and veal Boeuf et veau	350	333	317	317	301	309	352	* 301	..	..
Pork * Porc *	919	853	807	870	853	824	810	716	..	..
Mutton and lamb Mouton et agneau	58	62	63	65	70	69	67	* 57	..	..
Oceania **Océanie**	**3 274**	**3 423**	**3 434**	**3 672**	**3 780**	**3 632**	**3 752**	**3 969**	**4 061**	**3 984**
Beaf and veal **Boeuf et veau**	**1 798**	**1 816**	**1 872**	**2 095**	**2 180**	**2 069**	**2 176**	**2 322**	**2 359**	**2 399**
Pork **Porc**	**336**	**349**	**360**	**370**	**386**	**397**	**406**	**402**	**430**	**421**
Mutton and lamb **Mouton et agneau**	**1 140**	**1 257**	**1 201**	**1 207**	**1 214**	**1 165**	**1 170**	**1 245**	**1 272**	**1 164**
Australia **Australie**	**2 069**	**2 097**	**2 239**	**2 398**	**2 482**	**2 353**	**2 634**	**2 757**	**2 811**	**2 813**
Beaf and veal Boeuf et veau	1 345	1 310	1 385	1 521	1 588	1 491	1 677	1 760	1 791	1 826
Pork Porc	253	260	271	283	297	308	317	312	336	326
Mutton and lamb[1] Mouton et agneau[1]	470	527	583	594	597	553	640	685	684	661
Fiji **Fidji**	**13**	**13**	**14**	**14**	**15**	**15**	**15**	**15**	**16**	**17**
Beaf and veal[1] Boeuf et veau[1]	10	10	11	11	11	11	11	12	12	12
Pork[1] Porc[1]	3	3	3	3	3	3	3	3	3	3
Mutton and lamb Mouton et agneau	1	1	1	1	1	1	1	1	1	1[1]
French Polynesia **Polynésie française**	**1**	**1**	**1**	**1**	**1**	**1**	**1**	**1**	**1**	**1**
Pork Porc	1	1	1	1	1	1	1	1	1	1[1]
Kiribati **Kiribati**	**1**	**1**	**1**	**1**	**1**	**1**	**1**	**1**	**1**	**1**
Pork[1] Porc[1]	1	1	1	1	1	1	1	1	1	1
New Caledonia **Nouvelle-Calédonie**	**3**	**4**	**4**	**3**	**3**	**3**	**3**	**4**	**5**	**5**
Beaf and veal Boeuf et veau	3	3	3	2	2	2	2	3	3	3[1]
Pork Porc	1	1	1	1	1	1	1	1	1	1[1]
New Zealand **Nouvelle-Zélande**	**1 145**	**1 263**	**1 133**	**1 210**	**1 233**	**1 213**	**1 050**	**1 143**	**1 180**	**1 100**
Beaf and veal Boeuf et veau	433	487	468	555	572	557	479	540	545	550
Pork Porc	43	48	48	44	46	45	43	44	48	48
Mutton and lamb Mouton et agneau	668[1]	729[1]	617[1]	611	616	611	528	559	587	502
Palau[3] **Palaos[3]**	**1**	**1**	**1**	**1**	**1**	**1**	**1**	**1**	**1**	**1**
Pork[1] Porc[1]	1	1	1	1	1	1	1	1	1	1

42

Meat
Production: thousand metric tons [*cont.*]
Viande
Production : milliers de tonnes métriques [*suite*]

Country or area Pays ou zone	1984	1985	1986	1987	1988	1989	1990	1991	1992	1993
Papua New Guinea **Papouasie-Nvl-Guinée** Beaf and veal[1]	27	27	28	28	28	29	29	29	30	30
Boeuf et veau[1] Pork[1]	2	2	2	2	2	2	2	2	2	2
Porc[1]	25	25	26	26	26	27	27	27	28	28
Samoa **Samoa** Beaf and veal	4	4	4	4	4	5	5	5	4	5
Boeuf et veau Pork[1]	* 1	* 1	1	1	1	1	1	1[1]	1[1]	1[1]
Porc[1]	3	3	3	4	4	4	4	4	4	4
Solomon Islands **Iles Salomon** Beaf and veal[1]	2	2	2	2	3	3	3	3	3	3
Boeuf et veau[1] Pork[1]	1	1	1	1	1	1	1	1	1	0
Porc[1]	2	2	2	2	2	2	2	2	2	2
Tonga **Tonga** Pork[1]	2	2	2	2	2	2	2	2	2	2
Porc[1]	1	1	1	1	1	2	2	2	2	1
Vanuatu **Vanuatu** Beaf and veal	5	5	5	5	5	5	5	6	6	6
Boeuf et veau Pork[1]	2	2	2	3	3	3	3	3	3[1]	3[1]
Porc[1]	3	3	3	3	2	2	2	2	2	3
former USSR†* **ancienne URSS†*** Beaf and veal	14 037	14 050	14 781	15 480	16 211	16 494	16 476	15 174	..	..
Boeuf et veau Pork	7 244	7 370	7 840	8 281	8 616	8 800	8 814	8 261	..	..
Porc Mutton and lamb *	5 927	5 853	6 047	6 299	6 595	6 700	6 654	5 997	..	..
Mouton et agneau *	866	827	894	900	1 000	994	1 008	916	..	..

Source:
Food and Agriculture Organization of the United Nations
(Rome).

Source:
Organisation des Nations Unies pour l'alimentation et
l'agriculture (Rome).

† For detailed descriptions of data pertaining to
former Czechoslovakia, Germany, SFR Yugoslavia and former
USSR, see Annex I - Country or area nomenclature, regional
and other groupings.

† Pour les descriptions en détails des données
relatives à l'ancienne Tchécoslovaquie, l'Allemagne, la Rfs
Yougoslavie et l'ancienne URSS, voir l'Annexe I -
Nomenclature des pays ou zones, groupements régionaux et
autres groupements.

1 FAO estimate.
2 Prior to 1993, including data for Eritrea.
3 Including data for Federated States of Micronesia, Marshall
Is. and Northern Mariana Is.

1 Estimation de la FAO.
2 Avant de 1993, y compris les données pour Erythrée.
3 Y compris les données pour les Etats fédérés de Micronésie,
les îles Marshall et les îles Mariannes du Nord.

43
Beer
Bière
Production: thousand hectolitres
Production : milliers d'hectolitres

Country or area Pays ou zone	1983	1984	1985	1986	1987	1988	1989	1990	1991	1992
Total	963 123	948 796	957 526	976 526	994 237	1 032 408	1 054 153	1 093 511	1 128 357	1 147 096
Algeria Algérie	468	520	526	503	491	475	365	325	301	337
Angola Angola	690	* 650	653	583	466	...	...	...	...	...
Argentina Argentine	3 157	4 079	3 827	5 548	5 861	5 232	5 110	6 173	7 991	...
Armenia Arménie	..	..	..	..	..	..	..	..	419	...
Australia [12] Australie [12]	19 724	18 729	18 553	18 627	18 589	18 912	19 508	19 390	19 155	18 623
Austria Autriche	8 403	8 440	8 836	9 017	8 638	8 938	9 174	9 799	9 971	10 176
Azerbaijan Azerbaïdjan	..	..	..	..	..	..	..	..	5 159	1 855
Barbados Barbade	74	58	55	63	73	76	80	77	66	63
Belarus Bélarus	...	...	...	...		...		...	3 389	...
Belgium Belgique	14 224	14 311	13 931	13 715	13 988	13 792	13 164	14 141	13 799	14 259
Belize Belize	32	24	23	21	26	29	32	39	40	43
Bolivia Bolivie	717	701	606	803	...	...		...	127	...
Botswana Botswana	648	667	...	...	830	1 002	1 076	1 214	1 283	1 290
Brazil Brésil	25 861	25 980	27 092	34 005	33 893	36 445	42 343	43 849	47 107	...
Bulgaria Bulgarie	5 505	5 764	5 838	6 023	6 212	6 332	6 720	6 507	4 880	4 695
Burkina Faso Burkina Faso	701	701	525	422	389	401	398	350	394	71 .
Burundi Burundi	644	798	817	897	937	953	919	1 107	981	1 007
Cameroon Cameroun	3 354	3 976	4 904	5 308	5 857	5 105	...	...	...	...
Canada Canada	23 332	23 558	23 237	23 547	...	...	...	...	...	...
Central African Rep. Rép. centrafricaine	211	218	276	280	299	...	...	...	...	...
Chad Tchad	114	135	152	118	107	109	115	116	144	129
Chile Chili	1 760	1 781	1 892	2 050	2 548	2 650	2 765	2 653	2 788	3 349
China [3] Chine [3]	16 300	22 400	31 000	41 301	54 043	65 600	64 339	69 221	83 800	102 064

43

Beer
Production: thousand hectolitres [*cont.*]
Bière
Production : milliers d'hectolitres [*suite*]

Country or area Pays ou zone	1983	1984	1985	1986	1987	1988	1989	1990	1991	1992
Colombia Colombie	14 494	...	15 509	16 915	15 355	11 973	14 027	15 098	...	...
Congo Congo	788	906	882	897	762	744	618	566	686	708
Côte d'Ivoire Côte d'Ivoire	1 400	1 245	...	...	...	...	...	...	...	...
Croatia Croatie	..	..	..	..	..	..	..	..	2 248	2 720
Cuba Cuba	2 582	2 607	2 736	2 931	3 288	3 324	3 333	...	...	...
Cyprus Chypre	228	232	247	257	271	296	318	342	331	...
former Czechoslovakia† anc. Tchécoslovaquie†	24 957	23 768	22 354	22 789	22 228	22 670	23 333	21 966	..	..
Czech Republic République tchèque	..	..	..	..	..	..	..	..	17 902	18 899
Denmark[4] Danemark[4]	11 038	8 671	8 286	9 064	8 754	9 160	9 217	9 362	...	9 775
Dominican Republic Rép. dominicaine	999	945	1 038	1 099	1 269	...	...	...	...	...
Ecuador Equateur	626	...	...	...	...	...	...	...	...	1 826
Egypt Egypte	314	360	423	460	470	510	490	500	440	420
El Salvador El Salvador	344	...	...	...	...	...	...	...	...	...
Estonia Estonie	..	..	..	..	..	..	..	..	68	43
Ethiopia[5] Ethiopie[5]	655	613	818	797	797	877	777	499[6]	...	...
Fiji Fidji	191	185	178	160	148	157	175	194	183	...
Finland Finlande	2 840	2 968	3 110	3 238	3 509	3 752	3 947	4 151	4 418	4 253
France France	22 086	20 280	19 300	19 000	...	...	...	...	...	...
French Polynesia Polynésie française	96	94	95	118	120	125	121	121	124	129
Gabon Gabon	733	769	850	888	879	851	786	819	814	785
Germany † Allemagne†	..	..	..	..	..	..	..	..	112 071	114 089
F. R. Germany R. f. Allemagne	91 382	87 309	87 895	88 476	86 962	87 044	87 761	99 150	..	..
former German D. R. anc. R. d. allemande	25 313	24 500	24 288	24 316	24 128	24 521	24 843	15 885	..	..
Ghana Ghana	310	452	429	546	589	614	...	...	...	...

43

Beer
Production: thousand hectolitres [cont.]
Bière
Production : milliers d'hectolitres [suite]

Country or area Pays ou zone	1983	1984	1985	1986	1987	1988	1989	1990	1991	1992
Greece[3] Grèce[3]	2 850	2 970	3 262	3 103	3 461	4 524	3 869	3 961	3 772	...
Grenada Grenade	12	12	14	16	17	20	21	25	...	...
Guatemala Guatemala	645	672	715	613	795	869	...	...	...	...
Guyana Guyana	45	77	80	80	95	134	...	...	...	...
Honduras Honduras	469	511	* 500	* 548	...	...	...	...	...	...
Hungary Hongrie	7 830	7 962	8 740	8 963	9 045	9 425	9 722	9 918	9 570	9 162
Iceland Islande	30	36	34	31	38	51	39	31	...	...
India[7] Inde[7]	1 673	1 923	2 030	2 036	1 492	1 635	1 972	1 915	2 204	...
Indonesia Indonésie	942	847	797	718	833	957	953	1 042	1 044	...
Ireland[8] Irlande[8]	4 223	4 214	4 352	4 479	4 360	4 624	5 094	5 236	...	...
Israel Israël	408	415	511	523	536	556	511	567	...	...
Italy Italie	10 320	9 201	10 381	11 372	11 503	11 589	10 615	11 248	11 049	10 489
Jamaica Jamaïque	604	517	568	632	700	809	852	887	715	828
Japan[9] Japon[9]	50 534	45 978	48 522	50 754	54 922	58 572	62 869	65 636	69 157	70 106
Jordan Jordanie	63	50	46	45	46	45	46	...	...	...
Kazakhstan Kazakhstan	..	..	..	..	..	..	..	..	31 330	23 011
Kenya Kenya	2 200	2 280	2 630	3 010	3 070	3 140	3 150	3 311	...	...
Korea, Republic of Corée, République de	7 096	7 626	7 919	8 040	8 789	10 312	12 108	13 045	15 928	15 673
Kyrgyzstan Kirghizistan	..	..	..	..	..	..	..	..	445	310
Latvia Lettonie	..	..	..	..	..	..	..	..	1 295	859
Lebanon * Liban *	...	...	...	...	...	...	...	130	...	...
Liberia Libéria	117	111	97	106	136	158	...	...	...	...
Luxembourg Luxembourg	651	634	738	732	662	635	618	600	572	569
Madagascar Madagascar	236	229	241	255	240	201	232	298	236	226

43

Beer
Production: thousand hectolitres [cont.]
Bière
Production : milliers d'hectolitres [suite]

Country or area Pays ou zone	1983	1984	1985	1986	1987	1988	1989	1990	1991	1992
Malawi Malawi	889	622	658	627	675	659	757	752	763	...
Malaysia Malaisie	954	998	1 026	* 1 013	988	1 115	1 263	1 401	1 413	...
Mali Mali	9	...	...	...	...	...	...	...	...	...
Mauritius Maurice	150	166	172	188	238	261	254		...	...
Mexico Mexique	24 139	25 616	27 215	27 353	31 482	33 261	37 355	38 734	41 092	42 262
Mongolia Mongolie	91	97	89	65	50	49	67		...	...
Morocco[10] Maroc[10]	383	393	411	...	...		...	...	...	...
Mozambique Mozambique	445	375	228	230	214	297	413	337	367	...
Myanmar[11] Myanmar[11]	34	40	57	50	17	10	21	24	30	19
Nepal[12] Népal[12]	31	23	60	37	52	63	73	...	...	...
Netherlands[4] Pays-Bas[4]	17 330	17 050	17 530	17 990	17 550	17 437	18 908	20 055	19 893	...
New Zealand[4] Nouvelle-Zélande[4]	3 679	3 839	3 863	4 062	4 087	3 925	3 916	3 890	3 627	...
Nicaragua Nicaragua	503	522	...	...	...	...	...	...	...	...
Niger Niger	100	...	...	...	...	...	...	...	...	...
Nigeria Nigéria	7 516	7 355	8 052	10 160	6 695	...	...	...	...	...
Norway Norvège	1 911	1 944	1 991	...	2 198	2 238	2 229	2 281	...	...
Panama Panama	708	734	797	925	1 014	871	1 042	1 162	1 219	1 163
Papua New Guinea Papouasie-Nvl-Guinée	461	* 474	* 550	* 555	...	...	...	...	...	...
Paraguay Paraguay	717	753	768	887	918	903	1 055	1 076	1 126	1 140
Peru Pérou	5 230	5 388	5 724	7 477	8 559	7 018	5 548	5 732	...	...
Philippines Philippines	9 260	19 030	11 395	8 780	1 175	...	...	...	...	...
Poland Pologne	10 306	9 867	11 078	11 307	11 897	12 238	12 082	11 294	13 633	14 139
Portugal Portugal	4 265	3 682	3 795	4 128	4 977	5 619	6 874	6 919	6 309	6 454
Puerto Rico[1] Porto Rico[1]	377	347	260	261	266	360	525	628	644	545

43

Beer
Production: thousand hectolitres [*cont.*]
Bière
Production : milliers d'hectolitres [*suite*]

Country or area Pays ou zone	1983	1984	1985	1986	1987	1988	1989	1990	1991	1992
Republic of Moldova République de Moldova	..	..	..	..	..	..	..	..	660	410
Romania Roumanie	9 928	9 845	9 847	10 603	10 364	10 655	11 513	10 527	9 803	10 014
Russian Federation Fédération de Russie	..	..	..	..	..	..	..	..	32 700	...
Rwanda Rwanda	555	576	631	578	596	711	...	...	...	...
Saint Kitts and Nevis Saint-Kitts-et-Nevis	9	9	10	11	14	15	17	17	17	...
Sao Tome and Principe * Sao Tomé-et-Principe *	23	17	31	30	29	...	...	...	...	...
Senegal Sénégal	189	164	168	197	157	170	...	...	...	...
Seychelles Seychelles	39	38	41	42	47	50	52	53	59	70
Sierra Leone Sierra Leone	65	102	60	31	50	40	...	...	...	...
Slovakia Slovaquie	..	..	..	..	..	..	..	..	4 082	3 686
Slovenia Slovénie	..	..	..	..	..	..	..	..	2 203	1 783
South Africa Afrique du Sud	11 586	12 400	12 284	13 751	16 710	18 330	18 610	17 750	17 710	18 290
Spain Espagne	20 823	21 464	22 475	23 510	24 788	26 141	27 546	27 940	26 482	...
Sri Lanka Sri Lanka	55	61	74	85	92	* 92	60	...	...	...
Suriname Suriname	151	150	142	112	124	99	117	122	122	...
Sweden Suède	3 705	3 616	3 784	3 979	4 106	4 391	4 722	4 711	4 663	...
Switzerland[4] Suisse[4]	4 171	4 078	4 076	4 087	4 045	4 049	4 121	4 143	4 137	4 020
Syrian Arab Republic Rép. arabe syrienne	81	89	94	101	91	83	95	99	99	102
Thailand Thaïlande	1 456	1 639	1 052	863	973	1 303	1 801	2 635	2 840	3 252
Togo Togo	392	359	423	464	452	...	...	...	...	...
Trinidad and Tobago Trinité-et-Tobago	356	347	274	246	371	349	527	412	487	402
Tunisia Tunisie	423	396	373	355	347	405	...	...	...	...
Turkey Turquie	3 226	2 622	1 965	1 890	2 470	2 671	2 994	3 544	4 188	4 843
Uganda Ouganda	142	148	81	66	169	215	195	194	...	...

43
Beer
Production: thousand hectolitres [*cont.*]
Bière
Production : milliers d'hectolitres [*suite*]

Country or area Pays ou zone	1983	1984	1985	1986	1987	1988	1989	1990	1991	1992
Ukraine Ukraine	..	..	..	..	..	..	..	..	13 093	...
former USSR† ancienne URSS†	66 081	65 385	65 721	48 907	50 711	55 811	60 182	62 507	..	..
United Kingdom Royaume-Uni	60 324	60 105	59 655	59 439	59 895	60 156	54 950	70 800	...	...
United Rep.Tanzania Rép. Unie de Tanzanie	650	670	758	652	588	530	537	450	498	493
United States' Etats-Unis'	228 945	228 955	228 960	230 588	229 321	231 985	232 510	236 670	...	237 029
Uruguay Uruguay	...	...	468	678	709	603	...	...	...	...
Viet Nam Viet Nam	527	845	866	872	840	976	892	1 000	1 312	1 685
former Dem. Yemen anciennce Yémen dém.	61	66	70	59	50	...	...	...	...	...
Yugoslavia, SFR† Yougoslavie, Rfs†	12 398	10 925	10 656	11 643	12 054	11 970	11 286	...	..	...
Zaire Zaïre	3 069	3 699	4 222	4 284	...	...	...	...	...	...
Zambia Zambie	21 861	* 4 169	* 717[13]	* 782[13]	714[13]	825[13]	...	...	...	...
Zimbabwe Zimbabwe	1 235	920	1 100	...	...	...	...	...	...	...

Source:
Industrial statistics database of the Statistical Division
of the United Nations Secretariat.

† For detailed descriptions of data pertaining to
former Czechoslovakia, Germany, SFR Yugoslavia and former
USSR, see Annex I - Country or area nomenclature, regional
and other groupings.

1 Twelve months ending 30 June of year stated.

2 Excluding light beer containing less than 1.15% by volume of
alcohol.
3 Original data in metric tons.
4 Sales.
5 Twelve months ending 7 July of the year stated.

6 Beginning 1990, excluding Eritrea.
7 Production by large and medium scale establishments only.
8 Twelve months ending 30 September of year stated.

9 Twelve months beginning 1 April of year stated.

10 Including some carbonated drinks.
11 Government production only.
12 Twelve months beginning 16 July of year stated.

13 Incomplete coverage.

Source:
Base de données pour les statistiques industrielles de la
Division de statistique du Secrétariat de l'ONU.

† Pour les descriptions en détails des données
relatives à l'ancienne Tchécoslovaquie, l'Allemagne, la Rfs
Yougoslavie et l'ancienne URSS, voir l'Annexe I -
Nomenclature des pays ou zones, groupements régionaux et
autres groupements.

1 Période de douze mois finissant le 30 juin de l'année
indiquée.
2 Non compris la bière légère contenant moins de 1.15 p. 100
en poids d'alcohol.
3 Données d'origine exprimées en tonnes.
4 Ventes.
5 Période de douze mois finissant le 7 juillet de l'année
indiquée.
6 A partir de 1990, non compris Erythrée.
7 Production des grandes et moyennes entreprises seulement.
8 Période de douze mois finissant le 30 septembre de l'année
indiquée.
9 Période de douze mois commençant le 1er avril de l'année
indiquée.
10 Y compris certaines boissons gazeuses.
11 Production de l'Etat seulement.
12 Période de douze mois commençant le 16 juillet de l'année
indiquée.
13 Couverture incomplète.

44
Cigarettes
Cigarettes
Production: millions
Production : millions

Country or area Pays ou zone	1983	1984	1985	1986	1987	1988	1989	1990	1991	1992
Total	**4 463 541**	**4 611 937**	**4 758 379**	**4 880 989**	**5 019 687**	**5 121 037**	**5 139 872**	**5 289 032**	**4 988 363**	**5 005 247**
Albania Albanie	* 6 100[1]	6 200[1]	5 348	5 624	5 467	5 310	6 184	4 947	1 703	...
Algeria[2] Algérie[2]	17 500	18 000	18 500	19 000	17 699	17 016	15 950	18 775	17 848	14 423
Angola[1] Angola[1]	* 2 400	2 400	2 400	2 400	2 400	2 400	2 400	2 400	2 400	2 400
Argentina[2] Argentine[2]	28 241	30 843	31 092	32 112	29 745	32 000	33 722	33 140	34 560	* 37 000
Armenia Arménie	..	..	..	..	..	..	..	..	6 614	...
Australia[3] Australie[3]	35 385	32 667	32 453	...	...	34 106	34 736	36 263	34 977	* 34 000
Austria Autriche	15 625	14 907	16 051	15 354	15 067	14 324	14 402	14 961	16 406	15 836
Azerbaijan Azerbaïdjan	..	..	..	..	..	..	..	..	7 256	4 855
Bangladesh[4] Bangladesh[4]	14 031	14 843	14 393	14 365	14 762	14 031	14 088	12 289	13 604	12 535
Barbados[2] Barbade[2]	241	238	200	199	162	149	143	135	124	115
Belarus Bélarus	..	..	..	..	..	..	..	..	15 009	...
Belgium[5] Belgique[5]	29 738	29 422	30 184	28 619	28 953	28 683	27 489	27 758	27 303	29 576
Belize Belize	57	65	74	76	99	94	97	100	104	107
Bolivia[1] Bolivie[1]	628	368	533	899	900	1 200	1 200	1 200	1 200	1 200
Brazil[1] Brésil[1]	129 200	127 800	146 300	168 000	161 400	157 900	162 700	173 987	175 396	169 000
Bulgaria Bulgarie	91 296[3]	92 055[3]	93 975[3]	89 918[3]	90 300	86 800	82 600	73 000	73 300	71 000
Burkina Faso Burkina Faso	700	660	671	510	430	533	609	822	983	979
Burundi Burundi	294	334	293	257	284	333	286	384	450	453
Cambodia[1] Cambodge[1]	* 4 100	4 100	4 150	4 175	4 175	4 200	4 200	4 200	4 200	4 200
Cameroon Cameroun	1 954	2 319	2 128	3 800[1]	4 770[1]	4 800[1]	4 800[1]	4 900[1]	5 000[1]	5 000[1]
Canada Canada	63 900	61 600	63 486	55 632	54 030	53 858[1]	48 792[1]	46 111[1]	46 815[1]	45 500[1]
Central African Rep. Rép. centrafricaine	403	413	539	407	460	...	...	...	...	...
Chad Tchad	...	14 902	15 456	11 268	9 921	10 202	9 350	12 378	22 438	...
Chile Chili	7 680	8 107	8 053	8 296	8 183	9 061	9 686	10 011	10 000	10 790

44

Cigarettes
Production: millions [*cont.*]
Cigarettes
Production : millions [*suite*]

Country or area Pays ou zone	1983	1984	1985	1986	1987	1988	1989	1990	1991	1992
China[1] Chine[1]	968 800	1 066 000	1 180 000	1 296 500	1 440 500	1 545 000	1 597 800	1 648 765	1 599 700	1 650 000
Colombia[1] Colombie[1]	19 141	23 400	24 050	24 181	21 987	18 253	16 570	14 490	13 585	13 250
Congo[2] Congo[2]	895	903	1 027	888	787	770	709	* 1 000	581	431
Costa Rica[1] Costa Rica[1]	2 200	2 200	2 200	2 000	2 135	2 135	2 050	2 030	2 000	2 000
Côte d'Ivoire Côte d'Ivoire	3 640	3 210	4 000[1]	4 200[1]	4 400[1]	4 500[1]	4 500[1]	4 500[1]	4 500[1]	4 500[1]
Croatia Croatie	..	..	..	..	..	..	..	..	11 655	12 833
Cuba Cuba	16 802	18 697	17 961	16 841	15 398	16 885	16 520	...	...	...
Cyprus Chypre	2 878	2 586	2 366	1 914	3 717	4 412	3 935	4 601	5 497	* 4 500
former Czechoslovakia† anc. Tchécoslovaquie†	25 016	24 603	23 840	24 998	25 365	25 502	25 428	26 708	..	..
Denmark[6] Danemark[6]	9 846	10 583	10 966	11 246	11 162	11 144	11 209	11 387	11 407	11 439
Dominica Dominique	27	27	29	29	...	...	...	...	...	...
Dominican Republic Rép. dominicaine	3 604	3 696	3 826	4 164	4 473	4 802[1]	4 473[1]	4 805[1]	4 376[1]	4 582[1]
Ecuador Equateur	4 980	5 000	4 800	3 967	* 4 600	* 4 600	* 4 600	* 4 600	* 4 600	3 000
Egypt Egypte	35 571	44 599	47 520	44 394	47 500	43 846	43 208	39 837	40 154	42 516
El Salvador El Salvador	2 128	2 500[1]	2 300[1]	2 100[1]	2 100[1]	1 967[1]	1 970[1]	1 655[1]	1 620[1]	1 620[1]
Estonia Estonie	..	..	..	..	..	..	..	..	3 577	1 780
Ethiopia Ethiopie	1 946[7]	2 036[7]	2 229[7]	2 619[7]	3 040[7]	2 972[7]	2 711[7]	2 258[7,8]	* 2 300[1]	2 300[1]
Fiji Fidji	533	513	513	478	466	466	505	531	* 514	* 650
Finland Finlande	8 375	8 345	8 185	8 540	9 061	9 619	8 931	8 974	8 180	8 106
France France	62 147	60 729	67 376	59 122	54 120	* 53 307	* 54 225	53 000	* 50 311	* 49 000
Gabon Gabon	328	341	355	368	280	422	319	253	359	399
Germany † • Allemagne† F. R. Germany R. f. Allemagne	155 883	160 680	163 267	166 665	157 586	159 499	159 477	177 905	..	..
former German D. R. anc. R. d. allemande	27 387	28 018	26 909	27 364	27 625	28 576	28 625	22 469	..	..
Ghana Ghana	1 074	2 008	1 942	1 826	1 734	1 831	1 616	1 805	* 2 100	* 2 100

44

Cigarettes
Production: millions [*cont.*]
Cigarettes
Production : millions [*suite*]

Country or area Pays ou zone	1983	1984	1985	1986	1987	1988	1989	1990	1991	1992
Greece Grèce	24 286	25 699	27 635	27 118	28 859	26 244	26 123	26 175	27 700	* 29 250
Grenada Grenade	26	25	24	23	22	23	24	22	...	...
Guatemala Guatemala	2 156	1 968	1 988	1 926	2 270	1 858[1]	1 997[1]	1 955[1]	1 907[1]	2 001[1]
Guyana Guyana	408	373	467	481	477	470	600[1]	600[1]	600[1]	600[1]
Haiti Haïti	909	886	806	829	903	870[1]	870[1]	870[1]	870[1]	870[1]
Honduras Honduras	2 022	2 145	2 311	2 300[1]	2 278[1]	2 319[1]	2 582[1]	2 862[1]	2 300[1]	2 200[1]
Hong Kong[5] Hong-kong[5]	6 577	8 681	12 496	11 841	15 309	24 000	20 901	21 700	32 721	* 30 000
Hungary Hongrie	25 676	26 679	26 430	26 351	26 541	26 171	26 540	28 212	26 124	26 835
India[9] Inde[9]	86 900	86 217	80 600	74 991	64 782	54 013	58 066	61 162	65 270	...
Indonesia Indonésie	91 463	95 634	102 300	114 312	124 430	136 271	148 000	155 000	153 000	* 145 000
Iran, Islamic Rep. of[10] Iran, Rép. islamique d'[10]	15 104	16 154	16 168	15 239	15 068	13 790	9 923	12 319	* 15 000	* 15 000
Iraq Iraq	7 900[1]	7 000[1]	7 000[1]	10 000[1]	13 000[1]	20 250[1]	27 000[1]	26 000[1]	13 000[1]	5 794
Ireland Irlande	7 531	7 391	8 037	6 825	6 659	6 420	6 161	6 218	6 377	* 7 850
Israel Israël	6 373	6 714	6 709	6 723	6 888	5 882	5 245	5 440	* 5 650	* 5 650
Italy[2] Italie[2]	83 700	80 452	78 774	75 541	70 447	67 394	67 942	61 736	57 634	53 799
Jamaica Jamaïque	1 359	1 270	1 314	* 1 140	1 273	1 303	1 383	1 380	1 219	* 1 273
Japan[11] Japon[11]	306 320	306 867	303 000	309 200	273 700	267 000	268 400	268 100	275 000	279 000
Jordan Jordanie	4 535	5 027	3 905	3 732	4 378	3 678	2 926	4 100	3 800	3 105
Kenya Kenya	5 584	5 391	5 409	5 821	6 372	6 641	6 661	6 648	6 000	* 6 000
Korea, Republic of Corée, République de	75 275	77 984	75 532	78 694	81 816	86 014	86 759	91 923	94 336	96 648
Kyrgyzstan Kirghizistan	..	..	..	..	..	..	..	..	4 015	3 120
Lao People's Dem. Rep.[1] Rép. dém. pop. lao[1]	1 100	1 100	1 125	1 200	1 200	1 200	1 200	1 200	1 200	1 200
Latvia Lettonie	..	..	..	..	..	..	..	..	4 765	3 435
Lebanon[1] Liban[1]	200	200	1 800	* 1 800	2 500	3 500	4 000	4 000	4 000	4 000
Liberia Libéria	59	123	82	92	22[1]	22[1]	22[1]	22[1]	22[1]	22[1]

44
Cigarettes
Production: millions [*cont.*]
Cigarettes
Production : millions [*suite*]

Country or area Pays ou zone	1983	1984	1985	1986	1987	1988	1989	1990	1991	1992
Libyan Arab Jamah.[1] Jamah. arabe libyenne[1]	3 400	3 500	3 500	3 500	3 500	3 500	3 500	3 500	3 500	3 500
Madagascar[2] Madagascar[2]	1 780	2 137	2 368	2 188	2 669	1 817	2 341	1 955	1 950	2 223
Malawi Malawi	778	824	857	874	908	956	* 1 080	1 061	951	* 1 000
Malaysia[2] Malaisie[2]	13 502	14 671	13 839	13 706	13 729	15 904	16 169	17 331	17 498	16 574
Malta Malte	1 128	1 129[1]	1 300[1]	1 300[1]	1 400[1]	1 425[1]	1 450[1]	1 475[1]	1 475[1]	1 475[1]
Mauritius Maurice	982[2]	1 253[2]	1 418[2]	1 266[2]	1 392[2]	1 304[2]	1 318[2]	1 400[1]	1 400[1]	1 400[1]
Mexico Mexique	49 337	51 666	54 332	49 898	54 644	50 507	53 920	55 380	54 680	55 988
Morocco Maroc	586	548	593	663	671	684	615	640	602	515
Mozambique Mozambique	889	792	710	1 053	899	670	1 044	1 016	453	...
Myanmar[12] Myanmar[12]	2 949	2 515	3 506	1 574	553	952	505	979	682	396
Nepal Népal	3 741	4 252	4 741	5 600	6 046	5 664	6 706	...	...	...
Netherlands Pays-Bas	46 337	47 761	51 321	53 339	59 801	61 693	63 148	71 992	86 832	* 90 000
New Zealand Nouvelle-Zélande	6 196	6 274	5 767	5 471	5 400	6 250[1]	6 250[1]	6 250[1]	6 250[1]	6 250[1]
Nicaragua Nicaragua	1 919	2 318	2 400[1]	2 400[1]	2 400[1]	2 400[1]	2 400[1]	2 400[1]	2 400[1]	2 400[1]
Nigeria[1] Nigéria[1]	* 9 800	9 000	9 100	9 600	10 000	10 000	10 000	10 000	10 000	10 000
Norway[1] Norvège[1]	* 792	790	1 082	1 400	1 400	1 655	1 750	1 480	1 730	1 825
Pakistan[4] Pakistan[4]	38 199	40 096	38 921	39 593	39 929	40 697	31 567	32 279	29 887	29 673
Panama Panama	981	911	873	873	826	671	637	814	771	724
Paraguay Paraguay	931	878	834	750	1 134	* 2 730	2 730	992	827	777
Peru Pérou	3 582	3 489	3 102	3 741	3 420	2 671	2 439	2 672	...	...
Philippines[4] Philippines[4]	57 812	58 600	62 300	60 700	61 072	66 850[1]	69 700[1]	71 500[1]	70 710[1]	75 400[1]
Poland Pologne	82 823	86 363	90 021	94 212	98 666	89 681	81 342	91 497	90 407	86 571
Portugal[2] Portugal[2]	15 531	14 798	14 900	15 166	15 481	15 129	15 424	16 542	16 545	14 008
Republic of Moldova République de Moldova	..	..	..	..	..	..	..	..	9 164	8 582
Romania[3] Roumanie[3]	36 029[1]	32 888[1]	32 471[1]	31 013	32 918	33 349	22 121	18 090	17 722	17 781

44

Cigarettes
Production: millions [cont.]
Cigarettes
Production : millions [suite]

Country or area Pays ou zone	1983	1984	1985	1986	1987	1988	1989	1990	1991	1992
Russian Federation Fédération de Russie	..	..	..	..	..	..	..	..	144 000	...
Rwanda Rwanda	812	696	698	649	698	458	...	...	...	...
Senegal Sénégal	2 516[2]	2 501[2]	2 979[2]	3 300[2]	1 872[2]	1 348[2]	3 350[1]	3 350[1]	3 350[1]	3 350[1]
Seychelles Seychelles	52	65	56	60	68	61	58	67	69	62
Sierra Leone Sierra Leone	966	1 341	1 189	940	1 100	933	1 200[1]	1 200[1]	1 200[1]	1 200[1]
Singapore Singapour	3 165[2]	3 000[1]	2 400[1]	2 000[1]	2 300[1]	4 700[1]	5 982[1]	9 620[1]	10 500[1]	11 760[1 2]
Slovakia Slovaquie	..	..	..	..	..	..	..	..	8 721	...
Slovenia Slovénie	..	..	..	..	..	..	..	..	4 798	5 278
South Africa Afrique du Sud	31 101	32 501	31 704	30 680	17 773	28 002	36 665	40 792	40 163	35 563
Spain Espagne	77 396	83 884	80 495	81 705	87 362	85 612	75 817	75 995	81 843	* 87 500
Sri Lanka Sri Lanka	5 858	5 200	6 168	6 111	5 894	5 028	5 136	5 621	5 307	* 5 400
Sudan Soudan	1 354[2]	1 884[2]	2 700	2 200	2 200	1 800	750[1]	750[1]	750[1]	750[1]
Suriname Suriname	460	521	514	557	492	501	526	487	337	419
Sweden Suède	10 423	10 420	10 090	10 197	10 103	10 208	10 107	9 648	9 594	* 9 150
Switzerland Suisse	25 681	25 449	23 150	23 555	24 863	27 075	28 059	31 771	32 943	33 740
Syrian Arab Republic[2] Rép. arabe syrienne [2]	13 211	11 274	12 127	12 112	9 143	7 056	6 345	6 855	7 974	8 093
Thailand Thaïlande	28 941	29 170	29 192	29 530	31 407	33 992	37 365	38 316	39 359	* 40 500
Trinidad and Tobago[2] Trinité-et-Tobago[2]	1 148	1 064	987	920	583	766	726	701	881	656
Tunisia Tunisie	5 900	5 527	6 334	8 097	6 937	7 125	7 339[1]	7 530	7 920	* 79 200
Turkey Turquie	61 497	62 085	62 000	59 740	58 617	60 153	46 570	63 055	71 106	67 549
Uganda Ouganda	645	966	1 416	1 420	1 435	1 638	1 586	1 290	* 2 200	* 2 200
Ukraine Ukraine	..	..	..	..	..	..	..	..	66 645	...
former USSR† ancienne URSS†	369 193	373 793	381 274	383 878	378 475	358 218	343 288	313 082	..	..
United Kingdom[2 13] Royaume-Uni[2 13]	93 400	93 400	88 600	83 300	89 900	87 900	104 155	112 000	127 000	* 126 538
United Rep.Tanzania Rép. Unie de Tanzanie	3 800	4 000	2 666	2 748	2 635	2 785	2 845	3 742	3 870	* 4 150

44
Cigarettes
Production: millions [*cont.*]
Cigarettes
Production : millions [*suite*]

Country or area Pays ou zone	1983	1984	1985	1986	1987	1988	1989	1990	1991	1992
United States[4] Etats-Unis[4]	667 000	668 800	655 300	652 000	689 400	694 500	677 200	709 700	694 500	703 134
Uruguay Uruguay	* 3 750[1]	3 800[1]	3 098	3 583	3 383	3 446	3 900[1]	3 900[1]	3 900[1]	3 900[1]
Venezuela[1] Venezuela[1]	20 156	20 643	19 760	18 400	18 100	18 824	20 599	23 560	24 236	24 400
Viet Nam Viet Nam	18 476	21 236	21 012	22 364	19 630	17 754	23 288	24 990	25 960	* 24 600
former Dem. Yemen anciennce Yémen dém.	1 204	1 202	1 467	1 147	1 193	...	...	...	...	...
Yugoslavia, SFR† Yougoslavie, Rfs†	58 493	559 650	576 520	557 870	561 280	601 690	512 870	582 000	..	..
Zaire Zaïre	3 472[14]	3 500	3 525	3 600	4 500[1]	5 200	5 200[1]	5 200[1]	5 200[1]	5 200[1]
Zambia Zambie	* 1 400	1 400[1]	1 400[1]	1 450[1]	1 500[1]	1 500[1]	1 500[1]	1 500[1]	1 500[1]	1 500[1]
Zimbabwe[1] Zimbabwe[1]	2 400	2 300	2 318	2 300	2 420	2 560	2 518	2 600	3 240	3 025

Source:
Industrial statistics database of the Statistical Division
of the United Nations Secretariat.

† For detailed descriptions of data pertaining to
former Czechoslovakia, Germany, SFR Yugoslavia and former
USSR, see Annex I - Country or area nomenclature, regional
and other groupings.

1 Source: US Department of Agriculture, (Washington, DC).
2 Original data in units of weight. Computed on the basis of
  one million cigarettes per ton.
3 Including cigars.
4 Twelve months ending 30 June of year stated.

5 Including cigarillos.
6 Sales.
7 Twelve months ending 7 July of the year stated.

8 Beginning 1990, excluding Eritrea.
9 Production by large and medium scale establishments only.
10 Production by establishments employing 50 or more persons.
11 Twelve months beginning 1 April of year stated.

12 Government production only.
13 Sales by manufacturers employing 25 or more persons.
14 Including cut tobacco.

Source:
Base de données pour les statistiques industrielles de la
Division de statistique du Secrétariat de l'ONU.

† Pour les descriptions en détails des données
relatives à l'ancienne Tchécoslovaquie, l'Allemagne, la Rfs
Yougoslavie et l'ancienne URSS, voir l'Annexe I -
Nomenclature des pays ou zones, groupements régionaux et
autres groupements.

1 Source: "US Department of Agriculture," (Washington, DC).
2 Données d'origine exprimées en poids. Calcul sur la base
  d'un million cigarettes par tonne.
3 Y compris les cigares.
4 Période de douze mois finissant le 30 juin de l'année
  indiquée.
5 Y compris les cigarillos.
6 Ventes.
7 Période de douze mois finissant le 7 juillet de l'année
  indiquée.
8 A partir de 1990, non compris Erythrée.
9 Production des grandes et moyennes entreprises seulement.
10 Production des établissements occupant 50 personnes ou plus.
11 Période de douze mois commençant le 1er avril de l'année
   indiquée.
12 Production de l'Etat seulement.
13 Ventes des fabricants employant 25 personnes ou plus.
14 Y compris le tabac haché.

45
Fabrics
Tissus
Woven cotton and wool, cellulosic and non-cellulosic fibres: million square metres
Tissus de coton, laines, fibres cellulosiques et non cellulosiques : millions de mètres carrés

Country or area Pays ou zone	1983	1984	1985	1986	1987	1988	1989	1990	1991	1992
A. Cotton • Coton										
Total	71 186	70 678	73 563	89 676	81 410	91 387	100 034	87 991	92 753	96 612
Afghanistan[1] Afghanistan[1]	38	45	45	58	53	32	...	...	...	...
Algeria Algérie	* 105	* 105	* 103	* 87	* 88	80[2]	75[2]	75[2]	64[2]	83[2]
Armenia Arménie	..	..	..	..	..	..	..	..	10	
Australia[3 4] Australie[3 4]	32	33	37	39	38	39	36	40	36	39
Austria Autriche	77	64	85	85	93	89	88	108	100	86
Azerbaijan Azerbaïdjan	..	..	..	..	..	..	..	..	120	114
Bangladesh[3] Bangladesh[3]	* 71	* 70	* 75	* 70	* 70	63	63	63	63	63
Belarus Bélarus	..	..	..	..	..	..	..	..	144	...
Belgium Belgique	361[5]	361[5]	373[5]	388[5]	379[5]	340[5]	386	399	379	337
Bolivia * Bolivie *	3	1	1	2	...	...	...	...	...	...
Brazil[2] Brésil[2]	1 963	1 780	2 170	2 352	2 359	2 157	2 179	1 901	1 631	...
Bulgaria[6] Bulgarie[6]	371	374	356	352	358	370	367	254	138	102
Cameroon[2] Cameroun[2]	32	61	29	29	19	...	...	...	...	...
Central African Rep. * Rép. centrafricaine *	4	4	...	...	...	...	...	...	...	...
Chad Tchad	...	60	37	39	34	48	57	58	60	81
Chile[2] Chili[2]	34	49	54	62	61	39	38	31	31	25
China[2] Chine[2]	17 782	16 372	15 202	19 685	20 679	22 454	22 613	22 556	21 719	22 783
Côte d'Ivoire[7] Côte d'Ivoire[7]	129	118	...	...	...	...	...	...	...	...
Croatia Croatie	..	..	..	..	..	..	..	..	30	29
Cuba[8] Cuba[8]	156	156	177	183	202	203	184	...	...	...
former Czechoslovakia†[6] anc. Tchécoslovaquie†[6]	570	624	631	634	634	629	624	629		
Czech Republic République tchèque	..	..	..	..	..	..	..	..	381	309
Dominican Republic[9] Rép. dominicaine[9]	12	13	13	11	13	...	...	...	...	...

45

Fabrics
Woven cotton and wool, cellulosic and non-cellulosic fibres: million square metres [*cont.*]
Tissus
Tissus de coton, laines, fibres cellulosiques et non cellulosiques : millions de mètres carrés [*suite*]

Country or area Pays ou zone	1983	1984	1985	1986	1987	1988	1989	1990	1991	1992
Egypt Egypte	* 922[2]	* 899[2]	* 740[2]	* 726[2]	* 694[2]	589	601	603	609	613
El Salvador *[2 10] El Salvador *[2 10]	4	...	...	...	...	...	...	...	...	...
Estonia Estonie	..	..	..	..	..	..	..	..	168	111
Ethiopia[11 12] Ethiopie[11 12]	92	85	78	84	87	96	68	64	...	...
Finland[6] Finlande[6]	72	73	80	53	70	* 58	46	41	24	21
France[13] France[13]	920	907	861	841	843	820	841	808	751	736
Germany † Allemagne†	..	..	..	..	..	..	..	..	929	763
F. R. Germany[6] R. f. Allemagne[6]	884	943	990	946	1 014	962	943	896	..	..
former German D. R. anc. R. d. allemande	298	294	298	308	287	304	304	213	..	..
Ghana[6] Ghana[6]	8	12	15	...	...	...	...	...	...	...
Greece Grèce	300[13]	281[13]	288[13]	305[13]	* 381	* 382	* 394	...	...	...
Haiti Haïti	1	1	1	1	1	...	...	...	...	...
Honduras[10] Honduras[10]	9	11	10	...	...	...	...	...	...	...
Hong Kong Hong-kong	642	659	639	760	* 850	867	734	* 818	753	* 814
Hungary[6] Hongrie[6]	280	278	287	292	292	295	247	206	133	86
India Inde	10 315	10 547	12 610	12 367	12 912	12 269	12 497	16 224	* 15 340	* 16 428
Indonesia[2 14] Indonésie[2 14]	* 1 001	* 1 135	* 1 274	* 3 038	* 2 253	15 293	18 527	4 284	...	...
Ireland[6 15] Irlande[6 15]	36	34	...	...	...	...	...	...	...	...
Italy[13] Italie[13]	1 482	1 665	1 626	1 633	1 611	1 554	1 618	1 618	1 525	1 353
Jamaica[2] Jamaïque[2]	4	* 2	* 1	* 2	* 1	...	...	...	...	...
Japan[16] Japon[16]	2 079	2 090	2 061	1 974	1 837	1 885	1 915	1 765	1 603	1 465
Jordan[9] Jordanie[9]	1	2	* 2	1	2	...	...	...	...	...
Kenya Kenya	46	51	54	51	51	46	38	45	...	...
Korea, Republic of[6] Corée, République de[6]	442	395	482	559	567	623	648	620	599	469
Latvia Lettonie	..	..	..	..	..	..	..	..	45	22

45

Fabrics
Woven cotton and wool, cellulosic and non-cellulosic fibres: million square metres [cont.]
Tissus
Tissus de coton, laines, fibres cellulosiques et non cellulosiques : millions de mètres carrés [suite]

Country or area Pays ou zone	1983	1984	1985	1986	1987	1988	1989	1990	1991	1992
Madagascar Madagascar	91	82	78	81	75	68	72	59[2]	57[2]	38[2]
Malaysia Malaisie	241	...	...	...	...	...	...	...	...	...
Mali * [2] Mali * [2]	12	...	...	...	...	...	...	...	...	...
Mexico[10 13] Mexique[10 13]	476	460	511	504	626	424	...			
Myanmar[17] Myanmar[17]	107	109	68	* 53	* 36	* 31	21	47[2]	28[2]	27[2]
Nepal Népal	10 240	10 533	12 000	21 386	11 897	8 467	...	...	...	...
Netherlands[6 13] Pays-Bas[6 13]	107	114	105	101	93	79	86	79	72	...
Nicaragua[2] Nicaragua[2]	19	31	...	...	...	...	...	...	...	...
Niger * Niger *	7	7	6	31	39	20	...	...	...	...
Nigeria[10] Nigéria[10]	401	267	285	102	323	...	...	...	...	...
Norway Norvège	12[6]	19[6]	16[6]	16[6]	16[6]	14[6]	14[13]	* 14[13]	7[13]	...
Pakistan[10 18] Pakistan[10 18]	336	297	272	253	238	282	270	295	293	308
Paraguay Paraguay	7	7	* 11	* 14	* 7	14	16	18[2]	23[2]	23[2]
Philippines[2 6] Philippines[2 6]	* 216	184	191	* 217	* 304	* 478	* 535	...	...	...
Poland[6 19] Pologne[6 19]	797	868	887	878	801	839	811	474	332	290
Portugal[13] Portugal[13]	499	523	542	619	561	503	562	547	511	345
Republic of Moldova République de Moldova	..	..	..	..	..	..	..	..	165	150
Romania[6] Roumanie[6]	709	697	695	727	705	689	709	536	459	299
Russian Federation Fédération de Russie	..	..	..	..	..	..	..	..	5 949	...
Senegal Sénégal	10	* 8	* 3	...	* 3	...	...	...	...	...
Slovenia Slovénie	..	..	..	..	..	..	..	..	102[19]	78[19]
South Africa Afrique du Sud	147	147	139	158	172	163	188	175	170	130
Spain[13] Espagne[13]	633	576	604	583	640	662	705	698	705	...
Sri Lanka[2 10] Sri Lanka[2 10]	24	30	44	49	46	* 49	36	19	14	* 16
Sweden * Suède *	52[13]	56[13]	57[13]	54[13]	53[13]	67	65	...	...	...

45

Fabrics
Woven cotton and wool, cellulosic and non-cellulosic fibres: million square metres [cont.]
Tissus
Tissus de coton, laines, fibres cellulosiques et non cellulosiques : millions de mètres carrés [suite]

Country or area Pays ou zone	1983	1984	1985	1986	1987	1988	1989	1990	1991	1992
Switzerland[2] Suisse[2]	130	127	130	131	130	110	101	103	81	77
Syrian Arab Republic *[13] Rép. arabe syrienne *[13]	180	273	158	289	180	173	215	194	201	187
Thailand[6] Thaïlande[6]	743	784	825	881	...	...	...	...	...	...
Togo * Togo *	17	...	...	...	...	...	...	...	...	...
Turkey[2 20] Turquie[2 20]	659	685	589	390	399	422	660	714	626	710
Uganda[6 9] Ouganda[6 9]	...	...	10	10	10	11	12	8	...	...
Ukraine Ukraine	..	..	..	..	..	..	..	..	561	...
former USSR†[6] ancienne URSS†[6]	8 029	8 313	8 580	8 770	8 721	8 670	8 906	8 647	..	..
United Kingdom Royaume-Uni	290	305	315	308	280	240	* 246	* 198	* 185	...
United Rep.Tanzania Rép. Unie de Tanzanie	...	...	54	48	60	44	46	46	38	49
United States Etats-Unis	3 505	3 346	3 278	3 648	3 990	3 873	3 837	3 732	3 682	3 846
Viet Nam Viet Nam	343	435	439	427	432	458	403	380	335[2]	...
former Yemen Arab Rep.[2] anc. Yémen rép. arabe[2]	7	7	7	...	...	...	...	...	...	...
former Dem. Yemen anciennce Yémen dém.	4	4	3	3	3	...	...	...	...	...
Yugoslavia, SFR†[21] Yougoslavie, Rfs†[21]	379	318	344	358	366	351	339	..	..	...
Zaire[10] Zaïre[10]	61	* 58	* 50	...	...	...	...	...	...	...
Zambia Zambie	16	16	21	18	...	...	11	...	...	...

B. Wool • Laines

Total	3 279	3 388	3 382	3 407	3 366	3 508	3 442	3 184	2 869	2 866
Afghanistan *[1] Afghanistan *[1]	1	1	0	0	0	0	...	...	...	...
Algeria Algérie	15	15	18	18	12	7[2]	10[2]	13[2]	16[2]	13[2]
Armenia Arménie	..	..	..	..	..	..	..	..	4	...
Australia[3] Australie[3]	12	10	11	11	11	11	10	8	8	8
Austria Autriche	16	16	16	15	13	11	9	9	8	8
Azerbaijan Azerbaïdjan	..	..	..	..	..	..	..	..	10	7

45
Fabrics
Woven cotton and wool, cellulosic and non-cellulosic fibres: million square metres [cont.]
Tissus
Tissus de coton, laines, fibres cellulosiques et non cellulosiques : millions de mètres carrés [suite]

Country or area Pays ou zone	1983	1984	1985	1986	1987	1988	1989	1990	1991	1992
Belarus Bélarus	..	..	..	..	..	..	..	..	49	...
Belgium[22] Belgique[22]	12	11	10	9	6	5	5	5	6	4
Bulgaria[6] Bulgarie[6]	70	61	61	64	66	52	51	47	26	22
China Chine	236	298	360[2]	416[2]	438[2]	472[2]	461[2]	487[2]	514[2]	558[2]
Croatia Croatie	..	..	..	..	..	..	..	..	5	1
former Czechoslovakia†[6] anc. Tchécoslovaquie†[6]	93	85	87	85	85	86	87	77	..	..
Czech Republic République tchèque	..	..	..	..	..	..	..	..	49	48
Denmark[10 23] Danemark[10 23]	...	...	...	1	1	* 1	1	1	1	1
Ecuador *[2] Equateur *[2]	4	...	...	...	2	3	...	...	...	...
Egypt Egypte	19[2]	20[2]	22[2]	26[2]	24[2]	39[2]	* 28	23	23	23
Estonia Estonie	..	..	..	..	..	..	..	..	7	4
Finland[6] Finlande[6]	2	3	2	2	1	* 1	1	* 0	* 0	* 0
France France	41	41	45	39	29	19	21	16	14	13
Germany † Allemagne†	..	..	..	..	..	..	..	..	121	119
F. R. Germany[6] R. f. Allemagne[6]	79	87	104	102	100	104	109	108	..	..
former German D. R. anc. R. d. allemande	39	41	41	42	44	47	50	38	..	..
Greece[6 13] Grèce[6 13]	9	4	5	5	4	3	4	2	3	..
Hungary[6] Hongrie[6]	23	24	23	22	18	17	21	11	7	4
India *[2 24] Inde *[2 24]	...	221	140	117	107	139	149	..	...	...
Ireland[6] Irlande[6]	3	3	3	3	2	2	2	2	...	...
Italy[13] Italie[13]	453	480	497	448	436	457	457	418	428	445
Japan[10 16] Japon[10 16]	302	327	326	313	331	353	351	335	345	326
Kenya Kenya	4	3	4	3	4	3	1	0	...	...
Korea, Republic of[6] Corée, République de[6]	10	12	14	16	16	19	21	20	20	20
Kyrgyzstan Kirghizistan	..	..	..	..	..	..	..	..	13	11

45

Fabrics
Woven cotton and wool, cellulosic and non-cellulosic fibres: million square metres [*cont.*]
Tissus
Tissus de coton, laines, fibres cellulosiques et non cellulosiques : millions de mètres carrés [*suite*]

Country or area Pays ou zone	1983	1984	1985	1986	1987	1988	1989	1990	1991	1992
Latvia Lettonie	..	..	..	..	..	..	..	..	11	8
Mexico Mexique	12	13	13	11	9[13]	7[13]	8[13]	9[13]	8[13]	11[13]
Mongolia Mongolie	2	2	2	2	2	* 2	3[2]	...	...	...
Netherlands[6] Pays-Bas[6]	3	3	4	5	4	4	5	5	4	...
New Zealand Nouvelle-Zélande	3	3	3	2	2	2	...	...	...	...
Norway[6][25] Norvège[6][25]	2	3	3	3	2	2	1	...	...	...
Pakistan[3] Pakistan[3]	1	1	2	2	2	2	...	...	...	...
Poland[6][26] Pologne[6][26]	148	155	158	155	150	152	146	98	67	50
Portugal[13] Portugal[13]	27	27	34	34	35	31	32	34	29	33
Romania[6] Roumanie[6]	144	123	128	137	135	133	141	107	105	69
Russian Federation Fédération de Russie	..	..	..	..	..	..	..	..	492	...
Slovakia[6] Slovaquie[6]	..	..	..	..	..	..	..	..	12	11
Slovenia Slovénie	..	..	..	..	..	..	..	..	15	15
South Africa[25] Afrique du Sud[25]	4	7	5	4	5	4	5	4	4	4
Spain[13] Espagne[13]	25	27	30	32	35	34	40	31	32	...
Sweden *[13] Suède *[13]	1	1	1	1	1	1	1	1	...	...
Switzerland[25] Suisse[25]	14	14	14	14	12	11	9	9	10	8
Syrian Arab Republic[13] Rép. arabe syrienne[13]	4	4	5	4	1	2	1	2	0	1
Turkey[2][20] Turquie[2][20]	39	39	37	27	27	41	29	42	40	35
Ukraine Ukraine	..	..	..	..	..	..	..	..	79	...
former USSR† ancienne URSS†	911	872	841	842	854	876	890	865	..	..
United Kingdom[27] Royaume-Uni[27]	94	91	91	93	90	89	25	20	...	...
United States Etats-Unis	120	133	116	167	141	159	147	118	142	147
Yugoslavia, SFR†[21] Yougoslavie, Rfs†[21]	96	99	101	110	105	104	100	...	...	...
Zaire Zaïre	2	...	...	...	...	...	...	...	...	...

45

Fabrics
Woven cotton and wool, cellulosic and non-cellulosic fibres: million square metres [cont.]
Tissus
Tissus de coton, laines, fibres cellulosiques et non cellulosiques : millions de mètres carrés [suite]

Country or area Pays ou zone	1983	1984	1985	1986	1987	1988	1989	1990	1991	1992

C. Cellulosic and non-cellulosic fibres · Fibres cellulosiques et non cellulosiques

Country or area Pays ou zone	1983	1984	1985	1986	1987	1988	1989	1990	1991	1992
Total	**18 144**	**20 031**	**20 693**	**20 900**	**20 355**	**21 361**	**22 320**	**22 225**	**21 084**	**21 625**
Afghanistan[12] Afghanistan[12]	9	5	6	6	5	3	...	...	...	...
Algeria *[9] Algérie *[9]	21	20	20	...	...	...	...	...	...	...
Australia[3] Australie[3]	* 135	141	155	161	170	167	192	179	185	186
Austria Autriche	90	84	102	117	95	90	96	101	66	73
Belarus[28] Bélarus[28]	..	..	..	..	..	..	..	..	175	...
Belgium[29] Belgique[29]	2 547	2 641	2 630	2 762	2 828	3 154	3 639	3 955	3 883	4 259
Bulgaria[6 28] Bulgarie[6 28]	45	45	45	46	47	...	...	...	...	...
Côte d'Ivoire[2] Côte d'Ivoire[2]	3	2	2	...	...	...	...	...	...	...
Croatia Croatie	..	..	..	..	..	..	..	..	12	12
Cuba Cuba	14	16	28	...	...	...	...	...	...	...
former Czechoslovakia† anc. Tchécoslovaquie†	102	110	112	115	117	118	115	113	..	..
Czech Republic République tchèque	..	..	..	..	..	..	..	..	32	35
Ecuador[2] Equateur[2]	* 18	* 1	...	...	...	19	...	...	...	...
Egypt *[2] Egypte *[2]	13	...	...	...	...	...	...	...	...	...
El Salvador * El Salvador *	34	29	...	...	...	...	...	...	...	...
Ethiopia[11 12] Ethiopie[11 12]	6	6	6	6	6	6	5	5	...	...
Finland[6] Finlande[6]	34	35	# 42	36	34	* 35	* 34	27	* 17	* 14
France France	711	# 1 890	1 919	2 032	1 841	1 962	2 446	2 668	2 763	3 517
Germany † Allemagne†	..	..	..	..	..	..	..	..	1 477	1 359
F. R. Germany[6] R. f. Allemagne[6]	849	901	1 005	1 039	984	1 343	1 473	1 471	..	..
former German D. R. anc. R. d. allemande	244	253	251	247	256	258	249	...	...	...
Greece[10] Grèce[10]	98[13]	* 87[13]	* 9	...	...	...	...	...	...	...
Hong Kong[30] Hong-kong[30]	...	...	...	1	1	1	1	0	1	...

45

Fabrics
Woven cotton and wool, cellulosic and non-cellulosic fibres: million square metres [cont.]
Tissus
Tissus de coton, laines, fibres cellulosiques et non cellulosiques : millions de mètres carrés [suite]

Country or area Pays ou zone	1983	1984	1985	1986	1987	1988	1989	1990	1991	1992
Hungary[6] Hongrie[6]	91	90	91	86	82	75	67	45	35	22
India Inde	2 174	2 451	2 759	...	...	...	...	...	...	...
Iran, Islamic Rep. of[2] Iran, Rép. islamique d'[2]	247	...	...	...	...	...	...	...	...	...
Ireland[6] Irlande[6]	48	56	27	36	48	...	...	...	...	...
Japan[10 16] Japon[10 16]	3 927	3 994	3 787	3 560	3 307	3 340	3 365	3 376	3 263	3 175
Korea, Republic of Corée, République de	1 802	1 936	2 145	2 619	2 988	3 157	2 908	3 428	3 479	3 094
Mexico Mexique	527	530	584	499	# 368[13]	367[13]	4[13]	4[13]	5[13]	5[13]
Nepal[2 31] Népal[2 31]	5	8	8	14	18	16	...	...	...	...
Netherlands[6] Pays-Bas[6]	124	130	137[13]	* 179	* 172	* 170	18[13]	22[13]	26[13]	...
Norway Norvège	20[6]	25	...	...	...	...	...	...	...	...
Poland[6] Pologne[6]	154	154	154	146	138	154	148	113	88	98
Portugal *[13] Portugal *[13]	29	26	21	20	14	13	19	15	15	13
Republic of Moldova[28] République de Moldova[28]	..	..	..	..	..	..	..	..	44	22
Slovakia Slovaquie	..	..	..	..	..	..	..	..	26	...
Slovenia Slovénie	..	..	..	..	..	..	..	..	7	6
Spain[13 21] Espagne[13 21]	632	589	625	549	...	...	...	...	...	...
Sri Lanka[2 20] Sri Lanka[2 20]	2	1	1	...	...	...	...	...	...	...
Sweden Suède	67[32]	66[32]	70[32]	50	54	10[13 32]	9[13 32]	10[13 32]	6[13 32]	...
Thailand[32] Thaïlande[32]	708	757	812	875	...	...	...	...	...	...
Tunisia * Tunisie *	16	2	2	...	...	...	...	...	...	...
former USSR† ancienne URSS†	1 828	1 857	1 878	1 895	1 990	2 016	2 049	...	...	..
United Kingdom Royaume-Uni	368	388	415	436	407	423	...	...	...	...
Yugoslavia, SFR† Yougoslavie, Rfs†	24	26	31	33	26	30	37	...	...	...
Zaire * Zaïre *	4	...	...	...	...	...	...	...	...	...

45

Fabrics
Woven cotton and wool, cellulosic and non-cellulosic fibres: million square metres [cont.]

Tissus
Tissus de coton, laines, fibres cellulosiques et non cellulosiques : millions de mètres carrés [suite]

Source:
Industrial statistics database of the Statistical Division
of the United Nations Secretariat.

† For detailed descriptions of data pertaining to
former Czechoslovakia, Germany, SFR Yugoslavia and former
USSR, see Annex I - Country or area nomenclature, regional
and other groupings.

1 Twelve months beginning 21 March of year stated.

2 Original data in metres.
3 Twelve months ending 30 June of year stated.

4 Including pile and chenille fabrics of non-cellulosic
fibres.
5 Including cotton blankets and carpets.
6 After undergoing finishing processes.
7 Finished fabrics.
8 Including mixed cotton fabrics.
9 Including cellulosic fabrics.
10 Including finished fabrics and blanketing made of synthetic
fibers.
11 Twelve months ending 7 July of the year stated.

12 Beginning 1990, excluding Eritrea.
13 Original data in metric tons.
14 Including synthetic fabrics.
15 Beginning 1985, data are confidential.
16 Shipments.
17 Production by government-owned enterprises only.
18 Factory production only.
19 Including fabrics of cotton substitutes.
20 Government production only.
21 Including woven cellulosic fabrics.
22 Including woollen blankets and carpets.
23 Sales.
24 Mill production relating to wearable woollen fabrics only.

25 Pure woollen fabrics only.
26 Including fabrics of wool substitutes.
27 Deliveries of woollen & worsted fabrics (weight: more than
15% of wool or animal fibres).
28 Including silk fabrics.
29 Including blankets and carpets of cellulosic and
non-cellulosic fibres.
30 1982 and 1988 data are confidential (cellulosic fibres).

31 Twelve months beginning 16 July of year stated.

32 Including woven non-cellulosic fabrics.

Source:
Base de données pour les statistiques industrielles de la
Division de statistique du Secrétariat de l'ONU.

† Pour les descriptions en détails des données
relatives à l'ancienne Tchécoslovaquie, l'Allemagne, la Rfs
Yougoslavie et l'ancienne URSS, voir l'Annexe I -
Nomenclature des pays ou zones, groupements régionaux et
autres groupements.

1 Période de douze mois commençant le 21 mars de l'année
indiquée.
2 Données d'origine exprimées en mètres.
3 Période de douze mois finissant le 30 juin de l'année
indiquée.
4 Y compris les tissus bouclés et tissus chenille de fibres
non cellulosiques.
5 Y compris les couvertures et les tapis en coton.
6 Après opérations de finition.
7 Tissus finis.
8 Y compris les tissus de cotton mélangé.
9 Y compris les tissus en fibres cellulosiques.
10 Y compris les tissus finis et les couvertures en fibres
synthétiques.
11 Période de douze mois finissant le 7 juillet de l'année
indiquée.
12 A partir de 1990, non compris Erythrée.
13 Données d'origine exprimées en tonnes.
14 Y compris les tissus synthétiques.
15 A partir de 1985, les données sont confidentielles.
16 Expéditions.
17 Production des établissements d'Etat seulement.
18 Production des fabriques seulement.
19 Y compris les tissus de succédanés de coton.
20 Production de l'Etat seulement.
21 Y compris les tissus en fibres cellulosiques.
22 Y compris les couvertures et tapis en laine.
23 Ventes.
24 Production des usines correspondant aux tissus d'habillement
en laine seulement.
25 Tissus de laine pure seulement.
26 Y compris les tissus de succédanés de laine.
27 Quantités livrées de tissus de laine cardée et peignée
(poids : plus de 15% de laine ou fibres animales).
28 Y compris les tissus de soie.
29 Y compris les couvertures et les tapis en fibres
cellulosiques et non-cellulosiques.
30 1987 et 1988 les données sont confidentielles (fibres
cellulosiques).
31 Période de douze mois commençant le 16 juillet de l'année
indiquée.
32 Y compris les tissus en fibres non cellulosiques.

46
Leather footwear
Chaussures de cuir
Production: thousand pairs
Production : milliers de paires

Country or area Pays or zone	1983	1984	1985	1986	1987	1988	1989	1990	1991	1992
Total	3 971 793	4 150 214	4 360 925	4 479 965	4 226 629	4 579 123	4 528 067	4 539 906	4 111 383	4 398 463
Afghanistan Afghanistan	297	344	380	613	701	607	...	...	...	...
Algeria Algérie	* 16 780	18 135	17 909	18 421	18 100	14 689	14 943	16 376	11 824	9 040
Armenia Arménie	..	..	..			..		..	11 340	
Australia[1] Australie[1]	31 124	33 675	36 404	35 164	33 743	29 334	29 032	...	...	15 557
Austria Autriche	26 986	25 453	26 107	23 667	20 729	16 859	16 769	16 553	16 760	14 842
Azerbaijan Azerbaïdjan	..	..	..	..	..	..	..	..	10 262	5 221
Belarus Bélarus	..	..	..	..	..	..	..	..	45 343	...
Belgium Belgique	6 018	5 837	5 343	5 000	4 343	3 916	3 810	3 562	3 454	3 190
Bolivia Bolivie	1 739	1 142	1 302	1 312	...	...	...	...	...	...
Brazil Brésil	130 742	145 838	139 251	152 094	135 791	140 795	157 749	141 000	126 615	...
Bulgaria Bulgarie	26 057	27 997	29 623	29 932	30 668	32 284	33 417	35 052	17 048	16 166
Burkina Faso Burkina Faso	280	# 911	1 318	890	667	470	500	500	1 271	...
Cameroon[2] Cameroun[2]	6 108	3 572	3 964	2 415	1 725	1 733	1 800	1 800	...	...
Canada Canada	39 119	45 416	44 394	43 087	38 774	33 901	...	...	...	...
Central African Rep. Rép. centrafricaine	266	582	611	544	159	200	200	200	...	...
Chile Chili	7 091	6 847	6 357	6 318	6 277	6 614	7 601	6 314	8 481	8 783
China Chine	565 070	638 920	730 610	798 450	618 680	1 055 390	1 103 933	1 202 565	1 328 950	1 613 647
Colombia[3] Colombie[3]	12 299	...	9 555	12 377	10 731	17 200	12 496	11 284	...	...
Congo[4] Congo[4]	928	995	1 127	535	180	296	147	* 300	...	...
Côte d'Ivoire Côte d'Ivoire	6 380	5 900	1 600	1 700	1 700	1 800	1 800	1 800	...	...
Croatia Croatie	..	..	..	..	..	..	..	..	11 717	11 240
Cuba Cuba	13 172	12 473	12 396	13 492	14 183	13 300	11 004	13 400	...	...
Cyprus Chypre	11 571	12 524	12 277	12 273	13 097	13 377	11 643	10 447	10 591	...
former Czechoslovakia† anc. Tchécoslovaquie†	127 638	130 584	131 410	124 469	119 427	119 088	120 226	113 596	..	...

46
Leather footwear
Production: thousand pairs [cont.]
Chaussures de cuir
Production : milliers de paires [suite]

Country or area Pays or zone	1983	1984	1985	1986	1987	1988	1989	1990	1991	1992
Czech Republic République tchèque	..	..	..	..	..	..	..	..	41 400	36 572
Denmark[3] Danemark[3]	5 000	5 100	5 000	5 500	6 000	4 800	4 400	4 400	...	...
Dominican Republic Rép. dominicaine	9 275	9 636	9 458	9 611	* 2 500	...	...	...	...	...
Ecuador Equateur	1 424	* 1 400	* 1 400	1 567	1 400	1 500	1 500	1 500	...	1 936
Egypt Egypte	59 414	65 325	66 328	65 347	66 287	50 438	50 375	48 325	48 311	48 390
El Salvador El Salvador	2 732	# * 3 400	* 3 500	* 3 200	* 3 300	3 600	3 600	3 600	...	...
Estonia Estonie	..	..	..	..	..	..	..	..	6 301	3 208
Ethiopia[5 6] Ethiopie[5 6]	7 985	8 794	6 030	8 868	11 130	11 409	12 957	5 367[7]	...	...
Finland Finlande	6 666	7 867	7 407	9 414	8 635	6 860	5 243	4 752	3 683	3 419
France[2] France[2]	206 074	201 809	196 992	194 753	183 209	168 564	169 788	194 700	168 084	160 320
Germany † Allemagne†	..	..	..	..	..	..	..	..	84 435	63 544
F. R. Germany R. f. Allemagne	94 518	92 472	86 678	86 629	78 291	71 674	68 697	64 358	..	..
former German D. R. anc. R. d. allemande	81 683	82 425	83 551	85 518	87 980	91 382	91 518	61 822	..	..
Greece Grèce	12 599	11 180	11 305	11 430	13 900[4]	13 220	12 359	12 260	10 712	...
Haiti[3] Haïti[3]	519	528	500	600	500	500	500	500	...	...
Hong Kong Hong-kong	96 371	# 77 821	80 642	98 178	61 008	134 951	48 710	73 496	...	...
Hungary Hongrie	47 761	49 035	49 756	46 934	44 340	40 063	33 107	27 426	20 757	14 752
India[8] Inde[8]	13 735	59 212	167 938	193 385	194 455	200 320	198 761	198 404	201 449	...
Indonesia[9] Indonésie[9]	43 237	56 765	61 948	73 219	109 345	5 338	5 565	9 749	1 890	...
Iran, Islamic Rep. of[10] Iran, Rép. islamique d'[10]	25 285	25 602	24 717	21 129	21 946	21 005	21 849	23 374	...	...
Iraq Iraq	4 200[3]	4 500[3]	4 200[3]	3 305	4 900[3]	4 669	4 600	4 400	...	4 087
Ireland Irlande	2 609	2 716	1 914	2 492	2 202	2 097	1 913	1 403	...	...
Italy[3] Italie[3]	345 200	352 800	371 600	362 000	343 500	327 300	311 900	320 200	...	...
Jamaica[3] Jamaïque[3]	700	600	700	700	700	700	700	700	...	...
Japan[11 12] Japon[11 12]	48 387	55 707	53 387	51 975	51 984	56 023	53 819	54 054	53 351	52 455

46
Leather footwear
Production: thousand pairs [cont.]
Chaussures de cuir
Production : milliers de paires [suite]

Country or area Pays or zone	1983	1984	1985	1986	1987	1988	1989	1990	1991	1992
Kenya Kenya	1 303	1 182	1 361	1 465	1 306	1 381	1 693	1 605	...	...
Korea, Republic of Corée, République de	15 136	17 368	19 209	22 603	25 807	25 339	24 023	24 440	26 384	25 409
Kyrgyzstan Kirghizistan	..	..	..	..	..	..	..	..	9 507	5 751
Latvia Lettonie	..	..	..	..	..	..	..	..	7 778	8 885
Madagascar[9] Madagascar[9]	2 302	2 133	2 172	2 889	1 686	1 036	1 356	1 314	936	918
Malta Malte	1 524	1 691	...	...	...	...	...	...	...	...
Mexico Mexique	...	...	...	...	9 943	15 528	15 986	16 249	14 685	14 570
Mongolia Mongolie	2 226	2 677	2 883	3 149	3 517	3 921	4 140	3 000	...	...
Mozambique Mozambique	1 347	639	599	829	59	52	117	106	4	...
Nepal[13] Népal[13]	73	83	112	121	214	332	124	...	...	...
Netherlands[2] Pays-Bas[2]	10 599	9 585	9 452	9 349	8 556	6 249	6 197	5 598	5 137	...
New Zealand Nouvelle-Zélande	7 433	8 464	7 701	7 396	7 327	5 385	5 073	4 977	4 022	...
Nicaragua Nicaragua	2 812	2 549	* 1 000	* 1 000	* 1 100	...	...	...	...	...
Nigeria Nigéria	10 618	10 118	# 32 829	34 362	38 502	...	...	...	...	...
Norway Norvège	1 244	1 600[4]	1 400[4]	1 200[4]	1 000[4]	800[4]	800[4]	900[4]	...	...
Panama Panama	1 677	1 696	1 852	* 1 400	1 791	1 047	988	1 500	...	...
Paraguay Paraguay	4 887	5 034	4 790	4 936	5 308	5 053	5 094	5 300	...	...
Peru[3] Pérou[3]	2 187	2 107	2 179	2 423	1 623	943	904	636	...	...
Philippines[3] Philippines[3]	4 317	3 880	3 084	5 302	16 128	10 400	10 100	10 000	...	...
Poland Pologne	141 497	147 076	147 603	146 996	151 619	155 639	150 833	98 200	66 857	55 181
Portugal Portugal	24 157	28 790	31 333	34 167	35 173	37 762	43 667	75 378	39 658	40 419
Republic of Moldova République de Moldova	..	..	..	..	..	..	..	..	20 751	14 504
Romania Roumanie	95 221	96 210	99 732	105 093	102 871	99 751	110 733	80 670	63 196	41 237
Russian Federation Fédération de Russie	..	..	..	..	..	..	..	..	336 411	...
Rwanda Rwanda	4	5	16	21	24	25	...	...	...	...

46
Leather footwear
Production: thousand pairs [cont.]
Chaussures de cuir
Production : milliers de paires [suite]

Country or area Pays or zone	1983	1984	1985	1986	1987	1988	1989	1990	1991	1992
Saint Kitts and Nevis Saint-Kitts-et-Nevis	80	59	23	24	...	...	...	...	...	...
Senegal Sénégal	3 250	3 234	2 851	* 1 800	* 2 000	561	600	600	...	...
Singapore [4] Singapour [4]	3 914	4 755	3 789	* 2 600	* 2 700	2 700	2 800	2 800	...	...
Slovakia Slovaquie	..	..	..	..	..	..	..	..	26 744	22 875
Slovenia Slovénie	..	..	..	..	..	..	..	..	9 124	9 492
South Africa [2] Afrique du Sud [2]	45 542	46 316	45 117	44 094	44 508	42 727	54 461	55 524	52 089	52 029
Spain Espagne	114 190	125 250	113 888	114 621	109 714	117 707	114 089	117 199	115 190	...
Sri Lanka Sri Lanka	248	249	263	359	358	328	270	299	107	...
Sudan [3] Soudan [3]	# 8 500	8 000	8 500	6 000	5 000	4 000	4 000	4 000	...	...
Sweden Suède	5 618	4 490	3 826	3 977	4 263	2 518	2 088	1 904	...	...
Switzerland Suisse	5 418	5 781	5 813	5 462	4 832	4 544	4 285	4 039	3 353	3 035
Togo Togo	559	486	* 521	286	29	100	100	100	...	...
Tunisia [3] Tunisie [3]	3 200	3 600	4 000	5 000	5 300	7 000	6 900	7 000	...	...
Ukraine Ukraine	177 861	182 216	185 726	186 908	186 776	190 583	193 673	196 466	177 336	...
former USSR† ancienne URSS†	745 303	763 514	787 610	800 745	808 993	819 050	826 988	843 245	..	..
United Kingdom [14] Royaume-Uni [14]	121 152	123 807	128 698	129 017	124 298	120 589	101 318	92 673	...	...
United Rep.Tanzania Rép. Unie de Tanzanie	* 2 242	* 1 949	1 324	1 260	609	569	445	459	328	168
United States Etats-Unis	339 187	303 174	265 098	240 932	230 046	231 595	218 025	184 568	168 992	164 904
Viet Nam Viet Nam	...	...	...	...	...	5 290	...	5 827	6 188	5 672
former Yemen Arab Rep. anc. Yémen rép. arabe	...	...	240	217	...	...	...	...	...	...
former Dem. Yemen anciennce Yémen dém.	168	219	112	171	239	...	...	...	...	...
Yugoslavia, SFR† Yougoslavie, Rfs†	80 137	90 798	93 063	99 503	93 051	86 336	91 616	...	..	...
Zaire [3] Zaïre [3]	800	900	800	900	900	900	900	900	...	...

Source:
Industrial statistics database of the Statistical Division
of the United Nations Secretariat.

Source:
Base de données pour les statistiques industrielles de la
Division de statistique du Secrétariat de l'ONU.

46

Leather footwear
Production: thousand pairs [*cont.*]

Chaussures de cuir
Production : milliers de paires [*suite*]

† For detailed descriptions of data pertaining to former Czechoslovakia, Germany, SFR Yugoslavia and former USSR, see Annex I - Country or area nomenclature, regional and other groupings.

1 Twelve months ending 30 June of year stated.

2 Including rubber footwear.
3 Source: Food and Agriculture Organization (FAO), (Rome).

4 Including rubber and plastic footwear.

5 Twelve months ending 7 July of the year stated.

6 Including canvas, plastic and rubber footwear.

7 Beginning 1990, excluding Eritrea.
8 Production by large and medium scale establishments only.
9 Including plastic footwear.
10 Production by establishments employing 50 or more persons.
11 Production by establishments employing 10 or more persons.
12 Shipments.
13 Twelve months beginning 16 July of year stated.

14 Manufacturers' sales.

† Pour les descriptions en détails des données relatives à l'ancienne Tchécoslovaquie, l'Allemagne, la Rfs Yougoslavie et l'ancienne URSS, voir l'Annexe I - Nomenclature des pays ou zones, groupements régionaux et autres groupements.

1 Période de douze mois finissant le 30 juin de l'année indiquée.
2 Y compris les chaussures en caoutchouc.
3 Source: Organisation des Nations Unies pour l'alimentation et l'agriculture (FAO), (Rome).
4 Y compris les chaussures en caoutchouc et en matière plastique.
5 Période de douze mois finissant le 7 juillet de l'année indiquée.
6 Y compris les chaussures en toile, en matière plastique et en caoutchouc.
7 A partir de 1990, non compris Erythrée.
8 Production des grandes et moyennes entreprises seulement.
9 Y compris les chaussures en matière plastique.
10 Production des établissements occupant 50 personnes ou plus.
11 Production des établissements occupant 10 personnes ou plus.
12 Expéditions.
13 Période de douze mois commençant le 16 juillet de l'année indiquée.
14 Ventes des fabricants.

47
Sawnwood
Sciages

Production (sawn): thousand cubic metres
Production (sciés) : milliers de mètres cubes

Country or area Pays ou zone	1983	1984	1985	1986	1987	1988	1989	1990	1991	1992
World *Monde*	441 786	460 969	467 785	484 303	505 309	508 050	507 085	505 399	456 892	449 900
Africa Afrique	7 317	7 356	7 623	8 023	8 403	8 455	8 561	8 502	8 141	8 179
Algeria[1] Algérie[1]	13	13	13	13	13	13	13	13	13	13
Angola Angola	6	2	5	5	5	5	5 ·	5	5[1]	5[1]
Benin Bénin	9[1]	5	8	11	11[1]	11[1]	14	14[1]	27	24
Burkina Faso Burkina Faso	2[1]	2	2	2	2	2	2	2	2	2[1]
Burundi Burundi	1[1]	4	3	3[1]	3[1]	3[1]	3[1]	2	3	3[1]
Cameroon Cameroun	508	423	454	552	580	492	489	489[1]	489[1]	489[1]
Central African Rep. Rép. centrafricaine	61	58	56	54	52	52[1]	57	63	60	60[1]
Chad Tchad	1[1]	1[1]	1[1]	1[1]	1[1]	1[1]	1[1]	1	2	2[1]
Congo Congo	66	60	50	77	60	57	46	50	12	12[1]
Côte d'Ivoire Côte d'Ivoire	718	679	753	765	775	784	777	753	608	611
Equatorial Guinea Guinée équatoriale	23[1]	24[1]	39	51[1]	51[1]	51[1]	56[1]	52[1]	52[1]	52[1]
Ethiopia[2] Ethiopie[2]	45	45[1]	45[1]	45[1]	45[1]	39	34	22	12	12[1]
Gabon Gabon	108[1]	97	106	100[1]	* 46	* 37	* 30	* 37	* 32	32[1]
Gambia[1] Gambie[1]	1	1	1	1	1	1	1	1	1	1
Ghana Ghana	275	285	345	355	455	455[1]	537	472	400	410·
Guinea Guinée	90[1]	90[1]	90[1]	70	70	70	70	70	70	63
Guinea-Bissau[1] Guinée-Bissau[1]	16	16	16	16	16	16	16	16	16	16
Kenya Kenya	142	183	192	173	195	188	185	185[1]	185[1]	185[1]
Liberia Libéria	161	153	169	191	411	411[1]	411[1]	411[1]	411[1]	411[1]
Libyan Arab Jamah.[1] Jamah. arabe libyenne[1]	31	31	31	31	31	31	31	31	31	31
Madagascar Madagascar	234	234	234	234	234[1]	234	234	234	233[1]	233[1]
Malawi Malawi	23[1]	16	19	23	30	31	* 39	* 43	* 43	43
Mali Mali	6[1]	6	4	6	11	13	13[1]	13[1]	13[1]	13[1]

47
Sawnwood
Production (sawn): thousand cubic metres [*cont.*]
Sciages
Production (sciés) : milliers de mètres cubes [*suite*]

Country or area Pays ou zone	1983	1984	1985	1986	1987	1988	1989	1990	1991	1992
Mauritius Maurice	3	...	1	1	4	4	5	4	5	4
Morocco Maroc	130[1]	120[1]	100[1]	90[1]	80[1]	53	83	83[1]	83[1]	83[1]
Mozambique Mozambique	38	37	35	39	42	36	30	26	18	17
Nigeria Nigéria	2 402	2 512	2 712	2 712	2 712[1]	2 712	2 712	2 712	2 706[1]	2 706[1]
Réunion Réunion	1[1]	1[1]	2[1]	2	2	1	1	3	2	2[1]
Rwanda Rwanda	10[1]	12	13	13[1]	13[1]	13[1]	11	8	8	36
Sao Tome and Principe Sao Tomé-et-Principe	3	3[1]	3[1]	3[1]	4[1]	5	5[1]	5[1]	5[1]	5[1]
Senegal Sénégal	11[1]	11[1]	23	20	27	19	15	22	23	23[1]
Sierra Leone Sierra Leone	19	19[1]	14	12	12[1]	12[1]	12[1]	11	9	9[1]
Somalia[1] Somalie[1]	14	14	14	14	14	14	14	14	14	14
South Africa[3] Afrique du Sud[3]	1 620	1 635	1 510	* 1 734	* 1 734	* 1 873	1 873[1]	* 1 936	* 1 792	1 792
Sudan Soudan	13	13	13	13	13[1]	13	5	4	2[1]	3
Swaziland[1] Swaziland[1]	103	103	103	103	103	103	103	103	103	103
Togo Togo	1	1	3	2	4	4	3	6	2	3
Tunisia Tunisie	3	4	6	9	11[1]	12	20	16	17	6
Uganda Ouganda	23[1]	23[1]	23[1]	23[1]	23[1]	28	28[1]	28[1]	28[1]	28[1]
United Rep.Tanzania Rép. Unie de Tanzanie	91	102	109	154	156	156[1]	156[1]	156[1]	156[1]	156[1]
Zaire Zaïre	112	120	118	120	127	135	131	117	105	105[1]
Zambia Zambie	50	50[1]	50[1]	67	51	76	101	81	94	112
Zimbabwe Zimbabwe	131	149	138	114	175	190	190[1]	190[1]	250	250[1]
America, North **Amérique du Nord**	**127 191**	**141 747**	**146 633**	**158 011**	**173 163**	**171 993**	**167 164**	**168 279**	**158 235**	**167 630**
Bahamas[1] Bahamas[1]	1	1	1	1	1	1	1	1	1	1
Belize Belize	16[1]	19[1]	22	14	14[1]	14[1]	14[1]	14[1]	14[1]	14[1]
Canada Canada	42 070	49 869	54 586	54 853	61 775	60 737	59 245	54 906	52 040	56 318
Costa Rica Costa Rica	306[1]	412	412	412[1]	503	515	439	412	412	412[1]

47
Sawnwood
Production (sawn): thousand cubic metres [*cont.*]
Sciages
Production (sciés) : milliers de mètres cubes [*suite*]

Country or area Pays ou zone	1983	1984	1985	1986	1987	1988	1989	1990	1991	1992
Cuba Cuba	104	108	104	108	114	118	130	130[1]	130[1]	130[1]
El Salvador[1] El Salvador[1]	39	46	43	44	47	54	70	70	70	70
Guadeloupe[1] Guadeloupe[1]	1	1	1	1	1	1	1	1	1	1
Guatemala Guatemala	104	103	131	83	83[1]	21[1]	34[1]	14[1]	20[1]	20[1]
Haiti[1] Haïti[1]	14	14	14	14	14	14	14	14	14	14
Honduras Honduras	468	427	436	405	464	447	412	328	303	321
Jamaica Jamaïque	23	31	31[1]	26	30	44	40	40	32	28
Martinique Martinique	...	...	...	...	1	2	1	1[1]	1[1]	1
Mexico Mexique	1 827	1 975	2 205	2 143	2 410	2 528[1]	2 447[1]	2 366	* 2 696	2 696[1]
Nicaragua Nicaragua	222[1]	222[1]	222[1]	222[1]	180[1]	140[1]	110[1]	80	80[1]	80[1]
Panama Panama	53[1]	46	45	30[1]	25[1]	18	50[1]	48[1]	16	37
Trinidad and Tobago Trinité-et-Tobago	22	21	18	22	19	21	80	53	42	58
United States Etats-Unis	81 920	88 451	88 361	99 632	107 481	107 317	104 075	109 800	102 363	107 428
America, South Amérique du Sud	23 264	24 299	24 598	25 167	26 126	26 065	25 732	25 261	25 867	26 272
Argentina Argentine	1 121[1]	934[1]	901[1]	978[1]	1 139[1]	950[1]	950[1]	950[1]	950[1]	1 472
Bolivia Bolivie	52	46	40	55	56	51	102	102	120[1]	100[1]
Brazil[1] Brésil[1]	17 199	17 781	17 781	18 063	18 063	18 179	18 179	17 179	18 628	18 628
Chile Chili	1 610	2 001	2 194	2 026	2 677	2 710	2 684	3 327	3 218	3 019
Colombia Colombie	721[1]	603[1]	655[1]	813	813[1]	813[1]	813[1]	813[1]	813[1]	813[1]
Ecuador Equateur	1 142	1 212	1 215	1 258	1 265	1 280	1 492	1 641	865	900
French Guiana[1] Guyane française[1]	19	19	19	19	19	19	19	19	19	19
Guyana Guyana	70[1]	60	65	60	57	57[1]	57[1]	50[1]	10[1]	10[1]
Paraguay Paraguay	655[1]	834	758	766	862	906	510[1]	228	313	313[1]
Peru Pérou	390	484	535	615	628	546	392	499	486	486[1]
Suriname Suriname	59[1]	57	73	61[1]	42[1]	73[1]	44	44	40	43

47
Sawnwood
Production (sawn): thousand cubic metres [*cont.*]
Sciages
Production (sciés) : milliers de mètres cubes [*suite*]

Country or area Pays ou zone	1983	1984	1985	1986	1987	1988	1989	1990	1991	1992
Uruguay Uruguay	16	59	54	117	169	185	201	229	205	269
Venezuela Venezuela	210	210[1]	308	336	336[1]	296	* 290	180	200[1]	200[1]
Asia **Asie**	**95 828**	**97 156**	**99 740**	**100 848**	**105 646**	**105 989**	**107 854**	**104 647**	**99 789**	**97 346**
Afghanistan[1] Afghanistan[1]	400	400	400	400	400	400	400	400	400	400
Bangladesh Bangladesh	162[1]	154[1]	99	79	79[1]	79[1]	79[1]	79[1]	79[1]	79[1]
Bhutan[1] Bhoutan[1]	6	6	6	6	6	10	20	40	40	40
Brunei Darussalam Brunéi Darussalam	90	90[1]	90[1]	90[1]	90[1]	90[1]	90[1]	90[1]	90[1]	90[1]
Cambodia Cambodge	43[1]	43[1]	43[1]	43[1]	43[1]	43[1]	43[1]	71	122	122[1]
China[1] Chine[1]	23 905	25 791	27 234	26 702	26 577	26 522	25 177	23 160	20 521	19 317
Cyprus Chypre	55	59	63	59	57	55	60	22	16	14
Hong Kong Hong-kong	271[1]	248[1]	248[1]	248[1]	248[1]	248[1]	587	421	421[1]	421[1]
India Inde	14 495[1]	15 907[1]	17 460[1]	17 460[1]	17 460[1]	17 460[1]	17 460[1]	17 460[1]	17 460	17 460[1]
Indonesia Indonésie	6 315	6 620	7 118	7 549	9 887	10 290	10 357	9 145	8 638	8 438
Iran, Islamic Rep. of Iran, Rép. islamique d'	163[1]	172[1]	186[1]	202[1]	219[1]	239	262	169	173	173[1]
Iraq[1] Iraq[1]	8	8	8	8	8	8	8	8	8	8
Japan Japon	29 670	28 667	28 472	29 105	30 159	* 30 138	* 30 542	29 781[1]	28 264[1]	27 529[1]
Korea, Dem. P. R.[1] Corée, R. p. dém. de[1]	280	280	280	280	280	280	280	280	280	280
Korea, Republic of Corée, République de	3 518	2 974	3 018	3 563	4 145	4 014	4 194	3 897	4 041	3 513
Lao People's Dem. Rep. Rép. dém. pop. lao	25	26	16[1]	16	30[1]	40[1]	85[1]	78	71	110[1]
Lebanon Liban	22	22[1]	22[1]	22[1]	18	16	18	13	11	9
Malaysia Malaisie	7 282	5 933	5 494	5 525	6 285	6 662	8 275	8 849	8 993	9 369
Mongolia Mongolie	470[1]	470[1]	470[1]	470[1]	470[1]	470[1]	470[1]	470[1]	270	230
Myanmar Myanmar	674	664	615	568	392	283	375	296	239	282
Nepal Népal	220[1]	220[1]	220[1]	220[1]	220[1]	220[1]	220[1]	570	620	620[1]
Pakistan Pakistan	615	578	589	647	659	755	1 022	1 450	1 520	1 503

47
Sawnwood
Production (sawn): thousand cubic metres [*cont.*]
Sciages
Production (sciés) : milliers de mètres cubes [*suite*]

Country or area Pays ou zone	1983	1984	1985	1986	1987	1988	1989	1990	1991	1992
Philippines Philippines	1 222	1 234	1 062	978	* 1 233	1 033	975	841	726	452
Singapore[1] Singapour[1]	232	211	191	170	140	110	80	50	20	20
Sri Lanka Sri Lanka	29	23[1]	23[1]	20[1]	20[1]	20[1]	20[1]	10	5	5
Syrian Arab Republic[1] Rép. arabe syrienne[1]	9	9	9	9	9	9	9	9	9	9
Thailand Thaïlande	950	1 036	958	1 027	1 095	1 044	1 259	1 170	939	1 077
Turkey Turquie	4 343	4 923	4 923	4 923[1]	4 923[1]	4 923[1]	4 923[1]	4 923[1]	4 928	4 928[1]
Viet Nam Viet Nam	354[1]	389[1]	424[1]	459[1]	494[1]	529[1]	564[1]	896	885	849
Europe Europe	**85 796**	**87 749**	**85 199**	**84 381**	**84 213**	**85 259**	**87 412**	**88 100**	**80 184**	**79 674**
Albania Albanie	200[1]	200[1]	200[1]	200[1]	200[1]	200[1]	200[1]	382	382[1]	382[1]
Austria Autriche	6 269	6 315	6 001	5 818	5 944	6 478	6 920	7 509	7 239	7 020
Belgium-Luxembourg Belgique-Luxembourg	753	746	795	824	909	1 009	1 164	1 194	1 244	1 184
Bulgaria Bulgarie	1 504	1 483	1 515	1 338	1 490	1 459	1 426	1 126	1 132	1 132[1]
former Czechoslovakia† anc. Tchécoslovaquie†	5 143	5 227	5 219	5 251	5 186	5 128	4 860	4 764	3 621	3 621[1]
Denmark Danemark	829	829[1]	879	879[1]	861[1]	861[1]	861[1]	861[1]	861	620
Finland Finlande	8 023	8 265	7 333	7 143	7 563	7 823	7 763	7 503	5 983	6 983
France France	9 005	9 038	9 087	9 318	9 612	10 248	10 655	10 960	10 974	10 488
Germany † Allemagne†	..	..	..	..	..	..	..	..	13 322	13 496
F. R. Germany R. f. Allemagne	9 413	9 825	9 541	9 805	9 754	10 395	11 388	12 203	..	..
former German D. R. anc. R. d. allemande	2 449	2 491	2 491	2 431	2 465	2 489	2 521	2 521[1]	..	..
Greece Grèce	323	321	305	454	428	410	417	355	387	387[1]
Hungary Hongrie	1 204	1 322	1 284	1 277	1 225	1 237	1 257	1 068	936	428
Ireland Irlande	258	290	300	300[1]	300[1]	300[1]	356	386	386[1]	560
Italy Italie	2 025	2 234	2 599	1 919	1 905	2 095	1 998	1 950	1 850	1 853
Netherlands Pays-Bas	314	335	412	425	387	435	465	455	425	415
Norway Norvège	2 312	2 364	2 230	2 260	2 362	2 387	2 492	2 413	2 263	2 362

47
Sawnwood
Production (sawn): thousand cubic metres [*cont.*]
Sciages
Production (sciés) : milliers de mètres cubes [*suite*]

Country or area Pays ou zone	1983	1984	1985	1986	1987	1988	1989	1990	1991	1992
Poland Pologne	6 762	6 765	6 639	6 645	6 442	5 577	4 878	4 129	3 205	2 957
Portugal Portugal	2 360	2 606	1 860	2 070	2 095	2 088	2 090	2 090	1 670	1 460
Romania Roumanie	4 878	4 868	4 425	3 538	2 858	2 758	2 850	2 911	2 233	2 560
Spain Espagne	2 218	2 150	2 383	2 613	2 643	2 427	2 993	3 267	3 436	2 468
Sweden Suède	11 762	12 382	11 531	11 641	11 524	11 267	11 487	12 018	11 463	12 121
Switzerland Suisse	1 760	1 515	1 689	1 719	1 650	1 693	1 700	1 985	1 727	1 525
United Kingdom Royaume-Uni	1 619	1 520	1 717	1 807	1 823	1 919	2 191	2 271	2 241	2 448
Yugoslavia, SFR† Yougoslavie, Rfs†	4 413	4 658	4 764	4 706	4 587	4 577	4 481	3 779	3 204	3 204[1]
Oceania Océanie	**5 391**	**5 362**	**5 791**	**5 874**	**5 259**	**5 489**	**5 561**	**5 610**	**5 377**	**5 800**
Australia Australie	2 991	3 003	3 216	3 220	3 131	3 342	3 165	3 151	2 858	3 022
Fiji Fidji	82	80[1]	91[1]	78	88	103	94	94[1]	91[1]	91[1]
New Caledonia Nouvelle-Calédonie	4[1]	4[1]	4[1]	4[1]	4[1]	5	5	5[1]	3	2
New Zealand Nouvelle-Zélande	2 150	2 109	2 320	2 412	1 876	1 881	2 135	2 198	2 263	2 522
Papua New Guinea Papouasie-Nvl-Guinée	124[1]	124[1]	* 117	117[1]	117[1]	117[1]	117[1]	117[1]	117[1]	117[1]
Samoa[1] Samoa[1]	21	21	21	21	21	21	21	21	21	21
Solomon Islands Iles Salomon	15	17	17	15	13	12	16	16[1]	16[1]	16[1]
Tonga Tonga	1	1[1]	1	2	2[1]	2[1]	1[1]	1[1]	1[1]	1[1]
Vanuatu Vanuatu	3	3[1]	4[1]	6	7	7[1]	7[1]	7[1]	7[1]	7[1]
former USSR† ancienne URSS†	**97 000**	**97 300**	**98 200**	**102 000**	**102 500**	**104 800**	**104 800**	**105 000**	**79 300**	**65 000**

Source:
Food and Agriculture Organization of the United Nations (Rome).

† For detailed descriptions of data pertaining to former Czechoslovakia, Germany, SFR Yugoslavia and former USSR, see Annex I - Country or area nomenclature, regional and other groupings.

1 FAO estimate.
2 Including Eritrea.
3 Including data for Namibia.

Source:
Organisation des Nations Unies pour l'alimentation et l'agriculture (Rome).

† Pour les descriptions en détails des données relatives à l'ancienne Tchécoslovaquie, l'Allemagne, la Rfs Yougoslavie et l'ancienne URSS, voir l'Annexe I - Nomenclature des pays ou zones, groupements régionaux et autres groupements.

1 Estimation de la FAO.
2 Y compris Erythrée.
3 Y compris les données pour la Namibie.

48
Paper and paperboard
Papiers et cartons
Production: thousand metric tons
Production : milliers de tonnes métriques

Country or area Pays ou zone	1983	1984	1985	1986	1987	1988	1989	1990	1991	1992
World **Monde**	177 307	189 998	192 912	203 237	214 605	227 460	232 816	239 809	242 939	245 437
Africa **Afrique**	2 062	2 030	2 139	2 331	2 373	2 603	2 756	2 737	2 693	2 748
Algeria Algérie	107	135	110	120	120[1]	120[1]	120[1]	* 91	* 91	91[1]
Cameroon[1] Cameroun[1]	5	5	5	5	5	5	5 .	5	5	5
Egypt Egypte	110[1]	* 145	* 145	145[1]	* 160	160[1]	* 216	* 223	208	201
Ethiopia[2] Ethiopie[2]	9	10	10	10[1]	12[1]	10[1]	10[1]	* 8	6	3
Kenya Kenya	69	75	75	85	89	100	108	* 93	* 92	176
Libyan Arab Jamah. Jamah. arabe libyenne	5[1]	5[1]	* 6	* 6	6[1]	6[1]	6[1]	6[1]	6[1]	6[1]
Madagascar Madagascar	10	10	10	15	8	6[1]	6[1]	* 9	* 5	5[1]
Morocco Maroc	96	101	* 107	* 109	105	103[1]	102[1]	* 119	* 117	102
Mozambique Mozambique	2	1[1]	* 2	* 2	2[1]	2[1]	2[1]	* 1	* 1	1[1]
Nigeria Nigéria	18[1]	15	41	76	76[1]	95	73	79	65	65[1]
South Africa[3] Afrique du Sud[3]	1 520	1 422	1 489	* 1 611	1 600	1 800	1 899	* 1 904	* 1 905	1 905[1]
Sudan Soudan	9[1]	9[1]	9[1]	9[1]	* 10	10[1]	10[1]	* 4	* 3	3[1]
Tunisia Tunisie	30	28	48	53	62	70	82	78	72	64
Uganda Ouganda	...	...	* 2	* 2	2[1]	2[1]	2[1]	* 3	* 3	3[1]
United Rep.Tanzania Rép. Unie de Tanzanie	...	...	...	...	29	28	28[1]	25	25[1]	25[1]
Zaire Zaïre	3	2	2[1]	3	3	2[1]	1	1[1]	1[1]	3
Zambia Zambie	5[1]	3[1]	2	5	3	2	4	2	2	4
Zimbabwe Zimbabwe	65	64	76	* 75	81	82	82[1]	86	86[1]	86[1]
America, North **Amérique du Nord**	74 400	79 067	78 007	82 415	86 379	89 823	89 614	91 519	92 389	94 815
Canada Canada	13 353	14 222	14 448	15 259	16 044	16 639	16 555	16 466	16 559	16 585
Costa Rica Costa Rica	13	13	13	13[1]	13[1]	17	18	19	19[1]	19[1]
Cuba Cuba	109	122	132	150	148	141	168	* 123	* 118	118[1]
Dominican Republic Rép. dominicaine	9[1]	10	10[1]	10[1]	10[1]	10[1]	10[1]	10[1]	10[1]	10[1]

48
Paper and paperboard
Production: thousand metric tons [*cont.*]
Papiers et cartons
Production : milliers de tonnes métriques [*suite*]

Country or area Pays ou zone	1983	1984	1985	1986	1987	1988	1989	1990	1991	1992
El Salvador[1] El Salvador[1]	16	16	16	16	17	17	17	17	17	17
Guatemala Guatemala	16	18	14	17	14	14	14	14	14[1]	14[1]
Jamaica Jamaïque	18	18	15[1]	11[1]	2	3	4	4	4	5
Mexico Mexique	2 019	2 239	2 376	2 469	2 573	3 375	3 294[1]	2 873	2 896[1]	2 858
Panama Panama	43[1]	43[1]	24	26	26[1]	20	20	28	* 28	28[1]
United States Etats-Unis	58 804	62 366	60 959	64 444	67 532	69 587	69 514	71 965	72 724	75 161
America, South Amérique du Sud	**5 728**	**6 292**	**6 505**	**7 235**	**7 644**	**7 685**	**7 609**	**7 686**	**8 034**	**8 080**
Argentina Argentine	879	942	864	998	1 027	974[1]	917	891	* 963	976
Bolivia Bolivie	1[1]	1[1]	1[1]	1[1]	2	2[1]	2[1]	5	...	...
Brazil Brésil	3 426	3 768	4 022	4 525	4 712	4 685	4 806	4 844	4 888	4 913
Chile Chili	333	381	385	388	442	449	438	462	486	347
Colombia Colombie	366	413	* 446	457	488	501	501[1]	494	521	629
Ecuador Equateur	34[1]	34[1]	34	34	34	35	35	44	129	160
Paraguay Paraguay	13[1]	13[1]	8	8	10	11	11[1]	12	13	13
Peru Pérou	146	138	* 150	149	209	260	311	263	327	327
Uruguay Uruguay	43	45	45	64	66	59	63	61	75	83
Venezuela Venezuela	487	* 557	551	612	654	708	524	610	632	632[1]
Asia Asie	**32 452**	**35 038**	**38 153**	**40 984**	**45 066**	**49 279**	**53 014**	**56 920**	**60 146**	**62 172**
Bangladesh Bangladesh	148	149	* 104	* 113	* 114	96	97	92	97	97[1]
China Chine	8 414	9 514	11 197	12 601	14 231	15 700	16 487	17 328	18 525	19 942
Hong Kong Hong-kong	32	36	* 40	* 40	40[1]	40[1]	40[1]	* 80	* 115	115[1]
India Inde	1 481	1 557	* 1 530	* 1 746	* 1 812	1 861	1 902	* 2 060	* 2 240	* 2 400
Indonesia Indonésie	374	* 403	500	* 611	813	974	1 158	* 1 438	* 1 755	* 2 263
Iran, Islamic Rep. of Iran, Rép. islamique d'	78[1]	78[1]	78[1]	80[1]	90[1]	100	142[1]	* 211	* 235	235[1]
Iraq Iraq	28[1]	28[1]	28[1]	28[1]	28[1]	28[1]	28[1]	55	13	13[1]
Israel Israël	157	148	131	151	160	170	179	194	200	214

48
Paper and paperboard
Production: thousand metric tons [*cont.*]
Papiers et cartons
Production : milliers de tonnes métriques [*suite*]

Country or area Pays ou zone	1983	1984	1985	1986	1987	1988	1989	1990	1991	1992
Japan Japon	18 441	19 345	20 469	21 062	22 537	24 625	26 809	28 088	29 053	28 324
Jordan Jordanie	5	7	13	* 14	12	10	12	* 15	* 15	15[1]
Korea, Dem. P. R.[1] Corée, R. p. dém. de[1]	80	80	80	80	80	80	80	80	80	80
Korea, Republic of Corée, République de	1 982	2 207	2 312	2 773	3 163	3 659	4 018	4 524	4 922	5 504
Lebanon Liban	45[1]	45[1]	* 45	42	42[1]	37	37[1]	37[1]	42	42[1]
Malaysia Malaisie	* 45	50	* 53	* 73	97	120	251	* 275	* 293	* 407
Myanmar Myanmar	10[1]	10[1]	15[1]	23	9	8	10	11	11[1]	11[1]
Nepal Népal	2[1]	2[1]	...	* 2	2[1]	2[1]	2[1]	13	13	13[1]
Pakistan Pakistan	76	73	82	80	96	147	151	229	206	362
Philippines Philippines	280	290	268	218	358	314	239	245	395	570
Singapore Singapour	* 10	* 10	* 10	10[1]	10[1]	10[1]	10[1]	* 80	* 85	85[1]
Sri Lanka Sri Lanka	22[1]	23[1]	23	25	25[1]	28[1]	17	16	23	26
Syrian Arab Republic Rép. arabe syrienne	3[1]	3[1]	* 5	* 5	10[1]	19	19[1]	19[1]	1	1[1]
Thailand Thaïlande	300[1]	437	466	432[1]	465	514	520	877	958	577
Turkey Turquie	391	488	644	710	813	681	751	891	761	761[1]
Viet Nam Viet Nam	48[1]	55[1]	60[1]	65[1]	59	56	56[1]	62	108	115
Europe **Europe**	**51 008**	**55 523**	**55 762**	**57 610**	**60 407**	**64 697**	**66 483**	**67 462**	**67 253**	**67 579**
Albania Albanie	8[1]	8[1]	8[1]	8[1]	8[1]	26	26	44	44[1]	44[1]
Austria Autriche	1 789	1 922	2 127	2 183	2 396	2 650	2 754	2 932	3 090	3 252
Belgium-Luxembourg Belgique-Luxembourg	840	847	843	850	1 031	1 133	1 170	1 196	1 233	1 147
Bulgaria Bulgarie	442	445	454	458	456	477	438	322	258	153
former Czechoslovakia† anc. Tchécoslovaquie†	1 231	1 237	1 259	1 255	1 273	1 266	1 305	1 300	1 087	1 087[1]
Denmark Danemark	322	332	302	293	326	343[1]	328	335	356	341
Finland Finlande	6 388	7 318	7 447	7 549	8 011	8 652	8 579	8 777	8 505	8 823
France France	5 263	5 566	5 150	5 583	5 581	6 313	6 754	7 049	7 442	7 698

48

Paper and paperboard
Production: thousand metric tons [cont.]

Papiers et cartons
Production : milliers de tonnes métriques [suite]

Country or area Pays ou zone	1983	1984	1985	1986	1987	1988	1989	1990	1991	1992
Germany † Allemagne†	..	..	..	..	..	..	..	..	12 889	12 343
F. R. Germany R. f. Allemagne	8 273	9 145	9 178	9 409	9 938	10 576	11 259	11 873	..	..
former German D. R. anc. R. d. allemande	1 244	1 293	1 297	1 320	1 340	1 362	1 351	1 351[1]	..	..
Greece Grèce	277	294	282	283	280	280	280	361	387	387[1]
Hungary Hongrie	479	506	494	517	522	535	504	443	364	364[1]
Ireland Irlande	20[1]	20	* 22	37	29	33	34	35	36	36[1]
Italy Italie	4 259	4 722	4 587	4 631	4 882	5 512	5 640	5 587	5 795	5 959
Netherlands Pays-Bas	1 746	1 885	1 956	2 088	2 168	2 460	2 572	2 770	2 862	2 835
Norway Norvège	1 368	1 562	1 604	1 573	1 590	1 670	1 789	1 819	1 784	1 683
Poland Pologne	1 231	1 257	1 292	1 327	1 380	1 448	1 406	1 064	1 066	1 144
Portugal Portugal	592	671	706	590	628	681	740	780	877	959
Romania Roumanie	802[1]	806	801	811	816	819	* 709	547	359	359[1]
Spain Espagne	2 754	2 952	2 913	3 152	3 251	3 408	3 446	3 446	3 576	3 448
Sweden Suède	6 349	6 870	7 001	7 364	7 812	8 161	8 363	8 419	8 349	8 376
Switzerland Suisse	918	986	1 014	1 087	1 147	1 216	1 259	1 295	1 259[1]	1 305[1]
United Kingdom Royaume-Uni	3 208	3 591	3 712	3 941	4 184	4 295	4 475	4 824	4 951	5 152
Yugoslavia, SFR† Yougoslavie, Rfs†	1 205	1 288	1 313	1 301	1 358	1 381	1 302	894	685	685[1]
Oceania Océanie	**2 101**	**2 214**	**2 316**	**2 267**	**2 170**	**2 492**	**2 605**	**2 768**	**2 833**	**2 764**
Australia Australie	1 430	1 520	1 546	1 596	1 526	1 792	1 870	2 011	2 018	2 032
New Zealand Nouvelle-Zélande	671	694	* 770	671	644	700	735	757	815	732
former USSR† ancienne URSS†	**9 556**	**9 835**	**10 031**	**10 395**	**10 566**	**10 881**	**10 735**	**10 718**	**9 590**	**7 279**

Source:
Food and Agriculture Organization of the United Nations (Rome).

† For detailed descriptions of data pertaining to former Czechoslovakia, Germany, SFR Yugoslavia and former USSR, see Annex I - Country or area nomenclature, regional and other groupings.

1 FAO estimate.
2 Including Eritrea.
3 Including data for Namibia.

Source:
Organisation des Nations Unies pour l'alimentation et l'agriculture (Rome).

† Pour les descriptions en détails des données relatives à l'ancienne Tchécoslovaquie, l'Allemagne, la Rfs Yougoslavie et l'ancienne URSS, voir l'Annexe I - Nomenclature des pays ou zones, groupements régionaux et autres groupements.

1 Estimation de la FAO.
2 Y compris Erythrée.
3 Y compris les données pour la Namibie.

49

Tires
Pneumatiques : enveloppes
Production: thousands
Production : milliers

Country or area Pays ou zone	1983	1984	1985	1986	1987	1988	1989	1990	1991	1992
Total	**689 521**	**749 069**	**748 388**	**767 540**	**808 154**	**866 616**	**879 133**	**826 697**	**789 605**	**847 943**
Algeria * Algérie *	6	...	...	...	...	...	...	...	...	...
Argentina Argentine	4 463	5 357	3 904	4 865	5 209	5 615	4 826	4 955	4 739	5 627
Armenia Arménie	..	..	..	..	..	..	..	..	485	...
Australia [1 2] Australie [1 2]	5 821	6 165	5 846	6 600	5 576	7 450	7 450	7 600	...	...
Azerbaijan Azerbaïdjan	..	..	..	..	..	..	..	..	271	222
Belarus Bélarus	..	..	..	..	..	..	..	..	3 367	...
Brazil [3] Brésil [3]	18 230	21 020	22 827	24 376	28 224	29 255	29 215	29 162	28 926	30 306
Bulgaria Bulgarie	1 613	1 666	1 659	1 668	1 857	1 693	1 762	1 795	1 125	1 034
Canada [4] Canada [4]	23 533	25 519	26 655	28 765	27 200	25 032	20 664	21 692	...	...
Chile Chili	595	913	858	862	1 221	1 347	1 562	1 632	1 825	2 002
China [5] Chine [5]	12 710	15 690	19 260	19 243	23 332	29 910	32 262	32 091	38 723	51 834
Colombia Colombie	1 625	1 950	1 856	1 851	1 955	1 915	1 875	1 408	* 1 705	...
Côte d'Ivoire Côte d'Ivoire	12	...	...	...	...	...	...	...	...	...
Croatia Croatie	..	..	..	..	..	..	..	..	34	...
Cuba Cuba	255	328	345	319	232	320	230	...	...	...
Cyprus Chypre	113	119	105	91	* 91	* 92	89	* 99	* 72	...
former Czechoslovakia† anc. Tchécoslovaquie†	4 465	4 621	4 547	4 737	4 858	5 058	5 263	5 315	..	..
Czech Republic République tchèque	..	..	..	..	..	..	..	..	2 501	3 060
Ecuador Equateur	...	...	...	...	97	32	...	42	...	812
Egypt Egypte	730	775	845	1 011	1 038	1 006	1 126	1 171	1 243	1 186
Finland [6] Finlande [6]	1 352	1 427	1 339	...	...		...	...	...	...
France France	45 606	47 817	47 316	51 138	49 344	54 043	61 678	54 536	57 876	59 928
Germany † · Allemagne† F. R. Germany R. f. Allemagne	37 878	39 254	40 475	42 826	47 083	48 644	49 467	48 247	..	..

49
Tires
Production: thousands [cont.]
Pneumatiques : enveloppes
Production : milliers [suite]

Country or area Pays ou zone	1983	1984	1985	1986	1987	1988	1989	1990	1991	1992
former German D. R. anc. R. d. allemande	5 285	5 576	6 003	6 183	6 275	6 328	6 259	4 052	..	..
Hungary Hongrie	537	552	578	635	674	686	736	605	425	355
India Inde	5 529	5 952	5 042	6 383	5 616	7 537	8 066	8 460	8 312	8 172
Indonesia Indonésie	3 332	3 406	3 420	4 740	4 791	* 6 564	* 8 028	10 080	...	...
Iran, Islamic Rep. of Iran, Rép. islamique d'[7]	3 466	4 782	3 577	3 573	4 587	4 638	4 814	6 012	...	...
Israel Israël	1 045	1 009	930	744	920	573	751	778	...	...
Italy Italie	26 598	27 460	27 955	30 860	32 005	32 053	33 625	30 767	32 447	29 978
Jamaica Jamaïque	196	209	210	217	265	243	290	347	326	...
Japan Japon	123 094	131 199	135 745	135 323	139 086	150 562	155 038	153 226	153 677	154 900
Kenya[4] Kenya[4]	...	...	...	725	617	631	797	813	...	...
Korea, Republic of Corée, République de	12 033	15 126	15 207	18 214	20 060	24 250	24 535	29 776	35 307	38 048
Malaysia[5] Malaisie[5]	4 262	3 700	3 622	3 846	5 173	6 222	6 156	6 764	7 970	8 540
Mexico Mexique	8 722	9 779	10 472	9 330	10 164	10 474	11 038	11 855	12 148	12 568
Morocco Maroc	816	834	779	821	943	...	...	...	...	...
Mozambique Mozambique	111	51	80	60	22	17	29	23	5	...
New Zealand[9] Nouvelle-Zélande[9]	1 236	1 282	1 222	1 389	1 565	1 650	1 460	1 550	...	...
Nigeria Nigéria	2 205	1 604	2 154	* 2 941	...	...	...	...	...	...
Pakistan[2] Pakistan[2]	217	238	307	412	382	679	907	915	952	784
Panama Panama	49	49	37	...	22	23	23	...	...	...
Peru Pérou	676	782	784	914	946	868	710	637	561	...
Philippines Philippines	1 737	1 450	1 300	934	1 159	* 1 968	* 2 016	2 208	...	...
Poland Pologne	5 084	5 902	6 254	6 217	6 020	6 276	6 025	4 704	4 516	5 607
Portugal Portugal	2 011	2 154	2 286	2 385	2 754	3 261	3 292	2 976	2 184	...
Romania[9] Roumanie[9]	5 346	5 882	5 642	5 789	5 247	5 552	4 804	3 702	2 822	2 877
Russian Federation Fédération de Russie	..	..	..	..	..	..	..	..	33 522	...

49
Tires
Production: thousands [*cont.*]
Pneumatiques: enveloppes
Production: milliers [*suite*]

Country or area Pays ou zone	1983	1984	1985	1986	1987	1988	1989	1990	1991	1992
Slovakia Slovaquie	..	..	..	..	..	..	..	..	2 276	...
Slovenia Slovénie	..	..	..	..	..	..	..	..	3 289	4 133
South Africa Afrique du Sud	5 666	5 825	5 041	5 432	6 066	6 813	6 817	7 478	7 236	7 136
Spain Espagne	15 670	20 610	19 386	20 034	23 204	24 323	24 696	23 361	23 812	...
Sri Lanka Sri Lanka	197	164	180	200	340	343	339	382	392	...
Sweden Suède	2 616	2 741	2 996	3 001	2 852	* 2 980	* 2 915	* 2 162	2 158	...
Thailand Thaïlande	2 202	2 177	2 126	2 264	3 063	3 980	4 320	4 183	4 518	...
Tunisia Tunisie	172	205	261	358	374	471	504	576	...	528
Turkey Turquie	3 777	4 690	4 922	4 871	6 260	12 088	6 633	4 754	7 541	8 463
Ukraine Ukraine	..	..	..	..	..	..	..	..	7 859	...
former USSR†[10] ancienne URSS†[10]	62 025	63 737	65 171	66 023	67 802	69 125	69 705	...	..	..
United Kingdom Royaume-Uni	23 976	24 120	24 216	25 644	27 624	30 204	31 080	29 376	28 500	30 408
United Rep.Tanzania Rép. Unie de Tanzanie	...	...	113	139	197	188	213	208	185	158
United States Etats-Unis	186 924	209 375	195 972	190 296	202 980	211 356	212 868	210 660	202 391	230 250
Venezuela Venezuela	3 358	2 985	4 492	4 989	5 447	5 203	4 177	3 951	4 787	4 713
Yugoslavia, SFR†[1] Yougoslavie, Rfs†[1]	9 420	9 858	11 194	11 632	11 718	12 548	13 201	12 744	..	..

Source:
Industrial statistics database of the Statistical Division
of the United Nations Secretariat.

† For detailed descriptions of data pertaining to
former Czechoslovakia, Germany, SFR Yugoslavia and former
USSR, see Annex I - Country or area nomenclature, regional
and other groupings.

1 Including motorcycle tires.
2 Twelve months ending 30 June of year stated.

3 Including tires for motorcycles and bicycles.

4 Source: International Rubber Study Group, (London).
5 Tires of all types.
6 Beginning 1986, data are confidential.
7 Production by establishments employing 50 or more persons.
8 Including retreaded tires.
9 Including tires for vehicles operating off-the-road.
10 Including tires for agricultural vehicles, motorcycles and
   scooter.

Source:
Base de données pour les statistiques industrielles de la
Division de statistique du Secrétariat de l'ONU.

† Pour les descriptions en détails des données
relatives à l'ancienne Tchécoslovaquie, l'Allemagne, la Rfs
Yougoslavie et l'ancienne URSS, voir l'Annexe I -
Nomenclature des pays ou zones, groupements régionaux et
autres groupements.

1 Y compris les pneumatiques pour motocyclettes.
2 Période de douze mois finissant le 30 juin de l'année
   indiquée.
3 Y compris les pneumatiques pour motocyclettes et
   bicyclettes.
4 Source: "International Rubber Study Group", (Londres).
5 Pneumatiques de tous genres.
6 A partir de 1986, les données sont confidentielles.
7 Production des établissements occupant 50 personnes ou plus.
8 Y compris les pneumatiques rechapés.
9 Y compris les pneumatiques pour véhicules tous terrains.
10 Y compris les pneumatiques pour véhicules agricoles,
   motocyclettes et scooters.

50
Cement
Ciment

Production: thousand metric tons
Production : milliers de tonnes métriques

Country or area Pays ou zone	1983	1984	1985	1986	1987	1988	1989	1990	1991	1992
Total	906 745	933 899	949 972	995 875	1 043 524	1 110 405	1 146 182	1 153 849	1 109 590	1 189 177
Afghanistan[1] Afghanistan[1]	130	112	128	103	104	70	* 100	100	109	...
Albania Albanie	* 840 [2]	* 860 [2]	642	709	708	746	753	644	311	...
Algeria Algérie	4 776	5 538	6 096	6 448	7 541	7 195	6 819	6 337	6 323	7 093
Angola Angola	125	126	205	354 [2]	354 [2]	1 000 [2]	1 000 [2]	1 000 [2]	998 [2]	370 [2]
Argentina Argentine	5 882	5 220	4 795	5 558	6 302	6 024	4 439	3 612	4 399	...
Armenia Arménie	..	..	..	..	..	..	..	..	1 507	...
Australia Australie	5 351	4 655	5 680	6 106	5 920	6 158	6 901	7 075	6 110	5 731
Austria Autriche	4 907	4 899	4 560	4 569	4 518	4 763	4 749	4 903	5 017	5 029
Azerbaijan Azerbaïdjan	..	..	..	..	..	..	..	..	42	38
Bahamas[2] Bahamas[2]	26	...	...	...	...	...	...	...	...	...
Bangladesh[3] Bangladesh[3]	307	273	240	292	310	310	344	337	275	272
Barbados Barbade	...	64	148	199	147	185	225	213	144	71
Belarus Bélarus	..	..	..	..	..	..	..	..	2 402	...
Belgium Belgique	5 719	5 708	5 537	5 760	5 689	6 451	6 720	6 924	7 184	8 073
Benin[2] Bénin[2]	300	300	300	300	300	200	250	272	272	...
Bolivia Bolivie	310	220	241	343	386	448	499	560	583	...
Brazil Brésil	20 870	19 497	20 635	25 252	25 468	25 329	25 921	25 829	27 257	23 889
Bulgaria Bulgarie	5 644	5 717	5 296	5 702	5 494	5 535	5 036	4 710	2 374	2 132
Cameroon Cameroun	598	694	785	779	707	586	...	...	...	...
Canada Canada	7 871	9 387	10 192	10 611	12 603	12 350	12 591	11 745	9 372	8 592
Chile Chili	1 255	1 390	1 430	1 441	1 500	1 885	2 010	2 115	2 251	2 660
China Chine	108 250	123 020	145 950	166 060	186 249	210 140	210 295	209 711	244 656	308 217
Colombia Colombie	4 787	5 280	5 412	5 916	5 892	6 312	6 648	6 360	6 389	6 792
Congo Congo	28	* 45	62	2	38	86	121	21	* 103	...

50
Cement
Production: thousand metric tons [*cont.*]
Ciment
Production : milliers de tonnes métriques [*suite*]

Country or area Pays ou zone	1983	1984	1985	1986	1987	1988	1989	1990	1991	1992
Costa Rica[4] Costa Rica[4]	386	350	350	306	285	309	315	...	...	...
Côte d'Ivoire Côte d'Ivoire	636	552	535	775[2]	652[2]	700[2]	700[2]	500[2]	499[2]	...
Croatia Croatie	..	..	..	..	..	..	..	..	1 742	1 771
Cuba Cuba	3 231	3 347	3 182	3 305	3 535	3 566	3 759	3 696	2 599	...
Cyprus Chypre	943	853	659	857	854	868	1 042	1 133	1 134	...
former Czechoslovakia† anc. Tchécoslovaquie†	10 498	10 530	10 265	10 298	10 369	10 974	10 888	10 215	..	
Czech Republic République tchèque	..	..	..	..	..	..	..	..	5 610	6 111
Denmark[5] Danemark[5]	1 654	1 668	1 983	2 029	1 886	1 681	2 000	1 656	2 019	2 072
Dominican Republic Rép. dominicaine	1 057	1 109	1 001	952	1 209	1 235	* 1 269	* 1 189	* 998	...
Ecuador Equateur	1 494	1 755	2 008	2 118	2 875	2 126	* 1 548	1 792	1 774	2 072
Egypt Egypte	3 794	4 600	5 275	7 612	8 762	9 794	12 480	14 111	16 427	15 454
El Salvador[4] El Salvador[4]	435	399	450	460	480	455	447	444	448	...
Estonia Estonie	..	..	..	..	..	..	..	..	905	483
Ethiopia[6] Ethiopie[6]	120	165	228	270	350	406	* 412	324[7]	* 290	...
Fiji Fidji	110	98	93	92	59	44	58	78	79	84
Finland Finlande	1 979	1 656	1 695	1 495	1 579	1 619	1 693	1 649	1 343	1 133
France France	24 503	22 724	22 224	21 584	23 544	* 25 374	25 994	26 497	25 089	21 584
Gabon Gabon	194	207	244	210	141	132[2]	115[2]	115[2]	113[2]	120
Germany † • Allemagne† F. R. Germany R. f. Allemagne	30 466	28 909	25 758	26 580	25 268	26 215	28 499	30 456	..	..
former German D. R. anc. R. d. allemande	11 782	11 555	11 608	11 988	12 430	12 510	12 229	7 316	..	..
Ghana Ghana	278	235	356	219	294	412	* 565	* 675	* 675	...
Greece Grèce	14 196	13 521	12 855	12 542	11 869	12 777	12 319	13 142	13 151	10 668
Guadeloupe Guadeloupe	211	188	194	197	222	200[2]	200[2]	200[2]	200[2]	...
Guatemala Guatemala	491	455	574	1 392	1 260	* 880	* 591	* 611	* 623	...

50
Cement
Production: thousand metric tons [*cont.*]
Ciment
Production : milliers de tonnes métriques [*suite*]

Country or area Pays ou zone	1983	1984	1985	1986	1987	1988	1989	1990	1991	1992
Haiti Haïti	219	248	263	248	254	245	234	226	* 226	...
Honduras Honduras	485	368	348	360	375	268	321	326	402	...
Hong Kong Hong-kong	1 717	1 847	1 835	2 236	2 226	2 189	2 141	1 808	1 677	1 644
Hungary Hongrie	4 243	4 145	3 678	3 846	4 153	3 873	3 857	3 933	2 529	2 236
Iceland Islande	120	114	117	115	127	* 134	* 116	114	113	...
India Inde	25 261	29 541	31 971	34 983	37 135	41 136	44 197	46 170	51 660	53 700
Indonesia Indonésie	8 095	8 893	9 940	11 323	11 814	12 096	15 660	15 972	16 153	...
Iran, Islamic Rep. of[2] Iran, Rép. islamique d'[2]	10 655	12 064	11 954	12 148	12 852	11 926	12 587	14 429	15 000	...
Iraq Iraq	5 600[2]	8 000[2]	8 000[2]	7 992	9 780	9 162	12 500	13 000	* 5 000	2 453
Ireland Irlande	1 486	1 377	1 457	1 250	1 446	* 1 685	* 1 624	* 1 624	* 1 601	...
Israel Israël	2 058	1 889	1 596	1 624	2 226	2 326	2 289	2 868	* 2 902	...
Italy Italie	39 763	38 307	37 155	36 393	37 788	38 220	39 692	40 544	40 301	41 034
Jamaica Jamaïque	278	259	241	247	261	339	360	420	402	488
Japan Japon	80 891	78 860	72 847	71 264	71 551	77 554	79 717	84 445	89 564	88 252
Jordan Jordanie	1 269	2 026	2 022	1 837	2 472	1 780	1 930	1 780	1 800	2 575
Kazakhstan Kazakhstan	..	..	..	..	..	..	..	..	7 575	6 436
Kenya Kenya	1 180	1 134	1 115	1 174	1 243	1 201	1 316	1 512	1 497	...
Korea, Dem. P. R.[2] Corée, R. p. dém. de[2]	* 8 000	* 8 000	* 8 000	* 8 000	* 9 000	11 800	16 300	16 300	16 329	...
Korea, Republic of Corée, République de	21 282	20 413	20 509	23 530	25 946	29 611	30 821	33 914	39 167	44 444
Kuwait Koweït	1 124	1 184	1 066	1 014	888	984	1 108	800	* 299	...
Kyrgyzstan Kirghizistan	..	..	..	..	..	..	..	..	1 320	1 096
Latvia Lettonie	..	..	..	..	..	..	..	..	720	340
Lebanon[2] Liban[2]	1 500	1 250	* 1 000	* 900	* 900	* 900	* 900	907	907	...
Liberia Libéria	88	86	104	97	95	130	* 85	50	...	...
Libyan Arab Jamah. Jamah. arabe libyenne	3 093	6 000[2]	6 500[2]	2 077[2]	2 700[2]	2 700[2]	2 700[2]	2 700[2]	2 722[2]	...

50
Cement
Production: thousand metric tons [cont.]
Ciment
Production : milliers de tonnes métriques [suite]

Country or area Pays ou zone	1983	1984	1985	1986	1987	1988	1989	1990	1991	1992
Luxembourg Luxembourg	353	340	295	389	509	563	590	636	688	695
Madagascar Madagascar	36	37	28	32	44	33	24	29	32	30
Malawi Malawi	70	70	62	69	75	62	79	101	112	108
Malaysia Malaisie	3 231	3 469	3 128	3 569	3 316	3 861	4 794	5 881	7 451	8 366
Mali[2] Mali[2]	20	25	19	20	22	25	20	20	22	...
Martinique Martinique	207	189	191	209	221	247	244	277	291	...
Mexico Mexique	17 363	18 702	20 255	19 825	23 482	23 606	24 210	24 683	25 208	27 114
Mongolia Mongolie	165	141	151	425	541	502	513	510	* 399	...
Morocco Maroc	3 848	3 573	3 704	3 709	3 879	4 260	4 641	5 381	5 777	6 223
Mozambique Mozambique	188	105	77	73	73	64	80	78	63	...
Myanmar[9] Myanmar[9]	334	310	429	444	389	349	441	420	443	472
Nepal Népal	39	32	96	152	215	217	114	107	* 99	...
Netherlands Pays-Bas	3 108	3 176	2 911	3 099	2 929	3 418	3 546	3 682	3 546	3 300
New Caledonia Nouvelle-Calédonie	41	* 41	31	40	* 58	63	67	64	* 68	...
New Zealand Nouvelle-Zélande	760	823	863	895	880	793	729	681	581	599
Nicaragua[4] Nicaragua[4]	298	280	245	284	265	256	225	* 140	* 140	...
Niger[2] Niger[2]	30	38	38	26	29	40	27	27	28	...
Nigeria Nigéria	2 760	2 184	3 348	3 624	3 085	3 400[2]	3 500[2]	3 500[2]	3 501[2]	...
Norway Norvège	1 666	1 547	1 343	1 752	1 703	1 667	1 380	1 260	1 293	...
Pakistan[3] Pakistan[3]	3 938	4 503	4 732	5 773	6 508	7 072	7 125	7 488	7 762	8 321
Panama[2] Panama[2]	327	304	305	336	350	200	169	* 300	* 300	...
Paraguay Paraguay	153	109	46	179	269	256	256	344	343	476
Peru Pérou	1 967	1 947	1 757	2 207	2 584	2 514	2 105	2 184	2 124	2 052
Philippines Philippines	4 560	3 660	3 072	3 288	3 984	4 092	3 624	6 360	6 804	6 540
Poland Pologne	16 163	16 649	14 990	15 831	16 090	16 984	17 125	12 518	12 012	11 908

50
Cement
Production: thousand metric tons [*cont.*]
Ciment
Production : milliers de tonnes métriques [*suite*]

Country or area Pays ou zone	1983	1984	1985	1986	1987	1988	1989	1990	1991	1992
Portugal Portugal	6 060	5 514	5 279	5 425	5 853	6 471	6 673	7 188	7 342	7 730
Puerto Rico[3] Porto Rico[3]	855	896	865	896	1 094	1 163	1 257	1 302	1 296	1 266
Qatar Qatar	162	313	385	324	293	291	295	267	367	354
Republic of Moldova République de Moldova	..	..	..	..	..	..	..	..	1 809	705
Réunion Réunion	...	...	...	...	...	...	322	336	...	...
Romania Roumanie	13 027	12 991	11 189	13 054	12 435	13 124	12 225	9 468	6 692	6 271
Russian Federation Fédération de Russie	..	..	..	..	..	..	..	..	77 463	...
Rwanda Rwanda	...	8	32	47	57	51	* 68	* 60	* 60	...
Saudi Arabia Arabie saoudite	8 265	7 504	9 232	9 232	* 8 595	* 9 525	* 9 500	* 10 000	* 12 002	...
Senegal Sénégal	388	414	408	372	362	393	* 380	* 380	* 499	...
Sierra Leone Sierra Leone	42	0	17	34	24	...	9	...	...	...
Singapore Singapour	3 126	2 511	1 897	1 875	1 550	1 684	1 704	1 848	2 199	...
Slovakia Slovaquie	..	..	..	..	..	..	..	..	2 680	3 374
Slovenia Slovénie	..	..	..	..	..	..	..	..	1 801	1 568
South Africa Afrique du Sud	7 897	8 084	6 880	6 246	5 999	6 760	7 261	6 563	6 147	5 850
Spain Espagne	30 637	25 435	21 876	22 007	23 012	24 372	27 375	28 092	27 576	24 612
Sri Lanka Sri Lanka	480	401	380	558	619	633	596	579	620	588
Sudan Soudan	232	200	148	175	178	110	* 150	150	168	...
Suriname Suriname	70	50	79	61	40	34	52	55	24	14
Sweden Suède	2 240	2 393	2 101	2 044	2 238	4 427	4 541	5 000	4 493	...
Switzerland Suisse	4 138	4 181	4 254	4 393	4 617	4 965	5 461	5 206	4 716	4 260
Syrian Arab Republic Rép. arabe syrienne	3 719	4 279	4 357	4 316	3 870	3 330	3 976	3 049	2 843	3 246
Thailand Thaïlande	7 301	8 271	7 951	8 005	9 870	11 519	15 042	18 040	19 210	20 076
Togo Togo	232	243	284	338[2]	370[2]	378[2]	389[2]	400[2]	399[2]	...
Trinidad and Tobago Trinité-et-Tobago	393	406	329	338	326	360	384	438	486	482

50
Cement
Production: thousand metric tons [*cont.*]
Ciment
Production : milliers de tonnes métriques [*suite*]

Country or area Pays ou zone	1983	1984	1985	1986	1987	1988	1989	1990	1991	1992
Tunisia Tunisie	2 532	2 777	3 033	2 962	3 215	3 600	3 780	4 140	3 331	3 912
Turkey Turquie	13 595	15 738	17 581	20 004	21 980	22 568	23 704	24 404	25 842	28 331
Uganda Ouganda	31	25	12	16	16	15	17	27	24	36
Ukraine Ukraine	..	..	..	..		..	..	..	21 745	...
former USSR[†] ancienne URSS[†]	128 156	129 866	130 772	135 119	137 404	139 499	140 436	137 321	..	..
United Arab Emirates Emirats arabes unis	2 068	4 005[2]	4 205[2]	2 748[2]	3 106[2]	2 980[2]	3 112[2]	3 110[2]	3 012[2]	...
United Kingdom Royaume-Uni	13 396	13 481	13 339	13 413	14 311	16 506	16 849	14 736	12 002	...
United Rep.Tanzania Rép. Unie de Tanzanie	247	369	376	435	498	592	595	664	1 022	677
United States Etats-Unis	63 927	70 452	70 284	71 112	67 380	71 544	71 308	70 944	* 65 052	70 848
Uruguay Uruguay	401	374	317	329	420	435	* 465	469	458	552
Venezuela Venezuela	4 430	4 783	5 121	5 875	5 975	6 199	5 259	5 996	6 336	6 585
Viet Nam Viet Nam	907	1 296	1 503	1 526	1 665	1 954	2 088	2 534	3 127	3 926
former Yemen Arab Rep. anc. Yémen rép. arabe	623	709	698	708	* 760	* 646	* 700	* 700	* 798	...
Yugoslavia, SFR[†] Yougoslavie, Rfs[†]	9 588	9 315	9 028	9 128	8 963	8 840	8 560	7 956	..	...
Zaire Zaïre	513	534	444[2]	445[2]	492[2]	495[2]	460[2]	460[2]	449[2]	...
Zambia Zambie	392	220	316	334	375	405	385	432	376	...
Zimbabwe Zimbabwe	501	650	614	659	811	776	827	924	949	829

Source:
Industrial statistics database of the Statistical Division
of the United Nations Secretariat.

† For detailed descriptions of data pertaining to
former Czechoslovakia, Germany, SFR Yugoslavia and former
USSR, see Annex I - Country or area nomenclature, regional
and other groupings.

1 Twelve months beginning 21 March of year stated.

2 Source: US Bureau of Mines, (Washington, DC).
3 Twelve months ending 30 June of year stated.

4 Source: United Nations Economic Commission for Latin America
(ECLA), (Santiago).
5 Sales.
6 Twelve months ending 7 July of the year stated.

7 Beginning 1990, excluding Eritrea.
8 Production by establishments employing 50 or more persons.
9 Government production only.

Source:
Base de données pour les statistiques industrielles de la
Division de statistique du Secrétariat de l'ONU.

† Pour les descriptions en détails des données
relatives à l'ancienne Tchécoslovaquie, l'Allemagne, la Rfs
Yougoslavie et l'ancienne URSS, voir l'Annexe I -
Nomenclature des pays ou zones, groupements régionaux et
autres groupements.

1 Période de douze mois commençant le 21 mars de l'année
indiquée.
2 Source: "US Bureau of Mines," (Washington, DC).
3 Période de douze mois finissant le 30 juin de l'année
indiquée.
4 Source: Commission économique des Nations Unies pour
l'Amérique Latine (CEPAL), (Santiago).
5 Ventes.
6 Période de douze mois finissant le 7 juillet de l'année
indiquée.
7 A partir de 1990, non compris Erythrée.
8 Production des établissements occupant 50 personnes ou plus.
9 Production de l'Etat seulement.

51
Sulphuric acid
Acide sulfurique

Production: thousand metric tons
Production : milliers de tonnes métriques

Country or area Pays ou zone	1983	1984	1985	1986	1987	1988	1989	1990	1991	1992
Total	**129 473**	**137 230**	**132 564**	**131 421**	**134 938**	**142 970**	**141 182**	**135 918**	**124 849**	**126 799**
Albania Albanie	...	...	73	85	80	81	82	68	21	...
Algeria Algérie	156	249	...	...	30	61	60	39	46	52
Argentina Argentine	262	254	235	251	253	258	214	209	243	...
Australia[1] Australie[1]	1 734	1 706	1 783	1 788	1 680	1 818	1 904	1 464	986	...
Azerbaijan Azerbaïdjan	713	798	782	839	872	846	768	603	552	269
Bangladesh[1] Bangladesh[1]	3	4	4	6	8	5	6	5	6	4
Belarus Bélarus	1 228	1 258	1 194	1 171	1 186	1 185	1 179	1 177	998	...
Belgium[2] Belgique[2]	1 898	2 247	2 107	1 957	2 069	2 136	1 956	1 906	1 936	1 906
Bolivia Bolivie	1	1	0	0	...	...	...	...	1 174	...
Brazil Brésil	2 983	3 484	3 660	3 820	4 004	4 049	3 809	3 451	3 634	...
Bulgaria Bulgarie	861	908	810	807	689	840	846	522	356	404
Canada Canada	3 679	4 043	3 890	3 536	3 437	3 805	3 560	...	...	...
China Chine	8 696	8 172	6 764	7 631	9 833	11 113	11 533	11 969	13 329	14 087
Colombia Colombie	53	...	62	68	68	75	95	76	...	...
Côte d'Ivoire Côte d'Ivoire	1	1	...	...	...	...	...	...	...	...
Croatia Croatie	11	115	185	168	225	260	271	242	187	278
Cuba Cuba	356	336	373	395	370	391	377	...	...	...
Cyprus Chypre	13	0	0	0	17	2	1	0	0	...
former Czechoslovakia† anc. Tchécoslovaquie†	1 244	1 246	1 298	1 292	1 264	1 249	1 142	1 089	0	0
Czech Republic République tchèque	1 088	1 140	1 194	1 110	1 080	1 080	987	910	588	531
Denmark[34] Danemark[34]	20	27	36	23	33	91	122	90	37	38
Egypt Egypte	45	55	46	55	60	54	60	92	101	111
El Salvador El Salvador	4	...	...	...	...	...	...	...	...	...
Estonia Estonie	653	672	616	654	653	654	552	547	460	46

51
Sulphuric acid
Production: thousand metric tons [cont.]
Acide sulfurique
Production : milliers de tonnes métriques [suite]

Country or area Pays ou zone	1983	1984	1985	1986	1987	1988	1989	1990	1991	1992
Finland Finlande	1 145	1 165	1 207	1 100	870	1 179	1 129	1 010	1 015	1 087
France France	4 309	4 531	4 321	3 954	3 558	4 081	4 187	3 771	3 627	2 871
Germany † Allemagne†	..	..	..	..	..	..	..	..	3 064	...
F. R. Germany R. f. Allemagne	4 340	4 309	4 199	4 105	4 070	4 053	4 028	3 221	..	..
former German D. R. anc. R. d. allemande	926	885	883	883	867	799	835	431	..	..
Greece Grèce	1 016	1 155	1 086	952	912	987	1 023	950	841	...
Hungary[5] Hongrie[5]	645	587	556	564	592	529	502	263	141	99
India Inde	2 241	2 683	* 2 626	2 877	3 159	3 416	3 293	3 272	3 889	...
Indonesia Indonésie	35	44	58	46	71	...	...	40	52	
Iran, Islamic Rep. of * [6] Iran, Rép. islamique d' * [6]	150	200	200	200	200	200	...	...	...	...
Israel Israël	171	189	178	182	143	164	161	154	...	...
Italy Italie	2 540	2 636	2 724	2 605	2 724	2 499	2 212	2 038	1 853	1 733
Jamaica Jamaïque	17	...	...	...	...	...	...	...	...	...
Japan Japon	6 662	6 458	6 580	6 562	6 541	6 767	6 885	6 887	7 057	7 100
Kazakhstan Kazakhstan	3 327	3 173	3 098	3 283	3 440	3 605	3 455	3 151	2 815	2 349
Korea, Republic of Corée, République de	1 610	1 975	2 028	...	...	...	...	...	...	...
Kuwait Koweït	7	4	5	* 5	5	5	...	...	...	...
Mexico Mexique	2 996	3 196	2 222	2 149	# 582	534	414	455	362	195
Morocco[7] Maroc[7]	727	758	660	...	...	...	...	...	...	...
Netherlands Pays-Bas	1 451	1 609	1 508	1 209	1 043	1 144	...	...	...	...
Norway Norvège	440	460	448	484	516	796	...	...	...	...
Pakistan[1] Pakistan[1]	71	81	78	80	78	79	79	90	93	98
Peru Pérou	210	205	213	209	175	174	193	183	...	...
Philippines Philippines	56	40	62	25	...	37	...	...	...	...
Poland Pologne	2 786	2 769	2 863	2 965	3 149	3 154	3 115	1 721	1 088	1 244

51

Sulphuric acid
Production: thousand metric tons [cont.]
Acide sulfurique
Production : milliers de tonnes métriques [suite]

Country or area Pays ou zone	1983	1984	1985	1986	1987	1988	1989	1990	1991	1992
Portugal Portugal	431	553	497	448	331	293	289	260	...	...
Romania Roumanie	1 941	1 915	1 835	1 971	1 693	1 825	1 687	1 111	745	572
Russian Federation Fédération de Russie	11 064	11 510	12 017	12 761	12 785	12 935	12 422	12 767	11 597	...
Saudi Arabia Arabie saoudite	63	71	* 73	* 84	...	...	...	...	...	...
Slovakia Slovaquie	156	106	103	182	184	169	155	178	94	...
Slovenia Slovénie	203	217	232	216	185	204	189	125	86	121
Spain Espagne	2 995	3 683	3 391	3 577	3 318	3 440	3 325	2 848	1 628	...
Sweden Suède	890	927	958	1 014	992	962	902	855	928	
Syrian Arab Republic Rép. arabe syrienne	9	5	4	5	5	4	7	8	8	10
Thailand[7] Thaïlande[7]	56	63	66	78	81	60	64	72	82	...
Tunisia Tunisie	2 808	2 716	2 559	3 016	3 172	3 316	...	...	...	...
Turkey[8] Turquie[8]	338	353	514	583	102	313	313	295	205	248
Ukraine Ukraine	4 505	4 426	4 579	4 125	4 752	5 162	5 144	5 011	4 186	...
former USSR† ancienne URSS†	24 714	25 338	26 037	27 847	28 531	29 372	28 276	27 267	0	0
United Kingdom Royaume-Uni	2 629	2 654	2 525	2 330	2 335	2 257	1 977	...	...	...
United States[9] Etats-Unis[9]	33 975	37 914	35 964	32 650	35 612	38 628	39 282	40 222	39 432	40 697
Venezuela Venezuela	139	158	156	164	197	172	163	210	277	253
Viet Nam Viet Nam	...	...	...	14	14	17	10	8	9	7
Yugoslavia, SFR† Yougoslavie, Rfs†	1 300	1 471	1 489	1 595	1 592	1 713	1 617	...	0	0
Zaire Zaïre	160	* 153	* 150	...	...	...	...	...	...	...
Zambia Zambie	271	314	240	199	304	276	...	...	...	...

Source:
Industrial statistics database of the Statistical Division
of the United Nations Secretariat.

† For detailed descriptions of data pertaining to
former Czechoslovakia, Germany, SFR Yugoslavia and former
USSR, see Annex I - Country or area nomenclature, regional
and other groupings.

Source:
Base de données pour les statistiques industrielles de la
Division de statistique du Secrétariat de l'ONU.

† Pour les descriptions en détails des données
relatives à l'ancienne Tchécoslovaquie, l'Allemagne, la Rfs
Yougoslavie et l'ancienne URSS, voir l'Annexe I -
Nomenclature des pays ou zones, groupements régionaux et
autres groupements.

51

Sulphuric acid
Production: thousand metric tons [*cont.*]

Acide sulfurique
Production : milliers de tonnes métriques [*suite*]

1 Twelve months ending 30 June of year stated.

2 Production by establishments employing 5 or more persons.
3 Excluding quantities consumed by superphosphate industry.

4 Sales.
5 Including regenerated sulphuric acid.
6 Source: US Bureau of Mines, (Washington, DC).
7 Strength of acid not known.
8 Output of steel industry only.
9 Including data for government-owned, but privately-operated plants.

1 Période de douze mois finissant le 30 juin de l'année indiquée.
2 Production des établissements occupant 5 personnes ou plus.
3 Non compris les quantités utilisées par l'industrie des superphosphates.
4 Ventes.
5 Y compris l'acide sulfurique régénéré.
6 Source: "US Bureau of Mines," (Washington, DC).
7 Titre de l'acide inconnu.
8 Production de l'industrie sidérurgique seulement.
9 Y compris les données relatives à des usines appartenant à l'Etat mais exploitées par des entreprises privées.

52
Soap, washing powders and detergents
Savons, poudres pour lessives et détersifs

Production: thousand metric tons
Production : milliers de tonnes métriques

Country or area Pays ou zone	1983	1984	1985	1986	1987	1988	1989	1990	1991	1992
Total	18 817	19 192	20 497	20 756	21 440	21 736	22 751	23 113	20 539	19 862
Afghanistan Afghanistan	1	0	0	0	0	0	...	...	...	...
Albania [1] Albanie [1]	...	...	3	4	7	5	7	5	1	...
Algeria Algérie	104	111	111	151	193	225	223	263	235	184
Argentina [2] Argentine [2]	166	144	141	136	124	132	94	141	11 [3]	...
Armenia Arménie	..	..	..	..	..	..	..	..	12	...
Australia [4] Australie [4]	365	346	346	304	322	342	...	...	...	...
Austria Autriche	116	114	115	129	145	142	146	153	162	188
Azerbaijan Azerbaïdjan	..	..	..	..	..	..	..	..	101	69
Belarus Bélarus	..	..	..	..	..	..	..	..	71 [5]	...
Belgium Belgique	339	337	337	345	393	401	406	417	440	423
Bolivia Bolivie	7	3	7	1	...	...	...	...	4	...
Bulgaria Bulgarie	31	28	27	27	30	26 [3]	28 [3]	24 [3]	17 [3]	15 [3]
Burkina Faso [3] Burkina Faso [3]	30	12	13	10	13	11	10	14	* 26	14
Burundi [3] Burundi [3]	3	3	3	3	3	2	3	3	3	3
Cameroon Cameroun	25 [3]	31 [3]	26 [3]	31	39 [3]	23 [3]	...	...	...	...
Central African Rep. [3] Rép. centrafricaine [3]	1	1	...	...	0	...	...	...	...	...
Chad [3] Tchad [3]	...	...	1	1	4	3	4	4	4	5
Chile Chili	49	49	45	52	59	65	69	69	73	73
China Chine	1 542	1 752	2 000	2 271	2 310	2 514	2 576	2 581	2 367	1 006
Colombia Colombie	204	...	145 [3]	141 [3]	137 [3]	165 [3]	148 [3]	179 [3]	...	...
Congo [3] Congo [3]	5	6	2	2	2	1	1	3	5	3
Côte d'Ivoire Côte d'Ivoire	55	65	...	...	...	...	...	...	...	...
Croatia Croatie	..	..	..	..	..	..	..	..	67	52
Cuba [6] Cuba [6]	96	97	87	95	77	67	67	...	...	...

52
Soap, washing powders and detergents
Production: thousand metric tons [cont.]
Savons, poudres pour lessives et détersifs
Production : milliers de tonnes métriques [suite]

Country or area Pays ou zone	1983	1984	1985	1986	1987	1988	1989	1990	1991	1992
Cyprus Chypre	8	9	9	9	* 9	* 9	11	* 13	* 15	...
former Czechoslovakia† anc. Tchécoslovaquie†	110	112	118	122	128	131	134	138	..	
Czech Republic République tchèque	..	..	..	..		..	..	..	82	90
Denmark[7] Danemark[7]	201	212	216	209[1]	...	...	...	...	...	...
Dominica[3] Dominique[3]	7	4	6	6	7	...	...	...	...	...
Dominican Republic Rép. dominicaine	38	39	39	47	32[3]	...	...	...	...	...
Ecuador Equateur	65	...	...	...	47	47	...	95	...	57
Egypt Egypte	487	462	485	347	379	339	399	417	370	358
Estonia[1] Estonie[1]	..	..	..	..	..	..	..	..	28	4
Ethiopia[3][8] Ethiopie[3][8]	15	14	13	15	21	16	12	9[9]	...	...
Fiji[3] Fidji[3]	7	7	6	7	7	8	6	7	7	...
Finland Finlande	72	# 65	66	82	82	90	98	93	81	71
France France	753	752	760	818	784	749	808	811	781	822
Gabon Gabon	...	57	138	137	129	130	132	136	153	170
Germany †[3] Allemagne†[3]	..	..	..	..	..	..	..	..	122	113
F. R. Germany[11] R. f. Allemagne[11]	1 748[10]	1 724[10]	1 733[10]	106[3]	106[3]	101[3]	113[3]	126[3]	..	..
former German D. R. anc. R. d. allemande	461	486	498	499	518	476	527	291	..	..
Ghana[3] Ghana[3]	3	6	12	19	31	36	...	...	...	...
Greece Grèce	129	125	137	139	150	156	163	159	150	...
Guyana Guyana	1	1	0	0	1	1	...	...	...	...
Haiti Haïti	16	14[3]	13[3]	13[3]	...	...	...	...	...	
Hong Kong[1] Hong-kong[1]	19	29	26	43	36	35	62	84	56	...
Hungary Hongrie	119	116	114	125	127	126	130	144	103	92
India[12] Inde[12]	520	941	# 1 545	1 652	1 753	1 639	1 740	1 794	1 805	...
Indonesia Indonésie	136	131	193	260	248	250	...	502	455	...

52
Soap, washing powders and detergents
Production: thousand metric tons [*cont.*]
Savons, poudres pour lessives et détersifs
Production : milliers de tonnes métriques [*suite*]

Country or area Pays ou zone	1983	1984	1985	1986	1987	1988	1989	1990	1991	1992
Iran, Islamic Rep. of[13] Iran, Rép. islamique d'[13]	247	254	262	155	169[1]	149[1]	169[1]	212[1]	...	...
Israel[3] Israël[3]	7	6	6	6	6	5	4	4	...	...
Jamaica Jamaïque	13	13	12	13	12	13	11	11	9	...
Japan[14] Japon[14]	1 022	1 067	1 078	1 101	1 157	1 174	1 245	1 311	1 370	1 394
Jordan Jordanie	19	38	23	56	53	18[3]	26[3]	...	...	...
Kazakhstan[3] Kazakhstan[3]	..	..	..	..	..	..	..	..	38	24
Kenya Kenya	25[3]	22[3]	23[3]	55	55	71	79	74	...	...
Korea, Republic of Corée, République de	253	263	# 359	401	427	473	473	523	502	492
Kuwait[1] Koweït[1]	2	2	1	1	2	2	3	...	...	1
Latvia[5] Lettonie[5]	..	..	..	..	..	..	..	..	18	8
Liberia[3] Libéria[3]	4	5	4	3	...	...	...	...	...	...
Madagascar[3] Madagascar[3]	11	13	12	13	15	13	15	15	16	14
Malaysia[15] Malaisie[15]	35	34	42	48	51	63	67	80	88	102
Mexico Mexique	877	869	827	901	999	955	1 099	1 166	1 231	1 257
Mongolia[3] Mongolie[3]	5	5	4	4	4	3	3	...	...	...
Morocco[3] Maroc[3]	31	28	20	...	...	...	...	...	...	...
Mozambique Mozambique	20	13	9	9	14	16	20	9	10	...
Myanmar[3 16] Myanmar[3 16]	38	28	50	38	18	18	12	15	23	18
Nepal[3 17] Népal[3 17]	6	8	9	11	12	17	...	...	...	...
Netherlands[1] Pays-Bas[1]	262	266	274	...	...	...	...	...	...	...
New Zealand[18] Nouvelle-Zélande[18]	32	33	32	36	...	...	...	...	...	...
Nicaragua[3] Nicaragua[3]	32	40	...	...	...	...	...	...	...	...
Nigeria Nigéria	250	143	158	78	214	...	...	...	...	...
Norway Norvège	68	57	51	8	62	56	55	7[3]	...	...
Pakistan[15] Pakistan[15]	40	39	49	48	57	59	103	59	75	86

52
Soap, washing powders and detergents
Production: thousand metric tons [cont.]
Savons, poudres pour lessives et détersifs
Production : milliers de tonnes métriques [suite]

Country or area Pays ou zone	1983	1984	1985	1986	1987	1988	1989	1990	1991	1992
Panama Panama	13	14	15	...	17	14	17	...	...	...
Paraguay [3] Paraguay [3]	11	12	12	12	12	10	8	8	7	6
Peru [3] Pérou [3]	60	53	49	85	91	88	74	80	...	...
Philippines Philippines	223	186	198	220	221	...	...	...	...	...
Poland [19] Pologne [19]	248	241	273	309	315	339	331	215	199	174
Portugal Portugal	190	190	173	173	180	191	209	213	181	190
Republic of Moldova République de Moldova	..	..	..	..	..	..	..	..	23	12
Romania [20] Roumanie [20]	59	53	50	54	51	46	46	49	37	20
Russian Federation Fédération de Russie	..	..	..	..	..	..	..	..	1 106	...
Rwanda [3] Rwanda [3]	8	12	16	12	* 6	* 6	...	...	...	...
Saint Lucia [3] Sainte-Lucie [3]	1	0	0	0	0	0	0	0	0	...
Sao Tome and Principe [3] Sao Tomé-et-Principe [3]	0	0	0	0	1	...	...	...	...	...
Senegal [3] Sénégal [3]	38	35	30	...	33	...	...	...	...	...
Sierra Leone [3] Sierra Leone [3]	3	4	1	4	80	167	...	...	...	...
Singapore Singapour	27	25	24	24	26	28	28	...	...	...
Slovakia Slovaquie	..	..	..	..	..	..	..	..	23	...
Slovenia Slovénie	..	..	..	..	..	..	..	..	64	47
South Africa [21] Afrique du Sud [21]	185	173	187	192	198	207	199	216	231	233
Spain [21] Espagne [21]	1 102	1 145	1 137	1 078	1 170	1 262	1 282	1 376	1 424	...
Sri Lanka [3 16] Sri Lanka [3 16]	4	3	5	4	4	4	...	...	...	...
Sweden Suède	185	174	178	180	186	184 [1]	185 [1]	158 [1]	127 [1]	...
Switzerland [3 7 22] Suisse [3 7 22]	144	145	145	149	159	161	163	168	165	156
Syrian Arab Republic Rép. arabe syrienne	78	60	71	71	66	34	31	40	38	44
Thailand Thaïlande	87 [1]	93 [1]	109 [1]	107 [1]	111 [1]	178	197	190	196	...
Trinidad and Tobago [3] Trinité-et-Tobago [3]	2	2	3	3	4	4	3	2	2	2

52
Soap, washing powders and detergents
Production: thousand metric tons [cont.]

Savons, poudres pour lessives et détersifs
Production : milliers de tonnes métriques [suite]

Country or area Pays ou zone	1983	1984	1985	1986	1987	1988	1989	1990	1991	1992
Turkey Turquie	208	241	230	269	282	297	449	442	407	406
former USSR[5] ancienne URSS[5]	2 753	2 632	2 615	2 659	2 728	2 866	3 128	3 209	..	..
United Kingdom[23] Royaume-Uni[23]	1 212[24]	1 030[1]	1 128[1]	1 192[1]	1 166[1]	1 284[1]	1 415[1]	...	...	...
United Rep.Tanzania[3] Rép. Unie de Tanzanie[3]	...	...	14	15	19	21	20	23	24	20
Viet Nam[3] Viet Nam[3]	27	38	51	72	47	52	40	55	68	72
former Yemen Arab Rep.[3] anc. Yémen rép. arabe[3]	4	6	8	19	...	...	...	...	...	...
Yugoslavia, SFR† Yougoslavie, Rfs†	310	286	288	312	340	369	412	...	..	..
Zambia Zambie	10	10	10	11	...	...	...	...	...	...

Source:
Industrial statistics database of the Statistical Division
of the United Nations Secretariat.

† For detailed descriptions of data pertaining to
former Czechoslovakia, Germany, SFR Yugoslavia and former
USSR, see Annex I - Country or area nomenclature, regional
and other groupings.

1 Washing powders and detergents only.
2 Excluding liquid toilet soap.
3 Soap only.
4 Twelve months ending 30 June of year stated.

5 Soap data are in terms of 40 per cent fat content.

6 Excluding washing powder.
7 Sales.
8 Twelve months ending 7 July of the year stated.

9 Beginning 1990, excluding Eritrea.
10 Excluding surface-active organic agents.
11 Beginning 1986, washing powders and detergents data are
confidential.
12 Production by large and medium scale establishments only.
13 Soap production by establishments employing 10 or more
persons; washing powders/detergents prod. by estab.
employing 50 or more.
14 Surface-active agents only.
15 Toilet soap only.
16 Government production only.
17 Twelve months beginning 16 July of year stated.

18 Data refer to soap flakes and powder only.

19 Excluding detergents.
20 On the basis of 100 per cent active substances.
21 Synthetic detergents in powder form only.
22 Including other cleaning products.
23 Estimates of manufacturers' sales of finished detergents for
washing purposes, in terms of actual weight sold.

24 Including shipments of soap.

Source:
Base de données pour les statistiques industrielles de la
Division de statistique du Secrétariat de l'ONU.

† Pour les descriptions en détails des données
relatives à l'ancienne Tchécoslovaquie, l'Allemagne, la Rfs
Yougoslavie et l'ancienne URSS, voir l'Annexe I -
Nomenclature des pays ou zones, groupements régionaux et
autres groupements.

1 Poudres pour lessives et détersifs seulement.
2 Non compris le savon liquide de toilette.
3 Savons seulement.
4 Période de douze mois finissant le 30 juin de l'année
indiquée.
5 Les données des savons sont sur la base de 40 p. 100 de
matières grasses.
6 Non compris poudres pour lessives.
7 Ventes.
8 Période de douze mois finissant le 7 juillet de l'année
indiquée.
9 A partir de 1990, non compris Erythrée.
10 Non compris les produits organiques tensio-actifs.
11 A partir de 1986, les données des poudres pour lessives et
détersifs sont confidentielles.
12 Production des grandes et moyennes entreprises seulement.
13 Savons prod. des établissements occupant 10 personnes ou
plus; poudres pour lessives et détersifs prod. des étab.
occu.50 ou plus.
14 Produits tensio-actifs seulement.
15 Savons de toilette seulement.
16 Production de l'Etat seulement.
17 Période de douze mois commençant le 16 juillet de l'année
indiquée.
18 Données se rapportant aux savons en paillettes et en poudre
seulement.
19 Non compris les détersifs.
20 Sur la base de 100 p. 100 de substances actives.
21 Détersifs synthétiques en poudre seulement.
22 Y compris les autres produits de nettoyage.
23 Estimations des ventes de détersifs pour lessives par les
fabricants, sur la base du poids de produits effectivement
ecoulé.
24 Y compris les expéditions des savons.

53
Pig iron and crude steel
Fonte et acier brut
Production: thousand metric tons
Production : milliers de tonnes métriques

Country or area Pays ou zone	1983	1984	1985	1986	1987	1988	1989	1990	1991	1992
A. Pig-iron • Fonte										
Total	455 317	487 590	497 359	494 068	506 263	541 227	554 627	553 056	495 158	534 844
Algeria Algérie	1 126	1 191	1 477	1 262	1 493	1 515	1 315	1 054	893	944
Argentina Argentine	917	844	1 310	1 623	1 752	1 596	2 100	1 908	...	...
Australia[1] Australie[1]	4 990	5 258	5 341	5 925	5 783	5 455	5 875	6 188	5 600	6 394
Austria Autriche	3 319	3 745	3 704	3 349	3 451	3 665	3 823	3 452	3 439	...
Belgium Belgique	8 033	8 968	8 720	8 048	8 242	9 147	8 863	9 416	9 353	8 524
Brazil Brésil	12 945	17 230	18 961	20 163	20 944	23 454	24 363	21 141	22 695	23 152
Bulgaria Bulgarie	1 626	1 583	1 712	1 605	1 657	1 442	1 487	1 143	961	849
Canada Canada	8 567	9 643	9 660	9 246	9 720	9 490	10 200	7 344	8 828	...
Chile Chili	540	594	580	591	617	778	679	722	700	...
China Chine	37 380	40 010	43 840	50 638	55 032	57 040	58 200	62 380	67 000	75 890
Colombia Colombie	241	252	234	317	326	309	297	347	300	...
Croatia Croatie	..	..	..	..	..	..	..	..	69	...
former Czechoslovakia† anc. Tchécoslovaquie†	9 467	9 562	9 562	9 573	9 788	9 706	9 911	9 667	..	..
Czech Republic République tchèque	..	..	..	..	..	..	..	..	5 090	4 889
Egypt Egypte	196	159	87	121	147	132	112	108	113	60
Finland Finlande	1 898	2 034	# 1 890	1 871	2 063	2 173	2 284	2 283	2 332	2 452
France France	13 500	14 710	15 072	13 776	13 157	14 463	14 724	14 100	13 416	12 264
Germany † • Allemagne† F. R. Germany R. f. Allemagne	26 352	29 737	31 143	28 593	28 116	31 890	32 112	29 585	..	..
former German D. R. anc. R. d. allemande	2 196	2 343	2 565	2 726	2 743	2 774	2 722	2 128	..	..
Greece[2] Grèce[2]	138	138	140	160	160	160	160	160	160	...
Hungary Hongrie	2 047	2 096	2 095	2 054	2 108	2 093	1 954	1 697	1 314	1 179
India Inde	9 087	9 382	9 701	10 460	10 808	11 602	11 930	* 12 000	...	...

53
Pig iron and crude steel
Production: thousand metric tons [cont.]
Fonte et acier brut
Production : milliers de tonnes métriques [suite]

Country or area Pays ou zone	1983	1984	1985	1986	1987	1988	1989	1990	1991	1992
Iran, Islamic Rep. of Iran, Rép. islamique d'	838	1 208	1 122	850	976	1 012	* 1 000	* 1 000	...	...
Italy Italie	10 313	11 631	12 062	11 916	11 334	11 348	11 762	11 852	10 561	10 432
Japan Japon	72 937	80 403	80 569	74 651	73 418	79 295	80 196	80 229	79 985	80 229
Kazakhstan Kazakhstan	..	..	..	..	..	..		..	4 953	4 666
Korea, Dem. P. R.[2] Corée, R. p. dém. de[2]	* 5 500	* 5 700	* 5 800	* 5 800	* 5 800	* 6 500	* 6 500	* 6 500	6 500	...
Korea, Republic of Corée, République de	8 024[2]	8 763[2]	8 822	9 004	10 858	12 567	14 937	15 477	18 883	19 581
Luxembourg Luxembourg	2 316	2 768	2 754	2 650	2 305	2 520	2 684	2 645	2 463	2 255
Mexico Mexique	4 956	5 339	5 089	5 048	# 1 412	* 1 489	* 2 076	* 2 378	2 313	2 220
Morocco[2] Maroc[2]	15	15	15	15	15	15	15	15	15	...
Netherlands Pays-Bas	3 747	4 926	4 819	4 628	4 575	4 994	5 163	4 960	4 697	...
Norway Norvège	565	572	610	564	364	367	240	...	...	
Peru[2] Pérou[2]	113	66	163	216	185	202	227	225	230	...
Poland Pologne	9 286	9 533	9 335	10 096	10 024	9 837	9 075	8 352	6 297	6 315
Portugal Portugal	365	382	424	429	431	445	377	...	...	...
Romania Roumanie	8 190	9 557	9 212	9 330	8 673	8 941	9 052	6 355	4 525	3 110
Russian Federation Fédération de Russie	..	..	..	..	..	..	..	..	48 628	..
Slovakia Slovaquie	..	..	..	..	..	..	..	..	3 163	
South Africa Afrique du Sud	5 213	5 455	7 179	7 406	7 398	6 171[2]	6 543[2]	6 257[2]	6 968[2]	...
Spain Espagne	5 426	5 319	5 455	4 862	4 854	4 650	5 479	5 748	5 600	...
Sweden Suède	2 009	2 203	2 415	2 428	2 365	2 527	2 648	2 696	2 851	...
Switzerland Suisse	10[3]	54[3]	66[3]	147	140	134	141	129	105	102
Tunisia Tunisie	147	147	141	149	168	133	156	* 150	* 150	...
Turkey Turquie	2 953	3 015	3 371	3 926	4 438	12 894	18 868	32 829	32 865	58 978
Ukraine Ukraine	..	..	..	..	..	..	..	..	36 435	...
former USSR† ancienne URSS†	110 453	110 893	109 977	113 840	113 877	114 558	113 928	110 166	..	..

53
Pig iron and crude steel
Production: thousand metric tons [cont.]
Fonte et acier brut
Production : milliers de tonnes métriques [suite]

Country or area Pays ou zone	1983	1984	1985	1986	1987	1988	1989	1990	1991	1992
United Kingdom Royaume-Uni	9 548	9 420	10 167	9 632	11 916	12 943	12 551	12 320	12 100	...
United States Etats-Unis	44 186	47 090	45 763	40 176	43 851	50 571	50 687	49 668	44 123	47 377
Venezuela Venezuela	169	326	441	493	473	503	455	314	...	...
Yugoslavia, SFR† Yougoslavie, Rfs†	2 845	2 855	3 120	3 063	2 867	2 916	2 899	2 313	..	..
Zimbabwe Zimbabwe	584	400	674	644	575	600	520	521	* 525	...

B. Crude steel · Acier brut

Country or area Pays ou zone	1983	1984	1985	1986	1987	1988	1989	1990	1991	1992
Total	**675 507**	**721 122**	**730 056**	**724 282**	**744 577**	**785 630**	**787 367**	**772 941**	**715 189**	**727 872**
Albania Albanie	...	...	233	209	137	210	205	143	28	...
Algeria Algérie	893	1 130	1 215	1 123	1 380	* 1 303	* 945	* 769	797	760
Angola[4] Angola[4]	* 10	* 10	* 10	* 10	* 10	* 10	* 10	* 10	10	* 10
Argentina Argentine	2 828	2 508	2 775	3 116	3 463	3 527	* 3 874	* 3 624	2 972	...
Australia[1] Australie[1]	5 392	4 639	6 301	6 826	6 188	6 093	6 651	7 576	7 141	7 690
Austria Autriche	4 410	4 870	4 661	4 292	4 301[3]	4 376[3]	4 901[3]	4 395[3]	4 082	3 846
Azerbaijan Azerbaïdjan	..	..	..	..	..	..	..	..	1 127	809
Bangladesh[1] Bangladesh[1]	47	73	101	96	82	70	86	90	...	...
Belarus Bélarus	..	..	..	..	..	..	..	..	1 123	...
Belgium Belgique	10 266	11 413	10 781	9 803	9 844	11 306	11 053	11 546	11 419	...
Brazil Brésil	14 733	18 476	20 563	21 343	22 228	24 747	25 084	20 631	22 668	23 987
Bulgaria Bulgarie	2 831	2 878	2 944	2 965	3 045	2 875	2 899	2 184	1 615	1 551
Canada Canada	12 832	14 699	13 459	14 081	14 737	14 778	* 15 005	* 12 281	12 987	12 702
Chile[5] Chili[5]	618	692	689	706	726	899	800	772	805	...
China Chine	41 936	45 618	49 454	54 795	58 928	62 147	64 209	68 858	73 881	84 252
Colombia Colombie	482	507	525	632	691	777	711	733	700	...
Croatia Croatie	..	..	..	..	..	..	..	..	214	102
Cuba Cuba	352	325	401	412	402	321	314	270	270	...

53
Pig iron and crude steel
Production: thousand metric tons [cont.]
Fonte et acier brut
Production : milliers de tonnes métriques [suite]

Country or area Pays ou zone	1983	1984	1985	1986	1987	1988	1989	1990	1991	1992
former Czechoslovakia† anc. Tchécoslovaquie†	15 024	14 831	15 036	15 112	15 416	15 379	15 465	14 775	..	..
Czech Republic République tchèque	..	..	..	..	..	..	..	7 972	7 334
Denmark [3] Danemark [3]	492	548	528	632	606	650	624	610	633	...
Ecuador Equateur	22	17	18	17	25	24	23	20	20	...
Egypt [4] Egypte [4]	199	1 153	954	1 003	1 436	* 2 028	* 2 117	* 2 328	2 541	...
Finland Finlande	2 416	2 632	2 518	2 586	2 669	2 798	2 921	2 860	2 890	3 077
France France	17 582	19 000	18 808	17 857	17 689	19 108	19 535	19 304	18 708	18 190
Germany † Allemagne†	..	..	..	..	..	..	..	..	41 599	39 627
F. R. Germany R. f. Allemagne	35 728	39 389	40 488	37 134	36 248	41 023	41 078	38 433	..	..
former German D. R. anc. R. d. allemande	7 219	7 573	7 853	7 967	8 243	8 131	7 829	5 339	..	..
Greece Grèce	869	895	985	1 009	908	959	960	999	980	...
Hungary Hongrie	3 523	3 643	3 545	3 601	3 495	3 480	3 263	2 924	2 043	146 [6]
India Inde	10 139	10 348	11 187	11 427	12 262	13 019	15 522	16 479	17 577	18 723
Indonesia Indonésie	800	1 000	1 200	1 500	* 1 453	* 2 050	* 2 000	* 2 100	...	...
Iran, Islamic Rep. of Iran, Rép. islamique d'	703	784	764	741	783	883	* 1 000	* 1 000	...	...
Iraq Iraq	...	...	...	...	...	...	...	...	...	1 940
Ireland Irlande	141	166	203	208	220	200	324	325	293	...
Israel [4] Israël [4]	150	92	100	110	116	* 120	* 118	* 144	...	...
Italy Italie	21 810	24 110	23 898	22 882	22 858	23 943	25 507	25 647	25 270	24 924
Japan Japon	97 179	105 586	105 278	98 275	98 513	105 680	107 907	110 339	109 648	98 132
Kazakhstan Kazakhstan	..	..	..	..	..	..	..	..	6 377	6 063
Korea, Dem. P. R. * [4] Corée, R. p. dém. de * [4]	6 100	6 500	6 500	6 500	6 700	6 800	7 300	8 000	8 000	...
Korea, Republic of Corée, République de	16 356	17 833	18 650	18 793	19 481	21 318	23 705	24 459	27 864	29 557
Latvia Lettonie	..	..	..	..	..	..	..	..	374	246
Luxembourg Luxembourg	3 294	3 987	3 945	3 705	3 302	3 661	3 721	3 560	3 379	3 068

53
Pig iron and crude steel
Production: thousand metric tons [cont.]
Fonte et acier brut
Production : milliers de tonnes métriques [suite]

Country or area Pays ou zone	1983	1984	1985	1986	1987	1988	1989	1990	1991	1992
Mexico Mexique	6 747	7 293	7 174	6 960	7 211	7 312	7 329	8 221	7 462	7 848
Morocco Maroc	* 6	* 6	* 6	* 6	* 6	* 7	* 7	* 7	7	...
Netherlands Pays-Bas	4 477	5 739	5 517	5 283	5 083	5 518	5 681	5 412	5 171	5 439
New Zealand Nouvelle-Zélande	233	274	228	291	* 409[4]	* 460[4]	* 608[4]	* 765[4]	700[4]	...
Nigeria Nigéria	182	187	254	200	200	192	213	220	200	...
Norway Norvège	903	927	579	837	853	903	677	368	438	...
Peru Pérou	299	337	414	486	503	496	364	284	...	...
Philippines Philippines	200	250	260	260	* 250	300	* 300	* 300	...	...
Poland Pologne	16 236	15 180	15 361	16 283	16 263	* 15 943	12 466	11 501	9 311	8 448
Portugal Portugal	674	690	665	717	1 099	1 239	1 061	* 300[6]	...	...
Republic of Moldova République de Moldova	..	..	..	..	..	..	..	..	623	653
Romania Roumanie	13 302	15 229	14 587	15 026	14 631	15 057	15 165	10 624	7 509	5 614
Russian Federation Fédération de Russie	..	..	..	..	..	..	..	..	77 100	67 028
Saudi Arabia[4] Arabie saoudite[4]	* 400	* 842	* 1 106	* 1 100	1 365	1 614	* 1 810	1 833	...	...
Slovakia Slovaquie	..	..	..	..	..	..	..	..	4 107	4 498
Slovenia Slovénie	..	..	..	..	..	..	..	..	289	401
South Africa Afrique du Sud	7 190	7 827	8 582	8 127	9 123	8 837	9 337	8 691	9 358	...
Spain Espagne	13 262	13 379	14 679	* 12 137	* 11 629	* 11 679	* 12 564	* 12 818	12 700	...
Sweden Suède	4 204	4 674	4 851	4 719	4 683	4 779	4 692	4 435	...	...
Switzerland[3] Suisse[3]	835	978	987	1 075	866	989	1 064	1 105	955	...
Thailand[4] Thaïlande[4]	340	381	447	463	534	552	689	685	* 711	...
Tunisia Tunisie	163	166	160	181	188	159	192	200	200	...
Turkey Turquie	3 599[6]	4 113[6]	4 737[6]	5 989[6]	10 188	11 481	10 809	12 225	12 323	13 241
Uganda Ouganda	2	6	8	8	0	0	0	0	...	...
Ukraine Ukraine	..	..	..	..	..	..	..	..	46 743	...

53

Pig iron and crude steel
Production: thousand metric tons [*cont.*]
Fonte et acier brut
Production : milliers de tonnes métriques [*suite*]

Country or area Pays ou zone	1983	1984	1985	1986	1987	1988	1989	1990	1991	1992
former USSR† ancienne URSS†	161 372	163 097	163 808	170 098	171 417	172 217	168 421	162 326	..	..
United Kingdom Royaume-Uni	14 987	15 122	15 722	14 725	17 414	18 950	18 740	17 841	16 511	...
United States[7] Etats-Unis[7]	76 762	83 940	80 067	74 032	80 876	91 765	* 88 352	89 726	79 738	84 322
Uruguay Uruguay	46	41	39	31	30	29	37	34	...	...
Venezuela Venezuela	2 320	2 770	3 060	3 402	3 721	3 650	3 404	3 140	2 933	2 446
Viet Nam Viet Nam	...	...	...	64	70	74	85	101	149	196
Yugoslavia, SFR† Yougoslavie, Rfs†	4 140	4 236	4 476	4 524	4 368	4 488	4 542	3 608	..	..
Zimbabwe[4] Zimbabwe[4]	672	391	* 463	* 490	515	500	650	580	* 580	...

Source:
Industrial statistics database of the Statistical Division
of the United Nations Secretariat.

† For detailed descriptions of data pertaining to
former Czechoslovakia, Germany, SFR Yugoslavia and former
USSR, see Annex I - Country or area nomenclature, regional
and other groupings.

1 Twelve months ending 30 June of year stated.

2 Source: US Bureau of Mines, (Washington, DC) , including
crude and washed kaolin.

3 Source: Annual Bulletin of Steel Statistics for Europe,
United Nations Economic Commission of Europe (Geneva).

4 Source: US Bureau of Mines, (Washington, DC).
5 Source: "Instituto Latino Americano del Fierro y el Acero",
(Santiago).
6 Crude steel for casting only.
7 Excluding steel for castings made in foundries operated by
companies not producing ingots.

Source:
Base de données pour les statistiques industrielles de la
Division de statistique du Secrétariat de l'ONU.

† Pour les descriptions en détails des données
relatives à l'ancienne Tchécoslovaquie, l'Allemagne, la Rfs
Yougoslavie et l'ancienne URSS, voir l'Annexe I -
Nomenclature des pays ou zones, groupements régionaux et
autres groupements.

1 Période de douze mois finissant le 30 juin de l'année
indiquée.
2 Source: "US Bureau of Mines," (Wahington, DC), y compris le
kaolin brut et lavé.
3 Source: Bulletin annuel de statistiques de l'acier pour
l'Europe, Commission économique des Nations Unies pour
l'Europe (Genève).
4 Source: "US Bureau of Mines," (Washington, DC).
5 Source: "Instituto Latino Americano del Fierro y el Acero",
(Santiago).
6 L'acier brut pour moulages seulement.
7 Non compris l'acier pour les moulages fabriqués dans des
fonderies exploitées par des entreprises ne produisant pas
de lingots.

54
Aluminium
Aluminium

Production: thousand metric tons
Production : milliers de tonnes métriques

Country or area Pays ou zone	1983	1984	1985	1986	1987	1988	1989	1990	1991	1992
Total	18 006.2	19 931.9	19 665.5	19 709.8	21 300.6	22 504.9	23 171.7	21 846.1	20 484.3	21 294.9
Argentina[1] Argentine[1]	136.4	137.8	139.9	150.6	155.1	157.4	164.2[2]	163.0	166.2	155.6[2]
Australia[3] Australie[3]	441.6	658.9	867.3	* 925.0	* 960.0	* 1 120.5	* 1 289.2	* 1 268.0	1 264.6	...
Primary[3] 1re fusion[3]	403.9	617.9	822.3	870.0	921.0	1 074.0	1 240.8	1 235.1	* 1 235.0	...
Secondary[3] 2ème fusion[3]	37.7	41.0	45.0	55.0	39.0[4]	46.5[4]	48.4[4]	32.9	29.6	29.6[4]
Austria Autriche	482.6	380.7	285.0	283.7	322.3	391.1	334.1	246.3	205.9	...
Primary 1re fusion	94.2	151.9	152.7	148.8	156.1	165.9	169.0	159.1	89.0	...
Secondary[5] 2ème fusion[5]	388.4	228.8	132.4	134.9	166.1	225.2	165.2	87.2	* 116.9	...
Azerbaijan Azerbaïdjan	..	..	..	..	..	..	..	..	31.7	21.3
Primary 1re fusion	..	..	..	..	..	..	..	..	25.9	19.6
Secondary 2ème fusion	..	..	..	..	..	..	..	..	5.8	1.7
Bahrain[1 2] Bahreïn[1 2]	171.7	177.3	174.8	178.2	180.3	182.8	186.9	212.5	213.7	292.5
Belgium[2 6] Belgique[2 6]	1.0	2.0	2.0	2.0	3.2	3.0	3.0	3.0	3.0	3.0
Brazil Brésil	443.8	503.9	594.2	805.3	893.8[4]	938.4[4]	952.8[4]	982.7[4]	1 206.0[4]	1 260.4[4]
Primary 1re fusion	400.7	455.0	549.4	757.4	843.5[4]	873.5[4]	887.9[4]	930.6[4]	1 139.0[4]	1 193.3[4]
Secondary 2ème fusion	43.0	48.9	44.8	48.0	50.3[4]	64.9[4]	64.9[4]	52.1[4]	66.4[4]	67.1[4]
Cameroon[1] Cameroun[1]	53.6	125.4	81.8	51.0	79.0	80.0[2]	87.3[2]	87.5[2]	85.6[2]	82.5[2]
Canada *[7] Canada *[7]	1 154.2	1 286.0	1 347.3	1 420.2	1 582.2	1 647.5	1 631.8	1 635.1	1 889.3	2 017.7
Primary[7] 1re fusion[7]	1 091.2	1 222.0	1 282.3	1 355.2	1 540.4	1 534.5	1 554.8	1 567.4	1 821.6	1 950.0
Secondary[7] 2ème fusion[7]	63.0	64.0	65.0	65.0	41.7	113.0	77.0	67.7	67.7	67.7
China Chine	* 425.0[1 7]	* 400.0[1 7]	* 410.0[1 7]	* 410.0[1 7]	* 615.0[1 7]	718.4	758.4	854.3	900.0	1 096.4
Croatia[1] Croatie[1]	..	..	..	..	..	..	..	..	54.5	29.0
former Czechoslovakia† anc. Tchécoslovaquie†	66.3	66.0	66.0	66.1	66.4	67.4	69.3	69.8	..	..
Primary 1re fusion	36.2	31.6	31.7	33.1	32.4	31.4	32.6	30.1	..	..
Secondary 2ème fusion	* 30.1	* 34.4	* 34.3	* 33.0	* 34.0	* 36.0	* 36.7	* 39.7	..	..
Denmark[4] Danemark[4]	12.6	14.2	16.4	16.4	16.4	16.4	16.4	10.6	12.0	14.1
Egypt[1 8] Egypte[1 8]	140.2	128.4	135.5	114.0	148.7	142.6	146.2	141.1	141.0	139.4
Finland Finlande	14.9	# 1.6[6]	1.0[6]	0.2[6]	0.2[6]	* 2.0[6]	* 4.1[6]	* 4.9[6]	* 4.2[6]	* 4.5[6]

54
Aluminium
Production: thousand metric tons [*cont.*]
Aluminium
Production : milliers de tonnes métriques [*suite*]

Country or area Pays ou zone	1983	1984	1985	1986	1987	1988	1989	1990	1991	1992
France[9] **France**[9]	**522.9**	**504.0**	**457.3**	**495.2**	**568.2**	**541.0**	**554.8**	**533.4**	**471.8**	**636.7**
Primary 1re fusion	360.8	341.0	293.1	321.8	381.8	328.0	329.3	325.2	254.6	414.3
Secondary[9] 2ème fusion[9]	162.1	163.0	164.3	173.4	186.3	213.0	225.5	208.3	217.2	222.4
Germany † **Allemagne†**	..	..	..	..	..	..	..	..	**739.9**	**654.4**
Primary 1re fusion	..	..	..	..	..	..	..	..	690.3	602.8
Secondary 2ème fusion	..	..	..	..	..	..	..	..	49.5	51.9
F. R. Germany **R. f. Allemagne**	**781.6**	**815.1**	**789.7**	**805.3**	**773.7**	**787.0**	**784.7**	**759.6**	..	..
Primary 1re fusion	743.4	777.2	745.5	765.1	737.7	744.1	742.2	720.3[2]	..	..
Secondary 2ème fusion	38.2	38.0	* 44.2	* 40.3	* 36.0	* 42.9	* 42.6	39.4	..	..
former German D. R. **anc. R. d. allemande**	**59.0**[2]	**125.9**[9]	**127.1**[9]	**125.0**[9]	**122.3**[9]	**115.8**[9]	**107.7**[9]	**82.8**[9]	..	..
Primary 1re fusion	59.0[2]	67.8[9]	65.7[9]	66.0[9]	67.9[9]	61.2[9]	53.9[9]	41.2[9]	..	..
Secondary[9] 2ème fusion[9]	...	58.1	61.4	59.0	54.5	54.6	53.8	41.6	..	..
Ghana[1] **Ghana**[1]	**42.5**	**0.0**	**48.6**	**124.6**	**150.0**	**161.0**	**169.0**	**174.0**	**175.0**	**179.9**[2]
Greece **Grèce**	**157.9**	**189.0**	**175.5**	**175.9**	**223.6**	**207.9**	**209.1**	**225.9**	**175.1**	...
Hungary **Hongrie**	**87.2**	**89.0**	**86.9**	**86.4**	**85.2**	**83.8**	**85.1**	**81.1**	**63.3**	**26.9**
Primary 1re fusion	74.0	74.2	73.9	73.9	73.5	74.7	75.2	75.2	63.3	26.9
Secondary * 2ème fusion *	13.1	14.8	13.1	12.6	11.7	9.1	9.9	5.9	0.0	0.0
Iceland[1] **Islande**[1]	**76.1**	**80.4**	**73.4**	**75.9**	**83.5**	**82.0**	**88.7**	**86.8**	**88.8**	**89.9**[2]
India[1] **Inde**[1]	**203.9**	**268.3**	**259.9**	**234.9**	**245.3**	**289.7**	**425.3**	**427.6**	**504.3**	**496.3**[2]
Indonesia[1] **Indonésie**[1]	**138.6**	**206.9**	**216.8**	**220.0**	**219.9**	**180.0**	**196.9**[2]	**192.1**	**173.0**	**188.8**[2]
Iran, Islamic Rep. of[1] **Iran, Rép. islamique d'**[1]	**39.2**	**42.0**	**43.0**	**37.4**	**33.6**	**38.2**	***45.0**	***59.4**	**67.0**	**79.3**[2]
Italy **Italie**	**473.7**	**513.2**	**506.1**	**543.6**	**567.6**	**604.1**	**609.5**	**581.4**	**561.0**	**513.8**
Primary 1re fusion	195.7	230.2	224.1	242.6	232.6	226.3	219.5	231.8	218.0	160.7
Secondary 2ème fusion	278.0	283.0	282.0	301.0	335.0	377.8	390.0	349.6	348.0	353.1
Japan[9] **Japon**[9]	**1 096.3**	**1 131.4**	**1 098.5**	**1 013.6**	**1 085.1**	**1 053.3**	**1 081.5**	**1 140.6**	**1 148.5**	**1 112.2**
Primary 1re fusion	258.6	291.1	231.3	148.3	52.8	49.0	50.7	50.5	52.1	38.5
Secondary[9] 2ème fusion[9]	837.7	840.3	867.2	865.3	1 032.3	1 004.3	1 030.8	1 090.1	1 096.4	1 073.7
Korea, Republic of[1] **Corée, République de**[1]	**12.6**	**18.3**	**17.1**	**18.3**	**16.8**	**16.1**	**15.7**	**13.3**	**13.6**	...

54
Aluminium
Production: thousand metric tons [*cont.*]
Aluminium
Production : milliers de tonnes métriques [*suite*]

Country or area Pays ou zone	1983	1984	1985	1986	1987	1988	1989	1990	1991	1992
Mexico **Mexique**	**54.9**	**63.2**	**65.1**	**62.6**	**88.1**	**79.9**	**80.8**	**113.6**	**106.6**	**101.0**
Primary 1re fusion	39.8	43.6	42.8	46.2	79.3	75.4	67.6	56.8	42.8	17.4
Secondary [7] 2ème fusion [7]	15.1	19.6	22.3	16.4	8.8	4.5	13.2	56.8	63.8	83.6
Netherlands **Pays-Bas**	**293.6**	**309.1**	**312.9**	**362.6**	**377.3**	**394.1**	**409.4**	**392.1**	**367.8**	**385.3**
Primary 1re fusion	235.4	249.2	250.6	265.8	275.9	278.2	279.2	257.9	253.6	235.1 [2]
Secondary 2ème fusion	58.2	59.9	62.3	96.8	101.4	115.9	130.2	134.2	114.3	* 150.2
New Zealand [1] **Nouvelle-Zélande** [1]	**218.6**	**242.9**	**240.8**	**236.3**	**252.0** [4]	**255.6** [4]	**257.5** [4]	**259.7** [4]	**258.5** [4]	**242.9** [4]
Norway **Norvège**	**717.6**	**770.7**	**748.7**	**733.0**	**805.0**	**833.8**	**874.5**	**874.1**	**873.5**	**...**
Primary 1re fusion	713.0	765.1	742.7	725.8	797.8	826.6	867.3	867.1	866.5 [2]	...
Secondary 2ème fusion	4.6	5.6	6.0	7.2	7.2	7.2	7.2	7.0	7.0	40.0 [2]
Poland [1] **Pologne** [1]	**44.4**	**46.0**	**47.0**	**47.5**	**47.5**	**47.7**	**47.8**	**46.0**	**45.8**	**43.6**
Portugal [6] **Portugal** [6]	**2.4**	**2.4**	**2.5**	**2.1**	**2.3**	**2.5**	**4.0**	**9.3**	**6.0**	**6.0**
Romania [9] [10] **Roumanie** [9] [10]	**244.0**	**264.0**	**265.0**	**269.0**	**275.2**	**279.0**	**280.2**	**178.3**	**167.5**	**119.5**
Primary [10] 1re fusion [10]	223.0	244.0	247.0	253.0	260.5	265.6	269.1	168.0	158.2	112.0
Secondary 2ème fusion	21.0	20.0	18.0	16.0	14.7	13.4	11.2	10.3	9.2	7.5
Slovakia **Slovaquie**	..	..	..	..	..	..	..	..	**66.3**	**...**
Primary 1re fusion	..	..	..	..	..	..	..	..	49.4	...
Secondary 2ème fusion	..	..	..	..	..	..	..	..	* 16.9	...
Slovenia [1] **Slovénie** [1]	..	..	..	..	..	..	..	..	**90.2**	**84.8**
South Africa [17] **Afrique du Sud** [17]	**161.3**	**167.4**	**165.0**	*** 169.6**	**170.6**	**170.4**	**165.9**	**159.5**	**169.4**	**172.8**
Spain [4] **Espagne** [4]	**394.9**	**421.4**	**412.6**	**402.9**	**411.0**	**378.9**	**430.0**	**442.0**	**451.2**	**455.5**
Primary [4] 1re fusion [4]	357.6	380.8	370.1	354.7	341.0	293.9	352.4	355.3	355.2	359.0
Secondary [4] 2ème fusion [4]	37.4	40.6	42.5	48.2	70.0	85.0	77.6	86.7	96.0	96.5
Suriname [1] **Suriname** [1]	**33.6**	**23.0**	**28.8**	**28.7**	**1.9**	**9.8**	**28.4**	**31.3**	**30.7**	**32.4**
Sweden **Suède**	**109.7**	**113.6**	**114.1**	**109.9** [4]	**111.5** [4]	**130.6** [4]	**130.0** [4]	**126.3** [4]	**115.4** [4]	**93.7** [4]
Primary 1re fusion	82.0	82.8	83.7	77.1 [4]	81.5 [4]	98.6 [4]	97.0 [4]	96.3 [4]	96.9 [4]	77.2 [4]
Secondary [2] 2ème fusion [2]	27.7	30.8	30.4	32.8 [4]	30.0 [4]	32.0 [4]	33.0 [4]	30.0 [4]	18.5 [4]	16.5 [4]
Switzerland [1] **Suisse** [1]	**76.0**	**79.2**	**72.7**	**80.3**	**73.2**	**71.8**	**71.3**	**71.6**	**65.9**	**52.1**

54
Aluminium
Production: thousand metric tons [cont.]
Aluminium
Production : milliers de tonnes métriques [suite]

Country or area Pays ou zone	1983	1984	1985	1986	1987	1988	1989	1990	1991	1992
Turkey[1] **Turquie**[1]	30.4	37.9	54.1	60.0	41.7	56.7	57.2	61.0	56.0	61.3
former USSR†* [2] **ancienne URSS†*** [2]	3 100.0	3 200.0	3 250.0	3 300.0	3 400.0	3 500.0	3 450.0	3 523.0	..	..
United Kingdom[10] **Royaume-Uni**[10]	380.8	431.8	403.0	382.3	411.1	405.9	366.9	362.8	* 349.8[4]	...
Primary[10] 1re fusion[10]	252.5	287.9	275.4	275.9	294.4	300.2	297.3	289.8	293.5[4]	...
Secondary 2ème fusion	128.3	143.9	127.6	116.4	116.7	105.8	69.5	73.0	56.3	...
United States[9] **Etats-Unis**[9]	5 001.6	5 859.0	5 262.0	4 810.0	5 329.0	6 066.0	6 084.0	6 441.0	6 407.0	6 799.0
Primary 1re fusion	3 353.2	4 099.0	3 500.0	3 037.0	3 343.0	3 944.0	4 030.0	4 048.0	4 121.0	4 042.0
Secondary[9] 2ème fusion[9]	1 648.4	1 760.0	1 762.0	1 773.0	1 986.0	2 122.0	2 054.0	2 393.0	2 286.0	2 757.0
Venezuela[1] **Venezuela**[1]	335.3	385.2	402.8	421.4	430.2	443.3	565.6	598.8	610.0	507.5
Yugoslavia, SFR† **Yougoslavie, Rfs†**	283.6	301.6	316.1	319.7	281.1	313.3	331.7	291.0	..	..
Primary 1re fusion	258.2	292.3	306.3	309.2	280.6	312.7	331.0	* 290.0	..	..
Secondary 2ème fusion	25.4	9.3	9.7	10.5	0.5	0.7	0.6	* 1.0	..	..

Source:
Industrial statistics database of the Statistical Division of the United Nations Secretariat.

† For detailed descriptions of data pertaining to former Czechoslovakia, Germany, SFR Yugoslavia and former USSR, see Annex I - Country or area nomenclature, regional and other groupings.

1 Primary metal production only.
2 Source: "Metallgesellschaft Aktiengesellschaft", (Frankfurt).
3 Twelve months ending 30 June of year stated.

4 Source: World Metal Statistics (London).
5 Secondary aluminium produced from old scrap only.

6 Secondary metal production only.
7 Source: US Bureau of Mines, (Washington, DC).
8 Including aluminium plates, shapes and bars.
9 Including alloys.
10 Including pure content of virgin alloys.

Source:
Base de données pour les statistiques industrielles de la Division de statistique du Secrétariat de l'ONU.

† Pour les descriptions en détails des données relatives à l'ancienne Tchécoslovaquie, l'Allemagne, la Rfs Yougoslavie et l'ancienne URSS, voir l'Annexe I - Nomenclature des pays ou zones, groupements régionaux et autres groupements.

1 Production du métal de première fusion seulement.
2 Source: "Metallgesellschaft Aktiengesellschaft", (Francfort).
3 Période de douze mois finissant le 30 juin de l'année indiquée.
4 Source: "World Metal Statistics," (Londres).
5 Aluminium de deuxième fusion obtenu à partir de vieux déchets seulement.
6 Production du métal de deuxième fusion seulement.
7 Source: "US Bureau of Mines," (Washington, DC).
8 Y compris les tôles, les profilés et les barres d'aluminium.
9 Y compris les alliages.
10 Y compris la teneur pure des alliages de première fusion.

55
Radio and television receivers
Radiodiffusion et télévision : postes récepteurs

Production: thousands
Production : milliers

Country or area Pays ou zone	Radio receivers Radiodiffusion : postes récepteurs					Television receivers Télévision : postes récepteurs				
	1980	1989	1990	1991	1992	1980	1989	1990	1991	1992
Total	**185 538**	**132 235**	**139 413**	**126 671**	**118 577**	**72 170**	**121 567**	**127 762**	**124 868**	**127 929**
Albania Albanie	...	30	...	...	...	21	23	18	5	...
Algeria Algérie	112	120	213	215	192	94	219	283	176	218
Argentina Argentine	...	...	...	...	...	454	674	687	698	...
Armenia Arménie	..	..	..	167	...	...	...	...	...	...
Australia[1] Australie[1]	199	...	...	...	...	332[2]	162[2]	158[2]	...	...
Azerbaijan Azerbaïdjan	...	...	..	7	0	...	..	..	...	5 913
Bangladesh Bangladesh	228	102	105	...	...	8	79	83	...	...
Belarus Bélarus	..	..	..	932	...	...	..	..	1 103	...
Belgium[3 4] Belgique[3 4]	2 246	859	986	556	416	746	979	1 084	886	620
Brazil Brésil	6 769	7 210	5 151	5 317	...	3 254	2 920	3 196	3 265	...
Bulgaria Bulgarie	51	52	43	16	5	91	185	219	108	64
Chad Tchad	...	9	13	7	7	...	...	...	...	...
Chile Chili	...	...	...	...	...	107	24	0	0	...
China Chine	30 038[5]	18 347[5]	21 030[5]	19 691[5]	16 489[5]	2 492	27 665	26 847	26 914	28 678
Colombia Colombie	3	...	...	...	...	103	130	135	...	...
Croatia Croatie	..	..	..	3	1	..	..	..	3	...
Cuba Cuba	200	173	...	...	...	40	71	...	...	...
former Czechoslovakia† anc. Tchécoslovaquie†	250[6]	173[6]	187[6]	..	..	389	524	504	..	..
Denmark Danemark	88	...	...	...	...	77	125[7]	122[7]	117[7]	...
Ecuador Equateur	...	...	...	...	...	36	...	...	...	11
Egypt Egypte	171	43	59	39	36	308	194	333	264	260
Finland Finlande	257	...	...	...	...	341	413	412	327	268
France France	2 141	2 039	2 059	1 865	1 679	1 928	2 447	2 838	2 549	2 799

55

Radio and television receivers
Production: thousands [*cont.*]

Radiodiffusion et télévision : postes récepteurs
Production : milliers [*suite*]

Country or area Pays ou zone	Radio receivers Radiodiffusion : postes récepteurs					Television receivers Télévision : postes récepteurs				
	1980	1989	1990	1991	1992	1980	1989	1990	1991	1992
Germany † • Allemagne† F. R. Germany R. f. Allemagne	3 707	4 975	5 955	..	..	4 425	3 236	3 595	..	..
former German D. R. anc. R. d. allemande	915	1 151	522	..	..	578	775	632	..	..
Greece Grèce	...	...	...	...	...	137	37	10	2	...
Hong Kong Hong-kong	67 478	...	8 182	6 145	...	108	683	1 177	749	...
Hungary Hongrie	271	145	83	14	0	417	502	492	308	274
India Inde	1 918[8]	672[8]	647[8]	470[8]	...	88	1 237	1 322	1 190	...
Indonesia Indonésie	1 530[9]	1 905[9]	4 436[9]	4 659[9]	...	607	525	700	725	...
Iran, Islamic Rep. of[10] Iran, Rép. islamique d'[10]	57[11 12]	217[11 12]	232[11 12]	...	...	246	407	598	...	...
Iraq Iraq	...	...	...	...	...	...	...	...	...	16
Italy Italie	...	...	...	...	...	1 984	2 391	2 325	2 434	2 151
Japan Japon	15 343	10 690	10 955	14 213	9 418	15 205	...	...	...	...
Kazakhstan Kazakhstan	...	..	..	86	97	...	...	...	...	...
Kenya Kenya	...	...	...	...	...	...	2 077	4 186	...	...
Korea, Republic of Corée, République de	3 972	1 112	1 468	836	619	6 819	15 469	16 201	16 129	16 311
Latvia Lettonie	...	..	..	1 230	630	...	...	...	...	...
Malaysia Malaisie	5 000[13]	28 450	37 019	31 920	31 360	157	2 375	3 238	4 838	5 553
Mexico Mexique	1 029	...	...	...	...	964	536	633	490	435
Mozambique Mozambique	164	97	12	19	...	...	...	...	...	...
Myanmar[14] Myanmar[14]	1	7	4	1	1	...	0	...	...	...
Pakistan Pakistan	...	...	...	...	...	74	211	200	182	145
Peru Pérou	...	...	...	...	...	75	55	...	...	...
Poland Pologne	2 695	2 523	1 433	589	334	900	772	748	438	652
Portugal Portugal	593	1 071	1 435	1 522	1 815	467	304	...	318	238
Republic of Moldova République de Moldova	..	..	..	5	3	..	..	..	173	176

55
Radio and television receivers
Production: thousands [cont.]
Radiodiffusion et télévision : postes récepteurs
Production : milliers [suite]

Country or area Pays ou zone	Radio receivers Radiodiffusion : postes récepteurs					Television receivers Télévision : postes récepteurs				
	1980	1989	1990	1991	1992	1980	1989	1990	1991	1992
Romania Roumanie	863	526	384	384	79	541	511	401	389	318
Russian Federation Fédération de Russie	..	..	..	5 562	...	..	..	..	4 439	...
Singapore Singapour	17 070[16]	...	...	...	...	1 889	3 040	...	...	...
Slovakia Slovaquie	...	..	..	65[6]	...	...	..	..	202	...
Slovenia Slovénie	...	...	...	...	...	...	..	..	101	55
South Africa Afrique du Sud	861	941	775	...	...	338	344	373	486	376
Spain Espagne	188	15	10	28	...	763	2 100	2 466	2 807	...
Sweden Suède	26	...	...	...	...	308	333	284	224	...
Syrian Arab Republic Rép. arabe syrienne	...	...	...	...	...	72	90	18	17	20
Thailand Thaïlande	1 113	933	1 105	1 062	...	248	1 476	2 351	2 426	...
Trinidad and Tobago Trinité-et-Tobago	8	6	2	3	3	13	2	7	13	16
Turkey Turquie	52[17]	653[17]	143[17]	74[17]	104[17]	327	999	1 994	2 567	2 320
Ukraine Ukraine	..	..	..	892	...	..	..	..	3 616	...
former USSR† ancienne URSS†	8 478	8 561	9 168	..	..	7 528	9 938	10 540	..	..
United Kingdom Royaume-Uni	439[18]	34[18]	...	...	...	2 364	...	...	...	...
United Rep.Tanzania Rép. Unie de Tanzanie	223	56	71	102	108	...	...	...	...	...
United States Etats-Unis	7 672[4]	3 408[4]	3 014[4]	3 504	4 006	10 320[4]	15 478[4]	13 982[4]	12 865	13 532
Yugoslavia, SFR† Yougoslavie, Rfs†	125	86	...	..	..	543	503	...	..	..

Source:
Industrial statistics database of the Statistical Division
of the United Nations Secretariat.

† For detailed descriptions of data pertaining to
former Czechoslovakia, Germany, SFR Yugoslavia and former
USSR, see Annex I - Country or area nomenclature, regional
and other groupings.

1 Twelve months ending 30 June of year stated.

2 Colour television receivers only.

3 Production by establishments employing 5 or more persons.

Source:
Base de données pour les statistiques industrielles de la
Division de statistique du Secrétariat de l'ONU.

† Pour les descriptions en détails des données
relatives à l'ancienne Tchécoslovaquie, l'Allemagne, la Rfs
Yougoslavie et l'ancienne URSS, voir l'Annexe I -
Nomenclature des pays ou zones, groupements régionaux et
autres groupements.

1 Période de douze mois finissant le 30 juin de l'année
indiquée.
2 Récepteurs de télévision en couleur seulement.
3 Production des établissements occupant 5 personnes ou plus.

55

**Radio and television receivers
Production: thousands [*cont.*]**

**Radiodiffusion et télévision : postes récepteurs
Production : milliers [*suite*]**

4 Shipments.
5 Portable battery sets only.
6 Including record players.
7 Sales.
8 Production by large and medium scale establishments only.
9 Including radio with tape recording unit.

10 Production by establishments employing 50 or more persons.
11 Including sound reproducers.
12 Including tape recorders.
13 Data refer to Peninsular Malaysia only.
14 Government production only.
15 Including radio with tape recording unit and clock.

16 Excluding radio-phonograph combinations.
17 Including car tape-players.

4 Expéditions.
5 Appareils portatifs à piles seulement.
6 Y compris les tourne-disques.
7 Ventes.
8 Production des grandes et moyennes entreprises seulement.
9 Y compris les récepteurs de radio avec appareil enregistreur à bande magnétique incorporés.

10 Production des établissements occupant 50 personnes ou plus.
11 Y compris les lecteurs de son.
12 Y compris les magnétophones.
13 Données se rapportant à la Malaisie péninsulaire seulement.
14 Production de l'Etat seulement.
15 Y compris les récepteurs de radio avec appareil enregistreur à bande magnétique et pendule incorporés.

16 Non compris les radios-phonos.
17 Y compris les lecteurs de cassettes pour automobiles.

56
Passenger cars
Voitures de tourisme
Production: thousands
Production : milliers

Country or area Pays ou zone	1983	1984	1985	1986	1987	1988	1989	1990	1991	1992
Total	29 965	30 404	31 987	32 430	32 820	33 848	35 122	35 316	33 762	34 246
Argentina[1] Argentine[1]	132	142	118	143	166	142	112	87	114	221
Australia[1,2] Australie[1,2]	363	341	376	365	302	315	333	386	311	269
Austria Autriche	6	7	7	7	7	7	7	15	14	* 23
Brazil Brésil	748	679	759	815	683	782	731	663	705	816
Canada Canada	971	1 022	1 075	1 061	810	1 008	984	940	890	901
China Chine	6	6	9	* 10	* 12	* 11	...	...	* 40	...
former Czechoslovakia† anc. Tchécoslovaquie†	178	180	184	185	172	164	189	191	..	
Egypt Egypte	25	20	21	19	18	19	13	10	9	7
Finland[1] Finlande[1]	33	32	39	43	46	45	37	30	39	15
France France	3 228	2 909	2 631	2 773	3 052	3 228	3 415	3 293	3 190	3 326
Germany † Allemagne†	..	..	..	..	..	..	..	..	4 647	4 894
F. R. Germany R. f. Allemagne	3 875	3 783	4 165	4 269	4 348	4 312	4 536	4 634	..	..
former German D. R. anc. R. d. allemande	188	202	210	218	217	218	217	145	..	..
India[3] Inde[3]	45	64	89	101	123	157	178	177	193	182
Indonesia Indonésie	...	...	...	...	...	15	16	18	25	...
Italy[3] Italie[3]	1 395	1 439	1 384	1 653	1 712	1 883	1 971	1 873	1 632	1 475
Japan Japon	7 152	7 073	7 647	7 810	7 891	8 198	9 052	9 948	9 753	9 379
Korea, Republic of[1] Corée, République de[1]	128	167	262	457	778	868	846	935	1 119	1 259
Mexico[1] Mexique[1]	214	247	285	198	278	345	448	611	730	799
Netherlands[1] Pays-Bas[1]	106	109	108	119	125	120	133	123	85	...
Poland Pologne	269	278	283	290	293	293	285	266	167	219
Romania Roumanie	90	125	134	124	129	141	144	100	84	74
Russian Federation Fédération de Russie	..	..	..	..	..	..	..	..	1 030	...

56
Passenger cars
Production: thousands [*cont.*]

Voitures de tourisme
Production : milliers [*suite*]

Country or area Pays ou zone	1983	1984	1985	1986	1987	1988	1989	1990	1991	1992
Slovakia Slovaquie	..	..	..	..	..	..	..		4	...
Slovenia Slovénie	..	..	..	..	..	..	..		79	84
Spain Espagne	1 110	1 137	1 220	1 290	1 444	* 1 498	1 651	1 696	1 787	1 795
Sweden Suède	374	378	402	415	416	259	240	216	178	...
Ukraine Ukraine	..	..	..	..	..	..	..		156	...
former USSR† ancienne URSS†	1 316	1 327	1 333	1 326	1 332	1 262	1 217	1 259	..	..
United Kingdom Royaume-Uni	1 045	909	1 048	1 019	1 143	1 227	1 308	1 302	1 340	1 291
United States[4] Etats-Unis[4]	6 781	7 622	8 002	7 516	7 085	7 105	6 808	6 081	5 441	5 684
Yugoslavia, SFR† Yougoslavie, Rfs†	168	187	177	185	219	226	227	289	..	..

Source:
Industrial statistics database of the Statistical Division
of the United Nations Secretariat.

† For detailed descriptions of data pertaining to
former Czechoslovakia, Germany, SFR Yugoslavia and former
USSR, see Annex I - Country or area nomenclature, regional
and other groupings.

1 Including assembly.
2 Twelve months ending 30 June of year stated.

3 Excluding ordinance production.
4 Factory sales.

Source:
Base de données pour les statistiques industrielles de la
Division de statistique du Secrétariat de l'ONU.

† Pour les descriptions en détails des données
relatives à l'ancienne Tchécoslovaquie, l'Allemagne, la Rfs
Yougoslavie et l'ancienne URSS, voir l'Annexe I -
Nomenclature des pays ou zones, groupements régionaux et
autres groupements.

1 Y compris le montage.
2 Période de douze mois finissant le 30 juin de l'année
indiquée.
3 Non compris la production d'artillerie.
4 Ventes des fabriques.

57
Refrigerators for household use
Réfrigérateurs ménagers
Production: thousands
Production : milliers

Country or area Pays or zone	1983	1984	1985	1986	1987	1988	1989	1990	1991	1992
Total	**40 635**	**41 027**	**42 962**	**45 532**	**49 430**	**55 842**	**54 556**	**52 738**	**48 499**	**49 921**
Algeria Algérie	105	100	102	104	225	382	381	387	388	...
Antigua and Barbuda Antigua-et-Barbuda	2	1	1	1	1	1	...	...	...	...
Argentina Argentine	252	253	162	209	207	162	125	195	355	...
Australia[1] Australie[1]	277	265	252	328	289	386	380	346	363	372
Azerbaijan Azerbaïdjan	252	272	302	314	333	357	354	330	313	223
Belarus Bélarus	621	627	657	666	682	704	718	728	743	...
Brazil Brésil	1 742	1 491	1 706	2 220	2 366	1 875	2 381	2 441	2 453	...
Bulgaria Bulgarie	133	135	122	118	111	111	101	82	65	106
Canada Canada	501	439	510	569	...	...	...	...	...	...
Chile Chili	32	42	32	46	60	59	79	89	86	136
China Chine	189	587	1 448	2 250	4 013	7 576	6 708	4 631	4 699	4 858
Colombia Colombie	155	...	...	...	...	...	...	...	...	...
Cuba Cuba	13	22	26	17	5	7	9	...	...	...
former Czechoslovakia† anc. Tchécoslovaquie†	402	446	480	524	526	551	502	449	0	0
Denmark[2] Danemark[2]	817	814	915	902	817	* 217	257	278	269	294
Dominican Republic Rép. dominicaine	29	16	11	16	...	...	...	...	...	...
Ecuador Equateur	38	30	45	54	57	125	...	...	...	66
Egypt Egypte	422	457	514	536	601	693	477	246	260	232
Finland Finlande	...	...	# 149	141	165	194	198	166	150	144
France France	697	469	436	490	612	650	614	596	556	566
Germany † Allemagne†	0	0	0	0	0	0	0	0	4 226	4 298
F. R. Germany R. f. Allemagne	2 808	2 667	2 788	3 009	3 024	3 411	3 614	4 037	0	0
former German D. R. anc. R. d. allemande	763	895	973	1 018	1 075	1 124	1 140	1 005	0	0
Greece Grèce	150	118	139	148	130	114	130	119	85	...

57
Refrigerators for household use
Production: thousands [*cont.*]
Réfrigérateurs ménagers
Production : milliers [*suite*]

Country or area Pays or zone	1983	1984	1985	1986	1987	1988	1989	1990	1991	1992
Hungary Hongrie	453	473	457	438	462	469	390	438	443	483
India Inde	429	553	661	582	595	961	991	1 220	1 124	...
Indonesia Indonésie	115	97	108	137	100	98	115	196	194	...
Iran, Islamic Rep. of Iran, Rép. islamique d'[3]	686	726	721	435	354	326	351	651	...	...
Iraq Iraq	...	...	...	220	...	90	...	...	...	35
Italy Italie	3 900	3 576	3 357	3 590	3 794	3 942	4 082	3 972	4 155	4 011
Japan Japon	4 541	4 936	5 354	4 497	5 008	5 177	5 018	5 048	5 212	4 425
Kenya Kenya	...	...	...	13	15	19	21	21	...	...
Korea, Republic of Corée, République de	1 407	1 819	1 864	2 336	3 123	3 931	2 803	2 827	3 228	3 296
Kyrgyzstan Kirghizistan	...	...	...	...	...	...	...	...	...	1
Malaysia Malaisie	224	157	149	* 154	145	197	185	212	266	288
Mexico Mexique	451	357	336	329	282	264	372	396	487	541
Mozambique Mozambique	...	...	2	2	2	3	...	...	...	...
Nigeria Nigéria	...	...	...	0	0	...	...	...	...	...
Peru Pérou	68	52	54	60	78	54	32	45	...	...
Philippines Philippines	602	146	146	140	236	...	...	...	...	...
Poland Pologne	534	543	578	569	506	484	516	604	553	500
Portugal Portugal	180	227	209	192	311	382	376	464	529	550
Republic of Moldova République de Moldova	195	162	200	194	162	156	204	133	118	55
Romania Roumanie	440	402	400	404	420	442	470	393	389	402
Singapore Singapour	66	45	37	...	...	...	...	...	...	...
Slovakia Slovaquie	402	446	480	524	526	551	502	449	515	552
Slovenia Slovénie	762	911	877	673	722	751	797	844	720	661
South Africa [4] Afrique du Sud[4]	87	172	146	205	211	306	338	352	356	318
Spain Espagne	1 014	1 105	1 082	1 169	1 136	1 210	1 268	1 285	1 410	...

57
Refrigerators for household use
Production: thousands [*cont.*]
Réfrigérateurs ménagers
Production : milliers [*suite*]

Country or area Pays or zone	1983	1984	1985	1986	1987	1988	1989	1990	1991	1992
Sweden Suède	628	513	510	559	580	626	631	584	562	...
Syrian Arab Republic Rép. arabe syrienne	141	111	76	43	32	22	89	40	85	129
Thailand Thaïlande	...	...	...	...	...	548	730	855	789	...
Trinidad and Tobago Trinité-et-Tobago	20	22	10	14	17	14	12	14	13	10
Tunisia Tunisie	70	46	48	44	39	37	...	...	...	...
Turkey Turquie	413	501	488	659	834	862	815	986	1 019	1 093
Ukraine Ukraine	695	693	743	752	745	843	882	903	883	...
former USSR†[5] ancienne URSS†[5]	5 700	5 667	5 860	5 948	5 984	6 231	6 465	6 499	0	0
United Kingdom Royaume-Uni	1 315	1 228	1 266	1 254	1 281	1 405	1 244	1 312	...	...
United States Etats-Unis	5 902[6]	6 317[6]	6 419[6]	6 940[6]	7 231[6]	7 968[6]	8 013[6]	7 015[6]	7 599	9 396
Yugoslavia, SFR†[7] Yougoslavie, Rfs†[7]	636	746	732	885	738	1 164	971	...	0	0

Source:
Industrial statistics database of the Statistical Division
of the United Nations Secretariat.

† For detailed descriptions of data pertaining to
former Czechoslovakia, Germany, SFR Yugoslavia and former
USSR, see Annex I - Country or area nomenclature, regional
and other groupings.

1 Twelve months ending 30 June of year stated.

2 Sales.
3 Production by establishments employing 50 or more persons.
4 Including deep freezers and deep freeze-refrigerator
combinations.
5 Including freezers.
6 Shipments. Electric domestic refrigerators only.

7 Including refrigerators other than domestic.

Source:
Base de données pour les statistiques industrielles de la
Division de statistique du Secrétariat de l'ONU.

† Pour les descriptions en détails des données
relatives à l'ancienne Tchécoslovaquie, l'Allemagne, la Rfs
Yougoslavie et l'ancienne URSS, voir l'Annexe I -
Nomenclature des pays ou zones, groupements régionaux et
autres groupements.

1 Période de douze mois finissant le 30 juin de l'année
indiquée.
2 Ventes.
3 Production des établissements occupant 50 personnes ou plus.
4 Y compris congélateurs-conservateurs et congélateurs
combinés avec un réfrigérateur.
5 Y compris les congélateurs.
6 Expéditions. Réfrigérateurs électriques de ménage
seulement.
7 Y compris les réfrigérateurs autres que ménagers.

58
Washing machines for household use
Machines et appareils à laver, à usage domestique

Production: thousands
Production : en milliers

Country or area Pays or zone	1983	1984	1985	1986	1987	1988	1989	1990	1991	1992
Total	**32 823**	**35 779**	**40 104**	**41 899**	**44 466**	**47 704**	**45 894**	**45 033**	**37 500**	**37 006**
Argentina Argentine	145	171	103	188	156	137	63	153	178	...
Armenia Arménie	1	6	83	116	118	89	118	110	74	...
Australia [1] Australie [1]	...	...	276	282	400	394	397	330	326	296
Austria Autriche	72	71	72	83	80	80	72	77	...	
Belarus Bélarus	...	...	...	...	...	6	22	33	57	...
Belgium [2 3] Belgique [2 3]	237	158	124	113	109	109	81	60	99	138
Brazil Brésil	480	344	419	576	630	559	578	552	394	...
Bulgaria Bulgarie	130	141	156	159	172	169	177	90	74	69
Canada Canada	390	368	407	417	...	...	...	...	...	...
Chile Chili	33	50	43	62	71	81	108	124	134	189
China Chine	3 659	5 781	8 872	8 934	9 902	10 468	8 254	6 627	6 872	7 079
Croatia Croatie	6	2	5	11	14	17	13	3	1	1
former Czechoslovakia† anc. Tchécoslovaquie†	399	423	445	452	460	463	454	451	-	
Ecuador Equateur	6	4	1	...	2	2	...	...	...	...
Egypt Egypte	278	313	268	194	178	248	212	179	202	198
France [3] France [3]	1 418	1 370	1 262	1 444	1 288	1 561	1 670	1 636	1 645	1 713
Germany † • Allemagne† F. R. Germany [4] R. f. Allemagne [4]	1 606	1 692	1 827	1 986	2 113	2 194	2 430	...	...	
former German D. R. anc. R. d. allemande	504	525	502	495	497	503	521	556	-	
Greece Grèce	51	46	59	81	61	46	48	...	...	...
Hungary Hongrie	364	392	385	404	404	436	383	315	220	219
Indonesia Indonésie	10	10	12	13	13	16	10	17	19	...
Iran, Islamic Rep. of [5] Iran, Rép. islamique d' [5]	...	...	...	49	62	8	41	36	...	...
Israel [6] Israël [6]	12	11	11	13	11	9	8	13	...	...

58
Washing machines for household use
Production: thousands [*cont.*]
Machines et appareils à laver, à usage domestique
Production : en milliers [*suite*]

Country or area Pays or zone	1983	1984	1985	1986	1987	1988	1989	1990	1991	1992
Italy Italie	3 307	3 393	3 692	3 991	4 140	4 368	4 338	4 349	5 029	5 132
Japan Japon	4 981	5 277	5 092	4 661	4 772	5 118	5 141	5 576	5 587	5 225
Korea, Republic of Corée, République de	440	646	635	902	1 303	1 903	1 864	2 163	2 157	1 896
Kyrgyzstan Kirghizistan	195	220	221	225	218	234	250	234	209	94
Latvia Lettonie	610	634	647	659	660	657	612	570	427	18
Mexico Mexique	485	314	322	268	# 345	463	442	558	611	646
Myanmar [7] Myanmar [7]	...	2	0	...	...	...	...	...	...	...
Peru Pérou	12	11	10	18	28	17	10	11	...	...
Philippines Philippines	21	0	48	9	17	...	...	...	...	...
Poland Pologne	706	730	739	773	779	761	811	482	336	363
Portugal Portugal	...	...	...	...	...	...	16	34	7	...
Republic of Moldova République de Moldova	263	278	291	301	325	310	280	298	194	102
Romania Roumanie	352	208	210	263	242	236	204	205	188	159
Slovakia Slovaquie	171	176	189	198	203	200	199	202	144	122
Slovenia Slovénie	349	400	333	377	401	388	322	315	318	188
South Africa Afrique du Sud	62	66	54	60	45	52	68	109	87	44
Spain Espagne	1 198	1 100	1 083	1 217	1 407	1 552	1 378	1 425	1 522	...
Sweden Suède	64	69	68	77	76	109	104	107	103	...
Syrian Arab Republic Rép. arabe syrienne	44	52	49	33	65	42	34	44	29	41
Turkey Turquie	231	213	345	643	408	665	621	743	836	870
Ukraine Ukraine	310	330	372	390	457	533	651	788	830	...
former USSR† ancienne URSS†	4 250	4 534	5 068	5 383	5 779	6 104	6 698	7 818	-	-
United Kingdom [8] Royaume-Uni [8]	1 371	1 349	1 430	1 317	1 343	1 351	...	...	...	...
United States [3] Etats-Unis [3]	4 615	5 006	5 456	5 783	6 100	6 441	6 375	6 428	6 404	6 566
Yugoslavia, SFR† [9] Yougoslavie, Rfs† [9]	480	533	487	534	593	604	529	...	-	-

58

Washing machines for household use
Production: thousands [*cont.*]

Machines et appareils à laver, à usage domestique
Production : en milliers [*suite*]

Source:
Industrial statistics database of the Statistical Division
of the United Nations Secretariat.

† For detailed descriptions of data pertaining to
former Czechoslovakia, Germany, SFR Yugoslavia and former
USSR, see Annex I - Country or area nomenclature, regional
and other groupings.

1 Twelve months ending 30 June of year stated.

2 Production by establishments employing 5 or more persons.
3 Shipments.
4 Automatic washing machines only.
5 Production by establishments employing 50 or more persons.
6 Marketed local production.
7 Data refer to assembly. Government production only.

8 Deliveries of U.K. manufactured goods.
9 Including drying machines.

Source:
Base de données pour les statistiques industrielles de la
Division de statistique du Secrétariat de l'ONU.

† Pour les descriptions en détails des données
relatives à l'ancienne Tchécoslovaquie, l'Allemagne, la Rfs
Yougoslavie et l'ancienne URSS, voir l'Annexe I -
Nomenclature des pays ou zones, groupements régionaux et
autres groupements.

1 Période de douze mois finissant le 30 juin de l'année
indiquée.
2 Production des établissements occupant 5 personnes ou plus.
3 Expéditions.
4 Machines à laver automatiques seulement.
5 Production des établissements occupant 50 personnes ou plus.
6 Production locale commercialisée.
7 Données se rapportant à l'assemblage. Production de l'Etat
seulement.
8 Quantités livrées de marchandises fabriquées au Royaume-Uni.
9 Y compris les machines à sécher.

59
Machine tools
Machines-outils
Production: number
Production : nombre

Country or area Pays ou zone	1983	1984	1985	1986	1987	1988	1989	1990	1991	1992
A. Drilling and boring machines • Perceuses										
Total	158 911	173 848	162 845	157 817	151 674	159 500	154 573	144 850	117 864	100 747
Algeria Algérie	235	211	209	352	347	261	32	412	210	122
Austria Autriche	1 746	1 881	1 593	1 780	1 114	1 201	1 443	...	...	...
Azerbaijan Azerbaïdjan	...	..	..	..	..	..	..	..	326	237
Bangladesh Bangladesh	...	...	...	...	...	117	125	131	...	...
Belgium Belgique	...	37 528	...	36 893	...	...	...	...	...	...
Bulgaria Bulgarie	4 122	4 731	6 563	7 565	7 092	7 095	6 496	4 729	1 959	996
Croatia Croatie	..	..	..	..	..	..	..	..	458	346
former Czechoslovakia† anc. Tchécoslovaquie†	1 631	1 492	1 514	1 280	987	974	871	901	..	..
Czech Republic République tchèque	..	..	..	..	..	..	..	..	1 131	...
Denmark[1] Danemark[1]	43	...	185	165	125	* 305	338	348	394	223
Finland Finlande	50	...	...	51	...	103	74	69	42	10
France[2] France[2]	3 396	3 081	3 769	3 406	2 441	2 459	...	2 500	1 460	1 212
Germany † • Allemagne† F. R. Germany[3] R. f. Allemagne[3]	...	...	12 997	14 790	16 196	14 504	14 336	14 158	..	..
Hungary Hongrie	2 925	4 147	5 744	8 188	5 557	4 108	2 027	1 929	1 257	75
Indonesia Indonésie	...	...	...	...	...	112	139	89	527	...
Iran, Islamic Rep. of Iran, Rép. islamique d'	3 237	3 184	2 200	...	...	...	...	...	...	...
Japan Japon	31 617	41 224	36 807	27 968	24 077	32 954	39 152	40 171	33 929	22 973
Korea, Republic of Corée, République de	...	...	3 971	5 514	7 909	8 974	8 347	7 662	8 150	7 336
Mexico Mexique	...	...	...	...	5 760	6 312	4 518	2 138	2 187	855
Poland Pologne	8 300	11 130	10 439	12 636	12 394	12 193	5 339	2 238	2 995	1 858
Slovakia Slovaquie	..	..	..	..	..	..	..	..	10	...
Slovenia Slovénie	..	..	..	..	...	...	..	..	114	60

59
Machine tools
Production: number [*cont.*]
Machines-outils
Production : nombre [*suite*]

Country or area Pays ou zone	1983	1984	1985	1986	1987	1988	1989	1990	1991	1992
Spain Espagne	5 937	5 012	7 423	4 193	6 004	4 702	5 043	3 688	2 567	...
Sweden Suède	4 968	7 371	7 717	7 278	6 796	...	...	...	...	...
Turkey Turquie	...	...	...	...	140	172	22	8	57	...
former USSR† ancienne URSS†	64	50	48	90	302	358	571	304	..	..
United Kingdom Royaume-Uni	2 257	...	...	...	...	...	...	...	...	...
United States Etats-Unis	13 915 [4]	12 526 [4]	10 259 [4]	8 515 [4]	7 708 [4]	10 236 [4]	10 508 [4]	8 828 [4]	7 603	7 542 [4]
Viet Nam Viet Nam	...	...	...	50	30	...	...	...	...	...
Yugoslavia, SFR† Yougoslavie, Rfs†	1 760	2 279	2 507	3 057	2 045	2 098	2 355	...	..	..

B. Lathes • Tours

	1983	1984	1985	1986	1987	1988	1989	1990	1991	1992
Total	**146 038**	**155 237**	**149 690**	**141 712**	**135 747**	**143 730**	**151 444**	**149 390**	**126 486**	**105 928**
Algeria Algérie	140	138	149	161	120	110	150	270	273	310
Armenia Arménie	..	..	..	..	..	..	..	..	2 633	...
Austria Autriche	1 251	1 613	1 939	1 719	1 709	2 143	4 556	1 726	1 647	1 421
Bangladesh Bangladesh	...	...	...	...	...	202	214	223	...	...
Bulgaria Bulgarie	6 310	5 564	5 477	5 912	4 886	4 953	5 438	5 014	4 744	3 587
China Chine	49 694	55 675	...	...	...	...	...	...	...	...
Colombia Colombie	49	...	...	...	...	...	...	...	...	...
Croatia Croatie	..	..	..	..	..	..	..	..	584	463
former Czechoslovakia† anc. Tchécoslovaquie†	6 896	7 254	7 157	6 011	6 185	6 055	6 393	6 020	..	..
Czech Republic République tchèque	..	..	..	..	..	..	..	..	1 405	833
Denmark[1] Danemark[1]	...	...	...	...	18	* 5	...	545	384	279
Finland Finlande	2	...	...	...	...	2	2	4	1	0
France[4] France[4]	1 488	1 077	1 284	1 356	1 273	1 012	...	1 400	988	942
Germany† • Allemagne† F. R. Germany R. f. Allemagne	6 516	6 772	6 864	7 011	7 046	6 637	7 155	7 612	..	..

59
Machine tools
Production: number [*cont.*]
Machines-outils
Production : nombre [*suite*]

Country or area Pays ou zone	1983	1984	1985	1986	1987	1988	1989	1990	1991	1992
former German D. R.[5] anc. R. d. allemande[5]	2 629	3 225	2 598	2 522	2 466	2 482	2 481	...	..	
Hungary Hongrie	1 526	1 113	952	971	894	894	943	663	304	63
Indonesia Indonésie	...	...	...	...	...	36	19	2	45	...
Japan Japon	21 816	31 455	33 527	25 424	21 066	28 257	32 748	32 659	26 216	16 155
Korea, Republic of Corée, République de	5 141	6 020	5 206	6 209	7 898	8 605	9 156	10 597	11 324	6 643
Poland Pologne	4 348	4 701	4 474	4 676	4 104	3 970	3 824	5 178	2 184	1 105
Portugal Portugal	177	141	157	167	169	147	124	101	87	60
Romania Roumanie	...	...	5 148	5 083	5 038	4 958	4 747	3 702	2 883	1 583
Slovakia Slovaquie	..	..	..	..	..	..	..	..	4 417	...
Spain Espagne	1 158	1 674	1 264	1 336	1 599	1 484	2 074	1 353	1 265	...
Sweden[6] Suède[6]	571	500	431	349	296	113	...	...	...	...
Turkey Turquie	777	180	71	144	114	91	47	34	23	2
Ukraine Ukraine	..	..	..	..	..	..	..	..	3 300	...
former USSR† ancienne URSS†	8 915	8 239	6 881	8 033	6 396	6 933	5 991	6 018	-	-
United Kingdom Royaume-Uni	7 348	...	...	...	...	...	5 275	5 742	...	...
United States[4] Etats-Unis[4]	9 692	4 300	3 836	3 057	3 045	3 317	3 789	3 247	2 658	2 409
Yugoslavia, SFR† Yougoslavie, Rfs†	2 578	2 887	3 176	2 429	2 331	2 470	2 059	...	..	..

C. Milling machines • Fraiseuses

Total	*64 394*	*64 126*	*66 045*	*61 051*	*56 912*	*58 329*	*61 163*	*64 507*	*41 975*	*37 050*
Algeria Algérie	107	92	97	162	119	92	44	267	150	103
Armenia Arménie	..	..	..	..	..	..	..	..	759	...
Austria Autriche	552	682	713	554	564	828	1 142	458	526	844
Bangladesh Bangladesh	...	...	...	...	...	134	139	144	...	...
Belarus Bélarus	..	..	..	..	..	..	..	..	150	...
Bulgaria Bulgarie	1 082	1 209	1 981	1 320	1 307	1 453	1 450	1 240	961	432

59
Machine tools
Production: number [*cont.*]
Machines-outils
Production : nombre [*suite*]

Country or area Pays ou zone	1983	1984	1985	1986	1987	1988	1989	1990	1991	1992
China Chine	10 781	11 134	...	...	...	...	...	...	...	...
Colombia Colombie	79	...	...	...	...	...	...	...	...	...
Croatia Croatie	..		..		..	..		..	212	168
former Czechoslovakia† anc. Tchécoslovaquie†	2 617	2 548	2 260	2 278	2 299	2 208	1 596	1 470	..	..
Czech Republic République tchèque	..	..	..	..	..	..	..	..	1 706	1 355
Denmark[1] Danemark[1]	413	...	...	320	252	* 103	292	...	201	200
Finland Finlande	...	...	...	...	...	100	100	...	...	...
France[4] France[4]	1 959	1 665	1 434	1 451	966	570	...	700	496	401
Germany † · Allemagne† F. R. Germany[7] R. f. Allemagne[7] former German D. R. anc. R. d. allemande	... 3 447	... 3 429	... 3 335	... 3 081	... 3 007	... 3 298	... 3 135	12 150 3 051	.. ..	.. ..
Greece Grèce	...	853	594	618	1 404	1 674	1 890	1 805	1 452	...
Hungary Hongrie	79	53	39	39	42	22	66	136	297	256
Indonesia Indonésie	...	...	...	...	...	...	...	21	...	...
Japan Japon	9 713	10 723	11 399	7 258	4 859	6 959	8 612	8 492	7 584	3 913
Korea, Republic of Corée, République de	1 638	2 261	2 428	2 994	3 740	2 802	3 062	3 673	3 855	2 394
Poland Pologne	1 431	1 442	1 500	1 546	1 462	1 387	1 448	1 196	834	500
Portugal Portugal	135	107	95	79	98	148	134	...	...	...
Romania Roumanie	...	...	2 618	2 924	2 733	2 209	2 073	1 196	1 355	764
Slovenia Slovénie	..	..	..	..	..	..	..	..	13	12
Spain Espagne	7 459	5 041	6 823	6 519	6 984	6 533	8 057	7 062	4 169	...
Sweden Suède	388	412	293	...	...	...	...	98	...	...
Turkey Turquie	...	...	...	...	211	157	52	169	86	39
Ukraine Ukraine	..			..	..	..	..	..	1 208	...
former USSR† ancienne URSS†	3 700	3 399	3 491	3 130	2 110	1 989	2 421	2 435	..	..

59
Machine tools
Production: number [cont.]
Machines-outils
Production : nombre [suite]

Country or area Pays ou zone	1983	1984	1985	1986	1987	1988	1989	1990	1991	1992
United Kingdom Royaume-Uni	1 341	1 596	...	...	...	...	...	...	...	...
United States[4] Etats-Unis[4]	7 349	7 981	8 308	8 133	6 277	7 717	6 872	4 787	2 772	2 581
Yugoslavia, SFR† Yougoslavie, Rfs†	1 342	1 145	1 260	1 320	1 367	1 074	921	...	..	..

D. Metal-working presses • Presses pour le travail des métaux

	1983	1984	1985	1986	1987	1988	1989	1990	1991	1992
Total	**59 129**	**66 893**	**59 375**	**58 928**	**54 487**	**59 940**	**60 812**	**59 185**	**51 532**	**82 095**
Armenia Arménie	..	..	..	..	..	..	..	..	206	...
Austria Autriche	51	69	79	31	31	27	52	...	...	...
Brazil Brésil	1 543	1 384	1 824	2 136	2 312	1 499	1 415	1 182	1 142	...
Bulgaria Bulgarie	...	...	824	863	871	...	...	...	...	...
Colombia Colombie	12	...	...	...	...	...	...	...	...	...
former Czechoslovakia† anc. Tchécoslovaquie†	1 753	1 540	1 581	1 614	1 689	1 410	1 486	1 424	..	..
Czech Republic République tchèque	..	..	..	..	..	..	..	..	82	106
Finland Finlande	2 682	2 803	...	...	...	* 118	121	126	98	17
France[4] France[4]	1 171	1 166	1 082	1 015	787	1 883	781	1 300	1 235	1 385
Germany † Allemagne†	..	..	..	..	..	..	..	..	17 726	55 996
F. R. Germany R. f. Allemagne	12 538	14 153	13 289	15 804	13 620	14 372	16 460	16 643	..	..
former German D. R. anc. R. d. allemande	...	...	...	2 513	2 671	2 136	2 366	1 781	..	..
Greece Grèce	...	475	324	947	475	296	405	328	320	...
Japan Japon	15 358	21 351	19 023	16 185	15 395	20 449	21 574	22 571	19 173	12 458
Latvia Lettonie	..	..	..	..	..	..	..	..	3	...
Mexico Mexique	...	...	...	...	11	5	...	...	...	...
Netherlands[89] Pays-Bas[89]	284	187	280	328	...	...	...	...	...	...
Poland Pologne	44	67	64	49	170	66	63	50	44	42
Portugal Portugal	222	132	110	146	240	295	315	255	102	101
Slovakia Slovaquie	..	..	..	..	..	..	..	..	829	261

59
Machine tools
Production: number [*cont.*]
　　Machines-outils
　　Production : nombre [*suite*]

Country or area Pays ou zone	1983	1984	1985	1986	1987	1988	1989	1990	1991	1992
Slovenia Slovénie	..	..	..	..	..	..	..	..	119	92
Spain Espagne	2 447	3 332	1 647	2 239	1 846	2 483	2 035	1 600	2 152	...
Ukraine Ukraine	..	..	..	..	..	..	..	..	666	...
former USSR† ancienne URSS†	2 586	2 213	2 096	1 669	1 910	1 747	802	882	..	-
United Kingdom Royaume-Uni	562	552	...	...	...	...	...	...	...	...
United States[4] Etats-Unis[4]	9 452	10 814	9 137	9 439	8 134	9 593	8 959	7 285	5 912	5 822
Yugoslavia, SFR† Yougoslavie, Rfs†	2 492	1 783	1 961	1 773	2 236	1 871	2 280	...	..	...

Source:
Industrial statistics database of the Statistical Division
of the United Nations Secretariat.

† For detailed descriptions of data pertaining to
former Czechoslovakia, Germany, SFR Yugoslavia and former
USSR, see Annex I - Country or area nomenclature, regional
and other groupings.

1 Sales.
2 Limited coverage.
3 1982-1984, data are confidential (drilling and boring
　machines).
4 Shipments.
5 Excluding turning lathes for clock-makers.
6 Excluding automatic lathes not numerically controlled.
7 1983-1989, data are confidential.
8 Industrial sales.
9 Excluding mechanical presses.

Source:
Base de données pour les statistiques industrielles de la
Division de statistique du Secrétariat de l'ONU.

† Pour les descriptions en détails des données
relatives à l'ancienne Tchécoslovaquie, l'Allemagne, la Rfs
Yougoslavie et l'ancienne URSS, voir l'Annexe I -
Nomenclature des pays ou zones, groupements régionaux et
autres groupements.

1 Ventes.
2 Couverture limitée.
3 1982-1984, les données sont confidentielles (perceuses).
4 Expéditions.
5 Non compris les tours d'horloger.
6 Non compris les tours automatiques sans contrôle numérique.
7 1983-1989, les données sont confidentielles.
8 Ventes industrielles.
9 Non compris les presses à commande mécanique.

60
Lorries (trucks)
Camions
Production: number
Production : nombre

Country or area Pays ou zone	1983	1984	1985	1986	1987	1988	1989	1990	1991	1992
A. Assembled · Assemblés										
Total	479 701	529 141	456 347	407 882	416 424	517 542	580 467	641 462	623 309	647 584
Algeria[1] Algérie[1]	6 500	6 619	5 722	6 671	5 785	3 326	3 946	3 564	3 164	2 434
Bangladesh Bangladesh	...	...	...	551	457	809	806	504	528	...
Belgium[2 3] Belgique[2 3]	33 261	38 951	37 455	54 245	54 374	69 396	75 357	65 667	88 937	70 532
Chile Chili	1 510	2 556	3 528	3 016	* 4 548	6 012[4]	8 377[4]	8 028[4]	9 400[4]	14 352[4]
Colombia[4] Colombie[4]	6 775	11 220	5 878	8 056	8 869	13 718	13 572	12 660	8 900	10 116
Cuba Cuba	415	562	870	688	475	469	542	...	...	...
Denmark[5] Danemark[5]	46	109	237	...	...	...	...	...	...	...
Ghana *[4] Ghana *[4]	4 300	...	...	...	...	...	...	...	...	...
Greece Grèce	7 552	4 685	3 081	3 152	3 459	3 322	2 897	1 715	1 534	...
Indonesia Indonésie	570	770	1 000	920	925	...	...	...	...	...
Iran, Islamic Rep. of[6] Iran, Rép. islamique d'[6]	77 678	83 576	57 025	29 481	16 744	5 477	8 086	22 198	15 000	15 000
Iraq Iraq	...	...	297	557	...	...	...	...	...	...
Israel Israël	2 936	2 146	1 130	1 152	971	615	1 174	1 074	...	...
Kenya Kenya	...	...	...	2 120	1 947	2 086	1 679	1 701	1 296	...
Malaysia[7] Malaisie[7]	20 789	26 740	35 772	* 17 318	12 594	18 999	38 597	17 638	18 794	21 916
Morocco[4] Maroc[4]	2 707	2 847	3 373	3 768	2 256	8 256	9 624	7 800	8 000	7 500
Myanmar[8] Myanmar[8]	736	567	798	...	...	...	...	...	...	...
Netherlands Pays-Bas	1 869	2 709	2 605	3 135	3 959	4 036	4 092	...	...	...
New Zealand[7] Nouvelle-Zélande[7]	19 030	24 997	24 185	15 964	12 873	...	...	* 13 500	* 13 500	* 10 210
Nigeria Nigéria	2 551	1 958	2 091	6 150	3 747	...	...	...	...	...
Pakistan[9] Pakistan[9]	15 095	14 601	15 255	13 851	12 666	12 326	13 756	13 324	13 911	13 270
Peru[4] Pérou[4]	1 737	2 671	3 246	5 637	8 378	4 736	3 019	2 921	2 000	2 000
Philippines Philippines	3 864	# 271	44	111	780	1 008	...	...	...	...

60
Lorries (trucks)
Production: number [*cont.*]
Camions
Production : nombre [*suite*]

Country or area Pays ou zone	1983	1984	1985	1986	1987	1988	1989	1990	1991	1992
Portugal Portugal	13 955	7 076	9 174	14 068	23 481	55 291	61 996	59 939	53 309	54 431
Slovenia Slovénie	..	..	..	..	..	..	..	..	611	0
South Africa Afrique du Sud	102 903	123 070	94 461	77 929	93 534	108 663	119 591	107 922	97 178	93 599
Thailand[4] Thaïlande[4]	70 440	81 850	58 244	53 116	68 815	97 722	155 094	236 221	206 172	* 235 000
Trinidad and Tobago[4] Trinité-et-Tobago[4]	3 090	3 631	3 173	...	911	1 356	381	1 124	1 711	1 698
Tunisia Tunisie	8 464	9 798	7 328	2 890	2 607	1 115	948	1 560	1 068	768
Turkey Turquie	23 335	23 217	25 025	19 935	20 966	20 048	19 045	27 014	29 967	37 195
United Kingdom Royaume-Uni	1 181	1 315	1 168	1 142	1 170	1 628	...	...	...	...
United Rep.Tanzania[1] Rép. Unie de Tanzanie[1]	...	...	481	353	333	341	470	637	479	171
Venezuela Venezuela	40 000	40 000	44 000	52 000	41 000	46 000	6 000	12 000	24 000	30 000
Yugoslavia, SFR† Yougoslavie, Rfs†	...	48	...	...	79	918	434	451	..	..
Zaire Zaïre	1 994	2 335	...	...	...	...	...	...	...	...

B. Produced • Fabriqués

Total	9 550 192	11 353 223	11 953 944	11 840 052	12 433 392	13 503 575	12 972 566	11 479 678	11 216 930	13 890 131
Argentina[10][11] Argentine[10][11]	25 052	15 842	12 767	15 350	12 801	13 334	10 550	9 524	9 028	13 862
Armenia Arménie	..	..	..	..	..	..	..	..	6 823	...
Australia[9][12] Australie[9][12]	20 496	10 747	28 947	25 912	20 462	22 303	29 557	25 958	17 666	14 550
Austria Autriche	1 722	776	487	147	416	798	557	736	5 168	4 089
Azerbaijan Azerbaïdjan	..	..	..	..	..	..	..	..	3 246	402
Belarus Bélarus	..	..	..	..	..	..	..	..	38 178	...
Brazil[13] Brésil[13]	125 943	161 323	181 469	212 860	206 523	244 076	247 641	217 318	218 250	217 147
Bulgaria[10] Bulgarie[10]	6 445	6 456	6 860	6 748	7 346	6 811	7 888	7 285	2 778	945
Canada[14] Canada[14]	554 400	807 848	855 609	784 892	825 187	1 018 317	949 200	789 932	789 600	965 239
China Chine	137 100	181 800	269 000	229 100	298 400	403 300	363 400	289 700	382 500	476 700
former Czechoslovakia† anc. Tchécoslovaquie†	38 454	42 739	44 680	47 002	47 669	45 468	46 177	49 006	..	..

60
Lorries (trucks)
Production: number [cont.]
Camions
Production : nombre [suite]

Country or area Pays ou zone	1983	1984	1985	1986	1987	1988	1989	1990	1991	1992
Czech Republic République tchèque	..	..	..	..	..	..	..	..	20 419	12 287
Egypt Egypte	2 736	3 278	3 082	3 345	2 580	1 745	1 475	1 371	1 127	1 529
Finland[10] Finlande[10]	58	20	# 315	360	390	771	852	910	545	578
France France	451 185	412 757	* 448 692	495 372	* 502 901	535 571	577 495	539 796	461 640	494 124
Germany † Allemagne†	..	..	..	..	..	..	..	..	356 059	325 901
F. R. Germany R. f. Allemagne	268 600	236 876	262 199	267 147	243 760	258 314	274 496	315 010	..	..
former German D. R. anc. R. d. allemande	39 557	43 105	45 305	44 887	41 897	39 572	38 786	31 360		
Hungary Hongrie	358	201	736	1 025	1 406	748	162	154	11	30
India[14] Inde[14]	69 509	75 739	85 352	76 992	105 996[11]	119 868[11]	115 200[11]	145 200[11]	146 400[11]	141 600[11]
Indonesia Indonésie	...	...	...	...	...	20	1 145	1 280	26	...
Italy Italie	166 543	150 550	171 245	169 985	191 096	219 805	240 807	234 393	229 860	194 616
Japan Japon	3 897 366	4 308 257	4 535 947	4 399 699	4 297 737	4 431 099	3 918 400	3 486 618	3 433 790	3 053 477
Korea, Republic of[10] Corée, République de[10]	51 807	63 117	71 217	86 385	109 427	129 956	169 703	238 520	246 522	294 550
Kyrgyzstan Kirghizistan	..	..	..	..	..	..	..	..	23 621	14 818
Mexico[10] Mexique[10]	67 053	94 591	132 389	115 611	112 002	145 895	184 725	199 123	238 804	269 591
Netherlands Pays-Bas	6 269	7 360	7 799	8 086	9 407	9 753	11 679	29 800	26 100	23 982
Poland[15] Pologne[15]	42 933	46 903	49 114	46 088	45 550	46 834	43 853	38 956	20 100	17 657
Romania Roumanie	15 846	16 485	20 788	14 485	12 715	16 556	13 515	8 457	7 592	4 456
Russian Federation Fédération de Russie	..	..	..	..	..	..	..	..	615 868	...
Slovenia Slovénie	..	..	..	..	..	..	..	..	1 513	377
Spain Espagne	117 172	107 828	155 744	204 999	254 250	313 200[14]	343 899[14]	302 400[14]	219 600[14]	2 121 887[14]
Sweden Suède	* 41 121	* 53 775	* 54 648	* 59 618	62 335	76 507[14]	81 670[14]	74 400[14]	75 000[14]	...
Ukraine Ukraine	..	..	..	..	..	..	..	..	25 096	...
former USSR†[16] ancienne URSS†[16]	* 860 400	* 849 600	* 926 000	* 942 000	* 958 000	* 974 000	* 900 000	...	..	..
United Kingdom[15 17] Royaume-Uni[15 17]	222 980	202 965	243 792	210 994	227 389	294 633	326 590[11]	273 600[11]	222 000[11]	240 000[11]

60

Lorries (trucks)
Production: number [*cont.*]

Camions
Production : nombre [*suite*]

Country or area Pays ou zone	1983	1984	1985	1986	1987	1988	1989	1990	1991	1992
United States[18] Etats-Unis[18]	2 302 032	3 435 890	3 323 372	3 355 863	3 821 410	4 120 574	4 061 950	3 720 000	3 372 000	4 118 578
Yugoslavia, SFR† Yougoslavie, Rfs†	16 438	15 778	15 772	14 483	13 723	13 747	11 194	9 989	..	..

Source:
Industrial statistics database of the Statistical Division
of the United Nations Secretariat.

† For detailed descriptions of data pertaining to
former Czechoslovakia, Germany, SFR Yugoslavia and former
USSR, see Annex I - Country or area nomenclature, regional
and other groupings.

1 Including buses.
2 Production by establishments employing 5 or more persons.
3 Shipments.
4 Including motor coaches and buses.
5 Sales.
6 Production by establishments employing 50 or more persons.
7 Including vans and buses.
8 Government production only.
9 Twelve months ending 30 June of year stated.

10 Including assembly.
11 Including buses and motor coaches.
12 Finished and partly finished.
13 Incomplete coverage.
14 Excluding ordinance production.
15 Including special-purpose vehicles.
16 Excluding production for armed forces.
17 Including electrically-powered vehicles.
18 Factory sales.

Source:
Base de données pour les statistiques industrielles de la
Division de statistique du Secrétariat de l'ONU.

† Pour les descriptions en détails des données
relatives à l'ancienne Tchécoslovaquie, l'Allemagne, la Rfs
Yougoslavie et l'ancienne URSS, voir l'Annexe I -
Nomenclature des pays ou zones, groupements régionaux et
autres groupements.

1 Y compris les autobus.
2 Production des établissements occupant 5 personnes ou plus.
3 Expéditions.
4 Y compris les autocars et autobus.
5 Ventes.
6 Production des établissements occupant 50 personnes ou plus.
7 Y compris les autobus et les camionnettes.
8 Production de l'Etat seulement.
9 Période de douze mois finissant le 30 juin de l'année
indiquée.

10 Y compris le montage.
11 Y compris les autocars et autobus.
12 Finis et semi-finis.
13 Couverture incomplète.
14 Non compris la production d'artillerie.
15 Y compris véhicules à usages spéciaux.
16 Non compris production destinée aux forces armées.
17 Y compris les véhicules à propulsion électrique.
18 Ventes des fabriques.

Technical notes, tables 41-60

Industrial activity comprises mining and quarrying, manufacturing and the production of electricity, gas and water. These activities correspond to the major divisions 2, 3 and 4 respectively of the *International Standard Industrial Classification of All Economic Activities*.[42]

Many of the tables are based on data compiled for the United Nations *Industrial Commodity Statistics Yearbook*. [21] Exceptions are indicated in notes at the end of the tables.

The methods used by countries for the computation of industrial output are, as a rule, consistent with the recommendations on this subject by the United Nations and provide a satisfactory basis for comparative analysis.[41] In some cases, however, the definitions and procedures underlying computations of output differ from approved guidelines. The differences, where known, are indicated in the footnotes to each table.

A. *Food, beverages and tobacco*

Table 41 covers the production of centrifugal sugar from both beet and cane and the figures are expressed as far as possible in terms of raw sugar. However, where exact information about polarisation is lacking, data are expressed in terms of sugar "tel quel" and are footnoted accordingly. Unless otherwise stated, the data refer to calendar years.

The consumption data relate to the apparent consumption of centrifugal sugar in the country concerned, including sugar used for the manufacture of sugar-containing products whether exported or not and sugar used for purposes other than human consumption as food. Unless otherwise specified the statistics are expressed in terms of raw value (i.e. sugar polarizing at 96 degrees). However, where exact information is lacking, data are expressed in terms of sugar "tel quel" and are footnoted accordingly. The world and regional totals also include data for countries whose sugar consumption was less than 10 thousand metric tons.

Table 42 refers to meat from animals slaughtered within the national boundaries irrespective of the origin of the animals. Production figures of beef and veal (including buffalo meat), pork (including bacon and ham) and mutton and lamb (including goat meat), are in terms of carcass weight, excluding edible offals, tallow and lard. All data refer to total meat production, i.e from both commercial and farm slaughter.

Table 43 refers to beer made from malt, including ale, stout, porter.

Table 44 refers to cigarettes only.

Notes techniques, tableaux 41-60

L'activité industrielle comprend les industries extractives (mines et carrières), les industries manufacturières et la production d'électricité, de gaz et d'eau. Ces activités correspondent aux grandes divisions 2, 3 et 4, respectivement, de la *Classification internationale type par industrie de toutes les branches d'activité économique* [42].

Un grand nombre de ces tableaux sont établis sur la base de données compilées pour l'*Annuaire de statistiques industrielles par produit* des Nations Unies [21]. Les exceptions sont indiquées dans des notes au bas des tableaux.

En règle générale, les méthodes employées par les pays pour le calcul de leur production industrielle sont conformes aux recommandations des Nations Unies à ce sujet et offrent une base satisfaisante pour une analyse comparative [41]. Toutefois, dans certains cas, les définitions des méthodes sur lesquelles reposent les calculs de la production diffèrent des directives approuvées. Lorsqu'elles sont connues, les différences sont indiquées dans les notes au bas des tableaux.

A. *Alimentation, boisson et tabacs*

Le *Tableau 41* porte sur la production de sucre centrifugé à partir de la betterave et de la canne à sucre, et les chiffres sont exprimés autant que possible en sucre brut. Toutefois, en l'absence d'informations exactes sur la polarisation, les données sont exprimées en sucre tel quel, accompagnées d'une note au bas du tableau. Sauf indication contraire, les chiffres se rapportent à des années civiles.

Les données de la consommation se rapportent à la consommation apparente de sucre centrifugé dans le pays en question, y compris le sucre utilisé pour la fabrication de produits à base de sucre, exportés ou non, et le sucre utilisé à d'autres fins que pour la consommation alimentaire humaine. Sauf indication contraire, les statistiques sont exprimés en valeur brute (sucre polarisant à 96°). Toutefois, en l'absence d'informations exactes, les données sont exprimées en sucre tel quel, accompagnées d'une note au bas du tableau. Les totaux mondiaux et régionaux comprennent également les données relatives aux pays où la consommation de sucre est inférieure à 10.000 tonnes métriques.

Le *Tableau 42* indique la production de viande provenant des animaux abattus à l'intérieur des frontières nationales, quelle que soit leur origine. Les chiffres de production de viande de boeuf et de veau (y compris la viande de buffle), de porc (y compris le bacon et le jambon) et de mouton et d'agneau (y compris la viande de chèvre) se rapportent à la production en poids de carcasses et ne comprennent pas le saindoux, le suif et les abats comestibles. Toutes les données se rapportent à la production totale de viande, c'est-à-dire à la fois aux animaux abattus à des fins commerciales et des animaux sacrifiés à la ferme.

B. *Textile, wearing apparel and leather industries*

In *table 45*, *Fabrics of cotton and of wool*: data refer to woollen and worsted fabrics, woven fabrics of cotton at loom stage before undergoing finishing processes such as bleaching, dyeing, printing, mercerizing, lazing, etc. Fabrics of fine hair are excluded.

Woven fabrics of cellulosic and non-cellulosic fibres: fabrics of continuous and discontinuous rayon and acetate fibres and non-cellulosic fibres other than textile glass fibres, including pile and chenille fabrics at loom stage.

In *table 46*, data refer to total production of leather footwear for children, men and women and all other footwear such as footwear with outer soles of wood or cork, sports footwear and orthopedic leather footwear. House slippers and sandals of various types are included. Rubber footwear, however, are excluded.

C. *Wood and wood products; paper and paper products*

Table 47 refers to aggregate of sawnwood-coniferous, non-coniferous and sleepers. Data cover wood planed, unplaned, grooved, tongued and the like, sawn lengthwise or produced by a profile-chipping process, and planed wood which may also be finger-jointed, tongued or grooved, chamfered, rabbeted, V-jointed, beaded and so on. Wood flooring is excluded. Sleepers may be sawn or hewn.

Table 48 refers to the production of all paper and paper board. Data cover newsprint, printing and writing paper, construction paper and paperboard, household and sanitary paper, special thin paper, wrapping and packaging paper and paperboard.

D. *Chemicals and related products*

Table 49 refers to the production of rubber tires for passenger cars and commercial vehicles. Unless otherwise stated, data do not cover tires for vehicles operating off the road, motorcycles, bicycles and animal-drawn road vehicles, or the production of inner tubes.

Table 50 refers to all hydraulic cements used for construction (portland, metallurgic, aluminous, natural, and so on).

In *table 51*, data refer to H_2SO_4 in terms of pure monohydrate sulphuric acid, including the sulphuric acid equivalent of oleum or fuming sulphuric acid.

In *table 52*, data on soaps refer to normal soaps of commerce, including both hard and soft soaps. Included are, in particular, household soaps, toilet soaps, transparent soaps, shaving soaps, medicated soaps, disinfectant soaps, abrasive soaps, resin and naphthenate soaps, and industrial soaps.

Washing powders and detergents refer to organic surface—active agents, surface—active preparations and washing preparations whether or not containing soap.

Tableau 43 : Bière produite à partir du malte, y compris ale, stout et porter (bière anglaise, blonde et brune).

Le *Tableau 44* se rapporte seulement aux cigarettes.

B. *Textile, habillement et cuir*

Au *tableau 45*, *Tissus de coton et de laine* : données se rapportent aux tissus de laine cardée ou peignée, tissus de coton, avant les operations de finition, c'est-à-dire avant d'être blanchis teints, imprimés, mercerisés, glacés, etc. A l'exclusion des tissus de poils fins.

Tissus de fibres cellulosiques et non-cellulosiques : tissus sortant du metier à tisser de fibres de rayonne et d'acetate et tissus composés de fibres non cellulosiques, autres que les fibres de verre, continues ou discontinues; cette rubrique comprend les velours, peluches, tissus boucles et tissus chenille.

Au *tableau 46*, les données se rapportent à la production totale de chaussures de cuir pour enfants, hommes et dames et toutes les autres chaussures telles que chaussures à semelles en bois ou en liège, chaussures pour sports et orthopediques en cuir. Chaussures en caoutchouc ne sont pas compris.

C. *Bois et produits dérivés; papier et produits dérivés*

Les données du *Tableau 47* sont un agrégat des sciages de bois de cônifères et de non-cônifères et des traverses de chemins de fer. Elles comprennent les bois rabotés, non rabotés, rainés, languetés, etc. sciés en long ou obtenus à l'aide d'un procédé de profilage par enlèvement de copeaux et les bois rabotés qui peuvent être également à joints digitiformes languetés ou rainés, chanfreinés, à feuillures, à joints en V, à rebords, etc. Cette rubrique ne comprend pas les éléments de parquet en bois. Les traverses de chemin de fer comprennent les traverses sciées ou équaries à la hache.

Le *Tableau 48* se rapporte à la production de tout papier et carton. Les données comprennent le papier journal, les papiers d'impression et d'écriture, les papiers et cartons de construction, les papiers de ménage et les papiers hygiéniques, les papiers minces spéciaux, les papiers d'empaquetage et d'emballage et carton.

D. *Produits chimiques et apparentés*

Le *Tableau 49* se rapporte à la production de pneus en caoutchouc pour voitures particulières et véhicules utilitaires. Sauf indication contraire, elles ne couvrent pas les pneus pour véhicules non routiers, motocyclettes, bicyclettes et véhicules routiers à traction animale, ni la production de chambres à air.

Le *Tableau 50* se rapporte à tous les ciments hydrauliques utilisés dans la construction (portland, métallurgique, alumineux, naturel, etc.).

Au *tableau 51*, les données se rapportent au H_2SO_4 sur la base de l'acide sulfurique monohydraté, y compris l'équivalent en acide sulfurique de l'oléum ou acide sulfurique fumant.

E. *Basic metal industries*

Table 53 includes foundry and steel making pig-iron. Figures on crude steel include both ingots and steel for castings. In selected cases data are obtained from the United States of America Bureau of Mines (Washington, DC), Instituto Latino-Americano del Ferro y el Acero (Santiago) and the United Nations Economic Commission for Europe. Detailed references to sources of data are given in the United Nations *Industrial Commodity Statistics Yearbook*.[21]

Table 54 refers to aluminium obtained by electrolytic reduction of alumina (primary) and remelting metal waste or scrap (secondary).

F. *Non-metallic mineral products and fabricated metal products, machinery and equipment*

Table 55 refers to total production of radio and television receivers of all kinds.

In *table 56*, passenger cars include three-and four-wheeled road motor vehicles other than motor-cycle combinations intended for the transport of passengers and seating not more than nine persons (including the driver), which are manufactured wholly or mainly from domestically-produced parts and passenger cars shipped in "knocked-down" form for assembly abroad.

In *table 57*, data refer to refrigerators of the compression type or of the absorption type, of the sizes commonly used in private households. Insulated cabinets to contain an active refrigerating element (block ice) but no machine are excluded.

In *table 58*, these washing machines usually include electrically-driven paddles or rotating cylinders (for keeping the cleaning solution circulating through the fabrics) or alternative devices. Washing machines with attached wringers or cetrifugal spin driers, and centrifugal spin driers designed as independent units, are included.

Table 59, *machine-tools* presented in this table include drilling and boring machines, lathes, milling machines, and metal-working presses. Drilling and boring machines refer to metal-working machines fitted with a baseplate, stand or other device for mounting on the floor, or on a bench, wall or another machine. Lathes refer to metal-working lathes of all kinds, whether or not automatic, including slide lathes, vertical lathes, capstan and turret lathes, production (or copying) lathes. Milling machines refer to metal-working machines designed to work a plane or profile surface by means of rotating tools, known as milling cutters. Metal-working presses are mechanical, hydraulic and pneumatic presses used for forging, stamping, cutting out etc. Forge hammers are excluded. Detailed product definitions are given in the United Nations *Industrial Commodity Statistics Yearbook* [21].

Au *tableau 52*, les savons se rapportent aux produits commercialemenmt désignés sous le nom de savon, y compris les savons durs et les savons mous. Cette rubrique comprend notamment: les savons de ménage, les savons de toilette, les savons translucides, les savons à barbe, les savons médicinaux, les savons désinfectants, les savons abrasifs, les savons de résines ou de naphténates et les savons industriels.

Les poudres pour lessives et les détersifs se rapportent aux produits organiques tensio—actifs, préparations tensio—actives et préparations pour lessives contenant ou non du savon.

E. *Industries métallurgiques de base*

Les données du *tableau 53* se rapportent à la production de fonte et d'acier. Les données sur l'acier brut comprennent les lingots et l'acier pour moulage. Dans certains cas, les données proviennent du United States of America Bureau of Mines (Washington, DC), de l'Instituto Latino-Americano del Ferro y el Acero (Santiago) et de la United States Economic Commission for Europe. Pour plus de détails sur les sources de données, se reporter à *l'Annuaire des statistiques industrielles par produit* des Nations Unies [21].

Le *Tableau 54* se rapporte à la production d'aluminium obtenue par réduction électrolytique de l'alumine (production primaire) et par refusion de déchets métalliques (production secondaire).

F. *Produits minéraux non métalliques et fabrications métallurgiques, machines et équipements*

Tableau 55 : Production totale de postes récepteurs de radiodiffusion et de télévision de toutes sortes.

Tableau 56 : Les voitures de tourime comprennent les véhicules automobiles routiers à trois ou quatre roues, autres que les motocycles, destinés au transport de passagers, dont le nombre de places assises (y compris celle du conducteur) n'est pas supérieur à neuf et qui sont construits entièrement ou principalement avec des pièces fabriqués dans le pays, et les voitures destinées au transport de passagers exportées en pièces détachées pour être montées à l'étranger.

Au *tableau 57*, les données se rapportent aux appareils frigorifiques du type à compression ou à absorption de la taille des appareils communément utilisés dans les ménages. Cette rubrique ne comprend pas les glacières conçues pour contenir un élément frigorifique actif (glace en bloc) mais non un équipement frigorifique.

Au *tableau 58*, ces machines à laver comprennent généralement des pales ou des cylindres rotatifs (destinés à assurer le brassage continu du liquide et du linge) ou des dispositifs à mouvements alternés, mus électriquement. Cette rubrique comprend les machines à laver avec essoreuses à rouleau ou essoreuses centrifuges et les essoreuses centrifuges conçues comme des appareils indépendants.

In *table 60, lorries assembled from imported parts* include road motor vehicles designed for the conveyance of goods, including vehicles specially equipped for the transport of certain goods, which are assembled wholly or mainly from imported parts. Articulated vehicles (that is, units made up of a road motor vehicle and a semi-trailer) are included. Ambulances, prison vans and special purpose lorries and vans, such as fire-engines are excluded.

Lorries produced include road motor vehicles designed for the conveyance of goods, including vehicles specially equipped for the transport of certain goods, which are manufactured wholly or mainly from domestically-produced parts. Articulated vehicles (that is, units made up of a road motor vehicle and a semi-trailer) are included. Ambulances, prison vans, and special purpose lorries and vans, such as fire-engines, are excluded.

Tableau 59, machines-outils présentés dans ce tableau comprennent les perceuses, tours, fraiseuses, et presses pour le travail des métaux. Perceuses se rapportent aux machines-outils pour le travail des métaux, munies d'un socle, d'un pied ou d'un autre dispositif permettant de les fixer au sol, à un établi, à une paroi ou à une autre machine. Tours se rapportent aux tours à métaux, de tous types, automatiques ou non, y compris les tours parallèles, les tours verticaux, les tours à revolver, les tours à reproduire. Fraiseuses se rapportent aux machines-outils pour le travail des métaux conçues pour usiner une surface plane ou un profil au moyen d'outils tournants appelés fraises. Presses pour le travail des métaux se rapportent aux presses à commande mécanique, hydraulique et penumatique servant à forger, à estamper, à matricer etc. Cette rubrique ne comprend pas les outils agissant par chocs. Pour plus de détails sur les description des produits se reporter à l'*Annuaire des statistiques industrielles par produit* [21] des Nations Unies.

Au *tableau 60, camions assemblés à partir de pièces importées* comprennent les véhicules automobiles routiers conçus pour le transport des marchandises, y compris les véhicules spécialement équipés pour le transport de certaines marchandises, et qui sont montés entièrement ou principalement avec des pièces importées. Les véhicules articulés (c'est-à-dire les ensembles composés d'un véhicule automobile routier et d'une semi-remorque) sont compris dans cette rubrique. Cette rubrique ne comprend pas les ambulances, les voitures cellulaires et les camions à usages spéciaux, tels que les voitures-pompes à incendie.

Camions fabriqués comprennent les véhicules automobiles routiers conçus pour le transport des marchandises, y compris les véhicules spécialement équipés pour le transport de certaines marchandises, et qui sont montés entièrement ou principalement avec des pièces importées. Les véhicules articulés (c'est-à-dire les ensembles composés d'un véhicule automobile routier et d'une semi-remorque) sont compris dans cette rubrique. Cette rubrique ne comprend pas les ambulances, les voitures cellulaires et les camions à usages spéciaux, tels que les voitures-pompes à incendie.

61
Railways: traffic
Chemins de fer : trafic
Passenger and net ton-kilometres: millions
Voyageurs et tonnes-kilomètres : millions

Country or area Pays ou zone	1984	1985	1986	1987	1988	1989	1990	1991	1992	1993
Albania Albanie										
Passenger-kilometres										
Voyageurs-kilomètres	581	564	619	662	703	753	779	...	...	...
Net ton-kilometres										
Tonnes-kilomètres	636	605	622	629	626	674	584	...	...	...
Algeria Algérie										
Passenger-kilometres										
Voyageurs-kilomètres	1 835	2 011	2 035	1 972	2 439	2 724	2 991	3 192	...	...
Net ton-kilometres										
Tonnes-kilomètres	2 631	3 048	2 934	2 937	2 814	2 698	2 690	2 710	...	...
Angola Angola										
Passenger-kilometres										
Voyageurs-kilomètres	300	331	326	...	...	...	...	...	...	...
Net ton-kilometres										
Tonnes-kilomètres	1 700	1 615	1 720	...	...	...	...	...	...	...
Argentina Argentine										
Passenger-kilometres										
Voyageurs-kilomètres	10 469	10 743	12 459	12 475	10 271	10 533	10 512	8 045	6 749	5 836
Net ton-kilometres										
Tonnes-kilomètres	11 208	9 501	8 761	7 952	8 983	8 237	7 578	5 460	4 388	4 477
Armenia Arménie										
Passenger-kilometres										
Voyageurs-kilomètres	485	490	486	493	417	381	316	320	446	435
Net ton-kilometres										
Tonnes-kilomètres	5 117	5 140	4 958	5 139	4 803	5 120	4 884	4 177	1 280	451
Austria Autriche										
Passenger-kilometres										
Voyageurs-kilomètres	7 210	7 499	7 542	7 568	7 994	8 663	9 017	9 428	9 799	9 599
Net ton-kilometres										
Tonnes-kilomètres	11 398	12 066	11 436	11 263	11 331	11 962	12 796	13 181	12 448	12 030
Bangladesh Bangladesh										
Passenger-kilometres[1]										
Voyageurs-kilomètres[1]	6 284	6 031	6 005	6 027	5 052	4 338	5 070	4 587	5 348	...
Net ton-kilometres[1]										
Tonnes-kilomètres[1]	779	813	612	503	678	666	643	651	718	...
Belarus Bélarus										
Passenger-kilometres[2]										
Voyageurs-kilomètres[2]	13 154	13 731	14 199	14 965	15 989	16 525	16 852	15 795	18 017	19 500
Net ton-kilometres										
Tonnes-kilomètres	71 603	73 213	77 943	79 862	82 231	81 734	75 430	65 551	56 441	42 919
Belgium Belgique										
Passenger-kilometres										
Voyageurs-kilomètres	6 444	6 572	6 069	6 270	6 348	6 400	6 539	6 771	6 798	6 694
Net ton-kilometres										
Tonnes-kilomètres	7 905	8 254	7 423	7 266	7 694	8 049	8 354	8 153	8 074	7 568
Benin Bénin										
Passenger-kilometres										
Voyageurs-kilomètres	162	150	167	120	...	...	...	...	...	...
Net ton-kilometres										
Tonnes-kilomètres	177	179	186	191	...	...	...	...	...	...
Bolivia Bolivie										
Passenger-kilometres										
Voyageurs-kilomètres	684	748	657	500	369	386	388	350	334	288
Net ton-kilometres										
Tonnes-kilomètres	548	494	464	505	424	512	541	683	710	692
Botswana Botswana										
Net ton-kilometres[3]										
Tonnes-kilomètres[3]	1 337	1 297	1 328	1 401	...	...	...	...	...	...

61

Railways: traffic
Passenger and net ton-kilometres: millions [*cont.*]
Chemins de fer : trafic
Voyageurs et tonnes-kilomètres : millions [*suite*]

Country or area Pays ou zone	1984	1985	1986	1987	1988	1989	1990	1991	1992	1993
Brazil Brésil										
Passenger-kilometres										
Voyageurs-kilomètres	15 578	16 362	15 782	15 273	13 891[4]	18 813[4]	18 202	18 374	15 118	...
Net ton-kilometres[5]										
Tonnes-kilomètres[5]	92 167	99 881	103 877	109 433	119 754	125 046	120 439	121 431	116 569	...
Bulgaria Bulgarie										
Passenger-kilometres										
Voyageurs-kilomètres	7 538	7 785	8 004	8 075	8 143	7 601	7 793	4 866	5 393	5 837
Net ton-kilometres[5]										
Tonnes-kilomètres[5]	18 134	18 172	18 327	17 842	17 585	17 034	14 132	8 685	7 758	7 702
Cameroon Cameroun										
Passenger-kilometres										
Voyageurs-kilomètres	492	440	433	466	495	458	442	530	445	352
Net ton-kilometres										
Tonnes-kilomètres	868	999	758	622	685	743	684	679	613	653
Canada Canada										
Passenger-kilometres										
Voyageurs-kilomètres	2 915	3 040	2 831	2 709	2 989	3 178	2 004	1 426	...	...
Net ton-kilometres[5]										
Tonnes-kilomètres[5]	257 561	245 284	246 722	272 122	274 571	252 075	250 117	262 425	...	...
Chile Chili										
Passenger-kilometres										
Voyageurs-kilomètres	1 424	1 522	1 274	1 176	1 013	1 058	1 077	1 125	1 010	938
Net ton-kilometres										
Tonnes-kilomètres	2 315	2 577	2 555	2 657	2 809	2 946	2 787	2 717	2 715	2 464
China Chine										
Passenger-kilometres[6]										
Voyageurs-kilomètres[6]	204 600	241 600	258 696	284 304	326 000	303 700	261 263	282 810	315 224	348 330
Net ton-kilometres[6]										
Tonnes-kilomètres[6]	724 800	812 600	876 504	947 196	987 740	1 039 423	1 062 238	1 097 200	1 157 555	1 195 464
Colombia Colombie										
Passenger-kilometres										
Voyageurs-kilomètres	189	228	178	171	148	152	141	79	16	0
Net ton-kilometres[5]										
Tonnes-kilomètres[5]	733	777	691	563	464	361	391	298	243	459
Congo Congo										
Passenger-kilometres										
Voyageurs-kilomètres	408	437	456	400	419	434	410	435	...	...
Net ton-kilometres										
Tonnes-kilomètres	477	518	536	449	477	467	421	397	...	...
Côte d'Ivoire Côte d'Ivoire										
Passenger-kilometres[7]										
Voyageurs-kilomètres[7]	996	1 008	1 015	* 1 021	...	...	...	...	...	...
Net ton-kilometres[5][7]										
Tonnes-kilomètres[5][7]	530	544	562	578	...	...	...	...	...	...
Croatia Croatie										
Passenger-kilometres										
Voyageurs-kilomètres	..	..	..	..	..	..	..	* 1 500	* 984	* 1 056
Net ton-kilometres										
Tonnes-kilomètres	..	..	..	..	..	..	..	* 3 612	* 1 776	* 1 680
Cuba Cuba										
Passenger-kilometres										
Voyageurs-kilomètres	2 360	2 257	2 200	2 189	2 627	2 891	...	...	...	...
Net ton-kilometres										
Tonnes-kilomètres	2 304	2 409	2 155	2 105	2 087	2 048	...	...	...	...
Czech Republic République tchèque										
Passenger-kilometres										
Voyageurs-kilomètres	..	..	..	..	..	..	..	..	11 753	8 548

61
Railways: traffic
Passenger and net ton-kilometres: millions [*cont.*]
Chemins de fer : trafic
Voyageurs et tonnes-kilomètres : millions [*suite*]

Country or area Pays ou zone	1984	1985	1986	1987	1988	1989	1990	1991	1992	1993
Net ton-kilometres Tonnes-kilomètres	..	..	..	..	..	..	..	..	31 116	25 579
former Czechoslovakia† anc. Tchécoslovaquie†										
Passenger-kilometres Voyageurs-kilomètres	19 323	19 839	19 935	20 029	19 408	19 669	19 335	19 263	..	..
Net ton-kilometres Tonnes-kilomètres	74 015	73 598	75 152	73 525	75 294	71 985	64 326	49 933	..	..
Denmark Danemark										
Passenger-kilometres[8] Voyageurs-kilomètres[8]	4 618	4 910	4 876	4 860	4 850	4 733	4 851	4 777	4 600	...
Net ton-kilometres[9] Tonnes-kilomètres[9]	1 670	1 768	1 800	1 644	1 671	1 723	1 787	1 858	1 870	...
Ecuador Equateur										
Passenger-kilometres Voyageurs-kilomètres	49	53	55	63	77	83	82	53	53	39
Net ton-kilometres Tonnes-kilomètres	6	9	7	8	8	6	5	2	3	3
Egypt Egypte										
Passenger-kilometres[3] Voyageurs-kilomètres[3]	15 627	16 853	18 485	23 796	24 929	27 083	28 684	20 950	36 644	...
Net ton-kilometres[3] Tonnes-kilomètres[3]	2 597	2 756	2 908	3 021	4 348	2 853	3 045	3 162	3 229	...
El Salvador El Salvador										
Passenger-kilometres Voyageurs-kilomètres	5	5	5	6	6	5	6	8	6	...
Net ton-kilometres Tonnes-kilomètres	24	25	24	39	36	22	38	35	38	...
Estonia Estonie										
Passenger-kilometres Voyageurs-kilomètres	1 672	1 649	1 721	1 764	1 722	1 562	1 510	1 273	950	722
Net ton-kilometres Tonnes-kilomètres	6 515	6 446	6 736	7 134	7 989	9 609	6 977	6 545	3 646	4 152
Ethiopia Ethiopie										
Passenger-kilometres[10 11] Voyageurs-kilomètres[10 11]	268	275	276	315	342	298	277	291	204	230
Net ton-kilometres Tonnes-kilomètres	117	144	166	150	141	129	126	122	84	112
Finland Finlande										
Passenger-kilometres Voyageurs-kilomètres	3 276	3 224	2 676	3 106	3 147	3 208	3 331	3 230	3 057	3 007
Net ton-kilometres[12] Tonnes-kilomètres[12]	7 979	8 066	6 951	7 402	7 815	7 958	8 357	7 634	7 848	9 259
France France										
Passenger-kilometres Voyageurs-kilomètres	60 390	62 070	59 860	59 970	63 290	64 490	63 740	62 300	62 980	58 430
Net ton-kilometres[13] Tonnes-kilomètres[13]	57 470	55 780	51 690	51 330	52 290	53 270	51 530	51 480	50 400	45 900
Georgia Géorgie										
Passenger-kilometres Voyageurs-kilomètres	3 678	3 724	3 684	3 614	3 442	2 858	2 497	2 135	1 210	1 003
Net ton-kilometres Tonnes-kilomètres	13 316	13 487	13 132	12 500	13 020	12 671	12 355	9 916	3 677	1 750
Germany † Allemagne† F. R. Germany R. f. Allemagne										
Passenger-kilometres Voyageurs-kilomètres	39 575	43 451	42 129	39 965	41 760	42 023	44 588	46 711	47 389	48 150
Net ton-kilometres Tonnes-kilomètres	59 835	63 873	57 916	58 947	59 922	61 981	61 864	63 558	57 819	53 071

61
Railways: traffic
Passenger and net ton-kilometres: millions [*cont.*]
Chemins de fer : trafic
Voyageurs et tonnes-kilomètres : millions [*suite*]

Country or area Pays ou zone	1984	1985	1986	1987	1988	1989	1990	1991	1992	1993
former German D. R. anc. R. d. allemande										
Passenger-kilometres										
Voyageurs-kilomètres	22 919	22 451	22 402	22 563	22 775	23 588	17 397	10 323	9 851	9 853
Net ton-kilometres										
Tonnes-kilomètres	55 422	57 582	57 916	58 096	59 374	58 027	40 229	18 662	15 029	13 575
Ghana Ghana										
Passenger-kilometres										
Voyageurs-kilomètres	157	244	294	318	...	...	...	...	...	...
Net ton-kilometres										
Tonnes-kilomètres	44	80	101	114	...	...	...	...	...	...
Greece Grèce										
Passenger-kilometres										
Voyageurs-kilomètres	1 652	1 732	1 950	1 973	1 963	2 011	1 978	1 995	2 004	1 726
Net ton-kilometres[14]										
Tonnes-kilomètres[14]	770	733	702	599	604	657	647	606	563	523
Guatemala Guatemala										
Passenger-kilometres										
Voyageurs-kilomètres	47 826	27 967	19 874	15 596	9 094	10 213	15 960	12 531	...	...
Net ton-kilometres										
Tonnes-kilomètres	51 895	72 725	78 218	74 859	48 185	51 546	44 134	47 233	...	...
Hong Kong Hong-kong										
Passenger-kilometres										
Voyageurs-kilomètres	1 480	1 781	1 933	2 136	2 360	2 469	2 533	2 912	3 121	3 269
Net ton-kilometres										
Tonnes-kilomètres	47	52	69	72	70	69	70	65	61	51
Hungary Hongrie										
Passenger-kilometres										
Voyageurs-kilomètres	10 511	10 463	10 452	10 486	10 758	10 414	11 403[15]	9 861	9 184	8 432
Net ton-kilometres										
Tonnes-kilomètres	22 308	21 814	22 095	21 253	20 573	19 364	16 781[16]	11 938	10 015	7 708
India Inde										
Passenger-kilometres[17]										
Voyageurs-kilomètres[17]	226 582	240 614	256 535	269 389	263 731	280 848	295 644	314 564	...	...
Net ton-kilometres[17]										
Tonnes-kilomètres[17]	172 632	196 600	214 096	222 528	222 374	229 602	235 785	250 238	...	...
Indonesia Indonésie										
Passenger-kilometres										
Voyageurs-kilomètres	6 379	6 774	7 327	7 516	7 863	8 426	9 290	9 514	10 458	12 337
Net ton-kilometres										
Tonnes-kilomètres	1 173	1 333	1 465	1 759	2 359	2 921	3 190	3 470	3 779	3 955
Iran, Islamic Rep. of Iran, Rép. islamique d'										
Passenger-kilometres										
Voyageurs-kilomètres	6 130	5 585	4 638	3 674	4 661	4 752	4 573	4 585	5 298	6 422
Net ton-kilometres										
Tonnes-kilomètres	7 566	6 888	7 316	8 625	8 047	7 963	9 409	7 701	8 002	9 124
Iraq Iraq										
Passenger-kilometres										
Voyageurs-kilomètres	1 227	1 118	1 005	1 150	1 570	...	...	...	...	...
Net ton-kilometres[18]										
Tonnes-kilomètres[18]	1 245	1 245	1 294	1 534	2 023	...	...	...	...	...
Ireland Irlande										
Passenger-kilometres										
Voyageurs-kilomètres	903	1 023	1 075	1 191	1 181	1 226	1 223	1 243	1 222	1 079
Net ton-kilometres										
Tonnes-kilomètres	601	601	574	563	545	556	582	600	633	575
Israel Israël										
Passenger-kilometres										
Voyageurs-kilomètres	215	209	179	173	166	151	168	192	...	...

61
Railways: traffic
Passenger and net ton-kilometres: millions [*cont.*]
Chemins de fer : trafic
Voyageurs et tonnes-kilomètres : millions [*suite*]

Country or area Pays ou zone	1984	1985	1986	1987	1988	1989	1990	1991	1992	1993
Net ton-kilometres Tonnes-kilomètres	970	942	952	1 062	1 032	1 016	1 044	1 092	...	...
Italy Italie										
Passenger-kilometres Voyageurs-kilomètres	39 045	39 194	40 500	41 395	43 343	44 443	45 513	46 427	48 361	...
Net ton-kilometres[18] Tonnes-kilomètres[18]	17 870	18 024	17 410	18 625	19 567	20 587	21 217	21 680	21 830	...
Jamaica Jamaïque										
Passenger-kilometres[19] Voyageurs-kilomètres[19]	28	25	42	29	22	24	-	-	-	-
Net ton-kilometres[19] Tonnes-kilomètres[19]	...	62	117	123	72	18	-	-	-	-
Japan Japon										
Passenger-kilometres Voyageurs-kilomètres	324 991	328 450	333 425	341 439	356 468	369 642	383 735	396 472	403 245	401 864
Net ton-kilometres Tonnes-kilomètres	23 191	22 099	20 917	20 307	22 911	24 767	26 656	27 292	26 899	25 619
Jordan Jordanie										
Passenger-kilometres Voyageurs-kilomètres	1	1	1	1	1	1	2	2	2	2
Net ton-kilometres Tonnes-kilomètres	870	692	771	720	625	610	711	791	797	711
Kazakhstan Kazakhstan										
Passenger-kilometres Voyageurs-kilomètres	15 019	15 749	16 922	17 888	18 637	18 921	19 734	19 365	19 700	20 500
Net ton-kilometres Tonnes-kilomètres	373 074	382 507	397 907	404 583	416 875	409 573	406 963	374 230	286 100	190 000
Kenya Kenya										
Passenger-kilometres Voyageurs-kilomètres	4 842	4 842	1 478	1 558	828	732	699	658	563	464
Net ton-kilometres Tonnes-kilomètres	2 034	1 858	1 831	1 702	1 755	1 910	1 808	1 865	1 627	1 312
Korea, Republic of Corée, République de										
Passenger-kilometres Voyageurs-kilomètres	21 884	22 595	23 563	24 457	25 978	27 390	29 864	33 470	34 787	33 693
Net ton-kilometres Tonnes-kilomètres	12 033	12 296	12 813	13 061	13 784	13 605	13 663	14 494	14 256	14 658
Latvia Lettonie										
Passenger-kilometres[20] Voyageurs-kilomètres[20]	5 310	5 214	5 573	5 679	5 761	5 449	5 366	3 930	3 656	2 359
Net ton-kilometres[9] Tonnes-kilomètres[9]	20 799	19 933	20 691	21 380	21 689	21 132	18 538	16 739	10 115	9 852
Lithuania Lituanie										
Passenger-kilometres Voyageurs-kilomètres	3 377	3 417	3 494	3 565	3 665	3 470	3 640	3 225	2 740	2 700
Net ton-kilometres Tonnes-kilomètres	19 999	20 927	21 076	21 205	22 595	21 749	19 258	17 748	11 337	11 030
Luxembourg Luxembourg										
Passenger-kilometres Voyageurs-kilomètres	286	283	278	269	277	280	261	...	...	...
Net ton-kilometres Tonnes-kilomètres	584	645	604	593	639	704	709	713	672	647
Madagascar Madagascar										
Passenger-kilometres Voyageurs-kilomètres	202	178	208	209	242	204	198	...	...	...
Net ton-kilometres[5] Tonnes-kilomètres[5]	226	208	188	174	174	207	209	...	...	...

61

Railways: traffic
Passenger and net ton-kilometres: millions [cont.]
Chemins de fer : trafic
Voyageurs et tonnes-kilomètres : millions [suite]

Country or area Pays ou zone	1984	1985	1986	1987	1988	1989	1990	1991	1992	1993
Malawi Malawi										
Passenger-kilometres [17] Voyageurs-kilomètres [17]	109	123	112	113	113	112	107	92	* 72	...
Net ton-kilometres [17] Tonnes-kilomètres [17]	121	94	125	107	77	65	79	59	* 60	...
Malaysia Malaisie										
Passenger-kilometres [21] Voyageurs-kilomètres [21]	1 512	1 409	1 369	1 425	1 518	1 701	1 830	1 763	1 618	1 543
Net ton-kilometres [21] Tonnes-kilomètres [21]	1 077	1 018	1 042	1 119	1 326	1 361	1 405	1 262	1 081	1 157
Mali Mali										
Passenger-kilometres Voyageurs-kilomètres	149	173	177	196	177	184	184	...	...	...
Net ton-kilometres Tonnes-kilomètres	239	241	225	199	227	279	273	...	...	...
Mauritania Mauritanie										
Net ton-kilometres Tonnes-kilomètres	14 217	13 929	13 396	13 434	14 931	16 623	...	...	...	
Mexico Mexique										
Passenger-kilometres Voyageurs-kilomètres	5 951	6 015	5 874	5 828	5 619	5 383	5 336	4 725	4 794	3 219
Net ton-kilometres Tonnes-kilomètres	44 592	45 306	40 608	40 475	41 177	38 570	36 408	32 986	34 229	35 901
Mongolia Mongolie										
Passenger-kilometres Voyageurs-kilomètres	420	436	467	486	531	579	...	...	...	...
Net ton-kilometres Tonnes-kilomètres	5 121	5 960	6 333	6 180	6 241	5 956	...	...	...	...
Morocco Maroc										
Passenger-kilometres [22] Voyageurs-kilomètres [22]	1 620	1 933	1 958	2 069	2 092	2 168	2 237	2 345	2 233	1 904
Net ton-kilometres [22] Tonnes-kilomètres [22]	4 517	4 562	4 953	4 880	5 706	4 519	5 107	4 523	5 001	4 415
Mozambique Mozambique										
Passenger-kilometres Voyageurs-kilomètres	284	161	183	105	75	74	...	...	...	...
Net ton-kilometres Tonnes-kilomètres	538	290	301	353	306	403	...	...	...	...
Myanmar Myanmar										
Passenger-kilometres Voyageurs-kilomètres	3 859	3 834	3 554	4 486	3 830	3 920	4 374	4 482	* 4 608	* 4 680
Net ton-kilometres [5] Tonnes-kilomètres [5]	640	600	507	545	356	458	524	579	* 600	* 648
Netherlands Pays-Bas										
Passenger-kilometres Voyageurs-kilomètres	8 790	9 007	8 919	9 396	9 664	10 235	11 060	12 796	15 350	15 245
Net ton-kilometres Tonnes-kilomètres	3 103	3 274	3 050	3 010	3 194	3 108	3 070	3 033	2 802	2 703
New Zealand Nouvelle-Zélande										
Net ton-kilometres Tonnes-kilomètres	3 165 [23]	3 192 [23]	3 051 [23]	2 912 [23]	2 924 [23]	2 682 [23]	2 744 [3]	2 364 [3]	2 475 [3]	2 629 [3]
Nicaragua Nicaragua										
Passenger-kilometres Voyageurs-kilomètres	...	26	22	15	14	12	3	3	6	...
Net ton-kilometres * Tonnes-kilomètres *	...	4	...	...	...	...	...	...	...	...

61
Railways: traffic
Passenger and net ton-kilometres: millions [*cont.*]
Chemins de fer : trafic
Voyageurs et tonnes-kilomètres : millions [*suite*]

Country or area Pays ou zone	1984	1985	1986	1987	1988	1989	1990	1991	1992	1993
Nigeria Nigéria										
Passenger-kilometres										
Voyageurs-kilomètres	3 310	3 555	3 117	2 813	1 902	997	453	...	434	555
Net ton-kilometres										
Tonnes-kilomètres	1 644	1 709	1 725	1 743	...	...	...	...	120	161
Norway Norvège										
Passenger-kilometres										
Voyageurs-kilomètres	2 198	2 241	2 225	2 187	2 110	2 136	2 136	2 153	2 201	2 316
Net ton-kilometres										
Tonnes-kilomètres	2 650	2 932	3 015	2 822	2 617	2 780	2 354	2 675	2 294	2 872
Pakistan Pakistan										
Passenger-kilometres[1]										
Voyageurs-kilomètres[1]	17 806	16 848	16 919	18 544	19 732	20 373	19 963	18 159	16 759	16 274
Net ton-kilometres[1]										
Tonnes-kilomètres[1]	7 203	8 272	7 819	8 033	8 364	7 226	5 704	5 964	5 860	5 940
Panama Panama										
Passenger-kilometres										
Voyageurs-kilomètres	...	...	...	...	...	41 416	44 260	62 951	54 234	46 440
Net ton-kilometres										
Tonnes-kilomètres	...	...	...	...	...	35 299	30 020	37 856	48 380	42 482
Paraguay Paraguay										
Passenger-kilometres										
Voyageurs-kilomètres	6	2	2	2	2	2	2	1	1	1
Net ton-kilometres										
Tonnes-kilomètres	17	13	17	17	18	14	4	3	3	3
Peru Pérou										
Passenger-kilometres[5]										
Voyageurs-kilomètres[5]	483	476	490	594	596	659	469	320	226	165
Net ton-kilometres[5]										
Tonnes-kilomètres[5]	1 061	1 049	1 022	1 148	971	929	848	825	817	844
Philippines Philippines										
Passenger-kilometres										
Voyageurs-kilomètres	230	146	173	219	230	230	264	228	121	102
Net ton-kilometres										
Tonnes-kilomètres	17	13	15	16	15	13	36	12	9	5
Poland Pologne										
Passenger-kilometres										
Voyageurs-kilomètres	53 179	51 978	48 526	48 285	52 134	55 888	50 373	40 115	32 571	30 865
Net ton-kilometres										
Tonnes-kilomètres	123 503	120 642	121 775	121 381	122 204	111 140	83 530	65 146	73 333	78 437
Portugal Portugal										
Passenger-kilometres										
Voyageurs-kilomètres	5 456	5 725	5 803	5 907	6 036	5 908	5 664	5 692	...	...
Net ton-kilometres										
Tonnes-kilomètres	1 239	1 306	1 448	1 615	1 708	1 719	1 588	1 784	...	...
Republic of Moldova République de Moldova										
Passenger-kilometres										
Voyageurs-kilomètres	1 587[2]	1 648[2]	1 685	1 766[2]	1 616	1 514	1 464	1 280	1 718[2]	1 661[2]
Net ton-kilometres										
Tonnes-kilomètres	16 838	16 614	16 890	15 820	15 989	15 632	15 007	11 883	7 861	4 965
Romania Roumanie										
Passenger-kilometres[24]										
Voyageurs-kilomètres[24]	28 785	31 082	32 304	33 520	34 643	35 456	30 582	25 429	24 269	19 402
Net ton-kilometres[5]										
Tonnes-kilomètres[5]	75 159	74 215	79 092	78 070	80 607	81 131	57 253	37 853	28 170	25 170
Russian Federation Fédération de Russie										
Passenger-kilometres										
Voyageurs-kilomètres	240 100	246 300	257 900	264 300	272 500	270 100	274 400	255 000	253 200	272 200

61
Railways: traffic
Passenger and net ton-kilometres: millions [*cont.*]
Chemins de fer : trafic
Voyageurs et tonnes-kilomètres : millions [*suite*]

Country or area Pays ou zone	1984	1985	1986	1987	1988	1989	1990	1991	1992	1993
Net ton-kilometres Tonnes-kilomètres	244 100	2 506 000	2 585 000	2 581 000	2 606 000	2 557 000	2 523 000	2 326 000	1 967 000	1 608 000
Saudi Arabia Arabie saoudite										
Passenger-kilometres Voyageurs-kilomètres	79	72	71	81	92	156	141	145	126	139
Net ton-kilometres Tonnes-kilomètres	525	415	321	491	470	797	351	480	453	451
Senegal Sénégal										
Passenger-kilometres Voyageurs-kilomètres	140	139	143	155	139	143	183	173	...	...
Net ton-kilometres Tonnes-kilomètres	400	468	492	524	478	535	612	485	...	...
South Africa Afrique du Sud										
Passenger-kilometres[23][25] Voyageurs-kilomètres[23][25]	27 416	24 009	23 032	21 413	...	138 555	120 542	103 781	89 466	71 573
Net ton-kilometres[23][25] Tonnes-kilomètres[23][25]	83 329	90 162	93 331	92 178	...	95 736	90 716	85 524	88 586	89 716
Spain Espagne										
Passenger-kilometres Voyageurs-kilomètres	16 571	17 066	16 776	16 602	16 983	15 999	16 733	16 333	* 16 368	...
Net ton-kilometres[5] Tonnes-kilomètres[5]	12 077	12 077	11 766	11 942	12 145	12 048	11 613	10 755	* 9 360	...
Sri Lanka Sri Lanka										
Passenger-kilometres[26] Voyageurs-kilomètres[26]	2 280	2 111	1 972	1 881	1 859	1 734	2 781	2 698	2 613	2 852
Net ton-kilometres[26] Tonnes-kilomètres[26]	263	247	203	195	197	162	164	170	177	168
Sudan Soudan										
Passenger-kilometres Voyageurs-kilomètres	836	849	...	...	...	...	...	...	...	...
Net ton-kilometres Tonnes-kilomètres	1 770	1 860	...	...	...	...	...	...	...	...
Sweden Suède										
Passenger-kilometres Voyageurs-kilomètres	6 690	6 803	6 363	6 215	6 289	6 361	6 353	5 745	5 583	5 975
Net ton-kilometres Tonnes-kilomètres	17 776	18 420	18 553	18 406	18 687	19 156	19 599	18 815	18 609	17 337
Switzerland Suisse										
Passenger-kilometres Voyageurs-kilomètres	9 780	10 163	10 095	11 695	11 998	12 283	12 678	13 834	13 209	...
Net ton-kilometres Tonnes-kilomètres	7 224	7 379	7 279	7 124	7 875	8 560	8 794	8 659	8 212	7 821
Syrian Arab Republic Rép. arabe syrienne										
Passenger-kilometres Voyageurs-kilomètres	757	944	904	1 029	1 133	1 113	1 140	1 314	1 254	855
Net ton-kilometres Tonnes-kilomètres	966	1 251	1 418	1 508	1 569	1 350	1 265	1 238	1 699	1 097
Tajikistan Tadjikistan										
Passenger-kilometres Voyageurs-kilomètres	..	..	..	..	..	..	..	..	103	...
Net ton-kilometres Tonnes-kilomètres	..	..	..	..	..	..	..	..	604	...
Thailand Thaïlande										
Passenger-kilometres[26] Voyageurs-kilomètres[26]	9 643	9 140	9 274	9 583	10 301	10 935	11 612	12 820	14 136	14 587
Net ton-kilometres[26] Tonnes-kilomètres[26]	2 618	2 718	2 583	2 729	2 867	3 065	3 291	3 365	3 075	3 059
Togo Togo										
Passenger-kilometres Voyageurs-kilomètres	100	102	109	117	...	...	...	...	...	...

61
Railways: traffic
Passenger and net ton-kilometres: millions [*cont.*]
Chemins de fer : trafic
Voyageurs et tonnes-kilomètres : millions [*suite*]

Country or area Pays ou zone	1984	1985	1986	1987	1988	1989	1990	1991	1992	1993
Net ton-kilometres[9] Tonnes-kilomètres[9]	10	10	11	12	...	...	...	...	...	...
Tunisia Tunisie										
Passenger-kilometres[14] Voyageurs-kilomètres[14]	742	744	750	798	1 014	1 039	1 019	1 020	* 1 200	...
Net ton-kilometres[5 27] Tonnes-kilomètres[5 27]	1 695	1 710	1 877	1 986	2 156	2 064	1 834	1 813	* 2 016	...
Turkey Turquie										
Passenger-kilometres Voyageurs-kilomètres	6 277	6 489	6 052	6 174	6 708	6 845	6 410	6 048	6 259	7 147
Net ton-kilometres Tonnes-kilomètres	7 679	7 959	7 396	7 403	8 149	7 707	8 031	8 093	8 383	8 517
Uganda Ouganda										
Passenger-kilometres Voyageurs-kilomètres	246	234	195	212	118	69	108	60	...	...
Net ton-kilometres Tonnes-kilomètres	79	60	71	77	83	90	103	139	...	...
Ukraine Ukraine										
Passenger-kilometres Voyageurs-kilomètres	64 799	66 954	68 580	71 425	72 859	73 218	76 038	70 968	76 196	75 890
Net ton-kilometres Tonnes-kilomètres	499 320	497 916	506 123	496 001	504 689	497 333	473 953	402 290	337 761	246 356
United Kingdom Royaume-Uni										
Passenger-kilometres[17 28] Voyageurs-kilomètres[17 28]	29 508	30 381	31 099	33 140	34 322	33 648	33 191	32 466	31 418	30 354
Net ton-kilometres[17 28] Tonnes-kilomètres[17 28]	11 774	16 000	16 600	17 451	18 100	16 400	16 000	15 300	15 486	13 765
United Rep.Tanzania Rép. Unie de Tanzanie										
Passenger-kilometres Voyageurs-kilomètres	1 192	1 194	1 024	647	855	832	809	990	...	...
Net ton-kilometres Tonnes-kilomètres	565	660	814	789	936	990	956	983	...	...
United States Etats-Unis										
Passenger-kilometres[29] Voyageurs-kilomètres[29]	16 564	17 649	8 069	8 639	9 154	9 402	9 726	10 101	9 800	9 974
Net ton-kilometres[30] Tonnes-kilomètres[30]	1 345 488	1 280 394	1 283 736	1 388 388	1 603 853	1 632 284	1 664 690	1 672 589	1 557 180	1 619 258
Uruguay Uruguay										
Passenger-kilometres[31] Voyageurs-kilomètres[31]	331	241	196	140	-	-	-	-	-	-
Net ton-kilometres Tonnes-kilomètres	273	185	210	210	213	243	204	203	215	178
Venezuela Venezuela										
Passenger-kilometres Voyageurs-kilomètres	12	8	17	22	29	38	64	55	47	44
Net ton-kilometres Tonnes-kilomètres	11	14	12	18	40	39	35	40	36	26
Yugoslavia, SFR† Yougoslavie, Rfs†										
Passenger-kilometres Voyageurs-kilomètres	4 063	4 198	4 542	4 356	4 213	4 654	4 794	2 935	..	..
Net ton-kilometres[5] Tonnes-kilomètres[5]	9 864	9 638	9 259	8 738	8 395	8 707	7 744	5 760	..	..
Zaire Zaïre										
Passenger-kilometres Voyageurs-kilomètres	450	504	511	522	465	467	469	...	...	...

61
Railways: traffic
Passenger and net ton-kilometres: millions [cont.]
Chemins de fer : trafic
Voyageurs et tonnes-kilomètres : millions [suite]

Country or area Pays ou zone	1984	1985	1986	1987	1988	1989	1990	1991	1992	1993
Net ton-kilometres Tonnes-kilomètres	1 509	1 599	1 624	1 681	1 670	1 687	1 655	...	...	...
Zambia Zambie										
Passenger-kilometres Voyageurs-kilomètres	475	488	496	510	512	...	...	...	...	...
Net ton-kilometres Tonnes-kilomètres	1 417	1 401	1 407	1 420	1 431	...	...	...	...	...
Zimbabwe Zimbabwe										
Net ton-kilometres [3][32] Tonnes-kilomètres [3][32]	6 408	6 200	6 574	5 451	5 551	5 287	5 590	5 413	5 887	4 581

Source:
Transport statistics database of the Statistical Division of
the United Nations Secretariat.

† For detailed descriptions of data pertaining to
former Czechoslovakia, Germany, SFR Yugoslavia and former
USSR, see Annex I - Country or area nomenclature, regional
and other groupings.

1 Twelve months beginning 1 July of year stated.

2 Including passengers carried without revenues.
3 Twelve months ending 30 June of year stated.
4 Including urban railways traffic.

5 Including service traffic.
6 May include service traffic.
7 Abidjan-Ouagadougou line, which lies in Burkina Faso.

8 Including ferry traffic.
9 Including passengers' baggage and parcel post (Denmark and
Latvia: also mail).
10 Including traffic of Djibouti portion of Djibouti-Addis
Ababa line.
11 Twelve months beginning 8 July of year stated.
12 Beginning 1984, excluding local transport.
13 Including passengers' baggage.
14 Including military traffic (Greece: also government
traffic).
15 Including military, government and railway personnel.

16 Excluding suburban railways.
17 Twelve months beginning 1 April of year stated.
18 Excluding livestock.
19 Beginning 1990, railway closed.
20 Including railway personnel.
21 Peninsular Malaysia only.
22 Principal railways.
23 Twelve months ending 31 March of year stated.
24 Including military and government personnel.
25 Including Namibia.
26 Twelve months ending 30 September of year stated.
27 Ordinary goods only.
28 Excluding Northern Ireland.
29 Excluding commuter railroads beginning 1986.

30 Class I railways only.
31 Beginning 1988, passenger transport suspended.
32 Including traffic in Botswana.

Source:
Base de données pour les statistiques des transports de la
Division de statistique du Secrétariat de l'ONU.

† Pour les descriptions en détails des données
relatives à l'ancienne Tchécoslovaquie, l'Allemagne, la Rfs
Yougoslavie et l'ancienne URSS, voir l'Annexe I -
Nomenclature des pays ou zones, groupements régionaux et
autres groupements.

1 Douze mois commençant le premier juillet de l'année
indiquée.
2 Y compris passagers transportés gratuitement.
3 Douze mois finissant le 30 juin de l'année indiquée.
4 Y compris les lignes situées a l'intérieur d'une
agglomération urbaine.
5 Y compris le trafic de service.
6 Le trafic de service peut être compris.
7 Ligne Abidjan-Ouagadougou dont un tronçon passe en Burkina
Faso.
8 Y compris le trafic par ferry.
9 Y compris les bagages des voyageurs et les colis postaux
(Danemark et Lettonie: courrier aussi).
10 Y compris le trafic de la ligne Djibouti-Addis Abéba en
Djibouti.
11 Douze mois commençant le 8 juillet de l'année indiquée.
12 A compter de 1984 non compris le transport local.
13 Y compris les bagages des voyageurs.
14 Y compris le trafic militaire (Grèce: et de l'Etat aussi).
15 Y compris le militaires, les fonctionnaires et le personnel
de chemin de fer.
16 Non compris les lignes de banlieues.
17 Douze mois commençant le premier avril de l'année indiquée.
18 Non compris le bétail.
19 A compter de 1990, les chemins de fer fermés.
20 Y compris le personnel de chemins de fer.
21 Malasie péninsulaire seulement.
22 Chemins de fer principaux.
23 Douze mois finissant le 31 mars de l'année indiquée.
24 Y compris les militaires et les fonctionnaires.
25 Y compris Namibie.
26 Douze mois finissant le 30 septembre de l'année indiquée.
27 Petite vitesse seulement.
28 Non compris l'Irlande du Nord.
29 A compter de 1986 non compris les chemins de fer de
banlieue.
30 Réseaux de catégorie 1 seulement.
31 Transport passager interrompu à partir de 1988.
32 Y compris le trafic en Botswana.

62
Motor vehicles in use
Véhicules automobiles en circulation
Passenger cars and commercial vehicles: thousand units
Voitures de tourisme et véhicules utilitaires : milliers de véhicules

Country or area Pays or zone	1984	1985	1986	1987	1988	1989	1990	1991	1992	1993
World Monde										
Passenger cars[1]										
Voitures de tourisme[1]	364 042.2	373 667.2	393 351.7	395 129.4	407 959.3	422 240.3	438 525.0	450 596.3	451 927.6	464 514.4
Commercial vehicles[1]										
Véhicules utilitaires[1]	109 891.4	115 164.7	120 859.7	124 036.9	129 592.7	133 831.8	139 856.6	140 969.3	141 929.8	144 680.9
Afghanistan Afghanistan										
Passenger cars[1]										
Voitures de tourisme[1]	32.8	32.0	31.1	31.0	31.0	31.0	31.0	31.0	31.0	...
Commercial vehicles[1]										
Véhicules utilitaires[1]	26.2	25.5	24.8	24.8	25.0	25.0	25.0	25.0	25.0	...
Albania Albanie										
Commercial vehicles										
Véhicules utilitaires	2.8	2.5	2.5	2.5	2.8	2.8	* 2.8	...	...	...
Algeria Algérie										
Passenger cars										
Voitures de tourisme	582.0	611.0	639.0	667.0	725.0[1]	725.0[1]	725.0[1]	725.0[1]	725.0[1]	...
Commercial vehicles										
Véhicules utilitaires	275.0	300.0	317.0	324.0	480.0[1]	480.0[1]	480.0[1]	480.0[1]	480.0[1]	...
American Samoa Samoa américaines										
Passenger cars										
Voitures de tourisme	3.4	4.0	...	...	...	4.9	...	3.9	5.0	4.6
Commercial vehicles										
Véhicules utilitaires	0.4	0.5	...	...	...	0.2	...	0.3	0.3	0.4
Angola Angola										
Passenger cars[1]										
Voitures de tourisme[1]	129.0	125.9	122.4	122.0	122.0	122.0	122.0	122.0	122.0	...
Commercial vehicles[1]										
Véhicules utilitaires[1]	42.8	42.3	41.5	41.0	41.0	41.0	41.0	41.0	42.2	...
Antigua and Barbuda Antigua-et-Barbuda										
Passenger cars										
Voitures de tourisme	9.6	11.0	12.4	14.2	15.4	17.1	18.1	19.2	13.5[2]	...
Commercial vehicles										
Véhicules utilitaires	1.7	2.0	2.4	2.7	3.3	3.9	4.3	3.8	3.5[2]	...
Argentina Argentine										
Passenger cars										
Voitures de tourisme	3 759.3	3 878.2	4 037.9	4 137.2	4 079.8	4 235.1	4 283.7	* 4 405.0	* 4 809.0	* 4 856.0
Commercial vehicles										
Véhicules utilitaires	1 409.4	1 432.0	1 459.2	1 483.5	1 470.6	1 483.8	1 500.8	* 1 554.0	* 1 648.0	* 1 664.0
Australia Australie										
Passenger cars[3]										
Voitures de tourisme[3]	6 469.6	6 636.2	6 842.5	6 985.4	7 072.8	7 243.6	7 442.2	7 672.2	7 734.1	7 913.2
Commercial vehicles[3]										
Véhicules utilitaires[3]	1 673.2	1 751.0	1 838.2	1 881.1	1 949.9	1 977.6	2 047.3	2 104.3	1 915.4	2 041.3
Austria Autriche										
Passenger cars[4]										
Voitures de tourisme[4]	2 468.5	2 530.8	2 609.4	2 684.8	2 784.8	2 902.9	2 991.3	3 100.0	3 244.9	3 367.6
Commercial vehicles[4 5]										
Véhicules utilitaires[4 5]	570.9	580.2	591.6	605.8	624.9	643.1	648.3	657.6	674.6	685.7
Bahamas Bahamas										
Passenger cars										
Voitures de tourisme	48.1	54.1	54.5	59.3	69.0[1]	69.0[1]	69.0[1]	69.0[1]	69.0[1]	...
Commercial vehicles										
Véhicules utilitaires	7.1	8.5	9.5	12.6	14.0[1]	14.0[1]	14.0[1]	14.0[1]	14.0[1]	...
Bahrain Bahreïn										
Passenger cars										
Voitures de tourisme	73.2	79.4	82.8	84.9	90.3	94.9	98.6	102.5	108.0[1]	123.9
Commercial vehicles										
Véhicules utilitaires	20.4	21.8	22.3	23.0	21.3	22.7	24.2	28.3	25.0[1]	27.1

62
Motor vehicles in use
Passenger cars and commercial vehicles: thousand units [cont.]
Véhicules automobiles en circulation
Voitures de tourisme et véhicules utilitaires : milliers de véhicules [suite]

Country or area Pays or zone	1984	1985	1986	1987	1988	1989	1990	1991	1992	1993
Bangladesh Bangladesh										
Passenger cars										
Voitures de tourisme	28.1	29.4	30.9	32.4	34.2	37.8	42.1	44.2	45.6	46.6
Commercial vehicles										
Véhicules utilitaires	40.7	42.4	44.2	45.9	47.7	49.5	51.9	54.9	57.4	59.5
Barbados Barbade										
Passenger cars[4]										
Voitures de tourisme[4]	31.0	32.8	34.8	37.0	38.7	43.1	39.0[1]	41.0[1]	41.0[1]	...
Commercial vehicles[4,6]										
Véhicules utilitaires[4,6]	4.9	4.8	4.7	5.1	4.3	4.8	10.0[1]	11.0[1]	11.0[1]	...
Belgium Belgique										
Passenger cars										
Voitures de tourisme	3 258.4	3 300.5	3 336.2	3 453.1	3 567.4	3 688.1	3 814.6	3 934.0	3 991.6	4 079.5
Commercial vehicles										
Véhicules utilitaires	310.7	318.9	326.9	344.1	358.7	377.7	396.0	414.3	420.0	427.7
Belize Belize										
Passenger cars[1]										
Voitures de tourisme[1]	3.1	3.1	3.5	3.0	3.0	1.6	1.7	2.0	2.0	...
Commercial vehicles[1]										
Véhicules utilitaires[1]	3.4	3.3	3.6	4.0	4.0	2.7	2.8	3.0	3.0	...
Benin Bénin										
Passenger cars										
Voitures de tourisme	25.0	25.0	25.0	26.0	22.0[1]	22.0[1]	22.0[1]	22.0[1]	22.0[1]	...
Commercial vehicles										
Véhicules utilitaires	11.0	12.0	12.0	13.0	12.0[1]	12.0[1]	12.0[1]	12.0[1]	12.2[1]	...
Bermuda Bermudes										
Passenger cars										
Voitures de tourisme	17.3	17.8	17.7	18.2	18.9	19.5	19.7	20.1	21.4[1]	...
Commercial vehicles										
Véhicules utilitaires	3.1	3.2	3.2	3.3	3.4	3.6	3.8	3.6	3.1[1]	...
Bolivia Bolivie										
Passenger cars[1]										
Voitures de tourisme[1]	32.3	33.2	33.0	74.7	93.1	101.6	109.0	261.0	261.0	...
Commercial vehicles[1]										
Véhicules utilitaires[1]	46.0	46.9	46.6	135.8	110.5	141.9	152.4	58.0	63.0	...
Botswana Botswana										
Passenger cars										
Voitures de tourisme	26.1	29.3	32.5	33.8	37.4	41.6	50.4	50.9	57.0	...
Commercial vehicles										
Véhicules utilitaires	13.1	13.7	14.3	14.9	16.8	17.9	19.7	20.8	23.4	...
Brazil Brésil										
Passenger cars										
Voitures de tourisme	9 198.4	9 527.3	* 9 885.2	* 10 035.5	* 10 274.4	10 475.3	10 597.5	12 128.0[1]	12 128.0[1]	...
Commercial vehicles										
Véhicules utilitaires	2 285.2	2 410.0	* 2 350.0	* 2 373.6	* 2 417.8	2 451.6	2 472.5	1 075.0[1]	1 170.8[1]	...
British Virgin Islands Iles Vierges britanniques										
Passenger cars[7]										
Voitures de tourisme[7]	3.3	3.6	4.0	4.3	4.8	5.5	6.2	6.5	6.9	8.0
Brunei Darussalam Brunéi Darussalam										
Passenger cars										
Voitures de tourisme	70.6	78.6	83.4	87.8	92.1	98.8	106.6	114.1	122.1	129.9
Commercial vehicles										
Véhicules utilitaires	9.3	9.7	10.0	10.2	10.4	10.8	11.4	11.9	13.7	14.5
Bulgaria Bulgarie										
Passenger cars										
Voitures de tourisme	1 014.8	1 064.0	1 117.7	1 173.6	1 220.8	1 270.0	1 317.4	1 359.0	1 411.3	1 505.5
Commercial vehicles										
Véhicules utilitaires	137.8	139.6	140.2	141.1	141.2	140.8	127.3	116.3	99.7	80.4

62
Motor vehicles in use
Passenger cars and commercial vehicles: thousand units [*cont.*]
Véhicules automobiles en circulation
Voitures de tourisme et véhicules utilitaires : milliers de véhicules [*suite*]

Country or area Pays or zone	1984	1985	1986	1987	1988	1989	1990	1991	1992	1993
Burkina Faso Burkina Faso										
Passenger cars										
Voitures de tourisme	21.7	22.3	23.2	24.1	25.6	27.5	29.2	30.9	...	...
Commercial vehicles										
Véhicules utilitaires	21.6	22.4	24.0	25.0	25.5	26.9	28.2	29.4	...	...
Burundi Burundi										
Passenger cars										
Voitures de tourisme	8.5	9.2	11.1	12.2	13.0	13.9	15.0	16.4	17.5	18.5
Commercial vehicles										
Véhicules utilitaires	5.1	5.6	7.5	8.4	9.0	9.7	10.4	11.3	11.8	12.3
Cameroon Cameroun										
Passenger cars										
Voitures de tourisme	72.4	77.1	80.8	78.3	73.7	68.7	63.4	57.2	...	...
Commercial vehicles										
Véhicules utilitaires	41.3	43.5	44.9	43.9	41.0	37.7	34.3	30.9	...	...
Canada Canada										
Passenger cars[4]										
Voitures de tourisme[4]	10 780.7	11 118.1	11 477.0	11 772.5	12 086.0	12 811.3	12 622.0	13 061.1	13 322.4[1]	...
Commercial vehicles[4]										
Véhicules utilitaires[4]	3 099.1	3 148.5	3 212.0	3 567.8	3 765.9	3 458.4	3 931.3	3 679.8	3 688.4[1]	...
Cape Verde Cap-Vert										
Passenger cars										
Voitures de tourisme	2.0	2.0	2.0	2.0	...	...	10.0[2]	10.0[2]	10.0[2]	...
Commercial vehicles										
Véhicules utilitaires	1.0	1.0	1.0	1.0	...	...	4.5[2]	5.0[2]	5.0[2]	...
Cayman Islands Iles Caïmanes										
Passenger cars										
Voitures de tourisme	...	7.9	8.0	9.8	9.1	9.7	10.7	10.8	8.8[2]	...
Commercial vehicles										
Véhicules utilitaires	...	1.6	1.7	2.1	1.9	2.1	2.4	2.6	1.3[2]	...
Central African Rep. Rép. centrafricaine										
Passenger cars										
Voitures de tourisme	...	19.1	15.1	12.1	12.4	11.1	12.2	9.2	8.0	10.4
Commercial vehicles										
Véhicules utilitaires	...	5.6	4.7	4.2	4.1	3.2	3.0	2.3	1.7	2.4
Chad Tchad										
Passenger cars										
Voitures de tourisme	10.0	11.0	11.0	11.0	...	...	8.5[2]	8.5[2]	9.0[2]	...
Commercial vehicles										
Véhicules utilitaires	3.0	3.0	3.0	3.0	...	...	6.5[2]	6.5[2]	7.0[2]	...
Chile Chili										
Passenger cars										
Voitures de tourisme	630.4	624.9	590.7	618.5	669.1	661.3	710.4	765.5	826.8	896.5
Commercial vehicles[8]										
Véhicules utilitaires[8]	235.4	257.9	242.1	267.9	297.0	332.8	361.3	398.9	437.5	479.5
Colombia Colombie										
Passenger cars[9]										
Voitures de tourisme[9]	767.8	805.5	842.0	...	645.0[1]	715.3[1]	715.0[1]	715.0[1]	715.0[1]	...
Commercial vehicles[9]										
Véhicules utilitaires[9]	381.1	390.9	400.6	...	617.0[1]	665.3[1]	665.0[1]	665.0[1]	615.0[1]	...
Comoros Comores										
Passenger cars										
Voitures de tourisme	1.0	1.0	1.0	1.0	...	...	...	...	...	...
Commercial vehicles										
Véhicules utilitaires	3.0	4.0	4.0	4.0	...	...	...	...	...	...
Congo Congo										
Passenger cars										
Voitures de tourisme	37.8[1]	25.8[1]	26.1[1]	26.0[1]	26.0[1]	26.0[1]	26.0[1]	26.0[1]	26.0	...
Commercial vehicles										
Véhicules utilitaires	22.1[1]	19.6[1]	20.0[1]	20.0[1]	20.0[1]	20.0[1]	20.0[1]	20.0[1]	20.1	...

62
Motor vehicles in use
Passenger cars and commercial vehicles: thousand units [*cont.*]
Véhicules automobiles en circulation
Voitures de tourisme et véhicules utilitaires : milliers de véhicules [*suite*]

Country or area Pays or zone	1984	1985	1986	1987	1988	1989	1990	1991	1992	1993
Costa Rica Costa Rica										
Passenger cars[4]										
Voitures de tourisme[4]	106.2	111.5	119.1	127.2	135.0	143.9	168.8	180.8	204.2	220.1
Commercial vehicles[4]										
Véhicules utilitaires[4]	66.7	69.5	76.3	84.2	89.6	94.6	95.1	96.3	110.2	114.9
Côte d'Ivoire Côte d'Ivoire										
Passenger cars										
Voitures de tourisme	176.0	176.0	178.0	178.0	168.0[1]	155.0[1]	155.0[1]	155.0[1]	155.3[1]	...
Commercial vehicles										
Véhicules utilitaires	97.0	99.0	90.0	90.0	91.0[1]	90.0[1]	90.0[1]	90.0[1]	90.3[1]	...
Cuba Cuba										
Passenger cars										
Voitures de tourisme	200.1	206.3	217.2	229.5	241.3	...	...	...	...	...
Commercial vehicles										
Véhicules utilitaires	164.5	172.8	184.2	194.9	208.4	...	...	...	...	...
Cyprus Chypre										
Passenger cars[4]										
Voitures de tourisme[4]	118.1	125.7	131.6	142.6	152.7	165.4	178.2	189.7	197.8	203.2
Commercial vehicles[4]										
Véhicules utilitaires[4]	43.0	47.3	50.7	56.5	61.8	68.8	76.6	84.3	89.2	92.6
Czech Republic République tchèque										
Passenger cars										
Voitures de tourisme	..	..	..	..	..	..	..	..	2 522.8	2 693.9
Commercial vehicles										
Véhicules utilitaires	..	..						..	336.2	330.0
former Czechoslovakia† anc. Tchécoslovaquie†										
Passenger cars										
Voitures de tourisme	2 639.6	2 726.3	2 812.4	2 904.0	3 000.0	3 122.3	3 242.3	3 341.8	..	..
Commercial vehicles										
Véhicules utilitaires	378.1	388.2	400.8	413.0	423.0	446.7	461.6	474.7	..	..
Denmark Danemark										
Passenger cars[4 10]										
Voitures de tourisme[4 10]	1 451.7	1 513.3	1 571.1	1 601.6	1 610.3	1 611.7	1 603.8	1 607.0	1 619.5	1 630.3
Commercial vehicles[4 10]										
Véhicules utilitaires[4 10]	255.3	270.3	286.2	298.0	305.0	306.5	305.9	312.7	319.9	329.4
Djibouti Djibouti										
Passenger cars										
Voitures de tourisme	6.0	6.0	7.0	7.0	...	...	13.0[2]	13.0[2]	13.0[2]	...
Commercial vehicles										
Véhicules utilitaires	1.0	1.0	1.0	1.0	...	...	2.0[2]	2.5[2]	3.0[2]	...
Dominica Dominique										
Passenger cars										
Voitures de tourisme	2.9	3.2	3.5	3.7	4.1	4.5	5.0	4.7[2]	4.7[2]	...
Commercial vehicles										
Véhicules utilitaires	1.6	2.0	2.3	2.7	3.1	3.7	4.3	5.5[2]	5.5[2]	...
Dominican Republic Rép. dominicaine										
Passenger cars[11]										
Voitures de tourisme[11]	102.0	100.0	132.9	151.7	...	129.7	139.1	135.9	126.5	160.1
Commercial vehicles										
Véhicules utilitaires	61.3	57.2	76.6	84.2	...	87.5	89.9	94.3	93.5	111.4
Ecuador Equateur										
Passenger cars										
Voitures de tourisme	120.2	121.3	136.5	140.7	146.5	176.2	165.6	181.2	194.5	202.4
Commercial vehicles										
Véhicules utilitaires	179.6	176.0	185.1	188.5	190.2	221.6	207.3	206.0	232.7	231.4
Egypt Egypte										
Passenger cars										
Voitures de tourisme	847.0	900.0	933.0	965.0	980.0	1 019.0	1 054.0	1 081.0	1 117.0	1 143.0
Commercial vehicles										
Véhicules utilitaires	264.0	293.0	315.0	332.0	350.0	353.0	380.0	389.0	408.0	422.0

62
Motor vehicles in use
Passenger cars and commercial vehicles: thousand units [*cont.*]
Véhicules automobiles en circulation
Voitures de tourisme et véhicules utilitaires : milliers de véhicules [*suite*]

Country or area Pays or zone	1984	1985	1986	1987	1988	1989	1990	1991	1992	1993
El Salvador El Salvador										
Passenger cars										
Voitures de tourisme	85.3	52.9[1]	52.1[1]	52.0[1]	52.0[1]	52.0[1]	52.0[1]	53.0[1]	55.6[1]	...
Commercial vehicles										
Véhicules utilitaires	65.2	64.8[1]	64.0[1]	65.0[1]	65.0[1]	65.0[1]	65.0[1]	65.0[1]	67.4[1]	...
Estonia Estonie										
Passenger cars										
Voitures de tourisme	164.0	173.0	182.0	194.0	206.0	222.0	242.0	261.1	283.5	317.4
Ethiopia Ethiopie										
Passenger cars										
Voitures de tourisme	47.5[12]	41.1[12]	41.2[12]	48.3[12]	43.5[12]	42.8[12]	38.5	40.2	40.1[1]	...
Commercial vehicles										
Véhicules utilitaires	11.1[12]	18.2[12]	18.7[12]	10.9[12]	21.8[12]	21.6[12]	20.3	20.2	18.8[1]	...
Fiji Fidji										
Passenger cars[4 13]										
Voitures de tourisme[4 13]	31.0	32.5	33.6	34.4	34.9	37.5	40.3	42.1	44.0	45.3
Commercial vehicles[4 13]										
Véhicules utilitaires[4 13]	25.8	27.2	28.2	28.8	29.4	30.5	32.4	34.3	36.0	36.9
Finland Finlande										
Passenger cars										
Voitures de tourisme	1 474.0	1 546.1	1 619.8	1 698.7	1 795.9	1 908.9	1 939.9	1 922.5	1 936.3	1 872.9
Commercial vehicles										
Véhicules utilitaires	182.9	188.7	196.6	207.4	222.9	254.0	273.5	273.4	271.2	261.4
France France										
Passenger cars										
Voitures de tourisme	20 800.0	21 090.0	21 500.0	21 970.0	22 520.0	23 010.0	23 550.0	23 810.0	24 020.0	24 385.0
Commercial vehicles										
Véhicules utilitaires	3 072.0	3 209.0	3 298.0	3 419.0	3 547.0	3 674.0	3 810.0	3 922.0	4 099.0	4 036.0
French Guiana Guyane française										
Passenger cars[1]										
Voitures de tourisme[1]	21.6	22.4	23.4	24.0	24.0	24.0	25.0	27.0	27.7	...
Commercial vehicles[1]										
Véhicules utilitaires[1]	6.4	6.7	6.9	7.0	7.0	8.0	9.0	10.0	10.4	...
Gabon Gabon										
Passenger cars										
Voitures de tourisme	16.0	16.0	16.0	17.0	...	...	22.0[2]	23.0[2]	23.0[2]	...
Commercial vehicles										
Véhicules utilitaires	11.0	11.0	11.0	11.0	...	...	16.0[2]	17.0[2]	17.0[2]	...
Gambia Gambie										
Passenger cars										
Voitures de tourisme	5.3	4.2	3.8	3.9	4.5	5.3	6.5[2]	6.0[2]	7.0[2]	...
Commercial vehicles										
Véhicules utilitaires	1.3	1.6	1.4	1.4	1.5	2.0	1.5[2]	2.5[2]	3.0[2]	...
Georgia Géorgie										
Passenger cars										
Voitures de tourisme	..	..	..	425.7	440.8	448.7	481.9	479.0	479.0	468.8
Commercial vehicles										
Véhicules utilitaires	..	..	..	74.8	75.7	75.9	86.3	94.0	66.7	56.0
Germany † Allemagne† **F. R. Germany R. f. Allemagne**										
Passenger cars										
Voitures de tourisme	25 217.8	25 844.5	26 917.4	27 908.2	28 878.2	29 755.4	30 684.8	31 321.7	32 007.0	32 652.0
Commercial vehicles										
Véhicules utilitaires	1 614.6	1 629.0	1 664.3	1 704.4	1 753.0	1 811.9	1 895.4	1 985.8	2 134.6	2 826.0
former German D. R. anc. R. d. allemande										
Passenger cars										
Voitures de tourisme	3 157.1	3 306.2	3 462.2	3 600.4	* 3 743.6	* 3 898.9	4 817.0	* 6 300.0	* 7 000.0	* 6 500.0
Commercial vehicles[5]										
Véhicules utilitaires[5]	647.0	656.8	668.5	685.3	* 706.0	* 732.3	774.8	* 1 000.0	* 700.0	* 500.0

62

Motor vehicles in use
Passenger cars and commercial vehicles: thousand units [*cont.*]
Véhicules automobiles en circulation
Voitures de tourisme et véhicules utilitaires : milliers de véhicules [*suite*]

Country or area Pays or zone	1984	1985	1986	1987	1988	1989	1990	1991	1992	1993
Ghana Ghana										
Passenger cars[1] Voitures de tourisme[1]	60.2	59.1	58.3	58.0	58.0	66.0	82.2	90.0	90.0	...
Commercial vehicles[1] Véhicules utilitaires[1]	45.0	45.5	45.7	46.0	46.0	40.0	42.1	43.0	44.2	...
Gibraltar Gibraltar										
Passenger cars Voitures de tourisme	8.5	10.6	13.0	13.1	15.5	17.7	19.8	...	24.0	...
Commercial vehicles Véhicules utilitaires	0.8	1.0	1.0	1.6	2.0	2.3	2.5	...	2.9	...
Greece Grèce										
Passenger cars Voitures de tourisme	1 154.9	1 263.4	1 359.2	1 428.5	1 503.9	1 605.1	1 735.5	1 777.5	1 829.1	1 958.5
Commercial vehicles Véhicules utilitaires	588.9	619.2	645.7	663.8	697.1	732.2	784.2	811.2	8 205.0	848.9
Greenland Groënland										
Passenger cars[4] Voitures de tourisme[4]	1.6	1.7	2.1	2.0	2.1	2.0	2.0	1.9	2.0	2.1
Commercial vehicles[4] Véhicules utilitaires[4]	1.4	1.4	1.3	1.6	1.5	1.6	1.2	1.5	1.5	1.4
Guadeloupe Guadeloupe										
Passenger cars Voitures de tourisme	82.7	89.4	71.3[1]	79.0[1]	79.0[1]	83.0[1]	86.0[1]	89.0[1]	94.7[1]	...
Commercial vehicles Véhicules utilitaires	26.3	28.3	28.1[1]	31.0[1]	31.0[1]	33.0[1]	34.0[1]	35.0[1]	36.0[1]	...
Guam Guam										
Passenger cars Voitures de tourisme	57.6	60.6	54.9	55.9	56.2	62.0	...	72.8	76.7	74.7
Commercial vehicles Véhicules utilitaires	15.3	16.4	15.1	19.5	16.7	22.8	...	29.7	30.2	30.6
Guatemala Guatemala										
Passenger cars[1] Voitures de tourisme[1]	96.0	96.7	95.1	95.0	95.0	95.0	95.0	95.0	98.7	...
Commercial vehicles[1] Véhicules utilitaires[1]	94.0	93.7	92.7	93.0	93.0	93.0	93.0	93.0	95.0	...
Guinea Guinée										
Passenger cars Voitures de tourisme	10.0	11.0	11.0	11.0	50.1	57.7	60.9	65.6	23.1[2]	...
Commercial vehicles Véhicules utilitaires	10.0	11.0	12.0	12.0	24.6	25.9	27.8	29.0	13.0[2]	...
Guinea-Bissau Guinée-Bissau										
Passenger cars Voitures de tourisme	3.0	4.0	4.0	4.0	...	...	3.3[2]	3.3[2]	3.5[2]	...
Commercial vehicles Véhicules utilitaires	3.0	3.0	3.0	3.0	...	...	2.4[2]	2.4[2]	2.5[2]	...
Guyana Guyana										
Passenger cars[1] Voitures de tourisme[1]	30.0	29.4	28.9	20.3	22.0	24.0	24.0	24.0	24.0	...
Commercial vehicles[1] Véhicules utilitaires[1]	12.0	11.7	11.7	8.5	9.0	9.0	9.0	9.0	9.0	...
Haïti Haïti										
Passenger cars Voitures de tourisme	24.6	26.1	19.5	19.7	20.6	27.7	* 25.8	32.0[1]	32.0[1]	...
Commercial vehicles Véhicules utilitaires	10.6	10.7	13.0	22.6	22.8	22.8	* 9.6	21.0[1]	21.0[1]	...
Honduras Honduras										
Passenger cars Voitures de tourisme	28.5[4]	34.2[4]	26.8[1]	27.0[1]	27.0[1]	27.0[1]	38.6	43.7	68.5	...
Commercial vehicles Véhicules utilitaires	...	...	50.2[1]	51.0[1]	52.0[1]	52.0[1]	80.4	92.9	102.0	...

62
Motor vehicles in use
Passenger cars and commercial vehicles: thousand units [*cont.*]
Véhicules automobiles en circulation
Voitures de tourisme et véhicules utilitaires : milliers de véhicules [*suite*]

Country or area Pays or zone	1984	1985	1986	1987	1988	1989	1990	1991	1992	1993
Hong Kong Hong-kong										
Passenger cars										
Voitures de tourisme	163.4	160.9	155.6	162.3	177.4	197.2	214.9	229.3	254.6	277.5
Commercial vehicles										
Véhicules utilitaires	73.9	81.4	91.7	107.3	118.4	126.1	131.8	132.4	134.4	135.6
Hungary Hongrie										
Passenger cars										
Voitures de tourisme	1 344.1	1 435.9	1 538.9	1 660.3	1 789.6	1 732.4	1 944.6	2 015.5	2 058.3	2 091.6
Commercial vehicles										
Véhicules utilitaires	213.4	223.4	238.3	254.2	259.6	267.8	288.5	289.6	259.3	266.8
Iceland Islande										
Passenger cars										
Voitures de tourisme	100.2	103.0	112.3	120.1	125.2	124.3	119.7	120.9	120.1	116.2
Commercial vehicles										
Véhicules utilitaires	13.0	14.2	13.1	12.9	13.2	13.5	14.5	16.0	6.0	15.6
India Inde										
Passenger cars										
Voitures de tourisme	1 455.2	1 606.5	1 780.0	2 006.9	2 295.3	2 486.3	2 736.0	3 013.3	2 806.5[1]	...
Commercial vehicles[14]										
Véhicules utilitaires[14]	1 846.6	2 113.3	2 101.8	2 543.5	2 855.2	3 073.0	3 425.2	3 740.6	2 396.7[1]	...
Indonesia Indonésie										
Passenger cars										
Voitures de tourisme	927.0[8]	990.7[8]	1 064.0[8]	1 170.1[8]	1 073.1[8]	1 182.2[8]	1 313.2[1]	1 494.6	1 590.8	1 700.5
Commercial vehicles										
Véhicules utilitaires	982.5	1 072.6	1 138.9	1 257.1	1 278.3	1 387.3	1 492.8[1]	1 592.7	1 666.0	1 729.0
Iran, Islamic Rep. of Iran, Rép. islamique d'										
Passenger cars[9][15]										
Voitures de tourisme[9][15]	1 745.0	1 851.0	1 919.0	1 958.0	1 981.0	2 000.0	1 557.0[1]	1 557.0[1]	1 557.0[1]	...
Commercial vehicles[9][15]										
Véhicules utilitaires[9][15]	695.0	760.0	819.0	853.0	862.0	870.0	533.0[1]	561.0[1]	584.1[1]	...
Iraq Iraq										
Passenger cars[1]										
Voitures de tourisme[1]	250.7	258.4	251.3	251.0	630.3	672.2	672.0	672.0	672.0	...
Commercial vehicles[1]										
Véhicules utilitaires[1]	265.3	273.4	269.2	269.0	344.8	368.0	368.0	368.0	368.0	...
Ireland Irlande										
Passenger cars[16][17]										
Voitures de tourisme[16][17]	716.8[3]	715.3[3]	717.2[3]	742.8[3]	755.7[3]	779.8[3]	802.7[3]	843.2	865.4	898.3
Commercial vehicles										
Véhicules utilitaires	89.3[3]	98.7[3]	106.9[3]	116.6[3]	124.4[3]	135.9[3]	149.5[3]	155.2	152.0	142.9
Israel Israël										
Passenger cars										
Voitures de tourisme	599.3	613.9	648.8	696.7	753.5	777.9	813.0[1]	862.0[1]	948.0[1]	...
Commercial vehicles										
Véhicules utilitaires	113.1	114.7	120.8	132.0	143.8	149.2	163.0[1]	179.0[1]	199.1[1]	...
Italy Italie										
Passenger cars										
Voitures de tourisme	20 888.2	22 494.6	23 495.5	24 320.2	25 290.3	26 267.4	27 415.8	28 519.0	29 497.0[1]	...
Commercial vehicles										
Véhicules utilitaires	2 182.1	2 308.8	2 428.2	2 002.7	2 120.5	2 350.0	2 416.7	2 529.6	2 763.0[1]	...
Jamaica Jamaïque										
Passenger cars										
Voitures de tourisme	42.0	42.9	44.5	52.9	63.1	64.8	68.5	77.8	73.0	...
Commercial vehicles										
Véhicules utilitaires	23.2	26.1	20.7	23.0	26.9	24.5	28.2	29.8	38.4	...
Japan Japon[18][19]										
Passenger cars[18][19]										
Voitures de tourisme[18][19]	27 144.0	27 844.0	28 654.0	29 478.0	30 776.0	32 621.0	34 924.0	37 076.0	38 964.0	40 772.0
Commercial vehicles[18]										
Véhicules utilitaires[18]	16 477.0	17 377.0	18 346.0	19 401.0	20 592.0	21 330.0	21 571.0	21 575.0	21 383.0	21 132.0

62
Motor vehicles in use
Passenger cars and commercial vehicles: thousand units [*cont.*]
Véhicules automobiles en circulation
Voitures de tourisme et véhicules utilitaires : milliers de véhicules [*suite*]

Country or area Pays or zone	1984	1985	1986	1987	1988	1989	1990	1991	1992	1993
Jordan Jordanie										
Passenger cars[4] Voitures de tourisme[4]	139.5	143.4	149.8	153.6	158.9	159.9	172.0	166.8	181.5	175.3
Commercial vehicles[4] Véhicules utilitaires[4]	56.2	55.8	57.8	60.9	63.2	63.4	68.3	61.4	51.5	63.8
Kenya Kenya										
Passenger cars Voitures de tourisme	122.0[4 9 20]	126.2[4 9]	127.4[4 9]	133.3[4 9]	141.8[1]	150.0[1]	157.7[1]	163.5	165.4	171.5
Commercial vehicles Véhicules utilitaires	96.6[4 9 21]	103.8[4 9]	102.9[4 9]	110.8[4 9]	118.9[1]	149.0	138.6	150.1	168.4	172.8
Korea, Republic of Corée, République de										
Passenger cars Voitures de tourisme	465.1	556.7	664.2	844.4	1 118.0	1 558.7	2 074.9	2 727.9	3 461.1	4 271.3
Commercial vehicles Véhicules utilitaires	468.4	541.0	627.2	746.9	895.0	1 092.3	1 308.4	1 505.1	1 745.1	1 976.6
Kuwait Koweït										
Passenger cars Voitures de tourisme	384.1	417.4	410.0	429.2	458.1	488.0	...	554.7	579.8	600.0
Commercial vehicles Véhicules utilitaires	134.3	140.4	133.6	135.2	132.9	134.3	...	149.8	151.1	147.0
Latvia Lettonie										
Passenger cars Voitures de tourisme	201.5	214.5	226.1	236.3	251.3	264.1	282.7	328.5	350.0	367.5
Commercial vehicles Véhicules utilitaires	70.0	71.5	73.1	74.1	74.9	76.9	79.0	83.3	93.0	72.1
Lesotho Lesotho										
Passenger cars Voitures de tourisme	5.1[4]	6.1[4]	6.7[4]	5.0	...	...	...	...	...	...
Commercial vehicles Véhicules utilitaires	16.1[4]	14.6[4]	16.3[4]	* 13.0	...	...	...	...	...	...
Liberia Libéria										
Passenger cars Voitures de tourisme	16.0	16.0	17.0	18.0	...	...	...	...	...	...
Commercial vehicles Véhicules utilitaires	14.0	14.0	15.0	15.0	...	...	...	...	...	...
Libyan Arab Jamah. Jamah. arabe libyenne										
Passenger cars Voitures de tourisme	408.0	420.0	428.0	433.0	448.0[2]	448.0[2]	448.0[2]	450.0[2]	448.0[2]	...
Commercial vehicles Véhicules utilitaires	207.0	210.0	216.0	223.0	322.0[2]	322.0[2]	322.0[2]	330.0[2]	322.0[2]	...
Lithuania Lituanie										
Passenger cars Voitures de tourisme	319.0	341.2	370.5	390.8	417.7	452.0	493.0	530.8	565.3	597.7
Commercial vehicles Véhicules utilitaires	93.2	94.2	95.7	97.1	38.3	99.7	105.9	107.7	112.5	115.1
Luxembourg Luxembourg										
Passenger cars Voitures de tourisme	145.8	151.6	156.0	162.5	168.5	177.0	183.4	191.6	200.7	208.8
Commercial vehicles[5] Véhicules utilitaires[5]	26.3	26.7	27.4	28.0	28.7	30.0	31.6	33.6	35.8	38.0
Macau Macao										
Passenger cars[8] Voitures de tourisme[8]	17.5	17.9	19.5	19.6	21.0	22.4	24.7	26.2	29.9	32.6
Commercial vehicles[8] Véhicules utilitaires[8]	4.6	4.4	4.8	4.9	5.3	6.3	6.4	6.5	6.5	6.6
Madagascar Madagascar										
Passenger cars Voitures de tourisme	45.0	46.0	46.0	46.0	...	45.0	46.6[1]	47.0[1]	47.0[1]	...
Commercial vehicles Véhicules utilitaires	44.0	45.0	48.0	47.0	...	32.0	33.1[1]	33.0[1]	33.3[1]	...

62
Motor vehicles in use
Passenger cars and commercial vehicles: thousand units [cont.]
Véhicules automobiles en circulation
Voitures de tourisme et véhicules utilitaires : milliers de véhicules [suite]

Country or area Pays or zone	1984	1985	1986	1987	1988	1989	1990	1991	1992	1993
Malawi Malawi										
Passenger cars[4] Voitures de tourisme[4]	13.7	13.6	15.5[1]	14.9	19.1	16.1	23.2	15.0[1]	15.0[1]	...
Commercial vehicles[4] Véhicules utilitaires[4]	15.1	15.4	15.6[1]	16.7	12.9	19.3	21.8	18.0[1]	18.9[1]	...
Malaysia Malaisie										
Passenger cars Voitures de tourisme	1 292.0	1 384.0	1 453.6	1 504.2	1 578.9	1 689.4	1 845.6	2 024.8	2 158.2	2 291.1
Commercial vehicles[22] Véhicules utilitaires[22]	286.3	311.1	330.1	339.0	351.9	374.6	407.1	446.8	447.7	501.0
Mali Mali										
Passenger cars Voitures de tourisme	18.0	19.0	19.0	19.0	20.0	21.0[1]	21.0[1]	21.0[1]	21.0[1]	...
Commercial vehicles Véhicules utilitaires	12.0	13.0	13.0	13.0	13.0	8.0[1]	8.0[1]	8.0[1]	8.4[1]	...
Malta Malte										
Passenger cars Voitures de tourisme	80.3	82.3	85.6	89.5	97.6	110.6	109.2	121.6	125.0	152.6
Commercial vehicles Véhicules utilitaires	17.4	18.2	17.8	17.2	19.3	19.6	20.5	21.7	35.4	50.9
Martinique Martinique										
Passenger cars[1] Voitures de tourisme[1]	72.7	74.6	77.0	80.0	84.0	80.0	92.0	96.0	102.6	...
Commercial vehicles[1] Véhicules utilitaires[1]	24.5	25.3	26.0	27.0	28.0	29.0	30.0	31.0	31.9	...
Mauritania Mauritanie										
Passenger cars Voitures de tourisme	10.0	11.0	11.0	12.0	8.0[1]	8.0[1]	8.0[1]	8.0[1]	8.0[1]	...
Commercial vehicles Véhicules utilitaires	5.0	5.0	6.0	6.0	5.0[1]	5.0[1]	5.0[1]	5.0[1]	5.5[1]	...
Mauritius Maurice										
Passenger cars Voitures de tourisme	32.7	33.1	34.2	36.5	39.4	42.1	45.5	50.0	53.9	57.4
Commercial vehicles Véhicules utilitaires	10.6	10.8	11.2	12.2	13.7	15.1	16.9	19.1	20.9	22.5
Mexico Mexique										
Passenger cars Voitures de tourisme	4 802.1	5 102.4	5 202.9	5 336.2	5 806.9	6 219.1	6 893.3	7 497.1	7 749.6	8 014.2
Commercial vehicles Véhicules utilitaires	1 933.1	2 033.4	2 213.0	2 292.0	2 435.9	2 704.1	2 982.0	3 501.0	3 505.7	3 661.5
Morocco Maroc										
Passenger cars[8] Voitures de tourisme[8]	491.1	508.3	527.4	554.0	588.9	634.4	669.6	707.1	778.9	849.3
Commercial vehicles[8] Véhicules utilitaires[8]	232.7	239.9	247.7	255.1	263.1	272.4	282.9	295.5	302.3	316.7
Mozambique Mozambique										
Passenger cars Voitures de tourisme	49.0	46.0	45.0	45.0	# 84.0[1]	84.0[1]	84.0[1]	84.0[1]	84.0[1]	...
Commercial vehicles Véhicules utilitaires	19.0	18.0	18.0	18.0	# 24.0[1]	24.0[1]	25.0[1]	26.0[1]	26.2[1]	...
Myanmar Myanmar										
Passenger cars[4] Voitures de tourisme[4]	54.8	59.8	60.9	60.4	64.9	71.3	78.1	88.6	...	...
Commercial vehicles[4] Véhicules utilitaires[4]	50.2	51.9	53.5	50.7	51.6	53.7	54.9	56.6	...	...
Netherlands Pays-Bas										
Passenger cars[4 23 24] Voitures de tourisme[4 23 24]	4 772.4	4 901.1	4 949.9	5 117.7	5 250.6	5 371.4	5 509.2	5 569.1	5 658.3	5 755.4
Commercial vehicles[4 23 24] Véhicules utilitaires[4 23 24]	381.0	428.2	463.8	506.5	538.2	556.8	582.1	604.6	644.9	679.2

62
Motor vehicles in use
Passenger cars and commercial vehicles: thousand units [*cont.*]
Véhicules automobiles en circulation
Voitures de tourisme et véhicules utilitaires : milliers de véhicules [*suite*]

Country or area Pays or zone	1984	1985	1986	1987	1988	1989	1990	1991	1992	1993
New Caledonia Nouvelle-Calédonie										
Passenger cars[1]										
Voitures de tourisme[1]	43.9	43.8	46.4	48.0	50.0	53.0	54.0	55.0	56.7	...
Commercial vehicles[1]										
Véhicules utilitaires[1]	16.5	16.3	16.7	17.0	18.0	19.0	19.0	20.0	21.2	...
New Zealand Nouvelle-Zélande										
Passenger cars										
Voitures de tourisme	1 466.2	1 495.1	1 531.4[25]	...	1 382.3[25]	1 438.7[25]	1 497.7[25]	1 548.1[25]	1 551.4[25]	1 571.8[25]
Commercial vehicles										
Véhicules utilitaires	294.2	302.1	373.8[25]	...	289.2[25]	289.2[25]	306.6[25]	309.6[25]	317.8[25]	332.3[25]
Nicaragua Nicaragua										
Passenger cars[1]										
Voitures de tourisme[1]	32.7	31.9	31.1	31.0	30.0	31.1	31.1	31.0	31.3	...
Commercial vehicles[1]										
Véhicules utilitaires[1]	28.5	28.3	27.9	28.0	42.0	43.0	43.0	43.0	43.6	...
Niger Niger										
Passenger cars										
Voitures de tourisme	34.0	35.0	35.0	35.0	...	...	17.0[2]	18.0[2]	31.4[2]	...
Commercial vehicles										
Véhicules utilitaires	9.0	9.0	9.0	9.0	...	...	18.0[2]	18.0[2]	9.7[2]	...
Nigeria Nigéria										
Passenger cars										
Voitures de tourisme	360.0	350.0	391.0	337.0	308.0	259.0	273.0	...	...	...
Commercial vehicles										
Véhicules utilitaires	33.0	34.0	132.0	126.0	103.0	81.0	84.0	...	...	...
Norway Norvège										
Passenger cars[4,26]										
Voitures de tourisme[4,26]	1 428.7	1 514.0	1 592.2	1 623.1	1 622.0	1 612.7	1 612.0	1 614.6	1 619.4	1 633.0
Commercial vehicles[4,26]										
Véhicules utilitaires[4,26]	215.1	249.7	282.8	303.2	313.9	320.4	330.6	334.3	341.6	352.5
Pakistan Pakistan										
Passenger cars[4]										
Voitures de tourisme[4]	404.4	452.1	500.2	540.4	590.5	660.0	715.0	746.1	855.4	933.1
Commercial vehicles[4]										
Véhicules utilitaires[4]	144.2	153.2	172.8	175.0	194.4	259.6	269.1	279.8	310.8	332.4
Panama Panama										
Passenger cars										
Voitures de tourisme	121.0	129.0	134.3	129.4	129.5	120.8	132.9	144.2	149.9	...
Commercial vehicles										
Véhicules utilitaires	41.2	40.3	41.9	44.6	41.2	40.0	42.2	47.3	50.4	...
Papua New Guinea Papouasie-Nvl-Guinée										
Passenger cars[4]										
Voitures de tourisme[4]	17.4	16.0	16.6	17.1	...	...	...	...	11.5[2]	...
Commercial vehicles[4,13]										
Véhicules utilitaires[4,13]	29.0	26.6	27.0	26.1	...	...	...	...	29.8[2]	...
Paraguay Paraguay										
Passenger cars										
Voitures de tourisme	85.0	104.9	111.9	122.5	80.2	149.2	165.2	190.9	221.1	250.7
Commercial vehicles										
Véhicules utilitaires	24.5	22.0	23.2	27.2	18.9	25.2	25.7	30.7	34.9	37.7
Peru Pérou										
Passenger cars										
Voitures de tourisme	374.0	376.0	377.2	377.4	376.8	372.8	368.2	379.1	402.4	418.6
Commercial vehicles										
Véhicules utilitaires	217.0	220.3	226.5	233.4	239.8	239.5	237.4	244.9	270.6	288.8
Philippines Philippines										
Passenger cars										
Voitures de tourisme	360.7	347.9	356.7	358.8	376.6	413.0	454.6[1]	456.6	483.6	531.2
Commercial vehicles										
Véhicules utilitaires	534.1	514.5	526.7	554.7	598.2	671.7	764.9[1]	829.7	916.7	1 024.0

62
Motor vehicles in use
Passenger cars and commercial vehicles: thousand units [*cont.*]
Véhicules automobiles en circulation
Voitures de tourisme et véhicules utilitaires : milliers de véhicules [*suite*]

Country or area Pays or zone	1984	1985	1986	1987	1988	1989	1990	1991	1992	1993
Poland Pologne										
Passenger cars										
Voitures de tourisme	3 425.8	3 671.4	3 964.0	4 231.7	4 519.1	4 846.4	5 260.6	6 112.2	6 504.7	6 770.6
Commercial vehicles[27]										
Véhicules utilitaires[27]	813.4	864.3	914.4	955.0	1 010.5	1 069.6	1 138.1	1 240.2	1 299.5	1 321.9
Portugal Portugal										
Passenger cars[3 28]										
Voitures de tourisme[3 28]	1 600.7	1 701.7	1 813.0	1 947.3	2 152.5	2 343.4	2 552.3	2 774.7	3 049.8	3 295.1
Commercial vehicles[3 28]										
Véhicules utilitaires[3 28]	642.5[22]	669.1[22]	705.9[22]	607.9	669.6	739.2	812.8	881.2[8]	964.0	1 050.1
Puerto Rico Porto Rico										
Passenger cars										
Voitures de tourisme	896.6	1 142.6	1 225.1	1 304.3	1 337.7	1 321.9	1 332.1	1 347.7	1 356.8	...
Commercial vehicles										
Véhicules utilitaires	121.5	168.0	178.8	190.6	196.1	192.1	206.7	215.3	219.0	...
Qatar Qatar										
Passenger cars										
Voitures de tourisme	74.8	80.0	85.4	91.0	97.3	100.2	105.8	114.5	123.6	132.1
Commercial vehicles										
Véhicules utilitaires	32.5	33.4	36.8	39.3	41.7	43.7	46.8	53.9	57.5	61.5
Republic of Moldova République de Moldova										
Passenger cars										
Voitures de tourisme	129.2	148.1	163.1	171.3	184.9	195.7	209.0	218.1	166.3	166.4
Commercial vehicles										
Véhicules utilitaires	16.3	16.1	16.3	16.1	15.6	15.2	15.0	14.3	10.1	9.1
Réunion Réunion										
Passenger cars										
Voitures de tourisme	92.9	101.6	106.6	114.6	125.9	139.0	146.4	155.9	173.2[1]	...
Commercial vehicles										
Véhicules utilitaires	41.8	45.0	48.5	53.3	50.0[1]	53.0[1]	55.0[1]	58.0[1]	61.6[1]	...
Rwanda Rwanda										
Passenger cars										
Voitures de tourisme	5.6	6.1	6.7	7.1	7.2	7.9	7.2	15.0[2]	7.9[2]	...
Commercial vehicles										
Véhicules utilitaires	5.6	6.3	7.0	7.3	7.9	7.0	7.0	10.0[2]	2.0[2]	...
Saint Kitts and Nevis Saint-Kitts-et-Nevis										
Passenger cars										
Voitures de tourisme	3.0	3.1	3.3	3.3	3.4	3.9	4.1	3.7	...	...
Commercial vehicles										
Véhicules utilitaires	0.9	0.9	1.0	1.0	1.2	1.5	1.6	2.2	...	...
Saint Lucia Sainte-Lucie										
Passenger cars										
Voitures de tourisme	5.5	5.4	5.5	6.0	6.5	7.2	8.1	8.6	9.9[2]	...
Commercial vehicles										
Véhicules utilitaires	3.5	3.7	4.2	5.1	5.5	5.9	6.3	...	9.1[2]	...
St. Vincent-Grenadines St. Vincent-Grenadines										
Passenger cars										
Voitures de tourisme	5.0	4.9	5.1	4.9	5.2	5.3	5.3	5.3	5.0[2]	...
Commercial vehicles										
Véhicules utilitaires	1.9	2.0	2.3	2.4	2.6	2.7	2.8	2.8	2.0[2]	...
Samoa Samoa										
Passenger cars										
Voitures de tourisme	1.8	2.0	2.1	1.8	1.9	2.0	...	...	...	...
Commercial vehicles										
Véhicules utilitaires	2.7	2.8	2.0	2.5	2.5	3.9	...	...	...	...
Sao Tome and Principe Sao Tomé-et-Principe										
Passenger cars										
Voitures de tourisme	2.0	2.2	2.4	2.6	...	...	...	...	...	...

62
Motor vehicles in use
Passenger cars and commercial vehicles: thousand units [*cont.*]
Véhicules automobiles en circulation
Voitures de tourisme et véhicules utilitaires : milliers de véhicules [*suite*]

Country or area Pays or zone	1984	1985	1986	1987	1988	1989	1990	1991	1992	1993
Commercial vehicles Véhicules utilitaires	0.3	0.3	0.3	0.3	...	...	...	...	...	...
Saudi Arabia Arabie saoudite										
Passenger cars[1] Voitures de tourisme[1]	1 150.0	1 300.0	1 310.1	1 337.0	1 378.0	1 420.0	1 468.0	1 534.0	1 602.8	...
Commercial vehicles[1] Véhicules utilitaires[1]	1 200.0	1 450.0	1 452.6	1 477.0	1 477.0	1 499.0	1 536.0	1 592.0	1 681.9	...
Senegal Sénégal										
Passenger cars Voitures de tourisme	62.0	63.0	63.0	63.0	...	...	...	...	...	* 65.4
Commercial vehicles Véhicules utilitaires	36.0	36.0	36.0	36.0	...	...	...	...	...	* 24.0
Seychelles Seychelles										
Passenger cars Voitures de tourisme	3.4	3.5	3.5	3.4	3.6	4.0	4.3	4.7	4.9	6.1
Commercial vehicles Véhicules utilitaires	1.2	1.3	1.0	1.0	1.0	1.0	1.2	1.3	1.5	1.8
Sierra Leone Sierra Leone										
Passenger cars Voitures de tourisme	23.5	31.0	33.0	33.0	23.0[1]	23.0[1]	35.9[1]	36.0[1]	36.0[1]	...
Commercial vehicles Véhicules utilitaires	10.0	14.0	15.0	15.0	7.0[1]	7.0[1]	11.8[1]	12.0[1]	12.0[1]	...
Singapore Singapour										
Passenger cars Voitures de tourisme	232.3	236.2	234.6	236.1	251.4	271.2	286.8	300.1	302.8	321.9
Commercial vehicles Véhicules utilitaires	119.5	118.3	114.3	113.7	117.4	122.8	126.9	130.1	131.5	135.2
Somalia Somalie										
Passenger cars Voitures de tourisme	6.0	6.0	5.0	5.0	...	...	20.0[2]	10.5[2]	10.5[2]	...
Commercial vehicles Véhicules utilitaires	7.0	7.0	8.0	8.0	...	...	12.0[2]	12.0[2]	11.5[2]	...
South Africa Afrique du Sud										
Passenger cars[16] Voitures de tourisme[16]	3 018.7	3 096.6	3 237.2	3 286.8	3 222.4	3 498.2	3 599.8	3 698.2	3 739.2	...
Commercial vehicles[29] Véhicules utilitaires[29]	1 430.4	1 461.6	1 433.9	1 454.9	1 456.7	1 462.5	1 486.9	1 519.9	1 551.4	...
Spain Espagne										
Passenger cars Voitures de tourisme	8 874.4	9 273.7	9 643.3	10 218.5	10 787.5	11 467.7	11 995.6	12 537.1	13 102.3	13 440.7
Commercial vehicles Véhicules utilitaires	1 523.2	1 610.3	1 762.5	1 911.9	2 073.4	2 269.3	2 446.9	2 615.0	2 773.4	2 859.6
Sri Lanka Sri Lanka										
Passenger cars Voitures de tourisme	141.7[4]	148.6[4]	155.2[4]	147.8[4]	155.2[4]	163.8[4]	173.5	180.1	189.5	197.3
Commercial vehicles Véhicules utilitaires	121.8[4]	132.4[4]	141.3[4]	135.4[4]	139.2[4]	142.1[4]	146.0	152.7	159.9	166.3
Sudan Soudan										
Passenger cars Voitures de tourisme	160.0	171.0	177.0	185.0	...	...	116.0[2]	116.0[2]	116.0[2]	...
Commercial vehicles Véhicules utilitaires	23.0	23.0	23.0	24.0	...	...	56.9[2]	57.0[2]	57.0[2]	...
Suriname Suriname										
Passenger cars[4] Voitures de tourisme[4]	28.8	31.6	32.1	32.1	35.1	36.6	36.2	38.7	27.0[1]	...
Commercial vehicles Véhicules utilitaires	11.6	12.8	13.2	13.0	13.4	14.0	14.4	15.5	11.1[1]	...

62
Motor vehicles in use
Passenger cars and commercial vehicles: thousand units [*cont.*]
Véhicules automobiles en circulation
Voitures de tourisme et véhicules utilitaires : milliers de véhicules [*suite*]

Country or area Pays or zone	1984	1985	1986	1987	1988	1989	1990	1991	1992	1993
Swaziland Swaziland										
Passenger cars										
Voitures de tourisme	18.5	19.6	20.9	22.6	23.5	25.3	26.9	21.3[2]	22.0[2]	...
Commercial vehicles										
Véhicules utilitaires	17.4	20.1	20.6	21.9	23.6	24.4	26.3	15.9[2]	16.0[2]	...
Sweden Suède										
Passenger cars										
Voitures de tourisme	3 081.0[4]	3 151.2[4]	3 253.6[4]	3 366.6[4]	3 482.7[4]	3 578.0[4]	3 600.5[4]	3 619.4	3 586.7	3 566.0
Commercial vehicles										
Véhicules utilitaires	540.5	552.1	567.1	584.9	611.6	643.8	658.0	656.9	649.1	643.0
Switzerland Suisse										
Passenger cars[34]										
Voitures de tourisme[34]	2 552.1	2 617.2	2 678.9	2 732.7	2 745.4	2 899.6	2 993.5	3 065.8	3 098.6	3 116.6
Commercial vehicles[34]										
Véhicules utilitaires[34]	203.6	211.3	217.8	228.7	252.1	277.6	303.7	311.0	310.5	307.5
Syrian Arab Republic Rép. arabe syrienne										
Passenger cars										
Voitures de tourisme	126.4	120.1	123.9	125.3	124.3	124.7	126.0	126.9	128.0	149.8
Commercial vehicles										
Véhicules utilitaires	111.3	109.2	116.8	117.3	126.4	118.0	118.5	121.9	136.7	162.3
Thailand Thaïlande										
Passenger cars[3]										
Voitures de tourisme[3]	689.8	732.6	770.4	...	1 147.0	1 000.0	1 222.0	1 279.2	1 396.6	1 598.2
Commercial vehicles[3]										
Véhicules utilitaires[3]	674.8	684.7	689.9	...	789.0	835.0	976.0	1 008.8	1 188.3	1 483.7
Togo Togo										
Passenger cars										
Voitures de tourisme	2.1	2.1	2.2	2.6	# 25.0[1]	25.0[1]	25.0[1]	25.0[1]	25.0[1]	...
Commercial vehicles										
Véhicules utilitaires	2.1	0.8	0.8	0.9	# 15.0[1]	15.0[1]	15.0[1]	16.0[1]	16.1[1]	...
Tonga Tonga										
Passenger cars										
Voitures de tourisme	0.8	0.9	1.1	1.4	1.6	1.4	2.0	2.8	3.3[2]	...
Commercial vehicles										
Véhicules utilitaires	1.6	1.7	2.1	2.4	3.5	2.2	2.6	3.1	3.7[2]	...
Trinidad and Tobago Trinité-et-Tobago										
Passenger cars										
Voitures de tourisme	229.4	241.6[1]	244.1[1]	244.0[1]	244.0[1]	244.0[1]	145.8[1]	150.1[1]	161.1[1]	...
Commercial vehicles[1]										
Véhicules utilitaires[1]	60.6	79.2	79.0	79.0	79.0	79.0	58.0	60.0	42.5	...
Tunisia Tunisie										
Passenger cars										
Voitures de tourisme	145.0	166.0	171.0	179.0	292.6[1]	320.1[1]	320.1[1]	320.0[1]	320.0[1]	...
Commercial vehicles										
Véhicules utilitaires	150.0	160.0	181.0	188.0	166.9[1]	174.1[1]	174.0[1]	177.0[1]	180.5[1]	...
Turkey Turquie										
Passenger cars										
Voitures de tourisme	919.6	983.4	1 087.2	1 193.0	1 310.3	1 434.8	1 649.9	1 864.3	2 181.4	2 619.9
Commercial vehicles[21]										
Véhicules utilitaires[21]	361.6	381.1	401.8	414.9	427.3	438.0	452.9	469.0	490.9	518.4
Uganda Ouganda										
Passenger cars										
Voitures de tourisme	10.0	10.0	12.0	12.0	13.0	13.0	13.0	17.0	17.8[1]	...
Commercial vehicles										
Véhicules utilitaires	10.0	10.0	11.0	12.0	13.0	14.0	15.0	20.0	25.2[1]	...

62
Motor vehicles in use
Passenger cars and commercial vehicles: thousand units [*cont.*]
Véhicules automobiles en circulation
Voitures de tourisme et véhicules utilitaires : milliers de véhicules [*suite*]

Country or area Pays or zone	1984	1985	1986	1987	1988	1989	1990	1991	1992	1993
Ukraine Ukraine										
Passenger cars										
Voitures de tourisme	2 269.0	2 447.0	2 619.0	2 816.0	3 012.0	3 195.0	3 362.0	3 657.0	3 884.8	...
former USSR† ancienne URSS†										
Passenger cars										
Voitures de tourisme	6 270.8	6 670.6	7 082.9	7 539.2	7 955.5	8 438.1	8 963.9	9 712.6	10 531.3	11 518.2
United Kingdom Royaume-Uni										
Passenger cars										
Voitures de tourisme	17 213.3	17 737.1	18 355.1	18 859.1	19 940.0	20 925.0	21 485.0	21 515.0	21 904.0	...
Commercial vehicles										
Véhicules utilitaires	1 974.5	1 983.8	2 011.1	2 053.4	2 193.7	2 563.0	2 520.0	2 438.0	2 452.0	...
United Rep.Tanzania Rép. Unie de Tanzanie										
Passenger cars										
Voitures de tourisme	49.0	49.0	49.0	49.0	49.0	49.0	44.0	44.0[1]	44.0[1]	...
Commercial vehicles										
Véhicules utilitaires	32.0	33.0	33.0	33.0	33.0	33.0	54.0	55.0[1]	57.2[1]	...
United States Etats-Unis										
Passenger cars										
Voitures de tourisme	128 158.0	131 864.0	135 431.0	137 323.0	141 251.7[2]	143 081.4[2]	143 549.6[2]	142 955.6	144 213.4	146 314.2
Commercial vehicles										
Véhicules utilitaires	37 507.0	39 196.0	40 166.0	41 119.0	43 145.0[2]	44 179.1[2]	45 105.8[2]	45 416.3	46 148.8	47 749.1
Uruguay Uruguay										
Passenger cars										
Voitures de tourisme	292.4	306.3	318.4	332.7	350.2	360.3	379.6	389.6	418.0	425.6
Commercial vehicles										
Véhicules utilitaires	45.1	46.5	47.0	46.6	49.8	50.2	49.9	48.6	45.0	44.4
Vanuatu Vanuatu										
Passenger cars										
Voitures de tourisme	2.3[4 30]	2.3	3.8[1]	4.0[1]	4.0[1]	4.0[1]	4.0[1]	4.2[1]	4.0[1]	...
Commercial vehicles										
Véhicules utilitaires	1.1[4 30]	1.2	2.4[1]	2.0[1]	2.0[1]	2.0[1]	2.0[1]	2.0[1]	2.2[1]	...
Venezuela Venezuela										
Passenger cars										
Voitures de tourisme	1 559.0	1 598.0	1 656.0	1 718.0	1 740.0	1 615.0	1 582.0	1 540.0	1 566.0	1 579.0
Commercial vehicles										
Véhicules utilitaires	405.0	418.0	430.0	448.0	421.0	459.0	464.0	449.0	456.0	460.0
Yemen Yémen										
Passenger cars[30]										
Voitures de tourisme[30]	24.7	25.6	26.5	27.6	# 138.1[1]	145.4[1]	145.0[1]	145.0[1]	145.0[1]	...
Commercial vehicles[30]										
Véhicules utilitaires[30]	27.2	29.2	31.2	32.4	# 202.0[1]	219.1[1]	219.0[1]	219.0[1]	219.0[1]	...
Yugoslavia, SFR† Yougoslavie, Rfs†										
Passenger cars[31]										
Voitures de tourisme[31]	1 190.7	# 1 080.1	1 133.1	1 177.3	1 225.0	1 309.7	1 405.5	..	..	..
Commercial vehicles[31]										
Véhicules utilitaires[31]	110.2	# 102.0	111.2	117.4	121.4	127.2	132.5	..	..	..
Zaire Zaïre										
Passenger cars										
Voitures de tourisme	92.0	92.0	92.0	92.0	130.7	137.9	145.1	...	...	...
Commercial vehicles										
Véhicules utilitaires	80.0	80.0	80.0	80.0	76.8	84.8	92.8	...	...	...
Zambia Zambie										
Passenger cars[9]										
Voitures de tourisme[9]	...	71.8	74.0	75.2	96.0[1]	96.0[1]	96.0[1]	96.0[1]	96.0[1]	...

62
Motor vehicles in use
Passenger cars and commercial vehicles: thousand units [cont.]
Véhicules automobiles en circulation
Voitures de tourisme et véhicules utilitaires : milliers de véhicules [suite]

Country or area Pays or zone	1984	1985	1986	1987	1988	1989	1990	1991	1992	1993
Commercial vehicles[9] Véhicules utilitaires[9]	33.7	35.2	37.7	39.1	67.0[1]	68.0[1]	68.0[1]	68.0[1]	68.0[1]	...
Zimbabwe Zimbabwe										
Passenger cars Voitures de tourisme	200.0	241.0	266.0	274.0	276.0	285.0	* 290.0	* 300.0	...	...
Commercial vehicles Véhicules utilitaires	75.0	78.0	80.0	81.0	81.0	82.0	83.0	* 85.0	...	...

Source:
Transport statistics database of the Statistical Division of
the United Nations Secretariat.

† For detailed descriptions of data pertaining to
former Czechoslovakia, Germany, SFR Yugoslavia and former
USSR, see Annex I - Country or area nomenclature, regional
and other groupings.

1 Source: World Automotive Market Report, Auto and Truck
   International (Illinois).
2 Source: AAMA Motor Vehicle Facts and figures, American
   Automobile Manufacturers Association (Michigan).
3 Twelve months ending 30 September of year indicated.
4 Including vehicles operated by police or other governmental
   security organizations.
5 Including farm tractors.
6 Including jeeps.
7 Including commercial vehicles.
8 Including special-purpose vehicles.
9 Including vehicles no longer in circulation.
10 Including Faeroe Islands.
11 Excluding jeeps.
12 Twelve months ending 7 July.
13 Including ambulances (Fiji: fire engines also).

14 Including 3-wheeled passengers and goods vehicles.

15 Twelve months ending 20 March of year indicated.
16 Passenger cars include mini-buses equipped for transport of
   nine to fifteen passengers.
17 Including school buses.
18 Excluding small vehicles.
19 Including cars with a seating capacity of up to 10 persons.
20 Including light commercial vehicles.
21 Excluding tractors and semi-trailer combinations.
22 Excluding tractors.
23 Excluding diplomatic corps vehicles.
24 Twelve months ending 31 July.
25 Twelve months ending 31 March.
26 Including hearses (Norway: registered before 1981).
27 Excluding buses and tractors, but including special lorries.

28 Excluding Madeira and Azores.
29 Commercial vehicles include hearses, ambulances,
   fire-engines and jeeps specifically registered as commercial
   vehicles.
30 Former Democratic Yemen only.
31 Beginning 1985, data are not strictly comparable to data of
   previous years.

Source:
Base de données pour les statistiques des transports de la
Division de statistique du Secrétariat de l'ONU.

† Pour les descriptions en détails des données
relatives à l'ancienne Tchécoslovaquie, l'Allemagne, la Rfs
Yougoslavie et l'ancienne URSS, voir l'Annexe I -
Nomenclature des pays ou zones, groupements régionaux et
autres groupements.

1 Source: "World Automotive Market Report, Auto and Truck
   International" (Illinois).
2 Source: "AAMA Motor Vehicle Facts and Figures", American
   Automobile Manufacturers Association (Michigan).
3 Douze mois finissant le 30 septembre de l'année indiquée.
4 Y compris véhicules de la police ou d'autres services
   gouvernementaux d'ordre public.
5 Y compris tracteurs agricoles.
6 Y compris jeeps.
7 Y compris véhicules utilitaires.
8 Y compris véhicules à usages spéciaux.
9 Y compris véhicules retirés de la circulation.
10 Y compris les Iles Féroé.
11 Non compris jeeps.
12 Douze mois finissant le 7 juillet de l'année indiquée.
13 Y compris ambulances (Fidji: aussi les voitures de
   pompiers).
14 Y compris véhicules à trois roues (passagers et
   marchandises).
15 Douze mois finissant le 20 mars de l'année indiquée.
16 Voitures de tourisme comprennent mini-buses ayant une
   capacité de neuf à quinze passagers.
17 Y compris l'autobus de l'école.
18 Non compris véhicules petites.
19 Y compris véhicules comptant jusqu'à 10 places.
20 Y compris véhicules utilitaires légers.
21 Non compris ensembles tracteur-remorque et semi-remorque.
22 Non compris tracteurs.
23 Non compris véhicules des diplomates.
24 Douze mois finissant le 31 juillet.
25 Douze mois finissant le 31 mars.
26 Y compris corbillards (Norvège: enregistrés avant de 1981).
27 Non compris autobus et tracteurs, mais y compris camions
   spéciaux.
28 Non compris Madère et Azores.
29 Véhicules utilitaires comprennent corbillards, ambulances,
   voitures de pompiers et jeeps spécifiquement immatriculés
   comme véhicules utilitaires.
30 L'ancien Yémen démocratique seulement.
31 A compter de 1985, les données ne sont plus exactement
   comparables à celles des années précédants.

63
Merchant shipping: fleets
Transports maritimes : flotte marchande
Total, Oil tankers and Ore and bulk carrier fleets: thousand gross registered tons
Total, Pétroliers et Minéraliers et transporteurs de vracs : milliers de tonneaux de jauge brute

Flag Pavillon	1984	1985	1986	1987	1988	1989	1990	1991	1992	1993
	A. Total • Totale									
World *Monde*	**418 682**	**416 269**	**404 910**	**403 498**	**403 406**	**410 481**	**423 627**	**436 027**	**445 169**	**457 915**
Steam **Vapeur**	**99 114**	**89 857**	**76 896**	**72 228**	**67 989**	**65 070**	**63 974**	**62 662**	**56 800**	**...**
Motor **Moteur**	**319 569**	**326 412**	**328 014**	**331 270**	**335 417**	**345 410**	**359 653**	**373 364**	**388 200**	**...**
	Africa • Afrique									
Algeria Algérie	1 372	1 347	882	893	897	848	906	921	921	921
Angola Angola	97	91	92	92	91	93	93	93	94	88
Benin Bénin	5	5	5	5	5	5	5	2	2	1
Cameroon Cameroun	74	76	77	58	57	33	33	34	35	36
Cape Verde Cap-Vert	14	14	14	15	17	18	21	22	22	23
Comoros Comores	1	1	1	2	1	2	2	3	2	2
Congo Congo	8	8	8	8	8	8	9	9	9	10
Côte d'Ivoire Côte d'Ivoire	143	142	121	119	119	83	82	82	75	103
Djibouti Djibouti	3	3	3	3	3	3	3	3	3	4
Egypt Egypte	779	953	1 063	1 074	1 227	1 230	1 257	1 257	1 122	1 149
Equatorial Guinea Guinée équatoriale	6	6	6	6	6	6	6	6	7	2
Ethiopia[1] Ethiopie[1]	30	57	67	73	74	77	75	84	70	69
Gabon Gabon	97	98	98	24	25	25	24	25	25	36
Gambia Gambie	3	3	3	4	4	2	2	3	2	2
Ghana Ghana	186	163	166	142	125	126	126	135	135	118
Guinea Guinée	7	7	7	7	7	8	9	9	5	6
Guinea-Bissau Guinée-Bissau	4	4	4	4	4	4	4	4	4	4
Kenya Kenya	7	8	9	8	8	8	7	13	14	16
Liberia Libéria	62 025	58 180	52 649	51 412	49 734	47 893	54 700	52 427	55 918	53 919
Libyan Arab Jamah. Jamah. arabe libyenne	855	854	825	817	830	831	835	840	720	721

63
Merchant shipping: fleets
Total, Oil tankers and Ore and bulk carrier fleets: thousand gross registered tons [*cont.*]
Transports maritimes : flotte marchande
Total, Pétroliers et Minéraliers et transporteurs de vracs : milliers de tonneaux de jauge brute [*suite*]

Flag Pavillon	1984	1985	1986	1987	1988	1989	1990	1991	1992	1993
Madagascar Madagascar	78	74	74	64	92	70	74	73	45	34
Mauritania Mauritanie	15	17	23	30	37	40	41	42	43	44
Mauritius Maurice	42	38	152	163	157	130	99	82	122	194
Morocco Maroc	434	461	416	418	437	454	488	483	479	393
Mozambique Mozambique	46	41	43	36	36	38	40	37	39	36
Namibia Namibie	...	...	...	...	...	...	...	0	17	36
Nigeria Nigéria	442	443	564	594	587	500	496	493	516	515
Réunion Réunion	...	...	...	21	21	21	21	21	21	...
Saint Helena Sainte-Hélène	3	4	4	4	4	3	3	...	...	...
Sao Tome and Principe Sao Tomé-et-Principe	1	1	1	1	1	1	1	1	3	3
Senegal Sénégal	44	51	50	46	49	51	52	55	58	66
Seychelles Seychelles	1	2	4	3	3	3	3	4	4	4
Sierra Leone Sierra Leone	6	6	7	9	14	18	21	21	26	26
Somalia Somalie	28	29	16	18	13	11	17	17	17	18
South Africa Afrique du Sud	712	632	600	533	486	397	352	340	336	346
Sudan Soudan	96	96	96	97	97	97	58	45	45	64
Togo Togo	30	54	55	60	515	43	52	22	12	12
Tunisia Tunisie	277	284	286	285	281	282	278	276	280	269
United Rep.Tanzania Rép. Unie de Tanzanie	59	51	51	32	32	32	32	39	41	43
Zaire Zaïre	85	85	66	56	56	56	56	56	29	15
America, North · Amérique du Nord										
Anguilla Anguilla	4	4	4	4	3	3	3	5	5	4
Antigua and Barbuda Antigua-et-Barbuda	1	1	1	52	323	392	359	811	802	1 063
Bahamas Bahamas	3 192	3 907	5 985	9 105	8 963	11 579	13 626	17 541	20 616	21 224
Barbados Barbade	8	8	8	8	8	8	8	8	51	49

63
Merchant shipping: fleets
Total, Oil tankers and Ore and bulk carrier fleets: thousand gross registered tons [*cont.*]
Transports maritimes : flotte marchande
Total, Pétroliers et Minéraliers et transporteurs de vracs : milliers de tonneaux de jauge brute [*suite*]

Flag Pavillon	1984	1985	1986	1987	1988	1989	1990	1991	1992	1993
Belize Belize	1	1	1	1	1	1	1	...	37	148
Bermuda Bermudes	822	981	1 208	1 925	3 774	4 076	4 258	3 037	3 338	3 140
British Virgin Islands Iles Vierges britanniques	8	9	8	8	7	7	7	7	7	6
Canada Canada	3 449	3 344	3 160	2 971	2 902	2 825	2 744	2 685	2 610	2 541
Cayman Islands Iles Caïmanes	348	414	1 390	706	477	411	415	395	363	383
Costa Rica Costa Rica	20	20	13	15	15	13	14	14	8	8
Cuba Cuba	959	965	959	966	912	900	836	770	671	626
Dominica Dominique	1	3	2	2	2	3	2	2	2	4
Dominican Republic Rép. dominicaine	37	47	42	44	48	44	36	12	12	13
El Salvador El Salvador	4	4	4	4	4	4	2	2	2	2
Greenland Groënland	...	...	...	40	50	56	57	57	55	...
Grenada Grenade	0	0	0	1	1	1	1	1	1	1
Guadeloupe Guadeloupe	...	...	...	2	3	4	4	5	6	...
Guatemala Guatemala	16	16	9	5	5	5	5	1	2	1
Haiti Haïti	2	3	3	1	1	1	1	1	1	1
Honduras Honduras	277	357	555	506	582	691	712	816	1 045	1 116
Jamaica Jamaïque	9	9	9	13	14	14	14	14	11	11
Martinique Martinique	...	...	...	7	7	8	8	1	1	...
Mexico Mexique	1 489	1 467	1 520	1 532	1 448	1 388	1 320	1 196	1 114	1 125
Montserrat Montserrat	1	1	1	1	1	1	1	1	1	...
Netherlands Antilles[2] Antilles néerlandaises[2]	...	...	...	395	432	421	455	568	841	1 039
Nicaragua Nicaragua	19	18	23	13	14	5	5	5	4	4
Panama Panama	37 244	40 674	41 305	43 255	44 604	47 365	39 298	44 949	52 486	57 619
Puerto Rico Porto Rico	...	...	...	76	58	57	21	16	9	...
Saint Kitts and Nevis Saint-Kitts-et-Nevis	1	1	0	1	0	0	0	0	0	0

63
Merchant shipping: fleets
Total, Oil tankers and Ore and bulk carrier fleets: thousand gross registered tons [*cont.*]
Transports maritimes : flotte marchande
Total, Pétroliers et Minéraliers et transporteurs de vracs : milliers de tonneaux de jauge brute [*suite*]

Flag Pavillon	1984	1985	1986	1987	1988	1989	1990	1991	1992	1993
Saint Lucia Sainte-Lucie	3	2	3	2	2	2	2	2	2	2
Saint Pierre and Miquelon Saint-Pierre-et-Miquelon	...	...	...	3	4	4	4	3	6	...
St. Vincent-Grenadines St. Vincent-Grenadines	101	235	510	700	900	1 486	1 937	4 221	4 698	5 287
Trinidad and Tobago Trinité-et-Tobago	19	19	19	19	24	22	22	22	24	23
Turks and Caicos Islands Iles Turques et Caiques	3	3	4	3	4	3	3	5	4	4
United States Etats-Unis	19 292	19 518	19 900	20 086	20 758	20 263	19 744	18 565	14 435	14 087
America, South · Amérique du Sud										
Argentina Argentine	2 422	2 457	2 117	1 901	1 877	1 833	1 890	1 709	873	773
Brazil Brésil	5 722	6 057	6 212	6 324	6 123	6 078	6 016	5 883	5 348	5 216
Chile Chili	473	454	567	547	604	590	616	619	580	624
Colombia Colombie	374	366	380	424	412	379	372	313	250	238
Ecuador Equateur	412	444	438	421	428	402	385	384	348	286
Falkland Is. (Malvinas) Iles Falkland (Malvinas)	7	7	7	7	7	8	10	10	14	15
Guyana Guyana	24	23	23	22	15	15	15	16	17	17
Paraguay Paraguay	38	43	43	42	39	39	37	35	33	31
Peru Pérou	788	818	754	788	675	638	617	605	433	411
Suriname Suriname	15	15	13	11	11	11	13	13	13	13
Uruguay Uruguay	190	173	150	144	170	100	104	105	127	149
Venezuela Venezuela	1 003	985	998	999	982	1 087	935	970	871	971
Asia · Asie										
Azerbaijan Azerbaïdjan	..	..	..	..	..	..	..	..	637	667
Bahrain Bahreïn	44	48	52	44	54	55	47	262	138	103
Bangladesh Bangladesh	367	358	379	411	432	439	464	456	392	388
Brunei Darussalam Brunéi Darussalam	1	1	2	352	354	355	358	348	364	365
Cambodia Cambodge	...	...	...	...	...	...	...	...	4	6
China Chine	9 300	10 568	11 557	12 341	12 920	13 514	13 899	14 299	13 899	14 945

63
Merchant shipping: fleets
Total, Oil tankers and Ore and bulk carrier fleets: thousand gross registered tons [cont.]
Transports maritimes : flotte marchande
Total, Pétroliers et Minéraliers et transporteurs de vracs : milliers de tonneaux de jauge brute [suite]

Flag Pavillon	1984	1985	1986	1987	1988	1989	1990	1991	1992	1993
Cyprus Chypre	6 728	8 196	10 617	15 650	18 390	18 134	18 336	20 298	20 487	22 842
Hong Kong Hong-kong	5 784	6 858	8 180	8 035	7 329	6 151	6 565	5 876	7 267	7 664
India Inde	6 415	6 605	6 540	6 726	6 161	6 315	6 476	6 517	6 546	6 575
Indonesia Indonésie	1 857	1 936	2 086	2 121	2 126	2 035	2 179	2 337	2 367	2 440
Iran, Islamic Rep. of Iran, Rép. islamique d'	2 106	2 380	2 911	3 977	4 337	4 733	4 738	4 583	4 571	4 444
Iraq Iraq	1 074	1 012	1 016	1 002	953	1 056	1 044	931	902	902
Israel Israël	563	550	557	515	546	505	530	604	664	652
Japan Japon	40 358	39 940	38 488	35 932	32 074	28 030	27 078	26 407	25 102	24 248
Jordan Jordanie	48	48	42	33	32	32	42	135	61	71
Korea, Dem. P. R. Corée, R. p. dém. de	460	513	407	407	406	396	442	511	602	671
Korea, Republic of Corée, République de	6 771	7 169	7 184	7 214	7 334	7 832	7 783	7 821	7 407	7 047
Kuwait Koweït	2 551	2 350	2 581	2 088	735	1 865	1 855	1 373	2 258	2 218
Lebanon Liban	458	505	485	461	405	384	307	274	293	249
Macau Macao	...	...	...	3	4	3	3	3	3	2
Malaysia Malaisie	1 664	1 773	1 744	1 689	1 608	1 668	1 717	1 755	2 048	2 166
Maldives Maldives	137	133	85	100	104	94	78	42	52	55
Myanmar Myanmar	108	116	126	239	273	582	827	1 046	947	711
Oman Oman	14	17	15	25	25	24	23	23	15	16
Pakistan Pakistan	507	451	434	394	366	366	354	358	380	360
Philippines Philippines	3 441	4 594	6 922	8 681	9 312	9 385	8 515	8 626	8 470	8 466
Qatar Qatar	333	353	307	306	309	306	359	485	392	431
Saudi Arabia Arabie saoudite	3 863	3 137	2 978	2 692	2 269	2 119	1 683	1 321	1 016	998
Singapore Singapour	6 512	6 505	6 268	7 098	7 209	7 273	7 928	8 488	9 905	11 035
Sri Lanka Sri Lanka	746	635	622	594	410	287	350	333	285	294
Syrian Arab Republic Rép. arabe syrienne	56	58	63	63	64	74	80	109	144	209

63
Merchant shipping: fleets
Total, Oil tankers and Ore and bulk carrier fleets: thousand gross registered tons [*cont.*]
Transports maritimes : flotte marchande
Total, Pétroliers et Minéraliers et transporteurs de vracs : milliers de tonneaux de jauge brute [*suite*]

Flag Pavillon	1984	1985	1986	1987	1988	1989	1990	1991	1992	1993
Thailand Thaïlande	517	586	533	511	515	539	615	725	917	1 116
Turkey Turquie	3 125	3 684	3 424	3 336	3 281	3 240	3 719	4 107	4 136	5 044
United Arab Emirates Emirats arabes unis	766	869	654	732	825	839	750	889	884	804
Viet Nam Viet Nam	279	299	339	360	338	358	470	574	616	728
Yemen Yémen	..	..	..	..	..	..	17	17	16	24
former Yemen Arab Rep. anc. Yémen rép. arabe	3	3	7	200	196	196	..	..	..	..
former Dem. Yemen anciennce Yémen dém.	12	12	13	12	11	11	..	..	..	..
Europe · Europe										
Albania Albanie	56	56	56	56	56	56	56	59	59	59
Austria Autriche	129	134	125	194	201	204	139	139	140	160
Belgium Belgique	2 407	2 400	2 420	2 268	2 118	2 044	1 954	314	241	218
Bulgaria Bulgarie	1 283	1 322	1 385	1 551	1 392	1 375	1 360	1 367	1 348	1 314
Channel Islands Iles Anglo-Normandes	...	...	...	23	15	10	8	4	3	3
Croatia Croatie	..	..	..	..	..	..	..	..	210	193
former Czechoslovakia† anc. Tchécoslovaquie†	184	184	198	157	158	191	326	361	238	..
Czech Republic République tchèque	..	..	..	..	..	..	..	..	..	228
Denmark Danemark	5 211	4 942	4 651	4 714	4 322	4 785	5 008	5 698	5 269	5 293
Estonia Estonie	..	..	..	..	..	..	..	..	680	686
Faeroe Islands Iles Féroé	91	103	115	119	130	121	124	115	111	100
Finland Finlande	2 168	1 974	1 470	1 122	838	944	1 069	1 053	1 197	1 354
France[3] France[3]	8 945	8 237	5 936	5 264	4 395	4 286	3 721	3 879	3 869	4 252
Germany † Allemagne†	..	..	..	..	..	..	..	5 971	5 360	4 979
F. R. Germany R. f. Allemagne	6 242	6 177	5 565	4 318	3 917	3 967	4 301	..	..	..
former German D. R. anc. R. d. allemande	1 422	1 434	1 519	1 494	1 443	1 500	1 437	..	..	..
Gibraltar Gibraltar	247	583	1 613	2 827	3 042	2 611	2 008	1 410	492	384
Greece Grèce	35 059	31 032	28 391	23 560	21 979	21 324	20 522	22 753	25 739	29 134

63
Merchant shipping: fleets
Total, Oil tankers and Ore and bulk carrier fleets: thousand gross registered tons [cont.]
Transports maritimes : flotte marchande
Total, Pétroliers et Minéraliers et transporteurs de vracs : milliers de tonneaux de jauge brute [suite]

Flag Pavillon	1984	1985	1986	1987	1988	1989	1990	1991	1992	1993
Hungary Hongrie	80	77	86	77	76	76	98	104	92	45
Iceland Islande	179	180	176	174	175	183	177	168	177	174
Ireland Irlande	221	194	149	154	173	167	181	195	199	184
Isle of Man Ile de Man	...	...	...	1 914	2 137	2 111	1 824	1 937	1 628	1 563
Italy Italie	9 158	8 343	7 897	7 817	7 794	7 602	7 991	8 122	7 513	7 030
Latvia Lettonie	..	..	..	..	..	..	..	..	1 207	1 155
Lithuania Lituanie	..	..	..	..	..	..	..	..	668	639
Luxembourg Luxembourg	...	...	...	...	2	4	3	1 703	1 656	1 327
Malta Malte	1 366	1 856	2 015	1 726	2 686	3 329	4 519	6 916	11 005	14 163
Netherlands Pays-Bas	4 586	4 301	4 324	3 514	3 294	3 234	3 330	3 305	3 346	3 086
Norway Norvège	17 663	15 339	9 295	6 359	9 350	15 597	23 429	23 586	22 231	21 536
Poland Pologne	3 267	3 315	3 457	3 470	3 489	3 416	3 369	3 348	3 109	2 646
Portugal Portugal	1 571	1 437	1 114	1 045	985	723	851	887	972	1 002
Romania Roumanie	2 667	3 024	3 234	3 264	3 561	3 783	4 005	3 828	2 981	2 867
Russian Federation Fédération de Russie	..	..	..	..	..	..	..	..	16 302	16 814
Slovenia Slovénie	..	..	..	..	..	..	..	..	2	2
Spain[4] Espagne[4]	7 005	6 256	5 422	4 949	4 415	3 962	3 807	3 617	2 643	1 752
Sweden Suède	3 520	3 162	2 517	2 270	2 116	2 167	2 775	3 174	2 884	2 439
Switzerland Suisse	319	342	346	355	259	220	287	286	346	300
Ukraine Ukraine	..	..	..	..	..	..	..	..	5 222	5 265
United Kingdom Royaume-Uni	15 874	14 344	11 567	6 568	6 108	5 525	4 887	4 670	4 081	4 117
Yugoslavia, SFR† Yougoslavie, Rfs†	2 682	2 699	2 873	3 165	3 476	3 681	3 816	3 293	..	..
Oceania · Océanie										
Australia Australie	2 173	2 088	2 368	2 405	2 366	2 494	2 512	1 709	2 689	2 862
Cook Islands Iles Cook	...	...	...	...	4	5	6	7	5	5
Micronesia,Federated States of Micron, Etats fédérés de	...	...	...	...	...	...	6	8	9	9

63
Merchant shipping: fleets
Total, Oil tankers and Ore and bulk carrier fleets: thousand gross registered tons [*cont.*]
Transports maritimes : flotte marchande
Total, Pétroliers et Minéraliers et transporteurs de vracs : milliers de tonneaux de jauge brute [*suite*]

Flag Pavillon	1984	1985	1986	1987	1988	1989	1990	1991	1992	1993
Fiji Fidji	29	31	30	35	37	62	55	50	64	39
French Polynesia Polynésie française	...	...	...	21	18	17	20	20	23	...
Guam Guam	...	...	...	3	3	3	4	1	1	...
Kiribati Kiribati	2	2	3	3	4	4	4	4	5	5
Marshall Islands Iles Marshall	...	...	...	...	...	...	1 551	1 698	1 676	2 198
Nauru Nauru	67	67	67	66	60	41	32	15	5	1
New Caledonia Nouvelle-Calédonie	...	...	...	12	12	13	14	14	14	...
New Zealand Nouvelle-Zélande	285	296	314	334	332	252	254	269	238	218
Papua New Guinea Papouasie-Nvl-Guinée	26	29	31	36	38	37	37	36	46	47
Samoa Samoa	26	26	26	26	26	27	27	6	6	6
Solomon Islands Iles Salomon	6	6	6	6	9	8	8	8	8	7
Tonga Tonga	16	17	16	18	14	35	52	40	11	12
Tuvalu Tuvalu	1	1	1	1	1	2	1	1	12	70
Vanuatu Vanuatu	90	138	165	540	790	920	2 164	2 173	2 064	1 946
Wallis and Futuna Islands Iles Wallis et Futuna	...	...	...	39	44	59	39	42	80	80
former USSR† ancienne URSS†	**former USSR† · ancienne URSS†** 24 492	24 745	24 961	25 232	25 784	25 854	26 737	26 405	..	..

B. Oil tankers · Pétroliers

World *Monde*	147 463	138 448	128 426	127 600	127 843	129 578	134 836	138 897	138 149	143 077
Algeria Algérie	**Africa · Afrique** 594	594	119	134	119	40	39	39	32	32
Angola Angola	2	2	2	2	2	2	2	2	2	2
Egypt Egypte	104	97	99	95	255	244	263	262	170	195
Ethiopia Ethiopie	3	3	1	1	1	1	4	4	4	4
Gabon Gabon	74	74	74	0	0	0	0	1	1	1
Kenya Kenya	...	...	...	...	...	...	...	4	4	4

63
Merchant shipping: fleets
Total, Oil tankers and Ore and bulk carrier fleets: thousand gross registered tons [cont.]
Transports maritimes : flotte marchande
Total, Pétroliers et Minéraliers et transporteurs de vracs : milliers de tonneaux de jauge brute [suite]

Flag Pavillon	1984	1985	1986	1987	1988	1989	1990	1991	1992	1993
Liberia Libéria	34 113	31 585	28 675	28 249	27 961	26 667	28 763	26 700	27 440	26 273
Libyan Arab Jamah. Jamah. arabe libyenne	745	745	708	708	708	581	707	581	581	581
Madagascar Madagascar	8	10	8	9	9	5	5	5	9	9
Morocco Maroc	62	62	10	10	10	10	10	10	14	14
Mozambique Mozambique	7	0	1	1	1	1	1	1	1	1
Nigeria Nigéria	157	154	223	227	227	225	225	225	235	236
Réunion Réunion	...	...	...	...	...	...	...	...	19	...
South Africa Afrique du Sud	38	38	39	21	21	20	1	1	1	1
Tunisia Tunisie	132	132	132	132	132	27	27	27	27	6
United Rep.Tanzania Rép. Unie de Tanzanie	3	4	4	4	4	3	3	3	4	5
America, North · Amérique du Nord										
Antigua and Barbuda Antigua-et-Barbuda	...	...	...	...	2	47	11	14	5	7
Bahamas Bahamas	2 588	3 010	4 201	5 357	4 556	6 110	6 780	8 738	9 812	9 680
Barbados Barbade	...	...	...	...	...	...	...	...	44	44
Belize Belize	...	...	...	...	...	...	...	...	4	9
Bermuda Bermudes	220	220	301	895	2 836	3 273	3 285	1 987	2 058	1 838
Canada Canada	303	281	279	272	259	260	242	233	162	153
Cayman Islands Iles Caïmanes	59	46	737	218	75	43	79	72	31	31
Cuba Cuba	68	68	68	68	68	80	80	78	71	67
Dominican Republic Rép. dominicaine	1	1	1	1	1	1	1	1	1	1
Honduras Honduras	48	51	67	58	51	87	112	149	144	119
Jamaica Jamaïque	...	...	...	...	2	2	2	2	2	2
Mexico Mexique	663	540	606	587	522	533	507	507	479	478
Netherlands Antilles Antilles néerlandaises	...	...	...	...	...	...	...	...	32	32
Panama Panama	8 206	8 414	9 192	9 966	10 659	11 418	10 080	13 976	16 454	18 273
St. Vincent-Grenadines St. Vincent-Grenadines	7	...	83	96	141	203	295	378	834	1 112

63
Merchant shipping: fleets
Total, Oil tankers and Ore and bulk carrier fleets: thousand gross registered tons [cont.]
Transports maritimes : flotte marchande
Total, Pétroliers et Minéraliers et transporteurs de vracs : milliers de tonneaux de jauge brute [suite]

Flag Pavillon	1984	1985	1986	1987	1988	1989	1990	1991	1992	1993
United States Etats-Unis	7 836	7 472	7 296	7 428	7 949	7 956	8 532	8 069	5 493	5 013
America, South · Amérique du Sud										
Argentina Argentine	860	842	654	585	586	543	568	542	221	107
Brazil Brésil	1 811	1 834	1 938	1 943	1 849	1 838	1 897	1 947	1 930	2 068
Chile Chili	27	15	15	15	19	28	26	26	4	4
Colombia Colombie	31	32	36	12	14	14	14	11	6	6
Ecuador Equateur	163	160	157	159	159	120	120	116	112	75
Paraguay Paraguay	3	3	3	3	3	1	1	1	2	2
Peru Pérou	166	183	147	197	197	197	190	177	131	131
Uruguay Uruguay	95	96	76	77	118	47	47	47	46	46
Venezuela Venezuela	562	498	470	470	463	463	463	478	455	437
Asia · Asie										
Azerbaijan Azerbaïdjan	..	..	..	..	..	..	..	..	197	222
Bahrain Bahreïn	3	3	3	3	2	2	2	2	2	2
Bangladesh Bangladesh	51	38	40	50	60	49	50	51	51	51
China Chine	1 331	1 476	1 701	1 751	1 826	1 790	1 810	1 840	1 721	2 117
Cyprus Chypre	3 180	3 327	4 481	4 984	5 618	5 639	5 390	5 996	4 677	4 960
Hong Kong Hong-kong	583	440	884	1 095	945	827	1 000	756	880	818
India Inde	1 321	1 717	1 814	1 828	1 772	1 718	1 734	1 805	2 009	2 112
Indonesia Indonésie	420	491	617	644	660	580	582	595	594	608
Iran, Islamic Rep. of Iran, Rép. islamique d'	918	918	1 242	2 347	2 717	3 102	3 101	2 945	2 944	2 765
Iraq Iraq	797	747	777	770	729	831	829	725	720	719
Israel Israël	1	1	1	1	1	1	0	0	1	1
Japan Japon	15 212	14 089	12 365	10 798	9 628	7 879	7 584	7 204	7 167	7 249
Jordan Jordanie	...	...	...	...	...	...	...	50	50	50

63

Merchant shipping: fleets
Total, Oil tankers and Ore and bulk carrier fleets: thousand gross registered tons [*cont.*]
Transports maritimes : flotte marchande
Total, Pétroliers et Minéraliers et transporteurs de vracs : milliers de tonneaux de jauge brute [*suite*]

Flag Pavillon	1984	1985	1986	1987	1988	1989	1990	1991	1992	1993
Korea, Dem. P. R. Corée, R. p. dém. de	171	171	59	59	13	13	13	13	112	115
Korea, Republic of Corée, République de	1 024	1 002	977	963	951	808	593	543	612	620
Kuwait Koweït	1 431	1 290	1 629	1 261	133	1 092	1 101	1 044	1 706	1 548
Lebanon Liban	14	14	14	25	25	14	14	2	2	2
Malaysia Malaisie	216	224	238	241	180	163	179	249	256	257
Maldives Maldives	1	1	2	3	3	5	5	5	6	6
Myanmar Myanmar	2	3	3	3	3	2	6	9	3	3
Pakistan Pakistan	43	44	43	43	43	43	43	43	50	50
Philippines Philippines	555	562	656	755	480	403	372	376	388	414
Qatar Qatar	77	112	112	112	112	108	160	160	125	125
Saudi Arabia Arabie saoudite	1 951	1 578	1 604	1 636	1 300	1 214	928	561	265	277
Singapore Singapour	2 081	2 049	1 653	2 305	2 443	2 553	3 165	3 543	4 182	4 684
Sri Lanka Sri Lanka	192	137	140	98	5	8	78	78	74	74
Thailand Thaïlande	140	147	62	69	77	73	85	100	172	184
Turkey Turquie	1 166	1 582	1 037	846	828	793	777	772	828	903
United Arab Emirates Emirats arabes unis	513	629	372	396	450	441	332	334	458	326
Viet Nam Viet Nam	37	38	40	41	15	18	18	91	15	91
Yemen Yémen	..	..	..	..	..	..	2	2	2	2
former Yemen Arab Rep. anc. Yémen rép. arabe	...	...	...	193	193	193	...	...	..	..
former Dem. Yemen anciennce Yémen dém.	2	2	2	2	2	2	...	...	..	..
Belgium Belgique	**Europe · Europe**	232	266	224	252	273	272	12	2	2
	186									
Bulgaria Bulgarie	312	312	317	441	292	285	288	293	283	284
Croatia Croatie	..	..	..	..	..	..	..	..	7	7
Denmark Danemark	2 407	2 199	2 044	2 166	2 053	2 041	2 024	2 061	834	787
Estonia Estonie	..	..	..	..	..	..	..	..	6	6
Finland Finlande	986	926	597	457	211	155	245	209	256	307

63
Merchant shipping: fleets
Total, Oil tankers and Ore and bulk carrier fleets: thousand gross registered tons [cont.]
Transports maritimes : flotte marchande
Total, Pétroliers et Minéraliers et transporteurs de vracs : milliers de tonneaux de jauge brute [suite]

Flag Pavillon	1984	1985	1986	1987	1988	1989	1990	1991	1992	1993
France France	4 800	4 346	2 603	2 465	1 953	1 944	1 717	1 674	1 696[3]	1 869[3]
Germany † Allemagne†	..	..	..	..	..	..	..	249	89	89
F. R. Germany R. f. Allemagne	1 580	1 394	750	317	265	283	228	..	..	..
former German D. R. anc. R. d. allemande	31	36	36	36	36	36	6	..	..	..
Gibraltar Gibraltar	...	167	856	1 834	2 318	2 060	1 552	1 123	319	276
Greece Grèce	10 896	9 366	10 259	9 247	8 492	8 229	7 856	9 095	10 876	13 273
Iceland Islande	2	3	3	2	2	2	1	1	0	2
Ireland Irlande	9	9	8	15	15	19	19	19	8	8
Isle of Man Ile de Man	...	...	...	...	...	...	...	...	801	744
Italy Italie	3 498	3 601	2 561	2 631	2 670	2 461	2 560	2 685	2 115	1 949
Latvia Lettonie	..	..	..	..	..	..	..	..	533	536
Lithuania Lituanie	..	..	..	..	..	..	..	..	17	12
Luxembourg Luxembourg	...	...	...	...	2	2	2	264	107	55
Malta Malte	70	329	515	312	977	1 405	1 646	2 412	3 086	5 176
Netherlands Pays-Bas	1 143	745	931	774	615	575	590	618	364	407
Norway Norvège	8 780	7 263	3 304	2 785	4 397	7 074	10 794	10 904	9 220	9 265
Poland Pologne	237	318	318	317	235	154	154	136	89	89
Portugal Portugal	1 023	860	533	533	486	323	393	460	657	723
Romania Roumanie	295	384	384	384	523	596	645	679	520	446
Russian Federation Fédération de Russie	..	..	..	..	..	..	..	..	2 436	2 507
Spain Espagne	3 650	2 906	2 372	2 104	1 616	1 486	1 472	1 488	917	452
Sweden Suède	1 356	894	504	370	191	211	534	862	692	368
Ukraine Ukraine	..	..	..	..	..	..	..	..	80	80
United Kingdom Royaume-Uni	6 653	5 937	4 394	2 822	2 835	2 604	2 367	2 316	1 179	1 176
Yugoslavia, SFR† Yougoslavie, Rfs†	231	217	308	317	312	312	306	264	..	..

63
Merchant shipping: fleets
Total, Oil tankers and Ore and bulk carrier fleets: thousand gross registered tons [*cont.*]
Transports maritimes : flotte marchande
Total, Pétroliers et Minéraliers et transporteurs de vracs : milliers de tonneaux de jauge brute [*suite*]

Flag Pavillon	1984	1985	1986	1987	1988	1989	1990	1991	1992	1993
	Oceania · Océanie									
Australia Australie	612	588	662	702	702	678	677	708	780	780
Fiji Fidji	5	5	5	5	5	5	4	4	4	3
Marshall Islands Iles Marshall	...	...	...	...	...	...	...	...	1 146	1 601
New Zealand Nouvelle-Zélande	73	73	73	73	80	80	80	80	73	54
Papua New Guinea Papouasie-Nvl-Guinée	2	2	2	2	2	1	2	2	3	3
Vanuatu Vanuatu	...	...	26	248	252	175	233	233	184	24
Wallis and Futuna Islands Iles Wallis et Futuna	...	...	...	...	...	...	...	...	50	50
former USSR† ancienne URSS†	former USSR† · ancienne URSS†									
	4 662	4 591	4 087	4 207	4 368	4 128	4 167	4 068	..	..

C. Ore and bulk carrier fleets · Minéraliers et transporteurs de vracs

World *Monde*	128 334	133 983	132 908	131 028	129 635	129 482	133 190	135 884	139 042	140 915
	Africa · Afrique									
Algeria Algérie	81	57	57	57	77	95	153	172	172	172
Egypt Egypte	121	273	343	369	356	343	343	343	343	343
Liberia Libéria	21 651	20 773	18 028	16 788	15 647	14 374	16 099	15 629	17 035	15 640
Mauritius Maurice	16	16	125	97	81	39	47	47	80	116
Morocco Maroc	103	125	125	92	92	92	92	92	92	...
South Africa Afrique du Sud	161	129	125	88	88	...	...	...	...	...
Tunisia Tunisie	37	37	37	37	37	37	37	37	37	37
	America, North · Amérique du Nord									
Antigua and Barbuda Antigua-et-Barbuda	...	...	...	...	...	3	3	3	43	90
Bahamas Bahamas	233	405	902	2 105	2 368	2 822	3 698	4 872	4 312	4 515
Bermuda Bermudes	326	258	319	432	294	157	161	199	213	147
Canada Canada	2 056	2 000	1 851	1 654	1 590	1 550	1 462	1 414	1 425	1 377
Cayman Islands Iles Caïmanes	51	117	283	184	103	111	89	54	69	100
Cuba Cuba	62	62	62	62	62	62	62	62	31	30

63
Merchant shipping: fleets
Total, Oil tankers and Ore and bulk carrier fleets: thousand gross registered tons [*cont.*]
Transports maritimes : flotte marchande
Total, Pétroliers et Minéraliers et transporteurs de vracs : milliers de tonneaux de jauge brute [*suite*]

Flag Pavillon	1984	1985	1986	1987	1988	1989	1990	1991	1992	1993
Dominican Republic Rép. dominicaine	9	9	11	11	11	11	11	...	...	...
Honduras Honduras	...	...	...	25	60	103	56	57	92	90
Mexico Mexique	213	310	311	328	271	226	178	48	...	...
Netherlands Antilles [2] Antilles néerlandaises[2]	...	...	...	...	...	...	...	...	113	131
Panama Panama	14 558	17 431	17 400	17 733	17 924	17 513	13 293	13 669	17 327	19 280
St. Vincent-Grenadines St. Vincent-Grenadines	78	171	298	406	391	540	651	1 004	1 816	1 798
United States Etats-Unis	2 130	2 075	2 020	1 989	1 935	1 884	2 140	2 168	1 550	1 539
America, South · Amérique du Sud										
Argentina Argentine	451	505	515	457	465	459	502	365	62	62
Brazil Brésil	2 405	2 748	2 802	2 894	2 859	2 943	2 971	2 855	2 378	2 199
Chile Chili	224	240	314	278	320	295	296	296	279	297
Colombia Colombie	47	47	29	92	92	81	81	81	63	63
Ecuador Equateur	11	11	11	.	22	22	27	27	22	22
Peru Pérou	185	192	190	160	134	129	129	129	64	49
Venezuela Venezuela	54	75	86	86	109	157	147	147	96	147
Asia · Asie										
Bahrain Bahreïn	17	17	17	12	12	12	...	...	...	8
China Chine	2 961	3 531	3 871	4 143	4 391	4 726	4 907	5 206	5 405	5 714
Cyprus Chypre	1 711	2 741	3 739	6 788	8 789	8 821	9 226	10 234	10 816	12 277
Hong Kong Hong-kong	4 367	5 428	6 090	5 748	5 355	4 363	4 396	3 925	4 864	5 466
India Inde	3 053	2 954	2 943	3 152	2 729	3 025	3 182	3 134	2 912	2 939
Indonesia Indonésie	149	128	129	129	129	145	138	149	170	170
Iran, Islamic Rep. of Iran, Rép. islamique d'	527	777	1 109	1 080	1 068	1 059	1 059	1 059	1 047	1 049
Israel Israël	74	64	74	42	42	32	32	22	22	22
Japan Japon	13 227	13 900	13 895	12 611	10 847	9 234	8 788	8 652	8 509	7 336
Jordan Jordanie	26	26	26	26	25	25	25	13	...	10

63

Merchant shipping: fleets
Total, Oil tankers and Ore and bulk carrier fleets: thousand gross registered tons [*cont.*]
Transports maritimes : flotte marchande
Total, Pétroliers et Minéraliers et transporteurs de vracs : milliers de tonneaux de jauge brute [*suite*]

Flag Pavillon	1984	1985	1986	1987	1988	1989	1990	1991	1992	1993
Korea, Dem. P. R. Corée, R. p. dém. de	53	64	64	54	89	68	79	79	84	130
Korea, Republic of Corée, République de	3 856	4 258	4 269	4 140	4 227	4 492	4 708	4 463	3 931	3 575
Lebanon Liban	63	131	131	116	84	91	44	44	55	46
Malaysia Malaisie	428	468	456	378	378	347	347	319	449	556
Maldives Maldives	52	54	45	43	43	43	32	21	19	11
Myanmar Myanmar	...	...	...	80	112	355	518	569	553	415
Pakistan Pakistan	12	12	12	...	...	...	...	...	17	17
Philippines Philippines	1 595	2 756	4 805	6 388	6 906	6 987	6 344	6 262	6 040	5 999
Saudi Arabia Arabie saoudite	455	387	373	193	170	170	26	...	12	12
Singapore Singapour	2 044	2 258	2 478	2 465	2 287	2 083	2 190	2 132	2 626	2 889
Sri Lanka Sri Lanka	281	229	242	256	135	37	103	93	93	93
Syrian Arab Republic Rép. arabe syrienne	...	...	...	...	...	...	...	14	24	32
Thailand Thaïlande	12	13	28	16	...	10	10	32	69	158
Turkey Turquie	1 069	1 129	1 353	1 419	1 360	1 413	1 932	2 331	2 263	3 036
United Arab Emirates Emirats arabes unis	40	9	9	9	22	24	34	32	60	27
Viet Nam Viet Nam	14	14	14	14	14	14	14	21	21	21
Europe · Europe										
Austria Autriche	63	63	63	128	128	136	71	71	63	85
Belgium Belgique	1 424	1 365	1 416	1 350	1 164	1 091	979	...	...	...
Bulgaria Bulgarie	508	525	572	620	612	612	611	601	613	589
Croatia Croatie	..	..	..	..	..	..	..	..	32	3
former Czechoslovakia† anc. Tchécoslovaquie†	103	103	116	75	75	96	241	276	153	..
Czech Republic République tchèque	..	..	..	..	..	..	..	..	...	153
Denmark Danemark	446	423	290	251	163	326	353	525	563	498
Estonia Estonie	..	..	..	..	..	..	..	..	160	160
Finland Finlande	373	280	121	70	70	78	120	89	72	71

63
Merchant shipping: fleets
Total, Oil tankers and Ore and bulk carrier fleets: thousand gross registered tons [cont.]
Transports maritimes : flotte marchande
Total, Pétroliers et Minéraliers et transporteurs de vracs : milliers de tonneaux de jauge brute [suite]

Flag Pavillon	1984	1985	1986	1987	1988	1989	1990	1991	1992	1993
France France	1 625	1 400	957	858	698	641	357	406	343 [3]	463 [3]
Germany † Allemagne†	..	..	..	..	..	..	..	695	377	308
F. R. Germany R. f. Allemagne	842	768	559	402	344	318	397	..	..	..
former German D. R. anc. R. d. allemande	321	323	353	362	338	324	324	..	..	..
Gibraltar Gibraltar	59	...	...	716	560	398	344	192	73	58
Greece Grèce	16 439	15 324	13 202	10 557	10 060	9 987	9 783	10 802	11 638	12 482
Ireland Irlande	91	57	...	...	...	...	9	9	3	3
Isle of Man Ile de Man	...	...	...	...	...	...	...	...	304	171
Italy Italie	3 389	3 041	3 059	2 792	2 561	2 351	2 346	2 228	2 200	1 829
Lithuania Lituanie	..	..	..	..	..	..	..	..	112	116
Luxembourg Luxembourg	...	...	...	...	...	...	...	993	879	669
Malta Malte	623	806	952	882	1 019	1 172	1 774	2 772	5 103	5 854
Netherlands Pays-Bas	782	699	524	318	295	328	328	359	244	98
Norway Norvège	4 965	3 930	2 479	1 085	1 684	4 170	7 283	7 091	5 912	4 924
Poland Pologne	1 244	1 337	1 499	1 551	1 604	1 610	1 603	1 636	1 667	1 524
Portugal Portugal	133	158	231	286	267	178	223	197	75	27
Romania Roumanie	1 250	1 439	1 590	1 590	1 667	1 758	1 891	1 698	1 082	1 089
Russian Federation Fédération de Russie	..	..	..	..	..	..	..	..	1 903	1 821
Spain Espagne	1 278	1 321	1 175	1 060	1 105	955	850	732	504	223
Sweden Suède	442	394	271	134	127	176	383	415	209	119
Switzerland Suisse	202	247	264	284	210	183	252	252	312	268
Ukraine Ukraine	..	..	..	..	..	..	..	..	1 199	1 195
United Kingdom Royaume-Uni	3 374	3 014	2 150	1 492	1 286	1 249	749	743	122	104
Yugoslavia, SFR† Yougoslavie, Rfs†	1 045	1 115	1 182	1 369	2 212	1 915	2 018	1 707	..	..
Oceania · Océanie										
Australia Australie	984	932	1 185	1 173	1 178	1 106	1 112	1 004	987	1 030

63

Merchant shipping: fleets
Total, Oil tankers and Ore and bulk carrier fleets: thousand gross registered tons [*cont.*]
Transports maritimes : flotte marchande
Total, Pétroliers et Minéraliers et transporteurs de vracs : milliers de tonneaux de jauge brute [*suite*]

Flag Pavillon	1984	1985	1986	1987	1988	1989	1990	1991	1992	1993
Marshall Islands Iles Marshall	...	...	...	...	...	...	...	...	440	506
Nauru Nauru	37	37	37	37	37	17	17	...	...	...
New Zealand Nouvelle-Zélande	...	...	...	26	26	26	26	13	22	25
Vanuatu Vanuatu	57	106	77	208	361	482	1 133	1 132	1 052	1 117
former USSR† ancienne URSS†	2 744	2 974	3 433	3 535	3 805	4 115	4 183	3 902	..	..

Source:
Lloyd's Register of Shipping (London).

† For detailed descriptions of data pertaining to
former Czechoslovakia, Germany, SFR Yugoslavia and former
USSR, see Annex I - Country or area nomenclature, regional
and other groupings.

1 1993: excluding Eritrea.
2 Including Aruba.
3 Including French Antarctic Territory.
4 Including Canary Islands.

Source:
"Lloyd's Register of Shipping" (Londres).

† Pour les descriptions en détails des données
relatives à l'ancienne Tchécoslovaquie, l'Allemagne, la Rfs
Yougoslavie et l'ancienne URSS, voir l'Annexe I -
Nomenclature des pays ou zones, groupements régionaux et
autres groupements.

1 1993 : non compris Erythrée.
2 Y compris Aruba.
3 Y compris Territoire antarctiques française.
4 Y compris Iles Canaries.

64
International maritime transport
Transports maritimes internationaux
Vessels entered and cleared: thousand net registered tons
Navires entrés et sortis : milliers de tonneaux de jauge nette

Country or area Pays ou zone	1984	1985	1986	1987	1988	1989	1990	1991	1992	1993
Algeria Algérie										
Vessels entered Navires entrés	53 154	51 012	49 408	48 657	49 213	50 721	44 659	...	...	...
Vessels cleared Navires sortis	53 133	50 962	49 161	48 516	48 992	50 646	44 611	...	...	...
American Samoa Samoa américaines										
Vessels entered[1] Navires entrés[1]	944	851	...	...	...	...	...	...	...	...
Argentina Argentine										
Vessels entered[2] Navires entrés[2]	36 336	37 850	34 044	29 294	32 535	31 253	38 283	36 664[3]	11 379[4]	20 092[5]
Vessels cleared[2] Navires sortis[2]	53 133	50 962	49 161	48 516	48 992	50 646	44 611	...	...	...
Bangladesh Bangladesh										
Vessels entered[6] Navires entrés[6]	4 803	5 456	5 372	5 646	5 184	5 655	991	1 169	859	875
Vessels cleared[6] Navires sortis[6]	4 437	4 795	3 833	3 254	3 323	3 796	664	695	629	571
Belgium Belgique										
Vessels entered Navires entrés	86 817	88 098	87 458	90 822	96 548	190 157	204 857	211 767	236 323	229 915
Vessels cleared Navires sortis	74 965	77 086	72 235	75 037	76 919	155 017	162 236	165 185	189 286	190 795
Benin Bénin										
Vessels entered Navires entrés	2 411	...	...	...	...	...	...	...	...	...
Brazil Brésil										
Vessels entered Navires entrés	...	52 206	63 499	65 519	61 992	59 643	61 852	65 461	65 794	...
Vessels cleared Navires sortis	...	143 333	134 303	145 425	159 468	169 512	170 917	166 047	164 152	...
Brunei Darussalam Brunéi Darussalam										
Vessels entered Navires entrés	23 929	24 542	1 215 694	938 919	1 056 590	1 266 208	1 310 080	1 376 212	...	...
Vessels cleared Navires sortis	13 548	10 524	1 211 898	930 574	1 056 590	1 265 100	1 306 281	1 371 748	...	...
Cameroon Cameroun										
Vessels entered[2][7] Navires entrés[2][7]	5 865	6 362	6 427	6 059	5 562	5 333	5 521	5 426	5 344	5 279
Canada Canada										
Vessels entered[8] Navires entrés[8]	67 728	67 643	62 523	61 975	67 688	68 940	64 163	58 690	...	...
Vessels cleared[8] Navires sortis[8]	118 980	116 796	109 616	114 546	120 442	116 769	114 272	116 974	...	...
Cape Verde Cap-Vert										
Vessels entered Navires entrés	4 154	3 539	4 054	4 493	...	...	...	...	...	...
Colombia Colombie										
Vessels entered[2] Navires entrés[2]	14 990	16 092	17 497	19 691	18 781	20 410	22 713	23 732	21 927	24 874
Vessels cleared Navires sortis	14 791	15 799	17 218	19 331	18 826	20 437	22 448	23 809	22 056	24 967
Congo Congo										
Vessels entered Navires entrés	7 592	7 564	7 504	7 181	7 793	7 386	7 148	6 480	...	...

64

International maritime transport
Vessels entered and cleared: thousand net registered tons [*cont.*]
Transports maritimes internationaux
Navires entrés et sortis : milliers de tonneaux de jauge nette [*suite*]

Country or area Pays ou zone	1984	1985	1986	1987	1988	1989	1990	1991	1992	1993
Costa Rica Costa Rica										
Vessels entered Navires entrés	...	...	...	...	...	...	...	...	3 178	3 556
Vessels cleared Navires sortis	...	...	...	...	...	...	...	...	2 486	2 713
Cuba Cuba										
Vessels entered Navires entrés	11 000	12 400	12 900	15 400	12 100	...	...	...	...	...
Vessels cleared Navires sortis	10 900	12 100	12 700	15 300	12 000	...	...	...	...	...
Cyprus Chypre										
Vessels entered Navires entrés	12 296	11 436	12 429	12 839	13 231	14 793	14 964	12 860	14 791	14 918
Djibouti Djibouti										
Vessels entered[2] Navires entrés[2]	4 273	4 197	...	...	...	...	...	...	...	...
Dominican Republic Rép. dominicaine										
Vessels entered[2] Navires entrés[2]	7 112	8 969	9 094	9 219	9 344	9 469	9 594	9 719	9 844	9 969
Vessels cleared Navires sortis	7 328	8 996	9 111	9 226	9 340	9 455	9 569	9 684	9 798	9 913
Ecuador Equateur										
Vessels entered Navires entrés	11 885	12 653	12 694	12 524	16 303	17 008	17 484	21 152	20 806	22 610
Egypt Egypte										
Vessels entered Navires entrés	42 962	37 805	37 865	30 976	20 966	23 252	23 629	21 457	18 488	22 543
Vessels cleared Navires sortis	38 588	34 537	34 751	27 729	11 132	11 157	10 184	12 259	13 716	13 628
El Salvador El Salvador										
Vessels entered Navires entrés	2 500	2 341	...	...	...	...	...	...	...	...
Vessels cleared Navires sortis	1 397	909	...	...	...	...	...	...	...	...
Estonia Estonie										
Vessels entered Navires entrés	..	..	..	..	..	..	..	..	..	3 419
Vessels cleared Navires sortis	..	..	..	..	..	..	..	..	..	3 087
Fiji Fidji										
Vessels entered Navires entrés	3 552	699	1 590	2 025	1 739	2 603	3 012	...	...	...
Finland Finlande										
Vessels entered Navires entrés	50 722	55 740	65 098	68 203	70 420	85 265	102 500	112 418	119 238	117 003
Vessels cleared Navires sortis	50 717	55 713	65 509	68 807	70 877	84 438	102 995	111 948	119 040	121 946
France France										
Vessels entered Navires entrés	1 474 968	1 475 319	1 597 747	1 619 267	1 554 890	1 564 075	1 681 491	1 751 943	1 825 276	1 758 276
Gambia Gambie										
Vessels entered[1] Navires entrés[1]	749	659	954	892	982	969	...	...	...	...
Germany † Allemagne†										
Vessels entered Navires entrés	..	..	..	..	..	..	..	..	225 984	221 741

64

International maritime transport
Vessels entered and cleared: thousand net registered tons [*cont.*]
Transports maritimes internationaux
Navires entrés et sortis : milliers de tonneaux de jauge nette [*suite*]

Country or area Pays ou zone	1984	1985	1986	1987	1988	1989	1990	1991	1992	1993
Vessels cleared Navires sortis	..	..	..	..	..	..	..	..	199 441	196 456
F. R. Germany R. f. Allemagne										
Vessels entered Navires entrés	136 819	138 149	147 513	155 938	162 784	169 542	171 906	181 086	..	..
Vessels cleared Navires sortis	119 243	119 678	123 942	135 987	141 305	150 792	150 514	156 178	..	..
former German D. R. anc. R. d. allemande										
Vessels entered Navires entrés	9 643	10 343	11 023	10 324	10 469	10 820	8 713	...		
Vessels cleared Navires sortis	15 052	14 778	14 486	14 480	15 076	14 303	8 327	...	..	..
Gibraltar Gibraltar										
Vessels entered Navires entrés	152	176	251	403	304	231	209	...	...	...
Greece Grèce										
Vessels entered Navires entrés	25 176	27 800	28 700	31 365	31 638	31 608	35 152	36 679	37 789	32 429
Vessels cleared Navires sortis	20 524	26 276	24 777	23 917	24 448	25 365	22 262	20 118	20 401	18 467
Guadeloupe Guadeloupe										
Vessels entered[2] Navires entrés[2]	1 325	1 235	1 099	1 255	1 458	1 618	2 557	2 579	...	...
Guatemala Guatemala										
Vessels entered Navires entrés	1 255	1 345	1 544	1 826	1 877	2 018	2 111	2 349	...	...
Vessels cleared Navires sortis	1 302	1 424	1 562	1 557	1 579	1 705	1 772	1 740	...	...
Haiti Haïti										
Vessels entered Navires entrés	714	728	725	707	734	709	1 975	...	...	...
Hong Kong Hong-kong										
Vessels entered[9] Navires entrés[9]	82 457	91 608	98 137	106 396	115 736	126 308	131 802	140 121	161 109	184 901
Vessels cleared[9] Navires sortis[9]	82 533	91 631	98 078	106 451	115 641	126 562	131 636	140 198	161 292	184 658
India Inde										
Vessels entered[10][11] Navires entrés[10][11]	31 036	35 566	33 563	30 171	27 894	28 593	34 260	18 892	...	...
Vessels cleared[10][11] Navires sortis[10][11]	26 766	26 924	28 386	26 253	31 031	34 529	35 666	23 423	...	...
Indonesia Indonésie										
Vessels entered Navires entrés	66 801	54 814	63 589	65 245	82 125	82 846	109 490	113 380	128 571	142 968
Vessels cleared Navires sortis	12 427	15 886	20 302	21 449	21 601	22 798	26 105	34 903	38 178	44 959
Ireland Irlande										
Vessels entered[2] Navires entrés[2]	20 804	20 455	20 915	22 557	25 630	25 940	31 769	32 892	33 857	36 408
Israel Israël										
Vessels entered[2][12] Navires entrés[2][12]	13 747	13 230	14 408	15 857	15 962	...	...	...	...	...
Italy Italie										
Vessels entered Navires entrés	149 118	152 426	155 299	161 403	160 251	161 145	173 360	184 693	175 940	168 591
Vessels cleared Navires sortis	61 807	63 165	65 509	65 863	67 927	67 170	74 000	80 303	79 600	84 089

64
International maritime transport
Vessels entered and cleared: thousand net registered tons [*cont.*]
Transports maritimes internationaux
Navires entrés et sortis : milliers de tonneaux de jauge nette [*suite*]

Country or area Pays ou zone	1984	1985	1986	1987	1988	1989	1990	1991	1992	1993
Jamaica Jamaïque										
Vessels entered										
Navires entrés	...	...	...	...	...	...	...	...	...	9 892
Japan Japon										
Vessels entered[2]										
Navires entrés[2]	347 907	352 589	345 284	347 605	361 530	378 291	385 110	402 190	398 240	397 582
Jordan Jordanie										
Vessels entered										
Navires entrés	1 900	2 122	2 012	1 908	1 981	1 834	1 678	1 690	2 041	2 143
Vessels cleared										
Navires sortis	429	549	665	647	602	612	544	385	392	347
Kenya Kenya										
Vessels entered[2 7]										
Navires entrés[2 7]	5 825	5 743	6 134	6 172	6 133	6 091	6 134	5 897	7 112	7 102
Korea, Republic of Corée, République de										
Vessels entered										
Navires entrés	166 217	179 760	201 857	238 186	252 390	262 791	279 004	317 046	348 767	383 311
Vessels cleared										
Navires sortis	163 723	179 286	204 090	240 505	254 627	264 114	278 423	316 670	350 906	381 545
Macau Macao										
Vessels entered										
Navires entrés	9 598	11 482	...	...	...	...	...	...	...	...
Madagascar Madagascar										
Vessels entered[2]										
Navires entrés[2]	1 202	1 302	1 383	1 400	1 494	1 654	2 161	...	...	...
Vessels cleared[2]										
Navires sortis[2]	1 173	1 301	1 355	1 430	1 500	1 639	2 160	...	...	...
Malaysia Malaisie										
Vessels entered[13]										
Navires entrés[13]	69 803	69 032	55 197[14]	59 900[14]	79 521	86 033	95 724	...	...	...
Vessels cleared[13]										
Navires sortis[13]	70 596	69 012	54 925[14]	59 709[14]	79 735	85 301	94 713	...	...	...
Malta Malte										
Vessels entered										
Navires entrés	1 885	2 016	2 370	2 327	2 726	2 681	4 068	5 087	7 049	6 802
Vessels cleared										
Navires sortis	747	763	888	880	942	897	1 670	2 377	3 160	3 534
Mauritius Maurice										
Vessels entered[2]										
Navires entrés[2]	2 515	2 715	2 988	3 733	4 064	4 195	4 364	6 157	5 277	5 271
Vessels cleared										
Navires sortis	2 538	2 696	3 051	3 464	3 677	3 978	4 357	6 188	5 447	5 219
Mexico Mexique										
Vessels entered										
Navires entrés	11 008	10 619	9 009	10 880	14 262	17 318	19 274	18 209	20 887	19 731
Vessels cleared										
Navires sortis	72 387	84 139	82 796	88 950	90 432	84 575	89 111	93 680	97 356	32 816
Myanmar Myanmar										
Vessels entered										
Navires entrés	875	821	866	636	597	248	610	698	...	...
Vessels cleared										
Navires sortis	1 109	960	1 243	1 311	750	683	577	667	...	...
Netherlands Pays-Bas										
Vessels entered										
Navires entrés	210 353	213 668	215 340	210 547	220 323	232 153	365 350[15]	374 428[15]	383 164[15]	375 906[15]
Vessels cleared										
Navires sortis	205 105	209 119	212 823	180 611	190 524	200 373	224 415[15]	232 349[15]	239 572[15]	237 817[15]

64
International maritime transport
Vessels entered and cleared: thousand net registered tons [*cont.*]
Transports maritimes internationaux
Navires entrés et sortis : milliers de tonneaux de jauge nette [*suite*]

Country or area Pays ou zone	1984	1985	1986	1987	1988	1989	1990	1991	1992	1993
New Caledonia Nouvelle-Calédonie										
Vessels entered										
Navires entrés	2 908	2 526	...	...	...	...	...	...	...	...
Vessels cleared										
Navires sortis	2 885	2 524	...	...	...	...	...	...	...	...
New Zealand Nouvelle-Zélande										
Vessels entered										
Navires entrés	14 001	14 607	13 388	14 113	27 844[15]	30 890[15]	32 592[15]	38 069[15]	27 983[15]	...
Vessels cleared										
Navires sortis	13 934	14 613	13 365	14 107	27 247[15]	29 753[15]	31 967[15]	36 158[15]	27 508[15]	...
Nicaragua Nicaragua										
Vessels entered										
Navires entrés	1 300	1 326	...	...	...	...	...	...	...	...
Vessels cleared										
Navires sortis	430	442	...	...	...	...	...	...	...	...
Nigeria Nigéria										
Vessels entered										
Navires entrés	48 299	50 879	47 038	50 879	42 852	40 354	50 787	52 849	53 907	...
Vessels cleared										
Navires sortis	48 830	...	...	...	...	...	...	...	...	...
Norway Norvège										
Vessels entered										
Navires entrés	...	48 734	...	...	...	...	...	...	...	...
Pakistan Pakistan										
Vessels entered[6]										
Navires entrés[6]	13 194	14 493	14 736	15 168	15 190	15 331	15 057	16 289	19 401	18 785
Vessels cleared[6]										
Navires sortis[6]	6 046	7 155	6 899	7 141	6 728	7 265	6 743	7 692	7 382	7 284
Panama Panama										
Vessels entered										
Navires entrés	1 049	1 204	1 322	1 362	1 019	1 141	1 337	1 583	1 917	2 180
Vessels cleared										
Navires sortis	1 111	1 247	1 134	1 228	1 159	1 210	1 377	1 412	1 469	1 560
Peru Pérou										
Vessels entered										
Navires entrés	3 303	3 038	4 241	5 547	5 989	4 075	4 363	6 115	7 012	6 066
Vessels cleared										
Navires sortis	8 692	10 386	11 501	10 478	8 854	9 873	9 076	9 815	8 852	9 187
Philippines Philippines										
Vessels entered										
Navires entrés	18 577	18 395	19 380	21 787	24 130	25 729	30 441	28 969	29 876	32 388
Vessels cleared										
Navires sortis	19 949	18 927	20 282	20 593	20 830	20 441	22 810	22 442	19 411	22 431
Poland Pologne										
Vessels entered										
Navires entrés	11 550	11 279	11 467	12 989	14 394	15 129	12 959	13 116	15 573	15 544
Vessels cleared										
Navires sortis	21 692	18 320	17 939	18 298	19 844	19 477	21 131	18 277	20 031	23 222
Portugal Portugal										
Vessels entered[2 16]										
Navires entrés[2 16]	24 324	25 860	30 333	29 202	32 404	32 836	35 757	34 909	36 415	32 654
Réunion Réunion										
Vessels entered[2]										
Navires entrés[2]	1 849	1 892	1 995	2 189	2 391	2 547	2 728	2 792	...	...
Saint Helena Sainte-Hélène										
Vessels entered										
Navires entrés	55	91	276	174	79	49	78	...	...	...

64

International maritime transport
Vessels entered and cleared: thousand net registered tons [cont.]
Transports maritimes internationaux
Navires entrés et sortis : milliers de tonneaux de jauge nette [suite]

Country or area Pays ou zone	1984	1985	1986	1987	1988	1989	1990	1991	1992	1993
Saint Lucia Sainte-Lucie										
Vessels entered										
Navires entrés	2 009	2 257	2 255	2 408	2 329	2 276	2 063	...	...	...
Vessels cleared										
Navires sortis	2 004	2 252	2 256	3 126	...	...	...	...	...	...
St. Vincent-Grenadines St. Vincent-Grenadines										
Vessels entered										
Navires entrés	1 532	1 179	1 224	1 309	1 374	1 499	1 185	1 143	...	...
Vessels cleared										
Navires sortis	1 501	1 150	1 182	1 184	1 278	1 428	1 120	1 083	...	...
Samoa Samoa										
Vessels entered										
Navires entrés	575	453	434	659	523	499	501	527	425	530
Senegal Sénégal										
Vessels entered										
Navires entrés	10 229	9 750	5 258	9 872	...	...	...	8 985	9 447	9 625
Vessels cleared										
Navires sortis	10 275	9 824	5 237	9 858	...	...	...	8 987	9 477	9 769
Seychelles Seychelles										
Vessels entered										
Navires entrés	515	578	796	909	916	959	953	769	778	872
Singapore Singapour										
Vessels entered										
Navires entrés	73 607	75 520	77 529	81 712	...	...	...	...	...	...
Vessels cleared										
Navires sortis	60 142	62 239	65 501	68 990	...	...	...	...	...	...
South Africa Afrique du Sud										
Vessels entered[15]										
Navires entrés[15]	13 777	13 803	14 864	13 172	13 992	14 510	15 138	13 396	13 309	13 437
Vessels cleared[15]										
Navires sortis[15]	342 878	343 381	348 288	333 902	351 390	371 946	389 880	402 011	439 645	441 053
Spain Espagne										
Vessels entered										
Navires entrés	88 696	93 553	100 658	104 011	108 637	119 595	125 881	132 398	134 847	130 171
Vessels cleared										
Navires sortis	46 720	48 992	47 549	43 967	45 349	43 301	43 208	43 226	43 255	46 942
Sri Lanka Sri Lanka										
Vessels entered										
Navires entrés	12 790	13 608	15 422	13 870	14 161	17 084	20 148	20 545	22 087	24 955
Syrian Arab Republic Rép. arabe syrienne										
Vessels entered[27]										
Navires entrés[27]	11 180	10 893	...	...	...	...	...	...	2 836	3 525
Vessels cleared[2]										
Navires sortis[2]	10 922	10 809	...	...	...	...	...	...	2 992	3 459
Thailand Thaïlande										
Vessels entered										
Navires entrés	13 304	12 643	12 826	15 259	17 706	21 511	24 486	26 863	...	...
Vessels cleared										
Navires sortis	12 066	12 531	13 682	12 937	14 677	17 624	16 210	15 580	...	...
Tonga Tonga										
Vessels entered										
Navires entrés	1 662	1 765	1 765	1 778	1 611	1 874	1 816	1 950	...	...
Turkey Turquie										
Vessels entered										
Navires entrés	54 653	50 004	52 736	63 457	64 447	83 749	62 689	47 818	46 990	56 687
Vessels cleared										
Navires sortis	49 546	44 814	51 152	60 477	63 695	81 515	60 651	45 719	45 961	55 329

64
International maritime transport
Vessels entered and cleared: thousand net registered tons [cont.]
Transports maritimes internationaux
Navires entrés et sortis : milliers de tonneaux de jauge nette [suite]

Country or area Pays ou zone	1984	1985	1986	1987	1988	1989	1990	1991	1992	1993
Ukraine Ukraine										
Vessels entered										
Navires entrés	..	..	..	..	..	..	..	19 236	16 640	5 072
Vessels cleared										
Navires sortis	..	..	..	..	..	..	..	40 142	45 374	29 120
United States Etats-Unis										
Vessels entered[8][17]										
Navires entrés[8][17]	304 170	294 110	321 761	344 710	364 140	379 130	379 612	320 439	321 169	340 507
Vessels cleared[8][17]										
Navires sortis[8][17]	273 505	273 444	265 475	289 767	318 264	325 520	323 183	298 837	293 452	277 520
Venezuela Venezuela										
Vessels entered										
Navires entrés	18 852	21 337	20 579	22 663	26 101	17 529	15 774	19 882	19 758	22 087
Vessels cleared										
Navires sortis	6 966	7 044	13 320	8 826	8 223	6 641	18 294	11 500	15 194	17 211
former Dem. Yemen anciennce Yémen dém.										
Vessels entered[2]										
Navires entrés[2]	10 157	9 315	8 344	8 788	...	...	...	...	...	...
Vessels cleared[2]										
Navires sortis[2]	2 088	1 921	1 737	1 892	...	...	...	...	...	...
Yugoslavia, SFR† Yougoslavie, Rfs†										
Vessels entered										
Navires entrés	1 415	1 754	1 445	1 215	2 289	3 318	3 190	1 180	..	..
Vessels cleared										
Navires sortis	1 176	1 265	1 170	1 410	2 137	2 893	3 244	1 128	..	..

Source:
Transport statistics database of the Statistical Division of
the United Nations Secretariat.

† For detailed descriptions of data pertaining to
former Czechoslovakia, Germany, SFR Yugoslavia and former
USSR, see Annex I - Country or area nomenclature, regional
and other groupings.

1 Twelve months ending 30 June of year stated.
2 Including vessels in ballast.
3 Comprises Buenos Aires, Bahía Blanca, La Plata, Quequén, Mar
del Plata, Paraná Inferior, Paraná Medio and Rosario.
4 Buenos Aires only.
5 Comprising Buenos Aires, Rosario, Lib. Gral. San Martín,
Quequén and La Plata.
6 Twelve months beginning 1 July of year stated.

7 All entrances counted.
8 Including Great Lakes international traffic (Canada: also
St. Lawrence).
9 Including ocean-going vessels and river vessels as well as
vessels in ballast.
10 Twelve months beginning 1 April of year stated.
11 Excluding minor and intermediate ports.
12 Excluding tankers.
13 Data for Sarawak include vessels in ballast and all
entrances counted.
14 Peninsular Malaysia and Sarawak only.
15 Gross registered tons.
16 Including traffic with Portuguese overseas provinces.

17 Excluding traffic with United States Virgin Islands.

Source:
Base de données pour les statistiques des transports de la
Division de statistique du Secrétariat de l'ONU.

† Pour les descriptions en détails des données
relatives à l'ancienne Tchécoslovaquie, l'Allemagne, la Rfs
Yougoslavie et l'ancienne URSS, voir l'Annexe I -
Nomenclature des pays ou zones, groupements régionaux et
autres groupements.

1 Douze mois finissant le 30 juin de l'année indiquée.
2 Y compris navires sur lest.
3 Buenos aires, Bahía Blanca, La Plata, Quequén, Mar del
Plata, Paraná Inferior, Paraná Medio et Rosario.
4 Buenos Aires seulement.
5 Buenos Aires, Rosario, Lib. Gral. San Martín, Quequén et La
Plata.
6 Douze mois commençant le premier juillet de l'année
indiquée.
7 Toutes entrées comprises.
8 Y compris trafic international des Grands Lacs (Canada: et
du St. Laurent).
9 Y compris les grandes navigations et les navigations
fluviales et aussi les navires sur lest.
10 Douze mois commençant le premier avril de l'année indiquée.
11 Non compris les ports petits et moyens.
12 Non compris les bateaux citernes.
13 Les données pour Sarawak comprennent navires sur lest et
toutes entrées comprises.
14 Malaisie péninsulaire et Sarawak seulement.
15 Tonneaux de jauge bruts.
16 Y compris le trafic avec les provinces portugaises
d'outre-mer.
17 Non compris le trafic avec les Iles Vierges américaines.

65
Civil Aviation
Aviation civile

Passengers on scheduled services (000); Kilometres (million)
Passagers sur les services réguliers (000); Kilomètres (millions)

Country or area and traffic	Total Totale 1980	1991	1992	1993	International Internationaux 1980	1991	1992	1993	Pays ou zone et trafic
World									**Monde**
Kilometres flown	9362	14377	15555	16922	3613	5986	6840	7232	Kilomètres parcourus
Passengers carried	748978	1134065	1140625	1138123	163222	266510	302356	321054	Passagers transportés
Passengers km	1089343	1842934	1919332	1950230	466532	859748	982211	1046684	Passagers−km
Total ton−km	130999	230499	241148	249535	64279	128119	143562	155161	Total tonnes−km
Africa [1]									**Afrique** [1]
Kilometres flown	351	377	411	428	236	268	296	313	Kilomètres parcourus
Passengers carried	21235	24845	25555	23652	9003	12176	12753	12296	Passagers transportés
Passengers km	29724	39362	42846	42552	22434	31377	34921	35289	Passagers−km
Total ton−km	3544	4672	5078	5046	2811	3890	4294	4318	Total tonnes−km
Algeria									**Algérie**
Kilometres flown	28	27	26	36	19	15	14	20	Kilomètres parcourus
Passengers carried	2950	3385	3551	3254	1500	1673	1711	1676	Passagers transportés
Passengers km	2300	3092	3234	2901	1600	2071	2161	1991	Passagers−km
Total ton−km	220	303	310	296	156	206	210	207	Total tonnes−km
Angola									**Angola**
Kilometres flown	8	10	10	9	4	8	8	7	Kilomètres parcourus
Passengers carried	635	456	440	334	51	110	105	84	Passagers transportés
Passengers km	553	1241	1196	948	295	1061	1020	816	Passagers−km
Total ton−km	69	151	147	113	46	136	132	102	Total tonnes−km
Benin [2]									**Bénin** [2]
Kilometres flown	2	2	2	2	2	2	2	2	Kilomètres parcourus
Passengers carried	61	66	66	68	61	66	66	68	Passagers transportés
Passengers km	178	207	201	207	178	207	201	207	Passagers−km
Total ton−km	35	36	34	33	35	36	34	33	Total tonnes−km
Botswana									**Botswana**
Kilometres flown	1	3	3	3	0	2	2	2	Kilomètres parcourus
Passengers carried	39	102	111	111	20	68	77	77	Passagers transportés
Passengers km	15	73	76	76	7	55	58	58	Passagers−km
Total ton−km	1	7	8	8	1	6	6	6	Total tonnes−km
Burkina Faso [2]									**Burkina Faso** [2]
Kilometres flown	2	3	3	3	2	2	2	3	Kilomètres parcourus
Passengers carried	73	127	127	128	65	102	102	104	Passagers transportés
Passengers km	183	239	233	239	181	231	225	231	Passagers−km
Total ton−km	36	38	37	36	35	38	36	35	Total tonnes−km
Burundi									**Burundi**
Kilometres flown	0	0	0	0	0	0	0	0	Kilomètres parcourus
Passengers carried	11	8	9	9	11	7	7	7	Passagers transportés
Passengers km	5	2	2	2	5	2	2	2	Passagers−km
Total ton−km	1	0	0	0	1	0	0	0	Total tonnes−km
Cameroon									**Cameroun**
Kilometres flown	7	5	5	5	5	4	4	4	Kilomètres parcourus
Passengers carried	480	357	363	275	126	175	182	150	Passagers transportés
Passengers km	477	301	477	402	358	226	405	335	Passagers−km
Total ton−km	72	37	53	58	60	30	45	52	Total tonnes−km
Cape Verde									**Cap−Vert**
Kilometres flown	1	3	3	3	0	1	1	1	Kilomètres parcourus
Passengers carried	75	96	100	100	0	23	24	24	Passagers transportés
Passengers km	12	159	169	169	0	137	145	145	Passagers−km
Total ton−km	1	15	16	16	0	13	13	13	Total tonnes−km
Central African Rep. [2]									**Rép. centrafricaine** [2]
Kilometres flown	3	3	3	3	2	2	2	2	Kilomètres parcourus
Passengers carried	136	120	120	122	61	66	66	68	Passagers transportés
Passengers km	190	220	213	219	178	207	201	207	Passagers−km
Total ton−km	36	37	35	35	35	36	34	33	Total tonnes−km
Chad [2]									**Tchad** [2]
Kilometres flown	3	2	2	2	2	2	2	2	Kilomètres parcourus
Passengers carried	106	83	83	85	61	69	69	71	Passagers transportés
Passengers km	210	215	208	214	178	209	202	208	Passagers−km
Total ton−km	39	36	35	34	35	36	34	33	Total tonnes−km

65
Civil Aviation
Passengers on scheduled services (000); Kilometres (million) [cont.]
Aviation civile
Passagers sur les services réguliers (000); Kilomètres (millions) [suite]

Country or area and traffic	Total Totale				International Internationaux				Pays ou zone et trafic
	1980	1991	1992	1993	1980	1991	1992	1993	
Comoros									**Comores**
Kilometres flown	0	0	0	0	0	0	0	0	Kilomètres parcourus
Passengers carried	0	26	26	26	0	5	5	5	Passagers transportés
Passengers km	0	3	3	3	0	1	1	1	Passagers−km
Total ton−km	0	0	0	0	0	0	0	0	Total tonnes−km
Congo [2]									**Congo** [2]
Kilometres flown	3	3	3	3	2	2	2	2	Kilomètres parcourus
Passengers carried	117	229	229	231	67	71	71	73	Passagers transportés
Passengers km	197	257	250	256	182	209	203	209	Passagers−km
Total ton−km	37	41	39	38	36	36	34	33	Total tonnes−km
Côte d'Ivoire [2]									**Côte d'Ivoire** [2]
Kilometres flown	3	4	4	4	2	3	3	3	Kilomètres parcourus
Passengers carried	151	178	195	186	61	124	152	149	Passagers transportés
Passengers km	215	290	297	295	178	266	277	277	Passagers−km
Total ton−km	39	43	43	41	35	41	41	40	Total tonnes−km
Djibouti									**Djibouti**
Kilometres flown	2	1	...	...	2	1	...	...	Kilomètres parcourus
Passengers carried	63	131	...	...	44	106	...	...	Passagers transportés
Passengers km	53	67	...	...	50	65	...	...	Passagers−km
Total ton−km	5	6	...	...	5	6	...	...	Total tonnes−km
Egypt									**Egypte**
Kilometres flown	30	35	44	44	27	32	38	39	Kilomètres parcourus
Passengers carried	2028	2595	3609	2881	1270	1743	2075	1836	Passagers transportés
Passengers km	2870	5234	6323	5277	2536	4841	5629	4786	Passagers−km
Total ton−km	299	608	687	606	268	572	625	562	Total tonnes−km
Equatorial Guinea									**Guinée équatoriale**
Kilometres flown	0	0	0	0	0	0	0	0	Kilomètres parcourus
Passengers carried	13	14	14	14	13	14	14	14	Passagers transportés
Passengers km	7	7	7	7	7	7	7	7	Passagers−km
Total ton−km	1	1	1	1	1	1	1	1	Total tonnes−km
Ethiopia									**Ethiopie**
Kilometres flown	10	21	26	23	10	18	22	19	Kilomètres parcourus
Passengers carried	243	636	756	752	186	352	447	435	Passagers transportés
Passengers km	647	1568	1725	1717	597	1431	1577	1571	Passagers−km
Total ton−km	85	224	264	259	78	209	248	244	Total tonnes−km
Gabon									**Gabon**
Kilometres flown	6	6	6	6	5	4	4	4	Kilomètres parcourus
Passengers carried	331	436	471	302	95	131	145	152	Passagers transportés
Passengers km	374	479	536	570	313	399	450	480	Passagers−km
Total ton−km	61	70	77	82	54	62	68	73	Total tonnes−km
Gambia									**Gambie**
Kilometres flown	...	...	1	1	...	...	1	1	Kilomètres parcourus
Passengers carried	...	...	19	19	...	...	19	19	Passagers transportés
Passengers km	...	...	· 50	50	...	...	50	50	Passagers−km
Total ton−km	...	...	5	5	...	...	5	5	Total tonnes−km
Ghana									**Ghana**
Kilometres flown	4	5	5	4	3	5	5	4	Kilomètres parcourus
Passengers carried	279	192	206	152	179	192	200	152	Passagers transportés
Passengers km	330	331	352	387	304	331	350	387	Passagers−km
Total ton−km	34	58	62	61	32	58	61	61	Total tonnes−km
Guinea									**Guinée**
Kilometres flown	1	2	1	1	1	1	1	1	Kilomètres parcourus
Passengers carried	80	42	23	24	24	30	20	21	Passagers transportés
Passengers km	34	69	33	35	12	59	31	32	Passagers−km
Total ton−km	3	3	4	4	1	2	3	4	Total tonnes−km
Guinea−Bissau									**Guinée−Bissau**
Kilometres flown	1	1	1	1	0	0	0	0	Kilomètres parcourus
Passengers carried	22	21	21	21	8	8	8	8	Passagers transportés
Passengers km	8	10	10	10	4	6	6	6	Passagers−km
Total ton−km	1	1	1	1	0	1	1	1	Total tonnes−km
Kenya									**Kenya**
Kilometres flown	12	13	14	15	10	10	12	13	Kilomètres parcourus
Passengers carried	393	760	721	770	204	389	380	413	Passagers transportés
Passengers km	863	1479	1333	1459	782	1343	1205	1325	Passagers−km
Total ton−km	98	175	174	191	90	162	162	178	Total tonnes−km

65
Civil Aviation
Passengers on scheduled services (000); Kilometres (million) [*cont.*]
Aviation civile
Passagers sur les services réguliers (000); Kilomètres (millions) [*suite*]

Country or area and traffic	Total Totale				International Internationaux				Pays ou zone et trafic
	1980	1991	1992	1993	1980	1991	1992	1993	
Lesotho									**Lesotho**
Kilometres flown	1	1	1	1	0	1	1	1	Kilomètres parcourus
Passengers carried	52	56	26	23	10	28	20	19	Passagers transportés
Passengers km	11	14	9	9	4	10	8	8	Passagers−km
Total ton−km	1	1	1	1	0	1	1	1	Total tonnes−km
Liberia									**Libéria**
Kilometres flown	1	0	0	...	...	...	...	...	Kilomètres parcourus
Passengers carried	50	32	32	...	...	...	...	...	Passagers transportés
Passengers km	17	7	7	...	...	...	...	...	Passagers−km
Total ton−km	2	1	1	...	...	...	...	...	Total tonnes−km
Libyan Arab Jamahiriya									**Jamah. arabe libyenne**
Kilometres flown	12	19	11	5	6	11	3	−	Kilomètres parcourus
Passengers carried	1169	1884	1350	853	468	685	178	−	Passagers transportés
Passengers km	1101	2045	1116	565	632	1251	344	−	Passagers−km
Total ton−km	116	183	96	44	70	120	35	−	Total tonnes−km
Madagascar									**Madagascar**
Kilometres flown	7	6	6	7	2	3	3	3	Kilomètres parcourus
Passengers carried	448	314	344	419	54	68	79	94	Passagers transportés
Passengers km	379	385	432	499	226	290	325	368	Passagers−km
Total ton−km	54	61	66	72	38	52	56	60	Total tonnes−km
Malawi									**Malawi**
Kilometres flown	2	3	2	2	1	1	1	2	Kilomètres parcourus
Passengers carried	94	120	121	132	52	58	52	63	Passagers transportés
Passengers km	68	89	79	266	56	59	53	234	Passagers−km
Total ton−km	7	10	9	15	6	6	6	13	Total tonnes−km
Mali [2]									**Mali** [2]
Kilometres flown	2	0	2	2	1	0	2	2	Kilomètres parcourus
Passengers carried	71	0	66	68	42	0	66	68	Passagers transportés
Passengers km	97	0	201	207	82	0	201	207	Passagers−km
Total ton−km	9	0	34	33	8	0	34	33	Total tonnes−km
Mauritania [2]									**Mauritanie** [2]
Kilometres flown	3	3	3	4	2	2	2	3	Kilomètres parcourus
Passengers carried	141	213	213	215	76	89	89	91	Passagers transportés
Passengers km	218	282	275	281	188	230	224	230	Passagers−km
Total ton−km	39	43	41	40	36	38	36	35	Total tonnes−km
Mauritius									**Maurice**
Kilometres flown	3	15	18	17	3	15	17	17	Kilomètres parcourus
Passengers carried	84	525	578	582	77	504	551	550	Passagers transportés
Passengers km	185	2457	2799	2677	181	2444	2782	2658	Passagers−km
Total ton−km	20	301	342	334	20	300	341	333	Total tonnes−km
Morocco									**Maroc**
Kilometres flown	20	27	42	45	20	26	40	42	Kilomètres parcourus
Passengers carried	947	1430	2169	2140	906	1227	1884	1775	Passagers transportés
Passengers km	1868	2533	4297	4395	1855	2449	4197	4264	Passagers−km
Total ton−km	209	257	399	364	208	249	390	352	Total tonnes−km
Mozambique									**Mozambique**
Kilometres flown	6	5	5	4	3	3	3	3	Kilomètres parcourus
Passengers carried	282	283	225	206	52	95	88	84	Passagers transportés
Passengers km	467	465	387	411	260	284	248	282	Passagers−km
Total ton−km	51	52	44	47	29	33	29	34	Total tonnes−km
Namibia									**Namibie**
Kilometres flown	...	5	6	8	...	3	4	5	Kilomètres parcourus
Passengers carried	...	455	163	179	...	400	134	149	Passagers transportés
Passengers km	...	423	535	687	...	395	508	659	Passagers−km
Total ton−km	...	40	59	84	...	38	57	81	Total tonnes−km
Niger [2]									**Niger** [2]
Kilometres flown	3	2	2	2	2	2	2	2	Kilomètres parcourus
Passengers carried	111	66	66	68	61	66	66	68	Passagers transportés
Passengers km	199	207	201	207	178	207	201	207	Passagers−km
Total ton−km	37	36	34	33	35	36	34	33	Total tonnes−km
Nigeria									**Nigéria**
Kilometres flown	24	14	12	11	10	9	7	6	Kilomètres parcourus
Passengers carried	1939	930	647	608	301	374	230	208	Passagers transportés
Passengers km	1877	1391	990	913	996	1012	672	608	Passagers−km
Total ton−km	179	159	103	96	99	122	72	67	Total tonnes−km

65

Civil Aviation
Passengers on scheduled services (000); Kilometres (million) [*cont.*]
Aviation civile
Passagers sur les services réguliers (000); Kilomètres (millions) [*suite*]

Country or area and traffic	Total Totale				International Internationaux				Pays ou zone et trafic
	1980	1991	1992	1993	1980	1991	1992	1993	
Rwanda									**Rwanda**
Kilometres flown	...	1	0	0	...	1	0	0	Kilomètres parcourus
Passengers carried	...	10	9	9	...	3	4	4	Passagers transportés
Passengers km	...	13	2	2	...	11	1	1	Passagers−km
Total ton−km	...	2	0	0	...	2	0	0	Total tonnes−km
Sao Tome and Principe									**Sao Tomé−et−Principe**
Kilometres flown	...	0	0	0	...	0	0	0	Kilomètres parcourus
Passengers carried	...	22	22	22	...	13	13	13	Passagers transportés
Passengers km	...	8	8	8	...	4	4	4	Passagers−km
Total ton−km	...	1	1	1	...	0	0	0	Total tonnes−km
Senegal [2]									**Sénégal** [2]
Kilometres flown	3	3	3	3	2	2	2	3	Kilomètres parcourus
Passengers carried	113	138	138	140	71	110	110	112	Passagers transportés
Passengers km	196	229	222	216	182	220	214	208	Passagers−km
Total ton−km	37	37	36	35	36	37	35	34	Total tonnes−km
Seychelles									**Seychelles**
Kilometres flown	0	3	4	7	...	3	4	6	Kilomètres parcourus
Passengers carried	15	243	239	289	...	51	68	97	Passagers transportés
Passengers km	2	351	438	626	...	342	430	618	Passagers−km
Total ton−km	0	42	57	70	...	41	56	69	Total tonnes−km
Sierra Leone									**Sierra Leone**
Kilometres flown	1	0	0	0	0	0	0	0	Kilomètres parcourus
Passengers carried	56	0	17	18	26	0	17	18	Passagers transportés
Passengers km	86	0	51	55	77	0	51	55	Passagers−km
Total ton−km	9	0	5	5	8	0	5	5	Total tonnes−km
Somalia									**Somalie**
Kilometres flown	3	1	...	...	2	1	...	...	Kilomètres parcourus
Passengers carried	90	46	...	...	70	40	...	...	Passagers transportés
Passengers km	140	131	...	...	120	125	...	...	Passagers−km
Total ton−km	13	17	...	...	11	16	...	...	Total tonnes−km
South Africa									**Afrique du Sud**
Kilometres flown	67	67	73	81	37	30	35	43	Kilomètres parcourus
Passengers carried	4116	4819	4685	4799	801	842	969	1130	Passagers transportés
Passengers km	8920	8413	9511	10790	6088	4887	6229	7552	Passagers−km
Total ton−km	1074	950	1116	1285	778	593	785	954	Total tonnes−km
Sudan									**Soudan**
Kilometres flown	10	6	12	12	7	4	5	6	Kilomètres parcourus
Passengers carried	519	363	480	408	311	165	303	225	Passagers transportés
Passengers km	710	426	511	580	530	302	396	454	Passagers−km
Total ton−km	76	50	85	91	58	32	55	59	Total tonnes−km
Swaziland									**Swaziland**
Kilometres flown	1	1	1	1	1	1	1	1	Kilomètres parcourus
Passengers carried	31	59	57	58	31	59	57	58	Passagers transportés
Passengers km	30	45	42	44	30	45	42	44	Passagers−km
Total ton−km	3	4	4	4	3	4	4	4	Total tonnes−km
Togo [2]									**Togo** [2]
Kilometres flown	2	2	2	2	2	2	2	2	Kilomètres parcourus
Passengers carried	65	66	66	68	65	66	66	68	Passagers transportés
Passengers km	179	207	201	207	179	207	201	207	Passagers−km
Total ton−km	35	36	34	33	35	36	34	33	Total tonnes−km
Tunisia									**Tunisie**
Kilometres flown	14	13	15	17	13	13	15	17	Kilomètres parcourus
Passengers carried	978	1201	1250	1351	898	1059	1219	1345	Passagers transportés
Passengers km	1241	1407	1673	1877	1213	1359	1663	1875	Passagers−km
Total ton−km	123	144	168	173	120	140	167	172	Total tonnes−km
Uganda									**Ouganda**
Kilometres flown	3	0	1	1	2	0	1	1	Kilomètres parcourus
Passengers carried	83	26	32	40	43	25	32	40	Passagers transportés
Passengers km	68	13	17	24	58	13	17	24	Passagers−km
Total ton−km	15	1	2	2	14	1	2	2	Total tonnes−km
United Rep.Tanzania									**Rép. Unie de Tanzanie**
Kilometres flown	8	5	4	3	3	3	2	2	Kilomètres parcourus
Passengers carried	388	290	216	188	94	97	56	57	Passagers transportés
Passengers km	284	284	174	157	180	195	99	83	Passagers−km
Total ton−km	28	30	18	16	18	21	10	8	Total tonnes−km

65
Civil Aviation
Passengers on scheduled services (000); Kilometres (million) [cont.]
Aviation civile
Passagers sur les services réguliers (000); Kilomètres (millions) [suite]

Country or area and traffic	Total Totale				International Internationaux				Pays ou zone et trafic
	1980	1991	1992	1993	1980	1991	1992	1993	
Zaire									**Zaïre**
Kilometres flown	10	5	4	4	4	3	3	3	Kilomètres parcourus
Passengers carried	439	150	116	84	92	73	53	39	Passagers transportés
Passengers km	834	384	295	218	464	269	200	150	Passagers—km
Total ton—km	110	68	56	42	64	50	40	31	Total tonnes—km
Zambia									**Zambie**
Kilometres flown	10	6	5	4	9	5	4	3	Kilomètres parcourus
Passengers carried	257	293	246	219	114	157	127	102	Passagers transportés
Passengers km	467	655	509	393	425	611	470	359	Passagers—km
Total ton—km	89	82	63	47	85	78	59	44	Total tonnes—km
Zimbabwe									**Zimbabwe**
Kilometres flown	6	13	14	13	3	10	11	10	Kilomètres parcourus
Passengers carried	412	740	678	558	180	297	278	215	Passagers transportés
Passengers km	362	940	880	735	286	789	736	600	Passagers—km
Total ton—km	33	170	171	154	27	157	159	143	Total tonnes—km
America, North [1]									**Amérique du Nord** [1]
Kilometres flown	5074	7700	8025	8116	812	1541	1762	1801	**Kilomètres parcourus**
Passengers carried	336928	491546	507090	509876	38722	61480	66013	67712	**Passagers transportés**
Passengers km	467284	788001	836382	842690	112763	230924	258832	264197	**Passagers—km**
Total ton—km	53930	90370	96173	97741	14695	30326	33683	34811	**Total tonnes—km**
Antigua and Barbuda									**Antigua—et—Barbuda**
Kilometres flown	...	10	10	11	...	10	10	11	Kilomètres parcourus
Passengers carried	...	870	914	955	...	870	914	955	Passagers transportés
Passengers km	...	201	213	225	...	201	213	225	Passagers—km
Total ton—km	...	18	19	20	...	18	19	20	Total tonnes—km
Bahamas									**Bahamas**
Kilometres flown	10	6	4	4	4	4	2	2	Kilomètres parcourus
Passengers carried	345	1090	835	862	160	635	380	397	Passagers transportés
Passengers km	539	346	188	191	459	290	132	132	Passagers—km
Total ton—km	52	31	17	18	44	26	12	12	Total tonnes—km
Barbados									**Barbade**
Kilometres flown	1	...	...	...	1	...	...	...	Kilomètres parcourus
Passengers carried	46	...	...	...	46	...	...	...	Passagers transportés
Passengers km	330	...	...	...	330	...	...	...	Passagers—km
Total ton—km	30	...	...	...	30	...	...	...	Total tonnes—km
Canada									**Canada**
Kilometres flown	337	372	383	369	114	155	159	166	Kilomètres parcourus
Passengers carried	22453	16586	16818	17516	5530	5739	6171	6601	Passagers transportés
Passengers km	36234	39082	41253	40426	16293	22042	24160	24936	Passagers—km
Total ton—km	4095	4908	5109	5151	1944	2956	3173	3355	Total tonnes—km
Costa Rica									**Costa Rica**
Kilometres flown	8	11	13	17	8	11	13	16	Kilomètres parcourus
Passengers carried	431	504	589	699	332	454	535	645	Passagers transportés
Passengers km	495	1050	1204	1432	485	1044	1197	1425	Passagers—km
Total ton—km	68	151	165	189	67	150	165	188	Total tonnes—km
Cuba									**Cuba**
Kilometres flown	15	17	12	11	7	11	7	7	Kilomètres parcourus
Passengers carried	676	831	733	624	120	195	177	183	Passagers transportés
Passengers km	932	1598	1370	1321	666	1226	1042	1069	Passagers—km
Total ton—km	95	172	149	138	72	141	123	118	Total tonnes—km
Dominican Republic									**Rép. dominicaine**
Kilometres flown	6	3	4	3	6	3	4	3	Kilomètres parcourus
Passengers carried	466	264	323	328	466	264	323	328	Passagers transportés
Passengers km	550	227	283	280	550	227	283	280	Passagers—km
Total ton—km	60	21	26	28	60	21	26	28	Total tonnes—km
El Salvador									**El Salvador**
Kilometres flown	6	12	29	17	6	12	29	17	Kilomètres parcourus
Passengers carried	265	577	683	723	265	577	683	723	Passagers transportés
Passengers km	289	1195	1563	1738	289	1195	1563	1738	Passagers—km
Total ton—km	41	128	175	203	41	128	175	203	Total tonnes—km
Guatemala									**Guatemala**
Kilometres flown	4	4	6	6	4	4	6	6	Kilomètres parcourus
Passengers carried	119	165	230	240	119	165	230	240	Passagers transportés
Passengers km	159	230	366	384	159	230	366	384	Passagers—km
Total ton—km	21	29	52	56	21	29	52	56	Total tonnes—km

65
Civil Aviation
Passengers on scheduled services (000); Kilometres (million) [*cont.*]
        Aviation civile
        Passagers sur les services réguliers (000); Kilomètres (millions) [*suite*]

Country or area and traffic	Total Totale				International Internationaux				Pays ou zone et trafic
	1980	1991	1992	1993	1980	1991	1992	1993	
Haiti									**Haïti**
Kilometres flown	1	1	...	...	1	1	...	...	Kilomètres parcourus
Passengers carried	0	0	...	...	0	0	...	...	Passagers transportés
Passengers km	0	0	...	...	0	0	...	...	Passagers–km
Total ton–km	2	4	...	...	2	4	...	...	Total tonnes–km
Honduras									**Honduras**
Kilometres flown	7	5	5	5	7	5	4	4	Kilomètres parcourus
Passengers carried	508	447	438	449	369	253	245	255	Passagers transportés
Passengers km	387	336	309	323	369	303	275	289	Passagers–km
Total ton–km	43	45	40	42	41	42	36	38	Total tonnes–km
Jamaica									**Jamaïque**
Kilometres flown	14	12	12	13	13	10	11	11	Kilomètres parcourus
Passengers carried	723	894	983	1038	660	804	900	942	Passagers transportés
Passengers km	1207	1311	1460	1488	1200	1294	1448	1474	Passagers–km
Total ton–km	115	139	151	155	115	137	150	154	Total tonnes–km
Mexico									**Mexique**
Kilometres flown	157	191	219	210	63	87	91	86	Kilomètres parcourus
Passengers carried	12890	14901	15532	14621	2777	3717	3976	3490	Passagers transportés
Passengers km	13870	18267	19553	18216	6594	8709	9119	8016	Passagers–km
Total ton–km	1383	1726	1851	1746	665	872	909	811	Total tonnes–km
Nicaragua									**Nicaragua**
Kilometres flown	2	2	3	1	1	2	2	1	Kilomètres parcourus
Passengers carried	115	130	159	34	50	84	110	34	Passagers transportés
Passengers km	76	111	186	58	60	97	170	58	Passagers–km
Total ton–km	8	14	24	4	7	12	22	4	Total tonnes–km
Panama									**Panama**
Kilometres flown	7	4	5	8	6	4	5	8	Kilomètres parcourus
Passengers carried	355	343	283	320	307	343	283	320	Passagers transportés
Passengers km	409	267	340	380	395	267	340	380	Passagers–km
Total ton–km	40	34	43	50	38	34	43	50	Total tonnes–km
Trinidad and Tobago									**Trinité–et–Tobago**
Kilometres flown	16	25	24	26	16	25	24	25	Kilomètres parcourus
Passengers carried	877	1345	1354	1389	633	957	950	985	Passagers transportés
Passengers km	1505	3129	3077	3232	1485	3104	3050	3205	Passagers–km
Total ton–km	160	351	346	364	158	349	344	362	Total tonnes–km
United States									**Etats–Unis**
Kilometres flown	4469	7013	7284	7405	543	1189	1384	1427	Kilomètres parcourus
Passengers carried	295281	451247	466157	468974	25744	45483	49180	50611	Passagers transportés
Passengers km	409520	719770	764118	772048	82678	189867	214628	219691	Passagers–km
Total ton–km	47644	82514	87917	89486	11319	25327	28353	29327	Total tonnes–km
America, South [1]									**Amérique du Sud** [1]
Kilometres flown	**498**	**673**	**698**	**739**	**199**	**262**	**308**	**341**	**Kilomètres parcourus**
Passengers carried	**34008**	**41814**	**44732**	**45375**	**6049**	**7967**	**12137**	**12929**	**Passagers transportés**
Passengers km	**38628**	**57618**	**61034**	**65757**	**19547**	**31764**	**36359**	**41622**	**Passagers–km**
Total ton–km	**4790**	**7880**	**8488**	**9207**	**2907**	**5087**	**5847**	**6511**	**Total tonnes–km**
Argentina									**Argentine**
Kilometres flown	94	70	93	78	39	33	37	35	Kilomètres parcourus
Passengers carried	5589	4542	6711	5104	1300	1670	1787	1661	Passagers transportés
Passengers km	8031	9232	11696	9231	4413	6473	7178	6072	Passagers–km
Total ton–km	939	1101	1354	1117	599	797	895	780	Total tonnes–km
Bolivia									**Bolivie**
Kilometres flown	13	12	11	11	9	7	7	8	Kilomètres parcourus
Passengers carried	1342	1200	1214	1117	268	409	412	423	Passagers transportés
Passengers km	944	1022	1069	1092	570	723	780	843	Passagers–km
Total ton–km	121	101	107	120	89	75	81	96	Total tonnes–km
Brazil									**Brésil**
Kilometres flown	203	347	326	333	57	85	106	112	Kilomètres parcourus
Passengers carried	13008	19153	16388	16599	1330	2128	2707	3062	Passagers transportés
Passengers km	15572	28599	27897	29555	6008	12560	15144	16866	Passagers–km
Total ton–km	1956	3662	3796	4084	945	1919	2320	2514	Total tonnes–km
Chile									**Chili**
Kilometres flown	25	50	65	72	19	36	45	48	Kilomètres parcourus
Passengers carried	669	1406	1906	2360	299	497	589	717	Passagers transportés
Passengers km	1875	3039	3854	4425	1362	2116	2495	2772	Passagers–km
Total ton–km	324	706	879	1018	267	606	739	844	Total tonnes–km

65
Civil Aviation
Passengers on scheduled services (000); Kilometres (million) [*cont.*]
Aviation civile
Passagers sur les services réguliers (000); Kilomètres (millions) [*suite*]

Country or area and traffic	Total Totale 1980	1991	1992	1993	International Internationaux 1980	1991	1992	1993	Pays ou zone et trafic
Colombia									**Colombie**
Kilometres flown	45	71	69	87	22	33	37	40	Kilomètres parcourus
Passengers carried	4808	5540	5414	6425	752	685	758	832	Passagers transportés
Passengers km	4198	4465	4590	5296	2203	2212	2268	2376	Passagers−km
Total ton−km	532	976	952	1014	339	690	733	746	Total tonnes−km
Ecuador									**Equateur**
Kilometres flown	21	19	19	36	11	9	9	25	Kilomètres parcourus
Passengers carried	701	752	756	1243	255	252	251	491	Passagers transportés
Passengers km	975	1201	1188	4935	851	964	948	4521	Passagers−km
Total ton−km	131	186	176	596	115	156	146	550	Total tonnes−km
Guyana									**Guyana**
Kilometres flown	1	2	2	2	...	2	2	2	Kilomètres parcourus
Passengers carried	28	112	115	115	...	51	54	54	Passagers transportés
Passengers km	6	214	224	224	...	198	208	208	Passagers−km
Total ton−km	1	22	23	23	...	20	21	21	Total tonnes−km
Paraguay									**Paraguay**
Kilometres flown	4	9	9	8	3	7	8	7	Kilomètres parcourus
Passengers carried	129	309	314	337	109	275	279	303	Passagers transportés
Passengers km	262	1073	1141	1273	246	1054	1121	1253	Passagers−km
Total ton−km	26	102	125	139	24	99	122	136	Total tonnes−km
Peru									**Pérou**
Kilometres flown	25	19	19	22	9	8	8	11	Kilomètres parcourus
Passengers carried	1980	1491	1218	1362	221	224	231	336	Passagers transportés
Passengers km	1974	1759	1528	1926	822	865	848	1205	Passagers−km
Total ton−km	217	184	179	214	98	97	111	143	Total tonnes−km
Suriname									**Suriname**
Kilometres flown	3	3	3	3	2	3	3	3	Kilomètres parcourus
Passengers carried	144	196	216	216	120	180	200	200	Passagers transportés
Passengers km	245	534	604	604	240	530	600	600	Passagers−km
Total ton−km	26	63	71	70	25	63	71	70	Total tonnes−km
Uruguay									**Uruguay**
Kilometres flown	4	5	5	5	3	5	5	5	Kilomètres parcourus
Passengers carried	478	318	3330	3330	433	318	3330	3330	Passagers transportés
Passengers km	178	471	490	490	160	471	490	490	Passagers−km
Total ton−km	17	44	46	46	15	44	46	46	Total tonnes−km
Venezuela									**Venezuela**
Kilometres flown	61	66	76	80	26	33	41	44	Kilomètres parcourus
Passengers carried	5133	6795	7149	7166	960	1277	1540	1520	Passagers transportés
Passengers km	4367	6010	6753	6708	2671	3598	4278	4416	Passagers−km
Total ton−km	501	733	779	767	392	521	560	565	Total tonnes−km
Asia [1]									**Asie** [1]
Kilometres flown	**1270**	**2153**	**2562**	**2773**	**777**	**1324**	**1611**	**1736**	**Kilomètres parcourus**
Passengers carried	**106119**	**215966**	**240753**	**247315**	**34301**	**68059**	**80867**	**86639**	**Passagers transportés**
Passengers km	**158192**	**353174**	**407368**	**429578**	**108019**	**241304**	**285073**	**303414**	**Passagers−km**
Total ton−km	**20458**	**50029**	**56739**	**61481**	**15977**	**39478**	**45540**	**49938**	**Total tonnes−km**
Afghanistan									**Afghanistan**
Kilometres flown	3	4	4	4	3	1	1	1	Kilomètres parcourus
Passengers carried	76	212	212	197	51	92	92	83	Passagers transportés
Passengers km	163	205	205	197	155	151	151	145	Passagers−km
Total ton−km	36	27	27	25	35	22	22	20	Total tonnes−km
Azerbaijan									**Azerbaïdjan**
Kilometres flown	..	..	0	0	..	..	0	0	Kilomètres parcourus
Passengers carried	..	..	1455	806	..	..	16	21	Passagers transportés
Passengers km	..	..	3511	2091	..	..	33	41	Passagers−km
Total ton−km	..	..	339	194	..	..	3	4	Total tonnes−km
Bahrain									**Bahreïn**
Kilometres flown	7	11	16	18	7	11	16	18	Kilomètres parcourus
Passengers carried	522	876	981	1080	522	876	981	1080	Passagers transportés
Passengers km	714	1676	1922	2210	714	1676	1922	2210	Passagers−km
Total ton−km	91	211	252	297	91	211	252	297	Total tonnes−km
Bangladesh									**Bangladesh**
Kilometres flown	12	13	15	17	10	11	13	15	Kilomètres parcourus
Passengers carried	614	1080	1051	1083	270	630	624	667	Passagers transportés
Passengers km	1179	2190	2303	2556	1113	2100	2215	2470	Passagers−km
Total ton−km	126	296	290	278	119	287	282	270	Total tonnes−km

65
Civil Aviation
Passengers on scheduled services (000); Kilometres (million) [*cont.*]
Aviation civile
Passagers sur les services réguliers (000); Kilomètres (millions) [*suite*]

Country or area and traffic	Total Totale				International Internationaux				Pays ou zone et trafic
	1980	1991	1992	1993	1980	1991	1992	1993	
Bhutan									**Bhoutan**
Kilometres flown	...	0	0	0	...	0	0	0	Kilomètres parcourus
Passengers carried	...	8	9	9	...	8	9	9	Passagers transportés
Passengers km	...	4	5	5	...	4	5	5	Passagers−km
Total ton−km	...	0	0	0	...	0	0	0	Total tonnes−km
Brunei Darussalam									**Brunéi Darussalam**
Kilometres flown	...	7	10	16	...	7	10	16	Kilomètres parcourus
Passengers carried	...	307	410	604	...	307	410	604	Passagers transportés
Passengers km	...	487	900	1623	...	487	900	1623	Passagers−km
Total ton−km	...	65	111	211	...	65	111	211	Total tonnes−km
China									**Chine**
Kilometres flown	46	193	289	366	10	43	72	81	Kilomètres parcourus
Passengers carried	2568	19520	27345	31312	360	1363	4500	4667	Passagers transportés
Passengers km	3578	30132	40605	45000	913	6446	11162	11171	Passagers−km
Total ton−km	443	3207	4284	4848	134	1038	1677	1842	Total tonnes−km
Cyprus									**Chypre**
Kilometres flown	9	14	18	18	9	14	18	18	Kilomètres parcourus
Passengers carried	441	820	947	1011	441	820	947	1011	Passagers transportés
Passengers km	798	1803	2095	2179	798	1803	2095	2179	Passagers−km
Total ton−km	92	198	223	230	92	198	223	230	Total tonnes−km
India									**Inde**
Kilometres flown	85	109	114	104	43	43	48	46	Kilomètres parcourus
Passengers carried	6603	10717	11127	9442	1668	2074	2457	2194	Passagers transportés
Passengers km	10765	15585	16718	14396	6765	7863	8932	7858	Passagers−km
Total ton−km	1353	1881	1919	1661	966	1134	1169	1032	Total tonnes−km
Indonesia									**Indonésie**
Kilometres flown	88	165	192	209	24	53	70	83	Kilomètres parcourus
Passengers carried	5059	10402	11177	8945	922	2341	2773	2925	Passagers transportés
Passengers km	5907	16011	18758	19472	2774	9586	12076	12758	Passagers−km
Total ton−km	625	1913	2213	2386	337	1261	1538	1714	Total tonnes−km
Iran, Islamic Rep. of									**Iran, Rép. islamique d'**
Kilometres flown	16	35	32	34	8	11	11	12	Kilomètres parcourus
Passengers carried	1998	5537	4896	5352	396	757	730	750	Passagers transportés
Passengers km	2071	5645	4867	5045	995	2211	1914	1868	Passagers−km
Total ton−km	210	604	510	519	109	284	235	223	Total tonnes−km
Iraq									**Iraq**
Kilometres flown	13	0	0	...	12	0	0	...	Kilomètres parcourus
Passengers carried	620	28	53	...	434	6	12	...	Passagers transportés
Passengers km	1161	17	35	...	1083	7	15	...	Passagers−km
Total ton−km	157	1	3	...	150	1	1	...	Total tonnes−km
Israel									**Israël**
Kilometres flown	30	50	55	59	28	47	50	54	Kilomètres parcourus
Passengers carried	1483	2047	2391	2569	1043	1624	1857	2014	Passagers transportés
Passengers km	4727	7527	8361	8747	4590	7421	8203	8581	Passagers−km
Total ton−km	724	1548	1603	1654	712	1537	1588	1638	Total tonnes−km
Japan									**Japon**
Kilometres flown	365	532	569	577	150	267	286	284	Kilomètres parcourus
Passengers carried	45145	78720	81378	79010	4499	10714	11589	11260	Passagers transportés
Passengers km	51217	100789	108082	106360	22254	49811	55324	53979	Passagers−km
Total ton−km	6184	14134	14542	14668	3778	9395	9882	10031	Total tonnes−km
Jordan									**Jordanie**
Kilometres flown	21	23	31	35	21	23	31	35	Kilomètres parcourus
Passengers carried	1113	797	1109	1186	1070	782	1076	1145	Passagers transportés
Passengers km	2607	2439	3572	4004	2595	2435	3564	3993	Passagers−km
Total ton−km	316	386	503	581	315	386	502	580	Total tonnes−km
Kazakhstan									**Kazakhstan**
Kilometres flown	..	..	0	0	..	..	0	0	Kilomètres parcourus
Passengers carried	..	..	5273	2883	..	..	10	13	Passagers transportés
Passengers km	..	..	9713	5731	..	..	10	13	Passagers−km
Total ton−km	..	..	913	517	..	..	1	1	Total tonnes−km
Korea, Dem. P. R.									**Corée, R. p. dém. de**
Kilometres flown	3	3	3	3	0	1	1	1	Kilomètres parcourus
Passengers carried	192	223	235	242	20	35	35	36	Passagers transportés
Passengers km	90	182	189	197	17	101	103	108	Passagers−km
Total ton−km	10	20	20	21	2	10	10	11	Total tonnes−km

65
Civil Aviation
Passengers on scheduled services (000); Kilometres (million) [*cont.*]
Aviation civile
Passagers sur les services réguliers (000); Kilomètres (millions) [*suite*]

Country or area and traffic	Total Totale 1980	1991	1992	1993	International Internationaux 1980	1991	1992	1993	Pays ou zone et trafic
Korea, Republic of									**Corée, République de**
Kilometres flown	61	137	188	200	55	108	150	157	Kilomètres parcourus
Passengers carried	3567	16908	19767	21798	2105	4992	5633	6644	Passagers transportés
Passengers km	10833	20716	28004	31786	10240	16765	22935	26766	Passagers−km
Total−km	1850	4623	5825	6867	1797	4251	5327	6379	Total tonnes−km
Kuwait									**Koweït**
Kilometres flown	19	14	29	33	19	14	29	33	Kilomètres parcourus
Passengers carried	1076	840	1409	1554	1076	840	1409	1554	Passagers transportés
Passengers km	2114	1908	3529	4054	2114	1908	3529	4054	Passagers−km
Total−km	265	289	538	632	265	289	538	632	Total tonnes−km
Kyrgyzstan									**Kirghizistan**
Kilometres flown	..	..	...	−	..	..	...	−	Kilomètres parcourus
Passengers carried	..	..	...	637	..	..	...	−	Passagers transportés
Passengers km	..	..	...	1713	..	..	...	−	Passagers−km
Total−km	..	..	...	156	..	..	...	−	Total tonnes−km
Lao People's Dem. Rep.									**Rép. dém. pop. lao**
Kilometres flown	0	1	1	1	0	1	1	1	Kilomètres parcourus
Passengers carried	13	115	119	119	13	28	30	30	Passagers transportés
Passengers km	7	44	46	46	7	18	19	19	Passagers−km
Total ton−km	1	4	4	4	1	2	2	2	Total tonnes−km
Lebanon									**Liban**
Kilometres flown	43	17	19	18	43	17	19	18	Kilomètres parcourus
Passengers carried	930	536	600	677	930	536	600	677	Passagers transportés
Passengers km	1571	1150	1285	1459	1571	1150	1285	1459	Passagers−km
Total ton−km	680	276	272	260	680	276	272	260	Total tonnes−km
Malaysia									**Malaisie**
Kilometres flown	41	98	111	115	22	60	69	74	Kilomètres parcourus
Passengers carried	4516	11837	12757	13077	1822	4663	5081	5597	Passagers transportés
Passengers km	4076	14226	15714	17445	2916	11285	12625	14431	Passagers−km
Total ton−km	501	1942	2113	2178	384	1658	1824	1898	Total tonnes−km
Maldives									**Maldives**
Kilometres flown	1	0	0	0	1	−	−	−	Kilomètres parcourus
Passengers carried	27	9	9	9	27	−	−	−	Passagers transportés
Passengers km	20	3	3	3	20	−	−	−	Passagers−km
Total ton−km	2	0	0	0	2	−	−	−	Total tonnes−km
Mongolia									**Mongolie**
Kilometres flown	0	12	11	12	0	2	2	2	Kilomètres parcourus
Passengers carried	0	616	591	630	0	40	40	43	Passagers transportés
Passengers km	0	475	438	491	0	107	107	115	Passagers−km
Total ton−km	0	43	42	45	0	10	10	12	Total tonnes−km
Myanmar									**Myanmar**
Kilometres flown	6	4	4	4	1	0	0	0	Kilomètres parcourus
Passengers carried	481	319	319	319	64	19	19	19	Passagers transportés
Passengers km	218	140	140	140	56	14	14	14	Passagers−km
Total ton−km	21	14	14	14	6	2	2	2	Total tonnes−km
Nepal									**Népal**
Kilometres flown	5	9	9	9	3	5	5	6	Kilomètres parcourus
Passengers carried	380	672	680	712	164	357	365	381	Passagers transportés
Passengers km	234	706	721	769	196	655	670	715	Passagers−km
Total ton−km	22	69	71	76	19	65	67	72	Total tonnes−km
Oman									**Oman**
Kilometres flown	7	12	17	19	7	11	16	18	Kilomètres parcourus
Passengers carried	522	958	1081	1180	522	876	981	1080	Passagers transportés
Passengers km	714	1729	1987	2275	714	1676	1922	2210	Passagers−km
Total ton−km	91	216	258	303	91	211	252	297	Total tonnes−km
Pakistan									**Pakistan**
Kilometres flown	50	60	68	68	37	37	44	44	Kilomètres parcourus
Passengers carried	3029	5198	5681	5647	1501	1973	2251	2269	Passagers transportés
Passengers km	5696	9062	10095	9898	4522	6711	7574	7533	Passagers−km
Total ton−km	763	1210	1332	1322	643	963	1067	1069	Total tonnes−km
Philippines									**Philippines**
Kilometres flown	42	60	68	70	26	39	46	48	Kilomètres parcourus
Passengers carried	3246	5438	6137	5801	997	1833	2113	2229	Passagers transportés
Passengers km	5959	11028	12882	13085	4880	9307	10918	11295	Passagers−km
Total ton−km	709	1433	1602	1650	612	1260	1409	1478	Total tonnes−km

65
Civil Aviation
Passengers on scheduled services (000); Kilometres (million) [*cont.*]
Aviation civile
Passagers sur les services réguliers (000); Kilomètres (millions) [*suite*]

Country or area and traffic	Total Totale				International Internationaux				Pays ou zone et trafic
	1980	1991	1992	1993	1980	1991	1992	1993	
Qatar									**Qatar**
Kilometres flown	7	11	16	18	7	11	16	18	Kilomètres parcourus
Passengers carried	522	876	981	1080	522	876	981	1080	Passagers transportés
Passengers km	714	1676	1922	2210	714	1676	1922	2210	Passagers–km
Total ton–km	91	211	252	297	91	211	252	297	Total tonnes–km
Saudi Arabia									**Arabie saoudite**
Kilometres flown	91	87	106	113	47	47	59	64	Kilomètres parcourus
Passengers carried	9241	9409	11155	11864	2348	2989	3602	3800	Passagers transportés
Passengers km	9938	14881	17563	18572	4958	10083	12003	12646	Passagers–km
Total ton–km	1069	1848	2283	2415	591	1363	1718	1816	Total tonnes–km
Singapore									**Singapour**
Kilometres flown	69	133	158	181	69	133	158	181	Kilomètres parcourus
Passengers carried	3827	7745	8477	9271	3827	7745	8477	9271	Passagers transportés
Passengers km	14719	33452	37045	41262	14719	33452	37045	41262	Passagers–km
Total ton–km	1959	4992	5783	6826	1959	4992	5783	6826	Total tonnes–km
Sri Lanka									**Sri Lanka**
Kilometres flown	8	20	24	22	8	20	24	22	Kilomètres parcourus
Passengers carried	235	893	1046	994	235	893	1046	994	Passagers transportés
Passengers km	691	3449	4104	3624	691	3449	4104	3624	Passagers–km
Total ton–km	72	418	485	436	72	418	485	436	Total tonnes–km
Syrian Arab Republic									**Rép. arabe syrienne**
Kilometres flown	10	10	9	10	10	9	8	9	Kilomètres parcourus
Passengers carried	465	661	552	485	383	548	492	443	Passagers transportés
Passengers km	948	1127	966	817	908	1086	928	801	Passagers–km
Total ton–km	101	118	101	86	98	115	97	85	Total tonnes–km
Tajikistan									**Tadjikistan**
Kilometres flown	..	..	...	0	..	..	...	0	Kilomètres parcourus
Passengers carried	..	..	...	783	..	..	...	1	Passagers transportés
Passengers km	..	..	...	2231	..	..	...	5	Passagers–km
Total ton–km	..	..	...	205	..	..	...	0	Total tonnes–km
Thailand									**Thaïlande**
Kilometres flown	42	107	115	115	37	91	98	98	Kilomètres parcourus
Passengers carried	2459	7709	8547	10197	1924	4742	5343	6203	Passagers transportés
Passengers km	6276	18246	20427	22874	5988	16554	18616	20609	Passagers–km
Total ton–km	812	2561	2815	3167	789	2397	2637	2941	Total tonnes–km
Turkey									**Turquie**
Kilometres flown	14	39	64	80	8	28	47	59	Kilomètres parcourus
Passengers carried	1254	3160	4959	6120	377	1221	2158	2540	Passagers transportés
Passengers km	1103	3672	6060	7563	689	2663	4653	5693	Passagers–km
Total ton–km	113	407	670	846	72	315	541	679	Total tonnes–km
Turkmenistan									**Turkménistan**
Kilometres flown	..	..	...	—	..	..	...	—	Kilomètres parcourus
Passengers carried	..	..	...	748	..	..	...	—	Passagers transportés
Passengers km	..	..	...	1562	..	..	...	—	Passagers–km
Total ton–km	..	..	...	143	..	..	...	—	Total tonnes–km
United Arab Emirates									**Emirats arabes unis**
Kilometres flown	7	34	49	64	7	34	49	64	Kilomètres parcourus
Passengers carried	522	2042	2509	2936	522	2042	2509	2936	Passagers transportés
Passengers km	714	4861	6298	7794	714	4861	6298	7794	Passagers–km
Total ton–km	91	672	899	1126	91	672	899	1126	Total tonnes–km
Uzbekistan									**Ouzbékistan**
Kilometres flown	..	..	2	2	..	..	2	2	Kilomètres parcourus
Passengers carried	..	..	4032	2217	..	..	24	32	Passagers transportés
Passengers km	..	..	8125	4855	..	..	102	127	Passagers–km
Total ton–km	..	..	774	447	..	..	16	18	Total tonnes–km
Viet Nam									**Viet Nam**
Kilometres flown	0	3	3	3	0	2	2	2	Kilomètres parcourus
Passengers carried	6	198	204	211	6	127	130	137	Passagers transportés
Passengers km	3	188	201	209	3	116	125	133	Passagers–km
Total ton–km	0	17	18	19	0	11	11	12	Total tonnes–km
Yemen									**Yémen**
Kilometres flown	6	9	13	13	5	8	11	11	Kilomètres parcourus
Passengers carried	310	541	800	848	255	326	519	551	Passagers transportés
Passengers km	291	762	1124	1217	280	690	1016	1099	Passagers–km
Total ton–km	27	77	114	124	26	71	104	113	Total tonnes–km

65
Civil Aviation
Passengers on scheduled services (000); Kilometres (million) [cont.]
Aviation civile
Passagers sur les services réguliers (000); Kilomètres (millions) [suite]

Country or area and traffic	Total Totale				International Internationaux				Pays ou zone et trafic
	1980	1991	1992	1993	1980	1991	1992	1993	
Europe [1]									**Europe** [1]
Kilometres flown	1892	2852	3136	3283	1488	2231	2486	2615	Kilomètres parcourus
Passengers carried	128802	202514	219681	228760	69391	106574	120186	128704	Passagers transportés
Passengers km	203021	322787	367787	393816	173908	271580	311222	336020	Passagers−km
Total ton−km	27042	47551	52502	56896	24200	42531	46953	51122	Total tonnes−km
Austria									**Autriche**
Kilometres flown	22	48	62	72	22	46	61	70	Kilomètres parcourus
Passengers carried	1284	2606	3024	3297	1267	2485	2935	3212	Passagers transportés
Passengers km	1120	3605	4867	5629	1115	3559	4832	5595	Passagers−km
Total ton−km	118	405	568	669	118	401	564	665	Total tonnes−km
Belarus									**Bélarus**
Kilometres flown	..	..	...	1	..	..	...	1	Kilomètres parcourus
Passengers carried	..	..	...	805	..	..	...	42	Passagers transportés
Passengers km	..	..	...	2604	..	..	...	98	Passagers−km
Total ton−km	..	..	...	237	..	..	...	9	Total tonnes−km
Belgium									**Belgique**
Kilometres flown	55	73	76	86	55	73	76	86	Kilomètres parcourus
Passengers carried	1974	3018	3146	3651	1974	3018	3146	3651	Passagers transportés
Passengers km	4852	6223	6203	6484	4852	6223	6203	6484	Passagers−km
Total ton−km	842	1071	965	1003	842	1071	965	1003	Total tonnes−km
Bulgaria									**Bulgarie**
Kilometres flown	13	18	22	26	7	16	20	24	Kilomètres parcourus
Passengers carried	1788	646	814	916	486	516	673	776	Passagers transportés
Passengers km	775	1171	1636	2241	515	1116	1575	2180	Passagers−km
Total ton−km	80	112	163	231	55	107	157	225	Total tonnes−km
Croatia									**Croatie**
Kilometres flown	..	1	2	5	..	..	1	3	Kilomètres parcourus
Passengers carried	..	113	215	432	..	..	84	191	Passagers transportés
Passengers km	..	45	123	255	..	..	78	171	Passagers−km
Total ton−km	..	4	12	26	..	..	8	17	Total tonnes−km
former Czechoslovakia †									**anc. Tchécoslovaquie †**
Kilometres flown	25	22	26	..	16	19	24	..	Kilomètres parcourus
Passengers carried	1461	837	974	..	585	653	810	..	Passagers transportés
Passengers km	1539	1736	2135	..	1190	1655	2059	..	Passagers−km
Total ton−km	154	180	219	..	124	172	212	..	Total tonnes−km
Czech Republic									**République tchèque**
Kilometres flown	..	..	..	25	..	..	..	24	Kilomètres parcourus
Passengers carried	..	..	..	1025	..	..	..	1016	Passagers transportés
Passengers km	..	..	..	1900	..	..	..	1897	Passagers−km
Total ton−km	..	..	..	196	..	..	..	196	Total tonnes−km
Denmark [3]									**Danemark** [3]
Kilometres flown	33	58	60	65	27	45	47	52	Kilomètres parcourus
Passengers carried	3330	4582	4695	5077	1354	2395	2547	2835	Passagers transportés
Passengers km	3296	4440	4495	4913	2637	3685	3759	4130	Passagers−km
Total ton−km	423	544	536	580	359	467	461	502	Total tonnes−km
Estonia									**Estonie**
Kilometres flown	..	..	4	4	..	..	3	4	Kilomètres parcourus
Passengers carried	..	..	146	128	..	..	112	109	Passagers transportés
Passengers km	..	..	102	86	..	..	96	82	Passagers−km
Total ton−km	..	..	10	8	..	..	9	8	Total tonnes−km
Finland									**Finlande**
Kilometres flown	35	62	62	63	24	42	42	44	Kilomètres parcourus
Passengers carried	2512	3999	3898	3947	969	1790	1841	2074	Passagers transportés
Passengers km	2139	4719	4639	5529	1603	3804	3755	4712	Passagers−km
Total ton−km	243	556	535	662	194	476	458	590	Total tonnes−km
France [4]									**France** [4]
Kilometres flown	276	411	441	464	213	285	311	315	Kilomètres parcourus
Passengers carried	19521	32817	33607	34777	9952	12463	13580	13697	Passagers transportés
Passengers km	34130	50198	56701	59455	25938	33056	37308	38043	Passagers−km
Total ton−km	5131	8456	9293	9753	4317	6733	7315	7517	Total tonnes−km
Germany †									**Allemagne †**
Kilometres flown	..	434	468	474	..	365	392	399	Kilomètres parcourus
Passengers carried	..	24830	27578	29363	..	13677	15435	16555	Passagers transportés
Passengers km	..	43270	48965	52941	..	38773	44063	47808	Passagers−km
Total ton−km	..	8445	9166	10109	..	7973	8653	9542	Total tonnes−km

65
Civil Aviation
Passengers on scheduled services (000); Kilometres (million) [*cont.*]
Aviation civile
Passagers sur les services réguliers (000); Kilomètres (millions) [*suite*]

Country or area and traffic	Total Totale				International Internationaux				Pays ou zone et trafic
	1980	1991	1992	1993	1980	1991	1992	1993	
F. R. Germany									**R. f. Allemagne**
Kilometres flown	196	..	..	..	168	..	..	..	Kilomètres parcourus
Passengers carried	13046	..	..	..	7458	..	..	..	Passagers transportés
Passengers km	21056	..	..	..	18932	..	..	..	Passagers–km
Total ton–km	3524	..	..	..	3299	..	..	..	Total tonnes–km
Greece									**Grèce**
Kilometres flown	40	49	58	62	30	36	43	47	Kilomètres parcourus
Passengers carried	4891	4937	5466	5478	1656	1733	2137	2290	Passagers transportés
Passengers km	5062	6193	7262	7899	4030	5231	6263	6964	Passagers–km
Total ton–km	521	684	772	848	427	589	673	755	Total tonnes–km
Hungary									**Hongrie**
Kilometres flown	16	19	22	24	16	19	22	24	Kilomètres parcourus
Passengers carried	874	911	1029	1217	874	911	1029	1217	Passagers transportés
Passengers km	1020	1017	1125	1484	1020	1017	1125	1484	Passagers–km
Total ton–km	111	99	110	147	111	99	110	147	Total tonnes–km
Iceland									**Islande**
Kilometres flown	11	19	18	19	9	17	15	17	Kilomètres parcourus
Passengers carried	542	773	767	801	300	526	510	552	Passagers transportés
Passengers km	1295	1789	1840	1968	1235	1728	1775	1905	Passagers–km
Total ton–km	142	202	203	219	136	196	196	212	Total tonnes–km
Ireland									**Irlande**
Kilometres flown	22	44	42	39	21	42	40	37	Kilomètres parcourus
Passengers carried	1830	4765	5006	4650	1636	4407	4662	4354	Passagers transportés
Passengers km	2049	4163	4461	4209	2009	4101	4400	4157	Passagers–km
Total ton–km	271	480	501	467	266	475	496	463	Total tonnes–km
Italy									**Italie**
Kilometres flown	139	200	236	243	96	134	161	171	Kilomètres parcourus
Passengers carried	9956	18847	21767	21804	4191	6961	8480	8530	Passagers transportés
Passengers km	14076	22642	28667	29659	11209	16419	21719	22734	Passagers–km
Total ton–km	1813	3298	3878	4041	1535	2706	3235	3391	Total tonnes–km
Latvia									**Lettonie**
Kilometres flown	..	..	0	0	..	..	0	0	Kilomètres parcourus
Passengers carried	..	..	809	526	..	..	3	2	Passagers transportés
Passengers km	..	..	1218	792	..	..	8	5	Passagers–km
Total ton–km	..	..	125	81	..	..	1	1	Total tonnes–km
Lithuania									**Lituanie**
Kilometres flown	..	..	11	5	..	..	11	5	Kilomètres parcourus
Passengers carried	..	..	557	150	..	..	554	150	Passagers transportés
Passengers km	..	..	756	154	..	..	755	154	Passagers–km
Total ton–km	..	..	72	15	..	..	72	15	Total tonnes–km
Luxembourg									**Luxembourg**
Kilometres flown	3	6	7	8	3	6	7	8	Kilomètres parcourus
Passengers carried	162	406	461	471	162	406	461	471	Passagers transportés
Passengers km	55	258	286	290	55	258	286	290	Passagers–km
Total ton–km	5	24	27	27	5	24	27	27	Total tonnes–km
Malta									**Malte**
Kilometres flown	6	10	11	13	6	10	11	13	Kilomètres parcourus
Passengers carried	401	649	693	797	401	649	693	797	Passagers transportés
Passengers km	602	1014	1111	1250	602	1014	1111	1250	Passagers–km
Total ton–km	59	95	104	117	59	95	104	117	Total tonnes–km
Monaco									**Monaco**
Kilometres flown	0	0	0	0	0	0	0	0	Kilomètres parcourus
Passengers carried	44	43	44	44	44	43	44	44	Passagers transportés
Passengers km	1	1	1	1	1	1	1	1	Passagers–km
Total ton–km	0	0	0	0	0	0	0	0	Total tonnes–km
Netherlands [5]									**Pays–Bas** [5]
Kilometres flown	109	178	220	230	107	177	219	229	Kilomètres parcourus
Passengers carried	4984	8893	10088	11775	4633	8500	10011	11704	Passagers transportés
Passengers km	14643	28197	33351	38544	14596	28147	33302	38495	Passagers–km
Total ton–km	2347	4891	5556	6512	2342	4886	5552	6507	Total tonnes–km
Norway [3]									**Norvège** [3]
Kilometres flown	58	90	96	102	27	43	46	49	Kilomètres parcourus
Passengers carried	4804	8857	9469	10383	1354	2382	2591	2842	Passagers transportés
Passengers km	4068	6291	6584	7266	2637	3630	3736	4092	Passagers–km
Total ton–km	493	710	718	791	359	462	459	498	Total tonnes–km

65
Civil Aviation
Passengers on scheduled services (000); Kilometres (million) [cont.]
Aviation civile
Passagers sur les services réguliers (000); Kilomètres (millions) [suite]

Country or area and traffic	Total Totale				International Internationaux				Pays ou zone et trafic
	1980	1991	1992	1993	1980	1991	1992	1993	
Poland									**Pologne**
Kilometres flown	35	30	31	37	26	29	29	34	Kilomètres parcourus
Passengers carried	1711	1051	1102	1270	931	963	985	1106	Passagers transportés
Passengers km	2232	2878	2873	3335	1934	2845	2827	3272	Passagers−km
Total ton−km	207	288	293	357	182	285	289	351	Total tonnes−km
Portugal									**Portugal**
Kilometres flown	39	59	69	67	30	48	56	55	Kilomètres parcourus
Passengers carried	1978	3572	.4109	4026	971	2230	2512	2553	Passagers transportés
Passengers km	3459	7072	7790	7917	2793	6048	6665	6833	Passagers−km
Total ton−km	424	812	882	897	349	700	760	778	Total tonnes−km
Republic of Moldova									**République de Moldova**
Kilometres flown	..	..	0	0	..	..	0	0	Kilomètres parcourus
Passengers carried	..	..	571	312	..	..	1	2	Passagers transportés
Passengers km	..	..	1826	1078	..	..	3	4	Passagers−km
Total ton−km	..	..	176	100	..	..	0	0	Total tonnes−km
Romania									**Roumanie**
Kilometres flown	20	24	25	24	13	18	20	19	Kilomètres parcourus
Passengers carried	1112	1149	919	979	383	585	615	615	Passagers transportés
Passengers km	1209	2048	2042	1810	916	1798	1898	1657	Passagers−km
Total ton−km	109	178	174	161	83	157	162	148	Total tonnes−km
Russian Federation									**Fédération de Russie**
Kilometres flown	0	143	134	974	0	143	134	162	Kilomètres parcourus
Passengers carried	103754	128274	62174	36229	2503	3761	2872	3856	Passagers transportés
Passengers km	160299	224648	116139	76683	8982	15569	12269	15362	Passagers−km
Total ton−km	17510	22953	12085	7825	1135	1752	1423	1777	Total tonnes−km
Slovakia									**Slovaquie**
Kilometres flown	..	..	..	1	..	..	..	1	Kilomètres parcourus
Passengers carried	..	..	..	18	..	..	..	10	Passagers transportés
Passengers km	..	..	..	10	..	..	..	7	Passagers−km
Total ton−km	..	..	..	1	...	..	..	1	Total tonnes−km
Slovenia									**Slovénie**
Kilometres flown	..	..	3	5	..	..	3	5	Kilomètres parcourus
Passengers carried	..	..	188	291	..	..	180	276	Passagers transportés
Passengers km	..	..	196	278	..	..	195	275	Passagers−km
Total ton−km	..	..	19	28	..	..	19	28	Total tonnes−km
Spain									**Espagne**
Kilometres flown	164	220	246	248	96	126	148	148	Kilomètres parcourus
Passengers carried	15089	20945	23386	22279	5137	5443	6657	6956	Passagers transportés
Passengers km	15517	23200	27480	27105	10290	14800	18198	18551	Passagers−km
Total ton−km	1808	2726	3081	3025	1267	1876	2144	2168	Total tonnes−km
Sweden [3]									**Suède** [3]
Kilometres flown	66	109	114	120	41	62	69	76	Kilomètres parcourus
Passengers carried	5209	9827	9924	9730	2031	3410	3703	4042	Passagers transportés
Passengers km	5342	8163	8247	8428	3955	5350	5518	5945	Passagers−km
Total ton−km	666	935	932	993	538	672	689	757	Total tonnes−km
Switzerland									**Suisse**
Kilometres flown	98	146	158	172	97	141	152	165	Kilomètres parcourus
Passengers carried	5930	7974	8256	9887	5221	7008	7281	8534	Passagers transportés
Passengers km	10831	15327	16472	17704	10773	15087	16236	17443	Passagers−km
Total ton−km	1419	2475	2704	3007	1413	2445	2674	2977	Total tonnes−km
TFYR Macedonia									**L'ex−R.y. Macédoine**
Kilometres flown	..	..	...	5	..	..	...	5	Kilomètres parcourus
Passengers carried	..	..	...	187	..	..	...	187	Passagers transportés
Passengers km	..	..	...	292	..	..	...	292	Passagers−km
Total ton−km	..	..	...	28	..	..	...	28	Total tonnes−km
Ukraine									**Ukraine**
Kilometres flown	..	..	134	35	..	..	4	8	Kilomètres parcourus
Passengers carried	..	..	4906	1278	..	..	117	178	Passagers transportés
Passengers km	..	..	7906	1790	..	..	334	511	Passagers−km
Total ton−km	..	..	760	171	..	..	32	49	Total tonnes−km
United Kingdom [6]									**Royaume−Uni** [6]
Kilometres flown	426	614	684	729	370	527	592	636	Kilomètres parcourus
Passengers carried	25551	42861	47563	50188	18489	31187	35831	38061	Passagers transportés
Passengers km	56750	99856	115199	124882	54026	95191	110462	119950	Passagers−km
Total ton−km	6742	13680	15710	17387	6503	13284	15308	16970	Total tonnes−km

65
Civil Aviation
Passengers on scheduled services (000); Kilometres (million) [*cont.*]
Aviation civile
Passagers sur les services réguliers (000); Kilomètres (millions) [*suite*]

Country or area and traffic	Total Totale				International Internationaux				Pays ou zone et trafic
	1980	1991	1992	1993	1980	1991	1992	1993	
Yugoslavia, SFR †									**Yougoslavie, Rfs †**
Kilometres flown	35	31	..	..	21	26	..	..	Kilomètres parcourus
Passengers carried	3087	1888	..	..	969	1103	..	..	Passagers transportés
Passengers km	2984	3078	..	..	2094	2802	..	..	Passagers–km
Total ton–km	305	367	..	..	228	342	..	..	Total tonnes–km
Oceania [1]									**Océanie** [1]
Kilometres flown	**276**	**479**	**454**	**571**	**100**	**216**	**238**	**252**	**Kilomètres parcourus**
Passengers carried	**18133**	**29107**	**24402**	**36447**	**3254**	**6494**	**7361**	**8630**	**Passagers transportés**
Passengers km	**32195**	**57344**	**56694**	**75499**	**20881**	**37231**	**43053**	**49982**	**Passagers–km**
Total ton–km	**3727**	**7044**	**7122**	**9169**	**2554**	**5055**	**5770**	**6602**	**Total tonnes–km**
Australia									**Australie**
Kilometres flown	200	335	298	417	59	130	144	156	Kilomètres parcourus
Passengers carried	13649	21860	16605	28213	1961	4180	4786	5705	Passagers transportés
Passengers km	25555	44396	41899	59094	15769	26503	30568	36015	Passagers–km
Total ton–km	2900	5397	5258	7102	1897	3635	4140	4781	Total tonnes–km
Cook Islands									**Iles Cook**
Kilometres flown	...	1	...	...	...	1	...	...	Kilomètres parcourus
Passengers carried	...	10	...	...	...	10	...	...	Passagers transportés
Passengers km	...	27	...	...	...	27	...	...	Passagers–km
Total ton–km	...	3	...	...	...	3	...	...	Total tonnes–km
Fiji									**Fidji**
Kilometres flown	7	11	13	9	3	8	8	5	Kilomètres parcourus
Passengers carried	322	414	436	448	90	255	255	305	Passagers transportés
Passengers km	250	594	948	1027	141	573	924	1007	Passagers–km
Total ton–km	26	77	119	115	14	75	116	113	Total tonnes–km
Kiribati									**Kiribati**
Kilometres flown	0	1	1	1	0	0	0	0	Kilomètres parcourus
Passengers carried	0	25	25	26	0	2	2	3	Passagers transportés
Passengers km	0	9	9	10	0	5	5	6	Passagers–km
Total ton–km	0	2	2	2	0	1	1	1	Total tonnes–km
Marshall Islands									**Iles Marshall**
Kilometres flown	0	1	2	2	0	1	1	1	Kilomètres parcourus
Passengers carried	0	66	96	44	0	40	40	16	Passagers transportés
Passengers km	0	52	53	49	0	47	43	39	Passagers–km
Total ton–km	0	15	9	10	0	15	8	9	Total tonnes–km
Nauru									**Nauru**
Kilometres flown	5	2	3	3	5	2	3	3	Kilomètres parcourus
Passengers carried	64	59	117	117	64	59	117	117	Passagers transportés
Passengers km	107	110	206	206	107	110	206	206	Passagers–km
Total ton–km	10	11	20	20	10	11	20	20	Total tonnes–km
New Zealand									**Nouvelle–Zélande**
Kilometres flown	52	107	116	121	29	69	77	81	Kilomètres parcourus
Passengers carried	3497	5371	5784	6291	1041	1780	1968	2244	Passagers transportés
Passengers km	5725	11299	12679	14163	4563	9536	10828	12194	Passagers–km
Total ton–km	731	1454	1618	1823	599	1272	1430	1622	Total tonnes–km
Papua New Guinea									**Papouasie–Nvl–Guinée**
Kilometres flown	11	17	17	13	4	4	4	4	Kilomètres parcourus
Passengers carried	559	911	954	866	99	147	165	181	Passagers transportés
Passengers km	520	676	699	733	301	358	390	409	Passagers–km
Total ton–km	57	77	82	82	33	45	49	50	Total tonnes–km
Solomon Islands									**Iles Salomon**
Kilometres flown	1	2	2	2	0	1	1	1	Kilomètres parcourus
Passengers carried	42	69	69	69	0	8	8	8	Passagers transportés
Passengers km	39	13	13	13	0	4	4	4	Passagers–km
Total ton–km	3	1	1	1	0	0	0	0	Total tonnes–km
Tonga									**Tonga**
Kilometres flown	...	1	1	1	...	–	–	–	Kilomètres parcourus
Passengers carried	...	35	35	35	...	–	–	–	Passagers transportés
Passengers km	...	7	7	7	...	–	–	–	Passagers–km
Total ton–km	...	1	1	1	...	–	–	–	Total tonnes–km
Vanuatu									**Vanuatu**
Kilometres flown	...	1	1	2	...	1	1	2	Kilomètres parcourus
Passengers carried	...	53	59	67	...	53	59	67	Passagers transportés
Passengers km	...	115	129	142	...	115	129	142	Passagers–km
Total ton–km	...	11	13	14	...	11	13	14	Total tonnes–km

65

Civil Aviation
Passengers on scheduled services (000); Kilometres (million) [*cont.*]
Aviation civile
Passagers sur les services réguliers (000); Kilomètres (millions) [*suite*]

Country or area and traffic	Total Totale				International Internationaux				Pays ou zone et trafic
	1980	1991	1992	1993	1980	1991	1992	1993	
former USSR †									**ancienne URSS †**
Kilometres flown	0	143	270	1012	0	143	140	174	Kilomètres parcourus
Passengers carried	103754	128274	78412	46698	2503	3761	3040	4144	Passagers transportés
Passengers km	160299	224648	147220	100337	8982	15569	12751	16161	Passagers—km
Total ton—km	17510	22953	15046	9995	1135	1752	1475	1859	Total tonnes—km

Source:
International Civil Aviation Organization (Montreal).

Source:
Organisation de l'aviation civile internationale (Montréal).

† For detailed descriptions of data pertaining to former Czechoslovakia, Germany, SFR Yugoslavia and former USSR USSR, see Annex I – Country or area nomenclature, regional and other groupings.

1 Regional totals add to world totals. However, individual country statistics do not add to regional totals because (i) not all countries are shown, and (ii) the stastistics of France, the Netherlands and the United Kingdom have been distributed between two or more regions; France (Europe, Asia and Pacific, Africa, America: North and South), Netherlands (Europe and America: North and South), United Kingdom (Europe, Asia and Pacific, America: North and South).

2 Includes apportionment (1/10) of the traffic of Air Afrique, a multinational airline with headquarters in Côte d'Ivoire and operated by 10 African States until 1991. From 1992 includes apportionment (1/11) of the traffic of Air Afrique and operated by 11 African States.

3 Includes an apportionment of international operations performed by Scandinavian Airlines System (SAS), Denmark (2/7), Norway (2/7), Sweden (3/7).

4 Including data for airlines based in the territories and dependencies of France.

5 Including data for airlines based in the territories and dependencies of Netherlands.

6 Including data for airlines based in the territories and dependencies of United Kingdom.

† Pour les descriptions en détails des données relatives à l'ancienne Tchécoslovaquie, l'Allemagne, la Rfs Yougoslavie et l'ancienne URSS, voir l'Annexe I – Nomenclature des pays ou zones, groupements régionaux et autres groupements.

1 Les totaux régionaux s'ajoutent pour donner les totaux mondiaux. En revanche, les statistiques de chaque pays ne s'additionnent pas pour donner des totaux régionaux car (i) tous les pays ne sont pas indiqués et (ii) les statistiques de la France, des Pays—Bas et du Royaume—Uni concernent deux régions ou plus; France (Europe, Asie et Pacifique, Afrique, Amérique : Nord et Sud), Pays—Bas (Europe et Amérique : Nord et Sud), Royaume—Uni (Europe, Asie et Pacifique, Amérique : Nord et Sud).

2 Ces chiffres comprennent une partie du trafic (1/10) assurée par Air Afrique, compagnie aérienne multinationale dont le siège est situé en Côte d'Ivoire et est exploitée conjointement par 10 Etats Africains jusqu'à 1991. A partir de 1992 ces chiffres comprennent une partie du trafic (1/11) assurée par Air Afrique et est exploitée conjointement par 11 Etats Africains

3 Y compris une partie des vols internationaux effectués par le SAS, Danemark (2/7), Norvège (2/7) et Suède (3/7).

4 Y compris les données relatives aux compagnies aériennes ayant des bases d'opérations dans les territoires et dépendances de France.

5 Y compris les données relatives aux compagnies aériennes ayant des bases d'opérations dans les territoires et dépendances des Pays—Bas.

6 Y compris les données relatives aux compagnies aériennes ayant des bases d'opération dans les territoires et dépendances du Royaume—Uni.

Technical notes, tables 61-65

Table 61: Data refer to domestic and international traffic on all railway lines within each country shown, except railways entirely within an urban unit, and plantation, industrial mining, funicular and cable railways. The figures relating to passenger-kilometres include all passengers except military, government and railway personnel when carried without revenue. Those relating to ton-kilometres are freight net ton-kilometres and include both fast and ordinary goods services but exclude service traffic, mail, baggage and non-revenue governmental stores.

Table 62: For years in which a census or registration took place the census or registration figure is shown; for other years, unless otherwise indicated, the officially estimated number of vehicles in use is shown. The time of year to which the figures refer is variable. Special purpose vehicles such as two- or three-wheeled cycles and motorcycles, trams, trolley-buses, ambulances, hearses, military vehicles operated by police or other governmental security organizations are excluded. Passenger cars includes vehicles seating not more than nine persons (including the driver), such as taxis, jeeps and station wagons. Commercial vehicles includes: vans, lorries (trucks), buses, tractor and semi-trailer combinations but excluding trailers and farm tractors.

Table 63: Data refer to merchant fleets registered in each country on 30 June of the year stated (1983-1991). Beginning 1992, data refer to end of the year merchant fleets. They are given in gross registered tons (100 cubic feet or 2.83 cubic metres) and represent the total volume of all the permanently enclosed spaces of the vessels to which the figures refer. Vessels without mechanical means of propulsion are excluded, but sailing vessels with auxiliary power are included.

Part A of the table refers to the total of merchant fleets registered. Part B shows data for oil tanker fleets and part C data for ore-oil and bulk carrier fleets.

Table 64: The figures for vessels entered and cleared, unless otherwise stated, represent the sum of the net registered tonnage of sea-going foreign and domestic merchant vessels (power and sailing) entered with cargo from or cleared with cargo to a foreign port and refer to only one entrance or clearance for each foreign voyage. The data where possible exclude vessels "in ballast", i.e. entering without unloading or clearing without loading goods.

Notes techniques, tableaux 61-65

Tableau 61 : Les données se rapportent au trafic intérieur et international de toutes les lignes de chemins de fer du pays indiqué, à l'exception des lignes situées entièrement à l'intérieur d'une agglomération urbaine ou desservant une plantation ou un complexe industriel minier, des funiculaires et des téléfériques. Les chiffres relatifs aux voyageurs-kilomètres se rapportent à tous les voyageurs sauf les militaires, les fonctionnaires et le personnel des chemins de fer, qui sont transportés gratuitement. Les chiffres relatifs aux tonnes-kilomètres se rapportent aux tonnes-kilomètres nettes de fret et comprennent les services rapides et ordinaires de transport de marchandises, à l'exception des transports pour les besoins du service, du courrier, des bagages et des marchandises transportées gratuitement pour les besoins de l'Etat.

Tableau 62 : Pour les années où a eu lieu un recensement ou un enregistrement des véhicules, le chiffre indiqué est le résultat de cette opération; pour les autres années, sauf indication contraire, le chiffre indiqué correspond à l'estimation officielle du nombre de véhicules en circulation. L'époque de l'année à laquelle se rapportent les chiffres varie. Les véhicules à usage spécial, tels que les cycles à deux ou trois roues et motocyclettes, les tramways, les trolley-bus, les ambulances, les corbillards, les véhicules militaires utilisés par la police ou par d'autres services publics de sécurité ne sont pas compris dans ces chiffres. Les voitures de tourisme comprennent les véhicules automobiles dont le nombre de places assises (y compris celle du conducteur) n'est pas supérieur à neuf, tels que les taxis, jeeps et breaks. Les véhicules utilitaires comprennent les fourgons, camions, autobus et autocars, les ensembles tracteurs-remorques et semi-remorques, mais ne comprennent pas les remorques et les tracteurs agricoles.

Tableau 63 : Les données se rapportent à la flotte marchande enregistrée dans chaque pays au 30 juin de l'année indiquée (1983-1991). A partir de 1992, les données se rapportent à la flotte marchande à la fin de l'année. Elles sont exprimées en tonneaux de jauge brute (100 pieds cubes ou 2,83 mètres cubes) et représentent le volume total de tous les espaces clos en permanence dans les navires auxquels elle s'appliquent. Elles excluent les navires sans moteur, mais pas les voiliers avec moteurs auxiliaires.

Les données de la Partie A du tableau se rapportent au total de la flotte marchande enregistrée. Celles de la Partie B se rapportent à la flotte des pétroliers, et celles de la Partie C à la flotte des minéraliers et des transporteurs de vrac et d'huile.

Table 65: Data for total services cover both domestic and international scheduled services operated by airlines registered in each country. Scheduled services include supplementary services occasioned by overflow traffic on regularly scheduled trips and preparatory flights for newly scheduled services. Freight means all goods, except mail and excess baggage, carried for remuneration.

Tableau 64 : sauf indication contraire, les données relatives aux navires entrés et sortis représentent la jauge nette totale des navires marchands de haute mer (à moteur ou à voile) nationaux ou étrangers, qui entrent ou sortent chargés, en provenance ou à destination d'un port étranger. On ne compte qu'une seule entrée et une seule sortie pour chaque voyage international. Dans la mesure du possible, le tableau exclut les navires sur lest (c'est-à-dire les navires entrant sans décharger ou sortant sans avoir chargé).

Tableau 65 : Les données relatives au total des services se rapportent aux services réguliers, intérieurs ou internationaux, des compagnies de transport aérien enregistrées dans chaque pays. Les services réguliers comprennent aussi les vols supplémentaires nécessités par un surcroît d'activité des services réguliers et les vols préparatoires en vue de nouveaux services réguliers. Par fret, on entend toutes les marchandises transportées contre paiement, mais non le courrier et les excédents de bagage.

Table 66 follows overleaf
      Le tableau 66 est présenté au verso

66
Production, trade and consumption of commercial energy
Production, commerce et consommation d'énergie commerciale

Thousand metric tons of coal equivalent and kilograms per capita
Milliers de tonnes métriques d'équivalent houille et kilogrammes par habitant

Country or area	Year	Primary energy production – Production d'energie primaire					Changes in stocks	Imports	Exports
		Total Totale	Solids Solides	Liquids Liquides	Gas Gaz	Electricity Electricité	Variations des stocks	Imports Importations	Exports Exportations
Africa	**1980**	**577 211**	**98 232**	**436 325**	**35 227**	**7 427**	**6 277**	**69 920**	**425 314**
	1990	**703 444**	**139 493**	**461 565**	**92 179**	**10 206**	**3 120**	**72 374**	**460 825**
	1991	**736 187**	**142 719**	**485 575**	**97 527**	**10 365**	**2 114**	**72 551**	**488 969**
	1992	**722 012**	**139 822**	**474 804**	**97 136**	**10 250**	**−3 120**	**76 110**	**486 577**
Algeria	1980	100 905	7	74 964	25 903	31	3 575	2 593	69 185
	1990	158 008	10	89 282	68 700	17	1 274	1 481	110 032
	1991	159 904	15	88 078	71 776	36	157	1 756	113 827
	1992	153 093	15	83 507	69 547	24	−670	1 858	112 222
Angola	1980	10 847	..	10 610	102	135	0	230	9 047
	1990	34 030	..	33 642	222	167	0	21	31 621
	1991	35 713	..	35 324	222	167	0	21	33 403
	1992	37 527	..	37 137	222	169	0	21	35 332
Benin	1980	14	..	14	..	..	15	232	38
	1990	416	..	416	..	..	3	254	424
	1991	421	..	421	..	..	1	259	432
	1992	426	..	426	..	..	1	266	436
Burkina Faso	1980	..	..	..	..	..	1	207	0
	1990	..	..	..	..	..	4	260	0
	1991	..	..	..	..	..	0	266	0
	1992	..	..	..	..	..	0	268	0
Burundi	1980	2	2	..	..	0	0	58	..
	1990	19	6	..	..	13	4	99	..
	1991	21	5	..	..	16	4	110	..
	1992	19	6	..	..	13	3	97	..
Cameroon	1980	4 024	0	3 856	..	168	549	881	2 302
	1990	12 437	1	12 112	..	324	544	21	10 600
	1991	11 338	1	11 012	..	324	−79	23	10 119
	1992	10 024	1	9 698	..	325	0	23	8 723
Cape Verde	1980	..	..	..	..	..	0	63	0
	1990	..	..	..	..	..	0	47	0
	1991	..	..	..	..	..	0	49	0
	1992	..	..	..	..	..	0	51	0
Central African Rep.	1980	8	..	..	..	8	1	60	..
	1990	9	..	..	..	9	6	119	..
	1991	9	..	..	..	9	3	120	..
	1992	10	..	..	..	10	3	123	..
Chad	1980	..	..	..	..	..	0	134	..
	1990	..	..	..	..	..	3	94	..
	1991	..	..	..	..	..	1	55	..
	1992	..	..	..	..	..	3	63	..
Comoros	1980	0	..	..	..	0	0	19	..
	1990	0	..	..	..	0	0	31	..
	1991	0	..	..	..	0	0	31	..
	1992	0	..	..	..	0	0	31	..
Congo	1980	4 720	...	4 706	2	12	136	148	4 577
	1990	11 515	0	11 450	3	62	−21	13	10 635
	1991	11 524	0	11 462	3	59	143	17	10 491
	1992	12 417	0	12 361	4	52	130	22	11 371
Côte d'Ivoire	1980	338	..	171	..	167	265	2 812	740
	1990	613	..	450	..	163	27	4 684	358
	1991	611	..	457	..	154	56	4 717	381
	1992	593	..	464	..	129	49	4 770	383
Djibouti	1980	..	..	..	..	..	0	708	..
	1990	..	..	..	..	..	0	750	..
	1991	..	..	..	..	..	0	753	..
	1992	..	..	..	..	..	0	762	..
Egypt	1980	45 949	..	42 611	2 134	1 204	714	1 918	22 918
	1990	75 520	..	65 267	9 028	1 225	714	2 270	35 023
	1991	78 995	..	67 597	10 339	1 060	786	1 934	37 358
	1992	80 737	..	68 510	11 179	1 049	857	1 847	38 039
Equatorial Guinea	1980	0	..	..	..	0	1	28	..
	1990	0	..	..	..	0	1	55	..
	1991	0	..	..	..	0	0	58	..
	1992	0	..	..	..	0	0	55	..

Bunkers – Soutes   Consumption – Consommation

Air Avion	Sea Maritime	Unallocated Nondistribué	Per capita Par habitant	Total Totale	Solids Solides	Liquids Liquides	Gas Gaz	Electricity Electricité	Année	Pays ou zone
3 577	**9 424**	**12 530**	**435**	**190 008**	**74 220**	**84 366**	**24 019**	**7 403**	**1980**	**Afrique**
3 268	**5 693**	**21 018**	**448**	**281 894**	**101 148**	**120 543**	**50 152**	**10 051**	**1990**	
3 162	**5 654**	**19 256**	**447**	**289 583**	**103 928**	**124 188**	**51 185**	**10 282**	**1991**	
3 233	**4 975**	**18 817**	**432**	**287 640**	**101 226**	**127 544**	**48 751**	**10 119**	**1992**	
376	606	4 983	1 322	24 773	607	6 462	17 672	32	1980	Algérie
236	250	8 122	1 587	39 575	1 220	11 363	26 987	5	1990	
236	287	6 599	1 588	40 554	1 325	12 619	26 655	−45	1991	
236	230	5 679	1 426	37 255	1 325	13 871	22 149	−90	1992	
96	456	437	149	1 042	0	805	102	135	1980	Angola
231	974	327	98	898	0	509	222	167	1990	
228	985	228	93	889	0	500	222	167	1991	
236	985	100	91	896	0	505	222	169	1992	
15	..	0	52	179	..	164	..	14	1980	Bénin
22	..	0	47	220	..	197	..	23	1990	
25	..	0	46	222	..	197	..	25	1991	
25	..	0	46	229	..	200	..	29	1992	
..	..	..	29	205	0	205	..	0	1980	Burkina Faso
..	..	..	28	256	0	256	..	0	1990	
..	..	..	29	266	0	266	..	0	1991	
..	..	..	28	268	0	268	..	0	1992	
1	..	..	14	59	2	52	..	5	1980	Burundi
3	..	..	20	110	6	90	..	15	1990	
3	..	..	22	124	5	100	..	19	1991	
3	..	..	19	110	6	88	..	16	1992	
100	..	92	215	1 861	0	1 694	..	168	1980	Cameroun
19	..	113	103	1 182	1	857	..	324	1990	
19	..	101	101	1 200	1	875	..	324	1991	
19	..	93	99	1 211	1	885	..	325	1992	
0	7	..	194	56	0	56	..	..	1980	Cap–Vert
0	7	..	117	40	0	40	..	..	1990	
0	7	..	117	41	0	41	..	..	1991	
0	0	..	142	51	0	51	..	..	1992	
7	..	..	26	60	..	52	..	8	1980	Rép. centrafricaine
18	..	..	36	105	..	95	..	9	1990	
18	..	..	36	109	..	100	..	9	1991	
18	..	..	36	112	..	103	..	10	1992	
33	..	..	23	101	0	101	..	..	1980	Tchad
24	..	..	12	67	0	67	..	..	1990	
24	..	..	5	29	0	29	..	..	1991	
24	..	..	6	37	0	37	..	..	1992	
0	..	..	50	19	..	19	..	0	1980	Comores
0	..	..	57	31	..	31	..	0	1990	
0	..	..	55	31	..	31	..	0	1991	
0	..	..	53	31	..	31	..	0	1992	
6	13	0	83	138	...	123	2	13	1980	Congo
0	7	12	401	895	0	828	3	64	1990	
0	8	12	386	887	0	818	3	66	1991	
0	11	123	339	803	0	734	4	65	1992	
76	3	168	232	1 898	..	1 731	..	167	1980	Côte d'Ivoire
101	13	1 584	268	3 214	..	3 051	..	163	1990	
104	14	1 402	272	3 371	..	3 217	..	154	1991	
109	7	1 128	287	3 687	..	3 559	..	129	1992	
112	451	..	512	144	..	144	..	..	1980	Djibouti
96	487	..	323	167	..	167	..	..	1990	
96	491	..	311	166	..	166	..	..	1991	
93	64	..	1 108	605	..	605	..	..	1992	
236	1 465	1 603	478	20 930	685	16 907	2 134	1 204	1980	Egypte
206	2 149	3 094	650	36 604	1 218	25 132	9 028	1 225	1990	
221	2 148	2 230	662	38 186	978	25 810	10 339	1 060	1991	
221	2 007	3 015	652	38 446	942	25 276	11 179	1 049	1992	
..	..	..	124	27	..	26	..	0	1980	Guinée équatoriale
..	..	..	153	54	..	54	..	0	1990	
..	..	..	164	59	..	58	..	0	1991	
..	..	..	152	56	..	55	..	0	1992	

66
Production, trade and consumption of commercial energy
Thousand metric tons of coal equivalent and kilograms per capita [cont.]
Production, commerce et consommation d'énergie commerciale
Milliers de tonnes métriques d'équivalent houille et kilogrammes par habitant [suite]

Country or area	Year	Primary energy production – Production d'energie primaire					Changes in stocks Variations des stocks	Imports Importations	Exports Exportations
		Total Totale	Solids Solides	Liquids Liquides	Gas Gaz	Electricity Electricité			
Ethiopia	1980	...	..	..	..	...	...	...	...
	1990	215	..	..	..	215	36	1 642	202
	1991	217	..	..	..	217	36	1 652	209
	1992	222	..	..	..	222	43	1 645	209
Gabon	1980	12 757	..	12 705	20	32	−141	84	11 962
	1990	19 492	..	19 272	133	87	186	58	18 003
	1991	21 510	..	21 316	107	87	100	58	20 102
	1992	22 174	..	21 951	136	87	14	61	20 774
Gambia	1980	..	..	..	..	..	0	80	4
	1990	..	..	..	..	..	0	94	3
	1991	..	..	..	..	..	0	97	3
	1992	..	..	..	..	..	0	97	3
Ghana	1980	774	..	126	..	648	13	1 568	518
	1990	713	..	0	..	713	29	1 772	119
	1991	750	..	0	..	750	32	1 775	123
	1992	751	..	0	..	751	53	1 817	127
Guinea	1980	18	..	..	..	18	0	452	..
	1990	21	..	..	..	21	6	498	..
	1991	22	..	..	..	22	6	507	..
	1992	23	..	..	..	23	6	507	..
Guinea−Bissau	1980	..	..	..	..	..	0	71	0
	1990	..	..	..	..	..	3	106	0
	1991	..	..	..	..	..	3	109	0
	1992	..	..	..	..	..	3	110	0
Kenya	1980	130	..	..	..	130	−119	4 849	1 639
	1990	724	..	..	..	724	0	3 482	774
	1991	706	..	..	..	706	0	3 144	922
	1992	677	..	..	..	677	0	3 519	933
Liberia	1980	41	..	..	..	41	14	1 026	21
	1990	30	..	..	..	30	0	249	1
	1991	20	..	..	..	20	0	156	1
	1992	20	..	..	..	20	0	156	1
Libyan Arab Jamah.	1980	132 809	..	128 224	4 585	..	547	921	122 456
	1990	106 097	..	97 842	8 255	..	4 852	9	80 495
	1991	113 963	..	105 256	8 707	..	−2 685	9	94 704
	1992	109 756	..	100 742	9 014	..	−4 155	9	92 674
Madagascar	1980	18	..	..	..	18	−398	520	160
	1990	42	..	..	..	42	−14	502	62
	1991	41	..	..	..	41	0	562	70
	1992	42	..	..	..	42	0	562	79
Malawi	1980	48	..	..	..	48	0	315	0
	1990	86	..	..	..	86	1	279	0
	1991	92	..	..	..	92	1	298	0
	1992	95	..	..	..	95	1	299	0
Maldives	1980	..	..	..	..	..	..	20	...
	1990	..	..	..	..	..	..	87	44
	1991	..	..	..	..	..	..	97	46
	1992	..	..	..	..	..	..	91	50
Mali	1980	10	..	..	..	10	...	204	..
	1990	22	..	..	..	22	...	218	..
	1991	23	..	..	..	23	...	224	..
	1992	26	..	..	..	26	...	230	..
Mauritania	1980	3	..	..	..	3	0	314	..
	1990	3	..	..	..	3	1	1 352	..
	1991	3	..	..	..	3	1	1 398	..
	1992	3	..	..	..	3	0	1 485	..
Mauritius	1980	10	..	..	..	10	−4	446	0
	1990	10	..	..	..	10	40	885	0
	1991	9	..	..	..	9	27	898	0
	1992	14	..	..	..	14	34	986	0
Morocco	1980	974	680	20	90	184	−351	6 025	141
	1990	773	526	21	76	150	−75	10 031	0
	1991	775	551	17	52	156	−277	9 989	0
	1992	744	576	16	32	121	289	11 456	0

Bunkers – Soutes			Consumption – Consommation							
Air Avion	Sea Maritime	Unallocated Nondistribué	Per capita Par habitant	Total Totale	Solids Solides	Liquids Liquides	Gas Gaz	Electricity Electricité	Année	Pays ou zone
...	...	...	...	...	...	...	..	...	1980	Ethiopie
80	11	32	32	1 495	0	1 281	..	215	1990	
84	11	12	31	1 517	0	1 300	..	217	1991	
85	13	0	30	1 516	0	1 294	..	222	1992	
30	90	74	1 026	827	..	775	20	32	1980	Gabon
39	100	203	889	1 019	..	800	133	87	1990	
40	111	200	861	1 015	..	821	107	87	1991	
42	36	242	928	1 126	..	903	136	87	1992	
..	..	..	117	75	..	75	..	..	1980	Gambie
..	..	..	99	91	..	91	..	..	1990	
..	..	..	98	94	..	94	..	..	1991	
..	..	..	94	94	..	94	..	..	1992	
56	43	42	155	1 669	2	1 073	..	594	1980	Ghana
31	30	108	144	2 168	3	1 476	..	689	1990	
37	28	88	143	2 218	3	1 491	..	724	1991	
38	0	85	142	2 264	3	1 545	..	716	1992	
9	...	..	103	461	..	443	..	18	1980	Guinée
18	...	..	86	496	..	474	..	21	1990	
18	...	..	85	505	..	483	..	22	1991	
18	...	..	83	506	..	483	..	23	1992	
7	..	..	81	64	..	64	..	..	1980	Guinée–Bissau
7	..	..	100	96	..	96	..	..	1990	
9	..	..	99	97	..	97	..	..	1991	
9	..	..	98	99	..	99	..	..	1992	
374	273	534	137	2 278	18	2 091	..	169	1980	Kenya
88	233	75	129	3 036	152	2 140	..	745	1990	
0	182	85	109	2 661	133	1 806	..	722	1991	
0	199	165	114	2 899	159	2 033	..	706	1992	
12	47	35	500	938	..	897	..	41	1980	Libéria
6	33	0	93	239	..	209	..	30	1990	
6	17	0	57	151	..	131	..	20	1991	
6	17	0	55	151	..	131	..	20	1992	
251	71	3 100	2 400	7 304	1	5 695	1 608	..	1980	Jamah.arabe libyenne
310	113	4 394	3 508	15 943	5	9 334	6 604	..	1990	
310	99	5 285	3 454	16 260	5	9 651	6 604	..	1991	
295	99	4 714	3 311	16 139	5	9 570	6 564	..	1992	
1	26	−3	83	752	21	713	..	18	1980	Madagascar
1	16	11	37	467	15	410	..	42	1990	
1	19	18	38	496	15	440	..	41	1991	
1	19	9	37	496	15	438	..	42	1992	
15	..	..	56	348	65	235	..	48	1980	Malawi
16	..	..	37	348	18	243	..	86	1990	
16	..	..	38	373	11	269	..	92	1991	
16	..	..	37	377	12	269	..	95	1992	
..	..	..	127	20	..	20	..	..	1980	Maldives
..	..	..	194	42	..	42	..	..	1990	
..	..	..	229	51	..	51	..	..	1991	
..	..	..	177	41	..	41	..	..	1992	
22	..	..	28	192	..	182	..	10	1980	Mali
21	..	..	24	220	..	197	..	22	1990	
21	..	..	24	226	..	203	..	23	1991	
22	..	..	24	233	..	208	..	26	1992	
15	10	0	188	292	6	283	..	3	1980	Mauritanie
16	10	152	587	1 175	6	1 166	..	3	1990	
18	13	155	591	1 214	6	1 204	..	3	1991	
18	14	125	632	1 331	6	1 322	..	3	1992	
78	92	..	300	290	1	279	..	10	1980	Maurice
205	115	..	506	535	75	450	..	10	1990	
200	107	..	538	574	66	498	..	9	1991	
215	102	..	601	649	62	573	..	14	1992	
170	56	473	336	6 511	642	5 595	90	184	1980	Maroc
100	4	1 304	389	9 470	1 764	7 468	76	163	1990	
102	4	898	404	10 038	1 960	7 791	52	234	1991	
103	64	1 538	402	10 206	1 814	8 125	32	235	1992	

66

Production, trade and consumption of commercial energy
Thousand metric tons of coal equivalent and kilograms per capita [*cont.*]
Production, commerce et consommation d'énergie commerciale
Milliers de tonnes métriques d'équivalent houille et kilogrammes par habitant [*suite*]

Country or area	Year	Primary energy production – Production d'energie primaire					Changes in stocks Variations des stocks	Imports Importations	Exports Exportations
		Total Totale	Solids Solides	Liquids Liquides	Gas Gaz	Electricity Electricité			
Mozambique	1980	1 871	207	..	..	1 664	0	1 437	1 296
	1990	46	40	..	..	6	...	532	0
	1991	48	42	..	..	6	...	540	0
	1992	46	40	..	..	6	...	538	0
Niger	1980	20	20	..	..	..	...	259	..
	1990	172	172	..	..	..	0	321	..
	1991	172	172	..	..	..	1	329	..
	1992	170	170	..	..	..	0	339	..
Nigeria	1980	151 096	176	148 818	1 764	338	1 785	3 228	139 304
	1990	129 137	90	123 315	5 339	393	−4 571	2 543	112 067
	1991	141 540	90	135 199	5 858	393	2 999	2 695	116 281
	1992	137 819	95	130 807	6 524	393	471	2 710	114 517
Réunion	1980	37	..	..	..	37	21	355	0
	1990	67	..	..	..	67	3	608	1
	1991	57	..	..	..	57	0	730	1
	1992	61	..	..	..	61	0	719	1
Rwanda	1980	15	..	..	1	14	0	162	0
	1990	29	..	..	0	29	0	251	0
	1991	29	..	..	0	29	0	226	0
	1992	30	..	..	0	30	0	229	0
Sao Tome – Principe	1980	1	..	..	..	1	..	19	..
	1990	1	..	..	..	1	..	32	..
	1991	1	..	..	..	1	..	34	..
	1992	1	..	..	..	1	..	35	..
Senegal	1980	..	..	..	..	..	3	1 985	222
	1990	..	..	..	..	..	7	1 653	51
	1991	..	..	..	..	..	3	1 716	49
	1992	..	..	..	..	..	0	1 701	49
Seychelles	1980	..	..	..	..	..	1	99	..
	1990	..	..	..	..	..	−9	208	..
	1991	..	..	..	..	..	0	220	..
	1992	..	..	..	..	..	1	230	..
Sierra Leone	1980	..	..	..	..	..	0	437	1
	1990	..	..	..	..	..	3	332	3
	1991	..	..	..	..	..	3	376	4
	1992	..	..	..	..	..	3	381	4
Somalia	1980	..	..	..	..	..	77	580	135
	1990	..	..	..	..	..	...	...	...
	1991	..	..	..	..	..	...	...	...
	1992	..	..	..	..	..	...	...	...
South Africa Customs Un.	1980	93 519	93 379	..	..	140	−775	24 076	29 358
	1990	136 506	133 290	..	..	3 216	−3	22 462	42 788
	1991	139 399	135 843	..	..	3 556	−3	22 603	42 870
	1992	136 506	132 950	..	..	3 556	0	23 613	42 880
St.Helena and Depend.	1980	0	0	..	..	..	..	1	..
	1990	0	0	..	..	..	..	3	..
	1991	0	0	..	..	..	..	3	..
	1992	0	0	..	..	..	..	3	..
Sudan	1980	83	..	..	..	83	0	1 662	21
	1990	115	..	..	..	115	14	1 862	46
	1991	115	..	..	..	115	7	1 843	53
	1992	115	..	..	..	115	17	1 866	50
Togo	1980	0	0	..	..	0	0	724	495
	1990	1	0	..	..	1	0	282	13
	1991	1	0	..	..	1	0	300	9
	1992	1	0	..	..	1	0	295	9
Tunisia	1980	8 666	..	8 037	626	3	410	2 835	6 593
	1990	6 912	..	6 483	423	5	25	4 884	5 337
	1991	7 953	..	7 476	464	13	556	4 291	5 346
	1992	7 710	..	7 223	480	8	−242	4 689	5 721
Uganda	1980	80	..	..	..	80	0	308	38
	1990	94	..	..	..	94	0	401	17
	1991	95	..	..	..	95	0	440	18
	1992	96	..	..	..	96	0	451	18

Bunkers – Soutes			Consumption – Consommation							
Air Avion	Sea Maritime	Unallocated Nondistribué	Per capita Par habitant	Total Totale	Solids Solides	Liquids Liquides	Gas Gaz	Electricity Electricité	Année	Pays ou zone
37	122	35	150	1 818	288	1 046	..	485	1980	Mozambique
56	36	...	34	486	58	382	..	46	1990	
53	36	...	35	498	62	390	..	46	1991	
52	36	...	34	496	60	390	..	46	1992	
12	..	..	48	268	20	237	..	11	1980	Niger
21	..	..	61	471	172	276	..	23	1990	
22	..	..	60	479	172	284	..	23	1991	
22	..	..	59	487	170	294	..	23	1992	
562	468	244	166	11 961	180	9 690	1 764	327	1980	Nigéria
523	570	86	239	23 005	62	17 223	5 339	381	1990	
523	570	635	234	23 227	62	16 925	5 858	381	1991	
531	577	535	234	23 897	67	16 925	6 524	381	1992	
1	1	..	729	369	..	331	..	37	1980	Réunion
15	1	..	1 084	655	..	587	..	67	1990	
16	1	..	1 249	767	..	710	..	57	1991	
18	1	..	1 216	759	..	698	..	61	1992	
6	..	..	33	171	..	150	1	20	1980	Rwanda
13	..	..	38	267	..	236	0	30	1990	
13	..	..	34	241	..	211	0	30	1991	
12	..	..	34	247	..	216	0	31	1992	
0	..	..	213	20	..	19	..	1	1980	Sao Tomé–et–Principe
0	..	..	277	33	..	32	..	1	1990	
0	..	..	289	35	..	34	..	1	1991	
0	..	..	290	36	..	35	..	1	1992	
171	358	16	219	1 215	..	1 215	..	..	1980	Sénégal
144	190	7	171	1 253	..	1 253	..	..	1990	
152	189	38	171	1 284	..	1 284	..	..	1991	
152	186	22	168	1 292	..	1 292	..	..	1992	
32	20	..	714	45	..	45	..	..	1980	Seychelles
25	133	..	829	58	..	58	..	..	1990	
28	126	..	929	65	..	65	..	..	1991	
46	112	..	1 000	71	..	71	..	..	1992	
30	111	33	81	261	0	261	..	..	1980	Sierra Leone
24	108	47	37	147	0	147	..	..	1990	
25	107	67	42	171	0	171	..	..	1991	
25	107	65	42	177	0	177	..	..	1992	
31	43	103	28	190	..	190	..	..	1980	Somalie
...	...	...	...	...	..	...	..	..	1990	
...	...	...	...	...	..	...	..	..	1991	
...	...	...	...	...	..	...	..	..	1992	
103	4 483	104	2 555	84 323	67 861	15 151	...	1 310	1980	Un.douan.d'Afr.mérid
...	0	1 023	2 727	115 160	90 693	21 406	...	3 060	1990	
...	0	886	2 736	118 249	93 196	21 686	...	3 367	1991	
...	0	847	2 631	116 391	90 300	22 725	...	3 367	1992	
1	..	..	0	0	0	0	..	..	1980	Ste–Hélène et dépend
0	..	..	500	3	0	3	..	..	1990	
0	..	..	500	3	0	3	..	..	1991	
0	..	..	500	3	0	3	..	..	1992	
65	7	132	81	1 520	0	1 437	..	83	1980	Soudan
46	10	249	66	1 613	0	1 498	..	115	1990	
47	10	227	64	1 613	0	1 498	..	115	1991	
47	10	223	63	1 634	0	1 519	..	115	1992	
..	..	46	70	183	0	161	..	22	1980	Togo
..	..	0	76	270	0	234	..	36	1990	
..	..	0	80	292	0	258	..	33	1991	
..	..	0	76	287	0	249	..	39	1992	
112	10	268	643	4 108	87	3 392	626	3	1980	Tunisie
195	6	−20	774	6 253	129	4 360	1 760	5	1990	
115	1	2	755	6 223	99	4 761	1 346	15	1991	
227	1	25	793	6 667	111	4 607	1 942	7	1992	
0	..	..	27	349	..	305	..	44	1980	Ouganda
0	..	..	27	479	..	398	..	81	1990	
0	..	..	28	517	..	436	..	82	1991	
0	..	..	27	528	..	446	..	82	1992	

66

Production, trade and consumption of commercial energy
Thousand metric tons of coal equivalent and kilograms per capita [*cont.*]
Production, commerce et consommation d'énergie commerciale
Milliers de tonnes métriques d'équivalent houille et kilogrammes par habitant [*suite*]

Country or area	Year	Primary energy production – Production d'energie primaire					Changes in stocks Variations des stocks	Imports Importations	Exports Exportations
		Total Totale	Solids Solides	Liquids Liquides	Gas Gaz	Electricity Electricité			
United Rep. Tanzania	1980	66	1	..	..	65	0	959	84
	1990	79	4	..	..	75	4	1 029	47
	1991	80	4	..	..	76	3	1 069	35
	1992	81	4	..	..	77	3	1 035	35
Western Sahara	1980	..	..	..	..	..	..	81	..
	1990	..	..	..	..	..	..	98	..
	1991	..	..	..	..	..	..	98	..
	1992	..	..	..	..	..	..	100	..
Zaire	1980	2 120	138	1 461	..	521	2	1 309	1 267
	1990	2 829	78	2 014	..	737	18	1 528	1 700
	1991	2 778	80	1 960	..	738	7	1 576	1 657
	1992	2 788	85	1 964	..	739	0	1 564	1 628
Zambia	1980	1 606	488		..	1 118	−65	1 108	454
	1990	1 267	318		..	950	0	793	245
	1991	1 270	320		..	950	0	793	247
	1992	1 284	333		..	951	0	800	247
Zimbabwe	1980	3 626	3 134	..	..	492	0	1 329	334
	1990	5 391	4 958	..	..	433	0	1 174	158
	1991	5 979	5 596	..	..	383	221	1 627	252
	1992	5 912	5 547	..	..	365	−37	1 615	109
America, North	**1980**	**2 608 282**	**659 448**	**972 674**	**792 627**	**183 532**	**23 245**	**750 517**	**347 489**
	1990	**2 979 788**	**830 208**	**963 881**	**840 951**	**344 747**	**58 513**	**778 512**	**435 174**
	1991	**2 998 634**	**807 006**	**982 558**	**844 630**	**364 440**	**−2 523**	**762 465**	**473 213**
	1992	**3 001 698**	**800 716**	**973 143**	**863 795**	**364 045**	**−13 855**	**800 890**	**482 250**
Antigua and Barbuda	1980	..	..	..	..	..	0	172	24
	1990	..	..	..	..	..	0	204	10
	1991	..	..	..	..	..	0	200	10
	1992	..	..	..	..	..	0	200	10
Aruba	1980	..	..	..	..	..	...	...	...
	1990	..	..	..	..	..	2	800	0
	1991	..	..	..	..	..	2	840	0
	1992	..	..	..	..	..	2	735	0
Bahamas	1980	..	..	..	..	..	362	16 514	11 430
	1990	..	..	..	..	..	269	5 672	4 159
	1991	..	..	..	..	..	129	4 862	3 569
	1992	..	..	..	..	..	−101	4 361	3 286
Barbados	1980	78	..	60	18	..	−29	511	0
	1990	127	..	89	39	..	−43	745	277
	1991	119	..	89	30	..	14	855	268
	1992	122	..	93	29	..	13	512	64
Belize	1980	..	..	..	..	..	0	106	0
	1990	..	..	..	..	..	42	180	0
	1991	..	..	..	..	..	39	182	0
	1992	..	..	..	..	..	0	141	0
Bermuda	1980	..	..	..	..	..	3	270	0
	1990	..	..	..	..	..	0	305	0
	1991	..	..	..	..	..	0	259	0
	1992	..	..	..	..	..	0	213	0
British Virgin Islds	1980	..	..	..	..	..	..	12	..
	1990	..	..	..	..	..	..	23	..
	1991	..	..	..	..	..	..	23	..
	1992	..	..	..	..	..	..	25	..
Canada	1980	288 606	30 403	117 602	95 213	45 389	1 448	56 051	76 956
	1990	386 638	52 777	129 938	140 370	63 552	6 825	62 031	150 701
	1991	403 277	55 294	130 302	148 233	69 447	−2 865	56 108	172 577
	1992	415 860	49 183	135 217	162 645	68 815	−8 757	54 599	184 453
Cayman Islands	1980	..	..	..	..	..	...	88	..
	1990	..	..	..	..	..	...	128	..
	1991	..	..	..	..	..	...	138	..
	1992	..	..	..	..	..	...	141	..
Costa Rica	1980	259	..	..	..	259	42	1 149	58
	1990	430	..	..	..	430	44	1 465	155
	1991	446	..	..	..	446	−43	1 511	120
	1992	437	..	..	..	437	14	1 789	141

Bunkers – Soutes			Consumption – Consommation							
Air Avion	Sea Maritime	Unallocated Nondistribué	Per capita Par habitant	Total Totale	Solids Solides	Liquids Liquides	Gas Gaz	Electricity Electricité	Année	Pays ou zone
12	57	9	47	865	1	799	..	65	1980	Rép.Unie de Tanzanie
7	35	–4	40	1 018	4	938	..	75	1990	
40	31	10	39	1 031	4	951	..	76	1991	
40	31	3	37	1 004	4	923	..	77	1992	
4	0	..	592	77	..	77	..	..	1980	Sahara occidental
6	0	..	404	93	..	93	..	..	1990	
6	0	..	388	93	..	93	..	..	1991	
6	0	..	376	94	..	94	..	..	1992	
167	35	0	72	1 958	316	1 133	..	509	1980	Zaïre
151	48	38	64	2 402	287	1 401	..	714	1990	
155	50	38	63	2 447	299	1 434	..	714	1991	
152	45	36	62	2 491	303	1 472	..	715	1992	
52	..	–0	396	2 274	510	982		782	1980	Zambie
52	..	61	209	1 702	314	620	..	768	1990	
52	..	42	205	1 722	316	638	..	768	1991	
52	..	45	201	1 740	329	642	..	769	1992	
81	..	...	637	4 540	2 907	805	...	828	1980	Zimbabwe
96		...	637	6 311	4 946	826	...	540	1990	
78	..	...	692	7 055	5 210	1 246	...	598	1991	
0		...	712	7 454	5 531	1 411	...	512	1992	
11 288	55 145	96 541	7 616	2 825 091	571 841	1 273 489	796 230	183 530	1980	**Amérique du Nord**
2 113	30 995	35 478	7 529	3 196 027	708 008	1 320 649	822 600	344 771	1990	
1 944	32 887	32 795	7 486	3 222 783	703 336	1 304 463	850 240	364 744	1991	
2 015	32 788	32 052	7 483	3 267 338	707 863	1 320 551	874 582	364 342	1992	
81	0	0	1 098	67	..	67	..	..	1980	Antigua–et–Barbuda
50	0	0	2 250	144	..	144	..	..	1990	
52	0	0	2 156	138	..	138	..	..	1991	
52	0	0	2 123	138	..	138	..	..	1992	
..	..	...	...	...	..	...	..	..	1980	Aruba
..	..	536	3 925	263	..	263	..	..	1990	
..	..	536	4 522	303	..	303	..	..	1991	
..	..	428	4 485	305	..	305	..	..	1992	
72	1 048	2 445	5 510	1 157	0	1 157	..	..	1980	Bahamas
46	277	0	3 598	921	1	920	..	..	1990	
46	277	0	3 235	841	1	840	..	..	1991	
46	270	0	3 258	860	1	859	..	..	1992	
140	147	16	1 261	314	0	297	18	..	1980	Barbade
159	...	23	1 770	455	0	417	39	..	1990	
159	...	16	2 000	516	0	486	30	..	1991	
131	...	–44	1 811	469	0	440	29	..	1992	
15	0	..	623	91	..	91	..	0	1980	Belize
16	0	..	646	122	..	122	..	0	1990	
16	0	..	655	127	..	127	..	0	1991	
15	0	..	638	127	..	127	..	0	1992	
55	10	..	3 759	203	0	203	..	..	1980	Bermudes
25	0	..	4 590	280	0	280	..	..	1990	
25	0	..	3 774	234	0	234	..	..	1991	
22	0	..	3 081	191	0	191	..	..	1992	
..	..	..	1 000	12	..	12	..	..	1980	Iles Vierges brit.
..	..	..	1 438	23	..	23	..	..	1990	
..	..	..	1 353	23	..	23	..	..	1991	
..	..	..	1 471	25	..	25	..	..	1992	
1 016	2 173	4 009	10 533	259 055	29 019	120 961	67 034	42 042	1980	Canada
1 334	913	–2 988	10 503	291 883	33 583	108 141	86 663	63 497	1990	
1 151	948	–5 727	10 421	293 300	35 278	103 168	87 670	67 184	1991	
1 257	839	–7 406	10 534	300 073	36 090	106 037	92 208	65 738	1992	
10	..	..	4 588	78	..	78	..	..	1980	Iles Caïmanes
9	..	..	4 577	119	..	119	..	..	1990	
9	..	..	4 778	129	..	129	..	..	1991	
9	..	..	4 714	132	..	132	..	..	1992	
..	0	19	564	1 289	..	1 030	..	259	1980	Costa Rica
..	...	–19	565	1 714	..	1 264	..	450	1990	
..	...	72	581	1 808	..	1 360	..	448	1991	
..	...	45	635	2 026	..	1 597	..	429	1992	

66

Production, trade and consumption of commercial energy
Thousand metric tons of coal equivalent and kilograms per capita [*cont.*]
Production, commerce et consommation d'énergie commerciale
Milliers de tonnes métriques d'équivalent houille et kilogrammes par habitant [*suite*]

Country or area	Year	Primary energy production – Production d'energie primaire					Changes in stocks Variations des stocks	Imports Importations	Exports Exportations
		Total Totale	Solids Solides	Liquids Liquides	Gas Gaz	Electricity Electricité			
Cuba	1980	426	..	391	22	12	−144	14 985	801
	1990	1 098	..	1 037	48	13	0	16 112	1 513
	1991	1 131	..	1 068	49	13	0	12 696	171
	1992	1 399	..	1 337	49	13	0	12 232	170
Dominica	1980	1	..	..	..	1	...	18	..
	1990	2	..	..	..	2	...	28	..
	1991	2	..	..	..	2	...	28	..
	1992	2	..	..	..	2	...	28	..
Dominican Republic	1980	71	..	..	..	71	151	2 879	..
	1990	55	..	..	..	55	16	4 205	..
	1991	78	..	..	..	78	−41	4 414	..
	1992	270	..	..	..	270	−33	4 888	..
El Salvador	1980	580	..	..	..	580	14	995	0
	1990	719	..	..	..	719	−26	1 146	75
	1991	680	..	..	..	680	11	1 404	19
	1992	657	..	..	..	657	−1	1 559	23
Greenland	1980	0	0	..	..	..	..	263	...
	1990	0	0	..	..	..	..	273	12
	1991	0	0	..	..	..	..	267	9
	1992	0	0	..	..	..	..	239	10
Grenada	1980	..	..	..	..	..	...	24	..
	1990	..	..	..	..	..	...	59	..
	1991	..	..	..	..	..	...	59	..
	1992	..	..	..	..	..	...	61	..
Guadeloupe	1980	..	..	..	..	..	−15	281	0
	1990	..	..	..	..	..	0	677	0
	1991	..	..	..	..	..	0	710	0
	1992	..	..	..	..	..	0	732	0
Guatemala	1980	324	..	290	...	34	13	2 067	168
	1990	582	..	307	12	264	−10	1 926	234
	1991	564	..	289	11	264	7	1 990	227
	1992	709	..	437	12	260	0	2 421	354
Haiti	1980	27	..	..	..	27	4	316	..
	1990	23	..	..	..	23	−22	453	..
	1991	23	..	..	..	23	−25	426	..
	1992	23	..	..	..	23	14	407	..
Honduras	1980	96	..	..	..	96	6	915	30
	1990	256	..	..	..	256	−6	1 097	73
	1991	260	..	..	..	260	33	1 164	46
	1992	260	..	..	..	260	0	1 299	0
Jamaica	1980	15	..	..	..	15	−47	3 975	63
	1990	14	..	..	..	14	−108	3 617	119
	1991	15	..	..	..	15	−74	3 659	96
	1992	13	..	..	..	13	−46	3 765	77
Martinique	1980	..	..	..	..	..	−109	525	290
	1990	..	..	..	..	..	61	1 182	248
	1991	..	..	..	..	..	9	1 089	254
	1992	..	..	..	..	..	40	1 189	229
Mexico	1980	194 129	2 594	152 996	35 339	3 199	915	1 718	67 760
	1990	261 939	5 564	213 518	32 590	10 267	459	8 437	101 997
	1991	271 680	5 065	223 328	32 350	10 937	515	11 955	108 277
	1992	271 663	4 801	223 146	31 910	11 805	−1 108	15 053	110 097
Montserrat	1980	..	..	..	..	..	...	9	..
	1990	..	..	..	..	..	...	18	..
	1991	..	..	..	..	..	...	18	..
	1992	..	..	..	..	..	...	19	..
Netherland Antilles	1980	..	..	..	..	..	...	...	...
	1990	..	..	..	..	..	−404	17 782	12 376
	1991	..	..	..	..	..	−364	19 272	12 745
	1992	..	..	..	..	..	705	20 408	12 289
Nicaragua	1980	63	..	..	..	63	−9	931	2
	1990	524	..	..	..	524	−12	989	41
	1991	604	..	..	..	604	28	964	26
	1992	606	..	..	..	606	−4	1 138	1

Bunkers – Soutes			Consumption – Consommation							
Air Avion	Sea Maritime	Unallocated Nondistribué	Per capita Par habitant	Total Totale	Solids Solides	Liquids Liquides	Gas Gaz	Electricity Electricité	Année	Pays ou zone
...	...	1 089	1 407	13 665	143	13 488	22	12	1980	Cuba
90	0	821	1 395	14 785	198	14 526	48	13	1990	
90	0	608	1 212	12 958	148	12 748	49	13	1991	
90	0	739	1 171	12 633	182	12 388	49	13	1992	
0	..	..	257	19	..	18	..	1	1980	Dominique
0	..	..	423	30	..	28	..	2	1990	
0	..	..	423	30	..	28	..	2	1991	
0	..	..	423	30	..	28	..	2	1992	
1	..	9	489	2 788	0	2 717	..	71	1980	Rép. dominicaine
...	..	−66	606	4 310	9	4 246	..	55	1990	
...	..	194	598	4 339	57	4 204	..	78	1991	
...	..	−67	710	5 257	139	4 848	..	270	1992	
3	82	117	300	1 359	..	779	..	580	1980	El Salvador
...	...	48	342	1 768	..	1 048	..	720	1990	
...	...	37	382	2 017	..	1 337	..	681	1991	
...	...	36	400	2 159	..	1 495	..	664	1992	
..	..	..	5 260	263	2	261	..	..	1980	Groënland
..	..	..	4 661	261	0	261	..	..	1990	
..	..	..	4 625	259	0	259	..	..	1991	
..	..	..	4 000	228	0	228	..	..	1992	
0	...	..	270	24	..	24	..	..	1980	Grenade
1	...	..	637	58	..	58	..	..	1990	
1	...	..	637	58	..	58	..	..	1991	
3	...	..	637	58	..	58	..	..	1992	
41	...	..	777	254	..	254	..	..	1980	Guadeloupe
121	...	..	1 422	556	..	556	..	..	1990	
127	...	..	1 467	584	..	584	..	..	1991	
117	...	..	1 515	615	..	615	..	..	1992	
15	184	39	285	1 973	..	1 938	...	34	1980	Guatemala
...	...	13	247	2 272	..	1 997	12	264	1990	
...	...	78	237	2 240	..	1 965	11	264	1991	
...	...	140	270	2 635	..	2 363	12	260	1992	
13	...	..	61	325	...	298	..	27	1980	Haïti
59	...	..	68	440	12	405	..	23	1990	
52	...	..	64	422	25	374	..	23	1991	
35	...	..	56	381	26	331	..	23	1992	
...	23	48	253	904	..	809	..	96	1980	Honduras
...	...	29	258	1 257	..	1 038	..	219	1990	
...	...	21	264	1 325	..	1 089	..	236	1991	
...	...	87	284	1 471	..	1 212	..	260	1992	
29	21	36	1 822	3 887	1	3 872	..	15	1980	Jamaïque
37	...	39	1 497	3 543	52	3 478	..	14	1990	
44	...	157	1 451	3 450	51	3 385	..	15	1991	
66	...	−38	1 553	3 718	66	3 639	..	13	1992	
..	28	−13	1 006	328	..	328	..	..	1980	Martinique
..	14	66	2 203	793	..	793	..	..	1990	
..	14	12	2 198	800	..	800	..	..	1991	
..	40	63	2 217	816	..	816	..	..	1992	
75	50	10 819	1 733	116 228	3 971	77 376	31 617	3 264	1980	Mexique
106	61	9 907	1 868	157 845	5 719	108 816	33 212	10 098	1990	
106	61	10 607	1 900	164 070	5 415	113 357	34 533	10 765	1991	
106	57	11 575	1 882	165 987	5 494	113 379	35 438	11 676	1992	
..	1	..	583	7	..	7	..	..	1980	Montserrat
..	1	..	1 455	16	..	16	..	..	1990	
..	1	..	1 455	16	..	16	..	..	1991	
..	1	..	1 636	18	..	18	..	..	1992	
...	...	...	...	...	..	...	..	..	1980	Antilles néerland.
59	2 379	2 066	6 874	1 306	..	1 306	..	..	1990	
66	2 451	3 095	6 661	1 279	..	1 279	..	..	1991	
66	2 451	3 578	6 829	1 318	..	1 318	..	..	1992	
0	..	22	349	978	..	916	..	62	1980	Nicaragua
...	..	39	393	1 445	..	913	..	532	1990	
...	..	26	391	1 488	..	877	..	611	1991	
...	..	95	418	1 653	..	1 042	..	611	1992	

66

Production, trade and consumption of commercial energy
Thousand metric tons of coal equivalent and kilograms per capita [*cont.*]
Production, commerce et consommation d'énergie commerciale
Milliers de tonnes métriques d'équivalent houille et kilogrammes par habitant [*suite*]

Country or area	Year	Primary energy production – Production d'energie primaire					Changes in stocks Variations des stocks	Imports Importations	Exports Exportations
		Total Totale	Solids Solides	Liquids Liquides	Gas Gaz	Electricity Electricité			
Panama	1980	118	..	..	..	118	−3	6 129	505
	1990	272	..	..	..	272	84	2 072	548
	1991	250	..	..	..	250	73	2 269	626
	1992	232	..	..	..	232	166	3 172	1 147
Puerto Rico	1980	17	..	..	..	17	−750	13 445	5 194
	1990	34	..	..	..	34	2	10 621	1 460
	1991	36 ·	..	..	..	36	520	10 802	1 657
	1992	37	..	..	..	37	683	10 933	1 284
Saint Lucia	1980	..	..	..	..	..	0	53	..
	1990	..	..	..	..	..	0	78	..
	1991	..	..	..	..	..	0	78	..
	1992	..	..	..	..	..	0	81	..
St.Kitts−Nevis	1980	..	..	..	..	..	...	...	
	1990	..	..	..	..	..	0	31	
	1991	..	..	..	..	..	0	35	
	1992	..	..	..	..	..	0	35	
St.Pierre−Miquelon	1980	..	..	..	..	..	..	49	
	1990	..	..	..	..	..	..	120	
	1991	..	..	..	..	..	..	113	
	1992	..	..	..	..	..	..	106	
S.Vincent−Grenadines	1980	2	..	..	..	2	..	18	..
	1990	5	..	..	..	5	..	38	..
	1991	5	..	..	..	5	..	37	..
	1992	5	..	..	..	5	..	40	..
Trinidad and Tobago	1980	20 697	..	15 701	4 996	..	−53	10 496	22 344
	1990	18 127	..	11 246	6 881	..	171	1 508	10 445
	1991	18 624	..	10 645	7 979	..	402	2 959	10 231
	1992	17 933	..	10 027	7 906	..	165	4 056	10 984
United States	1980	2 102 772	626 451	685 633	657 039	133 649	15 670	528 746	99 320
	1990	2 308 943	771 867	607 746	661 012	268 317	50 044	608 711	129 665
	1991	2 300 841	746 646	616 837	655 977	281 380	−1 746	597 178	143 416
	1992	2 291 470	746 731	602 886	661 243	280 609	−5 666	631 937	138 984
U.S. Virgin Islands	1980	..	..	..	..	..	2 916	39 448	28 458
	1990	..	..	..	..	..	1 124	25 778	21 065
	1991	..	..	..	..	..	852	23 899	18 871
	1992	..	..	..	..	..	58	22 377	18 644
America, South	**1980**	**338 030**	**9 210**	**260 467**	**43 759**	**24 593**	**−1 650**	**92 837**	**161 474**
	1990	**466 532**	**27 000**	**315 789**	**76 792**	**46 950**	**4 450**	**82 496**	**204 672**
	1991	**494 317**	**30 690**	**333 255**	**81 040**	**49 331**	**2 370**	**84 482**	**226 071**
	1992	**499 078**	**28 024**	**340 230**	**80 692**	**50 131**	**4 254**	**86 817**	**224 281**
Antarctic Fisheries	1980	..	..	..	..	..	..	87	..
	1990	..	..	..	..	..	..	97	..
	1991	..	..	..	..	..	..	101	..
	1992	..	..	..	..	..	..	101	..
Argentina	1980	51 040	329	36 827	11 155	2 730	245	7 235	2 059
	1990	69 351	228	37 127	27 064	4 932	−236	4 033	6 501
	1991	73 187	246	37 816	30 220	4 905	104	5 058	7 367
	1992	78 415	181	42 520	30 680	5 035	751	5 268	9 866
Bolivia	1980	4 790	..	1 777	2 880	133	−38	1	2 543
	1990	5 349	..	1 760	3 435	154	35	1	2 817
	1991	5 440	..	1 861	3 408	170	58	1	2 845
	1992	5 719	..	1 872	3 681	166	−33	15	2 852
Brazil	1980	34 056	3 655	13 260	1 306	15 834	−1 633	69 220	2 773
	1990	80 423	2 917	46 145	5 139	26 222	2 565	58 326	5 506
	1991	81 235	3 293	45 629	5 026	27 287	−448	57 817	4 164
	1992	82 816	3 003	46 312	5 413	28 088	−1 759	58 238	4 838
Chile	1980	5 901	1 156	2 919	924	902	−243	6 331	130
	1990	8 271	2 578	1 858	2 364	1 471	502	10 253	301
	1991	7 988	2 607	1 677	2 092	1 613	1 077	10 426	129
	1992	7 884	1 938	1 509	2 381	2 057	95	10 802	87
Colombia	1980	20 158	3 982	9 594	4 792	1 790	−9	3 004	2 155
	1990	59 496	19 006	31 879	5 231	3 380	246	1 806	32 100
	1991	61 546	21 914	30 844	5 414	3 373	680	1 677	32 933
	1992	60 145	20 336	31 973	5 085	2 751	2 285	2 632	31 313

Bunkers – Soutes			Consumption – Consommation							
Air Avion	Sea Maritime	Unallocated Nondistribué	Per capita Par habitant	Total Totale	Solids Solides	Liquids Liquides	Gas Gaz	Electricity Electricité	Année	Pays ou zone
1	4 038	73	837	1 632	2	1 512	0	118	1980	Panama
0	..	28	745	1 683	35	1 285	77	286	1990	
0	..	0	745	1 820	49	1 424	80	267	1991	
0	..	55	817	2 036	54	1 653	79	250	1992	
75	279	−1 292	3 106	9 957	0	9 940	..	17	1980	Porto Rico
...	214	563	2 384	8 417	213	8 169	..	34	1990	
...	214	140	2 333	8 307	192	8 080	..	36	1991	
...	214	627	2 274	8 162	154	7 971	..	37	1992	
0	..	..	461	53	..	53	..	..	1980	Sainte−Lucie
0	..	..	586	78	..	78	..	..	1990	
0	..	..	578	78	..	78	..	..	1991	
0	..	..	591	81	..	81	..	..	1992	
..	..	..	...	...	..	...	..	..	1980	St−Kitts−Nevis
..	..	..	738	31	..	31	..	..	1990	
..	..	..	833	35	..	35	..	..	1991	
..	..	..	833	35	..	35	..	..	1992	
..	32	..	2 833	17	0	17	..	..	1980	St.Pierre−Miquelon
..	77	..	7 333	44	0	44	..	..	1990	
..	65	..	8 000	48	0	48	..	..	1991	
..	61	..	7 500	45	0	45	..	..	1992	
0	..	..	204	20	..	18	..	2	1980	S.Vincent−Grenadines
0	..	..	402	43	..	38	..	5	1990	
0	..	..	389	42	..	37	..	5	1991	
0	..	..	413	45	..	40	..	5	1992	
41	717	676	6 901	7 467	0	2 471	4 996	..	1980	Trinité−et−Tobago
0	40	585	6 792	8 395	0	1 514	6 881	..	1990	
0	..	107	8 667	10 843	0	2 864	7 979	..	1991	
0	..	159	8 443	10 681	0	2 775	7 906	..	1992	
9 454	41 565	74 236	10 499	2 391 273	538 704	1 023 094	692 544	136 931	1980	Etats−Unis
0	26 825	24 251	10 751	2 686 868	668 112	1 054 527	695 669	268 560	1990	
0	28 661	23 016	10 711	2 704 672	661 997	1 038 671	719 887	284 116	1991	
0	28 661	21 790	10 736	2 739 637	665 451	1 051 234	738 861	284 092	1992	
74	1 405	1 381	53 753	5 214	...	5 214	..	..	1980	Iles Vierges amér.
...	194	−463	37 824	3 858	73	3 785	..	..	1990	
...	194	−201	40 612	4 183	123	4 060	..	..	1991	
...	194	188	31 981	3 294	205	3 089	..	..	1992	
1 524	**3 159**	**18 489**	**1 033**	**247 870**	**15 503**	**163 966**	**43 826**	**24 575**	**1980**	**Amérique du Sud**
1 338	**3 599**	**21 927**	**1 068**	**313 041**	**24 799**	**164 593**	**76 873**	**46 776**	**1990**	
1 383	**3 961**	**19 604**	**1 090**	**325 410**	**26 792**	**168 594**	**80 794**	**49 230**	**1991**	
1 341	**3 762**	**23 490**	**1 082**	**328 765**	**23 919**	**174 242**	**80 644**	**49 961**	**1992**	
..	87	..	0	0	..	0	..	..	1980	Pêcheries antarctiq.
..	93	..	4 000	4	..	4	..	..	1990	
..	94	..	7 000	7	..	7	..	..	1991	
..	94	..	7 000	7	..	7	..	..	1992	
6	633	5 654	1 767	49 677	1 080	32 293	13 572	2 733	1980	Argentine
0	1 051	4 817	1 882	61 252	1 157	25 086	29 974	5 035	1990	
0	687	5 069	1 972	65 018	1 013	26 194	32 798	5 013	1991	
0	696	5 498	2 004	66 873	1 108	26 908	33 503	5 353	1992	
9	..	−17	428	2 293	1	1 698	462	133	1980	Bolivie
0	..	13	378	2 486	0	1 688	643	155	1990	
0	..	−18	380	2 556	0	1 748	637	171	1991	
0	..	165	399	2 750	0	1 727	856	167	1992	
847	867	8 186	760	92 236	7 444	67 678	1 306	15 808	1980	Brésil
666	789	12 244	788	116 979	13 334	69 230	5 146	29 268	1990	
700	1 113	11 857	805	121 665	15 375	70 643	5 034	30 613	1991	
647	1 201	11 176	812	124 951	14 943	73 561	5 409	31 038	1992	
10	82	481	1 056	11 772	2 591	7 286	992	902	1980	Chili
0	0	223	1 330	17 498	4 412	9 295	2 320	1 471	1990	
0	106	464	1 244	16 638	3 542	9 453	2 032	1 613	1991	
0	42	427	1 326	18 034	3 145	10 495	2 338	2 057	1992	
22	300	446	763	20 248	3 882	9 780	4 792	1 793	1980	Colombie
...	185	2 251	821	26 520	5 233	12 642	5 231	3 414	1990	
...	187	1 549	848	27 873	6 020	13 066	5 414	3 373	1991	
...	181	1 878	811	27 120	4 087	15 156	5 085	2 793	1992	

66

Production, trade and consumption of commercial energy
Thousand metric tons of coal equivalent and kilograms per capita [*cont.*]
Production, commerce et consommation d'énergie commerciale
Milliers de tonnes métriques d'équivalent houille et kilogrammes par habitant [*suite*]

Country or area	Year	Primary energy production – Production d'energie primaire					Changes in stocks Variations des stocks	Imports Importations	Exports Exportations
		Total Totale	Solids Solides	Liquids Liquides	Gas Gaz	Electricity Electricité			
Ecuador	1980	15 055	..	14 894	51	109	113	836	9 704
	1990	22 284	..	21 508	165	611	0	381	14 742
	1991	23 362	..	22 575	165	622	0	487	15 342
	1992	24 918	..	24 144	165	608	0	555	16 575
Falkland Is. (Malvinas)	1980	4	4	..	..	..	..	3	..
	1990	5	5	..	..	..	..	12	..
	1991	5	5	..	..	..	..	12	..
	1992	5	5	..	..	..	..	12	..
French Guiana	1980	..	..	..	..	..	...	174	
	1990	..	..	..	..	..	...	408	
	1991	..	..	..	..	..	...	413	
	1992	..	..	..	..	..	...	433	
Guyana	1980	1	..	..	..	1	...	845	
	1990	1	..	..	..	1	...	553	
	1991	1	..	..	..	1	...	543	
	1992	1	..	..	..	1	...	510	
Paraguay	1980	83	..	..	..	83	−29	649	19
	1990	3 341	..	..	..	3 341	19	1 016	3 048
	1991	3 608	..	..	..	3 608	−44	940	3 292
	1992	3 331	..	..	..	3 331	−49	1 172	3 076
Peru	1980	15 558	41	13 944	713	861	−94	258	4 493
	1990	12 033	79	9 899	768	1 286	287	2 385	3 303
	1991	11 029	68	8 858	691	1 412	184	3 458	3 480
	1992	10 825	83	8 919	620	1 203	−31	3 193	3 459
Suriname	1980	110	..	0	..	110	11	1 107	...
	1990	460	..	311	..	149	0	628	100
	1991	517	..	380	..	137	0	665	71
	1992	522	..	384	..	138	0	671	76
Uruguay	1980	279	..	..	..	279	106	2 942	3
	1990	861	..	..	..	861	−54	2 000	318
	1991	751	..	..	..	751	91	2 487	220
	1992	973	..	..	..	973	73	2 646	417
Venezuela	1980	190 996	44	167 252	21 938	1 761	−79	233	137 595
	1990	204 658	2 188	165 301	32 626	4 543	1 086	694	135 936
	1991	225 648	2 556	183 615	34 024	5 454	666	498	156 227
	1992	223 524	2 479	182 596	32 668	5 782	2 922	671	151 722
Asia	1980	**2 461 006**	**619 155**	**1 653 384**	**115 251**	**73 217**	**19 986**	**761 730**	**1 447 021**
	1990	**3 250 565**	**1 083 702**	**1 681 705**	**312 250**	**172 908**	**66 013**	**1 041 979**	**1 357 143**
	1991	**3 274 194**	**1 117 926**	**1 662 512**	**313 487**	**180 269**	**35 522**	**1 090 254**	**1 343 189**
	1992	**3 741 999**	**1 262 193**	**1 828 612**	**461 716**	**189 478**	**32 438**	**1 279 504**	**1 574 765**
Afghanistan	1980	3 710	119	12	3 491	88	2	487	3 361
	1990	484	105	6	279	94	15	904	0
	1991	454	94	5	270	85	18	846	0
	1992	318	8	0	252	59	19	488	0
Bahrain	1980	7 674	..	3 793	3 880	..	396	13 638	13 632
	1990	10 748	..	3 497	7 251	..	−167	14 762	15 008
	1991	10 558	..	3 665	6 893	..	112	15 212	15 175
	1992	10 543	..	3 650	6 893	..	92	15 623	15 989
Bangladesh	1980	1 617	..	9	1 537	72	−20	2 654	25
	1990	5 544	..	180	5 255	109	−212	3 402	0
	1991	6 562	..	177	6 282	103	−123	2 804	0
	1992	6 943	..	182	6 664	98	−30	3 352	3
Bhutan	1980	1	...	..	..	1	..	11	...
	1990	193	2	..	..	191	..	57	173
	1991	195	2	..	..	193	..	57	174
	1992	201	2	..	..	199	..	59	180
Brunei Darussalam	1980	29 779	..	17 769	12 010	..	219	98	27 274
	1990	24 113	..	10 181	13 932	..	−851	48	21 228
	1991	24 418	..	10 996	13 422	..	−1 194	105	21 678
	1992	25 399	..	12 057	13 343	..	−1 194	0	22 514
Cambodia	1980	6	..	..	..	6	...	135	
	1990	8	..	..	..	8	...	214	
	1991	8	..	..	..	8	...	220	
	1992	8	..	..	..	8	...	226	

Bunkers – Soutes			Consumption – Consommation							
Air Avion	Sea Maritime	Unallocated Nondistribué	Per capita Par habitant	Total Totale	Solids Solides	Liquids Liquides	Gas Gaz	Electricity Electricité	Année	Pays ou zone
52	129	213	713	5 680	..	5 518	51	111	1980	Equateur
63	233	−55	748	7 682	..	6 905	165	611	1990	
57	256	20	778	8 174	..	7 387	165	622	1991	
68	265	58	792	8 507	..	7 734	165	608	1992	
..	1	..	2 500	5	4	1		..	1980	Il. Falkland (Malvinas)
..	0	..	8 500	17	5	12	..	..	1990	
..	0	..	8 500	17	5	12	..	..	1991	
..	0	..	8 500	17	5	12	..	..	1992	
...	..	..	2 559	174	..	174	..	..	1980	Guyane française
25	..	..	3 274	383	..	383	..	..	1990	
22	..	..	3 179	391	..	391	..	..	1991	
22	..	..	3 186	411	..	411	..	..	1992	
6	7	..	1 096	832	0	832	..	1	1980	Guyana
16	4	..	670	533	0	532	..	1	1990	
15	4	..	654	524	0	523	..	1	1991	
13	4	..	610	493	0	490	..	3	1992	
4	..	10	232	727	..	644	..	83	1980	Paraguay
3	..	−5	299	1 292	..	997	..	295	1990	
3	..	−1	292	1 298	..	982	..	316	1991	
3	..	−5	323	1 477	..	1 170	..	307	1992	
0	115	279	636	11 024	195	9 255	713	861	1980	Pérou
0	...	156	494	10 671	198	8 419	768	1 286	1990	
0	...	283	479	10 540	408	8 029	691	1 412	1991	
0	...	702	441	9 888	321	7 744	620	1 203	1992	
1	9	0	3 366	1 195	26	1 059	..	110	1980	Suriname
0	...	200	1 973	789	0	640	..	149	1990	
0	...	296	2 012	815	0	678	..	137	1991	
0	...	294	2 012	823	0	685	..	138	1992	
50	142	163	946	2 758	4	2 470		283	1980	Uruguay
38	200	173	706	2 185	1	1 635	..	549	1990	
53	251	238	766	2 385	1	1 853	..	531	1991	
57	342	130	830	2 599	2	2 040	..	557	1992	
516	874	3 074	3 263	49 249	276	25 277	21 938	1 757	1980	Venezuela
526	1 137	1 911	3 320	64 756	459	27 129	32 626	4 543	1990	
532	1 355	−152	3 380	67 517	429	27 635	34 024	5 429	1991	
531	1 031	3 168	3 170	64 822	308	26 109	32 668	5 737	1992	
10 649	44 305	118 588	607	1 582 189	699 951	695 707	113 133	73 397	1980	Asie
12 015	43 488	154 750	844	2 659 136	1 214 668	950 481	321 167	172 820	1990	
17 080	41 536	163 388	866	2 763 734	1 269 895	996 552	317 040	180 246	1991	
18 791	44 582	200 416	950	3 150 512	1 384 928	1 133 729	440 832	191 024	1992	
22	..	0	51	813	119	475	130	88	1980	Afghanistan
9	..	0	91	1 364	105	887	279	94	1990	
9	..	0	81	1 273	94	824	270	85	1991	
9	..	0	47	779	8	444	252	75	1992	
0	267	2 025	14 383	4 991	0	1 111	3 880	..	1980	Bahreïn
0	...	2 824	16 012	7 846	0	594	7 251	..	1990	
0	...	2 912	14 990	7 570	0	677	6 893	..	1991	
0	...	2 398	14 783	7 687	0	794	6 893	..	1992	
0	86	230	45	3 950	168	2 173	1 537	72	1980	Bangladesh
9	7	654	78	8 487	402	2 721	5 255	109	1990	
0	7	504	81	8 978	129	2 464	6 282	103	1991	
0	0	348	89	9 975	241	2 971	6 664	98	1992	
..	..	..	9	11	1	9	..	1	1980	Bhoutan
..	..	..	50	77	18	38	..	20	1990	
..	..	..	50	78	18	38	..	22	1991	
..	..	..	51	80	18	40	..	22	1992	
..	..	−552	15 212	2 936	2	922	2 012	..	1980	Brunéi Darussalam
..	..	−565	16 922	4 349	0	1 165	3 183	..	1990	
..	..	−496	17 243	4 535	0	1 244	3 291	..	1991	
..	..	−533	17 149	4 613	0	1 207	3 406	..	1992	
..	..	0	22	141	0	135	..	6	1980	Cambodge
..	..	0	25	222	0	214	..	8	1990	
..	..	0	25	228	0	220	..	8	1991	
..	..	0	25	234	0	226	..	8	1992	

66

Production, trade and consumption of commercial energy
Thousand metric tons of coal equivalent and kilograms per capita [cont.]
Production, commerce et consommation d'énergie commerciale
Milliers de tonnes métriques d'équivalent houille et kilogrammes par habitant [suite]

Country or area	Year	Primary energy production – Production d'energie primaire					Changes in stocks Variations des stocks	Imports Importations	Exports Exportations
		Total Totale	Solids Solides	Liquids Liquides	Gas Gaz	Electricity Electricité			
China	1980	612 383	434 959	151 326	18 948	7 149	...	2 782	29 415
	1990	1 004 010	770 574	197 547	20 324	15 566	31 983	12 812	56 046
	1991	1 014 037	775 942	201 383	21 347	15 365	8 626	19 490	55 453
	1992	1 037 003	796 609	202 962	20 975	16 458	422	32 390	58 456
Cyprus	1980	..	..	..	..	..	31	1 387	0
	1990	..	..	..	..	..	−53	2 309	0
	1991	..	..	..	..	..	95	2 316	0
	1992	..	..	..	..	..	63	2 600	0
Hong Kong	1980			..	..	..	−300	9 307	295
	1990			..	..	..	514	17 711	3 178
	1991			..	..	..	117	18 683	4 041
	1992	..		..	..	..	−17	22 857	5 273
India	1980	114 066	92 081	13 425	1 726	6 834	−2 420	32 829	95
	1990	245 363	172 186	48 827	13 263	11 087	1 182	44 565	615
	1991	264 694	193 004	45 738	14 956	10 996	7 091	50 593	564
	1992	267 355	199 137	41 578	15 551	11 090	3 827	64 995	619
Indonesia	1980	132 256	304	110 864	20 718	370	5 023	10 836	99 114
	1990	222 699	7 330	132 705	80 026	2 637	7 540	12 987	104 852
	1991	228 095	13 715	147 004	64 781	2 595	9 591	13 668	123 447
	1992	234 734	21 147	142 890	67 871	2 826	4 264	18 020	132 563
Iran, Islamic Rep. of	1980	116 046	900	104 949	9 506	690	−2 241	237	62 367
	1990	264 297	1 300	230 077	32 173	747	11 091	10 583	162 770
	1991	285 060	1 400	248 507	34 286	867	8 606	9 923	179 278
	1992	286 072	1 500	250 092	33 285	1 195	−1 428	10 951	185 266
Iraq	1980	188 237	..	186 445	1 707	85	27	7	175 595
	1990	149 748	..	145 919	3 755	75	5 713	9	124 558
	1991	22 163	..	19 677	2 449	38	−3 159	1	3 986
	1992	40 020	..	36 911	3 022	86	3 407	0	7 503
Israel	1980	220	..	29	191	...	1 762	12 800	788
	1990	57	..	19	38	0	−771	16 399	1 464
	1991	48	..	16	31	1	−551	17 281	1 832
	1992	45	..	13	28	4	1 274	21 830	1 909
Japan	1980	63 073	16 059	777	2 905	43 333	5 038	436 700	2 181
	1990	99 970	7 268	790	2 861	89 051	2 092	507 015	11 421
	1991	105 591	7 084	1 073	2 988	94 446	1 641	520 639	7 351
	1992	107 041	6 684	1 192	3 022	96 143	−491	529 865	9 550
Jordan	1980	...	..	...	..	...	98	2 505	0
	1990	25	..	23	..	2	145	4 950	0
	1991	11	..	10	..	1	−213	4 450	67
	1992	6	..	4	..	2	103	5 452	0
Korea,Dem.Ppl's.Rep.	1980	44 764	42 000	..	..	2 764	...	4 077	100
	1990	85 100	81 200	..	..	3 900	...	9 224	514
	1991	87 700	83 800	..	..	3 900	...	8 907	489
	1992	88 348	85 400	..	..	2 948	...	8 621	464
Korea, Republic of	1980	13 774	12 239	..	..	1 535	2 570	45 129	0
	1990	31 495	11 066	..	..	20 429	−352	102 717	4 858
	1991	31 218	9 678	..	..	21 540	−1 195	126 179	13 205
	1992	29 292	7 693	..	..	21 598	1 287	149 619	17 828
Kuwait [1]	1980	134 712	..	125 038	9 674	..	1 579	1	113 968
	1990	96 658	..	89 699	6 960	..	−1 213	2 731	79 468
	1991	16 786	..	14 227	2 559	..	58	1	10 121
	1992	81 980	..	78 492	3 488	..	14	1	67 523
Lao People's Dem. Rep.	1980	114	0	..	..	114	..	90	94
	1990	102	1	..	..	101	..	113	75
	1991	110	1	..	..	109	..	123	81
	1992	107	1	..	..	106	..	133	78
Lebanon	1980	104	..	..	..	104	81	3 124	145
	1990	37	..	..	..	37	0	4 159	0
	1991	43	..	..	..	43	0	4 354	0
	1992	43	..	..	..	43	0	5 103	0

| Bunkers – Soutes | | | Consumption – Consommation | | | | | | | |
Air Avion	Sea Maritime	Unallocated Nondistribué	Per capita Par habitant	Total Totale	Solids Solides	Liquids Liquides	Gas Gaz	Electricity Electricité	Année	Pays ou zone
..	...	40 251	556	545 499	432 174	87 189	18 948	7 187	1980	Chine
..	...	35 357	787	893 437	727 642	129 679	20 324	15 792	1990	
..	...	36 429	812	933 020	754 087	141 870	21 347	15 716	1991	
..	...	37 669	836	972 846	777 908	156 894	20 975	17 069	1992	
111	24	13	1 922	1 209	0	1 209	..	..	1980	Chypre
348	83	35	2 701	1 896	108	1 788	..	..	1990	
283	80	25	2 582	1 833	15	1 818	..	..	1991	
364	84	24	2 877	2 066	26	2 040	..	..	1992	
1 044	978	..	1 447	7 291	8	7 321	...	−38	1980	Hong–kong
2 397	1 942	..	1 697	9 681	6 826	3 075	...	−221	1990	
2 373	1 696	..	1 821	10 455	7 363	3 468	...	−376	1991	
3 100	2 265	..	2 118	12 236	7 807	5 039	...	−610	1992	
814	335	8 052	203	140 018	95 165	36 298	1 726	6 829	1980	Inde
929	331	17 660	316	269 212	179 377	65 316	13 263	11 256	1990	
892	344	16 809	334	289 589	193 530	69 928	14 956	11 175	1991	
899	356	22 498	344	304 151	203 013	74 349	15 551	11 238	1992	
133	362	2 358	239	36 101	267	30 067	5 398	370	1980	Indonésie
340	286	31 088	501	91 580	2 956	43 113	42 874	2 637	1990	
385	293	33 064	404	74 983	5 750	47 637	19 001	2 595	1991	
485	300	32 898	436	82 244	5 331	49 512	24 574	2 826	1992	
7	5 806	4 670	1 164	45 674	960	34 810	9 213	690	1980	Iran, Rép.islamique d'
9	512	7 471	1 578	93 027	1 754	60 349	30 176	747	1990	
9	597	10 106	1 586	96 386	1 854	62 707	30 959	867	1991	
7	682	9 812	1 643	102 684	2 054	66 150	33 285	1 195	1992	
6	170	1 791	819	10 654	1	8 862	1 707	85	1980	Iraq
6	212	3 233	887	16 035	2	14 933	1 025	75	1990	
0	0	1 569	1 065	19 768	1	17 281	2 449	38	1991	
0	0	−282	1 546	29 392	0	26 284	3 022	86	1992	
348	...	1 293	2 276	8 829	6	8 652	191	−20	1980	Israël
507	247	463	3 121	14 546	3 703	10 852	38	−48	1990	
481	309	860	2 980	14 398	4 050	10 360	31	−43	1991	
599	268	1 375	3 267	16 451	4 946	11 516	28	−39	1992	
2 177	16 561	29 685	3 802	444 131	85 750	281 802	33 246	43 333	1980	Japon
817	7 476	20 989	4 567	564 190	114 825	291 625	68 689	89 051	1990	
6 391	8 276	17 961	4 718	584 610	119 659	296 658	73 847	94 446	1991	
6 655	8 250	23 605	4 743	589 338	118 265	299 887	75 043	96 143	1992	
283	..	108	690	2 016	..	2 016	..	...	1980	Jordanie
336	..	157	1 019	4 338	..	4 335	..	2	1990	
159	..	174	962	4 273	..	4 273	..	1	1991	
290	..	216	1 037	4 848	..	4 846	..	2	1992	
...	...	199	2 658	48 541	42 620	3 157	..	2 764	1980	Corée,Rép.pop.dém.de
...	...	−301	4 322	94 110	83 570	6 640	..	3 900	1990	
...	...	−360	4 348	96 478	86 095	6 483	..	3 900	1991	
...	...	−646	4 296	97 151	87 720	6 483	..	2 948	1992	
142	138	3 762	1 372	52 291	18 301	32 454	..	1 535	1980	Corée, République de
772	2 339	7 476	2 779	119 118	35 489	58 882	4 318	20 429	1990	
733	3 494	10 679	3 015	130 483	35 662	68 279	5 003	21 540	1991	
892	4 533	13 478	3 224	140 893	34 215	78 536	6 544	21 598	1992	
498	2 801	3 856	8 735	12 011	..	2 337	9 674	..	1980	Koweït [1]
442	1 049	4 867	6 895	14 776	..	5 087	9 689	..	1990	
147	43	518	2 847	5 900	..	3 341	2 559	..	1991	
147	369	1 805	6 256	12 124	..	8 636	3 488	..	1992	
..	..	..	34	110	0	89	..	21	1980	Rép. dém. pop. lao
..	..	..	33	140	1	109	..	30	1990	
..	..	..	35	151	1	120	..	31	1991	
..	..	..	36	163	1	130	..	31	1992	
355	14	95	951	2 538	5	2 420	..	113	1980	Liban
44	14	−13	1 625	4 151	0	4 109	..	42	1990	
44	14	−8	1 666	4 347	0	4 299	..	48	1991	
44	14	42	1 870	5 045	0	4 997	..	48	1992	

66

Production, trade and consumption of commercial energy

Thousand metric tons of coal equivalent and kilograms per capita [*cont.*]

Production, commerce et consommation d'énergie commerciale

Milliers de tonnes métriques d'équivalent houille et kilogrammes par habitant [*suite*]

| Country or area | Year | Primary energy production – Production d'energie primaire | | | | | Changes in stocks Variations des stocks | Imports Importations | Exports Exportations |
		Total Totale	Solids Solides	Liquids Liquides	Gas Gaz	Electricity Electricité			
Macau	1980	..	..	..	..	..	...	234	...
	1990	..	..	..	..	..	8	501	0
	1991	..	..	..	..	..	−21	500	0
	1992	..	..	..	..	..	−1	518	0
Malaysia	1980	19 474	...	19 194	109	171	628	9 219	16 439
	1990	61 168	105	43 433	17 141	489	138	13 498	48 816
	1991	66 529	180	44 450	21 354	546	−412	14 064	52 716
	1992	71 055	190	45 251	25 079	535	489	14 634	54 859
Mongolia	1980	1 707	1 707	...	..	..	...	965	...
	1990	3 066	3 066	...	..	..	...	1 072	214
	1991	3 018	3 018	...	..	..	...	994	198
	1992	2 912	2 912	...	..	..	...	903	165
Myanmar	1980	2 801	21	2 228	454	98	216	37	76
	1990	2 674	46	1 180	1 295	153	−4	188	17
	1991	2 363	51	1 001	1 158	152	−3	52	7
	1992	2 413	49	1 073	1 124	167	−4	56	9
Nepal	1980	22	..	..	..	22	...	235	1
	1990	86	..	..	..	86	−27	288	38
	1991	106	..	..	..	106	−41	413	49
	1992	107	..	..	..	107	−50	526	49
Oman	1980	21 157	..	20 040	1 118	..	269	1 227	19 765
	1990	52 310	..	48 783	3 527	..	−493	18	46 479
	1991	54 005	..	50 375	3 630	..	−229	30	47 653
	1992	55 253	..	52 804	2 449	..	−640	150	50 947
Pakistan	1980	10 816	1 057	688	8 000	1 071	71	8 327	1 164
	1990	22 632	2 118	3 884	14 442	2 188	466	13 738	389
	1991	24 128	2 058	4 669	15 010	2 391	−146	13 054	487
	1992	25 140	2 071	4 438	15 945	2 686	−324	13 532	650
Philippines	1980	3 966	222	764	..	2 980	96	15 861	126
	1990	8 638	840	350	..	7 448	464	19 631	592
	1991	8 781	853	236	..	7 692	−503	19 480	355
	1992	9 232	1 122	597	..	7 513	344	21 914	574
Qatar	1980	38 746	..	32 755	5 991	..	−157	88	32 250
	1990	36 039	..	28 616	7 423	..	−734	0	27 901
	1991	38 300	..	28 328	9 972	..	87	0	26 679
	1992	48 338	..	33 041	15 297	..	237	0	31 008
Saudi Arabia [1]	1980	728 427	..	726 767	1 660	..	7 920	2 838	682 134
	1990	523 831	..	483 074	40 757	..	5 587	0	406 210
	1991	655 436	..	611 318	44 118	..	9 148	0	524 162
	1992	674 087	..	628 277	45 810	..	10 776	0	532 369
Singapore	1980	..	..	..	..	..	−3 887	51 872	30 171
	1990	..	..	..	..	..	872	85 768	47 130
	1991	..	..	..	..	..	599	89 218	47 080
	1992	..	..	..	..	..	1 742	95 165	47 489
Sri Lanka	1980	182	..	..	..	182	193	2 809	228
	1990	386	..	..	..	386	44	2 712	143
	1991	383	..	..	..	383	−43	2 670	149
	1992	356	..	..	..	356	−8	2 897	8
Syrian Arab Republic	1980	13 516	..	13 138	64	315	38	7 534	10 315
	1990	35 754	..	32 888	2 173	694	196	1 077	19 689
	1991	38 564	..	35 260	2 536	768	503	340	19 678
	1992	40 757	..	37 227	2 623	907	1 644	835	20 632
Thailand	1980	741	573	11	0	156	−642	16 790	20
	1990	16 986	4 631	4 029	7 714	611	120	26 879	1 060
	1991	20 291	5 477	4 670	9 580	563	161	28 795	1 289
	1992	21 661	5 740	5 132	10 269	521	474	31 320	1 364
Turkey	1980	14 085	9 334	3 328	29	1 394	896	20 145	246
	1990	25 339	16 819	5 302	277	2 941	−1 045	41 309	3 024
	1991	26 278	16 882	6 233	265	2 897	−2 011	39 279	3 913
	1992	25 316	15 626	6 108	233	3 349	339	41 908	2 609
United Arab Emirates	1980	128 789	..	119 649	9 140	..	214	4 725	122 101
	1990	183 091	..	153 341	29 749	..	640	4	147 065
	1991	198 566	..	164 202	34 364	..	857	4	158 157
	1992	185 827	..	156 310	29 517	..	571	4	148 999

Bunkers – Soutes			Consumption – Consommation							
Air Avion	Sea Maritime	Unallocated Nondistribué	Per capita Par habitant	Total Totale	Solids Solides	Liquids Liquides	Gas Gaz	Electricity Electricité	Année	Pays ou zone
..	..	..	967	234	4	230	..	...	1980	Macao
..	..	..	1 439	492	0	481	..	11	1990	
..	..	..	1 466	522	0	510	..	12	1991	
..	..	..	1 405	520	0	506		14	1992	
0	82	238	821	11 306	77	10 951	97	181	1980	Malaisie
0	127	−1 073	1 490	26 658	2 040	19 395	4 741	482	1990	
0	156	2	1 534	28 132	2 288	20 251	5 050	543	1991	
0	160	−231	1 618	30 411	2 313	20 910	6 654	533	1992	
...	..	...	1 607	2 672	1 780	856	..	36	1980	Mongolie
...	..	...	1 802	3 923	2 857	1 038	..	28	1990	
...	..	...	1 713	3 814	2 824	979	..	10	1991	
...	..	...	1 606	3 650	2 751	889	..	9	1992	
0	0	525	60	2 021	58	1 412	454	98	1980	Myanmar
0	0	496	56	2 352	59	845	1 295	153	1990	
0	0	186	52	2 224	63	851	1 158	152	1991	
0	0	240	51	2 225	62	872	1 124	167	1992	
..	..	..	17	256	78	152	..	26	1980	Népal
..	..	..	19	363	13	259	..	91	1990	
..	..	..	26	511	67	343	..	102	1991	
..	..	..	31	634	92	436	..	106	1992	
171	337	0	1 673	1 842	..	724	1 118	..	1980	Oman
413	425	54	3 113	5 450	..	1 924	3 527	..	1990	
419	354	21	3 180	5 817	..	2 187	3 630	..	1991	
267	156	14	2 440	4 658	..	2 209	2 449	..	1992	
304	226	754	195	16 624	1 313	6 240	8 000	1 071	1980	Pakistan
146	48	857	283	34 464	3 178	14 656	14 442	2 188	1990	
217	26	1 397	280	35 202	3 139	14 662	15 010	2 391	1991	
221	28	848	288	37 248	3 185	15 432	15 945	2 686	1992	
137	243	992	377	18 233	506	14 747	...	2 980	1980	Philippines
544	96	1 731	409	24 843	2 159	15 237	...	7 448	1990	
554	77	1 960	416	25 818	2 346	15 780	...	7 692	1991	
654	96	3 132	415	26 346	1 921	16 913	...	7 513	1992	
96	..	85	28 646	6 560	..	569	5 991	..	1980	Qatar
161	..	−119	18 206	8 830	..	1 407	7 423	..	1990	
127	..	113	22 453	11 294	..	1 322	9 972	..	1991	
103	..	162	32 551	16 829	..	1 532	15 297	..	1992	
2 027	7 597	6 253	2 638	25 335	..	23 674	1 660	..	1980	Arabie saoudite [1]
1 621	16 125	7 170	5 429	87 117	..	46 360	40 757	..	1990	
1 483	12 457	14 186	5 701	94 000	..	49 881	44 118	..	1991	
1 621	13 045	19 194	5 770	97 082	..	51 272	45 810	..	1992	
652	5 971	7 041	4 937	11 924	4	11 932	..	−11	1980	Singapour
995	10 248	7 963	6 862	18 561	30	18 531	..	0	1990	
1 105	11 708	8 530	7 384	20 195	17	20 179	..	0	1991	
1 179	12 488	8 721	8 518	23 545	25	23 521	..	0	1992	
137	507	289	110	1 636	2	1 453	...	182	1980	Sri Lanka
83	554	93	127	2 182	1	1 795	...	386	1990	
57	443	122	133	2 325	3	1 939	...	383	1991	
74	404	125	150	2 651	2	2 293	...	356	1992	
351	...	1 711	992	8 635	4	8 261	64	306	1980	Rép. arabe syrienne
703	...	1 085	1 228	15 159	0	12 292	2 173	694	1990	
463	...	1 396	1 319	16 864	2	13 558	2 536	768	1991	
389	...	1 746	1 298	17 180	2	13 648	2 623	907	1992	
3	207	630	371	17 313	625	16 439	0	249	1980	Thaïlande
0	0	878	752	41 807	4 866	28 538	7 714	688	1990	
0	0	1 467	820	46 169	5 778	30 179	9 580	632	1991	
0	0	1 482	872	49 662	5 897	32 921	10 269	575	1992	
0	85	1 946	699	31 057	9 585	19 884	29	1 559	1980	Turquie
0	172	5 704	1 048	58 794	22 561	28 848	4 532	2 852	1990	
335	192	3 139	1 048	59 988	24 277	27 246	5 537	2 928	1991	
376	158	3 297	1 035	60 445	22 028	29 628	5 455	3 334	1992	
771	526	−3	9 759	9 905	..	4 240	5 665	..	1980	Emirats arabes unis
295	726	−558	20 902	34 927	..	9 451	25 476	..	1990	
324	741	−476	22 643	38 968	..	9 352	29 616	..	1991	
324	712	799	19 450	34 426	..	9 510	24 916	..	1992	

66
Production, trade and consumption of commercial energy
Thousand metric tons of coal equivalent and kilograms per capita [*cont.*]
Production, commerce et consommation d'énergie commerciale
Milliers de tonnes métriques d'équivalent houille et kilogrammes par habitant [*suite*]

Country or area	Year	Primary energy production – Production d'energie primaire					Changes in stocks	Imports	Exports
		Total Totale	Solids Solides	Liquids Liquides	Gas Gaz	Electricity Electricité	Variations des stocks	Imports Importations	Exports Exportations
Viet Nam	1980	5 582	5 300	...	...	282	...	1 783	500
	1990	9 140	4 627	3 849	4	660	−272	4 332	4 581
	1991	10 677	4 329	5 650	3	694	−243	3 734	6 757
	1992	13 362	4 792	7 842	3	725	−185	4 167	9 352
Yemen	1980	...	..	...	..	...	...	...	...
	1990	...	..	...	..	...	...	...	...
	1991	13 471		13 471			32	3 664	13 413
	1992	11 631	..	11 631	..	..	33	5 862	12 661
Europe	**1980**	**1 311 800**	**641 619**	**200 004**	**325 052**	**145 126**	**31 176**	**1 518 579**	**459 435**
	1990	**1 511 367**	**549 493**	**306 734**	**291 935**	**363 205**	**6 689**	**1 527 573**	**579 108**
	1991	**1 485 721**	**487 561**	**319 480**	**308 734**	**369 947**	**5 244**	**1 529 046**	**583 768**
	1992	**3 277 814**	**819 751**	**922 991**	**1 046 242**	**488 830**	**32 732**	**1 871 449**	**1 166 524**
Albania	1980	3 588	709	1 981	536	362	..	594	61
	1990	3 355	1 035	1 524	324	473	..	626	178
	1991	2 354	544	1 207	172	432	..	773	317
	1992	1 539	183	836	124	396		1 172	119
Austria	1980	9 614	1 281	2 139	2 663	3 532	1 054	20 719	906
	1990	8 143	910	1 703	1 539	3 991	134	25 475	1 366
	1991	8 421	774	1 896	1 730	4 020	−82	26 965	2 176
	1992	8 719	659	1 747	1 881	4 432	972	26 980	1 584
Belgium	1980	10 537	5 784	..	52	4 701	582	86 877	25 461
	1990	16 889	891	..	15	15 982	−93	85 383	26 876
	1991	17 170	1 113	..	13	16 044	377	90 270	28 198
	1992	16 859	564	..	8	16 287	1 161	91 419	28 421
Bulgaria	1980	10 992	7 644	393	209	2 746	220	27 284	1 653
	1990	13 584	7 804	86	15	5 679	−1 475	26 657	1 278
	1991	12 299	7 006	83	12	5 198	24	20 141	143
	1992	12 219	7 555	76	43	4 545	−468	16 534	359
former Czechoslovakia †	1980	67 539	64 354	133	787	2 265	642	44 650	8 527
	1990	64 862	54 275	176	779	9 633	−1 928	40 058	5 755
	1991	60 724	50 645	200	642	9 237	376	37 666	5 598
	1992	..	..	..	..	..	..	..	..
Denmark	1980	431	0	426	...	5	97	30 566	2 513
	1990	12 818	0	8 561	4 176	81	−530	21 866	9 473
	1991	15 381	0	10 130	5 156	94	−813	25 350	12 934
	1992	16 551	0	11 078	5 359	114	597	24 569	14 465
Faeroe Islands	1980	6	...	..	..	6	...	199	...
	1990	9	...	..	..	9	...	293	...
	1991	9	...	..	..	9	...	272	...
	1992	9	...	..	..	9	...	305	...
Finland	1980	4 072	366			3 706	−496	29 227	2 829
	1990	10 400	1 927	..	..	8 473	352	27 064	2 124
	1991	10 819	1 950	..	..	8 869	−1 338	26 892	4 013
	1992	10 868	1 857	..	..	9 011	−1 552	25 743	5 348
France incl. Monaco	1980	64 570	21 729	3 002	9 854	29 985	5 913	232 879	19 917
	1990	142 331	11 141	4 864	2 641	123 686	2 933	194 455	25 303
	1991	149 408	10 687	4 907	3 169	130 645	3 926	206 061	26 343
	1992	163 293	9 767	4 783	3 052	145 690	4 627	203 291	26 975
Germany †	1980	..	..	..	..	..	..	..	..
	1990	..	..	..	..	..	..	..	..
	1991	235 124	151 742	4 863	21 480	57 039	2 446	277 783	24 000
	1992	227 538	140 006	4 684	21 255	61 594	11 186	290 394	26 092
F. R. Germany	1980	174 822	126 239	7 430	22 780	18 373	7 411	257 432	32 934
	1990	184 223	103 487	5 133	18 629	56 973	−1 433	231 036	23 810
	1991	..	..	..	..	..	..	..	..
	1992	..	..	..	..	..	..	..	..
former German D.R.	1980	85 419	78 277	77	2 445	4 620	99	48 927	6 575
	1990	90 801	84 240	56	1 877	4 628	200	46 143	11 157
	1991	..	..	..	..	..	..	..	..
	1992	..	..	..	..	..	..	..	..
Gibraltar	1980	..	..	..	..	..	5	235	...
	1990	..	..	..	..	..	49	716	0
	1991	..	..	..	..	..	49	740	0
	1992	..	..	..	..	..	78	1 322	0

Bunkers – Soutes			Consumption – Consommation							
Air Avion	Sea Maritime	Unallocated Nondistribué	Per capita Par habitant	Total Totale	Solids Solides	Liquids Liquides	Gas Gaz	Electricity Electricité	Année	Pays ou zone
..	..	...	128	6 865	4 807	1 775	...	282	1980	Viet Nam
..	..	2	137	9 161	4 167	4 330	4	660	1990	
..	..	2	116	7 894	3 428	3 769	3	694	1991	
..	..	2	120	8 360	3 437	4 194	3	725	1992	
...	...	...	...	...	..	...	..	..	1980	Yémen
...	...	...	...	...	..	...	..	..	1990	
90	229	73	278	3 298		3 298	..	..	1991	
90	214	381	329	4 114	..	4 114		..	1992	
20 473	44 331	44 786	4 604	2 230 179	716 175	971 678	395 237	147 088	1980	Europe
31 057	51 878	23 903	4 704	2 346 304	671 736	853 094	453 753	367 720	1990	
24 473	51 101	17 691	4 644	2 332 491	613 376	872 152	474 168	372 794	1991	
25 592	52 812	98 786	5 205	3 772 817	948 076	1 215 107	1 121 603	488 032	1992	
..	..	728	1 270	3 393	846	1 710	536	301	1980	Albanie
..	..	785	918	3 019	1 210	1 001	324	485	1990	
..	..	775	611	2 036	794	906	172	165	1991	
..	..	918	498	1 675	443	780	124	327	1992	
99	..	−2 967	4 139	31 242	5 182	16 395	6 620	3 044	1980	Autriche
254	..	−10	4 137	31 875	6 012	14 426	7 502	3 935	1990	
265	..	615	4 180	32 411	5 351	15 255	7 692	4 114	1991	
290	..	420	4 154	32 432	4 701	14 747	8 483	4 500	1992	
594	3 462	1 932	6 636	65 382	15 687	31 174	14 144	4 377	1980	Belgique
1 362	5 969	1 517	6 697	66 640	14 247	23 902	12 967	15 525	1990	
1 322	6 108	1 956	6 962	69 479	14 261	25 536	13 864	15 817	1991	
1 304	6 070	2 939	6 830	68 383	12 678	25 038	14 365	16 303	1992	
..	302	2 197	3 826	33 904	13 499	12 152	5 027	3 226	1980	Bulgarie
..	465	2 480	4 170	37 493	12 974	10 667	7 707	6 144	1990	
..	299	909	3 466	31 065	11 020	7 990	6 596	5 459	1991	
..	387	511	3 135	27 963	10 546	6 730	5 810	4 877	1992	
..	..	4 079	6 462	98 941	63 030	22 284	11 136	2 491	1980	anc. Tchécoslovaquie †
..	..	4 786	6 149	96 307	56 747	14 040	15 329	10 190	1990	
..	..	3 489	5 666	88 927	51 258	11 902	16 309	9 459	1991	
..	..	...	...	...	...	...	...	...	1992	
772	607	−653	5 399	27 659	9 437	18 369	...	−147	1980	Danemark
1 021	1 397	−540	4 642	23 862	8 814	11 435	2 667	946	1990	
884	1 248	317	5 082	26 160	11 808	11 599	2 901	−148	1991	
864	1 309	−124	4 657	24 010	9 656	10 739	3 041	574	1992	
..	..	...	4 767	205	0	199	..	6	1980	Iles Féroé
..	..	...	6 426	302	0	293	..	9	1990	
..	..	...	5 979	281	0	272	..	9	1991	
..	..	...	6 681	314	0	305	..	9	1992	
221	851	2 508	5 729	27 386	6 680	15 626	1 225	3 855	1980	Finlande
472	824	−571	6 872	34 263	7 798	13 097	3 588	9 780	1990	
442	786	−397	6 830	34 206	7 728	12 920	3 793	9 765	1991	
391	990	−1 346	6 513	32 779	7 029	11 777	3 931	10 042	1992	
2 745	5 707	5 825	4 774	257 341	50 120	142 802	34 054	30 365	1980	France y comp.Monaco
5 505	3 654	4 702	5 193	294 688	28 713	109 618	38 292	118 066	1990	
4 251	3 792	5 811	5 460	311 346	30 096	113 657	43 446	124 148	1991	
4 886	3 664	4 590	5 618	321 842	26 497	113 078	43 184	139 082	1992	
...	...	...	...	...	...	...	...	...	1980	Allemagne †
...	...	...	...	...	...	...	...	...	1990	
0	2 970	4 155	6 003	479 336	158 843	171 904	91 620	56 968	1991	
	2 509	6 800	5 865	471 346	149 266	171 384	89 756	60 941	1992	
3 743	4 103	1 574	6 213	382 490	116 710	175 765	70 935	19 081	1980	R. f. Allemagne
5 601	2 939	1 592	6 241	382 751	107 892	141 230	76 619	57 010	1990	
...	...	...	...	...	...	...	...	...	1991	
...	...	...	...	...	...	...	...	...	1992	
..	..	4 073	7 385	123 599	86 072	23 583	9 145	4 800	1980	anc. R.d. allemande
..	..	7 056	7 295	118 530	86 310	18 024	9 446	4 751	1990	
..	..	...	...	...	...	...	...	...	1991	
..	..	...	...	...	...	...	...	...	1992	
4	196	..	1 034	30	0	30	..	...	1980	Gibraltar
10	629	..	1 000	28	0	28	..	−0	1990	
10	649	..	1 143	32	0	32	..	−0	1991	
6	1 214	..	857	24	0	24	..	0	1992	

66

Production, trade and consumption of commercial energy
Thousand metric tons of coal equivalent and kilograms per capita [*cont.*]
Production, commerce et consommation d'énergie commerciale
Milliers de tonnes métriques d'équivalent houille et kilogrammes par habitant [*suite*]

Country or area	Year	Primary energy production – Production d'energie primaire					Changes in stocks	Imports	Exports
		Total Totale	Solids Solides	Liquids Liquides	Gas Gaz	Electricity Electricité	Variations des stocks	Imports Importations	Exports Exportations
Greece	1980	4 760	4 341	0	0	418	756	34 041	13 919
	1990	11 821	10 164	1 192	219	245	−871	30 642	9 158
	1991	11 657	9 854	1 199	214	390	−382	29 076	6 618
	1992	11 259	9 783	985	197	294	1 286	32 396	7 006
Hungary	1980	21 891	9 826	3 977	8 074	14	579	23 482	3 617
	1990	20 613	6 297	3 748	5 445	5 123	283	22 194	1 882
	1991	20 158	6 013	3 522	5 499	5 123	−733	19 395	1 017
	1992	19 053	5 567	3 116	5 163	5 207	−732	17 419	1 384
Iceland	1980	437	..	..	..	437	−39	836	..
	1990	884	..	..	..	884	0	1 015	..
	1991	863	..	..	..	863	−15	883	..
	1992	811	..	..	..	811	73	1 014	..
Ireland	1980	2 723	1 451		1 169	103	72	9 252	300
	1990	4 733	1 606	..	3 006	121	590	10 946	918
	1991	4 733	1 572	..	3 042	118	−2	10 759	1 062
	1992	4 778	1 635	..	3 014	130	−209	10 679	1 137
Italy and San Marino	1980	29 405	696	2 610	16 446	9 653	73	189 076	16 238
	1990	37 598	397	6 671	22 269	8 260	2 017	215 596	26 850
	1991	39 814	356	6 190	23 763	9 504	−2 823	215 649	29 167
	1992	40 046	362	6 431	23 387	9 866	3 469	224 443	29 290
Luxembourg	1980	12	..	..	..	12	32	5 091	96
	1990	101	..	..	..	101	18	5 013	112
	1991	98	..	..	..	98	−32	5 324	99
	1992	75	..	..	..	75	−41	5 439	67
Malta	1980	...	...	..	..	..	0	615	0
	1990	...	...	..	..	..	0	834	0
	1991	...	...	..	..	..	0	831	0
	1992	...	...	..	..	..	0	834	0
Netherlands	1980	119 793	0	2 272	115 960	1 560	1 604	110 821	117 406
	1990	93 739	0	5 729	86 688	1 322	−1 051	133 833	109 824
	1991	104 621	0	5 364	98 001	1 255	1 017	133 049	114 512
	1992	104 532	0	4 802	98 286	1 445	1 219	130 634	113 867
Norway, Svalbard, and Jan Mayen Is.	1980	81 927	276	35 067	36 331	10 252	348	15 606	70 844
	1990	171 055	291	117 525	38 330	14 910	2 552	6 449	142 710
	1991	185 726	316	134 123	37 704	13 583	−380	6 482	159 903
	1992	206 192	375	152 894	38 543	14 380	775	5 484	179 013
Poland	1980	173 830	166 329	508	6 589	403	−3 263	34 939	29 972
	1990	136 549	132 564	239	3 339	407	1 647	34 205	28 076
	1991	132 132	127 746	227	3 739	419	1 144	28 494	21 717
	1992	128 861	124 501	286	3 636	439	1 004	28 962	20 713
Portugal	1980	1 170	177	..	..	993	848	14 440	600
	1990	1 312	165	..	..	1 148	1 117	24 043	3 362
	1991	1 292	158	..	..	1 133	100	23 801	3 085
	1992	759	129	..	..	630	166	27 074	4 448
Romania	1980	84 192	14 662	16 442	51 536	1 552	...	30 029	12 407
	1990	56 817	11 415	11 325	32 728	1 349	654	39 447	12 540
	1991	49 406	9 713	9 700	28 243	1 750	−1 510	25 654	3 669
	1992	46 151	10 106	9 448	25 159	1 437	1 308	23 203	3 267
Spain	1980	22 331	14 489	2 288	1	5 553	1 249	82 132	2 991
	1990	44 521	17 443	1 674	2 021	23 384	502	97 815	14 412
	1991	44 800	16 716	2 071	1 890	24 123	−1 496	103 919	16 735
	1992	42 620	15 644	2 040	1 642	23 294	866	108 229	14 157
Sweden	1980	17 124	17	36	..	17 071	3 565	45 238	6 833
	1990	34 295	10	4	..	34 280	−260	38 610	13 691
	1991	36 369	26	4	..	36 338	−996	39 030	13 525
	1992	33 171	364	1	..	32 806	−993	38 798	11 307
Switzerland,Liechtenstein	1980	9 061	..	..	...	9 061	567	21 559	2 348
	1990	12 591	..	..	5	12 586	−160	24 050	3 049
	1991	12 645	..	..	14	12 631	476	25 604	4 131
	1992	12 899	..	..	4	12 895	−509	24 841	3 855
United Kingdom	1980	284 951	108 556	115 110	47 048	14 237	9 258	99 289	79 079
	1990	300 798	79 280	131 268	64 958	25 292	308	115 780	103 974
	1991	308 337	78 228	130 854	72 204	27 050	6 015	124 878	104 418
	1992	308 315	70 045	135 191	73 549	29 530	2 164	123 371	107 684

Bunkers – Soutes			Consumption – Consommation							
Air Avion	Sea Maritime	Unallocated Nondistribué	Per capita Par habitant	Total Totale	Solids Solides	Liquids Liquides	Gas Gaz	Electricity Electricité	Année	Pays ou zone
1 080	1 218	1 599	2 098	20 229	4 767	14 968	0	494	1980	Grèce
1 134	3 710	−1 900	3 050	31 231	11 518	19 161	219	333	1990	
1 017	3 359	−2 117	3 133	32 238	11 090	20 465	214	469	1991	
1 030	3 858	−2 522	3 193	32 997	11 476	20 956	197	369	1992	
..	..	1 818	3 676	39 359	12 728	13 292	12 418	921	1980	Hongrie
..	..	1 909	3 737	38 733	8 960	10 550	12 733	6 490	1990	
..	..	1 644	3 648	37 624	8 752	10 219	12 626	6 027	1991	
..	..	1 680	3 327	34 139	7 347	10 060	11 099	5 633	1992	
29	...	..	5 623	1 282	25	820	..	437	1980	Islande
105	...	..	7 039	1 795	85	825		884	1990	
91	...	..	6 473	1 670	87	719	..	863	1991	
108	28	..	6 215	1 616	62	744		811	1992	
292	109	−232	3 362	11 434	2 454	7 708	1 169	103	1980	Irlande
500	27	491	3 755	13 154	4 604	5 423	3 006	121	1990	
481	49	51	3 953	13 852	4 721	5 971	3 042	118	1991	
435	23	138	3 969	13 934	4 714	6 077	3 014	130	1992	
2 323	6 011	1 303	3 410	192 533	15 998	129 899	36 235	10 401	1980	Italie y compris Saint–Marin
2 687	4 072	−6 165	3 922	223 732	19 293	128 937	62 984	12 517	1990	
3 079	3 562	−8 466	4 046	230 945	18 826	130 585	67 720	13 814	1991	
3 088	3 491	−7 175	4 067	232 326	17 405	135 462	65 257	14 202	1992	
93	..		13 415	4 883	2 434	1 414	674	361	1980	Luxembourg
189	..		12 588	4 796	1 510	2 095	614	576	1990	
196	..		13 400	5 159	1 424	2 471	709	555	1991	
189	..		13 590	5 300	1 357	2 639	739	563	1992	
112	40	..	1 429	463	1	462	...	..	1980	Malte
74	43	..	2 025	717	256	461	...	..	1990	
74	43	..	2 006	714	253	461	...	..	1991	
74	43	..	1 997	717	256	461	...		1992	
1 319	13 550	−3 310	7 073	100 043	5 386	38 255	54 880	1 523	1980	Pays–Bas
2 087	15 791	−7 959	7 282	108 880	15 269	42 240	48 918	2 453	1990	
2 311	16 179	−6 272	7 299	109 923	11 509	41 347	54 687	2 380	1991	
2 746	16 343	−6 959	7 115	107 951	11 363	41 233	52 844	2 510	1992	
0/	397	−477	6 460	26 421	1 402	13 444	1 379	10 195	1980	Norvège, Savalbard et Ile Jan–Mayen
118	645	2 567	6 811	28 913	1 237	11 583	3 137	12 956	1990	
115	563	2 608	6 893	29 400	1 125	12 215	2 818	13 242	1991	
118	704	2 351	6 703	28 716	1 149	11 541	2 718	13 307	1992	
..	78	5 227	4 969	176 755	143 178	20 816	12 386	374	1980	Pologne
..	618	3 332	3 596	137 080	107 848	16 236	12 717	279	1990	
..	262	2 402	3 536	135 103	107 409	16 304	11 293	97	1991	
..	415	2 367	3 484	133 324	105 516	16 715	11 150	−57	1992	
427	615	606	1 281	12 514	664	10 633	..	1 217	1980	Portugal
414	884	368	1 947	19 210	3 961	14 097	..	1 152	1990	
721	895	959	1 961	19 332	3 871	14 316	..	1 145	1991	
778	878	736	2 115	20 827	4 089	15 943	..	795	1992	
..	..	3 082	4 447	98 732	19 789	24 045	53 289	1 610	1980	Roumanie
..	..	3 131	3 445	79 938	17 656	18 712	41 052	2 518	1990	
..	..	2 511	3 036	70 389	14 132	19 351	34 290	2 616	1991	
..	..	1 348	2 744	63 431	15 136	16 064	30 278	1 953	1992	
1 247	2 347	8 206	2 355	88 424	18 240	62 034	2 766	5 384	1980	Espagne
1 608	5 586	4 863	2 938	115 365	28 177	55 967	7 889	23 332	1990	
2 133	5 622	6 114	3 038	119 610	29 545	57 138	8 888	24 039	1991	
1 666	5 697	6 920	3 081	121 544	28 438	60 527	9 206	23 373	1992	
237	1 232	−397	6 124	50 891	2 390	31 364	...	17 137	1980	Suède
399	968	650	6 713	57 456	3 617	18 934	842	34 063	1990	
442	1 143	1 810	6 912	59 474	3 382	19 032	881	36 179	1991	
483	1 307	−760	7 008	60 623	3 456	23 629	996	32 541	1992	
977	0	−1 001	4 371	27 729	499	18 012	1 375	7 843	1980	Suisse, Liechtenstein
1 453	26	157	4 679	32 115	533	16 665	2 590	12 328	1990	
1 384	12	13	4 648	32 233	453	16 253	3 240	12 288	1991	
1 492	25	−137	4 710	33 014	297	17 293	3 056	12 368	1992	
4 157	3 505	7 198	4 972	281 043	101 594	103 879	61 332	14 238	1980	Royaume–Uni
6 064	3 631	−4 823	5 335	307 423	90 563	115 491	74 610	26 759	1990	
5 253	3 561	−4 217	5 505	318 186	91 078	117 360	80 682	29 065	1991	
5 632	3 633	−645	5 403	313 218	84 934	115 850	80 853	31 580	1992	

66

Production, trade and consumption of commercial energy
Thousand metric tons of coal equivalent and kilograms per capita [*cont.*]
Production, commerce et consommation d'énergie commerciale
Milliers de tonnes métriques d'équivalent houille et kilogrammes par habitant [*suite*]

Country or area	Year	Primary energy production – Production d'energie primaire					Changes in stocks Variations des stocks	Imports Importations	Exports Exportations
		Total Totale	Solids Solides	Liquids Liquides	Gas Gaz	Electricity Electricité			
Yugoslavia, SFR †	1980	26 605	14 416	6 113	2 573	3 503	0	22 544	1 407
	1990	36 525	24 150	5 257	2 933	4 185	1 134	27 332	1 229
	1991	21 363	12 400	2 937	2 047	3 979	−104	23 304	389
	1992	..	..	..	..	..			
Oceania	**1980**	**120 948**	**70 817**	**31 283**	**13 587**	**5 260**	**−2 920**	**31 150**	**40 408**
	1990	**227 174**	**144 702**	**42 215**	**33 338**	**6 918**	**4 653**	**29 299**	**102 754**
	1991	**235 102**	**150 721**	**42 118**	**34 578**	**7 685**	**219**	**28 543**	**115 783**
	1992	**242 128**	**159 558**	**37 554**	**37 597**	**7 418**	**−620**	**31 396**	**125 841**
American Samoa	1980	..	..	..	..	..	..	201	..
	1990	..	..	..	..	..	..	267	..
	1991	..	..	..	..	..	..	267	..
	1992	..	..	..	..	..	..	274	..
Australia	1980	113 770	69 009	30 810	12 257	1 693	−3 250	19 137	39 348
	1990	211 301	142 768	39 500	27 206	1 828	4 915	19 238	100 299
	1991	217 679	148 544	39 257	27 901	1 978	342	18 279	113 094
	1992	224 427	157 174	34 828	30 489	1 936	−372	21 216	123 342
Cook Islands	1980	..	..	..	..	..	..	15	..
	1990	..	..	..	..	..	..	22	..
	1991	..	..	..	..	..	..	22	..
	1992	..	..	..	..	..	..	22	..
Fiji	1980	..	..	..	..	...	−18	580	101
	1990	47	..	..	..	47	−45	572	126
	1991	47	..	..	..	47	−27	565	194
	1992	48	..	..	..	48	0	573	178
French Polynesia	1980	...	..	..	..	...	0	248	..
	1990	9	..	..	..	9	0	397	..
	1991	9	..	..	..	9	0	400	..
	1992	9	..	..	..	9	0	403	..
Guam	1980	..	..	..	..	..	...	2 585	864
	1990	..	..	..	..	..	...	861	0
	1991	..	..	..	..	..	...	861	0
	1992	..	..	..	..	..	...	861	0
Kiribati	1980	..	..	..	..	..	...	13	..
	1990	..	..	..	..	..	...	10	..
	1991	..	..	..	..	..	...	10	..
	1992	..	..	..	..	..	...	10	..
Nauru	1980	..	..	..	..	..	...	66	..
	1990	..	..	..	..	..	...	69	..
	1991	..	..	..	..	..	...	69	..
	1992	..	..	..	..	..	...	70	..
New Caledonia	1980	34	..	..	..	34	...	904	...
	1990	57	..	..	..	57	0	741	12
	1991	42	..	..	..	42	0	825	22
	1992	42	..	..	..	42	0	814	17
New Zealand	1980	7 104	1 808	473	1 330	3 493	348	5 614	95
	1990	15 697	1 934	2 715	6 133	4 915	−218	5 056	2 299
	1991	17 262	2 177	2 861	6 678	5 546	−97	5 188	2 451
	1992	17 538	2 384	2 726	7 108	5 320	−248	5 086	2 282
Niue	1980	..	..	..	..	..	...	1	..
	1990	..	..	..	..	..	...	1	..
	1991	..	..	..	..	..	...	1	..
	1992	..	..	..	..	..	...	1	..
Palau [2]	1980	...	..	..	..	...	..	81	..
	1990	4	..	..	..	4	..	133	..
	1991	4	..	..	..	4	..	133	..
	1992	4	..	..	..	4	..	135	..
Papua New Guinea	1980	39	..	..	..	39	...	892	...
	1990	57	..	..	..	57	0	1 137	17
	1991	57	..	..	..	57	0	1 128	22
	1992	57	..	..	..	57	0	1 127	22

| Bunkers – Soutes | | | Consumption – Consommation | | | | | | | |
Air Avion	Sea Maritime	Unallocated Nondistribué	Per capita Par habitant	Total Totale	Solids Solides	Liquids Liquides	Gas Gaz	Electricity Electricité	Année	Pays ou zone
0	0	1 867	2 057	45 874	17 364	20 544	4 513	3 453	1980	Yougoslavie, Rfs †
0	0	5 484	2 352	56 009	25 933	17 957	8 002	4 117	1990	
0	0	3 021	1 731	41 360	14 560	15 974	6 687	4 139	1991	
...	...	...	...	...	...	...	...	...	1992	
2 565	2 611	2 865	4 697	106 569	40 062	47 727	13 520	5 260	1980	Océanie
3 447	1 771	−5 541	5 653	149 390	56 118	56 745	29 609	6 918	1990	
3 406	1 637	−7 446	5 588	150 048	57 790	56 307	28 266	7 685	1991	
3 594	1 668	−11 167	5 655	154 208	60 260	56 964	29 565	7 418	1992	
..	...	..	6 281	201	..	201	..	..	1980	Samoa américaines
..	130	..	2 915	137	..	137	..	..	1990	
..	130	..	2 854	137	..	137	..	..	1991	
..	130	..	2 880	144	..	144	..	..	1992	
1 132	1 614	2 631	6 276	91 433	38 312	39 171	12 257	1 693	1980	Australie
2 000	939	−4 761	7 529	127 147	54 326	47 517	23 477	1 828	1990	
2 081	847	−6 602	7 365	126 195	55 749	46 880	21 588	1 978	1991	
2 208	901	−10 145	7 464	129 708	58 098	47 218	22 457	1 936	1992	
...	..	..	833	15	..	15	..	..	1980	Iles Cook
12	..	..	556	10	..	10	..	..	1990	
12	..	..	526	10	..	10	..	..	1991	
12	..	..	526	10	..	10	..	..	1992	
106	45	..	546	346	21	325	..	...	1980	Fidji
88	37	..	567	412	24	341	..	47	1990	
59	40	..	469	345	18	280	..	47	1991	
41	40	..	484	361	20	293	..	48	1992	
...	112	..	901	136	..	136	..	...	1980	Polynésie française
59	46	..	1 523	300	..	291	..	9	1990	
59	46	..	1 505	304	..	294	..	9	1991	
59	46	..	1 490	307	..	297	..	9	1992	
488	302	8	8 617	922	..	922	..	..	1980	Guam
12	144	...	5 261	705	..	705	..	..	1990	
12	144	...	5 146	705	..	705	..	..	1991	
12	144	...	5 036	705	..	705	..	..	1992	
..	..	..	213	13	..	13	..	..	1980	Kiribati
..	..	..	139	10	..	10	..	..	1990	
..	..	..	135	10	..	10	..	..	1991	
..	..	..	133	10	..	10	..	..	1992	
7	..	..	8 429	59	..	59	..	..	1980	Nauru
7	..	..	6 100	61	..	61	..	..	1990	
7	..	..	6 100	61	..	61	..	..	1991	
7	..	..	6 300	63	..	63	..	..	1992	
...	21	..	6 413	917	120	763	..	34	1980	Nouvelle−Calédonie
22	7	..	4 506	757	164	537	..	57	1990	
22	13	..	4 765	810	176	592	..	42	1991	
22	13	..	4 647	804	170	592	..	42	1992	
245	484	225	3 637	11 322	1 609	4 957	1 263	3 493	1980	Nouvelle−Zélande
635	463	−781	5 463	18 354	1 603	5 703	6 133	4 915	1990	
544	411	−844	5 883	19 984	1 845	5 915	6 678	5 546	1991	
618	389	−1 022	5 992	20 607	1 971	6 207	7 108	5 320	1992	
0	..	..	333	1	..	1	..	..	1980	Nioué
0	..	..	500	1	..	1	..	..	1990	
0	..	..	500	1	..	1	..	..	1991	
0	..	..	500	1	..	1	..	..	1992	
7	..	..	497	73	..	73	..	...	1980	Palaos [2]
22	..	..	542	115	..	111	..	4	1990	
22	..	..	520	115	..	111	..	4	1991	
22	..	..	518	117	..	113	..	4	1992	
22	7	..	292	902	0	863	..	39	1980	Papouasie−Nvl−Guinée
40	4	..	295	1 132	1	1 074	..	57	1990	
38	4	..	285	1 120	1	1 063	..	57	1991	
37	4	..	279	1 120	1	1 063	..	57	1992	

66

Production, trade and consumption of commercial energy
Thousand metric tons of coal equivalent and kilograms per capita [*cont.*]

Production, commerce et consommation d'énergie commerciale
Milliers de tonnes métriques d'équivalent houille et kilogrammes par habitant [*suite*]

Country or area	Year	Primary energy production – Production d'energie primaire					Changes in stocks Variations des stocks	Imports Importations	Exports Exportations
		Total Totale	Solids Solides	Liquids Liquides	Gas Gaz	Electricity Electricité			
Samoa	1980	1	..	..	..	1	...	47	..
	1990	2	..	..	..	2	...	60	..
	1991	2	..	..	..	2	...	60	..
	1992	2	..	..	..	2	...	62	..
Solomon Islands	1980	..	..	..	..	..	...	53	..
	1990	..	..	..	..	..	...	80	..
	1991	..	..	..	..	..	...	80	..
	1992	..	..	..	..	..	...	79	..
Tonga	1980	..	..	..	..	..	...	19	..
	1990	..	..	..	..	..	...	44	..
	1991	..	..	..	..	..	...	49	..
	1992	..	..	..	..	..	...	47	..
Vanuatu	1980	..	..	..	..	..	...	56	..
	1990	..	..	..	..	..	...	32	..
	1991	..	..	..	..	..	...	31	..
	1992	..	..	..	..	..	...	29	..
Wake Island	1980	..	..	..	..	..	...	592	..
	1990	..	..	..	..	..	...	577	..
	1991	..	..	..	..	..	...	575	..
	1992	..	..	..	..	..	...	586	..
former USSR former USSR †	1980	1 995 707	562 288	865 148	518 586	49 685	1 120	12 064	354 213
	1990	2 352 318	486 714	818 844	939 347	107 413	7 445	25 234	389 362
	1991	2 217 666	433 974	740 707	935 323	107 661	...	4 484	285 498
	1992	..	..	..	..	..	..	..	..

Source:
Energy statistics database of the Statistical Division of the
United Nations Secretariat.

† For detailed descriptions of data pertaining to
former Czechoslovakia, Germany, SFR Yugoslavia and former
USSR, see annex I – Country or area nomenclature, regional
and other groupings.

1   Part Neutral Zone.
2   Including data for Federated States of Micronesia, Marshall Is. and
Northern Mariana Is.

Source:
Base de données pour les statistiques énergétiques de la Division
de statistique du Secrétariat de l'ONU.

† Pour les descriptions en détails des données
relatives à l'ancienne Tchécoslovaquie, l'Allemagne, la Rfs
Yougoslavie et l'ancienne URSS, voir l'Annexe I –
Nomenclature des pays ou zones, groupements régionaux et
autres groupements.

1   Part Zone neutrel.
2   Y compris les données pour les Etats fédérés de Micronésie, les
iles Marshall et les iles Mariannes du Nord.

Bunkers – Soutes			Consumption – Consommation							
Air Avion	Sea Maritime	Unallocated Nondistribué	Per capita Par habitant	Total Totale	Solids Solides	Liquids Liquides	Gas Gaz	Electricity Electricité	Année	Pays ou zone
..	..	..	302	48	0	47	..	1	1980	Samoa
..	..	..	389	63	0	60	..	2	1990	
..	..	..	384	63	0	60	..	2	1991	
..	..	..	388	64	0	62	..	2	1992	
4	...	...	211	48	..	48		..	1980	Iles Salomon
3	...	...	241	77	..	77		..	1990	
3	...	...	233	77	..	77		..	1991	
3	...	...	222	76	..	76		..	1992	
...	..	..	207	19	0	19	..	..	1980	Tonga
7	..	..	385	37	0	37	..	..	1990	
4	..	..	454	44	0	44	..	..	1991	
4	..	..	443	43	0	43	..	..	1992	
..	27	..	248	29	0	29	..	..	1980	Vanuatu
..	0	..	215	32	0	32	..	..	1990	
..	0	..	203	31	0	31	..	..	1991	
..	0	..	185	29	0	29	..	..	1992	
553	..	..	38 000	38	..	38	..	..	1980	Ile de Wake
539	..	..	38 000	38	..	38	..	..	1990	
542	..	..	34 000	34	..	34	..	..	1991	
549	..	..	37 000	37	..	37	..	..	1992	
										ancienne URSS
..	...	60 473	5 995	1 591 965	532 954	559 180	452 383	47 448	1980	ancienne URSS †
..	...	61 919	6 632	1 918 827	462 061	544 673	809 021	103 072	1990	
..	...	69 353	6 415	1 867 299	411 963	536 627	813 424	105 286	1991	
..	...	...	...	...	...	...	...	...	1992	

67
Production of selected energy commodities
Production des principaux biens de l'énergie
Thousand metric tons of coal equivalent
Milliers de tonnes métriques d'équivalent houille

Country or area Pays ou zone	Year Anneé	Hard coal, lignite & peat Houille, lignite & tourbe	Briquettes & cokes Agglomérés & cokes	Crude petroleum & NGL Pétrole brut & GNL	Light petroleum products Produits pétroliers légers	Heavy petroleum products Produits pétroliers lourds	Other petroleum products Autres produits pétroliers	LPG & refinery gas GLP et gaz de raffinerie	Natural gas Gaz naturel	Electricity Electricité
World	1980	2 660 769	415 111	4 419 285	1 474 670	2 353 615	275 299	189 154	1 844 089	1 199 309
Monde	1990	3 261 311	401 289	4 590 734	1 682 728	2 236 040	324 340	247 502	2 586 795	1 980 331
	1992	3 210 064	337 948	4 577 334	1 693 459	2 216 953	328 581	259 970	2 587 179	2 056 565
	1993	3 130 503	329 125	4 574 234	1 729 799	2 229 122	362 387	255 708	2 666 408	2 095 116
Africa	1980	98 232	2 504	436 325	31 189	63 644	2 832	1 483	35 227	23 068
Afrique	1990	139 493	3 438	461 565	56 198	91 863	4 479	3 318	92 179	41 689
	1992	139 822	3 435	474 804	57 282	92 690	4 689	3 384	97 136	43 223
	1993	145 311	3 662	472 074	58 573	94 338	4 560	3 525	100 366	43 879
Algeria	1980	7	0	74 964	5 108	9 230	145	458	25 903	875
Algérie	1990	10	0	89 282	10 407	20 098	472	1 399	68 700	1 978
	1992	15	0	83 507	10 561	19 613	515	1 399	69 547	2 246
	1993	20	0	84 655	11 035	19 188	544	1 414	71 708	2 385
Angola	1980	..	..	10 610	209	1 083	14	19	102	184
Angola	1990	..	..	33 642	481	1 393	20	28	222	226
	1992	..	..	37 137	485	1 399	21	26	222	228
	1993	..	..	35 994	491	1 385	21	28	223	228
Benin	1980	..	..	14	..	..	..	..	..	1
Bénin	1990	..	..	416	..	..	..	..	..	1
	1992	..	..	426	..	..	..	..	..	1
	1993	..	..	431	..	..	..	..	..	1
Burkina Faso	1980	..	..	..	..	..	..	..	..	14
Burkina Faso	1990	..	..	..	..	..	..	..	..	24
	1992	..	..	..	..	..	..	..	..	24
	1993	..	..	..	..	..	..	..	..	24
Burundi	1980	2	0	..	..	..	..	..	..	0
Burundi	1990	6	0	..	..	..	..	..	..	14
	1992	6	0	..	..	..	..	..	..	13
	1993	6	0	..	..	..	..	..	..	14
Cameroon	1980	0	..	3 856	436	664	79	16	..	178
Cameroun	1990	1	..	12 112	283	572	79	28	..	332
	1992	1	..	9 698	297	585	75	30	..	334
	1993	1	..	8 870	302	587	81	33	..	335
Cape Verde	1980	..	..	..	..	..	..	..	..	2
Cap–Vert	1990	..	..	..	..	..	..	..	..	4
	1992	..	..	..	..	..	..	..	..	5
	1993	..	..	..	..	..	..	..	..	5
Central African Republic	1980	..	..	..	..	..	..	..	..	8
Rép. centrafricaine	1990	..	..	..	..	..	..	..	..	12
	1992	..	..	..	..	..	..	..	..	12
	1993	..	..	..	..	..	..	..	..	12
Chad	1980	..	..	..	..	..	..	..	..	6
Tchad	1990	..	..	..	..	..	..	..	..	10
	1992	..	..	..	..	..	..	..	..	10
	1993	..	..	..	..	..	..	..	..	11
Comoros	1980	..	..	..	..	..	..	..	..	1
Comores	1990	..	..	..	..	..	..	..	..	2
	1992	..	..	..	..	..	..	..	..	2
	1993	..	..	..	..	..	..	..	..	2
Congo	1980	...	..	4 706	0	0	...	0	2	19
Congo	1990	0	..	11 450	252	597	12	5	3	62
	1992	0	..	12 361	255	509	113	6	4	53
	1993	0	..	12 391	265	513	19	6	4	53
Côte d'Ivoire	1980	..	..	171	651	1 215	10	16	..	215
Côte d'Ivoire	1990	..	..	450	1 366	1 670	133	22	..	282
	1992	..	..	464	1 781	1 785	129	23	..	227
	1993	..	..	463	1 794	1 789	139	26	..	235

67

Production of selected energy commodities
Thousand metric tons of coal equivalent [*cont.*]
Production des principaux biens de l'énergie
Milliers de tonnes métriques d'équivalent houille [*suite*]

Country or area Pays ou zone	Year Anneé	Hard coal, lignite & peat Houille, lignite & tourbe	Briquettes & cokes Agglomérés & cokes	Crude petroleum & NGL Pétrole brut & GNL	Light petroleum products Produits pétroliers légers	Heavy petroleum products Produits pétroliers lourds	Other petroleum products Autres produits pétroliers	LPG & refinery gas GLP et gaz de raffinerie	Natural gas Gaz naturel	Electricity Electricité
Djibouti	1980	..	..	..	..	..	..	..	..	14
Djibouti	1990	..	..	..	..	..	..	..	..	22
	1992	..	..	..	..	..	..	..	..	22
	1993	..	..	..	..	..	..	..	..	22
Egypt	1980	..	677	42 611	5 397	13 143	499	214	2 134	2 326
Egypte	1990	..	1 053	65 267	9 544	21 739	1 650	524	9 028	4 843
	1992	..	1 035	68 510	8 998	22 223	1 559	510	11 179	5 541
	1993	..	1 314	69 201	9 708	23 500	1 614	598	12 839	5 831
Equatorial Guinea	1980	..	..	..	..	..	..	..	..	2
Guinée équatoriale	1990	..	..	..	..	..	..	..	..	2
	1992	..	..	..	..	..	..	..	..	2
	1993	..	..	..	..	..	..	..	..	2
Ethiopia	1980	..	..	..	...	...	...	...	..	...
Ethiopie	1990	..	..	..	301	737	11	10	..	224
	1992	..	..	..	316	744	14	11	..	233
	1993	..	..	..	316	743	14	11	..	234
Gabon	1980	..	..	12 705	519	1 375	20	8	20	65
Gabon	1990	..	..	19 272	476	744	22	8	133	112
	1992	..	..	21 951	478	743	23	11	136	113
	1993	..	..	21 522	496	732	28	12	131	113
Gambia	1980	..	..	..	..	..	..	..	..	6
Gambie	1990	..	..	..	..	..	..	..	..	9
	1992	..	..	..	..	..	..	..	..	9
	1993	..	..	..	..	..	..	..	..	9
Ghana	1980	..	..	126	575	843	56	11	..	653
Ghana	1990	..	..	0	551	692	107	25	..	714
	1992	..	..	0	570	711	103	26	..	756
	1993	..	..	0	582	718	77	26	..	756
Guinea	1980	..	..	..	..	..	..	..	..	47
Guinee	1990	..	..	..	..	..	..	..	..	64
	1992	..	..	..	..	..	..	..	..	65
	1993	..	..	..	..	..	..	..	..	66
Guinea–Bissau	1980	..	..	..	..	..	..	..	..	2
Guinée–Bissau	1990	..	..	..	..	..	..	..	..	5
	1992	..	..	..	..	..	..	..	..	5
	1993	..	..	..	..	..	..	..	..	5
Kenya	1980	..	..	..	1 199	2 623	75	37	..	183
Kenya	1990	..	..	..	1 228	1 765	51	44	..	745
	1992	..	..	..	1 193	1 791	215	44	..	695
	1993	..	..	..	1 121	1 707	169	42	..	717
Liberia	1980	..	..	..	201	684	7	2	..	109
Libéria	1990	..	..	..	0	0	0	0	..	69
	1992	..	..	..	0	0	0	0	..	57
	1993	..	..	..	0	0	0	0	..	59
Libyan Arab Jamahiriya	1980	..	..	128 224	2 492	5 279	0	73	4 585	594
Jamah. arabe libyenne	1990	..	..	97 842	6 777	12 742	128	249	8 255	2 064
	1992	..	..	100 742	7 058	12 756	143	280	9 014	2 082
	1993	..	..	95 745	6 815	12 738	128	311	8 468	2 088
Madagascar	1980	..	0	..	206	525	1	8	..	54
Madagascar	1990	..	0	..	68	154	10	2	..	73
	1992	..	0	..	89	164	12	2	..	73
	1993	..	0	..	94	170	12	2	..	74
Malawi	1980	..	..	..	..	..	..	..	..	51
Malawi	1990	..	..	..	..	..	..	..	..	88
	1992	..	..	..	..	..	..	..	..	97
	1993	..	..	..	..	..	..	..	..	98
Mali	1980	..	..	..	..	..	..	..	..	13
Mali	1990	..	..	..	..	..	..	..	..	37
	1992	..	..	..	..	..	..	..	..	40
	1993	..	..	..	..	..	..	..	..	41

67
Production of selected energy commodities
Thousand metric tons of coal equivalent [*cont.*]
Production des principaux biens de l'énergie
Milliers de tonnes métriques d'équivalent houille [*suite*]

Country or area Pays ou zone	Year Anneé	Hard coal, lignite & peat Houille, lignite & tourbe	Briquettes & cokes Agglomérés & cokes	Crude petroleum & NGL Pétrole brut & GNL	Light petroleum products Produits pétroliers légers	Heavy petroleum products Produits pétroliers lourds	Other petroleum products Autres produits pétroliers	LPG & refinery gas GLP et gaz de raffinerie	Natural gas Gaz naturel	Electricity Electricité
Mauritania Mauritanie	1980	..	..	..	0	0	0	...	..	11
	1990	..	..	..	381	638	124	44	..	17
	1992	..	..	..	423	719	140	59	..	18
	1993	..	..	..	432	716	139	56	..	18
Mauritius Maurice	1980	..	..	..	..	..	..	..	..	57
	1990	..	..	..	..	..	..	..	..	93
	1992	..	..	..	..	..	..	..	..	114
	1993	..	..	..	..	..	..	..	..	121
Morocco Maroc	1980	680	0	20	1 275	4 231	73	305	90	605
	1990	526	0	21	1 809	5 487	265	362	76	1 218
	1992	576	0	16	1 791	6 050	321	364	32	1 268
	1993	604	0	14	1 667	6 264	272	384	32	1 218
Mozambique Mozambique	1980	207	..	..	269	692	7	39	..	1 720
	1990	40	..	..	...	...	...	...	..	60
	1992	40	..	..	...	...	...	...	..	60
	1993	40	..	..	...	...	...	...	..	60
Niger Niger	1980	20	..	..	..	..	..	..	..	16
	1990	172	..	..	..	..	..	..	..	20
	1992	170	..	..	..	..	..	..	..	21
	1993	172	..	..	..	..	..	..	..	21
Nigeria Nigéria	1980	176	..	148 818	3 518	4 971	...	95	1 764	882
	1990	90	..	123 315	8 535	8 484	...	85	5 339	1 449
	1992	95	..	130 807	8 260	8 252	...	93	6 524	1 449
	1993	95	..	134 263	8 335	8 266	...	93	6 524	1 449
Réunion Réunion	1980	..	..	..	..	..	..	..	..	42
	1990	..	..	..	..	..	..	..	..	111
	1992	..	..	..	..	..	..	..	..	134
	1993	..	..	..	..	..	..	..	..	139
Rwanda Rwanda	1980	..	..	..	..	..	..	..	1	14
	1990	..	..	..	..	..	..	..	0	30
	1992	..	..	..	..	..	..	..	0	31
	1993	..	..	..	..	..	..	..	0	29
St.Helena and dep. Ste−Hélène et dép.	1980	0	..	..	..	..	..	..	..	0
	1990	0	..	..	..	..	..	..	..	1
	1992	0	..	..	..	..	..	..	..	1
	1993	0	..	..	..	..	..	..	..	1
Sao Tome and Principe Sao Tomé−et−Principe	1980	..	..	..	..	..	..	..	..	1
	1990	..	..	..	..	..	..	..	..	2
	1992	..	..	..	..	..	..	..	..	2
	1993	..	..	..	..	..	..	..	..	2
Senegal Sénégal	1980	..	..	..	391	651	24	12	..	78
	1990	..	..	..	458	722	16	3	..	90
	1992	..	..	..	473	741	16	5	..	94
	1993	..	..	..	485	748	16	5	..	94
Seychelles Seychelles	1980	..	..	..	..	..	..	..	..	6
	1990	..	..	..	..	..	..	..	..	12
	1992	..	..	..	..	..	..	..	..	13
	1993	..	..	..	..	..	..	..	..	14
Sierra Leone Sierra Leone	1980	..	..	..	73	151	16	..	..	25
	1990	..	..	..	73	152	30	..	..	28
	1992	..	..	..	81	166	33	..	..	28
	1993	..	..	..	88	168	36	..	..	29
Somalia Somalie	1980	..	..	..	113	78	11	0	..	15
	1990	..	..	..	...	...	...	...	..	32
	1992	..	..	..	...	...	...	...	..	32
	1993	..	..	..	...	...	...	...	..	32
S.Africa Customs Un. Un.douan.d'Afr.merid	1980	93 379	1 578	..	6 729	11 626	1 727	109	..	11 104
	1990	133 290	1 843	..	11 543	9 650	1 159	208	..	22 560
	1992	132 950	1 848	..	12 338	10 170	1 052	213	..	23 090
	1993	138 673	1 848	..	12 575	10 794	1 052	227	..	23 398

67
Production of selected energy commodities
Thousand metric tons of coal equivalent [cont.]
Production des principaux biens de l'énergie
Milliers de tonnes métriques d'équivalent houille [suite]

Country or area Pays ou zone	Year Anneé	Hard coal, lignite & peat Houille, lignite & tourbe	Briquettes & cokes Agglomérés & cokes	Crude petroleum & NGL Pétrole brut & GNL	Light petroleum products Produits pétroliers légers	Heavy petroleum products Produits pétroliers lourds	Other petroleum products Autres produits pétroliers	LPG & refinery gas GLP et gaz de raffinerie	Natural gas Gaz naturel	Electricity Electricité
Sudan	1980	..	..	..	318	1 100	54	8	..	108
Soudan	1990	..	..	..	310	910	138	11	..	163
	1992	..	..	..	341	917	148	9	..	163
	1993	..	..	..	351	919	148	9	..	163
Togo	1980	0	..	..	208	246	..	..	..	2
Togo	1990	0	..	..	...	...	..	..	..	7
	1992	0	..	..	...	...	..,	..	..	11
	1993	0	..	..	...	...	..	..	..	11
Tunisia	1980	..	..	8 037	607	1 592	..	44	626	344
Tunisie	1990	..	..	6 483	708	1 606	..	239	423	680
	1992	..	..	7 223	818	1 306	..	247	480	759
	1993	..	..	6 696	922	1 353	..	214	436	788
Uganda	1980	..	..	..	..	..	..	..	..	80
Ouganda	1990	..	..	..	..	..	..	..	..	95
	1992	..	..	..	..	..	..	..	..	97
	1993	..	..	..	..	..	..	..	..	97
United Rep. Tanzania	1980	1	..	..	139	572	3	6	..	94
Rép.−Unie de Tanzanie	1990	4	..	..	244	539	3	8	..	109
	1992	4	..	..	255	554	3	9	..	111
	1993	4	..	..	265	560	3	9	..	111
Western Sahara	1980	..	..	..	..	..	..	..	..	9
Sahara occidental	1990	..	..	..	..	..	..	..	..	10
	1992	..	..	..	..	..	..	..	..	10
	1993	..	..	..	..	..	..	..	..	10
Zaire	1980	138	..	1 461	190	308	...	2	..	546
Zaïre	1990	78	..	2 014	135	347	13	2	..	756
	1992	85	..	1 964	144	354	20	2	..	759
	1993	92	..	1 828	152	342	20	2	..	760
Zambia	1980	488	45	..	366	762	10	3	..	1 129
Zambie	1990	318	32	..	270	425	35	16	..	955
	1992	333	32	..	280	438	33	16	..	956
	1993	337	31	..	282	436	29	16	..	956
Zimbabwe	1980	3 134	204	..	..	..	..	..	..	558
Zimbabwe	1990	4 958	509	..	..	..	..	..	..	1 174
	1992	5 547	520	..	..	..	..	..	..	1 058
	1993	5 266	469	..	..	..	..	..	..	939
America, North	1980	659 448	48 952	972 674	667 192	520 860	105 627	70 710	792 627	428 283
Amérique du Nord	1990	830 208	29 244	963 881	692 874	430 672	132 576	91 207	840 951	637 225
	1992	800 716	25 401	973 143	695 018	425 121	129 087	95 872	863 795	665 333
	1993	748 900	25 268	956 366	714 671	433 697	152 694	94 126	902 621	681 267
Antigua and Barbuda	1980	..	..	..	0	0	0	0	..	7
Antigua−et−Barbuda	1990	..	..	..	0	0	0	0	..	12
	1992	..	..	..	0	0	0	0	..	12
	1993	..	..	..	0	0	0	0	..	12
Aruba	1980	..	..	..	..	..	..	..	..	...
Aruba	1990	..	..	..	..	..	..	..	..	42
	1992	..	..	..	..	..	..	..	..	43
	1993	..	..	..	..	..	..	..	..	43
Bahamas	1980	..	..	..	1 662	10 137	..	39	..	105
Bahamas	1990	..	..	..	0	0	..	0	..	117
	1992	..	..	..	0	0	..	0	..	120
	1993	..	..	..	0	0	..	0	..	120
Barbados	1980	..	..	60	70	194	9	...	18	41
Barbade	1990	..	..	89	93	292	10	2	39	57
	1992	..	..	93	94	236	3	3	29	66
	1993	..	..	90	96	269	7	3	35	67
Belize	1980	..	..	..	..	..	..	..	..	7
Belize	1990	..	..	..	..	..	..	..	..	13
	1992	..	..	..	..	..	..	..	..	14
	1993	..	..	..	..	..	..	..	..	14

67
Production of selected energy commodities
Thousand metric tons of coal equivalent [cont.]
Production des principaux biens de l'énergie
Milliers de tonnes métriques d'équivalent houille [suite]

Country or area Pays ou zone	Year Anneé	Hard coal, lignite & peat Houille, lignite & tourbe	Briquettes & cokes Agglomérés & cokes	Crude petroleum & NGL Pétrole brut & GNL	Light petroleum products Produits pétroliers légers	Heavy petroleum products Produits pétroliers lourds	Other petroleum products Autres produits pétroliers	LPG & refinery gas GLP et gaz de raffinerie	Natural gas Gaz naturel	Electricity Electricité
Bermuda Bermudes	1980	..	..	..	..	..		..	..	41
	1990	..	..	..	..	..		..	..	60
	1992	..	..	..	..	..		..	..	64
	1993	..	..	..	..	..		..	..	64
British Virgin Islands Iles Vierges brittaniques	1980	..	..	..	..	..		..	..	3
	1990	..	..	..	..	..		..	..	6
	1992	..	..	..	..	..		..	..	6
	1993	..	..	..	..	..		..	..	6
Canada Canada	1980	30 403	5 164	117 602	60 023	60 838	7 395	7 812	95 213	55 946
	1990	52 777	3 465	129 938	54 898	45 022	13 155	8 656	140 370	77 334
	1992	49 183	3 468	135 217	52 929	41 024	12 303	8 932	162 645	84 017
	1993	53 598	3 418	142 027	53 661	42 661	13 356	8 518	179 593	88 352
Cayman Islands Iles Caïmanes	1980	..	..	..	..	..		..	..	8
	1990	..	..	..	..	..		..	..	28
	1992	..	..	..	..	..		..	..	29
	1993	..	..	..	..	..		..	..	29
Costa Rica Costa Rica	1980	..	..	..	187	519	20	12	..	270
	1990	..	..	..	173	479	20	6	..	435
	1992	..	..	..	184	553	1	5	..	509
	1993	..	..	..	155	577	14	5	..	539
Cuba Cuba	1980	..	..	391	2 573	5 877	497	262	22	1 227
	1990	..	20	1 037	3 015	6 458	614	281	48	1 803
	1992	..	13	1 337	2 251	4 291	583	244	49	1 367
	1993	..	13	1 393	2 176	4 294	533	239	51	1 358
Dominica Dominique	1980	..	..	..	..	..		..	..	1
	1990	..	..	..	..	..		..	..	4
	1992	..	..	..	..	..		..	..	4
	1993	..	..	..	..	..		..	..	4
Dominican Republic Rép. dominicaine	1980	..	..	..	597	1 367	..	79	..	407
	1990	..	..	..	758	1 443		37	..	454
	1992	..	..	..	909	1 883		39	..	686
	1993	..	..	..	812	1 900		47	..	722
El Salvador El Salvador	1980	..	..	..	257	552	23	40	..	592
	1990	..	..	..	233	627	19	50	..	744
	1992	..	..	..	294	772	26	54	..	733
	1993	..	..	..	369	847	21	59	..	792
Greenland Groenland	1980	0	..	..	..	..	..	..	..	20
	1990	0	..	..	..	..	..	..	..	27
	1992	0	..	..	..	..	..	..	..	27
	1993	...	..	..	..	..	..	..	..	31
Grenada Grenade	1980	..	..	..	..	..	..	..	..	3
	1990	..	..	..	..	..	..	..	..	7
	1992	..	..	..	..	..	..	..	..	8
	1993	..	..	..	..	..	..	..	..	8
Guadeloupe Guadeloupe	1980	..	..	..	..	..	..	..	..	38
	1990	..	..	..	..	..	..	..	..	92
	1992	..	..	..	..	..	..	..	..	111
	1993	..	..	..	..	..	..	..	..	118
Guatemala Guatemala	1980	..	..	290	265	794	..	3	...	205
	1990	..	..	307	218	570	..	8	12	286
	1992	..	..	437	236	804	..	12	12	347
	1993	..	..	503	236	747	..	16	12	379
Haiti Haïti	1980	..	..	..	..	..	..	..	..	39
	1990	..	..	..	..	..	..	..	..	58
	1992	..	..	..	..	..	..	..	..	53
	1993	..	..	..	..	..	..	..	..	48
Honduras Honduras	1980	..	..	..	210	493	..	5	..	114
	1990	..	..	..	157	415	..	9	..	281
	1992	..	..	..	145	359	..	6	..	284
	1993	..	..	..	...	...	..	..	..	303

67
Production of selected energy commodities
Thousand metric tons of coal equivalent [*cont.*]
Production des principaux biens de l'énergie
Milliers de tonnes métriques d'équivalent houille [*suite*]

Country or area Pays ou zone	Year Anneé	Hard coal, lignite & peat Houille, lignite & tourbe	Briquettes & cokes Agglomérés & cokes	Crude petroleum & NGL Pétrole brut & GNL	Light petroleum products Produits pétroliers légers	Heavy petroleum products Produits pétroliers lourds	Other petroleum products Autres produits pétroliers	LPG & refinery gas GLP et gaz de raffinerie	Natural gas Gaz naturel	Electricity Electricité
Jamaica	1980	..	..	..	451	837	29	54	..	218
Jamaïque	1990	..	..	..	359	1 114	36	43	..	302
	1992	..	..	..	451	1 245	29	44	..	270
	1993	..	..	..	319	712	26	28	..	282
Martinique	1980	..	..	..	379	228	..	17	..	34
Martinique	1990	..	..	..	363	598	..	23	..	84
	1992	..	..	..	369	620	..	26	..	97
	1993	..	..	..	375	628	..	28	..	97
Mexico	1980	2 594	2 768	152 996	25 978	41 778	3 866	1 956	35 339	9 234
Mexique	1990	5 564	2 276	213 518	31 294	52 895	2 970	3 768	32 590	21 426
	1992	4 801	2 014	223 146	31 780	53 867	3 065	3 805	31 910	23 361
	1993	5 450	2 126	223 429	32 370	55 412	3 087	3 962	33 336	25 025
Montserrat	1980	..	..	..	..	..	..	..	..	1
Montserrat	1990	..	..	..	..	..	..	..	..	2
	1992	..	..	..	..	..	..	..	..	2
	1993	..	..	..	..	..	..	..	..	2
Netherlands Antilles	1980	..	..	..	..	..	..	..	..	...
Antilles néerlandaises	1990	..	..	..	3 406	9 806	3 108	70	..	97
	1992	..	..	..	3 784	9 700	3 182	78	..	105
	1993	..	..	..	3 920	10 154	3 707	90	..	108
Nicaragua	1980	..	..	..	254	473	29	55	..	131
Nicaragua	1990	..	..	..	167	666	19	39	..	598
	1992	..	..	..	208	661	33	52	..	711
	1993	..	..	..	212	614	36	45	..	775
Panama	1980	..	..	..	559	2 087	26	34	..	223
Panama	1990	..	..	..	360	1 237	10	58	..	340
	1992	..	..	..	415	2 065	19	70	..	370
	1993	..	..	..	402	1 958	23	69	..	404
Puerto Rico	1980	..	..	..	5 580	8 011	1 870	587	..	1 622
Porto Rico	1990	..	..	..	4 904	4 468	3 068	146	..	1 883
	1992	..	..	..	4 349	3 524	2 279	109	..	2 019
	1993	..	..	..	4 411	3 539	2 315	109	..	2 032
Saint Kitts and Nevis	1980	..	..	..	..	..	..	..	..	...
Saint−Kitts−et−Nevis	1990	..	..	..	..	..	..	..	..	5
	1992	..	..	..	..	..	..	..	..	5
	1993	..	..	..	..	..	..	..	..	5
Saint Lucia	1980	..	..	..	..	..	..	..	..	7
Sainte−Lucie	1990	..	..	..	..	..	..	..	..	13
	1992	..	..	..	..	..	..	..	..	13
	1993	..	..	..	..	..	..	..	..	13
St.Pierre and Miquelon	1980	..	..	..	..	..	..	..	..	2
St.−Pierre−et−Miquelon	1990	..	..	..	..	..	..	..	..	6
	1992	..	..	..	..	..	..	..	..	6
	1993	..	..	..	..	..	..	..	..	6
St. Vincent and Grenadines	1980	..	..	..	..	..	..	..	..	3
St.−Vincent−et−Grenad.	1990	..	..	..	..	..	..	..	..	6
	1992	..	..	..	..	..	..	..	..	6
	1993	..	..	..	..	..	..	..	..	6
Trinidad and Tobago	1980	..	..	15 701	4 149	11 910	458	714	4 996	253
Trinité−et−Tobago	1990	..	..	11 246	1 785	4 519	47	278	6 881	440
	1992	..	..	10 027	1 704	6 083	50	370	7 906	488
	1993	..	..	9 151	1 881	5 482	47	283	6 901	469
Turks and Caicos Islands	1980	..	..	..	..	..	..	..	..	1
Iles Turques et Caïques	1990	..	..	..	..	..	..	..	..	1
	1992	..	..	..	..	..	..	..	..	1
	1993	..	..	..	..	..	..	..	..	1
United States	1980	626 451	41 019	685 633	550 215	319 051	86 300	58 769	657 039	357 243
Etats−Unis	1990	771 867	23 483	607 746	584 887	288 219	102 456	77 344	661 012	530 045
	1992	746 731	19 905	602 886	589 583	288 109	101 517	81 729	661 243	549 258
	1993	689 852	19 711	579 773	607 840	294 462	120 863	80 314	682 693	558 907

67
Production of selected energy commodities
Thousand metric tons of coal equivalent [*cont.*]
Production des principaux biens de l'énergie
Milliers de tonnes métriques d'équivalent houille [*suite*]

Country or area Pays ou zone	Year Anneé	Hard coal, lignite & peat Houille, lignite & tourbe	Briquettes & cokes Agglomérés & cokes	Crude petroleum & NGL Pétrole brut & GNL	Light petroleum products Produits pétroliers légers	Heavy petroleum products Produits pétroliers lourds	Other petroleum products Autres produits pétroliers	LPG & refinery gas GLP et gaz de raffinerie	Natural gas Gaz naturel	Electricity Electricité
U.S. Virgin Islands	1980	..	..	..	6 837	25 429	2 849	70	..	98
Iles Vierges américaines	1990	..	..	..	5 807	11 843	7 047	388	..	120
	1992	..	..	..	5 334	9 325	5 997	295	..	125
	1993	..	..	..	5 435	9 441	6 147	311	..	128
America, South	**1980**	**9 210**	**4 857**	**260 467**	**61 420**	**142 096**	**9 458**	**9 537**	**43 759**	**33 576**
Amérique du Sud	**1990**	**27 000**	**8 310**	**315 789**	**80 259**	**130 666**	**11 979**	**13 514**	**76 792**	**57 167**
	1992	**28 024**	**8 785**	**340 230**	**81 205**	**131 843**	**11 646**	**15 250**	**80 692**	**61 977**
	1993	**28 805**	**9 206**	**357 192**	**87 123**	**130 741**	**11 319**	**15 888**	**84 457**	**65 415**
Argentina	1980	329	468	36 827	10 243	22 568	2 202	1 342	11 155	5 455
Argentine	1990	228	857	37 127	10 966	17 933	2 169	1 739	27 064	8 064
	1992	181	630	42 520	11 475	18 403	2 817	1 997	30 680	8 673
	1993	141	652	44 731	11 598	18 152	2 878	2 056	31 556	9 670
Bolivia	1980	..	..	1 777	956	592	126	87	2 880	192
Bolivie	1990	..	..	1 760	762	491	35	57	3 435	262
	1992	..	..	1 872	699	521	32	71	3 681	296
	1993	..	..	1 779	721	503	25	72	3 651	300
Brazil	1980	3 655	3 838	13 260	21 174	48 015	3 072	5 342	1 306	17 134
Brésil	1990	2 917	6 871	46 145	25 819	47 228	5 293	7 823	5 139	27 926
	1992	3 003	7 344	46 312	25 266	47 532	5 114	8 546	5 413	30 131
	1993	2 917	7 653	47 159	27 829	45 989	4 265	9 137	5 907	31 001
Chile	1980	1 156	274	2 919	2 151	4 063	9	537	924	1 443
Chili	1990	2 578	320	1 858	2 922	5 125	134	492	2 364	2 257
	1992	1 938	445	1 509	3 175	5 172	161	622	2 381	2 747
	1993	1 610	467	1 496	3 569	5 430	192	663	2 291	2 949
Colombia	1980	3 982	249	9 594	4 512	5 355	1 015	588	4 792	2 533
Colombie	1990	19 006	261	31 879	6 359	8 330	315	1 129	5 231	4 348
	1992	20 336	366	31 973	6 939	8 610	338	1 569	5 085	4 421
	1993	20 162	433	33 145	8 904	9 801	414	1 571	5 093	4 950
Ecuador	1980	..	..	14 894	2 098	4 338	152	113	51	412
Equateur	1990	..	..	21 508	2 685	5 760	233	132	165	777
	1992	..	..	24 144	2 667	5 954	208	236	165	880
	1993	..	..	25 874	2 516	5 938	207	247	165	915
Falkland Is. (Malvinas)	1980	4	..	..	..	..	..	..	..	0
Iles Falkland (Malvinas)	1990	5	..	..	..	..	..	..	..	1
	1992	5	..	..	..	..	..	..	..	1
	1993	5	..	..	..	..	..	..	..	1
French Guiana	1980	..	..	..	..	..	..	..	..	1
Guyane francaise	1990	..	..	..	..	..	..	..	..	14
	1992	..	..	..	..	..	..	..	..	43
	1993	..	..	..	..	..	..	..	..	55
Guyana	1980	..	..	..	..	..	..	..	..	55
Guyana	1990	..	..	..	..	..	..	..	..	51
	1992	..	..	..	..	..	..	..	..	38
	1993	..	..	..	..	..	..	..	..	29
Paraguay	1980	..	..	..	136	239	3	8	..	29
Paraguay	1990	..	..	..	176	269	0	17	..	89
	1992	..	..	..	124	328	0	3	..	3 344
	1993	..	..	..	154	257	0	3	..	3 334
Peru	1980	41	23	13 944	4 118	5 946	105	264	713	3 864
Pérou	1990	79	0	9 899	3 319	7 024	70	308	768	1 233
	1992	83	0	8 919	3 099	7 117	70	310	620	1 697
	1993	81	0	9 004	3 178	7 579	82	312	618	1 615
Suriname	1980	..	..	0	..	..	..	..	..	1 760
Suriname	1990	..	..	311	..	..	..	..	..	194
	1992	..	..	384	..	..	..	..	..	185
	1993	..	..	387	..	..	..	..	..	170
Uruguay	1980	..	5	..	570	1 882	101	102	..	171
Uruguay	1990	..	0	..	430	1 085	80	131	..	412
	1992	..	0	..	431	1 046	108	109	..	914
	1993	..	0	..	115	351	12	50	..	1 093
										981

67
Production of selected energy commodities
Thousand metric tons of coal equivalent [*cont.*]
Production des principaux biens de l'énergie
Milliers de tonnes métriques d'équivalent houille [*suite*]

Country or area Pays ou zone	Year Anneé	Hard coal, lignite & peat Houille, lignite & tourbe	Briquettes & cokes Agglomérés & cokes	Crude petroleum & NGL Pétrole brut & GNL	Light petroleum products Produits pétroliers légers	Heavy petroleum products Produits pétroliers lourds	Other petroleum products Autres produits pétroliers	LPG & refinery gas GLP et gaz de raffinerie	Natural gas Gaz naturel	Electricity Electricité
Venezuela	1980	44	0	167 252	15 463	49 098	2 674	1 153	21 938	4 414
Venezuela	1990	2 188	0	165 301	26 820	37 420	3 649	1 685	32 626	7 310
	1992	2 479	0	182 596	27 328	37 159	2 797	1 786	32 668	8 532
	1993	3 889	0	193 617	28 536	36 739	3 246	1 777	35 174	8 769
Asia	1980	619 155	110 670	1 653 384	238 279	508 030	37 766	22 589	115 251	192 569
Asie	1990	1 083 702	152 842	1 681 705	337 485	644 269	54 209	46 340	312 250	400 181
	1992	1 262 193	156 119	1 828 612	389 036	734 699	59 188	63 094	461 716	472 765
	1993	1 293 086	166 075	1 874 454	410 483	766 920	61 161	65 689	491 746	503 155
Afghanistan	1980	119	38	12	..	..	..	..	3 491	119
Afghanistan	1990	105	37	6	..	..	..	..	279	139
	1992	8	0	0	..	..	..	..	252	86
	1993	7	0	0	..	..	..	..	243	85
Armenia	1980	..	..	..	..	..	..	..	..	..
Arménie	1990	..	..	..	..	..	..	..	..	..
	1992	..	..	..	..	..	..	..	6	1 106
	1993	..	..	..	..	..	..	..	5	774
Azerbaijan	1980	..	..	..	..	..	..	..	..	..
Azerbaidjan	1990	..	..	..	..	..	..	..	..	..
	1992	..	..	17 752	1 376	..	681	4 943	9 118	2 428
	1993	..	..	16 166	1 505	..	374	4 391	7 883	2 340
Bahrain	1980	..	..	3 793	6 161	10 604	474	78	3 880	204
Bahreïn	1990	..	..	3 497	7 082	10 527	562	39	7 251	429
	1992	..	..	3 650	7 616	11 279	660	39	6 893	431
	1993	..	..	3 612	6 713	9 935	481	36	8 677	532
Bangladesh	1980	..	..	9	667	982	45	5	1 537	326
Bangladesh	1990	..	..	180	595	577	46	14	5 255	990
	1992	..	..	182	633	650	33	12	6 664	1 174
	1993	..	..	185	688	752	36	12	7 131	1 190
Bhutan	1980	...	..	..	..	..	..	..	..	2
Bhoutan	1990	2	..	..	..	..	..	..	..	192
	1992	2	..	..	..	..	..	..	..	200
	1993	2	..	..	..	..	..	..	..	200
Brunei Darussalam	1980	..	..	17 769	42	88	..	31	12 010	58
Brunéi Darussalam	1990	..	..	10 181	326	194	..	3	13 932	152
	1992	..	..	12 057	356	256	..	3	13 343	154
	1993	..	..	11 924	377	264	..	5	13 188	158
Cambodia	1980	..	..	..	0	0	..	..	..	12
Cambodge	1990	..	..	..	0	0	..	..	..	20
	1992	..	..	..	0	0	..	..	..	21
	1993	..	..	..	0	0	..	..	..	22
China	1980	434 959	33 026	151 326	21 591	70 997	6 845	...	18 948	36 927
Chine	1990	770 574	71 085	197 547	38 502	84 104	8 048	8 080	20 324	76 305
	1992	796 609	77 444	202 962	46 838	91 746	8 930	10 104	20 975	92 796
	1993	820 425	90 382	207 447	53 035	97 107	9 459	10 629	22 545	103 736
Cyprus	1980	..	..	..	234	549	23	48	..	127
Chypre	1990	..	..	..	229	580	54	65	..	243
	1992	..	..	..	208	739	34	68	..	295
	1993	..	..	..	206	810	48	66	..	317
Georgia	1980	..	..	...	..	..	..	..	...	...
Géorgie	1990	..	..	...	..	..	..	..	...	...
	1992	..	..	143	..	..	..	..	0	1 409
	1993	..	..	143	..	..	..	..	0	1 191
Hong Kong	1980	..	..	..	..	..	..	..	..	1 554
Hong-kong	1990	..	..	..	..	..	..	..	..	3 557
	1992	..	..	..	..	..	..	..	..	4 309
	1993	..	..	..	..	..	..	..	..	4 470
India	1980	92 081	10 298	13 425	10 396	20 447	2 850	572	1 726	15 396
Inde	1990	172 186	8 893	48 827	23 238	38 008	7 382	3 225	13 263	37 080
	1992	199 137	9 610	41 578	22 330	40 501	7 727	3 885	15 551	42 541
	1993	209 524	9 700	39 521	22 958	41 252	7 052	4 166	15 696	45 484

67
Production of selected energy commodities
Thousand metric tons of coal equivalent [*cont.*]
Production des principaux biens de l'énergie
Milliers de tonnes métriques d'équivalent houille [*suite*]

Country or area Pays ou zone	Year Anneé	Hard coal, lignite & peat Houille, lignite & tourbe	Briquettes & cokes Agglomérés & cokes	Crude petroleum & NGL Pétrole brut & GNL	Light petroleum products Produits pétroliers légers	Heavy petroleum products Produits pétroliers lourds	Other petroleum products Autres produits pétroliers	LPG & refinery gas GLP et gaz de raffinerie	Natural gas Gaz naturel	Electricity Electricité
Indonesia	1980	304	0	110 864	12 389	22 814	885	1 149	20 718	1 748
Indonésie	1990	7 330	0	132 705	18 869	29 691	1 907	4 970	80 026	7 248
	1992	21 147	0	142 890	19 605	33 243	2 050	11 811	67 871	7 880
	1993	27 584	0	143 039	19 555	33 994	2 086	12 019	70 355	8 433
Iran, Islamic Republic of	1980	900	360	104 949	14 455	32 488	1 778	870	9 506	2 749
Iran, Rép. islamique d'	1990	1 300	360	230 077	14 485	34 802	4 003	1 399	32 173	7 260
	1992	1 500	414	250 092	17 205	38 677	4 433	1 709	33 285	8 429
	1993	1 460	427	249 407	18 587	40 839	4 719	2 020	36 041	8 842
Iraq	1980	..	..	186 445	4 509	7 584	515	0	1 707	1 404
Iraq	1990	..	..	145 919	6 142	17 590	1 004	326	3 755	3 582
	1992	..	..	36 911	7 717	22 102	962	622	3 022	3 108
	1993	..	..	46 882	8 156	22 887	991	932	3 395	3 231
Israel	1980	..	..	29	2 958	5 408	274	278	191	1 539
Israël	1990	..	..	19	3 632	7 333	265	353	38	2 567
	1992	..	..	13	4 558	8 682	379	451	28	3 032
	1993	..	..	11	5 842	9 489	611	413	28	3 194
Japan	1980	16 059	46 655	777	96 503	168 330	12 937	14 837	2 905	92 681
Japon	1990	7 268	42 925	790	91 649	129 133	15 072	17 392	2 861	157 523
	1992	6 684	40 281	1 192	107 039	143 917	14 989	18 433	3 022	167 459
	1993	6 349	39 657	1 082	112 111	146 176	15 505	18 607	3 084	175 318
Jordan	1980	..	..	...	973	1 385	144	64	..	131
Jordanie	1990	..	..	23	1 293	2 173	188	232	..	453
	1992	..	..	4	1 379	2 368	207	263	..	543
	1993	..	..	0	1 350	2 638	230	271	..	585
Kazakhstan	1980	..	..	..	..	..	..	..	..	..
Kazakhstan	1990	..	..	..	..	..	..	..	..	..
	1992	107 866	2 849	38 240	7 985	14 870			9 398	10 159
	1993	95 861	2 245	32 799	5 889	13 454			7 744	9 513
Korea, Dem. People's Rep.	1980	42 000	2 700	..	1 077	1 295			..	4 299
Corée, Rép. pop. dém. de	1990	81 200	3 240	..	1 842	2 458			..	6 572
	1992	85 400	3 240	..	1 809	2 408			..	4 668
	1993	88 200	3 240	..	1 787	2 393			..	4 668
Korea, Republic of	1980	12 239	14 521	..	7 495	26 991	741	510	..	5 788
Corée, Rép. de	1990	11 066	19 373	..	16 114	40 982	2 068	1 601	..	27 736
	1992	7 693	15 464	..	26 563	68 967	3 214	1 956	..	32 217
	1993	6 069	13 711	..	28 536	74 139	3 244	1 865	..	34 534
Kuwait, part Neutral Zone	1980	..	..	125 038	7 083	16 727	565	0	9 674	1 157
Koweït et prt. Zone Neut.	1990	..	..	89 699	11 780	27 577	461	464	6 960	2 320
	1992	..	..	78 492	5 745	17 339	46	43	3 488	2 099
	1993	..	..	141 757	7 907	24 276	447	357	5 951	2 236
Kyrgyzstan	1980	..	..	..	..	..	..	..	..	..
Kirghizistan	1990	..	..	..	..	..	..	..	..	..
	1992	1 297	..	162	..	..	..	..	84	1 461
	1993	995	..	126	..	..	..	..	48	1 362
Lao People's Dem. Rep.	1980	0	..	..	..	..	..	..	..	120
Rép. dém. populaire Lao	1990	1	..	..	..	..	..	..	..	107
	1992	1	..	..	..	..	..	..	..	112
	1993	1	..	..	..	..	..	..	..	111
Lebanon	1980	..	...	..	993	1 661	..	93	..	338
Liban	1990	..	...	..	132	468	..	6	..	184
	1992	..	...	..	100	354	..	3	..	479
	1993	..	0	..	112	350	..	5	..	485
Macau	1980	..	..	..	..	..	..	..	..	32
Macao	1990	..	..	..	..	..	..	..	..	97
	1992	..	..	..	..	..	..	..	..	122
	1993	..	..	..	..	..	..	..	..	146
Malaysia	1980	...	..	19 194	2 089	5 681	303	209	109	1 240
Malaisie	1990	105	..	43 433	3 806	9 558	346	591	17 141	3 103
	1992	190	..	45 251	4 226	9 478	404	625	25 079	3 917
	1993	260	..	44 592	4 596	10 363	424	819	28 491	4 370

67

Production of selected energy commodities
Thousand metric tons of coal equivalent [*cont.*]
Production des principaux biens de l'énergie
Milliers de tonnes métriques d'équivalent houille [*suite*]

Country or area Pays ou zone	Year Anneé	Hard coal, lignite & peat Houille, lignite & tourbe	Briquettes & cokes Agglomérés & cokes	Crude petroleum & NGL Pétrole brut & GNL	Light petroleum products Produits pétroliers légers	Heavy petroleum products Produits pétroliers lourds	Other petroleum products Autres produits pétroliers	LPG & refinery gas GLP et gaz de raffinerie	Natural gas Gaz naturel	Electricity Electricité
Maldives	1980	..	..	..	..	..	..	..	..	0
Maldives	1990	..	..	..	..	..	..	..	..	3
	1992	..	..	..	..	..	..	..	..	4
	1993	..	..	..	..	..	..	..	..	5
Mongolia	1980	1 707	..	...	...	...	...	..	..	192
Mongolie	1990	3 066	..	...	...	...	...	..	..	430
	1992	2 912	..	...	...	...	..	..	..	405
	1993	2 869	..	...	...	...	...	..	..	393
Myanmar	1980	21	..	2 228	559	847	404	5	454	183
Myanmar	1990	46	..	1 180	253	585	160	6	1 295	322
	1992	49	..	1 073	271	593	150	6	1 124	345
	1993	47	..	1 030	304	590	143	6	1 295	372
Nepal	1980	..	..	..	..	..	..	..	..	26
Nepal	1990	..	..	..	..	..	..	..	..	90
	1992	..	..	..	..	..	..	..	..	111
	1993	..	..	..	..	..	..	..	..	115
Oman	1980	..	..	20 040			0	...	1 118	118
Oman	1990	..	..	48 783	1 143	3 273	0	56	3 527	657
	1992	..	..	52 804	977	3 135	0	61	2 449	766
	1993	..	..	55 324	792	2 580	0	48	3 351	866
Pakistan	1980	1 057		688	2 077	3 770	347	57	8 000	1 840
Pakistan	1990	2 118	601	3 884	2 752	4 746	564	64	14 442	5 463
	1992	2 071	663	4 438	3 412	5 738	584	73	15 945	6 490
	1993	2 073	644	4 345	3 516	5 229	586	75	16 911	6 906
Philippines	1980	222	..	764	3 373	9 299	254	296	..	4 504
Philippines	1990	840	..	350	4 334	10 412	84	361	..	9 266
	1992	1 122	..	597	4 644	10 597	153	388	..	8 965
	1993	1 132	..	714	4 956	11 177	163	404	..	9 022
Qatar	1980	..	..	32 755	295	220	..	12	5 991	299
Qatar	1990	..	..	28 616	1 417	2 489	..	145	7 423	595
	1992	..	..	33 041	1 150	2 639	..	78	15 297	637
	1993	..	..	30 548	1 088	2 185	..	115	17 974	683
Saudi Arabia, pt. Neutral Zone	1980	..	..	726 767	9 162	27 962	2 276	448	1 660	2 323
Arabie saoudie,p.Zone nuet.	1990	..	..	483 074	32 425	71 742	1 859	1 383	40 757	5 732
	1992	..	..	628 277	33 854	68 659	2 289	1 725	45 810	6 832
	1993	..	..	606 336	34 355	70 181	2 460	1 896	47 797	7 779
Singapore	1980	..	..	..	17 075	24 087	1 577	250	..	852
Singapour	1990	..	..	..	24 763	37 890	1 927	715	..	1 918
	1992	..	..	..	26 562	41 608	2 092	761	..	2 155
	1993	..	..	..	27 794	44 619	2 171	808	..	2 329
Sri Lanka	1980	..	..	..	746	1 752	70	11	..	205
Sri Lanka	1990	..	..	..	778	1 628	69	100	..	387,
	1992	..	..	..	577	1 242	67	69	..	435
	1993	..	..	..	850	1 655	93	88	..	489
Syrian Arab Republic	1980	..	..	13 138	2 220	6 266	502	82	64	471
Rép. arabe syrienne	1990	..	..	32 888	3 261	12 569	533	238	2 173	1 409
	1992	..	..	37 227	3 021	12 782	510	255	2 623	1 543
	1993	..	..	38 679	2 917	12 801	395	253	2 597	1 552
Tajikistan	1980	...	...	...	...	...	..	..	...	...
Tadjikistan	1990	...	...	...	...	...	..	..	...	...
	1992	167	10	89	84	..	..	..	83	2 066
	1993	167	4	60	59		..	..	57	2 179
Thailand	1980	573	..	11	3 569	6 908	170	194	0	1 856
Thaïlande	1990	4 631	..	4 029	6 739	9 229	282	328	7 714	5 672
	1992	5 740	..	5 132	6 365	13 327	398	502	10 269	7 333
	1993	5 806	..	5 280	7 259	15 412	432	640	11 578	8 145
Turkey	1980	9 334	2 013	3 328	4 560	11 958	654	629	29	2 859
Turquie	1990	16 819	3 232	5 302	8 068	21 872	1 568	1 678	277	7 157
	1992	15 626	3 265	6 108	7 894	22 161	1 697	1 751	233	8 349
	1993	16 531	3 103	5 559	8 597	23 689	2 833	1 935	235	9 152

67

Production of selected energy commodities
Thousand metric tons of coal equivalent [*cont.*]
Production des principaux biens de l'énergie
Milliers de tonnes métriques d'équivalent houille [*suite*]

Country or area / Pays ou zone	Year / Anneé	Hard coal, lignite & peat / Houille, lignite & tourbe	Briquettes & cokes / Agglomérés & cokes	Crude petroleum & NGL / Pétrole brut & GNL	Light petroleum products / Produits pétroliers légers	Heavy petroleum products / Produits pétroliers lourds	Other petroleum products / Autres produits pétroliers	LPG & refinery gas / GLP et gaz de raffinerie	Natural gas / Gaz naturel	Electricity / Electricité
Turkmenistan Turkmenistan	1980	..	..	..	..	..	..	..	..	..
	1990	..	..	..	..	..	..	..	..	..
	1992			7 490	1 396	5 176	..	..	69 620	1 619
	1993	..	..	7 206	1 090	3 776	..	..	75 698	1 552
United Arab Emirates Emirats arabes unis	1980	..		119 649	295	502	..	23	9 140	773
	1990	..	..	153 341	4 969	7 880	..	373	29 749	2 098
	1992		..	156 310	5 165	8 101	..	404	29 517	2 145
	1993		..	149 383	5 076	8 236		466	30 529	2 159
Uzbekistan Ouzbekistan	1980	..	..	..	..	..	..	..	..	..
	1990	..	..	..	..	..	..	..	..	..
	1992	1 883	..	4 879	2 232	4 884	636	..	49 582	6 254
	1993	1 534	..	5 808	2 418	6 003	...	..	52 167	6 037
Viet Nam Viet Nam	1980	5 300		...	..	..	..	..	..	639
	1990	4 627	..	3 849	16	39	1	..	4	1 071
	1992	4 792	..	7 842	16	39	1	..	3	1 204
	1993	5 899	..	9 016	16	39	1	..	3	1 996
Yemen Yémen	1980	..	..	..	..	..	..	..	..	..
	1990	..	..	..	..	..	..	..	..	..
	1992	..	..	11 631	2 114	5 983	..	78	..	222
	1993	..	..	16 377	2 032	5 883	..	93	..	227
former Yemen Arab Rep. anc. Yémen rép. arabe	1980	..	..	..	..	..	..	..	..	23
	1990	..	..	12 095	165	381	..	..	..	102
	1992	..	..	..	..	..	..	..	..	..
	1993	..	..	..	..	..	..	..	..	..
former Dem. Yemen ancienne Yémen dém	1980	..	..	...	1 110	2 821	..	16	..	39
	1990	..	..	1 143	964	4 216	..	78	..	112
	1992	..	..	..	..	..	..	..	..	..
	1993	..	..	..	..	..	..	..	..	..
Europe **Europe**	1980	641 619	160 588	200 004	281 807	696 092	63 066	49 293	325 052	328 319
	1990	549 493	129 450	306 734	340 347	518 110	77 400	57 272	291 935	549 204
	1992	819 751	139 973	922 991	442 905	813 478	120 025	79 120	1 046 242	786 676
	1993	753 778	120 954	876 911	429 694	784 115	128 533	73 070	1 047 583	775 087
Albania Albanie	1980	709	155	1 981	362	1 348	1 862	..	536	456
	1990	1 035	386	1 524	203	798	785	..	324	515
	1992	183	232	836	149	579	713	..	124	412
	1993	175	200	786	134	457	464	..	119	424
Austria Autriche	1980	1 281	1 681	2 139	2 918	9 744	1 191	726	2 663	5 113
	1990	910	1 660	1 703	4 415	6 462	1 325	597	1 539	6 193
	1992	659	1 414	1 747	4 288	7 284	1 546	512	1 881	6 287
	1993	629	1 363	1 711	4 076	7 603	1 541	457	1 949	6 470
Belarus Bélarus	1980	..	..	..	..	..	..	..	..	..
	1990	..	..	..	..	..	..	..	..	..
	1992	1 359	..	2 857	7 168	19 276	621	407	346	4 618
	1993	950	..	2 864	5 266	13 708	387	333	345	4 099
Belgium Belgique	1980	5 784	5 964	..	13 318	31 283	2 153	1 682	52	9 647
	1990	891	5 421	..	12 756	22 813	5 671	1 510	15	19 326
	1992	564	4 584	..	13 504	25 187	6 926	1 236	8	19 682
	1993	332	4 000	..	12 440	24 869	6 864	1 270	6	19 005
Bosnia Herzegovina Bosnie—Herzégovine	1980	..	..	..	..	..	..	..	..	..
	1990	..	..	..	..	..	..	..	..	..
	1992	3 120	1 350	..	538	1 713	229	47	..	1 434
	1993	...	...	...	...	...	...	...	...	1 351
Bulgaria Bulgarie	1980	7 644	2 165	393	2 846	8 796	1 000	123	209	5 812
	1990	7 804	2 046	86	2 488	6 248	759	597	15	8 823
	1992	7 555	1 603	76	842	1 867	198	241	43	7 247
	1993	7 250	1 805	71	2 854	6 368	792	602	78	8 142
Croatia Croatie	1980	..	..	..	..	..	..	..	..	..
	1990	..	..	..	..	..	..	..	..	..
	1992	120	366	2 985	1 430	3 610	320	368	2 092	1 092
	1993	109	362	3 083	1 965	4 228	496	533	2 377	1 150

67

Production of selected energy commodities
Thousand metric tons of coal equivalent [*cont.*]
Production des principaux biens de l'énergie
Milliers de tonnes métriques d'équivalent houille [*suite*]

Country or area Pays ou zone	Year Anneé	Hard coal, lignite & peat Houille, lignite & tourbe	Briquettes & cokes Agglomérés & cokes	Crude petroleum & NGL Pétrole brut & GNL	Light petroleum products Produits pétroliers légers	Heavy petroleum products Produits pétroliers lourds	Other petroleum products Autres produits pétroliers	LPG & refinery gas GLP et gaz de raffinerie	Natural gas Gaz naturel	Electricity Electricité
Czech Republic	1980	..	..	..	..	..	..	..	..	..
République tchéque	1990	..	..	..	..	..	..	..	..	..
	1992	44 731	5 741	117	1 962	4 696	2 381	126	200	10 329
	1993	43 773	5 365	159	1 784	4 351	2 449	119	269	10 373
former Czechoslovakia †	1980	64 354	10 708	133	3 054	19 315	2 405	729	787	10 059
anciene Tchécoslovaquie †	1990	54 275	9 817	176	2 624	11 332	4 214	507	779	16 763
	1992	..	..	..	..	..	..	..	..	..
	1993	..	..	..	..	..	..	..	..	..
Denmark	1980	0	59	426	1 891	7 040	200	456	...	3 332
Danemark	1990	0	0	8 561	2 834	7 914	39	614	4 176	3 164
	1992	0	0	11 078	2 752	8 900	29	609	5 359	3 789
	1993	0	0	11 805	2 751	9 039	0	631	5 983	4 144
Estonia	1980	..	..	..	..	..	..	..	..	..
Estonie	1990	..	..	..	..	..	..	..	..	..
	1992	5 226	27	..	..	387	..	..	..	1 453
	1993	4 119	29	..	..	252	..	..	..	1 120
Faeroe Islands	1980	...	...	..	..	..	..	...	..	20
Iles Féroe	1990	...	...	..	..	..	..	...	..	26
	1992	...	...	..	..	..	..	..	..	25
	1993	0	0	..	..	..	..	0	..	24
Finland	1980	366	0	..	4 446	12 224	1 151	286	..	6 404
Finlande	1990	1 927	438	..	5 708	7 591	790	1 023	..	11 458
	1992	1 857	448	..	6 308	7 476	560	1 201	..	11 880
	1993	1 996	787	..	6 026	7 164	406	1 170	..	12 460
France incl. Monaco	1980	21 729	11 763	3 002	42 311	108 157	9 344	8 422	9 854	44 591
France y compris Monaco	1990	11 141	8 003	4 864	39 643	58 029	10 939	7 535	2 641	129 601
	1992	9 767	7 559	4 783	38 818	60 139	10 769	7 233	3 052	151 991
	1993	8 982	6 836	4 629	39 299	63 548	10 326	7 439	3 212	149 466
Germany †	1980	..	..	..	..	..	..	..	..	..
Allemagne †	1990	..	..	..	..	..	..	..	..	..
	1992	140 006	26 955	4 684	52 937	86 068	14 772	9 426	21 255	105 468
	1993	125 401	22 295	4 376	56 910	88 894	13 603	10 142	21 352	102 741
Federal Rep. of Germany	1980	126 239	34 671	7 430	47 176	95 820	11 984	10 008	22 780	56 013
Rép. fédérale d'Allemagne	1990	103 487	22 312	5 133	45 204	63 201	12 100	7 755	18 629	92 128
	1992	..	..	..	..	..	..	..	..	..
	1993	..	..	..	..	..	..	..	..	..
former German Dem. Rep.	1980	78 277	38 512	77	5 025	21 475	1 522	751	2 445	15 093
ancienne R. d. allemande	1990	84 240	35 895	56	7 066	14 654	1 746	1 274	1 877	17 342
	1992	..	..	..	..	..	..	..	..	..
	1993	..	..	..	..	..	..	..	..	..
Gibraltar	1980	..	..	..	..	..	..	..	..	7
Gibraltar	1990	..	..	..	..	..	..	..	..	10
	1992	..	..	..	..	..	..	..	..	11
	1993	..	..	..	..	..	..	..	..	11
Greece	1980	4 341	387	0	5 130	14 481	280	536	0	2 782
Grèce	1990	10 164	66	1 192	8 582	13 554	927	1 170	219	4 299
	1992	9 783	32	985	8 179	12 972	865	1 205	197	4 595
	1993	10 741	30	806	7 289	10 983	1 023	1 215	146	4 716
Hungary	1980	9 826	1 800	3 977	2 901	9 781	1 217	325	8 074	2 933
Hongrie	1990	6 297	1 983	3 748	4 261	6 149	955	440	5 445	6 904
	1992	5 567	662	3 116	3 692	6 243	719	409	5 163	7 356
	1993	4 533	443	2 959	3 625	6 595	835	433	5 558	7 458
Iceland	1980	..	..	..	..	..	..	..	..	443
Islande	1990	..	..	..	..	..	..	..	..	885
	1992	..	..	..	..	..	..	..	..	812
	1993	..	..	..	..	..	..	..	..	865
Ireland	1980	1 451	0	..	762	2 053	0	102	1 169	1 298
Irlande	1990	1 606	0	..	615	1 498	0	105	3 006	1 783
	1992	1 635	0	..	603	2 120	6	119	3 014	1 967
	1993	1 621	0	..	599	2 033	0	106	3 423	2 016

67
Production of selected energy commodities
Thousand metric tons of coal equivalent [cont.]
Production des principaux biens de l'énergie
Milliers de tonnes métriques d'équivalent houille [suite]

Country or area Pays ou zone	Year Anneé	Hard coal, lignite & peat Houille, lignite & tourbe	Briquettes & cokes Agglomérés & cokes	Crude petroleum & NGL Pétrole brut & GNL	Light petroleum products Produits pétroliers légers	Heavy petroleum products Produits pétroliers lourds	Other petroleum products Autres produits pétroliers	LPG & refinery gas GLP et gaz de raffinerie	Natural gas Gaz naturel	Electricity Electricité
Italy and San Marino	1980	696	7 448	2 610	35 316	94 768	5 673	6 066	16 446	26 033
Italie y comp. Saint−Marin	1990	397	6 356	6 671	41 712	75 736	7 464	7 406	22 269	30 197
	1992	362	5 413	6 431	41 952	80 737	7 677	6 654	23 387	31 608
	1993	362	4 929	6 630	41 708	78 950	21 882	7 761	24 891	31 413
Latvia	1980	..	..	..	..	..	..	..	..	..
Lettonie	1990	..	..	..	..	..	..	..	..	..
	1992	133	..	..	..	..	..	..	..	..
	1993	109								471
Lithuania	1980	..	..	..	..	..	..	..	..	482
Lituanie	1990	..	..	..	..	..	..	..	..	0
	1992	35	..	91	1 638	3 387	397	191	..	5 938
	1993	17		104	2 241	4 546	505	249		4 783
Luxembourg	1980	..	0	..	..	..	..	..	..	113
Luxembourg	1990	..	0	..	..	..	..	..	..	169
	1992	..	0	..	..	..	..	..	..	147
	1993	..	0	..	..	..	..	..	..	131
Malta	1980	...	0	..	..	..	..	..	..	65
Malte	1990	...	0	..	..	..	..	..	..	141
	1992	...	0	..	..	..	..	..	..	174
	1993	0	0	..	..	..	..	..	..	184
Netherlands	1980	0	2 387	2 272	25 159	49 528	3 227	6 007	115 960	9 005
Pays−Bas	1990	0	2 661	5 729	42 089	40 455	8 171	8 817	86 688	9 698
	1992	0	2 841	4 802	41 962	47 389	6 105	9 281	98 286	10 428
	1993	0	2 800	4 705	43 162	48 214	7 555	9 495	99 990	10 439
Norway, Svalbard & Jan Mayen Is	1980	276	339	35 067	3 527	7 582	315	677	36 331	10 269
Norvège, Svalbd., île J. Mayen	1990	291	0	117 525	7 259	10 154	283	1 337	38 330	14 967
	1992	375	0	152 894	7 551	11 641	507	1 379	38 543	14 434
	1993	256	0	163 800	7 407	11 417	393	1 333	38 444	14 740
Poland	1980	166 329	19 883	508	6 215	12 951	2 686	309	6 589	14 970
Pologne	1990	132 564	13 128	239	5 137	10 888	1 866	542	3 339	16 744
	1992	124 501	10 736	286	5 846	10 605	1 537	667	3 636	16 306
	1993	126 891	9 943	336	6 460	11 914	1 400	684	4 659	16 444
Portugal	1980	177	194	..	2 673	7 623	474	382	..	1 876
Portugal	1990	165	220	..	5 166	9 730	354	959	..	3 505
	1992	129	257	..	5 254	10 307	451	925	..	3 701
	1993	115	256	..	4 764	10 186	576	888	..	3 837
Moldova, Republic of	1980	..	..	..	..	..	..	..	..	..
Moldova, République de	1990	..	..	..	..	..	..	..	..	..
	1992	..	..	..	..	..	..	..	..	1 382
	1993	..	..	..	..	..	..	..	..	1 274
Romania	1980	14 662	4 218	16 442	8 427	25 326	2 281	2 405	51 536	8 290
Roumanie	1990	11 415	3 900	11 325	7 667	20 778	1 797	2 552	32 728	7 899
	1992	10 106	2 770	9 448	4 952	10 827	1 570	1 402	25 159	6 657
	1993	10 846	2 410	9 536	4 759	10 675	1 733	1 541	23 938	6 814
Russian Federation	1980	..	..	..	..	..	..	..	..	..
Fédération de Russie	1990	..	..	..	..	..	..	..	..	..
	1992	241 872	30 275	570 386	84 607	220 835	40 423	21 371	713 740	153 652
	1993	215 254	26 053	505 494	70 472	203 476	34 458	13 326	699 375	147 171
Slovakia	1980	..	..	..	..	..	..	..	..	..
Slovaquie	1990	..	..	..	..	..	..	..	..	..
	1992	1 497	1 876	100	1 205	3 149	1 595	146	409	5 526
	1993	1 507	1 876	96	1 175	3 889	1 079	155	282	5 669
Slovenia	1980	..	..	..	..	..	..	..	..	..
Slovénie	1990	..	..	..	..	..	..	..	..	..
	1992	1 159	..	3	416	445	141	..	20	2 472
	1993	1 065	..	3	376	435	18	..	16	2 420
Spain	1980	14 489	4 302	2 288	15 373	48 106	4 478	2 252	1	14 703
Espagne	1990	17 443	3 307	1 674	23 901	42 775	5 723	4 446	2 021	32 140
	1992	15 644	3 047	2 040	23 664	46 868	6 179	5 189	1 642	33 341
	1993	14 569	3 157	1 600	22 449	43 224	7 521	4 462	927	33 167

67
Production of selected energy commodities
Thousand metric tons of coal equivalent [*cont.*]
Production des principaux biens de l'énergie
Milliers de tonnes métriques d'équivalent houille [*suite*]

Country or area Pays ou zone	Year Anneé	Hard coal, lignite & peat Houille, lignite & tourbe	Briquettes & cokes Agglomérés & cokes	Crude petroleum & NGL Pétrole brut & GNL	Light petroleum products Produits pétroliers légers	Heavy petroleum products Produits pétroliers lourds	Other petroleum products Autres produits pétroliers	LPG & refinery gas GLP et gaz de raffinerie	Natural gas Gaz naturel	Electricity Electricité
Sweden	1980	17	1 137	36	4 693	19 693	897	171	..	18 418
Suède	1990	10	1 038	4	6 650	16 188	1 626	342	..	34 929
	1992	364	1 097	1	6 820	17 485	1 314	452	..	33 790
	1993	336	1 088	0	6 657	18 429	1 614	463	..	32 993
Switzerland, Liechtenstein	1980	..	...	..	2 009	4 062	187	408	...	9 178
Suisse, Liechtenstein	1990	..	...	..	1 491	2 406	205	352	5	12 731
	1992	..	...	..	1 852	3 691	203	473	4	13 092
	1993	..	...	..	2 043	4 165	180	538	3	13 308
TFYR Macedonia	1980	..	..	..	..	..	..	..	..	..
L'ex – r.p. Macédonie	1990	..	..	..	..	..	..	..	..	..
	1992	2 695		..	255	986	13	23	..	745
	1993	2 810	..	..	300	972	14	28	..	735
Ukraine	1980	..	..	..	..	..	..	..	..	..
Ukraine	1990	..	..	..	..	..	..	..	..	..
	1992	110 285	23 941	6 390	8 221	36 764	2 831	1 270	24 154	49 358
	1993	95 612	18 833	6 068	6 775	20 849	1 640	902	22 574	46 951
United Kingdom	1980	108 556	10 152	115 110	41 300	70 701	6 794	5 381	47 048	44 061
Royaume–Uni	1990	79 280	8 020	131 268	58 043	56 344	7 755	6 505	64 958	55 530
	1992	70 045	6 745	135 191	61 745	57 940	7 987	6 481	73 549	58 523
	1993	65 578	6 094	143 652	63 419	61 684	8 472	6 780	86 550	61 898
Yugoslavia	1980	..	..	..	..	..	..	..	..	..
Yugoslavie	1990	..	..	..	..	..	..	..	..	..
	1992	8 414	..	1 664	1 795	1 906	443	78	981	4 482
	1993	7 838	..	1 640	511	1 001	306	16	1 116	4 196
Yugoslavia, SFR †	1980	14 416	2 661	6 113	4 976	14 235	1 744	1 091	2 573	7 336
Yugoslavie, Rfs †	1990	24 150	2 793	5 257	4 833	12 414	1 908	888	2 933	11 333
	1992	..	..	..	..	..	..	..	..	..
	1993	..	..	..	..	..	..	..	..	..
Oceania	1980	**70 817**	**5 291**	**31 283**	**23 809**	**19 047**	**3 964**	**3 267**	**13 587**	**16 420**
Océanie	1990	**144 702**	**4 403**	**42 215**	**26 671**	**18 581**	**3 579**	**3 153**	**33 338**	**25 361**
	1992	**159 558**	**4 236**	**37 554**	**28 014**	**19 122**	**3 947**	**3 251**	**37 597**	**26 590**
	1993	**160 623**	**3 959**	**37 237**	**29 255**	**19 311**	**4 119**	**3 410**	**39 635**	**26 313**
American Samoa	1980	..	..	..	..	..	..	..	..	9
Samoa américaines	1990	..	..	..	..	..	..	..	..	11
	1992	..	..	..	..	..	..	..	..	11
	1993	..	..	..	..	..	..	..	..	11
Australia	1980	69 009	5 242	30 810	20 724	15 981	3 814	3 231	12 257	11 809
Australie	1990	142 768	4 395	39 500	22 790	16 070	3 388	2 870	27 206	19 049
	1992	157 174	4 236	34 828	24 294	16 497	3 612	2 905	30 489	19 623
	1993	158 112	3 959	34 322	25 439	16 609	3 760	3 056	32 714	20 091
Cook Islands	1980	..	..	..	..	..	..	..	..	1
Iles Cook	1990	..	..	..	..	..	..	..	..	2
	1992	..	..	..	..	..	..	..	..	2
	1993	..	..	..	..	..	..	..	..	2
Fiji	1980	..	..	..	..	..	..	..	..	38
Fidji	1990	..	..	..	..	..	..	..	..	57
	1992	..	..	..	..	..	..	..	..	59
	1993	..	..	..	..	..	..	..	..	57
French Polynesia	1980	..	..	..	..	..	..	..	..	31
Polynésie française	1990	..	..	..	..	..	..	..	..	36
	1992	..	..	..	..	..	..	..	..	39
	1993	..	..	..	..	..	..	..	..	40
Guam	1980	..	..	..	1 164	1 006	7	35	..	141
Guam	1990	..	..	..	...	...	...	...	..	98
	1992	..	..	..	...	...	...	...	..	98
	1993	..	..	..	...	...	...	...	..	98
Kiribati	1980	..	..	..	..	..	..	..	..	1
Kiribati	1990	..	..	..	..	..	..	..	..	1
	1992	..	..	..	..	..	..	..	..	1
	1993	..	..	..	..	..	..	..	..	1

67
Production of selected energy commodities
Thousand metric tons of coal equivalent [cont.]
Production des principaux biens de l'énergie
Milliers de tonnes métriques d'équivalent houille [suite]

Country or area Pays ou zone	Year Anneé	Hard coal, lignite & peat Houille, lignite & tourbe	Briquettes & cokes Agglomérés & cokes	Crude petroleum & NGL Pétrole brut & GNL	Light petroleum products Produits pétroliers légers	Heavy petroleum products Produits pétroliers lourds	Other petroleum products Autres produits pétroliers	LPG & refinery gas GLP et gaz de raffinerie	Natural gas Gaz naturel	Electricity Electricité
Nauru	1980	..	..	..	..	..	..	..	..	3
Nauru	1990	..	..	..	..	..	..	..	..	4
	1992	..	..	..	..	..	..	..	..	4
	1993	..	..	..	..	..	..	..	..	4
New Caledonia	1980	..	..	..	..	..	..	..	..	164
Nouvelle–Caledonie	1990	..	..	..	..	..	..	..	..	141
	1992	..	..	..	..	..	..	..	..	144
	1993	..	..	..	..	..	..	..	..	144
New Zealand	1980	1 808	49	473	1 920	2 060	143	...	1 330	4 037
Nouvelle–Zélande	1990	1 934	7	2 715	3 881	2 511	191	283	6 133	5 702
	1992	2 384	0	2 726	3 720	2 625	336	345	7 108	6 348
	1993	2 512	0	2 915	3 816	2 701	359	354	6 922	5 604
Niue	1980	..	..	..	..	..	..	..	..	0
Nioué	1990	..	..	..	..	..	..	..	..	0
	1992	..	..	..	..	..	..	..	..	0
	1993	..	..	..	..	..	..	..	..	0
Palau [2]	1980	..	..	..	..	..	..	..	..	0
Palaos [2]	1990	..	..	..	..	..	..	..	..	17
	1992	..	..	..	..	..	..	..	..	25
	1993	..	..	..	..	..	..	..	..	25
Papua New Guinea	1980	..	..	..	..	..	..	..	..	25
Papouasie–Nouv.–Guinée	1990	..	..	..	..	..	..	..	..	154
	1992	..	..	..	..	..	..	..	..	220
	1993	..	..	..	..	..	..	..	..	220
Samoa	1980	..	..	..	..	..	..	..	..	220
Samoa	1990	..	..	..	..	..	..	..	..	5
	1992	..	..	..	..	..	..	..	..	6
	1993	..	..	..	..	..	..	..	..	6
Solomon Islands	1980	..	..	..	..	..	..	..	..	6
Iles Salomon	1990	..	..	..	..	..	..	..	..	3
	1992	..	..	..	..	..	..	..	..	4
	1993	..	..	..	..	..	..	..	..	4
Tonga	1980	..	..	..	..	..	..	..	..	1
Tonga	1990	..	..	..	..	..	..	..	..	3
	1992	..	..	..	..	..	..	..	..	3
	1993	..	..	..	..	..	..	..	..	3
Vanuatu	1980	..	..	..	..	..	..	..	..	2
Vanuatu	1990	..	..	..	..	..	..	..	..	3
	1992	..	..	..	..	..	..	..	..	4
	1993	..	..	..	..	..	..	..	..	4
former USSR †	1980	562 288	82 249	865 148	170 974	403 847	52 586	32 275	518 586	177 073
ancienne URSS †	1990	486 714	73 603	818 844	148 893	401 878	40 118	32 698	939 347	269 504
	1992	..	..	..	..	..	..	..	..	..
	1993	..	..	..	..	..	..	..	..	..

Source:
Energy statistics database of the Statistical Division of the
United Nations Secretariat.

† For detailed descriptions of data pertaining to
former Czechoslovakia, Germany, SFR Yugoslavia and former
USSR, see annex I – Country or area nomenclature, regional
and other groupings.

1 Including Aruba.
2 Including data for Federated States of Micronesia, Marshall Is.
and Northern Mariana Is.

Source:
Base de données pour les statistiques énergétiques de la Division
de statistique du Secrétariat de l'ONU.

† Pour les descriptions en détails des données
relatives à l'ancienne Tchécoslovaquie, l'Allemagne, la Rfs
Yougoslavie et l'ancienne URSS, voir l'Annexe I –
Nomenclature des pays ou zones, groupements régionaux et
autres groupements.

1 Y compris l'Aruba.
2 Y compris les données pour les Etats fédérés de Micronésie, les îles
Marshall et les îles Mariannes du Nord.

Technical notes, tables 66 and 67

Tables 66 and 67: Data are presented in metric tons of coal equivalent (TCE), to which the individual energy commodities are converted in the interests of international uniformity and comparability.

The procedure to convert from original units to TCE is as follows:

Data in original unit (metric tons, TJ, kWh,m^3) x special factors = TCE

For special factors used to convert the commodities into coal equivalent and detailed description of methods, see the United Nations *Energy Statistics Yearbook* and related methodological publications. [20, 37, 38]

Table 66: The data on production refer to the first stage of production. Thus, for hard coal the data refer to mine production; for briquettes to the output of briquetting plants; for crude petroleum and natural gas to production at oil and gas wells; for natural gas liquids to production at wells and processing plants; for refined petroleum products to gross refinery output; for cokes and coke-oven gas to the output of ovens; for other manufactured gas to production at gas works, blast furnaces or refineries; and for electricity to the gross production of generating plants.

International trade of energy commodities is based on the "general trade" system, that is, all goods entering and leaving the national boundary of a country are recorded as imports and exports.

Bunkers refers to fuels supplied to ships and aircraft engaged in international transportation, irrespective of the carrier's flag.

In general, data on stocks refer to changes in stocks of producers, importers and/or industrial consumers at the beginning and end of each year.

Data on consumption refer to "apparent consumption" and are derived from the formula "production + imports − exports − bunkers +/− stock changes". Accordingly, the series on apparent consumption may in some cases represent only an indication of the magnitude of actual gross inland availability.

Table 67: Definitions of the energy commodities are as follows:

— Hard coal: Coal with a high degree of coalification, and with a gross calorific value above 24 MJ/kg (5,700 kcal/kg) on an ash-free but moist basis, and with a reflectance index of vitrinite of 0.5 and above;

— Lignite: Coal with a low degree of coalification which has retained the anatomical structure of the vegetable matter from which it was formed. Its gross calorific value is less than 24 MJ/kg (5,700 kcal/kg) on an ash-free but moist basis, and its reflectance index of vitrinite is less than 0.5;

Notes techniques, tableaux 66 et 67

Tableaux 66 et 67 : Les données relatives aux divers produits énergétiques ont été converties en tonnes métriques d'équivalent houille (TEC), dans un souci d'uniformité et pour permettre les comparaisons entre la production de différents pays.

La méthode de conversion utilisée pour passer des unités de mesure d'origine à l'unité commune est la suivante :

Données en unités d'origine (tonnes métriques, TJ, kWh, m^3) x facteurs de conversion = TEC

Pour les facteurs spéciaux utilisés pour convertir les produits énergétiques en équivalent houille et pour des descriptions détaillées des méthodes appliquées, se reporter à l'*Annuaire des statistiques de l'énergie* des Nations Unies et aux publications méthodologiques connexes [20, 37, 38].

Tableau 66 : Les données relatives à la production se rapportent au premier stade de production. Ainsi, pour la houille, les données se rapportent à la production minière; pour les briquettes, à la production des briquetteries; pour le pétrole brut et le gaz naturel, à la production des gisements de pétrole et de gaz; pour les condensats de gaz naturel, à la production au puits et aux installations de traitement; pour les produits pétroliers raffinés, à la production brute des raffineries; pour les cokes et le gaz des fours à coke, à la production des fours; pour les autres gaz manufacturés, à la production des usines à gaz, des hauts fourneaux ou des raffineries; et pour l'électricité, à la production brute des centrales.

Le commerce international des produits énergétiques est fondé sur le système du "commerce général", c'est-à-dire que tous les biens entrant sur le territoire national d'un pays ou en sortant sont respectivement enregistrés comme importations et exportations.

Les soutages se rapportent aux carburants fournis aux navires et aux avions assurant des transports internationaux, quel que soit leur pavillon.

En général, les variations des stocks se rapportent aux différences entre les stocks des producteurs, des importateurs ou des consommateurs industriels au début et à la fin de chaque année.

Les données sur la consommation se rapportent à la "consommation apparente" et sont obtenues par la formule "production + importations - exportations - soutage +/- variations des stocks". En conséquence, les séries relatives à la consommation apparente peuvent occasionnellement ne donner qu'une indication de l'ordre de grandeur des disponibilités intérieures brutes réelles.

Tableau 67 : Les définitions des produits énergétiques sont données ci-après :

— Peat: Solid fuel formed from the partial decomposition of dead vegetation under conditions of high humidity and limited air access (initial stage of coalification). Included is only that portion of peat used as fuel;

— Briquettes include the following commodities:

Patent fuel (hard coal briquettes): In the briquetting process, coal fines are moulded into artifacts of even shape under the influence of pressure and temperature with the admixture of binders.

Lignite briquettes: Lignite, after crushing and drying, is moulded under high pressure and without the admixture of binders to form artifacts of even shape;

Peat briquettes: Raw peat, after crushing and drying, is moulded under high pressure and without the admixture of binders to form artifacts of even shape;

— Coke: The solid residue obtained from the distillation of hard coal or lignite in the total absence of air (carbonization);

— Crude petroleum: Mineral oil consisting of a mixture of hydrocarbons of natural origin, yellow to black in colour, of variable specific gravity and viscosity, including crude mineral oils extracted from bituminous minerals (shale, bituminous sand, etc.). Data for crude petroleum include lease (field) condensate (separator liquids) which is recovered from gaseous hydrocarbons in lease separation facilities;

— Natural gas liquids (NGL): Liquid or liquefied hydrocarbons produced in the manufacture, purification and stabilization of natural gas. Their characteristics vary, ranging from those of butane and propane to heavy oils. Specifically included are natural gasolene, liquefied petroleum gas (LPG) from plants and plant condensate;

— Light petroleum products: Light products are defined (from the technological point of view) as liquid products obtained by distillation of crude petroleum at temperatures between 30 and 350°C, and/or having a specific gravity within the range of 0.625 to around 0.830;

— Heavy petroleum products: Heavy products are defined (from the technological point of view) as products obtained by distillation of crude petroleum at temperatures above 350°C and having a specific gravity higher than 0.830. Excluded are products which are not used for energy purposes, such as insulating oils, lubricants, paraffin wax, bitumen and petroleum coke;

— Liquefied petroleum gas (LPG): Hydrocarbons which are gaseous under conditions of normal temperature and pressure but are liquefied by compression or cooling to facilitate storage, handling and transportation;

— Refinery gas: Non-condensable gas collected in petroleum refineries which is generally used wholly as refinery fuel. It is also known as still gas;

– Houille : Charbon à haut degré de houillification et de pouvoir calorifique brut supérieur à 24 MJ/kg (5.700 kcal/kg) mesuré sans cendre, mais sur base humide, pour lequel l'indice de réflectance du vitrain est égal ou supérieur à 0,5;

– Lignite : Charbon d'un faible degré de houillification qui a gardé la structure anatomique des végétaux dont il est issu. Son pouvoir calorifique brut est inférieur à 24 MK/kg (5.700 kcal/kg) mesuré sans cendre, mais sur base humide, et son indice de réflectance du vitrain est égal ou inférieur à 0,5;

— Tourbe : Combustible solide issu de la décomposition partielle de végétaux morts dans des conditions de forte humidité et de faible circulation d'air (phase initiale de la houillification). N'est prise en considération ici que la tourbe utilisée comme combustible;

— Briquettes comprennent les produits suivants :

Agglomérés (briquettes de houille) : Par briquetage, les fines de charbon sont moulées en pains de forme régulière par compression et chauffage avec adjonction de liants.

Briquettes de lignite : Le lignite, après broyage et séchage, est moulé par compression, sans addition de liant, en pains de forme régulière.

Briquettes de tourbe : La tourbe brute, après broyage et séchage, est moulée par compression, sans addition de liant, en pains de forme régulière;

– Coke : Résidu solide obtenu lors de la distillation de houille ou de lignite en l'absence totale d'air (carbonisation);

– Pétrole brut : Huile minérale constituée d'un mélange d'hydrocarbures d'origine naturelle, de couleur variant du jaune au noir, d'une densité et d'une viscosité variables. Figurent également dans cette rubrique les huiles minérales brutes extraites de minéraux bitumeux (schiste, sable, etc.). Les données relatives au pétrole brut comprennent les condensats directement récupérés sur les sites d'exploitation des hydrocarbures gazeux (dans les installations prévues pour la séparation des phases liquide et gazeuse);

— Condensats de gaz naturel (GNL) : Hydrocarbures liquides ou liquéfiés produits lors de la fabrication, de la purification et de la stabilisation de gaz naturel. Leurs caractéristiques varient et se classent entre celles des gaz butane et propane et celles des huiles lourdes. Sont inclus en particulier dans cette rubrique l'essence naturelle, les gaz de pétrole liquéfiés (GPL) obtenus en usine et les condensats d'usine;

— Produits pétroliers légers : Les produits légers sont définis (du point de vue technologique) comme des produits liquides obtenus par distillation du pétrole brut à des températures comprises entre 30 et 350°C et/ou ayant une densité comprise entre 0,625 et 0,830 environ;

— Natural gas: A mixture of hydrocarbon compounds and small quantities of non-hydrocarbons existing in the gaseous phase, or in solution with oil in natural underground reservoirs at reservoir conditions;

— Electricity production: Refers to gross production, which includes the consumption by station auxiliaries and any losses in the transformers that are considered integral parts of the station. Excluded is electricity produced from pumped storage.

— Produits pétroliers lourds : Les produits lourds sont définis (du point de vue technologique) comme des produits obtenus par distilation du pétrole brut à des températures supérieures à 350°C et ayant une densité supérieure à 0,830. En sont exclus les produits qui ne sont pas utilisés à des fins énergétiques, tels que les huiles isolantes, les lubrifiants, les paraffines, le bitume et le coke de pétrole;

– Gaz de pétrole liquéfié (GPL) : Hydrocarbures qui sont à l'état gazeux dans des conditions de température et de pression normales mais sont liquéfiés par compression ou refroidissement pour en faciliter l'entreposage, la manipulation et le transport;

— Gaz de raffinerie : Comprend les gaz non condensables obtenus dans les raffineries de pétrole et qui sont généralement utilisés en totalité comme combustible de raffinerie. Ce produit est également appelé gaz de distillation;

– Gaz naturel : Mélanges de composés d'hydrocarbures et de petites quantités de composants autres que des hydrocarbures existant en phase gazeuse ou en solution huileuse dans des roches réservoirs souterraines naturelles, dans les conditions du réservoir;

– Production d'électricité : Se rapporte à la production brute qui comprend la consommation des équipements auxiliaires des centrales et les pertes au niveau des transformateurs considérés comme faisant partie intégrante de ces centrales. Elle ne comprend pas l'électricité produite à partir d'une accumulation par pompage.

68
Number of scientists, engineers and technicians engaged in research and experimental development
Nombre de scientifiques, d'ingénieurs et de techniciens employés à des travaux de recherche et de développement expérimental

Full—time equivalent (FTE)
Equivalent plein temps (EPT)

Country or area Pays ou zone	Year Année	Total	Scientists and engineers Scientifiques et ingénieurs		Technicians Techniciens	
			Total	% F	Total	% F
Africa · Afrique						
Benin [1] Bénin [1]	1989	1036	794	12.6	242	26.4
Burundi [1] Burundi [1]	1989	338	170[2]	10.0	168	...
Central African Rep. Rép. centrafricaine	1990	254	162	9.9	92	5.4
Congo [3] Congo [3]	1984	2335	862[2]	...	1473	...
Egypt [3] Egypte [3]	1991	46022	26415	...	19607	...
Gabon Gabon	1987	...	199[1]	...	18[4]	...
Guinea Guinée	1984	1893	1282	...	611	...
Libyan Arab Jamahiriya Jamah. arabe libyenne	1980	2600	1100	...	1500	...
Madagascar [5] Madagascar [5]	1989	1225	269	31.2	956	...
Mauritius [1] Maurice [1]	1992	559	389	...	170	...
Nigeria [6] Nigéria [6]	1987	7380	1338	...	6042	...
Rwanda Rwanda	1985	*138	71	...	*67	...
Senegal Sénégal	1981	4610	1948	...	2662	...
Seychelles Seychelles	1983	24	18	...	6	...
South Africa Afrique du Sud	1991	17108	12102	...	5006	...
Tunisia [1] Tunisie [1]	1992	* 3860	* 3260	...	* 600	...
America, North · Amérique du Nord						
Canada [7] Canada [7]	1991	92870	65350	...	27520	...
Costa Rica Costa Rica	1992	...	1722	...	...	...
Cuba [3] Cuba [3]	1992	24235	14770	43.2	9465	55.5
El Salvador [5 8] El Salvador [5 8]	1992	1714	102	...	1612	...
Guatemala [9] Guatemala [9]	1988	1783	858[2]	...	925	...
Jamaica [10] Jamaïque [10]	1986	33	18	55.6	15	20.0
Mexico Mexique	1984	46146	16679	25.9	29467	25.9
Nicaragua Nicaragua	1987	1027	725	...	302	...
Saint Lucia Sainte—Lucie	1984	139	53	...	86	...
Trinidad and Tobago Trinité—et—Tobago	1984	529	275	21.1	254	17.3
United States [11] Etats—Unis [11]	1988	...	*949200	...	...	...

68

Number of scientists, engineers and technicians engaged in research and experimental development
Full—time equivalent (FTE) [cont.]

Nombre de scientifiques, d'ingénieurs et de techniciens employés à des travaux de recherche et de dévelopment expérimental
Equivalent plein temps (EPT) [suite]

Country or area Pays ou zone	Year Année	Total	Scientists and engineers Scientifiques et ingénieurs		Technicians Techniciens	
			Total	% F	Total	% F
America, South · Amérique du Sud						
Argentina Argentine	1988	* 17329	* 11088	* 43.3	* 6241	...
Bolivia Bolivie	1991	2720	1681	41.6	1039	48.1
Brazil [12] Brésil [12]	1985	...	52863	...	...	...
Chile [3] Chili [3]	1988	7570	4630	...	2940	...
Colombia [1] Colombie [1]	1982	2107	1083	...	1024	...
Ecuador Ecuador	1990	3936	1732	...	2204	...
Guyana [3][13] Guyana [3][13]	1982	267	89	...	178	...
Paraguay Paraguay	1981	...	807	...	...	...
Peru Pérou	1981	...	4858	...	...	...
Uruguay Uruguay	1987	...	2093	34.4	...	...
Venezuela Venezuela	1992	* 4908	* 4258	* 35.0	* 650	* 31.5
Asia · Asie						
Brunei Darussalam [14] Brunéi Darussalam [14]	1984	136	20	...	116	...
China [15] Chine [15]	1992	1841445	1335336	...	506109	...
Cyprus Chypre	1992	312	147	...	165	...
India [16] Inde [16]	1990	* 224773	* 128036	6.0	96737	6.3
Indonesia Indonésie	1988	...	32038	...	...	...
Iran, Islamic Rep. of Iran, Rép. islamique d'	1985	5048	3194	...	1854	...
Israel Israel	1984	24400	20100	51.7 [17]	4300	32.6
Japan [7] Japon [7]	1992	813360	705346	...	108014	...
Jordan [3] Jordanie [3]	1986	447	418	12.9	29	20.7
Korea, Republic of [3][18] Corée, République de [3][18]	1992	102206	86953	6.4	15253	...
Kuwait [15] Koweït [15]	1984	2072	1511[2]	22.1[2]	561[2]	20.1[2]
Lebanon [19] Liban [19]	1980	186	180	...	6	...
Malaysia [18] Malaisie [18]	1988	6707	5537	...	1170	...
Nepal [15][18] Népal [15][18]	1980	409	334	...	75	...
Pakistan [3][20] Pakistan [3][20]	1990	15940	6626	7.0	9314	...
Philippines [3] Philippines [3]	1984	6685	4830	48.0	1855	...
Qatar [21] Qatar [21]	1986	290	229[2]	25.3	61[2]	3.3
Singapore [18] Singapour [18]	1987	4887	3361	19.3	1526	...
Sri Lanka Sri Lanka	1985	3483	2790	23.9	693	27.1
Tajikistan Tadjikistan	1992	...	3974	28.8	...	...

68

Number of scientists, engineers and technicians engaged in research and experimental development
Full−time equivalent (FTE) [*cont.*]

Nombre de scientifiques, d'ingénieurs et de techniciens employés à des travaux de recherche et de dévelopment
expérimental
Equivalent plein temps (EPT) [*suite*]

Country or area Pays ou zone	Year Année	Total	Scientists and engineers Scientifiques et ingénieurs		Technicians Techniciens	
			Total	% F	Total	% F
Thailand Thaïlande	1991	12650	9752	...	2898	...
Turkey Turquie	1991	13277[5]	11948	...	1329[5]	...
Uzbekistan Ouzbékistan	1992	* 44312	* 37625	* 45.2	* 6687	...
Viet Nam [22] Viet Nam [22]	1985	...	20000	...	...	...
Europe · Europe						
Austria Autriche	1989	17216	8782	...	8434	...
Belarus Bélarus	1992	38939	33685	16.2	5254	...
Belgium Belgique	1990	38773[23]	18465	...	20308[23]	...
Bulgaria Bulgarie	1992	48577	37825	45.9	10752	61.0
Croatia Croatie	1992	12746	8928	37.4	3818	61.1
former Czechoslovakia † [24] anc. Tchécoslovaquie † [24]	1989	108351	65475	...	42876	...
Czech Republic République Tchèque	1992	33456	20084	...	13372	...
Denmark Danemark	1991	25756[23]	12049	...	13707[23]	...
Finland [5] Finlande [5]	1991	21912[23]	11428	...	10484[23]	...
France France	1991	298592[23]	129215	...	169377[23]	...
Germany †· Allemagne † Federal Republic of Germany [7] Rép. féd. d'Allemagne [7]	1989	296727	176401	...	120326	...
former German Dem. Rep. [25] ancienne Rép. dém. allemande [25]	1989	195073[23]	127449	...	67624[23]	...
Greece [26] Grèce [26]	1986	1022	534	...	488	...
Hungary [24] Hongrie [24]	1992	19463[27]	12311	...	7152[27]	...
Iceland Islande	1989	1177[23]	773	...	404[23]	...
Ireland Irlande	1988	7642	6351	...	1291	...
Italy Italie	1990	120180	77876	...	42304	...
Latvia Lettonie	1992	...	8935	45.0	...	...
Lithuania Lituanie	1992	...	4750	32.0	...	...
Malta [28] Malte [28]	1988	39	34	...	5	...
Netherlands Pays−Bas	1991	66710[23]	40000	...	26710[23]	...
Norway Norvège	1991	20252[23]	13460	...	6792[23]	...
Poland Pologne	1992	94250	41440	...	52810	...
Portugal Portugal	1990	9663	5908	...	3755	...
Romania [29] Roumanie [29]	1992	39577	28203	44.5	11374	54.9
Russian Federation Fédération de Russie	1991	1079100	878500	...	200600	...
Slovenia Slovénie	1992	10404	5789	30.1	4615	47.6

68

Number of scientists, engineers and technicians engaged in research and experimental development
Full−time equivalent (FTE) [*cont.*]

Nombre de scientifiques, d'ingénieurs et de techniciens employés à des travaux de recherche et de dévelopment expérimental
Equivalent plein temps (EPT) [*suite*]

Country or area Pays ou zone	Year Année	Total	Scientists and engineers Scientifiques et ingénieurs Total	% F	Technicians Techniciens Total	% F
Spain [30] Espagne [30]	1990	49269[5]	37534	26.4	11735[5]	17.1
Sweden Suède	1991	53604[23]	26515	...	27089[23]	...
Switzerland [26] Suisse [26]	1989	25600	16300	...	9300	
TFYR Macedonia [17] L'ex−R.y. Macédoine [17]	1991	3296	2605	38.7	691	55.9
Ukraine Ukraine	1989	...	348600[31]	...	...	...
Yugoslavia, SFR † [3] Yougoslavie, Rfs † [3]	1989	53550	34770	...	18780	
Yugoslavia [3] Yougoslavie [3]	1992	15429	11246	...	4183	44.4
Oceania · Océanie						
Australia Australie	1990	57759	41837	...	15922	...
Fiji [32] Fidji [32]	1986	126	36	11.1	90	11.1
French Polynesia [32] Polynésie française [32]	1983	33	17	...	16	...
Guam [28] Guam [28]	1991	34	23	21.7	11	45.5
Kiribati Kiribati	1980	3	2	−	1	−
New Caledonia [33] Nouvelle−Calédonie [33]	1985	148	77	9.1	71	15.5
New Zealand Nouvelle−Zélande	1990	* 7865	* 5225	...	2640	...
Tonga [2][32] Tonga [2][32]	1981	15	11	−	4	25.0
former USSR · ancienne URSS						
former USSR † ancienne URSS †	1990	...	1694400[31]	...	...	...

Source:
United Nations Educational, Scientific and Cultural Organization (Paris).

Source:
Organisation des Nations Unies pour l'éducation, la science et la culture (Paris).

† For detailed descriptions of data pertaining to former Czechoslovakia, Germany, SFR Yugoslavia and former USSR, see Annex I — Country or area nomenclature, regional and other groupings.

1 Not including data for the productive sector.
2 Data include foreigners.
3 Not including military and defence R&D.
4 Data relate only to those in the general service sector.
5 Not including data for the higher education sector.
6 Data relate only to 23 out of 26 national research institutes under the Federal Ministry of Science and Technology.
7 Not including social sciences and humanities in the productive sector. '
8 Data refer to scientists and engineers and technicians engaged in public enterprises.
9 Data relate to the productive sector and the higher education sector only.
10 Data refer to the Scientific Research Council only.
11 Not including data for law, humanities and education.
12 Not including data either for scientists and engineers engaged in private productive enterprises or for military and defense R&D.
13 Data for the general service sector and for medical sciences in the higher education sector are excluded.

† Pour les descriptions en détails des données relatives à l'ancienne Tchécoslovaquie, l'Allemagne, la Rfs Yougoslavie et l'ancienne URSS, voir l'Annexe I — Nomenclature des pays ou zones, groupements régionaux et autres groupements.

1 Non compris les données relatives au secteur de la production.
2 Les données comprennent les étrangers.
3 Non compris les activités de R−D de caractère militaire ou relevant de la défense nationale.
4 Les données ne se réfèrent qu'au secteur de service général.
5 Non compris les données relatives au secteur de l'enseignement supérieur.
6 Les données ne concernent que 23 des 26 instituts de recherche nationaux sous tutelle du Ministère fédéral de la Science et de la Technologie.
7 Non compris les sciences sociales et humaines dans le secteur de la production.
8 Les données se réfèrent aux scientifiques et ingénieurs et techniciens employés dans les entreprises publiques.
9 Les données ne concernent que le secteur de la production et le secteur de l'enseignement supérieur.
10 Les données se réfèrent au "Scientific Research Council" seulement.
11 Non compris les données pour le droit, les sciences humaines et l'éducation.
12 Non compris les données relatives aux scientifiques et ingénieurs

68
Number of scientists, engineers and technicians engaged in research and experimental development
Full—time equivalent (FTE) [*cont.*]
Nombre de scientifiques, d'ingénieurs et de techniciens employés à des travaux de recherche et de dévelopment expérimental
Equivalent plein temps (EPT) [*suite*]

14 Data relate to 2 research institutes only.
15 Data refer to scientific and technological activities (STA).
16 Data for scientists and engineers include personnel in the higher education sector. Data for women scientists and engineers and for technicians in the higher education sector are not included.
17 Data include part—time personnel.
18 Not including data for social sciences and humanities.
19 Data refer to the Faculty of Science at the University of Lebanon only.
20 Data relate to R&D activities concentrated mainly in government—financed research establishments only.
21 Not including social sciences and humanities in the higher education sector.
22 Not including data for the general service sector.
23 Including auxiliary personnel.
24 Not including scientists and engineers engaged in the administration of R & D; of military R&D, only that part carried out in civil establishments is included.
25 With the exception of economics and computer sciences, R & D in social sciences and humanities is excluded.
26 Data relate only to the productive sector.
27 Including skilled workers.
28 Data relate to the higher education sector only.
29 Due to methodological changes, data are not strictly comparable with the previous years.
30 Not including private non—profit organization.
31 Data refer to scientific workers, i.e. all persons with a higher scientific degree or scientific title, regardless of the nature of their work, persons undertaking research work in scientific establishments and scientific teaching staff in institutions of higher education; they also include persons undertaking scientific work in industrial enterprises.
32 Data relate to one research institute only.
33 Data refer only to 6 out of 11 institutes.

employés dans les entreprises privées ni les activités de R—D de caractère militaire ou relevant de la défense national.
13 Les données relatives au secteur de service général et les sciences médicales du secteur de l'enseignement supérieur sont aussi exclues.
14 Les données ne concernent que 2 instituts de recherche.
15 Les données se réfèrent aux activités scientifiques et techniques (AST).
16 Les données relatives aux scientifiques et ingénieurs comprennent le personnel dans le secteur de l'enseignement supérieur. Les données pour les femmes scientifiques et ingénieurs et pour les techniciens dans l'enseignement supérieur ne sont pas comprises.
17 Les données comprennent les personnels à temps partiel.
18 Compte non tenu des sciences sociales et humaines.
19 Les données ne se réfèrent qu'à la Faculté des Sciences de l'Université de Liban.
20 Les données se réfèrent aux activités de R—D se trouvant pour la plupart dans les étabalissements de recherche financés par le gouvernement.
21 Compte non tenue des sciences sociales et humaines dans le secteur de l'enseignement supérieur.
22 Non compris les données pour le secteur de service général.
23 Y compris le personnel auxiliare.
24 Non compris les scientifiques et ingénieurs employés dans les services administratifs de R—D; pour la R—D de caractère militaire, seule la partie effectuée dans les établissements civils a été considérée.
25 A l'exception des sciences économiques et de l'informatique, la R—D dans les sciences sociales et humaines est exclue.
26 Les données ne concernent que le secteur de la production.
27 Y compris les ouvriers qualifiés.
28 Les données ne concernent que le secteur de l'enseignement supérieur.
29 Suite à des changements de méthodologie, les données ne sont pas strictement comparables avec celles des années précédentes.
30 Non compris les organisations privées à but non lucratif.
31 Les données se réfèrent aux travailleurs scientifiques, c.à.d., à toutes les personnes ayant un diplôme scientifique supérieur ou un titre scientifique, sans considération de la nature de leur travail, aux personnes qui effectuent un travail de recherche dans des insitutions scientifiques et au personnel scientifique enseignant dans des établissements d'enseignement supérieur; sont incluses aussi les personnes qui effectuent des travaux scientifiques dans les entreprises industrielles.
32 Les données ne concernent qu'un institut de recherche.
33 Les données ne se rapportent qu'à 6 instituts de recherche sur 11.

69
Expenditure for research and experimental development
Dépenses consacrées à la recherche
et au développement expérimental
National currency
Monnaie nationale

Country or area (Monetary unit) Pays ou zone (Unité monétaire)	Years Années	Total expenditure Dépenses totales (000)	Capital expenditure Dépenses de capital (000)	Current expenditure Dépenses courantes (000)	Current expenditure Dépenses courantes %
Africa · Afrique					
Benin (franc) [1] Bénin (franc) [1]	1989	...	...	3347695	...
Burundi (franc) [2] Burundi (franc) [2]	1989	536187	...	...	...
Central African Rep. (CFA franc) [3] Rép. centrafricaine (franc CFA) [3]	1984	680791	...	...	...
Congo (CFA franc) [4] Congo (franc CFA) [4]	1984	25530	14263	11267	44.1
Egypt (pound) [5] Egypte (livre) [5]	1991	* 955273	* 281463	* 673810	* 70.5
Gabon (CFA franc) [1] Gabon (franc CFA) [1]	1986	380000	130000	250000	65.8
Libyan Arab Jamahiriya (dinar) Jamah. arabe libyenne (dinar)	1980	22875	...	...	...
Madagascar (franc) Madagascar (franc)	1988	14371515	13378000	993515	6.9
Mauritius (rupee) [1] Maurice (roupie) [1]	1992	177000	48000	129000	72.9
Nigeria (naira) [6] Nigéria (naira) [6]	1987	86270	16655	69615	80.7
Rwanda (franc) Rwanda (franc)	1985	918560	819280	99280	10.8
Seychelles (rupee) [4] Seychelles (roupie) [4]	1983	12854	6771	6083	47.3
South Africa (rand) Afrique du sud (rand)	1991	2786086	240764	2545322	91.4
Tunisia (dinar) [1] Tunisie (dinar) [1]	1992	45000000	11000000	34000000	75.6
America, North · Amérique du Nord					
Canada (dollar) [7] Canada (dollar) [7]	1992	10289000	...	...	...
Costa Rica (colon) Costa Rica (colon)	1986	612000	...	...	...
Cuba (peso) [4 8] Cuba (peso) [4 8]	1992	247925	91216	156709	63.2
El Salvador (colon) [9] El Salvador (colon) [9]	1992	1083559[10]	131377	382603	74.4
Guatemala (quetzal) [11] Guatemala (quetzal) [11]	1988	31859	...	...	...
Jamaica (dollar) [12] Jamaïque (dollar) [12]	1986	4016	130	3886	96.8
Mexico (peso) [13] Mexique (peso) [13]	1989	1050283	...	...	...
Nicaragua (córdoba) [4] Nicaragua (córdoba) [4]	1987	...	...	988970	...
Panama (balboa) [14] Panama (balboa) [14]	1986	173	—	173	100.0
Saint Lucia (dollar) Sainte−Lucie (dollar)	1992	...	...	* 449	...
Trinidad and Tobago (dollar) Trinité−et−Tobago (dollar)	1984	143257	33336	109921	76.7
Turks and Caicos Is. (US dollar) Iles Turques et Caïques (E−U dollar)	1984	—	—	—	—
United States (dollar) [15] Etats−Unis (dollar) [15]	1988	139255000	4024000	135231000	97.1

69

Expenditure for research and experimental development
National currency [cont.]
Dépenses consacrées à la recherche et au développement expérimental
Monnaie nationale [suite]

Country or area (Monetary unit) Pays ou zone (Unité monétaire)	Years Années	Total expenditure Dépenses totales (000)	Capital expenditure Dépenses de capital (000)	Current expenditure Dépenses courantes (000)	Current expenditure Dépenses courantes %
America, South · Amérique du Sud					
Argentina (austral) Argentine (austral)	1992	* 664700	* 142300	* 522400	* 78.6
Bolivia (boliviano) Bolivie (boliviano)	1991	282899000	200536000	82363000	29.1
Brazil (cruzado) [13] [16] Brésil (cruzado) [13] [16]	1985	5390540	...	...	...
Chile (peso) [4] Chili (peso) [4]	1992	* 102196000	...	...	...
Colombia (peso) [1] Colombie (peso) [1]	1982	2754273	...	...	...
Ecuador (sucre) Ecuador (sucre)	1990	8443000			
Guyana (dollar) [4] [17] Guyana (dollar) [4] [17]	1982	2800	...	...	...
Peru (sol) [18] Pérou (sol) [18]	1984	159024000	...	...	...
Venezuela (bolivar) [4] [19] Venezuela (bolivar) [4] [19]	1992	19622200	...	...	...
Asia · Asie					
Brunei Darussalam (dollar) [20] Brunéi Darussalam (dollar) [20]	1984	10880	2660	8220	75.6
China (yuan) Chine (yuan)	1992	16908000	3993000	12915000	76.4
Cyprus (pound) [1] Chypre (livre) [1]	1992	5578	772	4806	86.2
India (rupee) Inde (roupie)	1990	41864300	8475300	33389000	79.8
Indonesia (rupiah) [21] Indonésie (rupiah) [21]	1988	259283000	64645000	194638000	75.1
Iran, Islamic Rep. of (rial) [22] Iran, Rép. Islamique d' (rial) [22]	1985	22010713	9464315	12546398	57.0
Israel (shekel) [23] Israël (shekel) [23]	1990	2323400	146800	2176600	93.7
Japan (yen) [7] [13] Japon (yen) [7] [13]	1991	13771524	2149488	11622036	84.4
Jordan (dinar) [4] Jordanie (dinar) [4]	1986	5587	1287	4300	77.0
Korea, Republic of (won) [13] [24] Corée, République de (won) [13] [24]	1992	4989031	1395668	3593363	72.0
Kuwait (dinar) [19] Koweït (dinar) [19]	1984	71163	8147	63016	88.6
Lebanon (pound) [25] Liban (livre) [25]	1980	22000	...	...	...
Malaysia (ringgit) [22] Malaisie (ringgit) [22]	1989	97200	...	...	...
Pakistan (rupee) [4] [26] Pakistan (roupie) [4] [26]	1987	5582081	1926257	3655824	65.5
Philippines (peso) Philippines (peso)	1984	613410	98610	514800	83.9
Qatar (riyal) Qatar (riyal)	1986	6650	–	6650	100.0
Singapore (dollar) [27] Singapour (dollar) [27]	1987	374744	151021	223723	59.7
Sri Lanka (rupee) Sri Lanka (roupie)	1984	256799	82464	174335	67.9
Thailand (baht) Thaïlande (baht)	1991	3928100	828600	3099500	78.9

69

Expenditure for research and experimental development
National currency [cont.]

Dépenses consacrées à la recherche et au développement expérimental
Monnaie nationale [suite]

Country or area (Monetary unit) Pays ou zone (Unité monétaire)	Years Années	Total expenditure Dépenses totales (000)	Capital expenditure Dépenses de capital (000)	Current expenditure Dépenses courantes (000)	Current expenditure Dépenses courantes %
Turkey (lira) [13] Turquie (livre) [13]	1991	3330047	1293048	2036999	61.2
Viet Nam (dong) [22] Viet Nam (dong) [22]	1985	498000	...	...	...
Europe · Europe					
Austria (schilling) Autriche (schilling)	1989	22966910	3744713	19222197	83.7
Belarus (rouble) Bélarus (rouble)	1992	8590900	2258700	6332200	73.7
Belgium (franc) Belgique (franc)	1990	108545800	...	...	...
Bulgaria (leva) Bulgarie (leva)	1992	3103800	256100	2847700	91.7
Croatia (dinar) Croatie (dinar)	1992	...	...	21874940	...
former Czechoslovakia (koruny) † [28] anc. Tchécoslovaquie (couron) † [28]	1989	24721000	2621000	22100000	89.4
Czech Republic (koruny) Rép. Tchèque (couron)	1992	14499000	2295000	12204000	84.2
Denmark (kroner) Danemark (couron)	1991	14100000	1932000	12168000	86.3
Finland (markkaa) Finlande (markkaa)	1991	10172000	1191000	8981000	88.3
France (franc) France (franc)	1991	163092000	15510000	147582000	90.5
Germany †· Allemagne † Federal Republic of Germany (D.mark) [7][29] Rép. féd. d'Allemagne (D.mark) [7][29]	1989	63871000[10]	8034000	55460000	87.3
former German Dem. Rep. (DDR mark) [19] ancienne Rép. dém. allemande (DDR mark) [19]	1989	...	...	11880000	...
Greece (drachma) Grèce (drachme)	1986	18331000	...	...	...
Hungary (forint) [28][30] Hongrie (forint) [28][30]	1992	30988000	3359000	27629000	89.2
Iceland (krona) Islande (couronne)	1989	3123000	...	...	...
Ireland (pound) Irlande (livre)	1988	185800	...	...	...
Italy (lire) [13] Italie (lire) [13]	1990	17001221	2193980	14807241	87.1
Latvia (lat) [8] Lettonie (lat) [8]	1992	2380	...	...	...
Malta (lire) [31] Malte (lire) [31]	1988	10	1	9	90.0
Netherlands (guilder) [7] Pays−Bas (florin) [7]	1991	10381000	1059000	9322000	89.8
Norway (kroner) Norvège (couron)	1991	12603100	1445300	11157800	88.5
Poland (zloty) [4][13] Pologne (zloty) [4][13]	1992	9557064	743185	8813879	92.2
Portugal (escudo) Portugal (escudo)	1990	52032200	10483400	41548800	79.9
Romania (leu) Roumanie (leu)	1992	42057000	2530000	39527000	94.0
Russian Federation Fédération de Russie	1991	...	...	23269600	...
Slovenia (tolar) Slovénie (tolar)	1992	15050524	1365784	13684740	90.9
Spain (peseta) Espagne (peseta)	1990	425829225[10]	81429191	342072440	81.2

69
Expenditure for research and experimental development
National currency [*cont.*]

Dépenses consacrées à la recherche et au développement expérimental
Monnaie nationale [*suite*]

Country or area (Monetary unit) Pays ou zone (Unité monétaire)	Years Années	Total expenditure Dépenses totales (000)	Capital expenditure Dépenses de capital (000)	Current expenditure Dépenses courantes (000)	Current expenditure Dépenses courantes %
Sweden (krona) Suède (couron)	1991	41352000	3326000	38026000	92.0
Switzerland (franc) [32] Suisse (franc) [32]	1989	...	...	5488000	...
United Kingdom (pound) Royaume−Uni (livre)	1991	11906000	...	...	...
Yugoslavia (dinar) [4] Yougoslavie (dinar) [4]	1992	335	196	139	41.5
Oceania · Océanie					
Australia (dollar) Australie (dollar)	1990	5087600	698800	4388800	86.3
Fiji (dollar) [33] Fidji (dollar) [33]	1986	3800	800	3000	78.9
French Polynesia (CFP franc) [33] Polynésie française (franc CFP) [33]	1983	324720	16280	308440	95.0
Guam (US dollar) [31] Guam (E−U dollar) [31]	1991	2215	−	...	...
New Caledonia (CFP franc) [34] Nouvelle−Calédonie (franc CFP) [34]	1985	800820	93273	707547	88.4
New Zealand (dollar) Nouvelle−Zélande (dollar)	1990	638000	68400	569600	89.3
Tonga (pa'anga) [33] Tonga (pa'anga) [33]	1980	426	147	279	65.5
former USSR · ancienne URSS					
former USSR (rouble) † [35] ancienne URSS (rouble) † [35]	1988	37800000	...	...	...

Source:
United Nations Educational, Scientific and Cultural Organization (Paris).

Source:
Organisation des Nations Unies pour l'éducation, la science et la culture (Paris).

† For detailed descriptions of data pertaining to former Czechoslovakia, Germany, SFR Yugoslavia and former USSR, see Annex I − Country or area nomenclature, regional and other groupings.

1 Not including data for the productive sector.
2 Not including data for the productive sector nor labour costs at the Ministry of Public Health.
3 Not including data for the general service sector.
4 Not including military and defence R&D.
5 Data refer to estimated budget for R & D.
6 Data relate only to 23 out of 26 national research institutes under the Federal ministry of Science and Technology.
7 Not including social sciences and humanities in the productive sector.
8 Data relate to government funds only.
9 Data refer to the R&D activities performed in public enterprises.
10 Including data for which a distribution by type of expenditure is not available; this figure has been excluded from percentage calculation in the last column.
11 Data refer to the productive sector and the higher education sector only.
12 Data relate to the Scientific Research Council only.
13 Figures in millions.
14 Data refer to the central government only.
15 Not including data for law, humanities and education. Data do not include capital expenditure in the productive sector.
16 Not including either military and defence R&D nor private productive enterprises.
17 Data for the general service sector and for medical sciences in the higher education sector are also excluded.

† Pour les descriptions en détails des données relatives à l'ancienne Tchécoslovaquie, l'Allemagne, la Rfs Yougoslavie et l'ancienne URSS, voir l'Annexe I − Nomenclature des pays ou zones, groupements régionaux et autres groupements.

1 Non compris les données relatives au secteur de la production.
2 Non compris les données relatives au secteur de la production ni les dépenses de personnel au Ministre de la Santé Publique.
3 Non compris les données pour le secteur de service général.
4 Non compris les activités de R−D de caractère militaire ou relevant de la défense nationale.
5 Les données se réfèrent au budget estimé pour la R−D.
6 Les données ne concernent que 23 des 26 instituts de recherche nationaux sous tutelle du Ministère fédéral de la Science et de la Technologie.
7 Non compris les données pour les sciences sociales et humaines du secteur de la production.
8 Les données ne concernent que les fonds publics seulement.
9 Les données se réfèrent aux activités de R−D exercées dans les entreprises publiques.
10 Y compris les données dont la répartition par type de dépenses n'est pas disponible; on n'a pas tenu compte de ce chiffre pour calculer le pourcentage de la dernière colonne.
11 Les données ne concernent que le secteur de la production et le secteur de l'enseignement supérieur.
12 Les données se réfèrent au "Scientific Research Council" seulement.
13 Chiffres en millions.
14 Les données ne concernent que le gouvernement central.
15 Non compris les données pour le droit, les sciences humaines et l'éducation. Les données ne comprennent pas les dépenses en capital dans les secteurs de la production.

69

Expenditure for research and experimental development
National currency [*cont.*]

Dépenses consacrées à la recherche et au développement expérimental
Monnaie nationale [*suite*]

18   Data refer to the budget allotment for science and technology.
19   Data refer to government expenditure on scientific and technological activities (STA).
20   Data relate to 2 research institutes only.
21   Data relate to the general service sector only.
22   Data relate to government expenditure only.
23   Data refer to the civilian sector only.
24   Not including military and defence R&D nor social sciences and humanities.
25   Data refer to the Faculty of Science at the University of Lebanon only.
26   Data relate to R&D activities concentrated mainly in government–financed research establishments only.
27   Not including foreign funds nor social sciences and humanities.
28   Of military R&D, only that part carried out in civil establishments is included.
29   Due to methodological changes, data are not strictly comparable with the previous years.
30   The total current expenditure includes 5229 million forints for which breakdown by type of current expenditure is not known.
31   Data relate to the higher education sector only.
32   Not including foreign and other funds.
33   Data relate to one research institute only.
34   Data refer only to 6 out of 11 research institutes.
35   Expenditure on Science from the national budget and other sources.

16   Non compris les activités de R–D de caractère militaire ou relevant de la défense nationale ni les entreprises privées de production.
17   Non compris les données relatives au secteur de service général et les sciences médicales du secteur de l'enseignement supérieur.
18   Les données se rérèrent aux crédits budgétaires relatifs à la science et à la technologie.
19   Les données concernent les dépenses du gouvernement relatives aux activités scientifiques et techniques (AST).
20   Les données se réfèrent à 2 instituts de recherche seulement.
21   Les données ne concernent que le secteur de service général.
22   Les données ne concernent que les dépenses du gouvernement.
23   Les données ne concernent que le secteur civil.
24   Non compris les activités de R–D de caractère militaire ou relevant de la défense nationale ni les données pour les sciences sociales et humaines.
25   Les données ne se réfèrent qu'à la Faculté des Sciences de l'Université de Liban.
26   Les données se réfèrent aux activités de R–D se trouvant pour la plupart dans les établissements de recherche financés par le gouvernement.
27   Non compris les fonds étrangers ni les sciences sociales et humaines.
28   Pour la R–D de caractère militaire, seule la partie effectuée dans les établissements civils a été considérée.
29   Suite à des changements de méthodologie, les données ne sont pas strictement comparables avec celles des années précédentes.
30   Le total comprend 5229 millions de forints dont la répartition par type de dépenses courantes n'est pas disponible.
31   Les données ne concernent que le secteur de l'enseignement supérieur.
32   Non compris les fonds étrangers et les fonds divers.
33   Les dónnées ne concernent qu'un institut de recherche.
34   Les données ne se rapportent qu'à 6 instituts de recherche sur 11.
35   Montant total des sommes dépensées pour la science d'après le budget national et autres sources.

70
Patents
Brevets
Applications, grants, patents in force: number
Demandes, délivrances, brevets en vigueur : nombre

Country or area Pays ou zone	Applications for patents Demandes de brevets			Grants of patents Brevets délivrés			Patents in force Brevets en vigueur		
	1990	1991	1992	1990	1991	1992	1990	1991	1992
African Intellectual Prop. Org[1] Org. africaine de la prop. intel.[1]	3 315	4 463	6 102	60	150	392	1 860	1 853	1 791
Algeria Algérie	185	139	144	592	617	83	...	...	...
Angola Angola	...	...	6	...	...	...	...	...	...
Argentina Argentine	...	...	2 453	...	406	663	...	...	...
Australia Australie	26 507	27 672	28 927	12 682	12 636	12 899	64 070	67 845	70 516
Austria Autriche	43 571	43 535	48 352	12 584	13 354	14 956	25 738	23 798	21 818
Bahamas Bahamas	27	28	31	...	...	31	...	...	...
Bahrain Bahreïn	...	...	31	...	...	26	...	...	...
Bangladesh Bangladesh	108	113	161	94	78	61	580	597	595
Barbados Barbade	3 138	4 336	5 875	...	...	...	...	...	...
Belgium Belgique	43 725	42 047	45 260	13 170	14 291	17 026	85 935	86 075	100 081
Botswana Botswana	45	68	89	29	39	29	...	...	...
Brazil Brésil	12 434	12 769	14 180	3 355	2 419	1 822	...	...	...
Bulgaria Bulgarie	6 396	5 584	7 136	1 468	417	394	26 614	1 591	824
Burundi Burundi	3	...	2	2	...	2	129	129	14
Canada Canada	37 917	38 380	44 064	14 187	15 473	18 332	339 184	333 370	331 158
Chile Chili	833	1 000	1 139	607	506	460	7 878	7 879	7 756
China Chine	...	11 423	14 409	...	4 122	3 966	...	...	...
Colombia Colombie	...	612	695	...	425	248	...	...	...
Costa Rica Costa Rica	66	...	...	12	...	...	338	...	...
Croatia Croatie	...	...	1 489	...	...	...	...	...	...
Cuba Cuba	260	24	20	57	3	11	340	388	170
Cyprus Chypre	43	53	57	43	54	57	517	688	727
former Czechoslovakia† anc. Tchécoslovaquie†	7 071	5 934	10 715	5 089	1 401	1 751	124 611	...	18 582

70
Patents
Applications, grants, patents in force: number [*cont.*]
Brevets
Demandes, délivrances, brevets en vigueur : nombre [*suite*]

Country or area Pays ou zone	Applications for patents Demandes de brevets			Grants of patents Brevets délivrés			Patents in force Brevets en vigueur		
	1990	1991	1992	1990	1991	1992	1990	1991	1992
Denmark Danemark	39 447	39 764	44 351	2 376	2 609	3 773	11 759	12 213	14 206
Ecuador Equateur	...	88	107	...	102	60	...	342	372
Egypt Egypte	789	787	818	306	403	290	...	...	...
El Salvador El Salvador	54	36	108	12	6	24	...	...	...
European Patent Office[2] Office européen de brevets[2]	62 778	66 822	70 747	24 757	26 643	30 408	...	...	...
Finland Finlande	12 610	12 099	14 927	2 467	2 683	2 695	16 576	16 629	18 367
France France	81 884	79 075	82 038	35 149	35 581	38 215	250 051	114 249	296 316
Gambia Gambie	40	62	81	25	35	26	...	...	...
Georgia Géorgie	...	...	428	...	...	...	...	...	...
Germany † · Allemagne† F. R. Germany R. f. Allemagne	110 349	109 187	115 209	42 860	43 190	46 520	232 791	244 488	253 267
former German D. R. anc. R. d. allemande	7 525	..	..	8 372	..	..	118 099	..	..
Ghana Ghana	62	87	101	45	37	24	176	...	...
Greece Grèce	18 908	32 359	35 958	1 975	3 688	6 361	...	...	...
Guatemala Guatemala	81	95	73	72	123	65	1 416	1 390	1 317
Guyana Guyana	6	...	...	3	...	...	218	...	...
Haïti Haïti	...	8	...	...	8	...	...	...	...
Honduras Honduras	23	19	30	19	...	24	509	441	465
Hong Kong Hong-kong	1 081	1 092	1 259	1 095	1 079	1 069	...	...	...
Hungary Hongrie	9 133	9 950	10 931	2 619	2 305	2 112	...	19 433	...
Iceland Islande	130	133	149	37	30	48	324	317	339
India Inde	3 820	3 595	3 424	1 611	1 572	1 469	10 550	11 126	10 031
Indonesia Indonésie	...	1 336	4 027	...	...	...	...	...	...
Iran, Islamic Rep. of Iran, Rép. islamique d'	355	427	400	299	286	205	...	...	...
Iraq Iraq	...	...	174	...	...	66	...	...	1 119

70
Patents
Applications, grants, patents in force: number [*cont.*]
Brevets
Demandes, délivrances, brevets en vigueur : nombre [*suite*]

Country or area Pays ou zone	Applications for patents Demandes de brevets			Grants of patents Brevets délivrés			Patents in force Brevets en vigueur		
	1990	1991	1992	1990	1991	1992	1990	1991	1992
Ireland Irlande	4 735	4 580	14 681	852	860	764	6 505	6 279	6 109
Israel Israël	3 908	3 717	3 727	1 641	2 346	2 681	10 202	10 863	11 837
Italy Italie	55 569	53 300	64 664	17 794	19 503	27 228	...	...	...
Jamaica Jamaïque	81	41	65	21	11	3	491	422	418
Japan Japon	376 792	380 453	384 456	59 401	36 100	92 100	589 750	579 695	601 635
Kazakhstan Kazakhstan	...	...	445	...	...	...	...	...	...
Kenya Kenya	68	131	110	38	47	58	...	1 777	...
Korea, Dem. P. R. Corée, R. p. dém. de	10 033	4 549	6 170	2 677	37	27	31 972	121	141
Korea, Republic of Corée, République de	31 387	36 154	40 157	7 762	8 691	10 502	24 803	28 271	39 487
Latvia Lettonie	...	...	661	...	...	...	...	...	...
Lesotho Lesotho	35	59	82	23	29	25	...	...	...
Liberia Libéria	...	17	...	...	17	...	...	490	...
Libyan Arab Jamah. Jamah. arabe libyenne	57	47	...	...	...	...	...	...	...
Lithuania Lituanie	...	29	251	...	...	13	...	...	13
Luxembourg Luxembourg	34 777	35 978	39 965	6 749	7 337	8 680	21 899	23 133	...
Macau Macao	...	...	3	...	...	...	...	...	33
Madagascar Madagascar	3 144	4 343	5 880	...	...	...	...	...	...
Malawi Malawi	3 204	4 402	5 947	62	74	57	288	350	383
Malaysia Malaisie	2 305	2 427	2 410	512	...	1 134	...	...	2 840
Malta Malte	19	27	25	10	24	12	622	134	...
Mauritius Maurice	4	10	12	13	11	9	212	142	120
Mexico Mexique	5 289	5 271	7 695	1 752	1 360	3 160	23 219	18 155	18 732
Monaco Monaco	3 347	4 728	29 622	84	49	89	416	396	441
Mongolia Mongolie	73	1 163	4 940	38	38	33	497	61	122

70

Patents
Applications, grants, patents in force: number [*cont.*]
    Brevets
    Demandes, délivrances, brevets en vigueur : nombre [*suite*]

Country or area Pays ou zone	Applications for patents Demandes de brevets			Grants of patents Brevets délivrés			Patents in force Brevets en vigueur		
	1990	1991	1992	1990	1991	1992	1990	1991	1992
Morocco Maroc	329	356	...	311	303	372	6 623	...	6 593
Namibia Namibie	163	133	...	139	124	...	1 082	1 151	...
Nepal Népal	...	...	3	...	...	6	...	...	29
Netherlands Pays-Bas	53 514	51 412	55 885	16 841	17 610	20 346	76 127	81 602	88 750
New Zealand Nouvelle-Zélande	4 671	4 533	4 546	3 481	3 598	2 988	...	...	...
Nicaragua Nicaragua	...	...	42	...	...	4	...	...	87
Nigeria Nigéria	258	...	...	170	...	...	...	...	...
Norway Norvège	11 885	12 572	14 104	2 666	2 821	2 998	15 309	16 046	16 600
Pakistan Pakistan	547	524	622	380	524	436	...	...	...
Panama Panama	68	87	84	35	46	40	1 473	1 519	1 548
Paraguay Paraguay	52	...	64	35	...	24	780	...	811
Patent Cooperation Treaty[3] Traité de Coopération de brevets[3]	19 157	22 247	...	...	...	...	...	...	...
Peru Pérou	268	247	283	165	193	260	1 118	892	1 149
Philippines Philippines	1 969	1 921	...	1 092	944	...	17 698	17 044	...
Poland Pologne	5 421	8 817	11 371	3 647	3 788	3 851	...	...	...
Portugal Portugal	3 642	3 555	13 402	563	453	1 253	10 385	3 731	6 205
Romania Roumanie	6 742	7 184	8 318	1 428	2 127	2 100	26 254	23 820	22 898
Russian Federation Fédération de Russie	...	1 203	59 239	...	...	7 897	...	...	22 779
Rwanda Rwanda	4	1	2	4	1	2	81	82	69
Saint Lucia Sainte-Lucie	2	...	1	2	...	1	45	...	...
Saudi Arabia Arabie saoudite	455	519	626	...	...	...	...	...	...
Singapore Singapour	1 028	1 104	1 354	1 238	1 091	1 281	...	...	...
Slovenia Slovénie	...	...	420	...	...	33	...	...	...
South Africa Afrique du Sud	10 469	10 202	10 128	5 465	5 885	6 125	...	...	...

70
Patents
Applications, grants, patents in force: number [cont.]
Brevets
Demandes, délivrances, brevets en vigueur : nombre [suite]

Country or area Pays ou zone	Applications for patents Demandes de brevets			Grants of patents Brevets délivrés			Patents in force Brevets en vigueur		
	1990	1991	1992	1990	1991	1992	1990	1991	1992
Spain Espagne	49 026	48 929	53 605	7 597	9 781	14 021	...	...	106 536
Sri Lanka Sri Lanka	3 279	4 494	6 044	134	104	58	666	770	828
Sudan Soudan	3 202	4 411	5 986	36	37	56	...	...	...
Swaziland Swaziland	33	60	126	21	28	70	...	...	...
Sweden Suède	49 596	48 568	52 726	15 949	16 767	18 672	84 929	53 327	88 240
Switzerland Suisse	49 402	48 496	52 941	16 152	16 808	17 967	99 287	101 275	102 360
Thailand Thaïlande	1 940	1 987	1 962	141	153	199	...	...	655
Trinidad and Tobago Trinité-et-Tobago	...	...	67	...	...	67	...	...	873
Tunisia Tunisie	160	128	...	522	180	...	4 461	4 090	...
Turkey Turquie	1 228	1 205	1 252	486	694	674	6 281	6 351	6 369
Uganda Ouganda	59	74	96	41	42	42	1 675	1 679	...
Ukraine Ukraine	...	...	579	...	...	226	...	...	226
former USSR† ancienne URSS†	123 797	30 180	...	84 658	1 215	...	9 369	6 956	...
United Kingdom Royaume-Uni	97 891	95 533	99 241	32 179	34 074	37 827	...	...	...
United States Etats-Unis	176 100	177 388	187 291	90 366	96 514	97 443	1 154 204	...	1 160 613
Uruguay Uruguay	...	171	179	...	125	...	...	...	...
Venezuela Venezuela	1 352	1 361	1 540	787	593	481	...	...	...
Viet Nam Viet Nam	79	62	83	14	23	35	95	117	68
Yugoslavia Yougoslavie	..	..	1 125	..	..	375	..	..	5 143
Yugoslavia, SFR† Yougoslavie, Rfs†	2 481	2 010	..	546	1 400	..	7 498	10 900	..
Zambia Zambie	105	120	161	93	73	48	903	878	892
Zimbabwe Zimbabwe	260	270	302	174	222	194	...	1 549	1 616

70
Patents
Applications, grants, patents in force: number [*cont.*]
Brevets
Demandes, délivrances, brevets en vigueur : nombre [*suite*]

Source:
World Intellectual Property Organization (Geneva).

† For detailed descriptions of data pertaining to
former Czechoslovakia, Germany, SFR Yugoslavia and former
USSR, see Annex I - Country or area nomenclature, regional
and other groupings.

1 Members of the African Intellectual Property Organization
(OAPI), which includes Benin, Burkina Faso, Cameroon,
Central African Republic, Chad, Congo, Côte d'Ivoire, Gabon,
Guinea, Mali, Mauritania, Niger, Senegal, Togo.
2 In 1992, the European Patent Office (EPO) was constituted by
the following member countries: Austria, Belgium, Denmark,
France, Germany, Greece, Ireland, Italy, Liechtenstein,
Luxembourg, Monaco, Netherlands, Portugal, Spain, Sweden,
Switzerland, United Kingdom.
3 In 1992, the member countries of the Patent Cooperation
Treaty (PCT) were as follows: Armenia, Australia, Austria,
Barbados, Belarus, Belgium, Benin, Brazil, Bulgaria, Burkina
Faso, Cameroon, Canada, Central African Republic, Chad,
Congo, Côte d'Ivoire, former Czechoslovakia, Democratic
People's Republic of Korea, Denmark, Finland, France, Gabon,
Georgia, Germany, Greece, Guinea, Hungary, Ireland, Italy,
Japan, Kazakhstan, Kyrgyzstan, Liechtenstein, Luxembourg,
Madagascar, Malawi, Mali, Mauritania, Monaco, Mongolia,
Netherlands, New Zealand, Norway, Poland, Portugal, Republic
of Korea, Republic of Moldova, Romania, Russian Federation,
Senegal, Spain, Sri Lanka, Sudan, Sweden, Switzerland,
Tajikistan, Togo, Ukraine, United Kingdom, United States,
Uzbekistan.

Source:
Organisation mondiale de la propriété intellectuelle
(Genève).

† Pour les descriptions en détails des données
relatives à l'ancienne Tchécoslovaquie, l'Allemagne, la Rfs
Yougoslavie et l'ancienne URSS, voir l'Annexe I -
Nomenclature des pays ou zones, groupements régionaux et
autres groupements.

1 Les membres de l'Organisation africaine de la propriété
intellectuelle (OAPI): Bénin, Burkina Faso, Cameroun, Congo,
Côte d'Ivoire, Gabon, Guinée, Mali, Mauritanie, Niger,
République centrafricaine, Sénégal, Tchad, Togo.
2 En 1992, l'Office européen de brevets (OEB) comprenait les
pays membres suivants: Allemagne, Autriche, Belgique,
Danemark, Espagne, France, Grèce, Irlande, Italie,
Liechtenstein, Luxembourg, Monaco, Pays-Bas, Portugal,
Royaume-Uni, Suède, Suisse.
3 En 1992, les pays membres du Traité de Coopération en
matière de brevets (PCT) étaient les suivants: Allemagne,
Arménie, Australie, Autriche, Barbade, Bélarus, Belgique,
Bénin, Brésil, Bulgarie, Burkina Faso, Cameroun, Canada,
Congo, Côte d'Ivoire, Danemark, Espagne, Etats-Unis,
Fédération de Russie, Finlande, France, Gabon, Géorgie,
Grèce, Guinée, Hongrie, Irlande, Italie, Japon, Kazakhstan,
Kirghizistan, Liechtenstein, Luxembourg, Madagascar, Malawi,
Mali, Mauritanie, Monaco, Mongolie, Norvège,
Nouvelle-Zélande, Ouzbékistan, Pays-Bas, Pologne, Portugal,
République centrafricaine, République de Corée, République
de Moldova, République populaire démocratique de Corée,
Roumanie, Royaume-Uni, Sénégal, Sri Lanka, Soudan, Suède,
Suisse, Tadjikistan, Tchad, anc. Tchécoslovaquie, Togo,
Ukraine.

Technical notes, tables 68-70

Tables 68 and 69 present selected results of international data compilation by UNESCO in the field of science and technology, mostly obtained in response to the statistical surveys of scientific and technological activities. More comprehensive data can be found in the UNESCO *Statistical Yearbook*. [29] In utilizing these results the reader should keep in mind the factors which have an obvious bearing on the comparability and degree of accuracy of the data.

The absolute figures for R&D (research and experimental development) expenditure should not be compared country by country. Such comparisons would require the conversion of national currencies into a common currency by means of special R&D exchange rates. Official exchange rates do not always reflect the real costs of R&D activities and comparisons based on such rates can result in misleading conclusions, although they can be used to indicate a gross order of magnitude.

Abridged definitions suggested for use in the UNESCO surveys are given below, from the UNESCO *Yearbook*.

Type of personnel

Scientists and engineers are persons with scientific or technical training (usually completions of third-level education) in any field of science, including natural sciences, engineering and technology, the medical and agricultural sciences and the social sciences and humanities, who are engaged in professional work on R&D activities, administrators and other high-level personnel who direct the execution of R&D activities.

Technicians are persons who have received vocational or technical training in any branch of knowledge or technology of a specified standard (i.e. at least three years after the first stage of second-level education).

Full-time equivalent (FTE) is a measurement unit representing one person working full-time for a given period. This unit is used to convert figures relating to the number of part-time workers into the equivalent number of full-time workers. Data concerning personnel are normally calculated in FTE, especially in the case of scientists and engineers.

Research and experimental development (R&D)

In general R&D is defined as any creative systematic activity undertaken in order to increase the stock of knowledge, including knowledge of man, culture and society, and the use of this knowledge to devise new applications. It includes fundamental research, applied research in such fields as agriculture, medicine, industrial chemistry, and experimental development work leading to new devices, products or processes.

Notes techniques, tableaux 68 à 70

Les *Tableaux 68 et 69* présentent certains résultats d'une compilation internationale de données effectuée par l'UNESCO dans les domaines de la science et de la technologie, obtenues pour la plupart en réponse à des enquêtes statistiques sur des activités scientifiques et technologiques. On trouvera des données plus complètes dans l'*Annuaire statistique* de l'UNESCO [29]. Pour interpréter les résultats ainsi obtenus, le lecteur doit tenir compte des divers facteurs qui influent manifestement sur la comparabilité et l'exactitude des données.

Il faut éviter de comparer les chiffres absolus concernant les dépenses de R-D d'un pays à l'autre. On ne pourrait procéder à des comparaisons détaillées qu'en convertissant en une même monnaie les sommes libellées en monnaie na-tionale au moyen de taux de change spécialement.applicables aux activités de R-D. Les taux de change officiels ne reflè-tent pas toujours le coût réel des activités de R-D, et les comparaisons établies sur la base de ces taux peuvent con-duire à des conclusions trompeuses; toutefois, elles peuvent être utilisées pour donner une idée de l'ordre de grandeur.

On trouvera ci-dessous des définitions abrégées, tirées de l'*Annuaire* de l'UNESCO, dont l'utilisation est suggérée pour les enquêtes de cette organisation.

Classification du personnel

Scientifiques et ingénieurs : Personnes ayant une formation scientifique ou technique (ayant généralement terminé des études supérieures) dans n'importe quel domaine de la science, y compris les sciences naturelles, l'ingénierie et la technologie, les sciences médicales et agricoles et les sciences sociales et humaines, qui s'adonnent à des activités professionnelles de R-D; administrateurs et autre personnel de haut niveau qui dirigent l'exécution des activités de R-D.

Techniciens : Personnes qui ont reçu une formation professionnelle ou technique d'un niveau spécifié (à savoir au moins trois ans après achèvement du premier cycle de l'enseignement du second degré) dans n'importe quelle branche du savoir ou de la technologie.

Equivalent plein temps (EPT). Unité d'évaluation qui correspond à une personne travaillant à plein temps pendant une période donnée. Cette unité est utilisée pour convertir le nombre de travailleurs à temps partiel en un nombre de travailleurs à plein temps. En principe, les données concernant le personnel sont calculée en EPT, surtout dans le cas des scientifiques et des ingénieurs.

Recherche et développement expérimental (R-D)

En général, la recherche scientifique et le développement expérimental englobent tous les travaux systématiques et créateurs entrepris afin d'accroître la connaissance, y compris la connaissance de l'homme, de la culture et de la société, et l'utilisation de cette connaissance pour en tirer de nouvelles applications. Elle comprend la recherche fondamentale, la recherche appliquée dans des domaines tels que l'agriculture, la médecine, la chimie industrielle et le développement expérimental conduisant à la mise au point de nouveaux dispositifs, produits ou procédés.

Expenditure for research and experimental development

Total is defined as all expenditure (both current and capital) made for this purpose in the course of a reference year in institutions and installations established in the national territory, as well as installations physically situated abroad, land or experimental facilities rented or owned abroad and ships, vehicles, aircraft and satellites used by national institutions. Amounts spent on R&D activities carried out by international organizations established in the country in question are excluded from this total.

Current expenditure includes all payments made for the performance of R&D activities within units, institutions or sectors of performance, whatever the source or origin of funds, covering the cost of labor, minor equipment and expendable supplies and other current expenses including share of overheads.

Capital expenditure includes all payments made for the performance of R&D activities and relating to expenditure on major equipment and other capital expenditure. Depreciation on major instruments, equipment and buildings etc. should be excluded.

Table 70: Data on patents include patent applications filed directly with the office concerned and grants made on the basis of such applications; inventors' certificates; patents of importation, including patents of introduction, revalidation patents and "patentes précautionales"; petty patents; patents applied and granted under the Patent Cooperation Treaty (PCT), the European Patent Convention, the Havana Agreement, the Harare Protocol of the African Regional Industrial Property Organization (ARIPO) and the African Intellectual Property Organization (OAPI).

Dépenses consacrées à la recherche et au développement

Le total se définit comme l'ensemble des dépenses (tant courantes qu'en capital) effectuées à ce titre, au cours d'une année de référence, dans les institutions et installations situées sur le territoire national, y compris dans les installations qui sont géographiquement situées à l'étranger : terrains ou installations d'essai acquis ou loués à l'étranger, ainsi que navires, véhicules, aéronefs et satellites utilisés par les institutions nationales. Sont exclues de ce total les dépenses pour les activités de R-D effectuées par les organisations internationales installées dans le pays considéré.

Les dépenses courantes comprennent tous les paiements effectués pour l'exécution d'activités de R-D à l'intérieur des unités, institutions ou secteurs d'exécution, quelle que soit la source ou l'origine des fonds, pour couvrir les dépenses de personnel, de petit matériel et de fournitures fongibles et autres dépenses courantes, y compris une part des frais généraux.

Les dépenses en capital comprennent tous les paiements effectués pour l'exécution d'activités de R-D qui ont trait aux dépenses de gros équipement et autres dépenses en capital. L'amortissement du gros appareillage, de l'équipement et des bâtiments doit en être exclu.

Tableau 70 : Les données relatives aux brevets comprennent les demandes de brevet déposées directement au bureau intéressé et les délivrances effectuées sur la base de ces demandes; les brevets d'invention; les brevets d'importation; y compris les brevets d'introduction, les brevets de revalidation et les "patentes précautionales"; les petits brevets, les brevets demandés et délivrés en vertu du traité de coopération sur les brevets, de la Convention européenne relative aux brevets, de l'Accord de la Havane, du Protocole d'Hararé de l'Organisation régionale africaine de la propriété industrielle (ARIPO) et de l'Organisation africaine de la propriété intellectuelle (OAPI).

Part Four
International Economic Relations

XV
International merchandise trade (tables 71-74)
XVI
International tourism (tables 75-77)
XVII
Balance of payments (table 78)
XVIII
International finance (tables 79 and 80)
XIX
Development assistance (tables 81-83)

Part Four of the *Yearbook* presents statistics on international economic relations in areas of international merchandise trade, international tourism, balance of payments and assistance to developing countries. The series cover all countries or areas of the world for which data are available.

Quatrième partie
Relations économiques internationales

XV
Commerce international des marchandises (tableaux 71 à 74)
XVI
Tourisme international (tableaux 75 à 77)
XVII
Balance des paiements (tableau 78)
XVIII
Finances internationales (tableaux 79 et 80)
XIX
Aide au développement (tableaux 81 à 83)

La quatrième partie de l'*Annuaire* présente des statistiques sur les relations économiques internationales dans les domaines du commerce international des marchandises, du tourisme international, de la balance des paiements et de l'assistance aux pays en développement. Les séries couvrent tous les pays ou les zones du monde pour lesquels des données sont disponibles.

71
Total imports and exports
Importations et exportations totales
Value in million US dollars
Valeur en millions de dollars E-U

Region, country or area Region, pays ou zone	1984	1985	1986	1987	1988	1989	1990	1991	1992	1993
A. Imports c.i.f. • Importations c.i.f.										
World[1] *Monde[1]*	1 912 570	1 929 422	2 199 871	2 559 625	2 914 958	3 161 451	3 566 689	3 545 670	3 805 374	3 716 705
Developed economies[1 2 3] Econ. développées[1 2 3]	1 352 949	1 385 311	1 553 889	1 843 777	2 080 730	2 273 181	2 589 822	2 591 496	2 710 739	2 535 248
Developing economies[3] Econ. en dévelop.[3]	474 908	456 816	464 016	522 214	627 379	690 055	781 066	862 158	999 831	1 091 065
OPEC+ OPEP+	116 957	98 493	91 367	89 540	99 776	101 095	110 293	128 021	156 892	153 325
LDC+ PMA+	18 810	17 952	18 976	19 406	20 994	21 814	24 353	22 815	25 211	25 267
Other[4] Autres[4]	91 235	93 589	181 966	193 634	206 849	198 215	195 802	92 016	94 804	90 393
America • Amérique										
Developed economies Economies développées	421 534	431 392	454 036	505 296	554 997	595 229	615 676	605 871	653 774	709 578
Canada[5] Canada[5]	77 786	80 642	85 494	92 596	112 718	119 796	123 247	124 782	129 268	139 054
United States Etats-Unis	346 364	352 463	382 295	424 442	459 542	492 922	516 987	508 363	553 923	603 438
Developing economies Econ. en dévelop.	82 871	80 916	85 488	90 085	99 609	106 384	118 473	133 536	158 678	169 867
LAIA+[6] ALAI+[6]	53 830	53 664	57 027	62 895	71 967	75 532	86 336	104 563	128 610	137 835
Argentina[7] Argentine[7]	4 585	3 814	4 724	5 818	5 322	4 203	4 077	8 275	14 872	16 651
Bolivia[7] Bolivie[7]	489	691	674	766	591	611	687	970	1 090	1 206
Brazil[5 7] Brésil[5 7]	15 210	14 332	15 557	16 581	16 055	19 875	22 524	22 950	23 068	27 740
Chile[7] Chili[7]	3 574	3 072	3 436	4 396	5 292	7 144	7 678	8 094	10 129	11 125
Colombia[7] Colombie[7]	4 498	4 141	3 862	4 322	5 002	5 004	5 590	4 906	6 516	...
Ecuador Equateur	1 616	1 767	1 810	2 252	1 714	1 855	1 862	2 399	2 501	2 562
Mexico Mexique	11 788	13 762	11 918	12 761	18 954	23 633	29 993	38 184	48 138	50 147
Paraguay[5 7] Paraguay[5 7]	644	631	832	679	778	819	1 352	1 212	1 424	1 698
Peru[5 7] Pérou[5 7]	2 212	1 835	2 909	3 562	3 348	2 749	3 470	4 195	4 860	4 901
Uruguay[7] Uruguay[7]	777	708	870	1 142	1 157	1 203	1 343	1 637	2 045	2 324
Venezuela[5] Venezuela[5]	6 874	7 418	8 600	8 711	11 476	7 030	6 365	9 963	12 128	10 979
CACM+[8] MCAC+[8]	5 091	5 087	4 773	5 478	5 719	6 116	6 360	6 795	8 394	9 540
Costa Rica[7] Costa Rica[7]	1 094	1 098	1 148	1 383	1 410	1 717	1 990	1 877	2 458	2 907

Region, country or area Region, pays ou zone	1984	1985	1986	1987	1988	1989	1990	1991	1992	1993

B. Exports f.o.b. • Exportations f.o.b.

Region, country or area Region, pays ou zone	1984	1985	1986	1987	1988	1989	1990	1991	1992	1993
World[1] *Monde*[1]	1 935 165	1 935 209	2 129 410	2 490 806	2 826 983	3 044 013	3 437 400	3 421 330	3 656 145	3 632 090
Developed economies[1,2,3] Econ. développées[1,2,3]	1 241 881	1 285 630	1 491 832	1 746 007	1 990 887	2 151 854	2 467 895	2 491 048	2 642 032	2 538 789
Developing economies[3] Econ. en dévelop.[3]	505 610	481 998	454 068	544 018	627 210	701 519	797 052	840 034	923 883	998 254
OPEC+ OPEP+	170 483	153 005	113 870	124 069	122 237	141 926	180 036	169 682	173 439	182 809
LDC+ PMA+	10 085	9 443	10 117	11 264	11 713	12 398	12 654	12 205	12 377	12 903
Other[4] Autres[4]	187 674	167 582	183 510	200 781	208 886	190 640	172 453	90 247	90 230	95 047

America • Amérique

Region, country or area Region, pays ou zone	1984	1985	1986	1987	1988	1989	1990	1991	1992	1993
Developed economies **Economies développées**	**309 769**	**305 475**	**308 730**	**340 551**	**422 276**	**468 159**	**496 668**	**521 619**	**553 187**	**577 044**
Canada[5] Canada[5]	90 269	90 953	90 325	98 171	117 112	121 835	127 634	127 163	134 441	145 185
United States Etats-Unis	223 976	218 815	227 158	254 122	322 427	363 812	393 592	421 730	448 163	464 773
Developing economies **Econ. en dévelop.**	**110 494**	**103 256**	**87 882**	**96 557**	**109 813**	**120 679**	**131 190**	**127 804**	**130 386**	**138 180**
LAIA+[6] ALAI+[6]	**90 215**	**85 198**	**69 876**	**80 396**	**93 035**	**102 819**	**112 465**	**111 760**	**115 262**	**122 586**
Argentina[7] Argentine[7]	8 107	8 396	6 852	6 360	9 135	9 579	12 353	11 978	12 235	12 869
Bolivia[7] Bolivie[7]	725	623	638	570	600	822	926	849	710	728
Brazil[5,7] Brésil[5,7]	27 005	25 639	22 349	26 224	33 494	34 383	31 414	31 620	35 793	38 597
Chile[7] Chili[7]	3 651	3 804	4 191	5 224	7 052	8 080	8 310	8 929	9 986	9 202
Colombia[7] Colombie[7]	3 462	3 552	5 102	4 642	5 037	5 717	6 766	7 232	6 917	...
Ecuador Equateur	2 620	2 905	2 172	1 928	2 192	2 354	2 714	2 852	3 007	2 904
Mexico Mexique	24 196	21 664	16 031	20 656	20 658	22 819	26 524	27 120	27 531	30 241
Paraguay[5,7] Paraguay[5,7]	386	305	275	379	657	1 119	959	737	658	...
Peru[5,7] Pérou[5,7]	3 147	2 979	2 531	2 661	2 701	3 488	3 231	3 329	3 484	3 463
Uruguay[7] Uruguay[7]	934	909	1 088	1 189	1 405	1 599	1 693	1 605	1 703	1 645
Venezuela[5] Venezuela[5]	15 997	14 438	8 660	10 577	10 113	12 867	17 586	15 518	13 247	14 378
CACM+[5] MCAC+[5]	**3 963**	**3 810**	**4 021**	**3 801**	**3 951**	**4 188**	**4 361**	**4 439**	**4 703**	**5 140**
Costa Rica[7] Costa Rica[7]	1 006	976	1 121	1 158	1 246	1 415	1 448	1 598	1 834	2 085

71

Total imports and exports
Value in million US dollars [*cont.*]

Importations et exportations totales
Valeur en millions de dollars E-U [*suite*]

Region, country or area Region, pays ou zone	1984	1985	1986	1987	1988	1989	1990	1991	1992	1993
El Salvador[7] El Salvador[7]	977	961	935	994	1 007	1 161	1 149	1 360	1 544	1 919
Guatemala[7] Guatemala[7]	1 279	1 175	959	1 447	1 557	1 654	1 649	1 851	2 463	2 599
Honduras[7] Honduras[7]	893	888	875	827	940	969	935	955	1 037	1 130
Nicaragua Nicaragua	848	964	857	827	805	615	638	751	892	...
Other America **Autres pays d'Amérique**	**23 950**	**22 165**	**23 688**	**21 711**	**21 923**	**24 736**	**25 777**	**22 178**	**21 674**	**22 492**
Antigua and Barbuda Antigua-et-Barbuda	132	112	207	222	225	...	...	...	...	...
Aruba[7] Aruba[7]	2 126	184	192	236	336	387	...	...	...	...
Bahamas Bahamas	4 098	3 078	3 289	3 041	2 264	3 001	2 920	1 801	...	...
Barbados Barbade	659	602	587	515	579	673	700	695	521	574
Belize Belize	130	128	122	143	181	216	211	251	273	281
Bermuda[5] Bermudes[5]	456	402	492	420	488	535	595	510	562	535
Cayman Islands[7] Iles Caïmanes[7]	142	147	161	195	231	259	288	267	...	...
Cuba[7] Cuba[7]	7 207	8 758	11 139	7 584	7 579	8 124	6 745	3 690	2 185	...
Dominica Dominique	58	55	56	66	88	107	118	110	111	...
Dominican Republic[5] Rép. dominicaine[5]	1 446	1 487	1 433	1 830	1 849	2 258	2 062	1 988	2 505	2 443
French Guiana[7] Guyane française[7]	248	257	295	394	507	568	785	769	669	524
Greenland Groënland	274	296	360	507	519	399	447	408	457	...
Grenada[7] Grenade[7]	56	69	84	89	92	99	105	121	...	...
Guadeloupe[7] Guadeloupe[7]	604	620	762	1 038	1 243	1 251	1 650	1 644	1 509	1 393
Guyana[7] Guyana[7]	219	248	240	...	...	258	311	307	382	...
Haiti[9] Haïti[9]	472	442	360	399	344	291	295	374	...	...
Jamaica Jamaïque	1 146	1 111	971	1 238	1 455	1 853	1 924	1 719	1 675	2 118
Martinique[7] Martinique[7]	686	683	879	1 119	1 290	1 317	1 779	1 695	1 750	1 556
Monteserrat Montserrat	18	18	...	25	27	37	48	35	34	...
Netherlands Antilles[7 10] Antilles néerlandaises[7 10]	4 032	2 256	1 112	1 502	1 404	1 620	2 141	2 174	...	...
Panama[7 11] Panama[7 11]	1 430	1 392	1 284	1 306	815	965	1 489	1 695	2 019	2 187

Region, country or area Région, pays ou zone	1984	1985	1986	1987	1988	1989	1990	1991	1992	1993
El Salvador[7] El Salvador[7]	717	695	755	591	609	496	588	571	555	734
Guatemala[7] Guatemala[7]	1 129	1 057	1 044	987	1 022	1 108	1 163	1 202	1 295	1 291
Honduras[7] Honduras[7]	725	780	854	791	842	859	831	792	802	814
Nicaragua Nicaragua	386	302	247	273	233	311	331 .	275	218	...
Other America **Autres pays d'Amérique**	**16 316**	**14 249**	**13 985**	**12 360**	**12 827**	**13 672**	**14 364**	**11 606**	**10 421**	**10 454**
Antigua and Barbuda Antigua-et-Barbuda	18	13	20	17	22	...	...	...	...	...
Aruba[7] Aruba[7]	2 089	290	24	26	31	23	...	...	...	...
Bahamas Bahamas	3 393	2 728	2 701	2 728	2 164	2 786	2 678	1 517	...	...
Barbados Barbade	391	347	275	156	176	187	214	205	189	179
Belize Belize	93	90	93	103	116	125	129	126	141	132
Bermuda[5] Bermudes[5]	41	23	65	29	31	50	60	55	85	...
Cayman Islands[7] Iles Caïmanes[7]	1	2	3	2	2	3	4	3	...	...
Cuba[7] Cuba[7]	5 462	6 531	7 615	5 402	5 518	5 392	4 910	3 550	2 050	...
Dominica Dominique	26	28	43	48	54	45	55	54	56	...
Dominican Republic[5] Rép. dominicaine[5]	868	735	718	711	890	924	735	658	566	555
French Guiana[7] Guyane française[7]	37	38	32	54	51	57	93	70	96	95
Greenland Groënland	169	174	257	346	391	418	454	341	333	...
Grenada[7] Grenade[7]	18	23	31	32	33	28	27	23	...	...
Guadeloupe[7] Guadeloupe[7]	87	72	104	93	158	112	118	147	130	128
Guyana[7] Guyana[7]	210	166	222	267	230	227	256	251	293	414
Haiti[9] Haïti[9]	179	168	184	214	179	144	158	103	...	...
Jamaica Jamaïque	705	566	589	706	880	998	1 158	1 079	1 048	1 057
Martinique[7] Martinique[7]	154	162	216	194	196	200	278	216	247	191
Monteserrat Montserrat	3	3	...	3	2	1	2	1	2	...
Netherlands Antilles[7][10] Antilles néerlandaises[7][10]	3 733	1 679	924	1 308	1 133	1 457	1 790	1 599	...	...
Panama[7][11] Panama[7][11]	258	301	327	336	292	297	321	342	510	508

71
Total imports and exports
Value in million US dollars [cont.]
Importations et exportations totales
Valeur en millions de dollars E-U [suite]

Region, country or area Région, pays ou zone	1984	1985	1986	1987	1988	1989	1990	1991	1992	1993
free zone of Colon Zone libre de Colon	1 338	1 586	1 931	2 005	1 844	...	...	...	...	...
Saint Lucia[7] Sainte-Lucie[7]	119	125	155	179	220	274	271	295	313	...
Saint Kitts and Nevis Saint-Kitts-et-Nevis	52	51	63	79	93	102	111	...	...	...
St. Vincent-Grenadines[7] St. Vincent-Grenadines[7]	77	79	87	89	110	127	136	...	...	...
Suriname[7] Suriname[7]	392	338	327	294	...	...	472	...	...	...
Trinidad and Tobago[7] Trinité-et-Tobago[7]	1 919	1 538	1 350	1 219	1 128	1 221	1 262	1 663	1 429	1 448

Europe · Europe

Region, country or area Région, pays ou zone	1984	1985	1986	1987	1988	1989	1990	1991	1992	1993
Developed economies **Economies développées**	**737 150**	**769 754**	**916 514**	**1 121 834**	**1 262 757**	**1 382 397**	**1 654 459**	**1 662 913**	**1 733 052**	**1 489 862**
EU+[12] **UE+[12]**	**630 664**	**656 063**	**774 643**	**949 142**	**1 071 409**	**1 180 572**	**1 417 342**	**1 444 011**	**1 516 023**	**1 295 765**
Belgium-Luxembourg[7] Belgique-Luxembourg[7]	55 303	56 211	68 729	83 550	92 453	98 596	120 325	121 060	125 153	...
Denmark[7] Danemark[7]	16 585	18 072	22 822	25 429	25 944	26 725	31 744	32 411	35 176	29 513
France[7 13] France[7 13]	104 377	108 379	129 435	158 499	177 288	190 963	233 207	230 832	238 875	200 785
F. R. Germany†[7 14 15] R. f. Allemagne†[7 14 15]	151 246	158 549	190 852	228 202	250 443	293 092	356 841	389 206	408 305	329 702
Greece[7] Grèce[7]	9 435	10 140	11 353	13 164	12 324	16 149	19 780	21 582	23 232	1 958
Ireland Irlande	9 670	10 019	11 559	13 639	15 571	17 423	20 682	20 756	22 483	21 387
Italy[7] Italie[7]	85 160	87 720	99 404	125 693	138 582	153 011	181 983	182 753	188 521	...
Netherlands[7] Pays-Bas[7]	62 102	65 382	75 690	91 495	99 522	104 467	126 195	127 252	134 412	121 568
Portugal[7] Portugal[7]	7 964	7 655	9 648	13 966	17 941	19 205	25 358	26 405	30 348	24 319
Spain[7] Espagne[7]	28 835	29 965	35 086	49 119	60 533	70 957	87 722	93 330	99 765	81 876
United Kingdom[13] Royaume-Uni[13]	105 219	109 593	126 367	154 407	189 719	199 220	224 550	209 864	221 638	206 321
EFTA+ **AELE+**	**102 757**	**109 958**	**137 357**	**167 148**	**185 457**	**195 969**	**229 543**	**216 356**	**214 146**	**191 329**
Austria[7] Autriche[7]	19 632	20 996	26 852	32 683	36 564	38 935	49 092	50 804	54 122	48 616
Finland Finlande	12 443	13 233	15 333	18 922	21 843	24 613	27 005	21 717	20 746	18 069
Iceland[7] Islande[7]	840	904	1 115	1 590	1 598	1 402	1 679	1 761	1 684	1 342
Norway Norvège	13 890	15 560	20 307	22 641	23 223	23 680	27 219	25 576	25 916	23 892
Sweden Suède	26 428	28 553	32 697	40 709	45 858	49 131	54 857	49 996	49 914	42 687

Region, country or area Region, pays ou zone	1984	1985	1986	1987	1988	1989	1990	1991	1992	1993
free zone of Colon Zone libre de Colon	1 533	1 788	2 183	2 278	2 119	...	...	...	...	...
Saint Lucia[7] Sainte-Lucie[7]	49	57	87	80	116	109	127	110	123	...
Saint Kitts and Nevis Saint-Kitts-et-Nevis	20	20	25	28	27	29	28	...	...	...
St. Vincent-Grenadines[7] St. Vincent-Grenadines[7]	54	63	64	52	85	75	83	...	...	...
Suriname[7] Suriname[7]	364	329	335	306	...	...	472	...	...	...
Trinidad and Tobago[7] Trinité-et-Tobago[7]	2 173	2 147	1 386	1 462	1 413	1 578	2 080	1 983	1 850	1 612

Europe • Europe

Region, country or area Region, pays ou zone	1984	1985	1986	1987	1988	1989	1990	1991	1992	1993
Developed economies Économies développées	714 272	755 116	926 909	1 114 808	1 237 317	1 336 753	1 605 698	1 575 497	1 668 170	1 516 816
EU+[12] UE+[12]	607 379	642 211	789 864	950 099	1 054 441	1 143 784	1 366 040	1 357 166	1 444 090	1 309 763
Belgium-Luxembourg[7] Belgique-Luxembourg[7]	51 778	53 762	68 960	83 308	92 149	100 095	118 328	118 355	123 564	...
Denmark[7] Danemark[7]	15 958	16 941	21 214	25 598	27 657	28 113	35 135	36 011	41 064	35 915
France[7 13] France[7 13]	97 570	101 709	124 863	148 402	167 813	173 073	210 169	213 441	231 913	206 259
F. R. Germany†[7 14 15] R. f. Allemagne†[7 14 15]	169 784	184 009	243 303	294 045	323 277	365 029	422 041	402 845	430 272	365 296
Greece[7] Grèce[7]	4 812	4 543	5 649	6 531	5 430	7 544	8 106	8 675	9 509	...
Ireland Irlande	9 644	10 362	12 586	15 996	18 745	20 694	23 747	24 223	28 336	28 909
Italy[7] Italie[7]	74 562	76 742	97 631	116 413	127 886	140 624	170 386	169 538	178 165	...
Netherlands[7] Pays-Bas[7]	65 872	68 418	80 512	93 096	103 586	108 285	131 783	133 684	139 967	131 156
Portugal[7] Portugal[7]	5 203	5 684	7 242	9 321	10 990	12 801	16 419	16 332	18 302	15 370
Spain[7] Espagne[7]	23 564	24 249	27 231	34 199	40 340	43 463	55 646	60 197	64 342	62 872
United Kingdom[13] Royaume-Uni[13]	93 864	101 414	106 975	131 210	145 482	153 299	185 326	185 306	190 542	181 559
EFTA+ AELE+	104 112	109 581	132 810	159 586	177 662	187 396	224 728	216 654	222 178	205 437
Austria[7] Autriche[7]	15 742	17 247	22 529	27 175	31 062	32 485	41 138	41 126	44 377	40 200
Finland Finlande	13 505	13 617	16 336	19 560	22 151	23 270	26 650	23 103	23 529	23 501
Iceland[7] Islande[7]	738	816	1 099	1 374	1 424	1 386	1 591	1 549	1 528	1 400
Norway Norvège	18 893	19 989	18 095	21 493	22 425	27 112	34 044	34 116	35 193	31 778
Sweden Suède	29 381	30 467	37 267	44 509	49 890	51 607	57 512	55 229	56 149	49 864

71
Total imports and exports
Value in million US dollars [cont.]
Importations et exportations totales
Valeur en millions de dollars E-U [suite]

Region, country or area Region, pays ou zone	1984	1985	1986	1987	1988	1989	1990	1991	1992	1993
Switzerland[7] Suisse[7]	29 525	30 711	41 053	50 602	56 372	58 207	69 691	66 502	61 763	56 722
Other Europe **Autres pays d'Europe**	**1 035**	**1 115**	**1 351**	**1 832**	**2 024**	**2 002**	**2 445**	**2 547**	**2 883**	**2 768**
Faeroe Islands Iles Féroé	262	248	336	513	478	346	338	303	...	...
Gibraltar Gibraltar	91	147	164	231	...	...	...	...	...	...
Malta Malte	717	759	887	1 140	1 353	1 480	1 961	2 114	2 349	2 166
Yugoslavia, SFR†[7 16] **Yougoslavie, Rfs†[7 16]**	**11 956**	**12 164**	**11 750**	**12 603**	**13 154**	**14 802**	**18 890**	**11 804**	**15 772**	**13 991**
Eastern Europe[17] **Europe de l'est[17]**	**88 714**	**75 650**	**84 872**	**88 803**	**90 205**	**74 628**	**66 244**	**44 725**	**51 254**	**41 683**
Bulgaria[5] Bulgarie[5]	12 714	13 656	15 249	16 211	16 582	14 881	12 893	3 017	5 228	4 315
former Czechoslovakia†[5 18] l'anc. Tchécoslovaquie†[5 18]	27 091	11 152	13 358	14 883	14 593	14 277	13 106	10 014	12 530	...
former German D. R.†[5 19] anc. R. d. allemande†[5 19]	22 940	23 433	27 414	28 786	29 647	17 778	...	...	...	...
Hungary Hongrie	7 326	7 919	9 292	9 450	9 136	8 803	8 764	11 532	11 122	12 520
Poland[5] Pologne[5]	10 638	10 799	11 208	10 844	12 240	10 085	9 528	14 261	15 913	18 834
Romania[5 7] Roumanie[5 7]	7 731	8 401	8 082	8 311	7 641	8 436	9 115	5 600	6 147	5 683
former USSR†[5 20] **ancienne URSS†[5 20]**	**84 714**	**87 296**	**88 871**	**96 061**	**107 229**	**114 567**	**120 651**	**43 458**	..	..
former USSR-Europe[21] anc. URSS-Europe[21]	..	..	..	..	..	..	..	..	39 565	32 442
Belarus Bélarus	..	..	..	..	..	..	..	..	1 061	747
Republic of Moldova Rép. de Moldova	..	..	..	..	..	..	..	..	164	181
Russian Federation Fédération de Russie	..	..	..	..	..	..	..	..	34 981	26 807
Ukraine Ukraine	..	..	..	..	..	..	..	..	1 953	2 431

Africa · Afrique

Region, country or area Region, pays ou zone	1984	1985	1986	1987	1988	1989	1990	1991	1992	1993
South Africa[5 22 23] Afrique du Sud[5 22 23]	16 240	11 448	12 372	14 623	17 612	17 664	17 665	17 837	18 714	19 090
Developing economies **Econ. en dévelop.**	**59 990**	**52 407**	**55 795**	**54 791**	**60 068**	**61 769**	**71 715**	**69 619**	**75 914**	**74 812**
Northern Africa **Afrique du Nord**	**35 382**	**26 637**	**29 596**	**27 376**	**31 294**	**32 719**	**38 393**	**33 245**	**36 281**	**36 653**
Algeria[7] Algérie[7]	10 289	9 841	9 234	7 028	7 399	9 209	10 122	7 538	8 648	...
Egypt[7] Egypte[7]	10 766	5 495	8 680	7 596	8 657	7 434	9 202	7 754	8 357	8 176
Libyan Arab Jamah. Jamah. arabe libyenne	6 222	4 101	4 193	4 722	5 878	5 049	5 599	5 356	...	...

Region, country or area Région, pays ou zone	1984	1985	1986	1987	1988	1989	1990	1991	1992	1993
Switzerland[7] Suisse[7]	25 853	27 446	37 484	45 474	50 712	51 536	63 793	61 532	61 403	58 694
Other Europe **Autres pays d'Europe**	**551**	**603**	**774**	**983**	**1 098**	**1 223**	**1 549**	**1 677**	**1 902**	**1 616**
Faeroe Islands Iles Féroé	158	179	249	345	348	346	418	435	...	...
Gibraltar Gibraltar	35	62	65	85	...	...	...	...	...	...
Malta Malte	395	400	497	605	714	844	1 130	1 252	1 543	1 349
Yugoslavia, SFR†[7 16] Yougoslavie, Rfs†[7 16]	**10 254**	**10 642**	**10 298**	**11 425**	**12 597**	**13 363**	**14 312**	**9 548**	...	...
Eastern Europe[17] Europe de l'est[17]	**96 022**	**80 301**	**86 263**	**92 815**	**98 328**	**81 467**	**68 276**	**43 973**	**44 543**	**31 250**
Bulgaria[5] Bulgarie[5]	12 850	13 348	14 192	15 905	17 223	16 013	13 347	3 835	4 268	3 582
former Czechoslovakia†[5 18] l'anc. Tchécoslovaquie†[5 18]	28 543	11 510	13 227	14 723	14 894	14 440	11 882	10 878	11 656	...
former German D. R.†[5 19] anc. R. d. allemande†[5 19]	24 836	25 268	27 729	29 871	30 672	17 334	...	...	...	...
Hungary Hongrie	7 840	8 254	8 875	9 204	9 739	9 584	9 707	10 301	10 680	8 604
Poland[5] Pologne[5]	11 750	11 488	12 074	12 205	13 956	13 155	14 322	14 460	13 187	14 143
Romania[5 7] Roumanie[5 7]	9 899	10 173	9 761	10 491	11 391	10 486	5 870	4 124	4 372	4 536
former USSR†[5 20] ancienne URSS†[5 20]	**91 652**	**87 281**	**97 247**	**107 966**	**110 559**	**109 173**	**104 177**	**46 274**	..	..
former USSR-Europe[21] anc. URSS-Europe[21]	..	..	..	..	..	..	..	..	45 687	50 859
Belarus Bélarus	..	..	..	..	..	..	..	..	751	715
Republic of Moldova Rép. de Moldova	..	..	..	..	..	..	..	..	119	174
Russian Federation Fédération de Russie	..	..	..	..	..	..	..	..	39 967	44 297
Ukraine Ukraine	..	..	..	..	..	..	..	..	3 173	3 116

Africa · Afrique

Region, country or area Région, pays ou zone	1984	1985	1986	1987	1988	1989	1990	1991	1992	1993
South Africa[5 22 23] Afrique du Sud[5 22 23]	17 331	16 340	17 777	20 553	20 755	22 318	23 834	23 288	23 416	23 339
Developing economies **Econ. en dévelop.**	**62 491**	**59 919**	**46 697**	**51 254**	**50 742**	**58 143**	**76 289**	**73 126**	**71 781**	**71 322**
Northern Africa **Afrique du Nord**	**31 528**	**28 404**	**22 153**	**24 650**	**22 878**	**27 014**	**37 929**	**35 330**	**32 697**	**30 926**
Algeria[7] Algérie[7]	12 821	10 149	7 831	8 564	7 707	9 454	13 306	12 645	11 137	...
Egypt[7] Egypte[7]	3 140	1 838	2 214	2 037	2 120	2 565	2 582	3 618	3 072	2 243
Libyan Arab Jamah. Jamah. arabe libyenne	11 148	12 314	7 748	8 765	6 684	8 239	13 877	11 213	...	...

71
Total imports and exports
Value in million US dollars [cont.]
Importations et exportations totales
Valeur en millions de dollars E-U [suite]

Region, country or area Region, pays ou zone	1984	1985	1986	1987	1988	1989	1990	1991	1992	1993
Morocco[7] Maroc[7]	3 911	3 850	3 803	4 230	4 773	5 493	6 925	6 894	7 356	7 162
Sudan[24] Soudan[24]	1 147	757	961	929	1 060	...	...	...	...	...
Tunisia Tunisie	3 221	2 757	2 891	3 039	3 689	4 387	5 513	5 190	6 415	6 215
Other Africa **Autres Pays d'Afrique**	**24 608**	**25 770**	**26 199**	**27 415**	**28 774**	**29 050**	**33 322**	**36 375**	**39 633**	**38 159**
CACEU+[25][26] UDEAC+[25][26]	**2 541**	**2 718**	**3 334**	**3 188**	**3 008**	**2 696**	**3 258**	**2 867**	**3 267**	**3 912**
Cameroon[7][26] Cameroun[7][26]	1 112	1 152	1 704	1 723	1 452	1 261	1 564	1 351	1 175	...
Central African Rep.[7][26] Rép. centrafricaine[7][26]	87	113	167	204	201	150	154	159	...	...
Congo[7][26] Congo[7][26]	618	598	597	529	564	517	617	...	...	...
Gabon[7][26] Gabon[7][26]	724	855	866	732	791	767	922	884	...	...
ECOWAS+[27] CEDEAO+[27]	**11 612**	**12 498**	**11 158**	**11 481**	**11 394**	**11 248**	**12 669**	**18 522**	**19 910**	**17 724**
Benin[7] Bénin[7]	288	343	386	348	326	207	...	...	...	...
Burkina Faso[7] Burkina Faso[7]	253	333	405	434	454	391	540	536	...	...
Cape Verde[7] Cap-Vert[7]	83	84	107	100	106	112	136	147	180	...
Côte d'Ivoire Côte d'Ivoire	1 497	1 750	2 055	2 242	2 100	2 185	...	...	...	...
Gambia Gambie	100	93	105	127	137	161	200	222	...	...
Ghana Ghana	608	866	1 046	1 166	904	1 277	...	...	...	...
Liberia[7] Libéria[7]	363	284	259	308	272	...	...	...	...	...
Mali[7] Mali[7]	279	300	444	374	504	340	...	...	...	...
Mauritania[7] Mauritanie[7]	208	233	221	235	240	222	...	...	...	...
Niger[7] Niger[7]	288	369	368	311	387	363	389	355	...	...
Nigeria Nigéria	5 868	6 205	4 029	3 918	3 889	3 419	4 318	9 098	8 359	...
Senegal[7] Sénégal[7]	968	812	962	1 023	1 080	1 221	1 292	...	...	...
Sierra Leone Sierra Leone	157	154	132	137	156	182	155	163	156	147
Togo[7] Togo[7]	271	288	312	424	487	472	581	444	...	...
Rest of Africa **Afrique N.D.A.**	**10 455**	**10 555**	**11 707**	**12 745**	**14 373**	**15 106**	**17 396**	**14 986**	**16 456**	**16 523**
Angola[7] Angola[7]	713	671	619	443	989	1 140	...	...	...	...

Region, country or area / Region, pays ou zone	1984	1985	1986	1987	1988	1989	1990	1991	1992	1993
Morocco[7] / Maroc[7]	2 171	2 165	2 433	2 807	3 626	3 337	4 265	4 285	3 977	3 424
Sudan[24] / Soudan[24]	629	367	333	504	509	671	...	...	...	...
Tunisia / Tunisie	1 797	1 738	1 760	2 139	2 395	2 930	3 527	3 714	4 040	3 804
Other Africa / **Autres Pays d'Afrique**	**30 963**	**31 515**	**24 544**	**26 604**	**27 864**	**31 129**	**38 360**	**37 796**	**39 083**	**40 397**
CACEU+[25 26] / **UDEAC+[25 26]**	**4 165**	**3 853**	**2 896**	**2 740**	**3 636**	**4 569**	**5 604**	**5 343**	**5 279**	**5 256**
Cameroon[7 26] / Cameroun[7 26]	886	722	782	806	1 559	1 683	2 034	1 932	1 815	...
Central African Rep.[7 26] / Rép. centrafricaine[7 26]	85	92	66	130	130	134	120	109	...	...
Congo[7 26] / Congo[7 26]	1 183	1 087	777	517	752	1 153	975	...	...	...
Gabon[7 26] / Gabon[7 26]	2 012	1 952	1 271	1 286	1 195	1 599	2 474	2 273	...	...
ECOWAS+[27] / **CEDEAO+[27]**	**18 072**	**19 429**	**12 897**	**14 329**	**13 436**	**15 083**	**20 070**	**20 929**	**22 459**	**23 000**
Benin[7] / Bénin[7]	167	152	104	114	71	97	...	...	...	...
Burkina Faso[7] / Burkina Faso[7]	79	70	83	155	142	95	151	105	...	...
Cape Verde[7] / Cap-Vert[7]	3	6	4	8	3	7	6	6	5	...
Côte d'Ivoire / Côte d'Ivoire	2 707	2 946	3 355	3 110	2 792	2 931	...	...	...	...
Gambia / Gambie	49	43	35	40	58	27	40	42	...	...
Ghana / Ghana	540	623	876	957	1 008	1 020	...	...	...	...
Liberia[7] / Libéria[7]	452	436	408	382	396	...	...	...	...	...
Mali[7] / Mali[7]	133	124	212	179	214	247	...	...	...	...
Mauritania[7] / Mauritanie[7]	292	374	349	428	354	437	...	...	...	...
Niger[7] / Niger[7]	259	259	317	312	289	244	283	312	...	...
Nigeria / Nigéria	12 010	13 113	5 899	7 383	6 875	8 138	12 912	12 255	11 787	...
Senegal[7] / Sénégal[7]	633	554	625	606	591	693	741	...	...	...
Sierra Leone / Sierra Leone	131	130	142	129	106	138	136	145	149	121
Togo[7] / Togo[7]	191	190	204	244	242	245	268	253	...	...
Rest of Africa / **Afrique N.D.A.**	**8 726**	**8 232**	**8 751**	**9 536**	**10 792**	**11 476**	**12 686**	**11 524**	**11 345**	**12 142**
Angola[7] / Angola[7]	2 018	2 224	1 319	2 147	2 494	2 989	3 910	3 410	...	...

71
Total imports and exports
Value in million US dollars [*cont.*]
Importations et exportations totales
Valeur en millions de dollars E-U [*suite*]

Region, country or area Region, pays ou zone	1984	1985	1986	1987	1988	1989	1990	1991	1992	1993
Burundi[7] Burundi[7]	187	189	202	206	206	188	235	255	221	205
Chad[7] Tchad[7]	182	166	212	226	252	240	286	297	...	...
Comoros[7] Comores[7]	43	37	37	52	...	...	...	...	...	...
Djibouti Djibouti	222	201	184	205	201	196	215	214	...	...
Ethiopia Ethiopie	943	989	1 101	1 101	1 085	953	1 076	472	707	...
Kenya Kenya	1 469	1 436	1 613	1 755	1 975	2 148	2 124	1 799	1 828	1 711
Madagascar[7] Madagascar[7]	366	402	353	302	364	342	571	435	453	455
Malawi Malawi	271	295	258	295	420	507	575	703	735	545
Mauritius Maurice	471	529	684	1 013	1 289	1 326	1 618	1 558	1 624	1 919
Mozambique[7] Mozambique[7]	540	424	543	642	736	808	878	899	855	955
Réunion[7] Réunion[7]	791	841	1 137	1 464	1 662	1 732	2 160	2 129	2 315	2 057
Rwanda Rwanda	283	294	352	357	369	333	290	306	288	...
Seychelles Seychelles	88	99	105	114	159	165	187	172	192	192
Somalia[7] Somalie[7]	104	112	284	132	...	...	...	...	...	...
Uganda Ouganda	344	298	307	555	544	390	293	197	516	...
United Rep.Tanzania Rép. Unie de Tanzanie	862	862	892	904	805	998	1 363	1 170	1 431	1 127
Zaire[7] Zaïre[7]	675	791	875	756	771	849	897	711	438	...
Zambia[5] Zambie[5]	596	714	603	739	839	779	1 243	801	908	...
Zimbabwe[5] Zimbabwe[5]	1 103	1 031	1 133	1 205	1 304	1 627	1 850	2 037	2 213	...

Asia • Asie

Region, country or area Region, pays ou zone	1984	1985	1986	1987	1988	1989	1990	1991	1992	1993
Developed economies **Economies développées**	144 416	138 926	136 419	163 233	199 071	220 304	247 819	251 145	248 312	257 949
Israel[7 28 29] Israël[7 28 29]	9 828	10 136	10 813	14 348	15 018	14 347	16 790	18 658	20 253	22 621
Japan Japon	136 181	130 515	127 588	151 076	187 411	209 755	235 424	237 290	233 265	241 652
Developing economies[30] **Econ. en dévelop.**[30]	316 623	307 858	307 136	360 390	449 769	502 057	566 777	641 720	744 013	827 253
Asia Middle East **Moyen-Orient d'Asie**	101 186	86 893	77 039	76 921	83 945	87 478	98 387	107 119	135 943	141 862
Non petroleum exports . Petrole non compris	...	...	...	...	...	...	...	...	...	...

Region, country or area Region, pays ou zone	1984	1985	1986	1987	1988	1989	1990	1991	1992	1993
Burundi[7] Burundi[7]	103	112	154	90	133	78	75	91	73	69
Chad[7] Tchad[7]	131	62	99	109	159	155	188	194	...	...
Comoros[7] Comores[7]	7	16	20	12	...	...	...	...	...	...
Djibouti Djibouti	13	14	20	28	23	25	25	17	...	...
Ethiopia Ethiopie	416	334	464	370	421	452	294	189	169	...
Kenya Kenya	1 081	958	1 200	961	1 071	970	1 031	1 107	1 362	1 374
Madagascar[7] Madagascar[7]	333	274	315	333	274	316	308	305	268	262
Malawi Malawi	314	248	245	277	288	267	417	469	396	319
Mauritius Maurice	372	440	676	880	997	987	1 194	1 194	1 301	1 464
Mozambique[7] Mozambique[7]	96	77	79	97	103	105	126	162	139	132
Réunion[7] Réunion[7]	80	97	135	169	158	161	190	150	213	168
Rwanda Rwanda	156	98	118	130	101	121	131	93	68	...
Seychelles Seychelles	26	28	18	22	32	34	57	49	48	52
Somalia[7] Somalie[7]	55	91	89	104	...	...	...	...	...	...
Uganda Ouganda	385	387	436	319	274	250	152	201	143	179
United Rep.Tanzania Rép. Unie de Tanzanie	396	245	329	282	269	373	408	360	416	349
Zaire[7] Zaïre[7]	1 003	954	1 097	970	1 108	1 249	987	831	429	...
Zambia[5] Zambie[5]	661	547	704	873	1 178	1 344	1 414	1 128	1 095	...
Zimbabwe[5] Zimbabwe[5]	1 148	1 110	1 301	1 425	1 646	1 546	1 729	1 530	1 438	...

Asia • Asie

Region, country or area Region, pays ou zone	1984	1985	1986	1987	1988	1989	1990	1991	1992	1993
Developed economies **Economies développées**	**173 920**	**181 737**	**215 985**	**237 615**	**269 743**	**281 256**	**295 333**	**322 251**	**347 824**	**370 741**
Israel[7][28][29] Israël[7][28][29]	5 807	6 260	7 154	8 454	8 198	11 072	12 080	11 891	13 119	14 779
Japan Japon	169 706	177 202	210 813	231 352	264 903	273 982	287 648	315 163	339 911	362 286
Developing economies[30] **Econ. en dévelop.[30]**	**320 477**	**306 294**	**307 144**	**382 352**	**451 097**	**506 312**	**572 524**	**626 505**	**704 590**	**773 047**
Asia Middle East **Moyen-Orient d'Asie**	**109 059**	**97 687**	**78 681**	**87 017**	**88 209**	**100 883**	**120 018**	**110 793**	**116 449**	**125 994**
Non petroleum exports[38] Petrole non compris[38]	15 396	18 779	30 034	25 310	30 162	30 506	39 187	33 244	30 283	34 917

71
Total imports and exports
Value in million US dollars [cont.]
Importations et exportations totales
Valeur en millions de dollars E-U [suite]

Region, country or area Region, pays ou zone	1984	1985	1986	1987	1988	1989	1990	1991	1992	1993
Bahrain Bahreïn	3 480	3 107	2 405	2 714	2 593	3 134	3 712	4 115	4 263	3 825
Cyprus Chypre	1 364	1 247	1 274	1 484	1 857	2 285	2 569	2 632	3 301	2 534
Iran, Islamic Rep. of [7][31] Iran, Rép. islamique d' [7][31]	...	...	...	...	...	...	...	...	30 662	...
Iraq [7] Iraq [7]	6 693	7 619	6 360	3 854	5 960	6 956	4 834	...	...	...
Jordan Jordanie	2 784	2 733	2 432	2 703	2 705	2 126	2 603	2 508	3 255	3 539
Kuwait [7] Koweït [7]	6 901	6 188	5 687	5 496	6 146	6 297	3 923	4 761	7 251	7 042
Oman Oman	2 748	3 153	2 384	1 822	2 202	2 257	2 681	3 194	3 769	4 114
Qatar Qatar	1 162	1 139	1 099	1 134	1 267	1 326	1 695	1 720	...	...
Saudi Arabia [7] Arabie saoudite [7]	33 696	23 622	19 108	20 110	21 784	21 154	24 069	...	...	28 198
Syrian Arab Republic [7] Rép. arabe syrienne [7]	4 116	3 967	2 728	2 481	2 231	2 097	2 526	3 151	3 490	4 140
Turkey [7] Turquie [7]	10 757	11 342	11 107	14 157	14 320	15 797	22 247	21 204	22 579	29 066
United Arab Emirates Emirats arabes unis	6 936	6 548	6 422	7 226	8 521	10 010	11 199	13 746	17 410	19 520
Yemen [32] Yémen [32]	2 378	1 998	1 640	1 378	...	...	...	...	...	...
Other Asia [33] Autres pays d'Asie [33]	215 438	220 965	230 097	283 469	365 825	414 579	468 390	534 600	605 438	682 644
Afghanistan [34] Afghanistan [34]	1 390	1 194	1 404	996	900	822	936	616	...	...
Bangladesh Bangladesh	2 827	2 505	2 546	2 715	3 042	3 651	3 651	3 353	3 908	3 989
Brunei Darussalam [7] Brunéi Darussalam [7]	626	615	656	641	744	859	1 020	1 178	1 176	...
China Chine	27 409	42 252	42 909	43 368	55 213	58 428	53 819	63 843	80 597	103 881
Hong Kong Hong-kong	28 568	29 703	35 367	48 465	63 896	72 155	82 474	100 255	123 430	138 658
India Inde	15 248	15 935	15 413	16 678	19 103	20 550	23 583	20 445	23 594	22 763
Indonesia [7] Indonésie [7]	13 882	10 262	10 718	12 891	13 249	16 444	21 837	25 869	27 311	28 086
Korea, Republic of [7] Corée, République de [7]	30 635	31 070	31 599	41 157	51 969	61 477	69 740	81 496	81 795	83 784
Lao People's Dem. Rep. [7] Rép. dém. pop. lao [7]	48	131	131	146	162	...	...	...	...	...
Macau Macao	798	779	890	1 121	1 290	1 478	1 534	1 843	1 948	1 893
Malaysia Malaisie	14 051	12 253	10 806	12 681	16 507	22 481	29 259	36 649	39 853	45 657

Region, country or area Région, pays ou zone	1984	1985	1986	1987	1988	1989	1990	1991	1992	1993
Bahrain Bahreïn	3 204	2 897	2 199	2 430	2 411	2 831	3 761	3 513	3 464	3 689
Cyprus Chypre	575	476	504	621	711	795	948	958	1 005	883
Iran, Islamic Rep. of[7][31] Iran, Rép. islamique d'[7][31]	12 527	13 449	...	...	...	...	...	...	...	...
Iraq Iraq	...	...	...	...	...	...	...	...	...	...
Jordan Jordanie	752	791	733	932	1 000	1 107	1 063	1 130	1 219	1 246
Kuwait[7] Koweït[7]	12 275	10 479	7 221	8 357	7 655	11 502	6 956	1 088	6 567	10 491
Oman Oman	3 068	3 938	1 834	2 491	2 476	4 068	5 508	4 871	5 428	5 299
Qatar Qatar	4 498	3 542	...	...	...	2 687	3 529	3 107	...	...
Saudi Arabia[7] Arabie saoudite[7]	37 545	27 481	20 187	23 199	24 377	28 382	44 417	...	...	...
Syrian Arab Republic[7] Rép. arabe syrienne[7]	1 853	1 637	1 325	1 350	1 345	3 006	4 062	3 143	3 093	3 146
Turkey[7] Turquie[7]	7 134	7 957	7 458	10 189	11 618	11 626	12 921	13 588	14 878	15 410
United Arab Emirates Emirats arabes unis	13 827	13 124	15 837	...	...	...	...	...	...	...
Yemen[32] Yémen[32]	43	54	37	101	...	...	...	...	...	...
Other Asia[33] Autres pays d'Asie[33]	211 418	208 607	228 463	295 335	362 888	405 429	452 505	515 711	583 693	643 171
Afghanistan[34] Afghanistan[34]	633	567	552	512	395	236	235	188	...	...
Bangladesh Bangladesh	932	986	880	1 067	1 291	1 305	1 671	1 689	2 098	2 273
Brunei Darussalam[7] Brunéi Darussalam[7]	3 204	2 972	1 797	1 902	1 708	1 883	2 228	2 698	2 370	...
China Chine	26 143	27 349	30 940	39 492	47 466	51 952	62 423	71 892	84 858	91 737
Hong Kong Hong-kong	28 323	30 187	35 439	48 476	63 163	73 140	82 160	98 577	119 512	135 248
India Inde	9 469	9 144	9 391	11 299	13 235	15 872	17 970	17 727	19 641	21 554
Indonesia[7] Indonésie[7]	21 902	18 590	16 075	17 135	19 465	22 160	25 674	29 543	33 861	33 612
Korea, Republic of[7] Corée, République de[7]	29 218	30 164	34 747	47 358	60 993	62 384	64 880	71 736	76 575	82 189
Lao People's Dem. Rep.[7] Rép. dém. pop. lao[7]	12	54	60	62	81	...	...	...	...	...
Macau Macao	913	906	1 046	1 397	1 492	1 644	1 694	1 655	1 749	1 685
Malaysia Malaisie	16 483	15 316	13 688	17 958	21 082	25 048	29 453	34 350	40 772	47 122

71

Total imports and exports
Value in million US dollars [cont.]
Importations et exportations totales
Valeur en millions de dollars E-U [suite]

Region, country or area Region, pays ou zone	1984	1985	1986	1987	1988	1989	1990	1991	1992	1993
Maldives Maldives	53	53	45	81	90	113	138	161	189	191
Myanmar Myanmar	239	284	304	268	244	191	270	646	651	814
Nepal Népal	416	455	460	571	679	582	686	765	792	880
Pakistan Pakistan	5 873	5 891	5 377	5 829	6 620	7 119	7 356	8 461	9 365	9 481
Philippines Philippines	6 241	5 445	5 134	6 811	8 731	10 732	13 042	12 853	15 459	17 271
Singapore Singapour	28 669	26 288	25 511	32 566	43 862	49 657	60 774	66 093	72 132	85 229
Sri Lanka Sri Lanka	1 845	1 807	1 829	2 027	2 211	2 088	2 634	3 084	3 410	3 732
Thailand Thaïlande	10 398	9 242	9 181	12 994	20 285	25 770	33 065	37 569	40 686	46 058
former USSR-Asia [35] anc. URSS-Asie [35]	..	..	..	..	..	..	..	..	2 632	2 747
Kazakhstan Kazakhstan	..	..	..	..	..	..	..	..	469	358
Uzbekistan Ouzbékistan	..	..	..	..	..	..	..	..	929	947

Oceania · Océanie

Region, country or area Region, pays ou zone	1984	1985	1986	1987	1988	1989	1990	1991	1992	1993
Developed economies Economies développées	33 135	33 443	33 310	37 566	44 884	56 085	53 040	52 385	55 415	57 157
Australia [5] Australie [5]	25 922	25 900	26 109	29 321	36 101	44 944	42 024	41 651	43 808	45 478
New Zealand Nouvelle-Zélande	6 521	6 293	6 368	7 640	7 709	9 223	9 976	8 828	9 679	10 118
Developing economies Econ. en dévelop.	3 471	3 472	3 847	4 345	4 779	5 043	5 211	5 479	5 530	5 142
American Samoa [7][36] Samoa américaines [7][36]	284	296	313	346	339	378	360	372	418	...
Fiji Fidji	450	442	435	379	461	581	754	652	624	634
French Polynesia [7] Polynésie française [7]	539	549	736	827	808	791	929	915	894	847
Kiribati [37] Kiribati [37]	18	15	14	18	22	23	27	26	37	
New Caledonia [7] Nouvelle-Calédonie [7]	311	348	458	626	604	766	883	863	917	850
Papua New Guinea [5] Papouasie-Nvl-Guinée [5]	1 086	1 008	1 068	1 262	1 593	1 546	1 271	1 615	1 521	1 299
Samoa Samoa	50	51	47	62	76	75	81	94	110	105
Solomon Islands [7] Iles Salomon [7]	79	83	72	81	98	114	95	110	97	101
Tonga Tonga	41	41	41	48	56	54	62	59	63	61

Region, country or area Region, pays ou zone	1984	1985	1986	1987	1988	1989	1990	1991	1992	1993
Maldives Maldives	18	23	25	31	40	45	52	54	40	35
Myanmar Myanmar	301	304	288	219	166	210	325	419	531	585
Nepal Népal	127	161	142	151	191	159	209	263	374	391
Pakistan Pakistan	2 614	2 740	3 379	4 178	4 527	4 779	5 523	6 464	7 273	6 672
Philippines Philippines	5 391	4 629	4 842	5 720	7 035	7 747	8 186	8 793	9 752	12 930
Singapore Singapour	24 072	22 815	22 495	28 692	39 305	44 661	52 730	58 964	63 435	74 008
Sri Lanka Sri Lanka	1 454	1 291	1 216	1 348	1 481	1 529	1 912	1 965	2 443	2 792
Thailand Thaïlande	7 413	7 121	8 879	11 722	15 953	20 078	23 084	28 428	32 473	37 173
former USSR-Asia[35] anc. URSS-Asie[35]	..	..	..	..	..	..	..	..	4 447	3 883
Kazakhstan Kazakhstan	..	..	..	..	..	..	..	..	1 398	1 271
Uzbekistan Ouzbékistan	...	...	...	...	...	...	...	...	869	707

Oceania · Océanie

	1984	1985	1986	1987	1988	1989	1990	1991	1992	1993
Developed economies Economies développées	**27 152**	**27 491**	**27 432**	**32 481**	**40 796**	**44 366**	**47 362**	**49 393**	**50 436**	**50 848**
Australia[5] Australie[5]	23 114	22 613	22 573	26 624	33 238	37 134	39 760	41 855	42 839	42 715
New Zealand Nouvelle-Zélande	5 524	5 722	5 882	7 196	8 849	8 875	9 394	9 649	9 839	10 537
Developing economies Econ. en dévelop.	**1 898**	**1 887**	**2 046**	**2 429**	**2 961**	**3 022**	**2 737**	**3 050**	**3 319**	**4 035**
American Samoa[7][36] Samoa américaines[7][36]	212	202	254	288	368	308	311	327	318	...
Fiji Fidji	256	236	335	381	372	444	497	450	435	407
French Polynesia[7] Polynésie française[7]	32	41	41	83	75	89	111	127	107	148
Kiribati[37] Kiribati[37]	11	4	2	2	5	5	3	3	5	...
New Caledonia[7] Nouvelle-Calédonie[7]	207	271	191	225	464	675	449	444	409	356
Papua New Guinea[5] Papouasie-Nvl-Guinée[5]	917	928	1 031	1 241	1 452	1 300	1 177	1 460	1 790	2 484
Samoa Samoa	19	16	11	12	15	13	9	6	6	6
Solomon Islands[7] Iles Salomon[7]	93	70	65	64	81	74	70	83	101	96
Tonga Tonga	9	5	6	6	8	9	11	13	12	17

71

Total imports and exports
Value in million US dollars [cont.]

Importations et exportations totales
Valeur en millions de dollars E-U [suite]

Region, country or area Region, pays ou zone	1984	1985	1986	1987	1988	1989	1990	1991	1992	1993
Vanuatu Vanuatu	69	70	57	70	71	71	96	83	82	...

Source:
Trade statistics database of the Statistical Division of the
United Nations Secretariat.
+ For Member States of this grouping, see
   Annex I - Other groupings.

† For detailed descriptions of data pertaining to
former Czechoslovakia, Germany, SFR Yugoslavia and former
USSR, see Annex I – Country or area nomenclature, regional
and other groupings.

1 Including trade conducted in accordance with the
   supplementary protocol to the treaty on the basis of
   relations between the Federal Republic of Germany and the
   former German Democratic Republic.
2 United States, Canada, Developed market economies of Europe,
   Israel, Japan, Australia, New Zealand and South African
   customs union.
3 This classification is intended for statistical convenience
   and does not necessarily express a judgement about the stage
   reached by a particular country in the development process.
4 Prior to January 1992 includes Eastern Europe and the former
   USSR. Beginning January 1992, includes Eastern Europe and
   the European countries of the former USSR.
5 Imports F.O.B.
6 Latin American Integration Association. Formerly Latin
   American Free Trade Association.
7 Country or area using special trade system. See technical
   notes for explanation of trade systems.

8 Central American Common Market.
9 Prior to 1985, year ending 30 September of the year stated.

10 Prior to 1986, includes Aruba.
11 Excluding trade of the free zone of Colon.
12 The EU totals have been recalculated for the whole period
   covered by this table to include the data for Portugal and
   Spain which joined the community on 1 January 1986.

13 Including monetary gold.
14 Prior to 1991, excludes trade conducted in accordance with
   the supplementary protocol to the treaty on the basis of
   relations between the Federal Republic of Germany and the
   former German Democratic Republic.
15 Data prior to January 1991, pertain to territorial
   boundaries of the Federal Republic of Germany prior to 3
   October 1990.
16 Beginning 1992, data refer to Croatia, Slovenia, TFYR
   Macedonia and former Yugoslavia.
17 Beginning 1993, includes Czech Republic and Slovakia.
18 Beginning 1985 data are not comparable to those shown for
   prior periods due to revisions of the Koruna to US dollar
   exchange rate.

19 Beginning 1989 data are not comparable to those shown for

Source:
Base de données pour les statistiques du commerce extérieur
de la Division de statistique du Secrétariat de l'ONU.
+ Les Etats membres de ce groupement, voir
   annexe I - Autres groupements.

† Pour les descriptions en détails des données
relatives à l'ancienne Tchécoslovaquie, l'Allemagne, la Rfs
Yougoslavie et l'ancienne URSS, voir l'Annexe I –
Nomenclature des pays ou zones, groupements régionaux et
autres groupements.

1 Y compris le commerce effectué en accord avec le protocole
   additionnel au traité définissant la base des relations
   entre la République Fédérale d'Allemagne et l'ancienne
   République Démocratique Allemande.
2 Etats-Unis, Canada, Pays développés d'Europe à economie de
   marche, Israel, Japon, Australie, Nouvelle-Zéalande et
   l'union douaniere de l'Afrique du Sud.
3 Cette classification est utilisée pour plus de commodite
   dans la presentation des statistiques et n'implique pas
   necessairèment un jugement quant au stage de developpement
   auquel est parvenu un pays donne.
4 Avant 1992, y compris de l'Europe de l'est et l'ancienne
   URSS. A partir de 1992, y compris de l'Europe de l'est et
   les pays européennes de l'ancienne URSS.
5 Importations F.O.B.
6 Association Latino-américaine d'intégration. Antérieurement
   Association Latino-américaine de libre-échange
7 Pays ou zone utilisant un système spécial du commerce. Pour
   l'explication du système de commerce, voir les notes
   techniques.
8 Marché commun de l'Amérique centrale.
9 Avant 1985, année finissant le 30 septembre de l'année
   indiquée.
10 Avant 1986, comprend Aruba.
11 Non compris le commerce de la zone libre de Colon.
12 On a récalculé pour toute la période considerée les totaux
   relatifs à la UE à fin d'y inclure les données rélatives à
   l'Espagne et le Portugal qui sont devenus membres de la
   communauté le 1 er janvier 1986.
13 Y compris l'or monétaire.
14 Avant 1991, non compris le commerce effectué en accord avec
   le protocole additionnel au traite définissant la base des
   relations entre la République fédérale d'allemagne et
   l'ancienne République démocratique allemande.
15 Les données relatives à la période précedant janvier 1991
   correspondent aux limites territoriales de la république
   fédérale d'Allemagne antérieur au 3 octobre 1990.
16 Avant 1992, les données se rapportent aux Croatie, Slovénie,
   TARY Macedonie et la république fédérative de Yougoslavie.
17 Avant 1993, y compris République tchèque et Slovaquie.
18 A partir de l'année 1985 les chiffres ne sont pas
   comparables aux chiffres indiqués pour les années antérieurs
   à cause des révisions de taux de change de Koruna au dollar
   de E-U.
19 A partir de l'année 1989 les chiffres ne sont pas

Region, country or area Région, pays ou zone	1984	1985	1986	1987	1988	1989	1990	1991	1992	1993
Vanuatu Vanuatu	44	31	17	18	20	22	19	18	24	...

prior periods due to revisions of the Mark to US dollar exchange rate.

20 For 1991, rouble values are converted to US dollars using commercial exchange rate and are not comparable to those shown for prior periods.

21 Includes Estonia, Latvia and Lithuania.
22 Excluding exports of gold bullion and gold coin.
23 The South African customs union comprising Botswana, Lesotho, Namibia, South Africa and Swaziland. Trade between the component countries is excluded.
24 Excluding exports of camels to Egypt.
25 Customs and Economic Union of Central Africa.
26 Inter-trade among the members of Customs and Economic Union of Central Africa is excluded.
27 Economic Community of West African States.
28 Imports and exports net of returned goods. The figures also exclude Judea and Samaria and the Gaza area.

29 Excluding military imports.
30 Beginning 1992, includes Armenia, Azerbaijan, Georgia, Kazakhstan, Kyrgyzstan, Tajikistan, Turkmenistan and Uzbekistan.
31 Year ending 20 December of the year stated.
32 Comprises trade of the former Democratic Yemen and former Yemen Arab Republic including any intertrade between them.

33 Includes imports and exports of People's Democratic Republic of Korea, Mongolia and Viet Nam which comprised former Centrally Planned Economies of Asia.

34 Year beginning 21 March of the year stated.
35 Includes Armenia, Azerbaijan, Georgia, Kyrgyzstan, Tajikistan, and Turkmenistan.
36 Year ending 30 September of the year stated.
37 Including Tuvalu (formerly Ellice Islands).
38 Data refer to total exports less petroleum exports of Asia Middle East countries where petroleum, in this case, is the sum of SITC groups 333, 334 and 335.

comparables aux chiffres indiqués pour les années antérieurs à cause des révisions de taux de change de Mark au dollar de E-U.

20 A partir de l'année 1991 les valeurs enroubles sont convertis en dollars des E.U. en utilisant le taux de change commercial et ne sont pas comparables aux données des périodes antérieures.

21 Y compris Estonie, Lettonie et Lituanie.
22 Non compris les exportations de Lingots et pièces d'or.
23 L'Union Douaniere de l'Afrique Meridionale comprend Botswana, Lesotho, Namibie, Afrique du Sud et Swaziland. Non compris le commerce entre ces pays.
24 Non compris les exportations des chameaux en Egypte.
25 Union douanière et économique de l'Afrique centrale.
26 Non compris le commerce entre les pays membres du UDEAC.

27 Communauté économique des états de l'Afrique de l'Ouest.
28 Importations et exportations nets. Ne comprennent pas les marchandises retournées. Sont également exclués les données de la Judee et de Samaria et ainsi que la zone de Gaza.
29 Non compris les importations des economats militaires.
30 A partir de 1992, y compris Arménie, Azerbaidjan, Géorgie, Kazakhstan, Kirghizistan, Tadjikistan, Turkménistan et Ouzbékistan.
31 Année finissant le 20 décembre de l'année indiquée.
32 Y compris le commerce de l'ancienne République populaire Démocratique de Yémen, le commerce de l'ancienne République Arabe de Yémen et le commerce entre eux.
33 Y compris les importations et les exportations de la République Démocratique de Corée, de la Mongolie et du Viet Nam auxquelles comprenait l'ancienne economie planifiée de l'Asie.
34 Année commençant le 21 mars de l'année indiquée.
35 Y compris Arménie, Azerbaidjan, Géorgie, Kirghizistan, Tadjikistan, et Turkménistan.
36 Année finissant le 30 septembre de l'année indiquée.
37 Y compris Tuvalu (anciennement îles Ellice).
38 Les données se rapportent aux exportations totales moins les exportations pétroliers. Dans ce cas, le pétrole est la somme des groupes CTCI 333, 334 et 335.

72
World exports by commodity classes and by regions
Exportations mondiales par classes de marchandises et par régions

In million US dollars f.o.b.

Exports from	Year	World Monde /1,2,3	Developed economies Econ. développées /2,3,4	Developing economies Economies en voie de développement /1,3,4 Total	OPEC+ OPEP+	Eastern Europe and former USSR Europe de l'Est et l'ancienne URSS Total /2,3,7	fmr USSR anc URSS /10	Europe Total /2	EEC+ CEE+ /2	EFTA+ AELE+	Developed Economies South Africa Afrique du Sud
									Total trade (SITC, Rev. 2 and Rev. 3, 0-9) /6		
World /1,2,3	1980	2000946	1335949	504031	127667	143964	61924	875843	744446	129567	13667
	1989	3024780	2128584	698252	100076	172321	74856	1311633	1118356	188747	12555
	1990	3396509	2437512	769567	111882	142939	65498	1577399	1352846	219432	12661
	1991	3438564	2447701	850438	126558	109319	53144	1583976	1372054	207411	13828
	1992	3685963	2582114	976161	146465	95658	41540	1663885	1447550	211627	15030
Developed economies /2,3,4	1980	1258935	891453	316150	99886	42235	21562	650015	542054	106360	12566
	1989	2129315	1638558	423216	68885	49004	27961	1105547	938118	164670	11442
	1990	2445205	1894303	477705	80179	49905	26382	1325795	1129540	192626	11525
	1991	2507065	1901607	526522	92251	56794	27615	1343611	1155747	184044	12074
	1992	2668725	1993439	589798	103248	62770	25582	1408624	1216363	188112	12439
Developing economies /1,3,4	1980	586897	401228	155476	22844	23015	13426	186082	172526	13342	1100
	1989	701110	436163	230203	26883	28579	18678	157958	144342	13409	1112
	1990	779361	471782	257967	28744	28054	19051	187043	171542	15286	1134
	1991	840461	502815	307006	32632	23136	16310	200909	184614	15639	1735
	1992	927432	536384	369548	41725	14934	8405	208336	192448	15404	2559
OPEC+	1980	306770	231096	70072	3966	3716	856	113998	106075	7869	75
	1989	141018	96055	40915	4824	2554	678	39089	36550	2526	5
	1990	164512	105115	41709	4513	3053	860	47251	44586	2651	6
	1991	165735	114691	45920	4371	2499	688	52456	49240	3188	7
	1992	174587	117543	53935	8930	2117	453	48490	46332	2121	16
Eastern Europe and the former USSR /2,3,7	1980	155115	43269	32406	4936	78714	26937	39747	29866	9865	1
	1989	194355	53863	44834	4308	94739	28218	48129	35897	10667	1
	1990	171944	71420	33896	2959	64980	20064	64561	51765	11520	2
	1991	91038	43280	16910	1675	29390	9219	39456	31694	7729	19
	1992	89807	52292	16814	1493	17954	7553	46926	38739	8110	32
former USSR /3	1980	76449	24431	19740	1740	32278	.	22672	17260	5413	0
	1989	109227	26070	32626	1067	50531	.	22947	17624	5300	0
	1990	104177	43065	25520	946	35592	.	38589	32252	6300	0
	1991	46274	...	...	...	...	.	...	...	...	0
	1992	.	.	.	.	.	.	.	.	.	...
former USSR-Europe /8	1980	.	.	.	.	.	.	.	.	.	
	1989	.	.	.	.	.	.	.	.	.	
	1990	.	.	.	.	.	.	.	.	.	
	1991	.	.	.	.	.	.	.	.	.	
	1992	45612	28853	9312	245	6001	.	25101	21271	3805	0
Developed economies-Europe /2	1980	801865	615412	149325	60121	31447	14681	540357	446087	92771	7888
	1989	1318944	1094321	171148	41216	37827	19041	926822	780983	143260	7617
	1990	1578155	1313211	204720	48755	40466	19279	1124857	952286	169216	7911
	1991	1585982	1308446	211081	52765	47127	20187	1133562	969681	160282	7894
	1992	1683666	1377738	233000	58314	53747	19438	1194604	1026287	164441	7730
European Economic Community+ /2	1980	689597	527375	132681	54303	23888	10829	462282	384556	76305	7168
	1989	1130391	936872	150609	37138	27979	13637	794137	674408	117330	7144
	1990	1351043	1122944	179195	43351	30039	13640	962609	820585	138967	7360
	1991	1367722	1125231	185025	46926	38950	17324	975825	839691	132834	7392
	1992	1455381	1186722	205613	52508	44688	16788	1029996	889075	137297	7159
European Free Trade Association+	1980	111610	87481	16588	5777	7541	3852	77552	61041	16434	719
	1989	187321	156446	20423	4020	9822	5385	131783	105715	25889	473
	1990	225512	188908	25370	5330	10402	5625	160973	130477	30199	551
	1991	216495	181726	25876	5757	8147	2845	156338	128648	27391	503
	1992	226335	189343	27200	5720	9020	2612	163073	135747	27073	570
Other developed economies	1980	457070	276042	166825	39765	10788	6881	109658	95967	13589	4678
	1989	810371	544237	252068	27669	11177	8920	178726	157134	21411	3826
	1990	867050	581099	272985	31425	9440	7103	200938	177254	23410	3613
	1991	921083	593161	315441	39486	9667	7428	210050	186066	23762	4180
	1992	985059	615701	356798	44934	9023	6145	214021	190076	23671	4709

En millions de dollars E.-U. f.o.b.

← Exportations vers

Economies développées /2,3,4				Developing economies / Economies en voie de développement /1,3,4								
Canada	USA E-U	Japan Japon	Australia New Zealand Australie Nouvelle-Zélande	Africa Afrique	America Amerique Total	LAIA+ ALADI+	Asia Asie Mid.East Moyen Orient	Other Autres /9	Oceania Océanie	Année	Exportations en provenance de ↓	

Commerce total (CTCI, Rev. 2 et Rev. 3, 0-9) /6

Canada	USA E-U	Japan	Aust/NZ	Africa	America Total	LAIA	Mid.East	Other	Oceania	Année	Provenance
50740	240320	124484	24458	84367	126105	83358	96519	173558	2994	1980	Monde /1,2,3
107764	456020	183881	45542	70459	124310	81609	91461	381821	4639	1989	
111004	476936	202130	44497	81938	132160	87641	101513	428766	5036	1990	
112196	475857	204417	42824	81869	150012	106091	113860	485210	5153	1991	
116052	519256	204260	47046	82803	177717	132855	133460	562549	5371	1992	
43176	122805	40279	17304	65004	76244	60160	67087	97449	2118	1980	Economies dévelopées /2,3,4
97010	288125	91610	34808	50602	84473	62111	58919	215923	3440	1989	
100709	308202	102235	34222	58966	95230	70099	68028	238219	3716	1990	
101348	299479	100471	31534	57462	109533	82719	78729	266735	3742	1991	
104640	320426	98482	34082	61333	129860	100438	89301	296656	4065	1992	
7325	116139	82453	7051	15007	44698	22367	23604	68810	876	1980	Economies en voie de développement /1,3,4
10420	165665	89399	10540	15937	31733	18593	27375	152205	1199	1989	
9992	166075	96255	10128	19439	28699	16871	29156	177408	1318	1990	
10641	174874	102101	11185	22660	36707	22924	32498	212313	1410	1991	
11022	197059	102955	12740	20107	46750	31635	40597	256526	1303	1992	
4483	56536	53076	2925	4243	25893	10275	9282	29068	10	1980	OPEP+
1305	27473	26822	1344	3296	9141	4867	8646	18880	56	1989	
1413	23211	31861	1348	5823	4615	1350	9260	21231	69	1990	
1243	27656	31726	1565	6664	6618	2529	9181	22854	79	1991	
979	34026	32034	1955	3425	8334	5158	12538	29039	47	1992	
239	1376	1752	103	4356	5164	832	5829	7299	0	1980	Europe de l'Est et l'ancienne URSS /2,3,7
334	2230	2873	195	3920	8105	905	5167	13693	0	1989	
302	2659	3640	147	3533	8231	670	4328	13139	1	1990	
207	1504	1846	106	1747	3772	448	2634	6162	1	1991	
389	1771	2823	224	1363	1106	781	3562	9368	3	1992	
46	233	1464	14	1380	3679	130	2210	4574	0	1980	l'ancienne URSS /3
61	842	2134	53	1494	6659	249	1624	10784	0	1989	
111	1189	3114	62	1515	7406	305	2554	10910	0	1990	
...	...	...	...	...	...	...	...	...	...	1991	
.	.	.	.	.	.	.	.	.	.	1992	
.	.	.	.	.	.	.	.	.	.	1980	l'ancienne URSS-Europe /8
.	.	.	.	.	.	.	.	.	.	1989	
.	.	.	.	.	.	.	.	.	.	1990	
.	.	.	.	.	.	.	.	.	.	1991	
226	876	2490	137	262	514	427	1832	6473	0	1992	
5776	44041	8181	6129	49626	25480	19374	41359	25147	519	1980	Economies développées-Europe /2
14160	99306	27733	11987	38667	24939	17006	37948	59466	1052	1989	
14652	111269	34454	12229	45990	29621	19957	45365	69885	1283	1990	
13979	101407	32720	10454	44165	32370	22446	50603	73160	1111	1991	
13355	109492	31630	11331	45962	37483	26682	56820	83159	1407	1992	
4953	38619	6672	5309	45716	22214	16875	36698	21653	487	1980	Communauté Economique Européenne+ /2
11606	85168	23085	10145	35743	21797	14572	33824	50563	1022	1989	
11832	95908	28486	10149	42727	25706	16942	39878	58953	1241	1990	
11492	87418	27126	8750	40719	28328	19414	44732	61834	1073	1991	
10889	94531	26351	9376	42746	32921	23389	50802	70873	1378	1992	
821	5393	1509	819	3870	3260	2498	4653	3493	32	1980	Association Européenne de Libre Echange+
2553	14058	4629	1841	2880	3128	2432	4101	8868	29	1989	
2806	15310	5954	2078	3203	3903	3008	5462	10874	42	1990	
2486	13923	5577	1703	3380	4022	3025	5845	11261	38	1991	
2464	14846	5267	1952	3137	4553	3284	5991	12214	29	1992	
37400	78764	32098	11175	15378	50763	40786	25727	72302	1599	1980	Autres economies développées
82850	188819	63877	22821	11935	59534	45105	20971	156457	2389	1989	
86057	196933	67781	21993	12976	65609	50142	22663	168334	2433	1990	
87369	198072	67751	21080	13297	77163	60273	28125	193575	2631	1991	
91285	210934	66852	22751	15371	92376	73757	32481	213497	2658	1992	

72
World exports by commodity classes and by regions (continued)
Exportations mondiales par classes de marchandises et par régions (suite)

In million US dollars f.o.b.

Exports from	Year	World Monde /1,2,3	Developed economies Econ. dévelop-pées /2,3,4	Developing economies /1,3,4 Total	OPEC+ OPEP+	Eastern Europe and former USSR Europe de l'Est et l'ancienne URSS Total /2,3,7	fmr USSR anc URSS /10	Europe Total /2	EEC+ CEE+ /2	EFTA+ AELE+	Develop Econom South Africa Afrique du Su
						Total trade (SITC, Rev. 2 and Rev. 3, 0-9) /6(continue					
Canada	1980	64935	55286	7601	1869	1776	1317	9447	8565	880	1
	1989	116003	105585	9657	1661	761	581	11501	9824	1674	12
	1990	126897	115641	10162	1630	1081	964	12209	10269	1927	14
	1991	126762	114385	10981	1848	1396	1290	11737	10250	1475	1
	1992	134617	122615	10774	1762	1227	1059	11469	9576	1891	1
United States	1980	216592	128821	82805	17465	3850	1510	65064	57849	7185	248
	1989	349356	223890	11 793	12982	5278	4262	92594	82516	9988	166
	1990	374449	242222	127368	13418	4226	3072	103832	93049	10666	175
	1991	400984	249153	146869	18587	4673	3500	109122	97514	11524	213
	1992	424871	252611	166710	21324	5252	3584	107457	97340	10032	247
Japan	1980	129807	61602	64619	18482	3585	2778	21436	18120	3264	180
	1989	275174	168064	103356	10772	3755	3082	56283	48191	8018	178
	1990	286947	170164	113384	13575	3308	2563	62412	53846	8437	15
	1991	314525	177648	134015	16474	2862	2115	68325	59557	8651	167
	1992	339651	186080	151727	18995	1844	1128	71584	62952	8488	175
Australia, New Zealand	1980	26673	15513	8291	1809	1513	1276	4334	4161	165	12
	1989	43976	27244	14852	2145	1312	987	7164	6567	593	12
	1990	47735	29483	16701	2651	715	500	7863	6816	1041	10
	1991	49768	29794	18608	2487	591	505	7206	6313	888	17
	1992	50765	29122	20573	2578	196	82	7480	6759	694	23
Developing economies-Africa	1980	94942	78737	12962	1003	2450	733	46251	43864	2329	30
	1989	56330	44744	8668	1526	2224	977	32735	31806	922	37
	1990	66548	55113	8764	1666	2168	1072	40125	39200	917	31
	1991	70092	57087	10145	1985	2235	1145	41878	40173	1410	31
	1992	70120	57798	10627	1886	896	244	42329	40806	1322	43
Developing economies-America	1980	107878	69406	29644	3601	6972	5196	26095	23954	2041	16
	1989	122876	85291	28687	3424	7804	5471	28831	26448	2355	34
	1990	133596	83891	28565	3971	6481	4224	30218	28087	2087	32
	1991	136644	95039	33885	4539	4924	4214	35641	33813	1790	34
	1992	135429	90015	42927	4953	1059	439	30961	29115	1832	48
Developing economies-Europe /5	1980	8977	3286	1659	851	4033	2489	2807	2385	417	
	1989	13363	6705	2041	799	4617	2899	5916	4928	981	
	1990	14391	8444	1880	747	4059	2681	7572	6587	978	
	1991	9548	5601	1247	495	2700	1786	5024	4370	649	
	1992	13952	8774	4156	296	993	436	8288	7149	1134	
Developing economies-Middle East	1980	210976	150989	53525	6014	3330	988	82802	76742	6059	72
	1989	100233	55916	39450	8383	3702	2152	24941	22637	2266	3
	1990	110733	64802	40900	7874	4180	2240	30474	27426	3021	5
	1991	105129	60545	40006	6924	3284	1840	28522	25476	3026	5
	1992	116121	66676	45251	11792	2728	783	28837	26853	1944	5
Developing economies-Other Asia	1980	161939	96880	57454	11376	6229	4020	27363	24820	2493	56
	1989	405124	240936	150822	12744	10187	7178	64626	57635	6865	36
	1990	451273	257306	177355	14474	11124	8834	77982	69599	8252	44
	1991	515731	281991	221103	18688	9940	7310	89228	80208	8721	101
	1992	588570	310426	266062	22791	9254	6499	97192	87848	9121	1578
former USSR-Asia /9	1980	.	.	.	.	.	.	.	.	.	
	1989	.	.	.	.	.	.	.	.	.	
	1990	.	.	.	.	.	.	.	.	.	
	1991	.	.	.	.	.	.	.	.	.	
	1992	4453	502	2683	843	377	.	425	366	58	
Developing economies-Oceania	1980	2184	1930	232	0	0	0	764	761	3	
	1989	3184	2571	535	6	46	0	909	889	20	
	1990	2821	2225	504	12	43	0	673	642	31	
	1991	3317	2552	620	1	53	15	616	573	43	
	1992	3240	2695	526	7	4	4	729	677	52	

En millions de dollars E.-U. f.o.b.

← Exportations vers

Economies développées /2,3,4				Developing economies / Economies en voie de développement /1,3,4						Année	Exportations en provenance de ↓
Canada	USA E-U	Japan Japon	Australia New Zealand / Australie Nouvelle-Zélande	Africa Afrique	America Amerique Total	LAIA+ ALADI+	Asia Asie Mid.East Moyen Orient	Other Autres /9	Oceania Océanie		

Commerce total (CTCI, Rev. 2 et Rev. 3, 0-9) /6(suite)

	41183	3726	660	932	3288	2396	681	2628	13	1980	Canada
.	85347	7430	1074	803	2169	1518	1068	5535	39	1989	
.	95217	7038	909	903	2325	1675	947	5911	27	1990	
.	95501	6243	674	820	2329	1751	1007	6764	21	1991	
.	104056	6198	669	722	2769	2198	840	6408	16	1992	
34102	.	20574	4651	6357	38021	31668	10266	27224	184	1980	Etats-Unis
74971	.	42752	9215	5942	47475	37611	10267	55260	353	1989	
78212	.	46130	9406	6068	52280	41969	10124	58051	289	1990	
79055	.	46111	9181	6562	61326	50671	13719	64523	376	1991	
83217	.	45836	9973	7287	73118	61378	15399	70110	489	1992	
2437	31747	.	4065	5958	8537	5923	13114	36480	407	1980	Japon
6807	93716	.	9151	3383	8855	5098	7775	82405	823	1989	
6726	90893	.	8106	3835	9712	5354	9478	89255	853	1990	
7251	92091	.	7574	3973	12243	6791	11583	105257	784	1991	
7090	96489	.	8158	4648	15053	8936	14201	117071	708	1992	
539	2804	6053	1627	611	358	267	1507	4754	991	1980	Australie, Nouvelle-Zélande
616	4852	11289	3144	502	576	486	1577	10927	1169	1989	
763	5483	11877	3357	600	707	610	1754	12316	1258	1990	
738	5200	12991	3441	498	727	583	1470	14438	1443	1991	
799	4679	12195	3686	505	775	659	1494	16362	1433	1992	
180	29733	1998	72	2978	5879	1476	1801	1463	1	1980	Economies en voie de développement-Afrique
312	9428	1608	92	3865	756	516	1708	1893	3	1989	
705	12278	1444	60	3933	708	458	1717	1907	4	1990	
275	12894	1415	56	4867	1032	915	1766	2001	4	1991	
282	12730	1608	71	5229	1211	1099	1383	2275	7	1992	
2804	34890	4541	189	2399	22985	11936	1592	2170	108	1980	Economies en voie de développement-Amérique
2340	46005	6697	706	1658	18626	11507	1978	6185	24	1989	
1587	43596	7128	630	1948	18174	12374	2109	6037	32	1990	
2081	48545	7615	438	1647	23139	15923	2180	6685	54	1991	
1988	49001	6773	423	1907	28941	20145	2454	9529	33	1992	
28	393	31	12	735	65	36	628	228	2	1980	Economies en voie de développement-Europe 5
54	624	35	36	696	230	47	838	276	1	1989	
62	691	39	66	715	110	47	781	273	2	1990	
41	458	26	44	475	73	31	518	180	1	1991	
43	379	31	22	257	69	35	360	238	1	1992	
2546	20180	42724	2655	3524	11745	7167	11120	26020	112	1980	Economies en voie de développement-Moyen Orient
712	10565	18659	974	3131	6307	3647	12220	16953	44	1989	
798	9827	22594	1007	5698	2503	312	12914	19102	54	1990	
804	7264	22844	963	6165	2626	445	12266	18548	58	1991	
540	13205	22736	1198	2546	2815	2706	16535	23071	0	1992	
1743	30648	32588	3848	5370	4019	1747	8463	38786	570	1980	Economies en voie de développement-Autres Pays d'Asie
6984	98632	61559	8339	6587	5810	2874	10627	126533	962	1989	
6825	99254	64415	7889	7145	7201	3679	11634	149761	1056	1990	
7423	105247	69563	8867	9505	9833	5608	15766	184457	1119	1991	
8154	121236	71246	10145	10168	13715	7651	19858	220987	1169	1992	
.	.	.	.	.	.	.	.	.	.	1980	l'ancienne URSS asiatique /9
.	.	.	.	.	.	.	.	.	.	1989	
.	.	.	.	.	.	.	.	.	.	1990	
.	.	.	.	.	.	.	.	.	.	1991	
0	39	38	0	32	119	37	1803	729	0	1992	
24	296	570	275	0	6	5	0	143	83	1980	Economies en voie de développement-Océanie
18	411	841	392	0	3	0	2	364	165	1989	
15	429	634	475	0	3	1	2	327	171	1990	
16	466	638	816	1	3	2	1	442	174	1991	
16	508	560	882	0	0	0	7	426	93	1992	

72
World exports by commodity classes and by regions (continued)
Exportations mondiales par classes de marchandises et par régions (suite)

In million US dollars f.o.b.

Food and raw materials (SITC, Rev. 2 and Rev. 3, 0 -

Exports from	Year	World Monde /1,2,3	Developed economies Econ. dévelop- pées /2,3,4	Developing economies Economies en voie de développement /1,3,4 Total	OPEC+ OPEP+	Eastern Europe and former USSR Europe de l'Est et l'ancienne URSS Total /2,3,7	fmr USSR anc URSS /10	Europe Total /2	EEC+ CEE+ /2	EFTA+ AELE+	South Africa Afriqu du Su
World /1,2,3	1980	819447	580748	174760	25564	50797	19003	348914	313431	34785	14
	1989	735518	499761	168672	22551	58296	20210	302317	272608	28512	11
	1990	817932	572858	175144	22666	46218	17153	359000	325289	32414	11
	1991	792613	568110	179613	21884	33846	16453	354614	322623	30593	12
	1992	824654	589943	195950	26154	27966	12841	366045	332566	32162	19
Developed economies /2,3,4	1980	301824	225914	56766	15350	11949	6051	169430	151067	17838	9
	1989	371436	287603	68058	13012	11205	7219	203872	184153	19057	7
	1990	422941	334791	71873	12888	10367	6065	241571	217857	22865	7
	1991	425981	335121	73287	12414	12138	7275	246035	222143	22983	7
	1992	447630	349631	77756	12348	14240	8418	255502	230818	23681	12
Developing economies /1,3,4	1980	455393	326670	107229	9440	16152	9376	153013	142854	9993	5
	1989	292220	185743	86982	8696	16232	10188	74453	70478	3914	4
	1990	326295	202566	92521	9108	14514	8967	84980	81049	3854	3
	1991	330737	213769	100964	9184	11172	7664	90805	86251	4080	4
	1992	338746	214501	114108	13461	6222	2739	86881	82478	4145	68
OPEC+	1980	299719	227802	66397	2198	3666	829	111889	104157	7679	
	1989	121826	84766	33145	1821	2490	638	34845	33578	1257	
	1990	142522	92967	32257	1692	2720	760	40941	39965	966	
	1991	141465	101751	34984	1551	2159	496	45501	44126	1348	
	1992	144520	103316	38285	4324	1919	334	41630	40045	1551	
Eastern Europe and the former USSR /2,3,7	1980	62229	28165	10765	775	22696	3576	26471	19510	6954	
	1989	71862	26415	13632	843	30859	2802	23993	17977	5542	
	1990	68697	35502	10750	670	21336	2121	32450	26383	5696	
	1991	35896	19221	5362	287	10535	1515	17774	14229	3529	
	1992	38278	25811	4086	345	7504	1684	23663	19271	4336	
former USSR /3	1980	44273	20034	8081	91	16158	.	18839	14185	4654	
	1989	53843	17418	10912	163	25513	.	15752	12308	3432	
	1990	...	...	...	...	...	.	...	...	...	
	1991	...	...	...	...	...	.	...	...	...	
	1992	.	.	.	.	.	.	.	.	.	
former USSR-Europe /8	1980	.	.	.	.	.	.	.	.	.	
	1989	.	.	.	.	.	.	.	.	.	
	1990	.	.	.	.	.	.	.	.	.	
	1991	.	.	.	.	.	.	.	.	.	
	1992	26119	19134	2337	141	4648	.	17320	14528	2772	
Developed economies-Europe /2	1980	177706	145878	21732	9321	5769	2489	136295	120810	14996	4
	1989	222205	191597	21950	6395	5295	2247	172330	154792	16896	4
	1990	263813	228766	24706	6812	5804	2422	206847	185324	20722	4
	1991	268664	232310	24755	6864	7305	3006	212917	191264	20793	4
	1992	284201	243875	26076	6940	10165	4939	223242	200814	21486	48
European Economic Community+ /2	1980	152142	123281	19929	8479	4634	1974	114643	101852	12324	33
	1989	189965	162419	20299	6003	4079	1773	145922	132261	13076	4
	1990	223935	192527	22599	6244	4478	1874	173951	157382	15836	4
	1991	229817	196776	22774	6349	6140	2744	180410	163550	16061	4
	1992	243776	207027	24107	6442	8765	4564	189623	171967	16781	4
European Free Trade Association+	1980	25338	22406	1798	838	1135	515	21480	18798	2660	
	1989	31829	28846	1636	381	1213	471	26128	22273	3797	
	1990	39427	35847	2081	556	1325	548	32536	27607	4860	
	1991	38359	35107	1952	501	1164	260	32129	27362	4705	
	1992	39937	36438	1942	485	1398	374	33250	28502	4682	
Other developed economies	1980	124118	80035	35034	6029	6180	3562	33135	30257	2843	4
	1989	149232	96006	46108	6616	5910	4972	31542	29361	2161	2
	1990	159128	106024	47167	6076	4564	3643	34723	32534	2142	3
	1991	157317	102811	48532	5550	4833	4269	33118	30879	2190	3
	1992	163429	105757	51680	5408	4074	3479	32260	30005	2195	7

En millions de dollars E.-U. f.o.b.

← Exportations vers

Economies développées /2,3,4				Developing economies Economies en voie de développement /1,3,4						Année	Exportations en provenance de ↓
Canada	USA E-U	Japan Japon	Australia New Zealand Australie Nouvelle-Zélande	Africa Afrique	America Amerique Total	LAIA ALADI +	Asia Asie Mid.East Moyen Orient	Other Autres /9	Oceania Océanie		

Produits alimentaires et matières brutes (CTCI, Rev. 2 et Rev. 3, 0 - 4)

Canada	USA E-U	Japan	Australia NZ	Africa	America Total	LAIA ALADI	Mid.East	Other /9	Oceania	Année	Exportations en provenance de
14862	136863	148486	9290	26547	61699	32071	32575	97152	1353	1980	Monde /1,2,3
12525	100666	115318	7027	22048	38878	22650	32393	100910	1403	1989	
16043	104593	128413	7076	28266	31874	16503	33617	110908	1620	1990	
15354	103418	127564	7201	28649	34956	19733	32634	116685	1674	1991	
15097	119161	128481	7815	22799	38390	26786	40323	134360	1527	1992	
6757	24053	22450	1670	14015	13160	9578	9464	18274	677	1980	Economies développées /2,3,4
9046	34705	35492	2721	11316	13588	9330	10913	30240	861	1989	
12417	39841	36207	2804	11885	13997	9299	10720	32514	946	1990	
11753	37167	35484	2650	11304	15382	10655	10521	33825	1004	1991	
11988	41086	35886	2675	11787	17182	12353	10576	36099	965	1992	
8086	112404	124810	7610	11476	46690	22394	21969	76524	676	1980	Economies en voie de développement /1,3,4
3433	65116	78373	4279	9791	22685	13078	19966	67610	542	1989	
3575	63776	90220	4224	15169	15055	6983	21515	75625	674	1990	
3559	65792	91164	4537	16850	18230	8915	21330	81589	670	1991	
3033	77664	90969	5130	10618	20895	14190	28410	96934	561	1992	
7041	80409	103375	5779	6868	38212	17376	15489	56546	20	1980	OPEP +
1923	36240	50131	2520	5224	14728	8132	13465	33865	100	1989	
2136	32128	60421	2444	10129	6133	1056	14666	37372	121	1990	
1989	34657	59856	2812	11305	8332	2366	14258	39122	140	1991	
1523	46840	60468	3451	4654	9897	6979	19086	49224	49	1992	
19	407	1226	10	1056	1850	98	1143	2353	0	1980	Europe de l'Est et l'ancienne URSS 2,3,7
46	846	1453	28	941	2606	241	1513	3060	0	1989	
50	976	1987	19	1212	2823	221	1382	2769	0	1990	
41	459	916	14	495	1344	163	783	1270	0	1991	
77	411	1625	10	393	312	243	1337	1326	0	1992	
2	51	1137	4	327	1526	10	427	1885	0	1980	l'ancienne URSS 3
6	338	1282	8	432	2284	19	621	2552	0	1989	
13	403	1812	11	...	...	...	...	...	...	1990	
6	179	804	5	...	...	...	...	...	...	1991	
.	.	.	.	.	.	.	.	.	.	1992	
.	.	.	.	.	.	.	.	.	.	1980	l'ancienne URSS-Europe 8
.	.	.	.	.	.	.	.	.	.	1989	
.	.	.	.	.	.	.	.	.	.	1990	
.	.	.	.	.	.	.	.	.	.	1991	
57	234	1520	2	113	116	97	943	1012	0	1992	
804	6435	1266	405	9856	2967	1517	6031	2123	87	1980	Economies développées-Europe 2
2658	11091	3802	729	7441	3139	1646	5677	4678	161	1989	
2979	13114	3970	802	8206	3421	1767	6129	5290	186	1990	
2714	10697	4131	692	7737	3566	1981	6569	5645	174	1991	
2659	11883	4243	713	7966	3871	2118	6432	6594	176	1992	
744	5902	1096	367	9155	2798	1398	5516	1897	86	1980	Communauté Economique Européenne + 2
2128	9561	3321	651	6970	2936	1508	5354	4254	160	1989	
2173	11283	3462	735	7619	3189	1596	5664	4880	185	1990	
1858	9369	3530	643	7198	3289	1784	6107	5217	174	1991	
1783	10302	3715	669	7459	3638	1957	5980	6074	175	1992	
60	514	170	38	699	170	119	512	226	1	1980	Association Européenne de Libre Echange +
529	1496	462	77	469	203	138	312	422	1	1989	
806	1812	495	67	584	231	170	452	401	1	1990	
854	1297	586	48	537	277	197	447	418	0	1991	
876	1551	518	44	504	228	156	437	516	1	1992	
5953	17617	21183	1265	4159	10193	8061	3432	16151	590	1980	Autres economies développées
6388	23614	31690	1992	3875	10449	7685	5237	25562	700	1989	
9438	26727	32237	2032	3679	10576	7532	4591	27224	760	1990	
9040	26470	31352	1958	3567	11816	8674	3952	28180	830	1991	
9328	29203	31643	1962	3821	13311	10236	4144	29506	789	1992	

72
World exports by commodity classes and by regions (continued)
Exportations mondiales par classes de marchandises et par régions (suite)

In million US dollars f.o.b.

Food and raw materials (SITC, Rev. 2 and Rev. 3, 0 - 4)(continu…)

Exports from / Year	World Monde /1,2,3	Developed economies Econ. développées /2,3,4	Developing economies Economies en voie de développement /1,3,4 Total	OPEC+ OPEP+	Eastern Europe and former USSR Europe de l'Est et l'ancienne URSS Total /2,3,7	fmr USSR anc URSS /10	Europe Total 2	EEC+ CEE+ /2	EFTA+ AELE+	Develo… Econom… Sout… Afric… Afriqu du Su
Canada 1980	28989	23881	3500	664	1441	1146	5277	4892	384	
1989	37324	32390	4287	1136	647	540	5800	5028	771	
1990	40312	34808	4548	905	946	875	5642	4957	672	
1991	39616	33622	4756	798	1239	1190	5238	4552	676	
1992	41892	35728	5203	800	961	911	4720	4140	578	
United States 1980	64022	37540	22307	3314	3059	1085	19497	18311	1172	2
1989	73395	40270	28765	3733	3965	3495	15574	14655	906	1
1990	76749	45462	28092	3246	2915	2337	17537	16580	930	1
1991	75199	43754	28377	3024	2902	2553	17128	16137	959	1
1992	77865	44590	30168	2954	2807	2429	16472	15560	882	5
Japan 1980	3693	1003	2517	542	172	75	414	372	38	
1989	4645	1262	3242	213	140	98	439	409	30	
1990	4929	1184	3626	201	117	82	450	413	37	
1991	5294	1210	3993	215	91	58	467	432	35	
1992	5777	1280	4434	266	63	54	455	418	38	
Australia, New Zealand 1980	19822	11068	5709	1494	1469	1256	2992	2907	78	
1989	25530	15686	8377	1530	1131	838	4867	4677	187	
1990	26677	16446	9160	1719	538	349	4719	4569	145	
1991	27719	16871	9825	1508	539	466	4512	4310	198	
1992	28477	16674	10149	1309	125	40	4724	4477	231	
Developing economies-Africa 1980	88165	73706	11647	797	2224	573	42233	39951	2227	1
1989	44385	36279	5736	696	1783	657	25723	24984	736	2
1990	54104	46083	5748	727	1734	692	32537	31764	770	1
1991	57338	48358	6867	881	1630	594	34537	32996	1260	1
1992	56692	48017	7246	914	686	100	34251	32900	1161	2
Developing economies-America 1980	86043	58053	20106	2048	6645	5104	20850	19152	1601	
1989	73299	51966	13311	1645	7337	5233	19092	17528	1542	1
1990	83443	49436	13908	2077	5998	3955	19723	18338	1346	1
1991	85548	62304	16463	2115	4563	3915	25935	24732	1170	1
1992	78777	55889	21342	2223	888	337	20997	19750	1237	3
Developing economies-Europe 5 1980	1940	1068	278	101	594	338	990	852	137	
1989	2179	1336	285	60	558	311	1211	1008	199	
1990	2392	1558	266	53	564	333	1438	1198	239	
1991	1582	1031	177	36	374	221	953	795	157	
1992	2458	1047	1179	20	220	47	992	863	126	
Developing economies-Middle East 1980	203698	148082	49450	3354	3179	904	80600	74802	5797	
1989	81157	45761	32278	3502	2042	625	18301	17385	899	
1990	89199	53607	32740	3443	2137	589	21281	20268	1002	
1991	83679	49189	31717	2834	1678	578	19185	18270	908	
1992	92366	55318	33906	6178	1934	398	19501	18462	1033	
Developing economies-Other Asia 1980	73766	44198	25553	3140	3510	2458	7825	7584	228	2
1989	88902	48529	35022	2793	4467	3363	9455	8908	529	
1990	95164	50282	39541	2801	4039	3398	9480	8984	475	
1991	100474	51331	45299	3316	2876	2341	9725	9024	549	
1992	106553	52722	50053	4125	2495	1857	10659	10064	546	
former USSR-Asia 9 1980	.	.	.	.	.	.	.	.	.	
1989	.	.	.	.	.	.	.	.	.	
1990	.	.	.	.	.	.	.	.	.	
1991	.	.	.	.	.	.	.	.	.	
1992	2359	353	1689	703	207	0	326	283	43	
Developing economies-Oceania 1980	1781	1564	196	0	0	0	515	512	3	
1989	2298	1872	351	0	46	0	672	664	7	
1990	1993	1600	318	7	42	0	521	498	23	
1991	2117	1557	440	1	52	15	470	434	36	
1992	1901	1508	383	1	0	0	481	438	43	

En millions de dollars E.-U. f.o.b.

← Exportations vers

Economies développées /2,3,4				Developing economies / Economies en voie de développement /1,3,4							
~anada	USA E-U	Japan Japon	Australia New Zealand Australie Nouvelle-Zélande	Africa Afrique	America Amerique Total	LAIA + ALADI +	Asia Asie Mid.East Moyen Orient	Other Autres /9	Oceania Océanie	Année	Exportations en provenance de ↓

~duits alimentaires et matières brutes (CTCI, Rev. 2 et Rev. 3, 0 - 4)(suite)

.	15036	3230	229	580	1300	805	207	1370	2	1980	Canada
.	20261	5908	322	467	854	597	707	2234	1	1989	
.	23146	5636	251	511	902	659	505	2621	1	1990	
.	23120	4992	162	518	946	722	345	2932	1	1991	
.	25793	4948	174	417	1161	957	336	3287	1	1992	
5234	.	11791	406	2465	8419	6902	1565	9422	67	1980	Etats-Unis
6000	.	17481	662	2526	9041	6639	3142	13787	95	1989	
9026	.	17562	711	2162	9091	6406	2737	13748	94	1990	
8627	.	16660	733	2189	10148	7371	2523	13281	106	1991	
8938	.	17401	699	2422	11338	8604	2605	13619	102	1992	
86	346	.	81	218	174	131	296	1764	64	1980	Japon
72	606	.	97	53	103	89	108	2884	94	1989	
67	526	.	103	38	49	36	91	3354	89	1990	
63	544	.	94	42	127	110	85	3638	96	1991	
42	647	.	98	33	107	89	109	4082	98	1992	
446	1859	5214	501	482	235	160	1335	3133	455	1980	Australie, Nouvelle-Zélande
233	2354	7290	890	453	409	327	1252	5683	507	1989	
264	2664	7812	948	516	479	388	1227	6330	572	1990	
279	2454	8588	953	407	537	426	966	7269	621	1991	
313	2335	8216	966	236	603	497	851	7871	587	1992	
167	29242	1644	70	2217	5799	1431	1676	1161	0	1980	Economies en voie de développement-Afrique
240	8788	1049	78	2477	663	447	1128	1072	3	1989	
671	11570	981	47	2576	616	383	1038	1055	4	1990	
192	12226	1040	43	3197	951	853	1108	1173	4	1991	
228	11975	939	52	3468	1098	1007	800	1381	6	1992	
2517	30341	3476	74	1724	15085	5967	1315	1622	101	1980	Economies en voie de développement-Amérique
1280	27047	3853	198	814	8520	4830	1286	2480	14	1989	
805	23877	4333	180	1109	8389	5436	1417	2750	13	1990	
1140	29775	4794	167	1058	10737	6518	1229	3252	22	1991	
935	28698	4437	157	1312	12807	7088	1453	5690	23	1992	
3	61	9	2	122	17	5	107	31	1	1980	Economies en voie de développement-Europe /5
7	88	11	7	133	6	2	124	21	0	1989	
9	90	6	8	139	12	7	77	38	0	1990	
6	59	4	5	91	8	5	51	27	0	1991	
6	36	4	5	85	24	4	29	10	0	1992	
5080	39864	82584	5075	6142	23346	14304	16111	50318	116	1980	Economies en voie de développement-Moyen Orient
1262	17150	34809	1808	4456	12533	7257	14510	28962	87	1989	
1447	17143	42158	1760	9391	4851	507	15785	31455	104	1990	
1471	12465	42279	1723	10103	5109	784	14766	29977	116	1991	
925	24263	42802	2166	3303	5434	5294	20250	37286	0	1992	
296	12644	36568	2145	1271	2438	683	2759	23264	396	1980	Economies en voie de développement-Autres Pays d'Asie
628	11706	37934	2055	1911	961	541	2917	34799	365	1989	
628	10719	42214	2066	1953	1187	650	3197	40090	475	1990	
736	10874	42528	2437	2401	1425	755	4176	46795	453	1991	
926	12288	42313	2614	2449	1532	797	5872	52248	475	1992	
.	.	.	.	.	.	.	.	.	.	1980	l'ancienne URSS asiatique .9
.	.	.	.	.	.	.	.	.	.	1989	
.	.	.	.	.	.	.	.	.	.	1990	
.	.	.	.	.	.	.	.	.	.	1991	
0	10	17	0	17	66	25	1467	140	0	1992	
24	252	530	243	0	5	5	0	129	62	1980	Economies en voie de développement-Océanie
15	336	716	133	0	0	0	2	275	73	1989	
14	375	527	162	0	0	0	2	238	79	1990	
15	392	519	162	0	0	0	1	365	74	1991	
12	404	474	137	0	0	0	6	319	57	1992	

72
World exports by commodity classes and by regions (continued)
Exportations mondiales par classes de marchandises et par régions (suite)

In million US dollars f.o.b.

Manufactured goods (SITC, Rev. 2 and Rev. 3, 5

Exports from	Year	World Monde /1,2,3	Developed economies Econ. développées /2,3,4	Developing economies Economies en voie de développement /1,3,4 Total	OPEC+ OPEP+	Eastern Europe and former USSR Europe de l'Est et l'ancienne URSS Total /2,3,7	fmr USSR anc URSS /10	Europe Total /2	EEC+ CEE+ /2	EFTA+ AELE+	Develo Econor Sou Afric Afriq du Su
World /1,2,3	1980	1134986	727352	316711	99523	85768	41864	508927	417115	90701	119
	1989	2182905	1569057	504139	75614	100338	52437	987326	829425	154624	111
	1990	2481266	1808638	572859	86719	86958	46541	1185329	1002122	179489	112
	1991	2551155	1824491	647310	101549	70028	35194	1199188	1026424	169704	122
	1992	2776014	1943149	757474	117158	65649	27440	1272884	1095412	174164	127
Developed economies /2,3,4	1980	929579	644161	253064	82816	29903	15318	467195	380880	85259	114
	1989	1688594	1302793	342329	54395	36955	20189	885912	742373	141480	104
	1990	1958817	1520615	391041	65134	38456	19640	1063476	896374	164370	105
	1991	2010205	1524592	434342	76993	43511	19698	1075264	917122	155269	109
	1992	2150792	1602747	492829	88017	47173	16346	1131623	969153	159392	108
Developing economies /1,3,4	1980	126194	71484	46865	13299	6733	3986	31452	28267	3139	5
	1989	400601	244988	141429	18098	12315	8469	82225	72881	9201	6
	1990	445093	264738	163828	19542	13486	10049	100517	89340	11050	7
	1991	499360	282467	203782	23320	11878	8571	108004	96685	11141	12
	1992	578211	315781	252572	28022	8568	5643	119610	108304	11093	18
OPEC+	1980	6659	3087	3495	1761	50	28	2013	1823	189	
	1989	16694	9082	7479	3002	64	39	3956	2855	1100	
	1990	21283	11625	9270	2820	332	99	6017	4508	1505	
	1991	23590	12479	10718	2819	340	192	6664	5028	1634	
	1992	29545	13922	15358	4601	197	118	6739	6179	558	
Eastern Europe and the former USSR /2,3,7	1980	79214	11707	16783	3408	49131	22560	10281	7969	2303	
	1989	93711	21277	20382	3121	51068	23779	19190	14171	3943	
	1990	77357	23286	17990	2043	35016	16852	21335	16409	4069	
	1991	41591	17432	9187	1236	14639	6926	15920	12618	3294	
	1992	47011	24621	12073	1120	9908	5452	21651	17957	3679	
former USSR /3	1980	18920	1323	6994	971	10604	.	1130	860	270	
	1989	29154	3211	11157	739	14785	.	2835	2015	814	
	1990	...	...	...	...	...		...	...	...	
	1991	...	...	...	...	...		...	...	...	
	1992	.	.	.	.	.		.	.	.	
former USSR-Europe /8	1980	.	.	.	.	.		.	.	.	
	1989	.	.	.	.	.		.	.	.	
	1990	.	.	.	.	.		.	.	.	
	1991	.	.	.	.	.		.	.	.	
	1992	15893	8111	6462	98	1317	.	6405	5382	1018	
Developed economies-Europe /2	1980	610141	459222	125171	49841	25446	12131	395252	318806	75456	73
	1989	1075359	891793	146343	34130	32039	16516	745158	619138	124115	70
	1990	1289202	1071525	176664	40878	34008	16519	906822	758237	146061	73
	1991	1288939	1061662	181746	44604	39019	16738	908067	768972	136386	73
	1992	1372720	1120572	202941	49919	42787	14055	959678	816401	140389	71
European Economic Community+ /2	1980	524266	394373	110575	44884	19028	8794	339399	276666	61797	66
	1989	919905	764256	127562	30472	23424	11594	639550	535550	102215	66
	1990	1103021	918394	153381	36101	24930	11435	778249	655023	120929	68
	1991	1110987	915262	157835	39309	32024	14141	783942	667314	114153	68
	1992	1185508	966952	177610	44637	35141	11789	829193	708412	118076	66
European Free Trade Association+	1980	85444	64498	14546	4920	6400	3337	55515	41821	13639	6
	1989	154635	126866	18681	3612	8593	4906	104986	82986	21881	4
	1990	185033	152167	23154	4716	9054	5071	127661	102326	25108	5
	1991	176679	145339	23762	5227	6967	2580	123106	100670	22202	4
	1992	185805	152407	25171	5208	7613	2234	129362	106885	22295	5
Other developed economies	1980	319438	184939	127893	32975	4457	3187	71943	62074	9803	40
	1989	613235	411001	195986	20265	4916	3673	140755	123234	17365	34
	1990	669615	449090	214377	24255	4448	3120	156655	138137	18309	31
	1991	721266	462930	252595	32389	4492	2960	167197	148150	18883	36
	1992	778072	482175	289888	38097	4386	2291	171946	152752	19003	37

En millions de dollars E.-U. f.o.b.

← Exportations vers

articles manufacturés (CTCI, Rev. 2 et Rev. 3, 5 - 8)

Economies développées /2,3,4				Developing economies — Economies en voie de développement /1,3,4						Année	Exportations en provenance de ↓
Canada	USA E-U	Japan Japon	Australia New Zealand Australie Nouvelle-Zélande	Africa Afrique	America Amerique Total	LAIA ALADI +	Asia Asie Mid.East Moyen Orient	Other Autres /9	Oceania Océanie		
40745	171463	127479	23430	64900	99800	71832	84663	157396	1575	1980	Monde /1,2,3
75276	380622	140211	41259	54273	99918	67303	77974	321377	3144	1989	
93562	393674	159984	39737	66748	102887	69239	87646	363911	3332	1990	
96036	385796	162164	37964	67441	118556	84657	99244	415956	3376	1991	
99454	433632	162676	42530	63943	143760	111029	121150	490940	3620	1992	
33642	95516	17109	15324	49912	61480	49309	56344	77016	1359	1980	Economies dévelopées /2,3,4
66008	247561	52922	31216	38766	67567	50251	46696	178343	2365	1989	
84558	259157	62458	30336	46313	77609	58148	55056	197967	2530	1990	
86340	252174	60882	27673	44984	90298	69139	64663	222997	2453	1991	
89177	269148	58629	30247	48579	108542	84927	76108	249575	2795	1992	
6898	75041	110164	8016	11991	35434	21797	24480	75937	216	1980	Economies en voie de développement /1,3,4
9000	132096	86637	9894	12924	27830	16426	28003	135095	779	1989	
8795	133536	96963	9295	18444	20955	10689	30217	158296	801	1990	
9550	132928	100837	10211	21390	26320	15256	33003	189279	923	1991	
9968	163170	102922	12191	14415	34448	25581	42843	233893	823	1992	
5133	48331	102518	5831	6069	25387	14807	16688	56779	20	1980	OPEP +
1471	24456	52535	2649	4895	13291	7672	15864	36753	108	1989	
1747	24250	62717	2600	9908	5899	1125	17256	41536	135	1990	
1841	19709	62384	3008	10983	6277	1473	16875	43978	159	1991	
1547	33528	63089	3856	3840	7168	6381	24340	56133	95	1992	
205	906	206	90	2997	2886	725	3839	4444	0	1980	Europe de l'Est et l'ancienne URSS /2,3,7
267	965	651	148	2583	4521	626	3275	7940	0	1989	
209	981	564	105	1991	4323	402	2373	7647	1	1990	
146	695	445	81	1067	1938	263	1578	3681	0	1991	
309	1314	1125	91	949	770	521	2199	7473	3	1992	
34	132	18	8	790	1741	115	1004	2220	0	1980	l'ancienne URSS /3
37	110	198	32	801	3402	192	700	5564	0	1989	
...	...	...	...	...	...	...	...	...	...	1990	
...	...	...	...	...	...	...	...	...	...	1991	
.	.	.	.	.	.	.	.	.	.	1992	
.	.	.	.	.	.	.	.	.	.	1980	l'ancienne URSS-Europe /8
.	.	.	.	.	.	.	.	.	.	1989	
.	.	.	.	.	.	.	.	.	.	1990	
.	.	.	.	.	.	.	.	.	.	1991	
167	606	898	15	147	377	315	875	4999	0	1992	
4803	36811	6807	5553	39018	21989	17419	34803	22529	421	1980	Economies développées-Europe /2
11293	87353	23701	11117	30876	21225	14870	31576	53660	880	1989	
11442	97248	30218	11276	37261	25821	17909	37904	63614	1086	1990	
11037	89737	28241	9623	35816	28330	20070	41900	66269	928	1991	
10547	96730	27069	10508	37414	33103	24144	48920	75296	1223	1992	
4041	31931	5471	4777	35840	18898	15045	30670	19383	390	1980	Communauté Economique Européenne + /2
9271	74780	19553	9357	28444	18300	12587	27799	45223	851	1989	
9437	83786	24784	9269	34604	22155	15079	32923	53143	1045	1990	
9444	77180	23268	7975	32938	24582	17250	36500	55474	890	1991	
8959	83368	22331	8601	34723	28789	21027	43369	63569	1195	1992	
760	4871	1336	775	3138	3085	2374	4128	3145	32	1980	Association Européenne de Libre Echange +
2021	12528	4147	1759	2390	2911	2282	3765	8405	29	1989	
1991	13429	5433	2005	2600	3655	2823	4970	10422	41	1990	
1594	12521	4971	1647	2816	3728	2813	5388	10739	37	1991	
1587	13277	4736	1906	2616	4310	3114	5540	11659	28	1992	
28839	58706	10302	9771	10894	39491	31889	21541	54487	938	1980	Autres economies développées
54716	160207	29222	20099	7891	46342	35381	15119	124683	1485	1989	
73116	161909	32240	19061	9052	51787	40239	17152	134353	1444	1990	
75302	162436	32641	18050	9168	61968	49069	22763	156728	1525	1991	
78630	172418	31560	19739	11165	75439	60783	27187	174278	1571	1992	

72
World exports by commodity classes and by regions (continued)
Exportations mondiales par classes de marchandises et par régions (suite)

In million US dollars f.o.b.

Manufactured goods (SITC, Rev. 2 and Rev. 3, 5 - 8)(continued)

Exports from	Year	World Monde /1,2,3	Developed economies Econ. développées /2,3,4	Developing economies Economies en voie de développement /1,3,4 Total	OPEC+ OPEP+	Eastern Europe and former USSR Europe de l'Est et l'ancienne URSS Total /2,3,7	fmr USSR anc URSS /10	Europe Total /2	EEC+ CEE+ /2	EFTA+ AELE+	Developed Economies South Africa Afrique du Sud
Canada	1980	33888	29401	4050	1200	332	169	3728	3354	372	9
	1989	74465	69862	4489	511	113	40	4994	4453	540	6
	1990	78831	73854	4878	700	97	52	5161	4561	600	5
	1991	79324	73804	5390	1032	129	72	5674	5173	500	4:
	1992	84610	79250	5121	946	239	121	5597	5072	525	5
United States	1980	144093	84501	57529	13488	784	424	42242	36909	5319	211!
	1989	245165	157368	8(555	8877	1242	762	73383	65150	8160	144.
	1990	282461	186994	9(892	9613	1227	719	81125	72480	8574	149(
	1991	306785	194238	110839	14523	1586	896	85702	76478	9180	186(
	1992	329192	197684	129334	17647	2174	1047	85438	77281	8109	177(
Japan	1980	124501	59710	61493	17874	3298	2595	20600	17397	3154	173(
	1989	266590	164487	98621	10466	3482	2857	54937	47122	7742	174(
	1990	277438	166243	108068	13284	3039	2332	60859	52629	8102	147;
	1991	304188	173307	128219	16161	2663	1954	66676	58310	8249	161(
	1992	328521	181490	145293	18638	1737	1036	69896	61561	8196	170(
Australia, New Zealand	1980	5712	3159	2330	288	18	0	1008	967	40	72
	1989	9888	6487	3360	306	34	6	1192	1131	61	45
	1990	10817	6860	3929	512	25	13	1388	1302	85	44
	1991	11858	7044	4779	590	32	22	1371	1288	83	59
	1992	12552	7047	5389	671	30	18	1543	1431	111	84
Developing economies-Africa	1980	6105	4578	1145	199	225	160	3593	3490	101	142
	1989	11541	8118	2886	826	433	318	6744	6560	180	114
	1990	12017	8661	2975	935	423	378	7320	7173	143	124
	1991	12333	8362	3224	1101	555	501	7073	6917	145	113
	1992	13084	9418	3324	968	211	144	7872	7703	159	179
Developing economies-America	1980	20558	10799	9413	1508	322	88	4912	4563	346	8!
	1989	48545	32721	15324	1777	465	237	9435	8787	641	170
	1990	48978	33823	14603	1888	481	268	10181	9533	644	152
	1991	49652	31905	17330	2416	356	293	9252	8709	538	184
	1992	55454	33473	21499	2713	171	102	9688	9099	586	155
Developing economies-Europe /5	1980	7001	2185	1379	749	3437	2150	1813	1529	280	0
	1989	11158	5348	1754	739	4056	2586	4685	3912	769	0
	1990	11955	6848	1613	694	3490	2345	6100	5359	735	1
	1991	7941	4545	1070	459	2326	1565	4049	3556	489	0
	1992	11396	7667	2940	275	771	387	7245	6237	1005	2
Developing economies-Middle East	1980	7044	2703	4048	2652	150	85	2112	1850	261	0
	1989	17021	8129	7143	4877	1660	1527	6519	5145	1353	28
	1990	21013	10719	8116	4427	2043	1650	8919	7066	1839	38
	1991	20958	10926	8227	4069	1605	1262	9068	7139	1915	43
	1992	23425	11149	11221	5604	793	385	9224	8292	898	43
Developing economies-Other Asia	1980	85133	50897	30851	8191	2599	1503	18793	16605	2152	310
	1989	311632	190126	114167	9873	5701	3801	54607	48254	6246	341
	1990	350581	204312	136366	11595	7048	5408	67847	60066	7682	418
	1991	407906	226329	173784	15274	7035	4950	78415	70225	8046	94!
	1992	474204	253564	213455	18456	6618	4622	85333	76733	8436	1475
former USSR-Asia 9	1980	.	.	.	.	.	.	.	.	.	
	1989	.	.	.	.	.	.	.	.	.	
	1990	.	.	.	.	.	.	.	.	.	
	1991	.	.	.	.	.	.	.	.	.	
	1992	1203	148	978	138	58	.	99	83	16	0
Developing economies-Oceania	1980	352	322	30	0	0	0	229	229	0	
	1989	703	546	155	6	1	0	234	224	11	
	1990	548	375	156	3	1	0	151	143	8	
	1991	571	401	147	0	1	0	147	140	7	
	1992	648	510	131	6	4	4	248	239	9	

En millions de dollars E.-U. f.o.b.

← Exportations vers

Articles manufacturés (CTCI, Rev. 2 et Rev. 3, 5 - 8)(suite)

Economies développées /2,3,4				Developing economies — Economies en voie de développement /1,3,4						Année	Exportations en provenance de ↓
Canada	USA E-U	Japan Japon	Australia New Zealand Australie Nouvelle-Zélande	Africa Afrique	America Amerique Total	LAIA+ ALADI+	Asia Asie Mid.East Moyen Orient	Other Autres /9	Oceania Océanie		
.	24592	494	430	348	1966	1581	469	1237	10	1980	Canada
.	62870	1136	735	320	1237	878	354	2524	37	1989	
.	66823	1105	633	365	1321	954	429	2707	21	1990	
.	66379	1142	493	287	1313	992	653	3095	16	1991	
.	71895	1154	477	287	1562	1219	490	2754	12	1992	
26338	.	8589	4167	3628	28645	24027	8035	16737	104	1980	Etats-Unis
47595	.	24619	8145	3305	35994	29187	6773	39951	217	1989	
66127	.	27793	8147	3748	40291	33478	6883	42527	155	1990	
67804	.	28294	7780	3887	48045	40965	10176	48288	21?	1991	
71363	.	27448	8661	4726	58345	50186	12159	53615	31c	1992	
2346	30961	.	3972	5721	8324	5766	12771	34222	336	1980	Japon
6709	91830	.	8979	3318	8657	4928	7627	78193	713	1989	
6634	88862	.	7937	3786	9593	5270	9353	84338	752	1990	
7158	89777	.	7406	3923	12033	6617	11450	99973	669	1991	
7007	93969	.	7970	4604	14853	8769	14045	111155	595	1992	
48	523	426	1078	107	61	45	137	1537	488	1980	Australie, Nouvelle-Zélande
77	1029	2113	2026	31	38	31	108	2659	517	1989	
98	1309	1867	2151	51	53	46	158	3151	515	1990	
104	1384	1941	2181	51	95	64	170	3843	618	1991	
123	1321	1569	2398	57	120	111	191	4387	635	1992	
13	473	347	2	615	66	44	117	300	0	1980	Economies en voie de développement-Afrique
71	615	552	12	1350	92	68	576	819	0	1989	
34	633	452	9	1318	92	75	678	852	1	1990	
83	593	364	7	1624	80	62	657	823	0	1991	
53	695	590	13	1712	113	92	580	888	0	1992	
282	4351	1049	115	675	7779	5899	274	548	7	1980	Economies en voie de développement-Amérique
1041	18697	2826	507	843	10064	6658	693	3695	10	1989	
743	19473	2765	450	838	9741	6914	692	3278	19	1990	
911	18435	2802	280	585	12332	9356	952	3415	31	1991	
1031	19960	2328	264	594	16074	13011	1001	3817	9	1992	
25	304	22	10	613	48	31	520	197	2	1980	Economies en voie de développement-Europe 5
48	535	24	29	562	223	45	713	254	1	1989	
53	598	33	58	575	98	39	703	235	2	1990	
35	396	22	39	384	65	26	467	153	1	1991	
36	337	27	14	170	43	30	332	225	1	1992	
5075	39653	80695	4951	6044	23248	14311	17737	49487	7	1980	Economies en voie de développement-Moyen Orient
1301	17454	33648	1804	4966	12530	7238	17205	26978	87	1989	
1503	17372	40711	1657	10070	4819	456	18340	29298	106	1990	
1523	12808	40765	1722	10776	5098	755	17481	27757	116	1991	
1001	24857	41442	2306	3426	5404	5267	24623	36400	0	1992	
1503	30217	28011	2931	4045	4292	1511	5832	25394	181	1980	Economies en voie de développement-Autres Pays d'Asie
6537	94720	49482	7412	5203	4921	2418	8817	103265	609	1989	
6461	95409	52912	7039	5643	6205	3203	9804	124548	605	1990	
6997	100626	56784	8080	8021	8743	5056	13446	157060	701	1991	
7843	117232	58453	9506	8512	12815	7181	16307	192466	779	1992	
.	.	.	.	.	.	.	.	.	.	1980	l'ancienne URSS asiatique /9
.	.	.	.	.	.	.	.	.	.	1989	
.	.	.	.	.	.	.	.	.	.	1990	
.	.	.	.	.	.	.	.	.	.	1991	
0	29	20	0	15	52	12	334	576	0	1992	
0	43	40	8	0	1	1	0	10	19	1980	Economies en voie de développement-Océanie
3	75	105	130	0	0	0	0	83	72	1989	
1	51	90	83	0	1	1	0	85	69	1990	
2	70	101	82	0	1	1	0	72	73	1991	
4	88	82	88	0	0	0	1	97	33	1992	

72
World exports by commodity classes and by regions (continued)
Exportations mondiales par classes de marchandises et par régions (suite)

Source:
International Trade statistics database of the Statistical Division of the
United Nations Secretariat.

+ For member states of this grouping, see Annex I:- Other groupings.

1/ Excluding the inter-trade between China, the Democratic People's
Republic of Korea, Mongolia and Viet Nam in 1980. The figures
shown for other years are based on the import statistics of China.

2/ Excluding the trade conducted in accordance with the
supplementary protocol to the treaty on the basis of relations
between the Federal Republic of Germany (FRG) and the former
German Democratic Republic (GDR). Data reported by FRG are as
follows:

Value in million United States dollars

	1980	1989	1990
FRG to GDR	2911	4351	13379
FRG from GDR	3072	3854	5129

3/ Exports of the former USSR for which country of destination is not
available are included in the totals for the World, the 'Developed

economies', the 'Developing economies' and 'Eastern Europe and
former USSR' but are excluded from the regional components of
these groupings.

4/ This classification is intended for statistical convenience and does
not, necessarily, express a judgement about the stage reached by a
particular country in the development process.

5/ Includes the Socialist Federal Republic of Yugoslavia only.

6/ Section 9 of the SITC, which comprises commodities and
transactions not classified elsewhere, is included in the total trade
but is not shown separately in this table.

7/ Beginning 1992, includes Eastern Europe and the European
countries of the former USSR.

8/ Includes Belarus, Estonia, Latvia, Lithuania, Republic of Moldova,
the Russian Federation, and Ukraine.

9/ Beginning 1992, includes Armenia, Azerbaijan, Georgia,
Kazakhstan, Kyrgyzstan, Tajikistan, Turkmenistan, and Uzbekistan.

10/ Prior to 1992, data refer to the former USSR. Beginning 1992,
data refer to the European countries of the former USSR.

Source:
Base de données pour les statistiques du commerce international de la Division de statistique du Secrétariat de l'ONU.

+ Les etats membres de ce groupement, voir annexe I:- Autres groupements.

1/ Non compris le commerce entre la Chine, la République populaire démocratique de Corée, la Mongolie et le Viet Nam en 1980. Les chiffres indiqués pour les autres années sont bases sur les statistiques d'importations de la Chine.

2/ Non compris le commerce effectué en accord avec le protocole additionnel au traité définissant la base des relations entre la République fédérale d'Allemagne (RfA) et l'ancienne République démocratique allemande (Rda). Les données fournies par RfA sont les suivantes:

Valeur en millions de dollars des E.-U.

	1980	1989	1990
de RfA vers Rda	2911	4351	13379
de RfA en prov. de Rda	3072	3854	5129

3/ Les exportations en provenance de l'ancienne URSS dont le , pays de destination ne sont pas disponibles sont comprises dans les totaux du Monde, des 'Économies développées', des 'Economies en voie de développement' et de 'Europe de l'Est et l'ancienne URSS', mais ils ne sont pas comprises dans chaque partie composant ces régions.

4/ Cette classification est utilisée pour plus de commodité dans la présentation des statistiques et n'implique pas nécessairement un jugement quant au stade de développement auquel est parvenu un pays donné.

5/ Y compris la République fédérative socialiste de Yougoslavie seulement.

6/ Section 9 de la CTCI, qui représente les articles et transactions non classes ailleurs est comprise dans le commerce total mais n'est pas présentée séparément dans ce tableau.

7/ A partir de janvier 1992, y compris de l'Europe de l'est et les pays européens de l'ancienne URSS.

8/ Y compris Bélarus, Estonie. Lettonie, Lituanie, République de Moldova, Fédération de Russie, et Ukraine.

9/ A partir de l'année 1992, y compris Azerbaïdjan, Arménie, Géorgie, Kazakhstan, Kirghizistan, Tadjikistan, Turkménistan, et Ouzbékistan.

10/ Avant l'année 1992, les données se rapportent à l'ancienne URSS. A partir de l'année 1992. les données se rapportent aux pays européens de l'ancienne URSS.

73
Total imports and exports: index numbers
Importations et exportations totales: indices

1980 = 100

Country or area Pays ou zone	1984	1985	1986	1987	1988	1989	1990	1991	1992	1993
A. Imports: Quantum index • Importations: Indice du quantum										
Australia Australie	139	134	128	131	154	187	173	170	183	...
Austria Autriche	109	116	120	127	137	152	168	171	179	...
Belgium-Luxembourg Belgique-Luxembourg	100	104	114	125	126	134	142	147	150	...
Bolivia Bolivie	...	...	...	...	...	...	...	...	...	...
Brazil Brésil	64	60	80	79	70	80	87	96	...	...
Bulgaria Bulgarie	121	134	139	137	144	138	106	88	...	...
Burkina Faso Burkina Faso	89	111	166	...	...	...	...	...	...	...
Canada Canada	120	118	131	146	155	163	164	168	179	196
Cyprus Chypre	155	139	145	165	...	...	...	...	...	...
former Czechoslovakia† anc. Tchécoslovaquie†	98	102	105	110	114	117	...	...	...	...
Denmark Danemark	110	118	126	125	123	123	131	137	143	132
Dominica Dominique	114	105	117	139	148	171	185	166	...	...
Dominican Republic Rép. dominicaine	...	...	...	...	...	...	...	...	...	...
Ecuador Equateur	39	80	44	45	41	47	...	...	...	...
Ethiopia Ethiopie	...	...	...	...	...	...	...	...	...	...
Faeroe Islands Iles Féroé	116	122	148	159	159	136	109	109	...	...
Fiji Fidji	109	101	99	92	...	...	...	...	...	...
Finland Finlande	98	104	110	119	130	144	138	115	112	109
France France	96	101	108	116	126	134	140	143	146	...
Germany† • Allemagne† F. R. Germany R. f. Allemagne	105	110	116	123	131	140	156	177	181	169
former German D. R. anc. R. d. allemande	103	105	110	108	...	...	...	...	...	...
Greece Grèce	119	138	141	157	129	173	195	219	232	...

Country or area Pays ou zone	1984	1985	1986	1987	1988	1989	1990	1991	1992	1993

A. Exports: Quantum index • Exportations: Indice du quantum

Country or area Pays ou zone	1984	1985	1986	1987	1988	1989	1990	1991	1992	1993
Australia Australie	112	126	131	146	138	147	156	182	196	...
Austria Autriche	122	134	134	138	151	172	190	201	211	...
Belgium-Luxembourg Belgique-Luxembourg	111	116	125	134	135	141	150	156	159	...
Bolivia Bolivie	89	74	83	72	73	83	88	90	88	82
Brazil Brésil	154	159	130	152	179	179	165	172	...	...
Bulgaria Bulgarie	132	142	136	139	142	139	107	75	...	...
Burkina Faso Burkina Faso	95	99	152	...	...	...	...	...	...	...
Canada Canada	137	137	144	159	162	164	172	174	189	208
Cyprus Chypre	121	107	100	121	...	...	...	...	...	...
former Czechoslovakia† anc. Tchécoslovaquie†	123	127	128	132	130	127	...	...	...	...
Denmark Danemark	119	123	126	128	135	142	154	164	172	167
Dominica Dominique	216	228	308	353	395	326	369	347	...	...
Dominican Republic Rép. dominicaine	106	100	87	99	115	92	89	86	103	...
Ecuador Equateur	140	170	181	118	173	170	173	182	187	209
Ethiopia Ethiopie	117	93	107	100	102	111	96	55	48	...
Faeroe Islands Iles Féroé	131	139	139	133	144	165	161	167	...	...
Fiji Fidji	92	100	92	94	...	...	...	...	...	...
Finland Finlande	114	115	116	118	121	121	125	114	124	147
France France	110	113	112	116	127	136	143	148	156	...
Germany † • Allemagne† F. R. Germany R. f. Allemagne	120	127	129	132	142	153	155	157	160	158
former German D. R. anc. R. d. allemande	129	132	134	130	...	...	...	...	...	...
Greece Grèce	113	117	137	139	104	143	135	155	166	...

73
Total imports and exports: index numbers
[cont.]

Importations et exportations totales: indices
1980 = 100 [suite]

Country or area Pays ou zone	1984	1985	1986	1987	1988	1989	1990	1991	1992	1993
A. Imports: Quantum index [cont.] • Importations: Indice du quantum [suite]										
Guatemala Guatemala	54	80	65	99	71	104	113	132	141	150
Hong Kong Hong-kong	136	146	164	217	275	299	334	396	483	548
Hungary Hongrie	100	107	110	113	112	114	108	114	106	128
Iceland Islande	98	106	114	143	135	120	121	127	117	101
India Inde	124	140	184	194	208	211	216	198	259	...
Indonesia Indonésie	...	...	...	...	...	...	...	...	...	...
Ireland Irlande	112	115	120	127	132	149	159	160	168	179
Israel Israël	121	126	147	165	166	159	172	198	220	248
Italy Italie	102	111	116	128	137	148	155	162	167	...
Japan Japon	109	110	120	131	153	165	175	179	178	186
Jordan Jordanie	116	119	125	130	137	114	118	117	160	175
Kenya Kenya	62	57	67	70	79	83	79	74	...	...
Korea, Republic of Corée, République de	146	155	168	202	230	265	301	351	358	...
Malawi Malawi	62	102	75	98	109	111	...	...	...	...
Malaysia Malaisie	...	...	...	...	...	...	...	...	...	...
Malta Malte	93	101	103	116	131	146	...	...	...	...
Morocco Maroc	92	89	88	97	105	113	124	141	155	...
Myanmar Myanmar	...	...	...	...	...	...	...	...	...	...
Netherlands Pays-Bas	104	111	116	123	130	137	144	148	152	...
New Zealand Nouvelle-Zélande	123	123	121	137	123	150	161	146	169	168
Norway[1] Norvège[1]	109	122	140	138	124	118	130	132	137	139
Pakistan Pakistan	111	116	112	115	150	156	150	157	183	185

Country or area Pays ou zone	1984	1985	1986	1987	1988	1989	1990	1991	1992	1993
A. Exports: Quantum index [cont.] • Exportations: Indice du quantum [suite]										
Guatemala Guatemala	133	88	123	98	99	109	109	136	157	148
Hong Kong[4] Hong-kong[4]	155	164	189	249	315	348	378	443	530	603
Hungary Hongrie	125	131	128	132	141	142	136	130	131	114
Iceland Islande	96	106	117	122	121	125	123	114	113	118
India Inde	72	68	77	93	98	119	132	150	152	...
Indonesia Indonésie	89	99	107	103	101	111	109	139	127	137
Ireland Irlande	144	149	158	180	190	211	229	243	275	313
Israel Israël	121	132	146	162	154	161	161	157	170	191
Italy Italie	116	124	127	129	137	149	154	154	160	...
Japan Japon	135	142	141	141	147	153	162	165	167	164
Jordan Jordanie	174	181	186	221	256	260	251	221	243	272
Kenya Kenya	87	91	105	101	106	106	112	121	...	...
Korea, Republic of Corée, République de	168	181	205	251	282	287	304	334	362	...
Malawi[4] Malawi[4]	78	99	113	108	107	83	...	...	...	...
Malaysia Malaisie	111	117	126	132	141	150	153	149	147	132
Malta Malte	104	107	108	113	116	139	...	...	...	...
Morocco Maroc	124	125	130	138	161	158	177	188	179	...
Myanmar Myanmar	121	94	112	82	55	68	94	90	100	137
Netherlands Pays-Bas	112	117	122	127	138	144	152	160	164	...
New Zealand Nouvelle-Zélande	115	128	125	129	134	130	138	157	156	162
Norway[1] Norvège[1]	119	123	127	144	150	173	185	158	213	224
Pakistan Pakistan	86	110	146	160	176	184	195	221	241	219

73

Total imports and exports: index numbers
[*cont.*]

Importations et exportations totales: indices
1980 = 100 [*suite*]

Country or area Pays ou zone	1984	1985	1986	1987	1988	1989	1990	1991	1992	1993
A. Imports: Quantum index [cont.] • Importations: Indice du quantum [suite]										
Panama Panama	...	...	...	...	...	...	...	...	...	...
Peru Pérou	...	...	...	...	...	...	...	...	...	...
Philippines Philippines	72	62	80	98	118	141	152	...	...	...
Poland Pologne	82	88	93	97	106	108	78	...	...	164
Singapore Singapour	127	123	134	153	196	215	247	271	290	344
Solomon Islands Iles Salomon	...	...	...	...	...	...	...	...	...	...
South Africa Afrique du Sud	94	69	72	72	84	84	76	79	80	82
Spain Espagne	97	103	121	148	183	212	233	261	284	275
Sri Lanka Sri Lanka	103	98	112	114	117	114	111	106	117	135
Sweden Suède	104	114	118	128	134	143	143	135	135	139
Switzerland Suisse	105	110	120	129	135	143	146	144	138	135
Syrian Arab Republic Rép. arabe syrienne	115	121	114	86	74	69	75	122	108	...
Thailand Thaïlande	113	109	106	130	181	218	265	288	309	346
Trinidad and Tobago Trinité-et-Tobago	117	103	105	83	70	85	81	...	...	...
Tunisia Tunisie	109	84	100	74	92	109	110	103	113	...
Turkey Turquie	249	...	...	...	...	229	264	253	257	353
former USSR† ancienne URSS†	128	134	126	124	129	162	159	...	...	...
United Kingdom Royaume-Uni	122	127	135	144	117	126	127	120	128	...
United Rep. Tanzania Rép. Unie de Tanzanie	85	87	79	...	...	...	...	...	...	...
United States Etats-Unis	133	145	161	165	171	181	184	181	195	214
Yugoslavia, SFR†[2] Yougoslavie, Rfs†[2]	67	67	70	78	...	...	...	...	...	...
Zimbabwe Zimbabwe	111	77	97	...	...	...	...	...	...	...

Country or area Pays ou zone	1984	1985	1986	1987	1988	1989	1990	1991	1992	1993
A. Exports: Quantum index [cont.] • Exportations: Indice du quantum [suite]										
Panama Panama	65	78	76	76	63	75	80	76	78	94
Peru Pérou	96	98	88	80	71	83	83	86	...	...
Philippines Philippines	99	94	115	116	131	145	156	...	...	...
Poland Pologne	106	108	113	118	129	129	140	...	...	135
Singapore Singapour	135	134	153	181	242	268	290	329	354	419
Solomon Islands Iles Salomon	139	127	156	114	126	124	134	142	...	...
South Africa Afrique du Sud	93	111	115	107	108	117	109	116	117	119
Spain Espagne	137	144	136	147	161	169	189	211	222	256
Sri Lanka Sri Lanka	122	126	135	137	141	153	169	158	182	207
Sweden Suède	127	131	135	140	144	147	146	145	146	157
Switzerland Suisse	110	117	121	121	129	136	142	138	145	145
Syrian Arab Republic Rép. arabe syrienne	89	83	80	84	93	143	184	190	193	...
Thailand Thaïlande	137	142	117	137	173	215	241	287	322	360
Trinidad and Tobago Trinité-et-Tobago	103	108	107	105	108	107	114	...	...	...
Tunisia Tunisie	89	80	90	93	94	108	98	107	109	...
Turkey Turquie	235	...	...	...	...	481	508	542	559	594
former USSR† ancienne URSS†	112	110	121	125	131	139	123	...	...	...
United Kingdom Royaume-Uni	113	120	124	131	105	112	120	122	126	...
United Rep. Tanzania Rép. Unie de Tanzanie	79	81	77	...	...	...	...	...	...	...
United States[5] Etats-Unis[5]	86	85	85	96	112	125	134	143	151	156
Yugoslavia, SFR†[2] Yougoslavie, Rfs†[2]	105	110	107	112	...	...	...	...	...	...
Zimbabwe Zimbabwe	100	96	119	...	...	...	...	...	...	...

73
Total imports and exports: index numbers
[cont.]

Importations et exportations totales: indices
1980 = 100 [suite]

Country or area Pays ou zone	1984	1985	1986	1987	1988	1989	1990	1991	1992	1993
B. Imports: Unit value index • Importations: Indice du valeur unitaire										
Australia[3] Australie[3]	121	144	157	166	162	161	167	170	177	191
Austria Autriche	114	117	107	102	104	106	104	104	102	...
Bangladesh Bangladesh	156	177	198	220	205	207	231	278	...	...
Belgium-Luxembourg Belgique-Luxembourg	152	152	127	119	128	138	135	132	128	...
Bolivia Bolivie	...	...	...	...	...	...	...	...	...	...
Brazil[2] Brésil[2]	97	93	72	85	92	99	103	96	...	...
Burkina Faso Burkina Faso	166	174	140	...	...	...	...	...	...	...
Canada Canada	115	122	122	118	118	119	121	119	124	132
Colombia[2] Colombie[2]	233	311	395	491	628	827	39	257	350	...
Cyprus Chypre	121	125	107	102	...	...	...	...	...	...
former Czechoslovakia† anc. Tchécoslovaquie†	142	144	146	142	130	130	...	...	...	...
Denmark Danemark	145	145	134	124	126	136	130	130	126	123
Dominica Dominique	102	88	86	96	104	106	112	117	...	...
Dominican Republic Rép. dominicaine	...	...	...	...	...	...	...	...	...	...
Ecuador Equateur	178	144	226	208	245	225	...	...	...	...
Egypt Egypte	101	102	125	194	280	...	...	...	...	...
Ethiopia Ethiopie	...	...	...	...	...	...	...	...	...	...
Faeroe Islands Iles Féroé	144	148	138	159	136	141	137	137	...	...
Fiji Fidji	110	109	100	123	...	...	...	...	...	...
Finland Finlande	131	135	121	119	122	126	128	131	145	163
France France	160	162	137	135	138	148	146	145	139	...
Germany † • Allemagne† F. R. Germany R. f. Allemagne	121	124	104	98	98	106	103	105	102	97
Ghana Ghana	730	270	853	...	...	...	...	...	...	...

Country or area Pays ou zone	1984	1985	1986	1987	1988	1989	1990	1991	1992	1993
B. Exports: Unit value index • Exportations: Indice du valeur unitaire										
Australia[3] Australie[3]	115	129	131	136	152	160	162	148	151	152
Austria Autriche	114	118	114	112	112	109	108	105	103	...
Bangladesh Bangladesh	125	180	154	143	167	166	172	180	...	...
Belgium-Luxembourg Belgique-Luxembourg	142	145	130	123	133	145	136	133	132	...
Bolivia Bolivie	93	89	74	59	56	62	57	55	43	36
Brazil[2] Brésil[2]	85	78	84	84	92	91	90	87	...	...
Burkina Faso Burkina Faso	193	166	116	...	...	...	...	...	...	...
Canada Canada	107	112	108	107	112	115	114	110	112	119
Colombia[2] Colombie[2]	203	268	415	397	512	533	729	854	861	...
Cyprus Chypre	131	126	119	123	...	...	...	...	...	...
former Czechoslovakia† anc. Tchécoslovaquie†	115	118	119	119	118	123	...	...	...	...
Denmark Danemark	143	148	142	140	139	149	145	146	143	140
Dominica Dominique	102	111	126	126	136	128	137	144	...	...
Dominican Republic Rép. dominicaine	77	75	83	47	77	77	76	75	66	...
Ecuador Equateur	80	78	46	52	44	52	61	53	59	45
Egypt Egypte	94	101	92	102	141	...	...	...	...	...
Ethiopia Ethiopie	96	95	126	93	97	93	87	84	90	...
Faeroe Islands Iles Féroé	136	144	144	152	148	147	148	163	...	...
Fiji[4] Fidji[4]	101	91	122	167	147	...	...	...	...	...
Finland Finlande	134	138	135	138	144	156	154	154	164	172
France France	152	157	150	149	154	163	160	159	155	...
Germany † • Allemagne† F. R. Germany R. f. Allemagne	116	121	117	114	114	120	119	118	117	112
Ghana Ghana	829	115	490	692	...	...	...	...	...	...

73

Total imports and exports: index numbers
[*cont.*]

Importations et exportations totales: indices
1980 = 100 [*suite*]

Country or area Pays ou zone	1984	1985	1986	1987	1988	1989	1990	1991	1992	1993
B. Imports: Unit value index [cont.] · Importations: Indice du valeur unitaire [suite]										
Greece Grèce	209	243	270	269	324	363	395	436	466	...
Guatemala Guatemala	126	92	93	91	117	107	91	88	112	108
Hong Kong Hong-kong	146	142	149	155	162	168	172	175	175	174
Hungary Hongrie	125	130	136	139	145	163	178	261	286	313
Iceland Islande	567	739	833	894	66	401	686	736	745	897
India Inde	116	124	103	114	131	164	191	233	239	...
Indonesia Indonésie	...	...	...	...	...	...	...	...	...	...
Ireland Irlande	146	150	133	133	142	151	143	146	144	151
Israel[2] Israël[2]	85	83	82	91	98	104	112	107	107	103
Italy Italie	170	183	150	148	154	166	165	164	162	...
Japan Japon	93	89	56	52	49	55	61	55	52	45
Jordan Jordanie	108	106	79	83	87	127	170	171	161	163
Kenya Kenya	193	228	216	219	241	291	350	390	...	...
Korea, Republic of[2] Corée, République de[2]	94	90	94	91	100	103	103	103	101	97
Liberia Libéria	99	92	96	102	...	...	...	...	...	...
Malawi Malawi	157	212	250	362	385	459	...	...	...	...
Malaysia Malaisie	106	105	92	93	...	...	...	...	...	...
Malta Malte	110	108	104	105	106	109	...	...	...	...
Mauritius Maurice	145	160	129	134	147	179	187	195	...	...
Mexico Mexique	...	...	...	...	...	...	136	137	138	139
Myanmar Myanmar	...	...	...	...	...	...	...	...	...	...
Netherlands Pays-Bas	126	127	105	98	98	105	104	103	102	...
New Zealand Nouvelle-Zélande	158	175	170	163	162	174	176	178	190	188

Country or area Pays ou zone	1984	1985	1986	1987	1988	1989	1990	1991	1992	1993
B. Exports: Unit value index [cont.] · Exportations: Indice du valeur unitaire [suite]										
Greece Grèce	207	235	252	275	326	375	408	448	483	...
Guatemala Guatemala	64	76	57	66	68	74	70	56	54	60
Hong Kong[4] Hong-kong[4]	145	146	149	154	159	167	171	176	178	176
Hungary Hongrie	116	119	120	124	132	153	168	220	241	269
Iceland Islande	552	711	860	973	144	435	681	807	750	795
India Inde	167	176	171	191	231	275	295	326	419	...
Indonesia Indonésie	85	79	56	56	53	52	64	56	54	49
Ireland Irlande	151	155	144	144	154	164	149	148	144	150
Israel[2] Israël[2]	87	86	89	94	107	111	120	121	122	121
Italy Italie	167	181	172	174	183	195	199	204	206	...
Japan Japon	101	101	85	80	78	84	87	86	87	83
Jordan Jordanie	121	117	101	94	108	170	203	225	217	211
Kenya Kenya	173	171	183	152	175	188	204	260	...	...
Korea, Republic of[2] Corée, République de[2]	99	96	102	107	122	134	134	134	132	133
Liberia Libéria	83	84	81	77	...	...	...	...	...	...
Malawi[4] Malawi[4]	193	194	189	246	290	336	...	...	...	...
Malaysia Malaisie	91	82	61	71	79	81	80	82	...	...
Malta Malte	106	106	109	114	126	132	...	...	...	...
Mauritius Maurice	143	163	173	193	206	229	258	274	293	...
Mexico Mexique	...	...	...	...	...	...	77	71	71	69
Myanmar Myanmar	88	94	75	58	67	83	72	73	61	59
Netherlands Pays-Bas	129	133	112	102	102	109	109	108	104	...
New Zealand Nouvelle-Zélande	150	163	159	168	179	203	201	191	206	211

73

Total imports and exports: index numbers
[*cont.*]

Importations et exportations totales: indices
1980 = 100 [*suite*]

Country or area Pays ou zone	1984	1985	1986	1987	1988	1989	1990	1991	1992	1993
	B. Imports: Unit value index [cont.] · Importations: Indice du valeur unitaire [suite]									
Norway[1] Norvège[1]	120	127	127	131	135	143	144	177	138	140
Pakistan Pakistan	165	171	153	179	184	216	235	253	259	277
Panama Panama	...	...	...	...	...	...	...	...	...	...
Peru Pérou	...	...	...	...	...	...	...	...	...	...
Philippines[2] Philippines[2]	108	104	83	85	94	93	...	...	...	...
Poland[2][3] Pologne[2][3]	136	167	195	263	443	1 175	8 527	...	...	62
Singapore Singapour	103	91	81	87	88	88	87	82	79	78
Solomon Islands Iles Salomon	...	...	...	...	...	...	...	...	...	...
South Africa Afrique du Sud	169	230	258	276	324	366	401	421	453	498
Spain[3] Espagne[3]	198	204	168	161	161	165	159	155	148	158
Sri Lanka Sri Lanka	133	143	133	150	183	272	377	335	403	...
Sweden[3] Suède[3]	148	151	139	143	148	155	157	157	151	156
Switzerland Suisse	101	105	95	90	94	101	100	101	103	102
Syrian Arab Republic Rép. arabe syrienne	132	110	132	133	164	158	160	161	157	...
Thailand Thaïlande	115	127	94	100	110	118	124	130	130	131
Trinidad and Tobago Trinité-et-Tobago	129	128	173	198	220	253	257	...	...	...
Tunisia Tunisie	136	146	145	160	168	184	207	204	195	...
Turkey Turquie	368	...	...	...	...	404	532	455	407	258
United Kingdom Royaume-Uni	138	144	138	141	135	141	144	146	147	...
United Rep. Tanzania Rép. Unie de Tanzanie	143	184	325	...	...	...	...	...	...	...
United States Etats-Unis	102	99	95	102	107	111	114	114	115	114
Yugoslavia, SFR†[2] Yougoslavie, Rfs†[2]	115	117	106	111	...	...	...	...	...	...
Zimbabwe Zimbabwe	131	153	160	...	...	...	...	...	...	...

Country or area Pays ou zone	1984	1985	1986	1987	1988	1989	1990	1991	1992	1993
B. Exports: Unit value index [cont.] • Exportations: Indice du valeur unitaire [suite]										
Norway[1] Norvège[1]	141	147	110	106	107	119	125	119	110	110
Pakistan Pakistan	121	123	128	155	166	180	197	204	209	228
Panama Panama	92	86	75	84	75	77	81	82	...	...
Peru[2] Pérou[2]	68	62	51	63	69	76	73	64	64	54
Philippines[2] Philippines[2]	94	85	73	84	100	98	87	...	...	...
Poland[2 3] Pologne[2 3]	124	155	185	268	459	1 410	9 084	...	...	69
Singapore Singapour	102	90	77	80	79	79	79	75	70	69
Solomon Islands Iles Salomon	140	135	120	181	217	220	203	192	...	...
South Africa Afrique du Sud	148	191	225	245	292	351	367	386	409	449
Spain[3] Espagne[3]	178	190	187	192	194	204	197	196	197	206
Sri Lanka Sri Lanka	174	154	137	159	180	290	411	360	469	...
Sweden[3] Suède[3]	146	151	150	153	162	172	175	175	170	167
Switzerland Suisse	109	113	109	109	110	117	119	123	123	124
Syrian Arab Republic Rép. arabe syrienne	93	91	67	62	60	74	96	94	65	...
Thailand Thaïlande	97	103	104	111	121	124	127	131	132	134
Trinidad and Tobago Trinité-et-Tobago	104	99	97	102	97	123	146	...	...	...
Tunisia Tunisie	142	143	128	146	156	177	189	184	178	...
Turkey Turquie	359	...	...	...	...	330	570	544	579	507
United Kingdom Royaume-Uni	136	144	132	137	133	139	144	145	147	...
United Rep.Tanzania Rép. Unie de Tanzanie	161	154	286	...	...	...	...	...	...	...
United States[5] Etats-Unis[5]	114	112	113	115	123	127	128	129	129	130
Yugoslavia, SFR†[2] Yougoslavie, Rfs†[2]	107	106	103	109	...	...	...	...	...	...
Zimbabwe Zimbabwe	158	189	192	...	...	...	...	...	...	...

73

Total imports and exports: index numbers
[*cont.*]

Importations et exportations totales: indices
1980 = 100 [*suite*]

Country or area Pays ou zone	1984	1985	1986	1987	1988	1989	1990	1991	1992	1993
C. Terms of trade · Termes de l'échange										
Australia Australie	95	90	83	81	94	99	97	87	85	80
Austria Autriche	101	101	107	109	108	102	105	101	100	...
Bangladesh Bangladesh	80	102	78	65	81	80	75	65	...	...
Belgium-Luxembourg Belgique-Luxembourg	94	95	102	103	104	105	101	101	103	...
Brazil Brésil	88	84	116	99	100	91	87	90	...	...
Burkina Faso Burkina Faso	116	95	83	...	...	...	...	...	...	...
Canada Canada	93	91	89	91	95	97	94	92	91	90
Colombia Colombie	87	86	105	81	81	64	1 882	332	246	...
Cyprus Chypre	108	101	111	121	...	...	...	...	...	...
former Czechoslovakia† anc. Tchécoslovaquie†	81	82	81	83	91	95	...	...	...	...
Denmark Danemark	99	102	106	113	110	109	111	112	114	114
Dominica Dominique	100	126	145	131	131	121	122	123	...	...
Ecuador Equateur	45	54	21	25	18	23	...	...	...	...
Egypt Egypte	93	99	74	53	51	...	...	...	...	...
Faeroe Islands Iles Féroé	94	97	104	96	109	104	108	119	...	...
Fiji Fidji	92	84	121	136	...	...	...	...	...	...
Finland Finlande	102	102	112	116	118	124	120	118	113	106
France France	95	97	110	110	111	110	109	110	111	...
Germany† · Allemagne† F. R. Germany R. f. Allemagne	96	98	112	116	116	113	115	112	115	116
Ghana Ghana	114	43	57	...	...	...	...	...	...	...
Greece Grèce	99	97	93	102	101	103	103	103	104	...
Guatemala Guatemala	51	83	61	72	58	69	77	63	48	55
Hong Kong Hong-kong	99	102	100	100	98	99	100	101	101	101

Country or area Pays ou zone	1984	1985	1986	1987	1988	1989	1990	1991	1992	1993
D. Purchasing power of exports • Pouvoir d'achat des exportations										
Australia Australie	106	112	108	119	129	145	151	159	167	...
Austria Autriche	123	135	144	150	163	176	199	203	211	...
Bangladesh Bangladesh	...	...	...	...	...	...	...	...	...	...
Belgium-Luxembourg Belgique-Luxembourg	104	110	127	138	141	148	151	157	164	...
Brazil Brésil	136	134	151	151	179	164	144	155	...	...
Burkina Faso Burkina Faso	110	94	126	...	...	...	...	...	...	...
Canada Canada	128	125	127	145	154	159	162	160	171	187
Colombia Colombie	...	...	...	...	...	...	...	...	...	...
Cyprus Chypre	130	108	111	146	...	...	...	...	...	...
former Czechoslovakia† anc. Tchécoslovaquie†	100	104	104	110	118	120	...	...	...	...
Denmark Danemark	118	126	133	145	148	156	172	184	195	190
Dominica Dominique	216	288	447	462	516	393	451	427	...	...
Ecuador Equateur	63	91	37	30	31	39	...	...	...	...
Egypt Egypte	...	...	...	...	...	...	...	...	...	...
Faeroe Islands Iles Féroé	124	135	145	127	157	172	174	199	...	...
Fiji Fidji	85	84	112	128	...	...	...	...	...	...
Finland Finlande	117	118	129	136	143	150	150	134	140	155
France France	104	110	123	128	140	150	157	163	173	...
Germany † • Allemagne† F. R. Germany R. f. Allemagne	115	124	144	154	165	173	178	176	184	182
Ghana Ghana	...	...	...	...	...	...	...	...	...	...
Greece Grèce	112	113	128	142	104	148	140	159	172	...
Guatemala Guatemala	68	73	75	71	57	76	84	86	76	81
Hong Kong Hong-kong	154	168	189	248	310	346	376	445	537	609

73

Total imports and exports: index numbers
[cont.]

Importations et exportations totales: indices
1980 = 100 [suite]

Country or area Pays ou zone	1984	1985	1986	1987	1988	1989	1990	1991	1992	1993
C. Terms of trade [cont.] • Termes de l'échange [suite]										
Hungary Hongrie	93	92	88	89	91	94	94	84	84	86
Iceland Islande	97	96	103	109	219	109	99	110	101	89
India Inde	144	142	166	167	176	168	154	140	175	...
Ireland Irlande	103	104	108	108	109	109	104	101	100	99
Israel Israël	102	104	109	104	109	107	108	113	114	117
Italy Italie	98	99	115	118	119	117	121	124	127	...
Japan Japon	109	114	152	155	158	153	143	156	169	183
Jordan Jordanie	111	111	127	113	125	134	119	132	135	130
Kenya Kenya	90	75	85	69	72	65	58	67	...	...
Korea, Republic of Corée, République de	106	106	109	118	121	131	130	131	131	136
Liberia Libéria	84	91	84	76	...	...	...	...	...	...
Malawi Malawi	123	91	75	68	75	73	...	...	...	...
Malaysia Malaisie	86	78	66	76	...	...	...	...	...	...
Malta Malte	96	98	104	109	119	121	...	...	...	...
Mauritius Maurice	98	102	134	144	140	128	138	140	...	...
Mexico Mexique	...	...	...	...	...	...	56	52	52	49
Netherlands Pays-Bas	102	105	107	104	105	103	105	105	102	...
New Zealand Nouvelle-Zélande	95	93	93	103	111	116	114	107	108	112
Norway Norvège	118	116	86	81	79	84	86	68	80	78
Pakistan Pakistan	73	72	84	87	91	83	84	81	81	82
Philippines Philippines	87	82	88	99	106	105	0	...	...	...
Poland Pologne	91	93	95	102	104	120	107	...	...	111
Singapore Singapour	99	99	96	92	90	90	91	91	89	88

Country or area Pays ou zone	1984	1985	1986	1987	1988	1989	1990	1991	1992	1993
D. Purchasing power of exp.[cont.] • Pouvoir d'achat des exportations [suite]										
Hungary Hongrie	115	120	113	118	129	133	128	110	110	98
Iceland Islande	93	102	121	133	265	136	122	125	113	105
India Inde	103	97	128	155	172	200	203	209	266	...
Ireland Irlande	148	155	171	195	206	230	238	245	275	311
Israel Israël	124	137	158	168	169	171	173	178	194	223
Italy Italie	114	123	145	152	162	174	185	191	202	...
Japan Japon	148	161	214	219	232	234	230	258	282	300
Jordan Jordanie	194	201	236	250	320	349	298	291	326	353
Kenya Kenya	78	68	89	70	77	68	65	81	...	...
Korea, Republic of Corée, République de	177	192	223	296	342	374	394	436	472	...
Liberia Libéria	...	...	...	...	...	...	...	...	...	...
Malawi Malawi	95	91	85	74	81	61	...	...	...	...
Malaysia Malaisie	95	91	84	101	...	...	...	...	...	...
Malta Malte	100	105	113	122	138	168	...	...	...	...
Mauritius Maurice	...	...	...	...	...	...	...	...	...	...
Mexico Mexique	...	...	...	...	...	...	...	...	...	...
Netherlands Pays-Bas	115	123	130	132	145	149	159	168	167	...
New Zealand Nouvelle-Zélande	109	119	117	133	148	151	158	169	169	183
Norway Norvège	140	142	110	117	119	144	160	107	170	175
Pakistan Pakistan	63	79	122	139	159	153	163	178	195	181
Philippines Philippines	86	77	101	116	139	153	...	...	...	...
Poland Pologne	97	100	107	120	134	155	150	...	...	150
Singapore Singapour	134	133	147	166	218	240	264	301	314	371

73

Total imports and exports: index numbers
[cont.]

Importations et exportations totales: indices
1980 = 100 [suite]

Country or area Pays ou zone	1984	1985	1986	1987	1988	1989	1990	1991	1992	1993
C. Terms of trade [cont.] · Termes de l'échange [suite]										
South Africa Afrique du Sud	88	83	87	89	90	96	92	92	90	90
Spain Espagne	90	93	111	119	121	124	124	127	133	130
Sri Lanka Sri Lanka	131	108	103	106	99	107	109	107	116	...
Sweden Suède	99	100	108	108	109	111	112	112	113	107
Switzerland Suisse	108	108	115	121	118	116	119	122	119	121
Syrian Arab Republic Rép. arabe syrienne	71	83	51	47	36	47	60	58	42	...
Thailand Thaïlande	84	81	110	111	109	105	102	101	102	102
Trinidad and Tobago Trinité-et-Tobago	81	78	56	51	44	48	57	...	...	...
Tunisia Tunisie	105	98	89	91	93	96	91	90	91	...
Turkey Turquie	98	...	...	...	...	82	107	120	142	197
United Kingdom Royaume-Uni	98	100	96	97	98	99	100	99	100	...
United Rep.Tanzania Rép. Unie de Tanzanie	112	84	88	...	...	...	...	...	...	...
United States Etats-Unis	112	114	118	113	115	115	112	113	112	113
Yugoslavia, SFR† Yougoslavie, Rfs†	93	91	97	98	...	...	...	...	...	...
Zimbabwe Zimbabwe	121	123	120	...	...	...	...	...	...	...

Source:
Trade statistics database of the Statistical Division of the
United Nations Secretariat.

† For detailed descriptions of data pertaining to
former Czechoslovakia, Germany, SFR Yugoslavia and former
USSR, see Annex I – Country or area nomenclature, regional
and other groupings.

1 Excluding ships.
2 Calculated in terms of US dollars, for Poland beginning
1993.
3 Price index numbers.

Source:
Base de données pour les statistiques du commerce extérieur
de la Division de statistique du Secrétariat de l'ONU.

† Pour les descriptions en détails des données
relatives à l'ancienne Tchécoslovaquie, l'Allemagne, la Rfs
Yougoslavie et l'ancienne URSS, voir l'Annexe I –
Nomenclature des pays ou zones, groupements régionaux et
autres groupements.

1 Non compris les navires.
2 Calculés en dollars des Etats-Unis, pour la Pologne à partir
de 1993.
3 Indices des prix.

Country or area Pays ou zone	1984	1985	1986	1987	1988	1989	1990	1991	1992	1993
D. Purchasing power of exp.[cont.]• Pouvoir d'achat des exportations [suite]										
South Africa Afrique du Sud	81	92	100	95	98	112	100	106	106	107
Spain Espagne	124	134	152	175	194	209	234	268	296	334
Sri Lanka Sri Lanka	160	136	139	145	139	163	185	170	212	...
Sweden Suède	125	131	146	150	158	163	163	162	165	168
Switzerland Suisse	118	125	140	147	152	157	168	169	173	177
Syrian Arab Republic Rép. arabe syrienne	63	69	41	39	34	67	110	111	80	...
Thailand Thaïlande	115	115	129	152	189	226	246	289	327	368
Trinidad and Tobago Trinité-et-Tobago	83	84	60	54	48	52	65	...	...	...
Tunisia Tunisie	93	78	80	85	87	103	89	97	99	...
Turkey Turquie	229	...	...	...	...	392	544	648	795	1 168
United Kingdom Royaume-Uni	111	120	119	127	103	111	120	121	126	...
United Rep.Tanzania Rép. Unie de Tanzanie	89	67	68	...	...	...	...	...	...	...
United States Etats-Unis	97	97	100	108	129	144	150	161	170	177
Yugoslavia, SFR† Yougoslavie, Rfs†	98	100	104	110	...	...	...	...	...	...
Zimbabwe Zimbabwe	121	118	143	...	...	...	...	...	...	...

4 Domestic exports only.
5 Excludes military exports.

4 Seulement exportations domestiques.
5 Non compris les exportations militaires.

74
# Manufactured goods exports
# Exportations des produits manufacturés
1980 = 100

Region, country or area Région, pays ou zone	1984	1985	1986	1987	1988	1989	1990	1991	1992	1993
Unit value indices in US dollars · Indices de valeur unitaire en dollars des E-U										
Total[1]	87	87	101	113	122	121	131	132	136	...
Developed economies **Econ. développées**	86	86	103	116	124	123	136	135	139	137
America, North **Amérique du Nord**	118	118	120	124	132	137	139	141	138	137
Canada Canada	110	105	105	108	117	122	121	120	107	107
United States[2] Etats-Unis[2]	121	123	127	130	138	142	145	148	149	149
Europe **Europe**	76	77	97	113	120	118	135	132	137	131
EU+ **UE+**	76	78	97	113	119	117	135	132	138	134
Belgium-Luxembourg[3] Belgique-Luxembourg[3]	71	73	93	106	113	112	126	122	124	...
Denmark Danemark	80	83	105	124	129	124	147	141	151	...
France France	78	81	95	110	114	112	130	124	129	...
Germany † Allemagne†	73	75	100	118	121	119	137	133	141	...
Greece Grèce	78	74	83	93	99	103	118	112	111	...
Ireland[4] Irlande[4]	78	96	116	124	134	139	153	145	148	...
Italy[3] Italie[3]	81	81	102	119	126	128	149	148	151	...
Netherlands[3] Pays-Bas[3]	76	76	95	111	117	115	134	128	133	...
Portugal[4] Portugal[4]	71	73	80	100	107	103	121	122	129	...
Spain[4] Espagne[4]	77	75	97	93	113	113	136	129	148	...
United Kingdom Royaume-Uni	77	79	92	107	120	116	132	132	135	...
EFTA **AELE**	76	77	97	114	121	118	135	131	133	117
Austria[5] Autriche[5]	76	78	99	117	123	110	130	122	127	116
Finland Finlande	83	83	100	117	130	137	153	145	135	114
Iceland[4] Islande[4]	78	70	81	92	121	126	122	107	105	93
Norway Norvège	79	78	89	104	129	133	134	126	123	108
Sweden Suède	74	75	93	108	117	118	131	130	132	...

74
Manufactured goods exports
[*cont.*]

Exportations des produits manufacturés
1980 = 100 [*suite*]

Region, country or area Region, pays ou zone	1984	1985	1986	1987	1988	1989	1990	1991	1992	1993
Switzerland[3] Suisse[3]	76	75	100	120	120	114	136	134	140	...
Other Europe · Autres pays d'Europe										
Malta Malte	80	79	96	115	132	165	...	...	...	...
Other **Autres**	**95**	**93**	**110**	**121**	**135**	**134**	**133**	**142**	**149**	**159**
Australia Australie	83	73	73	83	105	110	109	102	94	...
Israel Israël	87	86	89	94	106	114	124	125	125	126
Japan Japon	97	95	115	126	139	138	136	147	157	169
New Zealand Nouvelle-Zélande	95	91	98	115	146	143	137	131	123	122
South Africa Afrique du Sud	62	54	60	74	78	84	88	85	90	...
Developing economies **Econ. en dévelop.**	**93**	**92**	**91**	**101**	**111**	**113**	**115**	**115**	**118**	...

Unit value indices in 'SDR'[6] · Indices de valeur unitaire en 'DTS'[6]

	1984	1985	1986	1987	1988	1989	1990	1991	1992	1993
***Total*[1]**	**110**	**112**	**112**	**114**	**118**	**123**	**126**	**125**	**125**	...
Developed economies **Econ. développées**	**109**	**111**	**114**	**117**	**120**	**125**	**130**	**129**	**129**	**128**
Developing economies[7] **Econ. en dévelop.[7]**	**118**	**118**	**101**	**102**	**108**	**115**	**110**	**110**	**109**	...

Unit value indices in national currency · Indices de valeur unitaire en monnaie nationale

America, North · Amérique du Nord	1984	1985	1986	1987	1988	1989	1990	1991	1992	1993
Canada Canada	122	123	125	123	122	123	120	117	111	118
United States[2] Etats-Unis[2]	121	123	127	130	138	142	145	148	149	149
Europe · Europe **EU+ · UE+**										
Belgium-Luxembourg[3] Belgique-Luxembourg[3]	141	146	141	136	...	...	...	...	...	...
Denmark Danemark	147	154	151	151	154	161	161	160	161	...
France France	161	170	155	156	161	169	167	165	162	...
Germany † Allemagne†	116	121	119	116	117	122	121	122	121	...
Greece Grèce	204	238	271	291	337	392	435	477	496	...
Italy[3] Italie[3]	166	179	176	179	191	...	...	...	...	...
Netherlands[3] Pays-Bas[3]	122	126	117	112	116	122	122	120	118	...

74

Manufactured goods exports
[*cont.*]

Exportations des produits manufacturés
1980 = 100 [*suite*]

Region, country or area Region, pays ou zone	1984	1985	1986	1987	1988	1989	1990	1991	1992	1993
United Kingdom Royaume-Uni	134	143	146	151	157	166	172	174	178	...
EFTA · AELE Austria[5] Autriche[5]	117	123	116	114	117	112	113	110	108	105
Finland Finlande	134	137	135	137	146	157	156	156	165	174
Norway Norvège	130	134	134	142	170	186	169	166	155	155
Sweden Suède	144	152	156	160	169	180	183	186	182	...
Switzerland[3] Suisse[3]	106	109	106	...	...	...	...	...	...	...
Other Europe · Autres pays d'Europe Malta Malte	107	107	109	115	126	164	...	...	...	...
Other · Autres Australia Australie	108	119	125	136	153	160	159	150	146	...
Japan Japon	101	101	85	81	79	84	87	87	88	83
New Zealand Nouvelle-Zélande	165	179	181	189	217	233	224	217	222	217
South Africa Afrique du Sud	116	152	176	194	229	283	293	302	328	...

Quantum indices · Indices de volume

Region, country or area Region, pays ou zone	1984	1985	1986	1987	1988	1989	1990	1991	1992	1993
Total[1]	119	124	129	139	151	164	174	184	189	...
Developed economies **Econ. développées**	114	120	122	128	138	148	156	162	167	166
America, North **Amérique du Nord**	93	95	96	106	122	131	146	155	169	178
Canada Canada	139	151	157	163	182	181	193	196	233	259
United States Etats-Unis	82	82	81	92	108	120	135	145	153	159
Europe **Europe**	116	122	125	131	140	151	157	162	166	161
EU+ **UE+**	115	121	124	130	139	151	157	162	166	159
Belgium-Luxembourg Belgique-Luxembourg	109	114	118	126	137	150	158	161	164	...
Denmark Danemark	122	125	127	130	139	144	155	164	177	...
France France	108	109	116	121	132	143	152	163	170	...

74
Manufactured goods exports
[*cont.*]

Exportations des produits manufacturés
1980 = 100 [*suite*]

Region, country or area Region, pays ou zone	1984	1985	1986	1987	1988	1989	1990	1991	1992	1993
Germany † Allemagne†	122	129	131	135	145	156	159	162	165	...
Greece Grèce	123	123	145	152	126	153	144	156	182	...
Ireland Irlande	167	151	153	180	199	218	234	254	286	...
Italy Italie	119	127	128	133	139	149	154	155	159	...
Netherlands Pays-Bas	116	122	129	134	142	149	156	165	167	...
Portugal Portugal	169	180	215	225	249	299	332	335	365	...
Spain Espagne	139	150	133	171	170	189	201	230	219	...
United Kingdom Royaume-Uni	99	106	108	116	122	134	142	142	144	...
EFTA **AELE**	**121**	**128**	**133**	**136**	**144**	**153**	**160**	**159**	**163**	**173**
Austria Autriche	122	130	137	139	153	178	196	206	214	211
Finland Finlande	120	128	132	137	136	140	146	132	144	171
Iceland Islande	132	117	125	139	139	131	122	111	114	124
Norway Norvège	115	120	124	130	118	127	142	147	151	152
Sweden Suède	128	134	138	141	149	151	152	148	149	...
Switzerland Suisse	115	123	129	130	145	154	160	157	160	...
Other Europe · Autres pays d'Europe **Malta** Malte	102	106	112	116	118	115	...	...	...	...
Other **Autres**	**134**	**143**	**142**	**144**	**149**	**157**	**165**	**169**	**174**	**174**
Australia Australie	107	114	118	140	141	152	171	202	235	...
Israel Israël	119	133	148	168	173	180	185	181	204	...
Japan Japon	137	144	143	143	148	156	164	167	168	166
New Zealand Nouvelle-Zélande	120	130	127	128	126	136	148	163	177	197
South Africa Afrique du Sud	116	149	159	150	160	168	175	188	233	...
Developing economies[7] **Econ. en dévelop.**[7]	**151**	**157**	**177**	**218**	**246**	**279**	**307**	**344**	**346**	...

74
Manufactured goods exports
[*cont.*]

Exportations des produits manufacturés
1980 = 100 [*suite*]

Region, country or area Region, pays ou zone	1984	1985	1986	1987	1988	1989	1990	1991	1992	1993
Value (thousand million US $) • Valeur (millards de dollars E-U)										
Total[¹]	1 083.40	1 138.30	1 363.80	1 648.70	1 927.00	2 088.80	2 402.30	2 366.30	2 511.10	...
Developed economies **Econ. développées**	906.60	955.64	1 159.00	1 370.80	1 579.50	1 688.20	1 957.20	2 014.50	2 150.30	2 097.60
America, North **Amérique du Nord**	195.36	199.19	205.06	232.91	286.97	319.63	361.29	389.25	413.80	435.11
Canada Canada	52.04	53.86	56.19	59.81	72.19	74.46	78.83	79.42	84.61	93.87
United States Etats-Unis	143.32	145.34	148.87	173.11	214.78	245.17	282.46	309.83	329.19	341.24
Europe **Europe**	532.42	571.06	734.12	894.89	1 011.50	1 075.30	1 289.10	1 290.20	1 372.70	1 276.30
EU+ **UE+**	453.46	486.97	624.17	762.34	861.55	919.91	1 103.00	1 112.10	1 185.50	1 101.30
Belgium-Luxembourg Belgique-Luxembourg	37.21	39.75	52.29	64.09	73.83	80.42	95.18	93.84	97.60	...
Denmark Danemark	9.00	9.61	12.35	14.92	16.63	16.61	21.14	21.40	24.66	...
France France	70.48	73.71	92.19	111.69	126.05	134.56	165.29	169.44	184.58	170.71
Germany † Allemagne†	148.63	161.58	217.32	264.48	293.10	309.59	362.01	361.85	387.53	329.17
Greece Grèce	2.51	2.37	3.15	3.69	3.25	4.12	4.45	4.59	5.31	...
Ireland Irlande	6.04	6.66	8.16	10.26	12.31	13.95	16.50	16.96	19.56	...
Italy Italie	63.06	67.77	86.09	104.04	115.83	125.92	150.63	150.98	158.35	159.62
Netherlands Pays-Bas	33.79	35.54	46.82	56.91	63.55	65.51	80.23	80.89	85.31	88.75
Portugal Portugal	3.92	4.32	5.62	7.34	8.72	10.09	13.20	13.39	15.38	...
Spain Espagne	16.68	17.57	20.11	24.62	29.90	33.18	42.63	46.08	50.61	52.89
United Kingdom Royaume-Uni	62.15	68.08	80.09	100.29	118.37	125.96	151.77	152.62	156.62	151.63
EFTA **AELE**	78.63	83.75	109.51	132.01	149.37	154.64	185.03	177.28	185.80	173.59
Austria Autriche	13.73	15.06	20.17	24.28	27.94	29.07	37.89	37.53	40.66	36.70
Finland Finlande	10.36	10.88	13.57	16.56	18.17	19.77	22.98	19.65	20.05	20.05
Iceland Islande	0.20	0.16	0.20	0.25	0.33	0.32	0.29	0.23	0.23	0.23
Norway Norvège	6.77	6.96	8.30	10.08	11.40	12.55	14.22	13.84	13.84	12.21
Sweden Suède	23.22	24.80	31.49	37.54	43.03	43.92	48.90	47.50	48.49	43.13
Switzerland Suisse	24.34	25.89	35.79	43.30	48.50	49.01	60.76	58.52	62.53	61.27

74
Manufactured goods exports
[*cont.*]

Exportations des produits manufacturés
1980 = 100 [*suite*]

Region, country or area Region, pays ou zone	1984	1985	1986	1987	1988	1989	1990	1991	1992	1993
Other Europe · Autres pays d'Europe										
Malta Malte	0.33	0.34	0.43	0.54	0.63	0.77	1.00	1.12	...	...
Other **Autres**	**178.82**	**185.39**	**219.85**	**243.03**	**280.94**	**293.25**	**306.87**	**335.08**	**363.76**	**386.16**
Australia Australie	3.94	3.69	3.83	5.23	6.60	7.44	8.28	9.19	9.83	...
Israel Israël	4.77	5.26	6.10	7.29	8.44	9.42	10.52	10.47	11.74	...
Japan Japon	164.91	170.78	203.36	222.81	257.07	266.59	277.44	304.34	328.52	348.56
New Zealand Nouvelle-Zélande	1.43	1.48	1.56	1.85	2.31	2.45	2.54	2.67	2.73	2.98
South Africa Afrique du Sud	3.76	4.19	4.99	5.85	6.52	7.35	8.10	...	...	...
Developing economies[7] Econ. en dévelop.[7]	**176.84**	**182.69**	**204.80**	**277.91**	**346.48**	**400.60**	**445.10**	**499.36**	**578.21**	...

Source:
Trade statistics database of the Statistical Division of the
United Nations Secretariat.
+　For Member States of this grouping, see
Annex I - Other groupings.

†　For detailed descriptions of data pertaining to
former Czechoslovakia, Germany, SFR Yugoslavia and former
USSR, see Annex I - Country or area nomenclature, regional
and other groupings.

1　Excludes trade of the countries of Eastern Europe and the
former USSR.
2　Beginning third quarter 1989 derived from price indices;
national unit value index discontinued.

3　Derived from sub-indices using current weights. Indices for
Belgium, beginning 1988, and for Switzerland, beginning
1987, are calculated by the United Nations Statistical
Division.

4　Indices are calculated by the United Nations Statistical
Division for the years beginning 1981.
5　Series linked at 1988 by a factor calculated by the United
Nations Statistical Division.
6　Special drawing right.
7　Includes the Socialist Federal Republic of Yugoslavia.

Source:
Base de données pour les statistiques du commerce extérieur
de la Division de statistique de la secrétariat de l'ONU.
+　Les Etats membres de ce groupement, voir
annexe I - Autres groupements.

†　Pour les descriptions en détails des données
relatives à l'ancienne Tchécoslovaquie, l'Allemagne, la Rfs
Yougoslavie et l'ancienne URSS, voir l'Annexe I -
Nomenclature des pays ou zones, groupements régionaux et
autres groupements.

1　Non compris le commerce des pays de l'Europe de l'Est et
l'ancienne URSS.
2　A partir du troisième trimestre de l'année 1989 calculés à
partir des indices des prix; l'indice de la valeur unitaire
nationale est discontinué.

3　Calculé à partir de sous-indices à coéfficients de
pondération correspondant à la période en cours. Les indices
pour la Belgique, à partir de 1988, et pour la Suisse, à
partir de 1987, sont calculés par la Division de statistique
des Nations Unies.

4　Les indices sont calculés par la Division de statistique des
Nations Unies pour les années à partir de 1981.
5　Les séries sont enchaînés à 1988 par un facteur calculé par
la Division de statistique des Nations Unies.
6　Le droit de tirage spécial.
7　Y compris la République socialiste fédérative de
Yougoslavie.

Technical notes, tables 71-74

Tables 71-74: Current data (annual, monthly and/or quarterly) for most of the series are published regularly by the Statistical Division in the United Nations *Monthly Bulletin of Statistics*. More detailed descriptions of the tables and notes on methodology appear in the United Nations 1977 *Supplement to the Statistical Yearbook and Monthly Bulletin of Statistics, International Trade Statistics: Concepts and Definitions* [43, 45] and the *International Trade Statistics Yearbook*.[22] More detailed data including series for individual countries showing the value in national currencies for imports and exports and notes on these series are to be found in the *International Trade Statistics Yearbook* and in the *Monthly Bulletin of Statistics*.[23]

Data are obtained from national published sources, from data supplied by the Governments for use in the following United Nations publications: *Commodity Trade Statistics* [17], *Monthly Bulletin of Statistics* and *Statistical Yearbook*; and from publications of other United Nations agencies.

Territory

The statistics reported by a country refer to the customs area of the country. In most cases, this coincides with the geographical area of the country.

System of trade

Two systems of recording trade are in common use, differing mainly in the way warehoused and re-exported goods are recorded:

(a) Special trade (S): special imports are the combined total of imports for direct domestic consumption (including transformation and repair) and withdrawals from bonded warehouses or free zones for domestic consumption. Special exports comprise exports of national merchandise, namely, goods wholly or partly produced or manufactured in the country, together with exports of nationalized goods. (Nationalized goods are goods which, having been included in special imports, are then exported without transformation);

(b) General trade (G): general imports are the combined total of imports for direct domestic consumption and imports into bonded warehouses or free zones. General exports are the combined total of national exports and re-exports. Re-exports, in the general trade system, consist of the outward movement of nationalized goods plus goods which, after importation, move outward from bonded warehouses or free zones without having been transformed;

(c) Semi-special trade (Sl): semi-special imports are general imports less all re-exports; semi-special exports are exports of domestic produce.

Notes techniques, tableaux 71-74

Tableaux 71-74 : La Division de statistique des Nations Unies publie régulièrement des données courantes (annuelles, mensuelles et/ou trimestrielles) pour la plupart des séries de ces tableaux dans le *Bulletin mensuel de statistique* des Nations Unies. Des descriptions plus détaillées des tableaux et des notes méthodologiques figurent dans *1977 Supplément à l'Annuaire statistique et au Bulletin mensuel de statistique* des Nations Unies, dans la publication *Statistiques du Commerce international, Concepts et définitions* [43, 45] et dans l'*Annuaire statistique du Commerce international* [22]. Des données plus détaillées, comprenant des séries indiquant la valeur en monnaie nationale des importations et des exportations des divers pays et les notes accompagnant ces séries figurent dans l'*Annuaire statistique du Commerce international* et dans le *Bulletin mensuel de statistique*.[23]

Les données proviennent de publications nationales et des informations fournies par les gouvernements pour les publications suivantes des Nations Unies : *"Commodity Trade Statistics"* [17], Bulletin mensuel de statistique et Annuaire statistique", ainsi que de publications d'autres institutions des Nations Unies.

Territoire

Les statistiques fournies par pays se rapportent au territoire douanier de ce pays. Le plus souvent, ce territoire coïncide avec l'étendue géographique du pays.

Système de commerce

Deux systèmes d'enregistrement du commerce sont couramment utilisés, qui ne diffèrent que par la façon dont sont enregistrées les marchandises entreposées et les marchandises réexposées :

(a) Commerce spécial (S) : les importations spéciales représentent le total combiné des importations destinées directement à la consommation intérieure (transformations et réparations comprises) et les marchandises retirées des entrepôts douaniers ou des zones franches pour la consommation intérieure. Les exportations spéciales comprennent les exportations de marchandises nationales, c'est-à-dire des biens produits ou fabriqués en totalité ou en partie dans le pays, ainsi que les exportations de biens nationalisés. (Les biens nationalisés sont des biens qui, ayant été inclus dans les importations spéciales, sont ensuite réexportés tels quels.)

(b) Commerce général (G) : les importations générales sont le total combiné des importations destinées directement à la consommation intérieure et des importations placées en entrepôt douanier ou destinées aux zones franches. Les exportations générales sont le total combiné des exportations de biens nationaux et des réexportations. Ces dernières, dans le système du commerce général, comprennent les exportations de biens nationalisés et de biens qui, après avoir été importés, sortent des entrepôts de douane ou des zones franches sans avoir été transformés.

Direct transit trade, i.e., goods merely being trans-shipped or moving through the country for the purpose of transport only, is excluded from the statistics of both special and general trade.

Valuation

Goods are, in general, valued according to the transaction value. In the case of imports, the transaction value is the value at which the goods were purchased by the importer plus the cost of transportation and insurance to the frontier of the importing country (a c.i.f. valuation). In the case of exports, the transaction value is the value at which the goods were sold by the exporter, including the cost of transportation and insurance, to bring the goods onto the transporting vehicle at the frontier of the exporting country (a f.o.b. valuation).

Currency conversion

Conversion of values from national currencies into United States dollars is done by means of external trade conversion factors which are generally weighted averages of exchange rates, the weight being the corresponding monthly or quarterly value of imports or exports.

Coverage

The statistics relate to merchandise trade. Merchandise trade is defined to include, as far as possible, all goods which add to or subtract from the material resources of a country as a result of their movement into or out of the country. Thus, ordinary commercial transactions, government trade (including foreign aid, war reparations and trade in military goods), postal trade and all kinds of silver (except silver coin after its issue), are included in the statistics. Since their movement affects monetary rather than material resources, monetary gold, together with currency and titles of ownership after their issue into circulation, are excluded.

Commodity classification

The commodity classification of trade is in accordance with the United Nations *Standard International Trade Classification* (SITC).[48]

World and regional totals

The regional, economic and world totals have been adjusted: (a) to include estimates for countries or areas for which full data are not available; (b) to include insurance and freight for imports valued f.o.b.; (c) to include countries or areas not listed separately; (d) to approximate special trade; (e) to approximate calendar years; and (f) where possible, to eliminate incomparabilities owing to geographical changes, by adjusting the figures for periods before the change to be comparable to those for periods after the change.

(c) Commerce semi-spécial (SI) : les importations semi-spéciales sont les importations générales moins l'ensemble des réexportations; les exportations semi-spéciales sont les exportations de produits du pays.

Le transit direct, c'est-à-dire les marchandises uniquement transbordées ou transportées à travers le pays, est exclu aussi bien du commerce général que du commerce spécial.

Evaluation

En général, les marchandises sont évaluées à la valeur de la transaction. Dans le cas des importations, cette valeur est celle à laquelle les marchandises ont été achetées par l'importateur plus le coût de leur transport et de leur assurance jusqu'à la frontière du pays importateur (valeur c.a.f.). Dans le cas des exportations, la valeur de la transaction est celle à laquelle les marchandises ont été vendues par l'exportateur, y compris le coût de transport et d'assurance des marchandises jusqu'à leur chargement sur le véhicule de transport à la frontière du pays exportateur (valeur f.o.b.).

Conversion des monnaies

Le conversion en dollars des Etats-Unis de valeurs exprimées en monnaie nationale se fait par application de coefficients de conversion du commerce extérieur, qui sont généralement les moyennes pondérées des taux de change, le poids étant la valeur mensuelle ou trimestrielle correspondante des importations ou des exportations.

Couverture

Les statistiques se rapportent au commerce des marchandises. Le commerce des marchandises se définit comme comprenant, dans toute la mesure du possible, toutes les marchandises qui ajoutent ou retranchent aux ressources matérielles d'un pays par suite de leur importation ou de de leur exportation par ce pays. Ainsi, les transactions commerciales ordinaires, le commerce pour le compte de l'Etat (y compris l'aide extérieure, les réparations pour dommages de guerre et le commerce des fournitures militaires), le commerce par voie postale et les transactions de toutes sortes sur l'argent (à l'exception des transactions sur les pièces d'argent après leur émission) sont inclus dans ces statistiques. La monnaie or ainsi que la monnaie et les titres de propriété après leur mise en circulation sont exclus, car leurs mouvements influent sur les ressources monétaires plutôt que sur les ressources matérielles.

Classification par marchandise

La classification par marchandise du commerce extérieur est celle adoptée dans la *Classification type por le commerce international* des Nations Unies (CTCI) [48].

Totaux mondiaux et régionaux

Les totaux économiques régionaux et mondiaux ont été ajustés de manière : (a) à inclure les estimations pour les

Quantum and unit value index numbers

These index numbers show the changes in the volume of imports or exports (quantum index) and the average price of imports or exports (unit value index).

Description of tables

Table 72: The purpose of this table is to provide data on the network of flows of broad groups of commodities within and between important economic and geographic areas of the world. The regional analysis in this table is in accordance with that of table 71.

Export data in this table are largely comparable to data shown in table 71 except that table 71 contains revised data for total exports which may not be available at the commodity/destination level needed for this table. Also, the regional totals shown in table 71 have been adjusted to exclude the re-exports of countries comprising each region. This adjustment is not made in this table since re-exports are often not available by commodity and by destination.

The commodity classification is in accordance with the United Nations *Standard International Trade Classification* (SITC), Revision 2 for 1980 through 1987 except for countries which report trade data only in terms of the SITC, Revised. Beginning in 1988 the commodity classification is in accordance with SITC, Revision 3 where data are available from countries. [48]

The data approximate total exports of all countries and areas of the world with the exception of the inter-trade of the centrally planned economies of Asia in 1980 and trade conducted in accordance with the supplementary protocol to the treaty on the basis of relations between the Federal Republic of Germany and the former German Democratic Republic. They are based on official export figures converted, where necessary, to US dollars. Where official figures are not available estimates based on the imports reported by partner countries and on other subsidiary data are used. Some official national data have been adjusted (a) to approximate the commodity groupings of SITC; and (b) to approximate calendar years.

The data include special category (confidential) exports, ships' stores and bunkers and exports of minor importance, the destination of which cannot be determined. These data are included in the world totals for each commodity group and in total exports, but are excluded from all regions of destination. Approximately 1 ½ percent of total exports are not distributed. All data are generally in accordance with the special trade system.

Table 73: These index numbers show the changes in the volume (quantum index) and the average price (unit value index) of total imports and exports. The terms of trade figures are calculated by dividing export unit value indices by the corresponding import unit value indices. The product of the net terms of trade and the quantum index of exports is called the index of the purchasing power of exports. The footnotes to countries appearing in table 71 also apply to the index numbers in this table.

pays ou régions pour lesquels on ne disposait pas de données complètes; (b) à inclure l'assurance et le fret dans la valeur f.o.b. des importations; (c) à inclure les pays ou régions non indiqués séparément; (d) à donner une approximation du commerce spécial; (e) à les ramener à des années civiles; et (f) à éliminer, dans la mesure du possible, les données non comparables par suite de changements géographiques, en ajustant les chiffres correspondant aux périodes avant le changement de manière à les rendre comparables à ceux des périodes après le changement.

Indices de quantum et de valeur unitaire

Ces indices indiquent les variations du volume des importations ou des exportations (indice de quantum) et du prix moyen des importations ou des exportations (indice de valeur unitaire).

Description des tableaux

Tableau 72 : Ce tableau a pour but de fournir des données sur l'ensemble des flux de grandes catégories de marchandises à l'intérieur des grandes régions économiques et géographiques du monde et entre ces régions. L'analyse régionale de ce tableau est conforme à celle de tableau 71.

Les données de ce tableau sur les exportations sont en grande partie comparables aux données fournies au tableau 71; toutefois, celui-ci présente des données révisées pour les totaux des exportations qui ne sont pas nécessairement disponibles au niveau des marchandises/destinations présentées au tableau 72. En outre, les totaux régionaux indiqués au tableau 71 ont été ajustés de manière à exclure les réexportations effectuées par les pays composant chaque région. Cet ajustement n'apparaît pas sur ce tableau, car il est fréquent que l'on ne dispose pas des chiffres des réexportations par marchandise et par destination.

La classification des marchandises est conforme à la *Classification type pour le commerce international* (CTCI), Révision 2 des Nations Unies pour les années 1980 à 1987, sauf pour les pays qui ne fournissent de statistiques commerciales que sur la base de la CTCI révisée. A partir de 1988, la classification des marchandises est conforme à la CTCI, Révision 3, pour les pays qui ont fourni des données [48].

Les données fournissent une approximation des exportations totales de tous les pays et régions du monde à l'exception du commerce intrarégional des économies à planification centrale d'Asie en 1980 et des échanges commerciaux effectués selon le protocole supplémentaire du traité sur la base des relations entre la République fédérale d'Allemagne et l'ancienne République démocratique allemande. Elles sont fondées sur les chiffres officiels des exportations convertis, le cas échéant, en dollars des Etats-Unis. En l'absence de chiffres officiels, on a utilisé des estimations fondées sur les importations notifiées par les pays partenaires et sur d'autres données subsidiaires. Certaines données nationales officielles ont été ajustées : a) sur la base des groupements de marchandises de la CTCI; et b) sur la base des années civiles.

Table 74: Manufactured goods are here defined to comprise sections 5 through 8 of the Standard International Trade Classification. These sections are: chemicals and related products, manufactured goods classified chiefly by material, machinery and transport equipment and miscellaneous manufactured articles. The economic and geographic groupings in this table are in accordance with those of table 71, although table 71 includes more detailed geographical sub-groups which make up the groupings "other developed market economies" and "developing market economies" of this table. In 1980 the exports of manufactured goods by countries included in the indices for "total market economies" accounted for approximately 91 per cent of world exports of manufactured goods.

The unit value indices are obtained from national sources, except those of a few countries which the United Nations Statistical Division compiles using their quantity and value figures. For countries that do not compiles indices for manufactured goods exports conforming to the above definition, sub-indices are aggregated to approximate an index of SITC sections 5-8. Unit value indices obtained from national indices are rebased, where necessary, so that 1980=100. Indices in national currency are converted into US dollars using conversion factors obtained by dividing the weighted average exchange rate of a given currency in the current period by the weighted average exchange rate in the base period. All aggregate unit value indices are current period weighted.

The indices in SDRs are calculated by multiplying the equivalent aggregate indices in United States dollars by conversion factors obtained by dividing the SDR/$US exchange rate in the current period by the rate in the base period.

The quantum indices are derived from the value data and the unit value indices. All aggregate quantum indices are base period weighted.

Ces données englobent les catégories spéciales (confidentielles) d'exportation, les marchandises à bord des navires et les exportations d'importance mineure, dont la destination ne peut être déterminée. Ces données sont comprises dans les totaux mondiaux pour chaque classe de marchandises et dans les exportations totales, mais sont exclues de ceux des régions de destination. Environ 1,5 % des exportations totales ne sont pas distribuées. Dans l'ensemble, les données sont conformes au système du commerce spécial.

Tableau 73 : Ces indices indiquent les variations du volume (indice de quantum) et du prix moyen (indice de valeur unitaire) des importations et des exportations totales. Les chiffres relatifs aux termes de l'échange se calculent en divisant les indices de valeur unitaire des exportations par les indices correspondants de valeur unitaire des importations. Le produit de la valeur nette des termes de l'échange et de l'indice du quantum des exportations est appelé indice du pouvoir d'achat des exportations. Les notes figurant au bas du tableau 71 concernant certains pays s'appliquent également aux indices du présent tableau.

Tableau 74 : Les produits manufacturés se définissent comme correspondant aux sections 5 à 8 de la Classification type pour le commerce international. Ces sections sont : produits chimiques et produits connexes, biens manufacturés classés principalement par matière première, machines et équipements de transport et articles divers manufacturés. Les groupements économiques et géographiques de ce tableau sont conformes à ceux du tableau 90; toutefois, le tableau 90 comprend des subdivisions géographiques plus détaillées qui composent les groupements "autres pays développés à économie de marché" et "pays en développement à économie de marché" du présent tableau. En 1980, les exportations de produits manufacturés des pays inclus dans les indices correspondant au "total économies de marché" représentaient environ 91 % des exportations mondiales de produits manufacturés.

Les indices de valeur unitaire sont obtenus de sources nationales, à l'exception de ceux de certains pays que la Division de statistique des Nations Unies compile en utilisant les chiffres de ces pays relatifs aux quantités et aux valeurs. Pour les pays qui n'établissent pas d'indices conformes à la définition ci-dessus pour leurs exportations de produits manufacturés, on fait la synthèse de sous-indices de manière à établir un indice proche de celui des sections 5-8 de la CTCI. Le cas échéant, les indices de valeur unitaire obtenus à partir des indices nationaux sont ajustés sur la base 1980=100. On convertit les indices en monnaie nationale en indices en dollars des Etats-Unis en utilisant des facteurs de conversion obtenus en divisant la moyenne pondérée des taux de change d'une monnaie donnée pendant la période courante par la moyenne pondérée des taux de change de la période de base. Tous les indices globaux de valeur unitaire sont pondérés pour la période courante.

On calcule les indices en DTS en multipliant les indices globaux équivalents en dollars des Etats-Unis par les facteurs de conversion obtenus en divisant le taux de change DTS/dollars EU de la période courante par le taux correspondant de la période de base.

On détermine les indices de quantum à partir des données de valeur et des indices de valeur unitaire. Tous les indices globaux de quantum sont pondérés par rapport à la période de base.

75
Tourist arrivals by region of origin
Arrivées de touristes par région de provenance

Country or area of destination and region of origin	1989	1990	1991	1992	1993	Pays ou zone de destination et région de provenance
Albania[1]	**13 962**	**29 997**	**12 892**	**23 430**	**45 152**	**Albanie**[1]
Africa	...	...	...	...	39	Afrique
Americas	17	401	510	...	3 736	Amériques
Europe	13 931	29 569	12 382	...	39 728	Europe
Asia, East and South East/Oceania	14	27	...	...	927	Asie, Est et Sud-Est et Océanie
Southern Asia	...	...	...	...	167	Asie du Sud
Western Asia	...	...	...	...	555	Asie occidentale
Algeria[2,3]	**1 206 865**	**1 136 918**	**1 193 210**	**1 119 548**	**1 127 551**	**Algérie**[2,3]
Africa	425 973	407 807	519 807	437 076	395 459	Afrique
Americas	3 788	5 764	4 606	5 146	5 121	Amériques
Europe	195 322	227 355	158 957	151 180	144 249	Europe
Asia, East and South East/Oceania	5 185	8 338	11 110	6 390	5 459	Asie, Est et Sud-Est et Océanie
Western Asia	30 891	36 551	28 202	24 304	21 711	Asie occidentale
American Samoa[4]	**47 188**	**47 337**	**39 746**	**31 444**	**...**	**Samoa américaines**[4]
Americas	13 470	10 623	8 418	6 334	...	Amériques
Europe	3 778	1 331	1 304	732	...	Europe
Asia, East and South East/Oceania	29 648	35 224	29 928	24 172	...	Asie, Est et Sud-Est et Océanie
Angola[2]	**...**	**...**	**...**	**...**	**20 582**	**Angola**[2]
Americas	...	...	...	...	669	Amériques
Europe	...	...	...	...	10 580	Europe
Anguilla	**28 761**	**31 181**	**31 002**	**32 076**	**37 658**	**Anguilla**
Americas	26 428	28 617	28 279	29 177	34 355	Amériques
Europe	1 798	2 002	2 077	2 105	2 405	Europe
Antigua and Barbuda[3,5]	**189 079**	**197 046**	**196 571**	**209 902**	**240 185**	**Antigua-et-Barbuda**[3,5]
Americas	134 573	136 918	128 771	136 255	149 743	Amériques
Europe	52 121	57 105	64 133	70 008	86 874	Europe
Argentina[3]	**2 491 719**	**2 727 987**	**2 870 346**	**3 030 913**	**3 532 053**	**Argentine**[3]
Americas	2 181 955	2 360 626	2 462 876	2 513 991	2 933 378	Amériques
Europe	223 092	260 848	290 447	365 963	443 665	Europe
Aruba	**344 336**	**432 762**	**501 324**	**541 714**	**562 034**	**Aruba**
Americas	305 204	387 253	443 338	485 525	506 134	Amériques
Europe	37 039	40 695	54 703	53 118	52 698	Europe
Asia, East and South East/Oceania	...	...	78	199	170	Asie, Est et Sud-Est et Océanie
Australia[3,6]	**2 080 300**	**2 214 900**	**2 370 400**	**2 603 300**	**2 996 100**	**Australie**[3,6]
Africa	17 300	18 400	18 300	24 300	38 200	Afrique
Americas	326 600	317 100	337 900	323 800	345 200	Amériques
Europe	532 000	549 900	530 100	578 400	638 500	Europe
Asia, East and South East/Oceania	1 169 300	1 293 600	1 451 600	1 640 900	1 933 800	Asie, Est et Sud-Est et Océanie
Southern Asia	19 100	19 200	16 400	16 000	15 900	Asie du Sud
Western Asia	14 400	14 300	12 100	14 700	18 000	Asie occidentale
Austria[7]	**18 201 763**	**19 011 397**	**19 091 828**	**19 098 478**	**18 256 766**	**Autriche**[7]
Africa	25 628	24 955	22 609	27 503	26 852	Afrique
Americas	843 895	1 067 420	617 796	785 532	684 482	Amériques
Europe	16 603 415	16 713 190	17 740 096	17 507 575	16 774 178	Europe
Asia, East and South East/Oceania	293 976	350 433	276 872	299 204	280 846	Asie, Est et Sud-Est et Océanie
Southern Asia	33 000	28 000	16 000	15 000	13 000	Asie du Sud
Western Asia	105 000	97 000	74 000	77 000	74 000	Asie occidentale
Bahamas	**1 575 070**	**1 561 600**	**1 427 035**	**1 398 895**	**1 488 680**	**Bahamas**
Americas	1 463 510	1 437 765	1 287 005	1 246 165	1 326 995	Amériques
Europe	91 320	96 625	112 045	122 140	133 085	Europe

75

Tourist arrivals by region of origin [*cont.*]

Arrivées de touristes par région de provenance [*suite*]

Country or area of destination and region of origin	1989	1990	1991	1992	1993	Pays ou zone de destination et région de provenance
Asia, East and South East/Oceania	7 615	10 635	16 320	17 350	15 580	Asie, Est et Sud-Est et Océanie
Bahrain	**1 341 792**	**1 375 997**	**1 674 169**	**1 419 282**	...	**Bahrein**
Americas	20 394	23 666	33 008	42 193	...	Amériques
Europe	88 144	75 913	68 031	61 184	...	Europe
Asia, East and South East/Oceania	170 223	188 469	205 167	146 733	...	Asie, Est et Sud-Est et Océanie
Western Asia	1 053 979	1 079 533	1 358 258	1 156 582	...	Asie occidentale
Bangladesh[3]	**128 064**	**115 369**	**113 242**	**110 475**	**126 785**	**Bangladesh**[3]
Africa	...	...	565	1 664	1 446	Afrique
Americas	9 282	8 616	7 322	9 663	10 414	Amériques
Europe	20 320	15 041	16 892	22 728	29 079	Europe
Asia, East and South East/Oceania	19 787	13 969	18 206	20 786	23 358	Asie, Est et Sud-Est et Océanie
Southern Asia	67 430	44 890	66 064	52 789	59 428	Asie du Sud
Western Asia	2 357	1 922	2 193	2 566	3 055	Asie occidentale
Barbados	**461 259**	**432 067**	**394 222**	**385 472**	**395 979**	**Barbade**
Americas	293 441	274 021	236 898	227 422	227 761	Amériques
Europe	162 480	152 803	153 954	154 277	161 942	Europe
Asia, East and South East/Oceania	1 157	1 126	...	...	...	Asie, Est et Sud-Est et Océanie
Belgium * [7]	**3 042 076**	**3 221 566**	**2 992 170**	**3 219 998**	**3 285 000**	**Belgique** * [7]
Africa	70 670	71 861	59 786	40 329	34 168	Afrique
Americas	199 559	213 293	162 442	196 041	175 282	Amériques
Europe	2 663 092	2 832 276	2 679 146	2 839 679	2 909 961	Europe
Asia, East and South East/Oceania	106 389	103 353	89 940	109 078	109 875	Asie, Est et Sud-Est et Océanie
Belize[2]	**172 829**	**216 395**	**215 442**	**247 346**	**284 487**	**Belize**[2]
Americas	81 047	89 867	84 033	91 046	101 959	Amériques
Europe	20 147	21 865	26 016	34 718	37 331	Europe
Benin	**75 000**[1]	**248 063**[2]	**388 949**[2]	**401 851**[2]	**650 502**[2]	**Bénin**
Africa	48 000	203 580	319 205	352 386	600 423	Afrique
Americas	3 510	4 313	6 758	4 794	4 905	Amériques
Europe	20 520	34 990	54 281	38 497	39 389	Europe
Asia, East and South East/Oceania	2 970	5 180	8 705	6 174	5 785	Asie, Est et Sud-Est et Océanie
Bermuda[4]	**416 891**	**433 776**	**384 695**	**374 497**	**412 473**	**Bermudes**[4]
Americas	383 103	396 776	350 268	341 671	378 660	Amériques
Europe	24 944	25 027	24 851	24 460	26 602	Europe
Asia, East and South East/Oceania	...	902	1 041	880	955	Asie, Est et Sud-Est et Océanie
Bhutan	**1 480**	**1 538**	**2 106**	**2 763**	**2 984**	**Bhoutan**
Americas	479	...	544	616	751	Amériques
Europe	621	...	807	1 229	1 323	Europe
Asia, East and South East/Oceania	335	...	722	728	725	Asie, Est et Sud-Est et Océanie
Southern Asia	...	...	33	...	66	Asie du Sud
Bolivia[1]	**193 557**	**217 071**	**220 902**	**244 583**	**268 968**	**Bolivie**[1]
Africa	668	650	733	621	460	Afrique
Americas	98 858	118 190	130 635	149 818	163 474	Amériques
Europe	78 920	83 375	77 140	82 602	93 660	Europe
Asia, East and South East/Oceania	15 111	14 856	12 394	11 542	11 374	Asie, Est et Sud-Est et Océanie
Botswana[2]	**691 041**	**844 295**	**899 005**	**916 126**	...	**Botswana**[2]
Africa	612 457	759 197	813 629	817 437	...	Afrique
Americas	11 206	10 854	10 527	12 673	...	Amériques
Europe	60 075	65 473	64 921	76 674	...	Europe
Asia, East and South East/Oceania	7 303	8 771	9 928	9 342	...	Asie, Est et Sud-Est et Océanie
Brazil	**1 402 897**	**1 091 067**	**1 192 216**	**1 474 864**	...	**Brésil**

75

Tourist arrivals by region of origin [cont.]

Arrivées de touristes par région de provenance [suite]

Country or area of destination and region of origin	1989	1990	1991	1992	1993	Pays ou zone de destination et région de provenance
Africa	27 453	27 956	21 047	23 697	...	Afrique
Americas	979 991	682 380	827 699	1 067 324	...	Amériques
Europe	348 592	330 741	307 351	341 532	...	Europe
Asia, East and South East/Oceania	40 991	43 553	31 987	36 587	...	Asie, Est et Sud-Est et Océanie
Western Asia	3 779	3 381	2 375	3 213	...	Asie occidentale
British Virgin Islands	**175 757**	**160 046**	**147 030**	**116 944**	**236 992**	**Iles Vierges britanniques**
Americas	159 567	141 164	135 169	96 468	211 742	Amériques
Europe	13 524	11 806	9 356	9 931	19 831	Europe
Brunei Darussalam	**392 751**	**376 636**	**...**	**500 259**	**...**	**Brunéi Darussalam**
Americas[2]	5 749	5 501	...	4 630	...	Amériques[2]
Europe	11 056	10 836	...	22 343	...	Europe
Asia, East and South East/Oceania	370 865	354 016	...	467 414	...	Asie, Est et Sud-Est et Océanie
Southern Asia	3 615	4 680	...	5 392	...	Asie du Sud
Bulgaria[2]	**8 220 860**	**10 329 537**	**6 818 449**	**6 123 844**	**8 302 472**	**Bulgarie[2]**
Africa	16 504	14 001	13 617	16 124	12 274	Afrique
Americas	20 217	17 195	18 153	26 394	26 591	Amériques
Europe	8 034 821	10 162 128	6 000 815	5 948 922	7 939 664	Europe
Asia, East and South East/Oceania	27 463	16 065	11 538	15 819	15 098	Asie, Est et Sud-Est et Océanie
Southern Asia	43 605	55 862	48 383	35 926	12 268	Asie du Sud
Western Asia	37 136	39 951	41 582	52 578	40 779	Asie occidentale
Burkina Faso[1]	**79 624**	**73 814**	**80 100**	**92 118**	**111 115**	**Burkina Faso[1]**
Africa	33 610	31 058	32 528	37 150	45 741	Afrique
Americas	4 553	4 158	5 557	5 861	6 650	Amériques
Europe	36 274	33 211	36 242	38 425	46 844	Europe
Asia, East and South East/Oceania	1 028	784	1 223	1 401	1 743	Asie, Est et Sud-Est et Océanie
Western Asia	267	210	321	640	595	Asie occidentale
Burundi[9]	**81 714**	**109 418**	**125 000**	**86 000**	**75 000**	**Burundi[9]**
Africa	38 406	51 426	59 000	41 000	36 000	Afrique
Americas	4 902	6 564	8 000	6 000	5 250	Amériques
Europe	31 052	41 579	47 000	32 000	27 750	Europe
Asia, East and South East/Oceania	7 354	9 849	10 000	7 000	6 000	Asie, Est et Sud-Est et Océanie
Cameroon[1]	**86 968**	**89 094**	**83 826**	**62 057**	**81 350**	**Cameroun[1]**
Africa	22 519	24 705	24 539	14 851	21 654	Afrique
Americas	8 171	8 538	6 642	4 595	6 218	Amériques
Europe	52 219	51 416	48 729	40 702	50 069	Europe
Asia, East and South East/Oceania	1 833	2 121	2 078	682	1 508	Asie, Est et Sud-Est et Océanie
Western Asia	1 200	1 519	975	583	1 069	Asie occidentale
Canada	**15 111 200**	**15 209 200**	**14 912 100**	**14 740 800**	**15 105 100**	**Canada**
Africa	46 900	49 600	46 100	47 600	48 700	Afrique
Americas	12 457 400	12 523 000	12 267 700	12 065 900	12 283 400	Amériques
Europe	1 679 900	1 685 600	1 695 200	1 723 900	1 856 400	Europe
Asia, East and South East/Oceania	778 200	817 200	781 800	777 000	788 000	Asie, Est et Sud-Est et Océanie
Southern Asia	70 100	63 400	55 500	52 200	49 700	Asie du Sud
Western Asia	8 100	7 800	7 100	8 200	9 200	Asie occidentale
Cayman Islands[4]	**209 804**	**253 158**	**237 351**	**241 843**	**287 277**	**Iles Caïmanes[4]**
Americas	197 321	238 103	218 712	217 261	259 553	Amériques
Europe	11 008	13 134	15 249	17 961	20 465	Europe
Asia, East and South East/Oceania	1 031	1 439	1 065	1 242	1 580	Asie, Est et Sud-Est et Océanie
Chad[2]	**12 345**	**9 156**	**20 501**	**16 991**	**21 227**	**Tchad[2]**
Africa	10 767	7 607	10 720	10 145	12 678	Afrique

75
Tourist arrivals by region of origin [*cont.*]
Arrivées de touristes par région de provenance [*suite*]

Country or area of destination and region of origin	1989	1990	1991	1992	1993	Pays ou zone de destination et région de provenance
Americas	189	120	1 229	1 143	1 428	Amériques
Europe	1 319	1 294	8 340	5 485	6 852	Europe
Asia, East and South East/Oceania	57	135	206	218	269	Asie, Est et Sud-Est et Océanie
Chile	**797 400**	**942 892**	**1 349 149**	**1 283 287**	**1 412 495**	**Chili**
Americas	678 900	817 808	1 220 250	1 151 810	1 251 008	Amériques
Europe	70 800	97 354	103 354	102 219	133 141	Europe
Asia, East and South East/Oceania	12 800	13 706	13 938	5 889	7 952	Asie, Est et Sud-Est et Océanie
China[10]	**1 460 970**	**1 747 315**	**2 710 103**	**4 006 427**	**4 655 857**	**Chine**[10]
Africa	14 349	12 582	17 384	26 740	27 675	Afrique
Americas	288 953	303 542	418 498	495 404	564 087	Amériques
Europe	394 484	446 260	789 534	1 519 890	1 592 046	Europe
Asia, East and South East/Oceania	713 528	930 166	1 395 108	1 864 503	2 356 995	Asie, Est et Sud-Est et Océanie
Southern Asia	43 682	48 503	75 463	89 035	103 059	Asie du Sud
Colombia[9]	**732 982**	**812 796**	**856 862**	**1 075 891**	**1 047 021**	**Colombie**[9]
Americas	682 780	766 276	799 442	993 512	972 573	Amériques
Europe	45 926	43 331	53 323	76 320	70 297	Europe
Asia, East and South East/Oceania	3 524	2 588	3 367	3 950	...	Asie, Est et Sud-Est et Océanie
Southern Asia	...	49	44	264	...	Asie du Sud
Western Asia	84	54	107	147	...	Asie occidentale
Comoros[4]	**13 031**	**7 627**	**16 942**	**18 921**	**23 671**	**Comores**[4]
Africa	6 148	1 526	6 641	7 656	9 545	Afrique
Americas	102	176	247	222	318	Amériques
Europe	5 935	5 520	9 684	10 381	13 012	Europe
Asia, East and South East/Oceania	774	297	370	662	796	Asie, Est et Sud-Est et Océanie
Congo[11]	**32 883**	**32 547**	**32 945**	**37 182**	**...**	**Congo**[11]
Africa	12 362	12 670	11 261	13 472	...	Afrique
Americas	2 354	1 665	1 910	1 857	...	Amériques
Europe	17 536	17 534	18 809	20 680	...	Europe
Cook Islands[12]	**32 907**	**34 218**	**39 984**	**50 009**	**52 868**	**Iles Cook**[12]
Americas	4 988	6 762	8 325	9 716	11 816	Amériques
Europe	5 649	6 015	8 417	13 702	17 524	Europe
Asia, East and South East/Oceania	22 208	21 330	23 126	26 418	23 348	Asie, Est et Sud-Est et Océanie
Costa Rica	**375 951**	**435 037**	**504 649**	**610 591**	**684 005**	**Costa Rica**
Africa	171	244	299	314	468	Afrique
Americas	321 519	367 964	425 505	509 852	555 616	Amériques
Europe	46 423	58 738	69 087	90 320	115 871	Europe
Asia, East and South East/Oceania	5 836	6 711	7 278	8 144	9 669	Asie, Est et Sud-Est et Océanie
Côte d'Ivoire[3]	**191 975**	**196 528**	**200 000**	**217 000**	**...**	**Côte d'Ivoire**[3]
Africa	99 476	104 752	100 587	118 982	...	Afrique
Americas	9 892	10 622	11 000	17 014	...	Amériques
Europe	75 951	73 212	80 990	73 391	...	Europe
Asia, East and South East/Oceania	2 729	3 912	3 407	3 954	...	Asie, Est et Sud-Est et Océanie
Southern Asia	1 000	444	393	803	...	Asie du Sud
Western Asia	2 927	3 586	3 623	2 856	...	Asie occidentale
Croatia[7]	**...**	**...**	**...**	**2 010 000**	**2 363 000**	**Croatie**[7]
Americas	...	...	...	30 000	143 000	Amériques
Europe	...	...	...	1 237 000	1 374 000	Europe
Asia, East and South East/Oceania	...	...	...	2 000	2 000	Asie, Est et Sud-Est et Océanie
Western Asia	...	...	...	1 000	2 000	Asie occidentale
Cuba[2]	**326 304**	**340 329**	**424 041**	**460 610**	**546 023**	**Cuba**[2]

75
Tourist arrivals by region of origin [cont.]
Arrivées de touristes par région de provenance [suite]

Country or area of destination and region of origin	1989	1990	1991	1992	1993	Pays ou zone de destination et région de provenance
Africa	1 904	2 395	1 880	2 979	1 674	Afrique
Americas	159 928	163 795	207 359	226 135	288 051	Amériques
Europe	161 443	168 788	196 767	218 363	246 084	Europe
Asia, East and South East/Oceania	1 068	1 858	1 505	1 661	1 876	Asie, Est et Sud-Est et Océanie
Cyprus	**1 377 636**	**1 561 479**	**1 385 129**	**1 991 000**	**1 841 000**	**Chypre**
Africa	...	19 000	17 000	20 810	14 290	Afrique
Americas	13 836	22 197	30 000	38 830	36 210	Amériques
Europe	1 144 968	1 354 342	1 200 313	1 764 750	1 653 575	Europe
Western Asia	144 204	158 994	130 000	114 230	105 000	Asie occidentale
former Czechoslovakia†[2]	**29 683 211**	**46 586 782**	**64 801 030**	**83 477 428**	**...**	**anc. Tchécoslovaquie†[2]**
Europe	29 221 568	40 002 463	47 105 508	63 492 132	...	Europe
Denmark *[1]	**1 425 555**	**1 508 167**	**1 656 417**	**1 716 110**	**1 642 500**	**Danemark *[1]**
Americas	118 111	110 972	80 944	84 778	81 111	Amériques
Europe	1 172 250	1 260 806	1 451 640	1 499 582	1 422 055	Europe
Asia, East and South East/Oceania	32 972	30 000	25 944	30 806	27 778	Asie, Est et Sud-Est et Océanie
Dominica	**36 727**	**45 087**	**46 312**	**46 959**	**51 937**	**Dominique**
Americas	28 434	34 919	35 651	35 864	39 637	Amériques
Europe	6 994	9 329	9 396	9 940	11 550	Europe
Asia, East and South East/Oceania	...	...	21	12	28	Asie, Est et Sud-Est et Océanie
Dominican Republic[2]	**1 400 000**	**1 533 217**	**1 320 976**	**1 523 762**	**1 718 686**	**Rép. dominicaine[2]**
Ecuador[2 3]	**334 557**	**362 072**	**364 585**	**403 242**	**471 367**	**Equateur[2 3]**
Africa	311	289	383	446	525	Afrique
Americas	266 011	292 433	295 819	323 233	378 568	Amériques
Europe	59 044	60 314	58 844	68 337	80 067	Europe
Asia, East and South East/Oceania	9 151	8 992	9 513	11 191	12 194	Asie, Est et Sud-Est et Océanie
Egypt[2]	**2 503 398**	**2 600 117**	**2 214 277**	**3 206 940**	**2 507 762**	**Egypte[2]**
Africa	238 512	326 584	172 572	204 138	187 148	Afrique
Americas	200 479	179 144	119 863	224 479	187 476	Amériques
Europe	1 188 783	1 123 161	889 950	1 664 906	1 205 740	Europe
Asia, East and South East/Oceania	118 770	125 905	87 522	162 156	131 389	Asie, Est et Sud-Est et Océanie
Southern Asia	24 651	18 275	12 238	25 148	26 465	Asie du Sud
Western Asia	731 408	826 218	931 248	924 897	767 307	Asie occidentale
El Salvador[3]	**130 602**	**194 268**	**198 918**	**314 482**	**267 425**	**El Salvador[3]**
Africa	...	...	7	88	86	Afrique
Americas	119 097	182 546	184 645	290 646	242 547	Amériques
Europe	9 862	10 002	12 333	20 712	21 421	Europe
Asia, East and South East/Oceania	1 489	1 463	1 931	3 027	3 358	Asie, Est et Sud-Est et Océanie
Western Asia	...	...	2	9	13	Asie occidentale
Ethiopia[13]	**76 844**	**79 346**	**81 581**	**83 213**	**93 072**	**Ethiopie[13]**
Africa	25 684	24 073	24 777	25 290	28 285	Afrique
Americas	4 943	8 144	8 352	8 513	9 521	Amériques
Europe	20 437	21 069	21 582	21 993	24 599	Europe
Asia, East and South East/Oceania	6 401	5 306	5 454	5 515	6 171	Asie, Est et Sud-Est et Océanie
Southern Asia	2 260	2 088	2 136	7 916	2 438	Asie du Sud
Western Asia	5 335	7 504	7 746	7 916	8 851	Asie occidentale
Fiji[3]	**250 565**	**278 996**	**259 350**	**278 534**	**287 462**	**Fidji[3]**
Americas	50 961	55 366	47 084	47 404	55 004	Amériques
Europe	35 320	43 984	42 820	46 308	50 019	Europe
Asia, East and South East/Oceania	159 199	178 369	168 705	183 873	181 306	Asie, Est et Sud-Est et Océanie
Finland *[1]	**662 447**	**649 509**	**579 168**	**590 199**	**680 980**	**Finlande *[1]**

75
Tourist arrivals by region of origin [cont.]
Arrivées de touristes par région de provenance [suite]

Country or area of destination and region of origin	1989	1990	1991	1992	1993	Pays ou zone de destination et région de provenance
Africa	4 900	5 100	3 000	3 010	1 611	Afrique
Americas	66 070	64 562	47 472	48 358	52 355	Amériques
Europe	549 799	528 089	482 972	480 425	561 042	Europe
Asia, East and South East/Oceania	22 914	24 055	21 067	21 145	18 822	Asie, Est et Sud-Est et Océanie
France[14]	**50 158 000**	**52 785 000**	**55 041 000**	**59 710 000**	**60 100 000**	**France**[14]
Africa	1 583 000	1 500 000	1 232 000	...	...	Afrique
Americas	3 005 000	3 300 000	2 748 000	2 604 000	2 665 000	Amériques
Europe	44 074 000	46 835 000	49 616 000	52 507 000	52 988 000	Europe
Asia, East and South East/Oceania	1 033 000	900 000	941 000	449 000	325 000	Asie, Est et Sud-Est et Océanie
Western Asia	213 000	250 000	...	...	...	Asie occidentale
French Polynesia[3]	**139 705**	**132 361**	**120 938**	**123 619**	**147 847**	**Polynésie française**[3]
Africa	468	150	132	145	195	Afrique
Americas	61 795	51 592	44 144	46 098	55 878	Amériques
Europe	47 596	46 615	44 508	46 078	58 157	Europe
Asia, East and South East/Oceania	29 462	33 655	31 827	30 865	33 246	Asie, Est et Sud-Est et Océanie
Southern Asia	47	37	27	28	18	Asie du Sud
Western Asia	144	103	108	115	193	Asie occidentale
Gabon	**113 000**	**108 000**	**128 000**	**133 000**	**115 000**	**Gabon**
Africa	14 700	14 000	19 900	21 400	22 000	Afrique
Americas	6 700	6 200	6 400	6 600	...	Amériques
Europe	84 900	81 400	97 600	99 700	93 000	Europe
Asia, East and South East/Oceania	6 700	6 400	4 100	5 300	...	Asie, Est et Sud-Est et Océanie
Gambia	**85 860**	**101 419**	**114 000**	**...**	**...**	**Gambie**
Africa	31 000	39 000	...	...	...	Afrique
Americas	433	581	...	...	...	Amériques
Europe	54 000	61 000	...	...	...	Europe
Asia, East and South East/Oceania	...	78	...	...	...	Asie, Est et Sud-Est et Océanie
Germany †						**Allemagne†**
F. R. Germany[7 15 16]	**14 653 201**	**15 626 858**	**14 294 604**	**15 913 214**	**14 347 710**	**R. f. Allemagne**[7 15 16]
Africa	191 610	180 461	140 803	149 816	142 338	Afrique
Americas	2 474 179	2 871 511	1 986 074	2 153 047	1 898 706	Amériques
Europe	10 498 001	10 934 518	10 700 494	11 884 139	10 692 563	Europe
Asia, East and South East/Oceania	1 310 294	1 411 521	1 214 491	1 378 983	1 263 462	Asie, Est et Sud-Est et Océanie
former German D. R.	**3 102 000**	**...**	**...**	**...**	**...**	**anc. R. d. allemande**
Europe	639 630	...	...	...	...	Europe
Ghana	**125 162**	**145 780**	**172 464**	**213 316**	**...**	**Ghana**
Africa	39 231	45 797	69 048	72 579	...	Afrique
Americas	15 595	11 823	15 573	17 915	...	Amériques
Europe	34 637	40 687	47 152	52 879	...	Europe
Asia, East and South East/Oceania	4 521	7 770	7 416	10 260	...	Asie, Est et Sud-Est et Océanie
Southern Asia	4 744	...	...	...	...	Asie du Sud
Western Asia	6 446	773	1 311	1 618	...	Asie occidentale
Greece[17]	**8 081 851**	**8 873 310**	**8 036 127**	**9 331 360**	**9 412 823**	**Grèce**[17]
Africa	50 112	44 000	31 475	31 604	28 929	Afrique
Americas	399 802	382 623	255 770	378 191	343 344	Amériques
Europe	7 257 154	8 109 849	7 525 649	8 628 436	8 782 981	Europe
Asia, East and South East/Oceania	292 135	274 932	177 816	242 840	207 407	Asie, Est et Sud-Est et Océanie
Southern Asia	5 007	4 451	4 226	6 794	7 615	Asie du Sud
Western Asia	77 641	57 455	41 191	43 495	42 547	Asie occidentale
Grenada	**65 722**	**76 447**	**85 000**	**87 554**	**93 919**	**Grenade**

75
Tourist arrivals by region of origin [cont.]
Arrivées de touristes par région de provenance [suite]

Country or area of destination and region of origin	1989	1990	1991	1992	1993	Pays ou zone de destination et région de provenance
Americas	31 780	40 225	44 093	44 974	48 805	Amériques
Europe	15 841	16 778	22 127	23 687	25 296	Europe
Asia, East and South East/Oceania	163	769	657	476	839	Asie, Est et Sud-Est et Océanie
Western Asia	277	28	20	32	31	Asie occidentale
Guadeloupe[1]	**123 038**	**125 663**	**132 253**	**121 278**	**124 321**	**Guadeloupe**[1]
Americas	18 962	16 608	11 263	10 576	12 918	Amériques
Europe	103 542	108 564	120 425	110 120	110 901	Europe
Guam[5]	**668 748**	**780 404**	**737 260**	**876 742**	**784 018**	**Guam**[5]
Americas	43 393	52 199	50 897	60 442	61 895	Amériques
Europe	2 793	2 643	1 998	2 643	2 278	Europe
Asia, East and South East/Oceania	610 237	710 627	671 837	798 831	709 983	Asie, Est et Sud-Est et Océanie
Guatemala	**437 019**	**508 514**	**513 620**	**541 025**	**561 917**	**Guatemala**
Americas	352 187	411 871	406 595	419 051	428 865	Amériques
Europe	74 043	84 585	93 630	105 747	115 883	Europe
Asia, East and South East/Oceania	7 079	8 202	9 400	10 741	11 821	Asie, Est et Sud-Est et Océanie
Southern Asia	176	246	241	546	536	Asie du Sud
Western Asia	307	281	268	186	206	Asie occidentale
Guyana	...	...	...	**74 879**	**107 127**	**Guyana**
Americas	...	...	...	62 972	97 336	Amériques
Europe	...	...	...	6 763	7 892	Europe
Haiti	**122 000**	**120 000**	**119 000**	...	...	**Haïti**
Americas	107 000	107 000	110 900	...	...	Amériques
Europe	10 000	10 000	7 600	...	...	Europe
Honduras[2]	**260 399**	**290 353**	**226 121**	**243 544**	**236 511**	**Honduras**[2]
Africa	94	101	125	92	269	Afrique
Americas	243 287	268 329	201 464	217 576	206 579	Amériques
Europe	13 874	18 049	20 277	21 049	24 311	Europe
Asia, East and South East/Oceania	3 144	3 874	4 255	4 827	5 332	Asie, Est et Sud-Est et Océanie
Hong Kong[2]	**5 984 501**	**6 580 850**	**6 795 413**	**8 010 524**	**8 937 500**	**Hong-kong**[2]
Africa	41 601	46 958	55 355	56 101	65 532	Afrique
Americas	812 919	807 649	822 394	924 223	1 008 289	Amériques
Europe	725 813	755 298	809 722	962 322	1 078 786	Europe
Asia, East and South East/Oceania	4 252 712	4 814 284	4 947 045	5 895 152	6 625 324	Asie, Est et Sud-Est et Océanie
Southern Asia	131 807	138 515	142 529	152 356	136 948	Asie du Sud
Western Asia	14 840	13 510	14 313	17 895	18 895	Asie occidentale
Hungary[3 18]	**14 490 000**	**20 510 000**	**21 860 000**	**20 188 000**	**22 804 000**	**Hongrie**[3 18]
Africa	26 000	23 000	17 000	20 000	20 000	Afrique
Americas	198 000	239 000	236 000	304 000	317 000	Amériques
Europe	14 195 000	20 158 000	21 431 000	19 688 000	22 333 000	Europe
Asia, East and South East/Oceania	60 000	90 000	175 000	176 000	134 000	Asie, Est et Sud-Est et Océanie
Iceland	**130 503**	**141 718**	**143 459**	**142 560**	**157 326**	**Islande**
Africa	363	329	344	297	251	Afrique
Americas	24 594	24 138	23 901	23 319	26 973	Amériques
Europe	102 876	114 364	116 332	115 608	125 759	Europe
Asia, East and South East/Oceania	2 242	2 343	2 668	3 029	3 922	Asie, Est et Sud-Est et Océanie
Southern Asia	127	116	123	138	176	Asie du Sud
Western Asia	45	35	62	78	105	Asie occidentale
India[3]	**1 736 093**	**1 707 158**	**1 677 508**	**1 867 651**	**1 764 830**	**Inde**[3]
Africa	59 228	58 154	59 616	68 154	59 526	Afrique
Americas	183 656	176 995	165 321	208 211	219 527	Amériques

75

Tourist arrivals by region of origin [cont.]

Arrivées de touristes par région de provenance [suite]

Country or area of destination and region of origin	1989	1990	1991	1992	1993	Pays ou zone de destination et région de provenance
Europe	634 336	630 049	562 731	658 496	683 579	Europe
Asia, East and South East/Oceania	199 387	206 958	176 705	214 242	214 269	Asie, Est et Sud-Est et Océanie
Southern Asia	536 755	520 920	591 210	579 079	466 399	Asie du Sud
Western Asia	121 890	112 470	118 924	135 957	118 430	Asie occidentale
Indonesia	**1 625 965**	**2 177 566**	**2 569 870**	**3 064 161**	**3 403 138**	**Indonésie**
Africa	1 312	1 973	3 095	3 263	6 558	Afrique
Americas	93 400	127 278	129 335	157 872	190 310	Amériques
Europe	379 625	484 383	481 684	561 657	659 726	Europe
Asia, East and South East/Oceania	1 122 739	1 536 557	1 926 663	2 302 369	2 490 260	Asie, Est et Sud-Est et Océanie
Southern Asia	9 240	14 415	16 324	23 825	26 688	Asie du Sud
Western Asia	10 674	12 960	12 769	15 175	29 596	Asie occidentale
Iran, Islamic Rep. of	**89 148**	**153 615**	**212 096**	**...**	**...**	**Iran, Rép. islamique d'**
Africa	1 316	1 442	1 804	...	...	Afrique
Americas	863	1 392	2 121	...	...	Amériques
Europe	19 700	37 073	61 205	...	...	Europe
Asia, East and South East/Oceania	8 717	11 767	12 237	...	...	Asie, Est et Sud-Est et Océanie
Southern Asia	44 323	83 308	97 755	...	...	Asie du Sud
Western Asia	14 221	18 620	36 965	...	...	Asie occidentale
Iraq[2]	**1 025 149**	**747 625**	**267 743**	**504 473**	**...**	**Iraq[2]**
Africa	15 364	10 470	3 430	15 124	...	Afrique
Americas	10 080	6 955	1 123	1 924	...	Amériques
Europe	85 369	55 104	4 737	6 086	...	Europe
Asia, East and South East/Oceania	11 016	11 159	3 955	2 062	...	Asie, Est et Sud-Est et Océanie
Southern Asia	45 651	33 107	1 316	2 924	...	Asie du Sud
Western Asia	798 053	597 013	194 646	474 291	...	Asie occidentale
Ireland[3]	**3 484 000**	**3 666 000**	**3 535 000**	**3 666 000**	**3 814 000**	**Irlande[3]**
Americas	427 000	443 000	356 000	417 000	422 000	Amériques
Europe	2 943 000	3 099 000	3 071 000	3 131 000	3 268 000	Europe
Asia, East and South East/Oceania	62 000	69 000	54 000	71 000	56 000	Asie, Est et Sud-Est et Océanie
Israel[3]	**1 176 536**	**1 063 406**	**943 259**	**1 509 520**	**1 655 642**	**Israël[3]**
Africa	25 836	26 571	20 712	32 271	36 687	Afrique
Americas	343 568	314 216	285 722	440 580	470 796	Amériques
Europe	705 511	625 178	547 453	906 468	997 521	Europe
Asia, East and South East/Oceania	40 760	36 628	31 750	57 221	62 722	Asie, Est et Sud-Est et Océanie
Southern Asia	4 708	4 018	2 289	3 834	5 663	Asie du Sud
Western Asia	52 589	52 287	50 350	60 747	71 342	Asie occidentale
Italy[7]	**20 584 562**	**20 862 965**	**20 241 217**	**20 424 982**	**21 025 353**	**Italie[7]**
Africa	45 194	52 300	42 592	39 958	46 538	Afrique
Americas	2 846 656	3 031 405	2 283 300	2 810 758	2 930 963	Amériques
Europe	15 723 856	15 686 006	16 078 258	15 350 782	15 624 015	Europe
Asia, East and South East/Oceania	1 016 072	1 084 923	924 599	1 201 218	1 344 342	Asie, Est et Sud-Est et Océanie
Western Asia	107 654	107 207	98 745	98 020	107 568	Asie occidentale
Jamaica	**714 771**	**840 777**	**844 607**	**909 010**	**978 713**	**Jamaïque**
Africa	511	461	676	908	999	Afrique
Americas	612 477	709 918	669 036	700 413	752 653	Amériques
Europe	96 550	121 049	159 849	187 874	202 362	Europe
Asia, East and South East/Oceania	3 649	7 091	12 638	17 052	19 778	Asie, Est et Sud-Est et Océanie
Southern Asia	253	359	372	304	445	Asie du Sud
Japan[23]	**2 835 064**	**3 235 860**	**3 532 651**	**3 581 540**	**3 410 447**	**Japon[23]**
Africa	10 065	10 392	11 510	10 816	11 096	Afrique

75
Tourist arrivals by region of origin [cont.]
Arrivées de touristes par région de provenance [suite]

Country or area of destination and region of origin	1989	1990	1991	1992	1993	Pays ou zone de destination et région de provenance
Americas	653 793	711 410	727 072	717 762	680 429	Amériques
Europe	464 654	527 810	532 995	545 403	527 931	Europe
Asia, East and South East/Oceania	1 637 646	1 907 546	2 164 477	2 246 314	2 142 289	Asie, Est et Sud-Est et Océanie
Southern Asia	63 944	73 762	92 585	57 074	44 325	Asie du Sud
Western Asia	2 206	2 178	1 893	2 340	2 503	Asie occidentale
Jordan[2]	**2 278 126**	**2 633 262**	**2 227 688**	**3 242 985**	**3 098 938**	**Jordanie**[2]
Africa	14 523	62 613	43 620	69 927	91 707	Afrique
Americas	48 257	38 538	23 978	39 250	51 512	Amériques
Europe	225 745	206 125	194 566	284 032	307 687	Europe
Asia, East and South East/Oceania	24 835	55 389	16 462	25 930	35 822	Asie, Est et Sud-Est et Océanie
Southern Asia	18 515	253 319	24 333	25 772	20 942	Asie du Sud
Western Asia	1 925 785	2 013 516	1 920 340	2 793 236	2 586 486	Asie occidentale
Kenya[3][18]	**734 700**	**814 400**	**817 550**	**698 540**	**...**	**Kenya**[3][18]
Africa	171 600	142 100	228 020	191 914	...	Afrique
Americas	97 700	77 700	67 830	59 936	...	Amériques
Europe	396 000	427 400	470 010	403 783	...	Europe
Asia, East and South East/Oceania	32 600	31 400	33 290	25 302	...	Asie, Est et Sud-Est et Océanie
Southern Asia	11 300	12 500	17 650	16 914	...	Asie du Sud
Kiribati[2]	**3 009**	**3 332**	**2 935**	**3 587**	**...**	**Kiribati**[2]
Americas	939	642	372	590	...	Amériques
Europe	107	10	64	185	...	Europe
Asia, East and South East/Oceania	1 583	2 261	2 202	2 323	...	Asie, Est et Sud-Est et Océanie
Korea, Dem. P. R.[2][9]	**2 728 054**	**2 958 839**	**3 196 340**	**3 231 081**	**3 331 226**	**Corée, R. p. dém. de**[2][9]
Africa	7 137	6 925	6 856	7 912	6 346	Afrique
Americas	359 790	370 143	358 570	375 706	370 688	Amériques
Europe	175 633	196 051	251 298	282 571	326 952	Europe
Asia, East and South East/Oceania	1 816 265	2 007 731	2 175 295	2 158 411	2 234 196	Asie, Est et Sud-Est et Océanie
Southern Asia	41 642	48 210	81 097	83 038	54 014	Asie du Sud
Western Asia	6 829	6 538	9 770	10 369	11 881	Asie occidentale
Kuwait[2]	**1 938 438**	**...**	**...**	**1 593 142**	**...**	**Koweït**[2]
Africa	2 500	...	...	488	...	Afrique
Americas	16 326	...	...	21 020	...	Amériques
Europe	61 250	...	...	22 672	...	Europe
Asia, East and South East/Oceania	230 964	...	...	238 537	...	Asie, Est et Sud-Est et Océanie
Western Asia	1 622 734	...	...	1 303 770	...	Asie occidentale
Lesotho[2]	**216 252**	**242 456**	**357 458**	**416 882**	**348 943**	**Lesotho**[2]
Africa	210 237	235 845	349 437	408 569	341 480	Afrique
Americas	1 185	1 570	1 608	1 609	1 236	Amériques
Europe	4 014	3 970	5 246	5 499	4 667	Europe
Asia, East and South East/Oceania	594	1 026	1 167	1 205	1 560	Asie, Est et Sud-Est et Océanie
Southern Asia	81	45	...	...	...	Asie du Sud
Liechtenstein[1]	**77 118**	**77 528**	**71 046**	**71 710**	**64 717**	**Liechtenstein**[1]
Africa	292	249	169	196	241	Afrique
Americas	11 014	10 984	6 824	7 377	6 047	Amériques
Europe	62 524	63 476	61 638	61 400	56 539	Europe
Asia, East and South East/Oceania	3 152	2 630	2 237	2 650	1 862	Asie, Est et Sud-Est et Océanie
Lithuania[1]	**...**	**...**	**...**	**319 211**	**284 172**	**Lituanie**[1]
Africa	...	...	...	120	139	Afrique
Americas	...	...	...	4 325	11 070	Amériques

75
Tourist arrivals by region of origin [*cont.*]
Arrivées de touristes par région de provenance [*suite*]

Country or area of destination and region of origin	1989	1990	1991	1992	1993	Pays ou zone de destination et région de provenance
Europe	...	...	...	313 601	269 185	Europe
Asia, East and South East/Oceania	...	...	...	1 165	3 778	Asie, Est et Sud-Est et Océanie
Luxembourg[7]	**874 652**	**820 476**	**861 284**	**796 051**	...	**Luxembourg**[7]
Africa	6 099	6 547	5 614	5 627	...	Afrique
Americas	47 138	50 026	37 296	39 132	...	Amériques
Europe	805 939	749 260	804 801	737 433	...	Europe
Asia, East and South East/Oceania	13 752	14 534	13 525	13 859	...	Asie, Est et Sud-Est et Océanie
Macau[2]	**5 619 289**	**5 942 210**	**7 488 610**	**7 699 178**	**7 701 007**	**Macao**[2]
Africa	3 444	2 896	3 404	3 412	5 687	Afrique
Americas	113 926	104 104	123 757	130 893	135 425	Amériques
Europe	187 279	181 147	207 096	230 269	235 989	Europe
Asia, East and South East/Oceania	5 290 255	5 631 024	7 103 936	7 197 386	7 171 955	Asie, Est et Sud-Est et Océanie
Southern Asia	13 337	10 636	8 214	10 532	10 864	Asie du Sud
Madagascar	**38 954**	**52 923**	**34 891**	**53 654**	...	**Madagascar**
Africa	7 974	10 149	7 151	5 923	...	Afrique
Americas	2 372	3 867	2 547	2 522	...	Amériques
Europe	27 597	36 702	23 880	42 709	...	Europe
Asia, East and South East/Oceania	1 011	2 205	1 262	2 500	...	Asie, Est et Sud-Est et Océanie
Malawi[18]	**117 069**	**129 912**	**127 004**	...	...	**Malawi**[18]
Africa	95 564	105 366	102 436	...	...	Afrique
Americas	3 810	4 077	3 650	...	...	Amériques
Europe	13 164	15 603	15 563	...	...	Europe
Asia, East and South East/Oceania	3 138	...	...	...	...	Asie, Est et Sud-Est et Océanie
Malaysia[19]	**4 846 320**	**7 445 908**	**5 847 213**	**6 016 209**	**6 503 860**	**Malaisie**[19]
Africa	...	...	...	13 365	23 106	Afrique
Americas	103 330	174 986	135 344	112 177	120 180	Amériques
Europe	282 744	455 061	420 833	343 897	373 214	Europe
Asia, East and South East/Oceania	4 253 953	6 481 732	4 967 844	5 313 828	5 782 170	Asie, Est et Sud-Est et Océanie
Southern Asia	39 396	108 411	71 411	94 488	42 153	Asie du Sud
Western Asia	17 431	33 876	47 815	24 214	24 413	Asie occidentale
Maldives[4]	**158 488**	**195 156**	**196 112**	**235 852**	**241 020**	**Maldives**[4]
Africa	372	296	321	480	975	Afrique
Americas	1 702	2 167	2 008	2 060	2 311	Amériques
Europe	123 343	152 041	147 045	165 961	174 403	Europe
Asia, East and South East/Oceania	18 941	24 011	28 632	44 601	37 194	Asie, Est et Sud-Est et Océanie
Southern Asia	14 092	16 632	19 093	22 750	26 000	Asie du Sud
Mali[1]	**32 103**	**43 913**	**37 962**	**37 843**	...	**Mali**[1]
Africa	9 140	13 669	16 589	17 907	...	Afrique
Americas	3 387	3 895	2 705	2 998	...	Amériques
Europe	15 316	22 330	16 452	14 103	...	Europe
Asia, East and South East/Oceania	558	954	517	462	...	Asie, Est et Sud-Est et Océanie
Western Asia	1 583	179	182	147	...	Asie occidentale
Malta[18]	**828 311**	**871 675**	**895 036**	**1 002 381**	...	**Malte**[18]
Africa	1 551	1 569	2 083	3 489	...	Afrique
Americas	15 160	15 456	13 934	15 891	...	Amériques
Europe	758 414	793 481	806 184	912 801	...	Europe
Asia, East and South East/Oceania	7 569	8 388	7 822	9 030	...	Asie, Est et Sud-Est et Océanie
Southern Asia	1 824	1 589	1 890	1 855	...	Asie du Sud
Western Asia	32 810	37 566	48 730	40 759	...	Asie occidentale
Marshall Islands	...	...	6 741	8 000	5 055	**Iles Marshall**

75
Tourist arrivals by region of origin [cont.]
Arrivées de touristes par région de provenance [suite]

Country or area of destination and region of origin	1989	1990	1991	1992	1993	Pays ou zone de destination et région de provenance
Americas	...	...	2 851	3 377	1 981	Amériques
Europe	...	...	449	410	238	Europe
Asia, East and South East/Oceania	...	...	3 379	4 124	2 752	Asie, Est et Sud-Est et Océanie
Martinique[4]	**311 584**	**281 517**	**315 132**	**320 693**	**366 353**	**Martinique**[4]
Americas	75 843	62 668	82 723	59 164	59 240	Amériques
Europe	232 320	216 746	231 464	259 117	303 480	Europe
Mauritius	**262 790**	**291 550**	**300 670**	**335 400**	**374 630**	**Maurice**
Africa	121 090	135 320	139 460	141 710	149 750	Afrique
Americas	2 960	3 460	2 920	3 050	3 640	Amériques
Europe	120 600	128 840	134 260	164 660	193 620	Europe
Asia, East and South East/Oceania	13 350	15 740	15 540	17 780	16 880	Asie, Est et Sud-Est et Océanie
Southern Asia	4 790	8 190	8 490	8 200	10 740	Asie du Sud
Mexico[9]	**14 964 000**	**17 176 000**	**16 281 000**	**17 273 000**	**16 534 000**	**Mexique**[9]
Americas	14 785 000	16 951 000	15 793 000	16 750 000	15 930 000	Amériques
Europe	157 000	189 000	328 000	363 000	473 000	Europe
Monaco[1]	**245 146**	**244 640**	**239 043**	**245 592**	**208 206**	**Monaco**[1]
Africa	455	637	430	375	447	Afrique
Americas	37 763	38 133	25 075	31 100	29 336	Amériques
Europe	187 152	181 601	190 443	189 870	154 524	Europe
Asia, East and South East/Oceania	6 660	10 048	8 438	9 411	8 903	Asie, Est et Sud-Est et Océanie
Western Asia	2 544	2 674	2 224	2 641	3 250	Asie occidentale
Mongolia	**236 540**	**147 236**	...	...	...	**Mongolie**
Africa	9	20	...	...	...	Afrique
Americas	782	1 117	...	...	...	Amériques
Europe	231 986	137 622	...	...	...	Europe
Asia, East and South East/Oceania	3 555	8 218	...	...	...	Asie, Est et Sud-Est et Océanie
Southern Asia	191	153	...	...	...	Asie du Sud
Western Asia	14	8	...	...	...	Asie occidentale
Montserrat[3]	...	**12 780**	**16 700**	**17 260**	**20 994**	**Montserrat**[3]
Americas	...	9 680	12 700	14 060	17 473	Amériques
Europe	...	1 900	2 500	2 800	2 981	Europe
Morocco[9]	**3 468 429**	**4 024 197**	**4 162 239**	**4 389 753**	**4 027 356**	**Maroc**[9]
Africa	989 870	1 538 478	2 133 777	1 731 985	1 305 879	Afrique
Americas	116 456	115 408	69 085	118 112	120 491	Amériques
Europe	1 272 058	1 189 981	853 125	1 284 411	1 396 133	Europe
Asia, East and South East/Oceania	17 790	18 315	14 768	17 990	19 117	Asie, Est et Sud-Est et Océanie
Western Asia	89 694	83 179	92 041	65 201	63 016	Asie occidentale
Nepal	**239 945**	**254 885**	**292 995**	**334 353**	**293 567**	**Népal**
Africa	606	611	956	1 263	985	Afrique
Americas	30 013	28 215	26 229	30 083	27 895	Amériques
Europe	111 664	117 607	118 065	139 655	128 978	Europe
Asia, East and South East/Oceania	48 521	42 217	45 910	50 307	44 501	Asie, Est et Sud-Est et Océanie
Southern Asia	48 119	65 629	100 077	112 351	90 219	Asie du Sud
Netherlands[7]	**5 206 000**	**5 795 100**	**5 841 900**	**6 082 900**	**5 756 700**	**Pays-Bas**[7]
Africa	60 200	57 700	55 500	52 700	50 500	Afrique
Americas	646 400	690 400	552 000	605 700	564 100	Amériques
Europe	4 168 300	4 690 900	4 922 000	5 053 100	4 801 800	Europe
Asia, East and South East/Oceania	331 100	356 100	312 400	371 400	340 300	Asie, Est et Sud-Est et Océanie
New Caledonia[4 9]	**82 161**	**86 870**	**80 930**	**78 264**	**80 753**	**Nouvelle-Calédonie**[4 9]
Africa	...	165	169	147	247	Afrique

75

Tourist arrivals by region of origin [cont.]

Arrivées de touristes par région de provenance [suite]

Country or area of destination and region of origin	1989	1990	1991	1992	1993	Pays ou zone de destination et région de provenance
Americas	2 004	1 663	1 326	1 329	1 381	Amériques
Europe	17 185	29 855	28 895	27 936	28 771	Europe
Asia, East and South East/Oceania	61 601	55 187	50 540	48 852	50 354	Asie, Est et Sud-Est et Océanie
Southern Asia	1 043	...	...	...	...	Asie du Sud
New Zealand[29]	**901 078**	**976 010**	**963 470**	**1 055 681**	**1 156 978**	**Nouvelle-Zélande**[29]
Africa	2 580	3 094	2 872	4 367	7 418	Afrique
Americas	173 076	178 743	168 035	163 081	178 926	Amériques
Europe	145 687	170 739	177 172	201 763	223 407	Europe
Asia, East and South East/Oceania	517 471	564 112	575 043	645 664	708 662	Asie, Est et Sud-Est et Océanie
Southern Asia	4 632	3 626	2 457	2 949	2 506	Asie du Sud
Western Asia	1 713	1 906	1 568	2 155	2 757	Asie occidentale
Nicaragua	**77 125**	**106 462**	**145 872**	**166 914**	**...**	**Nicaragua**
Africa	...	140	180	153	...	Afrique
Americas	52 816	80 333	118 312	140 390	...	Amériques
Europe	13 717	17 758	22 969	22 739	...	Europe
Asia, East and South East/Oceania	345	1 278	2 398	2 241	...	Asie, Est et Sud-Est et Océanie
Southern Asia	...	300	1 556	1 195	...	Asie du Sud
Western Asia	...	83	209	196	...	Asie occidentale
Niger[4]	**23 517**	**20 729**	**16 000**	**13 070**	**11 808**	**Niger**[4]
Africa	7 905	6 552	6 900	5 540	5 588	Afrique
Americas	1 915	1 814	1 574	1 189	1 076	Amériques
Europe	12 901	11 557	6 749	5 773	4 785	Europe
Asia, East and South East/Oceania	577	524	564	344	284	Asie, Est et Sud-Est et Océanie
Western Asia	203	174	...	111	45	Asie occidentale
Nigeria	**160 740**	**226 242**	**308 065**	**271 854**	**...**	**Nigéria**
Africa	136 949	186 664	272 104	225 847	...	Afrique
Americas	1 299	1 656	2 387	2 596	...	Amériques
Europe	16 133	28 318	25 913	33 687	...	Europe
Asia, East and South East/Oceania	2 882	3 810	3 643	3 834	...	Asie, Est et Sud-Est et Océanie
Southern Asia	2 350	4 191	2 661	4 443	...	Asie du Sud
Western Asia	1 106	1 574	1 320	1 402	...	Asie occidentale
Niue[4]	**481**	**1 047**	**993**	**1 668**	**3 358**	**Nioué**[4]
Americas	16	23	21	81	189	Amériques
Europe	7	17	35	42	124	Europe
Asia, East and South East/Oceania	458	1 004	934	1 542	3 030	Asie, Est et Sud-Est et Océanie
Northern Mariana Islands[2]	**333 638**	**435 454**	**429 745**	**505 295**	**545 803**	**Iles Marianas du Nord**[2]
Africa	...	17	28	35	23	Afrique
Americas	59 043	70 521	75 123	80 161	79 426	Amériques
Europe	1 240	1 611	1 776	1 840	2 190	Europe
Asia, East and South East/Oceania	273 039	362 986	352 265	422 776	463 577	Asie, Est et Sud-Est et Océanie
Southern Asia	...	280	490	378	236	Asie du Sud
Western Asia	...	30	63	105	102	Asie occidentale
Norway *[1]	**1 867 386**	**1 954 821**	**2 113 951**	**2 375 452**	**2 556 162**	**Norvège ***[1]
Americas	205 629	215 272	175 125	213 813	195 738	Amériques
Europe	1 353 530	1 408 293	1 659 428	1 785 513	1 903 914	Europe
Asia, East and South East/Oceania	40 047	42 248	42 500	49 920	60 434	Asie, Est et Sud-Est et Océanie
Oman[1]	**135 749**	**148 759**	**160 525**	**192 296**	**344 000**	**Oman**[1]
Africa	5 119	5 524	6 714	7 006	10 000	Afrique
Americas	6 771	7 956	7 648	9 725	14 000	Amériques
Europe	53 343	58 170	56 970	69 742	96 000	Europe

75
Tourist arrivals by region of origin [*cont.*]
Arrivées de touristes par région de provenance [*suite*]

Country or area of destination and region of origin	1989	1990	1991	1992	1993	Pays ou zone de destination et région de provenance
Asia, East and South East/Oceania	20 323	29 352	40 035	47 136	46 000	Asie, Est et Sud-Est et Océanie
Western Asia	22 810	20 506	27 732	30 641	45 000	Asie occidentale
Pakistan	**494 600**	**423 842**	**438 088**	**352 112**	**379 165**	**Pakistan**
Africa	8 500	6 495	8 925	10 013	12 975	Afrique
Americas	46 700	42 290	41 921	39 649	47 650	Amériques
Europe	143 200	145 560	127 934	133 499	176 341	Europe
Asia, East and South East/Oceania	31 900	33 597	35 431	34 901	38 974	Asie, Est et Sud-Est et Océanie
Southern Asia	238 800	177 500	200 629	111 438	79 748	Asie du Sud
Western Asia	25 500	18 400	23 171	22 566	23 418	Asie occidentale
Palau	...	...	**32 700**	**36 117**	**40 497**	**Palaos**
Africa	...	...	...	...	11	Afrique
Americas	...	...	6 411	8 032	8 232	Amériques
Europe	...	...	1 202	1 541	1 714	Europe
Asia, East and South East/Oceania	...	...	23 922	25 385	30 385	Asie, Est et Sud-Est et Océanie
Southern Asia	...	...	...	...	149	Asie du Sud
Western Asia	...	...	...	...	5	Asie occidentale
Panama[2]	**216 632**	**230 960**	**298 756**	**311 937**	**314 732**	**Panama**[2]
Africa	294	222	175	228	183	Afrique
Americas	187 565	206 198	267 871	281 216	279 961	Amériques
Europe	18 043	16 999	19 528	19 751	22 552	Europe
Asia, East and South East/Oceania	10 712	7 507	9 155	10 719	12 002	Asie, Est et Sud-Est et Océanie
Western Asia	18	34	27	23	34	Asie occidentale
Papua New Guinea	**48 918**	**40 742**	**37 346**	**42 816**	**40 476**	**Papouasie-Nvl-Guinée**
Africa	133	161	94	89	100	Afrique
Americas	7 203	5 286	6 011	5 866	4 495	Amériques
Europe	6 530	7 497	6 138	5 980	5 299	Europe
Asia, East and South East/Oceania	34 324	26 484	23 285	30 561	30 582	Asie, Est et Sud-Est et Océanie
Southern Asia	728	1 314	1 818	318	...	Asie du Sud
Paraguay[3 6 20]	**278 698**	**280 454**	**361 410**	**334 497**	**404 491**	**Paraguay**[3 6 20]
Americas	234 941	230 786	290 610	267 798	323 957	Amériques
Europe	15 886	30 317	41 634	38 668	48 053	Europe
Asia, East and South East/Oceania	20 234	12 144	15 830	14 952	18 202	Asie, Est et Sud-Est et Océanie
Peru	**333 594**	**316 873**	**232 012**	**216 534**	...	**Pérou**
Africa	608	566	531	392	...	Afrique
Americas	172 278	159 689	132 763	134 070	...	Amériques
Europe	137 648	131 718	82 102	66 147	...	Europe
Asia, East and South East/Oceania	22 015	23 701	15 659	14 885	...	Asie, Est et Sud-Est et Océanie
Southern Asia	781	904	760	845	...	Asie du Sud
Western Asia	169	220	152	178	...	Asie occidentale
Philippines[29]	**1 189 719**	**1 024 520**	**951 365**	**1 152 952**	**1 372 097**	**Philippines**[29]
Africa	...	...	387	766	1 219	Afrique
Americas	298 665	245 496	216 482	252 300	307 989	Amériques
Europe	224 187	168 321	131 825	154 620	183 194	Europe
Asia, East and South East/Oceania	488 228	412 389	452 119	588 860	705 128	Asie, Est et Sud-Est et Océanie
Southern Asia	17 532	16 039	15 826	16 829	18 357	Asie du Sud
Western Asia	13 788	12 441	17 249	18 991	17 775	Asie occidentale
Poland[2]	**8 232 600**	**18 210 747**	**36 845 777**	**49 015 200**	**60 951 100**	**Pologne**[2]
Africa	6 000	...	...	8 700	8 000	Afrique
Americas	120 200	...	149 200	166 000	173 100	Amériques
Europe	7 834 000	15 794 055	36 371 400	48 664 100	60 598 200	Europe

75

Tourist arrivals by region of origin [cont.]
Arrivées de touristes par région de provenance [suite]

Country or area of destination and region of origin	1989	1990	1991	1992	1993	Pays ou zone de destination et région de provenance
Asia, East and South East/Oceania	24 900	...	21 000	50 300	66 300	Asie, Est et Sud-Est et Océanie
Southern Asia	7 800			7 300	11 300	Asie du Sud
Western Asia	12 600	...	...	27 300	29 900	Asie occidentale
Portugal[3]	**7 115 901**	**8 019 919**	**8 656 956**	**8 884 143**	**8 433 900**	**Portugal**[3]
Africa	93 118	102 030	101 006	94 834	...	Afrique
Americas	405 000	413 000	347 000	363 428	286 800	Amériques
Europe	6 541 886	7 423 471	8 137 170	8 352 341	7 899 700	Europe
Asia, East and South East/Oceania	50 145	54 146	45 909	45 477	32 500	Asie, Est et Sud-Est et Océanie
Puerto Rico[4]	**2 443 785**	**2 559 737**	**2 612 991**	**2 658 797**	**2 856 629**	**Porto Rico**[4]
Americas	1 742 247	1 833 283	1 860 290	1 871 908	2 026 095	Amériques
Réunion	**181 769**	**200 276**	**186 026**	**217 350**	**241 691**	**Réunion**
Africa	30 436	33 927	37 133	43 182	44 575	Afrique
Americas	...	...	...	1 239	884	Amériques
Europe	139 373	153 246	136 735	163 950	190 290	Europe
Asia, East and South East/Oceania	...	...	...	506	1 187	Asie, Est et Sud-Est et Océanie
Southern Asia	...	...	...	708	150	Asie du Sud
Romania[2]	**4 852 200**	**6 533 315**	**5 360 179**	**6 280 027**	**5 785 575**	**Roumanie**[2]
Africa	...	15 017	9 289	6 375	5 000	Afrique
Americas	34 100	52 911	50 865	50 298	63 343	Amériques
Europe	4 710 500	6 326 809	5 088 953	6 070 430	5 625 022	Europe
Asia, East and South East/Oceania	86 900	29 790	33 918	31 339	30 808	Asie, Est et Sud-Est et Océanie
Southern Asia	...	22 770	26 606	20 073	9 991	Asie du Sud
Western Asia	...	32 679	34 747	39 428	39 879	Asie occidentale
Saint Kitts and Nevis[5]	**72 125**	**75 689**	**83 903**	**89 719**	**88 554**	**Saint-Kitts-et-Nevis**[5]
Americas	57 016	67 434	75 513	79 439	72 962	Amériques
Europe	6 149	7 356	7 530	9 865	10 739	Europe
Asia, East and South East/Oceania	...	30	51	42	...	Asie, Est et Sud-Est et Océanie
Saint Lucia[3 4]	**132 852**	**138 427**	**159 034**	**177 488**	**194 136**	**Sainte-Lucie**[3 4]
Americas	83 491	87 340	99 717	103 589	114 137	Amériques
Europe	47 672	49 393	57 548	72 022	77 964	Europe
St. Vincent-Grenadines	**49 883**	**53 913**	**51 629**	**53 316**	**56 691**	**St. Vincent-Grenadines**
Americas	35 513	39 139	37 088	36 596	38 820	Amériques
Europe	13 913	14 277	14 013	16 292	17 323	Europe
Samoa	**53 994**	**47 642**	**34 953**	**37 507**	**46 806**	**Samoa**
Americas	5 700	4 686	3 940	4 384	6 122	Amériques
Europe	3 244	5 936	3 416	3 754	5 064	Europe
Asia, East and South East/Oceania	43 810	33 879	25 311	27 720	34 325	Asie, Est et Sud-Est et Océanie
San Marino[2]	**2 845 201**	**2 912 264**	**3 112 995**	**3 208 290**	**3 072 030**	**Saint-Marin**[2]
Africa	131	...	...	...	...	Afrique
Americas	1 977	...	...	...	...	Amériques
Europe	2 819 669	2 330 290	...	...	...	Europe
Asia, East and South East/Oceania	355	...	...	...	...	Asie, Est et Sud-Est et Océanie
Saudi Arabia[2]	**1 178 038**	**1 982 149**	**2 290 213**	**2 688 643**	**2 738 304**	**Arabie saoudite**[2]
Africa	342 183	390 889	302 350	559 458	518 316	Afrique
Americas	16 576	16 706	21 493	25 086	25 679	Amériques
Europe	46 643	50 684	39 637	60 611	61 434	Europe
Asia, East and South East/Oceania	772 636	1 523 870	1 926 733	2 043 488	2 132 875	Asie, Est et Sud-Est et Océanie
Senegal[1]	**259 096**	**245 881**	**233 512**	**245 581**	**...**	**Sénégal**[1]
Africa	39 847	37 127	38 031	48 110	...	Afrique
Americas	13 845	11 005	8 514	11 150	...	Amériques

75
Tourist arrivals by region of origin [cont.]
Arrivées de touristes par région de provenance [suite]

Country or area of destination and region of origin	1989	1990	1991	1992	1993	Pays ou zone de destination et région de provenance
Europe	201 397	193 694	180 963	181 651	...	Europe
Asia, East and South East/Oceania	2 713	2 824	3 494	2 866	...	Asie, Est et Sud-Est et Océanie
Western Asia	988	967	2 064	1 461	...	Asie occidentale
Seychelles	**86 095**[4]	**103 770**[5]	**90 050**[5]	**98 547**[5]	**116 180**[5]	**Seychelles**
Africa	11 109	15 803	18 759	12 068	13 510	Afrique
Americas	2 597	2 614	2 406	2 497	3 779	Amériques
Europe	68 818	81 299	65 043	79 744	94 267	Europe
Asia, East and South East/Oceania	1 390	1 837	1 979	1 980	2 690	Asie, Est et Sud-Est et Océanie
Southern Asia	416	344	488	684	525	Asie du Sud
Western Asia	1 765	1 873	1 375	1 574	1 409	Asie occidentale
Singapore[21]	**4 829 950**	**5 322 854**	**5 414 651**	**5 989 940**	**6 425 778**	**Singapour**[21]
Africa	48 138	52 687	52 070	60 773	87 151	Afrique
Americas	344 564	370 551	354 021	399 293	425 515	Amériques
Europe	1 132 735	1 204 795	1 168 235	1 246 682	1 316 740	Europe
Asia, East and South East/Oceania	2 898 879	3 168 213	3 313 223	3 763 560	4 101 333	Asie, Est et Sud-Est et Océanie
Southern Asia	329 628	454 088	459 272	444 573	412 716	Asie du Sud
Western Asia	40 287	30 506	26 501	30 248	31 975	Asie occidentale
Slovenia[7]	**...**	**...**	**...**	**616 380**	**624 371**	**Slovénie**[7]
Americas	...	...	...	8 922	12 026	Amériques
Europe	...	...	...	598 056	602 810	Europe
Asia, East and South East/Oceania	...	...	...	1 930	2 581	Asie, Est et Sud-Est et Océanie
Solomon Islands	**9 860**	**9 195**	**11 105**	**12 446**	**11 576**	**Iles Salomon**
Americas	999	758	1 115	1 359	1 302	Amériques
Europe	657	791	1 327	1 580	1 500	Europe
Asia, East and South East/Oceania	7 314	6 929	8 636	9 467	8 715	Asie, Est et Sud-Est et Océanie
South Africa[3,22]	**930 393**	**1 029 094**	**1 709 554**	**2 891 721**	**3 358 193**	**Afrique du Sud**[3,22]
Africa	460 459	535 274	1 193 446	2 327 456	2 699 147	Afrique
Americas	63 863	66 972	67 104	75 013	91 722	Amériques
Europe	350 670	366 412	382 772	408 619	442 624	Europe
Asia, East and South East/Oceania	47 988	54 080	56 002	68 164	88 618	Asie, Est et Sud-Est et Océanie
Southern Asia	2 814	3 609	7 171	9 520	14 773	Asie du Sud
Western Asia	1 100	1 270	1 291	1 848	3 022	Asie occidentale
Spain[29]	**54 057 562**	**52 044 056**	**53 494 964**	**55 330 716**	**57 263 351**	**Espagne**[29]
Africa	2 873 586	2 384 172	2 058 685	2 251 149	2 379 669	Afrique
Americas	1 676 759	1 497 189	1 270 138	1 552 234	1 457 888	Amériques
Europe	45 873 185	44 335 013	46 236 764	47 322 066	49 474 766	Europe
Asia, East and South East/Oceania	391 160	435 061	377 701	429 123	457 025	Asie, Est et Sud-Est et Océanie
Southern Asia	40 536	44 003	41 268	36 946	42 726	Asie du Sud
Western Asia	46 811	40 459	30 424	31 930	31 997	Asie occidentale
Sri Lanka[3]	**184 732**	**297 888**	**317 703**	**393 669**	**392 250**	**Sri Lanka**[3]
Africa	362	490	279	573	480	Afrique
Americas	6 844	8 802	11 617	13 332	14 262	Amériques
Europe	120 064	198 132	198 413	256 485	257 883	Europe
Asia, East and South East/Oceania	33 494	58 488	63 668	73 605	61 890	Asie, Est et Sud-Est et Océanie
Southern Asia	21 900	29 142	41 122	46 257	54 813	Asie du Sud
Western Asia	2 068	2 834	2 604	3 417	2 922	Asie occidentale
Sudan	**22 715**[4]	**32 789**	**15 649**[4]	**...**	**...**	**Soudan**
Africa	2 115	2 036	1 423	...	...	Afrique
Americas	1 882	2 585	1 328	...	...	Amériques
Europe	9 350	10 678	4 513	...	...	Europe

75

Tourist arrivals by region of origin [cont.]

Arrivées de touristes par région de provenance [suite]

Country or area of destination and region of origin	1989	1990	1991	1992	1993	Pays ou zone de destination et région de provenance
Asia, East and South East/Oceania	1 724	1 858	1 111	...	...	Asie, Est et Sud-Est et Océanie
Southern Asia	965	1 174	726	...	...	Asie du Sud
Western Asia	4 318	5 701	2 818	...	...	Asie occidentale
Suriname[3][4]	**20 700**	**28 478**	**30 000**	**...**	**...**	**Suriname**[3][4]
Americas	3 204	3 725	...	...	...	Amériques
Europe	16 761	23 836	...	...	...	Europe
Swaziland[1]	**235 612**	**262 826**	**264 376**	**263 477**	**271 680**	**Swaziland**[1]
Africa	198 431	221 350	222 654	223 489	229 624	Afrique
Americas	4 931	5 500	5 533	4 616	5 224	Amériques
Europe	27 318	30 474	30 654	28 266	30 327	Europe
Asia, East and South East/Oceania	2 615	2 917	2 935	3 778	3 454	Asie, Est et Sud-Est et Océanie
**Sweden * **[7]	**842 668**	**730 519**	**622 241**	**646 059**	**674 954**	**Suède * **[7]
Americas	41 747	41 264	30 090	33 326	35 256	Amériques
Europe	739 643	629 243	537 447	561 338	580 150	Europe
Asia, East and South East/Oceania	9 313	10 792	9 982	9 546	11 173	Asie, Est et Sud-Est et Océanie
Switzerland[7]	**10 093 524**	**10 521 446**	**10 111 086**	**10 264 793**	**9 901 324**	**Suisse**[7]
Africa	127 509	135 085	112 510	101 435	99 180	Afrique
Americas	1 398 094	1 582 359	1 029 113	1 220 609	1 117 476	Amériques
Europe	7 610 501	7 800 483	8 108 576	7 943 683	7 690 118	Europe
Asia, East and South East/Oceania	637 274	671 244	551 031	634 125	604 843	Asie, Est et Sud-Est et Océanie
Southern Asia	48 292	52 021	41 916	39 939	40 534	Asie du Sud
Western Asia	24 220	20 043	18 412	18 645	17 158	Asie occidentale
Syrian Arab Republic[2][3]	**1 363 304**	**1 442 441**	**1 570 161**	**1 739 884**	**1 909 916**	**Rép. arabe syrienne**[2][3]
Africa	32 633	31 835	20 171	32 075	32 974	Afrique
Americas	14 976	14 034	12 185	15 739	15 688	Amériques
Europe	211 352	231 540	282 736	293 780	306 949	Europe
Asia, East and South East/Oceania	9 264	7 315	4 434	6 246	9 942	Asie, Est et Sud-Est et Océanie
Southern Asia	215 880	200 878	123 863	160 249	125 950	Asie du Sud
Western Asia	848 073	922 848	1 095 355	1 184 427	1 357 097	Asie occidentale
Thailand[3]	**4 809 508**	**5 298 860**	**5 086 899**	**5 136 443**	**5 760 533**	**Thaïlande**[3]
Africa	27 484	31 943	35 625	38 509	51 355	Afrique
Americas	366 016	381 894	326 812	354 910	359 726	Amériques
Europe	1 225 648	1 343 618	1 207 670	1 310 148	1 435 993	Europe
Asia, East and South East/Oceania	2 844 662	3 171 445	3 116 983	3 044 763	3 533 134	Asie, Est et Sud-Est et Océanie
Southern Asia	251 149	313 902	344 223	331 648	314 646	Asie du Sud
Western Asia	94 549	56 058	55 586	56 465	65 679	Asie occidentale
TFYR Macedonia	**...**	**...**	**...**	**105 676**	**121 000**	**L'ex-R.y. Macédoine**
Americas	...	...	...	1 717	4 000	Amériques
Europe	...	...	...	101 757	114 000	Europe
Asia, East and South East/Oceania	...	...	...	534	1 000	Asie, Est et Sud-Est et Océanie
Togo[1]	**115 218**	**103 246**	**65 098**	**48 559**	**...**	**Togo**[1]
Africa	51 238	48 399	37 582	28 533	...	Afrique
Americas	8 540	6 953	3 059	2 229	...	Amériques
Europe	51 374	44 660	22 198	16 192	...	Europe
Asia, East and South East/Oceania	3 931	3 173	2 194	1 550	...	Asie, Est et Sud-Est et Océanie
Tonga[4]	**21 029**	**20 917**	**22 007**	**23 020**	**25 513**	**Tonga**[4]
Americas	4 553	4 892	5 339	5 481	5 498	Amériques
Europe	3 708	3 175	3 617	3 725	5 374	Europe
Asia, East and South East/Oceania	12 660	12 793	12 996	13 740	14 569	Asie, Est et Sud-Est et Océanie
Southern Asia	73	39	34	41	64	Asie du Sud

75
Tourist arrivals by region of origin [*cont.*]
Arrivées de touristes par région de provenance [*suite*]

Country or area of destination and region of origin	1989	1990	1991	1992	1993	Pays ou zone de destination et région de provenance
Trinidad and Tobago	**194 228**	**194 521**	**220 206**	**234 759**	**248 815**	**Trinité-et-Tobago**
Americas	151 940	144 901	158 451	176 200	187 831	Amériques
Europe	32 865	38 226	49 342	46 459	46 971	Europe
Tunisia[3]	**3 222 236**	**3 203 787**	**3 224 015**	**3 539 950**	**3 655 698**	**Tunisie**[3]
Africa	497 858	590 922	911 836	973 650	845 529	Afrique
Americas	12 523	12 608	8 060	13 839	20 182	Amériques
Europe	1 670 538	1 705 451	1 086 564	1 770 827	2 037 350	Europe
Western Asia	1 003 546	841 523	1 179 395	674 808	582 512	Asie occidentale
Turkey[2]	**4 459 151**	**5 389 308**	**5 517 897**	**7 076 096**	**6 500 638**	**Turquie**[2]
Africa	74 840	56 265	31 150	46 575	...	Afrique
Americas	266 763	273 436	108 828	236 205	328 493	Amériques
Europe	3 493 009	4 438 965	4 839 064	6 295 747	5 407 530	Europe
Asia, East and South East/Oceania	93 604	106 635	59 980	82 050	85 531	Asie, Est et Sud-Est et Océanie
Southern Asia	272 286	280 425	267 638	182 826	...	Asie du Sud
Western Asia	254 800	229 917	208 448	230 308	...	Asie occidentale
Turks and Caicos Islands	**47 652**	**41 889**	**54 616**	**52 345**	**66 812**	**Iles Turques et Caiques**
Americas	37 393	31 991	43 390	45 140	53 112	Amériques
Europe	3 317	5 017	7 045	6 605	8 110	Europe
Tuvalu	**567**	**671**	**976**	**862**	**929**	**Tuvalu**
Americas	84	76	73	164	98	Amériques
Europe	110	129	97	176	135	Europe
Asia, East and South East/Oceania	371	433	805	472	637	Asie, Est et Sud-Est et Océanie
United Arab Emirates[1 23]	**628 508**	**632 903**	**716 642**	**944 350**	**1 087 733**	**Emirats arabes unis**[1 23]
Africa	25 559	25 617	21 843	34 619	39 280	Afrique
Americas	14 212	19 616	46 145	35 271	41 510	Amériques
Europe	146 132	151 280	147 250	192 426	295 746	Europe
Asia, East and South East/Oceania	94 647	90 780	100 178	121 474	136 581	Asie, Est et Sud-Est et Océanie
Southern Asia	136 438	144 631	139 544	211 020	182 320	Asie du Sud
Western Asia	211 520	200 979	261 682	349 540	392 296	Asie occidentale
former USSR†[2 24]	**7 752 326**	**7 203 635**	**6 894 711**	**3 009 488**	**5 895 917**	**ancienne URSS†**[2 24]
Africa	69 000	37 000	26 883	33 445	28 926	Afrique
Americas	253 658	282 658	233 340	245 542	331 572	Amériques
Europe	6 753 853	6 006 950	5 636 062	1 683 606	4 192 445	Europe
Asia, East and South East/Oceania	335 554	449 650	532 316	888 304	1 206 527	Asie, Est et Sud-Est et Océanie
Southern Asia	80 622	88 504	150 202	90 106	54 506	Asie du Sud
Western Asia	92 000	107 000	139 354	40 711	39 042	Asie occidentale
United Kingdom[2 18]	**17 338 000**	**18 013 000**	**17 125 000**	**18 535 000**	**19 186 000**	**Royaume-Uni**[2 18]
Africa	548 000	553 000	535 000	536 000	550 000	Afrique
Americas	3 730 000	3 936 000	3 125 000	3 664 000	3 561 000	Amériques
Europe	10 854 000	11 060 000	11 392 000	12 034 000	12 689 000	Europe
Asia, East and South East/Oceania	1 163 000	1 310 000	1 035 000	1 174 000	1 080 000	Asie, Est et Sud-Est et Océanie
Western Asia	457 000	466 000	447 000	481 000	539 000	Asie occidentale
United Rep.Tanzania[2]	**137 889**	**153 000**	**186 800**	**201 744**	**230 158**	**Rép. Unie de Tanzanie**[2]
Africa	54 138	59 691	72 878	78 708	89 790	Afrique
Americas	30 841	34 005	41 517	44 838	51 151	Amériques
Europe	49 071	55 072	67 238	72 618	82 834	Europe
Asia, East and South East/Oceania	3 839	4 232	5 167	5 580	6 383	Asie, Est et Sud-Est et Océanie
United States	**36 563 703**	**39 539 010**	**42 985 520**	**47 556 029**	**45 792 700**	**Etats-Unis**
Africa	136 952	137 140	138 601	149 835	168 969	Afrique
Americas	25 191 535	27 356 619	29 910 948	33 020 458	30 799 986	Amériques

75

Tourist arrivals by region of origin [*cont.*]
Arrivées de touristes par région de provenance [*suite*]

Country or area of destination and region of origin	1989	1990	1991	1992	1993	Pays ou zone de destination et région de provenance
Europe	6 436 816	6 857 674	7 568 226	8 463 801	8 864 733	Europe
Asia, East and South East/Oceania	4 489 862	4 872 916	5 076 571	5 614 004	5 642 339	Asie, Est et Sud-Est et Océanie
Southern Asia	153 897	165 808	155 591	149 893	144 344	Asie du Sud
Western Asia	154 641	148 853	135 583	158 038	172 329	Asie occidentale
United States Virgin Is. [1]	**451 764**	**370 008**	**376 371**	**385 575**	**448 950**	**Iles Vierges américaines** [1]
Africa	...	...	...	225	301	Afrique
Americas	439 674	354 218	357 881	366 366	421 273	Amériques
Europe	9 593	8 593	8 916	12 472	14 358	Europe
Asia, East and South East/Oceania	...	...	...	804	854	Asie, Est et Sud-Est et Océanie
Uruguay [2]	**1 240 431**	**1 267 040**	**1 509 962**	**1 801 672**	**2 003 000**	**Uruguay** [2]
Americas	993 643	954 038	1 211 452	1 498 913	1 629 132	Amériques
Europe	42 372	45 283	44 971	47 779	52 542	Europe
Asia, East and South East/Oceania	1 452	1 918	1 667	1 680	2 647	Asie, Est et Sud-Est et Océanie
Western Asia	1 452	1 449	702	943	...	Asie occidentale
Vanuatu	**22 986**	**34 728**	**39 548**	**42 673**	**45 074**	**Vanuatu**
Americas	808	1 082	1 300	1 343	1 475	Amériques
Europe	949	1 301	1 236	2 282	2 422	Europe
Asia, East and South East/Oceania	21 031	30 601	35 246	38 445	40 600	Asie, Est et Sud-Est et Océanie
Venezuela	**411 849**	**524 533**	**598 328**	**433 524**	**396 141**	**Venezuela**
Africa	214	571	446	1 056	610	Afrique
Americas	236 104	296 412	335 016	238 064	208 340	Amériques
Europe	169 983	219 319	254 796	187 096	179 857	Europe
Asia, East and South East/Oceania	3 487	4 411	4 611	4 009	4 154	Asie, Est et Sud-Est et Océanie
Southern Asia	398	610	891	714	568	Asie du Sud
Western Asia	864	1 809	879	1 147	1 064	Asie occidentale
Viet Nam	...	...	...	**440 000**	**601 527**	**Viet Nam**
Americas	...	...	...	14 563	23 361	Amériques
Europe	...	...	...	25 866	64 959	Europe
Asia, East and South East/Oceania	...	...	...	116 542	162 205	Asie, Est et Sud-Est et Océanie
Yemen [1]	**65 188**	**51 849**	**43 656**	**72 164**	...	**Yémen** [1]
Africa	1 991	2 052	2 653	3 346	...	Afrique
Americas	2 826	3 370	3 806	7 117	...	Amériques
Europe	40 494	30 081	21 689	43 777	...	Europe
Asia, East and South East/Oceania	6 617	4 366	5 021	6 922	...	Asie, Est et Sud-Est et Océanie
Western Asia	13 260	11 980	10 487	11 002	...	Asie occidentale
Yugoslavia, SFR† [7]	**8 644 080**	**7 879 529**	**1 459 600**	...	...	**Yougoslavie, Rfst†** [7]
Americas	299 342	279 455	46 800	...	...	Amériques
Europe	8 157 573	7 406 184	1 309 800	...	...	Europe
Asia, East and South East/Oceania	56 639	57 834	15 600	...	...	Asie, Est et Sud-Est et Océanie
Yugoslavia [1]	...	...	...	**139 091**	**64 996**	**Yougoslavie** [1]
Americas	...	...	...	4 640	2 616	Amériques
Europe	...	...	...	124 850	57 687	Europe
Asia, East and South East/Oceania	...	...	...	1 532	776	Asie, Est et Sud-Est et Océanie
Western Asia	...	...	...	...	...	Asie occidentale
Zambia [2]	**113 182**	**141 004**	**171 507**	**158 759**	...	**Zambie** [2]
Africa	79 073	105 343	141 327	98 656	...	Afrique
Americas	5 596	5 441	3 594	1 332	...	Amériques
Europe	22 461	22 588	19 257	24 086	...	Europe
Asia, East and South East/Oceania	3 006	3 875	3 481	28 782	...	Asie, Est et Sud-Est et Océanie
Southern Asia	2 899	2 679	2 883	3 930	...	Asie du Sud
Western Asia	121	62	79	253	...	Asie occidentale

75
Tourist arrivals by region of origin [*cont.*]
Arrivées de touristes par région de provenance [*suite*]

Country or area of destination and region of origin	1989	1990	1991	1992	1993	Pays ou zone de destination et région de provenance
Zimbabwe[25]	474 000	606 000	664 000	737 533	914 824	Zimbabwe[25]
Africa	337 974	436 220	513 580	641 004	759 981	Afrique
Americas	17 089	18 585	14 650	13 658	26 337	Amériques
Europe	67 758	81 827	63 556	67 396	105 017	Europe
Asia, East and South East/Oceania	13 427	16 054	15 775	15 380	23 489	Asie, Est et Sud-Est et Océanie

Source:
World Tourism Organization (Madrid).

† For detailed descriptions of data pertaining to
former Czechoslovakia, Germany, SFR Yugoslavia and former
USSR, see Annex I - Country or area nomenclature, regional
and other groupings.

1 International tourist arrivals in hotels and similar
  establishments.
2 International visitor arrivals at frontiers (including
  tourists and same-day visitors).
3 Excluding nationals of the country residing abroad.
4 Air arrivals.
5 Air and sea arrivals.
6 Excluding crew members.
7 International tourist arrivals at collective tourism
  establishments.
8 International tourist arrivals in hotels of regional
  capitals.
9 Including nationals of the country residing abroad.
10 Excluding ethnic Chinese arriving from Hong Kong, Macau and
   Taiwan.
11 International visitor arrivals at all means of accommodation
   in Brazzaville, Pointe Noire, Loubomo, Owando and Sibiti.

12 Air arrivals at Rarotonga.
13 International tourist arrivals at Addis Ababa, Asmara and
   Assab.
14 Since 1989 "survey at frontiers and car study realized by
   SOFRES".
15 Excluding camping sites.
16 As of 1990, tourists from the former German Democratic
   Republic will be regarded as domestic tourists.

17 Data based on surveys.
18 Departures.
19 Foreign tourist departures; includes Singapore residents
   crossing the frontier by road through Johore Causeway.

20 Arrivals by air and land.
21 Excluding Malaysian citizens arriving by land and cruise
   passengers, but including excursionists (same-day visitors).

22 Beginning Jan. 1992, contract and border traffic concession
   workers are included.
23 Dubai only.
24 1992-1993: Data refer to the Commonwealth of Independent
   States; excluding the Baltic States.
25 Excluding transit passengers.

Source:
Organisation mondiale du tourisme (Madrid).

† Pour les descriptions en détails des données
relatives à l'ancienne Tchécoslovaquie, l'Allemagne, la Rfs
Yougoslavie et l'ancienne URSS, voir l'Annexe I -
Nomenclature des pays ou zones, groupements régionaux et
autres groupements.

1 Arrivées de touristes internationaux dans les hôtels et
  établissements assimilés.
2 Arrivées de visiteurs internationaux aux frontières (y
  compris touristes et visiteurs de la journée).
3 A l'exclusion des nationaux du pays résidant à l'étranger.
4 Arrivées par voie aérienne.
5 Arrivées par voie aérienne et maritime.
6 A l'exclusion des membres des équipages.
7 Arrivées de touristes internationaux dans les établissements
  d'hébergement collectifs.
8 Arrivées de trousistes internationaux dans les hôtel des
  capitales de département.
9 Y compris les nationaux du pays résidant à l'étranger.
10 A l'exclusion des arrivées de personnes d'ethnie chinoise en
   provenance de Hong-kong, Macao et Taïwan.
11 Arrivées des touristes internationaux dans l'ensemble des
   moyens d'hébergement de Brazzaville, Pointe Noire, Loubomo,
   Owando et Sibiti.
12 Arrivées par voie aérienne à Rarotonga.
13 Arrivées de touristes internationaux à Addis Abeba, Asmara
   et Assab.
14 A partir de 1989 "enquête aux frontières et étude autocar
   réalisée par la SPFRES".
15 A l'exclusion des terrains de camping.
16 A partir de 1990, les touristes en provenance de l'ancienne
   République Démocratique Allemande seront considérés comme de
   touristes nationaux.
17 Données obtenues au moyen d'enquêtes.
18 Départs.
19 Départs de touristes étrangers; y compris les résidents de
   Singapour traversant la frontière par voie terrestre a
   travers le Johore Causeway.
20 Arrivées par voies aérienne et terrestre.
21 A l'exclusion des arrivées de malaisiens par voie terrestre
   et de passagers en croisière, mais y compris les
   excursionnistes (visiteurs de la journée).
22 A partir de jan. 1992 les données incluent les travailleurs
   contractuels et ceux de la zone frontière.
23 Dubai seulement.
24 1992-1993: Données se rapportent à la Communauté des Etats
   Indépendants à l'exlusion des Etats Baltes.
25 A l'exclusion des passagers en transit.

76
Tourist arrivals and international tourism receipts
Arrivées de touristes et recettes touristiques internationales

Country or area Pays ou zone	Number of tourist arrivals (000) Nombre d'arrivées de touristes (000)					Tourist receipts (million US dollars) Recettes touristiques (millions de dollars E-U)				
	1989	1990	1991	1992	1993	1989	1990	1991	1992	1993
World *Monde*	**429 025**	**455 664**	**463 209**	**502 899**	**512 939**	**215 569**	**261 001**	**267 491**	**303 989**	**305 816**
Africa **Afrique**	**16 248**	**17 573**	**18 181**	**20 768**	**20 478**	**6 597**	**7 237**	**6 991**	**8 648**	**7 248**
Algeria Algérie	1 207	1 137	1 193	1 120	1 128	64	64	84	75	55
Angola Angola	40	46	46	40	21	...	13	...	...	20
Benin Bénin	75	110	117	130	140	20	28	29	32	38
Botswana Botswana	448	543	592	590	607	54	65	79	65	74
Burkina Faso Burkina Faso	80	74	80	92	111	6	8	8	9	8
Burundi Burundi	82	109	125	86	75	3	4	4	3	3
Cameroon Cameroun	87	89	84	62	81	94	53	73	59	47
Cape Verde Cap-Vert	...	...	...	...	...	3	6	8	7	7
Central African Rep. Rép. centrafricaine	5	6	6	6	6	3	3	3	3	3
Chad Tchad	12	9	21	17	21	9	12	10	21	23
Comoros Comores	13	8	17	19	24	3	2	9	8	8
Congo Congo	33	33	33	37	35	7	8	8	8	2
Côte d'Ivoire Côte d'Ivoire	192	196	200	217	200	65	48	46	60	64
Djibouti Djibouti	41	33	33	28	25	5	6	6	10	13
Egypt Egypte	2 351	2 411	2 112	2 944	2 112	2 058	1 994	2 029	2 730	1 332
Equatorial Guinea Guinée équatoriale	...	...	...	...	...	1	1	2	2	2
Ethiopia Ethiopie	77	79	82	83	93	21	26	20	23	20
Gabon Gabon	113	108	128	133	115	4	3	4	5	5
Gambia Gambie	86	100	66	64	90	22	26	30	27	26
Ghana Ghana	125	146	172	213	233	72	81	118	167	288
Guinea Guinée	49	100	28	33	93	24	30	13	11	6
Kenya Kenya	735	814	805	782	826	420	466	432	442	413
Lesotho Lesotho	169	171	182	155	130	13	17	18	19	17

76
Tourist arrivals and international tourism receipts
[*cont.*]

Arrivées de touristes et recettes touristiques internationales
[*suite*]

Country or area Pays ou zone	Number of tourist arrivals (000) Nombre d'arrivées de touristes (000)					Tourist receipts (million US dollars) Recettes touristiques (millions de dollars E-U)				
	1989	1990	1991	1992	1993	1989	1990	1991	1992	1993
Libyan Arab Jamah. Jamah. arabe libyenne	95	96	90	89	63	5	6	5	6	5
Madagascar Madagascar	39	53	35	54	55	28	40	27	39	41
Malawi Malawi	117	130	127	135	137	13	11	13	11	12
Mali Mali	32	44	38	38	24	39	47	46	11	11
Mauritania Mauritanie	...	...	...	...	...	13	9	12	8	15
Mauritius Maurice	263	292	301	335	375	182	244	252	299	301
Morocco Maroc	3 468	4 024	4 162	4 390	4 027	1 146	1 259	1 052	1 360	1 243
Namibia Namibie	...	...	...	...	...	...	61	76	91	95
Niger Niger	24	21	16	13	11	14	17	16	17	16
Nigeria Nigéria	161	190	214	237	192	21	25	39	29	31
Réunion Réunion	182	200	186	217	242	...	...	...	...	...
Rwanda Rwanda	19	16	3	5	2	10	10	4	4	2
Sao Tome and Principe Sao Tomé-et-Principe	4	4	3	3	3	1	2	2	2	2
Senegal Sénégal	259	246	234	246	168	142	167	171	182	173
Seychelles Seychelles	86	104	90	99	116	91	120	99	117	116
Sierra Leone Sierra Leone	86	98	96	89	91	17	19	18	17	18
Somalia Somalie	40	46	46	20	20	...	...	...	...	...
South Africa Afrique du Sud	930	1 029	1 710	2 892	3 358	709	992	1 131	1 226	1 190
Sudan Soudan	23	33	16	17	15	45	21	8	5	3
Swaziland Swaziland	248	294	279	258	285	25	25	26	32	30
Togo Togo	115	103	65	49	24	50	58	49	39	18
Tunisia Tunisie	3 222	3 204	3 224	3 540	3 656	933	953	685	1 074	1 114
Uganda Ouganda	41	69	69	70	74	9	10	15	10	30
United Rep.Tanzania Rép. Unie de Tanzanie	138	153	187	202	230	60	65	95	120	147
Zaire Zaïre	51	55	33	22	22	6	7	7	7	6

76
Tourist arrivals and international tourism receipts
[*cont.*]

Arrivées de touristes et recettes touristiques internationales
[*suite*]

Country or area Pays ou zone	Number of tourist arrivals (000) Nombre d'arrivées de touristes (000)					Tourist receipts (million US dollars) Recettes touristiques (millions de dollars E-U)				
	1989	1990	1991	1992	1993	1989	1990	1991	1992	1993
Zambia Zambie	113	141	171	159	167	12	41	35	51	52
Zimbabwe Zimbabwe	474	606	664	738	955	55	64	75	105	103
America, North Amérique du Nord	**78 926**	**85 078**	**87 462**	**93 587**	**92 665**	**54 743**	**63 680**	**70 082**	**76 607**	**80 937**
Anguilla Anguilla	29	31	31	32	38	28	35	35	35	43
Antigua and Barbuda Antigua-et-Barbuda	189	197	197	210	240	271	298	314	329	372
Aruba Aruba	344	433	501	542	562	310	353	401	443	464
Bahamas Bahamas	1 575	1 562	1 427	1 399	1 489	1 310	1 333	1 193	1 244	1 304
Barbados Barbade	461	432	394	385	396	528	494	453	463	502
Belize Belize	70	88	94	105	102	44	51	53	65	73
Bermuda Bermudes	418	435	386	375	413	451	490	454	444	505
British Virgin Islands Iles Vierges britanniques	176	160	147	117	237	125	132	109	109	122
Canada Canada	15 111	15 209	14 912	14 741	15 105	5 047	5 612	5 886	5 712	5 897
Cayman Islands Iles Caïmanes	210	253	237	242	287	175	234	220	228	252
Costa Rica Costa Rica	376	435	505	610	684	207	275	331	431	577
Cuba Cuba	315	327	418	455	544	204	243	387	567	720
Dominica Dominique	37	45	46	47	52	16	20	24	25	29
Dominican Republic Rép. dominicaine	1 400	1 533	1 321	1 524	1 691	818	890	877	1 054	1 234
El Salvador El Salvador	131	194	199	314	267	63	145	157	128	121
Grenada Grenade	66	76	85	88	94	31	38	42	38	45
Guadeloupe Guadeloupe	284	288	370	341	453	203	197	234	269	370
Guatemala Guatemala	437	509	513	541	562	152	185	211	243	265
Haiti Haïti	122	120	119	120	120	50	46	46	46	46
Honduras Honduras	176	202	198	230	225	28	29	31	32	32
Jamaica Jamaïque	715	841	845	909	979	593	740	764	858	942
Martinique Martinique	312	282	315	321	366	272	240	255	282	332

76
Tourist arrivals and international tourism receipts
[*cont.*]

Arrivées de touristes et recettes touristiques internationales
[*suite*]

Country or area Pays ou zone	Number of tourist arrivals (000) Nombre d'arrivées de touristes (000)					Tourist receipts (million US dollars) Recettes touristiques (millions de dollars E-U)				
	1989	1990	1991	1992	1993	1989	1990	1991	1992	1993
Mexico Mexique	14 964	17 176	16 281	17 273	16 534	4 766	5 467	5 881	6 085	6 167
Montserrat Montserrat	17	13	17	17	21	11	14	11	14	15
Netherlands Antilles Antilles néerlandaises	771	850	842	882	840	371	454	474	524	721
Nicaragua Nicaragua	77	106	146	167	198	7	12	16	21	30
Panama Panama	192	214	277	291	298	161	172	202	222	228
Puerto Rico Porto Rico	2 444	2 560	2 613	2 659	2 857	1 254	1 366	1 436	1 520	1 629
Saint Kitts and Nevis Saint-Kitts-et-Nevis	72	76	84	90	89	60	63	74	67	69
Saint Lucia Sainte-Lucie	133	138	159	177	194	145	154	173	208	221
St. Vincent-Grenadines St. Vincent-Grenadines	50	54	52	53	57	43	54	53	54	55
Trinidad and Tobago Trinité-et-Tobago	190	195	220	235	250	85	95	101	111	80
Turks and Caicos Islands Iles Turques et Caiques	48	42	55	52	67	39	37	50	48	53
United States Etats-Unis	36 564	39 539	42 986	47 556	45 793	36 250	43 007	48 384	53 861	56 501
United States Virgin Is. Iles Vierges américaines	450	463	470	487	561	625	705	750	827	921
America, South Amérique du Sud	**8 307**	**8 635**	**9 658**	**10 423**	**11 594**	**4 936**	**5 781**	**6 644**	**7 358**	**8 532**
Argentina Argentine	2 492	2 728	2 870	3 031	3 532	1 671	1 976	2 336	3 090	3 614
Bolivia Bolivie	194	217	221	245	269	75	84	91	107	115
Brazil Brésil	1 403	1 091	1 192	1 475	1 650	1 225	1 444	1 559	1 307	1 449
Chile Chili	797	943	1 349	1 283	1 412	407	540	700	706	824
Colombia Colombie	733	813	857	1 076	1 047	335	406	468	705	780
Ecuador Equateur	335	362	365	403	471	187	188	189	192	230
Guyana Guyana	67	64	73	93	107	28	27	30	31	36
Paraguay Paraguay	279	280	361	334	404	112	112	165	154	204
Peru Pérou	334	317	232	217	273	271	259	252	237	268
Suriname Suriname	21	28	30	30	30	8	11	11	11	11

76

Tourist arrivals and international tourism receipts
[*cont.*]

Arrivées de touristes et recettes touristiques internationales
[*suite*]

Country or area Pays ou zone	Number of tourist arrivals (000) Nombre d'arrivées de touristes (000)					Tourist receipts (million US dollars) Recettes touristiques (millions de dollars E-U)				
	1989	1990	1991	1992	1993	1989	1990	1991	1992	1993
Uruguay Uruguay	1 240	1 267	1 510	1 802	2 003	228	238	333	381	447
Venezuela Venezuela	412	525	598	434	396	389	496	510	437	554
Asia Asie	56 277	63 496	65 123	75 804	82 058	38 393	42 666	41 710	50 270	56 237
Afghanistan Afghanistan	8	8	8	6	6	1	1	1	1	1
Bahrain Bahreïn	1 342	1 376	1 674	1 419	1 450	116	135	162	177	213
Bangladesh Bangladesh	128	115	113	110	127	18	11	9	8	15
Bhutan Bhoutan	1	2	2	3	3	2	2	2	3	3
Brunei Darussalam Brunéi Darussalam	393	377	344	500	590	32	35	35	35	36
Cambodia Cambodge	20	17	25	88	118	...	...	...	50	48
China Chine	9 361	10 484	12 464	16 512	18 982	1 861	2 218	2 845	3 947	4 683
Cyprus Chypre	1 378	1 561	1 385	1 991	1 841	990	1 258	1 026	1 539	1 396
Hong Kong Hong-kong	5 985	6 581	6 795	8 011	8 938	4 731	5 032	5 078	6 037	7 562
India Inde	1 736	1 707	1 678	1 868	1 765	1 535	1 437	1 351	1 415	1 487
Indonesia Indonésie	1 626	2 178	2 570	3 064	3 403	1 285	2 105	2 522	3 278	3 988
Iran, Islamic Rep. of Iran, Rép. islamique d'	89	154	212	185	205	15	28	57	31	39
Iraq Iraq	1 025	748	268	504	400	59	55	20	20	15
Israel Israël	1 177	1 063	943	1 509	1 654	1 468	1 382	1 306	1 876	2 110
Japan Japon	1 499	1 879	2 104	2 103	1 925	3 143	3 578	3 435	3 588	3 557
Jordan Jordanie	639	572	436	661	765	547	512	317	462	563
Korea, Republic of Corée, République de	2 728	2 959	3 196	3 231	3 331	3 556	3 559	3 426	3 272	3 510
Kuwait Koweït	89	15	4	65	73	143	132	253	273	83
Lao People's Dem. Rep. Rép. dém. pop. lao	25	25	25	25	25	2	3	8	18	34
Macau Macao	2 304	2 513	3 047	3 180	3 850	1 164	1 473	1 761	2 234	2 500
Malaysia Malaisie	4 846	7 446	5 847	6 016	6 504	1 038	1 667	1 530	1 768	1 876

76
Tourist arrivals and international tourism receipts
[*cont.*]

Arrivées de touristes et recettes touristiques internationales
[*suite*]

Country or area Pays ou zone	Number of tourist arrivals (000) Nombre d'arrivées de touristes (000)					Tourist receipts (million US dollars) Recettes touristiques (millions de dollars E-U)				
	1989	1990	1991	1992	1993	1989	1990	1991	1992	1993
Maldives Maldives	158	195	196	236	241	75	89	95	138	146
Mongolia Mongolie	237	147	147	140	150	...	...	...	...	...
Myanmar Myanmar	14	21	21	27	55	4	9	13	16	19
Nepal Népal	240	255	293	334	293	107	109	126	110	157
Oman Oman	136	149	161	192	344	56	69	63	85	95
Pakistan Pakistan	495	424	438	352	379	159	156	163	120	111
Philippines Philippines	1 076	893	849	1 043	1 246	1 465	1 306	1 281	1 674	2 122
Qatar Qatar	110	136	143	141	160	...	...	...	...	...
Saudi Arabia Arabie saoudite	775	827	720	750	993	2 050	1 884	1 000	1 000	1 121
Singapore Singapour	4 397	4 842	4 913	5 446	5 804	3 307	4 593	4 557	5 250	5 793
Sri Lanka Sri Lanka	185	298	318	394	392	76	132	157	201	208
Syrian Arab Republic Rép. arabe syrienne	411	562	622	684	703	297	300	410	600	700
Thailand Thaïlande	4 810	5 299	5 087	5 136	5 761	3 753	4 326	3 923	4 829	5 014
Turkey Turquie	3 921	4 799	5 158	6 549	5 904	2 557	3 225	2 654	3 639	3 959
United Arab Emirates Emirats arabes unis	629	633	717	944	1 088	...	...	...	...	...
Viet Nam Viet Nam	215	250	300	440	670	59	85	85	80	85
Yemen Yémen	65	52	44	72	70	26	20	21	47	45
Europe **Europe**	**256 812**	**268 530**	**270 659**	**289 605**	**291 960**	**104 596**	**134 278**	**134 212**	**152 716**	**144 650**
Albania Albanie	14	30	13	23	45	3	4	3	2	8
Austria Autriche	18 202	19 011	19 092	19 098	18 257	10 717	13 410	13 800	14 526	13 566
Belgium Belgique	3 007	3 163	2 944	3 220	3 285	3 057	3 718	3 606	4 053	4 071
Bulgaria Bulgarie	4 316	4 500	4 000	3 750	3 827	495	320	44	215	307
Croatia Croatie	..	..	..	2 010	2 363	..	..	..	...	...
Czech Republic République tchèque	7 079	7 278	7 565	10 900	11 500	..	400	690	1 126	1 558

76

Tourist arrivals and international tourism receipts
[*cont.*]

Arrivées de touristes et recettes touristiques internationales
[*suite*]

Country or area Pays ou zone	Number of tourist arrivals (000) Nombre d'arrivées de touristes (000)					Tourist receipts (million US dollars) Recettes touristiques (millions de dollars E-U)				
	1989	1990	1991	1992	1993	1989	1990	1991	1992	1993
Denmark Danemark	1 218	1 275	1 429	1 543	1 569	2 313	3 322	3 475	3 784	3 052
Finland Finlande	882	866	786	790	798	1 017	1 170	1 247	1 360	1 239
France France	49 549	52 497	55 041	59 710	60 100	16 245	20 185	21 375	25 051	23 410
Germany † · Allemagne† F. R. Germany R. f. Allemagne	16 115	17 045	15 648	15 913	14 348	8 463	10 531	10 424	11 055	10 509
former German D. R. anc. R. d. allemande	3 102	..	..	..	..	...	...	...	...	..
Gibraltar Gibraltar	162	132	97	88	89	91	112	98	82	83
Greece Grèce	8 082	8 873	8 036	9 331	9 413	1 976	2 587	2 567	3 268	3 335
Hungary Hongrie	14 490	20 510	21 860	20 188	22 804	542	824	1 002	1 231	1 181
Iceland Islande	131	142	143	143	157	108	139	136	129	132
Ireland Irlande	3 484	3 666	3 535	3 666	3 814	1 070	1 447	1 511	1 620	1 639
Italy Italie	25 935	26 679	25 878	26 113	26 379	11 938	20 016	18 421	21 450	20 521
Liechtenstein Liechtenstein	77	78	71	72	65	...	...	...	...	...
Luxembourg Luxembourg	875	820	861	796	831	286	290	285	287	290
Malta Malte	828	872	895	1 002	1 063	372	496	574	568	653
Monaco Monaco	245	245	239	246	208	...	...	...	...	...
Netherlands Pays-Bas	5 206	5 795	5 842	6 083	5 757	3 049	3 636	4 246	5 237	4 690
Norway Norvège	1 867	1 955	2 114	2 375	2 556	1 347	1 570	1 646	1 975	1 849
Poland Pologne	3 293	3 400	11 350	16 200	17 000	202	358	2 800	4 100	4 500
Portugal Portugal	7 116	8 020	8 657	8 884	8 434	2 685	3 555	3 710	3 721	4 176
Romania Roumanie	1 857	3 009	3 000	3 798	2 911	167	106	145	262	197
San Marino Saint-Marin	437	582	582	583	585	...	...	...	...	...
Slovakia Slovaquie	957	822	635	579	650	..	70	135	213	390
Slovenia Slovénie	..	..	..	616	624	..	..	..	671	734
Spain Espagne	38 867	37 441	38 539	39 638	40 085	16 174	18 593	19 004	22 181	19 425

76
Tourist arrivals and international tourism receipts
[cont.]

Arrivées de touristes et recettes touristiques internationales
[suite]

Country or area Pays ou zone	Number of tourist arrivals (000) Nombre d'arrivées de touristes (000)					Tourist receipts (million US dollars) Recettes touristiques (millions de dollars E-U)				
	1989	1990	1991	1992	1993	1989	1990	1991	1992	1993
Sweden Suède	837	731	623	650	659	2 536	2 916	2 704	3 055	2 650
Switzerland Suisse	12 600	13 200	12 600	12 800	12 400	5 543	6 789	7 026	7 463	7 001
TFYR Macedonia L'ex-R.y. Macédoine	..	..	..	106	121	..	..	..	11	12
United Kingdom Royaume-Uni	17 338	18 013	17 125	18 535	19 186	11 389	14 940	13 070	13 932	34 449
Yugoslavia Yougoslavie	..	..	..	156	77	..	..	..	88	23
Yugoslavia, SFR† Yougoslavie, Rfs†	8 644	7 880	1 459	..	..	2 230	2 774	468		..
Oceania Océanie	**4 703**	**5 148**	**5 231**	**5 812**	**6 315**	**6 054**	**7 089**	**7 602**	**8 145**	**7 973**
American Samoa Samoa américaines	47	47	40	31	40	9	10	10	10	10
Australia Australie	2 080	2 215	2 370	2 603	2 996	3 448	4 088	4 484	4 405	4 655
Cook Islands Iles Cook	33	34	40	50	53	15	16	21	27	33
Fiji Fidji	251	279	259	279	287	182	199	194	218	236
French Polynesia Polynésie française	140	132	121	124	148	138	171	150	170	182
Guam Guam	669	780	737	877	784	804	936	1 093	1 579	950
Kiribati Kiribati	3	3	3	4	5	1	2	2	2	2
Marshall Islands Iles Marshall	4	5	7	8	5	...	...	3	4	4
New Caledonia Nouvelle-Calédonie	82	87	81	78	81	95	80	65	94	110
New Zealand Nouvelle-Zélande	901	976	963	1 056	1 157	1 005	1 019	1 021	1 032	1 165
Niue Nioué	1	1	1	2	3	...	...	...	...	...
Palau Palaos	...	...	33	36	40	...	...	...	...	...
Northern Mariana Islands Iles Marianas du Nord	326	426	422	496	534	261	455	450	483	515
Papua New Guinea Papouasie-Nvl-Guinée	49	41	37	43	40	40	41	41	49	45
Samoa Samoa	54	48	35	38	47	19	20	18	19	20
Solomon Islands Iles Salomon	10	9	11	12	12	5	4	5	6	6
Tonga Tonga	21	21	22	23	26	9	9	10	9	10

76

Tourist arrivals and international tourism receipts
[*cont.*]

Arrivées de touristes et recettes touristiques internationales
[*suite*]

Country or area Pays ou zone	Number of tourist arrivals (000) Nombre d'arrivées de touristes (000)					Tourist receipts (million US dollars) Recettes touristiques (millions de dollars E-U)				
	1989	1990	1991	1992	1993	1989	1990	1991	1992	1993
Tuvalu Tuvalu	1	1	1	1	1	...	...	...	...	...
Vanuatu Vanuatu	23	35	40	43	45	23	39	35	38	30
former USSR† ancienne URSS†	**7 752**	**7 204**	**6 895**	**6 900**	**7 869**	**250**	**270**	**250**	**245**	**239**

Source:
World Tourism Organization (Madrid).

† For detailed descriptions of data pertaining to
former Czechoslovakia, Germany, SFR Yugoslavia and former
USSR, see Annex I - Country or area nomenclature, regional
and other groupings.

Source:
Organisation mondiale du tourisme (Madrid).

† Pour les descriptions en détails des données
relatives à l'ancienne Tchécoslovaquie, l'Allemagne, la Rfs
Yougoslavie et l'ancienne URSS, voir l'Annexe I -
Nomenclature des pays ou zones, groupements régionaux et
autres groupements.

77
International tourism expenditures
Dépenses provenant du tourisme international
Million US dollars
Millions de dollars E-U

Country or area Pays ou zone	1984	1985	1986	1987	1988	1989	1990	1991	1992	1993
World *Monde*	95 678	100 762	125 036	155 448	185 686	198 576	243 119	242 985	277 590	269 873
Africa **Afrique**	3 156	2 852	2 752	3 282	3 896	4 206	4 643	5 567	5 531	5 652
Algeria Algérie	503	606	440	239	294	212	149	140	163	163
Angola Angola	...	...	...	15	15	37	38	65	75	75
Benin Bénin	3	8	10	12	13	10	12	10	12	13
Botswana Botswana	20	17	19	26	32	34	39	40	51	51
Burkina Faso Burkina Faso	20	20	19	28	30	30	35	34	37	35
Burundi Burundi	9	9	14	17	15	14	17	18	21	20
Cameroon Cameroun	111	130	205	244	283	321	279	414	228	225
Cape Verde Cap-Vert	...	...	...	4	3	3	5	3	6	6
Central African Rep. Rép. centrafricaine	23	24	31	41	45	45	51	43	55	55
Chad Tchad	18	20	30	47	32	33	36	32	30	12
Comoros Comores	7	7	13	5	5	5	6	7	6	6
Congo Congo	55	63	67	79	126	86	113	106	95	81
Côte d'Ivoire Côte d'Ivoire	112	106	195	233	235	168	240	207	220	199
Djibouti Djibouti	...	...	...	...	...	...	...	...	7	7
Egypt Egypte	146	106	52	78	43	87	129	225	918	1 048
Equatorial Guinea Guinée équatoriale	...	...	...	...	...	8	8	9	9	9
Ethiopia Ethiopie	4	4	5	6	6	10	11	7	10	10
Gabon Gabon	83	83	131	132	134	124	137	112	143	132
Gambia Gambie	3	2	2	3	5	5	8	15	13	13
Ghana Ghana	10	10	11	12	12	13	13	14	17	17
Guinea Guinée	...	...	...	...	...	23	30	27	17	28
Kenya Kenya	14	15	22	24	23	27	38	24	29	48

77
International tourism expenditures
Million US dollars [cont.]
Dépenses provenant du tourisme international
Millions de dollars E-U [suite]

Country or area Pays ou zone	1984	1985	1986	1987	1988	1989	1990	1991	1992	1993
Lesotho Lesotho	5	5	8	9	11	10	12	11	11	7
Libyan Arab Jamah. Jamah. arabe libyenne	503	409	149	322	428	512	424	877	154	206
Madagascar Madagascar	21	24	31	26	30	38	40	32	37	37
Malawi Malawi	7	8	7	6	3	3	3	3	3	3
Mali Mali	17	37	49	57	58	56	62	62	65	61
Mauritania Mauritanie	18	17	18	25	23	31	23	26	31	31
Mauritius Maurice	18	19	26	51	64	79	94	110	142	128
Morocco Maroc	70	88	100	132	164	153	184	190	242	245
Namibia Namibie	...	...	...	...	...	...	63	72	81	81
Niger Niger	10	9	9	34	35	32	44	40	30	29
Nigeria Nigéria	229	200	90	55	41	416	576	839	348	234
Rwanda Rwanda	11	11	14	15	16	17	23	17	17	17
Sao Tome and Principe Sao Tomé-et-Principe	1	1	1	1	1	2	2	2	2	2
Senegal Sénégal	42	38	49	55	76	72	105	105	112	106
Seychelles Seychelles	7	9	10	69	13	17	20	12	16	16
Sierra Leone Sierra Leone	5	1	6	17	7	4	4	4	3	4
South Africa Afrique du Sud	651	413	603	838	958	936	1 117	1 148	1 544	1 598
Sudan Soudan	48	43	32	35	99	144	51	12	33	33
Swaziland Swaziland	19	11	12	14	18	10	14	20	17	17
Togo Togo	21	22	27	29	32	34	43	45	48	30
Tunisia Tunisie	142	126	107	95	119	134	179	129	166	203
Uganda Ouganda	...	...	9	13	11	10	8	18	18	40
United Rep.Tanzania Rép. Unie de Tanzanie	15	16	15	23	23	23	22	68	82	102
Zaire Zaïre	64	35	45	22	16	17	16	16	16	16

77
International tourism expenditures
Million US dollars [cont.]
Dépenses provenant du tourisme international
Millions de dollars E-U [suite]

Country or area Pays ou zone	1984	1985	1986	1987	1988	1989	1990	1991	1992	1993
Zambia Zambie	24	22	31	46	49	98	54	87	56	56
Zimbabwe Zimbabwe	66	58	38	48	58	63	66	70	95	97
America, North Amérique du Nord	29 000	30 956	32 540	36 957	42 426	47 542	55 196	54 524	59 488	59 926
Antigua and Barbuda Antigua-et-Barbuda	13	13	14	15	16	16	18	20	23	23
Aruba Aruba	23	17	12	22	23	28	40	47	50	57
Bahamas Bahamas	106	123	132	153	172	184	196	200	187	195
Barbados Barbade	23	23	29	36	37	45	47	44	41	41
Belize Belize	6	5	5	5	7	8	7	8	14	21
Bermuda Bermudes	64	76	81	77	109	121	119	126	134	139
Canada Canada	3 985	4 130	4 294	5 304	6 460	8 044	10 401	11 367	11 289	10 629
Costa Rica Costa Rica	50	53	60	77	72	114	148	149	223	267
Dominica Dominique	2	3	2	2	2	4	4	5	6	5
Dominican Republic Rép. dominicaine	89	84	89	95	75	85	101	109	115	118
El Salvador El Salvador	74	89	74	76	75	93	55	57	58	61
Grenada Grenade	3	3	4	4	5	4	5	5	4	4
Guatemala Guatemala	61	61	82	95	109	126	100	67	103	116
Haiti Haïti	39	43	37	42	34	33	32	33	25	25
Honduras Honduras	25	27	30	35	37	38	38	37	38	39
Jamaica Jamaïque	21	32	35	44	57	54	54	54	64	64
Mexico Mexique	868	903	871	896	3 202	4 248	5 519	5 814	6 107	5 562
Nicaragua Nicaragua	...	6	4	6	2	1	15	28	30	34
Panama Panama	67	65	83	90	88	86	99	109	120	129
Puerto Rico Porto Rico	446	411	416	488	533	589	630	689	735	774
Saint Kitts and Nevis Saint-Kitts-et-Nevis	5	2	2	3	3	3	4	4	5	5

77

International tourism expenditures
Million US dollars [cont.]

Dépenses provenant du tourisme international
Millions de dollars E-U [suite]

Country or area Pays ou zone	1984	1985	1986	1987	1988	1989	1990	1991	1992	1993
Saint Lucia Sainte-Lucie	39	44	15	15	21	13	17	18	21	21
St. Vincent-Grenadines St. Vincent-Grenadines	6	7	4	4	4	5	4	4	4	4
Trinidad and Tobago Trinité-et-Tobago	276	219	165	158	168	119	112	113	115	115
United States Etats-Unis	22 709	24 517	26 000	29 215	32 114	33 418	37 349	35 322	39 872	41 260
America, South **Amérique du Sud**	**3 879**	**3 551**	**4 562**	**4 204**	**4 231**	**4 089**	**5 560**	**6 115**	**7 009**	**8 526**
Argentina Argentine	682	671	888	890	975	1 014	1 171	1 739	2 211	2 445
Bolivia Bolivie	30	38	27	56	55	106	130	129	141	151
Brazil Brésil	939	1 145	1 464	1 249	1 084	750	1 505	1 214	1 221	1 842
Chile Chili	327	269	319	353	423	397	426	409	459	568
Colombia Colombie	308	169	611	454	538	494	454	509	641	665
Ecuador Equateur	155	196	156	170	167	169	175	177	178	190
Paraguay Paraguay	44	47	48	51	54	50	58	118	135	138
Peru Pérou	160	244	320	335	344	292	495	477	480	304
Suriname Suriname	17	13	12	8	10	10	12	16	11	11
Uruguay Uruguay	154	162	174	129	139	167	111	100	104	129
Venezuela Venezuela	1 063	597	543	509	509	640	1 023	1 227	1 428	2 083
Asia **Asie**	**13 778**	**15 254**	**17 931**	**22 507**	**32 947**	**39 469**	**44 721**	**46 787**	**53 558**	**56 014**
Afghanistan Afghanistan	3	2	1	1	1	1	1	1	1	1
Bahrain Bahreïn	121	125	59	80	77	77	94	98	141	103
Bangladesh Bangladesh	33	45	54	53	99	123	78	83	111	153
Cambodia Cambodge	...	...	...	...	...	...	...	...	...	4
China Chine	150	314	308	387	633	429	470	511	812	555
Cyprus Chypre	71	39	54	67	81	78	111	113	132	133
India Inde	306	354	302	352	397	416	393	394	394	400

77
International tourism expenditures
Million US dollars [cont.]
Dépenses provenant du tourisme international
Millions de dollars E-U [suite]

Country or area Pays ou zone	1984	1985	1986	1987	1988	1989	1990	1991	1992	1993
Indonesia Indonésie	516	591	570	790	592	722	836	969	1 166	1 539
Iran, Islamic Rep. of Iran, Rép. islamique d'	488	508	247	246	69	129	340	734	1 109	578
Israel Israël	726	549	801	1 043	1 161	1 261	1 442	1 551	1 674	2 313
Japan Japon	4 607	4 814	7 229	10 760	18 682	22 490	24 928	23 983	26 837	26 860
Jordan Jordanie	381	424	445	444	475	422	336	282	350	345
Korea, Republic of Corée, République de	576	606	613	704	1 354	2 602	3 166	3 784	3 794	4 105
Kuwait Koweït	1 540	1 988	1 941	2 257	2 358	2 250	1 837	2 012	1 797	1 888
Lao People's Dem. Rep. Rép. dém. pop. lao	...	...	...	...	1	2	1	6	10	9
Macau Macao	...	399	419	24	26	32	39	49	57	71
Malaysia Malaisie	1 141	1 158	1 669	1 234	1 306	1 365	1 450	1 584	1 770	1 960
Maldives Maldives	4	5	6	6	8	10	15	19	22	29
Myanmar Myanmar	1	19	1	1	1	2	2	1	1	1
Nepal Népal	28	29	29	35	44	48	45	38	52	93
Oman Oman	...	...	47	47	47	47	47	47	47	47
Pakistan Pakistan	198	202	221	248	338	337	440	555	679	633
Philippines Philippines	19	37	56	88	76	77	111	61	102	130
Singapore Singapour	601	613	645	795	930	1 334	1 803	1 942	2 403	3 022
Sri Lanka Sri Lanka	47	46	55	63	68	69	74	97	111	121
Syrian Arab Republic Rép. arabe syrienne	311	302	160	131	170	194	223	256	260	300
Thailand Thaïlande	305	280	296	381	602	750	854	1 266	1 590	2 092
Turkey Turquie	277	324	314	448	358	565	520	592	776	863
Yemen Yémen	63	52	18	37	47	81	81	81	81	81
Europe Europe	43 008	45 566	64 193	84 818	97 313	98 132	127 416	124 661	146 630	134 552
Albania Albanie	...	...	...	...	...	...	...	...	1	5

77
International tourism expenditures
Million US dollars [cont.]

Dépenses provenant du tourisme international
Millions de dollars E-U [suite]

Country or area Pays ou zone	1984	1985	1986	1987	1988	1989	1990	1991	1992	1993
Austria Autriche	2 624	2 723	4 026	5 592	6 307	6 266	7 723	7 392	8 393	8 180
Belgium Belgique	1 953	2 050	2 889	3 881	4 577	4 254	5 471	5 528	6 603	6 363
Bulgaria Bulgarie	...	...	...	69	74	113	189	128	313	257
former Czechoslovakia† anc. Tchécoslovaquie†	280	300	349	409	399	431	636	393	..	
Czech Republic République tchèque	..	..	..	..	..	..	..	..	467	525
Denmark Danemark	1 227	1 410	2 119	2 860	3 087	2 932	3 676	3 377	3 779	3 214
Finland Finlande	681	777	1 060	1 506	1 842	2 040	2 740	2 742	2 449	1 617
France France	4 271	4 557	6 513	8 493	9 715	10 031	12 424	12 321	13 914	12 805
Germany † · Allemagne† F. R. Germany R. f. Allemagne	12 423	12 809	18 000	23 341	24 564	23 553	29 509	31 027	36 626	37 514
Greece Grèce	339	368	494	508	735	816	1 090	1 015	1 186	1 003
Hungary Hongrie	157	208	225	250	647	947	477	443	640	741
Iceland Islande	85	94	129	213	200	176	278	295	288	264
Ireland Irlande	411	429	685	839	961	989	1 159	1 125	1 361	1 256
Italy Italie	2 098	2 283	2 910	4 536	5 929	6 774	14 045	11 648	16 530	13 053
Malta Malte	50	50	69	102	120	107	134	140	138	211
Netherlands Pays-Bas	3 277	3 448	4 901	6 408	6 701	6 461	7 376	8 149	9 649	8 974
Norway Norvège	1 488	1 722	2 511	3 067	3 532	2 986	3 679	3 413	3 870	3 565
Poland Pologne	225	184	186	203	251	215	423	143	132	181
Portugal Portugal	222	235	329	421	533	583	867	1 024	1 165	1 846
Romania Roumanie	85	64	22	30	33	35	103	143	260	195
Slovakia Slovaquie	..	..	..	..	..	..	..	..	155	262
Slovenia Slovénie	..	..	..	..	..	..	..	..	282	304
Spain Espagne	835	1 010	1 513	1 938	2 440	3 080	4 254	4 530	5 542	4 706

77
International tourism expenditures
Million US dollars [cont.]
Dépenses provenant du tourisme international
Millions de dollars E-U [suite]

Country or area Pays ou zone	1984	1985	1986	1987	1988	1989	1990	1991	1992	1993
Sweden Suède	1 713	1 967	2 821	3 784	4 565	4 961	6 134	6 291	6 969	4 464
Switzerland Suisse	2 282	2 399	3 368	4 339	5 019	4 907	5 817	5 682	6 068	5 803
United Kingdom Royaume-Uni	6 197	6 369	8 942	11 939	14 510	15 344	19 063	17 609	19 850	17 244
Yugoslavia, SFR† Yougoslavie, Rfs†	85	110	132	90	109	131	149	103	..	...
Oceania **Océanie**	**2 857**	**2 583**	**3 058**	**3 680**	**4 873**	**5 138**	**5 583**	**5 331**	**5 374**	**5 203**
Australia Australie	2 125	1 918	2 058	2 427	2 965	4 103	4 535	4 247	4 301	4 100
Fiji Fidji	18	18	24	53	35	31	31	36	35	39
Kiribati Kiribati	...	...	...	...	2	2	2	2	2	2
New Zealand Nouvelle-Zélande	455	389	637	852	988	944	958	987	977	1 003
Papua New Guinea Papouasie-Nvl-Guinée	29	21	22	32	48	42	42	42	42	42
Samoa Samoa	1	1	2	2	1	2	2	2	2	2
Solomon Islands Iles Salomon	3	4	6	6	9	10	11	12	11	11
Tonga Tonga	3	3	3	3	3	3	1	2	3	3
Vanuatu Vanuatu	2	2	2	1	2	1	1	1	1	1

Source:
World Tourism Organization (Madrid).

† For detailed descriptions of data pertaining to
former Czechoslovakia, Germany, SFR Yugoslavia and former
USSR, see Annex I - Country or area nomenclature, regional
and other groupings.

Source:
Organisation mondiale du tourisme (Madrid).

† Pour les descriptions en détails des données
relatives à l'ancienne Tchécoslovaquie, l'Allemagne, la Rfs
Yougoslavie et l'ancienne URSS, voir l'Annexe I -
Nomenclature des pays ou zones, groupements régionaux et
autres groupements.

Technical notes, tables 75-77

Tables 75 and 76: For statistical purposes, the term "international visitor" describes "any person who travels to a country other than that in which he/she has his/her usual residence but outside his/her usual environment for a period not exceeding 12 months and whose main purpose of visit is other than the exercise of an activity remunerated from within the country visited".

International visitors include:

(a) *Tourists*, (overnight visitor): "a visitor who stays at least one night in a collective or private accommodation in the country visited"; and

(b) *Same-day visitors*: "a visitor who does not spend the night in a collective or private accommodation in the country visited".

Unless otherwise stated, same-day visitors are not included in these tables.

The figures do not therefore include immigrants, residents in a frontier zone, persons domiciled in one country or area and working in an adjoining country or area, members of the armed forces and diplomats and consular representatives when they travel from their country of origin to the country in which they are stationed and vice-versa.

The figures also exclude persons in transit who do not formally enter the country through passport control, such as air transit passengers who remain for a short period in a designated area of the air terminal or ship passengers who are not permitted to disembark. This category would include passengers transferred directly between airports or other terminals. Other passengers in transit through a country are classified as visitors.

These data are based generally on a frontier check. In the absence of frontier check figures, data based on arrivals at accommodation establishments are given but these are not strictly comparable with frontier check data as they exclude certain types of tourists such as campers and tourists staying in private houses, while on the other hand they may contain some duplication when a tourist moves from one establishment to another.

Unless otherwise stated, table 75 shows the number of tourist arrivals at frontiers classified by their region of origin. Totals correspond to the total number of arrivals from the regions indicated in the table. However, these totals may not correspond to the number of tourist arrivals shown in table 76. The later excludes same-day visitors whereas they may be included in table 75. More detailed information will be found in *Yearbook of Tourism Statistics*, published by the World Tourism Organization. [32]

Notes techniques, tableaux 75-77

Tableaux 75 et 76 : A des fins statistiques, l'expression "*visiteur international*" désigne "toute personne qui se rend dans un pays autre que celui où elle a son lieu de résidence habituelle, mais différent de son environnement habituel, pour une période de 12 mois au maximum, dans un but principal autre que celui d'y exercer une profession rémunérée".

Entrent dans cette catégorie :

(a) Les *touristes* (visiteurs passant la nuit), c'est à dire "les visiteurs qui passent une nuit au moins en logement collectif ou privé dans le pays visité";

(b) Les *visiteurs ne restant que la journée*, c'est à dire "les visiteurs qui ne passent pas la nuit en logement collectif ou privé dans le pays visité".

Sauf indication contraire, les visiteurs ne restant que la journée ne sont pas inclus dans ces tableaux.

Par conséquent, ces chiffres ne comprennent pas les immigrants, les résidents frontaliers, les personnes domiciliées dans une zone ou un pays donné et travaillant dans une zone ou pays limitrophe, les membres des forces armées et les membres des corps diplomatique et consulaire lorsqu'ils se rendent de leur pays d'origine au pays où ils sont en poste, et vice versa.

Ne sont pas non plus inclus les voyageurs en transit, qui ne pénètrent pas officiellement dans le pays en faisant contrôler leurs passeports, tels que les passagers d'un vol en escale, qui demeurent pendant un court laps de temps dans une aire distincte de l'aérogare, ou les passagers d'un navire qui ne sont pas autorisés à débarquer. Cette catégorie com-prend également les passagers transportés directement d'une aérogare à l'autre ou à un autre terminal. Les autres passa-gers en transit dans un pays sont classés parmi les visiteurs.

Ces données reposent en général sur un contrôle à la frontière. A défaut, les données sont tirées des établissements d'hébergement, mais elles ne sont alors pas strictement comparables à celles du contrôle à la frontière, en ce qu'elles excluent, d'une part, certaines catégories de touristes telles que les campeurs et les touristes séjournant dans des maisons privées, et que, d'autre part, elles peuvent compter deux fois le même touriste si celui-ci change d'établissement d'hébergement.

Sauf indication contraire, le tableau 75 indique le nombre d'arrivées de touristes par région de provenance. Les totaux correspondent au nombre total d'arrivées de touristes des régions indiquées sur le tableau. Les chiffres totaux peuvent néanmoins, ne pas coïncider avec le nombre des arrivées de touristes indiqué dans le tableau 76, qui ne comprend pas les visiteurs ne restant que la journée, lesquels peuvent au contraire être inclus dans les chiffres du tableau 75. Pour plus de renseignements, consulter l'*Annuaire des statistiques du tourisme* publié par l'Organisation mondiale du tourisme [32].

Unless otherwise stated the data on tourist receipts have been supplied by the World Tourism Organization. Tourist receipts are defined as "expenditure of international inbound visitors including their payments to national carriers for international transport. They should also include any other prepayments made for goods/services received in the destination country. They should in practice also include receipts from same-day visitors, except in cases when these are so important as to justify a separate classification. It is also recommended that, for the sake of consistancy with the Balance of Payments recommendations of the International Monetary Fund, international fare receipts be classified separately.

Table 77: International tourism expenditure are defined as "expenditure of outbound visitors in other countries including their payments to foreign carriers for international transport. They should in practice also include expenditure of residents travelling abroad as same-day visitors, except in cases when these are so important as to justify a separate classification. It is also recommended that, for the sake of consistency with the Balance of Payments recommendations of the International Monetary Fund, international fare expenditure be classified separately".

For more detailed statistics on the number of tourists and expenditures, see *Yearbook of Tourism Statistics* published by the World Tourism Organization.[32] For detailed definitions of tourist receipts, see *Balance of Payments Yearbook* published by the International Monetary Fund.[10]

For detailed information on methods of collection for frontier statistics, accommodation statistics and foreign exchange statistics see *Methodological Supplement to World Travel and Tourism Statistics* published by the World Tourism Organization.[54; see also 47 and 51.]

Sauf indication contraire, les données sur les recettes du tourisme ont été fournies par l'Organisation mondiale du tourisme et sont définies comme "les sommes dépensées par les visiteurs internationaux arrivant dans le pays, y compris les sommes versées aux transporteurs nationaux en paiement de transports internationaux. Il faut également y inclure tout autre versement effectué à l'avance pour des biens ou services à recevoir dans le pays de destination. Dans la pratique, il faut également y inclure les recettes provenant de visiteurs ne restant que la journée, sauf dans les cas où elles sont suffisamment importantes pour justifier une classification distincte. Il est recommandé aussi, dans un souci de cohérence avec les recommandations du Fonds monétaire international visant la balance des paiements, de classer à part les recettes au titre des transports internationaux".

Tableau 77 : Les dépenses du tourisme international sont définies comme étant les dépenses effectuées par les résidents du pays en visite à l'étranger, y compris les sommes versées aux transporteurs étrangers en règlement des transports internationaux. En pratique, ce poste devrait également inclure les dépenses des résidents qui voyagent à l'étranger pour la journée, sauf dans le cas où ces dépenses sont suffisamment importantes pour justifier une classification distincte. Il est recommandé aussi, dans un souci de cohérence avec les recommandations du Fonds monétaire international visant la balance des paiements, de classer à part les dépenses au titre des transports internationaux.

Pour des statistiques plus détaillées sur le nombre de touristes et les dépenses, voir l'*Annuaire des statistiques du tourisme* publié par l'Organisation mondiale du tourisme [32]. Pour des définitions détaillées des recettes touristiques, voir *"Balance of Payments Yearbook"* publié par le Fonds monétaire international [10].

Pour plus de renseignements sur les méthodes de collecte de données statistiques sur les contrôles aux frontières, l'hébergement et les mouvements de devises, voir *"Supplément méthodologique aux statistiques des voyages et du tourisme mondiaux"* publié par l'Organisation mondiale du tourisme [54; voir aussi 47 et 51].

78
Summary of balance of payments
Résumé des balances des paiements
Millions of US dollars
Millions de dollars des E—U

Country or area	1987	1988	1989	1990	1991	1992	1993	Pays ou zone
Africa· Afrique								
Algeria								**Algérie**
Merchandise: Exp. fob	9 029.0	7 620.0	9 534.0	12 964.0	12 330.0	...	...	Marchandises : exp. fob
Merchandise: Imp. fob	−6 616.0	−6 675.0	−8 372.0	−8 777.0	−6 852.0	...	...	Marchandises : imp. fob
Serv. & Income: Credit	675.0	542.0	607.0	571.0	463.0	...	...	Serv. & revenu : crédit
Serv. & Income: Debit	−3 464.0	−3 917.0	−3 390.0	−3 671.0	−3 790.0	...	...	Serv. & revenu : débit
Private Unrequited Transfers	522.0	385.0	535.0	332.0	239.0	...	...	Transf. priv.sans contrep.
Offic. Unrequited Trans.,nie	−5.0	5.0	6.0	1.0	−23.0	...	...	Tr. offic. sans contrep.nia
Direct Investment, nie	−11.0	8.0	4.0	−4.0	−39.0	...	...	Investissements direct, nia
Portfolio Investments, nie	0.0	2.0	0.0	0.0	0.0	...	...	Invest. portefeuille, nia
Other Capital, nie	321.0	734.0	751.0	−1 090.0	−982.0	...	...	Autres capitaux, nia
Net Errors and Omissions	−802.0	337.0	−448.0	−336.0	−299.0	...	...	Erreurs et omissions nettes
Reserves and Related Items	352.0	959.0	774.0	10.0	−1 047.0	...	...	Rés. et postes assimilables
Angola								**Angola**
Merchandise: Exp. fob	2 322.0	2 492.0	3 014.0	3 883.9	3 449.3	3 833.0	...	Marchandises : exp. fob
Merchandise: Imp. fob	−1 303.0	−1 372.0	−1 338.0	−1 578.2	−1 347.2	−1 988.0	...	Marchandises : imp. fob
Serv. & Income: Credit	93.0	128.0	150.0	119.1	186.3	158.7	...	Serv. & revenu : crédit
Serv. & Income: Debit	−717.0	−1 749.0	−1 954.0	−2 583.2	−2 896.2	−2 840.6	...	Serv. & revenu : débit
Private Unrequited Transfers	−8.0	−6.0	−68.0	−139.7	−30.0	55.1	...	Transf. priv.sans contrep.
Offic. Unrequited Trans.,nie	60.0	38.0	64.0	62.6	58.2	47.2	...	Tr. offic. sans contrep.nia
Direct Investment, nie	119.0	131.0	200.0	−335.7	664.5	288.0	...	Investissements direct, nia
Portfolio Investments, nie	0.0	0.0	0.0	0.0	0.0	0.0	...	Invest. portefeuille, nia
Other Capital, nie	−64.0	−330.0	−320.0	−618.6	−1 611.7	−733.8	...	Autres capitaux, nia
Net Errors and Omissions	−834.0	−257.0	−678.0	−19.1	26.9	42.8	...	Erreurs et omissions nettes
Reserves and Related Items	332.0	925.0	930.0	1 208.9	1 499.9	1 137.7	...	Rés. et postes assimilables
Benin								**Bénin**
Merchandise: Exp. fob	363.4	379.1	178.4	287.2	328.6	362.3	332.7	Marchandises : exp. fob
Merchandise: Imp. fob	−483.8	−511.0	−316.6	−427.9	−482.4	−560.7	−571.4	Marchandises : imp. fob
Serv. & Income: Credit	111.8	118.9	87.8	114.6	122.6	141.7	134.6	Serv. & revenu : crédit
Serv. & Income: Debit	−228.9	−238.7	−149.8	−174.1	−169.1	−219.1	−195.3	Serv. & revenu : débit
Private Unrequited Transfers	50.9	51.7	65.8	86.3	87.9	99.0	87.2	Transf. priv.sans contrep.
Offic. Unrequited Trans.,nie	112.8	117.2	121.3	112.4	101.7	132.2	160.0	Tr. offic. sans contrep.nia
Direct Investment, nie	0.0	0.0	0.0	0.0	0.0	0.0	0.0	Investissements direct, nia
Portfolio Investments, nie	0.0	0.0	0.0	0.0	0.0	0.0	0.0	Invest. portefeuille, nia
Other Capital, nie	44.3	−7.7	18.6	50.4	52.1	−40.7	45.0	Autres capitaux, nia
Net Errors and Omissions	−41.6	−7.8	−71.3	−39.0	20.4	7.7	−32.8	Erreurs et omissions nettes
Reserves and Related Items	71.3	98.4	65.9	−10.0	−61.9	77.5	40.1	Rés. et postes assimilables
Botswana								**Botswana**
Merchandise: Exp. fob	1 586.6	1 468.9	1 819.7	1 753.2	...	...	...	Marchandises : exp. fob
Merchandise: Imp. fob	−803.9	−986.9	−1 185.1	−1 606.2	...	...	...	Marchandises : imp. fob
Serv. & Income: Credit	296.9	330.4	355.3	497.3	...	...	...	Serv. & revenu : crédit
Serv. & Income: Debit	−589.6	−791.9	−711.5	−782.1	...	...	...	Serv. & revenu : débit
Private Unrequited Transfers	6.8	−17.5	−30.6	−40.8	...	...	...	Transf. priv.sans contrep.
Offic. Unrequited Trans.,nie	166.7	184.6	250.5	316.2	...	...	...	Tr. offic. sans contrep.nia
Direct Investment, nie	113.6	39.9	42.2	38.2	...	...	...	Investissements direct, nia
Portfolio Investments, nie	0.0	0.0	0.0	0.0	...	...	...	Invest. portefeuille, nia
Other Capital, nie	−203.3	−65.2	70.8	153.3	...	...	...	Autres capitaux, nia
Net Errors and Omissions	−12.2	220.0	−34.8	−21.7	...	...	...	Erreurs et omissions nettes
Reserves and Related Items	−561.5	−382.3	−576.5	−307.2	...	...	...	Rés. et postes assimilables
Burkina Faso								**Burkina Faso**
Merchandise: Exp. fob	229.9	249.1	229.8	272.2	283.2	287.5	276.5	Marchandises : exp. fob
Merchandise: Imp. fob	−475.2	−486.8	−501.6	−593.2	−601.5	−642.3	−643.4	Marchandises : imp. fob
Serv. & Income: Credit	48.9	49.0	52.7	65.0	63.5	72.9	66.2	Serv. & revenu : crédit
Serv. & Income: Debit	−213.3	−235.4	−237.3	−272.9	−277.6	−291.7	−289.5	Serv. & revenu : débit
Private Unrequited Transfers	123.1	113.4	96.0	113.9	106.4	106.9	96.8	Transf. priv.sans contrep.
Offic. Unrequited Trans.,nie	235.6	261.4	393.1	312.9	321.8	369.5	375.4	Tr. offic. sans contrep.nia
Direct Investment, nie	0.0	0.0	0.0	0.0	0.0	0.0	0.0	Investissements direct, nia
Portfolio Investments, nie	0.0	0.0	0.0	0.0	0.0	0.0	0.0	Invest. portefeuille, nia
Other Capital, nie	73.7	76.0	−199.9	89.5	133.5	124.3	113.5	Autres capitaux, nia
Net Errors and Omissions	−5.3	−2.4	19.2	−5.2	18.8	−9.0	21.9	Erreurs et omissions nettes
Reserves and Related Items	−17.5	−24.3	148.0	17.7	−48.1	−18.3	−17.4	Rés. et postes assimilables

78
Summary of balance of payments
Millions of US dollars
Résumé des balances des paiements
Millions de dollars des E-U

Country or area	1987	1988	1989	1990	1991	1992	1993	Pays ou zone
Burundi								**Burundi**
Merchandise: Exp. fob	98.3	124.4	93.2	72.9	90.7	79.3	75.0	Marchandises : exp. fob
Merchandise: Imp. fob	−159.2	−166.1	−151.4	−189.0	−195.9	−181.8	−172.8	Marchandises : imp. fob
Serv. & Income: Credit	14.8	14.8	24.2	24.8	35.1	31.3	25.3	Serv. & revenu : crédit
Serv. & Income: Debit	−163.4	−140.7	−119.1	−148.5	−157.7	−161.3	−134.6	Serv. & revenu : débit
Private Unrequited Transfers	7.2	9.9	8.6	10.0	13.2	12.8	17.4	Transf. priv.sans contrep.
Offic. Unrequited Trans.,nie	105.7	87.2	132.4	163.6	182.5	164.9	163.4	Tr. offic. sans contrep.nia
Direct Investment, nie	1.4	1.2	0.5	1.2	0.9	0.6	0.3	Investissements direct, nia
Portfolio Investments, nie	0.0	0.0	0.0	0.0	0.0	0.0	0.0	Invest. portefeuille, nia
Other Capital, nie	129.7	83.1	63.9	76.8	69.6	98.3	52.5	Autres capitaux, nia
Net Errors and Omissions	−37.3	−6.6	−14.3	−15.1	−5.8	−18.7	−16.3	Erreurs et omissions nettes
Reserves and Related Items	2.6	−7.1	−38.0	3.2	−32.6	−25.5	−10.2	Rés. et postes assimilables
Cameroon								**Cameroun**
Merchandise: Exp. fob	1 688.7	1 841.2	1 853.8	2 125.4	1 957.5	1 934.1	1 144.2	Marchandises : exp. fob
Merchandise: Imp. fob	−1 434.8	−1 220.8	−1 136.8	−1 347.2	−1 173.1	−983.3	−927.5	Marchandises : imp. fob
Serv. & Income: Credit	423.8	473.0	487.8	389.4	424.3	432.7	367.8	Serv. & revenu : crédit
Serv. & Income: Debit	−1 471.4	−1 404.4	−1 463.3	−1 611.6	−1 572.9	−1 729.3	−1 321.7	Serv. & revenu : débit
Private Unrequited Transfers	−125.7	−134.9	−63.6	−51.9	−65.8	−94.4	−79.8	Transf. priv.sans contrep.
Offic. Unrequited Trans.,nie	26.6	30.8	32.4	28.9	28.4	134.9	75.3	Tr. offic. sans contrep.nia
Direct Investment, nie	0.4	38.6	−113.1	−72.1	−38.3	−50.4	−103.1	Investissements direct, nia
Portfolio Investments, nie	0.0	0.0	0.0	0.0	0.0	0.0	0.0	Invest. portefeuille, nia
Other Capital, nie	333.3	−14.1	432.0	−154.7	−324.0	−296.0	−192.9	Autres capitaux, nia
Net Errors and Omissions	89.3	228.2	−156.0	−165.4	34.9	−626.0	210.9	Erreurs et omissions nettes
Reserves and Related Items	469.7	162.3	126.6	859.2	729.1	1 277.8	826.8	Rés. et postes assimilables
Cape Verde								**Cap-Vert**
Merchandise: Exp. fob	7.8	3.3	6.7	5.6	4.1	4.4	...	Marchandises : exp. fob
Merchandise: Imp. fob	−92.8	−101.8	−106.9	−119.5	−132.2	−173.3	...	Marchandises : imp. fob
Serv. & Income: Credit	43.3	45.0	57.3	61.3	54.3	59.1	...	Serv. & revenu : crédit
Serv. & Income: Debit	−25.5	−24.2	−32.5	−36.9	−23.2	−31.1	...	Serv. & revenu : débit
Private Unrequited Transfers	34.0	39.5	43.2	52.1	57.8	69.9	...	Transf. priv.sans contrep.
Offic. Unrequited Trans.,nie	46.9	38.7	27.6	25.5	31.3	67.4	...	Tr. offic. sans contrep.nia
Direct Investment, nie	2.8	0.4	−0.6	−0.1	1.2	−0.8	...	Investissements direct, nia
Portfolio Investments, nie	0.0	0.0	0.0	0.0	0.0	0.0	...	Invest. portefeuille, nia
Other Capital, nie	5.4	−2.2	−0.5	5.1	1.5	6.8	...	Autres capitaux, nia
Net Errors and Omissions	0.9	−0.7	−1.2	5.2	−7.3	5.5	...	Erreurs et omissions nettes
Reserves and Related Items	−22.8	1.9	6.9	1.7	12.6	−8.0	...	Rés. et postes assimilables
Central African Rep.								**Rép. centrafricaine**
Merchandise: Exp. fob	128.9	133.7	148.1	150.5	125.6	123.5	...	Marchandises : exp. fob
Merchandise: Imp. fob	−197.7	−179.1	−186.0	−241.6	−178.7	−165.1	...	Marchandises : imp. fob
Serv. & Income: Credit	70.6	62.3	66.2	69.8	56.0	60.1	...	Serv. & revenu : crédit
Serv. & Income: Debit	−176.4	−172.2	−165.9	−190.9	−155.6	−169.3	...	Serv. & revenu : débit
Private Unrequited Transfers	−23.7	−27.6	−24.9	−32.9	−29.8	−32.1	...	Transf. priv.sans contrep.
Offic. Unrequited Trans.,nie	124.8	148.3	129.0	155.9	120.7	125.4	...	Tr. offic. sans contrep.nia
Direct Investment, nie	9.3	−8.6	−2.5	−3.1	−8.4	−8.7	...	Investissements direct, nia
Portfolio Investments, nie	0.0	0.0	0.0	0.0	0.0	0.0	...	Invest. portefeuille, nia
Other Capital, nie	54.1	18.2	22.7	72.6	32.9	41.6	...	Autres capitaux, nia
Net Errors and Omissions	−1.7	11.6	1.3	1.1	−1.9	0.7	...	Erreurs et omissions nettes
Reserves and Related Items	11.6	13.3	11.9	18.5	39.1	23.9	...	Rés. et postes assimilables
Chad								**Tchad**
Merchandise: Exp. fob	109.4	145.9	155.4	230.3	193.5	182.3	135.8	Marchandises : exp. fob
Merchandise: Imp. fob	−225.9	−228.4	−240.3	−259.5	−249.9	−243.0	−201.3	Marchandises : imp. fob
Serv. & Income: Credit	73.3	80.9	43.6	44.0	39.8	44.3	37.6	Serv. & revenu : crédit
Serv. & Income: Debit	−211.0	−233.4	−220.7	−252.0	−219.3	−239.1	−214.6	Serv. & revenu : débit
Private Unrequited Transfers	−9.8	−17.1	−20.2	−12.9	−19.8	−34.5	−21.2	Transf. priv.sans contrep.
Offic. Unrequited Trans.,nie	238.5	277.7	226.3	204.5	189.9	204.3	180.0	Tr. offic. sans contrep.nia
Direct Investment, nie	0.2	−12.6	6.2	0.0	−6.3	−11.8	0.0	Investissements direct, nia
Portfolio Investments, nie	0.0	0.0	0.0	0.0	0.0	0.0	0.0	Invest. portefeuille, nia
Other Capital, nie	8.5	36.9	68.3	56.1	66.2	51.8	67.3	Autres capitaux, nia
Net Errors and Omissions	16.5	−83.7	11.1	−33.3	−13.0	9.2	−34.8	Erreurs et omissions nettes
Reserves and Related Items	0.3	33.8	−29.7	22.9	18.6	36.5	51.2	Rés. et postes assimilables
Comoros								**Comores**
Merchandise: Exp. fob	11.6	21.5	18.1	17.9	24.4	...	...	Marchandises : exp. fob
Merchandise: Imp. fob	−44.2	−44.3	−35.7	−45.2	−53.6	...	...	Marchandises : imp. fob
Serv. & Income: Credit	16.2	18.6	21.8	20.2	27.5	...	...	Serv. & revenu : crédit
Serv. & Income: Debit	−44.7	−46.2	−42.6	−46.9	−48.1	...	...	Serv. & revenu : débit
Private Unrequited Transfers	0.9	3.1	2.7	5.5	3.7	...	...	Transf. priv.sans contrep.

78
Summary of balance of payments
Millions of US dollars
Résumé des balances des paiements
Millions de dollars des E−U

Country or area	1987	1988	1989	1990	1991	1992	1993	Pays ou zone
Offic. Unrequited Trans.,nie	38.7	40.8	41.1	39.2	37.2	...	...	Tr. offic. sans contrep.nia
Direct Investment, nie	7.6	3.8	3.3	−0.7	2.5	...	...	Investissements direct, nia
Portfolio Investments, nie	0.0	0.0	0.0	0.0	0.0	...	...	Invest. portefeuille, nia
Other Capital, nie	22.0	0.4	4.3	14.4	1.6	...	...	Autres capitaux, nia
Net Errors and Omissions	0.7	−1.4	−7.6	−9.2	1.7	...	...	Erreurs et omissions nettes
Reserves and Related Items	−8.8	3.7	−5.4	4.9	3.1	...	...	Rés. et postes assimilables
Congo								**Congo**
Merchandise: Exp. fob	876.7	843.2	1 160.5	1 388.7	1 107.7	1 178.7	1 107.5	Marchandises : exp. fob
Merchandise: Imp. fob	−419.9	−522.7	−532.0	−512.7	−494.5	−438.2	−490.9	Marchandises : imp. fob
Serv. & Income: Credit	127.7	99.7	97.5	113.9	118.0	78.6	65.0	Serv. & revenu : crédit
Serv. & Income: Debit	−818.3	−873.5	−857.3	−1 244.0	−1 188.2	−1 117.9	−1 130.5	Serv. & revenu : débit
Private Unrequited Transfers	−56.8	−56.6	−46.7	−62.8	−59.9	−63.1	−83.3	Transf. priv.sans contrep.
Offic. Unrequited Trans.,nie	67.9	64.4	93.1	65.7	55.3	44.6	24.7	Tr. offic. sans contrep.nia
Direct Investment, nie	43.4	9.1	0.0	0.0	0.0	0.0	0.0	Investissements direct, nia
Portfolio Investments, nie	0.0	0.0	0.0	0.0	0.0	0.0	0.0	Invest. portefeuille, nia
Other Capital, nie	−336.9	−71.1	−325.4	−72.0	9.6	−153.8	90.4	Autres capitaux, nia
Net Errors and Omissions	27.6	40.6	8.5	−40.6	−6.3	41.2	−35.4	Erreurs et omissions nettes
Reserves and Related Items	488.5	466.8	401.8	363.9	458.2	429.9	452.4	Rés. et postes assimilables
Côte d'Ivoire								**Côte d'Ivoire**
Merchandise: Exp. fob	2 949.7	2 691.3	2 696.8	3 027.9	2 686.2	2 880.0	2 734.1	Marchandises : exp. fob
Merchandise: Imp. fob	−1 863.3	−1 769.4	−1 777.1	−1 700.6	−1 706.8	−1 885.6	−1 662.3	Marchandises : imp. fob
Serv. & Income: Credit	613.2	627.5	541.7	581.8	553.0	588.2	564.0	Serv. & revenu : crédit
Serv. & Income: Debit	−2 309.9	−2 335.1	−2 226.3	−2 792.5	−2 788.0	−2 764.0	−2 644.1	Serv. & revenu : débit
Private Unrequited Transfers	−501.1	−514.4	−398.7	−553.5	−464.4	−492.3	−394.1	Transf. priv.sans contrep.
Offic. Unrequited Trans.,nie	141.4	58.9	196.3	227.3	230.1	245.2	173.4	Tr. offic. sans contrep.nia
Direct Investment, nie	87.5	51.7	18.5	31.6	80.8	77.1	30.4	Investissements direct, nia
Portfolio Investments, nie	−8.0	−14.1	1.3	0.0	0.0	0.0	0.0	Invest. portefeuille, nia
Other Capital, nie	−55.2	−175.3	−324.1	−290.5	92.2	−74.0	−53.7	Autres capitaux, nia
Net Errors and Omissions	12.6	−24.4	−38.8	−97.1	53.0	35.0	61.9	Erreurs et omissions nettes
Reserves and Related Items	933.1	1 403.1	1 310.6	1 565.6	1 263.8	1 390.4	1 190.4	Rés. et postes assimilables
Djibouti								**Djibouti**
Merchandise: Exp. fob	...	...	...	...	...	192.0	211.2	Marchandises : exp. fob
Merchandise: Imp. fob	...	...	...	...	...	−383.4	−402.3	Marchandises : imp. fob
Serv. & Income: Credit	...	...	...	...	...	174.4	183.9	Serv. & revenu : crédit
Serv. & Income: Debit	...	...	...	...	...	−95.8	−94.8	Serv. & revenu : débit
Private Unrequited Transfers	...	...	...	...	...	−89.6	−88.7	Transf. priv.sans contrep.
Offic. Unrequited Trans.,nie	...	...	...	...	...	113.8	102.4	Tr. offic. sans contrep.nia
Direct Investment, nie	...	...	...	...	...	2.0	3.0	Investissements direct, nia
Portfolio Investments, nie	...	...	...	...	...	0.0	0.0	Invest. portefeuille, nia
Other Capital, nie	...	...	...	...	...	58.0	4.0	Autres capitaux, nia
Net Errors and Omissions	...	...	...	...	...	11.4	73.3	Erreurs et omissions nettes
Reserves and Related Items	...	...	...	...	...	17.0	8.0	Rés. et postes assimilables
Egypt								**Egypte**
Merchandise: Exp. fob	3 115.0	2 770.0	2 907.0	3 604.0	3 856.0	3 400.0	3 243.0	Marchandises : exp. fob
Merchandise: Imp. fob	−8 095.0	−9 378.0	−8 841.0	−10 303.0	−9 831.0	−8 901.0	−9 923.0	Marchandises : imp. fob
Serv. & Income: Credit	4 130.0	4 982.0	5 123.0	7 148.0	7 951.0	8 901.0	9 307.0	Serv. & revenu : crédit
Serv. & Income: Debit	−3 725.0	−3 858.0	−4 672.0	−5 667.0	−5 507.0	−7 664.0	−7 334.0	Serv. & revenu : débit
Private Unrequited Transfers	3 604.0	3 770.0	3 293.0	4 284.0	4 054.0	6 104.0	5 664.0	Transf. priv.sans contrep.
Offic. Unrequited Trans.,nie	725.0	666.0	880.0	1 119.0	1 380.0	972.0	1 342.0	Tr. offic. sans contrep.nia
Direct Investment, nie	929.0	1 178.0	1 228.0	722.0	191.0	455.0	493.0	Investissements direct, nia
Portfolio Investments, nie	2.0	0.0	0.0	15.0	21.0	6.0	4.0	Invest. portefeuille, nia
Other Capital, nie	−1 263.0	130.0	−867.0	−11 776.0	−4 918.0	−629.0	−1 259.0	Autres capitaux, nia
Net Errors and Omissions	892.0	−362.0	414.0	630.0	730.0	716.0	−1 519.0	Erreurs et omissions nettes
Reserves and Related Items	−315.0	102.0	533.0	10 224.0	2 073.0	−3 360.0	−18.0	Rés. et postes assimilables
Equatorial Guinea								**Guinée équatoriale**
Merchandise: Exp. fob	38.5	44.7	32.7	37.8	35.8	...	...	Marchandises : exp. fob
Merchandise: Imp. fob	−47.9	−56.5	−43.6	−53.2	−59.6	...	...	Marchandises : imp. fob
Serv. & Income: Credit	6.1	5.9	5.8	4.5	6.2	...	...	Serv. & revenu : crédit
Serv. & Income: Debit	−47.2	−56.6	−39.8	−46.0	−52.4	...	...	Serv. & revenu : débit
Private Unrequited Transfers	−2.5	−3.8	−13.0	−17.1	−16.6	...	...	Transf. priv.sans contrep.
Offic. Unrequited Trans.,nie	27.6	45.8	36.8	54.9	62.0	...	...	Tr. offic. sans contrep.nia
Direct Investment, nie	0.0	0.0	−0.4	9.7	42.2	...	...	Investissements direct, nia
Portfolio Investments, nie	0.0	0.0	0.0	0.0	0.0	...	...	Invest. portefeuille, nia
Other Capital, nie	−1.0	4.9	10.5	2.0	−10.0	...	...	Autres capitaux, nia
Net Errors and Omissions	0.8	−1.7	−4.5	−2.4	−30.7	...	...	Erreurs et omissions nettes
Reserves and Related Items	25.6	17.4	15.5	9.7	23.2	...	...	Rés. et postes assimilables

78
Summary of balance of payments
Millions of US dollars
Résumé des balances des paiements
Millions de dollars des E-U

Country or area	1987	1988	1989	1990	1991	1992	1993	Pays ou zone
Ethiopia								**Ethiopie**
Merchandise: Exp. fob	355.2	400.0	443.8	292.0	167.6	169.9	198.8	Marchandises : exp. fob
Merchandise: Imp. fob	−932.7	−956.0	−817.9	−912.1	−470.8	−992.7	−706.0	Marchandises : imp. fob
Serv. & Income: Credit	318.7	288.9	301.8	313.8	282.7	290.1	299.6	Serv. & revenu : crédit
Serv. & Income: Debit	−366.4	−402.4	−408.4	−436.6	−381.1	−472.4	−377.8	Serv. & revenu : débit
Private Unrequited Transfers	129.6	180.5	145.7	229.1	222.4	341.5	251.8	Transf. priv.sans contrep.
Offic. Unrequited Trans.,nie	278.0	261.6	190.5	220.0	353.0	543.9	279.6	Tr. offic. sans contrep.nia
Direct Investment, nie	0.0	0.0	0.0	0.0	0.0	0.0	0.0	Investissements direct, nia
Portfolio Investments, nie	0.0	0.0	0.0	0.0	0.0	0.0	0.0	Invest. portefeuille, nia
Other Capital, nie	292.8	299.6	222.0	230.0	−204.1	−65.4	79.4	Autres capitaux, nia
Net Errors and Omissions	−182.8	−94.0	−32.0	−134.6	−254.9	−81.1	66.3	Erreurs et omissions nettes
Reserves and Related Items	107.4	22.3	−45.5	198.3	285.1	266.2	−91.7	Rés. et postes assimilables
Gabon								**Gabon**
Merchandise: Exp. fob	1 286.4	1 195.6	1 626.0	2 488.8	2 227.9	2 259.2	2 149.7	Marchandises : exp. fob
Merchandise: Imp. fob	−731.8	−791.2	−751.7	−805.1	−861.0	−886.3	−845.1	Marchandises : imp. fob
Serv. & Income: Credit	130.8	228.1	308.0	261.7	352.0	394.8	318.2	Serv. & revenu : crédit
Serv. & Income: Debit	−1 011.1	−1 103.7	−1 248.6	−1 643.3	−1 524.3	−1 793.8	−1 765.8	Serv. & revenu : débit
Private Unrequited Transfers	−147.8	−155.5	−135.2	−158.5	−125.5	−154.1	−140.9	Transf. priv.sans contrep.
Offic. Unrequited Trans.,nie	24.4	11.2	9.3	24.2	5.7	12.1	15.2	Tr. offic. sans contrep.nia
Direct Investment, nie	82.2	122.8	−38.6	44.7	−69.5	101.3	91.5	Investissements direct, nia
Portfolio Investments, nie	0.0	0.0	0.0	0.0	0.0	0.0	0.0	Invest. portefeuille, nia
Other Capital, nie	282.7	593.8	99.7	−411.1	−234.3	−321.9	−238.0	Autres capitaux, nia
Net Errors and Omissions	−51.0	−101.9	35.0	−38.0	8.6	−55.1	−5.8	Erreurs et omissions nettes
Reserves and Related Items	135.2	0.8	96.1	236.8	220.4	443.8	421.2	Rés. et postes assimilables
Gambia [2]								**Gambie** [2]
Merchandise: Exp. fob	74.5	83.1	100.2	110.6	142.9	147.0	...	Marchandises : exp. fob
Merchandise: Imp. fob	−95.0	−105.9	−125.4	−140.5	−185.0	−177.8	...	Marchandises : imp. fob
Serv. & Income: Credit	50.2	63.7	67.7	71.4	84.3	85.7	...	Serv. & revenu : crédit
Serv. & Income: Debit	−61.9	−61.4	−66.4	−64.9	−83.5	−74.2	...	Serv. & revenu : débit
Private Unrequited Transfers	13.5	12.8	6.7	14.1	15.1	13.3	...	Transf. priv.sans contrep.
Offic. Unrequited Trans.,nie	24.8	34.3	32.2	43.2	39.4	43.2	...	Tr. offic. sans contrep.nia
Direct Investment, nie	1.5	1.2	14.8	0.0	10.2	6.2	...	Investissements direct, nia
Portfolio Investments, nie	0.0	0.0	0.0	0.0	0.0	0.0	...	Invest. portefeuille, nia
Other Capital, nie	−2.1	8.5	−5.3	−6.1	10.6	12.6	...	Autres capitaux, nia
Net Errors and Omissions	5.5	−11.3	−20.8	−24.0	−16.7	−36.7	...	Erreurs et omissions nettes
Reserves and Related Items	−11.0	−24.8	−3.7	−3.8	−17.3	−19.2	...	Rés. et postes assimilables
Ghana								**Ghana**
Merchandise: Exp. fob	826.8	881.0	807.2	890.6	997.6	986.4	...	Marchandises : exp. fob
Merchandise: Imp. fob	−951.5	−993.4	−1 002.2	−1 198.9	−1 318.7	−1 456.7	...	Marchandises : imp. fob
Serv. & Income: Credit	79.3	77.7	81.9	93.1	110.3	128.9	...	Serv. & revenu : crédit
Serv. & Income: Debit	−376.3	−399.6	−407.8	−429.0	−462.8	−505.1	...	Serv. & revenu : débit
Private Unrequited Transfers	201.6	172.4	202.1	201.9	219.5	254.9	...	Transf. priv.sans contrep.
Offic. Unrequited Trans.,nie	123.2	196.1	220.2	213.8	201.4	213.8	...	Tr. offic. sans contrep.nia
Direct Investment, nie	4.7	5.0	15.0	14.8	20.0	22.5	...	Investissements direct, nia
Portfolio Investments, nie	0.0	0.0	0.0	0.0	0.0	0.0	...	Invest. portefeuille, nia
Other Capital, nie	251.0	204.0	198.6	310.2	318.1	299.1	...	Autres capitaux, nia
Net Errors and Omissions	−18.7	37.9	40.6	8.8	23.8	−0.4	...	Erreurs et omissions nettes
Reserves and Related Items	−140.1	−181.1	−155.6	−105.3	−109.2	56.6	...	Rés. et postes assimilables
Guinea								**Guinée**
Merchandise: Exp. fob	544.6	511.9	595.6	671.2	687.1	517.1	561.1	Marchandises : exp. fob
Merchandise: Imp. fob	−380.3	−510.6	−531.5	−585.7	−694.9	−608.4	−582.6	Marchandises : imp. fob
Serv. & Income: Credit	58.2	63.8	112.3	170.4	160.2	167.6	317.8	Serv. & revenu : crédit
Serv. & Income: Debit	−282.8	−373.2	−438.2	−528.7	−529.2	−471.4	−427.3	Serv. & revenu : débit
Private Unrequited Transfers	−13.3	1.7	−0.9	−32.5	−26.0	−1.5	64.0	Transf. priv.sans contrep.
Offic. Unrequited Trans.,nie	35.3	84.9	281.8	102.7	114.0	141.7	137.1	Tr. offic. sans contrep.nia
Direct Investment, nie	12.8	15.7	12.3	17.9	38.8	19.7	2.7	Investissements direct, nia
Portfolio Investments, nie	0.0	0.0	0.0	0.0	0.0	0.0	0.0	Invest. portefeuille, nia
Other Capital, nie	−1.7	59.1	−95.4	49.9	7.6	86.7	123.8	Autres capitaux, nia
Net Errors and Omissions	−9.3	−5.1	−44.9	52.1	112.3	18.6	−229.3	Erreurs et omissions nettes
Reserves and Related Items	36.5	151.9	109.0	82.8	130.2	129.7	32.9	Rés. et postes assimilables
Guinea−Bissau								**Guinée−Bissau**
Merchandise: Exp. fob	15.4	15.9	14.2	19.3	20.4	6.5	16.0	Marchandises : exp. fob
Merchandise: Imp. fob	−44.7	−58.9	−68.9	−68.1	−67.5	−83.5	−53.8	Marchandises : imp. fob
Serv. & Income: Credit	0.0	0.0	0.0	0.0	0.0	0.0	0.0	Serv. & revenu : crédit
Serv. & Income: Debit	−33.2	−37.0	−49.0	−35.4	−47.3	−43.8	−40.4	Serv. & revenu : débit
Private Unrequited Transfers	−2.0	1.5	1.2	1.0	−4.1	−0.6	−1.6	Transf. priv.sans contrep.

78
Summary of balance of payments
Millions of US dollars
Résumé des balances des paiements
Millions de dollars des E−U

Country or area	1987	1988	1989	1990	1991	1992	1993	Pays ou zone
Offic. Unrequited Trans.,nie	40.2	37.0	51.3	51.8	52.1	45.8	51.0	Tr. offic. sans contrep.nia
Direct Investment, nie	0.0	0.0	0.0	0.0	0.0	0.0	0.0	Investissements direct, nia
Portfolio Investments, nie	0.0	0.0	0.0	0.0	0.0	0.0	0.0	Invest. portefeuille, nia
Other Capital, nie	1.5	−3.4	−7.0	1.2	−8.8	2.1	−13.6	Autres capitaux, nia
Net Errors and Omissions	−7.7	3.5	−11.6	−1.5	−16.3	22.0	−16.0	Erreurs et omissions nettes
Reserves and Related Items	30.5	41.4	69.8	31.7	71.3	51.6	58.4	Rés. et postes assimilables
Kenya								**Kenya**
Merchandise: Exp. fob	908.7	1 017.5	926.1	1 010.5	1 053.8	1 004.0	1 185.6	Marchandises : exp. fob
Merchandise: Imp. fob	−1 622.6	−1 802.2	−1 963.4	−2 005.3	−1 697.3	−1 594.3	−1 492.8	Marchandises : imp. fob
Serv. & Income: Credit	829.9	874.3	1 008.5	1 222.8	1 151.9	1 148.3	1 143.9	Serv. & revenu : crédit
Serv. & Income: Debit	−824.6	−895.5	−933.3	−1 122.9	−1 067.4	−937.7	−925.2	Serv. & revenu : débit
Private Unrequited Transfers	72.0	89.0	101.5	167.8	144.4	68.3	147.1	Transf. priv.sans contrep.
Offic. Unrequited Trans.,nie	141.9	256.4	280.9	206.9	204.5	214.2	93.9	Tr. offic. sans contrep.nia
Direct Investment, nie	45.0	−1.8	60.8	57.1	18.8	6.4	1.5	Investissements direct, nia
Portfolio Investments, nie	0.0	0.0	0.0	0.0	0.0	0.0	0.0	Invest. portefeuille, nia
Other Capital, nie	316.8	384.1	573.0	303.8	77.7	−276.5	0.3	Autres capitaux, nia
Net Errors and Omissions	107.9	34.7	67.7	66.9	69.6	110.3	257.5	Erreurs et omissions nettes
Reserves and Related Items	25.1	43.4	−122.0	92.5	43.9	256.9	−411.8	Rés. et postes assimilables
Lesotho								**Lesotho**
Merchandise: Exp. fob	46.5	63.7	66.4	59.5	67.2	109.2	134.0	Marchandises : exp. fob
Merchandise: Imp. fob	−451.5	−559.4	−592.6	−672.6	−803.5	−932.6	−911.6	Marchandises : imp. fob
Serv. & Income: Credit	392.3	417.0	412.9	495.6	517.7	537.6	493.7	Serv. & revenu : crédit
Serv. & Income: Debit	−75.7	−84.0	−91.2	−103.3	−104.4	−115.3	−95.1	Serv. & revenu : débit
Private Unrequited Transfers	0.3	4.4	4.2	4.8	2.7	3.9	2.6	Transf. priv.sans contrep.
Offic. Unrequited Trans.,nie	111.8	133.7	210.7	281.1	403.5	434.9	398.0	Tr. offic. sans contrep.nia
Direct Investment, nie	5.7	21.0	13.4	17.1	7.5	2.7	15.0	Investissements direct, nia
Portfolio Investments, nie	0.0	0.0	0.0	0.0	0.0	0.0	0.0	Invest. portefeuille, nia
Other Capital, nie	−2.4	−29.0	−33.5	−62.1	−68.2	−65.2	49.8	Autres capitaux, nia
Net Errors and Omissions	−26.0	26.5	1.9	−2.8	20.1	74.8	16.1	Erreurs et omissions nettes
Reserves and Related Items	−0.9	6.1	7.9	−17.2	−42.4	−49.9	−102.4	Rés. et postes assimilables
Liberia								**Libéria**
Merchandise: Exp. fob	374.9	...	...	...	...	...	...	Marchandises : exp. fob
Merchandise: Imp. fob	−311.7	...	...	...	...	...	...	Marchandises : imp. fob
Serv. & Income: Credit	57.7	...	...	...	...	...	...	Serv. & revenu : crédit
Serv. & Income: Debit	−262.5	...	...	...	...	...	...	Serv. & revenu : débit
Private Unrequited Transfers	−21.4	...	...	...	...	...	...	Transf. priv.sans contrep.
Offic. Unrequited Trans.,nie	45.4	...	...	...	...	...	...	Tr. offic. sans contrep.nia
Direct Investment, nie	38.5	...	...	...	...	...	...	Investissements direct, nia
Portfolio Investments, nie	0.0	...	...	...	...	...	...	Invest. portefeuille, nia
Other Capital, nie	−223.8	...	...	...	...	...	...	Autres capitaux, nia
Net Errors and Omissions	30.3	...	...	...	...	...	...	Erreurs et omissions nettes
Reserves and Related Items	272.6	...	...	...	...	...	...	Rés. et postes assimilables
Libyan Arab Jamah.								**Jamah. arabe libyenne**
Merchandise: Exp. fob	6 292.0	5 653.0	7 274.0	11 352.0	...	...	...	Marchandises : exp. fob
Merchandise: Imp. fob	−5 820.0	−5 762.0	−6 509.0	−7 575.0	...	...	...	Marchandises : imp. fob
Serv. & Income: Credit	912.0	891.0	565.0	783.0	...	...	...	Serv. & revenu : crédit
Serv. & Income: Debit	−1 944.0	−2 074.0	−1 869.0	−1 878.0	...	...	...	Serv. & revenu : débit
Private Unrequited Transfers	−508.0	−497.0	−472.0	−446.0	...	...	...	Transf. priv.sans contrep.
Offic. Unrequited Trans.,nie	−60.0	−37.0	−16.0	−35.0	...	...	...	Tr. offic. sans contrep.nia
Direct Investment, nie	−230.0	42.0	90.0	54.0	...	...	...	Investissements direct, nia
Portfolio Investments, nie	−3 213.0	−222.0	−52.0	−115.0	...	...	...	Invest. portefeuille, nia
Other Capital, nie	3 305.0	343.0	1 150.0	−945.0	...	...	...	Autres capitaux, nia
Net Errors and Omissions	184.0	271.0	130.0	−37.0	...	...	...	Erreurs et omissions nettes
Reserves and Related Items	1 082.0	1 392.0	−292.0	−1 158.0	...	...	...	Rés. et postes assimilables
Madagascar								**Madagascar**
Merchandise: Exp. fob	327.0	284.0	321.0	319.0	338.0	328.0	...	Marchandises : exp. fob
Merchandise: Imp. fob	−315.0	−319.0	−320.0	−566.0	−440.0	−466.0	...	Marchandises : imp. fob
Serv. & Income: Credit	105.0	132.0	161.0	209.0	151.0	177.0	...	Serv. & revenu : crédit
Serv. & Income: Debit	−410.0	−443.0	−437.0	−450.0	−416.0	−411.0	...	Serv. & revenu : débit
Private Unrequited Transfers	34.0	38.0	42.0	49.0	52.0	88.0	...	Transf. priv.sans contrep.
Offic. Unrequited Trans.,nie	120.0	158.0	161.0	188.0	127.0	148.0	...	Tr. offic. sans contrep.nia
Direct Investment, nie	0.0	0.0	13.0	22.0	14.0	21.0	...	Investissements direct, nia
Portfolio Investments, nie	0.0	0.0	0.0	0.0	0.0	0.0	...	Invest. portefeuille, nia
Other Capital, nie	−11.0	−22.0	−68.0	−40.0	−56.0	−109.0	...	Autres capitaux, nia
Net Errors and Omissions	−12.0	53.0	−46.0	−9.0	−4.0	−52.0	...	Erreurs et omissions nettes
Reserves and Related Items	163.0	118.0	174.0	278.0	235.0	276.0	...	Rés. et postes assimilables

78
Summary of balance of payments
Millions of US dollars
Résumé des balances des paiements
Millions de dollars des E−U

Country or area	1987	1988	1989	1990	1991	1992	1993	Pays ou zone
Malawi								**Malawi**
Merchandise: Exp. fob	278.5	297.0	...	...	...	...	...	Marchandises : exp. fob
Merchandise: Imp. fob	−177.6	−253.0	...	...	...	...	...	Marchandises : imp. fob
Serv. & Income: Credit	43.7	37.8	...	...	...	...	...	Serv. & revenu : crédit
Serv. & Income: Debit	−243.8	−230.3	...	...	...	...	...	Serv. & revenu : débit
Private Unrequited Transfers	13.8	15.1	...	...	...	...	...	Transf. priv.sans contrep.
Offic. Unrequited Trans.,nie	30.1	80.4	...	...	...	...	...	Tr. offic. sans contrep.nia
Direct Investment, nie	0.1	0.0	...	...	...	...	...	Investissements direct, nia
Portfolio Investments, nie	4.2	0.8	...	...	...	...	...	Invest. portefeuille, nia
Other Capital, nie	72.3	130.6	...	...	...	...	...	Autres capitaux, nia
Net Errors and Omissions	24.2	−18.0	...	...	...	...	...	Erreurs et omissions nettes
Reserves and Related Items	−45.4	−60.2	...	...	...	...	...	Rés. et postes assimilables
Mali								**Mali**
Merchandise: Exp. fob	255.9	251.5	269.3	337.9	354.5	339.3	343.6	Marchandises : exp. fob
Merchandise: Imp. fob	−335.4	−359.1	−338.8	−432.4	−447.1	−484.0	−463.5	Marchandises : imp. fob
Serv. & Income: Credit	85.6	88.0	75.5	99.5	106.3	99.0	102.8	Serv. & revenu : crédit
Serv. & Income: Debit	−355.1	−372.1	−353.7	−456.0	−437.6	−461.3	−441.4	Serv. & revenu : débit
Private Unrequited Transfers	36.6	45.3	51.4	66.9	70.0	88.0	84.8	Transf. priv.sans contrep.
Offic. Unrequited Trans.,nie	210.0	256.8	216.6	250.8	312.9	312.8	271.2	Tr. offic. sans contrep.nia
Direct Investment, nie	−6.0	0.7	15.0	−6.6	3.5	−7.6	0.0	Investissements direct, nia
Portfolio Investments, nie	0.0	0.0	0.0	0.0	0.0	0.0	0.0	Invest. portefeuille, nia
Other Capital, nie	92.6	138.1	81.8	84.3	49.3	10.6	−5.7	Autres capitaux, nia
Net Errors and Omissions	1.5	−1.4	−1.5	1.1	29.0	−22.7	22.7	Erreurs et omissions nettes
Reserves and Related Items	14.3	−47.7	−15.8	54.6	−40.7	125.8	85.5	Rés. et postes assimilables
Mauritania								**Mauritanie**
Merchandise: Exp. fob	402.4	437.6	447.9	443.9	435.8	406.8	...	Marchandises : exp. fob
Merchandise: Imp. fob	−359.2	−348.9	−349.3	−382.9	−399.1	−461.3	...	Marchandises : imp. fob
Serv. & Income: Credit	37.0	39.4	39.2	30.6	33.2	21.2	...	Serv. & revenu : crédit
Serv. & Income: Debit	−306.3	−306.7	−251.7	−187.1	−185.8	−208.9	...	Serv. & revenu : débit
Private Unrequited Transfers	−20.6	−22.1	−25.0	−15.6	−17.2	24.5	...	Transf. priv.sans contrep.
Offic. Unrequited Trans.,nie	99.2	104.6	120.3	101.5	103.3	100.6	...	Tr. offic. sans contrep.nia
Direct Investment, nie	1.4	1.0	3.5	6.7	2.3	5.0	...	Investissements direct, nia
Portfolio Investments, nie	0.0	0.0	0.0	0.0	0.0	0.0	...	Invest. portefeuille, nia
Other Capital, nie	90.1	38.5	13.4	−7.2	24.4	70.4	...	Autres capitaux, nia
Net Errors and Omissions	−101.5	−16.0	−3.6	−62.3	19.5	58.8	...	Erreurs et omissions nettes
Reserves and Related Items	157.3	72.6	5.3	72.5	−16.3	−17.0	...	Rés. et postes assimilables
Mauritius								**Maurice**
Merchandise: Exp. fob	902.3	1 000.5	986.5	1 205.2	1 215.1	1 302.6	1 304.4	Marchandises : exp. fob
Merchandise: Imp. fob	−918.7	−1 166.1	−1 196.0	−1 474.8	−1 419.1	−1 473.4	−1 558.6	Marchandises : imp. fob
Serv. & Income: Credit	336.4	404.8	451.0	572.6	649.2	700.6	666.3	Serv. & revenu : crédit
Serv. & Income: Debit	−322.8	−393.3	−420.0	−519.7	−544.7	−623.3	−605.5	Serv. & revenu : débit
Private Unrequited Transfers	41.1	71.6	67.6	82.3	79.5	86.6	91.7	Transf. priv.sans contrep.
Offic. Unrequited Trans.,nie	25.4	21.2	7.2	14.5	1.9	5.4	8.3	Tr. offic. sans contrep.nia
Direct Investment, nie	17.3	23.6	34.9	40.4	8.1	−28.6	−18.5	Investissements direct, nia
Portfolio Investments, nie	0.0	0.0	0.0	−2.2	−0.4	0.0	−2.2	Invest. portefeuille, nia
Other Capital, nie	42.7	101.5	14.9	100.3	34.2	13.8	39.9	Autres capitaux, nia
Net Errors and Omissions	97.1	121.6	198.5	213.2	167.2	59.7	81.2	Erreurs et omissions nettes
Reserves and Related Items	−220.8	−185.4	−144.7	−231.9	−190.8	−43.3	−7.0	Rés. et postes assimilables
Morocco								**Maroc**
Merchandise: Exp. fob	2 781.0	3 608.0	3 312.0	4 210.0	4 277.0	3 956.0	3 682.0	Marchandises : exp. fob
Merchandise: Imp. fob	−3 850.0	−4 360.0	−4 992.0	−6 282.0	−6 253.0	−6 692.0	−6 062.0	Marchandises : imp. fob
Serv. & Income: Credit	1 411.0	1 798.0	1 700.0	2 111.0	1 975.0	2 643.0	2 609.0	Serv. & revenu : crédit
Serv. & Income: Debit	−1 938.0	−2 185.0	−2 431.0	−2 572.0	−2 688.0	−2 873.0	−3 045.0	Serv. & revenu : débit
Private Unrequited Transfers	1 579.0	1 303.0	1 356.0	2 012.0	2 013.0	2 179.0	2 138.0	Transf. priv.sans contrep.
Offic. Unrequited Trans.,nie	191.0	303.0	265.0	320.0	280.0	360.0	154.0	Tr. offic. sans contrep.nia
Direct Investment, nie	60.0	85.0	167.0	165.0	320.0	424.0	493.0	Investissements direct, nia
Portfolio Investments, nie	0.0	0.0	0.0	0.0	0.0	0.0	0.0	Invest. portefeuille, nia
Other Capital, nie	24.0	−311.0	655.0	1 723.0	1 155.0	947.0	473.0	Autres capitaux, nia
Net Errors and Omissions	38.0	22.0	−11.0	9.0	88.0	2.0	−372.0	Erreurs et omissions nettes
Reserves and Related Items	−297.0	−263.0	−21.0	−1 697.0	−1 167.0	−946.0	−70.0	Rés. et postes assimilables
Mozambique								**Mozambique**
Merchandise: Exp. fob	97.0	103.0	105.0	126.0	162.0	139.0	...	Marchandises : exp. fob
Merchandise: Imp. fob	−578.0	−662.0	−727.0	−790.0	−809.0	−799.0	...	Marchandises : imp. fob
Serv. & Income: Credit	137.0	157.0	167.0	173.0	203.0	223.0	...	Serv. & revenu : crédit
Serv. & Income: Debit	−349.0	−333.0	−392.0	−374.0	−402.0	−444.0	...	Serv. & revenu : débit
Private Unrequited Transfers	0.0	0.0	0.0	0.0	0.0	0.0	...	Transf. priv.sans contrep.

78
Summary of balance of payments
Millions of US dollars
Résumé des balances des paiements
Millions de dollars des E−U

Country or area	1987	1988	1989	1990	1991	1992	1993	Pays ou zone
Offic. Unrequited Trans.,nie	304.0	377.0	387.0	448.0	502.0	499.0	...	Tr. offic. sans contrep.nia
Direct Investment, nie	6.0	5.0	3.0	9.0	23.0	25.0	...	Investissements direct, nia
Portfolio Investments, nie	0.0	0.0	0.0	0.0	0.0	0.0	...	Invest. portefeuille, nia
Other Capital, nie	−98.0	−131.0	−58.0	−93.0	−210.0	−148.0	...	Autres capitaux, nia
Net Errors and Omissions	40.0	85.0	57.0	66.0	−4.0	32.0	...	Erreurs et omissions nettes
Reserves and Related Items	441.0	400.0	458.0	433.0	536.0	472.0	...	Rés. et postes assimilables
Namibia								**Namibie**
Merchandise: Exp. fob	...	...	...	1 101.0	1 252.0	1 288.0	...	Marchandises : exp. fob
Merchandise: Imp. fob	...	...	...	−1 117.0	−1 108.0	−1 177.0	...	Marchandises : imp. fob
Serv. & Income: Credit	...	...	...	308.0	378.0	346.0	...	Serv. & revenu : crédit
Serv. & Income: Debit	...	...	...	−547.0	−605.0	−619.0	...	Serv. & revenu : débit
Private Unrequited Transfers	...	...	...	25.0	24.0	24.0	...	Transf. priv.sans contrep.
Offic. Unrequited Trans.,nie	...	...	...	192.0	311.0	280.0	...	Tr. offic. sans contrep.nia
Direct Investment, nie	...	...	...	36.0	100.0	53.0	...	Investissements direct, nia
Portfolio Investments, nie	...	...	...	11.0	−26.0	4.0	...	Invest. portefeuille, nia
Other Capital, nie	...	...	...	−244.0	−285.0	−173.0	...	Autres capitaux, nia
Net Errors and Omissions	...	...	...	275.0	−54.0	−33.0	...	Erreurs et omissions nettes
Reserves and Related Items	...	...	...	−39.0	14.0	7.0	...	Rés. et postes assimilables
Niger								**Niger**
Merchandise: Exp. fob	411.9	369.0	311.0	303.4	283.9	265.6	238.4	Marchandises : exp. fob
Merchandise: Imp. fob	−409.6	−392.5	−368.6	−337.5	−273.3	−266.3	−244.0	Marchandises : imp. fob
Serv. & Income: Credit	60.6	50.0	57.7	71.3	58.5	52.1	49.8	Serv. & revenu : crédit
Serv. & Income: Debit	−245.2	−218.6	−202.5	−256.4	−195.7	−185.5	−175.5	Serv. & revenu : débit
Private Unrequited Transfers	−49.9	−34.9	−40.4	−48.8	−37.9	−39.7	−33.9	Transf. priv.sans contrep.
Offic. Unrequited Trans.,nie	142.1	143.7	132.0	184.4	157.4	129.2	136.3	Tr. offic. sans contrep.nia
Direct Investment, nie	0.0	0.0	0.0	0.0	0.0	0.0	0.0	Investissements direct, nia
Portfolio Investments, nie	0.0	0.0	0.0	0.0	0.0	0.0	0.0	Invest. portefeuille, nia
Other Capital, nie	71.7	26.1	35.7	22.8	−22.3	12.5	−25.0	Autres capitaux, nia
Net Errors and Omissions	−15.4	43.4	−4.1	−25.2	−40.4	15.6	−9.4	Erreurs et omissions nettes
Reserves and Related Items	33.9	13.7	79.3	86.2	69.8	16.5	63.4	Rés. et postes assimilables
Nigeria								**Nigéria**
Merchandise: Exp. fob	7 545.0	6 897.0	7 870.0	13 585.0	12 254.0	11 791.0	...	Marchandises : exp. fob
Merchandise: Imp. fob	−4 097.0	−4 271.0	−3 692.0	−4 932.0	−7 813.0	−7 181.0	...	Marchandises : imp. fob
Serv. & Income: Credit	270.0	404.0	704.0	1 176.0	1 097.0	1 208.0	...	Serv. & revenu : crédit
Serv. & Income: Debit	−3 762.0	−3 212.0	−3 918.0	−4 926.0	−5 080.0	−4 304.0	...	Serv. & revenu : débit
Private Unrequited Transfers	−19.0	−33.0	−19.0	1.0	12.0	22.0	...	Transf. priv.sans contrep.
Offic. Unrequited Trans.,nie	−5.0	21.0	145.0	84.0	732.0	731.0	...	Tr. offic. sans contrep.nia
Direct Investment, nie	603.0	377.0	1 882.0	588.0	712.0	897.0	...	Investissements direct, nia
Portfolio Investments, nie	−535.0	−65.0	−220.0	−197.0	−61.0	1 884.0	...	Invest. portefeuille, nia
Other Capital, nie	−4 188.0	−4 857.0	−5 307.0	−4 573.0	−3 284.0	−10 565.0	...	Autres capitaux, nia
Net Errors and Omissions	−306.0	−215.0	−110.0	235.0	−92.0	−122.0	...	Erreurs et omissions nettes
Reserves and Related Items	4 495.0	4 954.0	2 664.0	−1 042.0	1 523.0	5 638.0	...	Rés. et postes assimilables
Rwanda								**Rwanda**
Merchandise: Exp. fob	121.4	117.9	104.7	102.6	95.6	68.5	...	Marchandises : exp. fob
Merchandise: Imp. fob	−267.0	−278.6	−254.1	−227.7	−228.1	−240.4	...	Marchandises : imp. fob
Serv. & Income: Credit	56.4	56.9	52.3	46.6	46.5	36.1	...	Serv. & revenu : crédit
Serv. & Income: Debit	−171.3	−165.0	−142.0	−152.0	−128.8	−132.0	...	Serv. & revenu : débit
Private Unrequited Transfers	7.5	10.5	7.9	5.8	20.9	22.1	...	Transf. priv.sans contrep.
Offic. Unrequited Trans.,nie	118.6	113.8	108.7	115.8	159.8	161.1	...	Tr. offic. sans contrep.nia
Direct Investment, nie	17.5	21.0	15.5	7.7	4.6	2.2	...	Investissements direct, nia
Portfolio Investments, nie	0.0	0.0	0.0	−0.3	−0.1	0.0	...	Invest. portefeuille, nia
Other Capital, nie	104.7	72.7	38.4	48.3	94.6	60.2	...	Autres capitaux, nia
Net Errors and Omissions	1.5	0.4	1.9	30.3	0.2	18.2	...	Erreurs et omissions nettes
Reserves and Related Items	10.6	50.4	66.7	22.9	−65.2	4.0	...	Rés. et postes assimilables
Sao Tome and Principe								**Sao Tomé−et−Principe**
Merchandise: Exp. fob	6.5	9.5	4.9	4.2	...	...	...	Marchandises : exp. fob
Merchandise: Imp. fob	−13.6	−14.1	−13.3	−13.0	...	...	...	Marchandises : imp. fob
Serv. & Income: Credit	1.7	2.0	4.6	3.9	...	...	...	Serv. & revenu : crédit
Serv. & Income: Debit	−9.3	−9.0	−8.8	−9.3	...	...	...	Serv. & revenu : débit
Private Unrequited Transfers	−0.3	0.0	−0.2	0.1	...	...	...	Transf. priv.sans contrep.
Offic. Unrequited Trans.,nie	1.9	0.9	1.5	−0.2	...	...	...	Tr. offic. sans contrep.nia
Direct Investment, nie	0.0	0.2	0.0	0.0	...	...	...	Investissements direct, nia
Portfolio Investments, nie	0.0	0.0	0.0	0.0	...	...	...	Invest. portefeuille, nia
Other Capital, nie	8.9	6.1	6.7	7.6	...	...	...	Autres capitaux, nia
Net Errors and Omissions	0.0	0.0	−1.0	−2.7	...	...	...	Erreurs et omissions nettes
Reserves and Related Items	4.1	4.4	5.6	9.4	...	...	...	Rés. et postes assimilables

78
Summary of balance of payments
Millions of US dollars
Résumé des balances des paiements
Millions de dollars des E−U

Country or area	1987	1988	1989	1990	1991	1992	1993	Pays ou zone
Senegal								**Sénégal**
Merchandise: Exp. fob	670.9	678.6	758.6	911.6	824.2	831.9	722.6	Marchandises : exp. fob
Merchandise: Imp. fob	−955.8	−956.0	−998.4	−1 176.1	−1 114.1	−1 200.3	−1 105.4	Marchandises : imp. fob
Serv. & Income: Credit	453.2	491.3	498.0	585.8	584.5	618.5	588.4	Serv. & revenu : crédit
Serv. & Income: Debit	−708.9	−752.0	−722.5	−831.5	−793.3	−834.9	−790.4	Serv. & revenu : débit
Private Unrequited Transfers	6.3	5.9	6.3	29.4	28.4	36.6	40.3	Transf. priv.sans contrep.
Offic. Unrequited Trans.,nie	227.7	270.7	259.7	265.2	265.1	279.9	239.8	Tr. offic. sans contrep.nia
Direct Investment, nie	−2.0	0.7	0.0	0.0	0.0	0.0	0.0	Investissements direct, nia
Portfolio Investments, nie	0.7	1.1	0.0	0.0	0.0	0.0	0.0	Invest. portefeuille, nia
Other Capital, nie	210.6	148.0	33.4	4.2	51.0	79.0	42.2	Autres capitaux, nia
Net Errors and Omissions	−0.2	−3.3	1.6	−16.7	−26.2	82.6	114.8	Erreurs et omissions nettes
Reserves and Related Items	97.4	114.8	163.3	228.1	180.4	106.7	147.8	Rés. et postes assimilables
Seychelles								**Seychelles**
Merchandise: Exp. fob	8.1	17.3	14.5	28.1	18.8	19.6	...	Marchandises : exp. fob
Merchandise: Imp. fob	−96.3	−135.0	−139.6	−158.4	−146.3	−162.9	...	Marchandises : imp. fob
Serv. & Income: Credit	147.6	167.5	189.9	233.2	239.7	251.8	...	Serv. & revenu : crédit
Serv. & Income: Debit	−101.7	−102.1	−114.1	−130.1	−131.3	−131.2	...	Serv. & revenu : débit
Private Unrequited Transfers	−2.3	−4.9	−5.2	−2.5	−2.5	−2.8	...	Transf. priv.sans contrep.
Offic. Unrequited Trans.,nie	23.5	28.9	31.7	29.3	25.3	23.8	...	Tr. offic. sans contrep.nia
Direct Investment, nie	14.0	18.9	16.6	21.4	18.9	18.6	...	Investissements direct, nia
Portfolio Investments, nie	0.0	0.0	0.0	0.0	0.0	0.0	...	Invest. portefeuille, nia
Other Capital, nie	4.9	2.4	17.9	−7.0	9.3	−6.2	...	Autres capitaux, nia
Net Errors and Omissions	6.2	2.8	−8.2	−10.0	−21.5	−6.9	...	Erreurs et omissions nettes
Reserves and Related Items	−4.0	4.3	−3.5	−4.0	−10.5	−3.9	...	Rés. et postes assimilables
Sierra Leone								**Sierra Leone**
Merchandise: Exp. fob	138.9	104.5	139.5	139.8	149.5	...	...	Marchandises : exp. fob
Merchandise: Imp. fob	−114.8	−138.2	−160.4	−140.4	−138.3	...	...	Marchandises : imp. fob
Serv. & Income: Credit	44.1	52.1	38.5	70.5	68.4	...	...	Serv. & revenu : crédit
Serv. & Income: Debit	−105.2	−29.9	−84.6	−146.2	−78.7	...	...	Serv. & revenu : débit
Private Unrequited Transfers	0.0	0.3	0.1	0.1	2.7	...	...	Transf. priv.sans contrep.
Offic. Unrequited Trans.,nie	6.8	8.5	7.2	6.8	7.1	...	...	Tr. offic. sans contrep.nia
Direct Investment, nie	39.4	−23.1	22.4	32.4	7.5	...	...	Investissements direct, nia
Portfolio Investments, nie	0.0	0.0	0.0	0.0	0.0	...	...	Invest. portefeuille, nia
Other Capital, nie	−40.6	16.4	−40.3	−33.2	−9.8	...	...	Autres capitaux, nia
Net Errors and Omissions	−21.9	−62.5	29.2	49.2	11.1	...	...	Erreurs et omissions nettes
Reserves and Related Items	53.4	71.9	48.4	20.9	−19.4	...	...	Rés. et postes assimilables
Somalia								**Somalie**
Merchandise: Exp. fob	94.0	58.4	67.7	...	...	...	...	Marchandises : exp. fob
Merchandise: Imp. fob	−358.5	−216.0	−346.3	...	...	...	...	Marchandises : imp. fob
Serv. & Income: Credit	0.0	0.0	0.0	...	...	...	...	Serv. & revenu : crédit
Serv. & Income: Debit	−179.7	−164.6	−206.4	...	...	...	...	Serv. & revenu : débit
Private Unrequited Transfers	−13.1	6.4	−2.9	...	...	...	...	Transf. priv.sans contrep.
Offic. Unrequited Trans.,nie	343.3	217.3	331.2	...	...	...	...	Tr. offic. sans contrep.nia
Direct Investment, nie	0.0	0.0	0.0	...	...	...	...	Investissements direct, nia
Portfolio Investments, nie	0.0	0.0	0.0	...	...	...	...	Invest. portefeuille, nia
Other Capital, nie	−24.6	−103.3	−31.9	...	...	...	...	Autres capitaux, nia
Net Errors and Omissions	40.2	21.7	−0.8	...	...	...	...	Erreurs et omissions nettes
Reserves and Related Items	98.4	180.2	189.4	...	...	...	...	Rés. et postes assimilables
South Africa								**Afrique du Sud**
Merchandise: Exp. fob	21 088.0	22 432.0	22 399.0	23 560.0	23 289.0	23 645.0	24 068.0	Marchandises : exp. fob
Merchandise: Imp. fob	−13 925.0	−17 210.0	−16 810.0	−16 778.0	−17 156.0	−18 216.0	−18 287.0	Marchandises : imp. fob
Serv. & Income: Credit	3 110.0	3 244.0	3 544.0	4 391.0	4 487.0	4 669.0	4 443.0	Serv. & revenu : crédit
Serv. & Income: Debit	−7 549.0	−7 424.0	−7 755.0	−9 168.0	−8 433.0	−8 816.0	−8 549.0	Serv. & revenu : débit
Private Unrequited Transfers	97.0	90.0	129.0	−33.0	−28.0	32.0	68.0	Transf. priv.sans contrep.
Offic. Unrequited Trans.,nie	116.0	86.0	72.0	104.0	100.0	74.0	62.0	Tr. offic. sans contrep.nia
Direct Investment, nie	−163.0	98.0	10.0	−5.0	−8.0	−5.0	−8.0	Investissements direct, nia
Portfolio Investments, nie	−181.0	−54.0	−138.0	−50.0	78.0	1 496.0	225.0	Invest. portefeuille, nia
Other Capital, nie	−1 011.0	−1 683.0	−989.0	398.0	406.0	−1 305.0	−1 881.0	Autres capitaux, nia
Net Errors and Omissions	−220.0	−965.0	−575.0	−1 016.0	−1 228.0	−1 443.0	−2 936.0	Erreurs et omissions nettes
Reserves and Related Items	−1 361.0	1 386.0	113.0	−1 405.0	−1 506.0	−131.0	2 795.0	Rés. et postes assimilables
Sudan								**Soudan**
Merchandise: Exp. fob	265.0	427.0	544.4	326.5	302.5	213.4	...	Marchandises : exp. fob
Merchandise: Imp. fob	−694.8	−948.5	−1 051.0	−648.8	−1 138.2	−810.2	...	Marchandises : imp. fob
Serv. & Income: Credit	192.2	171.6	279.7	184.9	79.7	155.5	...	Serv. & revenu : crédit
Serv. & Income: Debit	−323.1	−341.3	−495.7	−376.0	−326.4	−297.6	...	Serv. & revenu : débit
Private Unrequited Transfers	133.7	216.3	412.4	59.8	45.2	123.7	...	Transf. priv.sans contrep.

78
Summary of balance of payments
Millions of US dollars
Résumé des balances des paiements
Millions de dollars des E−U

Country or area	1987	1988	1989	1990	1991	1992	1993	Pays ou zone
Offic. Unrequited Trans.,nie	194.6	117.0	160.0	81.4	82.5	109.0	...	Tr. offic. sans contrep.nia
Direct Investment, nie	0.0	0.0	3.5	0.0	0.0	0.0	...	Investissements direct, nia
Portfolio Investments, nie	0.0	0.0	0.0	0.0	0.0	0.0	...	Invest. portefeuille, nia
Other Capital, nie	85.8	67.5	114.4	116.9	584.1	316.4	...	Autres capitaux, nia
Net Errors and Omissions	−195.3	3.3	−160.0	9.3	97.8	31.0	...	Erreurs et omissions nettes
Reserves and Related Items	341.8	287.2	192.4	246.0	272.9	158.8	...	Rés. et postes assimilables
Swaziland								**Swaziland**
Merchandise: Exp. fob	423.6	466.2	493.8	556.6	596.6	637.7	649.8	Marchandises : exp. fob
Merchandise: Imp. fob	−369.4	−441.0	−515.4	−586.5	−632.3	−764.8	−774.4	Marchandises : imp. fob
Serv. & Income: Credit	148.2	184.6	216.8	273.1	266.9	268.4	225.0	Serv. & revenu : crédit
Serv. & Income: Debit	−189.1	−217.3	−278.7	−292.7	−295.5	−259.9	−252.3	Serv. & revenu : débit
Private Unrequited Transfers	2.8	4.2	2.4	0.0	−1.3	2.2	0.4	Transf. priv.sans contrep.
Offic. Unrequited Trans.,nie	46.1	68.6	85.5	98.3	90.8	121.5	114.4	Tr. offic. sans contrep.nia
Direct Investment, nie	54.2	38.3	57.3	31.2	46.5	47.2	29.0	Investissements direct, nia
Portfolio Investments, nie	1.2	6.1	6.9	−8.9	−0.5	0.0	−1.1	Invest. portefeuille, nia
Other Capital, nie	−64.9	−102.8	−76.8	−60.4	−37.9	−12.5	−64.8	Autres capitaux, nia
Net Errors and Omissions	−31.2	7.6	59.6	0.4	−20.0	51.8	26.3	Erreurs et omissions nettes
Reserves and Related Items	−21.5	−14.6	−51.3	−11.1	−13.4	−91.7	47.6	Rés. et postes assimilables
Togo								**Togo**
Merchandise: Exp. fob	397.5	435.3	411.7	395.2	393.1	322.3	214.7	Marchandises : exp. fob
Merchandise: Imp. fob	−437.1	−504.5	−470.1	−513.1	−452.3	−417.8	−248.6	Marchandises : imp. fob
Serv. & Income: Credit	137.7	126.0	152.0	180.7	172.3	160.6	88.1	Serv. & revenu : crédit
Serv. & Income: Debit	−262.7	−270.7	−259.7	−288.0	−273.3	−261.8	−193.2	Serv. & revenu : débit
Private Unrequited Transfers	7.0	6.9	9.2	9.9	14.2	8.7	9.5	Transf. priv.sans contrep.
Offic. Unrequited Trans.,nie	97.1	119.7	106.0	113.9	91.5	81.2	31.1	Tr. offic. sans contrep.nia
Direct Investment, nie	7.4	13.0	7.4	0.0	0.0	0.0	0.0	Investissements direct, nia
Portfolio Investments, nie	1.4	0.7	0.0	0.0	0.0	0.0	0.0	Invest. portefeuille, nia
Other Capital, nie	−63.0	19.1	14.1	62.2	46.4	10.6	−46.9	Autres capitaux, nia
Net Errors and Omissions	−15.7	−31.7	1.8	10.8	−33.7	−62.1	−42.2	Erreurs et omissions nettes
Reserves and Related Items	130.4	86.2	27.6	28.4	41.9	158.5	187.5	Rés. et postes assimilables
Tunisia								**Tunisie**
Merchandise: Exp. fob	2 101.0	2 399.0	2 931.0	3 515.0	3 696.0	4 014.0	3 804.0	Marchandises : exp. fob
Merchandise: Imp. fob	−2 829.0	−3 496.0	−4 139.0	−5 193.0	−4 895.0	−6 077.0	−5 872.0	Marchandises : imp. fob
Serv. & Income: Credit	1 267.0	1 905.0	1 618.0	1 782.0	1 473.0	2 083.0	2 138.0	Serv. & revenu : crédit
Serv. & Income: Debit	−1 119.0	−1 256.0	−1 230.0	−1 396.0	−1 448.0	−1 643.0	−1 688.0	Serv. & revenu : débit
Private Unrequited Transfers	481.0	548.0	485.0	591.0	574.0	570.0	595.0	Transf. priv.sans contrep.
Offic. Unrequited Trans.,nie	37.0	115.0	213.0	225.0	131.0	88.0	112.0	Tr. offic. sans contrep.nia
Direct Investment, nie	91.0	63.0	74.0	77.0	122.0	364.0	239.0	Investissements direct, nia
Portfolio Investments, nie	8.0	5.0	−6.0	2.0	21.0	46.0	13.0	Invest. portefeuille, nia
Other Capital, nie	39.0	131.0	126.0	302.0	195.0	648.0	651.0	Autres capitaux, nia
Net Errors and Omissions	48.0	27.0	−5.0	−28.0	77.0	5.0	16.0	Erreurs et omissions nettes
Reserves and Related Items	−126.0	−441.0	−65.0	123.0	55.0	−97.0	−7.0	Rés. et postes assimilables
Uganda								**Ouganda**
Merchandise: Exp. fob	333.6	266.3	277.7	177.8	173.2	151.2	196.7	Marchandises : exp. fob
Merchandise: Imp. fob	−475.6	−523.5	−588.3	−491.0	−377.1	−421.9	−474.7	Marchandises : imp. fob
Serv. & Income: Credit	0.0	0.0	0.0	0.0	23.6	38.6	100.0	Serv. & revenu : crédit
Serv. & Income: Debit	−236.2	−260.4	−260.5	−243.1	−318.5	−336.0	−354.3	Serv. & revenu : débit
Private Unrequited Transfers	0.0	0.0	0.0	0.0	103.4	207.3	163.3	Transf. priv.sans contrep.
Offic. Unrequited Trans.,nie	266.2	322.4	311.6	293.0	225.6	261.2	261.7	Tr. offic. sans contrep.nia
Direct Investment, nie	0.0	0.0	0.0	0.0	1.0	3.0	3.4	Investissements direct, nia
Portfolio Investments, nie	0.0	0.0	0.0	0.0	0.0	0.0	0.0	Invest. portefeuille, nia
Other Capital, nie	31.2	3.6	213.0	211.8	136.6	111.8	156.6	Autres capitaux, nia
Net Errors and Omissions	26.4	154.9	−38.0	9.5	0.6	9.0	5.5	Erreurs et omissions nettes
Reserves and Related Items	54.4	36.7	84.5	41.9	31.7	−24.2	−58.1	Rés. et postes assimilables
United Rep.Tanzania								**Rép. Unie de Tanzanie**
Merchandise: Exp. fob	287.9	386.5	415.1	407.8	362.2	400.7	462.0	Marchandises : exp. fob
Merchandise: Imp. fob	−1 000.5	−1 033.0	−1 070.1	−1 186.4	−1 284.7	−1 313.6	−1 299.9	Marchandises : imp. fob
Serv. & Income: Credit	111.3	120.6	122.7	141.1	150.0	155.6	290.2	Serv. & revenu : crédit
Serv. & Income: Debit	−427.6	−471.1	−487.2	−481.1	−502.2	−569.6	−580.4	Serv. & revenu : débit
Private Unrequited Transfers	314.3	231.9	182.4	158.5	269.2	325.0	193.2	Transf. priv.sans contrep.
Offic. Unrequited Trans.,nie	268.7	389.3	469.8	535.1	554.2	580.0	526.5	Tr. offic. sans contrep.nia
Direct Investment, nie	0.0	0.0	0.0	0.0	0.0	12.0	20.0	Investissements direct, nia
Portfolio Investments, nie	0.0	0.0	0.0	0.0	0.0	0.0	0.0	Invest. portefeuille, nia
Other Capital, nie	60.4	33.9	21.7	126.5	108.1	76.9	55.0	Autres capitaux, nia
Net Errors and Omissions	93.7	−42.5	18.8	217.0	−20.2	44.6	−18.6	Erreurs et omissions nettes
Reserves and Related Items	292.0	384.4	326.8	81.6	363.4	288.4	352.1	Rés. et postes assimilables

78
Summary of balance of payments
Millions of US dollars
Résumé des balances des paiements
Millions de dollars des E−U

Country or area	1987	1988	1989	1990	1991	1992	1993	Pays ou zone
Zaire								**Zaïre**
Merchandise: Exp. fob	1 731.0	2 178.0	2 201.0	2 138.0	...	...	...	Marchandises : exp. fob
Merchandise: Imp. fob	−1 376.0	−1 645.0	−1 683.0	−1 539.0	...	...	...	Marchandises : imp. fob
Serv. & Income: Credit	262.0	185.0	165.0	171.0	...	...	...	Serv. & revenu : crédit
Serv. & Income: Debit	−1 412.0	−1 458.0	−1 461.0	−1 549.0	...	...	...	Serv. & revenu : débit
Private Unrequited Transfers	−70.0	−67.0	−109.0	−81.0			...	Transf. priv.sans contrep.
Offic. Unrequited Trans.,nie	220.0	226.0	276.0	217.0			...	Tr. offic. sans contrep.nia
Direct Investment, nie	0.0	0.0	0.0	0.0			...	Investissements direct, nia
Portfolio Investments, nie	0.0	0.0	0.0	0.0			...	Invest. portefeuille, nia
Other Capital, nie	70.0	−8.0	−60.0	−222.0			...	Autres capitaux, nia
Net Errors and Omissions	13.0	−134.0	113.0	105.0			...	Erreurs et omissions nettes
Reserves and Related Items	561.0	723.0	558.0	761.0	...	...	...	Rés. et postes assimilables
Zambia								**Zambie**
Merchandise: Exp. fob	852.0	1 189.0	1 340.0	1 254.0	1 172.0	...	...	Marchandises : exp. fob
Merchandise: Imp. fob	−585.0	−687.0	−774.0	−1 511.0	−752.0	...	...	Marchandises : imp. fob
Serv. & Income: Credit	49.0	61.0	87.0	108.0	93.0	...	...	Serv. & revenu : crédit
Serv. & Income: Debit	−552.0	−893.0	−953.0	−825.0	−1 059.0	...	...	Serv. & revenu : débit
Private Unrequited Transfers	−20.0	−25.0	−30.0	−18.0	−22.0		...	Transf. priv.sans contrep.
Offic. Unrequited Trans.,nie	8.0	59.0	109.0	395.0	261.0		...	Tr. offic. sans contrep.nia
Direct Investment, nie	75.0	93.0	164.0	203.0	34.0		...	Investissements direct, nia
Portfolio Investments, nie	0.0	0.0	0.0	0.0	0.0		...	Invest. portefeuille, nia
Other Capital, nie	−238.0	−70.0	1 664.0	285.0	−24.0		...	Autres capitaux, nia
Net Errors and Omissions	153.0	40.0	−1 712.0	322.0	110.0		...	Erreurs et omissions nettes
Reserves and Related Items	258.0	232.0	105.0	−213.0	187.0	...	...	Rés. et postes assimilables
Zimbabwe								**Zimbabwe**
Merchandise: Exp. fob	1 452.0	1 664.9	1 693.5	1 747.9	1 693.8	1 527.6	1 609.1	Marchandises : exp. fob
Merchandise: Imp. fob	−1 071.0	−1 163.6	−1 318.3	−1 505.2	−1 645.7	−1 782.1	−1 487.0	Marchandises : imp. fob
Serv. & Income: Credit	197.6	207.8	267.9	287.1	299.6	331.1	407.1	Serv. & revenu : crédit
Serv. & Income: Debit	−578.2	−644.9	−693.1	−781.8	−905.4	−963.2	−850.9	Serv. & revenu : débit
Private Unrequited Transfers	−32.1	−13.2	−19.2	−3.0	3.4	39.7	26.5	Transf. priv.sans contrep.
Offic. Unrequited Trans.,nie	79.7	65.7	78.5	108.1	94.6	241.8	179.0	Tr. offic. sans contrep.nia
Direct Investment, nie	−30.5	4.1	−10.2	−12.2	2.8	15.0	28.0	Investissements direct, nia
Portfolio Investments, nie	0.9	−60.9	−36.7	−21.7	7.3	−9.5	−5.1	Invest. portefeuille, nia
Other Capital, nie	89.5	104.8	94.7	276.5	526.4	367.9	304.3	Autres capitaux, nia
Net Errors and Omissions	16.6	−63.0	−103.8	−9.9	−31.4	37.2	14.9	Erreurs et omissions nettes
Reserves and Related Items	−124.5	−101.6	46.8	−85.8	−45.2	194.6	−225.9	Rés. et postes assimilables

America, North· Amérique du Nord

Country or area	1987	1988	1989	1990	1991	1992	1993	Pays ou zone
Antigua and Barbuda								**Antigua−et−Barbuda**
Merchandise: Exp. fob	16.8	17.0	15.7	19.0	35.4	54.7	...	Marchandises : exp. fob
Merchandise: Imp. fob	−218.9	−201.1	−242.3	−230.7	−252.6	−260.9	...	Marchandises : imp. fob
Serv. & Income: Credit	199.4	237.2	274.5	318.8	324.2	339.9	...	Serv. & revenu : crédit
Serv. & Income: Debit	−120.1	−108.4	−139.4	−155.7	−150.5	−148.6	...	Serv. & revenu : débit
Private Unrequited Transfers	9.2	9.4	9.0	7.9	6.6	4.9	...	Transf. priv.sans contrep.
Offic. Unrequited Trans.,nie	2.5	2.8	3.3	2.5	2.7	1.4	...	Tr. offic. sans contrep.nia
Direct Investment, nie	38.6	31.1	41.3	58.8	52.4	19.6	...	Investissements direct, nia
Portfolio Investments, nie	0.0	0.0	0.0	0.0	0.0	0.0	...	Invest. portefeuille, nia
Other Capital, nie	38.0	−6.1	−5.3	−44.5	−47.1	−15.3	...	Autres capitaux, nia
Net Errors and Omissions	14.6	0.2	4.2	−23.6	−4.7	−14.4	...	Erreurs et omissions nettes
Reserves and Related Items	19.9	17.9	39.0	47.5	33.5	18.9	...	Rés. et postes assimilables
Aruba								**Aruba**
Merchandise: Exp. fob	45.1	87.4	107.5	155.5	878.8	1 069.2	1 154.4	Marchandises : exp. fob
Merchandise: Imp. fob	−249.8	−354.6	−397.4	−580.8	−1 402.8	−1 446.7	−1 546.5	Marchandises : imp. fob
Serv. & Income: Credit	269.0	337.1	364.7	425.8	490.6	585.6	617.4	Serv. & revenu : crédit
Serv. & Income: Debit	−84.0	−113.2	−124.9	−157.5	−173.5	−181.6	−193.7	Serv. & revenu : débit
Private Unrequited Transfers	−2.9	−0.9	3.4	−8.9	−14.1	8.4	2.5	Transf. priv.sans contrep.
Offic. Unrequited Trans.,nie	0.0	0.0	0.0	7.8	8.6	7.4	5.8	Tr. offic. sans contrep.nia
Direct Investment, nie	0.0	0.0	0.0	130.5	184.7	−37.0	−17.9	Investissements direct, nia
Portfolio Investments, nie	0.0	0.0	0.0	−6.4	−12.2	−6.8	−3.9	Invest. portefeuille, nia
Other Capital, nie	24.6	56.8	47.8	48.2	56.3	19.7	13.4	Autres capitaux, nia
Net Errors and Omissions	10.9	−12.8	20.4	−2.4	6.5	4.4	2.0	Erreurs et omissions nettes
Reserves and Related Items	−12.9	0.4	−21.5	−11.7	−22.8	−22.6	−33.4	Rés. et postes assimilables
Bahamas								**Bahamas**
Merchandise: Exp. fob	273.1	270.5	259.2	307.6	319.8	310.2	256.8	Marchandises : exp. fob
Merchandise: Imp. fob	−998.1	−982.9	−1 136.7	−1 190.2	−1 045.6	−1 069.2	−1 080.9	Marchandises : imp. fob
Serv. & Income: Credit	1 359.4	1 354.8	1 531.3	1 570.3	1 395.4	1 420.7	1 479.3	Serv. & revenu : crédit
Serv. & Income: Debit	−663.3	−673.4	−715.1	−773.6	−775.3	−708.8	−740.4	Serv. & revenu : débit

78
Summary of balance of payments
Millions of US dollars
Résumé des balances des paiements
Millions de dollars des E−U

Country or area	1987	1988	1989	1990	1991	1992	1993	Pays ou zone
Private Unrequited Transfers	−17.8	−28.9	−17.9	−10.6	−7.8	−12.8	−12.6	Transf. priv.sans contrep.
Offic. Unrequited Trans.,nie	14.2	14.4	18.9	21.2	27.4	26.2	25.3	Tr. offic. sans contrep.nia
Direct Investment, nie	10.8	36.7	25.0	−17.2	0.0	7.4	−24.1	Investissements direct, nia
Portfolio Investments, nie	0.0	0.0	0.0	0.0	0.0	0.0	0.0	Invest. portefeuille, nia
Other Capital, nie	−25.7	36.5	69.8	74.3	176.9	5.5	21.4	Autres capitaux, nia
Net Errors and Omissions	−12.5	−28.4	−61.1	27.5	−77.8	−7.9	94.2	Erreurs et omissions nettes
Reserves and Related Items	59.9	0.7	26.6	−9.3	−13.0	28.7	−19.0	Rés. et postes assimilables
Barbados								**Barbade**
Merchandise: Exp. fob	131.4	144.8	147.0	151.0	143.6	158.0	152.4	Marchandises : exp. fob
Merchandise: Imp. fob	−459.0	−518.7	−600.3	−624.1	−617.4	−464.7	−511.3	Marchandises : imp. fob
Serv. & Income: Credit	553.9	638.0	740.8	685.3	688.4	664.3	729.7	Serv. & revenu : crédit
Serv. & Income: Debit	−254.9	−238.2	−268.6	−271.2	−272.8	−254.2	−327.3	Serv. & revenu : débit
Private Unrequited Transfers	19.0	29.7	31.9	34.8	32.0	39.6	26.0	Transf. priv.sans contrep.
Offic. Unrequited Trans.,nie	−13.1	−13.2	−27.1	7.8	1.0	0.6	−5.2	Tr. offic. sans contrep.nia
Direct Investment, nie	4.6	10.5	5.4	9.7	6.0	13.5	6.8	Investissements direct, nia
Portfolio Investments, nie	−1.6	−0.2	−4.7	−6.0	−8.8	−4.1	−8.9	Invest. portefeuille, nia
Other Capital, nie	82.8	37.1	−23.1	44.3	20.6	−103.4	3.4	Autres capitaux, nia
Net Errors and Omissions	−56.0	−50.4	−44.1	−70.6	−32.6	−21.5	−44.7	Erreurs et omissions nettes
Reserves and Related Items	−7.2	−39.5	42.8	38.9	39.9	−28.3	−20.9	Rés. et postes assimilables
Belize								**Belize**
Merchandise: Exp. fob	102.8	119.4	124.4	129.2	126.1	140.6	132.0	Marchandises : exp. fob
Merchandise: Imp. fob	−126.9	−161.2	−188.5	−188.4	−223.6	−244.5	−250.5	Marchandises : imp. fob
Serv. & Income: Credit	61.8	82.4	95.5	125.9	131.0	149.3	156.4	Serv. & revenu : crédit
Serv. & Income: Debit	−59.3	−69.0	−81.6	−80.8	−87.3	−104.4	−115.9	Serv. & revenu : débit
Private Unrequited Transfers	15.2	15.2	20.7	16.3	15.4	17.6	15.4	Transf. priv.sans contrep.
Offic. Unrequited Trans.,nie	15.7	10.6	10.4	13.0	12.6	12.8	14.1	Tr. offic. sans contrep.nia
Direct Investment, nie	6.8	14.0	18.6	17.2	13.6	15.6	9.2	Investissements direct, nia
Portfolio Investments, nie	0.0	0.0	0.0	0.0	0.0	0.2	7.0	Invest. portefeuille, nia
Other Capital, nie	−4.2	13.3	6.8	7.9	8.6	6.6	16.6	Autres capitaux, nia
Net Errors and Omissions	−0.2	−2.9	9.1	−25.0	−12.8	6.3	1.5	Erreurs et omissions nettes
Reserves and Related Items	−11.9	−21.8	−15.5	−15.4	16.4	−0.1	14.2	Rés. et postes assimilables
Canada								**Canada**
Merchandise: Exp. fob	98 052.0	115 432.0	122 971.0	128 438.0	126 003.0	132 351.0	144 030.0	Marchandises : exp. fob
Merchandise: Imp. fob	−89 092.0	−107 274.0	−116 985.0	−120 108.0	−122 308.0	−126 370.0	−136 418.0	Marchandises : imp. fob
Serv. & Income: Credit	18 986.0	22 471.0	24 544.0	25 901.0	26 187.0	25 095.0	24 661.0	Serv. & revenu : crédit
Serv. & Income: Debit	−39 691.0	−47 980.0	−53 498.0	−55 727.0	−53 950.0	−53 232.0	−56 323.0	Serv. & revenu : débit
Private Unrequited Transfers	394.0	663.0	763.0	731.0	656.0	590.0	545.0	Transf. priv.sans contrep.
Offic. Unrequited Trans.,nie	−449.0	−416.0	−523.0	−783.0	−640.0	−494.0	−363.0	Tr. offic. sans contrep.nia
Direct Investment, nie	−501.0	2 571.0	442.0	3 130.0	−2 951.0	988.0	−1 233.0	Investissements direct, nia
Portfolio Investments, nie	8 615.0	8 277.0	14 676.0	8 568.0	15 379.0	8 407.0	21 034.0	Invest. portefeuille, nia
Other Capital, nie	8 868.0	14 377.0	7 416.0	11 876.0	11 452.0	5 462.0	8 903.0	Autres capitaux, nia
Net Errors and Omissions	−2 403.0	−561.0	486.0	−1 401.0	−2 314.0	1 397.0	−5 327.0	Erreurs et omissions nettes
Reserves and Related Items	−2 778.0	−7 558.0	−293.0	−625.0	2 486.0	5 807.0	492.0	Rés. et postes assimilables
Costa Rica								**Costa Rica**
Merchandise: Exp. fob	1 106.7	1 180.7	1 333.4	1 354.2	1 498.1	1 739.1	1 944.6	Marchandises : exp. fob
Merchandise: Imp. fob	−1 245.2	−1 278.6	−1 572.0	−1 796.7	−1 697.6	−2 210.9	−2 610.4	Marchandises : imp. fob
Serv. & Income: Credit	385.6	478.1	617.8	739.3	802.8	954.1	1 137.1	Serv. & revenu : crédit
Serv. & Income: Debit	−729.5	−814.1	−985.5	−912.7	−820.1	−1 026.0	−1 094.2	Serv. & revenu : débit
Private Unrequited Transfers	38.7	40.0	39.2	55.4	50.3	88.3	85.9	Transf. priv.sans contrep.
Offic. Unrequited Trans.,nie	67.3	90.4	87.2	66.5	67.3	75.0	67.0	Tr. offic. sans contrep.nia
Direct Investment, nie	75.8	121.4	95.2	160.4	172.8	221.6	275.0	Investissements direct, nia
Portfolio Investments, nie	0.0	−6.0	−13.2	−28.2	−13.0	−16.9	−5.1	Invest. portefeuille, nia
Other Capital, nie	−517.0	−387.3	−264.5	−254.5	−4.8	−5.3	65.3	Autres capitaux, nia
Net Errors and Omissions	131.2	224.6	208.9	56.4	99.9	201.9	19.7	Erreurs et omissions nettes
Reserves and Related Items	686.4	350.8	453.5	559.8	−155.7	−20.9	115.1	Rés. et postes assimilables
Dominica								**Dominique**
Merchandise: Exp. fob	49.3	57.0	46.3	56.1	55.6	54.6	48.3	Marchandises : exp. fob
Merchandise: Imp. fob	−58.9	−77.2	−94.4	−104.0	−96.5	−97.5	−98.8	Marchandises : imp. fob
Serv. & Income: Credit	19.8	24.9	27.6	35.1	40.0	44.9	49.3	Serv. & revenu : crédit
Serv. & Income: Debit	−27.7	−28.9	−34.7	−40.0	−42.5	−42.9	−41.4	Serv. & revenu : débit
Private Unrequited Transfers	7.6	8.4	11.7	12.8	12.7	12.2	13.7	Transf. priv.sans contrep.
Offic. Unrequited Trans.,nie	6.9	10.1	11.0	9.1	9.5	5.8	6.1	Tr. offic. sans contrep.nia
Direct Investment, nie	9.7	6.9	8.1	7.6	10.7	13.6	9.7	Investissements direct, nia
Portfolio Investments, nie	0.0	0.0	0.0	−0.4	0.0	0.0	−0.1	Invest. portefeuille, nia
Other Capital, nie	−0.8	−1.4	24.5	22.8	15.1	11.2	8.3	Autres capitaux, nia
Net Errors and Omissions	2.5	−0.8	0.1	6.0	−0.3	1.5	5.6	Erreurs et omissions nettes

78
Summary of balance of payments
Millions of US dollars
Résumé des balances des paiements
Millions de dollars des E–U

Country or area	1987	1988	1989	1990	1991	1992	1993	Pays ou zone
Reserves and Related Items	−8.4	1.0	−0.2	−5.1	−4.2	−3.4	−0.6	Rés. et postes assimilables
Dominican Republic								**Rép. dominicaine**
Merchandise: Exp. fob	711.3	889.7	924.4	734.5	658.3	562.5	511.5	Marchandises : exp. fob
Merchandise: Imp. fob	−1 591.5	−1 608.0	−1 963.8	−1 792.8	−1 728.8	−2 174.3	−2 118.4	Marchandises : imp. fob
Serv. & Income: Credit	863.6	1 021.9	1 259.9	1 356.9	1 407.5	1 585.5	1 880.3	Serv. & revenu : crédit
Serv. & Income: Debit	−678.1	−676.1	−820.6	−775.4	−759.2	−850.7	−876.0	Serv. & revenu : débit
Private Unrequited Transfers	273.1	288.8	300.5	314.8	329.5	346.6	361.8	Transf. priv.sans contrep.
Offic. Unrequited Trans.,nie	57.5	64.8	83.9	55.8	57.0	85.2	79.8	Tr. offic. sans contrep.nia
Direct Investment, nie	89.0	106.1	110.0	132.8	145.0	179.7	182.8	Investissements direct, nia
Portfolio Investments, nie	0.0	0.0	0.0	0.0	0.0	0.0	0.0	Invest. portefeuille, nia
Other Capital, nie	−46.1	−121.7	50.0	−150.0	−282.7	−103.4	81.6	Autres capitaux, nia
Net Errors and Omissions	248.9	35.6	−185.2	−294.1	426.7	305.8	−69.1	Erreurs et omissions nettes
Reserves and Related Items	72.3	−1.1	240.9	417.5	−253.3	63.1	−34.3	Rés. et postes assimilables
El Salvador								**El Salvador**
Merchandise: Exp. fob	589.6	610.6	497.8	580.2	588.0	597.5	731.7	Marchandises : exp. fob
Merchandise: Imp. fob	−938.7	−966.5	−1 089.5	−1 180.0	−1 294.1	−1 558.8	−1 766.9	Marchandises : imp. fob
Serv. & Income: Credit	361.1	352.2	336.7	323.2	341.9	408.4	437.2	Serv. & revenu : crédit
Serv. & Income: Debit	−415.0	−471.0	−463.8	−428.8	−474.9	−493.1	−524.6	Serv. & revenu : débit
Private Unrequited Transfers	180.5	202.1	207.8	324.0	469.9	708.8	823.2	Transf. priv.sans contrep.
Offic. Unrequited Trans.,nie	154.4	143.5	180.9	146.3	156.4	142.3	181.0	Tr. offic. sans contrep.nia
Direct Investment, nie	18.3	17.0	12.9	1.7	25.3	15.3	16.4	Investissements direct, nia
Portfolio Investments, nie	0.0	0.0	0.0	0.0	0.0	0.0	0.0	Invest. portefeuille, nia
Other Capital, nie	−77.3	35.3	92.7	−12.0	−86.5	−19.6	70.2	Autres capitaux, nia
Net Errors and Omissions	6.9	−107.1	126.3	270.3	125.9	65.5	90.3	Erreurs et omissions nettes
Reserves and Related Items	120.3	184.1	98.2	−24.9	148.2	133.7	−58.7	Rés. et postes assimilables
Grenada								**Grenade**
Merchandise: Exp. fob	31.9	32.8	28.2	26.6	23.2	19.9	...	Marchandises : exp. fob
Merchandise: Imp. fob	−89.1	−92.2	−99.0	−106.3	−113.6	−103.2	...	Marchandises : imp. fob
Serv. & Income: Credit	49.6	56.1	57.4	67.7	76.3	80.0	...	Serv. & revenu : crédit
Serv. & Income: Debit	−36.7	−38.2	−43.3	−48.6	−47.8	−47.4	...	Serv. & revenu : débit
Private Unrequited Transfers	14.9	15.6	16.7	17.0	18.7	17.5	...	Transf. priv.sans contrep.
Offic. Unrequited Trans.,nie	9.5	9.4	9.6	15.5	8.4	8.2	...	Tr. offic. sans contrep.nia
Direct Investment, nie	14.7	15.0	10.5	12.9	15.3	22.6	...	Investissements direct, nia
Portfolio Investments, nie	0.0	0.2	0.0	0.0	0.1	−0.2	...	Invest. portefeuille, nia
Other Capital, nie	6.2	−4.5	18.0	−1.1	9.8	1.8	...	Autres capitaux, nia
Net Errors and Omissions	0.1	−2.1	−5.3	12.4	9.6	7.9	...	Erreurs et omissions nettes
Reserves and Related Items	−1.1	8.0	7.3	3.8	0.1	−7.1	...	Rés. et postes assimilables
Guatemala								**Guatemala**
Merchandise: Exp. fob	977.9	1 073.3	1 126.1	1 211.4	1 230.0	1 283.7	1 363.2	Marchandises : exp. fob
Merchandise: Imp. fob	−1 333.2	−1 413.2	−1 484.4	−1 428.0	−1 673.0	−2 327.8	−2 384.0	Marchandises : imp. fob
Serv. & Income: Credit	189.4	227.4	328.7	377.0	522.7	683.1	721.5	Serv. & revenu : crédit
Serv. & Income: Debit	−469.9	−525.8	−587.3	−600.3	−523.1	−735.2	−765.6	Serv. & revenu : débit
Private Unrequited Transfers	101.0	141.7	178.8	205.3	257.7	338.8	362.0	Transf. priv.sans contrep.
Offic. Unrequited Trans.,nie	92.3	82.6	71.0	1.7	2.0	51.7	1.2	Tr. offic. sans contrep.nia
Direct Investment, nie	150.2	329.7	76.2	47.6	90.7	94.1	142.5	Investissements direct, nia
Portfolio Investments, nie	−16.0	−372.2	−63.9	−21.3	71.1	11.4	85.4	Invest. portefeuille, nia
Other Capital, nie	52.5	123.0	213.0	−72.5	571.0	505.0	561.3	Autres capitaux, nia
Net Errors and Omissions	−72.7	−2.4	54.7	36.2	83.3	81.8	85.2	Erreurs et omissions nettes
Reserves and Related Items	328.5	336.0	87.1	242.9	−632.4	13.6	−172.7	Rés. et postes assimilables
Haiti [4]								**Haïti** [4]
Merchandise: Exp. fob	210.1	180.4	148.3	265.8	202.0	75.6	81.6	Marchandises : exp. fob
Merchandise: Imp. fob	−311.2	−283.9	−259.3	−442.6	−448.6	−214.1	−266.6	Marchandises : imp. fob
Serv. & Income: Credit	115.5	100.7	93.1	59.1	59.6	39.5	37.8	Serv. & revenu : crédit
Serv. & Income: Debit	−216.6	−230.5	−219.1	−163.5	−170.5	−97.4	−104.1	Serv. & revenu : débit
Private Unrequited Transfers	56.2	63.4	59.3	61.0	69.5	70.0	73.4	Transf. priv.sans contrep.
Offic. Unrequited Trans.,nie	114.8	129.5	114.9	131.9	164.7	85.0	100.0	Tr. offic. sans contrep.nia
Direct Investment, nie	4.7	10.1	9.4	8.0	13.6	0.0	0.0	Investissements direct, nia
Portfolio Investments, nie	0.0	0.0	0.0	0.0	0.0	0.0	0.0	Invest. portefeuille, nia
Other Capital, nie	48.7	16.2	50.8	25.0	12.3	−20.6	−43.7	Autres capitaux, nia
Net Errors and Omissions	−16.4	14.4	−10.7	20.1	110.1	55.7	98.6	Erreurs et omissions nettes
Reserves and Related Items	−5.9	−0.3	13.2	35.2	−12.7	6.3	23.0	Rés. et postes assimilables
Honduras								**Honduras**
Merchandise: Exp. fob	821.8	881.1	903.2	886.9	834.7	833.1	846.0	Marchandises : exp. fob
Merchandise: Imp. fob	−871.4	−923.4	−955.7	−907.0	−912.5	−990.2	−943.9	Marchandises : imp. fob
Serv. & Income: Credit	143.5	166.7	182.0	166.3	220.7	269.6	295.2	Serv. & revenu : crédit
Serv. & Income: Debit	−450.3	−470.8	−492.4	−477.3	−512.6	−586.6	−587.0	Serv. & revenu : débit

78
Summary of balance of payments
Millions of US dollars
Résumé des balances des paiements
Millions de dollars des E-U

Country or area	1987	1988	1989	1990	1991	1992	1993	Pays ou zone
Private Unrequited Transfers	26.0	28.0	34.2	48.8	53.0	61.1	61.3	Transf. priv.sans contrep.
Offic. Unrequited Trans.,nie	85.2	157.4	148.4	95.9	103.3	114.8	72.6	Tr. offic. sans contrep.nia
Direct Investment, nie	38.6	48.3	51.0	43.5	52.1	47.6	34.8	Investissements direct, nia
Portfolio Investments, nie	0.6	−0.2	0.1	0.1	0.1	0.1	0.1	Invest. portefeuille, nia
Other Capital, nie	73.3	7.7	−116.3	−48.4	−149.2	−25.2	119.5	Autres capitaux, nia
Net Errors and Omissions	−33.4	−93.1	−138.9	−107.4	152.0	29.2	−81.8	Erreurs et omissions nettes
Reserves and Related Items	166.1	198.3	384.3	298.6	158.4	246.5	183.2	Rés. et postes assimilables
Jamaica								**Jamaïque**
Merchandise: Exp. fob	709.2	883.0	1 000.4	1 157.5	1 150.7	1 053.6	1 044.5	Marchandises : exp. fob
Merchandise: Imp. fob	−1 061.1	−1 240.3	−1 606.4	−1 679.6	−1 575.0	−1 529.2	−1 858.8	Marchandises : imp. fob
Serv. & Income: Credit	925.0	888.7	1 008.6	1 167.2	1 097.8	1 241.9	1 415.9	Serv. & revenu : crédit
Serv. & Income: Debit	−886.0	−1 005.8	−1 187.2	−1 248.1	−1 182.1	−1 095.2	−1 199.9	Serv. & revenu : débit
Private Unrequited Transfers	117.2	436.5	299.5	159.0	153.3	248.2	306.4	Transf. priv.sans contrep.
Offic. Unrequited Trans.,nie	53.5	69.1	187.5	116.0	99.5	91.6	65.5	Tr. offic. sans contrep.nia
Direct Investment, nie	53.4	−12.0	57.1	137.9	133.2	142.4	77.9	Investissements direct, nia
Portfolio Investments, nie	0.0	0.0	0.0	0.0	0.0	0.0	0.0	Invest. portefeuille, nia
Other Capital, nie	307.1	100.0	41.1	266.6	138.2	212.1	218.5	Autres capitaux, nia
Net Errors and Omissions	84.6	−46.0	10.0	29.3	−20.4	−59.9	79.2	Erreurs et omissions nettes
Reserves and Related Items	−302.9	−73.2	189.4	−105.8	4.8	−305.5	−149.2	Rés. et postes assimilables
Mexico								**Mexique**
Merchandise: Exp. fob	20 495.0	20 547.0	22 842.0	26 838.0	26 855.0	27 516.0	30 033.0	Marchandises : exp. fob
Merchandise: Imp. fob	−13 306.0	−20 273.0	−25 438.0	−31 271.0	−38 184.0	−48 193.0	−48 924.0	Marchandises : imp. fob
Serv. & Income: Credit	9 431.0	11 470.0	13 369.0	14 919.0	16 442.0	16 807.0	17 469.0	Serv. & revenu : crédit
Serv. & Income: Debit	−14 292.0	−16 373.0	−19 141.0	−21 912.0	−22 747.0	−23 957.0	−24 656.0	Serv. & revenu : débit
Private Unrequited Transfers	1 655.0	2 085.0	2 391.0	2 679.0	2 639.0	2 908.0	2 591.0	Transf. priv.sans contrep.
Offic. Unrequited Trans.,nie	264.0	170.0	152.0	1 296.0	107.0	113.0	96.0	Tr. offic. sans contrep.nia
Direct Investment, nie	1 184.0	2 011.0	2 785.0	2 549.0	4 742.0	4 393.0	4 901.0	Investissements direct, nia
Portfolio Investments, nie	−1 399.0	121.0	298.0	−3 985.0	12 138.0	19 175.0	27 867.0	Invest. portefeuille, nia
Other Capital, nie	−2 852.0	−6 627.0	−1 973.0	9 877.0	8 259.0	3 440.0	−709.0	Autres capitaux, nia
Net Errors and Omissions	2 954.0	−3 193.0	4 504.0	1 228.0	−2 278.0	−457.0	−1 436.0	Erreurs et omissions nettes
Reserves and Related Items	−4 134.0	10 062.0	211.0	−2 218.0	−7 973.0	−1 745.0	−7 232.0	Rés. et postes assimilables
Netherlands Antilles								**Antilles néerlandaises**
Merchandise: Exp. fob	106.1	172.8	253.8	208.3	210.9	242.6	226.3	Marchandises : exp. fob
Merchandise: Imp. fob	−771.7	−879.4	−1 017.8	−1 112.3	−1 118.9	−1 168.5	−1 130.3	Marchandises : imp. fob
Serv. & Income: Credit	981.7	1 240.5	1 299.6	1 512.9	1 566.8	1 693.9	1 641.0	Serv. & revenu : crédit
Serv. & Income: Debit	−340.6	−419.4	−480.5	−627.8	−653.8	−753.5	−747.3	Serv. & revenu : débit
Private Unrequited Transfers	−62.2	−61.9	−47.2	−48.5	−61.0	−46.3	−61.6	Transf. priv.sans contrep.
Offic. Unrequited Trans.,nie	34.4	19.6	27.0	21.7	49.1	43.5	56.0	Tr. offic. sans contrep.nia
Direct Investment, nie	2.4	3.8	12.6	5.7	32.3	38.5	13.2	Investissements direct, nia
Portfolio Investments, nie	13.9	−57.1	−75.8	−49.1	−30.7	−18.8	−12.5	Invest. portefeuille, nia
Other Capital, nie	−28.3	−5.1	−30.4	52.7	−43.1	22.1	21.5	Autres capitaux, nia
Net Errors and Omissions	34.7	19.5	14.6	6.5	6.2	5.8	37.8	Erreurs et omissions nettes
Reserves and Related Items	29.6	−33.4	44.2	29.8	42.2	−59.2	−44.0	Rés. et postes assimilables
Nicaragua								**Nicaragua**
Merchandise: Exp. fob	295.1	235.7	318.7	332.4	268.1	223.1	267.0	Marchandises : exp. fob
Merchandise: Imp. fob	−734.4	−718.3	−547.3	−569.7	−688.0	−770.8	−659.4	Marchandises : imp. fob
Serv. & Income: Credit	30.9	39.5	28.8	71.6	79.9	93.7	105.6	Serv. & revenu : crédit
Serv. & Income: Debit	−405.8	−402.3	−330.8	−341.1	−509.2	−650.6	−591.1	Serv. & revenu : débit
Private Unrequited Transfers	0.0	0.0	0.0	0.0	0.0	10.0	25.0	Transf. priv.sans contrep.
Offic. Unrequited Trans.,nie	135.4	130.0	168.9	201.6	844.4	260.6	208.6	Tr. offic. sans contrep.nia
Direct Investment, nie	0.0	0.0	0.0	0.0	0.0	15.0	38.8	Investissements direct, nia
Portfolio Investments, nie	0.0	0.0	0.0	0.0	0.0	0.0	0.0	Invest. portefeuille, nia
Other Capital, nie	125.9	303.5	−89.3	−161.1	−543.6	−553.3	−541.6	Autres capitaux, nia
Net Errors and Omissions	−78.9	51.9	−69.2	−181.2	84.7	60.2	128.1	Erreurs et omissions nettes
Reserves and Related Items	631.8	360.0	520.2	647.5	463.7	1 312.0	1 019.0	Rés. et postes assimilables
Panama								**Panama**
Merchandise: Exp. fob	2 593.0	2 452.0	2 681.0	3 316.0	4 146.0	5 012.0	5 299.0	Marchandises : exp. fob
Merchandise: Imp. fob	−3 058.0	−2 531.0	−3 084.0	−3 804.0	−4 960.0	−5 891.0	−6 152.0	Marchandises : imp. fob
Serv. & Income: Credit	3 235.0	2 059.0	2 101.0	2 203.0	2 247.0	2 414.0	2 347.0	Serv. & revenu : crédit
Serv. & Income: Debit	−2 515.0	−1 440.0	−1 613.0	−1 765.0	−1 699.0	−1 736.0	−1 604.0	Serv. & revenu : débit
Private Unrequited Transfers	−51.0	−40.0	−36.0	−22.0	−16.0	−27.0	−27.0	Transf. priv.sans contrep.
Offic. Unrequited Trans.,nie	114.0	111.0	106.0	118.0	125.0	146.0	138.0	Tr. offic. sans contrep.nia
Direct Investment, nie	57.0	−52.0	37.0	−18.0	−30.0	2.0	−41.0	Investissements direct, nia
Portfolio Investments, nie	−71.0	212.0	−89.0	−62.0	−16.0	−149.0	−559.0	Invest. portefeuille, nia
Other Capital, nie	−159.0	−405.0	−514.0	−705.0	−783.0	−526.0	−5.0	Autres capitaux, nia
Net Errors and Omissions	−646.0	−1 370.0	−421.0	377.0	565.0	444.0	196.0	Erreurs et omissions nettes

78
Summary of balance of payments
Millions of US dollars
Résumé des balances des paiements
Millions de dollars des E−U

Country or area	1987	1988	1989	1990	1991	1992	1993	Pays ou zone
Reserves and Related Items	501.0	1 004.0	832.0	361.0	422.0	314.0	407.0	Rés. et postes assimilables
Saint Kitts and Nevis								**Saint−Kitts−et−Nevis**
Merchandise: Exp. fob	28.0	27.4	28.6	27.7	29.3	31.9	...	Marchandises : exp. fob
Merchandise: Imp. fob	−70.0	−82.0	−86.8	−97.4	−97.1	−92.2	...	Marchandises : imp. fob
Serv. & Income: Credit	44.0	51.8	54.3	58.0	71.8	82.2	...	Serv. & revenu : crédit
Serv. & Income: Debit	−26.7	−34.3	−43.4	−44.7	−46.9	−56.1	...	Serv. & revenu : débit
Private Unrequited Transfers	10.5	10.5	14.2	12.8	12.4	13.1	...	Transf. priv.sans contrep.
Offic. Unrequited Trans.,nie	5.1	5.0	2.1	−0.5	0.8	0.6	...	Tr. offic. sans contrep.nia
Direct Investment, nie	16.7	13.1	40.8	48.8	21.4	13.7	...	Investissements direct, nia
Portfolio Investments, nie	0.0	0.0	0.0	0.0	0.0	0.0	...	Invest. portefeuille, nia
Other Capital, nie	−22.8	4.5	15.6	−1.4	8.2	13.4	...	Autres capitaux, nia
Net Errors and Omissions	15.7	3.9	−19.1	−3.1	0.7	3.3	...	Erreurs et omissions nettes
Reserves and Related Items	−0.5	0.1	−6.3	−0.1	−0.7	−9.8	...	Rés. et postes assimilables
Saint Lucia								**Sainte−Lucie**
Merchandise: Exp. fob	79.5	119.1	112.0	127.3	110.3	122.8	...	Marchandises : exp. fob
Merchandise: Imp. fob	−156.7	−194.5	−240.9	−238.7	−261.4	−275.4	...	Marchandises : imp. fob
Serv. & Income: Credit	104.8	126.7	146.4	160.2	186.0	204.8	...	Serv. & revenu : crédit
Serv. & Income: Debit	−60.7	−83.4	−92.8	−121.0	−124.5	−129.9	...	Serv. & revenu : débit
Private Unrequited Transfers	19.9	14.2	13.7	15.2	17.3	16.1	...	Transf. priv.sans contrep.
Offic. Unrequited Trans.,nie	9.0	5.4	5.3	0.3	4.5	6.4	...	Tr. offic. sans contrep.nia
Direct Investment, nie	15.0	16.4	26.6	44.8	57.7	46.4	...	Investissements direct, nia
Portfolio Investments, nie	0.0	0.0	0.0	0.0	0.0	−0.5	...	Invest. portefeuille, nia
Other Capital, nie	−0.9	−6.5	27.2	10.1	4.1	23.7	...	Autres capitaux, nia
Net Errors and Omissions	−0.7	5.0	8.2	8.3	13.7	−7.8	...	Erreurs et omissions nettes
Reserves and Related Items	−9.1	−2.5	−5.6	−6.5	−7.7	−6.5	...	Rés. et postes assimilables
Saint Vincent−Grenadines								**Saint Vincent−Grenadine**
Merchandise: Exp. fob	52.3	85.3	74.7	82.7	67.3	77.5	...	Marchandises : exp. fob
Merchandise: Imp. fob	−89.5	−110.0	−112.2	−119.6	−119.7	−118.6	...	Marchandises : imp. fob
Serv. & Income: Credit	31.3	34.4	45.4	52.0	50.5	52.0	...	Serv. & revenu : crédit
Serv. & Income: Debit	−30.9	−35.2	−47.8	−49.6	−52.9	−51.5	...	Serv. & revenu : débit
Private Unrequited Transfers	17.8	20.0	12.6	13.1	12.8	12.9	...	Transf. priv.sans contrep.
Offic. Unrequited Trans.,nie	6.6	6.3	8.9	16.1	18.3	11.1	...	Tr. offic. sans contrep.nia
Direct Investment, nie	5.5	9.5	10.6	7.7	8.8	18.9	...	Investissements direct, nia
Portfolio Investments, nie	0.0	0.0	0.0	0.0	0.0	0.0	...	Invest. portefeuille, nia
Other Capital, nie	3.9	−5.8	7.6	−5.1	13.0	7.9	...	Autres capitaux, nia
Net Errors and Omissions	−2.2	−2.2	1.6	6.8	−1.8	1.2	...	Erreurs et omissions nettes
Reserves and Related Items	5.2	−2.3	−1.4	−4.0	3.8	−11.3	...	Rés. et postes assimilables
Trinidad and Tobago								**Trinité−et−Tobago**
Merchandise: Exp. fob	1 396.9	1 453.3	1 534.6	1 935.2	1 751.3	1 661.9	1 477.2	Marchandises : exp. fob
Merchandise: Imp. fob	−1 057.6	−1 064.2	−1 045.2	−947.6	−1 210.3	−995.6	−952.9	Marchandises : imp. fob
Serv. & Income: Credit	243.6	307.7	329.5	393.1	477.1	512.0	416.5	Serv. & revenu : crédit
Serv. & Income: Debit	−785.2	−776.2	−849.9	−915.6	−1 025.1	−1 039.7	−832.4	Serv. & revenu : débit
Private Unrequited Transfers	−20.2	−23.0	−19.2	−21.0	−15.9	−15.7	−7.0	Transf. priv.sans contrep.
Offic. Unrequited Trans.,nie	−16.6	−6.6	−5.5	−4.4	2.1	−0.4	0.2	Tr. offic. sans contrep.nia
Direct Investment, nie	35.0	62.9	148.9	109.4	169.3	177.9	379.2	Investissements direct, nia
Portfolio Investments, nie	0.0	0.0	0.0	0.0	0.0	0.0	0.0	Invest. portefeuille, nia
Other Capital, nie	43.4	−204.4	−315.4	−615.7	−396.2	−332.1	−280.4	Autres capitaux, nia
Net Errors and Omissions	−94.8	21.1	45.4	−112.0	−29.0	−72.6	−41.8	Erreurs et omissions nettes
Reserves and Related Items	255.6	229.4	176.8	178.5	276.5	104.4	−158.6	Rés. et postes assimilables
United States [1]								**Etats−Unis** [1]
Merchandise: Exp. fob	250.2	320.2	362.1	389.3	416.9	440.4	456.9	Marchandises : exp. fob
Merchandise: Imp. fob	−409.8	−447.2	−477.4	−498.3	−491.0	−536.5	−589.4	Marchandises : imp. fob
Serv. & Income: Credit	197.1	237.4	279.4	307.5	300.2	291.0	298.6	Serv. & revenu : crédit
Serv. & Income: Debit	−181.5	−213.7	−240.7	−256.6	−239.7	−230.7	−237.8	Serv. & revenu : débit
Private Unrequited Transfers	−10.6	−12.0	−12.7	−13.0	−13.8	−13.3	−13.7	Transf. priv.sans contrep.
Offic. Unrequited Trans.,nie	−12.5	−13.0	−13.4	−20.6	20.5	−18.8	−18.4	Tr. offic. sans contrep.nia
Direct Investment, nie	31.0	41.8	30.9	18.0	−5.2	−31.1	−36.5	Investissements direct, nia
Portfolio Investments, nie	31.1	40.3	43.5	−33.0	8.6	16.6	−17.6	Invest. portefeuille, nia
Other Capital, nie	52.6	22.4	−7.7	37.0	21.5	57.3	68.2	Autres capitaux, nia
Net Errors and Omissions	−4.5	−12.6	53.1	40.0	−39.7	−17.2	21.1	Erreurs et omissions nettes
Reserves and Related Items	56.9	36.3	−16.9	29.8	21.8	42.2	68.7	Rés. et postes assimilables

America, South· Amérique du Sud

	1987	1988	1989	1990	1991	1992	1993	
Argentina								**Argentine**
Merchandise: Exp. fob	6 360.0	9 134.0	9 573.0	12 354.0	11 978.0	12 235.0	13 117.0	Marchandises : exp. fob
Merchandise: Imp. fob	−5 343.0	−4 892.0	−3 864.0	−3 726.0	−7 559.0	−13 685.0	−15 545.0	Marchandises : imp. fob
Serv. & Income: Credit	2 046.0	2 226.0	2 469.0	4 300.0	4 154.0	3 929.0	4 158.0	Serv. & revenu : crédit

78
Summary of balance of payments
Millions of US dollars
Résumé des balances des paiements
Millions de dollars des E−U

Country or area	1987	1988	1989	1990	1991	1992	1993	Pays ou zone
Serv. & Income: Debit	−7 290.0	−8 040.0	−9 491.0	−9 374.0	−10 013.0	−9 774.0	−9 628.0	Serv. & revenu : débit
Private Unrequited Transfers	−8.0	0.0	8.0	998.0	793.0	749.0	535.0	Transf. priv.sans contrep.
Offic. Unrequited Trans.,nie	0.0	0.0	0.0	0.0	0.0	0.0	−89.0	Tr. offic. sans contrep.nia
Direct Investment, nie	−19.0	1 147.0	1 028.0	1 836.0	2 439.0	4 179.0	6 305.0	Investissements direct, nia
Portfolio Investments, nie	−572.0	−718.0	−1 098.0	−1 346.0	−34.0	−680.0	−9 035.0	Invest. portefeuille, nia
Other Capital, nie	565.0	2.0	−7 938.0	−6 340.0	−2 245.0	5 339.0	−6 896.0	Autres capitaux, nia
Net Errors and Omissions	−112.0	−165.0	−249.0	715.0	−341.0	137.0	87.0	Erreurs et omissions nettes
Reserves and Related Items	4 373.0	1 306.0	9 562.0	583.0	828.0	−2 429.0	16 991.0	Rés. et postes assimilables
Bolivia								**Bolivie**
Merchandise: Exp. fob	518.7	542.5	723.5	830.8	760.3	608.4	...	Marchandises : exp. fob
Merchandise: Imp. fob	−646.3	−590.9	−729.5	−775.6	−804.2	−1 040.8	...	Marchandises : imp. fob
Serv. & Income: Credit	147.7	146.5	167.2	164.7	181.6	182.3	...	Serv. & revenu : crédit
Serv. & Income: Debit	−564.1	−537.8	−581.2	−578.0	−582.8	−526.4	...	Serv. & revenu : débit
Private Unrequited Transfers	18.0	12.7	20.6	21.6	23.0	22.7	...	Transf. priv.sans contrep.
Offic. Unrequited Trans.,nie	99.2	123.9	135.2	138.4	160.0	220.5	...	Tr. offic. sans contrep.nia
Direct Investment, nie	36.4	−12.0	−25.4	26.1	50.0	91.1	...	Investissements direct, nia
Portfolio Investments, nie	0.0	0.0	0.0	0.0	0.0	0.0	...	Invest. portefeuille, nia
Other Capital, nie	−178.1	−12.7	−78.0	66.9	64.6	282.8	...	Autres capitaux, nia
Net Errors and Omissions	174.6	46.6	−32.1	−11.4	53.3	34.3	...	Erreurs et omissions nettes
Reserves and Related Items	393.9	281.2	399.7	116.5	94.2	125.1	...	Rés. et postes assimilables
Brazil								**Brésil**
Merchandise: Exp. fob	26 210.0	33 773.0	34 375.0	31 408.0	31 619.0	35 793.0	38 783.0	Marchandises : exp. fob
Merchandise: Imp. fob	−15 052.0	−14 605.0	−18 263.0	−20 661.0	−21 041.0	−20 554.0	−25 711.0	Marchandises : imp. fob
Serv. & Income: Credit	2 520.0	3 050.0	4 442.0	4 919.0	4 223.0	5 206.0	5 163.0	Serv. & revenu : crédit
Serv. & Income: Debit	−15 198.0	−18 153.0	−19 773.0	−20 288.0	−17 765.0	−16 545.0	−20 525.0	Serv. & revenu : débit
Private Unrequited Transfers	113.0	107.0	226.0	813.0	1 521.0	2 240.0	1 682.0	Transf. priv.sans contrep.
Offic. Unrequited Trans.,nie	−43.0	−13.0	18.0	21.0	35.0	3.0	−29.0	Tr. offic. sans contrep.nia
Direct Investment, nie	1 087.0	2 794.0	744.0	236.0	−42.0	1 443.0	−292.0	Investissements direct, nia
Portfolio Investments, nie	−428.0	−498.0	−421.0	512.0	3 808.0	7 366.0	12 872.0	Invest. portefeuille, nia
Other Capital, nie	−10 941.0	−11 506.0	−12 848.0	−6 315.0	−7 895.0	−2 293.0	−4 200.0	Autres capitaux, nia
Net Errors and Omissions	−802.0	−827.0	−819.0	−296.0	852.0	−1 393.0	−853.0	Erreurs et omissions nettes
Reserves and Related Items	12 534.0	5 878.0	12 319.0	9 651.0	4 685.0	−11 266.0	−6 890.0	Rés. et postes assimilables
Chile								**Chili**
Merchandise: Exp. fob	5 224.0	7 052.0	8 080.0	8 310.0	8 928.0	9 986.0	9 202.0	Marchandises : exp. fob
Merchandise: Imp. fob	−3 994.0	−4 833.0	−6 502.0	−7 037.0	−7 354.0	−9 236.0	−10 181.0	Marchandises : imp. fob
Serv. & Income: Credit	1 268.0	1 399.0	1 736.0	2 260.0	2 593.0	2 873.0	3 131.0	Serv. & revenu : crédit
Serv. & Income: Debit	−3 432.0	−3 962.0	−4 233.0	−4 381.0	−4 494.0	−4 797.0	−4 631.0	Serv. & revenu : débit
Private Unrequited Transfers	65.0	63.0	58.0	54.0	40.0	74.0	61.0	Transf. priv.sans contrep.
Offic. Unrequited Trans.,nie	61.0	114.0	157.0	146.0	299.0	357.0	325.0	Tr. offic. sans contrep.nia
Direct Investment, nie	923.0	1 011.0	1 279.0	582.0	400.0	321.0	410.0	Investissements direct, nia
Portfolio Investments, nie	0.0	0.0	87.0	359.0	225.0	452.0	747.0	Invest. portefeuille, nia
Other Capital, nie	−1 959.0	−287.0	−155.0	2 110.0	202.0	2 111.0	1 616.0	Autres capitaux, nia
Net Errors and Omissions	−78.0	−109.0	−120.0	−32.0	400.0	359.0	−95.0	Erreurs et omissions nettes
Reserves and Related Items	1 922.0	−448.0	−387.0	−2 371.0	−1 239.0	−2 500.0	−585.0	Rés. et postes assimilables
Colombia								**Colombie**
Merchandise: Exp. fob	5 661.0	5 343.0	6 031.0	7 079.0	7 507.0	7 263.0	...	Marchandises : exp. fob
Merchandise: Imp. fob	−3 793.0	−4 516.0	−4 557.0	−5 108.0	−4 548.0	−6 030.0	...	Marchandises : imp. fob
Serv. & Income: Credit	1 368.0	1 665.0	1 578.0	1 947.0	1 984.0	2 432.0	...	Serv. & revenu : crédit
Serv. & Income: Debit	−3 901.0	−3 672.0	−4 151.0	−4 402.0	−4 292.0	−4 487.0	...	Serv. & revenu : débit
Private Unrequited Transfers	1 009.0	975.0	912.0	1 041.0	1 712.0	1 747.0	...	Transf. priv.sans contrep.
Offic. Unrequited Trans.,nie	−8.0	−11.0	−14.0	−15.0	−14.0	−13.0	...	Tr. offic. sans contrep.nia
Direct Investment, nie	293.0	159.0	547.0	484.0	433.0	740.0	...	Investissements direct, nia
Portfolio Investments, nie	48.0	0.0	179.0	−4.0	81.0	60.0	...	Invest. portefeuille, nia
Other Capital, nie	−353.0	781.0	−319.0	−454.0	−1 299.0	−517.0	...	Autres capitaux, nia
Net Errors and Omissions	67.0	−530.0	157.0	70.0	269.0	14.0	...	Erreurs et omissions nettes
Reserves and Related Items	−391.0	−194.0	−363.0	−638.0	−1 834.0	−1 209.0	...	Rés. et postes assimilables
Ecuador								**Equateur**
Merchandise: Exp. fob	2 021.0	2 202.0	2 354.0	2 714.0	2 851.0	3 008.0	2 903.0	Marchandises : exp. fob
Merchandise: Imp. fob	−2 054.0	−1 583.0	−1 693.0	−1 711.0	−2 207.0	−2 048.0	−2 325.0	Marchandises : imp. fob
Serv. & Income: Credit	449.0	457.0	536.0	563.0	587.0	652.0	680.0	Serv. & revenu : crédit
Serv. & Income: Debit	−1 672.0	−1 709.0	−1 808.0	−1 839.0	−1 808.0	−1 722.0	−1 748.0	Serv. & revenu : débit
Private Unrequited Transfers	0.0	0.0	0.0	0.0	0.0	0.0	0.0	Transf. priv.sans contrep.
Offic. Unrequited Trans.,nie	132.0	97.0	97.0	107.0	110.0	120.0	130.0	Tr. offic. sans contrep.nia
Direct Investment, nie	75.0	80.0	80.0	82.0	85.0	95.0	115.0	Investissements direct, nia
Portfolio Investments, nie	0.0	0.0	0.0	0.0	0.0	0.0	0.0	Invest. portefeuille, nia
Other Capital, nie	−352.0	−570.0	−608.0	−854.0	−677.0	−989.0	−317.0	Autres capitaux, nia

78
Summary of balance of payments
Millions of US dollars
Résumé des balances des paiements
Millions de dollars des E−U

Country or area	1987	1988	1989	1990	1991	1992	1993	Pays ou zone
Net Errors and Omissions	−244.6	−119.2	72.6	71.1	197.7	81.8	132.3	Erreurs et omissions nettes
Reserves and Related Items	1 645.6	1 145.2	969.4	866.9	861.3	802.2	429.7	Rés. et postes assimilables
Paraguay								**Paraguay**
Merchandise: Exp. fob	597.4	871.0	1 180.0	1 382.3	1 120.8	1 081.5	1 653.0	Marchandises : exp. fob
Merchandise: Imp. fob	−918.7	−1 030.1	−1 015.9	−1 635.8	−1 867.6	−1 950.6	−2 671.6	Marchandises : imp. fob
Serv. & Income: Credit	222.4	352.6	483.6	604.3	1 012.2	955.2	1 109.2	Serv. & revenu : crédit
Serv. & Income: Debit	−417.8	−438.8	−415.6	−578.7	−661.8	−720.1	−735.6	Serv. & revenu : débit
Private Unrequited Transfers	2.0	2.0	1.6	7.2	6.7	2.7	4.6	Transf. priv.sans contrep.
Offic. Unrequited Trans.,nie	25.0	33.2	22.3	48.4	65.6	31.2	37.4	Tr. offic. sans contrep.nia
Direct Investment, nie	5.3	8.4	12.8	76.3	83.5	136.6	111.0	Investissements direct, nia
Portfolio Investments, nie	0.0	0.0	0.0	0.0	0.0	0.0	0.0	Invest. portefeuille, nia
Other Capital, nie	86.7	−207.7	−186.7	−152.1	131.5	55.4	198.2	Autres capitaux, nia
Net Errors and Omissions	338.3	198.3	−90.6	362.4	472.0	457.7	483.2	Erreurs et omissions nettes
Reserves and Related Items	59.4	211.1	8.5	−114.3	−362.9	−49.6	−189.4	Rés. et postes assimilables
Peru								**Pérou**
Merchandise: Exp. fob	2 661.0	2 691.0	3 488.0	3 231.0	3 330.0	3 485.0	3 463.0	Marchandises : exp. fob
Merchandise: Imp. fob	−3 182.0	−2 790.0	−2 291.0	−2 892.0	−3 495.0	−4 050.0	−4 043.0	Marchandises : imp. fob
Serv. & Income: Credit	998.0	1 038.0	971.0	917.0	994.0	983.0	1 045.0	Serv. & revenu : crédit
Serv. & Income: Debit	−2 297.0	−2 396.0	−2 328.0	−2 432.0	−2 478.0	−2 561.0	−2 682.0	Serv. & revenu : débit
Private Unrequited Transfers	0.0	0.0	0.0	0.0	0.0	0.0	0.0	Transf. priv.sans contrep.
Offic. Unrequited Trans.,nie	180.0	211.0	235.0	275.0	318.0	433.0	417.0	Tr. offic. sans contrep.nia
Direct Investment, nie	32.0	26.0	59.0	41.0	−7.0	127.0	349.0	Investissements direct, nia
Portfolio Investments, nie	0.0	0.0	0.0	0.0	0.0	0.0	222.0	Invest. portefeuille, nia
Other Capital, nie	−1 180.0	−909.0	−1 193.0	−923.0	−959.0	−191.0	−596.0	Autres capitaux, nia
Net Errors and Omissions	−34.0	−187.0	−195.0	312.0	1 618.0	1 348.0	1 278.0	Erreurs et omissions nettes
Reserves and Related Items	2 822.0	2 316.0	1 254.0	1 471.0	679.0	426.0	547.0	Rés. et postes assimilables
Suriname								**Suriname**
Merchandise: Exp. fob	338.8	358.4	549.2	465.9	345.9	341.0	...	Marchandises : exp. fob
Merchandise: Imp. fob	−274.3	−239.4	−330.9	−374.4	−347.1	−272.5	...	Marchandises : imp. fob
Serv. & Income: Credit	82.1	24.0	24.6	23.0	23.7	23.4	...	Serv. & revenu : crédit
Serv. & Income: Debit	−73.8	−86.1	−98.0	−106.7	−110.3	−106.8	...	Serv. & revenu : débit
Private Unrequited Transfers	−0.4	−4.6	−5.6	−7.5	−7.3	−7.3	...	Transf. priv.sans contrep.
Offic. Unrequited Trans.,nie	2.4	10.3	23.5	34.4	19.3	33.4	...	Tr. offic. sans contrep.nia
Direct Investment, nie	−72.6	−95.8	−167.9	−43.0	10.4	−30.4	...	Investissements direct, nia
Portfolio Investments, nie	−0.1	0.0	0.0	0.5	−2.3	1.5	...	Invest. portefeuille, nia
Other Capital, nie	22.1	29.9	−5.0	27.6	24.4	−19.6	...	Autres capitaux, nia
Net Errors and Omissions	−33.6	−1.8	9.9	−9.4	−0.5	25.4	...	Erreurs et omissions nettes
Reserves and Related Items	9.4	5.2	0.1	−10.3	43.9	12.0	...	Rés. et postes assimilables
Uruguay								**Uruguay**
Merchandise: Exp. fob	1 182.3	1 404.5	1 599.0	1 692.9	1 604.7	1 801.4	1 731.6	Marchandises : exp. fob
Merchandise: Imp. fob	−1 079.9	−1 112.2	−1 136.2	−1 266.9	−1 543.7	−1 923.2	−2 118.3	Marchandises : imp. fob
Serv. & Income: Credit	510.4	463.0	636.5	723.9	830.9	1 055.3	1 166.5	Serv. & revenu : crédit
Serv. & Income: Debit	−761.6	−754.4	−973.8	−972.1	−889.6	−970.9	−1 031.4	Serv. & revenu : débit
Private Unrequited Transfers	0.0	0.0	0.0	0.0	0.0	0.0	0.0	Transf. priv.sans contrep.
Offic. Unrequited Trans.,nie	8.0	21.3	8.0	8.1	40.1	28.6	24.8	Tr. offic. sans contrep.nia
Direct Investment, nie	55.0	44.5	0.0	0.0	0.0	0.0	75.8	Investissements direct, nia
Portfolio Investments, nie	13.1	37.4	49.9	17.5	109.4	229.1	158.3	Invest. portefeuille, nia
Other Capital, nie	221.2	104.6	−55.8	−103.4	−538.6	−320.6	−39.6	Autres capitaux, nia
Net Errors and Omissions	−104.6	−247.1	−62.6	35.7	468.8	238.3	220.8	Erreurs et omissions nettes
Reserves and Related Items	−43.9	38.4	−65.0	−135.7	−82.0	−138.0	−188.5	Rés. et postes assimilables
Venezuela								**Venezuela**
Merchandise: Exp. fob	10 437.0	10 082.0	12 915.0	17 444.0	14 968.0	13 988.0	14 019.0	Marchandises : exp. fob
Merchandise: Imp. fob	−8 870.0	−12 080.0	−7 283.0	−6 807.0	−10 131.0	−12 714.0	−11 117.0	Marchandises : imp. fob
Serv. & Income: Credit	2 446.0	2 623.0	2 695.0	4 032.0	3 605.0	3 149.0	3 274.0	Serv. & revenu : crédit
Serv. & Income: Debit	−5 312.0	−6 287.0	−5 979.0	−6 107.0	−6 357.0	−7 823.0	−8 082.0	Serv. & revenu : débit
Private Unrequited Transfers	−73.0	−123.0	−171.0	−259.0	−316.0	−347.0	−310.0	Transf. priv.sans contrep.
Offic. Unrequited Trans.,nie	−18.0	−24.0	−16.0	−24.0	−33.0	−6.0	−7.0	Tr. offic. sans contrep.nia
Direct Investment, nie	−16.0	21.0	77.0	96.0	1 769.0	429.0	−55.0	Investissements direct, nia
Portfolio Investments, nie	0.0	0.0	−158.0	13 579.0	192.0	61.0	731.0	Invest. portefeuille, nia
Other Capital, nie	283.0	−1 923.0	−5 148.0	−18 231.0	−420.0	1 841.0	532.0	Autres capitaux, nia
Net Errors and Omissions	−505.0	3 117.0	1 418.0	−1 742.0	−1 516.0	−295.0	407.0	Erreurs et omissions nettes
Reserves and Related Items	1 628.0	4 594.0	1 650.0	−1 981.0	−1 761.0	1 717.0	608.0	Rés. et postes assimilables
Asia · Asie								
Afghanistan								**Afghanistan**
Merchandise: Exp. fob	538.7	453.8	252.3	...	...	...	...	Marchandises : exp. fob
Merchandise: Imp. fob	−904.5	−731.8	−623.5	...	...	...	...	Marchandises : imp. fob

78
Summary of balance of payments
Millions of US dollars
Résumé des balances des paiements
Millions de dollars des E−U

Country or area	1987	1988	1989	1990	1991	1992	1993	Pays ou zone
Serv. & Income: Credit	54.8	92.9	28.3	...	...	...	...	Serv. & revenu : crédit
Serv. & Income: Debit	−167.6	−131.5	−111.3	...	...	...	...	Serv. & revenu : débit
Private Unrequited Transfers	0.0	0.0	−1.2	...	...	...	...	Transf. priv.sans contrep.
Offic. Unrequited Trans.,nie	311.7	342.8	312.1	...	...	...	...	Tr. offic. sans contrep.nia
Direct Investment, nie	0.0	0.0	0.0	...	...	...	...	Investissements direct, nia
Portfolio Investments, nie	0.0	0.0	0.0	...	...	...	...	Invest. portefeuille, nia
Other Capital, nie	−33.9	−4.1	−59.6	...	...	...	...	Autres capitaux, nia
Net Errors and Omissions	211.6	−47.7	182.8	...	...	...	...	Erreurs et omissions nettes
Reserves and Related Items	−10.8	25.6	20.1	...	...	...	...	Rés. et postes assimilables
Bahrain								**Bahreïn**
Merchandise: Exp. fob	2 429.5	2 411.4	2 831.1	3 760.6	3 513.0	3 417.3	...	Marchandises : exp. fob
Merchandise: Imp. fob	−2 442.3	−2 334.0	−2 820.2	−3 339.9	−3 703.5	−3 730.3	...	Marchandises : imp. fob
Serv. & Income: Credit	1 146.3	1 162.5	1 250.5	1 196.0	1 216.5	1 264.1	...	Serv. & revenu : crédit
Serv. & Income: Debit	−1 203.5	−1 223.4	−1 294.4	−1 557.7	−1 620.5	−1 773.7	...	Serv. & revenu : débit
Private Unrequited Transfers	−243.6	−193.1	−198.9	−272.3	−303.5	−270.7	...	Transf. priv.sans contrep.
Offic. Unrequited Trans.,nie	113.3	366.5	102.1	458.8	101.9	100.0	...	Tr. offic. sans contrep.nia
Direct Investment, nie	−35.9	222.1	180.9	−3.5	−6.9	−8.5	...	Investissements direct, nia
Portfolio Investments, nie	0.0	0.0	0.0	−80.6	−34.6	0.0	...	Invest. portefeuille, nia
Other Capital, nie	−17.8	−436.4	−446.5	540.4	−305.6	374.7	...	Autres capitaux, nia
Net Errors and Omissions	−88.7	117.0	207.0	−519.6	1 424.7	545.3	...	Erreurs et omissions nettes
Reserves and Related Items	342.7	−92.5	188.5	−182.2	−281.6	81.9	...	Rés. et postes assimilables
Bangladesh								**Bangladesh**
Merchandise: Exp. fob	1 076.9	1 291.0	1 304.8	1 672.4	1 688.7	2 097.9	2 277.9	Marchandises : exp. fob
Merchandise: Imp. fob	−2 445.6	−2 734.4	−3 300.1	−3 259.4	−3 074.5	−3 353.8	−3 560.9	Marchandises : imp. fob
Serv. & Income: Credit	295.4	332.5	423.1	455.8	501.0	583.4	630.7	Serv. & revenu : crédit
Serv. & Income: Debit	−666.7	−793.9	−923.4	−880.2	−862.2	−954.7	−1 095.8	Serv. & revenu : débit
Private Unrequited Transfers	788.3	827.1	806.8	828.3	901.8	1 019.8	1 132.4	Transf. priv.sans contrep.
Offic. Unrequited Trans.,nie	713.9	804.3	589.1	785.2	909.8	788.2	812.9	Tr. offic. sans contrep.nia
Direct Investment, nie	3.2	1.8	0.2	3.2	1.4	3.7	14.0	Investissements direct, nia
Portfolio Investments, nie	−0.1	0.0	1.7	0.3	2.2	8.7	8.4	Invest. portefeuille, nia
Other Capital, nie	555.0	396.8	831.3	694.3	464.0	526.0	465.8	Autres capitaux, nia
Net Errors and Omissions	−123.8	6.6	−43.1	−75.7	−98.4	−84.0	12.0	Erreurs et omissions nettes
Reserves and Related Items	−196.5	−132.0	309.5	−224.2	−433.8	−635.2	−697.6	Rés. et postes assimilables
Bhutan								**Bhoutan**
Merchandise: Exp. fob	55.0	74.9	73.4	...	...	...	...	Marchandises : exp. fob
Merchandise: Imp. fob	−88.7	−125.1	−105.1	...	...	...	...	Marchandises : imp. fob
Serv. & Income: Credit	11.5	13.2	15.9	...	...	...	...	Serv. & revenu : crédit
Serv. & Income: Debit	−41.4	−45.8	−37.1	...	...	...	...	Serv. & revenu : débit
Private Unrequited Transfers	3.4	4.5	4.4	...	...	...	...	Transf. priv.sans contrep.
Offic. Unrequited Trans.,nie	75.2	63.4	57.9	...	...	...	...	Tr. offic. sans contrep.nia
Direct Investment, nie	0.0	0.0	0.0	...	...	...	...	Investissements direct, nia
Portfolio Investments, nie	0.0	0.0	0.0	...	...	...	...	Invest. portefeuille, nia
Other Capital, nie	18.5	32.4	3.7	...	...	...	...	Autres capitaux, nia
Net Errors and Omissions	−9.0	5.5	−23.2	...	...	...	...	Erreurs et omissions nettes
Reserves and Related Items	−24.5	−23.0	10.1	...	...	...	...	Rés. et postes assimilables
Cambodia								**Cambodge**
Merchandise: Exp. fob	...	...	...	...	...	265.0	284.0	Marchandises : exp. fob
Merchandise: Imp. fob	...	...	...	...	...	−443.0	−479.0	Marchandises : imp. fob
Serv. & Income: Credit	...	...	...	...	...	50.0	64.0	Serv. & revenu : crédit
Serv. & Income: Debit	...	...	...	...	...	−110.0	−155.0	Serv. & revenu : débit
Private Unrequited Transfers	...	...	...	...	...	9.0	3.0	Transf. priv.sans contrep.
Offic. Unrequited Trans.,nie	...	...	...	...	...	262.0	326.0	Tr. offic. sans contrep.nia
Direct Investment, nie	...	...	...	...	...	33.0	37.0	Investissements direct, nia
Portfolio Investments, nie	...	...	...	...	...	0.0	0.0	Invest. portefeuille, nia
Other Capital, nie	...	...	...	...	...	−19.0	−9.0	Autres capitaux, nia
Net Errors and Omissions	...	...	...	...	...	−39.0	−43.0	Erreurs et omissions nettes
Reserves and Related Items	...	...	...	...	...	−7.0	−28.0	Rés. et postes assimilables
China								**Chine**
Merchandise: Exp. fob	34 734.0	41 054.0	43 220.0	51 519.0	58 919.0	69 568.0	75 659.0	Marchandises : exp. fob
Merchandise: Imp. fob	−36 395.0	−46 369.0	−48 840.0	−42 354.0	−50 176.0	−64 385.0	−86 313.0	Marchandises : imp. fob
Serv. & Income: Credit	5 413.0	6 327.0	6 497.0	8 872.0	10 698.0	14 844.0	15 583.0	Serv. & revenu : crédit
Serv. & Income: Debit	−3 676.0	−5 233.0	−5 575.0	−6 314.0	−7 000.0	−14 781.0	−17 710.0	Serv. & revenu : débit
Private Unrequited Transfers	249.0	416.0	238.0	222.0	444.0	804.0	883.0	Transf. priv.sans contrep.
Offic. Unrequited Trans.,nie	−25.0	3.0	143.0	52.0	387.0	351.0	289.0	Tr. offic. sans contrep.nia
Direct Investment, nie	1 669.0	2 344.0	2 613.0	2 657.0	3 453.0	7 156.0	23 115.0	Investissements direct, nia
Portfolio Investments, nie	1 051.0	876.0	−180.0	−241.0	235.0	−57.0	3 049.0	Invest. portefeuille, nia

78
Summary of balance of payments
Millions of US dollars
Résumé des balances des paiements
Millions de dollars des E−U

Country or area	1987	1988	1989	1990	1991	1992	1993	Pays ou zone
Other Capital, nie	3 281.0	3 913.0	1 290.0	839.0	4 344.0	−7 349.0	−2 690.0	Autres capitaux, nia
Net Errors and Omissions	−1 518.0	−957.0	115.0	−3 205.0	−6 767.0	−8 211.0	−10 096.0	Erreurs et omissions nettes
Reserves and Related Items	−4 783.0	−2 374.0	479.0	−12 047.0	−14 537.0	2 060.0	−1 769.0	Rés. et postes assimilables
Cyprus								**Chypre**
Merchandise: Exp. fob	566.4	645.5	717.3	846.7	875.0	903.2	...	Marchandises : exp. fob
Merchandise: Imp. fob	−1 326.8	−1 666.8	−2 072.1	−2 308.9	−2 363.1	−2 989.9	...	Marchandises : imp. fob
Serv. & Income: Credit	1 375.8	1 628.3	1 878.6	2 345.1	2 156.6	2 854.0	...	Serv. & revenu : crédit
Serv. & Income: Debit	−567.5	−650.6	−696.6	−859.4	−891.8	−1 039.9	...	Serv. & revenu : débit
Private Unrequited Transfers	26.4	24.6	22.9	21.9	20.6	22.0	...	Transf. priv.sans contrep.
Offic. Unrequited Trans.,nie	17.5	27.4	17.0	20.8	16.2	9.0	...	Tr. offic. sans contrep.nia
Direct Investment, nie	52.0	62.1	69.9	126.7	66.7	106.7	...	Investissements direct, nia
Portfolio Investments, nie	0.0	0.0	92.8	−37.4	124.3	54.0	...	Invest. portefeuille, nia
Other Capital, nie	28.9	86.8	247.6	371.9	54.5	162.9	...	Autres capitaux, nia
Net Errors and Omissions	−109.0	−86.6	−48.9	−230.1	−148.9	−299.0	...	Erreurs et omissions nettes
Reserves and Related Items	−63.7	−70.7	−228.5	−297.3	89.8	217.1	...	Rés. et postes assimilables
India								**Inde**
Merchandise: Exp. fob	11 884.0	13 510.0	16 144.0	18 286.0	...	...	...	Marchandises : exp. fob
Merchandise: Imp. fob	−17 661.0	−20 091.0	−22 254.0	−23 437.0	...	...	...	Marchandises : imp. fob
Serv. & Income: Credit	3 813.0	4 218.0	4 586.0	5 061.0	...	...	...	Serv. & revenu : crédit
Serv. & Income: Debit	−6 235.0	−7 537.0	−8 372.0	−9 783.0	...	...	...	Serv. & revenu : débit
Private Unrequited Transfers	2 636.0	2 295.0	2 567.0	2 337.0	...	...	...	Transf. priv.sans contrep.
Offic. Unrequited Trans.,nie	370.0	457.0	503.0	499.0	...	...	...	Tr. offic. sans contrep.nia
Direct Investment, nie	0.0	0.0	0.0	0.0	...	...	...	Investissements direct, nia
Portfolio Investments, nie	0.0	0.0	0.0	0.0	...	...	...	Invest. portefeuille, nia
Other Capital, nie	5 734.0	7 243.0	7 349.0	5 670.0	...	...	...	Autres capitaux, nia
Net Errors and Omissions	−409.0	−112.0	−285.0	−571.0	...	...	...	Erreurs et omissions nettes
Reserves and Related Items	−133.0	16.0	−238.0	1 937.0	...	...	...	Rés. et postes assimilables
Indonesia								**Indonésie**
Merchandise: Exp. fob	17 206.0	19 509.0	22 974.0	26 807.0	29 635.0	33 796.0	36 607.0	Marchandises : exp. fob
Merchandise: Imp. fob	−12 532.0	−13 831.0	−16 310.0	−21 455.0	−24 834.0	−26 774.0	−28 376.0	Marchandises : imp. fob
Serv. & Income: Credit	1 626.0	1 861.0	2 437.0	2 897.0	3 739.0	4 209.0	4 923.0	Serv. & revenu : crédit
Serv. & Income: Debit	−8 655.0	−9 190.0	−10 548.0	−11 655.0	−13 062.0	−14 582.0	−15 798.0	Serv. & revenu : débit
Private Unrequited Transfers	86.0	99.0	167.0	166.0	130.0	229.0	346.0	Transf. priv.sans contrep.
Offic. Unrequited Trans.,nie	171.0	155.0	172.0	252.0	132.0	342.0	282.0	Tr. offic. sans contrep.nia
Direct Investment, nie	385.0	576.0	682.0	1 093.0	1 482.0	1 777.0	2 004.0	Investissements direct, nia
Portfolio Investments, nie	−88.0	−98.0	−173.0	−93.0	−12.0	−88.0	−201.0	Invest. portefeuille, nia
Other Capital, nie	3 184.0	1 739.0	2 409.0	3 495.0	4 227.0	4 440.0	3 878.0	Autres capitaux, nia
Net Errors and Omissions	−753.0	−933.0	−1 315.0	744.0	91.0	−1 279.0	−3 078.0	Erreurs et omissions nettes
Reserves and Related Items	−630.0	113.0	−495.0	−2 251.0	−1 528.0	−2 070.0	−587.0	Rés. et postes assimilables
Iran, Islamic Rep. of [3]								**Iran, Rép. islamique d' [3]**
Merchandise: Exp. fob	11 916.0	10 709.0	13 081.0	19 305.0	18 661.0	19 868.0	...	Marchandises : exp. fob
Merchandise: Imp. fob	−12 005.0	−10 608.0	−13 448.0	−18 330.0	−25 190.0	−23 274.0	...	Marchandises : imp. fob
Serv. & Income: Credit	437.0	467.0	798.0	892.0	881.0	846.0	...	Serv. & revenu : crédit
Serv. & Income: Debit	−2 438.0	−2 437.0	−3 122.0	−4 040.0	−5 800.0	−5 940.0	...	Serv. & revenu : débit
Private Unrequited Transfers	0.0	0.0	2 500.0	2 500.0	2 000.0	1 996.0	...	Transf. priv.sans contrep.
Offic. Unrequited Trans.,nie	0.0	0.0	0.0	0.0	0.0	0.0	...	Tr. offic. sans contrep.nia
Direct Investment, nie	0.0	0.0	0.0	0.0	0.0	0.0	...	Investissements direct, nia
Portfolio Investments, nie	0.0	0.0	0.0	0.0	0.0	0.0	...	Invest. portefeuille, nia
Other Capital, nie	1 711.0	320.0	3 261.0	295.0	6 033.0	4 703.0	...	Autres capitaux, nia
Net Errors and Omissions	155.0	539.0	−770.0	−946.0	1 321.0	1 636.0	...	Erreurs et omissions nettes
Reserves and Related Items	224.0	1 010.0	−2 300.0	324.0	2 094.0	165.0	...	Rés. et postes assimilables
Israel								**Israël**
Merchandise: Exp. fob	9 249.0	10 286.0	11 067.0	12 139.0	12 029.0	13 314.0	14 804.0	Marchandises : exp. fob
Merchandise: Imp. fob	−12 870.0	−13 099.0	−12 900.0	−15 120.0	−16 946.0	−18 260.0	−20 411.0	Marchandises : imp. fob
Serv. & Income: Credit	4 708.0	5 198.0	5 686.0	6 196.0	6 531.0	7 466.0	7 339.0	Serv. & revenu : crédit
Serv. & Income: Debit	−6 666.0	−7 280.0	−7 717.0	−8 564.0	−8 704.0	−9 186.0	−9 852.0	Serv. & revenu : débit
Private Unrequited Transfers	1 491.0	1 486.0	1 853.0	2 123.0	2 282.0	2 722.0	2 853.0	Transf. priv.sans contrep.
Offic. Unrequited Trans.,nie	3 407.0	3 369.0	3 246.0	3 783.0	4 392.0	4 163.0	3 894.0	Tr. offic. sans contrep.nia
Direct Investment, nie	152.0	173.0	87.0	−65.0	−73.0	−112.0	−374.0	Investissements direct, nia
Portfolio Investments, nie	158.0	4 172.0	1 023.0	−210.0	548.0	−740.0	1 750.0	Invest. portefeuille, nia
Other Capital, nie	1 012.0	−4 935.0	−2 174.0	−40.0	−564.0	−1 268.0	624.0	Autres capitaux, nia
Net Errors and Omissions	9.0	−539.0	1 227.0	272.0	332.0	445.0	854.0	Erreurs et omissions nettes
Reserves and Related Items	−661.0	1 170.0	−1 398.0	−515.0	173.0	1 457.0	−1 481.0	Rés. et postes assimilables
Japan [1]								**Japon [1]**
Merchandise: Exp. fob	224.6	259.8	269.6	280.4	306.6	330.9	351.3	Marchandises : exp. fob
Merchandise: Imp. fob	−128.2	−164.8	−192.7	−216.8	−203.5	−198.5	−209.7	Marchandises : imp. fob

78
Summary of balance of payments
Millions of US dollars
Résumé des balances des paiements
Millions de dollars des E−U

Country or area	1987	1988	1989	1990	1991	1992	1993	Pays ou zone
Serv. & Income: Credit	79.7	111.8	143.9	166.0	188.6	194.1	203.7	Serv. & revenu : crédit
Serv. & Income: Debit	−85.4	−123.1	−159.5	−188.2	−206.3	−204.2	−207.6	Serv. & revenu : débit
Private Unrequited Transfers	−1.0	−1.1	−1.0	−1.0	−0.7	−1.3	−2.3	Transf. priv.sans contrep.
Offic. Unrequited Trans.,nie	−2.7	−3.0	−3.3	−4.5	−11.8	−3.3	−3.8	Tr. offic. sans contrep.nia
Direct Investment, nie	−18.4	−34.7	−45.2	−46.3	−29.4	−14.5	−13.6	Investissements direct, nia
Portfolio Investments, nie	−91.3	−52.8	−32.5	−14.5	35.5	−28.4	−65.7	Invest. portefeuille, nia
Other Capital, nie	64.3	21.3	29.8	39.2	−77.9	−63.6	−24.3	Autres capitaux, nia
Net Errors and Omissions	−3.7	3.1	−21.8	−20.9	−7.7	−10.5	−0.3	Erreurs et omissions nettes
Reserves and Related Items	−37.9	−16.5	12.8	6.6	6.6	−0.6	−27.7	Rés. et postes assimilables
Jordan								**Jordanie**
Merchandise: Exp. fob	933.1	1 007.4	1 109.4	1 063.8	1 129.5	1 218.9	1 246.3	Marchandises : exp. fob
Merchandise: Imp. fob	−2 400.1	−2 418.7	−1 882.5	−2 300.7	−2 302.2	−2 998.7	−3 145.2	Marchandises : imp. fob
Serv. & Income: Credit	1 350.0	1 461.3	1 278.2	1 514.5	1 465.5	1 561.7	1 672.8	Serv. & revenu : crédit
Serv. & Income: Debit	−1 576.9	−1 695.2	−1 298.9	−1 549.6	−1 570.2	−1 714.7	−1 597.3	Serv. & revenu : débit
Private Unrequited Transfers	742.9	799.8	565.4	457.4	408.1	781.3	997.1	Transf. priv.sans contrep.
Offic. Unrequited Trans.,nie	599.1	551.7	613.2	587.6	475.7	386.3	356.6	Tr. offic. sans contrep.nia
Direct Investment, nie	38.3	23.8	−18.1	69.1	−25.6	44.1	19.5	Investissements direct, nia
Portfolio Investments, nie	0.0	0.0	0.0	0.0	0.0	0.0	0.0	Invest. portefeuille, nia
Other Capital, nie	426.9	350.4	97.6	503.6	2 122.9	952.2	−699.3	Autres capitaux, nia
Net Errors and Omissions	27.9	123.4	0.3	75.4	321.4	160.9	777.6	Erreurs et omissions nettes
Reserves and Related Items	−141.3	−203.9	−464.7	−421.0	−2 025.2	−392.0	372.0	Rés. et postes assimilables
Korea, Republic of								**Corée, République de**
Merchandise: Exp. fob	46 244.0	59 648.0	61 408.0	63 123.0	69 581.0	75 169.0	80 950.0	Marchandises : exp. fob
Merchandise: Imp. fob	−38 585.0	−48 203.0	−56 811.0	−65 127.0	−76 561.0	−77 315.0	−79 090.0	Marchandises : imp. fob
Serv. & Income: Credit	10 011.0	11 252.0	12 643.0	14 269.0	15 531.0	16 010.0	18 253.0	Serv. & revenu : crédit
Serv. & Income: Debit	−9 034.0	−9 984.0	−12 432.0	−14 712.0	−17 124.0	−18 625.0	−20 220.0	Serv. & revenu : débit
Private Unrequited Transfers	1 199.0	1 404.0	200.0	266.0	20.0	257.0	633.0	Transf. priv.sans contrep.
Offic. Unrequited Trans.,nie	19.0	44.0	48.0	9.0	−173.0	−25.0	−142.0	Tr. offic. sans contrep.nia
Direct Investment, nie	418.0	720.0	453.0	−105.0	−241.0	−497.0	−540.0	Investissements direct, nia
Portfolio Investments, nie	−113.0	−482.0	−29.0	811.0	3 116.0	5 742.0	10 725.0	Invest. portefeuille, nia
Other Capital, nie	−9 239.0	−4 492.0	−3 050.0	2 263.0	3 950.0	1 909.0	−6 845.0	Autres capitaux, nia
Net Errors and Omissions	1 184.0	−591.0	690.0	−2 005.0	753.0	1 099.0	−715.0	Erreurs et omissions nettes
Reserves and Related Items	−2 104.0	−9 316.0	−3 120.0	1 208.0	1 148.0	−3 724.0	−3 009.0	Rés. et postes assimilables
Kuwait								**Koweït**
Merchandise: Exp. fob	8 221.0	7 709.0	11 396.0	6 989.0	869.0	6 548.0	10 413.0	Marchandises : exp. fob
Merchandise: Imp. fob	−4 726.0	−5 394.0	−5 470.0	−3 411.0	−4 053.0	−6 292.0	−6 040.0	Marchandises : imp. fob
Serv. & Income: Credit	7 145.0	8 974.0	10 498.0	9 665.0	7 078.0	7 404.0	6 262.0	Serv. & revenu : crédit
Serv. & Income: Debit	−4 547.0	−4 691.0	−5 038.0	−4 129.0	−5 745.0	−5 416.0	−2 932.0	Serv. & revenu : débit
Private Unrequited Transfers	−1 102.0	−1 179.0	−1 283.0	−770.0	−426.0	−829.0	−1 229.0	Transf. priv.sans contrep.
Offic. Unrequited Trans.,nie	−158.0	−140.0	−211.0	−4 181.0	−23 372.0	−1 098.0	−129.0	Tr. offic. sans contrep.nia
Direct Investment, nie	−571.0	−462.0	−841.0	−208.0	−243.0	−1 067.0	−775.0	Investissements direct, nia
Portfolio Investments, nie	50.0	−459.0	−568.0	−381.0	−612.0	263.0	−46.0	Invest. portefeuille, nia
Other Capital, nie	−4 823.0	−6 458.0	−6 907.0	1 234.0	39 628.0	12 063.0	−2 329.0	Autres capitaux, nia
Net Errors and Omissions	−1 337.0	172.0	−301.0	−5 706.0	−11 849.0	−9 723.0	−4 679.0	Erreurs et omissions nettes
Reserves and Related Items	1 847.0	1 928.0	−1 275.0	897.0	−1 276.0	−1 851.0	1 485.0	Rés. et postes assimilables
Lao People's Dem. Rep.								**Rép. dém. pop. lao**
Merchandise: Exp. fob	64.3	57.8	63.3	78.7	96.6	132.6	231.8	Marchandises : exp. fob
Merchandise: Imp. fob	−216.2	−149.4	−193.8	−185.5	−197.9	−232.8	−379.5	Marchandises : imp. fob
Serv. & Income: Credit	26.3	18.2	23.3	25.9	41.1	67.0	91.8	Serv. & revenu : crédit
Serv. & Income: Debit	−37.9	−33.0	−35.5	−33.5	−63.4	−84.1	−95.6	Serv. & revenu : débit
Private Unrequited Transfers	3.5	6.7	8.3	10.9	10.4	8.6	9.5	Transf. priv.sans contrep.
Offic. Unrequited Trans.,nie	27.0	24.9	54.8	55.6	61.0	60.7	98.8	Tr. offic. sans contrep.nia
Direct Investment, nie	0.0	2.0	4.0	6.0	8.0	9.0	29.9	Investissements direct, nia
Portfolio Investments, nie	0.0	0.0	0.0	0.0	0.0	0.0	0.0	Invest. portefeuille, nia
Other Capital, nie	111.4	55.7	95.8	79.4	88.2	52.8	−1.1	Autres capitaux, nia
Net Errors and Omissions	13.5	21.2	−24.6	−36.3	−28.8	−9.4	30.2	Erreurs et omissions nettes
Reserves and Related Items	8.1	−4.1	4.4	−1.2	−15.2	−4.4	−15.8	Rés. et postes assimilables
Malaysia								**Malaisie**
Merchandise: Exp. fob	17 754.0	20 852.0	24 633.0	28 636.0	33 534.0	39 613.0	45 984.0	Marchandises : exp. fob
Merchandise: Imp. fob	−11 918.0	−15 306.0	−20 251.0	−26 014.0	−33 007.0	−36 238.0	−42 801.0	Marchandises : imp. fob
Serv. & Income: Credit	3 229.0	3 597.0	4 185.0	5 878.0	5 978.0	7 042.0	7 594.0	Serv. & revenu : crédit
Serv. & Income: Debit	−6 567.0	−7 484.0	−8 390.0	−9 472.0	−10 776.0	−12 385.0	−13 402.0	Serv. & revenu : débit
Private Unrequited Transfers	−25.0	70.0	−17.0	3.0	29.0	65.0	84.0	Transf. priv.sans contrep.
Offic. Unrequited Trans.,nie	69.0	81.0	97.0	51.0	8.0	67.0	75.0	Tr. offic. sans contrep.nia
Direct Investment, nie	423.0	719.0	1 668.0	2 332.0	3 998.0	5 183.0	5 206.0	Investissements direct, nia
Portfolio Investments, nie	140.0	−448.0	−107.0	−255.0	170.0	−1 108.0	−984.0	Invest. portefeuille, nia

78
Summary of balance of payments
Millions of US dollars
Résumé des balances des paiements
Millions de dollars des E−U

Country or area	1987	1988	1989	1990	1991	1992	1993	Pays ou zone
Other Capital, nie	−2 100.0	−2 245.0	−230.0	−292.0	1 454.0	4 708.0	6 785.0	Autres capitaux, nia
Net Errors and Omissions	114.0	−267.0	−358.0	1 085.0	−151.0	−292.0	2 802.0	Erreurs et omissions nettes
Reserves and Related Items	−1 119.0	430.0	−1 230.0	−1 953.0	−1 238.0	−6 655.0	−11 343.0	Rés. et postes assimilables
Maldives								**Maldives**
Merchandise: Exp. fob	34.9	44.6	51.3	58.1	59.2	51.1	38.5	Marchandises : exp. fob
Merchandise: Imp. fob	−66.5	−87.3	−111.3	−121.2	−141.8	−167.9	−177.8	Marchandises : imp. fob
Serv. & Income: Credit	77.6	85.3	102.6	124.4	128.9	171.1	181.2	Serv. & revenu : crédit
Serv. & Income: Debit	−45.3	−40.2	−45.2	−56.5	−60.8	−69.4	−77.8	Serv. & revenu : débit
Private Unrequited Transfers	−2.4	−5.0	−5.1	−7.4	−16.6	−18.9	−20.0	Transf. priv.sans contrep.
Offic. Unrequited Trans.,nie	9.7	11.5	18.3	11.2	22.1	14.3	8.3	Tr. offic. sans contrep.nia
Direct Investment, nie	5.1	1.2	4.4	5.6	6.5	6.6	6.9	Investissements direct, nia
Portfolio Investments, nie	0.0	0.0	0.0	0.0	0.0	0.0	0.0	Invest. portefeuille, nia
Other Capital, nie	−10.5	−2.7	7.4	1.7	−1.1	18.9	18.0	Autres capitaux, nia
Net Errors and Omissions	−1.4	6.2	−20.1	−17.0	2.6	−1.0	22.6	Erreurs et omissions nettes
Reserves and Related Items	−1.2	−13.6	−2.3	1.1	1.0	−4.8	0.1	Rés. et postes assimilables
Mongolia								**Mongolie**
Merchandise: Exp. fob	817.1	829.1	795.8	468.1	...	...	...	Marchandises : exp. fob
Merchandise: Imp. fob	−1 827.6	−1 849.9	−1 913.6	−1 051.0	...	...	...	Marchandises : imp. fob
Serv. & Income: Credit	89.2	94.5	43.9	53.2	...	...	...	Serv. & revenu : crédit
Serv. & Income: Debit	−69.0	−106.7	−160.5	−121.3	...	...	...	Serv. & revenu : débit
Private Unrequited Transfers	−0.3	−0.3	0.0	0.0	...	...	...	Transf. priv.sans contrep.
Offic. Unrequited Trans.,nie	0.0	0.0	3.9	7.4	...	...	...	Tr. offic. sans contrep.nia
Direct Investment, nie	0.0	0.0	0.0	0.0	...	...	...	Investissements direct, nia
Portfolio Investments, nie	0.0	0.0	0.0	0.0	...	...	...	Invest. portefeuille, nia
Other Capital, nie	1 143.4	1 019.4	1 318.7	546.8	...	...	...	Autres capitaux, nia
Net Errors and Omissions	−76.5	14.6	41.5	−4.8	...	...	...	Erreurs et omissions nettes
Reserves and Related Items	−76.3	−0.7	−129.7	101.6	...	...	...	Rés. et postes assimilables
Myanmar								**Myanmar**
Merchandise: Exp. fob	219.6	165.7	222.8	222.6	...	...	...	Marchandises : exp. fob
Merchandise: Imp. fob	−452.7	−370.2	−304.3	−524.3	...	...	...	Marchandises : imp. fob
Serv. & Income: Credit	75.4	49.6	59.3	95.9	...	...	...	Serv. & revenu : crédit
Serv. & Income: Debit	−129.3	−113.5	−101.4	−264.4	...	...	...	Serv. & revenu : débit
Private Unrequited Transfers	8.4	8.6	14.3	10.2	...	...	...	Transf. priv.sans contrep.
Offic. Unrequited Trans.,nie	98.7	83.9	41.4	28.8	...	...	...	Tr. offic. sans contrep.nia
Direct Investment, nie	0.0	0.0	7.8	161.1	...	...	...	Investissements direct, nia
Portfolio Investments, nie	0.0	0.0	84.0	232.9	...	...	...	Invest. portefeuille, nia
Other Capital, nie	202.9	139.7	74.2	24.6	...	...	...	Autres capitaux, nia
Net Errors and Omissions	14.7	116.7	52.7	21.3	...	...	...	Erreurs et omissions nettes
Reserves and Related Items	−37.7	−80.5	−150.6	−8.7	...	...	...	Rés. et postes assimilables
Nepal								**Népal**
Merchandise: Exp. fob	162.2	193.8	161.2	217.9	274.5	376.3	397.0	Marchandises : exp. fob
Merchandise: Imp. fob	−512.4	−664.9	−568.1	−666.6	−756.9	−752.1	−858.6	Marchandises : imp. fob
Serv. & Income: Credit	224.5	240.0	224.3	229.5	266.9	307.3	362.1	Serv. & revenu : crédit
Serv. & Income: Debit	−137.9	−165.3	−158.2	−178.6	−200.2	−241.8	−275.5	Serv. & revenu : débit
Private Unrequited Transfers	67.2	60.1	48.5	60.4	53.7	45.7	74.3	Transf. priv.sans contrep.
Offic. Unrequited Trans.,nie	73.1	64.8	49.0	48.2	57.5	83.2	78.2	Tr. offic. sans contrep.nia
Direct Investment, nie	0.0	0.0	0.0	0.0	0.0	0.0	0.0	Investissements direct, nia
Portfolio Investments, nie	0.0	0.0	0.0	0.0	0.0	0.0	0.0	Invest. portefeuille, nia
Other Capital, nie	190.7	252.7	196.1	304.5	457.1	335.9	283.5	Autres capitaux, nia
Net Errors and Omissions	−3.6	12.5	5.2	4.9	10.7	0.8	4.6	Erreurs et omissions nettes
Reserves and Related Items	−63.8	6.3	42.1	−20.2	−163.4	−155.4	−65.6	Rés. et postes assimilables
Oman								**Oman**
Merchandise: Exp. fob	3 805.0	3 342.0	4 068.0	5 508.0	4 871.0	5 555.0	5 365.0	Marchandises : exp. fob
Merchandise: Imp. fob	−1 769.0	−2 107.0	−2 225.0	−2 623.0	−3 112.0	−3 627.0	−4 030.0	Marchandises : imp. fob
Serv. & Income: Credit	533.0	270.0	397.0	443.0	418.0	341.0	434.0	Serv. & revenu : crédit
Serv. & Income: Debit	−1 091.0	−1 095.0	−1 159.0	−1 347.0	−1 547.0	−1 569.0	−1 528.0	Serv. & revenu : débit
Private Unrequited Transfers	−702.0	−762.0	−791.0	−817.0	−871.0	−1 181.0	−1 329.0	Transf. priv.sans contrep.
Offic. Unrequited Trans.,nie	8.0	42.0	16.0	−57.0	−3.0	−16.0	18.0	Tr. offic. sans contrep.nia
Direct Investment, nie	35.0	92.0	112.0	141.0	149.0	87.0	99.0	Investissements direct, nia
Portfolio Investments, nie	0.0	0.0	0.0	0.0	0.0	0.0	0.0	Invest. portefeuille, nia
Other Capital, nie	−225.0	130.0	−127.0	−640.0	372.0	226.0	−49.0	Autres capitaux, nia
Net Errors and Omissions	−486.0	−379.0	33.0	−472.0	253.0	462.0	−39.0	Erreurs et omissions nettes
Reserves and Related Items	−108.0	467.0	−324.0	−135.0	−530.0	−280.0	1 058.0	Rés. et postes assimilables
Pakistan								**Pakistan**
Merchandise: Exp. fob	3 938.0	4 405.0	4 796.0	5 380.0	6 381.0	6 880.0	6 760.0	Marchandises : exp. fob
Merchandise: Imp. fob	−6 254.0	−7 097.0	−7 366.0	−8 094.0	−8 642.0	−9 671.0	−9 312.0	Marchandises : imp. fob

78
Summary of balance of payments
Millions of US dollars
Résumé des balances des paiements
Millions de dollars des E−U

Country or area	1987	1988	1989	1990	1991	1992	1993	Pays ou zone
Serv. & Income: Credit	1 083.0	946.0	1 323.0	1 518.0	1 596.0	1 625.0	1 577.0	Serv. & revenu : crédit
Serv. & Income: Debit	−2 187.0	−2 393.0	−2 808.0	−3 238.0	−3 559.0	−4 148.0	−4 237.0	Serv. & revenu : débit
Private Unrequited Transfers	2 440.0	2 101.0	2 207.0	2 276.0	2 344.0	3 068.0	1 942.0	Transf. priv.sans contrep.
Offic. Unrequited Trans.,nie	418.0	615.0	513.0	504.0	483.0	377.0	334.0	Tr. offic. sans contrep.nia
Direct Investment, nie	110.0	173.0	167.0	242.0	261.0	346.0	348.0	Investissements direct, nia
Portfolio Investments, nie	132.0	126.0	15.0	87.0	92.0	370.0	292.0	Invest. portefeuille, nia
Other Capital, nie	127.0	1 248.0	957.0	843.0	538.0	1 840.0	2 309.0	Autres capitaux, nia
Net Errors and Omissions	17.0	23.0	−242.0	−103.0	−78.0	121.0	−91.0	Erreurs et omissions nettes
Reserves and Related Items	177.0	−147.0	438.0	585.0	584.0	−808.0	78.0	Rés. et postes assimilables
Philippines								**Philippines**
Merchandise: Exp. fob	5 720.0	7 074.0	7 821.0	8 186.0	8 840.0	9 824.0	11 375.0	Marchandises : exp. fob
Merchandise: Imp. fob	−6 737.0	−8 159.0	−10 419.0	−12 206.0	−12 051.0	−14 519.0	−17 597.0	Marchandises : imp. fob
Serv. & Income: Credit	3 454.0	3 592.0	4 586.0	4 842.0	5 623.0	7 497.0	7 528.0	Serv. & revenu : crédit
Serv. & Income: Debit	−3 454.0	−3 672.0	−4 274.0	−4 231.0	−4 273.0	−4 618.0	−5 294.0	Serv. & revenu : débit
Private Unrequited Transfers	376.0	500.0	473.0	357.0	473.0	473.0	398.0	Transf. priv.sans contrep.
Offic. Unrequited Trans.,nie	197.0	275.0	357.0	357.0	354.0	344.0	301.0	Tr. offic. sans contrep.nia
Direct Investment, nie	307.0	936.0	563.0	530.0	544.0	228.0	763.0	Investissements direct, nia
Portfolio Investments, nie	19.0	50.0	280.0	−50.0	110.0	40.0	−164.0	Invest. portefeuille, nia
Other Capital, nie	−8.0	−415.0	511.0	1 577.0	2 273.0	2 940.0	2 687.0	Autres capitaux, nia
Net Errors and Omissions	68.0	493.0	402.0	593.0	−138.0	−520.0	292.0	Erreurs et omissions nettes
Reserves and Related Items	58.0	−674.0	−300.0	45.0	−1 755.0	−1 689.0	−289.0	Rés. et postes assimilables
Saudi Arabia								**Arabie saoudite**
Merchandise: Exp. fob	23 138.0	24 315.0	28 312.0	44 246.0	47 623.0	47 049.0	44 918.0	Marchandises : exp. fob
Merchandise: Imp. fob	−18 283.0	−19 805.0	−19 231.0	−21 490.0	−25 968.0	−30 248.0	−25 897.0	Marchandises : imp. fob
Serv. & Income: Credit	13 113.0	12 809.0	13 015.0	12 398.0	11 728.0	10 855.0	9 680.0	Serv. & revenu : crédit
Serv. & Income: Debit	−19 506.0	−15 650.0	−20 892.0	−23 634.0	−40 737.0	−33 726.0	−26 262.0	Serv. & revenu : débit
Private Unrequited Transfers	−4 935.0	−6 510.0	−8 542.0	−11 236.0	−13 746.0	−13 397.0	−15 717.0	Transf. priv.sans contrep.
Offic. Unrequited Trans.,nie	−3 300.0	−2 499.0	−2 200.0	−4 401.0	−6 489.0	−1 501.0	−940.0	Tr. offic. sans contrep.nia
Direct Investment, nie	−1 175.0	−328.0	−654.0	1 864.0	160.0	−79.0	−79.0	Investissements direct, nia
Portfolio Investments, nie	6 150.0	3 057.0	−1 786.0	−3 340.0	470.0	−3 646.0	8 448.0	Invest. portefeuille, nia
Other Capital, nie	7 438.0	3 092.0	8 470.0	218.0	27 008.0	19 028.0	7 345.0	Autres capitaux, nia
Net Errors and Omissions	0.0	0.0	0.0	0.0	0.0	0.0	0.0	Erreurs et omissions nettes
Reserves and Related Items	−2 640.0	1 519.0	3 508.0	5 376.0	−49.0	5 664.0	−1 496.0	Rés. et postes assimilables
Singapore								**Singapour**
Merchandise: Exp. fob	27 464.0	37 993.0	43 572.0	51 095.0	57 156.0	62 068.0	71 959.0	Marchandises : exp. fob
Merchandise: Imp. fob	−29 910.0	−40 338.0	−45 687.0	−55 812.0	−60 948.0	−67 850.0	−80 025.0	Marchandises : imp. fob
Serv. & Income: Credit	10 348.0	13 162.0	16 414.0	21 907.0	24 285.0	27 200.0	29 970.0	Serv. & revenu : crédit
Serv. & Income: Debit	−7 825.0	−9 629.0	−11 148.0	−14 652.0	−16 003.0	−17 075.0	−19 169.0	Serv. & revenu : débit
Private Unrequited Transfers	−170.0	−209.0	−242.0	−274.0	−340.0	−405.0	−482.0	Transf. priv.sans contrep.
Offic. Unrequited Trans.,nie	−64.0	−90.0	−125.0	−169.0	−159.0	−190.0	−215.0	Tr. offic. sans contrep.nia
Direct Investment, nie	2 630.0	3 537.0	2 004.0	4 004.0	4 444.0	5 982.0	6 062.0	Investissements direct, nia
Portfolio Investments, nie	252.0	−293.0	−76.0	−1 140.0	−802.0	−819.0	−944.0	Invest. portefeuille, nia
Other Capital, nie	−2 412.0	−2 256.0	−1 473.0	1 794.0	−2 708.0	397.0	4 335.0	Autres capitaux, nia
Net Errors and Omissions	782.0	−217.0	−503.0	−1 322.0	−728.0	−3 208.0	−3 913.0	Erreurs et omissions nettes
Reserves and Related Items	−1 095.0	−1 659.0	−2 738.0	−5 431.0	−4 198.0	−6 100.0	−7 578.0	Rés. et postes assimilables
Sri Lanka								**Sri Lanka**
Merchandise: Exp. fob	1 393.9	1 477.1	1 505.1	1 853.0	2 003.3	2 301.4	2 785.7	Marchandises : exp. fob
Merchandise: Imp. fob	−1 866.0	−2 017.5	−2 055.1	−2 325.6	−2 808.0	−3 016.5	−3 527.8	Marchandises : imp. fob
Serv. & Income: Credit	397.5	407.9	404.2	532.6	601.1	689.5	745.1	Serv. & revenu : crédit
Serv. & Income: Debit	−744.0	−787.9	−787.1	−898.9	−995.0	−1 069.3	−1 103.6	Serv. & revenu : débit
Private Unrequited Transfers	312.8	320.0	330.7	362.4	401.3	461.7	559.8	Transf. priv.sans contrep.
Offic. Unrequited Trans.,nie	179.9	206.1	188.6	178.1	202.4	182.6	160.2	Tr. offic. sans contrep.nia
Direct Investment, nie	58.2	43.6	17.7	42.5	43.8	121.0	187.6	Investissements direct, nia
Portfolio Investments, nie	0.0	0.0	0.0	0.0	32.1	25.7	67.2	Invest. portefeuille, nia
Other Capital, nie	336.7	212.4	559.3	435.6	613.2	354.6	669.6	Autres capitaux, nia
Net Errors and Omissions	−122.5	37.4	−115.0	−115.1	225.6	173.3	131.1	Erreurs et omissions nettes
Reserves and Related Items	53.6	101.2	−48.3	−64.8	−319.9	−223.9	−674.9	Rés. et postes assimilables
Syrian Arab Republic								**Rép. arabe syrienne**
Merchandise: Exp. fob	1 357.0	1 348.0	3 013.0	4 156.0	3 438.0	3 100.0	3 153.0	Marchandises : exp. fob
Merchandise: Imp. fob	−2 226.0	−1 986.0	−1 821.0	−2 062.0	−2 354.0	−2 941.0	−3 475.0	Marchandises : imp. fob
Serv. & Income: Credit	625.0	689.0	915.0	919.0	1 129.0	1 349.0	1 578.0	Serv. & revenu : crédit
Serv. & Income: Debit	−1 148.0	−1 097.0	−1 537.0	−1 723.0	−2 098.0	−2 316.0	−2 503.0	Serv. & revenu : débit
Private Unrequited Transfers	334.0	360.0	430.0	385.0	350.0	550.0	600.0	Transf. priv.sans contrep.
Offic. Unrequited Trans.,nie	760.0	536.0	223.0	88.0	234.0	313.0	40.0	Tr. offic. sans contrep.nia
Direct Investment, nie	0.0	0.0	0.0	0.0	0.0	0.0	0.0	Investissements direct, nia
Portfolio Investments, nie	0.0	0.0	0.0	0.0	0.0	0.0	0.0	Invest. portefeuille, nia

78
Summary of balance of payments
Millions of US dollars
Résumé des balances des paiements
Millions de dollars des E−U

Country or area	1987	1988	1989	1990	1991	1992	1993	Pays ou zone
Other Capital, nie	399.0	85.0	−1 708.0	−1 836.0	−515.0	−50.0	571.0	Autres capitaux, nia
Net Errors and Omissions	−23.0	34.0	420.0	110.0	−112.0	70.0	100.0	Erreurs et omissions nettes
Reserves and Related Items	−79.0	32.0	66.0	−36.0	−72.0	−76.0	−64.0	Rés. et postes assimilables
Thailand								**Thaïlande**
Merchandise: Exp. fob	11 595.0	15 781.0	19 834.0	22 811.0	28 232.0	32 100.0	36 410.0	Marchandises : exp. fob
Merchandise: Imp. fob	−12 019.0	−17 856.0	−22 750.0	−29 561.0	−34 222.0	−36 261.0	−40 556.0	Marchandises : imp. fob
Serv. & Income: Credit	4 168.0	5 944.0	7 046.0	8 478.0	9 526.0	10 769.0	12 355.0	Serv. & revenu : crédit
Serv. & Income: Debit	−4 334.0	−5 760.0	−6 874.0	−9 222.0	−11 369.0	−13 340.0	−15 449.0	Serv. & revenu : débit
Private Unrequited Transfers	100.0	47.0	47.0	26.0	163.0	323.0	281.0	Transf. priv.sans contrep.
Offic. Unrequited Trans.,nie	125.0	189.0	199.0	187.0	98.0	54.0	32.0	Tr. offic. sans contrep.nia
Direct Investment, nie	182.0	1 081.0	1 726.0	2 303.0	1 847.0	1 969.0	1 493.0	Investissements direct, nia
Portfolio Investments, nie	346.0	530.0	1 486.0	−38.0	−81.0	927.0	5 455.0	Invest. portefeuille, nia
Other Capital, nie	534.0	2 228.0	3 387.0	6 833.0	9 993.0	6 900.0	7 495.0	Autres capitaux, nia
Net Errors and Omissions	248.0	411.0	928.0	1 419.0	431.0	−517.0	−347.0	Erreurs et omissions nettes
Reserves and Related Items	−945.0	−2 596.0	−5 029.0	−3 235.0	−4 618.0	−2 925.0	−7 169.0	Rés. et postes assimilables
Turkey								**Turquie**
Merchandise: Exp. fob	10 322.0	11 929.0	11 780.0	13 026.0	13 672.0	14 892.0	15 610.0	Marchandises : exp. fob
Merchandise: Imp. fob	−13 551.0	−13 706.0	−15 999.0	−22 581.0	−20 998.0	−23 082.0	−29 772.0	Marchandises : imp. fob
Serv. & Income: Credit	4 195.0	6 026.0	7 098.0	8 933.0	9 315.0	10 451.0	11 843.0	Serv. & revenu : crédit
Serv. & Income: Debit	−4 162.0	−4 812.0	−5 476.0	−6 496.0	−6 816.0	−7 262.0	−7 829.0	Serv. & revenu : débit
Private Unrequited Transfers	2 066.0	1 827.0	3 135.0	3 349.0	2 854.0	3 147.0	3 035.0	Transf. priv.sans contrep.
Offic. Unrequited Trans.,nie	324.0	332.0	423.0	1 144.0	2 245.0	912.0	733.0	Tr. offic. sans contrep.nia
Direct Investment, nie	106.0	354.0	663.0	700.0	783.0	779.0	622.0	Investissements direct, nia
Portfolio Investments, nie	282.0	1 178.0	1 386.0	547.0	623.0	2 411.0	3 917.0	Invest. portefeuille, nia
Other Capital, nie	1 503.0	−2 490.0	−1 269.0	2 790.0	−3 803.0	458.0	4 424.0	Autres capitaux, nia
Net Errors and Omissions	−505.0	515.0	969.0	−469.0	926.0	−1 222.0	−2 275.0	Erreurs et omissions nettes
Reserves and Related Items	−580.0	−1 153.0	−2 710.0	−943.0	1 199.0	−1 484.0	−308.0	Rés. et postes assimilables

Europe· Europe

Country or area	1987	1988	1989	1990	1991	1992	1993	Pays ou zone
Albania								**Albanie**
Merchandise: Exp. fob	311.2	344.6	393.7	322.1	73.0	70.0	111.6	Marchandises : exp. fob
Merchandise: Imp. fob	−315.6	−382.3	−455.8	−455.9	−281.0	−540.5	−601.5	Marchandises : imp. fob
Serv. & Income: Credit	23.4	30.5	40.6	31.5	10.0	22.9	142.5	Serv. & revenu : crédit
Serv. & Income: Debit	−20.7	−26.9	−28.4	−31.0	−59.3	−126.8	−192.9	Serv. & revenu : débit
Private Unrequited Transfers	6.9	7.0	9.1	15.0	8.1	149.7	273.1	Transf. priv.sans contrep.
Offic. Unrequited Trans.,nie	0.0	0.0	1.5	0.0	81.2	374.0	282.1	Tr. offic. sans contrep.nia
Direct Investment, nie	0.0	0.0	0.0	0.0	0.0	20.0	58.0	Investissements direct, nia
Portfolio Investments, nie	0.0	0.0	0.0	0.0	0.0	0.0	0.0	Invest. portefeuille, nia
Other Capital, nie	−5.9	139.3	359.4	−117.7	−181.2	−52.2	−13.9	Autres capitaux, nia
Net Errors and Omissions	12.3	22.0	4.8	−2.0	125.2	47.4	−10.3	Erreurs et omissions nettes
Reserves and Related Items	−11.6	−134.2	−324.9	238.0	224.0	35.5	−48.7	Rés. et postes assimilables
Austria								**Autriche**
Merchandise: Exp. fob	26 582.0	30 108.0	31 901.0	40 336.0	40 285.0	43 386.0	39 257.0	Marchandises : exp. fob
Merchandise: Imp. fob	−31 406.0	−34 889.0	−37 482.0	−47 348.0	−48 882.0	−52 228.0	−47 082.0	Marchandises : imp. fob
Serv. & Income: Credit	19 644.0	22 923.0	25 225.0	32 502.0	35 172.0	38 661.0	38 464.0	Serv. & revenu : crédit
Serv. & Income: Debit	−15 010.0	−18 353.0	−19 280.0	−24 319.0	−26 383.0	−29 582.0	−30 593.0	Serv. & revenu : débit
Private Unrequited Transfers	−10.0	37.0	−57.0	110.0	26.0	−742.0	−684.0	Transf. priv.sans contrep.
Offic. Unrequited Trans.,nie	−70.0	−73.0	−71.0	−108.0	−102.0	−198.0	−236.0	Tr. offic. sans contrep.nia
Direct Investment, nie	96.0	126.0	−280.0	−1 047.0	−933.0	−1 056.0	−625.0	Investissements direct, nia
Portfolio Investments, nie	344.0	1 924.0	1 547.0	1 368.0	1 180.0	1 970.0	8 738.0	Invest. portefeuille, nia
Other Capital, nie	594.0	−1 635.0	100.0	−339.0	−259.0	129.0	−4 254.0	Autres capitaux, nia
Net Errors and Omissions	−433.0	322.0	−613.0	−1 170.0	731.0	2 253.0	−782.0	Erreurs et omissions nettes
Reserves and Related Items	−333.0	−491.0	−990.0	15.0	−836.0	−2 593.0	−2 202.0	Rés. et postes assimilables
Belgium−Luxembourg								**Belgique−Luxembourg**
Merchandise: Exp. fob	76 088.0	85 496.0	89 988.0	107 654.0	106 019.0	113 638.0	103 837.0	Marchandises : exp. fob
Merchandise: Imp. fob	−76 268.0	−84 273.0	−89 020.0	−107 064.0	−106 085.0	−112 307.0	−99 905.0	Marchandises : imp. fob
Serv. & Income: Credit	48 264.0	56 112.0	71 850.0	96 971.0	108 251.0	125 144.0	120 527.0	Serv. & revenu : crédit
Serv. & Income: Debit	−43 864.0	−51 986.0	−67 902.0	−90 646.0	−101 703.0	−117 520.0	−109 283.0	Serv. & revenu : débit
Private Unrequited Transfers	−114.0	41.0	47.0	−597.0	−280.0	−503.0	−602.0	Transf. priv.sans contrep.
Offic. Unrequited Trans.,nie	−1 313.0	−1 796.0	−1 765.0	−1 369.0	−1 470.0	−1 984.0	−1 986.0	Tr. offic. sans contrep.nia
Direct Investment, nie	−427.0	1 428.0	245.0	1 794.0	3 213.0	27.0	6 627.0	Investissements direct, nia
Portfolio Investments, nie	−2 429.0	−4 572.0	−2 900.0	−7 532.0	−7 120.0	−3 872.0	−8 196.0	Invest. portefeuille, nia
Other Capital, nie	2 346.0	−614.0	−2 549.0	4 084.0	673.0	−3 805.0	−12 162.0	Autres capitaux, nia
Net Errors and Omissions	−11.0	59.0	−86.0	−2 844.0	−992.0	1 847.0	−938.0	Erreurs et omissions nettes
Reserves and Related Items	−2 273.0	104.0	2 094.0	−451.0	−504.0	−665.0	2 081.0	Rés. et postes assimilables
Bulgaria								**Bulgarie**
Merchandise: Exp. fob	10 297.0	9 283.0	8 268.0	6 113.0	3 737.0	3 956.0	3 971.0	Marchandises : exp. fob

78
Summary of balance of payments
Millions of US dollars
Résumé des balances des paiements
Millions de dollars des E−U

Country or area	1987	1988	1989	1990	1991	1992	1993	Pays ou zone
Merchandise: Imp. fob	−11 308.0	−9 889.0	−8 960.0	−7 427.0	−3 769.0	−4 169.0	−4 301.0	Marchandises : imp. fob
Serv. & Income: Credit	1 273.0	1 268.0	1 350.0	957.0	455.0	1 195.0	1 276.0	Serv. & revenu : crédit
Serv. & Income: Debit	−1 090.0	−1 167.0	−1 504.0	−1 478.0	−569.0	−1 386.0	−1 507.0	Serv. & revenu : débit
Private Unrequited Transfers	108.0	103.0	77.0	125.0	50.0	40.0	37.0	Transf. priv.sans contrep.
Offic. Unrequited Trans.,nie	0.0	0.0	0.0	0.0	19.0	3.0	0.0	Tr. offic. sans contrep.nia
Direct Investment, nie	0.0	0.0	0.0	4.0	56.0	41.0	55.0	Investissements direct, nia
Portfolio Investments, nie	0.0	0.0	0.0	0.0	0.0	0.0	0.0	Invest. portefeuille, nia
Other Capital, nie	480.0	1 545.0	−40.0	−2 818.0	39.0	674.0	106.0	Autres capitaux, nia
Net Errors and Omissions	−257.0	−486.0	375.0	70.0	0.0	−85.0	28.0	Erreurs et omissions nettes
Reserves and Related Items	497.0	−657.0	434.0	4 454.0	−17.0	−271.0	334.0	Rés. et postes assimilables
former Czechoslovakia †								**anc. Tchécoslovaquie †**
Merchandise: Exp. fob	15 183.0	15 027.0	14 217.0	11 635.0	10 596.0	11 463.0	..	Marchandises : exp. fob
Merchandise: Imp. fob	−15 454.0	−14 642.0	−14 074.0	−13 057.0	−10 717.0	−13 297.0	..	Marchandises : imp. fob
Serv. & Income: Credit	3 220.0	3 301.0	3 363.0	3 171.0	3 583.0	4 811.0	..	Serv. & revenu : crédit
Serv. & Income: Debit	−2 647.0	−2 659.0	−2 678.0	−3 182.0	−2 614.0	−3 149.0	..	Serv. & revenu : débit
Private Unrequited Transfers	95.0	94.0	130.0	258.0	75.0	51.0	..	Transf. priv.sans contrep.
Offic. Unrequited Trans.,nie	−27.0	−29.0	−22.0	−52.0	−14.0	90.0	..	Tr. offic. sans contrep.nia
Direct Investment, nie	0.0	0.0	257.0	187.0	586.0	1 073.0	..	Investissements direct, nia
Portfolio Investments, nie	0.0	0.0	0.0	0.0	0.0	−43.0	..	Invest. portefeuille, nia
Other Capital, nie	−108.0	−893.0	−548.0	455.0	−1 565.0	−1 276.0	..	Autres capitaux, nia
Net Errors and Omissions	−3.0	6.0	−81.0	−543.0	861.0	−144.0	..	Erreurs et omissions nettes
Reserves and Related Items	−260.0	−207.0	−563.0	1 127.0	−789.0	422.0	..	Rés. et postes assimilables
Denmark								**Danemark**
Merchandise: Exp. fob	25 695.0	27 537.0	28 728.0	36 072.0	36 783.0	40 650.0	37 070.0	Marchandises : exp. fob
Merchandise: Imp. fob	−24 900.0	−25 654.0	−26 304.0	−31 197.0	−32 035.0	−33 446.0	−29 258.0	Marchandises : imp. fob
Serv. & Income: Credit	10 475.0	13 299.0	14 288.0	18 841.0	23 119.0	30 064.0	35 411.0	Serv. & revenu : crédit
Serv. & Income: Debit	−14 051.0	−16 303.0	−17 688.0	−21 936.0	−25 019.0	−32 120.0	−38 004.0	Serv. & revenu : débit
Private Unrequited Transfers	−56.0	−88.0	80.0	−46.0	−150.0	−131.0	−133.0	Transf. priv.sans contrep.
Offic. Unrequited Trans.,nie	−164.0	−131.0	−223.0	−362.0	−715.0	−749.0	−375.0	Tr. offic. sans contrep.nia
Direct Investment, nie	−534.0	−217.0	−976.0	−350.0	−299.0	−1 219.0	340.0	Investissements direct, nia
Portfolio Investments, nie	3 683.0	1 231.0	−2 749.0	2 900.0	1 854.0	10 127.0	12 661.0	Invest. portefeuille, nia
Other Capital, nie	4 210.0	2 261.0	1 368.0	1 860.0	−4 657.0	−13 046.0	−15 080.0	Autres capitaux, nia
Net Errors and Omissions	85.0	−619.0	−347.0	−2 407.0	−2 183.0	−357.0	1 220.0	Erreurs et omissions nettes
Reserves and Related Items	−4 443.0	−1 316.0	3 821.0	−3 374.0	3 303.0	226.0	−3 851.0	Rés. et postes assimilables
Estonia								**Estonie**
Merchandise: Exp. fob	..	..	..	..	..	458.6	802.7	Marchandises : exp. fob
Merchandise: Imp. fob	..	..	..	..	..	−448.2	−900.0	Marchandises : imp. fob
Serv. & Income: Credit	..	..	..	..	..	202.1	370.5	Serv. & revenu : crédit
Serv. & Income: Debit	..	..	..	..	..	−185.1	−340.3	Serv. & revenu : débit
Private Unrequited Transfers	..	..	..	..	..	−0.3	−0.4	Transf. priv.sans contrep.
Offic. Unrequited Trans.,nie	..	..	..	..	..	96.9	105.6	Tr. offic. sans contrep.nia
Direct Investment, nie	..	..	..	..	..	157.6	160.4	Investissements direct, nia
Portfolio Investments, nie	..	..	..	..	..	0.0	−0.2	Invest. portefeuille, nia
Other Capital, nie	..	..	..	..	..	−97.2	32.0	Autres capitaux, nia
Net Errors and Omissions	..	..	..	..	..	−127.5	−65.6	Erreurs et omissions nettes
Reserves and Related Items	..	..	..	..	..	−56.8	−164.6	Rés. et postes assimilables
Finland								**Finlande**
Merchandise: Exp. fob	19 079.0	21 826.0	22 882.0	26 101.0	22 516.0	23 571.0	23 135.0	Marchandises : exp. fob
Merchandise: Imp. fob	−17 700.0	−20 686.0	−23 101.0	−25 376.0	−20 195.0	−19 619.0	−16 743.0	Marchandises : imp. fob
Serv. & Income: Credit	5 057.0	6 363.0	6 611.0	8 193.0	6 919.0	6 374.0	5 991.0	Serv. & revenu : crédit
Serv. & Income: Debit	−7 675.0	−9 686.0	−11 425.0	−14 905.0	−14 929.0	−14 455.0	−12 812.0	Serv. & revenu : débit
Private Unrequited Transfers	−162.0	−87.0	−252.0	−341.0	−308.0	−286.0	−99.0	Transf. priv.sans contrep.
Offic. Unrequited Trans.,nie	−328.0	−425.0	−510.0	−634.0	−768.0	−532.0	−452.0	Tr. offic. sans contrep.nia
Direct Investment, nie	−885.0	−2 092.0	−2 620.0	−2 501.0	−1 289.0	−20.0	−1 234.0	Investissements direct, nia
Portfolio Investments, nie	1 413.0	3 162.0	3 401.0	5 806.0	9 333.0	7 995.0	5 822.0	Invest. portefeuille, nia
Other Capital, nie	6 746.0	1 073.0	2 681.0	9 957.0	−2 840.0	−3 886.0	−4 258.0	Autres capitaux, nia
Net Errors and Omissions	−1 523.0	807.0	1 276.0	−2 366.0	−328.0	−1 306.0	927.0	Erreurs et omissions nettes
Reserves and Related Items	−4 022.0	−255.0	1 058.0	−3 935.0	1 889.0	2 163.0	−276.0	Rés. et postes assimilables
France								**France**
Merchandise: Exp. fob	141 658.0	160 188.0	170 761.0	206 670.0	207 129.0	225 318.0	195 114.0	Marchandises : exp. fob
Merchandise: Imp. fob	−150 325.0	−168 726.0	−181 412.0	−220 341.0	−217 305.0	−223 563.0	−188 117.0	Marchandises : imp. fob
Serv. & Income: Credit	79 344.0	89 530.0	102 552.0	134 410.0	151 966.0	181 508.0	187 542.0	Serv. & revenu : crédit
Serv. & Income: Debit	−69 713.0	−79 114.0	−89 031.0	−122 502.0	−140 840.0	−170 426.0	−178 197.0	Serv. & revenu : débit
Private Unrequited Transfers	−2 297.0	−2 437.0	−2 668.0	−4 011.0	−3 020.0	−2 816.0	−729.0	Transf. priv.sans contrep.
Offic. Unrequited Trans.,nie	−3 114.0	−4 237.0	−5 863.0	−9 461.0	−4 961.0	−5 684.0	−5 413.0	Tr. offic. sans contrep.nia
Direct Investment, nie	−4 071.0	−6 010.0	−9 189.0	−21 639.0	−8 783.0	−9 426.0	151.0	Investissements direct, nia

78
Summary of balance of payments
Millions of US dollars

Résumé des balances des paiements
Millions de dollars des E−U

Country or area	1987	1988	1989	1990	1991	1992	1993	Pays ou zone
Portfolio Investments, nie	5 442.0	7 798.0	25 392.0	34 809.0	13 818.0	34 035.0	3 756.0	Invest. portefeuille, nia
Other Capital, nie	−6 104.0	1 999.0	−7 323.0	12 464.0	−8 718.0	−44 150.0	−14 593.0	Autres capitaux, nia
Net Errors and Omissions	850.0	940.0	−5 568.0	1 425.0	4 897.0	2 109.0	2 686.0	Erreurs et omissions nettes
Reserves and Related Items	8 329.0	67.0	2 350.0	−11 823.0	5 816.0	13 094.0	−2 201.0	Rés. et postes assimilables
Germany †								**Allemagne †**
F. R. Germany [1]								**R. f. Allemagne [1]**
Merchandise: Exp. fob	278.5	308.6	325.0	391.3	378.6	406.7	364.2	Marchandises : exp. fob
Merchandise: Imp. fob	−208.3	−228.9	−247.2	−320.2	−355.4	−373.9	−318.9	Marchandises : imp. fob
Serv. & Income: Credit	82.5	87.5	100.5	134.2	145.5	160.5	153.2	Serv. & revenu : crédit
Serv. & Income: Debit	−90.1	−98.4	−102.6	−136.3	−151.8	−182.4	−181.8	Serv. & revenu : débit
Private Unrequited Transfers	−5.7	−6.4	−5.7	−7.2	−7.0	−8.6	−8.1	Transf. priv.sans contrep.
Offic. Unrequited Trans.,nie	−10.6	−11.7	−12.3	−15.5	−28.9	−23.4	−23.4	Tr. offic. sans contrep.nia
Direct Investment, nie	−7.7	−11.9	−7.5	−19.5	−15.0	−10.6	−15.6	Investissements direct, nia
Portfolio Investments, nie	−1.9	−43.8	−4.6	−2.4	23.1	39.4	122.7	Invest. portefeuille, nia
Other Capital, nie	−15.3	−16.0	−60.7	−33.8	5.0	28.6	−94.2	Autres capitaux, nia
Net Errors and Omissions	−1.0	2.5	4.5	15.0	7.9	6.8	−11.5	Erreurs et omissions nettes
Reserves and Related Items	−20.3	18.4	10.6	−5.5	−2.1	−43.0	13.3	Rés. et postes assimilables
Greece								**Grèce**
Merchandise: Exp. fob	5 612.0	5 933.0	5 994.0	6 365.0	6 797.0	6 009.0	5 035.0	Marchandises : exp. fob
Merchandise: Imp. fob	−11 112.0	−12 005.0	−13 377.0	−16 543.0	−16 909.0	−17 612.0	−15 592.0	Marchandises : imp. fob
Serv. & Income: Credit	4 604.0	5 445.0	5 191.0	6 968.0	7 757.0	9 319.0	9 218.0	Serv. & revenu : crédit
Serv. & Income: Debit	−3 362.0	−3 980.0	−4 352.0	−5 045.0	−5 402.0	−6 331.0	−5 907.0	Serv. & revenu : débit
Private Unrequited Transfers	1 370.0	1 713.0	1 381.0	1 817.0	2 149.0	2 417.0	2 414.0	Transf. priv.sans contrep.
Offic. Unrequited Trans.,nie	1 665.0	1 936.0	2 602.0	2 901.0	4 034.0	4 058.0	4 085.0	Tr. offic. sans contrep.nia
Direct Investment, nie	683.0	907.0	752.0	1 005.0	1 135.0	1 144.0	977.0	Investissements direct, nia
Portfolio Investments, nie	0.0	0.0	0.0	0.0	0.0	0.0	0.0	Invest. portefeuille, nia
Other Capital, nie	1 291.0	947.0	1 999.0	2 997.0	2 826.0	1 475.0	3 840.0	Autres capitaux, nia
Net Errors and Omissions	223.0	41.0	−538.0	−185.0	−183.0	−853.0	−631.0	Erreurs et omissions nettes
Reserves and Related Items	−974.0	−937.0	348.0	−280.0	−2 204.0	374.0	−3 439.0	Rés. et postes assimilables
Hungary								**Hongrie**
Merchandise: Exp. fob	9 967.0	9 989.0	10 493.0	9 151.0	9 688.0	10 097.0	8 119.0	Marchandises : exp. fob
Merchandise: Imp. fob	−9 887.0	−9 406.0	−9 450.0	−8 617.0	−9 330.0	−10 108.0	−12 140.0	Marchandises : imp. fob
Serv. & Income: Credit	1 227.0	1 287.0	1 522.0	3 164.0	2 847.0	3 829.0	3 301.0	Serv. & revenu : crédit
Serv. & Income: Debit	−2 088.0	−2 559.0	−3 283.0	−4 107.0	−3 669.0	−4 325.0	−4 275.0	Serv. & revenu : débit
Private Unrequited Transfers	105.0	117.0	130.0	794.0	834.0	843.0	711.0	Transf. priv.sans contrep.
Offic. Unrequited Trans.,nie	0.0	0.0	0.0	−7.0	34.0	15.0	21.0	Tr. offic. sans contrep.nia
Direct Investment, nie	0.0	0.0	0.0	0.0	1 462.0	1 479.0	2 339.0	Investissements direct, nia
Portfolio Investments, nie	0.0	0.0	0.0	0.0	0.0	0.0	3 918.0	Invest. portefeuille, nia
Other Capital, nie	235.0	680.0	901.0	−801.0	12.0	−1 063.0	−174.0	Autres capitaux, nia
Net Errors and Omissions	160.0	50.0	−141.0	10.0	−82.0	2.0	724.0	Erreurs et omissions nettes
Reserves and Related Items	281.0	−158.0	−172.0	413.0	−1 795.0	−770.0	−2 545.0	Rés. et postes assimilables
Iceland								**Islande**
Merchandise: Exp. fob	1 376.1	1 425.4	1 401.5	1 588.6	1 551.5	1 523.1	1 399.2	Marchandises : exp. fob
Merchandise: Imp. fob	−1 428.2	−1 439.4	−1 267.3	−1 509.1	−1 598.9	−1 521.7	−1 217.9	Marchandises : imp. fob
Serv. & Income: Credit	569.7	557.8	549.4	639.3	642.4	684.6	682.2	Serv. & revenu : crédit
Serv. & Income: Debit	−707.8	−764.1	−764.5	−852.9	−899.1	−887.8	−866.2	Serv. & revenu : débit
Private Unrequited Transfers	0.9	0.7	0.9	6.1	2.8	1.6	5.9	Transf. priv.sans contrep.
Offic. Unrequited Trans.,nie	−1.7	−1.9	−3.9	−6.3	−8.0	−7.8	−8.0	Tr. offic. sans contrep.nia
Direct Investment, nie	1.7	−15.9	−35.6	−3.1	24.1	9.2	3.3	Investissements direct, nia
Portfolio Investments, nie	0.0	0.0	0.6	0.3	26.5	−15.8	−40.3	Invest. portefeuille, nia
Other Capital, nie	230.0	242.7	160.0	213.7	249.3	282.8	−12.4	Autres capitaux, nia
Net Errors and Omissions	−58.8	−4.1	13.5	−2.3	19.9	1.9	−5.9	Erreurs et omissions nettes
Reserves and Related Items	18.1	−1.2	−54.6	−74.3	−10.5	−70.1	60.1	Rés. et postes assimilables
Ireland								**Irlande**
Merchandise: Exp. fob	15 566.0	18 389.0	20 356.0	23 356.0	23 660.0	27 905.0	28 729.0	Marchandises : exp. fob
Merchandise: Imp. fob	−12 952.0	−14 567.0	−16 352.0	−19 387.0	−19 493.0	−21 092.0	−20 557.0	Marchandises : imp. fob
Serv. & Income: Credit	3 153.0	3 824.0	4 328.0	5 958.0	6 265.0	6 657.0	5 874.0	Serv. & revenu : crédit
Serv. & Income: Debit	−7 167.0	−9 117.0	−10 417.0	−12 499.0	−12 175.0	−13 981.0	−13 217.0	Serv. & revenu : débit
Private Unrequited Transfers	−160.0	−119.0	−93.0	−63.0	−56.0	−59.0	−51.0	Transf. priv.sans contrep.
Offic. Unrequited Trans.,nie	1 464.0	1 662.0	1 670.0	2 680.0	3 242.0	3 022.0	2 867.0	Tr. offic. sans contrep.nia
Direct Investment, nie	89.0	92.0	85.0	99.0	97.0	102.0	89.0	Investissements direct, nia
Portfolio Investments, nie	−207.0	990.0	650.0	−199.0	−1 070.0	−3 189.0	2 451.0	Invest. portefeuille, nia
Other Capital, nie	749.0	−881.0	−2 308.0	−1 805.0	−2 229.0	−3 377.0	−2 930.0	Autres capitaux, nia
Net Errors and Omissions	350.0	320.0	1 145.0	2 608.0	2 221.0	470.0	659.0	Erreurs et omissions nettes
Reserves and Related Items	−886.0	−592.0	937.0	−748.0	−464.0	3 542.0	−3 915.0	Rés. et postes assimilables

78
Summary of balance of payments
Millions of US dollars
Résumé des balances des paiements
Millions de dollars des E−U

Country or area	1987	1988	1989	1990	1991	1992	1993	Pays ou zone
Italy								**Italie**
Merchandise: Exp. fob	116 712.0	127 859.0	140 556.0	170 304.0	169 465.0	178 155.0	168 456.0	Marchandises : exp. fob
Merchandise: Imp. fob	−116 629.0	−128 782.0	−142 219.0	−168 931.0	−169 911.0	−175 070.0	−136 178.0	Marchandises : imp. fob
Serv. & Income: Credit	39 045.0	41 914.0	47 309.0	69 370.0	71 521.0	90 362.0	87 864.0	Serv. & revenu : crédit
Serv. & Income: Debit	−40 607.0	−46 197.0	−54 916.0	−85 245.0	−89 046.0	−115 694.0	−103 584.0	Serv. & revenu : débit
Private Unrequited Transfers	1 769.0	1 614.0	1 091.0	1 152.0	−292.0	−468.0	450.0	Transf. priv.sans contrep.
Offic. Unrequited Trans.,nie	−2 720.0	−3 006.0	−3 720.0	−3 478.0	−5 798.0	−5 192.0	−5 832.0	Tr. offic. sans contrep.nia
Direct Investment, nie	1 809.0	1 218.0	6.0	−1 174.0	−4 821.0	−2 786.0	−3 660.0	Investissements direct, nia
Portfolio Investments, nie	−7 318.0	294.0	3 147.0	−357.0	−6 037.0	−4 356.0	71 764.0	Invest. portefeuille, nia
Other Capital, nie	14 419.0	15 198.0	21 585.0	44 171.0	35 072.0	19 066.0	−64 082.0	Autres capitaux, nia
Net Errors and Omissions	−987.0	−1 695.0	−1 568.0	−14 190.0	−6 871.0	−8 009.0	−18 334.0	Erreurs et omissions nettes
Reserves and Related Items	−5 493.0	−8 416.0	−11 270.0	−11 623.0	6 718.0	23 992.0	3 135.0	Rés. et postes assimilables
Malta								**Malte**
Merchandise: Exp. fob	631.3	758.3	866.3	1 154.2	1 283.6	1 557.1	1 351.1	Marchandises : exp. fob
Merchandise: Imp. fob	−1 024.6	−1 222.3	−1 327.7	−1 753.0	−1 897.3	−2 104.0	−1 953.2	Marchandises : imp. fob
Serv. & Income: Credit	718.4	831.5	860.0	1 065.2	1 129.2	1 207.5	1 191.5	Serv. & revenu : crédit
Serv. & Income: Debit	−384.7	−465.8	−506.5	−609.5	−638.5	−724.0	−717.8	Serv. & revenu : débit
Private Unrequited Transfers	63.1	97.6	54.8	32.4	32.5	9.4	7.3	Transf. priv.sans contrep.
Offic. Unrequited Trans.,nie	23.4	68.6	49.6	55.1	83.2	84.3	53.6	Tr. offic. sans contrep.nia
Direct Investment, nie	19.4	40.8	51.7	45.6	77.0	39.5	...	Investissements direct, nia
Portfolio Investments, nie	−7.5	−38.4	−57.7	−1.9	−241.3	−209.5	...	Invest. portefeuille, nia
Other Capital, nie	−14.5	14.8	−40.7	−87.2	181.0	247.1	...	Autres capitaux, nia
Net Errors and Omissions	−28.8	−50.4	64.4	24.2	−93.9	−62.8	...	Erreurs et omissions nettes
Reserves and Related Items	4.5	−34.8	−14.2	74.9	84.6	−44.8	...	Rés. et postes assimilables
Netherlands								**Pays−Bas**
Merchandise: Exp. fob	86 158.0	97 442.0	101 317.0	122 071.0	122 625.0	129 195.0	120 303.0	Marchandises : exp. fob
Merchandise: Imp. fob	−80 991.0	−88 966.0	−93 162.0	−111 741.0	−111 885.0	−117 855.0	−107 388.0	Marchandises : imp. fob
Serv. & Income: Credit	37 350.0	41 588.0	49 211.0	57 614.0	61 144.0	65 021.0	64 805.0	Serv. & revenu : crédit
Serv. & Income: Debit	−36 275.0	−41 082.0	−45 328.0	−55 772.0	−59 945.0	−64 960.0	−63 135.0	Serv. & revenu : débit
Private Unrequited Transfers	−1 050.0	−909.0	−944.0	−1 159.0	−1 222.0	−1 619.0	−1 662.0	Transf. priv.sans contrep.
Offic. Unrequited Trans.,nie	−1 234.0	−1 159.0	−1 309.0	−2 084.0	−3 189.0	−3 354.0	−3 553.0	Tr. offic. sans contrep.nia
Direct Investment, nie	−5 626.0	−2 334.0	−6 480.0	−3 103.0	−7 262.0	−6 751.0	−4 355.0	Investissements direct, nia
Portfolio Investments, nie	1 864.0	3 372.0	7 253.0	−5 173.0	−859.0	−9 208.0	1 674.0	Invest. portefeuille, nia
Other Capital, nie	2 644.0	−1 623.0	−8 797.0	3 624.0	1 789.0	8 671.0	−8 496.0	Autres capitaux, nia
Net Errors and Omissions	23.0	−4 704.0	−1 312.0	−4 001.0	−1 121.0	7 286.0	8 348.0	Erreurs et omissions nettes
Reserves and Related Items	−2 861.0	−1 626.0	−450.0	−277.0	−77.0	−6 427.0	−6 542.0	Rés. et postes assimilables
Norway								**Norvège**
Merchandise: Exp. fob	21 191.0	23 075.0	27 171.0	34 313.0	34 212.0	35 162.0	31 989.0	Marchandises : exp. fob
Merchandise: Imp. fob	−21 951.0	−23 284.0	−23 401.0	−26 552.0	−25 516.0	−25 860.0	−23 974.0	Marchandises : imp. fob
Serv. & Income: Credit	11 585.0	12 994.0	14 195.0	16 661.0	16 871.0	16 683.0	15 257.0	Serv. & revenu : crédit
Serv. & Income: Debit	−13 948.0	−15 544.0	−16 619.0	−18 954.0	−18 994.0	−21 246.0	−19 426.0	Serv. & revenu : débit
Private Unrequited Transfers	−194.0	−168.0	−222.0	−237.0	−338.0	−490.0	−313.0	Transf. priv.sans contrep.
Offic. Unrequited Trans.,nie	−788.0	−962.0	−910.0	−1 208.0	−1 185.0	−1 288.0	−1 082.0	Tr. offic. sans contrep.nia
Direct Investment, nie	−687.0	−699.0	161.0	−467.0	−2 180.0	305.0	1 232.0	Investissements direct, nia
Portfolio Investments, nie	2 283.0	4 226.0	3 043.0	561.0	−3 107.0	862.0	−228.0	Invest. portefeuille, nia
Other Capital, nie	3 637.0	1 373.0	−1 148.0	−854.0	−2 294.0	−1 542.0	6 052.0	Autres capitaux, nia
Net Errors and Omissions	−1 349.0	−1 149.0	−1 305.0	−2 848.0	−219.0	−3 442.0	−1 659.0	Erreurs et omissions nettes
Reserves and Related Items	220.0	138.0	−965.0	−414.0	2 751.0	855.0	−7 850.0	Rés. et postes assimilables
Poland								**Pologne**
Merchandise: Exp. fob	12 026.0	13 846.0	12 869.0	15 837.0	14 393.0	13 929.0	13 582.0	Marchandises : exp. fob
Merchandise: Imp. fob	−11 236.0	−12 757.0	−12 822.0	−12 248.0	−15 104.0	−14 060.0	−17 087.0	Marchandises : imp. fob
Serv. & Income: Credit	2 433.0	2 743.0	3 611.0	3 803.0	4 260.0	5 501.0	4 780.0	Serv. & revenu : crédit
Serv. & Income: Debit	−5 160.0	−5 630.0	−6 676.0	−6 836.0	−6 463.0	−8 940.0	−7 823.0	Serv. & revenu : débit
Private Unrequited Transfers	1 558.0	1 691.0	1 521.0	2 206.0	723.0	213.0	621.0	Transf. priv.sans contrep.
Offic. Unrequited Trans.,nie	0.0	0.0	88.0	305.0	45.0	253.0	139.0	Tr. offic. sans contrep.nia
Direct Investment, nie	4.0	−7.0	−7.0	89.0	298.0	665.0	1 697.0	Investissements direct, nia
Portfolio Investments, nie	0.0	0.0	0.0	0.0	0.0	0.0	0.0	Invest. portefeuille, nia
Other Capital, nie	−3 322.0	−10 654.0	−1 789.0	−8 820.0	−4 481.0	−1 710.0	644.0	Autres capitaux, nia
Net Errors and Omissions	91.0	−267.0	−110.0	133.0	−767.0	−148.0	−106.0	Erreurs et omissions nettes
Reserves and Related Items	3 606.0	11 035.0	3 315.0	5 531.0	7 096.0	4 297.0	3 553.0	Rés. et postes assimilables
Portugal								**Portugal**
Merchandise: Exp. fob	9 266.0	10 874.0	12 720.0	16 311.0	16 231.0	18 195.0	15 444.0	Marchandises : exp. fob
Merchandise: Imp. fob	−12 847.0	−16 392.0	−17 585.0	−23 141.0	−24 079.0	−27 735.0	−22 330.0	Marchandises : imp. fob
Serv. & Income: Credit	3 646.0	4 037.0	4 630.0	6 603.0	6 941.0	7 718.0	9 142.0	Serv. & revenu : crédit
Serv. & Income: Debit	−3 401.0	−3 906.0	−4 152.0	−5 461.0	−5 784.0	−6 188.0	−8 024.0	Serv. & revenu : débit
Private Unrequited Transfers	3 404.0	3 597.0	3 726.0	4 509.0	4 593.0	4 794.0	3 842.0	Transf. priv.sans contrep.

78
Summary of balance of payments
Millions of US dollars
Résumé des balances des paiements
Millions de dollars des E−U

Country or area	1987	1988	1989	1990	1991	1992	1993	Pays ou zone
Offic. Unrequited Trans.,nie	368.0	725.0	814.0	998.0	1 381.0	3 032.0	2 874.0	Tr. offic. sans contrep.nia
Direct Investment, nie	476.0	842.0	1 653.0	2 447.0	1 985.0	1 186.0	1 136.0	Investissements direct, nia
Portfolio Investments, nie	816.0	1 814.0	1 050.0	961.0	1 895.0	−3 064.0	−530.0	Invest. portefeuille, nia
Other Capital, nie	−604.0	−2 363.0	1 302.0	−845.0	656.0	928.0	−4 229.0	Autres capitaux, nia
Net Errors and Omissions	653.0	1 640.0	497.0	1 160.0	1 893.0	978.0	−171.0	Erreurs et omissions nettes
Reserves and Related Items	−1 777.0	−867.0	−4 654.0	−3 542.0	−5 713.0	156.0	2 847.0	Rés. et postes assimilables
Romania								**Roumanie**
Merchandise: Exp. fob	10 491.0	11 392.0	10 487.0	5 770.0	4 266.0	4 364.0	4 892.0	Marchandises : exp. fob
Merchandise: Imp. fob	−8 313.0	−7 642.0	−8 437.0	−9 114.0	−5 372.0	−5 558.0	−6 020.0	Marchandises : imp. fob
Serv. & Income: Credit	908.0	1 023.0	1 015.0	785.0	784.0	713.0	862.0	Serv. & revenu : crédit
Serv. & Income: Debit	−1 043.0	−851.0	−551.0	−801.0	−908.0	−1 090.0	−1 118.0	Serv. & revenu : débit
Private Unrequited Transfers	0.0	0.0	0.0	0.0	20.0	19.0	103.0	Transf. priv.sans contrep.
Offic. Unrequited Trans.,nie	0.0	0.0	0.0	106.0	198.0	46.0	119.0	Tr. offic. sans contrep.nia
Direct Investment, nie	0.0	0.0	0.0	−18.0	37.0	73.0	87.0	Investissements direct, nia
Portfolio Investments, nie	0.0	0.0	0.0	0.0	0.0	0.0	−73.0	Invest. portefeuille, nia
Other Capital, nie	−1 083.0	−4 223.0	−1 376.0	1 631.0	283.0	1 307.0	955.0	Autres capitaux, nia
Net Errors and Omissions	81.0	16.0	114.0	147.0	15.0	−12.0	139.0	Erreurs et omissions nettes
Reserves and Related Items	−1 041.0	285.0	−1 252.0	1 494.0	677.0	138.0	54.0	Rés. et postes assimilables
Slovenia								**Slovénie**
Merchandise: Exp. fob	..	..	..	..	..	6 683.0	6 082.8	Marchandises : exp. fob
Merchandise: Imp. fob	..	..	..	..	..	−5 891.9	−6 237.2	Marchandises : imp. fob
Serv. & Income: Credit	..	..	..	..	..	1 289.3	1 518.3	Serv. & revenu : crédit
Serv. & Income: Debit	..	..	..	..	..	−1 200.1	−1 216.1	Serv. & revenu : débit
Private Unrequited Transfers	..	..	..	..	..	35.4	62.2	Transf. priv.sans contrep.
Offic. Unrequited Trans.,nie	..	..	..	..	..	10.4	−16.7	Tr. offic. sans contrep.nia
Direct Investment, nie	..	..	..	..	..	113.0	111.6	Investissements direct, nia
Portfolio Investments, nie	..	..	..	..	..	−8.9	3.0	Invest. portefeuille, nia
Other Capital, nie	..	..	..	..	..	−117.2	−174.5	Autres capitaux, nia
Net Errors and Omissions	..	..	..	..	..	−280.5	−68.6	Erreurs et omissions nettes
Reserves and Related Items	..	..	..	..	..	−632.5	−64.7	Rés. et postes assimilables
Spain								**Espagne**
Merchandise: Exp. fob	33 561.0	39 652.0	43 301.0	53 888.0	58 901.0	63 921.0	60 232.0	Marchandises : exp. fob
Merchandise: Imp. fob	−46 547.0	−57 650.0	−67 797.0	−83 454.0	−89 654.0	−94 954.0	−75 950.0	Marchandises : imp. fob
Serv. & Income: Credit	23 867.0	27 440.0	28 875.0	34 496.0	38 175.0	45 709.0	39 321.0	Serv. & revenu : crédit
Serv. & Income: Debit	−13 749.0	−17 729.0	−19 919.0	−26 006.0	−30 201.0	−38 967.0	−32 808.0	Serv. & revenu : débit
Private Unrequited Transfers	2 277.0	3 018.0	3 163.0	3 053.0	2 201.0	2 613.0	1 708.0	Transf. priv.sans contrep.
Offic. Unrequited Trans.,nie	358.0	1 485.0	1 444.0	1 204.0	3 859.0	3 197.0	2 857.0	Tr. offic. sans contrep.nia
Direct Investment, nie	3 825.0	5 786.0	6 955.0	10 904.0	6 919.0	6 758.0	4 173.0	Investissements direct, nia
Portfolio Investments, nie	3 799.0	2 291.0	7 989.0	5 361.0	18 725.0	−790.0	24 687.0	Invest. portefeuille, nia
Other Capital, nie	6 605.0	6 538.0	3 398.0	12 037.0	6 332.0	448.0	−27 552.0	Autres capitaux, nia
Net Errors and Omissions	−1 291.0	−2 414.0	−2 693.0	−4 521.0	−1 117.0	−5 407.0	−1 474.0	Erreurs et omissions nettes
Reserves and Related Items	−12 706.0	−8 416.0	−4 716.0	−6 962.0	−14 141.0	17 472.0	4 807.0	Rés. et postes assimilables
Sweden								**Suède**
Merchandise: Exp. fob	44 013.0	49 367.0	51 071.0	56 835.0	54 543.0	55 366.0	49 347.0	Marchandises : exp. fob
Merchandise: Imp. fob	−39 528.0	−44 487.0	−47 056.0	−53 433.0	−48 184.0	−48 643.0	−41 679.0	Marchandises : imp. fob
Serv. & Income: Credit	13 318.0	15 697.0	18 331.0	23 414.0	24 171.0	24 357.0	19 721.0	Serv. & revenu : crédit
Serv. & Income: Debit	−16 569.0	−19 732.0	−23 699.0	−31 218.0	−33 176.0	−37 259.0	−29 661.0	Serv. & revenu : débit
Private Unrequited Transfers	−379.0	−472.0	−709.0	−645.0	−395.0	−439.0	−162.0	Transf. priv.sans contrep.
Offic. Unrequited Trans.,nie	−1 016.0	−1 140.0	−1 338.0	−1 645.0	−1 653.0	−2 168.0	−1 624.0	Tr. offic. sans contrep.nia
Direct Investment, nie	−4 141.0	−5 798.0	−8 387.0	−12 594.0	−909.0	−345.0	2 356.0	Investissements direct, nia
Portfolio Investments, nie	−1 065.0	−1 359.0	−1 250.0	2 467.0	6 546.0	981.0	1 290.0	Invest. portefeuille, nia
Other Capital, nie	5 976.0	10 054.0	19 564.0	29 444.0	−6 969.0	9 833.0	7 895.0	Autres capitaux, nia
Net Errors and Omissions	153.0	−1 190.0	−5 303.0	−5 130.0	6 159.0	5 269.0	−4 955.0	Erreurs et omissions nettes
Reserves and Related Items	−762.0	−938.0	−1 223.0	−7 495.0	−133.0	−6 952.0	−2 530.0	Rés. et postes assimilables
Switzerland								**Suisse**
Merchandise: Exp. fob	55 219.0	62 725.0	65 366.0	77 488.0	73 745.0	79 353.0	74 932.0	Marchandises : exp. fob
Merchandise: Imp. fob	−60 647.0	−67 301.0	−69 690.0	−83 878.0	−77 550.0	−78 863.0	−72 695.0	Marchandises : imp. fob
Serv. & Income: Credit	32 484.0	35 886.0	38 204.0	47 226.0	46 980.0	46 973.0	44 988.0	Serv. & revenu : crédit
Serv. & Income: Debit	−19 266.0	−20 754.0	−24 155.0	−31 563.0	−30 229.0	−30 277.0	−27 670.0	Serv. & revenu : débit
Private Unrequited Transfers	−1 550.0	−1 711.0	−1 665.0	−2 183.0	−2 275.0	−2 373.0	−2 227.0	Transf. priv.sans contrep.
Offic. Unrequited Trans.,nie	45.0	−2.0	−18.0	−146.0	−346.0	−624.0	−633.0	Tr. offic. sans contrep.nia
Direct Investment, nie	1 047.0	−8 289.0	−5 023.0	−1 409.0	−3 363.0	−4 422.0	−5 730.0	Investissements direct, nia
Portfolio Investments, nie	−1 733.0	−7 415.0	−3 023.0	−1 127.0	−11 978.0	−6 127.0	−17 308.0	Invest. portefeuille, nia
Other Capital, nie	−7 377.0	963.0	429.0	−8 921.0	3 665.0	−4 820.0	5 595.0	Autres capitaux, nia
Net Errors and Omissions	4 983.0	3 473.0	995.0	5 688.0	2 322.0	5 599.0	1 157.0	Erreurs et omissions nettes
Reserves and Related Items	−3 205.0	2 426.0	−1 420.0	−1 172.0	−970.0	−4 420.0	−411.0	Rés. et postes assimilables

78

Summary of balance of payments
Millions of US dollars
Résumé des balances des paiements
Millions de dollars des E−U

Country or area	1987	1988	1989	1990	1991	1992	1993	Pays ou zone
United Kingdom								**Royaume−Uni**
Merchandise: Exp. fob	129 847.0	143 078.0	150 696.0	181 729.0	182 579.0	188 451.0	182 084.0	Marchandises : exp. fob
Merchandise: Imp. fob	−148 866.0	−181 237.0	−191 239.0	−214 471.0	−200 853.0	−211 879.0	−201 953.0	Marchandises : imp. fob
Serv. & Income: Credit	123 214.0	148 414.0	168 929.0	197 527.0	190 274.0	180 188.0	165 167.0	Serv. & revenu : crédit
Serv. & Income: Debit	−106 954.0	−133 491.0	−157 791.0	−189 082.0	−184 180.0	−165 504.0	−154 294.0	Serv. & revenu : débit
Private Unrequited Transfers	−205.0	−480.0	−491.0	−535.0	−531.0	−485.0	−405.0	Transf. priv.sans contrep.
Offic. Unrequited Trans.,nie	−5 340.0	−5 857.0	−6 964.0	−8 206.0	−1 931.0	−8 621.0	−7 441.0	Tr. offic. sans contrep.nia
Direct Investment, nie	−15 639.0	−15 873.0	−4 931.0	13 103.0	−344.0	−2 856.0	−11 079.0	Investissements direct, nia
Portfolio Investments, nie	43 615.0	8 286.0	−31 200.0	−9 130.0	−24 449.0	−9 251.0	−66 551.0	Invest. portefeuille, nia
Other Capital, nie	−23 990.0	25 720.0	53 713.0	26 093.0	54 259.0	10 458.0	83 934.0	Autres capitaux, nia
Net Errors and Omissions	−2 606.0	10 041.0	3 697.0	438.0	−1 216.0	12 142.0	4 306.0	Erreurs et omissions nettes
Reserves and Related Items	6 923.0	1 399.0	15 582.0	2 533.0	−13 608.0	7 356.0	6 233.0	Rés. et postes assimilables
			Oceania· Océanie					
Australia								**Australie**
Merchandise: Exp. fob	27 014.0	33 182.0	36 893.0	39 332.0	42 005.0	42 375.0	42 240.0	**Marchandises : exp. fob**
Merchandise: Imp. fob	−26 739.0	−33 898.0	−40 311.0	−38 964.0	−38 491.0	−40 820.0	−42 363.0	Marchandises : imp. fob
Serv. & Income: Credit	8 964.0	11 837.0	12 901.0	14 121.0	14 583.0	15 089.0	15 822.0	Serv. & revenu : crédit
Serv. & Income: Debit	−17 623.0	−22 771.0	−28 740.0	−31 145.0	−29 737.0	−28 353.0	−26 392.0	Serv. & revenu : débit
Private Unrequited Transfers	1 219.0	1 732.0	2 138.0	1 965.0	2 075.0	1 524.0	738.0	Transf. priv.sans contrep.
Offic. Unrequited Trans.,nie	−201.0	−258.0	−194.0	−158.0	−245.0	−361.0	−414.0	Tr. offic. sans contrep.nia
Direct Investment, nie	−1 193.0	3 073.0	4 617.0	6 892.0	1 776.0	5 379.0	1 540.0	Investissements direct, nia
Portfolio Investments, nie	4 323.0	5 832.0	471.0	2 475.0	4 497.0	−328.0	1 799.0	Invest. portefeuille, nia
Other Capital, nie	5 183.0	10 418.0	11 510.0	4 456.0	6 328.0	4 980.0	5 553.0	Autres capitaux, nia
Net Errors and Omissions	−576.0	−3 896.0	1 344.0	2 754.0	−3 107.0	−4 223.0	1 423.0	Erreurs et omissions nettes
Reserves and Related Items	−371.0	−5 251.0	−628.0	−1 727.0	316.0	4 737.0	55.0	Rés. et postes assimilables
Fiji								**Fidji**
Merchandise: Exp. fob	307.3	356.6	417.5	469.6	427.4	417.1	422.5	Marchandises : exp. fob
Merchandise: Imp. fob	−345.2	−389.9	−487.9	−644.6	−549.5	−539.5	−653.5	Marchandises : imp. fob
Serv. & Income: Credit	262.8	308.2	414.0	467.3	493.1	516.1	578.3	Serv. & revenu : crédit
Serv. & Income: Debit	−227.6	−244.8	−301.5	−320.7	−354.5	−371.5	−376.4	Serv. & revenu : débit
Private Unrequited Transfers	−20.8	−3.6	−14.5	−22.2	−24.6	−16.3	−11.1	Transf. priv.sans contrep.
Offic. Unrequited Trans.,nie	10.3	32.2	26.6	16.7	30.0	25.9	27.1	Tr. offic. sans contrep.nia
Direct Investment, nie	−14.4	30.2	−18.5	75.5	19.4	48.6	23.1	Investissements direct, nia
Portfolio Investments, nie	0.0	0.0	0.0	0.0	0.0	0.0	0.0	Invest. portefeuille, nia
Other Capital, nie	−54.3	16.1	−53.9	−26.9	−17.0	−11.8	−37.0	Autres capitaux, nia
Net Errors and Omissions	18.9	7.3	4.8	21.1	−14.6	−15.2	−15.2	Erreurs et omissions nettes
Reserves and Related Items	62.9	−112.1	13.5	−35.8	−9.6	−53.4	42.1	Rés. et postes assimilables
Kiribati								**Kiribati**
Merchandise: Exp. fob	4.0	5.0	5.0	4.0	3.0	...	...	Marchandises : exp. fob
Merchandise: Imp. fob	−18.0	−22.0	−23.0	−27.0	−26.0	...	...	Marchandises : imp. fob
Serv. & Income: Credit	19.0	21.0	22.0	23.0	24.0	...	...	Serv. & revenu : crédit
Serv. & Income: Debit	−13.0	−15.0	−15.0	−17.0	−16.0	...	...	Serv. & revenu : débit
Private Unrequited Transfers	0.0	−1.0	0.0	0.0	0.0	...	...	Transf. priv.sans contrep.
Offic. Unrequited Trans.,nie	17.0	17.0	17.0	23.0	27.0	...	...	Tr. offic. sans contrep.nia
Direct Investment, nie	0.0	0.0	0.0	0.0	0.0	...	...	Investissements direct, nia
Portfolio Investments, nie	0.0	0.0	0.0	0.0	0.0	...	...	Invest. portefeuille, nia
Other Capital, nie	−1.0	−2.0	−4.0	−4.0	−1.0	...	...	Autres capitaux, nia
Net Errors and Omissions	−4.0	0.0	5.0	5.0	−1.0	...	...	Erreurs et omissions nettes
Reserves and Related Items	−3.0	−3.0	−7.0	−7.0	−10.0	...	...	Rés. et postes assimilables
New Zealand								**Nouvelle−Zélande**
Merchandise: Exp. fob	7 245.0	8 831.0	8 846.0	9 191.0	9 555.0	9 781.0	10 463.0	Marchandises : exp. fob
Merchandise: Imp. fob	−6 663.0	−6 667.0	−7 873.0	−8 294.0	−7 483.0	−8 108.0	−8 749.0	Marchandises : imp. fob
Serv. & Income: Credit	2 819.0	3 092.0	2 885.0	3 421.0	2 564.0	2 463.0	2 892.0	Serv. & revenu : crédit
Serv. & Income: Debit	−5 313.0	−5 842.0	−5 599.0	−5 590.0	−5 979.0	−5 690.0	−6 067.0	Serv. & revenu : débit
Private Unrequited Transfers	178.0	310.0	491.0	650.0	728.0	740.0	575.0	Transf. priv.sans contrep.
Offic. Unrequited Trans.,nie	−40.0	−43.0	−27.0	−46.0	−43.0	−56.0	−46.0	Tr. offic. sans contrep.nia
Direct Investment, nie	48.0	289.0	−427.0	−169.0	0.0	0.0	0.0	Investissements direct, nia
Portfolio Investments, nie	0.0	0.0	30.0	11.0	0.0	0.0	0.0	Invest. portefeuille, nia
Other Capital, nie	−839.0	−3 134.0	−534.0	−316.0	−1 239.0	−2 352.0	−2 774.0	Autres capitaux, nia
Net Errors and Omissions	317.0	244.0	981.0	1 321.0	386.0	1 743.0	1 933.0	Erreurs et omissions nettes
Reserves and Related Items	2 248.0	2 921.0	1 226.0	−178.0	1 511.0	1 477.0	1 773.0	Rés. et postes assimilables
Papua New Guinea								**Papouasie−Nvl−Guinée**
Merchandise: Exp. fob	1 243.9	1 475.3	1 318.5	1 173.8	1 482.6	1 950.9	2 504.7	Marchandises : exp. fob
Merchandise: Imp. fob	−1 129.6	−1 384.5	−1 341.3	−1 106.8	−1 403.8	−1 321.7	−1 134.8	Marchandises : imp. fob
Serv. & Income: Credit	155.2	240.9	254.9	311.9	374.1	389.0	338.4	Serv. & revenu : crédit
Serv. & Income: Debit	−583.9	−761.3	−674.0	−612.5	−863.0	−1 113.7	−1 204.9	Serv. & revenu : débit

78
Summary of balance of payments
Millions of US dollars
Résumé des balances des paiements
Millions de dollars des E−U

Country or area	1987	1988	1989	1990	1991	1992	1993	Pays ou zone
Private Unrequited Transfers	−104.5	−124.6	−130.7	−107.2	−64.1	−62.6	−129.8	Transf. priv.sans contrep.
Offic. Unrequited Trans.,nie	204.3	217.7	217.3	225.5	323.6	255.0	172.7	Tr. offic. sans contrep.nia
Direct Investment, nie	115.4	119.7	221.3	156.3	202.8	290.9	0.6	Investissements direct, nia
Portfolio Investments, nie	0.0	0.0	0.0	0.0	0.0	0.0	0.0	Invest. portefeuille, nia
Other Capital, nie	61.7	125.4	43.7	60.4	−140.0	−442.6	−658.0	Autres capitaux, nia
Net Errors and Omissions	39.5	37.9	31.6	−78.6	2.2	−17.5	29.1	Erreurs et omissions nettes
Reserves and Related Items	−1.9	53.5	58.7	−22.9	85.5	72.2	82.0	Rés. et postes assimilables
Solomon Islands								**Iles Salomon**
Merchandise: Exp. fob	63.2	81.9	74.7	70.1	83.4	101.7	...	Marchandises : exp. fob
Merchandise: Imp. fob	−69.5	−104.6	−94.3	−77.3	−92.0	−87.4	...	Marchandises : imp. fob
Serv. & Income: Credit	24.3	25.6	30.4	27.7	33.0	37.0	...	Serv. & revenu : crédit
Serv. & Income: Debit	−60.8	−78.3	−87.0	−86.3	−98.7	−88.9	...	Serv. & revenu : débit
Private Unrequited Transfers	3.0	5.2	6.1	4.7	1.9	1.8	...	Transf. priv.sans contrep.
Offic. Unrequited Trans.,nie	22.6	32.2	36.8	33.3	36.3	34.0	...	Tr. offic. sans contrep.nia
Direct Investment, nie	10.5	1.7	11.6	10.4	14.5	14.2	...	Investissements direct, nia
Portfolio Investments, nie	0.0	0.0	0.0	0.0	0.0	0.0	...	Invest. portefeuille, nia
Other Capital, nie	−2.8	42.1	13.5	12.4	0.6	8.3	...	Autres capitaux, nia
Net Errors and Omissions	2.8	−10.9	−5.2	−8.6	8.4	−6.2	...	Erreurs et omissions nettes
Reserves and Related Items	6.6	5.1	13.5	13.7	12.6	−14.4	...	Rés. et postes assimilables
Tonga								**Tonga**
Merchandise: Exp. fob	7.0	6.4	9.9	9.0	10.6	14.3	11.9	Marchandises : exp. fob
Merchandise: Imp. fob	−35.1	−44.1	−48.1	−49.5	−51.4	−50.7	−49.5	Marchandises : imp. fob
Serv. & Income: Credit	22.1	24.7	25.4	35.0	26.4	21.3	21.3	Serv. & revenu : crédit
Serv. & Income: Debit	−17.0	−25.4	−24.0	−23.1	−24.8	−23.8	−21.4	Serv. & revenu : débit
Private Unrequited Transfers	22.4	21.3	24.9	30.6	30.2	28.0	34.9	Transf. priv.sans contrep.
Offic. Unrequited Trans.,nie	6.4	5.7	10.0	10.5	7.1	9.6	6.9	Tr. offic. sans contrep.nia
Direct Investment, nie	0.2	0.1	0.1	0.1	0.2	1.2	0.4	Investissements direct, nia
Portfolio Investments, nie	0.0	0.0	−2.2	−13.6	−3.7	−0.5	−0.1	Invest. portefeuille, nia
Other Capital, nie	0.1	4.8	3.2	0.0	5.5	5.4	3.0	Autres capitaux, nia
Net Errors and Omissions	−4.9	7.2	−1.3	1.9	−0.9	−1.2	0.3	Erreurs et omissions nettes
Reserves and Related Items	−1.1	−0.7	2.0	−0.9	1.0	−3.6	−7.7	Rés. et postes assimilables
Vanuatu								**Vanuatu**
Merchandise: Exp. fob	13.7	15.4	13.7	13.7	14.9	17.8	17.4	Marchandises : exp. fob
Merchandise: Imp. fob	−57.1	−57.9	−57.9	−79.3	−74.0	−66.8	−64.7	Marchandises : imp. fob
Serv. & Income: Credit	67.3	63.4	63.7	92.1	91.0	87.0	78.6	Serv. & revenu : crédit
Serv. & Income: Debit	−67.5	−58.8	−48.3	−57.2	−76.1	−74.3	−75.2	Serv. & revenu : débit
Private Unrequited Transfers	5.8	9.8	4.6	11.1	17.7	7.0	10.8	Transf. priv.sans contrep.
Offic. Unrequited Trans.,nie	32.4	30.8	20.7	29.9	32.1	32.9	32.1	Tr. offic. sans contrep.nia
Direct Investment, nie	12.9	10.8	9.2	13.1	25.5	26.5	26.7	Investissements direct, nia
Portfolio Investments, nie	0.0	0.0	0.0	0.0	0.0	0.0	0.0	Invest. portefeuille, nia
Other Capital, nie	6.5	−0.9	15.0	0.7	−53.3	−2.8	−10.6	Autres capitaux, nia
Net Errors and Omissions	−15.1	−17.3	−13.0	−19.4	19.3	−26.5	−11.7	Erreurs et omissions nettes
Reserves and Related Items	1.1	4.7	−7.8	−4.7	3.1	−0.8	−3.5	Rés. et postes assimilables

Source:
International Monetary Fund (Washington, DC).

† For detailed descriptions of data pertaining to
former Czechoslovakia, Germany, SFR Yugoslavia and former
USSR, see Annex I − Country or area nomenclature, regional
and other groupings.

1 Billions of US Dollars.
2 Fiscal year ending June 30.
3 Fiscal year beginning March 21.
4 Fiscal year ending September 30.

Sources:
Fonds montaire international (Washington, DC).

† Pour les descriptions en détails des données relatives à l'ancienne
Tchécoslovaquie, l'Allemagne, la Rfs Yougoslavie et l'ancienne
URSS, voir l'Annexe I − Nomenclature des pays ou zones,
groupements régionaux et autres groupements.

1 Milliards de dollars des E−U.
2 L'années fiscales finissant la 30 juin.
3 L'années fiscales commençant le 21 mars.
4 L'années fiscales finissant le 30 septembre.

Technical notes, table 78

A balance of payments can be broadly described as the record of an economy's international economic transactions. It shows (a) transactions in goods, services and income between an economy and the rest of the world, (b) changes of ownership and other changes in that economy's monetary gold, special drawing rights (SDRs) and claims on and liabilities to the rest of the world, and (c) unrequited transfers and counterpart entries needed to balance in the accounting sense any entries for the foregoing transactions and changes which are not mutually offsetting.

The detailed definitions concerning the content of the basic categories of the balance of payments are given in the *Balance of Payments Manual (Fourth Edition)*.[35] Brief explanatory notes are given below to clarify the scope of the major items.

Merchandise covers exports and imports of goods and related distributive services at the customs frontiers of the exporting economy, i.e. f.o.b. value. International transactions in non-monetary gold are included in this item. Most merchandise exports and imports data are based on customs records.

Services and income covers transactions in real resources between residents and non-residents other than those classified as merchandise, including (a) shipment and other transportation services, including freight, insurance and other distributive services in connection with the movement of commodities, (b) travel, i.e. goods and services acquired by non-resident travellers in a given country and similar acquisitions by resident travellers abroad, and (c) investment income which covers income of non-residents from their financial assets invested in the compiling economy (debit) and similar income of residents from their financial assets invested abroad (credit).

Private unrequited transfers covers unrequited transfers between residents other than the general government and central bank of the compiling economy on the one hand, and non-residents other than general government and central banks on the other hand.

Official unrequited transfers covers inter-official transfers together with other transfers in which either the transferor or the transferee is the general government or central bank whether of the compiling country or of a foreign country.

Direct investment covers all capital transactions between direct investment enterprises and the direct investors themselves or any of those investors' other direct investment enterprises.

Portfolio investment covers investment in long-term bonds and corporate equities other than "direct investment" and "reserves".

Notes techniques, tableau 78

La balance des paiements peut se définir d'une façon générale comme le relevé des transactions économiques internationales d'une économie. Elle indique (a) les transactions sur biens, services et revenus entre une économie et le reste du monde, (b) les transferts de propriété et autres variations intervenues dans les avoirs en or monétaire de cette économie, dans ses avoirs en droits de tirages spéciaux (DTS) ainsi que dans ses créances financières sur le reste du monde ou dans ses engagements financiers envers lui et (c) les "inscriptions de transferts sans contrepartie" et de "contrepartie" destinées à équilibrer, d'un point de vue comptable, les transactions et changements précités qui ne se compensent pas réciproquement.

Les définitions détaillées relatives au contenu des postes fondamentaux de la balance des paiements figurent dans le *Manuel de la balance des paiements (Quatrième édition)* [35]. De brèves notes explicatives sont présentées ci-après pour clarifier la portée de ces principales rubriques.

Marchandise : exportations et importations de biens et de services de distribution y afférents à la frontière douanière de l'économie exportatrice, c'est-à-dire leur valeur f.o.b. Les transactions internationales sur or non monétaire sont comprises sous ce poste. La plupart des données relatives aux exportations et importations de marchandises sont tirées des registres douaniers.

Services et revenus : transactions en ressources effectuées entre résidents et non résidents, autres que celles qui sont considérées comme des marchandises, notamment : (a) expéditions et autres services de transport, y compris le fret, l'assurance et les autres services de distribution liés aux mouvements de marchandises; (b) voyages, à savoir les biens et services acquis par des voyageurs non résidents dans un pays donné et achats similaires faits par des résidents voyageant à l'étranger; et (c) revenus des investissements, qui correspondent aux revenus que les non-résidents tirent de leurs avoirs financiers placés dans l'économie déclarante (débit) et les revenus similaires que les résidents tirent de leurs avoirs financiers placés à l'étranger (crédit).

Transferts privés sans contrepartie : transferts sans contrepartie entre les résidents de l'économie déclarante autres que le gouvernement et la banque centrale, d'une part, et les non-résidents autres que les gouvernements et les banques centrales, d'autre part.

Transferts officiels sans contrepartie : transferts entre secteurs officiels, ainsi que les transferts où l'une des parties est le gouvernement ou la banque centrale de l'économie déclarante ou d'une économie étrangère.

Investissements directs : toutes les transactions sur capitaux entre, d'une part, les entreprises d'investissement direct et, d'autre part, les investisseurs directs eux-mêmes ou toute entreprise où ils ont effectué des investissements directs.

Other capital (long-term and short-term) covers transactions in all financial assets and liabilities other than those specified in the table.

Reserves and related items covers those assets (together with the use of Fund credit) that are conceived of as available for use by an economy's central authorities in meeting balance of payments needs.

Investissements de portefeuille : investissements en obligations à long terme et actions de société autres que les "investissements directs" et les "réserves".

Autres capitaux (à long et à cour terme): transactions sur tous les avoirs et engagements financiers autres que ceux spécifiés dans le tableau.

Réserves et postes assimilables : variations de ces actifs (ainsi que l'utilisation des crédits du Fonds) qui sont considérés comme disponibles par les responsables de l'économie pour répondre aux besoins de la balance des paiements.

79
Exchange rates
Cours des changes
National currency per US dollar
Valeur du dollar des Etats-Unis en monnaie nationale

Country (monetary unit) Pays (unité monétaire)	1985	1986	1987	1988	1989	1990	1991	1992	1993	1994
Afghanistan: afghani Afghanistan : afghani										
End of period										
Fin de période	50.600	50.600	50.600	50.600	50.600	50.600	50.600	50.600	50.600	50.600
Period average										
Moyenne sur période	50.600	50.600	50.600	50.600	50.600	50.600	50.600	50.600	50.600	50.600
Albania: lek Albanie : lek										
End of period										
Fin de période	7.000	7.000	...	6.000	6.400	15.000	25.000	97.000	95.000	...
Algeria: dinar Algérie : dinar										
End of period										
Fin de période	4.773	4.823	4.936	6.731	8.032	12.191	21.392	22.781	24.123	42.892
Period average										
Moyenne sur période	5.028	4.702	4.850	5.915	7.609	8.957	18.473	21.836	23.345	35.058
Angola: kwanza Angola : kwanza										
End of period										
Fin de période	29.918	29.918	29.918	29.918	29.918	29.918	...	...	...	...
Antigua and Barbuda: EC dollar Antigua-et-Barbuda : EC dollar										
End of period										
Fin de période	2.700	2.700	2.700	2.700	2.700	2.700	2.700	2.700	2.700	2.700
Argentina: peso Argentine : peso										
End of period										
Fin de période	0.000	0.000	0.000	0.001	0.179	0.558	0.998	0.990	0.998	0.999
Period average										
Moyenne sur période	0.000	0.000	0.000	0.001	0.042	0.488	0.954	0.991	0.999	0.999
Aruba: florin Aruba : florin										
End of period										
Fin de période	...	1.790	1.790	1.790	1.790	1.790	1.790	1.790	1.790	1.790
Period average										
Moyenne sur période	...	1.790	1.790	1.790	1.790	1.790	1.790	1.790	1.790	1.790
Australia: dollar Australie : dollar										
End of period										
Fin de période	1.469	1.504	1.384	1.169	1.262	1.293	1.316	1.452	1.477	1.287
Period average										
Moyenne sur période	1.432	1.496	1.428	1.280	1.265	1.281	1.284	1.362	1.471	1.368
Austria: schilling Autriche : schilling										
End of period										
Fin de période	17.280	13.710	11.250	12.565	11.815	10.677	10.689	11.354	12.143	11.095
Period average										
Moyenne sur période	20.689	15.267	12.642	12.348	13.231	11.370	11.676	10.989	11.632	11.422
Bahamas: dollar Bahamas : dollar										
End of period										
Fin de période	1.000	1.000	1.000	1.000	1.000	1.000	1.000	1.000	1.000	1.000
Bahrain: dinar Bahreïn : dinar										
End of period										
Fin de période	0.376	0.376	0.376	0.376	0.376	0.376	0.376	0.376	0.376	0.376
Period average										
Moyenne sur période	0.376	0.376	0.376	0.376	0.376	0.376	0.376	0.376	0.376	0.376
Bangladesh: taka Bangladesh : taka										
End of period										
Fin de période	31.000	30.800	31.200	32.270	32.270	35.790	38.580	39.000	39.850	40.250
Period average										
Moyenne sur période	27.995	30.407	30.950	31.733	32.270	34.569	36.596	38.951	39.567	40.212
Barbados: dollar Barbade : dollar										
End of period										
Fin de période	2.011	2.011	2.011	2.011	2.011	2.011	2.011	2.011	2.011	2.011

79
Exchange rates
National currency per US dollar [*cont.*]
Cours des changes
Valeur du dollar des Etats-Unis en monnaie nationale [*suite*]

Country (monetary unit) Pays (unité monétaire)	1985	1986	1987	1988	1989	1990	1991	1992	1993	1994
Belgium: franc Belgique : franc										
End of period										
Fin de période	50.360	40.410	33.153	37.345	35.760	30.982	31.270	33.180	36.110	31.837
Period average										
Moyenne sur période	59.378	44.672	37.334	36.768	39.404	33.418	34.148	32.149	34.596	33.456
Belize: dollar Belize : dollar										
End of period										
Fin de période	2.000	2.000	2.000	2.000	2.000	2.000	2.000	2.000	2.000	2.000
Benin: CFA franc Bénin : franc CFA										
End of period[1]										
Fin de période[1]	378.050	322.750	267.000	302.950	289.400	256.450	259.000	275.320	294.770	# 534.600
Period average[1]										
Moyenne sur période[1]	449.260	346.300	300.540	297.850	319.010	272.260	282.110	264.690	283.160	# 555.200
Bhutan: ngultrum Bhoutan : ngultrum										
End of period										
Fin de période	12.165	13.122	12.877	14.949	17.035	18.073	25.834	26.200	31.380	31.380
Period average										
Moyenne sur période	12.369	12.611	12.961	13.917	16.225	17.505	22.742	25.918	30.493	31.374
Bolivia: boliviano Bolivie : boliviano										
End of period										
Fin de période	1.692	1.923	2.210	2.470	2.980	3.400	3.745	4.095	4.475	4.695
Period average										
Moyenne sur période	0.440	1.922	2.055	2.350	2.692	3.173	3.581	3.900	4.265	4.620
Botswana: pula Botswana : pula										
End of period										
Fin de période	2.101	1.838	1.566	1.936	1.872	1.871	2.072	2.257	2.565	2.717
Period average										
Moyenne sur période	1.903	1.879	1.679	1.829	2.015	1.860	2.022	2.110	2.423	2.685
Brazil: real Brésil : real										
End of period[2]										
Fin de période[2]	0.000	0.000	0.000	0.280	4.130	64.390	0.190	2.200	0.050	# 0.846
Period average[2]										
Moyenne sur période[2]	0.000	0.000	0.000	0.100	1.030	24.840	0.150	1.640	0.032	...
Bulgaria: lev Bulgarie : lev										
End of period										
Fin de période	1.030	0.940	0.870	0.820	0.820	2.840	21.810	24.490	32.710	65.020
Period average										
Moyenne sur période	1.030	0.940	0.870	0.830	0.840	2.190	17.790	23.340	27.590	54.130
Burkina Faso: CFA franc Burkina Faso : franc CFA										
End of period[1]										
Fin de période[1]	378.050	322.750	267.000	302.950	289.400	256.450	259.000	275.320	294.770	# 534.600
Period average[1]										
Moyenne sur période[1]	449.260	346.300	300.540	297.850	319.010	272.260	282.110	264.690	283.160	# 555.200
Burundi: franc Burundi : franc										
End of period										
Fin de période	111.965	124.165	114.470	149.940	175.430	165.350	191.100	236.550	264.380	246.940
Period average										
Moyenne sur période	120.691	114.171	123.564	140.395	158.667	171.255	181.513	208.303	242.780	252.662
Cameroon: CFA franc Cameroun : franc CFA										
End of period[1]										
Fin de période[1]	378.050	322.750	267.000	302.950	289.400	256.450	259.000	275.320	294.770	# 534.600
Period average[1]										
Moyenne sur période[1]	449.260	346.300	300.540	297.850	319.010	272.260	282.110	264.690	283.160	# 555.200
Canada: dollar Canada : dollar										
End of period										
Fin de période	1.397	1.380	1.300	1.193	1.158	1.160	1.156	1.271	1.324	1.403

79
Exchange rates
National currency per US dollar [*cont.*]
Cours des changes
Valeur du dollar des Etats-Unis en monnaie nationale [*suite*]

Country (monetary unit) Pays (unité monétaire)	1985	1986	1987	1988	1989	1990	1991	1992	1993	1994
Period average Moyenne sur période	1.365	1.389	1.326	1.231	1.184	1.167	1.146	1.209	1.290	1.366
Cape Verde: escudo Cap-Vert : escudo										
End of period Fin de période	85.375	76.565	65.775	73.665	73.045	66.085	66.470	73.089	85.992	81.140
Period average Moyenne sur période	91.632	80.145	72.466	72.067	77.978	70.031	71.408	68.018	80.427	81.891
Central African Rep.: CFA franc Rép. centrafricaine : franc CFA										
End of period[1] Fin de période[1]	378.050	322.750	267.000	302.950	289.400	256.450	259.000	275.320	294.770	# 534.600
Period average[1] Moyenne sur période[1]	449.260	346.300	300.540	297.850	319.010	272.260	282.110	264.690	283.160	# 555.200
Chad: CFA franc Tchad : franc CFA										
End of period[1] Fin de période[1]	378.050	322.750	267.000	302.950	289.400	256.450	259.000	275.320	294.770	# 534.600
Period average[1] Moyenne sur période[1]	449.260	346.300	300.540	297.850	319.010	272.260	282.110	264.690	283.160	# 555.200
Chile: peso Chili : peso										
End of period Fin de période	183.860	204.730	238.140	247.200	297.370	337.090	374.510	382.120	428.470	402.920
Period average Moyenne sur période	161.081	193.016	219.540	245.047	267.155	305.062	349.372	362.588	404.349	420.077
China: yuan renminbi Chine : yuan renminbi										
End of period Fin de période	3.201	3.722	3.722	3.722	4.722	5.222	5.434	5.752	5.800	8.446
Period average Moyenne sur période	2.937	3.453	3.722	3.722	3.765	4.783	5.323	5.515	5.762	8.619
Colombia: peso Colombie : peso										
End of period Fin de période	172.200	219.000	263.700	335.860	433.920	568.730	706.860	811.770	917.330	831.270
Period average Moyenne sur période	142.312	194.261	242.607	299.174	382.568	502.259	633.045	759.282	863.065	844.836
Comoros: franc Comores : franc										
End of period[3] Fin de période[3]	378.046	322.746	266.997	302.947	289.397	256.447	258.997	275.322	294.772	# 400.950
Period average[3] Moyenne sur période[3]	449.265	346.308	300.538	297.850	319.010	272.266	282.108	264.693	283.164	# 416.403
Congo: CFA franc Congo : franc CFA										
End of period[1] Fin de période[1]	378.050	322.750	267.000	302.950	289.400	256.450	259.000	275.320	294.770	# 534.600
Period average[1] Moyenne sur période[1]	449.260	346.300	300.540	297.850	319.010	272.260	282.110	264.690	283.160	# 555.200
Costa Rica: colon Costa Rica : colon										
End of period Fin de période	53.700	58.875	69.250	79.500	84.350	103.550	135.425	137.430	151.440	165.070
Period average Moyenne sur période	50.453	55.986	62.776	75.805	81.504	91.579	122.432	134.506	142.172	157.067
Côte d'Ivoire: CFA franc Côte d'Ivoire : franc CFA										
End of period[1] Fin de période[1]	378.050	322.750	267.000	302.950	289.400	256.450	259.000	275.320	294.770	# 534.600
Period average[1] Moyenne sur période[1]	449.260	346.300	300.540	297.850	319.010	272.260	282.110	264.690	283.160	# 555.200
Croatia: new dinar Croatie : nouveau dinar										
End of period Fin de période	..	..	..	..	..	..	..	0.798	6.562	5.629

79

Exchange rates
National currency per US dollar [*cont.*]
Cours des changes
Valeur du dollar des Etats-Unis en monnaie nationale [*suite*]

Country (monetary unit) Pays (unité monétaire)	1985	1986	1987	1988	1989	1990	1991	1992	1993	1994
Period average Moyenne sur période	..	..	..	..	..	..	0.020	0.740	3.580	6.000
Cuba: peso Cuba : peso										
End of period Fin de période	0.910	0.793	0.773	0.776	0.791	0.700	0.700	0.700	0.700	...
Cyprus: pound Chypre : livre										
End of period Fin de période	0.543	0.512	0.439	0.466	0.479	0.435	0.439	0.483	0.520	0.476
Period average Moyenne sur période	0.613	0.518	0.481	0.467	0.494	0.458	0.458	0.450	0.497	0.492
former Czechoslovakia†: koruna anc. Tchécoslovaquie† : couronne										
End of period Fin de période	16.000	14.380	13.000	14.310	14.290	28.000	27.840	28.900	..	..
Period average Moyenne sur période	17.400	14.900	13.690	14.360	15.050	17.950	29.480	28.260	..	..
Czech Republic: koruna République tchèque : couronne										
End of period Fin de période	..	..	..	..	..	..	..	29.955	28.049
Period average Moyenne sur période	..	..	..	..	..	..	..	29.153	28.785
Denmark: krone Danemark : couronne										
End of period Fin de période	8.969	7.342	6.096	6.874	6.607	5.776	5.913	6.255	6.772	6.083
Period average Moyenne sur période	10.596	8.091	6.840	6.731	7.310	6.189	6.396	6.036	6.484	6.361
Djibouti: franc Djibouti : franc										
End of period Fin de période	177.721	177.721	177.721	177.721	177.721	177.721	177.721	177.721	177.721	177.721
Dominica: EC dollar Dominique : EC dollar										
End of period Fin de période	2.700	2.700	2.700	2.700	2.700	2.700	2.700	2.700	2.700	2.700
Dominican Republic: peso Rép. dominicaine : peso										
End of period Fin de période	2.940	3.077	4.960	6.340	6.340	11.350	12.660	12.575	12.767	13.064
Period average Moyenne sur période	3.113	2.904	3.845	6.112	6.340	8.525	12.692	12.774	12.679	13.160
Ecuador: sucre Equateur : sucre										
End of period Fin de période	95.750	146.500	221.500	432.510	648.420	878.200	1 270.580	1 844.250	2 043.780	2 269.000
Period average Moyenne sur période	69.560	122.780	170.460	301.610	526.350	767.750	1 046.250	1 533.960	1 919.100	2 196.730
Egypt: pound Egypte : livre										
End of period Fin de période	0.700	0.700	0.700	0.700	1.100	2.000	3.330	3.330	3.370	3.392
El Salvador: colon El Salvador : colon										
End of period Fin de période	2.500	5.000	5.000	5.000	5.000	8.030	8.080	9.170	8.670	8.750
Equatorial Guinea: CFA franc Guinée équatoriale : franc CFA										
End of period[1] Fin de période[1]	378.050	322.750	267.000	302.950	289.400	256.450	259.000	275.320	294.770	# 534.600
Period average[1] Moyenne sur période[1]	449.260	346.300	300.540	297.850	319.010	272.260	282.110	264.690	283.160	# 555.200
Estonia: kroon Estonie : couronne										
End of period Fin de période	..	..	..	..	..	..	..	12.910	13.880	...

79
Exchange rates
National currency per US dollar [*cont.*]
Cours des changes
Valeur du dollar des Etats-Unis en monnaie nationale [*suite*]

Country (monetary unit) Pays (unité monétaire)	1985	1986	1987	1988	1989	1990	1991	1992	1993	1994
Period average Moyenne sur période	...	...	...	...	...	...	...	...	13.220	...
Ethiopia: birr Ethiopie : birr										
End of period Fin de période	2.070	2.070	2.070	2.070	2.070	2.070	2.070	5.000	5.000	5.950
Fiji: dollar Fidji : dollar										
End of period Fin de période	1.120	1.145	1.441	1.405	1.494	1.459	1.473	1.564	1.541	1.409
Period average Moyenne sur période	1.154	1.133	1.244	1.430	1.483	1.481	1.476	1.503	1.542	1.464
Finland: markka Finlande : markka										
End of period Fin de période	5.417	4.794	3.946	4.169	4.059	3.634	4.133	5.245	5.784	4.743
Period average Moyenne sur période	6.198	5.069	4.396	4.183	4.291	3.823	4.044	4.479	5.712	5.223
France: franc France : franc										
End of period Fin de période	7.561	6.455	5.340	6.059	5.788	5.129	5.180	5.506	5.895	5.346
Period average Moyenne sur période	8.985	6.926	6.011	5.957	6.380	5.445	5.642	5.294	5.663	5.552
Gabon: CFA franc Gabon : franc CFA										
End of period[i] Fin de période[i]	378.050	322.750	267.000	302.950	289.400	256.450	259.000	275.320	294.770	# 534.600
Period average[i] Moyenne sur période[i]	449.260	346.300	300.540	297.850	319.010	272.260	282.110	264.690	283.160	# 555.200
Gambia: dalasi Gambie : dalasi										
End of period Fin de période	3.461	7.426	6.439	6.659	8.315	7.495	8.957	9.217	9.535	9.579
Period average Moyenne sur période	3.894	6.938	7.074	6.709	7.585	7.883	8.803	8.887	9.129	9.576
Germany † Allemagne†										
End of period Fin de période	..	..	..	..	..	..	1.516	1.614	1.726	1.549
Period average Moyenne sur période		..	..	..	..	..	1.660	1.562	1.653	1.623
F. R. Germany: deutsche mark R. f. Allemagne : deutsche mark										
End of period Fin de période	2.461	1.941	1.581	1.780	1.698	1.494	..	..	..	..
Period average Moyenne sur période	2.944	2.171	1.797	1.756	1.880	1.616	..	..	..	..
former German D. R.: mark anc. R. d. allemande : mark										
End of period Fin de période	2.600	2.000	1.650	1.870	1.790	1.500	..	..	..	..
Ghana: cedi Ghana : cedi										
End of period Fin de période	59.988	90.009	176.056	229.885	303.030	344.828	390.625	520.833	819.672	1 052.630
Period average Moyenne sur période	54.365	89.204	153.733	202.346	270.000	326.332	367.831	437.087	649.061	956.711
Greece: drachma Grèce : drachme										
End of period Fin de période	147.760	138.760	125.925	148.100	157.790	157.625	175.280	214.580	249.220	240.100
Period average Moyenne sur période	138.119	139.981	135.429	141.860	162.417	158.514	182.266	190.624	229.250	242.603
Grenada: EC dollar Grenade : EC dollar										
End of period Fin de période	2.700	2.700	2.700	2.700	2.700	2.700	2.700	2.700	2.700	2.700

79
Exchange rates
National currency per US dollar [*cont.*]
Cours des changes
Valeur du dollar des Etats-Unis en monnaie nationale [*suite*]

Country (monetary unit) Pays (unité monétaire)	1985	1986	1987	1988	1989	1990	1991	1992	1993	1994
Guatemala: quetzal Guatemala : quetzal										
End of period										
Fin de période	1.000	2.500	2.500	2.705	3.400	5.015	5.043	5.274	5.815	5.649
Period average										
Moyenne sur période	1.000	1.875	2.500	2.620	2.816	4.486	5.029	5.171	5.635	5.751
Guinea: franc Guinée : franc										
End of period										
Fin de période	22.473	235.630	440.000	550.000	620.000	680.000	802.950	922.410	972.414	981.024
Period average										
Moyenne sur période	24.333	333.452	428.402	474.396	591.646	660.167	753.858	902.001	955.490	961.636
Guinea-Bissau: peso Guinée-Bissau : peso										
End of period										
Fin de période	176.280	238.650	851.320	1 362.760	1 987.200	2 508.620	4 959.150	8 655.560	11 464.000	15 369.000
Period average										
Moyenne sur période	159.270	203.630	559.010	1 109.710	1 810.140	2 185.460	3 658.610	6 933.910	10 082.000	12 892.000
Guyana: dollar Guyana : dollar										
End of period										
Fin de période	4.150	4.400	10.000	10.000	33.000	45.000	122.000	126.000	130.750	142.500
Period average										
Moyenne sur période	4.252	4.272	9.756	10.000	27.159	39.533	111.811	125.002	126.730	138.290
Haiti: gourde Haïti : gourde										
End of period										
Fin de période	5.000	5.000	5.000	5.000	5.000	6.807	8.240	10.950	12.800	12.950
Honduras: lempira Honduras : lempira										
End of period										
Fin de période	2.000	2.000	2.000	2.000	2.000	2.000	5.400	5.830	7.260	9.400
Hong Kong: dollar Hong-kong : dollar										
End of period										
Fin de période	7.811	7.795	7.760	7.808	7.807	7.801	7.781	7.743	7.726	7.738
Period average										
Moyenne sur période	7.791	7.803	7.798	7.806	7.800	7.790	7.771	7.741	7.736	7.728
Hungary: forint Hongrie : forint										
End of period										
Fin de période	47.347	45.927	46.387	52.537	62.543	61.449	75.620	83.970	100.700	110.690
Period average										
Moyenne sur période	50.119	45.832	46.971	50.413	59.066	63.206	74.735	78.988	91.933	105.160
Iceland: krona Islande : couronne										
End of period										
Fin de période	42.060	40.240	35.660	46.220	61.170	55.390	55.620	63.920	72.730	68.300
Period average										
Moyenne sur période	41.508	41.104	38.677	43.014	57.042	58.284	58.996	57.546	67.603	69.944
India: rupee Inde : roupie										
End of period										
Fin de période	12.165	13.122	12.877	14.949	17.035	18.073	25.834	26.200	31.380	31.380
Period average										
Moyenne sur période	12.369	12.611	12.961	13.917	16.225	17.503	22.742	25.918	30.493	31.374
Indonesia: rupiah Indonésie : rupiah										
End of period										
Fin de période	1 125.000	1 641.000	1 650.000	1 731.000	1 797.000	1 901.000	1 992.000	2 062.000	2 110.000	2 200.000
Period average										
Moyenne sur période	1 110.580	1 282.560	1 643.850	1 685.700	1 770.060	1 842.810	1 950.320	2 029.920	2 087.100	2 160.750
Iran, Islamic Rep. of: rial Iran, Rép. islamique d' : rial										
End of period										
Fin de période	84.228	75.644	65.622	68.589	70.235	65.307	64.591	67.039	1 758.560	1 735.970

79
Exchange rates
National currency per US dollar [*cont.*]
Cours des changes
Valeur du dollar des Etats-Unis en monnaie nationale [*suite*]

Country (monetary unit) Pays (unité monétaire)	1985	1986	1987	1988	1989	1990	1991	1992	1993	1994
Period average Moyenne sur période	91.052	78.760	71.460	68.683	72.015	68.096	67.505	65.552	1 267.770	1 748.750
Iraq: dinar Iraq : dinar										
End of period Fin de période	0.311	0.311	0.311	0.311	0.311	0.311	0.311	0.311	0.311	0.311
Ireland: pound Irlande : livre										
End of period Fin de période	0.804	0.715	0.597	0.663	0.643	0.563	0.571	0.614	0.709	0.646
Period average Moyenne sur période	0.946	0.743	0.673	0.656	0.706	0.605	0.621	0.588	0.677	0.669
Israel: new shekel Israël : nouveau shekel										
End of period Fin de période	1.499	1.486	1.539	1.685	1.963	2.048	2.283	2.764	2.986	3.018
Period average Moyenne sur période	1.179	1.488	1.595	1.599	1.916	2.016	2.279	2.459	2.830	3.011
Italy: lira Italie : lire										
End of period Fin de période	1 678.500	1 358.130	1 169.250	1 305.770	1 270.500	1 130.150	1 151.060	1 470.860	1 703.970	1 629.740
Period average Moyenne sur période	1 909.440	1 490.810	1 296.070	1 301.630	1 372.090	1 198.100	1 240.610	1 232.410	1 573.670	1 612.440
Jamaica: dollar Jamaïque : dollar										
End of period Fin de période	5.480	5.480	5.500	5.480	6.480	8.038	21.492	22.185	32.474	33.201
Period average Moyenne sur période	5.559	5.478	5.487	5.489	5.745	7.184	12.116	22.960	24.948	33.086
Japan: yen Japon : yen										
End of period Fin de période	200.500	159.100	123.500	125.850	143.450	134.400	125.200	124.750	111.850	99.740
Period average Moyenne sur période	238.540	168.520	144.640	128.150	137.960	144.790	134.710	126.650	111.200	102.210
Jordan: dinar Jordanie : dinar										
End of period Fin de période	0.368	0.344	0.329	0.477	0.648	0.665	0.675	0.691	0.704	0.701
Period average Moyenne sur période	0.395	0.350	0.338	0.374	0.575	0.664	0.681	0.680	0.693	0.699
Kenya: shilling Kenya : shilling										
End of period Fin de période	16.284	16.042	16.515	18.599	21.601	24.084	28.074	36.216	68.163	44.839
Period average Moyenne sur période	16.432	16.226	16.454	17.747	20.572	22.915	27.508	32.217	58.001	56.051
Kiribati: dollar Kiribati : dollar										
End of period Fin de période	1.469	1.504	1.384	1.169	1.261	1.293	1.316	1.452	1.477	1.287
Period average Moyenne sur période	1.432	1.496	1.428	1.280	1.265	1.281	1.284	1.362	1.471	1.368
Korea, Republic of: won Corée, République de : won										
End of period Fin de période	890.200	861.400	792.300	684.100	679.600	716.400	760.800	788.400	808.100	788.700
Period average Moyenne sur période	870.020	881.450	822.570	731.470	671.460	707.760	733.350	780.650	802.670	803.440
Kuwait: dinar Koweït : dinar										
End of period Fin de période	0.289	0.292	0.270	0.283	0.292	...	0.284	0.303	0.298	0.300
Period average Moyenne sur période	0.301	0.291	0.279	0.279	0.294	...	...	0.293	0.302	0.297

79
Exchange rates
National currency per US dollar [*cont.*]
Cours des changes
Valeur du dollar des Etats-Unis en monnaie nationale [*suite*]

Country (monetary unit) Pays (unité monétaire)	1985	1986	1987	1988	1989	1990	1991	1992	1993	1994
Lao People's Dem. Rep.: new kip Rép. dém. pop. lao : nouveau kip										
End of period										
Fin de période	95.000	95.000	367.500	452.500	713.500	695.500	711.500	717.000	716.000	...
Period average										
Moyenne sur période	45.000	95.000	175.117	392.012	583.015	708.570	702.500	716.000	716.000	...
Lebanon: pound Liban : livre										
End of period										
Fin de période	18.100	87.000	455.000	530.000	505.000	842.000	879.000	1 838.000	1 711.000	1 647.000
Period average										
Moyenne sur période	16.420	38.370	224.600	409.230	496.690	695.090	928.230	1 712.790	1 741.360	1 680.070
Lesotho: loti Lesotho : maloti										
End of period										
Fin de période	2.558	2.183	1.930	2.378	2.536	2.563	2.743	3.053	3.397	3.543
Period average										
Moyenne sur période	2.229	2.285	2.036	2.273	2.623	2.587	2.761	2.852	3.268	3.551
Liberia: dollar Libéria : dollar										
End of period										
Fin de période	1.000	1.000	1.000	1.000	1.000	1.000	1.000	1.000	1.000	1.000
Libyan Arab Jamah.: dinar Jamah. arabe libyenne : dinar										
End of period										
Fin de période	* 0.296	* 0.314	* 0.271	* 0.285	* 0.292	* 0.270	* 0.268	* 0.301	0.325	0.360
Luxembourg: franc Luxembourg : franc										
End of period										
Fin de période	50.360	40.410	33.153	37.345	35.760	30.982	31.270	33.180	36.110	31.837
Period average										
Moyenne sur période	59.378	44.672	37.334	36.768	39.404	33.418	34.148	32.149	34.596	33.456
Madagascar: franc Madagascar : franc										
End of period										
Fin de période	635.790	769.810	1 234.270	1 526.430	1 532.540	1 465.830	1 832.660	1 910.170	1 962.670	...
Period average										
Moyenne sur période	662.480	676.340	1 069.210	1 407.110	1 603.440	1 494.150	1 835.360	1 863.970	1 913.780	...
Malawi: kwacha Malawi : kwacha										
End of period										
Fin de période	1.679	1.952	2.054	2.535	2.679	2.647	2.664	4.396	4.494	...
Period average										
Moyenne sur période	1.719	1.861	2.209	2.561	2.760	2.729	2.803	3.603	4.403	...
Malaysia: ringgit Malaisie : ringgit										
End of period										
Fin de période	2.426	2.603	2.493	2.715	2.703	2.701	2.724	2.612	2.701	2.560
Period average										
Moyenne sur période	2.483	2.581	2.520	2.619	2.709	2.705	2.750	2.547	2.574	2.624
Maldives: rufiyaa Maldives : rufiyaa										
End of period										
Fin de période	7.129	7.244	9.395	8.525	9.205	9.620	10.320	10.535	11.105	11.770
Period average										
Moyenne sur période	7.098	7.151	9.223	8.785	9.041	9.552	10.253	10.569	10.957	11.586
Mali: CFA franc Mali : franc CFA										
End of period[1]										
Fin de période[1]	378.040	322.740	267.000	302.940	289.400	256.450	259.000	275.320	294.770	# 534.600
Period average[1]										
Moyenne sur période[1]	449.260	346.300	300.540	297.850	319.010	272.260	282.110	264.690	283.160	# 555.200
Malta: lira Malte : lire										
End of period[1]										
Fin de période	0.424	0.369	0.312	0.332	0.337	0.301	0.306	0.374	0.395	0.368

79
Exchange rates
National currency per US dollar [cont.]
Cours des changes
Valeur du dollar des Etats-Unis en monnaie nationale [suite]

Country (monetary unit) Pays (unité monétaire)	1985	1986	1987	1988	1989	1990	1991	1992	1993	1994
Period average Moyenne sur période	0.469	0.393	0.345	0.331	0.348	0.318	0.323	0.319	0.382	0.378
Mauritania: ouguiya Mauritanie : ouguiya										
End of period Fin de période	77.070	74.080	71.600	75.730	83.550	77.840	77.820	115.100	124.160	128.370
Period average Moyenne sur période	77.085	74.375	73.878	75.261	83.051	80.609	81.946	87.027	120.806	123.575
Mauritius: rupee Maurice : roupie										
End of period Fin de période	14.310	13.137	12.175	13.834	14.996	14.322	14.794	16.998	18.656	17.863
Period average Moyenne sur période	15.442	13.466	12.878	13.438	15.250	14.863	15.652	15.563	17.648	17.960
Mexico: new peso Mexique : nouveau peso										
End of period Fin de période	0.372	0.923	2.210	2.281	2.641	2.945	3.071	3.115	3.106	5.325
Period average Moyenne sur période	0.257	0.612	1.378	2.273	2.461	2.813	3.018	3.095	3.116	3.375
Mongolia: tugrik Mongolie : tughrik										
End of period Fin de période	...	...	...	...	...	14.000	39.400	105.070	# 396.510	414.090
Period average Moyenne sur période	...	...	...	...	...	...	9.520	42.560	...	# 412.720
Morocco: dirham Maroc : dirham										
End of period Fin de période	9.621	8.712	7.800	8.211	8.122	8.043	8.150	9.049	9.651	8.960
Period average Moyenne sur période	10.062	9.104	8.359	8.209	8.488	8.242	8.706	8.538	9.299	9.203
Mozambique: metical Mozambique : metical										
End of period Fin de période	45.980	43.940	451.270	699.460	1 027.310	1 159.610	2 033.000	2 951.400	5 343.160	6 651.000
Period average Moyenne sur période	48.940	45.820	329.530	594.670	844.340	1 053.090	1 763.990	2 550.400	3 874.240	6 038.590
Myanmar: kyat Myanmar : kyat										
End of period Fin de période	7.842	7.039	6.110	6.410	6.494	6.080	6.014	6.241	6.246	5.903
Period average Moyenne sur période	8.475	7.330	6.653	6.395	6.705	6.339	6.284	6.105	6.157	5.975
Namibia: rand Namibie : rand										
End of period Fin de période	2.558	2.183	1.930	2.378	2.536	2.562	2.743	3.053	3.397	3.543
Period average Moyenne sur période	2.229	2.285	2.036	2.273	2.623	2.587	2.761	2.852	3.268	3.551
Nepal: rupee Népal : roupie										
End of period Fin de période	20.700	22.000	21.600	25.200	28.600	30.400	42.700	43.200	49.240	49.880
Period average Moyenne sur période	18.246	21.230	21.819	23.289	27.189	29.369	37.255	42.717	48.607	49.397
Netherlands: guilder Pays-Bas : florin										
End of period Fin de période	2.772	2.192	1.777	1.999	1.915	1.690	1.710	1.814	1.941	1.735
Period average Moyenne sur période	3.321	2.450	2.026	1.977	2.121	1.821	1.870	1.758	1.857	1.820
Netherlands Antilles: guilder Antilles néerlandaises : florin										
End of period Fin de période	1.800	1.800	1.800	1.800	1.790	1.790	1.790	1.790	1.790	1.790

79
Exchange rates
National currency per US dollar [*cont.*]
Cours des changes
Valeur du dollar des Etats-Unis en monnaie nationale [*suite*]

Country (monetary unit) Pays (unité monétaire)	1985	1986	1987	1988	1989	1990	1991	1992	1993	1994
New Zealand: dollar Nouvelle-Zélande : dollar										
End of period										
Fin de période	2.006	1.910	1.521	1.592	1.674	1.701	1.848	1.944	1.790	1.556
Period average										
Moyenne sur période	2.023	1.913	1.695	1.526	1.672	1.676	1.734	1.862	1.851	1.687
Nicaragua: gold córdoba Nicaragua : córdoba or										
End of period[4]										
Fin de période[4]	5.600	14.000	14.000	184.000	7.630	600.000	5.000	5.000	6.350	7.112
Period average[4]										
Moyenne sur période[4]	7.770	19.500	20.530	53.140	3.120	140.921	4.270	5.000	5.620	6.723
Niger: CFA franc Niger : franc CFA										
End of period[1]										
Fin de période[1]	378.050	322.750	267.000	302.950	289.400	256.450	259.000	275.320	294.770	# 534.600
Period average[1]										
Moyenne sur période[1]	449.260	346.300	300.540	297.850	319.010	272.260	282.110	264.690	283.160	# 555.200
Nigeria: naira Nigéria : naira										
End of period										
Fin de période	1.000	3.317	4.141	5.353	7.651	9.001	9.862	19.646	21.882	21.997
Period average										
Moyenne sur période	0.894	1.754	4.016	4.537	7.365	8.038	9.909	17.298	22.065	21.996
Norway: krone Norvège : couronne										
End of period										
Fin de période	7.582	7.400	6.232	6.570	6.615	5.907	5.973	6.924	7.518	6.762
Period average										
Moyenne sur période	8.597	7.395	6.737	6.517	6.904	6.260	6.483	6.214	7.094	7.058
Oman: rial Omani Oman : rial omani										
End of period										
Fin de période	0.345	0.384	0.384	0.384	0.384	0.384	0.384	0.384	0.384	0.384
Pakistan: rupee Pakistan : roupie										
End of period										
Fin de période	15.980	17.250	17.450	18.650	21.420	21.900	24.720	25.700	30.120	30.800
Period average										
Moyenne sur période	15.928	16.647	17.399	18.003	20.541	21.707	23.801	25.083	28.107	30.567
Panama: balboa Panama : balboa										
End of period										
Fin de période	1.000	1.000	1.000	1.000	1.000	1.000	1.000	1.000	1.000	1.000
Papua New Guinea: kina Papouasie-Nvl-Guinée : kina										
End of period										
Fin de période	1.012	0.961	0.878	0.826	0.860	0.953	0.953	0.987	0.981	1.179
Period average										
Moyenne sur période	1.000	0.971	0.908	0.867	0.859	0.955	0.952	0.965	0.978	1.011
Paraguay: guarani Paraguay : guaraní										
End of period										
Fin de période	320.000	550.000	550.000	550.000	1 218.000	1 258.000	1 352.300	1 600.200	1 880.000	1 940.000
Period average										
Moyenne sur période	306.670	339.170	550.000	550.000	1 056.220	1 229.810	1 325.180	1 500.260	1 744.350	1 911.540
Peru: new sol Pérou : nouveau sol										
End of period[5]										
Fin de période[5]	13 945.000	13 950.000	33 000.000	500.000	5 261.400	516.900	960.000	1 630.000	2 160.000	2 180.000
Period average[5]										
Moyenne sur période[5]	10 974.900	13 947.500	16 835.800	128.800	2 666.200	187.900	772.500	1 245.800	1 988.300	2 195.000
Philippines: peso Philippines : peso										
End of period										
Fin de période	19.032	20.530	20.800	21.335	22.440	28.000	26.650	25.096	27.699	24.418

79
Exchange rates
National currency per US dollar [*cont.*]
      Cours des changes
      Valeur du dollar des Etats-Unis en monnaie nationale [*suite*]

Country (monetary unit) Pays (unité monétaire)	1985	1986	1987	1988	1989	1990	1991	1992	1993	1994
Period average Moyenne sur période	18.607	20.386	20.568	21.095	21.737	24.310	27.479	25.512	27.120	26.417
Poland: zloty Pologne : zloty										
End of period Fin de période	0.015	0.020	0.032	0.050	0.650	0.950	1.096	1.577	2.134	2.437
Period average Moyenne sur période	0.015	0.017	0.026	0.043	0.144	0.950	1.058	1.363	1.811	2.272
Portugal: escudo Portugal : escudo										
End of period Fin de période	157.487	146.117	129.865	146.371	149.841	133.600	134.184	146.758	176.812	159.093
Period average Moyenne sur période	170.395	149.587	140.882	143.954	157.458	142.555	144.482	134.998	160.800	165.993
Qatar: riyal Qatar : riyal										
End of period Fin de période	3.640	3.640	3.640	3.640	3.640	3.640	3.640	3.640	3.640	3.640
Period average Moyenne sur période	3.640	3.640	3.640	3.640	3.640	3.640	3.640	3.640	3.640	3.640
Republic of Moldova: lei République de Moldova : lei										
End of period Fin de période	..	..	..	..	..	..	0.002	0.414	3.640	4.270
Romania: lei Roumanie : lei										
End of period Fin de période	15.730	15.280	13.740	14.370	14.440	34.710	# 189.000	460.000	1 276.000	1 767.000
Period average Moyenne sur période	17.141	16.153	14.557	14.277	14.922	22.432	76.387	# 307.953	760.051	1 655.090
Russian Federation: rouble Fédération de Russie : rouble										
End of period Fin de période	..	..	..	..	..	..	..	...	1 200.000	...
Rwanda: franc Rwanda : franc										
End of period Fin de période	93.490	84.180	73.020	76.710	77.620	121.120	119.790	146.270	146.370	...
Period average Moyenne sur période	101.262	87.640	79.672	76.445	79.977	82.597	125.140	133.350	144.248	...
Saint Kitts and Nevis: EC dollar Saint-Kitts-et-Nevis : EC dollar										
End of period Fin de période	2.700	2.700	2.700	2.700	2.700	2.700	2.700	2.700	2.700	2.700
Saint Lucia: EC dollar Sainte-Lucie : EC dollar										
End of period Fin de période	2.700	2.700	2.700	2.700	2.700	2.700	2.700	2.700	2.700	2.700
St. Vincent-Grenadines: EC dollar St. Vincent-Grenadines : EC dollar										
End of period Fin de période	2.700	2.700	2.700	2.700	2.700	2.700	2.700	2.700	2.700	2.700
Sao Tome and Principe: dobra Sao Tomé-et-Principe : dobra										
End of period Fin de période	41.196	36.993	72.827	98.176	140.366	140.982	280.021	375.540	516.700	...
Period average Moyenne sur période	44.604	38.589	54.211	86.343	124.672	143.331	201.816	321.337	429.854	732.628
Saudi Arabia: riyal Arabie saoudite : riyal										
End of period Fin de période	3.645	3.745	3.745	3.745	3.745	3.745	3.745	3.745	3.745	3.745
Period average Moyenne sur période	3.622	3.703	3.745	3.745	3.745	3.745	3.745	3.745	3.745	3.745
Senegal: CFA franc Sénégal : franc CFA										
End of period[1] Fin de période[1]	378.050	322.750	267.000	302.950	289.400	256.450	259.000	275.320	294.770	# 534.600

79

Exchange rates
National currency per US dollar [*cont.*]
Cours des changes
Valeur du dollar des Etats-Unis en monnaie nationale [*suite*]

Country (monetary unit) Pays (unité monétaire)	1985	1986	1987	1988	1989	1990	1991	1992	1993	1994
Period average[1] Moyenne sur période[1]	449.260	346.300	300.540	297.850	319.010	272.260	282.110	264.690	283.160	# 555.200
Seychelles: rupee Seychelles : roupie										
End of period Fin de période	6.602	5.929	5.143	5.397	5.467	5.119	5.063	5.254	5.258	4.969
Period average Moyenne sur période	7.134	6.177	5.600	5.384	5.646	5.337	5.289	5.122	5.182	5.056
Sierra Leone: leone Sierra Leone : leone										
End of period Fin de période	5.208	35.587	23.041	39.063	65.359	188.679	434.783	526.316	577.634	613.008
Period average Moyenne sur période	5.094	16.092	34.043	32.514	59.813	151.446	295.344	499.442	567.459	586.740
Singapore: dollar Singapour : dollar										
End of period Fin de période	2.105	2.175	1.998	1.946	1.894	1.744	1.630	1.645	1.608	1.461
Period average Moyenne sur période	2.200	2.177	2.106	2.012	1.950	1.813	1.728	1.629	1.616	1.527
Slovakia: koruna Slovaquie : couronne										
Period average Moyenne sur période	..	..	..	..	..	..	..	..	30.770	32.040
Slovenia: new dinar Slovénie : nouveau dinar										
End of period Fin de période	..	..	..	..	..	..	..	..	124.000	117.000
Period average Moyenne sur période	..	..	..	..	..	..	27.570	81.290	113.240	128.810
Solomon Islands: dollar Iles Salomon : dollar										
End of period Fin de période	1.613	1.986	1.974	2.118	2.397	2.614	2.795	3.100	3.248	...
Period average Moyenne sur période	1.481	1.741	2.003	2.083	2.293	2.529	2.715	2.928	3.188	...
Somalia: shilling Somalie : shilling										
End of period Fin de période	42.500	90.500	100.000	270.000	929.500	...	...	...	...	...
Period average Moyenne sur période	39.487	72.000	105.177	170.453	490.675	...	...	...	...	...
South Africa: rand Afrique du Sud : rand										
End of period Fin de période	2.558	2.183	1.930	2.378	2.536	2.563	2.743	3.053	3.397	3.543
Period average Moyenne sur période	2.229	2.285	2.036	2.273	2.623	2.587	2.761	2.852	3.268	3.551
Spain: peseta Espagne : peseta										
End of period Fin de période	154.150	132.395	109.000	113.450	109.720	96.909	96.688	114.623	142.214	131.739
Period average Moyenne sur période	170.044	140.048	123.478	116.487	118.378	101.934	103.912	102.379	127.260	133.958
Sri Lanka: rupee Sri Lanka : roupie										
End of period Fin de période	27.408	28.520	30.763	33.033	40.000	40.240	42.580	46.000	49.561	49.980
Period average Moyenne sur période	27.163	28.017	29.445	31.807	36.047	40.063	41.371	43.830	48.322	49.415
Sudan: pound Soudan : livre										
End of period Fin de période	2.500	2.500	4.500	4.500	4.500	4.500	14.993	135.135	217.391	400.000
Period average Moyenne sur période	2.304	2.500	3.000	4.500	4.500	4.500	6.956	97.432	159.314	289.609

79
Exchange rates
National currency per US dollar [*cont.*]
Cours des changes
Valeur du dollar des Etats-Unis en monnaie nationale [*suite*]

Country (monetary unit) Pays (unité monétaire)	1985	1986	1987	1988	1989	1990	1991	1992	1993	1994
Suriname: guilder Suriname : florin										
End of period										
Fin de période	1.785	1.785	1.785	1.785	1.785	1.785	1.785	1.785	1.785	1.785
Swaziland: lilangeni Swaziland : lilangeni										
End of period										
Fin de période	2.558	2.183	1.930	2.378	2.536	2.563	2.743	3.053	3.397	3.543
Period average										
Moyenne sur période	2.223	2.285	2.036	2.273	2.623	2.587	2.761	2.852	3.268	3.551
Sweden: krona Suède : couronne										
End of period										
Fin de période	7.615	6.819	5.848	6.157	6.227	5.698	5.529	7.043	8.303	7.461
Period average										
Moyenne sur période	8.604	7.124	6.340	6.127	6.447	5.919	6.047	5.824	7.783	7.716
Switzerland: franc Suisse : franc										
End of period										
Fin de période	2.076	1.623	1.278	1.504	1.546	1.295	1.355	1.456	1.479	1.311
Period average										
Moyenne sur période	2.457	1.799	1.491	1.463	1.636	1.389	1.434	1.406	1.478	1.368
Syrian Arab Republic: pound Rép. arabe syrienne : livre										
End of period										
Fin de période	3.925	3.925	3.925	11.225	11.225	11.225	11.225	11.225	11.225	11.225
Thailand: baht Thaïlande : baht										
End of period										
Fin de période	26.650	26.130	25.070	25.240	25.690	25.290	25.280	25.520	25.540	25.090
Period average										
Moyenne sur période	27.159	26.299	25.723	25.294	25.702	25.585	25.517	25.400	25.319	25.150
Togo: CFA franc Togo : franc CFA										
End of period [1]										
Fin de période[1]	378.050	322.750	267.000	302.950	289.400	256.450	259.000	275.320	294.770	# 534.600
Period average[1]										
Moyenne sur période[1]	449.260	346.300	300.540	297.850	319.010	272.260	282.110	264.690	283.160	# 555.200
Tonga: pa'anga Tonga : pa'anga										
End of period										
Fin de période	1.469	1.504	1.384	1.170	1.258	1.296	1.332	1.390	1.379	1.258
Period average										
Moyenne sur période	1.432	1.496	1.428	1.275	1.261	1.280	1.296	1.347	1.384	1.320
Trinidad and Tobago: dollar Trinité-et-Tobago : dollar										
End of period										
Fin de période	3.600	3.600	3.600	4.250	4.250	4.250	4.250	4.250	5.814	5.933
Period average										
Moyenne sur période	2.450	3.600	3.600	3.844	4.250	4.250	4.250	4.250	5.351	5.925
Tunisia: dinar Tunisie : dinar										
End of period										
Fin de période	0.757	0.840	0.778	0.898	0.905	0.837	0.864	0.951	1.047	0.991
Period average										
Moyenne sur période	0.834	0.794	0.829	0.858	0.949	0.878	0.925	0.884	1.004	1.012
Turkey: lira Turquie : livre										
End of period										
Fin de période	576.860	757.790	1 020.900	1 814.840	2 313.690	2 930.070	5 079.920	8 564.430	14 472.500	38 726.000
Period average										
Moyenne sur période	521.980	674.510	857.210	1 422.350	2 121.680	2 608.640	4 171.820	6 872.420	10 984.600	29 608.700
Uganda: shilling Ouganda : shilling										
End of period										
Fin de période	14.000	14.000	60.000	165.000	370.000	540.000	915.000	1 217.150	1 130.150	926.770

79
Exchange rates
National currency per US dollar [*cont.*]
  Cours des changes
  Valeur du dollar des Etats-Unis en monnaie nationale [*suite*]

Country (monetary unit) Pays (unité monétaire)	1985	1986	1987	1988	1989	1990	1991	1992	1993	1994
Period average Moyenne sur période	6.720	14.000	42.840	106.140	223.090	428.850	734.010	1 133.830	1 195.020	979.450
Ukraine: rouble Ukraine : rouble										
End of period Fin de période	..	..	..	..	..	..	637.700	25 000.000	...	
former USSR†: rouble ancienne URSS† : rouble										
End of period Fin de période	0.770	0.684	0.602	0.612	0.633	1.600		..	..	..
United Arab Emirates: dirham Emirats arabes unis : dirham										
End of period Fin de période	3.671	3.671	3.671	3.671	3.671	3.671	3.671	3.671	3.671	3.671
Period average Moyenne sur période	3.671	3.671	3.671	3.671	3.671	3.671	3.671	3.671	3.671	3.671
United Kingdom: pound Royaume-Uni : livre										
End of period Fin de période	0.692	0.678	0.534	0.553	0.623	0.519	0.535	0.661	0.675	0.640
Period average Moyenne sur période	0.779	0.682	0.612	0.562	0.611	0.563	0.567	0.570	0.667	0.653
United Rep.Tanzania: shilling Rép. Unie de Tanzanie : shilling										
End of period Fin de période	16.499	51.719	83.717	125.000	192.300	196.600	233.900	335.000	479.871	523.453
Period average Moyenne sur période	17.472	32.698	64.260	99.292	143.377	195.056	219.157	297.708	405.274	509.631
United States: dollar Etats-Unis : dollar										
End of period Fin de période	1.000	1.000	1.000	1.000	1.000	1.000	1.000	1.000	1.000	1.000
Period average Moyenne sur période	1.000	1.000	1.000	1.000	1.000	1.000	1.000	1.000	1.000	1.000
Uruguay: peso Uruguay : peso										
End of period Fin de période	0.125	0.181	0.281	0.451	0.805	1.594	2.490	3.481	# 4.418	5.615
Period average Moyenne sur période	0.101	0.152	0.227	0.359	0.605	1.171	2.019	3.027	# 3.948	5.053
Uzbekistan: rouble Ouzbékistan : rouble										
End of period Fin de période	...	...	...	...	...	...	1.670	414.500	1 200.000	...
Vanuatu: vatu Vanuatu : vatu										
End of period Fin de période	100.250	116.240	100.560	105.050	110.700	109.250	110.790	119.000	120.800	112.080
Period average Moyenne sur période	106.032	106.076	109.849	104.426	116.042	117.061	111.675	113.392	121.581	116.405
Venezuela: bolivar Venezuela : bolívar										
End of period Fin de période	7.500	14.500	14.500	14.500	43.079	50.380	61.554	79.450	105.640	# 170.000
Period average Moyenne sur période	7.500	8.083	14.500	14.500	34.681	46.900	56.816	68.376	90.826	148.503
Viet Nam: dong Viet Nam : dong										
End of period Fin de période	22.500	22.500	281.250	1 125.000	5 375.000	8 125.000	...	...	...	...
Period average Moyenne sur période	...	22.740	78.290	606.520	4 463.950	6 482.800	...	...	...	...
Yemen: rial Yémen : rial										
End of period Fin de période	..	..	..	..	..	12.010	12.010	12.010	12.010	12.010

79
Exchange rates
National currency per US dollar [cont.]
Cours des changes
Valeur du dollar des Etats-Unis en monnaie nationale [suite]

Country (monetary unit) Pays (unité monétaire)	1985	1986	1987	1988	1989	1990	1991	1992	1993	1994
former Dem. Yemen: dinar anciennce Yémen dém. : dinar										
End of period										
Fin de période	0.345	0.345	0.345	0.345	0.345	..	..	..	..	..
former Yemen Arab Rep.: rial anc. Yémen rép. arabe : rial										
End of period										
Fin de période	8.100	12.250	9.900	9.760	9.760	..	..	..	..	..
Period average										
Moyenne sur période	7.363	9.639	10.342	9.772	9.760	..	..	..	..	..
Yugoslavia, SFR†: new dinar Yougoslavie, Rfs† : nouveau dinar										
End of period										
Fin de période	0.031	0.046	0.124	0.521	11.816	10.657	19.735	..	..	..
Period average										
Moyenne sur période	0.027	0.038	0.073	0.252	2.876	11.318	19.638	..	..	..
Zaire: zaire Zaïre : zaïre										
End of period⁶										
Fin de période⁶	18.598	23.700	43.833	91.333	151.540	666.667	21.224	663.330	35.000	3 250.000
Period average⁶										
Moyenne sur période⁶	16.621	19.871	37.460	62.343	127.121	239.475	5.195	215.136	2.514	# 1 194.120
Zambia: kwacha Zambie : kwacha										
End of period										
Fin de période	5.700	12.710	8.000	10.004	21.650	42.753	88.968	359.712	500.000	...
Period average										
Moyenne sur période	3.140	7.788	9.519	8.266	13.814	30.289	64.640	172.214	452.763	...
Zimbabwe: dollar Zimbabwe : dollar										
End of period										
Fin de période	1.641	1.678	1.663	1.943	2.270	2.636	5.051	5.482	6.935	8.387
Period average										
Moyenne sur période	1.614	1.667	1.662	1.806	2.119	2.452	3.621	5.098	6.483	8.152

Source:
International Monetary Fund (Washington, DC).

† For detailed descriptions of data pertaining to
former Czechoslovakia, Germany, SFR Yugoslavia and former
USSR, see Annex I - Country or area nomenclature, regional
and other groupings.

1 The official rate is pegged to the French franc. Beginning
January 12, 1994, the CFA franc was devalued to CFAF 100 per
French franc from CFAF 50 at which it has been fixed since
1948.
2 Reais per million US dollar through 1990, per thousand US
dollar for 1991-1992 and per US dollar for 1993.
3 The official rate is pegged to the French franc. Beginning
January 12, 1994, the CFA franc was devalued to CFAF 75 per
French franc from CFAF 50 at which it has been fixed since
1948.
4 Gold córdoba per billion US dollar through 1987, per million
US dollar for 1988, per thousand US dollar for 1989-1990 and
per US dollar for 1991-1994.
5 New soles per billion US dollar through 1987, per million US
dollar for 1988-1989, and per thousand US dollar for
1990-1994.
6 New Zaires per US dollar through 1990, per thousand US
dollar for 1991-1992, and per US dollar for 1993.

Source:
Fonds monétaire international (Washington, DC).

† Pour les descriptions en détails des données
relatives à l'ancienne Tchécoslovaquie, l'Allemagne, la Rfs
Yougoslavie et l'ancienne URSS, voir l'Annexe I -
Nomenclature des pays ou zones, groupements régionaux et
autres groupements.

1 Le taux de change officiel est raccroché au taux de change
du franc français. Le 12 janvier 1994, le franc CFA a été
dévalué de 50 par franc français, valeur qu'il avait
conservée depuis 1948, à 100 par franc français.
2 Reais par million de dollars des États-Unis jusqu'en 1990,
par millier de dollars des États-Unis en 1991 et 1992 et par
dollar des État-Unis à compter de 1993.
3 Le taux de change officiel est raccroché au taux de change
du franc français. Le 12 janvier 1994, le CFA a été dévalué
de 50 par franc français, valeur qu'il avait conservée
depuis 1948, à 75 par franc français.
4 Cordobas or par milliard de dollars des États-Unis jusqu'en
1987, par million de dollars en 1988, par millier de dollars
en 1989-1990 et par dollar en 1991-1994.
5 Nouveaux soles par milliard de dollars des États-Unis
jusqu'en 1987, par million de dollars en 1988 et 1989, et
par millier de dollars en 1990-1994.
6 Nouveaux zaïres par dollar des États-Unis jusqu'en 1990, par
millier de dollars des États-Unis en 1991 et 1992, et par
dollar des États-Unis en 1993.

80
Total external and public/publicly guaranteed long-term debt of developing countries
Total de la dette extérieure et dette publique extérieure à long terme garantie par l'Etat des pays en développement

Million US dollars

Millions de dollars E-U

A. Total external debt • Total de la dette extérieure

Country or area[1]	1986	1987	1988	1989	1990	1991	1992	1993	Pays ou zone[1]
Total long-term debt (LDOD)	983432	1128199	1126897	1150567	1226315	1285902	1328263	1423993	**Total de la dette à long terme (LDOD)**
Public/publicly guaranteed	901032	1053167	1064586	1095122	1160351	1208598	1230712	1305042	Dette publique ou garantie par l'Etat
Official creditors	405571	497287	518601	551145	613958	663119	679858	729148	Créanciers publics
Multilateral	140548	177705	178167	186898	214736	232553	239530	257531	Multilatéraux
IBRD	68217	88947	84159	84684	95860	100302	98092	102991	BIRD
IDA	28067	33365	36179	39363	45103	49755	53607	58318	IDA
Bilateral	265023	319582	340434	364247	399221	430566	440328	471617	Bilatéraux
Private creditors	495462	555881	545985	543977	546394	545478	550854	575894	Créanciers privés
Bonds	41279	46499	51613	58248	120155	130475	144314	192974	Obligations
Commercial banks	332061	366024	360117	351782	277043	270425	257854	232754	Banques commerciales
Other private	122122	143358	134255	133947	149196	144579	148686	150167	Autres institutions privées
Private non-guaranteed	82400	75032	62311	55445	65964	77305	97552	118951	**Dette privé non garantie**
Undisbursed debt	186623	211073	201249	205092	217247	240915	239534	239342	**Dette (montants non versés)**
Official creditors	133222	151320	146902	154330	166558	181814	185180	184844	Créanciers publics
Multilateral	79932	87815	87654	94104	102881	114624	122739	119119	Multilatéraux
IBRD	38019	39484	37968	42552	43250	47613	50072	47836	BIRD
IDA	13716	15932	15418	16324	19206	21163	21425	21608	IDA
Bilateral	53290	63505	59248	60226	63677	67189	62441	65724	Bilatéraux
Private creditors	53401	59753	54348	50763	50689	59102	54354	54498	Créanciers privés
Commitments	91153	112342	115765	115714	123889	142586	128670	133657	**Engagements**
Official creditors	44247	52968	53315	61378	64704	76510	64606	57000	Créanciers publics
Multilateral	27855	29080	29237	32943	36164	44421	42275	33063	Multilatéraux
IBRD	14559	14116	12419	16820	15405	18271	15484	14187	BIRD
IDA	3221	4566	4573	4837	6310	7254	6162	5764	IDA
Bilateral	16391	23888	24079	28435	28539	32090	22332	23937	Bilatéraux
Private creditors	46907	59375	62450	54337	59186	66076	64064	76657	Créanciers privés
Disbursements	112825	120232	124862	122265	136022	137499	159593	179461	**Versements**
Public/publicly guaranteed	103918	111907	116386	109106	117072	118321	124199	136626	Dette publique ou garantie par l'Etat
Official creditors	46118	49031	49008	48275	55820	58183	50666	54568	Créanciers publics
Multilateral	22074	24772	23948	23504	28259	30127	28825	31199	Multilatéraux
IBRD	10197	11315	12163	10841	13628	12053	10434	13151	BIRD
IDA	3192	3915	3836	3591	4378	4605	5143	4870	IDA
Bilateral	24044	24259	25061	24772	27561	28056	21841	23369	Bilatéraux
Private creditors	57800	62876	67378	60831	61251	60139	73533	82058	Créanciers privés
Bonds	3707	4289	9487	8986	8957	12944	14177	35578	Obligations
Commercial banks	30301	36182	36016	25850	18483	20002	22204	21471	Banques commerciales
Other private	23792	22404	21875	25995	33811	27193	37152	25009	Autres institutions privées
Private non-guaranteed	8907	8325	8476	13158	18951	19177	35394	42835	**Dette privé non garantie**
Principal repayments	75133	83945	88062	86130	91760	89709	97687	110014	**Remboursements du principal**
Public/publicly guaranteed	64165	73208	76419	73558	82218	79218	83293	84441	Dette publique ou garantie par l'Etat
Official creditors	19316	22592	24950	24904	26619	28968	30237	30775	Créanciers publics
Multilateral	6993	9859	12813	11782	13059	15133	16127	16747	Multilatéraux
IBRD	4605	6553	9273	7952	8540	9535	10366	10417	BIRD
IDA	136	151	173	210	251	308	345	401	IDA
Bilateral	12323	12733	12136	13122	13561	13834	14110	14028	Bilatéraux
Private creditors	44849	50616	51469	48654	55599	50250	53056	53666	Créanciers privés
Bonds	2950	3239	6598	3829	6239	3513	9523	12615	Obligations
Commercial banks	24981	30778	25532	25443	27070	21669	22078	21848	Banques commerciales
Other private	16919	16599	19339	19381	22290	25068	21455	19202	Autres institutions privées
Private non-guaranteed	10968	10737	11643	12572	9542	10492	14394	25574	**Dette privé non garantie**

80

Total external and public/publicly guaranteed long-term debt of developing countries
Million US dollars [*cont.*]

Total de la dette extérieure et dette publique extérieure à long terme garantie par l'Etat des pays en développement
Millions de dollars E-U [*suite*]

A. Total external debt · Total de la dette extérieure

Country or area	1986	1987	1988	1989	1990	1991	1992	1993	Pays ou zone
Net flows	37692	36287	36800	36135	44262	47790	61906	69447	**Apports nets**
Public/publicly									**Dette publique ou**
guaranteed	39753	38699	39967	35548	34854	39104	40906	52186	**garantie par l'Etat**
Official creditors	26802	26438	24059	23371	29201	29216	20429	23793	Créanciers publics
Multilateral	15081	14912	11134	11722	15201	14994	12698	14451	Multilatéraux
IBRD	5592	4762	2890	2889	5088	2518	68	2734	BIRD
IDA	3056	3764	3664	3381	4127	4297	4797	4469	IDA
Bilateral	11721	11526	12924	11649	14001	14222	7731	9342	Bilatéraux
Private creditors	12951	12260	15909	12177	5653	9889	20477	28393	Créanciers privés
Bonds	758	1050	2889	5157	2718	9431	4654	22963	Obligations
Commercial banks	5320	5405	10484	407	−8587	−1667	126	−378	Banques commerciales
Other private	6873	5806	2536	6613	11522	2124	15697	5807	Autres institutions privées
Private non−guaranteed	−2061	−2412	−3167	587	9409	8686	21000	17261	**Dette privé non garantie**
Interest payments (LINT)	57020	57615	65196	60006	59440	60317	57394	57569	**Paiements d'intérets (LINT)**
Public/publicly									**Dette publique ou**
guaranteed	49525	50715	58591	54685	54454	54713	51415	51107	**garantie par l'Etat**
Official creditors	14712	16114	17968	18527	20706	22426	22601	24517	Créanciers publics
Multilateral	7788	9395	10371	9944	11389	12940	13042	13565	Multilatéraux
IBRD	5184	6241	6973	6380	7155	7964	7796	8029	BIRD
IDA	237	276	296	268	303	348	372	399	IDA
Bilateral	6924	6719	7597	8584	9317	9486	9559	10952	Bilatéraux
Private creditors	34813	34602	40623	36158	33748	32287	28814	26590	Créanciers privés
Bonds	2593	2837	3211	3599	5347	8638	8427	9526	Obligations
Commercial banks	25624	24444	29380	24750	20436	15187	13050	10601	Banques commerciales
Other private	6596	7321	8031	7809	7965	8462	7338	6464	Autres institutions privées
Private non−guaranteed	7495	6900	6606	5321	4986	5605	5980	6462	**Dette privé non garantie**
Net transfers	−19329	−21328	−28396	−23871	−15177	−12528	4512	11878	**Transferts nets**
Public/publicly									**Dette publique ou**
guaranteed	−9772	−12017	−18623	−19137	−19600	−15609	−10509	1079	**garantie par l'Etat**
Official creditors	12090	10325	6091	4844	8496	6790	−2171	−724	Créanciers publics
Multilateral	7293	5518	763	1778	3812	2054	−344	887	Multilatéraux
IBRD	408	−1478	−4083	−3491	−2067	−5447	−7728	−5295	BIRD
IDA	2818	3488	3368	3113	3824	3949	4425	4070	IDA
Bilateral	4797	4807	5327	3066	4684	4736	−1828	−1611	Bilatéraux
Private creditors	−21862	−22342	−24714	−23981	−28096	−22399	−8337	1803	Créanciers privés
Bonds	−1836	−1787	−322	1558	−2629	793	−3773	13438	Obligations
Commercial banks	−20304	−19039	−18897	−24344	−29023	−16854	−12924	−10979	Banques commerciales
Other private	277	−1516	−5495	−1195	3557	−6337	8359	−657	Autres institutions privées
Private non−guaranteed	−9557	−9311	−9773	−4734	4423	3081	15020	10799	**Dette privé non garantie**
Total debt service									**Total du service de la**
(LTDS)	132154	141560	153258	146136	151200	150027	155081	167584	**dette (LTDS)**
Public/publicly									**Dette publique ou**
guaranteed	113691	123924	135009	128243	136672	133930	134708	135548	**garantie par l'Etat**
Official creditors	34028	38706	42918	43432	47325	51393	52838	55292	Créanciers publics
Multilateral	14782	19254	23184	21726	24447	28073	29169	30312	Multilatéraux
IBRD	9789	12793	16247	14331	15695	17499	18162	18446	BIRD
IDA	373	427	468	477	554	656	717	799	IDA
Bilateral	19247	19452	19734	21706	22878	23320	23669	24980	Bilatéraux
Private creditors	79662	85218	92092	84811	89347	82537	81870	80256	Créanciers privés
Bonds	5543	6076	9809	7428	11586	12151	17950	22141	Obligations
Commercial banks	50604	55221	54912	50193	47507	36856	35128	32449	Banques commerciales
Other private	23515	23920	27370	27190	30254	33530	28793	25666	Autres institutions privées
Private non−guaranteed	18463	17637	18249	17893	14528	16096	20373	32036	**Dette privé non garantie**

80
Total external and public/publicly guaranteed long-term debt of developing countries
Million US dollars [*cont.*]

Total de la dette extérieure et dette publique extérieure à long terme garantie par l'Etat des pays en développement
Millions de dollars E-U [*suite*]

B. Public and publicly guaranteed long-term debt • Dette publique extérieure à long terme garantie par l'Etat

Country or area Pays ou zone	1984	1985	1986	1987	1988	1989	1990	1991	1992	1993
Albania Albanie	...	...	...	...	...	0.5	36.3	75.9	112.6	173.9
Algeria Algérie	14133.2	16397.9	19498.8	23095.1	24422.1	24637.9	26396.7	25964.4	25225.1	24586.7
Angola Angola	596.8	1830.3	2576.1	3958.4	4605.1	6032.7	7115.3	7213.5	7570.7	7727.2
Argentina Argentine	26700.1	37327.2	40958.1	49221.0	47546.3	51832.3	46905.6	47567.8	47036.0	55414.9
Armenia Arménie	..	..	..	..	..	..	..	1.0	2.9	140.0
Azerbaijan Azerbaïdjan	..	..	..	..	..	..	..	..	...	35.5
Bangladesh Bangladesh	5030.0	5965.1	7354.9	8958.2	9482.0	9916.6	11451.5	11960.7	12243.8	13047.9
Barbados Barbade	300.1	354.9	426.2	447.8	529.0	480.0	504.2	482.7	400.6	346.6
Belarus Bélarus	..	..	..	..	..	..	..	..	188.7	864.3
Belize Belize	74.7	94.8	97.3	110.9	118.5	127.9	135.0	149.6	155.2	162.2
Benin Bénin	581.1	662.3	769.4	932.2	913.9	1107.8	1156.6	1263.9	1322.7	1408.9
Bhutan Bhoutan	2.7	8.8	21.0	40.2	67.0	71.9	80.3	84.9	82.7	83.2
Bolivia Bolivie	3371.9	3511.3	4070.3	4621.3	4139.9	3428.8	3690.4	3534.2	3671.0	3687.1
Botswana Botswana	266.4	349.1	409.1	545.6	534.7	549.0	553.9	614.7	620.6	665.6
Brazil Brésil	70780.8	74708.4	84878.2	91765.2	89949.1	84361.7	83760.7	81476.7	86204.1	86649.6
Bulgaria Bulgarie	2583.9	3801.5	5806.0	7904.5	8304.9	9273.0	9816.5	9990.4	9957.1	9746.2
Burkina Faso Burkina Faso	369.1	456.2	576.7	744.0	767.4	648.2	749.9	882.6	986.6	1093.0
Burundi Burundi	320.0	409.0	524.2	711.6	755.6	832.2	851.4	901.2	947.0	999.2
Cambodia Cambodge	218.9	221.9	225.5	229.8	228.3	238.6	242.0	242.5	233.7	239.4
Cameroon Cameroun	1692.0	2005.0	2432.4	2826.8	2905.8	3763.3	4704.2	4912.7	5417.3	5436.0
Cape Verde Cap–Vert	76.2	96.4	108.6	121.3	122.1	125.8	139.9	141.3	144.1	148.8
Central African Rep. Rép. centrafricaine	216.5	290.5	399.3	544.5	589.9	636.9	641.8	739.4	751.8	797.2
Chad Tchad	153.6	159.6	200.2	266.9	308.5	325.2	445.6	541.5	658.8	704.6
Chile Chili	10616.8	12896.8	14689.4	15541.8	13696.4	10865.5	10426.1	10070.5	9577.7	8867.7
China Chine	6178.5	9936.8	16571.5	25923.6	32577.4	37041.6	45397.4	49342.3	58341.2	70023.5
Colombia Colombie	7733.7	9573.1	12180.6	13827.7	13846.0	13989.0	14670.8	14468.8	13238.5	12860.6
Comoros Comores	101.7	129.7	158.8	188.1	188.0	161.0	171.6	157.4	168.5	169.4
Congo Congo	1864.2	2362.2	2799.3	3392.7	3486.8	3502.7	4183.2	4021.9	3857.4	4097.1
Costa Rica Costa Rica	3183.3	3534.3	3625.4	3708.5	3560.7	3559.3	3076.3	3319.9	3207.4	3138.7
Côte d'Ivoire Côte d'Ivoire	4780.0	5772.7	6669.5	8282.2	7840.3	8304.4	10014.3	10634.6	10668.7	10550.5
Croatia Croatie	..	..	..	..	..	..	..	..	...	869.9

80
External debt of developing countries
Million US dollars [*cont.*]
Dette extérieure des pays en développement
Millions de dollars E–U [*suite*]

B. Public and publicly guaranteed long–term debt · Dette publique extérieure à long terme garantie par l'Etat

Country or area Pays ou zone	1984	1985	1986	1987	1988	1989	1990	1991	1992	1993
Czech Republic République tchèque	1807.4	1905.8	2100.9	2583.4	2727.7	3106.2	3953.8	4117.6	3883.0	5392.2
Djibouti Djibouti	61.6	96.0	119.2	154.3	158.3	131.3	156.1	170.9	176.8	192.6
Dominica Dominique	36.3	42.2	45.5	57.3	59.3	66.5	79.1	86.2	85.8	85.6
Dominican Republic Rép. dominicaine	2363.6	2690.9	2929.7	3220.2	3282.3	3333.7	3433.6	3751.1	3724.4	3762.8
Ecuador Equateur	6555.2	7198.7	8259.6	8991.1	9027.7	9427.0	9866.8	9950.9	9831.4	9935.1
Egypt Egypte	29149.2	34981.2	38052.1	44318.1	44409.4	42513.8	34337.6	36110.6	35808.6	36603.4
El Salvador El Salvador	1476.2	1555.3	1585.9	1675.4	1684.8	1825.6	1911.9	2056.9	2147.3	1897.0
Equatorial Guinea Guinée équatoriale	80.3	112.8	139.5	173.2	184.6	205.3	209.2	215.5	214.4	218.7
Estonia Estonie	..	..	..	..	..	..	..	..	33.8	85.8
Ethiopia Ethiopie	1450.2	1864.6	2221.3	2737.3	3070.4	3325.7	3627.9	3975.6	4184.8	4530.3
Fiji Fidji	279.3	302.3	311.7	336.4	335.6	301.6	305.9	270.7	226.8	199.3
Gabon Gabon	668.3	943.8	1427.9	2146.3	2287.6	2596.1	3135.0	3180.0	3001.8	2889.4
Gambia Gambie	151.8	176.6	212.1	265.4	276.8	289.1	309.1	322.8	346.4	348.8
Georgia Géorgie	..	..	..	..	..	..	..	..	79.3	568.0
Ghana Ghana	1157.8	1306.7	1730.1	2262.1	2203.4	2356.6	2700.7	2986.5	3123.5	3340.5
Grenada Grenade	42.1	46.6	51.8	65.6	69.9	72.1	91.0	101.3	97.7	96.2
Guatemala Guatemala	1950.0	2169.1	2293.9	2337.9	2141.9	2120.6	2240.0	2234.4	2105.8	2300.9
Guinea Guinée	1117.9	1291.9	1620.6	1883.3	2031.0	1958.1	2243.7	2402.7	2471.3	2675.4
Guinea–Bissau Guinée–Bissau	187.2	261.7	309.1	402.7	428.9	474.4	557.3	596.9	607.9	633.5
Guyana Guyana	697.3	780.5	860.0	958.4	991.4	1262.5	1741.4	1755.2	1666.0	1726.8
Haiti Haïti	475.1	522.3	575.7	673.5	683.3	684.4	745.6	609.5	625.9	617.6
Honduras Honduras	1774.4	2138.0	2378.9	2700.8	2757.5	2866.2	3416.3	3082.9	3211.4	3479.2
Hungary Hongrie	7062.6	9966.9	12383.8	15673.5	15612.3	16634.0	18006.4	18931.2	17843.2	20357.0
India Inde	24356.7	30283.7	37175.7	44379.2	48039.3	62774.4	69339.5	71886.0	77487.1	80984.9
Indonesia Indonésie	22268.5	26777.5	32625.9	40853.0	41200.3	41029.9	44963.0	48735.2	49416.8	52451.0
Iran, Islamic Rep. of Iran, Rép.islamic d'	2456.7	2390.0	2412.8	2279.9	2055.1	1861.6	1796.9	2064.5	1715.5	8879.6
Jamaica Jamaïque	2635.2	3140.4	3286.7	3761.0	3741.3	3755.8	3930.6	3806.2	3685.6	3604.1
Jordan Jordanie	2689.8	3256.2	4049.9	4876.2	5351.9	6221.6	7023.4	7447.0	6913.8	6824.6
Kazakhstan Kazakhstan	..	..	..	..	..	..	..	..	25.8	1552.3
Kenya Kenya	2296.5	2690.3	3430.5	4311.7	4267.4	4218.9	4849.8	4991.7	4978.0	5121.2
Korea, Republic of Corée, République de	23833.0	28279.1	29350.9	23889.7	20024.6	17038.1	18786.6	22481.5	24050.8	24566.4
Kyrgyzstan Kirghizistan	..	..	..	..	..	..	..	..	0.0	248.1
Lao People's Dem. Rep. Rép. dém. populaire lao	505.0	606.0	856.7	1151.4	1322.8	1463.2	1757.7	1849.6	1886.9	1948.4

80
External debt of developing countries
Million US dollars [cont.]
Dette extérieure des pays en développement
Millions de dollars E-U [suite]

B. Public and publicly guaranteed long-term debt · Dette publique extérieure à long terme garantie par l'Etat

Country or area Pays ou zone	1984	1985	1986	1987	1988	1989	1990	1991	1992	1993
Latvia Lettonie	..	..	..	..	..	..	..	..	26.0	118.8
Lebanon Liban	391.8	390.4	391.0	393.6	363.6	352.4	360.9	339.7	304.2	375.1
Lesotho Lesotho	129.8	167.4	190.0	251.6	274.9	312.7	374.9	414.0	446.1	472.0
Liberia Libéria	773.9	885.0	987.6	1114.5	1076.8	1064.2	1116.0	1106.3	1081.2	1070.3
Lithuania Lituanie	..	..	..	..	..	..	..	..	9.5	163.5
Madagascar Madagascar	2010.5	2427.8	2955.0	3560.5	3677.1	3609.6	3782.4	3971.7	3909.4	3919.9
Malawi Malawi	708.6	791.7	947.5	1160.9	1205.6	1271.3	1405.0	1527.6	1562.1	1723.5
Malaysia Malaisie	13169.3	14506.0	16276.6	17884.1	14632.0	12627.7	12684.2	14013.8	13468.4	13862.7
Maldives Maldives	49.9	49.0	58.7	61.9	59.3	54.4	64.0	78.0	92.7	111.7
Mali Mali	1092.0	1301.2	1578.5	1905.9	1910.5	2043.8	2345.0	2454.8	2470.4	2506.3
Malta Malte	90.8	99.4	95.2	95.6	84.8	78.5	124.6	146.4	129.4	127.9
Mauritania Mauritanie	1147.9	1332.9	1584.2	1851.1	1831.1	1763.8	1827.4	1858.3	1866.8	1960.2
Mauritius Maurice	336.4	398.0	449.2	582.5	643.1	633.8	746.3	806.5	744.6	717.4
Mexico Mexique	69726.0	72703.0	75826.2	84357.8	80598.2	76114.0	77557.5	79119.3	72263.8	74450.4
Mongolia Mongolie	...	...	...	...	...	...	...	...	288.0	344.3
Morocco Maroc	11811.8	13822.0	15840.6	18609.8	19498.8	20310.8	22142.6	20300.7	20510.4	20310.1
Mozambique Mozambique	1112.4	2538.4	3025.3	3701.5	3770.0	3937.2	4170.0	4122.7	4513.9	4650.1
Myanmar Myanmar	2196.5	2896.5	3617.8	4245.1	4220.2	4044.9	4444.2	4556.8	4974.0	5134.4
Nepal Népal	430.8	545.0	709.7	936.2	1099.8	1294.9	1571.8	1708.1	1754.8	1937.6
Nicaragua Nicaragua	4114.8	4970.8	5806.7	6447.3	7020.0	7660.5	8244.9	8762.5	8991.1	8773.1
Niger Niger	674.4	832.5	995.2	1244.0	1286.3	1136.6	1320.3	1252.5	1291.3	1353.6
Nigeria Nigéria	11392.6	13139.1	19086.1	28464.1	29057.9	30995.6	32586.5	33183.9	28457.5	28236.7
Oman Oman	1339.8	1908.1	2461.1	2441.4	2479.4	2619.2	2398.7	2471.6	2340.0	2319.1
Pakistan Pakistan	9741.0	10579.2	11781.2	13412.5	13837.2	14418.9	16409.9	17620.4	18454.8	20305.6
Panama Panama	3180.7	3323.2	3494.7	4026.5	4005.3	3935.4	3988.1	3918.2	3771.3	3709.3
Papua New Guinea Papouasie-Nvl-Guinée	960.5	1069.3	1233.0	1429.7	1252.9	1314.1	1501.8	1591.6	1533.7	1515.5
Paraguay Paraguay	1247.5	1533.8	1825.6	2223.9	2093.0	2095.3	1713.5	1685.4	1365.2	1282.8
Peru Pérou	9148.6	9815.8	10999.8	12635.0	12337.2	12616.8	13634.1	15300.7	15417.4	16123.2
Philippines Philippines	11300.2	13714.1	19264.5	22893.5	22439.6	22401.3	24073.5	25063.2	25605.0	27471.2
Poland Pologne	16332.3	29734.1	31903.5	36038.4	33627.0	34519.3	39262.7	45048.9	42940.9	41425.9
Portugal Portugal	10770.4	12825.5	13950.2	15039.1	14006.2	14691.3	17132.4	19542.6	21432.8	25173.0
Republic of Moldova République de Moldova	..	..	..	..	..	..	..	..	38.5	201.6
Romania Roumanie	6254.8	5805.1	5652.5	5342.7	2116.5	198.8	263.4	355.5	1334.5	2080.0

80
External debt of developing countries
Million US dollars [*cont.*]
Dette extérieure des pays en développement
Millions de dollars E–U [*suite*]

B. Public and publicly guaranteed long–term debt · Dette publique extérieure à long terme garantie par l'Etat

Country or area Pays ou zone	1984	1985	1986	1987	1988	1989	1990	1991	1992	1993
Russian Federation Fédération de Russie	17557.9	21396.5	23328.4	29718.8	30972.8	35741.7	48003.1	54942.3	64703.4	72768.6
Rwanda Rwanda	243.8	329.8	416.5	559.9	609.2	598.9	687.9	768.9	804.6	835.8
Saint Kitts and Nevis Saint–Kitts–et–Nevis	10.6	12.8	16.9	21.1	26.6	32.0	36.6	39.6	39.3	39.6
Saint Lucia Sainte–Lucie	20.3	23.1	28.4	41.9	52.5	61.2	72.0	75.0	89.5	96.9
St. Vincent and Grenadines St. Vincent–Grenadines	22.7	24.7	28.3	38.7	45.4	51.3	57.3	60.3	61.0	62.3
Samoa Samoa	63.0	63.7	64.6	71.4	71.1	71.9	91.0	113.3	117.8	139.2
Sao Tome and Principe Sao Tomé–et–Pricipe	53.8	62.0	76.7	91.1	100.9	114.8	135.7	180.4	195.6	225.7
Senegal Sénégal	1691.1	2058.6	2623.2	3321.8	3253.4	2659.4	2938.7	2862.7	2960.9	3010.8
Seychelles Seychelles	51.5	74.0	108.2	138.7	131.3	130.8	149.1	151.0	146.6	138.0
Sierra Leone Sierra Leone	322.4	387.5	465.8	543.9	537.7	529.7	604.4	656.9	680.4	728.3
Slovakia Slovaquie	864.6	783.6	776.2	998.5	1146.9	1222.1	1506.0	1784.4	1658.9	2058.3
Slovenia Slovénie	..	..	..	..	..	..	..	..	...	1255.7
Solomon Islands Iles Salomon	36.9	53.9	71.1	94.9	101.5	99.4	104.3	99.4	93.4	94.9
Somalia Somalie	1282.7	1411.9	1554.3	1743.3	1779.2	1813.3	1925.8	1945.2	1897.9	1897.0
Sri Lanka Sri Lanka	2349.3	2838.6	3452.3	4081.4	4151.3	4280.8	4938.2	5654.0	5617.6	5936.2
Sudan Soudan	6191.0	6601.7	7121.6	8043.1	8002.7	8468.9	9155.3	9220.1	8983.6	8993.7
Swaziland Swaziland	175.9	214.0	250.1	283.6	254.3	247.5	257.1	256.5	231.3	217.7
Syrian Arab Republic Rép. arabe syrienne	7551.2	9508.6	11387.5	14347.2	15094.9	15693.6	14917.0	16353.2	15912.4	16234.1
Tajikistan Tadjikistan	..	..	..	..	..	..	..	..	9.7	41.2
TFYR Macedonia L'ex–R.y. Macédoine	..	..	..	..	..	..	..	..	...	528.2
Thailand Thaïlande	7186.4	9860.3	11488.0	13831.8	13195.4	12427.9	12629.9	13418.2	13426.6	14561.9
Togo Togo	681.4	791.6	890.5	1052.3	1062.8	945.1	1084.9	1141.9	1135.2	1128.4
Tonga Tonga	19.4	23.8	28.4	35.6	37.0	38.1	44.4	44.2	42.6	43.7
Trinidad and Tobago Trinité–et–Tobago	1062.9	1299.2	1582.0	1636.2	1816.4	1785.3	1778.9	1737.2	1698.1	1703.5
Tunisia Tunisie	3703.4	4454.3	5281.0	6065.7	5953.1	6102.8	6662.2	7109.1	7201.7	7424.1
Turkey Turquie	16570.1	19551.3	24909.9	31537.4	33559.8	34714.4	38511.8	39040.5	39513.3	43320.5
Turkmenistan Turkménistan	..	..	..	..	..	..	..	..	...	9.0
Uganda Ouganda	700.6	893.1	1110.1	1606.5	1640.0	1917.1	2234.0	2358.7	2520.8	2616.7
Ukraine Ukraine	..	..	..	..	..	..	..	..	456.5	3456.8
United Rep. Tanzania Rép. Unie de Tanzanie	3027.8	3417.1	4457.1	5229.7	5394.9	5506.9	6298.1	6482.6	6578.8	6734.2
Uruguay Uruguay	2527.8	2695.0	2895.3	3126.4	2950.6	4018.3	4523.9	4355.7	4497.6	4628.7
Uzbekistan Ouzbékistan	..	..	..	..	..	..	..	..	9.9	736.0
Vanuatu Vanuatu	4.9	6.9	8.1	13.7	15.3	20.9	30.6	38.1	39.6	39.4

80
External debt of developing countries
Million US dollars [*cont.*]
Dette extérieure des pays en développement
Millions de dollars E-U [*suite*]

B. Public and publicly guaranteed long-term debt · Dette publique extérieure à long terme garantie par l'Etat

Country or area Pays ou zone	1984	1985	1986	1987	1988	1989	1990	1991	1992	1993
Venezuela Venezuela	18778.2	17737.7	25329.1	25008.4	25180.9	25166.2	24508.7	24938.6	25829.7	26855.7
Viet Nam Viet Nam	...	...	...	...	...	18299.9	20600.0	20598.6	21252.5	21553.9
Yemen Yémen	2583.3	2966.6	3439.0	4058.8	4393.8	4643.2	5153.8	5255.5	5253.3	5341.0
Yugoslavia Yougoslavie	..		..	..		..	..	..	..	8198.7
Yugoslavia, SFR † Yougoslavie, Rfs †	8484.8	11149.3	12236.1	14272.9	14053.0	14109.5	12986.4	11640.5	11015.5	..
Zaire Zaïre	4284.4	4957.8	5917.5	7207.6	6941.3	7965.8	9006.1	9271.1	8947.7	8769.1
Zambia Zambie	2596.2	3100.0	3836.1	4465.4	4435.8	4231.5	4852.0	4990.9	4719.5	4665.6
Zimbabwe Zimbabwe	1502.0	1777.5	2046.8	2375.5	2227.5	2275.4	2464.1	2611.0	2787.5	3020.7

Source:
World Debt Tables 1994-95 (volumes 1 and 2), The World Bank (Washington, DC).

† For detailed descriptions of data pertaining to former Czechoslovakia, Germany, SFR Yugoslavia and former USSR, see Annex I – Country or area nomenclature, regional and other groupings.

1 The following abbreviations have been used in the table:
LDOD: Long-term debt outstanding and disbursed
IBRD: International Bank for Reconstruction and Development
IDA: International Development Association
LINT: Loan interest
LTDS: Long-term debt service

Source:
"World Debt Tables 1994-95 (volumes 1 and 2)", La Banque mondiale (Washington, DC).

† Pour les descriptions en détails des données relatives à l'ancienne Tchécoslovaquie, l'Allemagne, la Rfs Yougoslavie et l'ancienne URSS, voir l'Annexe I – Nomenclature des pays ou zones, groupements régionaux et autres groupements.

1 Les abbréviations ci-après ont été utilisées dans le tableau:
LDOD : Dette à long terme
BIRD : Banque internationale pour la réconstruction et le développement
IDA : Association internationale de développement
LINT : Paiement des intérêts
LTDS : Service de la dette

Technical notes, tables 79 and 80

Table 79: Foreign exchange rates are shown in units of national currency per US dollar and refer to end-of-period and period-average quotations. Unless otherwise stated, the table refers to the midpoint (average of selling and buying rates) market rates for all dates so far as data are available. For dates prior to those for which market rates are available, the series represent fixed par value of central rates agreed with the International Monetary Fund. For further information see *International Financial Statistics*.[11]

Table 80: Data are extracted from *World Debt Tables 1994-1995, External Debt for Developing Countries*, published by the World Bank.[30]

Long term external debt is defined as debt that has an original or extended maturity of more than one year and is owed to non-residents and repayable in foreign currency, goods, or services. A distinction is made between:

— Public debt which is an external obligation of a public debtor, which could be a national government, a political sub-division, an agency of either of the above or, in fact, any autonomous public body;

— Publicly guaranteed debt, which is an external obligation of a private debtor that is guaranteed for repayment by a public entity;

— Private non-guaranteed external debt, which is an external obligation of a private debtor that is not guaranteed for repayment by a public entity.

The data referring to public and publicly guaranteed debt do not include data for (a) transactions with International Monetaty Fund, (b) debt repayable in local currency, (c) direct investment and (d) short-term debt (that is, debt with an original maturity of less than a year).

The data referring to private non-guaranteed debt also exclude the above items but include contractual obligations on loans to direct-investment enterprises by foreign parent companies or their affiliates.

Data are aggregated by type of creditor. The breakdown is as follows:

Official creditors:

(a) Loans from international organizations (multilateral loans), excluding loans from funds administered by an international organization on behalf of a single donor government. The latter are classified as loans from governments;

(b) Loans from governments (bilateral loans) and from autonomous public bodies;

Private creditors:

(a) Suppliers: Credits from manufacturers, exporters, or other suppliers of goods;

(b) Financial markets: Loans from private banks and other private financial institutions as well as publicly issued and privately placed bonds;

Notes techniques, tableaux 79 et 80

Tableau 79 : Les cours des changes sont exprimés par nombre d'unités de monnaie nationale pour un dollar des Etats-Unis et se rapportent aux cotations en fin de période et pour les moyennes sur la période. Sauf indication contraire, le tableau indique les cours moyens (moyenne des cours à la vente et à l'achat) pour toutes les dates pour lesquelles des données sont disponibles. En ce qui concerne les dates antérieures à celles pour lesquelles on dispose de données, les séries portent sur les parités fixes des taux centraux convenus avec le Fonds monétaire international. Pour plus de renseignements, voir *Satistiques financières internationales* [11].

Tableau 80 : les données sont extraites des *Tableaux de la dette internationale 1994-1995, Dette extérieure des pays pour développement*, publiés par la Banque mondiale [30].

La dette extérieure à long terme désigne la dette dont l'échéance initiale ou reportée est de plus d'un an, due à des non-résidents et remboursable en devises, biens ou services. On établit les distinctions suivantes :

– La dette publique, qui est une obligation extérieure d'un débiteur public, pouvant être un gouvernement, un organe politique, une institution de l'un ou l'autre ou, en fait, tout organisme public autonome.

– La dette garantie par l'Etat, qui est une obligation extérieure d'un débiteur privé, dont le remboursement est garanti par un organisme public.

– La dette extérieure privée non garantie, qui est une obligation extérieure d'un débiteur privé, dont le remboursement n'est pas garanti par un organisme public.

Les statistiques relatives à la dette publique ou à la dette garantie par l'Etat ne comprennent pas les données concernant : (a) les transactions avec le Fonds monétaire international; (b) la dette remboursable en monnaie nationale; (c) les investissements directs; et (d) la dette à court terme (c'est-à-dire la dette dont l'échéance initiale est inférieure à un an).

Les statistiques relatives à la dette privée non garantie ne comprennent pas non plus les éléments précités, mais comprennent les obligations contractuelles au titre des prêts consentis par des sociétés mères étrangères ou leurs filiales à des entreprises créées dans le cadre d'investissements directs.

Les données sont groupées par type de créancier, comme suit :

Créanciers publics :

(a)   Les prêts obtenus auprès d'organisations internationales (prêts multilatéraux), à l'exclusion des prêts au titre de fonds administrés par une organisation internationale pour le compte d'un gouvernement donateur précis, qui sont classés comme prêts consentis par des gouvernements.

(b)   Les prêts consentis par des gouvernements (prêts bilatéraux) et par des organisations publiques autonomes.

Créanciers privés :

(a)   Fournisseurs : Crédits consentis par des fabricants exportateurs et autre fournisseurs de biens;

(c) Other: External liabilities on account of nationalized properties and unclassified debts to private creditors.

A distinction is made between the following categories of external public debt:

— Debt outstanding (including undisbursed) is the sum of disbursed and undisbursed debt and represents the total outstanding external obligations of the borrower at year-end;

— Debt outstanding (disbursed only) is total outstanding debt drawn by the borrower at year end;

— Commitments are the total of loans for which contracts are signed in the year specified;

— Disbursements are drawings on outstanding loan commitments during the year specified;

— Service payments are actual repayments of principal amortization and interest payments made in foreign currencies, goods or services in the year specified;

— Net flows (or net lending) are disbursements minus principal repayments;

— Net transfers are net flows minus interest payments or disbursements minus total debt-service payments.

The countries included in the table are those for which data are sufficiently reliable to provide a meaningful presentation of debt outstanding and future service payments as of the end of 1994.

(b)   Marchés financiers : prêts consentis par des banques privées et autres institutions financières privées, et émissions publiques d'obligations placées auprès d'investisseurs privés;

(c)   Autres créanciers : engagements vis-à-vis de l'extérieur au titre des biens nationalisés et dettes diverses à l'égard de créanciers privés.

On fait une distinction entre les catégories suivantes de dette publique extérieure :

– L'encours de la dette (y compris les fonds non décaissés) est la somme des fonds décaissés et non décaissés et représente le total des obligations extérieures en cours de l'emprunteur à la fin de l'année;

– L'encours de la dette (fonds décaissés seulement) est le montant total des tirages effectués par l'emprunteur sur sa dette en cours à la fin de l'année;

– Les engagements représentent le total des prêts dont les contrats ont été signés au cours de l'année considérée;

– Les décaissements sont les sommes tirées sur l'encours des prêts pendant l'année considérée;

– Les paiements au titre du service de la dette sont les remboursements effectifs du principal et les paiements d'intérêts effectués en devises, biens ou services pendant l'année considérée;

– Les flux nets (ou prêts nets) sont les décaissements moins les remboursements de principal;

– Les transferts nets désignent les flux nets moins les paiements d'intérêts, ou les décaissements moins le total des paiements au titre du service de la dette.

Les pays figurant sur ce tableau sont ceux pour lesquels les données sont suffisamment fiables pour permettre une présentation significative de l'encours de la dette et des paiements futurs au titre du service de la dette à la fin de 1994.

81
Disbursements to developing countries or areas of bilateral and multilateral official development assistance
Versements mises à la disposition des pays ou zones en développement au titre de l'aide publique bilatérale et multilatérale au développement

Region, country or area Région, pays ou zone	Year Année	Disbursements ($US) Versements ($E-U)			
		Bilateral Bilatérale (millions)	Multilateral [1] Multilatérale [1] (millions)	Total (millions)	Per capita [2] Par habitant [2]
Total	**1990**	**38667.9**	**23667.7**	**62335.6**	...
	1991	**43118.6**	**23938.7**	**67057.3**	...
	1992	**43067.1**	**22118.4**	**65185.5**	...
Africa	**1990**	**15964.1**	**8435.3**	**24399.4**	**41.1**
Afrique	**1991**	**16693.9**	**9115.4**	**25809.3**	**42.2**
	1992	**16339.2**	**9995.3**	**26334.5**	**41.8**
Algeria	1990	232.2	329.5	561.7	22.5
Algérie	1991	306.9	332.8	639.7	25.1
	1992	374.1	269.8	643.9	24.6
Angola	1990	164.5	104.7	269.2	29.3
Angola	1991	158.9	135.2	294.1	30.9
	1992	194.7	166.4	361.1	36.5
Benin	1990	125.7	144.7	270.4	58.4
Bénin	1991	160.0	108.0	268.0	56.1
	1992	171.2	97.5	268.7	54.5
Botswana	1990	121.2	46.7	167.9	131.6
Botswana	1991	104.2	42.9	147.1	111.7
	1992	93.4	36.0	129.4	95.2
Burkina Faso	1990	238.7	85.2	323.9	36.0
Burkina Faso	1991	270.1	140.6	410.7	44.5
	1992	267.7	166.9	434.6	45.7
Burundi	1990	157.6	103.2	260.8	47.4
Burundi	1991	122.9	134.8	257.7	45.4
	1992	148.7	161.4	310.1	53.0
Cameroon	1990	339.1	201.0	540.1	46.9
Cameroun	1991	377.1	252.3	629.4	53.1
	1992	579.0	212.1	791.1	64.9
Cape Verde	1990	76.7	32.9	109.6	321.4
Cap-Vert	1991	80.5	24.7	105.2	300.6
	1992	82.0	41.6	123.6	343.3
Central African Rep.	1990	99.9	147.8	247.7	84.6
Rép. centrafricaine	1991	98.1	73.3	171.4	57.1
	1992	106.7	64.1	170.8	55.5
Chad	1990	183.3	130.9	314.2	56.6
Tchad	1991	137.6	126.0	263.6	46.3
	1992	148.4	92.4	240.8	41.2
Comoros	1990	30.6	15.6	46.2	85.1
Comores	1991	30.5	29.0	59.5	105.5
	1992	23.1	24.2	47.3	80.9
Congo	1990	202.0	12.0	214.0	95.9
Congo	1991	117.7	13.7	131.4	57.1
	1992	101.7	12.6	114.3	48.2
Cote d'Ivoire	1990	530.6	459.9	990.5	82.7
Côte d'Ivoire	1991	434.7	385.9	820.6	66.1
	1992	527.4	315.3	842.7	65.5
Djibouti	1990	88.3	18.2	106.5	206.0
Djibouti	1991	82.9	22.6	105.5	197.6
	1992	92.0	22.4	114.4	209.5
Egypt	1990	3171.8	59.4	3231.2	57.4
Egypte	1991	4157.0	283.4	4440.4	77.0
	1992	3000.9	355.1	3356.0	56.9
Equatorial Guinea	1990	43.6	18.7	62.3	177.0
Guinée équatoriale	1991	35.2	20.5	55.7	154.7
	1992	35.9	24.2	60.1	162.9
Ethiopia	1990	509.7	442.5	952.2	19.1
Ethiopie	1991	464.4	642.6	1107.0	21.5
	1992	457.7	738.0	1195.7	22.3
Gabon	1990	126.9	62.1	189.0	164.9
Gabon	1991	140.5	51.0	191.5	162.4
	1992	64.8	42.8	107.6	88.7

81
Disbursements to developing countries or areas of bilateral and multilateral
official development assistance [cont.]
Versements mises à la disposition des pays ou zones en développement au titre de l'aide publique bilatérale
et multilatérale au développement [suite]

Region, country or area Région, pays ou zone	Year Année	Disbursements ($US) Versements ($E-U)			
		Bilateral Bilatérale (millions)	Multilateral[1] Multilatérale[1] (millions)	Total (millions)	Per capita[2] Par habitant[2]
Gambia Gambie	1990	56.9	41.8	98.7	106.9
	1991	55.0	44.8	99.8	103.6
	1992	50.3	65.3	115.6	115.4
Ghana Ghana	1990	264.9	427.7	692.6	46.1
	1991	448.6	530.0	978.6	63.2
	1992	332.3	388.4	720.7	45.2
Guinea Guinée	1990	139.0	141.7	280.7	48.8
	1991	173.3	200.2	373.5	63.0
	1992	233.5	236.7	470.2	76.9
Guinea-Bissau Guinée-Bissau	1990	77.7	52.2	129.9	134.8
	1991	64.9	51.3	116.2	118.1
	1992	59.2	45.5	104.7	104.1
Kenya Kenya	1990	735.2	392.1	1127.3	47.7
	1991	608.4	234.0	842.4	34.4
	1992	516.7	321.7	838.4	33.0
Lesotho Lesotho	1990	85.2	59.7	144.9	80.9
	1991	74.1	49.2	123.3	67.0
	1992	67.9	75.5	143.4	75.8
Liberia Libéria	1990	42.3	35.1	77.4	30.1
	1991	56.7	99.9	156.6	58.9
	1992	26.1	92.9	119.0	43.3
Libyan Arab Jamahiriya Jamahiriya arabe libyenne	1990	7.6	6.7	14.3	3.1
	1991	3.0	89.8	92.8	19.7
	1992	1.5	−6.8	−5.3	−1.1
Madagascar Madagascar	1990	268.2	135.5	403.7	32.1
	1991	274.2	184.6	458.8	35.3
	1992	215.5	145.3	360.8	26.9
Malawi Malawi	1990	216.2	280.8	497.0	53.1
	1991	208.7	311.9	520.6	53.3
	1992	207.9	356.6	564.5	55.5
Mali Mali	1990	312.5	158.8	471.3	51.2
	1991	279.7	172.4	452.1	47.6
	1992	239.1	190.2	429.3	43.7
Mauritania Mauritanie	1990	106.4	95.7	202.1	100.9
	1991	110.3	102.0	212.3	103.4
	1992	116.4	112.2	228.6	108.5
Mauritius Maurice	1990	75.7	15.2	90.9	86.0
	1991	61.5	2.6	64.1	60.1
	1992	34.7	3.4	38.1	35.3
Morocco Maroc	1990	595.4	521.0	1116.4	45.9
	1991	610.8	626.9	1237.7	49.8
	1992	733.6	663.1	1396.7	55.0
Mozambique Mozambique	1990	751.5	255.0	1006.5	70.9
	1991	771.6	303.8	1075.4	74.5
	1992	1010.0	458.7	1468.7	99.7
Namibia Namibie	1990	39.4	83.9	123.3	91.4
	1991	95.1	89.1	184.2	133.0
	1992	97.9	45.5	143.4	100.8
Niger Niger	1990	254.6	138.8	393.4	50.9
	1991	264.4	107.0	371.4	46.5
	1992	261.5	109.1	370.6	44.8
Nigeria Nigéria	1990	172.7	391.0	563.7	5.9
	1991	171.6	300.0	471.6	4.8
	1992	137.7	388.5	526.2	5.2
Rwanda Rwanda	1990	183.2	98.9	282.1	40.4
	1991	232.9	126.8	359.7	50.1
	1992	187.5	165.1	352.6	47.9
Saint Helena Sainte-Hélène	1990	23.3	1.4	24.7	4116.7
	1991	14.8	0.5	15.3	2550.0
	1992	15.2	0.6	15.8	2633.3
Sao Tome and Principe Sao Tomé-et-Principe	1990	32.2	24.0	56.2	472.3
	1991	25.1	27.2	52.3	432.2
	1992	27.0	30.2	57.2	461.3

81
Disbursements to developing countries or areas of bilateral and multilateral
official development assistance [cont.]
Versements mises à la disposition des pays ou zones en développement au titre de l'aide publique bilatérale
et multilatérale au développement [suite]

Region, country or area Région, pays ou zone	Year Année	Disbursements ($US) Versements ($E–U)			
		Bilateral Bilatérale (millions)	Multilateral [1] Multilatérale [1] (millions)	Total (millions)	Per capita [2] Par habitant [2]
Senegal	1990	589.2	215.4	804.6	109.8
Sénégal	1991	421.4	186.3	607.7	80.8
	1992	454.0	258.1	712.1	92.4
Seychelles	1990	32.7	3.0	35.7	510.0
Seychelles	1991	16.8	7.2	24.0	342.9
	1992	15.4	5.6	21.0	295.8
Sierra Leone	1990	39.9	21.9	61.8	15.5
Sierra Leone	1991	67.9	35.6	103.5	25.3
	1992	74.1	53.4	127.5	30.4
Somalia	1990	269.6	142.7	412.3	47.5
Somalie	1991	116.0	70.2	186.2	21.2
	1992	495.9	146.2	642.1	72.4
Sudan	1990	420.0	388.9	808.9	32.9
Soudan	1991	368.8	495.2	864.0	34.2
	1992	188.0	351.5	539.5	20.8
Swaziland	1990	36.1	4.7	40.8	54.8
Swaziland	1991	31.1	19.7	50.8	66.4
	1992	26.6	22.0	48.6	61.8
Togo	1990	155.0	100.2	255.2	72.3
Togo	1991	124.5	74.8	199.3	54.7
	1992	134.9	90.0	224.9	59.8
Tunisia	1990	214.4	318.9	533.3	66.0
Tunisie	1991	263.9	478.1	742.0	90.0
	1992	296.8	247.9	544.7	64.8
Uganda	1990	244.4	383.5	627.9	35.0
Ouganda	1991	285.3	316.2	601.5	32.3
	1992	254.6	445.9	700.5	36.4
United Rep. Tanzania	1990	844.1	298.9	1143.0	44.6
Rép. Unie de Tanzanie	1991	763.8	276.9	1040.7	39.4
	1992	815.2	495.7	1310.9	48.2
Zaire	1990	632.7	331.1	963.8	25.7
Zaïre	1991	342.7	216.2	558.9	14.5
	1992	162.7	101.4	264.1	6.6
Zambia	1990	408.9	90.8	499.7	61.3
Zambie	1991	582.8	122.4	705.2	83.8
	1992	698.6	282.5	981.1	113.1
Zimbabwe	1990	295.9	129.3	425.2	42.9
Zimbabwe	1991	359.2	180.7	539.9	53.0
	1992	535.2	507.3	1042.5	99.6
Other and unallocated	1990	898.9	233.9	1132.8	...
Autres et non–ventilés	1991	1066.3	162.7	1229.0	...
	1992	1146.3	257.5	1403.8	...
Americas	**1990**	**4188.4**	**5805.4**	**9993.8**	**23.0**
Amériques	**1991**	**4856.7**	**3460.1**	**8316.8**	**18.7**
	1992	**4293.7**	**1876.2**	**6169.9**	**13.6**
Anguilla	1990	2.4	1.5	3.9	557.1
Anguilla	1991	5.1	1.8	6.9	985.7
	1992	3.5	1.6	5.1	637.5
Antigua and Barbuda	1990	2.9	1.7	4.6	71.9
Antigua–et–Barbuda	1991	5.8	1.5	7.3	114.1
	1992	4.0	0.9	4.9	75.4
Argentina	1990	166.2	340.6	506.8	15.6
Argentine	1991	270.6	337.8	608.4	18.5
	1992	254.7	−114.4	140.3	4.2
Aruba	1990	28.9	1.1	30.0	447.8
Aruba	1991	24.8	0.3	25.1	374.6
	1992	29.9	0.6	30.5	448.5
Bahamas	1990	0.4	30.2	30.6	119.5
Bahamas	1991	0.2	58.3	58.5	225.0
	1992	0.4	34.1	34.5	130.7
Barbados	1990	1.4	9.5	10.9	42.4
Barbade	1991	1.5	−0.1	1.4	5.4
	1992	0.8	−0.5	0.3	1.2

81

Disbursements to developing countries or areas of bilateral and multilateral official development assistance [*cont.*]

Versements mises à la disposition des pays ou zones en développement au titre de l'aide publique bilatérale et multilatérale au développement [*suite*]

Region, country or area Région, pays ou zone	Year Année	Disbursements ($US) Versements ($E-U)			
		Bilateral Bilatérale (millions)	Multilateral [1] Multilatérale [1] (millions)	Total (millions)	Per capita [2] Par habitant [2]
Belize	1990	18.8	15.6	34.4	182.0
Belize	1991	16.5	7.5	24.0	123.7
	1992	13.9	13.6	27.5	138.2
Bermuda	1990	42.1	0.1	42.2	691.8
Bermudes	1991	−4.8	–	−4.8	−77.4
	1992	−5.1	0.1	−5.0	−80.6
Bolivia	1990	364.7	209.5	574.2	87.4
Bolivie	1991	356.1	184.4	540.5	80.3
	1992	501.9	264.5	766.4	111.2
Brazil	1990	142.1	−309.4	−167.3	−1.1
Brésil	1991	165.8	−254.1	−88.3	−0.6
	1992	−278.7	−517.0	−795.7	−5.2
British Virgin Islands	1990	3.0	4.9	7.9	49.3
Iles Vierges britanniques	1991	3.3	6.7	10.0	58.8
	1992	2.1	9.1	11.2	65.9
Cayman Islands	1990	2.1	7.3	9.4	361.5
Iles Caïmanes	1991	−0.8	5.5	4.7	174.1
	1992	−1.4	8.9	7.5	267.9
Chile	1990	83.3	500.9	584.2	44.4
Chili	1991	105.5	305.5	411.0	30.7
	1992	117.7	259.8	377.5	27.8
Colombia	1990	86.6	12.6	99.2	3.1
Colombie	1991	117.3	155.8	273.1	8.3
	1992	221.0	−44.0	177.0	5.3
Costa Rica	1990	206.6	12.7	219.3	72.3
Costa Rica	1991	167.4	24.8	192.2	61.7
	1992	134.0	30.1	164.1	51.4
Cuba	1990	33.6	17.4	51.0	4.8
Cuba	1991	19.2	18.4	37.6	3.5
	1992	11.8	12.9	24.7	2.3
Dominica	1990	10.8	8.8	19.6	276.1
Dominique	1991	7.7	10.1	17.8	250.7
	1992	8.3	6.1	14.4	202.8
Dominican Republic	1990	72.7	61.9	134.6	18.9
Rép. dominicaine	1991	64.8	28.8	93.6	12.9
	1992	52.0	23.2	75.2	10.2
Ecuador	1990	122.2	121.5	243.7	23.7
Equateur	1991	169.2	100.3	269.5	25.7
	1992	204.5	50.2	254.7	23.7
El Salvador	1990	312.0	−3.1	308.9	59.7
El Salvador	1991	241.6	74.7	316.3	59.9
	1992	315.0	76.9	391.9	72.6
Falkland Islands	1990	1.8	–	1.8	900.0
Iles Falkland	1991	3.2	0.6	3.8	1900.0
	1992	0.1	0.3	0.4	200.0
Grenada	1990	5.0	9.0	14.0	153.8
Grenade	1991	9.5	7.3	16.8	184.6
	1992	5.6	9.6	15.2	167.0
Guatemala	1990	149.5	67.8	217.3	23.6
Guatemala	1991	154.5	−4.4	150.1	15.9
	1992	177.1	−52.0	125.1	12.8
Guyana	1990	35.8	114.4	150.2	188.7
Guyana	1991	35.2	94.0	129.2	161.3
	1992	26.0	57.4	83.4	103.2
Haiti	1990	117.1	54.8	171.9	26.5
Haïti	1991	139.6	41.8	181.4	27.4
	1992	76.9	25.9	102.8	15.2
Honduras	1990	383.5	52.8	436.3	89.4
Honduras	1991	225.0	59.8	284.8	56.6
	1992	206.9	122.0	328.9	63.5
Jamaica	1990	251.9	38.2	290.1	122.6
Jamaïque	1991	150.6	35.9	186.5	78.4
	1992	112.7	−4.8	107.9	45.1

81

Disbursements to developing countries or areas of bilateral and multilateral
official development assistance [cont.]

Versements mises à la disposition des pays ou zones en développement au titre de l'aide publique bilatérale
et multilatérale au développement [suite]

Region, country or area Région, pays ou zone	Year Année	Disbursements ($US) Versements ($E-U)			
		Bilateral Bilatérale (millions)	Multilateral [1] Multilatérale [1] (millions)	Total (millions)	Per capita [2] Par habitant [2]
Mexico Mexique	1990	144.7	2853.2	2997.9	35.5
	1991	260.9	1128.5	1389.4	16.1
	1992	300.7	635.2	935.9	10.6
Montserrat Montserrat	1990	7.8	0.5	8.3	754.5
	1991	8.2	1.2	9.4	854.5
	1992	4.7	2.6	7.3	663.6
Netherlands Antilles Antilles néerlandaises	1990	53.0	8.2	61.2	322.1
	1991	78.4	3.3	81.7	425.5
	1992	88.9	7.2	96.1	497.9
Nicaragua Nicaragua	1990	288.5	45.0	333.5	90.7
	1991	700.3	17.9	718.2	188.7
	1992	472.8	196.9	669.7	169.3
Panama Panama	1990	96.0	−38.2	57.8	24.1
	1991	107.8	−55.2	52.6	21.5
	1992	193.7	−230.1	−36.4	−14.6
Paraguay Paraguay	1990	47.6	−20.3	27.3	6.3
	1991	128.0	−17.6	110.4	24.8
	1992	65.5	−3.9	61.6	13.5
Peru Pérou	1990	350.4	54.1	404.5	18.7
	1991	600.1	29.2	629.3	28.6
	1992	377.8	−53.6	324.2	14.4
Saint Kitts and Nevis Saint−Kitts−et−Nevis	1990	5.0	3.0	8.0	190.5
	1991	2.6	4.9	7.5	178.6
	1992	3.8	4.5	8.3	197.6
Saint Lucia Sainte−Lucie	1990	6.2	9.8	16.0	120.3
	1991	13.9	12.0	25.9	191.9
	1992	15.0	19.0	34.0	248.2
St. Vincent and Grenadines St. Vincent−et−Grenadines	1990	5.2	9.8	15.0	140.2
	1991	7.2	11.1	18.3	169.4
	1992	6.5	9.7	16.2	148.6
Suriname Suriname	1990	51.2	14.5	65.7	164.3
	1991	31.6	14.8	46.4	114.6
	1992	73.0	8.1	81.1	198.3
Trinidad and Tobago Trinité−et−Tobago	1990	6.1	40.3	46.4	37.5
	1991	−5.0	34.8	29.8	23.8
	1992	−3.3	72.3	69.0	54.5
Turks and Caicos Islands Iles Turques et Caiques	1990	8.9	3.9	12.8	1066.7
	1991	15.8	1.9	17.7	1475.0
	1992	14.1	0.9	15.0	1153.8
Uruguay Uruguay	1990	41.7	45.3	87.0	28.1
	1991	38.2	160.2	198.4	63.8
	1992	59.9	148.6	208.5	66.6
Venezuela Venezuela	1990	75.7	1056.7	1132.4	58.1
	1991	23.1	556.5	579.6	29.0
	1992	30.7	595.5	626.2	30.6
Other and unallocated Autres et non−ventilés	1990	352.8	337.6	690.4	...
	1991	391.3	253.8	645.1	...
	1992	394.6	188.0	582.6	...
Asia **Asie**	**1990**	**11399.7**	**8106.7**	**19506.4**	**6.7**
	1991	**13234.9**	**9688.8**	**22923.7**	**7.7**
	1992	**13552.4**	**8440.6**	**21993**	**7.3**
Afghanistan Afghanistan	1990	100.4	38.2	138.6	9.2
	1991	101.8	34.8	136.6	8.7
	1992	126.7	74.2	200.9	12.1
Bahrain Bahreïn	1990	1.9	4.2	6.1	12.4
	1991	0.9	0.2	1.1	2.2
	1992	1.2	15.8	17.0	32.7
Bangladesh Bangladesh	1990	1103.3	986.9	2090.2	19.3
	1991	811.8	948.2	1760.0	16.0
	1992	853.2	946.6	1799.8	16.0
Bhutan Bhoutan	1990	20.1	28.6	48.7	31.5
	1991	37.5	26.7	64.2	41.0
	1992	34.1	23.9	58.0	36.7

81

Disbursements to developing countries or areas of bilateral and multilateral official development assistance [*cont.*]

Versements mises à la disposition des pays ou zones en développement au titre de l'aide publique bilatérale et multilatérale au développement [*suite*]

Region, country or area Région, pays ou zone	Year Année	Disbursements ($US) Versements ($E–U)			
		Bilateral Bilatérale (millions)	Multilateral[1] Multilatérale[1] (millions)	Total (millions)	Per capita[2] Par habitant[2]
Brunei Darussalam	1990	3.7	0.1	3.8	14.8
Brunéi Darussalam	1991	3.6	0.0	3.6	13.7
	1992	5.3	0.1	5.4	20.1
Cambodia	1990	28.5	13.1	41.6	4.7
Cambodge	1991	49.9	41.1	91.0	10.0
	1992	94.8	111.4	206.2	21.9
China	1990	1511.7	1014.3	2526.0	2.2
Chine	1991	1252.5	1446.4	2698.9	2.3
	1992	2077.2	1491.1	3568.3	3.1
Hong Kong	1990	19.4	18.8	38.2	6.7
Hong–kong	1991	13.5	22.7	36.2	6.3
	1992	−62.3	23.3	−39.0	−6.8
India	1990	752.2	1900.6	2652.8	3.1
Inde	1991	1770.2	2300.9	4071.1	4.7
	1992	1198.0	1952.5	3150.5	3.6
Indonesia	1990	1541.1	1212.2	2753.3	15.1
Indonésie	1991	1770.6	1353.3	3123.9	16.8
	1992	1971.4	902.4	2873.8	15.2
Iran, Islamic Rep. of	1990	34.8	−16.5	18.3	0.3
Iran, Rép. islamique d'	1991	83.2	61.2	144.4	2.4
	1992	64.4	125.9	190.3	3.0
Iraq	1990	−8.6	25.5	16.9	0.9
Iraq	1991	431.5	124.6	556.1	30.0
	1992	65.9	74.1	140.0	7.4
Israel	1990	1370.7	27.4	1398.1	300.0
Israël	1991	1715.4	23.8	1739.2	360.0
	1992	2058.6	7.2	2065.8	410.2
Jordan	1990	435.0	85.9	520.9	122.3
Jordanie	1991	682.7	212.6	895.3	201.5
	1992	313.0	111.8	424.8	90.9
Korea, Dem. People's Rep.	1990	0.9	7.5	8.4	0.2
Corée, Rép. pop. dém. de	1991	1.4	7.6	9.0	0.2
	1992	4.7	7.5	12.2	0.3
Korea, Republic of	1990	54.8	−377.0	−322.2	−14.8
Corée, République de	1991	52.7	−269.4	−216.7	−9.8
	1992	2.8	−149.1	−146.3	−6.5
Kuwait	1990	2.2	4.9	7.1	3.3
Koweït	1991	2.9	1.6	4.5	2.2
	1992	1.6	0.8	2.4	1.2
Lao People's Dem. Rep.	1990	51.2	100.0	151.2	36.0
Rép. dém. populaire lao	1991	66.8	76.5	143.3	33.1
	1992	76.9	89.1	166.0	37.1
Lebanon	1990	71.7	30.5	102.2	40.0
Liban	1991	59.0	5.6	64.6	24.8
	1992	68.8	16.4	85.2	31.6
Macau	1990	0.1	0.1	0.2	0.6
Macao	1991	0.1	–	0.1	0.3
	1992	0.1	0.0	0.1	0.3
Malaysia	1990	458.6	89.6	548.2	30.6
Malaisie	1991	272.9	102.0	374.9	20.4
	1992	195.0	7.5	202.5	10.8
Maldives	1990	11.6	10.1	21.7	100.5
Maldives	1991	13.9	17.0	30.9	138.6
	1992	15.0	22.6	37.6	162.8
Mongolia	1990	6.3	6.8	13.1	6.0
Mongolie	1991	59.5	10.0	69.5	31.2
	1992	67.5	55.4	122.9	54.1
Myanmar	1990	83.1	80.6	163.7	3.9
Myanmar	1991	105.9	73.1	179.0	4.2
	1992	82.7	32.0	114.7	2.6
Nepal	1990	239.0	197.0	436.0	22.6
Népal	1991	290.4	168.8	459.2	23.2
	1992	275.6	170.8	446.4	22.0

81

Disbursements to developing countries or areas of bilateral and multilateral
official development assistance [cont.]

Versements mises à la disposition des pays ou zones en développement au titre de l'aide publique bilatérale
et multilatérale au développement [suite]

Region, country or area Région, pays ou zone	Year Année	Disbursements ($US) Versements ($E–U)			
		Bilateral Bilatérale (millions)	Multilateral [1] Multilatérale [1] (millions)	Total (millions)	Per capita [2] Par habitant [2]
Oman Oman	1990	11.3	2.7	14.0	8.0
	1991	12.7	12.4	25.1	13.7
	1992	19.2	28.3	47.5	24.9
Pakistan Pakistan	1990	653.5	979.8	1633.3	13.4
	1991	471.5	1336.1	1807.6	14.4
	1992	469.4	1062.1	1531.5	11.8
Philippines Philippines	1990	1105.1	666.6	1771.7	29.1
	1991	860.1	450.2	1310.3	21.1
	1992	1538.6	613.7	2152.3	33.9
Qatar Qatar	1990	1.3	0.9	2.2	4.5
	1991	1.3	1.0	2.3	4.6
	1992	1.3	0.8	2.1	4.1
Saudi Arabia Arabie saoudite	1990	12.8	30.8	43.6	2.7
	1991	12.6	32.1	44.7	2.7
	1992	48.1	7.2	55.3	3.3
Singapore Singapour	1990	−3.2	−37.8	−41.0	−15.2
	1991	6.4	−11.6	−5.2	−1.9
	1992	18.6	−11.7	6.9	2.5
Sri Lanka Sri Lanka	1990	403.8	321.3	725.1	42.1
	1991	457.5	425.0	882.5	50.6
	1992	248.8	387.6	636.4	36.0
Syrian Arab Republic Rép. arabe syrienne	1990	69.4	19.6	89.0	7.2
	1991	236.4	−16.3	220.1	17.2
	1992	50.4	26.3	76.7	5.8
Thailand Thaïlande	1990	731.7	−35.2	696.5	12.5
	1991	639.2	−52.5	586.7	10.4
	1992	696.0	−405.9	290.1	5.1
United Arab Emirates Emirats arabes unis	1990	2.8	2.4	5.2	3.1
	1991	−7.8	2.2	−5.6	−3.3
	1992	−10.4	2.1	−8.3	−4.7
Viet Nam Viet Nam	1990	107.9	81.7	189.6	2.8
	1991	135.2	103.0	238.2	3.5
	1992	473.9	109.3	583.2	8.4
Yemen Yémen	1990	168.8	118.3	287.1	25.4
	1991	220.8	64.9	285.7	24.1
	1992	149.0	92.6	241.6	19.3
Other and unallocated Autres et non−ventilés	1990	240.8	466.2	709.2	...
	1991	538.4	553.3	1088.8	...
	1992	257.2	414.8	666.0	...
Oceania **Oceanie**	**1990**	**1214.7**	**222.5**	**1437.2**	**247.8**
	1991	**1212.9**	**225.6**	**1438.5**	**242.8**
	1992	**1297.3**	**190.5**	**1487.8**	**245.9**
Cook Islands Iles Cook	1990	10.1	2.2	12.3	683.3
	1991	11.2	2.2	13.4	705.3
	1992	11.8	5.4	17.2	905.3
Fiji Fidji	1990	43.5	−7.3	36.2	49.9
	1991	39.6	5.2	44.8	61.0
	1992	54.1	10.7	64.8	86.9
French Polynesia Polynésie française	1990	258.0	1.4	259.4	1316.8
	1991	309.3	7.4	316.7	1567.8
	1992	323.9	5.8	329.7	1600.5
Kiribati Kiribati	1990	17.7	2.8	20.5	284.7
	1991	15.9	4.3	20.2	273.0
	1992	22.0	4.8	26.8	357.3
Nauru Nauru	1990	0.2	−	0.2	20.0
	1991	0.4	−	0.4	40.0
	1992	0.2	−	0.2	20.0
New Caledonia Nouvelle−Calédonie	1990	300.2	1.6	301.8	1796.4
	1991	308.7	1.3	310.0	1823.5
	1992	354.5	3.6	358.1	2069.9
Niue Nioué	1990	7.0	0.2	7.2	3600.0
	1991	8.9	0.5	9.4	4700.0
	1992	4.5	0.3	4.8	2400.0

81

Disbursements to developing countries or areas of bilateral and multilateral
official development assistance [cont.]

Versements mises à la disposition des pays ou zones en développement au titre de l'aide publique bilatérale
et multilatérale au développement [suite]

Region, country or area Région, pays ou zone	Year Année	Disbursements ($US) Versements ($E–U)			
		Bilateral Bilatérale (millions)	Multilateral [1] Multilatérale [1] (millions)	Total (millions)	Per capita [2] Par habitant [2]
Palau [3]	1990	61.9	1.1	63.0	4200.0
Palau [3]	1991	26.5	3.6	30.1	1881.3
	1992	7.2	0.2	7.4	462.5
Papua New Guinea	1990	320.0	171.3	491.3	128.0
Papouasie–Nvl–Guinée	1991	321.7	137.2	458.9	116.9
	1992	348.9	103.3	452.2	112.6
Samoa	1990	27.7	20.0	47.7	294.4
Samoa	1991	25.8	30.9	56.7	345.7
	1992	29.5	20.5	50.0	303.0
Solomon Islands	1990	31.1	14.4	45.5	142.2
Iles Salomon	1991	28.1	7.7	35.8	108.2
	1992	27.4	17.8	45.2	131.8
Tokelau	1990	4.4	0.4	4.8	2400.0
Tokélaou	1991	4.0	0.4	4.4	2200.0
	1992	4.0	0.3	4.3	2150.0
Tonga	1990	24.2	5.5	29.7	309.4
Tonga	1991	14.4	5.5	19.9	205.2
	1992	19.7	5.0	24.7	254.6
Tuvalu	1990	4.8	0.2	5.0	555.6
Tuvalu	1991	4.4	1.0	5.4	600.0
	1992	7.5	0.9	8.4	933.3
Vanuatu	1990	42.1	7.8	49.9	334.9
Vanuatu	1991	35.7	16.9	52.6	343.8
	1992	32.4	8.2	40.6	258.6
Wallis and Futuna Islands	1990	0.0	0.9	0.9	64.3
Iles Wallis et Futuna	1991	–	1.5	1.5	107.1
	1992	0.3	0.5	0.8	57.1
Other and unallocated	1990	61.9	0.1	62.0	...
Autres et non–ventilés	1991	58.2	–	58.2	...
	1992	49.6	3.3	52.9	...
Unspecified	**1990**	**5116.8**	**769.8**	**5886.6**	...
Non–specifiés	**1991**	**5643.3**	**1139.7**	**6783.0**	...
	1992	**6111.4**	**1155.0**	**7266.4**	...

Source:
Organization for Economic Co–operation and Development (Paris).

1   As reported by OECD/DAC, IDA, agencies of the United Nations
    family and the European Development Fund. Excluding
    non–concessional flows (i.e., less than 25% grant element).
2   Population based on estimates of midyear population.
3   Including data for Federated States of Micronesia,
    Marshall Is. and Northern Mariana Is.

Source:
Organisation de coopération et de développement économiques (Paris).

1   Chiffre communiqués par le Comité d'aide au développement de
    l'OCDE, l'IDA, les agences des Nations Unies et le Fonds Européen
    de Développement. Non compris les apports non libéraux
    (c'est–à–dire) dont l'élément de libéralité est intérieur à 25 p.100).
2   Population d'après des estimations de la population au milieu
    de l'année.
3   Y compris les données pour les Etats fédérés de Micronésie,
    les îles Marshall et les îles Mariannes du Nord.

82
Net official development assistance from DAC countries to developing countries and multilateral organizations
Aide publique au développement nette de pays du CAD aux pays en développement et aux organisations multilatérales

Net disbursements: million US dollars and as % of GNP
Versements nets: millions de dollars E–U et en % de PNB

Country or area Pays ou zone	1988 Million US $ Millions E–U $	1988 As % of GNP En % de PNB	1989 Million US $ Millions E–U $	1989 As % of GNP En % de PNB	1990[1] Million US $ Millions E–U $	1990[1] As % of GNP En % de PNB	1991[1] Million US $ Millions E–U $	1991[1] As % of GNP En % de PNB	1992[1] Million US $ Millions E–U $	1992[1] As % of GNP En % de PNB	1993 Million US $ Millions E–U $	1993 As % of GNP En % de PNB
Total	**47045**	**0.34**	**45732**	**0.32**	**52955**	**0.33**	**56678**	**0.33**	**60850**	**0.33**	**55963**	**0.30**
Australia Australie	1101	0.46	1020	0.38	955	0.34	1050	0.38	1015	0.37	953	0.35
Austria Autriche	301	0.24	282	0.23	394	0.25	547	0.34	556	0.30	544	0.30
Belgium Belgique	601	0.39	703	0.46	889	0.46	831	0.41	870	0.39	808	0.39
Canada Canada	2347	0.50	2320	0.44	2470	0.44	2604	0.45	2515	0.46	2373	0.45
Denmark Danemark	922	0.89	937	0.93	1171	0.94	1200	0.96	1392	1.02	1340	1.03
Finland Finlande	608	0.59	706	0.63	846	0.63	930	0.78	644	0.64	355	0.45
France France	5463	0.58	5802	0.61	7163	0.60	7386	0.62	8270	0.63	7915	0.63
Germany † Allemagne †	4731	0.39	4948	0.41	6320	0.42	6890	0.40	7583	0.39	6954	0.37
Ireland Irlande	57	0.20	49	0.17	57	0.16	72	0.19	70	0.16	81	0.20
Italy Italie	3193	0.39	3613	0.42	3395	0.31	3347	0.30	4122	0.34	3043	0.31
Japan Japon	9134	0.32	8965	0.31	9069	0.31	10952	0.32	11151	0.30	11259	0.26
Luxembourg Luxembourg	19	0.20	18	0.19	25	0.21	42	0.33	38	0.26	50	0.35
Netherlands Pays–Bas	2231	0.98	2094	0.94	2538	0.92	2517	0.88	2753	0.86	2525	0.82
New Zealand Nouvelle–Zélande	104	0.27	87	0.22	95	0.23	100	0.25	97	0.26	98	0.25
Norway Norvège	985	1.13	917	1.05	1205	1.17	1178	1.13	1273	1.16	1014	1.01
Portugal Portugal	84	0.21	113	0.25	148	0.25	213	0.31	302	0.36	246	0.29
Spain Espagne	231	0.07	537	0.14	959	0.20	1261	0.24	1518	0.27	1213	0.25
Sweden Suède	1534	0.86	1799	0.96	2007	0.91	2116	0.90	2460	1.03	1769	0.98
Switzerland Suisse	617	0.32	558	0.30	750	0.32	863	0.36	1139	0.45	793	0.33
United Kingdom Royaume–Uni	2645	0.32	2587	0.31	2638	0.27	3201	0.32	3243	0.31	2908	0.31
United States Etats–Unis	10141	0.21	7676	0.15	11394	0.21	11262	0.20	11709	0.20	9721	0.15

Source:
Organization for Economic Co–operation and Development (Paris).

Source:
Organisation de coopération et de développement économiques (Paris).

† For detailed descriptions of data pertaining to former Czechoslovakia, Germany, SFR Yugoslavia and former USSR, see Annex I – Country or area nomenclature, regional and other groupings.

† Pour les descriptions en détails des données relatives à l'ancienne Tchécoslovaquie, l'Allemagne, la Rfs Yougoslavie et l'ancienne URSS, voir l'Annexe I – Nomenclature des pays ou zones, groupements régionaux et autres groupements.

1 Except for total including debt forgiveness of non–official development assistance claims.

1 Sauf pour le total, ces chiffres incluent l'annulation des créances au titre de l'assistance autre que l'aide publique au développement.

83
Socio-economic development assistance through the United Nations system
Assistance en matière de développement socio-économique fournie par le système des Nations Unies

Thousand US dollars

Milliers de dollars E-U

A. Development grant expenditures [1] · Aide au développement [1]

Country or area Pays ou zone	Year Année	UNDP PNUD — UNDP programme Programme de PNUD	UNDP PNUD — Special funds Fonds gérés	UNFPA FNUAP	UNICEF FISE	WFP PAM	Other UN system Autres organis. –ONU — Regular budget Budget ordinaire	Other UN system Autres organis. –ONU — Extra- budgetary Extra- budgétaire	Total	Gov't self- supporting Auto- assistance gouverne- mentale
Total	1992	1026782	137558	128234	743812	1575229	241602	727191	4580408	67714
Total	1993	1031000	173393	134321	803701	1487716	345810	891889	4867830	73090
Regional programmes	1992	60491	17461	24357	105825	4255	68959	203414	484762	6404
Totaux régionaux	1993	70882	44453	29715	117158	0	90254	477706	830168	4313
Africa	1992	0	7931	4140	5753	4255	15331	82680	120090	1242
Afrique	1993	0	10873	3932	1781	0	21451	89828	127865	1191
Asia	1992	0	2545	2747	0	0	10600	39241	55133	338
Asie	1993	0	2786	3292	2391	0	13961	43702	66132	15
Latin America	1992	0	3118	1821	19563	0	14491	28587	67580	111
Amérique latine	1993	0	2593	1802	8013	0	17293	30973	60674	77
Middle East	1992	0	407	233	3	0	4165	4382	9190	176
Moyen orient	1993	0	457	844	5	0	7741	4372	13419	223
Interregional	1992	40851	0	15416	0	0	19183	0	75450	0
Interrégional	1993	41775	19746	19845	0	0	28133	271225	380724	472
Global	1992	19640	3460	0	80506	0	5189	48524	157319	4537
Global	1993	29107	7998	0	104968	0	1675	37606	181354	2335
Country programmes	1992	811960	105148	101752	636376	1570974	160027	289720	3675957	58578
Programmes, pays	1993	859068	122131	103484	684456	1462427	240474	309980	3782020	66248
Afghanistan	1992	14814	2091	200	6343	24572	1113	500	49633	57
Afghanistan	1993	18298	756	185	11201	39532	2781	3805	76558	1158
Albania	1992	764	0	885	389	0	588	177	2803	24
Albanie	1993	1446	0	262	922	0	711	1297	4638	24
Algeria	1992	2277	0	904	884	6379	886	350	11680	47
Algérie	1993	1793	0	592	1406	5168	1365	788	11112	244
Angola	1992	11230	1922	833	12361	46580	1218	1507	75651	0
Angola	1993	3746	115	858	14357	57362	1544	906	78888	0
Anguilla	1992	378	40	0	0	0	0	0	418	0
Anguilla	1993	318	18	0	0	0	0	0	336	0
Antigua and Barbuda	1992	65	3	24	0	150	61	104	407	0
Antigua-et-Barbuda	1993	107	5	1	0	38	16	42	209	0
Argentina	1992	31587	11	10	1309	0	1152	1390	35459	899
Argentine	1993	59429	736	6	1559	0	1497	1920	65147	1417
Armenia	1992	0	0	0	0	0	76	0	76	0
Arménie	1993	0	0	0	779	844	71	0	1694	0
Aruba	1992	209	0	0	0	0	7	0	216	0
Aruba	1993	485	0	0	0	0	21	2	508	0
Bahamas	1992	15	0	0	0	0	421	153	589	−24
Bahamas	1993	−17	0	0	0	0	529	288	800	6
Bahrain	1992	120	0	−2	0	0	137	23	278	0
Bahreïn	1993	63	0	−7	0	0	260	26	342	0
Bangladesh	1992	24749	3847	2587	40947	76551	3327	2523	154531	21
Bangladesh	1993	21450	8923	3157	31618	25004	6594	3408	100154	38
Barbados	1992	391	0	69	0	0	121	374	955	0
Barbade	1993	9	7	2	0	0	180	458	656	0
Belize	1992	249	8	−1	868	0	444	322	1890	0
Belize	1993	387	159	43	821	0	589	504	2503	109
Benin	1992	8327	1111	792	3035	3678	867	570	18380	38
Bénin	1993	6601	1527	605	3377	7290	1423	933	21756	331
Bermuda	1992	...	...	...	...	...	...	...	...	...
Bermudes	1993	0	0	0	0	0	3	0	3	0
Bhutan	1992	5879	458	313	2059	3090	623	1109	13531	0
Bhoutan	1993	4576	1087	184	2601	2425	1011	1293	13177	0
Bolivia	1992	10463	807	911	5192	6306	1348	3970	28997	0
Bolivie	1993	13284	3329	1162	4648	10622	1704	5375	40124	0
Botswana	1992	2788	512	689	1025	6209	719	826	12768	120
Botswana	1993	3654	701	476	1427	4708	1227	916	13109	38

83
Socio-economic development assistance through the United Nations system
Thousand US dollars [cont.]

Assistance en matière de développement socio-économique fournie par le système des Nations Unies
Milliers de dollars E-U [suite]

A. Development grant expenditures [1] · Aide au développement [1]

Country or area Pays ou zone	Year Année	UNDP programme Programme de PNUD	Special funds Fonds gérés	UNFPA FNUAP	UNICEF FISE	WFP PAM	Regular budget Budget ordinaire	Extra-budgetary Extra-budgétaire	Total	Auto-assistance gouverne-mentale
Brazil	1992	22264	200	1339	6276	9672	2547	4875	47173	960
Brésil	1993	37836	2813	938	9023	9208	3021	3689	66528	2219
British Virgin Islands	1992	86	0	7	0	0	1	0	94	0
Iles Vierges britanniques	1993	54	0	1	0	0	1	21	77	0
Brunei Darussalam	1992	5	0	0	0	0	18	0	23	0
Brunéi Darussalam	1993	0	0	0	0	0	35	34	69	26
Bulgaria	1992	308	0	36	47	0	973	383	1747	0
Bulgarie	1993	420	0	66	0	0	721	429	1636	5
Burkina Faso	1992	10572	4972	923	2935	3170	901	770	24243	4
Burkina Faso	1993	7979	3011	779	4358	7409	2175	985	26696	0
Burundi	1992	8783	1194	1129	2718	2152	1012	1748	18736	379
Burundi	1993	5885	643	826	2914	4968	1726	2104	19066	573
Cambodia	1992	7742	752	1	11937	23177	1230	5051	49890	3
Cambodge	1993	24215	4548	20	15101	20094	1664	5405	71047	35
Cameroon	1992	2323	424	333	1444	844	1509	1037	7914	188
Cameroun	1993	1612	642	685	1603	1872	2133	1757	10304	540
Cape Verde	1992	1544	3059	357	978	7878	1220	2669	17705	17
Cap-Vert	1993	1226	337	363	1709	5606	1624	2553	13418	36
Cayman Islands	1992	12	0	0	0	0	945	0	957	0
Iles Caïmanes	1993	11	0	0	0	0	0	15	26	0
Central African Rep.	1992	4802	1431	921	2058	4205	0	1355	14772	0
Rép. centrafricaine	1993	2891	2767	853	1717	5744	1466	1325	16763	0
Chad	1992	8797	1380	968	3133	6971	1451	2669	25369	218
Tchad	1993	8180	1907	954	2597	5485	1719	1478	22320	202
Chile	1992	5063	224	179	1097	0	1630	2141	10334	107
Chili	1993	8283	110	116	1081	0	1588	2106	13284	146
China	1992	41625	3252	10406	21849	28120	3990	5285	114527	0
Chine	1993	38410	6360	10319	17647	23782	6746	8411	111675	35
Colombia	1992	13418	21	509	2060	5864	1342	1983	25197	26
Colombie	1993	24348	601	576	1566	1348	1887	2063	32389	162
Comoros	1992	3154	579	564	504	1597	726	267	7391	0
Comores	1993	2611	1348	384	695	1649	1430	274	8391	7
Congo	1992	538	27	318	1381	854	1186	1091	5395	339
Congo	1993	784	11	213	866	1083	1554	1344	5855	356
Cook Islands	1992	327	41	85	0	0	160	148	761	0.
Iles Cook	1993	495	21	104	0	0	271	103	994	0
Costa Rica	1992	2594	108	165	721	1479	1050	4034	10151	2.
Costa Rica	1993	2871	435	268	741	1015	1439	5355	12124	45
Côte d'Ivoire	1992	4067	4	412	1919	4381	1039	1228	13050	1
Côte d'Ivoire	1993	2504	1	833	1306	6335	1818	1386	14183	50
Cuba	1992	2700	61	582	712	5683	1745	428	11911	3
Cuba	1993	2183	28	1040	1314	9300	1993	1339	17197	971
Cyprus	1992	415	18	28	0	0	396	204	1061	22
Chypre	1993	637	0	27	0	0	435	168	1267	3
Czech Republic	1992	519	0	0	0	0	453	608	1580	1
République tchèque	1993	270	0	0	0	0	484	785	1539	1
Djibouti	1992	1232	485	26	1363	1875	604	340	5925	0
Djibouti	1993	1427	717	61	1037	2839	1132	472	7685	0
Dominica	1992	193	41	0	0	210	7	186	637	0
Dominique	1993	207	13	6	0	223	98	130	677	0
Dominican Republic	1992	5202	183	734	1115	332	1219	337	9122	0
Rép. dominicaine	1993	9753	13	288	1023	244	1359	588	13268	0
Ecuador	1992	3603	38	554	1541	3800	2100	3137	14773	17
Equateur	1993	5838	94	662	1956	835	2422	2994	14801	215
Egypt	1992	9498	736	1197	7004	12372	2610	5732	39149	1211
Egypte	1993	13692	2517	576	5444	10461	3144	8038	43872	3375
El Salvador	1992	6089	2120	657	1061	15709	823	432	26891	104
El Salvador	1993	11368	3532	668	1403	5831	1361	602	24765	32

83
Socio-economic development assistance through the United Nations system
Thousand US dollars [cont.]
Assistance en matière de développement socio-économique fournie par le système des Nations Unies
Milliers de dollars E-U [suite]

A. Development grant expenditures [1] · Aide au développement [1]

Country or area Pays ou zone	Year Année	UNDP PNUD UNDP programme Programme de PNUD	Special funds Fonds gérés	UNFPA FNUAP	UNICEF FISE	WFP PAM	Other UN system Autres organis. -ONU Regular budget Budget ordinaire	Extra-budgetary Extra-budgétaire	Total	Gov't self-supporting Auto-assistance gouverne-mentale
Equatorial Guinea	1992	2735	294	627	1259	1469	681	273	7338	249
Guinée équatoriale	1993	1638	280	273	1010	2832	1000	207	7240	-7
Eritrea	1992	..	..	..	..	..	..	..		..
Erythrée	1993	1645	1389	180	448	15243	326	83	19314	0
Ethiopia	1992	12580	4696	1509	30356	146263	2284	14359	212047	296
Ethiopia	1993	14614	2394	2753	20069	101280	2752	8511	152373	414
Fiji	1992	1180	95	368	0	0	1185	472	3300	62
Fidji	1993	877	337	253	0	0	1602	538	3607	329
French Guiana	1992	0	0	0	0	0	36	0	36	0
Guyane française	1993	0	0	0	0	0	39	0	39	0
French Polynesia	1992	0	0	0	0	0	53	0	137	0
Polynésie française	1993	0	0	0	0	0	85	84	169	0
Gabon	1992	1296	2	343	0	0	707	619	2967	156
Gabon	1993	1149	0	258	313	0	1137	723	3580	281
Gambia	1992	4014	2444	807	692	3974	847	1761	14539	0
Gambie	1993	4176	1381	354	735	2640	1351	1176	11813	0
Ghana	1992	9080	296	1303	4080	6315	1305	3555	25934	1343
Ghana	1993	6851	538	1036	4431	15011	1515	3069	32451	914
Greece	1992	0	0	0	0	0	396	23	419	0
Grèce	1993	0	0	0	0	0	265	167	432	3
Grenada	1992	286	20	15	0	416	194	128	1059	0
Grenade	1993	264	3	3	0	270	83	92	715	0
Guam	1992	0	0	0	0	0	9	28	37	0
Guam	1993	0	0	0	0	0	19	30	49	0
Guatemala	1992	4279	74	941	1377	2363	737	765	10536	0
Guatemala	1993	5129	19	504	2438	7379	1139	791	17399	0
Guinea	1992	6322	1908	628	3138	3491	1578	1669	18734	493
Guinée	1993	4780	1374	677	3654	3466	2066	1951	17968	501
Guinea-Bissau	1992	4939	1000	626	1427	3577	1192	3056	15817	1891
Guinée-Bissau	1993	5075	4117	22	1989	4229	1479	1396	18307	727
Guyana	1992	4678	205	49	656	505	472	262	6827	0
Guyana	1993	4169	280	23	721	1190	791	495	7669	0
Haiti	1992	3036	541	277	2850	3279	420	1194	11597	0
Haïti	1993	4154	711	713	4390	3676	784	2474	16902	0
Honduras	1992	9044	89	811	878	5826	542	1340	18530	0
Honduras	1993	9360	137	886	1266	4368	731	2856	19604	912
Hong Kong	1992	61	13	0	0	0	157	0	231	0
Hong-kong	1993	39	0	0	0	0	133	18	190	18
Hungary	1992	32	0	250	0	0	861	1944	3087	122
Hongrie	1993	55	0	-10	0	0	418	1732	2195	146
India	1992	19521	67	10289	63352	43979	3043	10912	151163	3862
Inde	1993	34000	93	7006	57322	26281	9203	15364	149269	4536
Indonesia	1992	18598	448	2638	14763	14738	5780	7524	64489	2173
Indonésie	1993	15277	535	2132	13778	3562	8652	5722	49658	1939
Iran, Islamic Rep. of	1992	2916	0	161	2074	11223	1813	876	19063	254
Iran, Rép. islamique d'	1993	2867	10	1961	1791	12388	2446	1176	22639	804
Iraq	1992	473	239	0	28014	26159	989	1645	57519	34
Iraq	1993	-12	327	1	55855	26631	1703	13343	97848	2
Jamaica	1992	4881	81	168	1020	1627	1247	462	9486	7
Jamaïque	1993	3556	104	215	1900	5098	1335	471	12679	9
Jordan	1992	1943	52	372	2146	4409	1371	797	11090	12
Jordanie	1993	1535	543	465	1909	4503	1713	809	11477	9
Kenya	1992	11714	512	1329	11312	71769	1759	2819	101214	8
Kenya	1993	10669	583	829	18041	88918	2326	3560	124926	46
Kiribati	1992	416	127	184	0	0	225	125	1077	0
Kiribati	1993	303	140	110	0	0	465	117	1135	0
Korea, Dem. P. R.	1992	3833	31	752	636	0	980	1312	7544	0
Corée, R. p. dém. de	1993	3445	39	955	545	0	1763	534	7281	136

83

Socio-economic development assistance through the United Nations system

Thousand US dollars [*cont.*]

Assistance en matière de développement socio-économique fournie par le système des Nations Unies

Milliers de dollars E-U [*suite*]

A. Development grant expenditures [1] · Aide au développement [1]

Country or area Pays ou zone	Year Année	UNDP PNUD UNDP programme Programme de PNUD	Special funds Fonds gérés	UNFPA FNUAP	UNICEF FISE	WFP PAM	Other UN system Autres organis. -ONU Regular budget Budget ordinaire	Extra-budgetary Extra-budgétaire	Total	Gov't self-supporting Auto-assistance gouverne-mentale
Korea, Republic of	1992	996	47	86	281	0	1435	250	3095	0
Corée, République de	1993	596	0	63	744	0	1872	300	3575	152
Kuwait	1992	336	0	0	0	0	92	29	457	0
Koweït	1993	1566	0	0	23	0	227	141	1957	82
Lao People's Dem. Rep.	1992	8082	0	132	2683	3433	1066	1496	16892	0
Rép. dém. pop. lao	1993	6512	3695	152	2642	926	1858	1830	17615	0
Lebanon	1992	1776	0	5	2533	2154	1028	6656	14152	10
Liban	1993	3060	0	7	1902	1885	1283	2150	10287	125
Lesotho	1992	3975	1123	605	1344	12211	1340	2650	23248	362
Lesotho	1993	2738	436	779	1907	7100	1545	2275	16780	1029
Liberia	1992	1974	124	80	10908	69550	1414	108	84158	0
Libéria	1993	1953	151	17	9107	65976	1951	203	79358	0
Libyan Arab Jamahiriya	1992	843	0	4	0	0	527	14501	15875	12185
Jamah. arabe libyenne	1993	2336	23	7	0	0	790	11207	14363	10460
Madagscar	1992	13342	46	753	5453	11936	1146	1321	33997	102
Madagascar	1993	11801	287	934	6673	2881	1950	1941	26467	309
Malawi	1992	11307	851	974	6422	139929	942	1225	161650	0
Malawi	1993	10985	1740	1135	5910	68909	1606	1327	91612	0
Malaysia	1992	2486	4	437	626	0	928	527	5008	75
Malaisie	1993	2122	759	60	546	0	1638	381	5506	71
Maldives	1992	2406	1956	208	1043	0	642	461	6716	0
Maldives	1993	1593	393	274	1111	0	1088	456	4915	18
Mali	1992	12096	3333	905	6550	4768	1636	2406	31694	0
Mali	1993	8344	2993	855	6788	3156	2550	2302	26988	0
Malta	1992	153	0	0	0	0	99	271	523	96
Malte	1993	100	1	0	0	0	118	130	349	146
Marshall Islands	1992	205	18	34	0	0	21	46	324	8
Iles Marshall	1993	124	70	168	0	0	74	43	479	0
Mauritania	1992	4114	1836	419	1387	5436	1506	588	15286	0
Mauritanie	1993	3397	2565	642	1882	21131	2329	480	32426	0
Mauritius	1992	628	0	179	600	1419	300	569	3695	0
Maurice	1993	255	0	196	654	1457	735	215	3512	28
Mexico	1992	2619	34	2918	3581	4340	1922	2501	17915	881
Mexique	1993	2915	114	2505	3132	9858	2456	3147	24127	1174
Mongolia	1992	2029	171	605	531	0	1739	396	5471	0
Mongolie	1993	1470	655	813	719	1583	2490	2949	10679	0
Montserrat	1992	231	16	0	0	0	0	0	247	0
Montserrat	1993	149	50	0	0	0	1	0	200	0
Morocco	1992	6270	29	1227	3455	23100	1598	699	36378	0
Maroc	1993	2985	15	2706	2888	23149	2367	968	35078	110
Mozambique	1992	16802	4896	1010	25998	100567	1321	3641	154235	116
Mozambique	1993	12256	3766	1437	26570	58820	2034	3583	108466	35
Myanmar	1992	12618	0	122	6600	0	2128	1702	23170	468
Myanmar	1993	8622	0	484	7453	0	4330	2742	23631	1252
Namibia	1992	3802	1006	130	3776	7017	982	1154	17867	161
Namibie	1993	3698	757	339	4438	3991	1437	1689	16349	87
Nepal	1992	14850	1085	2090	8229	6043	2413	3014	37724	51
Népal	1993	9844	965	2251	8319	9970	4269	4154	39772	344
Netherlands Antilles	1992	317	0	0	0	0	61	42	420	28
Antilles néerlandaises	1993	209	0	0	0	0	85	66	360	14
Nicaragua	1992	7164	223	528	1366	8695	1105	4126	23207	1
Nicaragua	1993	10520	1614	563	1973	2931	1396	5001	23998	38
Niger	1992	12117	5095	721	2959	14569	1241	8682	45384	0
Niger	1993	9911	3411	1095	3487	6087	1958	5506	31455	0
Nigeria	1992	10049	5	1013	15719	0	2129	6272	35187	3110
Nigéria	1993	9452	3	870	20273	0	2979	5261	38838	3900
Niue	1992	227	36	0	0	0	0	0	263	0
Nioué	1993	96	15	0	0	0	10	0	121	0

83

Socio–economic development assistance through the United Nations system
Thousand US dollars [cont.]

Assistance en matière de développement socio–économique fournie par le système des Nations Unies
Milliers de dollars E–U [suite]

A. Development grant expenditures [1] · Aide au développement [1]

Country or area Pays ou zone	Year Année	UNDP PNUD UNDP programme Programme de PNUD	Special funds Fonds gérés	UNFPA FNUAP	UNICEF FISE	WFP PAM	Other UN system Autres organis. –ONU Regular budget Budget ordinaire	Extra-budgetary Extra-budgétaire	Total	Gov't self–supporting Auto–assistance gouverne-mentale
Oman	1992	2762	100	95	759	0	725	491	4932	167
Oman	1993	1187	37	184	751	0	1051	1702	4912	1231
Pakistan	1992	22267	154	1637	16338	86882	2634	2886	132798	0
Pakistan	1993	16018	259	854	12603	13035	4497	4208	51474	525
Panama	1992	7675	18	272	778	1346	851	241	11181	54
Panama	1993	11620	14	314	743	697	1080	333	14801	8
Papua New Guinea	1992	6252	151	140	987	0	1576	1728	10834	512
Papouasie–Nvl–Guinée	1993	5009	489	172	989	0	2318	1584	10561	327
Paraguay	1992	9045	114	454	1235	1808	631	1271	14558	962
Paraguay	1993	12670	24	208	1308	3842	841	−621	18272	−962
Peru	1992	9018	168	683	2793	8762	1822	2577	25823	619
Pérou	1993	23634	496	1587	2899	11321	2318	4445	46700	1813
Philippines	1992	4573	474	4865	10634	2618	1777	5764	30705	166
Philippines	1993	4237	467	6570	7689	785	2087	4952	26787	585
Poland	1992	562	5	11	0	0	873	2636	4087	141
Pologne	1993	500	15	6	0	0	528	1831	2880	451
Portugal	1992	595	0	149	0	0	585	71	1400	26
Portugal	1993	403	0	43	0	0	281	146	873	28
Qatar	1992	738	4	1	0	0	78	219	1040	219
Qatar	1993	1723	6	0	0	0	154	28	1911	22
Réunion	1992	...	0	...	...	...	...	...	...	...
Réunion	1993	0	0	0	0	0	44	0	44	0
Romania	1992	1257	0	211	1144	0	751	1015	4378	220
Roumanie	1993	1141	0	199	876	0	759	1101	4076	91
Rwanda	1992	8227	1929	609	4500	7304	1241	2553	26363	36
Rwanda	1993	5110	1556	717	4638	53939	2242	2137	70339	0
Saint Kitts and Nevis	1992	489	4	3	0	402	117	62	1077	0
Saint–Kitts–et–Nevis	1993	311	10	−1	0	153	124	77	674	0
Saint Lucia	1992	232	0	16	0	3	74	102	427	0
Sainte–Lucie	1993	345	24	19	0	0	143	112	643	0
Saint Vincent/Grenadines	1992	174	52	4	0	548	153	62	993	0
Saint Vincent/Grenadines	1993	302	50	22	0	315	121	77	887	0
Sao Tome and Principe	1992	1149	410	380	822	5697	509	140	9107	119
Sao Tomé–et–Principe	1993	782	228	60	640	2608	885	298	5501	129
Saudi Arabia	1992	5882	4	0	0	0	712	12145	18743	11844
Arabie saoudite	1993	6699	4	0	0	0	981	11357	19041	10935
Senegal	1992	7724	4829	1171	4529	9809	1579	8173	37814	0
Sénégal	1993	5402	3444	1234	3584	8582	2130	7806	32182	102
Seychelles	1992	42	15	109	0	126	567	133	992	0
Seychelles	1993	222	11	177	0	125	842	165	1542	0
Sierra Leone	1992	4173	1513	476	3317	2934	1063	708	14184	298
Sierra Leone	1993	5044	823	342	3795	4764	1494	1473	17735	58
Singapore	1992	−8	0	0	0	0	229	25	246	25
Singapour	1993	0	0	0	0	0	395	0	395	0
Solomon Islands	1992	875	22	134	0	0	304	191	1526	6
Iles Salomon	1993	694	4	191	0	0	667	309	1865	44
Somalia	1992	2665	−13	6	28380	63904	1453	319	96714	25
Somalie	1993	8511	753	19	31786	76451	2264	2628	122412	0
Sri Lanka	1992	12264	27	746	4385	5645	1834	2173	27074	42
Sri Lanka	1993	7298	781	701	3177	3301	3508	1912	20678	27
Sudan	1992	17504	5186	550	28247	94852	2316	5888	154543	825
Soudan	1993	11594	4314	2215	32414	90853	4242	6989	152621	597
Suriname	1992	184	40	17	0	0	407	227	875	0
Suriname	1993	119	4	12	0	0	434	395	964	12
Swaziland	1992	1869	46	247	935	7622	783	487	11989	0
Swaziland	1993	1459	94	249	905	5993	1254	498	10452	0
Syrian Arab Republic	1992	2394	14	1195	1006	10761	1443	705	17518	47
Rép. arabe syrienne	1993	1655	24	1626	815	14328	2214	613	21275	162

83
Socio-economic development assistance through the United Nations system
Thousand US dollars [cont.]
Assistance en matière de développement socio-économique fournie par le système des Nations Unies
Milliers de dollars E-U [suite]

A. Development grant expenditures [1] · Aide au développement [1]

Country or area Pays ou zone	Year Année	UNDP PNUD		Special funds Fonds gérés	UNFPA FNUAP	UNICEF FISE	WFP PAM	Other UN system Autres organis. -ONU		Total	Gov't self-supporting Auto-assistance gouverne-mentale
		UNDP programme Programme de PNUD						Regular budget Budget ordinaire	Extra-budgetary Extra-budgétaire		
Thailand	1992	3423	697	474	4548	23414	2425	2589	37570	31
Thaïlande	1993	2372	806	186	4572	4969	4392	2766	20063	0
TFYR Macedonia	1992	..	..	..	..	..	..	0	..	..
L'ex-R.y. Macédoine	1993	0	0	0	0	187271	0	0	187271	0
Togo	1992	4674	1257	557	2221	1288	746	694	11437	0
Togo	1993	2904	847	119	1994	416	987	1010	8277	0
Tokelau	1992	222	11	38	0	0	11	0	282	0
Tokélaou	1993	302	0	34	0	0	39	2	377	0
Tonga	1992	229	32	37	0	0	460	135	893	0
Tonga	1993	429	22	146	0	0	828	120	1545	0
Trinidad and Tobago	1992	1812	43	0	0	0	589	757	3201	415
Trinité-et-Tobago	1993	573	51	0	0	0	828	497	1949	37
Tunisia	1992	1351	0	1044	1213	2605	1446	700	8359	71
Tunisie	1993	2016	0	1111	970	3722	1960	1377	11156	58
Turkey	1992	3035	1	692	5037	3529	1334	2884	16512	1686
Turquie	1993	3326	17	577	3277	1031	1579	3150	12957	2205
Turks and Caicos Islands	1992	161	0	16	0	0	0	35	212	0
Iles Turques et Caiques	1993	169	0	-2	0	0	0	62	229	0
Tuvalu	1992	342	382	85	0	0	62	7	878	0
Tuvalu	1993	294	48	43	0	0	99	12	496	0
Uganda	1992	14747	3284	2599	16487	10841	1324	2245	51527	808
Ouganda	1993	12224	2344	2115	17209	16981	2149	1401	54423	486
United Arab Emirates	1992	1868	5	0	0	0	84	42	1999	26
Emirats arabes unis	1993	1580	6	0	0	0	299	69	1954	55
United Rep. Tanzania	1992	18801	4642	2596	18199	3944	1558	8229	57969	3264
Rép. Unie de Tanzanie	1993	17916	3418	2964	11934	7681	2252	8180	54345	2901
Uruguay	1992	5484	2	98	262	0	828	803	7477	209
Uruguay	1993	8973	433	58	728	0	812	1623	12627	70
Vanuatu	1992	426	584	1181	0	0	582	157	2930	0
Vanuatu	1993	147	153	105	0	0	992	168	1565	0
Venezuela	1992	7914	0	9	860	0	1203	863	10849	117
Venezuela	1993	8758	111	32	1097	0	1160	628	11786	40
Viet Nam	1992	17589	2261	4685	11206	12214	2124	8659	58738	161
Viet Nam	1993	12967	6195	3750	16596	15891	4375	1239	61013	-951
Yemen	1992	7784	2252	855	2908	9430	2643	3097	28969	61
Yémen	1993	2859	2177	761	2834	10684	3880	3007	26202	753
Yugoslavia, SFR †	1992	219	12	27	13495	18669	135	186	32743	53
Yougoslavie, Rfs †	1993	243	7	15	19773	0	12	67	20117	2
Zaire	1992	7038	84	267	9120	5869	1427	892	24697	116
Zaïre	1993	6471	1	12	9026	5918	1955	716	24099	29
Zambia	1992	4074	634	667	3861	23883	1705	4765	39589	18
Zambie	1993	3645	212	587	6435	9042	2383	4913	27217	52
Zimbabwe	1992	3446	570	1149	3520	59140	1341	2975	72141	0
Zimbabwe	1993	5524	1530	1532	6506	5210	1847	2812	24961	0
Other countries	1992	13287	5801	219	2311	4716	3294	10641	40269	1789
Autres pays	1993	14844	178	2516	10330	5420	6029	13756	53073	834
Not elsewhere classified	1992	154331	14949	2125	1611	0	12616	234057	419689	2732
Non-classé ailleurs	1993	101050	6809	1122	2087	25289	15082	104203	255642	2529

83
Socio–economic development assistance through the United Nations system
Thousand US dollars [cont.]
Assistance en matière de développement socio–économique fournie par le système des Nations Unies
Milliers de dollars E–U [suite]

B. Development loan, relief and other expenditures [1] · Prêts au développement, secours et autres dépenses [1]

Country or area / Pays ou zone	Year / Année	Loans Prêts IFAD / FIDA	IDA / IDA	IBRD / BIRD	IFC / SFI	Total develop. grants / Aide totale au développement	Grand total / Total général	IBRD/IDA technical assistance / BIRD/IDA assistance technique	Relief and related grants / Secours et aide connexe
Total	1992	130752	4144000	−6865000	958000	4580408	2948160	1196986	1240420
Total	1993	160777	4319000	−4317000	1910290	4867830	6940897	1447443	1347689
Regional programmes	1992	0	0	0	0	484762	484762	0	31991
Totaux régionaux	1993	0	0	0	0	830168	830168	0	34435
Africa	1992	0	0	0	0	120090	120090	0	908
Afrique	1993	0	0	0	0	127865	127865	0	4359
Asia	1992	0	0	0	0	55133	55133	0	0
Asie	1993	0	0	0	0	66132	66132	0	12723
Latin America	1992	0	0	0	0	67580	67580	0	0
Amérique latine	1993	0	0	0	0	60674	60674	0	1474
Middle East	1992	0	0	0	0	9190	9190	0	31083
Moyen orient	1993	0	0	0	0	13419	13419	0	15721
Interregional	1992	0	0	0	0	75450	75450	0	0
Interrégional	1993	0	0	0	0	380724	380724	0	158
Global	1992	0	0	0	0	157319	157319	0	0
Global	1993	0	0	0	0	181354	181354	0	0
Country programmes	1992	130752	4144000	−6860000	958000	3675957	2048709	1172885	99242
Programmes, pays	1993	160777	4318000	−4313000	1910290	3782020	5858087	1447308	1175346
Afghanistan	1992	0	0	0	0	49633	49633	0	45177
Afghanistan	1993	0	0	0	0	76558	76558	0	16623
Albania	1992	0	2000	0	0	2803	4803	109	0
Albanie	1993	0	26000	0	0	4638	30638	1360	23
Algeria	1992	519	0	−11000	0	11680	1199	8718	0
Algérie	1993	1611	0	−107000	4460	11112	−89817	15592	5567
Angola	1992	0	6000	0	0	75651	81651	2055	2
Angola	1993	0	10000	0	0	78888	88888	6750	4333
Anguilla	1992	0	0	0	0	418	418	0	0
Anguilla	1993	0	0	0	0	336	336	0	0
Antigua and Barbuda	1992	0	0	0	0	407	407	0	0
Antigua−et−Barbuda	1993	0	0	0	0	209	209	0	0
Argentina	1992	426	0	−439000	82000	35459	−321115	39654	0
Argentine	1993	2541	0	940000	261020	65147	1268708	89111	1482
Armenia	1992	0	0	0	0	76	76	0	0
Arménie	1993	0	0	1000	0	1694	2694	5	6865
Aruba	1992	0	0	0	0	216	216	0	0
Aruba	1993	0	0	0	0	508	508	0	0
Bahamas	1992	0	0	0	0	589	589	468	0
Bahamas	1993	0	0	0	0	800	800	72	0
Bahrain	1992	0	0	0	0	278	278	0	0
Bahreïn	1993	0	0	0	0	342	342	0	0
Bangladesh	1992	7303	271000	0	2000	154531	434834	21811	−10
Bangladesh	1993	5565	241000	0	500	100154	347219	26111	18964
Barbados	1992	0	0	−7000	0	955	−6045	286	0
Barbade	1993	0	0	−8000	0	656	−7344	179	0
Belize	1992	401	0	0	0	1890	2291	259	0
Belize	1993	375	0	−1000	430	2503	2308	582	2557
Benin	1992	2056	28000	0	0	18380	48436	7030	0
Bénin	1993	1772	25000	0	0	21756	48528	5876	3535
Bermuda	1992	...	...	...	...		...	...	...
Bermudes	1993	0	0	0	0	3	3	0	
Bhutan	1992	892	1000	0	0	13531	15423	0	0
Bhoutan	1993	1432	1000	0	0	13177	15609	0	0
Bolivia	1992	3245	51000	−35000	41000	28997	89242	12921	0
Bolivie	1993	4348	60000	0	15080	40124	119552	11454	167
Botswana	1992	166	0	−27000	0	12768	−14066	780	0
Botswana	1993	1381	0	0	0	13109	14490	3546	712
Brazil	1992	0	0	−1332000	36000	47173	−1248827	39520	945
Brésil	1993	0	0	−1387000	186370	66528	−1134102	56899	945

83
Socio-economic development assistance through the United Nations system
Thousand US dollars [cont.]
Assistance en matière de développement socio-économique fournie par le système des Nations Unies
Milliers de dollars E-U [suite]

B. Development loan, relief and expenditures [1] · Prêts au développement, secours et dépenses [1]

Country or area Pays ou zone	Year Année	Loans Prêts				Total develop. grants Aide totale au dévelop- pement	Grand total Total général	IBRD/IDA technical assistance BIRD/IDA assistance technique	Relief and related grants Secours et aide connexe
		IFAD FIDA	IDA IDA	IBRD BIRD	IFC SFI				
British Virgin Islands	1992	0	0	0	0	94	94	0	0
Iles Vierges britanniques	1993	0	0	0	0	77	77	0	0
Brunei Darussalam	1992	0	0	0	0	23	23	0	0
Brunéi Darussalam	1993	0	0	0	0	69	69	0	0
Bulgaria	1992	0	0	85000	0	1747	86747	584	0
Bulgarie	1993	0	0	−9000	0	1636	−7364	735	64
Burkina Faso	1992	1577	46000	0	0	24243	71820	4643	0
Burkina Faso	1993	0	58000	0	0	26696	84696	6747	717
Burundi	1992	406	45000	0	0	18736	64142	8065	0
Burundi	1993	776	32000	0	0	19066	51842	9168	3435
Cambodia	1992	0	0	0	0	49890	49890	0	28751
Cambodge	1993	0	0	0	0	71047	71047	0	31396
Cameroon	1992	179	0	−8000	19000	7914	19093	16035	0
Cameroun	1993	0	0	0	10000	10304	20304	14391	966
Cape Verde	1992	788	1000	0	0	17705	19493	712	0
Cap−Vert	1993	1257	3000	0	0	13418	17675	1116	0
Cayman Islands	1992	0	0	0	0	957	957	0	0
Iles Caïmanes	1993	0	0	0	0	26	26	0	0
Central African Rep.	1992	1965	18000	0	0	14772	34737	10322	0
Rép. centrafricaine	1993	2667	21000	0	0	16763	40430	9596	1947
Chad	1992	0	30000	0	0	25369	55369	11501	0
Tchad	1993	0	22000	0	0	22320	44320	8923	78
Chile	1992	0	0	−113000	30000	10334	−72666	11649	0
Chili	1993	0	0	−208000	22000	13284	−172716	23092	313
China	1992	17025	777000	93000	12000	114527	1013552	46696	44
Chine	1993	18501	831000	433000	4000	111675	1398176	50116	3330
Colombia	1992	0	0	−697000	50000	25197	−621803	21602	0
Colombie	1993	0	0	−395000	19720	32389	−342891	16522	40
Comoros	1992	262	2000	0	0	7391	9653	972	0
Comores	1993	155	5000	0	0	8391	13546	1634	0
Congo	1992	249	0	0	0	5395	5644	313	0
Congo	1993	0	0	0	0	5855	5855	0	1152
Cook Islands	1992	0	0	0	0	761	761	0	0
Iles Cook	1993	0	0	0	0	994	994	0	0
Costa Rica	1992	450	0	−49000	1000	10151	−37399	84	0
Costa Rica	1993	525	0	−65000	7900	12124	−44451	341	3966
Côte d'Ivoire	1992	1292	74000	−213000	1000	13050	−123658	7017	0
Côte d'Ivoire	1993	398	2000	−314000	4860	14183	−292559	6219	9502
Cuba	1992	299	0	0		11911	12210	0	103
Cuba	1993	0	0	0	0	17197	17197	0	812
Cyprus	1992	0	0	3000	2000	1061	6061	1363	0
Chypre	1993	0	0	0	0	1267	1267	3936	9985
Czech Republic	1992	0	0	116000	0	1580	117580	0	0
République tchèque	1993	0	0	0	32160	1539	33699	0	218
Djibouti	1992	55	3000	0	0	5925	8980	780	0
Djibouti	1993	120	3000	0	0	7685	10805	949	5293
Dominica	1992	71	0	0	0	637	708	456	0
Dominique	1993	144	0	0	0	677	821	56	1223
Dominican Republic	1992	344	0	−9000	0	9122	466	2459	0
Rép. dominicaine	1993	1932	0	−19000	7340	13268	3540	2502	16
Ecuador	1992	487	0	−74000	10000	14773	−48740	6538	18
Equateur	1993	275	0	−108000	10000	14801	−82924	11723	112
Egypt	1992	11035	−10000	−66000	0	39149	−25816	3336	6
Egypte	1993	7783	1000	−160000	9500	43872	−97845	3436	2016
El Salvador	1992	0	0	−15000	0	26891	11891	1469	0
El Salvador	1993	800	0	19000	0	24765	44565	5471	2192
Equatorial Guinea	1992	1198	2000	0	0	7338	10536	1544	0
Guinée équatoriale	1993	1326	4000	0	0	7240	12566	1486	0
Eritrea	1992	..	..	..	..	..	..	..	131
Erythrée	1993	0	0	0	0	19314	19314	0	131

83
Socio–economic development assistance through the United Nations system
Thousand US dollars [*cont.*]

Assistance en matière de développement socio–économique fournie par le système des Nations Unies
Milliers de dollars E–U [*suite*]

B. Development loan, relief and expenditures [1] • Prêts au développement, secours et dépenses [1]

Country or area Pays ou zone	Year Année	Loans Prêts IFAD FIDA	IDA IDA	IBRD BIRD	IFC SFI	Total develop. grants Aide totale au dévelop- pement	Grand total Total général	IBRD/IDA technical assistance BIRD/IDA assistance technique	Relief and related grants Secours et aide connexe
Ethiopia	1992	1560	99000	0	0	212047	312607	9008	68
Ethiopie	1993	0	218000	0	0	152373	370373	7647	34221
Fiji	1992	0	0	−15000	0	3300	−11700	243	0
Fidji	1993	0	0	−15000	4200	3607	−7193	355	0
French Guiana	1992	0	0	0	0	36	36	0	0
Guyane française	1993	0	0	0	0	39	39	0	22
French Polynesia	1992	0	0	0	0	137	137	0	0
Polynésie française	1993	0	0	0	0	169	169	0	0
Gabon	1992	0	0	−4000	0	2967	−1033	4589	0
Gabon	1993	0	0	−1000	0	3580	2580	4152	170
Gambia	1992	672	20000	0	0	14539	35211	5066	0
Gambie	1993	326	6000	0	840	11813	18979	1823	0
Ghana	1992	2282	157000	0	14000	25934	199216	26968	0
Ghana	1993	3191	189000	0	113960	32451	338602	27516	3940
Greece	1992	0	0	0	0	419	419	0	0
Grèce	1993	0	0	0	0	432	432	0	1661
Grenada	1992	0	0	0	0	1059	1059	0	0
Grenade	1993	0	0	0	0	715	715	0	0
Guam	1992	0	0	0	0	37	37	0	0
Guam	1993	0	0	0	0	49	49	0	0
Guatemala	1992	2488	0	−110000	0	10536	−96976	90	0
Guatemala	1993	1660	0	−55000	81000	17399	45059	119	4686
Guinea	1992	37	83000	0	0	18734	101771	21053	0
Guinée	1993	2060	128000	0	0	17968	148028	30388	16405
Guinea−Bissau	1992	627	10000	0	0	15817	26444	4115	0
Guinée−Bissau	1993	353	12000	0	0	18307	30660	4434	1150
Guyana	1992	423	3000	0	0	6827	10250	1003	0
Guyana	1993	784	28000	0	0	7669	36453	3652	0
Haiti	1992	202	0	0	0	11597	11799	183	0
Haïti	1993	0	0	0	0	16902	16902	0	256
Honduras	1992	534	58000	0	0	18530	77064	996	0
Honduras	1993	412	53000	0	0	19604	73016	1001	496
Hong Kong	1992	0	0	0	0	231	231	0	0
Hong−kong	1993	0	0	0	0	190	190	0	17195
Hungary	1992	0	0	54000	23000	3087	80087	22014	0
Hongrie	1993	0	0	−62000	28150	2195	−31655	8699	4515
India	1992	2345	634000	−507000	107000	151163	387508	58308	0
Inde	1993	2589	661000	−293000	134090	149269	653948	60934	5209
Indonesia	1992	3499	0	−511000	76000	64489	−367012	271769	0
Indonésie	1993	7143	0	−425000	155040	49658	−213159	267221	3150
Iran, Islamic Rep. of	1992	0	0	72000	0	19063	91063	0	64
Iran, Rép. islamique d'	1993	0	0	48000	0	22639	70639	1922	23767
Iraq	1992	0	0	0	0	57519	57519	0	6579
Iraq	1993	0	0	0	0	97848	97848	0	2186
Jamaica	1992	0	0	−110000	0	9486	−100514	3194	0
Jamaïque	1993	338	0	−45000	5000	12679	−26983	5527	0
Jordan	1992	3579	0	34000	0	11090	48669	731	0
Jordanie	1993	80	0	−31000	5000	11477	−14443	513	566
Kenya	1992	2340	75000	0	1000	101214	179554	15605	0
Kenya	1993	646	208000	0	500	124926	334072	15089	59099
Kiribati	1992	0	0	0	0	1077	1077	0	0
Kiribati	1993	0	0	0	0	1135	1135	0	0
Korea, Dem. P. R.	1992	0	0	0	0	7544	7544	0	0
Corée, R. p. dém. de	1993	0	0	0	0	7281	7281	0	0
Korea, Republic of	1992	0	0	−428000	0	3095	−424905	0	0
Corée, République de	1993	0	0	−378000	0	3575	−374425	6956	40
Kuwait	1992	0	0	0	0	457	457	0	0
Koweït	1993	0	0	0	0	1957	1957	0	0
Lao People's Dem. Rep.	1992	0	37000	0	0	16892	53892	2866	0
Rép. dém. pop. lao	1993	0	36000	0	0	17615	53615	5943	3230

83
Socio-economic development assistance through the United Nations system
Thousand US dollars [cont.]
Assistance en matière de développement socio-économique fournie par le système des Nations Unies
Milliers de dollars E-U [suite]

B. Development loan, relief and expenditures [1] · Prêts au développement, secours et dépenses [1]

Country or area Pays ou zone	Year Année	Loans Prêts IFAD FIDA	IDA IDA	IBRD BIRD	IFC SFI	Total develop. grants Aide totale au dévelop- pement	Grand total Total général	IBRD/IDA technical assistance BIRD/IDA assistance technique	Relief and related grants Secours et aide connexe
Lebanon	1992	0	0	−7000	0	14152	7152	0	1634
Liban	1993	0	0	15000	8650	10287	33937	0	1339
Lesotho	1992	526	8000	0	0	23248	31774	3906	0
Lesotho	1993	431	5000	0	0	16780	22211	25653	256
Liberia	1992	0	0	0	0	84158	84158	0	109
Libéria	1993	0	0	0	0	79358	79358	0	4865
Libyan Arab Jamahiriya	1992	0	0	0	0	15875	15875	0	0
Jamah. arabe libyenne	1993	0	0	0	0	14363	14363	0	1055
Madagascar	1992	2503	26000	0	0	33997	62500	16621	0
Madagascar	1993	2002	35000	0	2850	26467	66319	14437	366
Malawi	1992	2952	76000	0	0	161650	240602	11415	9
Malawi	1993	1950	134000	0	0	91612	227562	14133	28082
Malaysia	1992	306	0	−130000	0	5008	−124686	12538	0
Malaisie	1993	0	0	−92000	0	5506	−86494	10159	3677
Maldives	1992	0	8000	0	0	6716	14716	2598	0
Maldives	1993	482	3000	0	0	4915	8397	2087	0
Mali	1992	2196	55000	0	0	31694	88890	4503	0
Mali	1993	1707	38000	0	16720	26988	83415	8133	1102
Malta	1992	0	0	0	0	523	523	0	0
Malte	1993	0	0	0	0	349	349	0	142
Marshall Islands	1992	0	0	0	0	324	324	0	0
Iles Marshall	1993	0	0	0	0	479	479	0	37
Mauritania	1992	263	10000	0	0	15286	25549	6450	1371
Mauritanie	1993	880	25000	0	0	32426	58306	4209	5843
Mauritius	1992	319	0	−24000	1000	3695	−18986	799	0
Maurice	1993	0	0	−16000	0	3512	−12488	1976	0
Mexico	1992	1481	0	−525000	77000	17915	−428604	58088	0
Mexique	1993	3818	0	−800000	55950	24127	−716105	80307	10882
Mongolia	1992	0	−36000	0	0	5471	−30529	1602	0
Mongolie	1993	0	3000	0	0	10679	13679	1161	179
Montserrat	1992	0	0	0	0	247	247	0	0
Montserrat	1993	0	0	0	0	200	200	0	0
Morocco	1992	2462	0	−40000	54000	36378	52840	11855	0
Maroc	1993	6243	0	−175000	42090	35078	−91589	17962	290
Mozambique	1992	0	104000	0	0	154235	258235	16535	1473
Mozambique	1993	0	90000	0	3500	108466	201966	15938	43048
Myanmar	1992	0	5000	0	0	23170	28170	43	0
Myanmar	1993	0	0	0	0	23631	23631	1071	259
Namibia	1992	0	0	0	0	17867	17867	0	0
Namibie	1993	0	0	0	0	16349	16349	0	784
Nepal	1992	467	61000	0	0	37724	99191	10089	0
Népal	1993	1039	60000	0	750	39772	101561	10105	7983
Netherlands Antilles	1992	0	0	0	0	420	420	0	0
Antilles néerlandaises	1993	0	0	0	0	360	360	0	0
Nicaragua	1992	0	72000	0	0	23207	95207	0	169
Nicaragua	1993	1803	13000	−30000	0	23998	8801	22	704
Niger	1992	755	15000	0	0	45384	61139	5859	0
Niger	1993	1710	16000	0	0	31455	49165	4425	172
Nigeria	1992	2301	23000	−287000	34000	35187	−192512	35795	0
Nigéria	1993	2550	36000	−310000	72380	38838	−160232	46249	674
Niue	1992	0	0	0	0	263	263	0	0
Nioué	1993	0	0	0	0	121	121	0	0
Oman	1992	0	0	0	0	4932	4932	2789	0
Oman	1993	0	0	0	0	4912	4912	1009	0
Pakistan	1992	6396	182000	86000	26000	132798	433194	0	232
Pakistan	1993	5060	203000	7000	19000	51474	285534	28884	26683
Panama	1992	0	0	−240000	0	11181	−228819	0	0
Panama	1993	0	0	−75000	970	14801	−59229	278	202
Papua New Guinea	1992	355	0	−22000	0	10834	−10811	8596	0
Papouasie−Nvl−Guinée	1993	69	0	−23000	0	10561	−12370	6533	987

83
Socio–economic development assistance through the United Nations system
Thousand US dollars [cont.]

Assistance en matière de développement socio–économique fournie par le système des Nations Unies
Milliers de dollars E–U [suite]

B. Development loan, relief and expenditures [1] · Prêts au développement, secours et dépenses [1]

Country or area Pays ou zone	Year Année	Loans Prêts IFAD FIDA	IDA IDA	IBRD BIRD	IFC SFI	Total develop. grants Aide totale au dévelop- pement	Grand total Total général	IBRD/IDA technical assistance BIRD/IDA assistance technique	Relief and related grants Secours et aide connexe
Paraguay	1992	0	0	−53000	0	14558	−38442	1665	0
Paraguay	1993	1175	0	−53000	0	18272	−33553	3400	13
Peru	1992	1561	0	0	5000	25823	32384	92	30
Pérou	1993	4140	0	−69000	26340	46700	8180	0	155
Philippines	1992	729	32000	−94000	28000	30705	−2566	30600	0
Philippines	1993	2081	0	2000	31310	26787	62178	45375	8109
Poland	1992	0	0	310000	25000	4087	339087	6541	110
Pologne	1993	0	0	253000	11140	2880	267020	16095	315
Portugal	1992	0	0	0	0	1400	1400	645	0
Portugal	1993	0	0	0	0	873	873	499	629
Qatar	1992	0	0	0	0	1040	1040	0	0
Qatar	1993	0	0	0	0	1911	1911	0	0
Réunion	1992	...	...	...	...	...	...	...	...
Réunion	1993	0	0	0	0	44	44	0	0
Romania	1992	0	0	209000	0	4378	213378	1621	1698
Roumanie	1993	0	0	170000	1870	4076	175946	3510	198
Rwanda	1992	2587	27000	0	0	26363	55950	5748	0
Rwanda	1993	2869	34000	0	0	70339	107208	8326	4695
Saint Kitts and Nevis	1992	0	0	0	0	1077	1077	704	0
Saint–Kitts–et–Nevis	1993	0	0	0	0	674	674	852	0
Saint Lucia	1992	227	0	0	0	427	654	872	0
Sainte–Lucie	1993	104	0	0	0	643	747	581	0
Saint Vincent/Grenadines	1992	0	0	0	0	993	993	80	0
St. Vincent/Grenadines	1993	277	0	0	0	887	1164	115	0
Sao Tome and Principe	1992	640	5000	0	0	9107	14747	2192	0
Sao Tomé–et–Principe	1993	820	5000	0	0	5501	11321	1487	0
Saudi Arabia	1992	0	0	0	0	18743	18743	0	0
Arabie saoudite	1993	0	0	0	0	19041	19041	0	0
Senegal	1992	9	94000	0	0	37814	131823	7465	0
Sénégal	1993	0	36000	0	190	32182	68372	12192	3977
Seychelles	1992	0	0	−1000	2000	992	1992	0	0
Seychelles	1993	215	0	−1000	2300	1542	3057	35	0
Sierra Leone	1992	1958	25000	0	10000	14184	51142	117	140
Sierra Leone	1993	1152	34000	0	3000	17735	55887	4432	1554
Singapore	1992	0	0	0	0	246	246	0	0
Singapour	1993	0	0	0	0	395	395	0	0
Solomon Islands	1992	49	0	0	0	1526	1575	0	442
Iles Salomon	1993	181	1000	0	0	1865	3046	0	28
Somalia	1992	0	0	0	0	96714	96714	−91	4611
Somalie	1993	0	0	0	0	122412	122412	0	29169
Sri Lanka	1992	2467	60000	0	0	27074	89541	10227	4739
Sri Lanka	1993	2783	114000	0	0	20678	137461	13848	1511
Sudan	1992	1788	47000	0	0	154543	203331	14444	15963
Soudan	1993	2473	0	0	0	152621	155094	12513	605
Suriname	1992	0	0	0	0	875	875	0	0
Suriname	1993	0	0	0	0	964	964	0	3143
Swaziland	1992	764	0	0	0	11989	12753	0	1080
Swaziland	1993	593	0	−6000	1570	10452	6615	0	0
Syrian Arab Republic	1992	0	0	0	0	17518	17518	0	18221
Rép. arabe syrienne	1993	585	0	0	0	21275	21860	0	533206
Thailand	1992	3154	0	−638000	7000	37570	−590276	5379	0
Thaïlande	1993	2542	0	−172000	46020	20063	−103375	15429	144
TFYR Macedonia	1992	..	..	..	..	..	..	..	..
L'ex–R.y. Macédoine	1993	0	0	0	0	187271	187271	0	0
Togo	1992	1259	28000	0	0	11437	40696	8122	0
Togo	1993	606	3000	0	0	8277	11883	1900	0
Tokelau	1992	0	0	0	0	282	282	0	0
Tokélaou	1993	0	0	0	0	377	377	0	0
Tonga	1992	315	0	0	0	893	1208	0	0
Tonga	1993	0	0	0	0	1545	1545	0	0

83
Socio-economic development assistance through the United Nations system
Thousand US dollars [cont.]

Assistance en matière de développement socio-économique fournie par le système des Nations Unies
Milliers de dollars E-U [suite]

B. Development loan, relief and expenditures [1] · Prêts au développement, secours et dépenses [1]

Country or area Pays ou zone	Year Année	Loans Prêts IFAD FIDA	IDA IDA	IBRD BIRD	IFC SFI	Total develop. grants Aide totale au dévelop-pement	Grand total Total général	IBRD/IDA technical assistance BIRD/IDA assistance technique	Relief and related grants Secours et aide connexe
Trinidad and Tobago	1992	0	0	−5000	0	3201	−1799	687	0
Trinité−et−Tobago	1993	0	0	17000	0	1949	18949	975	0
Tunisia	1992	2372	0	−156000	5000	8359	−140269	6146	0
Tunisie	1993	3020	0	−16000	11000	11156	9176	7396	103
Turkey	1992	1769	0	−921000	107000	16512	−795719	22588	1035
Turquie	1993	5151	0	−830000	210470	12957	−601422	44694	5383
Turks and Caicos Islands	1992	0	0	0	0	212	212	0	0
Iles Turques et Caiques	1993	0	0	0	0	229	229	0	0
Tuvalu	1992	0	0	0	0	878	878	0	0
Tuvalu	1993	0	0	0	0	496	496	0	0
Uganda	1992	2267	143000	0	0	51527	196794	21428	0
Ouganda	1993	3213	127000	0	300	54423	184936	30275	7584
United Arab Emirates	1992	0	0	0	0	1999	1999	0	0
Emirats arabes unis	1993	0	0	0	0	1954	1954	0	0
United Rep. Tanzania	1992	779	216000	0	2000	57969	276748	28813	0
Rép. Unie de Tanzanie	1993	3757	123000	0	300	54345	181402	24568	3896
Uruguay	1992	0	0	98000	6000	7477	111477	4593	0
Uruguay	1993	1072	0	−48000	2800	12627	−31501	5608	100
Vanuatu	1992	0	3000	0	0	2930	5930	744	0
Vanuatu	1993	0	0	0	0	1565	1565	12	0
Venezuela	1992	195	0	71000	21000	10849	103044	5507	0
Venezuela	1993	1581	0	−91000	129690	11786	52057	4027	822
Viet Nam	1992	0	−1000	0	0	58738	57738	0	42
Viet Nam	1993	951	−1000	0	0	61013	60964	0	15025
Yemen	1992	2672	44000	0	0	28969	75641	6000	0
Yémen	1993	8183	36000	0	0	26202	70385	7948	3484
Yugoslavia, SFR †	1992	0	0	−172000	2000	32743	−137257	416	55
Yougoslavie, Rfs †	1993	0	0	0	0	20117	20117	12	25
Zaire	1992	564	40000	0	3000	24697	68261	16113	0
Zaïre	1993	57	0	0	0	24099	24156	14424	6275
Zambia	1992	2280	167000	0	0	39589	208869	2531	0
Zambie	1993	3673	168000	0	0	27217	198890	15649	3271
Zimbabwe	1992	1748	75000	8000	36000	72141	192889	4575	0
Zimbabwe	1993	2512	61000	39000	26010	24961	153483	4099	7174
Other countries	1992	2042	7000	−4000	4211	0	40269	45311	1789
Autres pays	1993	0	32000	636000	2714	60000	53073	781073	33567
Not elsewhere classified	1992	0	0	−5000	0	419689	414689	24101	1109187
Non−classé ailleurs	1993	0	1000	−4000	0	255642	252642	135	137908

Source:
1995 Triennial Policy Review of operational activities of the
United Nations system (E/1995/98) and Comprehensive statistical
data on operational activities for development for the years
1992 (E/1994/64/Add.2) and 1993 (A/50/202/Add.2).

† For detailed descriptions of data pertaining to former
Czechoslovakia, Germany, SFR Yugoslavia and former USSR,
see Annex I – Country or area nomenclature, regional and
other groupings.

1 The following abbreviations have been used in the table:
UNDP: United Nations Development Programme
UNFPA: United Nations Population Fund
UNICEF: United Nations Children's Fund
WFP: World Food Programme
IDA: International Development Association
IFAD: International Fund for Agricultural Development
IBRD: International Bank for Reconstruction and Development
IFC: International Finance Corporation

Source:
"1995 Triennial Policy Review of operational activities of the United
Nations system (E/1995/98)" et Information statistique détail concernant
les activités opérationelles du développement, 1992 (E/1994/64/Add.2) et
1993 (A/50/202/Add.2).

† Pour les descriptions en détails des données relatives à l'ancienne
Tchécoslovaquie, l'Allemagne, la Rfs Yougoslavie et l'ancienne URSS,
voir l'Annexe I – Nomenclature des pays ou zones, groupements
et régionaux autres groupements.

1 Les abbréviations ci−après ont été utilisées dans le tableau:
PNUD : Programme des Nations Unies pour le développement
FNUAP : Fonds des Nations Unies pour la population
FISE : Fonds des Nations Unies pour l'enfance
PAM : Programme alimentaire mondiale
IDA : Association internationale de développement
FIDA : Fonds international de développement agricole
BIRD : Banque internationale pour la réconstruction et le
développement
SFI : Société financière internationale

Technical notes, tables 81-83

Table 81 presents estimates of flows of financial resources to countries in the developing regions either directly (bilaterally) or through multilateral institutions (multilaterally).

For the purpose of this table, the developing regions include all Africa, except South Africa; Latin America and the Caribbean, including the British Virgin Islands; Asia, except Japan; and Oceania, except Australia, New Zealand and United States possessions and territories "Developed market economies" comprise 21 members of the Development Assistance Committee of the Organisation for Economic Co-operation and Development (OECD): Australia, Austria, Belgium, Canada, Denmark, Finland, the Federal Republic of Germany, France, Ireland, Italy, Japan, Luxembourg (since December 1992), Netherlands, New Zealand, Norway, Portugal (since December 1991), Spain (since December 1991), Sweden, Switzerland, the United Kingdom of Great Britain and Northern Ireland and the United States of America.

The multilateral institutions include the World Bank Group, regional banks, financial institutions of the European Community and a number of United Nations institutions, programmes and trust funds.

The main source of data is the Development Assistance Committee of OECD to which member countries reported data on their flow of resources to developing countries and multilateral institutions. Data in the *Statistical Yearbook* do not include the less developed countries in Europe as recipients.

Additional information on definitions, methods and sources can be found in OECD's *Geographical Distribution of Financial Flows to Developing Countries*.[16]

Table 82 presents the development assistance expenditures of donor countries. This table includes donors contributions to multilateral agencies, so the overall totals differ from those in table 81, which include disbursements by multilateral agencies.

Table 83: includes data on expenditures on operational activities for development undertaken by the organizations of the United Nations system. Operational activities encompass, in general, those activities of a development cooperation character that seek to mobilize or increase the potential and capacity of countries to promote economic and social development and welfare, including the transfer of resources to developing countries or regions in a tangible or intangible form. The table also covers, as a memo item, expenditures on activities of an emergency character, the purpose of which is immediate relief in crisis situations, such as assistance to refugees, humanitarian work and activities in respect of disasters.

Notes techniques, tableaux 81 à 83

Le *Tableau 81* présente les estimations des flux de ressources financières mises à la disposition des pays des régions en développement soit directement (aide bilatérale) soit par l'intermédiaire d'institutions multilatérales (aide multilatérale).

Aux fins de ce tableau, les régions en développement comprennent l'Afrique, sauf l'Afrique du Sud; l'Amérique latine et les Caraïbes, y compris les Iles vierges britanniques; l'Asie, sauf le Japon; et l'Océanie, sauf l'Australie, la Nouvelle-Zélande et les possessions et territoires des Etats-Unis. Les "pays développés à économie de marché" comprennent les 21 membres du Comité d'aide au développement de l'Organisation de coopération et de développement économiques (OCDE), à savoir : l'Allemagne (République fédérale d'), l'Australie, l'Autriche, la Belgique, le Canada, le Danemark, l'Espagne (à partir de décembre 1991), les Etats-Unis d'Amérique, la Finlande, la France, l'Irelande, l'Italie, le Japon, le Luxembourg (à partir de décembre 1992), la Norvège, la Nouvelle-Zélande, les Pays-Bas, le Portugal (à partir de décembre 1991), le Royaume-Uni de Grande-Bretagne et d'Irlande du Nord, la Suède et la Suisse.

Les institutions multilatérales comprennent le Groupe de la Banque mondiale, les banques régionales, les institutions financières de la Communauté européenne et un certain nombre d'institutions, de programmes et de fonds d'affectation spéciale des Nations Unies.

La principale source de données est le Comité d'aide au développement de l'OCDE, auquel les pays membres ont communiqué des données sur les flux de ressources qu'ils mettent à la disposition des pays en développement et des institutions multilatérales. Les données présentées dans l'*Annuaire statistique* ne comprennent pas l'aide fournie aux pays moins développés d'Europe.

Pour plus de renseignements sur les définitions, méthodes et sources, se reporter à la *Répartition géographique des ressources financières de l'OCDE* [16].

Le *Tableau 82* présente les dépenses que les pays donateurs consacrent à l'aide publique au développement (APD). Ces chiffres incluent les contributions des donateurs à des agences multilatérales, de sorte que les totaux diffèrent de ceux du tableau 81, qui incluent les dépenses des agences multilatérales.

Le *Tableau 83* présente des données sur les dépenses consacrées à des activités opérationnelles pour le développement par les organisations du système des Nations Unies. Par "activités opérationnelles", on entend en général les activités ayant trait à la coopération au développement, qui visent à mobiliser ou à accroître les potentialités et aptitudes que présentent les pays pour promouvoir le développement et le bien-être économiques et sociaux, y

Expenditures on operational activities for development are financed from contributions from governments and other official and non-official sources to a variety of funding channels in the United Nations system. These include United Nations funds and programmes such as contributions to the United Nations Development Programme; contributions to funds administered by United Nations Development Programme; regular (assessed) and other extrabudgetary contributions to specialized agencies, International Atomic Energy Agency, and other organizations for the purposes of operational activities; contributions to the International Development Association (IDA) and to the International Fund for Agricultural Development (IFAD).

Data are taken from the 1992 and 1993 reports of the Secretary-General to the assembly on operational activities for development [18].

compris les transferts de ressources vers les pays ou régions en développement sous forme tangible ou non. Ce tableau indique également, pour mémoire, les dépenses liées à des activités revêtant un caractère d'urgence, qui ont pour but d'apporter un secours immédiat dans les situations de crise, telles que l'aide aux réfugiés, l'assistance humanitaire et les secours en cas de catastrophe.

Les dépenses consacrées aux activités opérationnelles pour le développement sont financées au moyen de contributions que les gouvernements et d'autres sources officielles et non officielles apportent à divers organes de financement, tels que fonds et programmes, du système des Nations Unies. On peut citer notamment les contributions au Programme des Nations Unies pour le développement; les contributions aux fonds gérés par le Programme des Nations Unies pour le développement; les contributions régulières (budgétaires) et les contributions extrabudgétaires aux institutions spécialisées, à l'Agence internationale de l'énergie atomique et à d'autres organismes aux fins d'activités opérationnelles; les contributions à l'Association internationale de développement (IDA) et au Fonds international de développement agricole (FIDA).

Les données sont extraites des rapports annuels de 1992 et 1993 du Secrétaire général à la session de l'Assemblée générale sur les activités opérationnelles pour le développement [18].

Country and area nomenclature, regional and other groupings

A. *Changes in country or area names*

In the periods covered by the statistics in the present issue of the *Statistical Yearbook* (in general, 1983-1992, 1984-1993), and as indicated at the end of each table, the following major changes in designation have taken place:

Former *Czechoslovakia*: Since 1 January 1993, data for the Czech Republic and Slovakia, where available, are shown separately under the appropriate country name. For periods prior to 1 January 1993, where no separate data are available for the Czech Republic and Slovakia, unless otherwise indicated, data for the former Czechoslovakia are shown under the country name "former Czechoslovakia".

Germany: Through the accession of the German Democratic Republic to the Federal Republic of Germany with effect from 3 October 1990, the two German States have united to form one sovereign State. As from the date of unification, the Federal Republic of Germany acts in the United Nations under the designation "Germany". All data shown which pertain to Germany prior to 3 October 1990 are indicated separately for the Federal Republic of Germany and the former German Democratic Republic based on their respective territories at the time indicated;

SFR Yugoslavia: Data provided for Yugoslavia prior to 1 January 1992 refer to the Socialist Federal Republic of Yugoslavia which was composed of six republics. Data provided for Yugoslavia after that date refer to the Federal Republic of Yugoslavia which is composed of two republics (Serbia and Montenegro);

Former *USSR*: In 1991, the Union of Soviet Socialist Republics formally dissolved into fifteen independent countries (Armenia, Azerbaijan, Belarus, Estonia, Georgia, Kazakhstan, Kyrgyzstan, Latvia, Lithuania, Republic of Moldova, Russian Federation, Tajikistan, Turkmenistan, Ukraine and Uzbekistan). Whenever possible, data are shown for the individual countries. Otherwise, data are shown for the former USSR.

Other changes in designation during the periods are listed below:

Brunei Darussalam was formerly listed as Brunei;
Burkina Faso was formerly listed as Upper Volta;
Cambodia was formerly listed as Democratic Kampuchea;
Cameroon was formerly listed as United Republic of Cameroon;
Côte d'Ivoire was formerly listed as Ivory Coast;
Myanmar was formerly listed as Burma;

Nomenclature des pays ou zones, groupements régionaux et autres groupements

A. *Changements dans le nom des pays ou zones*

Au cours des périodes sur lesquelles portent les statistiques, dans cette édition de l'*Annuaire Statistique* (1983-1992 et 1984-1993, en générale), et comme indiqués à la fin de chaque tableau les changements principaux de désignation suivants ont eu lieu :

L'ancienne *Tchécoslovaquie*: Depuis le 1er janvier 1993, les données relatives à la République tchèque, et à la Slovaquie, lorsqu'elles sont disponibles, sont présentées séparément sous le nom de chacun des pays. En ce qui concerne la période précédant le 1er janvier 1993, pour laquelle on ne possède pas de données séparées pour les deux Républiques, les données relatives à l'ancienne Tchécoslovaquie sont, sauf indication contraire, présentées sous le titre "ancienne Tchécoslovaquie".

Allemagne: En vertu de l'adhésion de la République démocratique allemande à la République fédérale d'Allemagne, prenant effet le 3 octobre 1990, les deux Etats allemands se sont unis pour former un seul Etat souverain. A compter de la date de l'unification, la République fédérale d'Allemagne est désigné à l'ONU sous le nom d'"Allemagne". Toutes les données se rapportant à l'Allemagne avant le 3 octobre figurent dans deux rubriques séparées basées sur les territoires respectifs de la République fédérale d'Allemagne et l'ancienne République démocratique allemande selon la période indiquée;

Rfs Yougoslavie : Les données fournies pour la Yougoslavie avant le 1er janvier 1992 se rapportent à la République fédérative socialiste de Yougoslavie, qui était composée de six républiques. Les données fournies pour la Yougoslavie après cette date se rapportent à la République fédérative de Yougoslavie, qui est composée de deux républiques (Serbie et Monténégro);

Ancienne *URSS* : En 1991, l'Union des républiques socialistes soviétiques s'est séparé en 15 pays distincts (Arménie, Azerbaïdjan, Bélarus, Estonie, Géorgie, Kazakhstan, Kirghizistan, Lettonie, Lituanie, la République de Moldova, Fédération de Russie, Tadjikistan, Turkménistan, Ukraine, Ouzbékistan). Les données sont présentées pour ces pays pris séparément quand cela est possible. Autrement, les données sont présentées pour l'ancienne URSS.

Les autres changements de désignation couvrant les périodes mentionnées sont énumérés ci-dessous :

Le *Brunéi Darussalam* apparaissait antérieurement sous le nom de Brunéi;

Le *Burkino Faso* apparaissait antérieurement sous le nom de la Haute-Volta;

Le *Cambodge* apparaissait antérieurement sous le nom de la Kampuchea démocratique;

Le *Cameroun* apparaissait antérieurement sous le nom de République-Unie du Cameroun;

Myanmar apparaissait antérieurement sous le nom de Birmanie;

Palau was formerly listed as Pacific Islands and includes data for Federated States of Micronesia, Marshall Islands and Northern Mariana Islands;

Saint Kitts and Nevis was formerly listed as Saint Christopher and Nevis;

Yemen comprises the former Republic of Yemen and the former Democratic Yemen;

Data relating to the People's Republic of China generally include those for Taiwan Province in the field of statistics relating to population, area, natural resources, natural conditions such as climate. In other fields of statistics, they do not include Taiwan Province unless otherwise stated.

B. *Regional groupings*

The scheme of regional groupings given below presents seven regions based mainly on continents. Five of the seven continental regions are further subdivided into 21 regions that are so drawn as to obtain greater homogeneity in sizes of population, demographic circumstances and accuracy of demographic statistics.[19, 26] This nomenclature is widely used in international statistics and is followed to the greatest extent possible in the present *Yearbook* in order to promote consistency and facilitate comparability and analysis. However, it is by no means universal in international statistical compilation, even at the level of continental regions, and variations in international statistical sources and methods dictate many unavoidable differences in particular fields in the present *Yearbook*. General differences are indicated in the footnotes to the classification presented below. More detailed differences are given in the footnotes and technical notes to individual tables.

Neither is there international standardization in the use of the terms "developed" and "developing" countries, areas or regions. These terms are used in the present publication to refer to regional groupings generally considered as "developed": these are Europe and former USSR, the United States of America and Canada in northern America, and Australia, Japan and New Zealand in Asia and Oceania. These designations are intended for statistical convenience and do not necessarily express a judgement about the stage reached by a parlicular country or area in the development process. Differences from this usage are indicated in the notes to individual tables.

Palaos apparaissait antétieurement sous le nom de Iles du Pacifique y compris les données pour les Etats fédérés de Micronésie, les îles Marshall et îles Mariannes du Nord;

Saint-Kitts-et-Nevis apparaissait antérieurement sous le nom de Saint-Christophe-et-Nevis;

Le *Yémen* comprend l'ancienne République de Yémen et l'ancien Yémen démocratique;

Les données relatives à la République populaire de Chine comprennent en général les données relatives à la province de Taïwan lorqu'il s'agit de statistiques concernant la population, la superficie, les ressources naturelles, les conditions naturelles telles que le climat, etc. Dans les statistiques relatives à d'autres domaines, la province de Taïwan n'est pas comprise, sauf indication contraire.

B. *Groupements régionaux*

Le système de groupements régionaux présenté ce-dessous comporte sept régions basés principalement sur les continents. Cinq des sept régions continentales sont elles-mêmes subdivisées, formant ainsi 21 régions délimitées de manière à obtenir une homogénéité accrue dans les effectifs de population, les situations démographiques et la précision des statistiques démographiques [19, 26]. Cette nomenclature est couramment utilisée aux fins des statistiques internationales et a été appliquée autant qu'il a été possible dans le présent *Annuaire* en vue de renforcer la cohérence et de faciliter la comparaison et l'analyse. Son utilisation pour l'établissement des statistiques internationales n'est cependant rien moins qu'universelle, même au niveau des régions continentales, et les variations que présentent les sources et méthodes statistiques internationales entraînent inévitablement de nombreuses différences dans certains domaines de cet *Annuaire*. Les différences d'ordre général sont indiquées dans les notes figurant au bas de la classification présentée ci-dessous. Les différences plus spécifiques sont mentionées dans les notes techiques et notes infrapaginales accompagnant les divers tableaux.

L'application des expressions "développés" et "en développement" aux pays, zones ou régions n'est pas non plus normalisée à l'échelle internationale. Ces expressions sont utilisées dans la présente publication en référence aux groupements régionaux généralement considérés comme "développés", à savoir l'Europe et l'ancienne URSS, les Etats-Unis d'Amérique et le Canada en Amérique du Nord, et l'Australie, le Japon et la Nouvelle-Zélande dans la région de l'Asie et du Pacifique. Ces appellations sont employées pour des raisons de commodité statistique et n'expriment pas nécessairement un jugement sur le stade de développement atteint par tel ou tel pays ou zone. Les cas différant de cet usage sont signalés dans les notes accompagnant les tableaux concernés.

Africa

Eastern Africa
British Indian Ocean Territory
Burundi
Comoros
Djibouti
Ethiopia
Kenya
Madagascar
Malawi
Mauritius
Mozambique
Reunion
Rwanda
Seychelles
Somalia
Uganda
United Republic of Tanzania
Zambia
Zimbabwe

Middle Africa
Angola
Cameroon
Central African Republic
Chad
Congo
Equatorial Guinea
Gabon
Sao Tome and Principe
Zaire

Northern Africa
Algeria
Egypt
Libyan Arab Jamahiriya
Morocco
Sudan
Tunisia
Western Sahara

Southern Africa
Botswana
Lesotho
Namibia
South Africa
Swaziland

Western Africa
Benin
Burkina Faso
Cape Verde
Côte d'Ivoire
Gambia

Afrique

Afrique orientale
Territoire britannique de l'océan indien
Burundi
Comores
Djibouti
Ethiopie
Kenya
Madagascar
Malawi
Maurice
Mozambique
Réunion
Rwanda
Seychelles
Somalie
Ouganda
République-Unie de Tanzanie
Zambie
Zimbabwe

Afrique centrale
Angola
Cameroun
République centrafricaine
Tchad
Congo
Guinée équatoriale
Gabon
Sao Tomé-et-Principe
Zaïre

Afrique septentrionale
Algérie
Egypte
Jam. arabe libyenne
Maroc
Soudan
Tunisie
Sahara occidental

Afrique méridionale
Botswana
Lesotho
Namibie
Afrique du Sud
Swaziland

Afrique occidentale
Bénin
Burkina Faso
Cap-Vert
Côte d'Ivoire
Gambie

Ghana	Ghana
Guinea	Guinée
Guinea-Bissau	Guinée-Bissau
Liberia	Libéria
Mali	Mali
Mauritania	Mauritanie
Niger	Niger
Nigeria	Nigéria
St. Helena	Sainte-Hélène
Senegal	Sénégal
Sierra Leone	Sierra Leone
Togo	Togo

Northern America / **Amérique septentrionale**

Bermuda	Bermudes
Canada	Canada
Greenland	Groenland
St. Pierre and Miquelon	Saint-Pierre-et-Miquelon
United States of America	Etats-Unis d'Amérique

Latin America and Caribbean / **Amérique latine et Caraïbes**

Caribbean / *Caraïbes*

Anguilla	Anguilla
Antigua and Barbuda	Antigua-et-Barbuda
Aruba	Aruba
Bahamas	Bahamas
Barbados	Barbade
British Virgin Islands	Iles Vierges britanniques
Cayman Islands	Iles Caïmanes
Cuba	Cuba
Dominica	Dominique
Dominican Republic	République dominicaine
Grenada	Grenade
Guadeloupe	Guadeloupe
Haiti	Haïti
Jamaica	Jamaïque
Martinique	Martinique
Montserrat	Montserrat
Netherlands Antilles	Antilles néerlandaises
Puerto Rico	Porto Rico
St. Kitts and Nevis	St. Christophe/Nevis
St. Lucia	Sainte-Lucie
St. Vincent/Grenadines	St. Vincent/Grenadines
Trinidad and Tobago	Trinité-et-Tobago
Turks and Caicos Islands	Iles Turques et Caiques
US Virgin Islands	Iles Vierges américaines

Central America / *Amérique centrale*

Belize	Belize
Costa Rica	Costa Rica
El Salvador	El Salvador
Guatemala	Guatemala
Honduras	Honduras
Mexico	Mexique
Nicaragua	Nicaragua
Panama	Panama

South America	*Amérique du Sud*
Argentina	Argentine
Bolivia	Bolivie
Brazil	Brésil
Chile	Chili
Colombia	Colombie
Ecuador	Equateur
Falkland Islands (Malvinas)	Iles Falkland (Malvinas)
French Guiana	Guyane française
Guyana	Guyana
Paraguay	Paraguay
Peru	Pérou
Suriname	Suriname
Uruguay	Uruguay
Venezuela	Venezuela
Asia	**Asie**
Central Asia	*Asie centrale*
Kazakhstan	Kazakhstan
Kyrgyzstan	Kirghizistan
Tajikistan	Tadjikistan
Turkmenistan	Turkménistan
Uzbekistan	Ouzbékistan
Eastern Asia	*Asie orientale*
China	Chine
Hong Kong	Hong-kong
Japan	Japon
Korea, Democratic People's Republic	Corée, république populaire démocratique de
Korea, Republic of	Corée, République de
Macau	Macao
Mongolia	Mongolie
South-eastern Asia	*Asie méridionale orientale*
Brunei Darussalam	Brunéi Darussalam
Cambodia	Cambodge
East Timor	Timor oriental
Indonesia	Indonésie
Lao People's Democratic Republic	République démocratique populaire Lao
Malaysia	Malaisie
Myanmar	Myanmar
Philippines	Philippines
Singapore	Singapour
Thailand	Thaïlande
Viet Nam	Viet Nam
Southern Asia	*Asie méridionale*
Afghanistan	Afghanistan
Bangladesh	Bangladesh
Bhutan	Bhoutan
India	Inde
Iran (Islamic Republic of)	Iran, République islamique d'
Maldives	Maldives
Nepal	Népal
Pakistan	Pakistan
Sri Lanka	Sri Lanka

Western Asia
Armenia
Azerbaijan
Bahrain
Cyprus
Georgia
Iraq
Israel
Jordan
Kuwait
Lebanon
Oman
Palestine
Qatar
Saudi Arabia
Syrian Arab Republic
Turkey
United Arab Emirates
Yemen

Europe
Eastern Europe
Belarus
Bulgaria
Czech Republic
Germany: [a]
  former German Democratic Republic
Hungary
Poland
Republic of Maldova
Romania
Russian Federation
Slovakia
Ukraine

Western Europe [b]
Austria
Belgium
France
Germany: [a]
  Federal Republic of Germany
Liechtenstein
Luxembourg
Monaco
Netherlands
Switzerland

  Northern Europe
Channel Islands
Denmark
Estonia
Faeroe Islands
Finland
Iceland
Ireland

Asie occidentale
Arménie
Azerbaïjan
Bahreïn
Chypre
Géorgie
Iraq
Israël
Jordanie
Koweït
Liban
Oman
Palestine
Qatar
Arabie saoudite
République arabe syrienne
Turquie
Emirats arabes unis
Yémen

Europe
Europe orientale
Bélarus
Bulgarie
République tchèque
Allemagne: [a]
  ancienne République démocratique allemande
Hongrie
Pologne
République de Maldova
Roumanie
Fédération de Russie
Slovaquie
Ukraine

Europe occidentale [b]
Autriche
Belgique
France
Allemagne: [a]
  République fédérale d'Allemagne
Liechtenstein
Luxembourg
Monaco
Pays-Bas
Suisse

  Europe septentrionale
Iles Anglo-Normandes
Danemark
Estonie
Iles Féroé
Finlande
Islande
Irlande

Isle of Man	Iles de Man
Latvia	Lettonie
Lithuania	Lithuanie
Norway	Norvège
Svalbard and Jan Mayen Islands	Svalbard et îles Jan Mayen
Sweden	Suède
United Kingdom	Royaume-Uni

Southern Europe	Europe méridionale
Albania	Albanie
Andorra	Andorre
Bosnia and Herzegovina	Bosnie-Herzégovine
Croatia	Croatie
Gibraltar	Gibraltar
Greece	Grèce
Holy See	Saint-Siège
Italy	Italie
Malta	Malte
Portugal	Portugal
San Marino	Saint-Marin
Slovenia	Slovénie
Spain	Espagne
The Former Yougoslav Rep. of Macedonia	L'ex-République yougoslave de Macédoine
Yugoslavia	Yougoslavie

Oceania / **Océanie**

Australia and New Zealand / *Australie et Nouvelle Zélande*

Australia	Australie
Christmas Islands	Iles Christmas
Cocos (Keeling) Islands	Iles des Cocos (Keeling)
New Zealand	Nouvelle-Zélande
Norfolk Island	Ile Norfolk

Pacific / *Pacifique*

Melanesia	Melenésie
Fiji	Fidji
New Caledonia	Nouvelle-Calédonie
Papua New Guinea	Papouasie-Nouv.-Guinée
Solomon Islands	Iles Salomon
Vanuatu	Vanuatu

Micronesia	Micronésie
Guam	Guam
Kiribati	Kiribati
Marshall Islands	Iles Marshall
Micronesia, Federated States of	Micronésie, Etats fédératives de
Nauru	Nauru
Northern Marianna Islands	Iles Mariannes du Nord
Palau	Palaos
Wake Island	Ile de Wake

Polynesia	Polynésie
American Samoa	Samoa américaine
Cook Islands	Iles Cook
French Polynesia	Polynésie française

Johnston Island	Ile Johnston
Midway Islands	Iles Midway
Niue	Nioué
Pitcairn	Pitcairn
Samoa	Samoa
Tokelau	Tokélau
Tonga	Tonga
Tuvalu	Tuvalu
Wallis and Futuna Islands	Iles Wallis et Futuna
former **Union of Soviet Socialist Republics**	ancienne **Union des républiques Socialistes Soviétiques**
former USSR	ancienne URSS

---

a   Through the accession of the German Democratic Republic to the Federal Republic of Germany with effect from 3 October 1990, the two German States have united to form one sovereign State. As from the date of unification, the Federal Republic of Germany acts in the United Nations under the designation of "Germany". All data shown which pertain to Germany prior to 3 October 1990 are indicated separately for the Federal Republic of Germany and the former German Democratic Republic based on their respective territories at the time indicated.

b   Where the term "western Europe" is used in the present publication in distinction to "eastern Europe", it refers to all regions of Europe except eastern Europe (that is, it is comprised of northern and southern as well as western Europe).

---

a   En vertu de l'adhésion de la République démocratique allemande à la République fédérale d'Allemagne, prenant effet le 3 octobre 1990, les deux Etats allemands se sont unis pour former un seul Etat souverain. A compter de la date de l'unification, la République fédérale d'Allemagne est désigné à l'ONU sous le nom d'Allemagne'. Toutes les données se rapportant à l'Allemagne avant le 3 octobre figurent dans deux rubriques séparées basées sur les territoires respectifs de la République fédérale d'Allemagne et l'ancienne République démocratique allemande selon la période indiquée.

b   Lorsque l'expression "Europe occidentale" est utilisée dans la présente publication par opposition à l'expression "Europe orientale", elle s'applique à toutes les régions de l'Europe à l'exception de l'Europe orientale (c'est-à-dire qu'elle englobe l'Europe septentrionale et l'Europe méridionale aussi bien que l'Europe occidentale proprement dite).

C. *Other groupings*

Following is a list of other groupings and their compositions presented in the *Yearbook*. These groupings are organized mainly around economic and trade interests in regional associations.

Central American Common Market (CACM)
- Costa Rica
- El Salvador
- Guatemala
- Honduras
- Nicaragua

Central African Customs and Economic Union (CACEU)
- Cameroon
- Central African Republic
- Chad
- Congo
- Equatorial Guinea
- Gabon

Economic Community of West African States (ECOWAS)
- Benin
- Burkina Faso
- Cape Verde
- Côte d'Ivoire
- Gambia
- Ghana
- Guinea
- Guinea-Bissau
- Liberia
- Mali
- Mauritania
- Niger
- Nigeria
- Senegal
- Sierra Leone
- Togo

European Union (EU) [a]
- Belgium
- Denmark
- France
- Germany
- Greece
- Ireland
- Italy
- Luxembourg
- Netherlands
- Portugal
- Spain
- United Kingdom

C. *Autres groupements*

On trouvera ci-après une liste des autres groupements et de leur composition, présentée dans l'*Annuaire*. Ces groupements correspondent essentiellement à des intérêts économiques, et commerciaux d'après les associations régionales.

Marché commun de l'Amérique centrale (MCAC)
- Costa Rica
- El Salvador
- Guatemala
- Honduras
- Nicaragua

Union douanière et économique de l'Afrique centrale (UDEAC)
- Cameroun
- République centrafricaine
- Tchad
- Congo
- Guinée équatoriale
- Gabon

Communauté économique des états de l'Afrique de l'Ouest (CEDEAO)
- Bénin
- Burkina Faso
- Cap-Vert
- Côte d'Ivoire
- Gambie
- Ghana
- Guinée
- Guinée-Bissau
- Libéria
- Mali
- Mauritanie
- Niger
- Nigéria
- Sénégal
- Sierra Leone
- Togo

Union européenne (UE) [a]
- Belgique
- Danemark
- France
- Allemagne
- Grèce
- Irlande
- Italie
- Luxembourg
- Pays-Bas
- Portugal
- Espagne
- Poyaume-Uni

European Free Trade Association (EFTA) [b]
    Austria
    Finland
    Iceland
    Liechtenstein
    Norway
    Sweden
    Switzerland

Latin American Integration Association (LAIA)
    Argentina
    Bolivia
    Brazil
    Chile
    Colombia
    Ecuador
    Mexico
    Paraguay
    Peru
    Uruguay
    Venezuela

Least developed countries (LDC) [c]
    Afghanistan
    Bangladesh
    Benin
    Bhutan
    Botswana
    Burkina Faso
    Burundi
    Cambodia
    Cape Verde
    Central African Republic
    Chad
    Comoros
    Djibouti
    Equatorial Guinea
    Ethiopia
    Gambia
    Guinea
    Guinea-Bissau
    Haiti
    Kiribati
    Lao People's Democratic Republic
    Lesotho
    Liberia
    Madagascar
    Malawi
    Maldives
    Mali
    Mauritania
    Mozambique
    Myanmar

Association européenne de libre échange (AELE) [b]
    Autriche
    Finlande
    Islande
    Liechtenstein
    Norvège
    Suède
    Suisse

Association Latino-américaine d'intégration (LAIA)
    Argentine
    Bolivie
    Brésil
    Chili
    Colombie
    Equateur
    Mexique
    Paraguay
    Pérou
    Uruguay
    Venezuela

Les pays moins avancés (PMA) [c]
    Afghanistan
    Bangladesh
    Bénin
    Bhoutan
    Botswana
    Burkina Faso
    Burundi
    Cambodge
    Cap-Vert
    République centrafricaine
    Tchad
    Comores
    Djibouti
    Guinée équatoriale
    Ethiopie
    Gambie
    Guinée
    Guinée-Bissau
    Haïti
    Kiribati
    République démocratique populaire lao
    Lesotho
    Libéria
    Madagascar
    Malawi
    Maldives
    Mali
    Mauritanie
    Mozambique
    Myanmar

Nepal	Népal
Niger	Niger
Rwanda	Rwanda
Samoa	Samoa
Sao Tomé and Principe	Sao Tomé-et-Principe
Sierra Leone	Sierra Leone
Solomon Islands	Iles Salomon
Somalia	Somalie
Sudan	Soudan
Togo	Togo
Tuvalu	Tuvalu
Uganda	Ouganda
United Republic of Tanzania	République-Unie de Tanzanie
Vanuatu	Vanuatu
Yemen	Yémen
Zaire	Zaïre
Zambia	Zambie

Organization of Petroleum Exporting Countries (OPEC)	*Organisation des pays exportateurs de pétrole* (OPEP)
Algeria	Algérie
Ecuador	Equateur
Gabon	Gabon
Indonesia	Indonésie
Iran, Islamic Republic of	Iran, République islamique d'
Iraq	Iraq
Kuwait	Koweït
Libyan Arab Jamahiriya	Jamahiriya arabe libyenne
Nigeria	Nigéria
Qatar	Qatar
Saudi Arabia	Arabie saoudite
United Arab Emirates	Emirats arabes unis
Venezuela	Venezuela

---

a Beginning January 1995, includes Austria, Finland and Sweden.	a A partir de janvier 1995, y compris Autriche, Finlande et Suède.
b Beginning January 1995, excludes Austria, Finland and Sweden.	b A partir de janvier 1995, non compris Autriche, Finlande et Suède.
c As determined by the General Assembly in its resolution 46/206.	c Comme déterminés par l'Assemblée générale dans sa résolution 46/206.

Annex II

Conversion coefficients and factors

The metric system of weights and measures is employed in the *Statistical Yearbook*. In this system, the relationship between units of volume and capacity is: 1 litre = 1 cubic decimetre exactly (as decided by the 12th International Conference of Weights and Measures, New Delhi, November 1964).

Section A shows the equivalents of the basic metric, British imperial and United States units of measurements. According to an agreement between the national standards institutions of English-speaking nations, the British and United States units of length, area and volume are now identical, and based on the yard = 0.9144 metre exactly. The weight measures in both systems are based on the pound = 0.45359237 kilogram exactly (Weights and Measures Act 1963 (London), and *Federal Register* announcement of 1 July 1959: *Refinement of Values for the Yard and Pound* (Washington D.C.)).

Section B shows various derived or conventional conversion coefficients and equivalents.

Section C shows other conversion coefficients or factors which have been utilized in the compilation of certain tables in the *Statistical Yearbook*. Some of these are only of an approximate character and have been employed solely to obtain a reasonable measure of international comparability in the tables.

For a comprehensive survey of international and national systems of weights and measures and of units weights for a large number of commodities in different countries, see *World Weights and Measures* (United Nations publication, Sales No. E.66.XVII.3).

A. Equivalents of metric, British imperial and United States units of measure

Annexe II

Coefficients et facteurs de conversion

L'Annuaire statistique utilise le système métrique pour les poids et mesures. La relation entre unités métriques de volume et de capacité est: 1 litre = 1 décimètre cube (dm³) exactement (comme fut décidé à la Conférence internationale des poids et mesures, New Delhi, novembre 1964).

La section A fournit les équivalents principaux des systèmes de mesure métrique, britannique et américain. Suivant un accord entre les institutions de normalisation nationales des pays de langue anglaise, les mesures britanniques et américaines de longueur, superficie et volume sont désormais identiques, et sont basées sur le yard = 0:9144 mètre exactement. Les mesures de poids se rapportent, dans les deux systèmes, à la livre (pound) = 0.45359237 kilogramme exactement ("Weights and Measures Act 1963" (Londres), et *"Federal Register Announcement of 1 July 1959: Refinement of Values for the Yard and Pound"* (Washington, D.C.)).

La section B fournit divers coefficients et facteurs de conversion conventionnels ou dérivés.

La section C fournit d'autres coefficients ou facteurs de conversion utilisés dans l'élaboration de certains tableaux de *l'Annuaire statistique*. D'aucuns ne sont que des approximations et n'ont été utilisés que pour obtenir un degré raisonnable de comparabilité sur le plan international.

Pour une étude d'ensemble des systèmes internationaux et nationaux de poids et mesures, et d'unités de poids pour un grand nombre de produits dans différents pays, voir *"World Weights and Measures"* (publication des Nations Unies, No de vente E.66.XVII.3).

A. Equivalents des unités métriques, britanniques et des Etats-Unis

Metric units Unités métriques	British imperial and US equivalents Equivalents en mesures britanniques et des Etats-Unis	British imperial and US units Unités britanniques et des Etats-Unis	Metric equivalents Equivalents en mesures métriques	
Length–Longeur				
1 centimetre–centimètre (cm) ..	0.3937008 inch	1 inch.	2.540	cm
1 metre–mètre (m)	(3.280840 feet (1.093613 yard	1 foot . . . . 1 yard	30.480 0.9144	cm m
1 kilometre – kilomètre (km) . . .	(0.6213712 mile (0.5399568 int. naut. mile	1 mile . . . . 1 international nautical mile	1609.344 1852.000	m m
Area – Superficie				
1 square centimetre – cm².	0.1550003 square inch	1 square inch	6.45160	cm²
1 square metre – m²	(10.763910 square feet (1.195990 square yards	1 square foot . . . . 1 square yard	9.290304 0.83612736	dm² m²
1 hectare – ha	2.471054 acres	1 acre	0.4046856	ha
1 square kilometre – km²	0.3861022 square mile	1 square mile	2.589988	km²
Volume				
1 cubic centimetre – cm³	0.06102374 cubic inch	1 cubic inch	16.38706	cm³
1 cubic metre – m³	(35.31467 cubic feet (1.307951 cubic yards	1 cubic foot . . . . 1 cubic yard	28.316847 0.76455486	dm³ m³
Capacity – Capacité				
1 litre (l)	(0.8798766 imp. quart (1.056688 U.S. liq. quart (0.908083 U.S. dry quart	1 British imperial quart . . . . 1 U.S. liquid quart . . . . 1 U.S. dry quart	1.136523 0.9463529 1.1012208	l l l
1 hectolitre (hl)	(21.99692 imp. gallons (26.417200 U.S. gallons (2.749614 imp. bushels (2.837760 U.S. bushels	1 imperial gallon . . . . 1 U.S. gallon . . . . 1 imperial bushel . . . . 1 U.S. bushel	4.546092 3.785412 36.368735 35.239067	l l l l
Weight or mass – Poids				
1 kilogram (kg)	(35.27396 av. ounces (32.15075 troy ounces (2.204623 av. pounds	1 av. ounce . . . . 1 troy ounce . . . . 1 av. pound . . . . 1 cental (100 lb.) . . . . 1 hundredweight (112 lb.)	28.349523 31.10348 453.59237 45.359237 50.802345	g g g kg kg
1 ton – tonne (t)	(1.1023113 short tons (0.9842065 long tons	1 short ton (2 000 lb.) . . . . 1 long ton (2 240 lb.)	0.9071847 1.0160469	t t

B. Various conventional or derived coefficients

Railway and air transport

1 passenger-mile = 1.609344 voyageur (passager) - kilomètre
1 short ton-mile = 1.459972 tonne-kilomètre
1 long ton-mile = 1.635169 tonne kilomètre

Tonnage de navire

1 cubic metre – m³ = (0.353 register ton – tonne de jauge
(0.841 British shipping ton
(0.885 US shipping ton

1 metric ton – tonne métrique – 0.984 dwt ton

Electric energy

1 Kilowatt (kW) = (1.34102 British horsepower (hp)
(1.35962 cheval vapeur (cv)

C. Other coefficients or conversion factors employed in *Statistical Yearbook* tables

Roundwood

Equivalent in solid volume without bark.

Sugar

1 metric ton raw sugar = 0.9 metric ton refined sugar.
For the United States and its possessions:
1 metric ton refined sugar = 1.07 metric tons raw sugar

B. Divers coefficients conventionnels ou dérivés

Transport ferroviaire et aérien

1 voyageur (passager) - kilomètre = 0.621371) passenger-mile
1 tonne-kilomètre =   (0.684945 short ton-mile
(0.611558 long ton-mile**Ship tonnage**

1 register ton (100 cubic feet) – tonne de jauge = 2.83m³
1 British shipping ton (42 cubic feet) = 1.19m³
1 U.S. shipping ton (40 cubic feet) = 1.13m³
1 deadweight ton (dwt ton = long ton) = 1.016047 metric ton – tonne métrique

Energie électrique

1 British horsepower (hp) = 0.7457 kW
1 cheval vapeur (cv) = 0.735499 kW

C. Autres coefficients ou facteurs de conversion utilisés dans les tableaux de l'*Annuaire statistique*

Bois rond

Equivalences en volume solide sans écorce.

Sucre

1 tonne métrique de sucre brut = 0.9 tonne métrique de sucre raffiné.
Pour les Etats-Unis et leurs possessions:
1 tonne métrique de sucre raffiné = 1.07 t.m. de sucre brut.

<table>
<tr><td>

D. Selected energy conversion factors

Crude petroleum

    1 barrel = 42 U.S. gallons = 34.97 imperial gallons = 158.99 litres = 0.15899 cubic metres.
    1 cubic metre = 6.2898 barrels.
    The equivalent of barrels in metric tons depends on the specific gravity of the petroleum which varies from country to country. The average specific gravity for each producing country is indicated in the table on the production of crude petroleum in the *Energy Statistics Yearbook*.

Coal equivalent[1] (metric tons unless otherwise indicated):

</td><td>

D. Facteurs de conversion pour certains produits en matière d'énergie

Pétrole brut

    1 baril = 42 gallons E.U. = 34.97 gallons britanniques = 158.99 litres = 0.15899 m³.
    1m³ = 6.2898 barils.
    L'équivalent du baril en tonnes métriques dépend du poids spécifique du pétrole qui varie d'un pays à l'autre. Le poids spécifique moyen utilisé pour chaque pays producteur se trouve dans l'Annuaire des statistiques de l'énergie dans le tableau relatif à la production de pétrole brut.

Equivalent en houille[1] (tonnes métriques sauf indication contraire):

</td></tr>
</table>

Coal, anthracite and bituminous	1.0		Charbon, anthracite et la houille bitumineuse	1.0
Coal briquettes	1.0		Briquettes de charbon	1.0
Cokes of coal	0.9		Cokes de charbon	0.9
Lignite	0.385		Lignite	0.385
Cokes of brown coal or lignite	0.67		Cokes de charbon brun ou de lignite	0.67
Lignite briquettes	0.67		Briquettes de lignite	0.67
Peat for fuel	0.325		Tourbe pour combustible	0.325
Peat briquettes	0.5		Briquettes de tourbe	0.5
Crude petroleum	1.429		Pétrole brut	1.429
Natural gas liquids (weighted average)	1.542		Condensats provenant du gaz naturel (moyenne pondérée)	1.542
Liquefied petroleum gases	1.554		Gaz de pétrole liquéfié	1.554
Natural gas (terajoules[2])	34.121		Gaz naturel (terajoules[2])	34.121

<table>
<tr><td>

Coal equivalent (metric tons of hydro, nuclear and geothermal electricity:
1000 kWh = 0.123

</td><td>

Equivalent en houille (tonnes métriques) d'électricité, hydraulique, nucléaire et géothermique:
1000 kWh = 0.123

</td></tr>
</table>

---

[1] It should be noted that the base used for coal equivalency comprises 7000 calories/gramme.
[2] Under standard conditions of l5°C, 1013.25 mbar, dry.

[1] Veuillez noter que l'équivalence en houille est faite sur la base de 7000 calories/gramme.
[2] En volume standard (à l5°C, 1013.25 mbar, gaz sec).

Tables added and omitted

A. *Tables added*

In the present issue of the *Statistical Yearbook* (1993), the following tables have been added:

Table 22: Expenditure on gross domestic product at current prices;

Table 23: Gross domestic product by kind of economic activity at current prices.

B. *Tables omitted*

The following tables in the 1992 edition have been deleted from the present issue:

Table 8: Export price index numbers of primary commodities and non-ferrous base metals;

Table 10: Education preceding the first level;

Table 14: Vital statistics summary and expectation of life at birth;

Table 17: Newsprint consumption;

Table 27: Money supply;

Table 37: Roots and tubers;

Table 38: Pulses;

Table 40: Vegetables;

Table 41: Fruit;

Table 42: Tobacco;

Table 46: Tractors in use;

Table 48: Uranium;

Table 49: Gold;

Table 52: Wine;

Table 55: Non-alcoholic beverages;

Table 60: Wood-based panels;

Table 61: Rubber: synthetic and reclaimed;

Table 69: Merchant vessels;

Table 89: External trade conversion factors;

Table 94: Structure of world exports by commodity classes and regions;

Table 100: International reserves excluding gold.

C. The following tables in the *1992* edition are not presented in the present issue because of insufficient new data:

Table 15: Selected indicators of life expectancy, child-bearing and mortality;

Table 22 (in *Yearbook*, 38th issue): Food supply;

Table 21: Cinemas:number, seating capacity, annual attendance and box office receipts;

Table 82: Selected indicators of natural resources;

Table 83: Emissions of CO_2 and consumptions CFCs and halons;

Table 84: Concentration of suspended particulate matter at selected sites;

Tableaux ajoutés et supprimés

A. *Tableaux ajoutés*

Dans ce numéro de l'*Annuire statistique* (1993), les tableaux suivants ont été ajoutés :

Tableau 22: Dépenses imputées au produit intérieur brut aux prix courants;

Tableau 23: Produit intérieur brut par genre d'activité economique aux prix courants.

B. *Tableaux supprimés*

Les tableaux suivants de l'édition de 1992 ont été supprimés de la présente édition:

Tableau 8: Indices des prix des exportations des matières premières et des métaux non-férreux;

Tableau 10: Enseignement précédant le premier degré;

Tableau 14: Aperçu des statistiques de l'état civil et espérance de vie à la naissance;

Tableau 17: Consommation de papier journal;

Tableau 27: Disponibilités monétaires;

Tableau 37: Racines et tubercules;

Tableau 38: Légumineuses sèches;

Tableau 40: Légumes;

Tableau 41: Fruits;

Tableau 42: Tabac;

Tableau 46: Tracteurs en service;

Tableau 48: Uranium;

Tableau 49: Or;

Tableau 52: Vin;

Tableau 55: Boissons non alcooliques;

Tableau 60: Panneaux à base de bois;

Tableau 61: Caoutchouc: cynthétique et régénéré;

Tableau 69: Navires marchands;

Tableau 89: Facteurs de conversion pour le commerce extérieur;

Tableau 94: Structure des exportations mondiales par catégories de marchandises et par régions;

Tableau 100: Réserves internationales, l'or non compris.

C. Les tableaux suivants dans l'édition précédente *1992* de l'*Annuaire statistique* mais qui n'ont pas été repris dans la présente édition fautes de données nouvelles suffisantes :

Tableau 15: Choix d'indicateurs de l'espérance de vie, de maternité et de la mortalité;

Tableau 22 (dans l'*Annuire*, 38ème édition): Disponibilités alimentaires;

Tableau 21: Cinémas : nombre d'établissements, nombre de sièges, fréquentation annuelle et recettes guichet;

Tableau 82: Choix d'indicateurs concernant certaines ressources naturelles;

Table 85: Global water quality in selected rivers;

Table 86: Surface and land area and land use;

Table 87: Selected indicators of environmental protection.

These tables will be updated in future issues of the *Yearbook* when new data become available.

Tableau 83: Emissions de CO_2 et consommation de CFCs et halons;

Tableau 84: Concentration de particules en suspension en divers lieux;

Tableau 85: Qualité générale de l'eau de certains cours d'eau;

Tableau 86: Superficie totale, superficie des terres et utilisation des terres;

Tableau 87: Choix d'indicateurs de la protection de l'environnement.

Ces tableaux seront actualisés dans les futures livraisons de l'*Annuaire* à mesure que des données nouvelles deviendront disponibles.

Statistical sources and references

A. Statistical sources*
1. AAMA Motor Vehicle, *Facts and Figures 1994* (Detroit, USA) and previous issues.
2. Auto and Truck International, *1994-95 World Automotive Market Report* (Illinois, USA) and previous issues.
3. Food and Agriculture Organization of the United Nations, *FAO Yearbook: Fertilizer 1993* (Rome).
4. _____, *FAO Yearbook: Fishery Statistics, Catches and Landings 1992* (Rome).
5. _____, *FAO Yearbook: Forest Products 1992* (Rome).
6. _____, *FAO Yearbook: Production 1993* (Rome).
7. International Civil Aviation Organization, *Cvil Aviation Statistics of the World 1993* (Montreal).
8. _____, Digest of statistics, Traffic (Montreal).
9. International Labour Office, *Year Book of Labour Statistics 1994* (Geneva).
10. International Monetary Fund, *Balance of Payments Yearbook 1994* (Washington, DC).
11. _____, *International Financial Statistics* (Washington, DC, monthly).
12. International Sugar Organization, *Sugar Yearbook 1993* (London) and previous issues.
13. International Telecommunications Union, *Yearbook of Telecommunication Statistics 1982-1991* (Geneva).
14. Lloyd's Register of Shipping, *Annual Summary of Merchant Ships Completed 1993* (London) and previous issues.
15. Organisation for Economic Cooperation and Development, *Development Cooperation: Efforts and Policies of the Members of the Development Assistance Committee 1993* and *1994* (Paris).
16. _____, *Geographic Distribution of Financial Flows 1989/1992* (Paris).
17. United Nations, *Commodity Trade Statistics*, Series D (United Nations serial publication).
18. _____, "Comprehensive statistical data on operational activities for development for the years 1992 and 1993" (E/1994/64/Add.2 and A/50/202/add.2).
19. _____, *Demographic Yearbook 1992* (United Nations publication, Sales No. E/F.94.XIII.1) and previous issues.
20. _____, *Energy Statistics Yearbook 1992* (United Nations publication, Sales No. E/F.94.XVII.9).

---

* The following United Nations organization also provided data for the present issue of the *Statistical Yearbook*: World Health Organization, table 13.

Sources statistiques et références

A. Sources statistiques*
1. "AAMA Motor Vehicle, *Facts and Figures 1994*" (Detroit, USA) et les éditions précédentes.
2. "Auto and Truck International, *1994-1995 World Automotive Market Report*" (Illinois) et les éditions précédentes.
3. Organisation des Nations Unies pour l'alimentation et l'agriculture, *FAO Annuaire: Engrais 1993* (Rome).
4. _____, *FAO Annuaire: Statistiques des pêches, captures et quantités débarquées 1992* (Rome).
5. _____, *FAO Annuaire des produits forestiers 1992* (Rome).
6. _____, *FAO Annuaire: Production 1993* (Rome).
7. Organisation de l'aviation civile internationale, *Statistiques mondiales de l'aviation civile 1993* (Montréal).
8. _____, "Digest of statistics, Traffic" (Montréal).
9. Bureau international du Travail, *Annuaire des statistiques du Travail 1994* (Genève).
10. Fonds monétaire international, "*Balance of Payments Yearbook 1994*", (Washington DC).
11. _____, *Statistiques financières internationales* (Washington, DC, mensuel).
12. "International Sugar Organization, Sugar Yearbook 1993" (Londres) et éditions précédentes.
13. Union international des télécommunications, *Annuaire statistique des télécommunications 1982-1991* (Genève).
14. "Lloyd's Register of Shipping, *Annual Summary of Merchant Ships Completed 1993*" (Londres) et éditions précédentes.
15. Organisation de Coopération de Développement Economiques, *Coopération pour le développement : Efforts et politiques des membres du comité d'aide au développement 1993 et 1994* (Paris).
16. _____, *Répartition géographique des Ressources Financières 1989/1992* (Paris).
17. Organisation des Nations Unies, "*Commodity Trade Statistics*", (publication des Nations Unies, Série D).
18. _____, *Données statistiques détaillées sur les activités opérationnelles de développement pour les années 1992 et 1993* (E/1994/64/Add.2 et A/50/202/add.2).
19. _____, *Annuaire démographique 1992* (publication des Nations Unies, No de vente 94.XIII.1) et les éditions précédentes.
20. _____, *Annuaire des statistiques de l'énergie 1992* (publication des Nations Unies, No de vente E/F.94.XVII.9).

---

* L'organisation des Nations Unies suivante a envoyé aussi des données pour ce numéro de l'Annuaire statistique: Organisation mondiale de la santé, tableau 13.

21. _____, *Industrial Commodity Statistics Yearbook 1992* (United Nations publications, Sales No. E/F.94.XVII.13).

22. _____, *International Trade Statistics Yearbook 1993*, vols. I and II (United Nations publication, Sales No. E/F.95.XVII.3).

23. _____, *Monthly Bulletin of Statistics*, various issues up to March 1995 (United Nations publication, Series Q).

24. _____, *National Accounts Statistics: Main Aggregates and Detailed Tables, 1992* (United Nations publication, Sales No. E.95.XVII.4) Part I and II.

25. _____, *World Comparisons of Purchasing Power and Real Product for 1980, Parts one and two* (United Nations publication, Sales Nos. E.86.XVII.9 and E.86.XVII.10).

26. _____, *World Population Prospects: 1994* (United Nations publications, forthcoming).

27. _____, *World Urbanization 1992* (United Nations wall poster, Sales No. E.93.XIII.2).

28. _____, *World Urbanization Prospects 1994* (United Nations publication, Sales No. E.95.XIII.12).

29. United Nations Educational, Scientific and Cultural Organization, *Statistical Yearbook 1994* (Paris).

30. World Bank, *World Debt Tables 1994-95* (Washington, DC).

31. World Intellectual Property Organization, *Industrial Property Statistics 1992* (Geneva).

32. _____, *Yearbook of Tourism Statistics 1994* (Madrid).

B. *References*

33. Food and Agriculture Organization of the United Nations, *The Fifth World Food Survey 1985* (Rome 1985).

34. International Labour Office, *International Standard Classification of Occupations, Revised Edition 1968* (Geneva, 1969); revised edition, 1988, ... *ISCO-88* (Geneva, 1990).

35. International Monetary Fund, *Balance of Payments Manual, Fifth Edition* (Washington, DC, 1993).

36. United Nations, *Basic Methodological Principles Governing the Compilation of the System of Statistical Balances of the National Economy*, Studies in Methods, Series F, No. 17, Rev. 1, vols. 1 and 2 (United Nations publications, Sales No. E.89.XVII.5 and E.89.XVII.3).

37. _____, *Energy Statistics: Definitions, Units of Measure and Conversion Factors*, Series F, No. 44 (United Nations publication, Sales No. E.86.XVII.21).

38. _____, *Energy Statistics—A Manual for Developing Countries*, Series F, No. 56 (United Nations publication, Sales No. E.91.XVII.10).

21. _____, *Annuaire des statistiques industrielles par produit 1992* (publications des Nations Unies, No de vente E/F.94.XVII.13).

22. _____, *Annuaire statistique du Commerce international 1993*, Vols. I et II (publication des Nations Unies, No de vente E/F.95.XVII.3).

23. _____, *Bulletin mensuel de statistique*, différentes éditions, jusqu'en mars 1995 (publication des Nations Unies, Série Q).

24. _____, "*National Accounts Statistics: Main Aggregates and Detailed Tables 1992*" (publication des Nations Unies, No de vente E.95.XVII.4) partie I et II.

25. _____, "*World Comparisons of Purchasing Power and Real Product for 1980, Parts one and two*" (publications des Nations Unies, Nos. de vente E.86.XVII.9 et E.86.XVII.10).

26. _____, "*World Population Prospects: 1994*", (publications des Nations Unies, à paraître).

27. _____, "*World Urbanization 1992*" (tableau mural des Nations Unies, No de vente E.93.XIII.2)

28. _____, "*World Urbanization Prospects 1994*" (publication des Nations Unies, No de vente E.95.XIII.12).

29. Organisation des Nations Unies pour l'éducation, la science et la culture, *Annuaire statistique 1994* (Paris).

30. Banque mondiale, "*World Debt Tables 1994-1995*" (Washington, DC).

31. Organisation mondiale de la propriété intellectuelle, *Statistiques de propriété industrielle 1992* (Genève).

32. _____, *Annuaire des statistiques du tourisme 1994* (Madrid).

B. *Références*

33. Organisation des Nations Unies pour l'alimentation et l'agriculture, *Cinquième enquête mondiale sur l'alimentation 1985* (Rome, 1985).

34. Organisation internationale du Travail, *Classification internationale type des professions, édition révisée* 1968 (Genève, 1969); édition révisée 1988...*CITP-88* (Genève, 1990).

35. Fonds monétaire international, *Manuel de la balance des paiements, cinquième édition* (Washington, DC, 1993).

36. Organisation des Nations Unies, *Principes méthodologiques de base régissant l'établissement des balances statistiques de l'économie nationale*, Série F, No 17, Rev.1 Vol. 1 et Vol. 2 (publication des Nations Unies, No de vente F.89.XVII.5 et F.89.XVII.3).

37. _____, *Statistiques de l'énergie: définitions, unités de mesures et facteurs de conversion*, Série F, No 44 (publication des Nations Unies, No de vente F.86.XVII.21).

38. _____, *Statistiques de l'énergie - Manuel pour les pays en développement*, Série F, No 56 (publication des Nations Unies, No de vente F.91.XVII.10).

39. _____, *Handbook of Vital Statistics Systems and Methods*, vol. I, *Legal, Organization and Technical Aspects*, Series F, No. 35, vol. I (United Nations publication, Sales No. E.91.XVII.5).

40. _____, *Handbook on Social Indicators*, Studies in Methods, Series F, No. 49 (United Nations publication, Sales No. E.89.XVII.6).

41. _____, *International Recommendations for Industrial Statistics*, Series M, No. 48, Rev. 1 (United Nations publication, Sales No. E.83.XVII.8).

42. _____, *International Standard Industrial Classification of All Economic Activities*, Statistical Papers, Series M, No. 4, Rev. 2 (United Nations publication, Sales No. E.68.XVII.8); Rev. 3 (United Nations publication, Sales No. E.90.XVII.11).

43. _____, *International Trade Statistics: Concepts and Definitions*, Series M, No. 52, Rev. 1 (United Nations publication, Sales No. E.82.XVII.14).

44. _____, *Methods Used in Compiling the United Nations Price Indexes for External Trade*, volume 1, Statistical Papers, Series M, No. 82 (United Nations Publication, Sales No. E.87.XVII.4).

45. _____, *1977 Supplement to the Statistical Yearbook and the Monthly Bulletin of Statistics*, Series S and Series Q, Supplement 2 (United Nations publication, Sales No. E.78.XVII.10).

46. _____, *Principles and Recommendations for Population and Housing Censuses*, Statistical Papers, Series M, No. 67 (United Nations publication, Sales No. E.80.XVII.8).

47. _____, *Provisional Guidelines on Statistics of International Tourism*, Statistical Papers, Series M, No. 62 (United Nations publication, Sales No. E.78.XVII.6).

48. _____, *Standard International Trade Classification, Revision 3*, Statistical Papers, Series M, No. 34, Rev. 3 (United Nations publication, Sales No. E.86.XVII.12), *Revision 2*, Series M, No. 34, Rev. 2 (United Nations publication), *Revision*, Series M, No. 34, Revision (United Nations publication, Sales No. E.61.XVII.6).

49. _____, *A System of National Accounts, Studies in Methods*, Series F, No. 2, Rev. 3 (United Nations publication, Sales No. E.69.XVII.3).

50. _____, *A System of National Accounts 1993*, Studies in Methods, Series F, No. 2, Rev. 4 (United Nations publication, Sales No. E.94.XVII.4).

51. _____, and World Tourism Organizations *Recommendations on Tourism Statistics*, Statistical Papers, Series M, No. 83 (United Nations publication, Sales No. E.94.XVII.6).

39. _____, "*Handbook of Vital Statistics System and Methods*, Vol. 1, *Legal, Organization and Technical Aspects*", Série F, No 35, Vol. 1 (publication des Nations Unies, No de vente E.91.XVII.5).

40. _____, *Manuel des Indicateurs sociaux*, Série F, No 49 (publication des Nations Unies, No de vente F.89.XVII.6).

41. _____, *Recommandations internationales concernant les statistiques industrielles*, Série M, No 48, Rev. 1 (publication des Nations Unies, No de vente F.83.XVII.8).

42. _____, *Classification Internationale type, par Industrie, de toutes les branches d'activité économique*, Série M, No 4, Rev. 2 (publication des Nations Unies, No de vente F.68.XVII.8); Rev. 3 (publication des Nations Unies, No de vente F.90.XVII.11).

43. _____, *Statistiques du commerce International: Concepts et définitions*, Série M, No 52, Rev. 1 (publication des Nations Unies, No de vente F.82.XVII.14).

44. _____, *Méthodes utilisées par les Nations Unies pour établir les indices des prix des produits de base entrant dans le commerce international*, Série M, No 82, Vol. 1 (publication des Nations Unies, No de vente F.87.XVII.4).

45. _____, *1977 Supplément à l'Annuaire statistique et au bulletin mensuel de statistique*, Série S et Série Q, supplément 2 (publication des Nations Unies, No de vente F.78.XVII.10).

46. _____, *Principes et recommandations concernant les recensements de la population et de l'habitation*, Série M, No 67 (publication des Nations Unies, No de vente F.80.XVII.8).

47. _____, *Directives provisoires pour l'établissement des statistiques du tourisme International*, Série M, No 62 (publication des Nations Unies, No de vente 78.XVII.6).

48. _____, *Classification type pour le commerce International (troisième version révisée)*, Série M, No 34, Rev. 3 (publication des Nations Unies, No de vente F.86.XVII.12), *Révision 2*, Série M, No 34, Rev. 2 (publication des Nations Unies), *Révision*, Série M, No. 34, Révision (publication des Nations Unies, No de vente F.61.XVII.6).

49. _____, *Système de comptabilité nationale*, Série F, No 2, Rev. 3 (publication des Nations Unies, No de vente F.69.XVII.3).

50. _____, *Système de comptabilité nationale 1993*, Série F, No 2, Rev. 4 (publication des Nations Unies, No de vente F.94.XVII.4).

51. _____, et l'Organisation mondiale du tourisme "*Recommendations on Tourism Statistics*, Statistical Papers", Série M, No. 83 (publication des Nations Unies, No. de vente E.94.XVII.6).

52. _____, *Towards a System of Social and Demographic Statistics, Studies in Methods*, Series F, No. 18 (United Nations publication, Sales No. E.74.XVII.8).

53. World Health Organization, *Manual of the International Statistical Classification of Diseases, Injuries and Causes of Death*, vol. 1 (Geneva, 1977). See also *Demographic Yearbook*.[19]

54. World Tourism Organization, *Methodological Supplement to World Travel and Tourism Statistics* (Madrid, 1985).

52. _____, *Vers un système de statistiques démographiques et sociales, Etudes méthodologiques*, Série F, No 18 (publication des Nations Unies, No. de vente F.74.XVII.8).

53. Organisation mondiale de la santé, *Manuel de la classification statistique internationale des maladies, traumatismes et causes de décès*, Vol. 1 (Genève, 1977). Voir aussi *Annuaire démographique* [19].

54. Organisation mondiale du tourisme, *Supplément méthodologique aux statistiques des voyages et du tourisme mondiaux* (Madrid, 1985).

Index

Note: References to tables are indicated by **boldface** type. For citations of organizations, see Index of organizations.

Index of organizations

Litho in United Nations, New York
93722—September 1995—6,975
ISBN 92-1-061163-2
ISSN 0082-8459

United Nations publication
Sales No. E/F.95.XVII.1
ST/ESA/STAT/SER.S/16